Truck Engines

Fuel & Computerized Management Systems

Sean Bennett
Centennial College
Toronto, CANADA

CENGAGE Learning™

Andover • Melbourne • Mexico City • Stamford, CT • Toronto • Hong Kong • New Delhi • Seoul • Singapore • Tokyo

Truck Engines: Fuel & Computerized Management Systems
Sean Bennett

This edition is reprinted with license from Delmar, a part of Cengage Learning, for sale in India, Pakistan, Bangladesh, Nepal and Sri Lanka.

ISBN-13: 978-81-315-1128-2
ISBN-10: 81-315-1128-6

Cengage Learning India Private Limited
418, F.I.E., Patparganj
Delhi 110092
India

Tel: 91-11-43641111
Fax: 91-11-43641100
Email: asia.infoindia@cengage.com

Cengage Learning is a leading provider of customized learning solutions with office locations around the globe, including Andover, Melbourne, Mexico City, Stamford (CT), Toronto, Hong Kong, New Delhi, Seoul, Singapore, and Tokyo. Locate your local office at: www.cengage.com/global

Cengage Learning Products are represented in Canada by Nelson Education, Ltd.

For product information, visit our website at **www.cengage.co.in**

Printed in India
Second Indian Reprint 2012

Contents

Part 2: Hydromechanical Diesel and Alternate Fuel Systems

Preface to the India Edition

Innovation in diesel engine technology in recent years has been driven to a large extent by the EPA tiered introduction of tailpipe emissions. Most of the topics in this edition of the textbook are either directly or indirectly connected to the introduction of EPA Tier 3 and Tier 4 diesel emissions standards. Many would argue that the rate of technological change is such that rigid curriculum models do not best serve the learning requirements of today's diesel technicians. Immediately after graduation from postsecondary programs, diesel techs have to hit the shop floor with their feet running. They must function at a high level of efficiency and technical competence in a competitive environment very different from that of a generation ago.

This textbook attempts to be vocational in character. Along with addressing subject matter appearing in the curriculum of programs targeting eventual certification for graduates, the most current technology is explained in sufficient detail to help technicians understand what they have to contend with in the real world truck shop. Good programs of study should not only be as up to date as possible but should emphasize performance-based learning outcomes. There is little point in merely understanding concepts if they cannot be put into practice. Today's truck technicians must become lifelong learners if they are going to succeed. Although designed for entry-level diesel technicians, a secondary objective of this textbook is to cover some of the learning requirements of certified, practicing technicians in helping them maintain technical currency. It should be underlined that in order to maintain the technical expertise required of diesel engine specialists today, regular participation in OEM corporate training programs is essential.

Every person learns in different ways, but mechanical technicians tend to learn by doing rather than by reading. This book attempts to cover the concepts that will help the technician master repair technology, but this comes with a reminder that true mastery of its subject matter will not be achieved without some hands-on exposure. As both a teacher and an author, I am made constantly aware that for students to effectively learn the complex technologies built into today's trucks, both a conceptual and a practical approach are required. Teachers will commonly hear students inform them that they have not been able to fully understand the contents of a chapter until they have revisited it after the experience of working on it. As such it is my hope that this book will not merely serve as a learning guide for entry-level technicians, but earn a second life in their toolboxes as a reference text.

Although the primary focus is on on-highway commercial diesel engines, differences between these and their off-highway counterparts are rapidly disappearing. The EPA has now instituted off-highway diesel engine emissions standards identical to on-highway standards with the sole difference that they are implemented one calendar year later. For instance, the Tier 4 implementation date for on-highway diesel engines is 2010, while it is 2011 for off-highway diesels. This move by the

EPA has effectively ended the hydromechanical engine management era for off-highway diesel engines at the Tier 3 stage (2007/2008).

I would like to especially thank the many individuals who have contributed in big ways and small. Valuable input can be as brief as pointing out a simple typo, or as complex as submitting detailed suggestions for improvements. In this edition, I am indebted to Bernie Andringa of Skagit Valley College, Mount Vernon, Washington, for his detailed proofing and annotations: Most of his feedback is incorporated in this edition. I would also like to thank my colleague Al Thompson of Centennial College, Toronto, for not only teaching me some of the fundamentals of the photography I have used in this book, but also for taking some of the photographs. I am also indebted to the editorial team in Clifton Park, NY, who have done so much work behind the scenes to help me grow as a technical author.

In concluding this preface, I should say a few words about where diesel technology might be heading. In recent years some of the differences that distinguish gasoline- and diesel-fueled engines have been eroding. Since 2004 we have seen a significant increase in direct-injected (DI), gasoline-fueled engines accompanied by the increases in compression ratios that DI makes possible. DI gasoline fueling along with fast-response, piezo-actuated injectors enable the engine electronics to better control cylinder combustion using multipulse injection events. This combustion management strategy utilized for a generation of diesel engines has led to the development in Europe of a gasoline-fueled engine that uses a combination of spark and compression ignition depending on how the engine is operated. Similarly diesel technology has borrowed from its gasoline-fueled counterparts. With an ever-increasing number of diesel engines adopting common rail fueling in addition to comprehensive exhaust gas aftertreatment, similarities between gasoline and diesel engines are on the rise. In terms of fundamental operating principles, common rail fueling makes diesel fuel management almost as simple as gasoline fuel injection in an auto engine. This, along with the complexity of the electronically managed emissions controls required for many years in automobile engines, reinforces the fact that the challenge for diesel techs for the next decade will be in mastery of electronic control systems.

Sean Bennett, April 2008
Feedback and comments are welcomed: email@seanbennett.org

Acknowledgments

Individuals

Ray Amlung, Cummins Engineering, Columbus, Indiana

Bernie Andringa, Skagit Valley College, Mount Vernon, Washington

Jim Bardeau, Centennial College, Toronto

Kabeer Barudin, Harper Detroit Diesel, Toronto

Guy Bennett, Avolution Multimedia, Sarnia, Ontario

James Bennett, Avolution Multimedia, Sarnia, Ontario

Bev Blaine, Cummins Ontario, Toronto

Darrin Bruneau, Freightliner Service Operations, Toronto

Mike Cerato, Centennial College, Toronto

Howard Chesneau, Fuel Quality Services, Georgia

George Clark, Centennial College, Toronto

Dave Coffey, Caterpillar Engines, Peoria, Illinois

David Connant, Nashville Auto Diesel College

Cameron Cox, SAIT, Calgary, Alberta

Dara Cruickshank Bennett, Angharan, Millbrook, Ontario

Jeff Curtis, Bellingham Technical College, Bellingham, Washington

George Czata, Harper Detroit Diesel, Toronto

Pat Dillard, Nashville Auto Diesel College, Nashville, Tennessee

Enzo DiPietroantonio, Cummins Eastern Canada, Toronto

Owen Duffy, Centennial College, Toronto

Craig Fedder, Navistar International Trucks, Chicago, Illinois

David Fehling, Association of Diesel Specialists, Kansas City, MO

Donald Fetherolf, Mack Trucks Incorporated, Allentown, Pennsylvania

Scott Furr, Cummins Atlantic, Charlotte, North Carolina

Terry Harkness, Toromont Caterpillar, Toronto

Alan Hertzog, NA Institute, Volvo Trucks, Allentown, Pennsylvania

Bob Huzij, Cambrian College, Sudbury, Ontario

Serge Joncas, Volvo Training, Toronto

Gord King, Fanshawe College, London, Ontario

Marty Kubiak, Detroit Diesel Corporation, Detroit, Michigan

Steve Lamb, Harper Freightliner, Toronto

Bobby Leatherman, Nashville Auto Diesel College, Nashville, Tennessee

Patrick Leitner, Freightliner NE Training Center, New Jersey

George Liidemann, Centennial College, Toronto

Sam Lightowler, Toronto Transit Commission, Toronto

Jim Lonnie, Caterpillar, Peoria, Illinois

Roger Look, Caterpillar, Peoria, Illinois

Ben Macaro, Cummins Eastern Canada, Toronto

Bob Marshal, University of Northern Ohio, Lima, Ohio

Brad Martin, Detroit Diesel Corporation, London, Ontario

Kerry Matthews, Cummins Atlantic, Charlotte, North Carolina

Timothy Meyer, Cummins Fuels Division, Columbus, Indiana

Don Mitchell, Case Corporation, Moline, Illinois

John Montgomery, Volvo-Mack Trucks, Director of Customer Satisfaction, Toronto

Roger Morin, Volvo Training Center, Toronto

John Murphy, GM Sales, Toronto

Stephen Pang, Toronto Transit Commission, Toronto

George Parsons, Sault College, Sault Ste. Marie, Ontario

Mike Perreira, Cummins Ontario, Toronto

Martin Restoule, Algonquin College, Ottawa, Ontario

Jack Rosebro, Perfect Sky, Carpinteria, California

Ken Riley, Toromont Caterpillar Training Dept, Toronto

Fred Schmidt, Donaldson Company, Minneapolis, Minnesota

Wayne Scott, Freightliner LLC, Portland, Oregon

Craig Smith, Nova Freightliner, Truro, Nova Scotia

Darren Smith, Centennial College, Toronto

Angelo Spano, Centennial College, Toronto

Dan Sullivan, Sullivan Solutions, Durham, North Carolina

Stefanie Thiel, Mahle Gruppe, Stuttgart, Germany

Al Thompson, Centennial College and Toronto ATS

Pierre Valley, Mack Trucks Training, Toronto

Gary Van Nederynan, Orion Bus, Oriskany, New York

George Vass, Volvo Trucks Training, Toronto

Gus Wright, Centennial College, Toronto

Organizations and Corporations

Association of Diesel Specialists, Kansas City, Missouri

Avolution Multimedia, Sarnia, Ontario

Caterpillar Engines, Peoria, Illinois

Cummins Engine Company, Columbus, Indiana

Cummins Ontario, Toronto, Ontario

Davco Manufacturing, Saline, Michigan

Delphi Fuel Division, Troy, Michigan

Denso Corporation, Aichi, Japan

Detroit Diesel Corporation, Detroit, Michigan

Donaldson Corporation, Minneapolis, Minnesota

Harper Detroit Diesel, Toronto

Hino Toyota, Bloomfield Hills, Michigan

Imperial Oil, Toronto, Ontario

Institution of Mechanical Engineers, London, England

International Trucks, Chicago, Illinois

Isuzu America, Cerritos, California

Jacobs Manufacturing, Bloomfield, Connecticut

Kent Moore, Warren, Michigan

Mack Trucks Canada, Toronto

Mack Trucks Incorporated, Allentown, Pennsylvania

Mahle Gruppe, Stuttgart, Germany

Mercedes-Benz, Montvale, New Jersey

Metropolitan Transit Authority, Brooklyn, New York

Michigan Freightliner LLC, Portland, Oregon

Mid-Ontario Freightliner, Toronto

Nashville Auto Diesel College, Nashville, Tennessee

Robert Bosch Corporation, Chicago, Illinois

Society of Automotive Engineers, Warrendale, Pennsylvania

Stanadyne, Windsor, Connecticut

Sullivan Training Systems, Raleigh, North Carolina

Superflow Corporation, Colorado Springs, Colorado

Technical and Maintenance Council, American Trucking Association, Alexandria, Virginia

Toromont Caterpillar, Toronto

Toronto Transit Commission, Toronto

University of Northwestern Ohio, Lima, Ohio

Vibratech TVD, Springfield, New York

Volvo Trucks, Allentown, Pennsylvania

Williams Controls Incorporated, Portland, Oregon

Reviewers

Roy Benningfield, Universal Technical Institute, Exton, PA

Gregory Clouse, Montana State University, Havre, MT

Cameron Cox, Southern Alberta Institute of Technology, Calgary, AB

Pat Dillard, Nashville Auto-Diesel College, Nashville, TN

John C. Holland, Lincoln College of Technology, Denver, CO

Robert E. Marshal, University of Northwestern Ohio, Lima, OH 45805

Wilmer Martin, Universal Technical Institute, Exton, PA

David Miller, Lincoln College of Technology, Denver, CO

Adam Oyerbides, Lincoln Technical Institute, Grand Prairie, TX

James Potter, Universal Technical Institute, Houston, TX

Mike Slicker, Lincoln College of Technology, Denver, CO

Jason Speer, Lincoln Technical Institute, Indianapolis, IN

Lynn Stilger, Montana State University, Havre, MT

Murray Tamas, Thompson Rivers University, Kamloops, BC

PART 1

Diesel Engine Fundamentals

1

Introduction

INTRODUCTION

The primary objective of this textbook is to provide a basic understanding of the truck diesel engine and its fuel management circuits to the extent required by truck, bus, and heavy equipment technicians. Much of the book deals with electronically managed engines, but some coverage of older hydromechanical systems has been retained because it is still a requirement of curriculum, NATEF standards, and ASE tasks. The technology addressed in this textbook focuses on principles of operation, the objective being to help the technician understand the systems rather than address the detail of the procedure required to repair them. Some repair detail appears in the e-resource that accompanies this book, but students and technicians should develop the habit of using online service information systems whenever possible.

Most engine **original equipment manufacturers (OEMs)** mandate that technicians be certified to diagnose and repair their engines and the computerized systems that manage them. Certification requires that the technician attend proprietary courses. These product-specific courses usually last a week or less and deliver the information and procedures required to diagnose and repair one specific engine series. Because of the delivery time constraints, product-specific courses seldom touch on the principles of operation. This textbook is very concerned with why and how components and systems function and less concerned with reproducing procedure and specifications that are best obtained from OEM literature and training courses.

Shift Toward the Green *Diesel*

The focus of this book is the diesel engine and its hydromechanical and electronic fuel management systems. Because of the increasingly aggressive statutory noxious emissions controls required of commercial highway diesel engines and, more recently, their off-highway counterparts, diesel engines have undergone some radical changes in recent years. First the industry experienced a shift away from the **hydromechanical management** systems during the late 1980s. By the late 1990s, industry accepted that commercial diesel engines required management by computer, which we generally refer to as **electronic engine management**. Since 2004 the changes have been even more dramatic.

The drive toward producing a clean diesel engine has led to significant changes to the type of fuels used, more precise computerized control of combustion, and extensive exhaust gas aftertreatment devices. A very recent development has been a *green* state movement, led by ten states that have elected to adopt California's CARB standards: significantly, the ten states involved in this movement represent more than half of the gross national product (GNP) of the United States.

Today, the trucking industry has not only fully accepted computer controls on diesel engines but has become accustomed to a cycle of progressively tougher emissions controls every 3 or 4 years, not just to manage the diesel engine, but also to manage most other major chassis systems. Computer-controlled or electronically managed engines and other chassis systems are universal today because they fail less often, make vehicles much easier to drive, require less maintenance, and produce better fuel mileage. Trucks today use a technology known as **multiplexing** that allows the various chassis computers to network with each other. Multiplexing allows the engine, transmission, brakes, and other chassis systems to talk to each other and make possible hi-tech chassis features such as yaw control and antirollover electronics.

CARB and EPA

The **California Air Resources Board (CARB)** and **Environmental Protection Agency (EPA)** led the drive toward emissions control standards that made the polluting diesel engines of a generation ago a thing of the past. As each new set of emissions control standards is introduced, so also are new and exciting technological advances that make the diesel engine today a work in progress, and while the tendency may not be to embrace these changes, we have come to accept them. In anticipation of the introduction of **ultra-low sulfur (ULS)** fuel in 2007, engine manufacturers were reporting in 2004 that fuel economy could be impacted by up to 5%. By the time ULS legislation was effected in October 2006, engine performance technology had advanced so that the actual hit on fuel economy turned out to be 0% to 1% negative, depending on the application.

Objectives of this Book

This textbook attempts to make engine and fuel management technology easy to understand for the technician specializing in truck service and repair. As each year passes, the role of the truck technician changes with the technology. Technician skills today emphasize diagnostic abilities followed by a remove/replace procedure. That can be tough on novice technicians. Diagnosis is a culminating skill. A generation ago the individual responsible for diagnosis in a truck shop was widely experienced, usually with many years of field practice. In the real world of today's truck shop, the technician is required to perform less disassembly/reconditioning and much more troubleshooting and diagnosis. Most diagnosis of current diesel engines is performed using online **service information systems (SIS)** so the technician must possess first-rate computer skills. In fact, to get the most out of the approach this book uses toward diesel technology, it helps if the student has access to one of the SIS data hubs.

Core Skills

Computers are part of the way of life in the modern truck shop. They are used to diagnose engine malfunctions, program customer and proprietary data to engine control modules (ECMs), and track every aspect of working life in a business. A heavy emphasis is placed on computer literacy throughout this book. Change in the trucking industry is exponential. The specific systems addressed by this textbook will become obsolete within a relatively short time. However, the technical skill sets addressed in its pages should become the building blocks for a career that will necessarily have to adapt to change as each innovation is introduced. Figure 1–1 shows a 2008 Freightliner Cascadia with a late generation multiplexed chassis and Figure 1–2 is an overhead view of a post-2007, common rail fueled Cummins ISC. Note the clutter of electronic and emission control hardware on the ISC engine, the focal point of the learning objectives of this book.

ROLE OF THE TRUCKING INDUSTRY AND THE TRUCK TECHNICIAN

The trucking industry in North America grows by the year. In many ways, new truck sales provide a barometer of how the economy is doing in any given year. The saying, "If you got it, a truck brought it," is true for most of the consumer items we purchase. Even if that item was transported by train, boat, or plane for a portion of its journey, trucks would have played

Figure 1–1 *Freightliner Cascadia 2008 chassis.*

a role in pickup and delivery at stages of the journey.

Millions of people are directly and indirectly employed by the trucking industry. Trucks must be designed, built in factories, marketed, and then operated and maintained. The role of the technician in this industry is a small but crucial one. If all the truck technicians in the United States withheld their labor for 2 weeks, a large percentage of the economy would shut down. The trucking industry, besides employing mechanical technicians to maintain and repair equipment, employs drivers, dispatchers, warehouse personnel, people to market services, and people to manage operations.

Careers in Trucking

For some truck technicians, practicing as a technician will be just one of the career roles they play in the industry. A recent retiring **chief executive officer (CEO)** of a major truck OEM described his years with the company, beginning when he started as a janitor, apprenticed as a mechanic, practiced as a technician, and progressed with the company as a service manager, sales manager, and production operations manager. He moved through upper management and finally achieved the CEO position. While this is unlikely to be the objective of most aspiring truck technicians, a recent study indicated that within 5 years of achieving certification status, 50% of truck technicians no longer worked as hands-on technicians. Most were employed in other positions in the trucking industry. There are many opportunities in management, sales, training, field service operations, and information systems that the qualified truck technician can explore.

Many truck technicians have no desire to do anything but repair trucks. They will, over time, become the experts the industry needs to maintain the ever more sophisticated equipment on our highways. There are engine specialists in the truck garages of North America who know and understand a given engine model better than the engineers who designed

Figure 1–2 *Overhead view of a post-2007 common rail fueled Cummins ISC engine.*

it. Over a period of time they learn to diagnose every rattle and burp the engine can produce.

THE TRUCK TECHNICIAN OF THE TWENTY-FIRST CENTURY

For a technician to effectively work on modern electronically managed engines, considerable general experience, knowledge of mechanical principles, and product-specific training are required. Persons choosing a career as a truck technician in North America usually begin by attending a 1- or 2-year college program, followed by a period of hands-on training culminating in professional examinations conducted by the **National Institute for Automotive Service Excellence (ASE)**, the **Association of Diesel Specialists (ADS)**, or other licensing agencies. In Canada, most provinces mandate both the structure and duration of apprenticeship (3 to 5 years), and then license technicians by on-the-job standards of competency and their ability to pass certification examinations.

Cost-driven Change

In recent years, many labor-intensive procedures (such as the out-of-chassis overhaul of major components) have been moved out of the truck garage to remote and often offshore remanufacturing centers. Remanufacturing centers are often located in a jurisdiction where labor rates are low and the remanufacturing processes can be subdivided, allowing lesser skilled but specialist workers to perform most of the labor. This is a result of the service and repair industry becoming highly cost conscious.

A 2005 MIT study estimated that over 6% of the GNP of the United States (some $820 billion) was required annually to repair the damage caused by mechanical wear in general. Reducing this kind of dollar expenditure by fractions of a percent can represent massive savings. Every aspect of the repair procedure is analyzed for efficiency and cost-effectiveness. The truck engine technician of a generation ago who may have diagnosed, disassembled, reconditioned, and tested an engine is probably responsible today only for the diagnosis. That expert is followed by someone of lesser experience who will remove and replace the engine on a rebuilt/exchange basis. The rebuild facility will often be located in a jurisdiction where labor costs are lower, and, more often, this means in another country.

Telematics

Another factor that has powerfully influenced change in the trucking industry is the improvement in communications technology resulting in increasing use of **telematics**. Telematics is the long-distance transmission of data. Combinations of microwave and satellite telecommunications are used by many fleet operators to track vehicle location and communicate with onboard electronic systems. In other words, the various computer-controlled systems that are networked on the chassis are also communicating with remote devices.

Skills Required

What this tells us is that the changes the industry has undergone during the information technology (IT) revolution have produced a need for a different type of truck and a different kind of truck technician. The modern truck is a network of networks in itself. Chassis computers share data over a data backbone to optimize vehicle operation and make the vehicle safer. For instance, the response capability of the various system computers networked to a chassis data

Figure 1–3 *Chassis computer-controlled systems used on a modern truck. (Courtesy of Daimler Trucks North America)*

bus greatly exceeds that of the human driver in the event of a pending vehicle rollover. Figure 1–3 shows some of the many computer-controlled subsystems on a modern truck. Today's truck technician requires the following skills:

- Computer literacy
- Language comprehension accuracy
- Hands-on mechanical competence
- Ability to adapt to frequent technological change
- Ability to distill information in order to produce diagnoses

There is no doubt that the technician of today is required to know much more than the technician of a generation ago. A technician is also required to perform to higher levels of efficiency, competency, and ethics. The truck technician has always been expected to possess a wide variety of skills. Some of those skills are mechanical and incorporate many abilities that some might regard as specialties in their own right: machinist, plumber, electrician, welder, and truck driver, to name a few. And while we have already emphasized the technicians' requirement of sound computer skills to work with vehicle controls and office data management systems, the technician also plays a role in maintaining good customer relations and the basics of business management.

To Specialize or Not to Specialize

Ideally, the truck technician should have a broad base of training that accommodates at some point every vehicle system and component. It is a poor strategy to specialize in one vehicle system and to learn it to the exclusion of all others. Every system on a vehicle enables it to function and therefore is to some extent interdependent. This interdependence has increased with the introduction of data bus networking that optimizes the performance of all the chassis computer-controlled systems. Specialization is usually desirable only when the "specialist" has a broad base of experience and knowledge in all vehicle systems and then develops expertise in the subject of the specialty.

Changes in Types of Repair Work

The nature of the work in truck repair facilities is changing. The fuel injection pump shop specializing in the overhaul and testing of hydromechanical pumps has become nearly obsolete as far as the trucking industry is concerned. Truck and engine OEMs offer increasingly extended warranties. This means that fleets may employ technicians to perform little other than running repairs such as lighting repairs and vehicle servicing. The result is that proportionally a higher percentage of the major system repair work is performed by the OEM dealerships. This probably makes sense. The OEM dealership is more likely to have properly trained personnel and specialized equipment to perform the repairs. The per hour labor rates at such a facility are likely to be high, but then the financial investment in training and specialized tools is significant. Figure 1–4 shows a truck technician performing an emissions test.

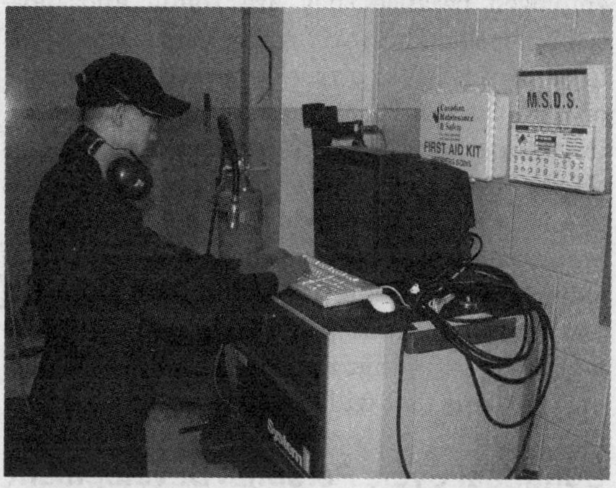

Figure 1–4 *Technician using software to conduct an emissions test.*

TRAINING AND THE TRUCK TECHNICIAN

Some of the training a truck technician requires will take place in college. However, much of the most valuable training is less formal. Often the best teachers are not found in classrooms, but are expert practitioners who take pride in helping others by sharing their expertise. Most people are born not knowing much about mechanical technology and, therefore, must learn. The apprentice technician will depend on those with greater experience and knowledge in order to learn. In most cases, knowledge cannot be hoarded or controlled.

The student who wishes to learn will usually find a means and will come to resent the individual who tries to block that goal.

It is often a measure of the health of a company to observe how it handles the learning and training of its employees. Sound organizations will develop a working atmosphere in which every person understands the benefits of a trained, knowledgeable colleague. Technicians who are disinclined to teach others what they have been taught themselves are usually motivated by insecurity. This type of insecurity is more often a fault of the employer rather than the individual, and it most often occurs in operations in which every person is somehow ranked according to their perceived ability and within which advancement is discouraged.

Many apprentice truck technicians are first employed because they are young and have strong backs. There are many heavy, dirty jobs to be performed in a truck garage, and there is no doubt that a 20-year-old can remove and reinstall a transmission in better average times than a 50-year-old. However, it is in the interest of any employer to develop the potential of apprentice technicians, from the employer point of view, to prepare for the future, and from the learning technicians' point of view, to give those persons a sense that they are gradually acquiring the skills that will result in expertise.

Continual Cycle Personal Development

A generation ago, the newly certified technician may have been able to consider himself or herself a finished product and exempt from further in-school training. This is not so today; the modern technician requires regular upgrading as the technology changes. One major truck OEM uses the Japanese word ***kaizan***, meaning "continuous improvement," to describe its corporate philosophy, and technicians would do well to adopt kaizan as their motto. The learning process for anyone working with today's technology is a lifelong one. Learning for a truck technician is seldom a process confined to classrooms, although this is an important element of it. The learning process for a technician really consists of three stages: formal, self-development, and experience.

Formal Training

Formal training involves some study in classrooms and is structured to produce a result such as knowledge of a practice or procedure, backed by certification or a diploma. Most technicians are required to successfully complete some formal training before they are certified to practice. Additionally, depending on the specialties of their employer's operation, technicians should attend formal upgrading on a regular basis. Formal upgrading courses are usually designed to address some very specific subject matter or procedures and usually last from 1 to 5 days; again, they often culminate in testing and the awarding of a certificate.

Self-Development

Self-driven training has to come from within, so many overlook it. Smart technicians read. It is a good idea to make reading a habit. Reading trade magazines is an excellent means of keeping up to date; most truck shops make these available to their technicians but few take advantage of them. There are many excellent publications, some of which cover very specific subject matter, while others cover the industry in general. Some recommended publications include:

TMC Publications The best way to receive American Trucking Association (ATA) **Technical and Maintenance Council (TMC)** publications is to join as a technician member. The TMC publishes bulletins such as the *TMC Fleet Advisor* and other technical periodicals. Once a year the TMC publishes a compendium of **Recommended Practices (RPs)** titled *Recommended Maintenance Practices Manual.* This is an essential reference for truck technicians because it benchmarks standards of accepted practice in the industry. Coverage is bumper to bumper. Check out the TMC at http://tmc.truckline.com. The address is:

American Trucking Association
Technology and Maintenance Council
950 N. Glebe Road, Suite 210
Arlington, VA, 22203-4181
Phone: (703) 838-1763

Automotive Engineering Published monthly by the **Society of Automotive Engineers (SAE)** and sent to all of its members, *Automotive Engineering* covers the whole automotive spectrum from an engineering/technological perspective focusing on innovation. Most innovations in the truck, bus, and heavy equipment industries are covered.

Automotive News *Automotive News* covers the whole automotive industry (including trucks), mostly from the sales perspective. It tracks the monthly and yearly sales data for all vehicles, including each category of truck by class size.

Commercial Carrier Journal (CCJ) *Commercial Carrier Journal* is published monthly and covers the trucking industry from the service and maintenance perspective. It is highly recommended as a means of keeping up to date with the changes in truck technology as they happen.

Diesel Progress *Diesel Progress* covers the diesel engine industry from the smallest to the largest engines. It is a great magazine for the engine and fuels specialist. Most new diesel engine technologies are profiled in *Diesel Progress* ahead of production. It is one of the best ways of keeping informed of upcoming emissions legislation, buying trends, and emerging technology.

Fleet Advisor *Fleet Advisor* is the TMC monthly newsletter that covers a broad range of issues relevant to the trucking industry. It publishes reports from the truck driver, technical, legal, and business perspectives of the industry. It announces the Recommended Practices (RPs) under discussion and invites participation from TMC members.

OEM Service Information Systems

OEM service literature is written with the objective of sequencing a procedure and seldom makes reading entertaining. However, many OEMs have their own in-house magazines to keep their employees up to date, and the technical service bulletins (TSBs) of today are usually presented in a more readable format than they were 10 years ago because OEMs realize their importance as an information tool. Many OEMs also provide learning packages deliverable in either CD-ROM or online formats. These are usually interactive and often are directed toward in-house certification at their conclusion. Often self-directed learning packages are integrated with the OEM online service information systems so technicians can certify on a specific system and learn at whatever pace they feel like. Technicians should never forget that self-development tends to be rewarding and has a way of stimulating interest. Those who do not bother, stagnate.

Experience

For the technician, there is no substitute for the hands-on experience of tackling a procedure. Employers should ensure that after investing in sending a technician to a 1-week course training for a specific procedure that the technician be given the opportunity to practice the newly learned skills at the first opportunity. If 6 months slip by between the training course and the hands-on application that the techniques target, much of the learned content will have been forgotten.

Be an Observer

Apprentice technicians should make a practice of using their eyes and ears a lot. They should watch successful technicians and learn from them: how they organize their work area, how they strategize each task they undertake, how they test the results of a repair procedure, and how they approach and deal with customers. Novice truck technicians have a steep learning curve to adjust to and during the process it will be important to learn how to think. There is little room for impulse, guesswork, or carelessness in working with today's technology, and the consequences of a mistake can be very costly.

Policy Adjustments

It would take a self-righteous truck technician indeed to claim that he or she had never made a mistake. Errors do happen on the truck shop floor. While making mistakes may be part of life, in service facilities it is essential that they be limited. Mistakes are costly to business practice and to the shop's reputation. They are known as **policy adjustments** by those administrating billing. When you are unfortunate enough to make a mistake, regard it as a learning experience that should stick in your mind, never to be repeated. Technicians who make frequent mistakes will probably be required to seek employment in a different industry.

CUSTOMER SERVICE AND PROFESSIONALISM

A generation ago, few truck diesel technicians had to concern themselves with what is today known as customer service. Most large operations had personnel on their payroll to deal with customer service issues and, in fact, many technicians worked with the expectation that the customer adapt to their way of working and certainly never the other way around. The catchphrase back then was *take it or leave it*. Many service facilities had signs directed at customers expressly forbidding them access to the working area of service garages, usually citing insurance regulations. Things have changed.

Edward Demmings

The pioneer of modern customer service philosophy was an American, Edward Demmings. Demmings's ideas on customer service initially met with little acceptance in North America, but he had an easier time convincing the Japanese. In his lifetime, Edward Demmings acquired almost heroic status in Japan. Japanese companies adopted his teachings and used them to step into the world arena and establish Japan as a major economic force second only to the United States. In applying Demmings's teachings to the production technology and customer service, the Japanese auto industry has, over the years, established itself as being the best positioned for world dominance over the foreseeable future.

Total Quality Management

Edward Demmings is credited with coining the phrase **total quality management (TQM)**. TQM is claimed to be used by many companies today. It is best described as a whole corporate philosophy that views customer service as being the very reason for that company being in business and includes every employee as a stakeholder. In other words, every person within a corporation is regarded as a member of a team, and every member of the team contributes significantly to the image that the organization projects. The notion that every person in a company had an essential role to play was alien to many North American and European organizations in the 1970s and 1980s, which maintained strictly hierarchical structures in which those at the base of the pyramid tended to be little valued. The Japanese exploited this and essentially, among their many imports to North America, reimported the teachings of Edward Demmings.

It should be noted that more than half of the main players in the truck manufacturing industry claim to practice some form of TQM, including the company with the largest market share, which has set many of the trends of the industry over the past two decades, forcing the rest to either keep up or lose market share. TQM has been dismissed as a trend by some and indeed many companies have treated Demmings's teachings as a sort of smorgasbord, picking and choosing those items that they feel might result in an increase in business profitability, and perhaps omitting those items that they feel might cost them money. This is a case of missing the point. But TQM has influenced the way in which all companies do business.

Kaizan

We first used this Japanese word a little earlier in the chapter to describe a model of continuous personal self-improvement. But Japanese manufacturing has used this "continuous improvement" model on a much broader scale. Imported Japanese automobiles were often derided during the 1960s with accusations of low quality, probably with some justification. A gradual reversal took place during the 1970s and into the 1980s. Japanese manufacturers built a solid base by listening to customer concerns and constantly adjusting to them. Listening to and responding to customer complaints laid down the building blocks of the modern Japanese auto industry, based around kaizan production cycles.

Today, we more often than not associate Japanese manufactured products with good quality backed by excellent customer service support. It matters little to the buying public that much of this Japanese product is now manufactured in America by American workers, and as a nation we have come to associate the product of Japanese companies as being superior because of their emphasis on customer service and quality.

The 24/7 Operation

Many truck facilities are offering extended service hours—often at 24-hour, 7-day-a-week facilities. As soon as one operation offers this type of service in a given area, others must too if they are to effectively compete. In addition, service operations are recognizing the crucial role played by their technicians in delivering customer service. Many are training their personnel, offering seminars with titles such as "How to Communicate with Customers" and "Handling the Difficult Customer." Operations that used to discourage customers from even entering the service garage are now welcoming the customer, encouraging them to talk to the technician performing the work, and providing them with waiting facilities more in keeping with a hotel lobby than with what used to be associated with a truck service garage.

Honesty

The modern truck shop makes for a working environment that can be difficult for the novice technician. A truck owner/operator with 20 years in the business is likely to have learned a thing or two about the technology he operates and can easily intimidate a first-year apprentice truck technician. It is essential to underline the fact that one of the core premises of TQM is honesty. Technicians will find that in dealing with customers, the relationship will always be made easier when the approach is open and sincere. Until recently, two of the worst traits of the automotive garage were the fact that many customers felt that either they were being talked down to or that their repairs were misrepresented either to inflate a bill or disguise a lack of competence in diagnosing a problem. Using an honest approach will probably not reduce the number of policy adjusted (write-off time) work orders, but it will have the effect of increasing the number of repeat customers.

Professionalism

All technicians should develop a sense of professionalism. You will never witness a doctor or dentist criticizing the handiwork of a colleague who previously worked on the body or mouth in question although there are surely cases where there is justification to do just that. Unfortunately, it is all too easy to hear a technician deriding the abilities of whoever previously worked on a piece of equipment. This is simply unprofessional and reflects as badly on the person doing the criticizing as the person supposedly at fault. Actually, it reflects badly on the whole industry. Mistakes happen. Remember that what goes around, comes around. Accept that and the fact that the very best technicians can have bad days, commit errors, and have lapses in concentration. Technicians should make a practice of learning from their own mistakes and those of their colleagues and actively work to improve the level of professionalism in the industry.

Six Sigma

If you have made a visit to any Caterpillar or Cummins dealership recently, you will have noticed more than a few posters advocating Six Sigma practices. **Six Sigma** was developed by Motorola Corporation as a corporate means of eliminating defects. A "defect" is defined as a "unit that is not a member of the intended population," so it is not entirely clear whether a defect might include employees who fail to pull their weight. The concept was pioneered by Bill Smith at Motorola in 1986 and since its introduction, Motorola claimed to have saved over $17 billion. Six Sigma has become an element of many companies' production improvement initiatives including Honeywell (Allied Signal), Raytheon, Cummins, and Caterpillar.

Key Concepts of Six Sigma

Six Sigma identifies the following six key concepts:

- Quality: the attributes a customer expects from a product.
- Defect: failure to deliver what a customer wants
- Process capability: what a company's systems and processes can deliver
- Variation: what a customer sees and feels about a product
- Stability: process of managing consistent, predictable processes to improve what a customer sees and feels
- Design for Six Sigma: adapting to meet customer needs (external) and process capability (internal)

Caterpillar practices Six Sigma in all its manufacturing facilities. It borrows from martial arts the color of belts to represent status: for instance, master black belts, black belts, green belts, and yellow belts represent diminishing levels of status within the Six Sigma hierarchy. Master black belts devote all their time to Six Sigma analysis and promotion, black belts report to the master black belts, while at a lower level, yellow belts are mere rookies into the system.

Fundamentally, Six Sigma is about teamwork, as is TQM. Six Sigma stratifies the team a little more than TQM. However, if employees can be motivated to buy into the program, it has been proven to produce improvement, so the key is establishing employee motivation. As a society we embrace trends and catchphrases and maybe with good reason: any program that promises some change for the better is likely to result in some improvement, even if short-lived. More recently Six Sigma has been integrated with the TRIZ methodology for problem solving and product design.

TRIZ

TRIZ comes from Russia. This is another much used term in the workplace these days and if your employer suddenly installs an employee suggestion box and then sincerely attempts to get you to use it, it may be due to TRIZ. The acronym **TRIZ** is sourced from the Russian phrase for "theory of inventive problem solving." The fundamental principles of TRIZ tell us that:

- If you have a problem, somebody someplace has already solved it (or one similar): being inventive means finding that solution and adapting it to your problem.
- Compromises should not be accepted because they reduce effectiveness: in other words, compromises should be eliminated where possible, a concept that might be difficult to accept in North America where compromise is the norm.

Customer Service Summary

Throughout a career, most truck technicians will have to participate in both small and big ways in customer service initiatives. You might feel like dismissing some of these as trends that will die a natural death. The core truth is that as a technician you will be, in effect, selling your expertise in the marketplace. And if you are selling anything, you need a buyer. The salespersons and companies that are selling your skills probably regard their skills in marketing just as highly as you might regard your technical skills. So when it comes to understanding and practicing customer service you have no choice. The best mechanical technician in the world cannot survive in the marketplace if no one is willing to pay for those skills.

DEFINING ENGINES BY DISPLACEMENT

In this textbook our objective is to primarily focus on highway engines used in truck and bus chassis. However, many of the engines referenced are used in a range of other applications, which include:

- Off-highway mechanical drive mobile
- Off-highway hydrostatic drive mobile
- Marine motive power
- Land-based genset
- Marine genset
- Land-based auxiliary power

Because this textbook is used in programs that do not address highway equipment, it is necessary to define terms such as **small bore**, **medium bore**, and **large bore** when referring to a diesel engine so that no confusion results. It should be stressed that this is a loose definition and each OEM will probably have opinions that vary somewhat.

- Small bore: diesel engines with displacements between 5.9 and 8 liters
- Medium bore: diesel engines with displacements between 8 and 12 liters
- Large bore: diesel engines with displacements between 12 and 16 liters

In reality these definitions only apply to the way we refer to engine displacement in the trucking industry. If our perspective was that of a railway locomotive diesel engine specialist, then a 16-liter engine would be considered small bore. It should also be recognized that as diesel engine technology develops, especially as it applies to power over weight, in many cases the small bore engine (by our definition) is often taking on tasks that required a medium bore engine a generation ago.

REVIEW EXERCISES

1. Log on to the Internet and search http://www.asecert.org, http://www.diesel.net, and http://www.dieselnet.com. Explore the Web pages of some of the major truck engine OEMs. Select a company that has an e-mail query page and solicit information on one of its products.
2. Log onto the TMC Website and bookmark it into your favorites.
3. Find out what it takes to become a TMC technician member. If you qualify, consider joining.
4. Contact the ASE online or by mail and request electronic or hard-copy information on its testing criteria and locations.
5. Investigate the role of the trucking industry in your community. Contact a local trucking association and request data on the business activity of its members.
6. Obtain copies of *Diesel Progress,* the *CCJ,* and other trade publications. Remember that many publications make electronic versions available and accessible via the Internet.
7. Contact some OEMs in the automotive and truck manufacturing fields and ask them about their customer service philosophy. A lack of response can sometimes be as informative as a response.
8. Check out the ATA Website at http://www.trucking.org. Bookmark this and http://www.dieselnet.com into your Favorites file and consult weekly. Do the same with at least a couple of the diesel engine OEMs.

2

Hand and Shop Tools, Precision Tools, and Units of Measurement

INTRODUCTION

This chapter is intended to provide a guide to tools for novice truck technicians. The tools are loosely divided into the categories of hand tools, precision measuring, and shop tools (Figure 2–1).

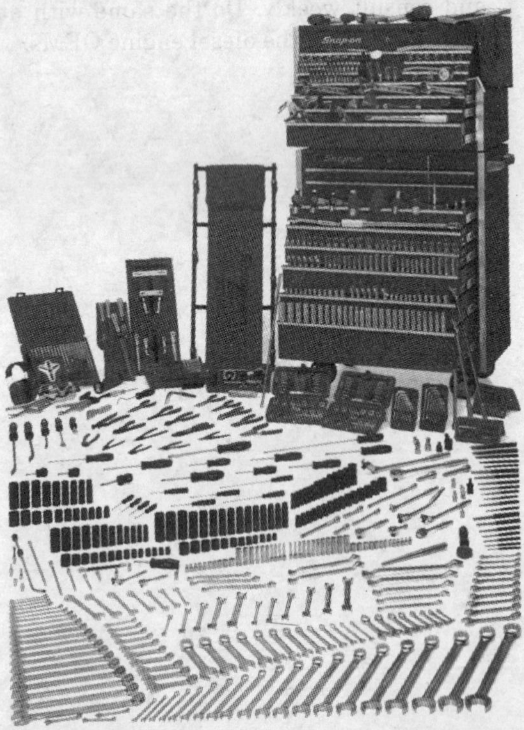

Figure 2–1 *A complete tool kit. (Courtesy of Snap-on Tools Corporation)*

There is also a guide to the contents of a truck technician's toolbox; however, the rookie technician should invest in a minimum of tools before obtaining employment and then develop a tool collection with the job requirements in mind. Some guidance in standard to metric conversion units is also provided. The modern truck technician is usually expected to work on engines engineered on metric systems using specifications generally presented in standard or English values, so it helps to be familiar with both systems.

HAND TOOLS

The technician will need to possess a basic set of hand tools and while some basic guidelines are provided later in this chapter, the tools selected and their quality will be largely determined by the nature of the work. Hand tools vary considerably in price and before spending large sums of money, the technician should determine whether the expenditure is justified by the amount of use they will get. Most better quality hand tools carry lifetime warranties; however, this may not cover a tool that wears out, so it is good to question the extent of the warranty offered. Diesel technicians seldom wear out tools and the main problem is usually loss. Because of the high price of hand tools, most technicians learn to check the contents of their toolboxes carefully after completing each job. Most loss of tools is usually the result of carelessness on the part of the technician. Thousands of wrenches are lost every week because they are left on a truck, bus, or car chassis.

Open-end Wrenches

Open-end wrenches have open jaws on either side of the wrench, usually with different sizes at either end and slightly offset (Figure 2–2A). The wrench should be of sufficient quality that the jaws do not spread when force is applied and that the jaws are not made too bulky to restrict access to difficult-to-get-at fasteners. They may damage softer fasteners (such as brass pipe nuts) because their design permits them to impart force to only two of the six flats of a hex nut.

Combination Wrenches

Most technicians usually own a couple of sets of combination wrenches including the common sizes. A combination wrench is manufactured with a box end and an open end, both of the same nominal size (Figure 2–2B and C). Most truck diesel technicians should only consider better quality combination wrenches that are guaranteed for life. There exists a wide range of prices between lower quality wrenches that tend to be heavy and clumsy (but often as strong as the best quality) and the best quality, which are light and smooth to handle. It is useful to own a set of top quality combination wrenches in sizes up to 3/4″ (19 mm) because these are able to access fasteners on engines that heavier, clumsier wrenches do not access. Less expensive but fully warranted wrenches can be considered for use on sizes larger than 3/4″ (19 mm). There are many cheaper, poor quality wrenches on the market, many of them imported. These are seldom guaranteed. Poor quality wrenches are dangerous and should be used by no technician, whether amateur or professional.

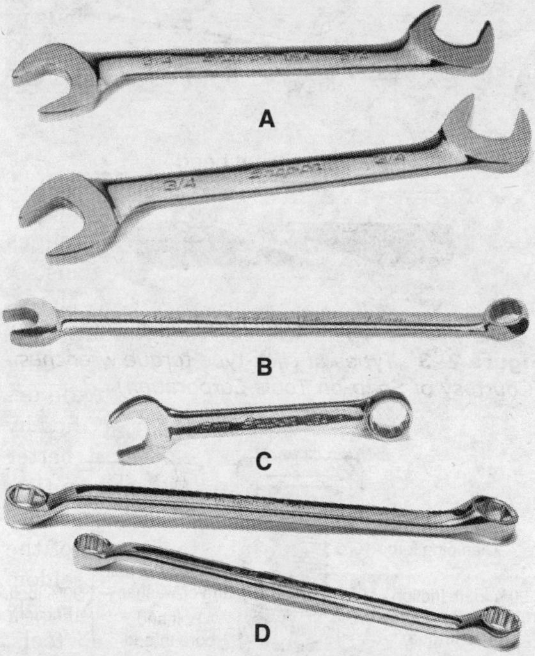

Figure 2–2 *Wrenches: A. Open-end; B. and C. Combination; D. Box-end. (Courtesy of Snap-on Tools Corporation)*

Box-end Wrenches

A box-end wrench surrounds the fastener and may be of hexagonal or double-hexagonal design (Figure 2–2D). As most bolts and nuts are of a hexagonal design, the hex box-end wrench will grip more securely; however, it will be less versatile where access is restricted as it can only fit on the fastener in 6 radial positions through a rotation rather than the 12 radial positions of the double-hex, box-end wrench.

Adjustable Wrenches

The adjustable wrench consists of a rigid jaw integral with the handle and an adjustable jaw moved by a worm adjuster screw. The truck technician should probably own a couple of these and then resolve to use them as little as possible. Their advantage is versatility and their ability to sometimes grip a worn fastener. Their disadvantage is that they cause wear because the adjustable jaw never fits tightly to the flats on a hex fastener and it tends to round them out. Never apply excessive force to an adjustable wrench.

Line Wrenches

A line wrench is designed to grip to a pipe or line hex nut and act on four of the six flats of its hex. It has the appearance of a box-end wrench with a small section removed so that it fits through the pipe to enclose the pipe nut. The line wrench should be used in place of the open-end wrench to avoid damaging pipe nuts.

Socket Wrenches

Diesel technicians will require complete socket sets in ¼″, ⅜″, and ½″ drive sizes and may consider a ¾″ drive. What constitutes a complete set of sockets varies a little by manufacturer but typically ¼″ drive sets are provided with sockets up to ½″ (12 mm), ⅜″ drives up to ¾″ (19 mm), and ½″ drives up to 1¼″ (30 mm). Sockets may be of the hex or double-hex design and enclose the fastener. The socket may be hand rotated by a ratchet or flex bar and power rotated by a pneumatically powered wrench or impact wrench.

Impact sockets are manufactured of softer alloys than those designed to be driven manually to prevent fracture. Technicians should purchase good quality sockets because the consequence of a failed socket is personal injury. Deep sockets permit access to a nut in which a greater length of the bolt or stud is exposed. A crowsfoot socket is essentially an open-end wrench that can be turned by a ratchet; it grips two of the six flats of a nut and is probably mostly used for final torquing a difficult-to-access nut. A line socket is the socket counterpart to the line wrench. It grips four of the six flats of a nut and its main use is to deliver final torque to a pipe nut.

Ratchets and Breaker/Flex Bars

Reversible ratchets used in conjunction with sockets are tools much used by any technician. They are used to rapidly turn fasteners by hand and should be of good quality because the consequence of failure is personal injury. However, they are not designed to accommodate high torque loads. The ratchet spur wheel is locked to one direction of rotation by a single or double cog. The spur and cog cannot be observed because they are enclosed in the ratchet head, but this determines the ultimate strength of the tool. A breaker bar (also known as a flex bar, power bar, and Johnson bar) has a grip bar and pivoting drive square to engage with a socket in the same way a ratchet does, so it is available in ¼″, ⅜″, ½″, ¾″, and 1″ sizes. A breaker/flex bar can be used to release fasteners that require considerably more force than could be safely applied to a ratchet; however, the use of "helpers" such as a pipe over the handle should be avoided.

Torque Wrenches

Torque wrenches measure resistance to turning effort (Figure 2–3). The objective of torquing fasteners is to ensure that a specified clamping force between two components is achieved. However, an estimated 90% of the applied force to a torque wrench is required to overcome the friction of the fastener thread surface area, with about 10% contributing to clamping force (Figure 2–4). Most technicians possess at least one torque wrench and probably do not use it as much as they should. In assembling engine and fuel system components, every fastener should be torqued to specification. Studies indicate that when technicians fail to use torque wrenches, they overtorque

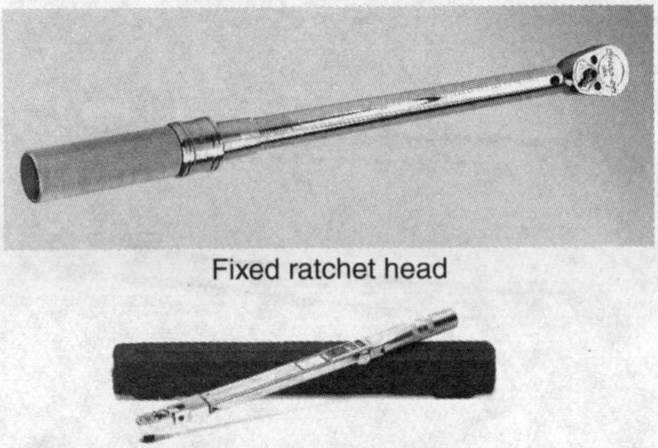

Figure 2–3 *Types of click-type torque wrenches. (Courtesy of Snap-on Tools Corporation)*

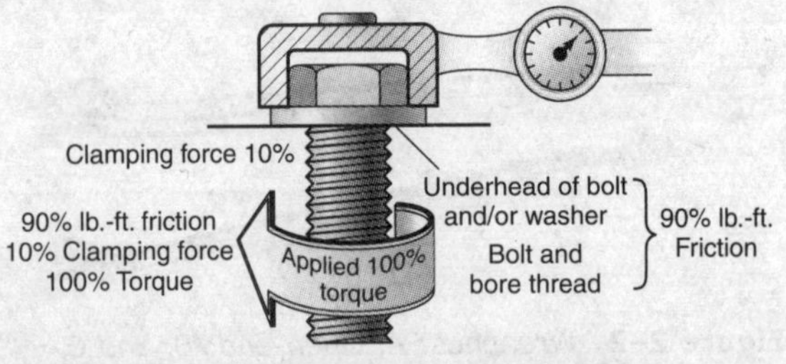

Figure 2–4 *Torque to overcome friction. (Courtesy of Navistar International Corp.)*

fasteners to values 50% to 100% over the specification. This action can damage fasteners and distort components, including cylinder blocks and heads. The commonly used torque wrench is the sensory or click type: when the selected torque value is attained, the wrench produces an audible click. Diesel technicians are often required to torque large numbers of fasteners to the same specification such as when torquing cylinder heads to a cylinder block; it makes sense to use the click-type torque wrench when performing this type of procedure.

Stiction

Stiction is stationary friction. Referencing Figure 2-4, the student can see that starting to turn a bolt requires a lot more force than keeping it turning. The tighter the bolt is torqued, the greater the stiction, and this can affect torque readings. Lubricating bolts can help but the type of lubricant used is important. This means that the lube suggested by the service literature should always be used. This can range from nothing at all to engine lube to locking adhesives. Failure to observe this can result in either high or low clamping pressures.

Types of Torque Wrenches

Click-type torque wrenches should always be backed off to a zero reading after use and their calibration should be routinely checked. Dial-type torque wrenches have circular dial scales with a needle that indicates the applied torque value. These tend to be more useful when torquing sequences of fasteners at different torque values; dial-type torque wrenches are usually higher priced than click-type torque wrenches. Beam-type torque wrenches use a flexible, middle alloy steel shaft (beam) that deflects when torque is applied: a needle pointer indicates the applied torque effort. These tend to be the most economical but have good accuracy and seldom require calibrating. They should be stored carefully as the needle pointer is vulnerable.

Technicians should probably own a ³/₈″ and ½″ drive torque wrench of the sensory or click type. The torque wrench calibration should be checked annually with a torque wrench tester. Torque specifications are provided by OEMs in order to obtain reasonably consistent clamping pressures between mated components. Torque values for fuel injection and engine component fasteners are usually factored for lubricated threads.

Torque Wrench Extensions

Torque wrenches are mainly used with sockets, line sockets, or crowsfoot jaw wrenches. Torque readings are a product of the length of the handle measured in feet or inches and the force applied to the handle. For example, a 2-foot (ft.)-long torque wrench when applied with 10 pounds (lb) of force would produce 20 lb-ft. of torque. To express any torque value specified in lb-ft. in lb-in. simply multiply by 12. So 20 lb-ft. is equivalent to 240 lb-in. When a standard extension (one that is at right angles to the plane of the wrench) is used with a torque wrench, no torque reading adjustment is required. If a linear extension (one on the same linear plane as the wrench) is used, the mechanical advantage is increased, and the torque reading should be adjusted using the following formula:

$$TS = \frac{T \times L^1}{(L^1 + L^2)}$$

T = torque in lb-ft.

TS = torque scale in lb

L^1 = torque wrench frame length in inches

L^2 = torque wrench extension length in inches

So if a torque specification is 50 lb-ft. and an 18″ torque wrench is to be used in conjunction with a 12″ linear extension, the following calculation should be used to determine what the correct reading on the torque wrench scale should be:

$$TS = \frac{T \times L^1}{(L^1 + L^2)} = \frac{50 \times 18}{(18 + 12)}$$

$$TS = \frac{900}{30}$$

TS = 30 lb-ft.

Torque wrenches may be calibrated in the standard or metric systems: pound-feet or Newton-meters.

1 lb-ft. = 1.356 Newton-meters

1 Newton-meter = 0.7375 lb-ft.

Torque-to-Yield Bolts

Torque-to-yield bolts are designed to stretch to their yield point as they are tightened. They are often used where more precise clamping load forces are required so they are typically seen on cylinder heads, connecting rods, and crankshaft main bearings. As a torque-to-yield bolt is tightened it is stretched to its elastic limit so it is designed for one-off use. When service literature on a repair procedure indicates that a bolt should be discarded and replaced with a new one, usually a torque-to yield fastener is used.

Tightening Torque-to-Yield Fasteners

Tightening torque-to-yield fasteners is usually performed progressively using a combination of torque and turning angle. The first step is to torque the fastener or set of fasteners to a low specification so that an even clamping load is applied to the mating components. This is usually followed by a second pass at an incrementally higher torque value. After this, each fastener is turned a specified number of degrees measured with a protractor or template placed over the fastener hex: this may occur in two or three additional steps. For example, each fastener might be turned 90 degrees, followed by an additional 45 degrees. Torque-to-yield ensures a more accurate and even clamping force. The tools used for torque-to-yield fasteners are:

- Protractors with a movable pointer
- Electronic (they measure the turn angle with more precision)

Hammers

Mechanical technicians mostly use ball peen hammers in various different weights. The specified weight of a hammer is the head weight, which starts at ½ lb and should go up to about 4 lb in weight. While there is no place for the carpenter's claw hammer in the technician's toolbox, a 4 lb cross peen hammer can be a useful addition, and the engine specialist should also own a 5 lb rubber (dead blow) mallet and a couple of soft-faced or fiberglass hammers. Safety glasses should be worn whenever using any striking tool. The impact faces of hammers should be inspected regularly and discarded when the face becomes damaged. Properly heat-treated hammers should possess highest hardness at the contact face and be relatively soft behind it to buffer the shock loads. Hammer handles are also important, and a hammer should not be used when its handle is damaged. The handle may be made of hickory, in which case it is susceptible to damage, or steel and integral with the head with a rubber-cushioned grip. Some examples are shown in Figure 2–5A and B.

Pliers

Most technicians will require a large selection of pliers. They are used for gripping and cutting. Pliers used for working on electrical circuits should have insulated handles. Each type of pliers is named, and some examples are needle nose, slip joint, lineman, and side-cutter. Figure 2–5C shows some examples.

Screw Extractors

Fasteners occasionally fail when the fastener head shears. When methods such as welding a nut onto a fastener that has had its hex head sheared off have failed, a screw extractor must be used. First, the fastener must be drilled as centrally as possible to 75% of its depth and half its diameter, then the appropriately sized screw extractor must be inserted. Two types exist: the taper square screw extractor is designed to bite into and grab the bore of the drilled hole; it can be progressively driven into the hole if the edges round out. The left-hand twist screw extractor works by tapping its way into the drilled hole in the fastener as it is turned counterclockwise.

Stud Extractors

Stud extractors may be used to extract a fastener only when not sheared flush. Two types are used: the collet type fits over the exposed length of the stud and locks to it as it is rotated counterclockwise. The wedge type consists of a splined circular wedge that locks to the stud as it is rotated. Wedge-type stud extractors should only be used with hand tools as they are driven eccentrically.

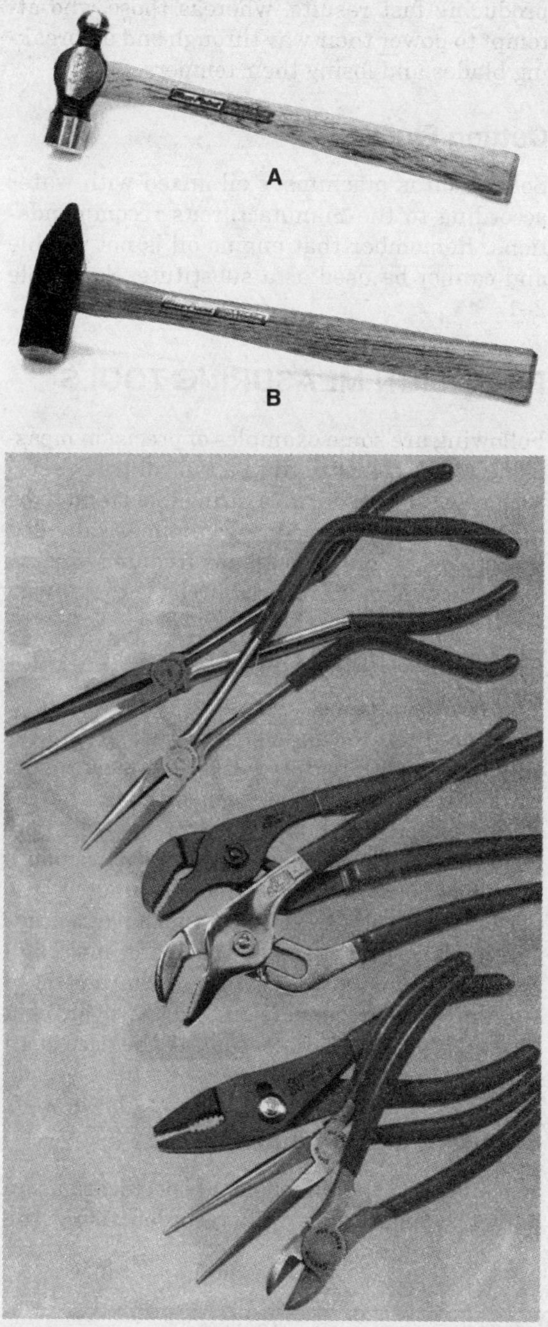

Figure 2–5 *Types of hammers and pliers: A. Ball peen hammer; B. Cross peen hammer; C. Selection of different types of pliers. (Courtesy of Snap-on Tools Corporation)*

Taps and Dies

Taps and dies are designed to cut threads in both standard and metric specifications. Taps cut internal threads in bores; quite often the technician will use a tap to repair damaged threads. Three types are used:

1. Taper tap: used to cut threads to a virgin bore
2. Plug tap: used to finish cut or repair threads
3. Bottom tap: used to cut the final threads in a blind hole

Dies cut external threads. Most are designed with graduated teeth and a taper, enabling them to cut threads to a shaft. It is important to use an appropriate cutting medium. Again, technicians are more likely to use these to repair damaged threads. In fact, most toolbox quality taps and dies should not be used to cut virgin threads in hardened steels, especially if the fastener is critical.

Thread Chasers

A thread chaser is a die designed for the sole purpose of repairing minor damage to an existing thread—it cuts in much the same way a die does, but it is not designed to cut new threads.

Thread Pitch Gauges

Thread pitch gauges are designed to measure the thread pitch (angle) and number of threads per inch (tpi) of any thread pattern. They are used to determine which tap, die, or fastener is appropriate.

Reamers

Reamers are rotary cutting tools that enlarge an existing hole to an exact dimension. They are used where a greater degree of accuracy is required than a drill is capable of delivering. Reamers are available in several types: adjustable reamers can be set to cut holes through a range of dimensions, and spiral fluted, taper reamers are designed to be driven by air or electric power tools and enlarge a hole accurately to a specific size.

Drill Bits

Drill bits are driven by an electric or pneumatic power tool. Twist drills must be machined to the correct cutting pitch (angle) normally by grinding on a fine abrasive wheel. The point angle of a drill bit is the combined angle of the dressed cutting edges, usually about 120 degrees for drilling most steels and cast irons. This would produce a cutting pitch of about 30 degrees. The point angle is increased when harder steels must be drilled. The two cutting edges of the twist drill must be angled identically from the center point of the drill and have the same radial dimension; a slight difference will result in a larger hole than the drill shank specification. Drills sharpened by hand seldom produce exactly sized holes. Drill speeds should be adjusted for the material being cut and the size of the hole.

Hacksaws

Most technicians will be required to use a hacksaw from time to time. A hacksaw is designed specifically to cut metals. A hacksaw should have a rigid frame and the blade selected should have the appropriate number of teeth per inch (tpi) for the metal to be cut; generally, the harder the metal, the more tpi required. Metal thickness is also a factor and a rule of thumb is that at least two teeth must be in contact with the material to be cut. Better quality hacksaw blades tend to be cost effective as they last until they wear out, while cheaper blades tend to break. Inspect the blades and replace them when dulled or missing teeth are evident. A hacksaw should be used with a light but firm grip. An even horizontal stroke with no rocking will produce the fastest cutting rates. A relaxed, calm approach to the cutting task has a way of producing fast results, whereas those who attempt to power their way through end up breaking blades and losing their tempers.

Cutting Fluids

Soluble oil is machinist's oil mixed with water according to the manufacturer's recommendations. Remember that engine oil is not soluble and cannot be used as a substitute. See Table 2–1.

PRECISION MEASURING TOOLS

Following are some examples of precision measuring instruments used in a typical truck service garage. Precision measuring tools tend to be high-cost items. They are sometimes provided by employers, but technicians frequently using a specific tool may wish to purchase their own.

Electronic Digital Calipers

Electronic digital calipers (EDCs) are a great addition to the toolbox. EDCs are an invaluable tool for the diesel technician who works extensively in the engines and fuels area. These dimensional measuring tools will perform inside, outside, and depth measurements to half a thousandth or 0.0005″ accuracy. EDCs perform with good accuracy and have the advantage of being easier to read than micrometers. Additionally, they will perform metric to standard linear conversions at the push of a button. Figure 2–6 shows some of the uses of an EDC.

Standard Micrometers

Since most of the dimensional specifications on a metric engine are still recorded using the

Table 2–1 *Recommended Cutting Fluid Processes*

Metal	*Drilling*	*Reaming*	*Tapping*
Mild steel	Soluble oil/water	Lard	Lard
Middle alloy	Soluble oil/water	Lard	Lard
Cast iron	Dry	Dry	Lard
Brass	Dry	Water	Dry
Copper	Soluble oil/water	Soluble oil/water	Soluble oil/water
Aluminum	Kerosene	Kerosene	Soluble oil/water

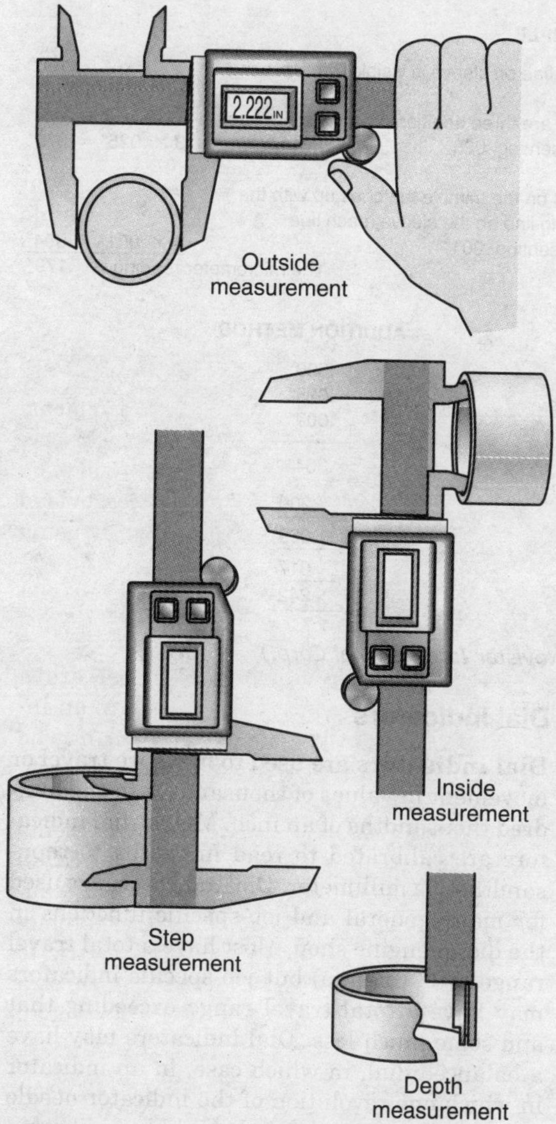

Figure 2–6 *Various uses of a set of digital calipers.*

English system, every technician must know how to read a standard micrometer. A standard **outside micrometer** consists of a frame in the shape of the letter *G*: on one side of the frame is a fixed anvil and on the other is a spindle assembly. The spindle assembly consists of an accurately machined screw that rotates in the spindle nut. The spindle is rotated by a thimble: the point at which the thimble registers on the micrometer sleeve calibration scale indicates the reading.

From the point at which the spindle contacts the anvil and the micrometer dimensional readout is zero, the thimble must be rotated through 40 complete revolutions to produce a reading of 1 inch. This means that the inch is divided by 40. Therefore, each complete rotation of the thimble equals:

$$\frac{1.00}{40}\text{ inch} = 0.025''$$

So each full rotation of the thimble is equivalent to 0.025″. The calibration scale on the sleeve is calibrated in units of 0.025″ and every fourth stroke equals 0.100″, which is usually indicated by single digits 1 through 10. The leading edge of the thimble is divided into 25 calibration strokes, each representing 0.001″. Understanding exactly how the micrometer is calibrated enables the technician to read it easily. All standard micrometers measure dimensions through 1 inch and are read in the same manner. Changing the frame and anvil assembly permits the micrometer to be used for measuring dimensions larger than 1 inch but the readings are still confined to the 1-inch window of spindle travel. An outside micrometer is used to measure the outside dimensions of a component such as a shaft or set of shims, whereas an **inside micrometer** is used to measure internally such as a bore dimension. Figure 2–7 shows both outside and inside micrometers and the terminology used to describe them. Study Figure 2–8. This shows how to read a standard micrometer.

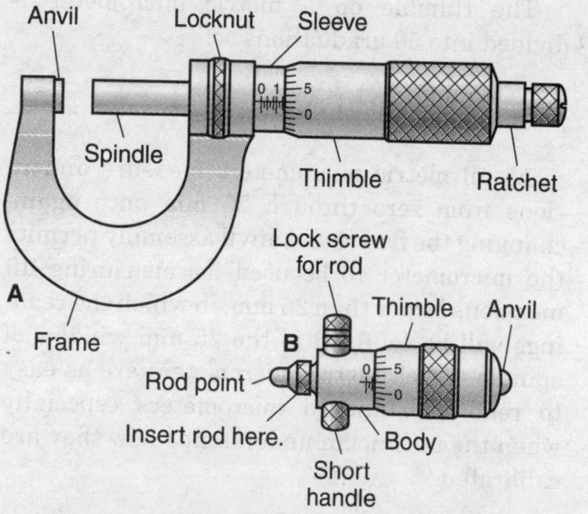

Figure 2–7 *Nomenclature and components of a micrometer: A. An outside micrometer; B. An inside micrometer.*

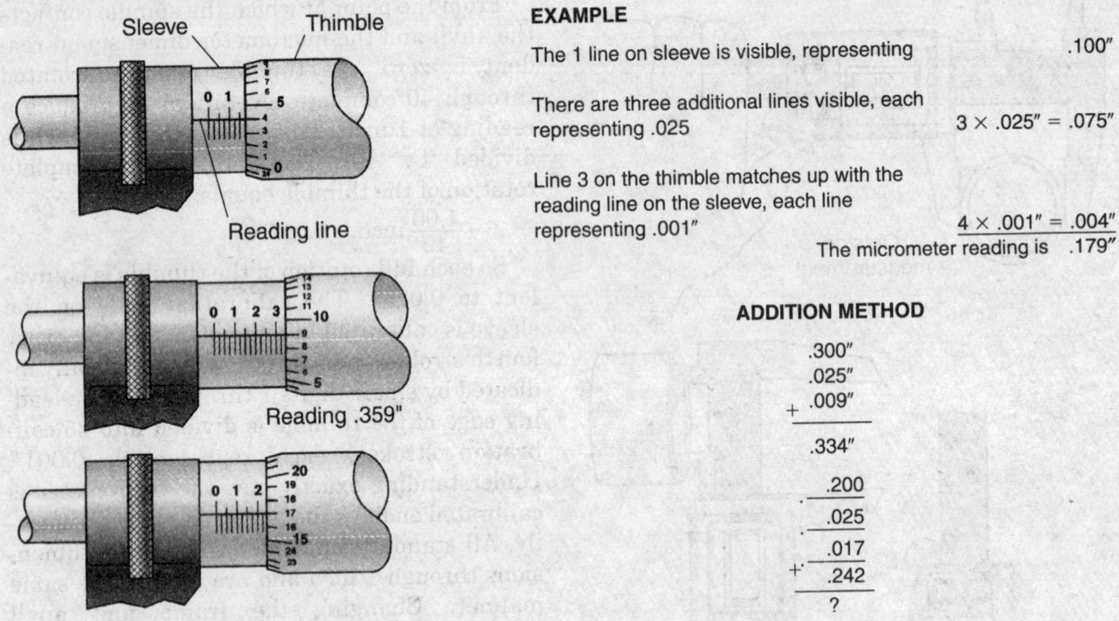

Figure 2–8 *How to read a micrometer. (Courtesy of Navistar International Corp.)*

Metric Micrometers

All metric micrometers have a scale that reads 25 mm and, from the zero measurement when the spindle contacts the anvil, must be turned through 50 rotations. Therefore, each revolution of the thimble equals:

$$\frac{25.0\ \text{mm}}{50} = 0.5\ \text{mm}$$

The thimble on a metric micrometer is divided into 50 graduations so:

$$\frac{0.5\ \text{mm}}{50} = 0.01\ \text{mm}$$

As all metric micrometers measure dimensions from zero through 25 mm once again, changing the frame and anvil assembly permits the micrometer to be used for measuring dimensions larger than 25 mm, in which the readings will be confined to the 25 mm window of spindle travel. Metric micrometers are as easy to read as standard micrometers especially when the technician understands how they are calibrated.

Dial Indicators

Dial indicators are used to measure travel or movement in values of thousandths to one hundred thousandths of an inch. Metric dial indicators are calibrated to read in tenths to thousandths of a millimeter. Dial indicators are used for many general and job-specific functions in the diesel engine shop. Most have a total travel range of 1″ (25 mm) but job-specific indicators may have a total travel range exceeding that and some much less. Dial indicators may have a balanced dial, in which case, in an indicator in which one revolution of the indicator needle represents 0.100″, the dial would be calibrated in units of 0.001″ to 0.050″ through 180 degrees of the dial and then graduated back down to 0.001″ through the other side of the dial; the 0 at the center of the dial face would be marked with a + on one side and a – on the other. This type of dial indicator is useful in determining the **total indicated runout (TIR)** of a rotating component. Figure 2–9 demonstrates dial indicator terminology.

For instance, when measuring flywheel concentricity, a magnetic base dial indicator would

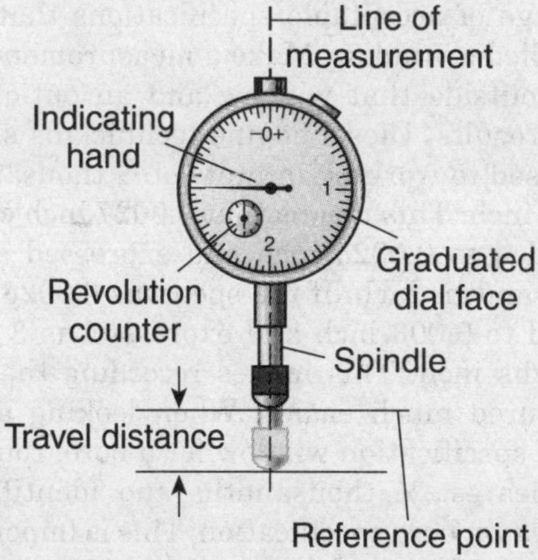

Figure 2–9 *Dial indicator terminology.*

be placed on the engine crankshaft and the engine would be rotated manually through 360 degrees. Using chalk on the flywheel housing face, the indicator dial should be set to zero at the start point, and as the engine is rotated through 360 degrees, readings both above and below zero should be added to indicate the TIR. A secondary dial gauge is used to indicate complete revolutions of the primary face gauge. Dial indicators should always be mounted so that the plunger is at right angles to the component being measured. Figure 2–10 shows the various positions of a universal dial indicator.

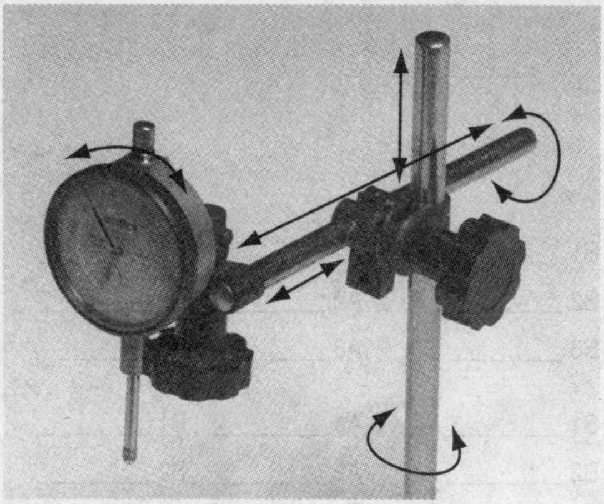

Figure 2–10 *Adjustment features of a dial indicator.*

Dial Bore Gauges

Dial bore gauges are used to measure bore dimensions such as **inside diameter (id)**, taper, and out-of-round. They are calibrated in 0.001″ or 0.0001″ and are a fast way of rapidly assessing cylinder bore dimensions. The typical dial bore gauge consists of a shaft, on top of which is the dial indicator; at the base of the shaft is a measuring sled consisting of guides and an actuating plunger. One of the three guides is located diametrically opposite to the actuating plunger. The actuating plunger is responsible for producing the indicator readings. Figure 2–11 shows a dial bore gauge being used to measure cylinder bore.

Figure 2–11 *Dial bore gauge being used to measure cylinder bore.*

Set-up Options

The guide located diametrically opposite the actuating plunger is both removable (to permit the dial bore gauge to measure different bore dimensions) and adjustable within a range. There are a number of different methods of setting up a dial bore gauge and each technician will in time develop a preference for a particular method; in addition, some OEMs have very specific methods that they want used. For instance, one OEM prefers that a C-tram be used: the dimension of the C-tram is exactly that of the maximum permitted bore size,meaning that every within- specification measurement should read on the *minus* side of the indicator scale.

Dial Bore Gauge Procedure

Many students experience some difficulty in setting up and making measurements using a dial bore gauge: this is because so many different methods are used. I find that the following method works well if students do not have a lot of experience in setting up dial bore gauges. The steps are:

1. Identifying the specification window
2. Setting the micrometer
3. Setting the dial bore gauge
4. Making bore measurements
5. Recording the measurement

Identifying the Specification Window

Most OEMs use units of 0.0001 inch to express cylinder bore specifications. They will provide a range of acceptable specifications that may be called a window. Make a measurement that falls outside that window and an out-of-spec bore results. Diesel engine technicians should get used to working in units of ½ thousandths of an inch. This means that 0.0027 inch can be rounded to 0.0025 inch and expressed as 2½ thousandths inch. If the spec was 0.0028 inch, round to 0.003 inch and express it as 3 thousandths inch. This makes recording the data measured much easier. When looking at the OEM specification window for a bore, round to the nearest ½ thousandths and identify the *maximum* bore specification. This is important. It means that at the start of making measurements, every acceptable specification will read on the *minus* side of the dial bore gauge zero.

Setting the Micrometer

Set the micrometer reading to the maximum permitted bore specification. Lock the micrometer. Record the maximum specification on the bore data chart shown in Figure 2–12. Now wrap the micrometer frame in a wiper and gently but firmly clamp it into a vise.

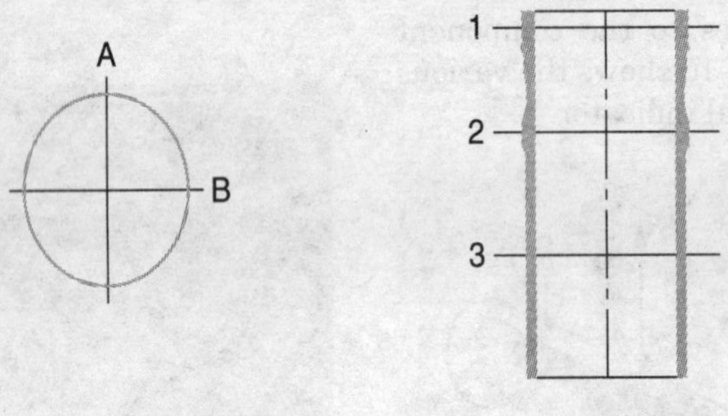

1 A1 ______	B1 ______	2 A1 ______	B1 ______	3 A1 ______	B1 ______
A2 ______	B2 ______	A2 ______	B2 ______	A2 ______	B2 ______
A3 ______	B3 ______	A3 ______	B3 ______	A3 ______	B3 ______
4 A1 ______	B1 ______	5 A1 ______	B1 ______	6 A1 ______	B1 ______
A2 ______	B2 ______	A2 ______	B2 ______	A2 ______	B2 ______
A3 ______	B3 ______	A3 ______	B3 ______	A3 ______	B3 ______

Maximum specification ______ (set micrometer to this dimension).

Minimum specification ______

Figure 2–12 *Cylinder bore data recording chart for a six-cylinder engine. Note the locations for taking longitudinal and transverse measurements.*

Setting the Dial Bore Gauge

Now set the dial bore gauge to the mike dimension. Select the appropriate length adjustable guide and screw inward into the sled until the dimension between the adjustable guide and the measuring plunger is less than the micrometer setting. Supporting the measuring plunger against the micrometer anvil, begin to screw the adjustable guide counterclockwise (CCW) until it makes contact with the micrometer spindle. Now while continuing to rotate the adjustable guide CCW the indicator reading needle will begin to rotate: make sure that it rotates at least one full revolution (on a typical indicator, this will be 100 thousandths). Remove the dial bore gauge. Lock the jam nut on the adjustable guide. It does not matter if the adjustable guide moves slightly as the jam nut is engaged. Next, insert the dial bore gauge back into the mike. Make sure the dial bore indicator turns through approximately one rotation of travel, then zero the indicator and lock the setting. Having set the dial bore gauge, remove and reinstall it a couple of times to check the measurement.

Each time the dial bore gauge is installed into the mike it should read exactly zero. When it does, the zero corresponds to the *maximum* permitted bore specification. This means that *any* positive reading on the indicator is out of spec. This also means that when recording the data, every in-spec reading should be a minus reading in a within-spec bore.

Making Bore Measurements

If making measurements to determine the serviceability of cylinder liners, they should be made in the following locations:

1. Top of the ring belt sweep
2. Bottom of the ring belt sweep
3. Midway between measurement 2 and the bottom of the liner

These locations are shown in Figure 2–12. To make a bore measurement, gently hold the dial bore gauge between two fingers on the grip of the handle above the indicator. Allow the measuring sled to pivot in the bore using the dial bore handle to sweep the device through an arc. Watch the needle as the dial bore gauge moves through each sweep, and look for the stroke-over point—in other words, the point at which the needle reverses.

Recording the Measurements

This is where technicians should keep things as simple as possible. Think in terms of relative thousandths. Avoid recording the specification and complete the chart as discussed above. On this engine, the OEM defines the specification tolerance as 4.8744 inch to 4.8768 inch. First, round these values to the nearest half thousandths of an inch. Now record the following specs onto the bore chart shown in Figure 2–12:

- Minimum specified bore size: 4.874½
- Maximum specified bore size: 4.877

The specification window within which this cylinder sleeve should measure is 2 ½ thousandths. Observe the way in which the bore chart shown in Figure 2–13 has been completed, and observe the out-of-spec cylinder.

Depth Gauges

Depth gauges may be of the micrometer or dial gauge type. In each case, the instrument consists of a block to which either a micrometer assembly or dial indicator mechanism is attached. The micrometer depth gauge is read opposite from the standard micrometer: When the spindle is flush with the block (that is, it is in its most retracted position), the micrometer reads zero. As the thimble is rotated, the plunger extends beyond the flush position on the block to produce readings on the sleeve calibration scale. The calibration scales on both standard and metric depth micrometers are read in the same way as regular micrometers. The dial-type depth gauge is simply a dial indicator mounted on a block. It is read in a manner exactly similar to a dial indicator.

Combination Square

The combination square consists of a right-angle square, a protractor, and a center gauge assembled on a steel ruler. A good quality, precision combination square may be used as a square, protractor, center gauge, depth gauge, height gauge, level, straightedge, and ruler. It is a valuable addition to the technician's toolbox.

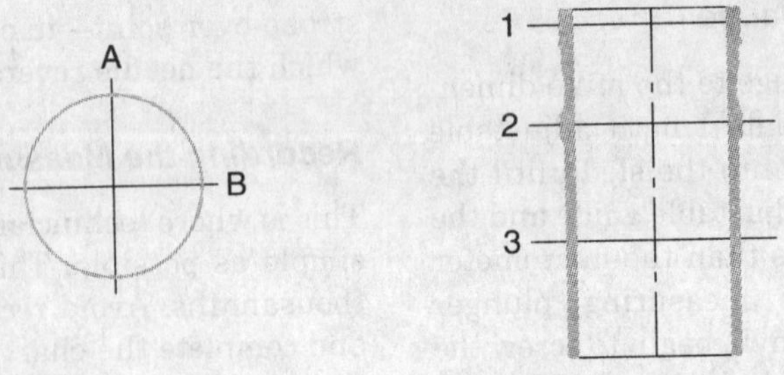

Cylinder	A1	A2	A3	B1	B2	B3
1	−1	−1½	−2	−½	−1½	−2
2	−1	−1½	−2	−1½	−2	−2
3	(+1½)	−½	−1	(+1)	−½	−1½
4	−½	−1½	−2	−1	−1	−2
5	−½	−1	−1½	−½	−1½	−2
6	−½	−1	−1½	−1	−1	−2

Maximum specification 4.877" (set micrometer to this dimension).

Minimum specification 4.874½"

Figure 2–13 *Completed cylinder bore data recording chart using the method outlined in this chapter: Note the out-of-spec data on #3 cylinder.*

Telescoping Gauges (Snap Gauges)

Telescoping gauges are used to measure internal dimensions. They have no integral calibration and must be used in conjunction with a standard outside micrometer. In other words, they are a comparison measuring instrument. A set of telescoping gauges is usually capable of measuring dimensions from ½″ up to 6″. The gauge has the appearance of a T: the T bar is equipped with a spring-loaded plunger, which, when released by the locking handle, expands to the dimension to be measured because it is spring loaded. The gauge may then be locked by the locking handle and removed from the bore; an outside micrometer is then used to measure the T bar dimension.

The diesel engine rebuilder should generally attempt to avoid the use of telescoping gauges. Because it is an indirect measuring instrument, the chances of making inaccurate measurements increase. Telescoping gauges should not be used to measure cylinder bores, firstly because of the potential to make inaccurate measurements, and secondly because it takes much longer than when a dial bore gauge is used.

Small-hole Gauges

Small-hole gauges, like telescoping gauges, are comparison measuring instruments used to measure small cylindrical bores in conjunction with an outside micrometer. In the typical small-hole gauge, a tapered spindle is rotated by a handle to spread split ball halves, moving them outward to contact the bore walls being measured. The gauge is adjusted for minimal drag in the bore, then measured with an outside micrometer. A typical application of a split ball gauge would be the measuring of valve guide bores.

Plastigage

Plastigage™ is used to check friction-bearing clearances. It consists of a cylindrical plastic thread enclosed in an envelope calibrated in the dimensions that the Plastigage is designed to measure. To measure bearing-to-shaft clearance, a small strip should be cut and placed across the width of the bearing shell. Next, the cap should be torqued to specification, which results in flattening the Plastigage. The cap is then removed and the width to which the

Plastigage has been flattened should be measured against the calibration scale on the envelope. The wider the flattened dimension of the plastigage, the narrower the bearing clearance. Plastigage is used to measure rod and main bearing clearance on engines. When crankshaft main journal clearance is measured, the engine must be inverted so that the weight of the crankshaft is fully supported by the cylinder block and not by the main caps. Chapter 6 goes into a little more detail in how to use Plastigage and the method is shown in that chapter in Figure 6–23 and Figure 6–24. Plastigage is available in four size ranges, but in engine technology the following is generally used:

Green: clearance range –0.001″ to 0.003″
Red: clearance range –0.002″ to 0.006″
Blue: clearance range –0.004″ to 0.009″

- Carefully remove Plastigage from the shaft when the measurement is complete.
- A Plastigage test strip that is flattened irregularly indicates journal taper.

Dividers and Calipers

Dividers are used for measuring dimensions between lines or points and scribing reference points and arcs. **Calipers** are designed with internally or externally arced legs to perform internal and external measurements. Dividers and calipers are comparison measuring instruments that require the use of a calibrated measuring instrument such as a micrometer or ruler to produce a specific dimension.

Precision Straightedge

A precision straightedge is manufactured from a middle alloy carbon steel. It should be encased in a protective wood or plastic cover and hung vertically when stored. A precision straightedge is used for such tasks as measuring cylinder block wear and warpage in conjunction with a set of thickness gauges.

Thickness Gauges (Feeler Gauges)

Thickness gauges are precisely machined blades of tool steel usually packaged in sets. They are available in standard and metric dimensions and tend to be one of the most used items in the technician's toolbox. Thickness gauges are used for adjusting valve lash; checking connecting rod endplay; checking backlash on gear sets; checking ring end gap; and, used with a precision straightedge, checking cylinder head and cylinder block warpage and wear.

The technician should be aware that thickness gauges wear with frequent use and should be measured from time to time with a micrometer. Individual blades can be easily replaced in a thickness gauge set.

Truck Technician's Toolbox

Not every technician's toolbox will be equipped as shown in Figure 2–1, but the following information might help as a guideline. Begin with the safety items:

Safety glasses

Table 2–2 *Ratchet Drives and Socket Sizes*

	Double Hex	*Hex*	*Hex Deep*	*Soft Impact*
¼0 drive		4 mm–12 mm 5/16″–½″	4 mm–12 mm 5/16″–½″	
3/8″ drive	10 mm–19 mm 3/8″–¾″	10 mm–19 mm 3/8″–¾″	10 mm–19 mm 3/8″–¾″	
½″ drive	½″–15/16″ 12 mm–19 mm	½″–1¼″ 12 mm–24 mm	½″–5/16″ 12 mm–19mm	½″–1¼″ 12 mm–24 mm
¾″ drive		7/8″–1½″ 20 mm–30 mm		

Hearing protection muffs and plugs
Uncured leather gloves (use for heavy lifting)
Latex gloves (these provide the hands with protection against oil and grease)

¼″ drive ratchet
3/8″ drive ratchet
½″ drive ratchet ½″ drive breaker/flex bar
¾″ drive ratchet ¾″ drive breaker/flex bar
Sockets: Purchasing in sets is usually more economical. See Table 2–2.
Allen sockets—an assortment of sizes
½″ drive torque wrench—click or dial type to 250 ft.-lb (swivel/flex head useful for engine work in tight locations)
3/8″ drive torque wrench—preferably dial type to 120 in.-lb
¾″ and 1″ drive torque wrenches—usually provided by shops
4:1 torque multiplier—usually provided by shops
Combination wrenches—standard: 5/32″–1¼″; Metric: 4 mm–30 mm
Open- and box-end wrenches—standard: 5/32″–1¼″; Metric: 4 mm–30 mm
Note that purchasing wrenches in sets, rather than individually, is usually significantly more economical.
Allen keys—standard and metric sets
Screwdrivers—purchase in sets: slotted, Phillips, TORX (to #30)
Digital multimeter (DDM)—2½- or 3½-digit resolution: Consider purchasing a rubber holster to protect the tool.
Circuit test light
Circuit testing clips and cables
Breakout Ts
Breakout boxes—specialty diagnostic breakout boxes are usually provided by shops.
Electronic Service Tools (ESTs)—usually provided by shops: avoid investing in costly ESTs and software because it rapidly becomes obsolete.
Coolant hydrometer (not a recommended instrument)
Refractometer (for coolant and battery electrolyte)— usually provided by shops. Ensure that it is calibrated for electrolyte, ethylene glycol (EG), and propylene glycol (PG).
4 lb cross peen hammer (optional)
1½ lb ball peen hammer (wood handle)
1½ lb nylon/rubber head
Hacksaw—Selecting a good quality frame with high rigidity and using the best blades can pay off in saved frustration and sweat.
Prybar set—to 18″ size (length)
Cold chisel set—to 1″
Set of punches—to 1″
Brass drift—1″ × 8″
Mild steel drift—1″ × 12″
Stud extractor set + wheel stud extractor collar
Nut splitter
Bolt cutter
Seal and bearing drivers—normally provided by shops
Lineman pliers 8″ and 12″
Terminal crimpers—bent nose electronic pliers
Specialty terminal crimpers/connector disassembly tools—usually provided by shops
Wire strippers—needle nose pliers
Sidecutters
Tin snips—straight cut
Slip joint/waterpump pliers—12″, 24″, 36″
Vise grips (Never let anyone kid you that these are the tools of an amateur. There are thousands of valid uses for them in the truck shop.)
10″ pipe jaw/10″ straight jaw/8″ needle nose vise grips
18″ pipe wrench
Adjustable wrenches—6″, 8″, 12″
½″ chuck pneumatic drill
HS drill bits to 3/8″ size
3/8″ drive air ratchet
½″ drive impact gun (see Figure 2–13)
Pneumatic chisel/hammer
Truck tire air chuck

Tire gauge to 150 psi–1 MPa
Air blower nozzle
Air hose—often provided by shops (see Figure 2–13)
Hand primer pump coupled into #8 hydraulic hose and fittings
Variable focus flashlight
Soldering gun
Telescoping mirror
Telescoping magnet
0–1″ micrometer
0–1″ depth micrometer
0–25 mm micrometer
0–25 mm depth micrometer
Micrometer sets exceeding 1″ are provided by shops.
6″–150 mm vernier caliper—digital preferred
12″ tape measure
Dial bore gauge—telescoping gauges (usually provided by shops)
Stethoscope (usually provided by shops)
Infrared thermometer (usually provided by shops)
Fluorescent trouble light—often provided by shops
Heavy-duty hand tools—wrenches sized over 1¼″ or 30 mm, ¾″ and 1″ drive sockets, and other heavy-duty specialty tools are normally provided by shops.
Roller cabinet—many shallow drawers preferable to fewer deep ones
Top box—also consider a side cabinet for shop manuals and fluid containers
Creeper—often provided by shops

SHOP TOOLS

Shop tools are those tools generally provided by the employer. Tools that are too large to fit into a toolbox, high in cost, or highly specialized to a specific procedure should be provided by the service garage.

Sledgehammers

Sledgehammers are designed so that the weight of the head (and perhaps the length of the arms holding the hammer) defines the force imparted. Sledgehammers have a variety of functions in the truck shop and while they are unlikely to be called on in engine reconditioning, they may be required in some of the procedures required to remove components from a chassis. They are usually manufactured in 8, 12, and 16 lb weights. Before using a sledgehammer, the handle should be inspected for damage and the head to handle securement checked. When swinging a sledgehammer, the hand grip should be firm but relaxed and the weight of the head allowed to define the amount of force delivered. On no account should the operator attempt to amplify this force with muscle power as it usually results in missing the target. If the force is insufficient to achieve the objective, select a heavier hammer.

Presses

Most service garages will have at least one power press. Extreme caution is required when operating a power press: components should be properly supported, and mandrels/drivers should be used when required. Arbor presses are hand actuated. Whenever pressing components using any kind of power press, always consider both the consequences of component slippage and where separated components will fall. Personal safety and that of those working in the vicinity must always be considered. Figure 2–14 shows a press driving a bearing onto a shaft.

Scissor Jacks

Scissor jacks are designed to quickly raise one end of a truck to heights of up to 8 feet above the shop floor. Clevises on the jack fit under each frame rail and the truck is raised by an air-actuated piston. Scissor jacks are an invaluable shop tool but they must be safely used. Ensure that the lift clevises are positioned at a safe location on the frame rails and allowance is made for the relative movement between the truck and jack during raising and lowering. When the truck has been raised, engage the mechanical stops and make sure that the weight is supported on them and not the power piston. Also check the weight of the vehicle to be lifted and the load capacity of the

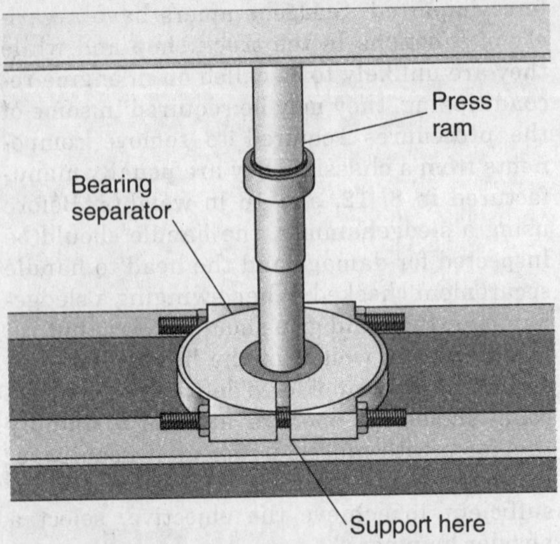

Figure 2–14 *Press and bearing puller used to install bearing on shaft.*

scissor jack. It is good practice to double block a raised truck by providing two mechanical supports. Chock the wheels on the end of the chassis not being raised after the unit has been raised to the desired height: Do *not* attempt to do this before hoisting because the vehicle must be allowed to roll unhindered during the lift.

"A" Frame Hoists

"A" frame hoists are often large enough to pass over the top of a truck and use a block and tackle (chain falls) lift mechanism. They can be used to lift components such as an engine or a cab from the chassis. They may also be capable of hoisting one end of the chassis, but extreme caution should be exercised because of the tendency of anything lifted by chains to swing. The hoist mechanism should be inspected annually whether or not local regulations require the inspection.

Cherry Pickers

These are portable, hydraulically actuated, one arm boom hoists that have many uses in the truck and bus garage. They are available in a variety of sizes and load-carrying ratings. The pickup arm is usually adjustable in length, and the longer the adjustment arm setting, the less the load lift potential. If the load-carrying ability is exceeded, they will topple. Figure 2–15 shows a cherry picker.

Figure 2–15 *Cherry picker.*

Transmission and Clutch Jacks

Transmission and clutch jacks are usually hydraulically actuated and designed to fit under the truck frame and support the transmission/clutch. It is important to ensure that transmissions are chained securely to the jack especially when the mass of the unit is top heavy such as in triple countershaft units. When it is expected that the jack must support the transmission stationary such as when removing an engine from a chassis, ensure that the transmission jack load is supported mechanically. Clutch jacks should be used when installing heavy-duty clutches: a 15½ in. clutch pack can weigh somewhere around 175 lb (80 kilograms [kg]) and should never be handled without some kind of assistance.

Spreader Bars

A **spreader bar** is a rigid bar, usually adjustable in length, used for lifting engines out of a chassis. The spreader bar should be adjusted to the length of the engine; it can then be attached to the engine at three or four points: the bar should be close to level before a lift is attempted. Chains should be installed so that the chain length is at a minimum with the spreader bar clear of the engine. The chains should be attached to the engine by means of hooks to lifting eyes located on either the cylinder head or cylinder block assembly. Never fit lifting eyes to the rocker housing fasteners. Some engines have just two permanently fitted lifting eyes. It is usually safe to lift these engines using the spreader bar on a two-point lift.

Load Rotor

The load rotor is alternate to the spreader bar when it comes to hoisting an engine: it permits a limited ability to tilt the load during the lift. Load rotors consist of a ratcheting chain block with a single chain equipped with hooks at either end. The hooks fit to the engine lift eyes and the load rotor chain block locks the chain, permitting a different length of chain on either side of the block. This feature often permits the chains to be fitted to the engine and clear the upper engine components.

Chains

Chains are rated by working load limit—a value that is normally equivalent to about 25% of the tensile strength of the chain material. Many jurisdictions require that chains be inspected annually. Additionally, technicians should visually inspect all chains before using them. The saying that every chain is only as strong as its weakest link bears true. Truck diesel engines can weigh more than a ton and while relying on a chain to support this kind of weight, the technician should try to avoid working under the load. Apart from inspecting the **chain hoists** and checking their load rating, the technician should check the hooks, lifting eyes, lifting eye fasteners, and connecting links.

Slings

Slings may have to be used to hoist engines from the chassis in some applications. These are normally manufactured from synthetic fibers or braided steel wire. Again, the load capacity must be checked and the sling integrity inspected. Avoid using steel cable slings unless also using the engine lift eyes.

Air Tools

Air tools are extensively used in any truck service location. Some of the air tools are owned by technicians; others are provided by the shop. Technicians working around pneumatic equipment sometimes forget that it can be dangerous. It makes sense to wear eye protection and be aware that dusts driven into the air by pneumatic tools can cause breathing problems. Most shops provide heavy-duty pneumatic tools such as 1″ drive air guns. A ½″ drive impact wrench such as that shown in Figure 2–16 will be one of the most frequently used tools in the truck technician's cabinet. The engine technician primarily uses this for disassembly. Buying a good quality air gun and properly maintaining it by keeping it moisture-free and oiled (observe the oiler and filter in Figure 2–16) will help ensure that it functions well for a number of years.

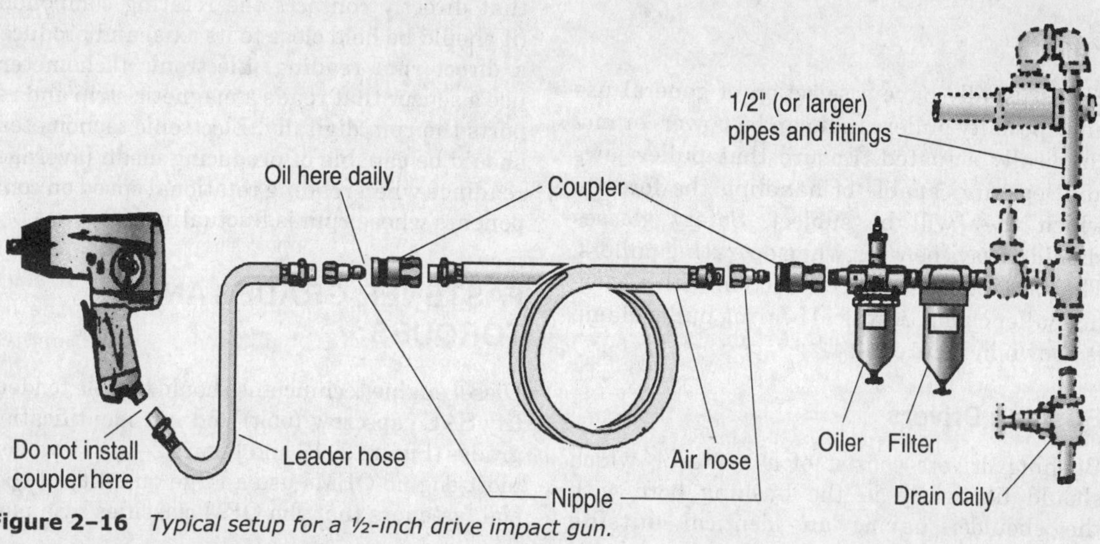

Figure 2–16 *Typical setup for a ½-inch drive impact gun.*

Oxyacetylene Equipment

Technicians use oxyacetylene for heating and cutting probably on a daily basis. Less commonly, this equipment is used for brazing and welding. Technicians using oxyacetylene stations require some basic instruction in the safety requirements and handling of this equipment. Despite the fact that oxyacetylene equipment is used by every technician, seldom is the novice technician provided with adequate instruction.

Steam and High-pressure Washers

Hot water, high-pressure washers have generally replaced the steam cleaners more commonly used a decade ago. Hot water, high-pressure washers are safer and usually require less maintenance than steam washers. The technician should be aware of the potential for damage when using any type of high-temperature, high-pressure washers; eye protection and gloves should be worn when operating this equipment. Environmental regulations in most jurisdictions mandate that the runoff from this type of cleaning operation not be permitted to directly enter sewage systems. Power wash runoff should be filtered through a water separator system, and the separator tanks should be pumped out regularly. Heavy fines may be imposed when sewage is contaminated with oil and road dirt washed off trucks, buses, and their engines.

Pullers

Shops usually have a selection of general use and specialty pullers that can be power- or mechanically actuated. Ensure that puller jaws and legs are capable of handling the force to which they will be subject. Safety glasses should always be worn when operating pullers. They are often used in conjunction with a bearing puller clamp. A typical bearing puller clamp is shown in Figure 2–14.

Bushing Drivers

Bushing drivers consist of a mandrel, which should fit tightly in the bushing bore with the shoulder having an identical **outside diameter (od)** to the bore that the bushing is pressed into. This should enable bushings to be removed and installed without damaging either the bushing or the bore to which it is fitted. Bushings may be installed using direct mechanical force such as a hammer or slide hammer, or by using hydraulic or pneumatic power drivers. Always wear safety glasses when using bushing drivers. Some bushings must be reamed for final fit after installation.

Glass Bead Blasters/Sandblasters

Most shops rebuilding engines are equipped with a glass bead blaster or sandblaster. Usually these are encased in an enclosed housing and powered pneumatically. They are an ideal method of cleaning up components, especially when the components are coated with high-tack adhesives or gasket remains that can be difficult to remove. In enclosed housing glass bead blasters, protective gloves are integral with the unit and in most cases they can only be actuated when the component is placed inside and the cover sealed. Armored glass permits the object being blasted to be observed. Technicians should consider using hearing protection as these units are capable of producing high noise levels.

Tachometers

Tachometers measure rotational speed. Mechanical tachometers consist of a pickup button that directly contacts the rotating component (it should be held close to its axis) and produces a direct rpm reading. Electronic tachometers use a sensor that reads a magnetic strip and reports the rpm digitally. Electronic tachometers should be capable of producing mean (average) readings when reading rotational speed on components whose rpm is fluctuating.

FASTENER GRADES AND TORQUES

Diesel engine technicians should be able to identify SAE capscrew (bolt) and nut identification grades (Figure 2–17 and Figure 2–18). However, most engine OEMs use a large quantity of special fasteners that the OEM classifies by a part

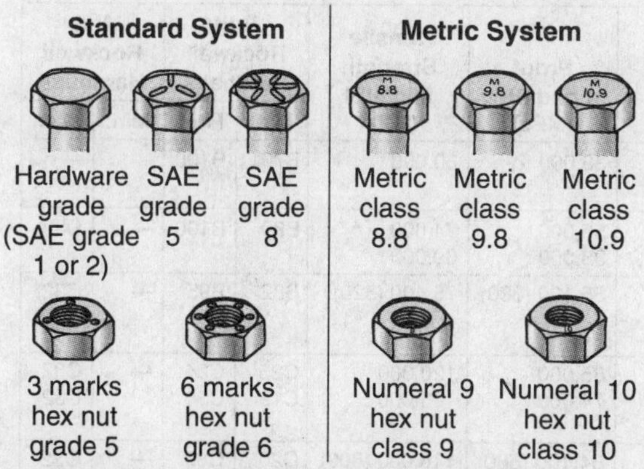

Figure 2–17 *Fastener grade identification. (Courtesy of Mack Inc.)*

number. This is often done with the specific objective of discouraging crossover to any other than OEM fasteners either because of some very specific metallurgical properties of the fastener, which make crossover impossible, or simply to prevent the use of substandard quality fasteners. In recent years, the North American market has been subject to bogus fasteners with the appropriate SAE markings, mostly originating in Asia, that have found their way onto aircraft and highway equipment at both manufacturing and repair facilities. Unfortunately, these dangerous and illegal fasteners are not usually discovered until analysis by accident investigators. It is important to purchase fasteners from reputable suppliers and perform destructive testing on fastener samples at random intervals.

The SAE fastener grades commonly used on trucks and their engines are:

Grade 5 category—manufactured from medium carbon steels and heat treated to provide:

	Proof Loads	*Tensile Strength*
to ¾″	85,000 psi	120,000 psi
¾″–1″	78,000 psi	115,000 psi
1″–1½″	74,000 psi	105,000 psi

Grade 8 category—manufactured from medium carbon alloyed steels and heat treated and roll threaded to provide:

	Proof Loads	*Tensile Strength*
up to 1½″	120,000 psi	150,000 psi

SAE Grade 5 fasteners are identified by three radial strokes.

SAE Grade 8 fasteners are identified by six radial strokes.

Clamping Force

Fasteners are designed to provide clamping force. When the fasteners responsible for clamping a cylinder head to a cylinder block are torqued to a common torque value, the objective is to ensure a consistent clamping force throughout the assembly. Fasteners such as Huck™ fasteners provide a more reliable clamping force value because they are set by defining the clamping force and eliminate the variables encountered when using a torque value to set clamping force.

Tensile Strength and Yield Strength

Fasteners are rated by **tensile strength**, the amount of force that must be exacted on a round bar of 1″ sectional area to make it fracture. The **yield strength** is the amount of force that must be exerted using the same test to cause the bar to permanently deform; typically this force is about 10% lower than the tensile strength value in steels. The difference between the tensile strength and yield strength is sometimes used to denote the elasticity of a material.

Shear Strength

Shear strength is a measure of a fastener's ability to withstand force applied at a 90-degree angle to the axis of the bolt. Some bolts are made to withstand shear forces, such as body-bound bolts, which are manufactured with an interference fit shank. The interference fit shank consists of an extended shoulder that should be driven home with a hammer before applying torque to the nut.

THE METRIC SYSTEM AND ENGLISH/METRIC CONVERSION

The metric system has been authorized by an act of Congress in the United States and by federal legislation in Canada. The metric system is used by most countries in the world, and has been generally adopted by industry for use in the United States as a replacement for the standard system. For example, every truck diesel engine engineered in the United States since 1980 has been a metric engine. However, the specifications reproduced in service literature are usually

Grade Marking	Specification	Material	Nominal Size Dia. In.	Proof Load PSI (MPs)	Tensile Strength Min. PSI (MPa)	Bolt Rockwell Hardness		Nut Rockwell Hardness	
						Min.	Max.	Min.	Max.
	ASTM A307 Grade A SAE J429 Grade 1	Low carbon steel	¼ thru 1½	33,000	60,000	B70	B100	—	—
	SAE J429 Grade 2	Low carbon steel	¼ thru ¾ over ¾ to 1½	55,000 33,000	74,000 60,000	B80	B100	—	C32
	ISO SAE J1199 Property Class 5.8	Low or medium carbon steel	M5 thru M24	55,100 (380)	75,400 (520)	B82	B95	—	C32
	ASTM A449 Type 1 SAE J429 Grade 5	Medium carbon steel, quenched and tempered	¼ thru 1 over 1 to 1½	85,000 74,000	120,000 105,000	C25 C19	C34 C30	—	C32 C32
	ISO/DIN SAE J1199 Property Class 8.8	Medium carbon steel, quenched and tempered	M3 thru M16 M17 thru M36	84,100 (580) 87,000 (600)	116,000 (800) 120,350 (830)	C20 C23	C30 C34	—	C32
	SAE J429 Grade 5.1 (SEMS)	Low or medium carbon steel, quenched and tempered with assembled washer	No. 6 thru 5/8	85,000	120,000	C25	C40	—	—
	ISO SAE J1199 Property Class 9.6	Medium carbon steel, quenched and tempered	M1.6 thru M16	94,250 (650)	130,500 (900)	C27	C36	—	C32
	SAE J429 Grade 7	Medium carbon allow steel, quenched and tempered, roll threaded after heat treatment	¼ thru 1½	105,000	133,000	C28	C34	—	
	ASTM A354 Grade BD Bowma-Torq®/Grade 8 SAE J429 Grade 8	Medium carbon alloy steel, quenched and tempered	¼ thru 1½	120,000	150,000	C33	C39	C24	C36
	SAE J429 Grade 8.2	Low carbon boron martensite steel, fully killed, fine grain, quenched and tempered	¼ thru 1	120,000	150,000	C35	C42	—	—
	ISO SAE J1199 Property Class 10.9	Medium carbon alloy steel, quenched and tempered	M6 thru M36	120,350 (830)	150,800 (1040)	C33	C39	C26	C36
	ISO Property Class 12.9	Medium carbon alloy steel, quenched and tempered	M1.6 thru M36	140,650 (970)	176,900 (1220)	C38	C44	C26	C36
	Bowmalloy®	Proprietary medium carbon alloy steel, quenched and tempered	¼ thru 1½	156,000	180,000 Min. 2000,000 Max.	C38	C42	C26	C36

*Manufacturer's identification symbols are required per ASTM, ISO, or SAE

Figure 2–18 *Bolt head identification chart. (Courtesy of Navistar International Corp.)*

presented in both standard and metric systems with the consequence that most mechanical technicians are forced to have an understanding of both. It should be noted that the British replaced the "English" system of weights and measures with the metric system more than 40 years ago.

The metric system is a decimal system, the meter being the basis of all measures, whether of length, surface, capacity, volume, or weight. The meter measures 39.37 inches and is theoretically one ten-millionth of the distance from the Equator to either the North or the South Pole. The unit of weight is the gram (15.432 grains), and is the weight of a cubic centimeter of water at its greatest density at about 39°F.

Multiples of the units are expressed by the Greek prefixes "deca," "hecto," "kilo," "mega," and "giga," indicating, respectively, tens, hundreds, thousands, millions, and billions. Decimal parts of the units are indicated by the Latin prefixes "deci," "centi," "milli," "micro," "nano," and "pico," meaning, respectively, tenth, hundredth, thousandth, millionth, thousand millionth, and billionth.

Metric Weights and Measures

Listings of the commonly used metric weights and measures as well as a conversion table for common metric measurements to standard units follow.

Metric Weights

Milligram	(1/1000 gm)	= 0.0154 gr.
Centigram	(1/100 gm)	= 0.1543 gr.
Decigram	(1/10 gm)	= 1.5432 gr.
Gram		= 15,432 gr.
Decagram	(10 gm)	= 0.3527 oz.
Hectogram	(100 gm)	= 3.5274 oz.
Kilogram	(1000 gm)	= 2.2046 lb
Myriagram	(10,000 gm)	= 22.046 lb

Metric Dry Measures

Milliliter	(1/1000 L)	= 0.061 cu. in.
Centiliter	(1/100 L)	= 0.6102 cu. in.
Deciliter	(1/10 L)	= 6.1022 cu. in.
Liter		= 0.908 qt.
Decaliter	(10 L)	= 9.08 qt.
Hectoliter	(100 L)	= 2.838 bu.
Kiloliter	(1000 L)	= 1.308 cu. yd.

Metric Liquid Measures

Milliliter	(1/1000 L)	= 0.0338 fl. oz.
Centiliter	(1/100 L)	= 0.338 fl. oz.
Deciliter	(1/10 L)	= 0.845 gill
Liter		= 1.0567 qt.
Decaliter	(10 L)	= 2.6418 gal.
Hectoliter	(100 L)	= 26.417 gal.
Kiloliter	(1000 L)	= 264.18 gal.

Metric Measures of Length

Millimeter	(1/1000 m)	= 0.0394 in.
Centimeter	(1/100 m)	= 0.3937 in.
Decimeter	(1/10 m)	= 3.937 in.
Meter		= 39.37 in.
Decameter	(10 m)	= 393.7 in.
Hectometer	(100 m)	= 328.1 ft.
Kilometer	(1000 m)	= 0.62137 mi.
	(1 mile = 1.6093 km)	
Myriameter	(10,000 m)	= 6.2137 mi.

Metric Surface Measures

Centare	(1 sq. m)	= 1,550 sq. in.
Are	(100 sq. m)	= 119.6 sq. yd.
Hectare	(10,000 sq. m)	= 2.471 acre

Metric to Standard Conversion Formulae

The following formulae can be used to translate metric values to standard and vice versa.

Linear Measurements

Centimeters × 0.3937 = in.
Meters = 39.37 in.
Kilometers = 0.621 mi.
Kilometers × 3280.89 = ft.
Square centimeters × 0.155 = sq. in.
Cubic centimeters × 0.06102 = cu. ins.
Cubic meters × 35.3144 = cu. ft.

Liters × 0.2642 = gal. (231 cu. in.)
Kilograms × 2.2046 = lb
Kilograms per square millimeter × 1422.3 = lb per sq. in.
Kilograms per square centimeter × 14,223 = lb per sq. in.

Torque Conversion

1 lb-ft. = 1.355 Newton-meters (N·m)
1 N·m = 0.738 lb-ft.

Temperature Conversion

$$\text{Degrees Fahrenheit} = \frac{9 \times C}{5} + 32$$

$$\text{Degrees Celsius} = \frac{5 \times (F - 32)}{9}$$

Pressure Conversions

Fuel injection test instruments are often calibrated in **units of atmosphere (atms)**. Technicians should become familiar with the process of converting units of pressure into the metric and standard systems. This is an easy way:

Atmospheric pressure @ sea level = 14.7 psi
= 101.3 kPa = 1 unit of atmosphere or 1 atms

Remembering the following is not as mathematically accurate as the previous equivalent, but it is a fast method of converting pressure values that is accurate enough for quick conversions:

15 psi = 100 kPa = 1 atms

So, to convert 45 psi to kPa:

$\frac{45}{15}$ = 3 units of atmosphere or 3 atms

3 atms = 100 kPa = 300 kPa

In Europe, a unit of pressure measurement known as a **bar** is used. Be familiar with bar, because Bosch uses this unit for specifications in the many diesel fuel systems it manufactures. One bar is equivalent to 105 newtons per square meter, which is not precisely equivalent to one atmosphere. Although units of bar and atms are often used as if they were exactly equivalent, this is not so, and where precise values are required they should not be confused. Some fuel injection comparator equipment is calibrated in bar.

1 atms = 14.7 psi = 101.3 kPa = 1.033 bar
= 29.92″ Hg = 407.19″ H_2O

Power Conversion

1 hp = 550 lb-ft. per second = 0.746 kW = 42.4 Btus per minute

REVIEW QUESTIONS

1. When the spindle contacts the anvil on a standard 0–1″ micrometer, it should read:
   a. 0
   b. 0.001″
   c. 0.025″
   d. 1″
2. How many complete rotations must the thimble of a standard micrometer be turned to travel through a reading of zero to a reading of 1 inch?
   a. 25
   b. 40
   c. 50
   d. 100
3. How many complete rotations must the thimble of a metric micrometer be turned to travel through a reading of zero to a reading of 25 mm?
   a. 25
   b. 40
   c. 50
   d. 100
4. When the thimble of a metric micrometer is turned through one full revolution the dimension between the anvil and the spindle has changed by:
   a. 0.1 mm
   b. 0.5 mm
   c. 2.5 mm
   d. 5.0 mm
5. When using a dial indicator to check the concentricity of a flywheel housing, during a single rotation of the flywheel the reading on the positive side of the zero on the dial peaks at 0.003″ while the reading on the negative side peaks at 0.006″. What is the TIR?
   a. 0.003″
   b. 0.006″
   c. 0.009″
   d. 0.018″

6. Which of the following precision measuring instruments would be required to measure a valve guide bore?
   a. Dial indicator
   b. Inside micrometer
   c. Split ball gauge and micrometer
   d. Dial bore gauge
7. If 300 kPa is converted to pounds per square inch, the result would be closest to which of the following values?
   a. 15 psi
   b. 30 psi
   c. 45 psi
   d. 300 psi
8. To which of the following values would 1.5 mm be closest?
   a. 0.0625″
   b. 0.125″
   c. 0.250″
   d. 1.50″
9. Convert 550° Fahrenheit into Celsius.
   a. 240
   b. 288
   c. 385
   d. 550
10. A wrench with box and open ends at either end, both of the same nominal dimension, is known as a(n):
    a. Torque wrench
    b. Combination wrench
    c. Box-end wrench
    d. Adjustable wrench
11. Which of the following is used to identify an SAE grade 8 bolt?
    a. 3 radial strokes on the capscrew head
    b. 5 radial strokes on the capscrew head
    c. 6 radial strokes on the capscrew head
    d. 8 radial strokes on the capscrew head
12. If a 12″ linear extension is used on a torque wrench with a 24″ bar, calculate the reading on the torque wrench scale required to produce an actual torque value of 250 lb-ft.
    a. 36 lb-ft.
    b. 120 lb-ft.
    c. 167 lb-ft.
    d. 323 lb-ft.
13. Convert 250 lb-ft. to Newton-meters and select the closest value from the answers below.
    a. 167 N·m
    b. 340 N·m
    c. 410 N·m
    d. 500 N·m
14. Convert 600 hp into kW and select the closest value from the answers below.
    a. 350 kW
    b. 450 kW
    c. 550 kW
    d. 650 kW
15. The cutting fluid recommended for use when cutting threads in mild steel is:
    a. Soluble oil and water
    b. Lard
    c. Dry
    d. Kerosene
16. When drilling into cast iron, the correct method calls for the procedure to be performed:
    a. With kerosene
    b. With lard
    c. Preheated
    d. Dry
17. A bolt designed so that the shoulder has a small interference fit with the bore it is to be fitted to is called a(an):
    a. SAE #5
    b. SAE #8
    c. Body bound bolt
    d. Huck fastener
18. The working load of a chain is normally what percentage of the rated tensile strength of the chain material?
    a. 10%
    b. 25%
    c. 50%
    d. 75%
19. Which Plastigage color code should be selected to measure a main bearing clearance that the manufacturer specifies must be between 0.0023″ and 0.0038″?
    a. Red
    b. Green
    c. Blue
20. After Plastigage checking a main bearing, using red Plastigage, the measuring strip has not been deformed. *Technician A* states that the bearing clearance must be greater than 0.006″. *Technician B* states that a strip of green coded Plastigage must be used to perform the measurement. Who is correct?
    a. Technician A only
    b. Technician B only
    c. Both A and B
    d. Neither A nor B

3

History of the Heat Engine

INTRODUCTION

The modern vehicle has reached a stage of technological development in which an operator rarely has to consider exactly what happens when the ignition key is turned and energizes the series of circuits that make it run. It is an action that many of us perform several times a day in a world in which a person may travel more in a single day than our forebearers of just 200 years ago did in a lifetime.

The modern vehicle is a result of humanity's pursuit of constant improvement. We may be able to identify the individual credited with the invention of the diesel engine, but it should always be remembered that he had the cumulative practical and technical achievements of all those who preceded him as a starting base for his research. The first diesel engines were crude and unreliable machines. In a hundred or so years the observations, research, and sometimes lifelong efforts of thousands of engineers, technicians, and designers, most of whom are "nameless," have resulted in the technical excellence of the modern, highly reliable diesel engine.

Thousands of components are integral in the modern truck chassis, each with roles in various mechanical, hydraulic, pneumatic, electrical, and electronic circuits. Each had to be invented, manufactured, and systematically improved to its present stage of development. Every invention will surely be further improved as each year goes by. And as each year goes by, the rate of improvement accelerates. Consider the truck manufactured just 20 years ago and performance match it head to head with today's version. Aside from subjective considerations such as appearance, the contemporary vehicle will in most cases significantly outperform its predecessors just as surely as the 2030 model will outperform today's model.

Progressive Development

In inventing the diesel engine, Rudolph Diesel took a significant step forward in the progress of technology as applied to the modern truck. There were significant steps both before and after him. The argument has been made that the invention of the Otto cycle and diesel engines was inevitable after the theoretical foundation provided by the French scientist Beau de Rochas, but it would be difficult to make the argument that Diesel's engine of 1892 was not a pretty significant event.

This chapter identifies some of the individuals in the history of technology whose accomplishments were important to the development of both the **heat engine** (any machine that converts the heat energy of a fuel to mechanical outcomes) and the modern diesel engine. In examining the history of technology until fairly recent times, it was usually possible to match an accomplishment with the name of the inventor. Today the thrust toward technological advance is something that large corporations must finance, and "inventions" are usually the work of dozens of individuals and are *owned* by the funding company.

Within this list of innovators, there are people who may be described as mathematicians, scientists, engineers, chemists, technicians, physicists, and diesel mechanics with many different levels of education but with certainly one label in common—they were inventors. Most inventions result from an individual's refusal to accept the impossible.

In many ways, the history of technology is the history of modern man. Take any current truck, put it on an empty asphalt parking lot, and completely disassemble it. Disassemble every component and subcomponent until further disassembly is impossible. There will be 100,000 or so manufactured components that will seem to have little obvious connection to a truck. Think about how each of those components was designed and manufactured and how each evolved. How many countries were sourced for the raw materials? How many individuals were involved in the manufacture? Now think about the number of people through the ages who were involved in the evolution of that truck. Start with the invention of the wheel. There is no chance that the first wheel was as remotely round as a wheel on a vehicle today; the progressive efforts of thousands of minds helped make that happen.

Technology is Rational

Everything in technology is rational. In other sciences such as physics or astronomy, the more people learn, the more they discover they do not know. There are no mysteries in technology, just challenges that take a little longer to figure out. So when next stymied with a troubleshooting problem on an engine, technicians must remember that the cause will be rational. They should make it a challenge to discover the cause and then explain it in terms that make sense to them. Technicians should think a little about what James Watt and Rudolf Diesel did and then refuse to accept that the concept of the unexplained can exist in technology.

HISTORY OF THE MODERN DIESEL ENGINE

This collection of minibiographies and events is not intended to be a comprehensive history of technology but rather a listing of significant innovations. The innovators and some critical events are generally listed chronologically although some overlapping occurs in periods of hectic innovation such as at the end of the nineteenth century.

Hero of Alexandria (Around A.D. 60) Aka Heron, Greece

Noted for:

The aeolipile
Hero's formula (area of a triangle)

Hero was a Greek mathematician and inventor who authored a number of books, mostly on the subject of mathematics. His importance to technology is as the inventor of the first known heat engine somewhere around the year A.D. 60. Hero called his machine an **aeolipile** (Figure 3–1) but in our age it is described as a **reaction turbine**. The aeolipile was a spherical vessel with an axial shaft through its center

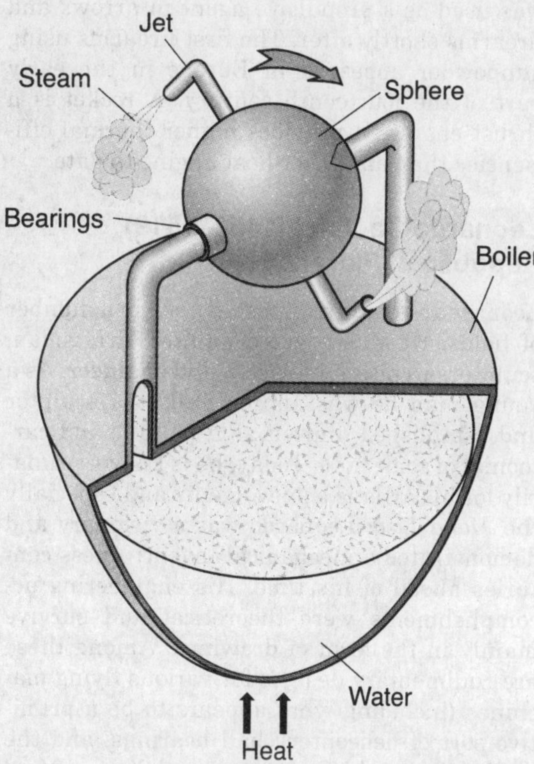

Figure 3–1 *Hero's aeolipile or reaction turbine. (Courtesy of Facts on File)*

supported by bearings and mounted over a heat source. The vessel was fitted with two canted (angled) nozzle jets. When filled with water and heated, escaping steam from the jets produced thrust and resulted in rotary motion. This engine was rated at less than 1% thermal efficiency. It was the forerunner of all modern engines, but very especially the jet turbine, because it developed thrust. In one of his books, *Pneumatica,* Hero describes a number of devices including fire engines, water fountains, coin-operated machines, water clocks, and various other steam-powered machines; it is doubtful whether any of these were actually built. In many ways, Hero was a visionary and thinker rather than a doer, much as Leonardo da Vinci a few centuries later.

Gunpowder and the Rocket (Thirteenth Century, somewhere in China)

The invention of both gunpowder and the rocket is attributed to the Chinese. Fireworks were known to exist in China in 1232 and gunpowder was used as a propulsive agent in arrows and firearms shortly after. The first firearms using gunpowder appeared in Europe in the early part of the fourteenth century. A rocket is a thrust engine. It provides higher thermal efficiencies than any other heat engine to date.

Leonardo Da Vinci (1452–1519) Republic of Florence

Leonardo's genius manifested itself in a number of fields. He was a noted painter, draftsman, sculptor, architect, biologist, and engineer. As a young man he was apprenticed to a sculptor and established himself as a painter and cartoonist of some note. Today, he is known primarily for his artistic achievements and especially the *Mona Lisa.* Leonardo was a visionary and demonstrated a mechanical inventiveness centuries ahead of his time. His engineering accomplishments were theoretical and survive mainly in the form of drawings. Among these are rudimentary designs for various flying machines (including what appears to be a primitive sort of helicopter), ball bearings, and the smoke jack (chimney smoke driven, vaned rotor). He is included in here for his theoretical work on ball bearings and vaned rotors.

Denis Papin (1647–1712) France/United Kingdom

Noted for:

Designing the first pressure cooker

Conceiving the first cylinder and piston steam engine

Both of Papin's inventions are, of course, closely related. The pressure cooker works by fitting a saucepan with a tightly fitting sealed lid, permitting the vessel's contents to rise to a temperature above normal boiling temperature, thus enabling shorter cooking times. A pressure relief valve prevents explosions (usually!). Anyone who has watched water boil in a saucepan with a lid will probably sooner or later figure out that steam might be used to move a piston in a cylinder. Papin was the first to realize this and he sketched some rudimentary designs for steam engines, none of them as it turned out, very practical. He never actually built an engine, but it is known that Savery referenced his work.

Thomas Savery (1650–1715) United Kingdom

Noted for:

The first *useful* steam engine

The Marine Odometer

There is no doubt that a number of steam engines were built in the 1,600 plus years between Hero's reaction turbine and 1698, but not one of them was capable of doing any real work. Thomas Savery was a military engineer, and in 1698 he patented an engine designed to pump water out of mines. This consisted of a closed vessel filled with water into which pressurized steam was piped, forcing the water to a higher level at which the steam was condensed by a sprinkler. The vacuum created by condensing the steam to a liquid again was used to draw still more water upward. The action was made more or less continuous by having two containing vessels in the same apparatus or engine. Savery built and sold his engines both for pumping out mines and supplying water to buildings. He later worked with Newcomen in developing his atmospheric engine.

Thomas Newcomen (1663–1729) United Kingdom

Noted for:

- Inventing and building the atmospheric steam
- The (Newcomen) engine

Thomas Newcomen was an ironmonger who tackled the high cost of using horses to draw water out of mines. With his assistant John Cawley, a plumber, he developed the Savery engine by making it double acting; that is, steam was used both to produce work first by pressure and subsequently as it condensed, by vacuum. Newcomen formed a partnership with Savery in his later years and together they built engines that pumped out mines all over Europe. Note the size of the engine in Figure 3–2.

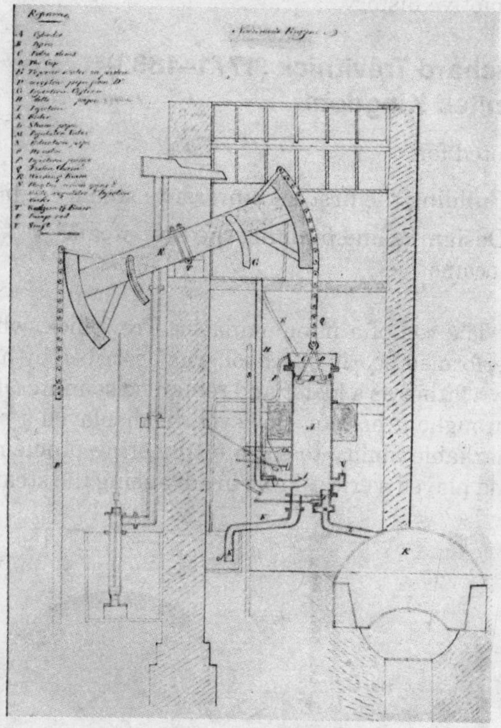

Figure 3–2 *Newcomen's 1712 engine. (Courtesy of Institute of Mechanical Engineers)*

James Watt (1736–1819) United Kingdom

Noted for:

- The Watt steam engine
- The centrifugal governor
- The physics of power analysis
- The units of power (BHP and watt)
- Planetary gearing
- Crank and throw principle
- The pressure gauge

James Watt was born just outside Glasgow, Scotland, the son of a successful ship and house builder (Figure 3–3). In childhood he was educated mostly by his mother and in his father's workshops, where at an early age he was entrusted with making models and repairing instruments. He was trained formally as an instrument technician for 1 year in London, following which he obtained employment at the University of Glasgow repairing mathematical instruments such as quadrants, compasses, and scales. This job was significant as it enabled him to continue his education and, in an informal way, meet many of the leading scientists and mathematicians of the day. One of these was Joseph Black, who invented the concept of latent heat. Watt first focused on the steam engines of the day when asked to repair a Newcomen single chamber steam engine. He was puzzled by the inefficiency of the machine, realizing that the loss of

Figure 3–3 *James Watt. (Courtesy of Institute of Mechanical Engineers)*

latent heat (the heat energy required to change the state of a substance from a liquid to a gas) prevented it from achieving better than 1% thermal efficiency.

Watt analyzed **power** production from a mathematical perspective. He then designed and patented a steam engine in 1765 with a separate condenser and expansion chamber (Figure 3–4). Until James Watt's 1765 engine, no heat engine had achieved better than 1% thermal efficiency; within 10 years, engines were built with thermal efficiencies of 4%. To put this achievement into perspective, it took another 100 years before a further quadrupling of thermal efficiency took place.

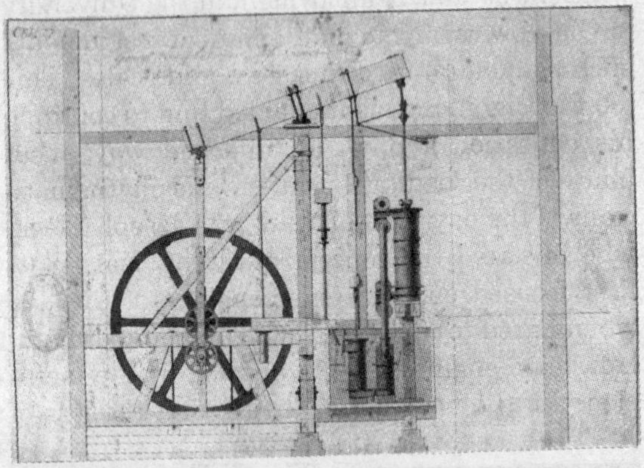

Figure 3–4 *Watt's condensing engine. (Courtesy of Institute of Mechanical Engineers)*

Watt worked for 7 years as a land surveyor after patenting his steam engine and then moved to Birmingham, England, in 1744 where he spent the rest of his life. He formed a business partnership with Matthew Boulton, an iron works manufacturer, and together they made steady improvements to the steam engine. Financially successful, Watt devoted most of the rest of his long life to inventing. Among his patents are the double-acting piston, the centrifugal governor, crank throw principle, sun and planet gearing, and the pressure gauge. Perhaps his most important contribution to technology from the twenty-first century perspective was that he wrote the language and mathematics required to understand power production, which gave his followers the tools to develop the modern engine.

Nicolas-Joseph Cugnot (1723–1804) France

Noted for:

Designing and building the world's first automobile

Cugnot was a French military engineer who designed a two-piston, steam engine referencing the work of Papin but independently from Newcomen and Watt. His engine was the first to use high-pressure steam expansively without condensation. It was mounted in a tricycle-type carriage in which a single front wheel was responsible for driving and steering the vehicle. The vehicle was severely handicapped by its need for an on-board water supply and maintaining the high-pressure steam and therefore had little practical value. However, it was the first automobile and as such has an honored place in automotive history.

Richard Trevithick (1771–1833) United Kingdom

Noted for:

Building the first high-pressure steam engine

Designing and building the first steam locomotive

The son of a mine manager, Trevithick, who performed poorly in school, was described by his own father as a loafer and remained semiliterate throughout his life. However, he displayed a remarkable ability to solve engineering problems and played a critical role in advancing the steam

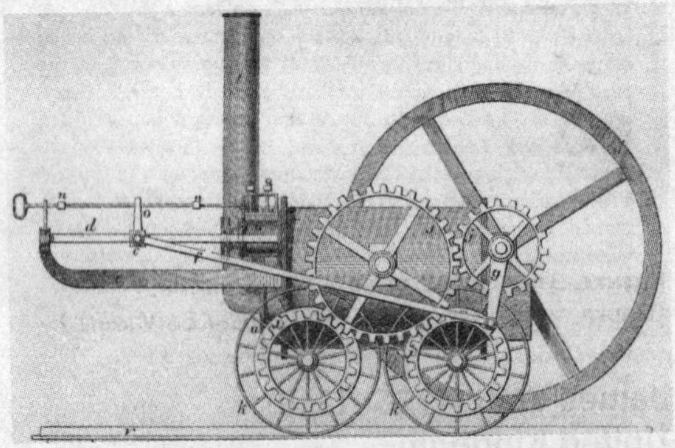

Figure 3–5 *Trevithick's locomotive of 1804. (Courtesy of Institute of Mechanical Engineers)*

engine. He reportedly had little business sense and never profited from his genius. Trevithick's success with building steam engines and locomotives occurred largely because he focused on the materials used to construct pressure vessels, rails, and drivetrain components. His locomotives were designed primarily for on-site mine operations (Figure 3–5). He died in poverty and was buried in an unmarked grave.

George Stephenson (1781–1848) United Kingdom

Noted for:

- Inventing and building the first updraft steam engine
- Building the world's first commercial railway
- Achieving the first land speed record—58 km/h (36 mph)

Stephenson was the son of a mechanic who operated a coal mine pump, atmospheric engine; he evidently had no formal education and started work at an early age. At 19 years of age he was operating a Newcomen engine, and he taught himself to read and write in his 20s. He was later taught mathematics by his only son who studied the subject in college. Stephenson first sought to improve the power of the steam engines of the day and he achieved this by inventing *steam blast* venting in which exhaust steam was redirected up the chimney, pulling air after it, and increasing the breathing draft. This innovation is said to have made the steam locomotive truly practical and began a countdown on the days the horse was to be used as a railway "engine."

Next, Stephenson turned his attention to establishing a functional railway, connecting centers located some distance apart and capable of carrying both cargo and passengers. In 1825, the commercial railway age effectively began when 450 passengers were carried from Darlington to Stockton in England at a speed of 24 km/h (15 mph). Stephenson followed this up by connecting the cities of Liverpool and Manchester, which are 64 km (40 miles) apart, by a railway line. A little later, he designed and built a locomotive named the *Rocket* to run on this track and take part in a challenge competing against three other entries; it won, achieving a then record-breaking speed of 58 km/h (36 mph). Perhaps Stephenson's greatest achievement was demonstrating the viability of reliable engine-powered transportation; his locomotives *Locomotion, Planet,* and *Rocket* (Figure 3–6) heralded a boom in railway construction throughout the world.

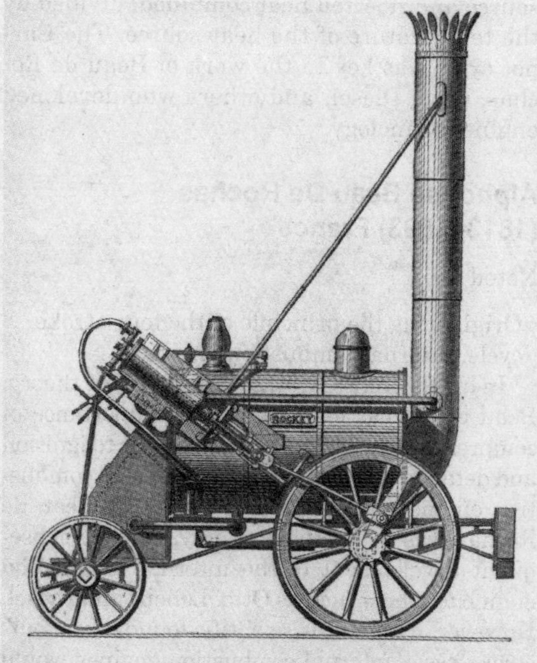

Stephenson's "Rocket."

Figure 3–6 *Stephenson's locomotive. (Courtesy of Institute of Mechanical Engineers)*

Sadi Carnot (1796–1832) France

Noted for:

- The Carnot cycle relating to heat engines

Carnot was the eldest son of the French revolutionary figure Lazare Carnot and was educated by his father and at the Ecole Polytechnique in Paris. However, as a result of the Napoleonic wars, he spent most of his working life in the French military. He developed what is described today as the Carnot cycle in the latter part of his short life, which he devoted to the study of physics.

The **Carnot cycle** relates to the ideal cyclical sequence of changes in fluids in a heat engine. During an operating cycle an engine develops heat from a heat source, delivers the results of the heat expansion to a powertrain, rejects the heat not converted to mechanical force (to a heat sink/atmosphere), and then must produce work during compression at low temperature. Carnot asserted that the ratio of work output to the heat input (represented by the heat energy potential of the fuel) is equal to the difference between the temperatures of the heat source and rejected heat combined, divided by the temperature of the heat source. The Carnot cycle was key to the work of Beau de Rochas, Otto, Diesel, and others who developed engine technology.

Alphonse Beau De Rochas (1813–1893) France

Noted for:

- Originating the principle of the four-stroke cycle, internal combustion engine

In his theoretical works on heat engine theory, Beau de Rochas emphasized the importance of compressing the air-fuel mixture prior to ignition and detailed a four-stroke cycle, internal combustion engine that he patented in 1862. Beau de Rochas was a theoretician only. He left subsequent development of the internal combustion engine to others such as Otto, Lenoir, and Diesel. However, without his scientific foundation, four-stroke cycle, internal combustion engines would not have replaced steam engines by the end of the nineteenth century.

Etienne Lenoir (1822–1900) Belgium

Noted for:

- Building the first commercially successful internal combustion engine
- The two-stroke cycle
- The electric locomotive brake

Lenoir's engine was a converted double-acting steam engine with side valves to admit air-fuel mixture and discharge combustion end gases. The engine was fueled with coal gas (essentially methane) and could achieve about a 4% thermal efficiency. Lenoir also built the first automobile with an internal combustion engine, adapting his two-stroke cycle engine to run on liquid fuels extracted from coal tar.

Allessandro Volta (1743–1827) Italy

Noted for:

- Inventing the battery
- Discovering methane gas
- Identifying static electricity

The first known battery was that invented by the Italian physicist Allessandro Volta in about 1800. Volta was an academic who taught at the University of Padua, where he was known as a philosopher. He is said to have demonstrated his battery to Napoleon. The voltaic cell was later developed by the English chemist John Daniell in 1836.

Nikolaus Otto (1832–1891) Germany

Noted for:

- Building the first four-stroke cycle, internal combustion engine

Otto built and attempted to patent the first four-stroke cycle engine in 1876, using the principles employed in the typical automobile engine of today. He was initially granted a patent for his engine but later had it revoked when it was contested by Beau de Rochas. Notwithstanding this setback, Otto built thousands of engines, and his name is used today to describe the four-stroke cycle, spark-ignited engine principle: the Otto cycle (Figure 3–7).

Rudolph Diesel (1858–1913) France/Germany

Noted for:

- Inventing the four-stroke cycle, compression ignition engine (diesel cycle)

Rudolph Diesel was born in Paris and educated in France, England, and, finally, Germany, where at the Technische Hochschule in Munich he was known as a brilliant scholar (Figure 3–8). His first job was working on thermodynamics with the refrigeration engineer Carl von Linde, and he joined the Paris-based firm in 1880. However, Diesel devoted much personal time to developing an internal combustion engine that would theoretically produce thermal efficiency values much greater than the Otto cycle engines.

Figure 3–7 *1885 ad for an Otto engine. (Courtesy of Cummins Inc.)*

In 1890 he moved to Berlin, and in 1892 he patented the compression ignition engine that today bears his name (Figure 3–9). From 1892 to 1897, Diesel worked with the MANN and Krupp companies, producing a series of increasingly successful diesel engines and culminating in a single-cylinder, coal dust-fueled version that would produce 25 brake horsepower (BHP). From the moment that Diesel introduced his patent in 1892, his engine was an international success, and versions were built in France, the United Kingdom, and the United States before the turn of the century.

Diesel was both a great linguist and salesman. In 1897, he decided to leave the development of the diesel engine to others and instead devoted himself to popularizing it throughout the world. He lectured in many different countries and became extremely wealthy on the royalty proceeds of his 1892 patent. His economic success did not, however, translate into any personal happiness. He suffered from bouts of depression and alcoholism and in 1913 mysteriously disappeared from the deck of a ship traveling from Dresden to London in what may have been a suicide.

Figure 3–8 *Rudolf Diesel.*

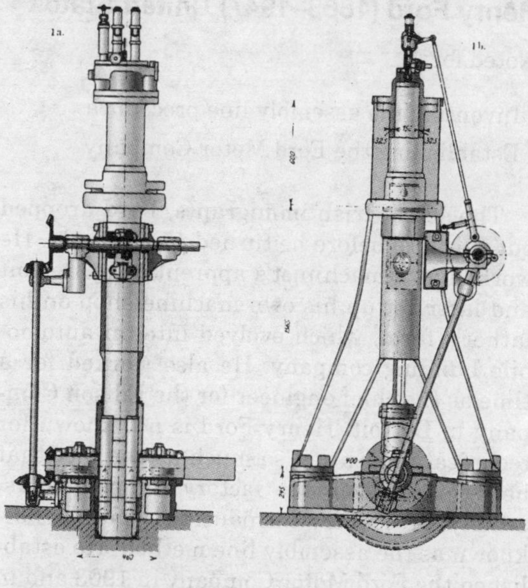

Figure 3–9 *Line drawing of Diesel's 1893 engine. (Courtesy of Cummins Inc.)*

Robert Bosch (1861–1942) Germany

Noted for:

The invention of the spark plug
The invention of the magneto
The invention of high-pressure oil injection for diesel engines

Robert Bosch was educated in the United States, where he worked with Thomas Edison and his colleague, Siegmund Bergmann. Bosch returned to Stuttgart, Germany, where he founded Bosch GmbH in 1886, a firm that became Europe's largest auto parts manufacturer. While continuing to develop the company that bore his name, Bosch was active as an engineer and inventor, primarily in the auto electrical field. However, in 1927 he overcame a problem that had stunted the progress of the diesel engine for a number of years, namely that of introducing the fuel charge to a high-compression diesel engine. Bosch devised his port-helix metering, high-pressure liquid injection pump to overcome the problem of precise fueling of diesel engines running at higher speeds. His system and competitors' copies of it remain in use today with surprisingly few hydromechanical modifications and electronic management.

Henry Ford (1863–1947) United States

Noted for:

Inventing the assembly line procedure
Establishing the Ford Motor Company

The son of Irish immigrants, Ford dropped out of school before he turned 15 years old. He worked as a machinist's apprentice in Detroit and later set up his own machine shop on his father's farm, which evolved into an automobile building company. He also worked for a time as the chief engineer for the Edison Company in Detroit. Henry Ford is not known for technical innovation so much as the fact that he revolutionized the factory component assembly procedure, introducing what is now known as the assembly line method. He established the Ford Motor Company in 1903 and in 1908 the Model T appeared. By 1913, mass production enabled him to sell it for $500. His business philosophy was to reduce unit cost and maximize sales.

Harry Ricardo (1883–1974) United Kingdom

Noted for:

The Ricardo combustion chamber—side valve design
Fuel research and theories of ignition, combustion, and detonation
The indirect injected diesel engine
Stratified charge combustion
Smoke emission controls

Ricardo was born in London and studied as a mechanical engineer at Cambridge University. Although he is best known for the SI engine side valve combustion chamber that bears his name, Ricardo is important in the history of diesel technology for his research in the nature of combustion in the diesel engine; many of his conclusions are still in practice to this day. During the Second World War, Ricardo worked extensively on controlling smoke emission from engines powering British tanks, which was making them conspicuous to enemy gunners. He was the pioneer of emissions technology and his work on diesel fuel combustion dynamics still has relevance today.

Hans List (1896–1996) Austria

Noted for:

Researching the scavenging of two-stroke cycle engines
Developing high-speed, direct-injected diesel engines
Authoring *The Internal Combustion Engine,* 1935

List was born in Graz in 1896, the son of a railway design engineer. He studied thermodynamics at the Graz Technical University and wrote his PhD thesis on the control of diesel engines. When he was 50 years old, List started a consultancy engineering firm called AVL in his hometown of Graz, which rapidly became world renowned in addressing the combustion dynamics of diesel engines. Many current high-speed, direct-injected diesel engines had at least some of their engineering undertaken by List's AVL GmbH.

Clessie Cummins (1888–1968) United States

Noted for:

- Inventing the mechanically actuated, liquid fuel injector
- Entering the first diesel-powered car in the Indianapolis 500
- Building the Cummins engine company
- Being known as the "father" of the truck engine
- Inventing the engine compression brake
- Inventing the Cummins cycle engine—a four-stroke, three-barrel design

Clessie Cummins was the first of five children born to an Indiana farmer/cooper (barrelmaker). He attended 13 different schools because his parents moved around. He left school after completing grade 8, and he remained contemptuous of formal education throughout his life. Cummins began his working life employed as a chauffeur for an Indiana banking family called the Irwins. He turned the Irwins' garage into a machine shop and with the financial help of his employers

Figure 3–10 *Cummins 6 BHP, four-stroke cycle engine of 1919. (Courtesy of Cummins Inc.)*

introduced his first diesel engine to the marketplace in 1919, a 6 BHP, single-cylinder, kerosene-fueled engine (Figure 3–10). Over the next few years, he and his company became associated with a number of publicity stunts, all of which had the objective of furthering Cummins's firm conviction that the diesel engine was the engine of the future. The first of these was staged in 1930 when a 1925 Packard, powered by a 50 BHP, four-cylinder Cummins U model, drove 792 miles from Indianapolis to the New York auto show on $1.38

Figure 3–11 *Cummins at Daytona Beach in 1931. (Courtesy of Cummins Inc.)*

of fuel. The following year he set the speed record for a diesel engine-powered car in Daytona Beach (Figure 3–11) using a U model engine now rated at 86 BHP @1,700 rpm of 101.4 mph; by 1934, Cummins had increased this record to 138 mph using a Model H engine powering a Duesenberg. Cummins also entered Cummins-powered Duesenbergs in the Indianapolis 500 with moderate success on the track but with great success in promoting diesel engines.

The model H (Figure 3–12) had been designed for a truck application, however, and Cummins adopted the same publicity stunt methods of marketing the diesel as a truck engine, first hauling a 3½ ton load from coast to coast for $11.38 of fuel in the middle of the Depression and next setting the world endurance record for any vehicle driving 14,600 miles around the Indy 500 track. The math was simple: The truck averaged 43.4 mph at a cost of ½ cent of fuel per mile and only one refuel was required at the 10,000-mile point. Clessie Cummins spent his retirement years doing what he enjoyed most—inventing. One invention was the engine compression brake, variations of which are widely used in current trucks, and another was a four-stroke cycle, three- or five-barrel, opposed piston engine that never saw production but did serve as a testament to his creativity.

Figure 3–12 *Cummins H engine. (Courtesy of Cummins Inc.)*

Vernon Roosa (1912–1989) United States

Noted for:

Inventing the inlet metering, rotary distributor diesel injection pump

The pencil injector (diesel nozzle type)

Holding over 350 U.S. patents

Vernon Roosa was born on a dairy farm in Alligerville, New York. He left school at the age of 15 and his first paid job was that of a gas pump attendant. He later became a mechanic with a bus company where he first acquired some proficiency with diesel engines. Roosa subsequently worked for a Long Island manufacturing company that made diesel-powered gensets; it was here that he first conceived the principle of a low-cost, low-maintenance rotary distributor injection pump. Roosa attempted to interest American Bosch (now Ambac), Robert Bosch GmBh, and CAV (now Delphi Lucas CAV) in his new concept for fueling diesel engines but initially had no takers. He then took his design to the Hartford Machine and Screw Company (this later became absorbed by Stanadyne Corp.) and persuaded them to invest in it. Almost simultaneously CAV became interested, resulting in both Hartford Machine and Screw (DB-series pumps) and CAV (DPA) releasing almost identical versions of his invention in 1952. The CAV DPA diesel fuel injection pump would become the biggest selling diesel fuel injection pump worldwide for reasons that are discussed in Chapter 22.

Evolution of the Computer (500 B.C.–Today)

The abacus is considered to be the earliest computing device. It is thought to be Babylonian (now modern Iraq) in origin (beginning around 500 B.C.) but its use spread throughout Europe and the Far East. It is still used in remote areas of the Far East today in place of a cash register. The abacus is a simple arithmetic calculator with racks of counting beads that represent numeric values. Expert practitioners can use them at speeds that compete with cash register input/output transactions.

The first real analog computer, the slide rule, was invented by Edmund Gunter, an English mathematician, in 1620. The slide rule was used extensively in mathematics and engineering until it was replaced by electronic digital computer technology in the 1970s. Blaise Pascal, the French scientist and philosopher, is credited with conceiving the first mechanical digital calculating machine, which was built in 1642. Various forms of Pascal's digital calculator were used until the introduction of the modern electronic digital calculator.

Eckert and Mauchly (see next section) built the first automatic electronic digital computer and launched the computer revolution that has resulted in the affordable personal computer that in a mere 20 years has changed the world more rapidly than any invention in history. At least one computer is required to manage every highway truck diesel engine sold in America today.

Presper Eckert (1919–1995) and John Mauchly (1907–1980) United States

Noted for:

Inventing the electronic digital computer

During the Second World War, Eckert and Mauchly were contracted by the United States Army to accelerate the recomputation of artillery firing tables. The result was the world's first electronic **digital computer** that could handle data in coded formats. The research took place at the University of Pennsylvania. Eckert and Mauchly called their machine an electronic numeric integrator and calculator (ENIAC) and, far from being a mere prototype of the modern computer, it actually contained all of the high-speed circuitry used in today's computers. These were the men who made today's technology possible.

Well, sort of. With them and any of the previously mentioned names in this chapter, the only real question is: How much longer would the world have waited before another person would have made the invention? The answer is, in most cases, not very long. Such is the progression of technology.

Tim Berners-Lee (1955–present) United Kingdom

Noted for:

Inventing the World Wide Web in 1990

Founding the www Consortium at MIT in 1991

Most experts agree that this was the most important invention of the twentieth century because it changed the way people communicate. Simply, it opened the doors for *anyone* to contribute to an information stream. It has revolutionized education and the speed at which technology develops: Nothing has democratized progress and thought in the way the Internet has. In trucking, generally, and in truck repair technology, the Internet is used daily.

REVIEW QUESTIONS

1. Who invented the first heat engine?
   a. Otto
   b. da Vinci
   c. Hero
   d. Diesel
2. Which of the following properly describes the first heat engine?
   a. Internal combustion
   b. Two-stroke cycle
   c. Four-stroke cycle
   d. Reaction turbine
3. Who invented the term *horsepower*?
   a. Papin
   b. Watt
   c. Stephenson
   d. Diesel
4. Who built the world's first useful steam engine?
   a. Savery
   b. Newcomen
   c. Watt
   d. Carnot
5. Who designed the first centrifugal governor?
   a. da Vinci
   b. Watt
   c. Bosch
   d. Diesel
6. In what year did Diesel patent his compression ignition engine?
   a. 1864
   b. 1892
   c. 1896
   d. 1912
7. Who patented the first high-pressure liquid fuel injection pump for a diesel engine?
   a. Otto
   b. Diesel
   c. Ricardo
   d. Bosch

8. Who is known as the "father of the highway diesel engine"?
   a. Diesel
   b. Bosch
   c. Cummins
   d. Roosa
9. What are the U.S. inventors Eckert and Mauchly known for?
   a. The first electronic digital computer
   b. The first turbocharger
   c. The first internal combustion engine
   d. The first dry cell battery
10. The world's first automobile was built by:
    a. Trivithick
    b. Watt
    c. Carnot
    d. Cugnot

4

Engine Basics

INTRODUCTION

This chapter begins by introducing some basic engine terminology: Before technicians can properly understand how an engine functions they have to become familiar with some of the language used to describe its operation. After introducing the basic terminology, the text goes on to describe the diesel cycle, two-stroke diesel cycle, and Otto cycle. At the end of the chapter, some more complex terms that build on the first batch of basic terms are discussed. The engine terminology introduced in this chapter is used repeatedly throughout the textbook; many of the terms are expanded on in later chapters.

KEY ENGINE TERMS

The terms explained here are the building blocks required to understand engine technology. Mostly the technically correct terms are used in this textbook, but remember that the terminology used on the shop floor might differ. When referencing manufacturers' service literature, a basic understanding of the key terms that are introduced in this chapter is expected. The following text simply interprets many of the words that are later used to describe the various engine cycles and key events within those cycles.

Basic Definitions

The following terms have to be understood before the basics of engine technology can be grasped. The good news is that you probably are familiar with many of them. But be careful because these definitions may be slightly different from those you are familiar with.

Engine. A machine that converts one energy form to another form of energy.

Heat engine. An engine that converts the potential **heat energy** of a combustible fuel into mechanical work. The diesel is a heat engine that functions by converting the heat energy of diesel fuel into mechanical work at its flywheel.

Internal combustion engine. A heat engine in which the combustion of fuel is contained within a cylinder. Differentiated from a steam engine in which the fuel is combusted outside of the *engine*.

Diesel engine. An internal combustion engine in which the cylinder fuel air charge is ignited by the heat of compression.

Kinetic energy. The energy of motion. When a heat engine attempts to convert the energy available in a fuel to mechanical work some heat losses result. Kinetic energy describes that percentage of the potential energy of the fuel that actually gets converted to usable mechanical energy.

Top dead center (TDC). The uppermost point of the piston travel in an engine cylinder.

Bottom dead center (BDC). The lowest point of piston travel in the engine cylinder.

Before top dead center (BTDC). Any point of piston travel through its upstroke.

After top dead center (ATDC). A point of piston travel through its downstroke.

Bore. The diameter of a cylinder. Bore is expressed as the piston sectional area over which cylinder pressures act. The bore dimension in an engine cylinder is identified in Figure 4–1.

Stroke. The distance through which a piston travels from BDC to TDC. Stroke is established by the crank throw offset; that is, the distance from the crankshaft centerline to the throw centerline multiplied by 2 equals the stroke dimension. Figure 4–1 identifies the bore and stroke dimensions in an engine cylinder.

Swept volume. The volume displaced by the piston in the cylinder as it moves from BDC to TDC. It can be calculated if both stroke and bore are known.

Clearance volume. The remaining volume in an engine cylinder when the piston is at the top of its travel or TDC. Clearance volume influences actual compression temperatures and cylinder breathing efficiencies. The clearance volume on older indirect injection (IDI) diesel engines was considerable but it is much less on today's direct-injected (DI) engines.

Cylinder volume. The total volume in the cylinder when the piston is at BDC: swept volume plus clearance volume.

Engine displacement. The *swept volume* of all the engine cylinders expressed in cubic inches or cubic centimeters/liters.

Displacement – bore × bore × stroke × 0.7854 × number of cylinders

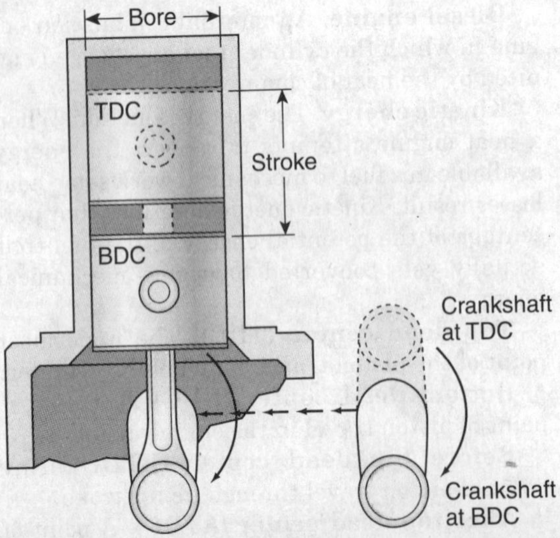

Figure 4–1 *Bore and stroke.*

Examples:

Navistar International MaxxForce 13

Engine displacement calculation data: 6-cylinder engine, bore 126 mm, stroke 166 mm

Engine displacement = 126 × 126 × 166 × 0.7854 × 6
= 12,419,134 cubic millimeters
= 12.4 liters (rounded)

Caterpillar ACERT post-2007 C15

Engine displacement calculation data: 6-cylinder engine, bore 5.4″, stroke 6.75″

Engine displacement = 5.4 × 5.4 × 6.75 × 0.7854 × 6
= 927.54 cubic inches
= 928 cubic inches (rounded)

Square engine. The term used to describe an engine in which the cylinder bore diameter is exactly equal to the piston stroke dimension. When bore and stroke values are expressed, bore always appears before stroke.

Oversquare engine. The term used to describe an engine in which the cylinder bore diameter is larger than the stroke dimension. A majority of spark-ignited, gasoline-fueled engines fall into this category, though this could change as direct-injected, gasoline-fueled engines become popular.

Undersquare engine. The term used to describe an engine in which the cylinder bore diameter is smaller than the stroke dimension. Most high-compression diesel engines are undersquare as shown by the displacement that was calculated earlier.

Compression ignition (CI). CI is the acronym commonly used to describe any diesel engine in which the fuel charge is ignited by heat generated on the compression stroke.

Spark ignited (SI). An engine in which the air-fuel charge is ignited by a timed electrical spark. When diesel engine platforms are converted to run natural gas (NG), they require an SI system.

Direct injection (DI). Either a CI or an SI engine in which the fuel charge is injected directly into the engine cylinder rather than to a precombustion chamber or part of the intake manifold. Almost all current diesel engines use DI as well as a new generation of gasoline-fueled, SI engines.

Indirect injection (IDI). A CI or an SI engine in which the fuel charge is introduced outside the engine cylinder to a precombustion chamber, cylinder head intake tract, or intake manifold.

Ratio. The quantitative relationship between two values expressed by the number of times one contains the other. The term is commonly used in automotive technology to describe the drive/driven relationships of two meshed gears, the mechanical advantage of levers, and the cylinder compression ratio.

Compression ratio. A measure of the cylinder volume when the piston is at BDC versus the cylinder volume when the piston is at TDC. Compression ratios in diesel engines fall between 14:1 and 24:1. Current, high-speed, turbocharged truck and bus diesel engines have compression ratios typically between 16:1 and 17:1. Figure 4–2 shows a diesel engine with a compression ratio of 17:1.

Compression pressure. The actual cylinder pressure developed on the compression stroke. Actual compression pressures developed range from 2.41 MPa (350 psi) to 4.82 MPa (700 psi) in CI engines. The higher the compression pressure, the more heat developed in the

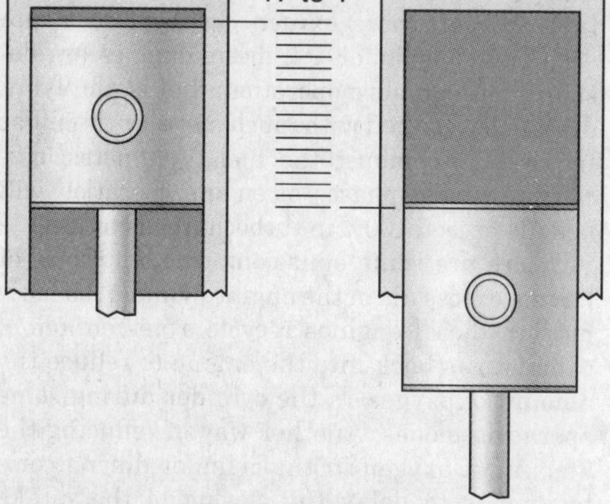

Figure 4–2 *Diesel engine compression ratio.*

cylinder. Truck diesel engines typically produce compression pressures of ± 600 psi.

Combustion pressure. The peak pressure developed during the power stroke. In today's electronically controlled truck diesel engines, combustion pressures may peak at up to five times the compression pressure.

Naturally aspirated (NA). Describes an engine whose only means of inducing air (or air-fuel mixture) into its cylinders is the low cylinder pressure created by the downstroke of the piston.

Manifold boost. Describes the extent of charge pressure above atmospheric delivered to the cylinders in a turbocharged engine. Most current truck and bus diesel engines are boosted; that is, they are turbocharged.

Volumetric efficiency. A measure of an engine's breathing efficiency. It is usually defined as the ratio between the volume of actual fresh air taken into the cylinder before the intake valve(s) close versus the cylinder swept volume. However, properly expressed, volumetric efficiency is a ratio of *masses*, not volumes. If the example of a liquid filling a pump cylinder is used to explain volumetric efficiency, it is the weight of liquid drawn into the pump cylinder in a cycle versus the maximum weight of the liquid the cylinder could contain. In a diesel engine the fluid being charged to the cylinder is air, which happens to be compressible. This means that in a turbocharged engine volumetric efficiency can often exceed 100%. Therefore, another way of defining the term is to say that it is the amount of air charged to the engine cylinder in an actual cycle versus the amount it would contain if it were at atmospheric pressure. Volumetric efficiency is usually expressed in percentage terms.

Fire point. The temperature at which a flammable liquid gives off sufficient vapor for continuous combustion to take place. This is also known as *ignition temperature*. The fire point or ignition temperature of a diesel fuel is specified by a cetane number (CN), which is explained in some detail in Chapter 16.

THE DIESEL CYCLE

A cycle is a recurring sequence of events. The **diesel cycle** is usually described by the four

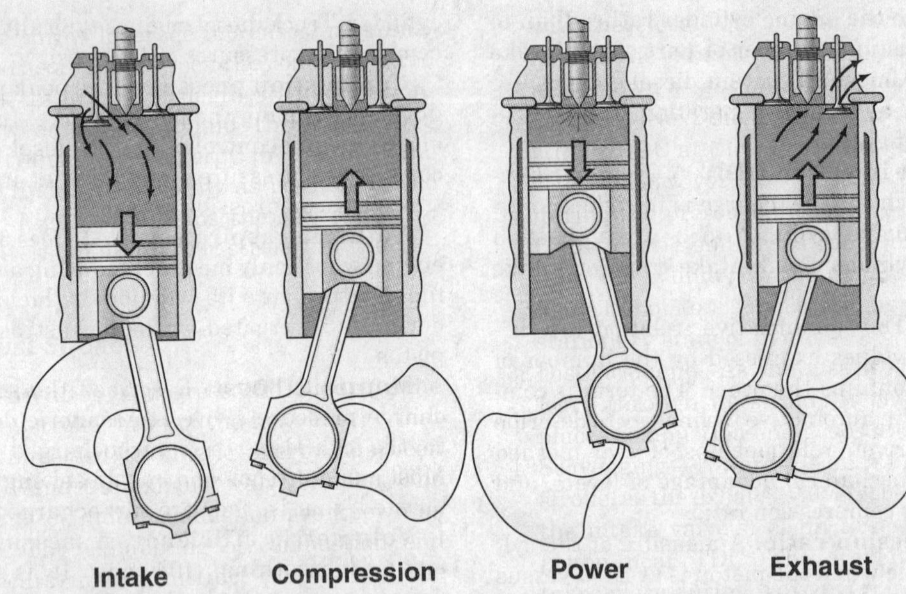

Figure 4–3 *The four-stroke diesel cycle.*

strokes of the pistons made as an engine is turned through two revolutions. A complete cycle of a diesel engine requires two full rotations and this translates into 720 crankshaft degrees. Each of the four strokes that comprise the cycle involves moving a piston either from the top of its travel to its lowest point of travel or vice versa; each stroke of the cycle therefore translates into 180 crankshaft degrees. The four strokes that comprise the four-stroke cycle are: intake, compression, power, and exhaust (Figure 4–3). The diesel cycle by definition is a four-stroke cycle. Two-stroke cycle compression ignition engines exist but, correctly, these should be qualified as two-stroke cycle diesel engines.

Direct-injection, Compression Ignition Engine

The events that make up the four-stroke cycle diesel engine take place in sequence as follows:

1. Intake Stroke

The piston is drawn from TDC to BDC with the cylinder head intake valve(s) held open. The downstroke of the piston creates lower-than-atmospheric pressure in the cylinder and in a naturally aspirated engine, this induces a charge of fresh, filtered air into the cylinder. Because most current truck and bus engines are turbocharged, the cylinder will actually be filled with charged air (that is, at a pressure above atmospheric) when the intake valve(s) open and the piston travels downward. The actual cylinder charge depends on the manifold boost value (extent of turboboost: this varies with how the engine is operated).

The *air* that is taken into the cylinder is a mixture of gaseous elements: approximately four-fifths nitrogen and one-fifth oxygen. The oxygen is required to combust the fuel. Note that the fuel is not introduced into the cylinder until later in the cycle. By pressurizing the air charge using a turbocharger, more oxygen can be forced into each engine cylinder. All diesel engines are designed for lean burn operation; that is, the cylinder will be charged with much more air than that required to combust the fuel. Volumetric efficiency in most phases of engine operation will usually exceed 100% in turbocharged engines.

There are some emissions disadvantages of having excess air in the engine cylinder, so most modern diesel engines recycle a percentage of exhaust gas back into the engine to reduce the amount of oxygen in the cylinder during some operating modes. Another way of reducing the amount of oxygen in the cylinder during combustion is to delay the closing of the intake valve on the compression stroke. This is called variable valve timing (VVT) and it has the effect of lowering the compression ratio. Figure 4–4A demonstrates the intake stroke on a modern diesel engine.

2. Compression Stroke

The piston is now driven from BDC to TDC with the intake and exhaust valves closed, compressing the charge of air in the cylinder and in doing so, heating it. Compression pressures in diesel engines vary from 400 psi (2,750 kPa) to 800 psi (5,500 kPa). The actual amount of heat generated from these compression pressures also varies, but it usually substantially exceeds the minimum ignition temperature values of the fuel. Compression ratios used to achieve the compression pressure required of diesel engines generally vary from a low of 14:1 to a high of 25:1. However, in modern turbocharged, highway diesel engines, compression ratios are typically around 16:1 to 17:1. Figure 4–4B illustrates the compression stroke on a modern diesel engine.

3. Expansion or Power Stroke

Shortly before the completion of the compression stroke, atomized fuel is introduced directly into the engine cylinder by a multi-orifii (multiple-hole) nozzle assembly. The fuel exits the injector nozzle in the liquid state in droplets appropriately sized for combustion in a DI engine. Once exposed to the heated air charge in the cylinder these liquid droplets are first vaporized then ignited. The ignition point is designed to usually occur just before the piston is positioned at TDC with the objective of peaking the gas pressure acting on the piston during the expansion stroke at 10 to 20 degrees ATDC. However, because this is the objective at all engine speeds and loads, it is difficult to uniformly achieve it with hydromechanical fuel systems. Noxious emissions requirements of newer electronically managed engines often mean that the power stroke may not be managed to produce optimum mechanical efficiency because of the requirement to remain within legal emission specifications. Figure 4–4C illustrates the power stroke in a typical diesel engine.

In managing the power stroke it is desirable to have little pressure acting on the piston at TDC. Cylinder gas pressure should peak at 15 to 20 degrees ATDC when throw leverage exists but is close to minimum. As gas pressure acts on the piston and forces it through its stroke, the cylinder pressure will decrease, but

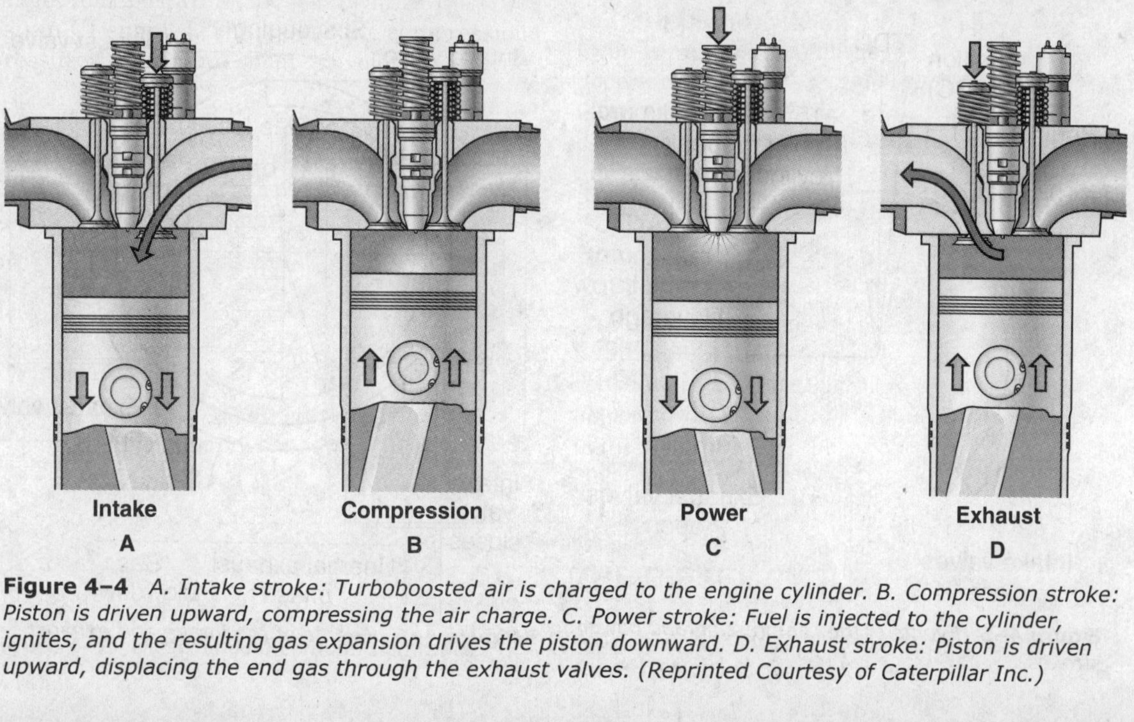

Figure 4–4 *A. Intake stroke: Turboboosted air is charged to the engine cylinder. B. Compression stroke: Piston is driven upward, compressing the air charge. C. Power stroke: Fuel is injected to the cylinder, ignites, and the resulting gas expansion drives the piston downward. D. Exhaust stroke: Piston is driven upward, displacing the end gas through the exhaust valves. (Reprinted Courtesy of Caterpillar Inc.)*

as it does throw leverage increases, maximizing when the angle between the connecting rod and crank throw is at 90 degrees. This relationship between pressure and throw leverage helps to transmit the energy produced in the engine cylinder as smoothly as possible to the flywheel. Figure 4–5 maps the key events of the compression and power strokes in the diesel cycle.

4. Exhaust Stroke

Somewhere after 90 degrees ATDC during the expansion stroke, most of the heat energy that can be converted to kinetic energy has been converted and the exhaust valve(s) open. The products of cylinder combustion are known as end gas. The exhausting of combustion end gases occurs in four distinct phases and the process begins during the latter portion of the power stroke (see Figure 4–4D):

1. Pressure differential—At the moment the exhaust valves open during the latter portion of the power stroke, pressure is higher in the cylinder than in the exhaust manifold. High-pressure end gas in the cylinder will therefore flow to the lower pressure zone in the exhaust manifold. This phase is also known as *gas blowdown*.
2. Inertial—Next the piston comes to a standstill at BDC at the completion of the power stroke. However, gas inertia established during the pressure differential phase will result in the end gases continuing to flow from the cylinder to the exhaust manifold while the piston is in a stationary and near-stationary state of motion.
3. Displacement—As the piston is forced upward through its stroke, it positively displaces combustion end gases above it.
4. Scavenging—Toward the end of the exhaust stroke, as the exhaust valve(s) begin to close, the intake valve(s) begin to open with the piston near TDC. The scavenging phase takes place during valve overlap and can be highly effective in expelling end gases and providing some piston crown cooling. The efficiency of the scavenging process is greatest with turbocharged engines. Figure 4–6 shows the events of the two breathing strokes, intake and exhaust.

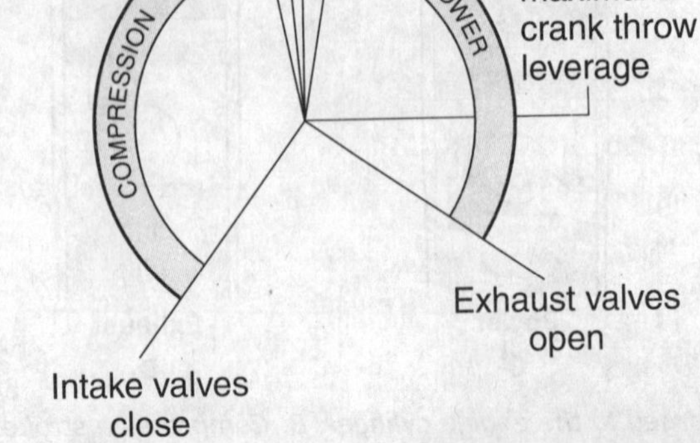

Figure 4–5 *Events of the compression and power strokes.*

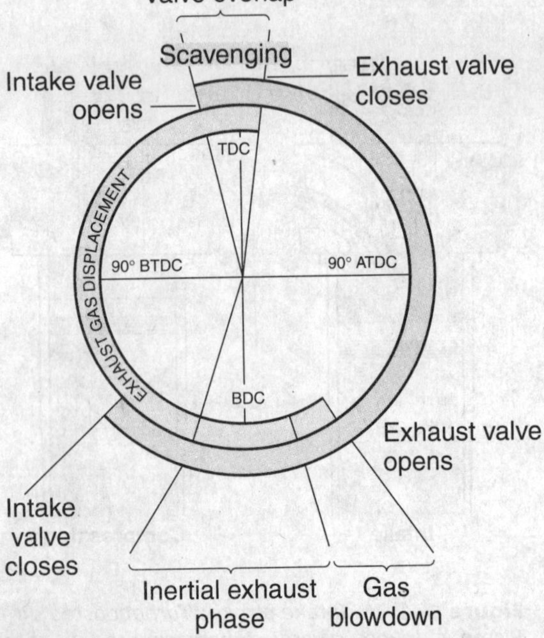

Figure 4–6 *Events of the intake and exhaust strokes.*

THE TWO-STROKE CYCLE DIESEL ENGINE

The intake and exhaust strokes are eliminated, meaning that in 360 degrees of rotation, each engine cylinder has fired. Theoretically, the two-stroke cycle diesel engine should develop twice as much power as a four-stroke cycle engine of the same displacement, but in reality this is not achieved mainly due to reduced cylinder breathing efficiency. The two-stroke cycle diesel engine principle used to be widely used by the diesel industry generally but not commonly in on-highway diesels. The only examples in the truck and bus industry were manufactured by Detroit Diesel Corporation and not too many of these remain because they cannot meet current emissions standards. Many of the on-highway, two-stroke cycle engines still in existence today are managed electronically but even these are disappearing fast. Refer to Figure 4–7 to follow the explanation of the two-stroke cycle in the next sections.

Two-stroke Cycle Construction

The physical characteristics of the two-stroke cycle engine differ from those of the four-stroke cycle diesel engine mainly because cylinder breathing must take place in less than one-fifth of the time. All the valves in the cylinder head, usually four per cylinder, are exhaust valves. After combustion, the cylinder end gases must be expelled and to enable this, air must be pumped through the cylinder from an air box charged by a Roots blower, which is sometimes aided by a turbocharger. In on-highway applications, because of widely variable speed and load variations, engine breathing requires the use of a Roots blower. A Roots blower is a positive displacement pump that works efficiently at all rotational speeds. Its disadvantage is that it is gear driven and leeches some engine power.

A turbocharger may completely replace the Roots blower in applications such as a genset designed to run at continuously high output, resulting in sufficient exhaust gas heat to drive the turbine with sufficient efficiency to scavenge the cylinders. But in on-highway applications the Roots blower must be used and a turbocharger is used in addition to the blower. The cylinder liners are machined with ports designed to be exposed when the piston is in the lower portion of its downstroke. When these ports are exposed to the air box by the downward traveling piston, the cylinder is charged with air for scavenging and breathing. The ports are usually canted (angled) to encourage a vortex (cyclonic) airflow dynamic.

Two-stroke Cycle Events

The two-stroke cycle sequence begins with the piston at BDC when cylinder *scavenging* takes place. At this moment the piston has fully exposed the canted intake ports and the exhaust valves are fully opened. Air from the air box rushes into the cylinder and displaces the combustion end gases, spiraling them upward to exit through the exhaust valves. Air from the air box continues to charge the cylinder until the piston

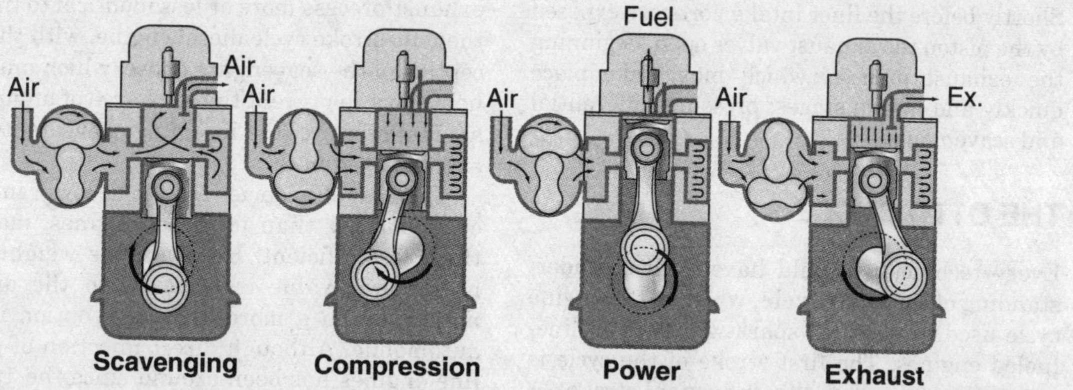

Figure 4–7 *The two-stroke diesel cycle.*

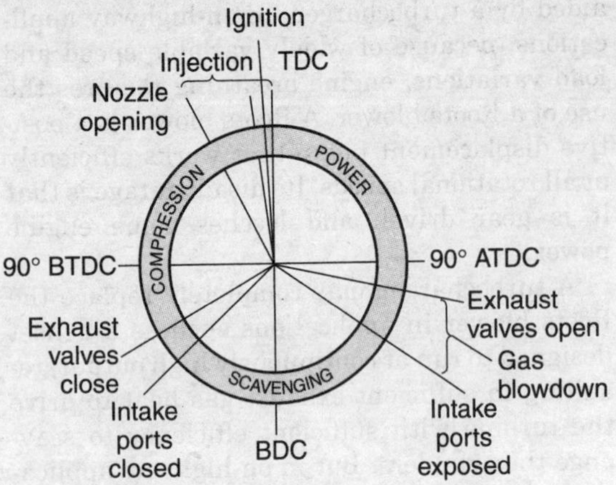

Figure 4–8 *Two-stroke diesel cycle events.*

reverses and its upward travel closes off the intake ports; the exhaust valves close almost simultaneously. Every upward stroke of the piston is therefore a compression stroke. Figure 4–8 maps the events of the two-stroke cycle diesel engine.

Scavenging Efficiency

If all the end gases were effectively expelled, only air would be compressed during the compression stroke. However, scavenging efficiency is a problem with these engines and any end gases remaining in the cylinder after the exhaust valves close will dilute the incoming air charge. Shortly before TDC the fueling of the cylinder begins directly into the engine cylinder. After a short delay, ignition occurs and expanding combustion gases act on the piston and drive it downward through the power stroke. Every downward stroke of the piston is a power stroke. Shortly before the liner intake ports are exposed by the piston the exhaust valves open, beginning the exhaust process, which must take place quickly and in two stages: pressure differential and scavenging.

THE OTTO CYCLE

Every technician should have a clear understanding of the **Otto cycle**, which is the engine cycle used in most SI (spark-ignited) gasoline-fueled engines. The first stroke of the cycle is appropriately called the induction stroke. A charge of air-fuel mixture is induced into the engine cylinder by the low pressure created by the downstroke of the piston in the engine cylinder as it moves from TDC to BDC. The air-fuel charge is mixed outside the engine cylinder, either by a carburetor located upstream from the intake plenum or by injectors located at the intake tract of the cylinder head. When the piston stops at BDC and the intake valve(s) close, the cylinder contains all of the ingredients, that is, the fuel and air required for the subsequent power stroke. The mixture of fuel and air is compressed on the second stroke of the cycle, which is the compression stroke. Compression ratios must be somewhat lower on SI engines because the compression ignition of the more volatile fuel used (gasoline) must be avoided. Compression ratios on gasoline engines have crept upward over the years. Something around 8:1 to 10:1 is typical. This produces compression pressures around 150 psi (10 bar). The cylinder charge is ignited just before the completion of the compression stroke.

The power stroke can produce peak combustion pressures approximately five times the compression pressure. The gasoline fuels used are chemically more complex than diesel fuels and must be formulated with a mixture of fractions (hydrocarbon compounds derived from crude petroleum) that ignite in a sequential chain reaction, expanding to apply pressure to the piston evenly through the power stroke. As with the diesel engine, the cylinder pressure and the crank throw angle must be phased to transfer the power smoothly and evenly to the flywheel. At some point after the piston has passed 90 degrees ATDC on the power stroke, the exhaust valves open and begin a four-phase exhaust process more or less identical to that of the four-stroke cycle diesel engine, with the exception of the scavenging phase, which must be held short to prevent the discharge of unburned air-fuel mixture (the incoming charge) into the exhaust system.

Thermal efficiencies tend to be lower in Otto cycle engines than in diesel engines, making them less efficient. However, the weight-per-horsepower value is better than the diesel, making them a more attractive option in an automobile. Although direct injection of gasoline engines has been around since the 1920s, recent innovations have made this an option on a range of current automobile engines. DI has the potential to greatly improve the efficiency of gasoline-fueled engines.

Atkinson Cycle

The **Atkinson cycle** is a variation on the Otto cycle. By connecting the piston to the crankshaft with a double pivot linkage, an Atkinson cycle enables the four piston strokes of the Otto cycle to take place in one crankshaft revolution. It also permits a longer power stroke than compression stroke, promising theoretically higher efficiencies than the Otto cycle. Today, the term "Atkinson cycle engine" is being misused to describe a number of VVT engines in which the intake valves can simply be held open longer. Delaying intake valve closure promotes the reverse flow of air back into the intake manifold because the beginning of the compression stroke is retarded. Diesel technicians should understand the term because it has been used to describe diesel VVT engines.

Miller Cycle

The **Miller cycle** is another variation on the Otto cycle engine. All Miller cycle engines use a supercharger. As with the Atkinson cycle, the intake valves are held open for the first part of the compression stroke. The result is that the piston compresses both the cylinder and intake manifold up to the supercharger during the first part of the compression stroke. This is said to increase efficiency.

ENGINE SYSTEMS AND CIRCUITS

Many of the components in the diesel (compression ignition or CI) engine are identical to those in the Otto (spark-ignited or SI) cycle engine. For study purposes, the engine components are divided as follows:

1. Engine housing components—the cylinder block, cylinder head(s), oil sump, rocker covers, timing gear covers, and flywheel housing.
2. Engine powertrain—the components directly responsible for power delivery including the piston assemblies, connecting rods, crankshaft assembly, vibration damper, and flywheel.
3. Engine feedback assembly—the engine's self-management components also known as the valvetrain assembly. This term is used to describe the diesel engine timing geartrain, camshaft, valvetrains, valves, fueling apparatus, and accessory drive components.
4. Engine lubrication circuit—the oil pump, relief valve, lubrication circuitry, full flow filter(s), bypass filters, and heat exchangers.
5. Engine cooling circuit—the coolant pump, thermostat(s), water jacket, coolant manifold, filter, shutters, fan assembly, radiator, and other heat exchangers.
6. Engine breathing system—the engine intake and exhaust system components including precleaners, air cleaners, ducting, turbocharger, Roots blower, charge air heat exchangers, tip turbine assemblies, intake and exhaust manifolds, pyrometer, exhaust gas recirculation (EGR) system, exhaust piping, engine silencer, catalytic converter, and other external emission control apparatus.
7. Engine Fuel Management System—the fuel storage, pumping, metering, and quantity control apparatus including a management computer, sensors and actuators, hydraulic injectors, mechanical unit injectors (MUIs), electronic unit injectors (EUIs), hydraulically actuated electronic unit injectors (HEUIs), electronic unit pumps (EUPs), common rail (CR) injection, hydromechanical injection pumps, fuel tanks, filters, and transfer pumps.

Diesel Fuel

Diesel fuel is covered in some detail in later sections of this book, so this is just a quick introduction. The fuel used in modern diesel engines on North American highways is composed of roughly 85% carbon and 12% to 16% hydrogen, much the same as the chemical composition of gasolines. Unlike gasolines, diesel fuel does not vaporize readily at ambient temperatures, so it is less likely to form combustible mixtures of fuel and air. The heat required to ignite the fuel oil is defined by the most volatile fractions of the fuel. This temperature is usually around 250°C; it would be higher, around 290°C (550°F), using fuel with the poorest ignition quality that can be sold legally for use on North American highways.

Using diesel fuel oils with a fire point of around 250°C, at least this temperature must be achieved in the engine cylinder during the compression stroke of the piston if the fuel is to be ignited. In fact, actual cylinder temperatures generated on the compression stroke tend to be considerably higher than the minimum required to ignite the fuel. The greater the difference between these two temperature values, the shorter the ignition lag. **Ignition lag** is the time between the entry of the first droplets of fuel into the engine cylinder and the moment of ignition that begins combustion.

ADVANCED ENGINE TERMS

After building an understanding of some basic engine terms and learning how the different engine cycles function, be ready for some of the more complex terms required to fully understand the technology of the diesel engine. These terms are used later on in the textbook and usually are defined in more detail.

Mean effective pressure (MEP). A way of describing the cylinder pressure that can actually be converted into torque. Simply it is combustion expansion pressure (pressure developed in the engine cylinder during the power stroke) minus the compression pressure. In a four-stroke cycle CI engine it is assumed that cylinder pressure through the intake and exhaust strokes is zero for purposes of calculating MEP, because through both there is no significant amount of pressure. MEP describes the relationship between the *work performed* by the piston (in compressing the air charge) and the *work received* by the piston (through its downstroke on the power stroke). If the engine is going to continue to rotate, there has to be a net gain in terms of work. MEP is an important definition. For instance, it is important to understand MEP to explain how an internal engine compression brake functions. The principle of the internal engine compression brake is to apply braking force to the drivetrain by making the piston do all of the work it normally does through the compression stroke but then eliminate the force it would normally be subject to receive during the powerstroke.

Constant pressure cycle. The theoretical diesel cycle engine presumes that the fuel supplied to the cylinder during the expansion stroke will be at a rate permitting combustion pressure to remain constant through a large portion of the stroke. The theoretical diesel cycle is neither practical nor desirable (see the following section).

Cylinder pressure and throw leverage. The objective of any engine is to transfer the power developed in its cylinders as smoothly and evenly as possible to the power take-off mechanism, which is usually a flywheel. The relationship between the crankshaft throw (to which the piston assembly is connected) and the crankshaft centerline is that of a lever. A lever is a device that provides a mechanical advantage. The amount of leverage (mechanical advantage) depends on the rotational position of the throw, which ranges from zero or no leverage when the throw is positioned at TDC to maximum leverage when the throw is positioned at a 90-degree angle with the connecting rod after TDC. This makes the relationship between cylinder pressure (gas pressure acting on the piston) and throw leverage (the position of the piston) critical in meeting the objective of a smooth/even transfer of power from the engine cylinders to the drivetrain.

Let us take a look at this principle in operation. When the piston is at TDC beginning a power stroke, it is desirable to have minimum cylinder pressure because in this position throw leverage is zero; therefore, no power transfer is possible. A properly setup fuel system attempts to manage cylinder pressure so that in any given performance mode it peaks somewhere between 10 degrees and 20 degrees ATDC when there is some but nevertheless a small amount of throw leverage. As the piston is forced down through the power stroke, the gas pressure acting on the piston diminishes, but as it does throw leverage increases. Ideally, this relationship between cylinder pressure and crank throw leverage should be managed in a way that results in consistent torque delivery from an engine cylinder through the power stroke until the throw forms a 90-degree angle with the connecting rod: This occurs a little before true 90 degrees ATDC.

Boyle's law (Robert Boyle, U.K., 1627–1691). States that the absolute pressure that a given quantity of gas at constant temperature exerts against the walls of a container is universally proportional to the volume occupied. In other words, assuming a constant temperature, the pressure of a specific quantity of gas depends on the volume of the vessel it is contained in. So to use an example of this law as it applies to a diesel engine, it means that a constant temperature mass of gas (air) in a cylinder as its volume is reduced by moving the piston will exert pressure on the cylinder walls because the number of molecules will remain the same, but they will have less room to move in, thereby causing the pressure rise.

Charles's law (Jacques Charles, France, 1746–1823). States that the increase in temperature in gases produces the same increase in volume if the pressure remains constant. In other words, heating a gas must result in an increase in volume if the pressure is to remain unchanged. Using the Celsius scale, it can be proved that the volume of a gas increases by 0.003663 of its volume at zero Celsius for every one degree of temperature rise. On the Fahrenheit scale, the volume increases by 0.002174 for every one degree of temperature rise above 32°. By graphing this equation negatively, a point would be reached at which the gas would have no volume; the vibration of the molecules would cease and the gas would contain no heat energy and would cease to exist as a substance. This would occur at absolute zero or – 273°C (–460°F). To conclude, if the volume of a gas is changed by increasing its temperature while keeping its pressure constant, then its volume will increase proportionally with temperature rise.

The first law of thermodynamics. States that heat energy and mechanical energy are naturally convertible. This is predicated by the law of conservation of energy, which states that energy can neither be created nor destroyed. This means that the total energy available remains constant. However, energy can change its form. Heat energy can be changed into mechanical energy (the operating principle of any internal combustion engine) or vice versa. Similarly, heat energy can be changed into electrical energy and vice versa.

The second law of thermodynamics. States that heat will not flow from a cool body to a warmer body without some kind of assistance, but that it will flow from a warm body to a cooler body. If the objective is to force heat from a cool body to a hot body, such as in an air-conditioning system, some form of external assistance must be applied.

Friction. Force is required to move an object over the surface of another. Friction is the resistance to motion between two objects in contact with each other. Friction is factored by both load and surface condition. Smooth surfaces produce less friction than rough surfaces, and if a lubricant such as water or oil is added, friction diminishes. Lubricants coat and separate two surfaces from each other and reduce friction, but the lubricant itself provides some resistance to movement, which is known as viscous (fluid) friction. A friction bearing such as a crankshaft main bearing provides a sliding friction dynamic, whereas ball bearings provide a rolling friction dynamic that usually offers less resistance to motion.

Static friction. Describes the characteristic of an object at rest to attempt to stay that way. For example, the engine piston at its travel limit stops momentarily before the crankshaft and connecting rod reverse its movement. When the piston is momentarily stopped, the crankshaft must overcome the static friction of the stationary piston, which places both the connecting rod and a portion of the crankshaft under tension. This tensile loading of the connecting rod and crankshaft is amplified as rotational speed increases.

Inertia. Describes the tendency of an object in motion to stay in motion or, conversely, an object at rest to remain that way. Kinetic inertia describes the characteristic of an object in motion to stay in motion. For example, an engine piston moving in one direction must be stopped at its travel limit, and its kinetic inertia must be absorbed by the crankshaft and connecting rod. The inertia principle is used by the engine vibration damper and the flywheel—the inertial mass represented by the flywheel would have to be greatest in a single-cylinder, four-stroke cycle engine. As the number of cylinders increases, the inertial mass represented by the flywheel can be reduced due to the greater mass of rotating components and the higher frequency of power strokes.

Joule's heat efficiency (James Prescott Joule, U.K., 1818–1889). **Joule** established the relationship between the units of heat and work that could be done. This relationship is used to describe the potential energy of a fuel and is known as Joule's mechanical heat equivalent.

1 Btu of potential heat energy = 778 ft.-lb of mechanical energy

1 J (joule) of potential heat energy = 1 J of mechanical energy

1 J = 1 Newton/meter (N·m) = 0.7374 ft.-lb

Calorific value. The potential heat energy of a fuel. A heat engine attempts to convert the potential heat energy of a fuel into kinetic energy: The thermal efficiency of the engine is a measure of how successful this is. The calorific value of fuels is measured in Btu (English system) or joules and calories (metric system).

British thermal unit (Btu). A Btu is a measure of heat energy. One Btu is the amount of heat required to raise the temperature of 1 pound of water 1 degree Fahrenheit.

Thermal efficiency. A measure of the combustion efficiency of an engine calculated by comparing the heat energy potential of the fuel (calorific value) with the amount of usable mechanical work produced. Electronically controlled CI engines can have thermal efficiency values exceeding 40%.

Rejected heat. That percentage of the heat potential of the fuel that is not converted into useful work by an engine. If a CI engine operating at optimum efficiency can be said to have a thermal efficiency of 40%, then 60% of the calorific value of the fuel has to be discharged as *rejected heat*. Half of the rejected heat is typically transferred to the engine hardware to be dissipated to the atmosphere by the engine cooling system, and the other half exits in the exhaust gas. A turbocharger makes use of rejected heat by compressing the intake air forced into the engine cylinders, thereby increasing the thermal efficiency of the engine. Figure 4–9 shows how the potential energy of the fuel (100%) is released in a typical diesel engine: 33% is converted into useful mechanical energy delivered to the flywheel. The remainder of that potential heat energy totaling 67% is released as rejected heat. The rejected heat is dispersed as radiated heat (7%) to engine coolant (30%), and into the exhaust (30%).

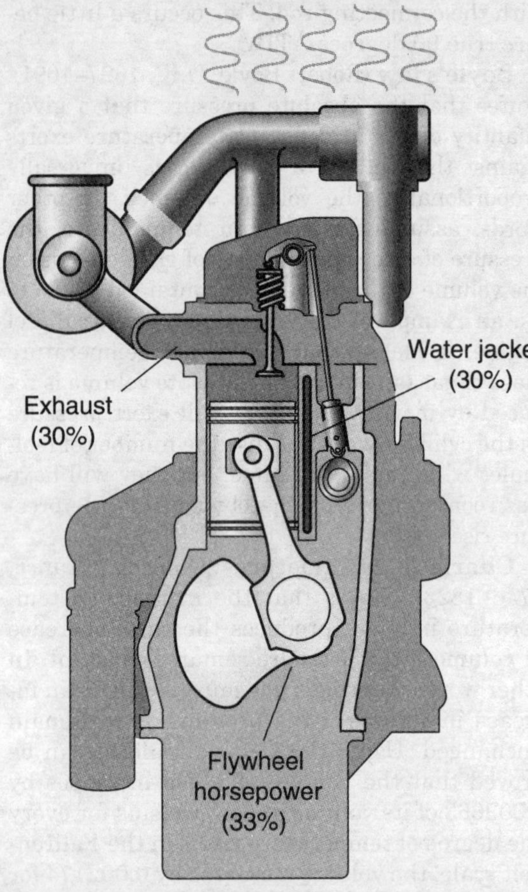

Figure 4–9 *How the potential energy of diesel fuel is released in a diesel engine.*

REVIEW QUESTIONS

1. A popular six-cylinder Navistar diesel engine has a bore of 120 mm and a stroke of 155 mm. What is the engine displacement?
   a. 9.4 liters
   b. 10.5 liters
   c. 12.4 liters
   d. 13.6 liters

2. Which of the following accurately describes engine displacement?
   a. Total piston swept volume
   b. Mean effective pressure
   c. Peak horsepower
   d. Peak torque
3. Engine breathing in the two-stroke cycle is usually referred to as:
   a. Inertial
   b. Scavenging
   c. Exhaust displacement
4. The tendency of an object in motion to want to stay in motion is known as:
   a. Kinetic energy
   b. Dynamic friction
   c. Inertia
   d. Mechanical force
5. When a modern diesel engine is running at optimum efficiency, the percentage of rejected heat would typically be:
   a. 20%
   b. 40%
   c. 60%
   d. 80%
6. What percentage of the potential heat energy of the fuel does the modern diesel engine convert to kinetic energy when it is operating close to optimum efficiency?
   a. 20%
   b. 40%
   c. 60%
   d. 80%
7. In which of the four strokes of the cycle is the cylinder pressure at its highest in a running diesel engine?
   a. Intake
   b. Compression
   c. Power
   d. Exhaust
8. Where does scavenging take place on a four-stroke cycle diesel engine?
   a. BDC after the power stroke
   b. TDC after the compression stroke
   c. Valve overlap
   d. 10 to 20 degrees ATDC on the power stroke
9. Ideally, where should peak cylinder pressure occur during the power stroke?
   a. TDC
   b. 10 to 20 degrees ATDC
   c. 90 degrees ATDC
   d. At gas blowdown
10. Which of the seven subcircuits of the engine deals with the camshaft, valvetrains, and mechanical actuation of injectors and fuel pumping apparatus?
    a. Powertrain
    b. Feedback assembly
    c. Lubrication circuit
    d. Engine housing circuit
11. A Cummins ISB engine has a displacement of 6.7 liters. Convert this to cubic inches.
    a. 365 cubic inches
    b. 409 cubic inches
    c. 444 cubic inches
    d. 530 cubic inches
12. A 2007 Caterpillar ACERT C13 engine has an actual total cylinder swept volume of 12.5 liters. Convert this to cubic inches.
    a. 454 cubic inches
    b. 555 cubic inches
    c. 610 cubic inches
    d. 762 cubic inches
13. A 2004 Mack Truck E-Tech engine has a total displacement of 728 cubic inches. Express this value in metric terms.
    a. 11.1 liters
    b. 11.9 liters
    c. 12.7 liters
    d. 14 liters
14. A six-cylinder engine has a bore of 4.875″ and a stroke of 6.5″. Calculate the total displacement and express it in cubic inches.
    a. 530 cubic inches
    b. 610 cubic inches
    c. 728 cubic inches
    d. 855 cubic inches
15. A Volvo six-cylinder engine has a bore of 144 mm and a stroke of 165 mm. Calculate the total displacement and express it in liters.
    a. 11.1 liters
    b. 12.7 liters
    c. 14.8 liters
    d. 16.1 liters
16. When running an engine at idle speed, which would typically be the optimum location to peak the cylinder pressure?
    a. 20 degrees BTDC
    b. 10 degrees BTDC
    c. TDC
    d. 15 degrees ATDC

17. When running an engine at rated speed and load, which would typically be the optimum location to peak the cylinder pressure?
    a. 30 degrees BTDC
    b. 15 degrees BTDC
    c. TDC
    d. 15 degrees ATDC
18. At which point in the engine cycle does the crank throw leverage peak?
    a. TDC
    b. Around 15 degrees ATDC
    c. Just before 90 degrees ATDC
    d. BTDC
19. Which law states that all energy forms are convertible?
    a. The first law of thermodynamics
    b. The second law of thermodynamics
    c. Boyle's law
    d. Charles's law
20. The energy of motion is known as:
    a. Inertia
    b. Kinetic
    c. Thermal
    d. Potential
21. A six-cylinder engine has a bore of 5.69 inches and a stroke of 6.5 inches. Calculate the total displacement and express it in liters.
    a. 11.1 liters
    b. 12.7 liters
    c. 14.8 liters
    d. 16.1 liters
22. A six-cylinder engine has a bore of 4.875″ and a stroke of 6.5″. Calculate the total displacement and express it in liters.
    a. 10.3 liters
    b. 11.9 liters
    c. 12.5 liters
    d. 12.8 liters

5

Power

INTRODUCTION

In the 1,600 years between Hero and Watt (see Chapter 3), many heat engines had been designed and built, but not one had achieved better than 1% thermal efficiency and most a lot less. James Watt began a process of analyzing all the factors that contribute to an engine's ability to perform work and wrote the language and mathematics to describe it. It is important that the diesel technician acquire a basic understanding of power production. Although it is possible to operate diagnostic equipment such as **dynamometers** simply by matching specifications to readouts on test instruments, the technician will not be capable of properly analyzing test data without understanding some of the language and technology of power.

Measuring Power

The intent of this chapter is to simplify some of the concepts of power. Technicians can further develop their knowledge in this area by consulting more advanced reference texts and doing some online research. When it comes to understanding power, both standard and metric measurement systems are used by industry. In reality, however, when rating the power of engines it is far more likely to use the term **horsepower** rather than its metric equivalent kilowatts. The best solution is for the technician to become familiar with both standard and metric systems. Despite the fact that every diesel engine a technician is likely to encounter in a modern truck was engineered using the metric system, engine OEMs will most often publish power analysis specifications using only the standard system. This means that technicians should first begin to understand power using the standard system.

DEFINITIONS AND FORMULAE

To understand the basics of power, it is first necessary to understand the language used to describe it. The term *power* is very often misused, which adds to the difficulty in understanding it. We begin by introducing a series of definitions and then work toward developing them into the formulae required to calculate power. It is important that the contents of this section be properly understood before tackling the latter chapters on engine dynamometers and power analysis.

Energy

Energy is best defined as the capacity for producing work. Energy exists in a number of forms; kinetic, potential, electrical, thermal, chemical, and nuclear are its common forms. Kinetic energy is the energy of motion, and any body in a state of motion is described as possessing kinetic energy. Air in its compressed state and contained in a reservoir has the potential for creating motion and is therefore described as *potential energy*. A compressed spring does also, so it, too, has potential energy.

First Law of Thermodynamics

The first law of thermodynamics tells us that energy can be neither created nor destroyed; however, the way in which energy manifests itself can be changed. In an internal combustion engine, the potential heat energy of a fuel is released as it is combusted. This produces gas expansion resulting in pressure that acts on a piston and drives it through its power stroke. In this way, the heat potential of the fuel produces kinetic energy. Not all of the potential heat energy of the fuel can be successfully converted to kinetic energy. The energy released by combusting fuel in an engine cylinder that cannot be converted to kinetic energy is described as **rejected heat**. Rejected heat has to be dissipated to atmosphere.

Effects of Heat Energy

Heat is easily converted into mechanical energy and this can happen in many ways. The sun's heat raises millions of tons of water vapor high into the atmosphere daily, so all the mechanical energy of falling water, whether as rain or snow in rivers or glaciers, stems directly from the sun's heat. The term *heat engine* is used to describe a wide range of engines that convert the heat potential of fuels into mechanical energy. Included in the category of heat engines are rocket engines, steam engines, jet turbines, gasoline, and diesel engines.

Force

Force is generally defined in terms of the effects it produces, although it should always be remembered that force can be exerted with no result. If force is applied to a body at rest, it may be sufficient to cause the body to move; however, it might not. In other words, you can apply a lot of force to an object, but unless the object is moved no work results. Using the standard system, force is measured in pounds. In the Système Internationale (SI) metric system, the Newton is the unit by which force is measured.

1N = force that produces acceleration of 1 meter/second on a mass of 1 kg

How Force Is Produced in an Engine

In a diesel engine, force is represented by cylinder pressure. Cylinder pressure results from combusting fuel in the engine cylinder. This force acts on the sectional area of the piston crown and is transmitted to rotary motion by acting on the crank throw. In other words it produces **torque**. Torque is twisting force and because it is classified as force it does not necessarily result in any work accomplished. If you can recall attempting to ride your tricycle up a steep hill when you were 3 years old, there would have come a point when you simply could not apply enough force to the pedals (producing torque at the drive wheel) to keep moving. In a diesel engine, the amount of force is controlled by the amount of fuel delivered to the engine cylinders. This is because of the excess air factor in a diesel.

Work

Work is accomplished when the application of force produces a result. When the definition is applied to an engine, work is accomplished when force acting on the piston results in piston travel. In the following formula, standard values are always listed first:

$$\text{Work (watt-second/joules)} = \text{Force (pounds/Newtons)} \times \text{Distance (feet/meters)}$$

To use a nonengine analogy to define work, if two persons of identical weight run exactly 100 meters, each has accomplished the same amount of work.

Engine Torque

Torque is turning effort. In a typical engine, the force acting on the piston is transmitted to a crankshaft throw. The throw is a lever. The farther offset it is from the crankshaft centerline, the more leverage it can exert. Torque is the product of force applied to the crank throw. The ability to produce torque in an engine is directly related to cylinder pressure. Because of this we can say that peak torque will always occur when cylinder pressures are at their highest.

A bicycle is also an engine. The power source is the rider. A bicycle has a crankshaft driven by a pair of throws called pedals, offset 180 degrees. Peak torque is realized whenever muscle force acting on the pedals is at a maximum. High torque is required to propel a bicycle and the weight of its rider up a steep hill so gear selection must be made so that the required torque is within the capabilities of its power source, the cyclist.

Torque = Force (pounds/Newtons) × Leverage (feet/meters)

Power

Power is the rate of doing work. If two sprinters of equal weight run 100 meters, each will have accomplished the same amount of work. However, if one runs this distance in 20 seconds and the other in 10 seconds, the latter has twice the power.

$$\text{Power (watts)} = \frac{\text{Work (joules)}}{\text{Time (seconds)}}$$

Horsepower

James Watt coined the term horsepower before he developed his steam engine. Steam engines before Watt had at best 1% thermal efficiencies. That meant that only 1% of the heat potential of the fuel combusted was converted into useful mechanical energy. It also meant that the draft horse was more often the engine of choice. In fact, the draft horse was used widely enough in those days that most people had at least an idea of what the animal was capable of. Watt closely observed the horse at work with a mathematician's eye and came to this conclusion: A medium-sized draft horse could raise a 330 lb weight through a distance of 100 ft. in 60 seconds.

Horsepower Formula

Subsequent observation has concluded that this is actually about 50% more than the average draft horse could sustain over a working day, but Watt obviously did not observe the performance of the horse for a prolonged period. Truly, a draft horse can only produce about ½ horsepower continuously. But that does not matter today. The foregoing data, expressed as a formula, is used to calculate horsepower:

$$\begin{aligned} 330 \text{ lb} \times 100 \text{ ft.} &= 33{,}000 \text{ lb-ft./min.} \\ &= 550 \text{ lb-ft. per second} \\ &= 1 \text{ HP} \\ 1 \text{ HP} &= 33{,}000 \text{ lb-ft. work in } 1 \text{ minute} \\ &= \frac{33{,}000}{6} \text{ lb-ft.} = 550 \text{ lb-ft in } 1 \text{ second} \end{aligned}$$

Importance of Time

Power is always related to time—it is simply the rate of doing work. A cyclist in competition will usually select a gear ratio that permits the pumping out of power strokes at the fastest rate manageable, knowing that each downstroke of a pedal from top dead center (TDC) to bottom dead center (BDC) translates into an actual distance on the road. If a race were to be held among a group of runners of identical weight, every runner who completed the race would have accomplished the same amount of work. The work accomplished by each runner would be the act of transporting their mass over a defined distance. However, one of the runners will have won the race by crossing the finish line ahead of the other runners. This runner has accomplished the race distance in the least amount of time and can be said to have demonstrated the most power.

Power Calculations

In a diesel engine, as with the cyclist, the number of power strokes per second, in conjunction with the actual torque value, determines brake power. A dynamometer is a tool for testing engine power. It applies a resistance to turning effort and measures it; it therefore measures *torque.* When torque is factored with *time* (engine rpm), the brake power can be calculated. This is usually an automatic function of an engine dynamometer.

Brake Power

Brake power is power measured at the flywheel of an engine. It is always lower than indicated power (calculated power) because of the amount of power consumed by the engine in overcoming internal friction and pumping losses—usually about 10%. The term is used to describe actual engine power over calculated power. Almost always, diesel engine original equipment manufacturers (OEMs) rate the power of their engine using brake power, which is expressed as **brake horsepower (BHP)**.

Calculating Brake Power

Despite the fact that industry in both Canada and the United States has embraced the metric system, most of us still think using the standard system of weights and measurements. In the trucking industry especially, we have been especially reluctant to adopt the metric system. That means that manufacturers sell (in engine ratings) and express brake power as horsepower.

$$\text{Brake Horsepower} = \frac{\text{Force} \times \text{Distance} \times \text{Time}}{33{,}000 \text{ lb} \times 1 \text{ ft.} \times 1 \text{ min.}}$$

Distance must take into account that an internal combustion engine rotates and, therefore, the distance factor is not linear but circular (torque); to accommodate this, pi (3.1416) is used to construct the equation, which is simplified as follows:

$$2\pi = 2 \times 3.1416 = 6.2832$$

$$33{,}000 \div 6.2832 = 5252.10084$$

or, for calculation purposes, 5252

In the first brake power formula, the values time and distance are multiplied by each other and by the force value. In a diesel engine force is expressed as *torque*. However, time and distance are commonly expressed as one value, rpm. So for practical purposes, the brake power formula that is used becomes:

Brake Horsepower Power

$$= \frac{\text{Torque (1 lb-ft.)} \times \text{rpm (time)}}{5252}$$

$$\text{Brake Power (kW)} = \frac{\text{Torque} \times \text{rpm}}{9429}$$

This formula can be used to calculate brake horsepower providing the torque value and rotational speed values are known (Figure 5–1). For instance, a Cummins ISX is specified to produce a peak torque value of 2,050 lb-ft. at 1,200 rpm—to calculate the amount of power this engine produces at this speed and torque:

$$\text{BHP} = \frac{2050 \text{ lb-ft.} \times 1200 \text{ rpm}}{5252}$$

$$= 468 \text{ BHP}$$

Using the metric system, brake power is measured in watts, so using the formulae explained earlier in this section:

1 Newton-meter = 1 joule

1 joule per second = 1 watt

Because engine brake power has such a close relationship with speed, a sure way of increasing power is simply to allow the engine to turn faster. In the days of hydromechanical engines, the simple act of increasing governed engine rpm would usually increase engine power—though possibly at the expense of such factors as specified idle and peak torque rpm. Now you know enough of the math of power output to pick up a calculator and be able to check the accuracy of OEM power graphs such as the one reproduced in Figure 5–1: Pick any point on the torque curve, match it to the rpm directly below it, and using the formula just shown, the result should be close to that shown on the power band (at the same rpm) of the graph.

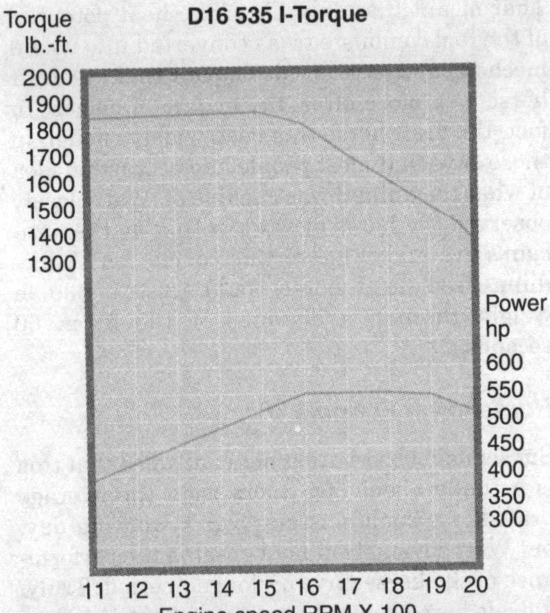

Figure 5–1 *Volvo D16 liter power and torque graph. (Courtesy of Volvo Trucks NA)*

Conversions

1 lb-ft.	= 1.356 N·m
1 N·m	= 0.737 lb-ft.
1 HP	= 0.746 kW 5 746 watts
1 kW	= 1.341 HP
1 HP	= 2,545 Btu per hour
1 metric HP	= 4,500 kilogram/meters per minute
	= 0.9863 HP

As mentioned before, truck, heavy equipment, and bus diesel engines are usually classified by rating their actual or brake power in horsepower (BHP), but most OEMs also provide a **kilowatt (kW)** power rating. In all cases, power is measured at the flywheel (see Figure 5–2 and Figure 5–3). The brake power of the engine measured at the chassis wheels, such as would be displayed by a chassis dynamometer, would be somewhat less. There are other methods of reckoning power values, some of which are listed later in this chapter.

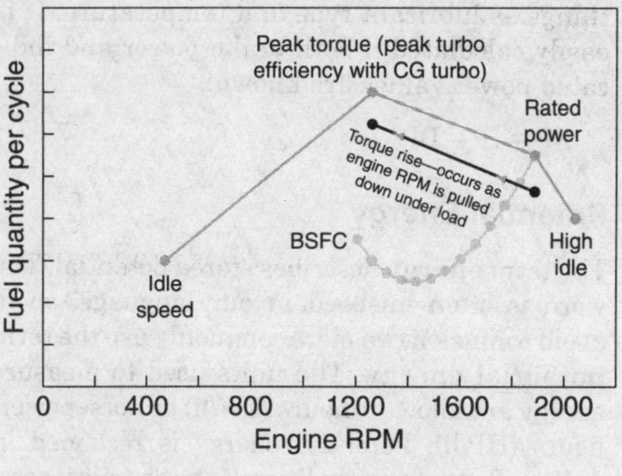

Figure 5–2 *Simple fuel map showing torque rise and BSFC curve.*

Indicated Power (IP)

Indicated power is calculated power. It is useful for comparison purposes but, obviously, it is not as reliable of an indicator of true power as brake power. The following formula is used to calculate indicated power:

P MEP (Pressure)
L Length of stroke
A Piston cross sectional area
N Number of power strokes per minute
C Number of engine cylinders

Performance curves

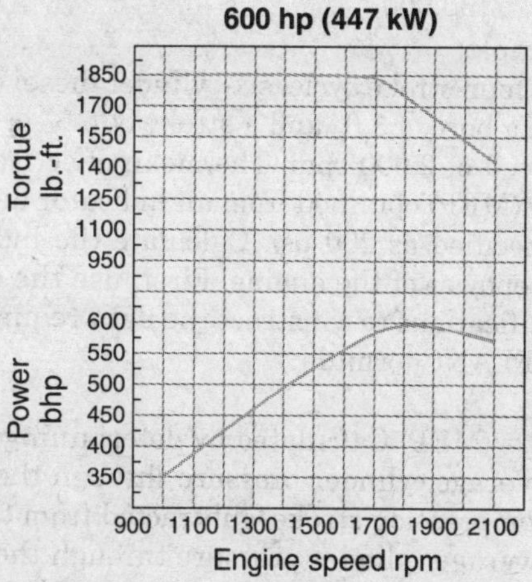

Performance data

Operating range (rpm)	1200–2100
Governed speed—rpm	**2100**
Advertised hp (kW)	600 (447)
Max hp (kW)	600 (447)
Peak torque—lb.-ft. (N·m)	**1850 (2509)**
Peak torque—rpm	1200
Torque rise (%)	29
Altitude capability—ft (m)	10,000 (3048)

Performance curves

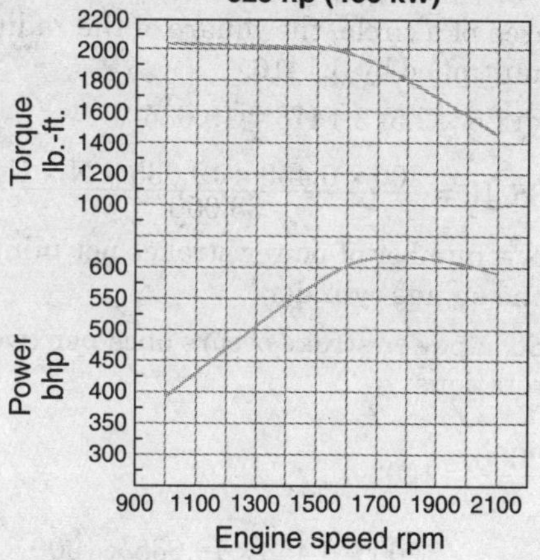

Performance data

Operating range (rpm)	1200–2100
Governed speed—rpm	**2100**
Advertised hp (kW)	625 (466)
Max hp (kW)	625 (466)
Peak torque—lb.-ft. (N·m)	**2050 (2779)**
Peak torque—rpm	1200
Torque rise (%)	38
Altitude capability—ft (m)	10,000 (3048)

Figure 5–3 *Caterpillar C15 performance curves. (Reprinted Courtesy of Caterpillar Inc.)*

$$IP = \frac{PLANC}{33{,}000}$$

Example:

A four-stroke cycle, six-cylinder diesel engine with a bore of 5.0″ and a stroke of 5.5″ is run at a speed of 2,000 rpm. The mean effective pressure (MEP) obtained from an indicator diagram is specified as 200 psi. Calculate the indicated horsepower of the engine. First, use the engine specifications to construct the data required for the PLANC formula:

P = (MEP) Calculated by determining the average cylinder pressure through the compression stroke, subtracted from the average cylinder pressure through the power stroke. This data is provided: 200 psi

$$\text{So, } IP = \frac{200 \times L \times A \times N \times C}{33{,}000}$$

$$L = \text{Stroke} = 5.5 \text{ Length in fee} = \frac{5.5}{12} = 0.458$$

$$\text{So, } IP = \frac{200 \times 0.458 \times A \times N \times C}{33{,}000}$$

A = Piston sectional area: Bore diameter = 5.0″, so radius = 2.5″. To calculate the area of a circle, the square of the radius is multiplied by 3.1416:

$$= (2.5 \times 2.5)\ 3.1416 = 19.635$$

$$\text{So, } IP = \frac{200 \times 0.458 \times 19.635 \times N \times C}{33{,}000}$$

N = number of power strokes per minute in one engine cylinder

So, a power stroke occurs once per two full rotations:

$$\frac{2000}{2} = 1000$$

$$\text{So, } IP = \frac{200 \times 0.458 \times 19{,}635 \times 1000 \times C}{33{,}000}$$

C = Number of engine cylinders = 6

$$\text{So, } IP = \frac{200 \times 0.458 \times 19{,}635 \times 1000 \times 6}{33{,}000}$$

Now the equation is complete, so solve it:

$$IP = \frac{200 \times 0.458 \times 19.635 \times 1000 \times 6}{33{,}000}$$

$$IP = 327 \text{ HP}$$

SAE Power

SAE power is a brake power rating known to be especially reliable because test criteria and conditions are rigorously detailed. The engine must be tested in a controlled temperature environment with all parasitic loads attached, including the primary transmission from which the torque load is applied and power measured. SAE power ratings can be expressed as gross power and net power. Both are measured at the flywheel. However, gross power is raw engine power with all potential parasitic loads removed (including emissions controls) while net SAE power is that produced by the engine operated with its normal emissions controls and parasitic loads.

Friction Power

The amount of power required to overcome the friction of all the moving parts in an engine—friction power increases as engine size and rotational speeds increase and is affected by such things as lubricant type and temperature. It is easily calculated if both brake power and indicated power values are known:

$$FP = IP - BP$$

Potential Energy

The term **energy** describes stored potential. The word is often misused in our language, so to avoid confusion, we more commonly use the term **potential energy**. The units used to measure energy are kilowatt/hours (kW/h) or horsepower/hours (HP/h). Potential energy is reckoned in units called calories or Btus. If the chemical composition of a fuel is known before it is combusted, we can state exactly how much potential energy that fuel has. As fuel is combusted, it releases its potentialenergy. If the fuel happens to be combusted in an engine cylinder, we can determine exactly how much of its"potential" has actually been converted into useful mechanical energy. This is expressed as a percentage and called **thermal efficiency**.

Mechanical Efficiency

Mechanical efficiency is a measure of the pumping and friction efficiency of an engine and, like *friction power*, is a term more likely to be found in a textbook than in a listing of engine specifications. Usually expressed in percentage terms:

Mechanical Efficiency = BP ÷ IP

For example, an engine that has been tested to produce 400 BHP and has its indicated power calculated at 475 BHP would have a mechanical efficiency rating of 84%.

400 ÷ 475 = 84%

Continuous Brake Horsepower

As mentioned earlier, Watt's definition of a unit of horsepower equation was flawed because the draft horse performance had been observed through a single pull rather than continuously. In the same way, to reliably spec engine power, we have to look at the power that engine can produce continuously. Continuous brake horsepower is a measure of the power an engine can produce continuously without compromising engine longevity. The term is used by one diesel engine OEM, but it means pretty much what its competitors refer to as brake power.

Brake Power at the Wheels

A chassis dynamometer is the most accurate way to test the power output of diesel engines. This tests power produced at the wheels, meaning that engine power must be transmitted through drivetrain and wheel assemblies to a floor-mounted roller assembly. Actual brake power readings on a chassis dynamometer are typically 10% to 20% below readings taken on an engine dynamometer (i.e., power at the flywheel). Chassis dyno testing is discussed in a later chapter, but it is important to know the powertrain of the specific chassis being tested before trying to make sense of chassis dynamometer horsepower readings.

Torque Rise

The term **torque rise** originated from the days of hydromechanically managed engines. These engines had rigid fuel maps and constant geometry turbochargers. However, although today's engines use "soft" fuel maps (computer-controlled) and variable geometry turbos, the term is still used. Torque rise occurs between peak torque rpm and rated speed rpm. Because diesel engines in mobile equipment are required to operate over a wide rpm and load range, high torque rise can be desirable.

Torque Rise Profile

The term **torque rise profile** comes from the graphic representation of the upper portion of the fuel map often shown by engine OEMs on their sales literature. This has been simplified in Figure 5–2. Torque rise profile is perhaps best understood as being the engine-operating range. For a truck with a mechanical transmission, this is important because it defines the upper and lower shift limits. An engine described as high torque rise when compared with the same engine with identical rated power but low torque rise:

- May result in poor fuel efficiency given a heavy driver's boot
- Requires a transmission with less gear ratios and/or less shifting
- Is described by truck drivers as being "more powerful"

Torque rise is often expressed as a percentage of rated speed. In other words, an engine with a nominal rated speed of 2,000 rpm and torque rise of 35% would produce peak torque at 1,300 rpm. This is what is shown in Figure 5–1. Diesel engine governing is discussed more in detail later in Chapter 26 but this should help to begin with. The performance curve shown in Figure 5–2 shows a performance curve that would be considered a high torque rise engine by today's standards of fuel economy.

Generally, a fleet truck engineered for a linehaul highway application requires fuel system programming (or governing, in older engines) for fuel economy and, therefore, does not require a high torque rise characteristic. Sure enough, the driver may be shifting gears a little more often while getting to a cruise speed, but once there it is desirable to operate the engine as efficiently as possible, in the sweet spot of the torque rise profile. This is usually somewhere close to the midpoint between peak torque rpm and rated rpm. However, a dump truck hauling aggregates in a constant stop-start application would require a much wider engine-operating rpm range, a high torque rise engine.

Torque Rise Programming

Terms like *high torque rise* and *low torque rise* are used less often today because there are so many options for programming engine operation according to the expected conditions. A diesel engine programmed with low torque rise for fuel economy in a linehaul application can be reprogrammed for high torque rise performance usually with no hardware changes. The key to this being possible is the variable geometry turbocharger.

In the days of hydromechanical and early generation electronically controlled diesel engines, the two primary variations on an engine described as high torque rise over the conventional version of the same engine had to do with:

- Fuel system management
- Turbocharger geometry

The features that defined the high torque rise engine were:

1. The fuel injected per cycle at peak torque was designed to significantly exceed that injected per cycle at rated speed.
2. The turbocharger geometry was tailored to provide a constant rate of airflow per minute from the peak torque rpm (base of the torque rise profile) through to the **rated speed** (peak BHP rpm). This meant that the mass of airflow delivered to the engine cylinder per cycle was greatest at peak torque.

In the days before programmable engine management and variable geometry turbochargers, making significant changes to the torque rise profile on a diesel engine usually meant costly fuel system recalibration and the replacing of the turbocharger.

Specific Fuel Consumption

Specific fuel consumption (SFC) is the fuel consumed per unit of work produced. This data described as **brake specific fuel consumption (BSFC)** is often diagrammed into power/fuel map/performance graphs to demonstrate fuel efficiency through the torque rise of the engine. If the data is constructed around brake power values (and it usually is when truck engine data is displayed) the specific fuel consumption is labeled as BSFC. When the data is constructed around indicated power specifications, the term *indicated specific fuel consumption* (IFSC) is used. A hypothetical BSFC is shown in Figure 5–2 mainly to get across the idea that in diesel engines, as rpm increases, fuel consumption increases with it; this explains why maximum engine rpms have tended to go down over recent years.

Load

Load is usually defined as the ratio of power developed to the normal **rated power** at a specific rpm of an engine and is expressed as a percentage. In some instances, load is used to express the ratio of torque developed to that of peak torque and, again, this is expressed as a percentage. This can be observed when using most OEM software during diagnostic testing. When performing dynamometer testing, after the rated power/speed value is established, the load data can be useful in establishing whether abnormalities exist in the fuel map.

Interpreting OEM Fuel Maps

Each engine OEM has differing methods of graphically representing power on a graph. However, you should now have the key tools to interpret these graphs. Figure 5–3, Figure 5–4, and Figure 5–5 show how Caterpillar, Detroit Diesel, and Navistar International use graphical data to sell their engines: The performance curves depicted are all post-2007 versions of the engines. As with Figure 5–1, you can do the math shown here to check the data shown in the graphs.

B-life and Power

Today, technicians who work in the trucking industry sooner or later will come across the terms *B10* and *B50-life* in relation to engine longevity. **B-life** is explained in more detail in Chapter 28, but for now B-ratings is defined as the projection in percentages of engines that will fail at various mileage intervals. In the trucking industry the miles are linehaul miles (truck vocational and off-highway

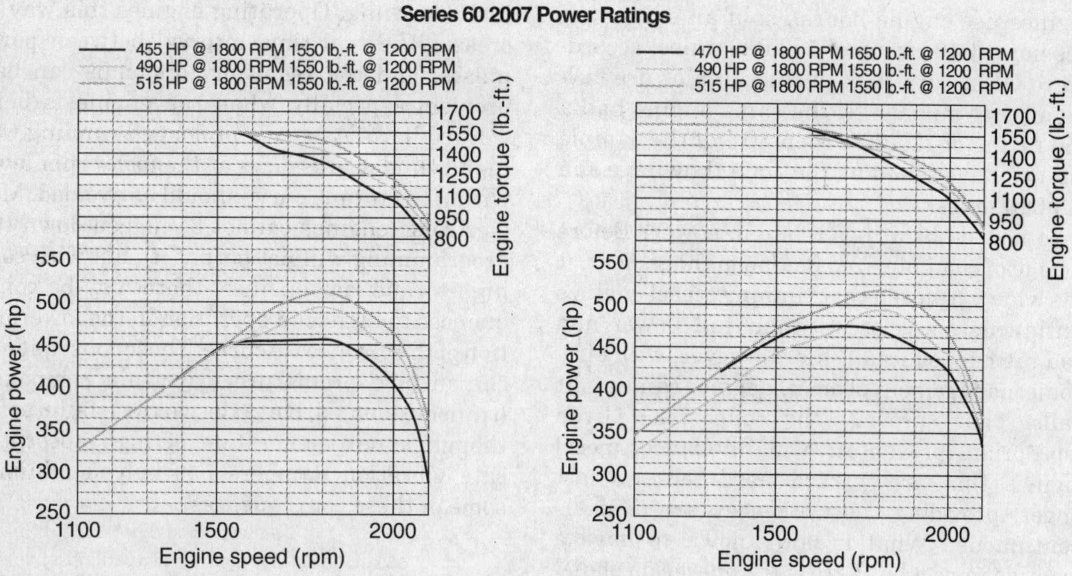

Figure 5–4 *Detroit Diesel Series 60 (14-liter) performance curves. (Courtesy of Detroit Diesel Corporation)*

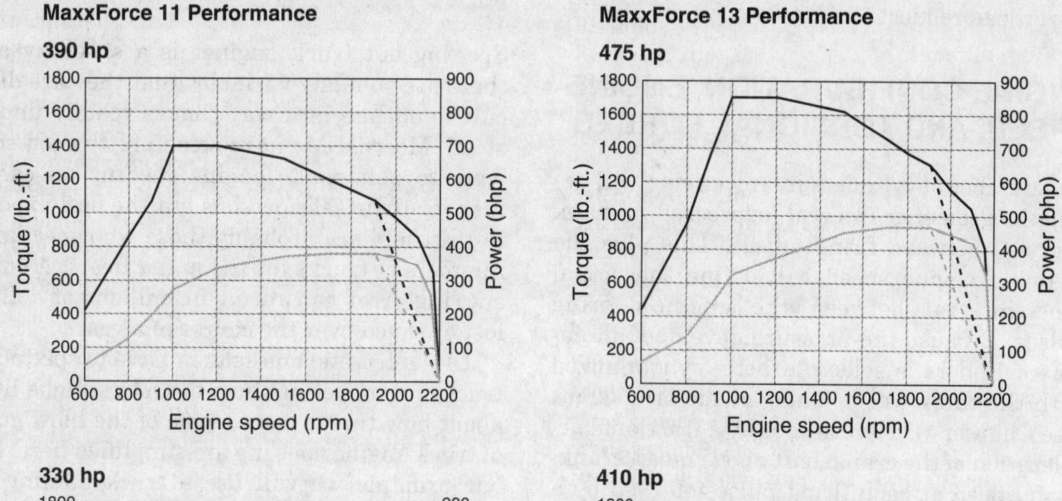

Figure 5–5 *Navistar MaxxForce (11- and 13-liter) performance curves. (Courtesy of Navistar International Corp.)*

engines use engine hours), so if an engine series has a B10 rating of 550,000 miles, according to OEM data, 10% of the samples are failing at that mileage. If the same engine had a B50 rating of 725,000, then 50% of the sample engines have failed by the time they have run 725,000 miles.

So how do B-ratings relate to power? Before the age of computerized engine management, it was widely believed that running diesel engines continuously at close to their full power and load ratings reduced B-life. However, with electronic management of diesel engines came audit trails that proved the opposite. Given appropriate preventative maintenance, diesel engines that are worked hard tend to last longer, providing the work they are doing is "continuous." What is now known to have a negative impact on B-life is the operation of diesel engines inconsistently alternating between periods of high power/load request and low power/load request.

ENGINE CONFIGURATION, ENGINE SPEED, AND TORSIONAL FORCES

The number of cylinders in any engine will define the geometric interval (expressed in crank degrees) between firing pulses. This plus the factor of engine speed will define the actual time interval between the engine's firing pulses. Because the pressure developed in engine cylinders is seldom exactly synchronized with the throw vector angle of the crankshaft (see Chapter 4), there is usually a fractional acceleration of the crankshaft above mean crankshaft speed at each firing pulse followed by a fractional slowdown just prior to the next firing pulse. These torsional forces (sometimes called **torsionals**) are amplified at slower engine rpm levels when the time interval between each power stroke is greatest and can affect the driven components in the drivetrain.

Run-slow, Gear-fast

Today's diesel engines, which are designed for fuel economy combined with low emissions, tend to wind out at much lower rpm levels than those of a generation ago. This type of engine programming is referred to as *run-slow, gear-fast* governing. Operating engines this way increases the real time interval between power pulses, and the resulting torsionals can be a problem especially when the engine is being worked hard. As a rule prolonged running with high cylinder pressures at the *lower* rpm levels within the torque curve should be avoided. Most electronic engines cannot be lugged down (the programming will not permit it), but with some high torque rise engines, there may be consequences to high load operation in the lower portion of the torque rise curve. These consequences can involve premature drivetrain component (transmission, drive axle carrier) failures although a new generation of high absorption rate vibration dampers has helped minimize some of these consequences.

HOW MUCH POWER DO YOU NEED?

Spec'ing out truck engines is a specialty and there are so many variables that they are difficult to outline in a way that is readily understood. Most OEMs have some pretty good software to help make engine spec'ing an easier task for their salespersons but the best experts in this field are probably those who spec large numbers of trucks for the major fleets. A small spec'ing error can result in millions of dollars lost or saved over the course of a year.

Diesel technicians seldom have to spec out a truck, but it does not hurt to understand a little about how this is done. Some of the hard math of truck engine spec'ing are simplified here. For our example we will use a truck hauling an 80,000 lb (36,290 kg) load and overlook some of the gearing and aerodynamic complexities that could arise:

- It takes 255 BHP to move the truck down a level asphalt road at 60 mph with no wind.
- On a 1% uphill grade to maintain the same speed, another 65 BHP will have to be added, so now the engine requires 320 BHP.
- Back on the level highway, if there is a 5 mph headwind, another 16 BHP will have to be added to maintain that road speed of 60 mph. Should the headwind be 10 mph, another 36 BHP has to added.

- Parasitic losses such as an air-conditioning system (up to 15 BHP) and engine fan (viscous type: averaging 4–6 BHP, on-off type up to 15 BHP but only on 2% of running time) all have to be factored into the BHP requirement for a specific truck.
- So running up a 3% grade (not that steep a hill) with a 20 mph headwind (not that strong a wind) is going to require a whopping 531 horsepower to maintain the same road speed. Throw in some of the common parasitic losses an engine has to sustain and you can see why some operators, especially brokers, are targeting 600+ BHP engines.

REVIEW QUESTIONS

1. Calculate the brake power of an engine producing 1,150 lb-ft. of torque at the flywheel when it is turning at 2,000 rpm.
2. Use the PLANC formula to calculate the indicated power of a six-cylinder, four-stroke cycle engine run at 1,800 rpm using the following data:
   stroke: 6.0 inches
   bore: 5.5 inches
   MEP: 220 psi
3. Which of the following is required to produce peak torque in a diesel engine?
   a. Peak brake power
   b. Peak cylinder pressure
   c. Peak volumetric efficiency
   d. Peak mechanical efficiency
4. In most hydromechanical fuel systems, when maximum engine rpm is increased, the engine brake power potential will also increase.
   a. True
   b. False
5. Which of the following statements helps explain why an engine develops peak brake power at a higher rpm than peak torque?
   a. Cylinder pressures are highest
   b. More power strokes per second occur
   c. More time for combustion
   d. Better volumetric efficiency at high rpm
6. Which of the following is true of an engine described as having high torque rise when comparing it with its standard version with the same nominal rated power?
   a. It will produce higher brake power.
   b. It will have a higher maximum speed.
   c. It will provide a wider operating rpm range.
   d. It will be more fuel efficient.
7. A group of cyclists all weighing 80 kg (175 lb) and riding on bicycles of equal weight ride a 10-kilometer (6.4-mile) race and all finish it. Which of the cyclists has performed the most work?
   a. The winner
   b. All have performed equal work
   c. The loser
8. Usable power at the flywheel of an engine is known as:
   a. BSFC
   b. Brake power
   c. Indicated power
   d. Torque
9. When engine power is calculated rather than tested, it is known as:
   a. Mean effective pressure
   b. BSFC
   c. Brake power
   d. Indicated power
10. Which of the following expresses 1 HP correctly?
    a. 33,000 lb-ft. of work per second
    b. 550 lb-ft. of work per second
    c. 5,252 lb-ft. of work per minute
11. Convertatorquevalueof25lb-ft.intoNewton-meters.
    a. 19.8
    b. 21.3
    c. 33.9
    d. 38.1
12. A Cummins ISM engine, programmed to 330 HP, produces a peak torque value of 1350 ft.-lb at 1,300 rpm. Convert the torque value to Newton-meters.
    a. 1,007 N·m
    b. 1,275 N·m
    c. 1,830 N·m
    d. 2,225 N·m
13. A Caterpillar C15 is rated at 600 HP. Express this in kilowatts.
    a. 400 kW
    b. 448 kW
    c. 464 kW
    d. 488 kW

14. A Navistar MaxxForce 13 produces 355 kW of power at 1,800 rpm. Convert the kW to an HP value.
    a. 276 HP
    b. 355 HP
    c. 455 HP
    d. 475 HP
15. When comparing truck engine brake power measured at the flywheel to that measured at the drive wheels, which one of the following should be true about brake power?
    a. There is no difference if the transmission and drive carrier are functioning properly.
    b. Power at the wheels exceeds that at the flywheel by 10% to 20%.
    c. Power can only be measured at the drive wheels.
    d. Power at the flywheel exceeds that at the drive wheels by 10% to 20%.

6

Engine Powertrain Components

INTRODUCTION

This chapter addresses the group of engine components responsible for transmitting the gas pressures developed in engine cylinders to the engine power takeoff mechanism, usually a flywheel. This group of components defined as the engine powertrain includes:

- Pistons
- Piston rings
- Wrist pins
- Connecting rods
- Crankshafts
- Friction bearings
- Flywheels
- Vibration dampers

This chapter takes a close look at all of the powertrain components in sequence. The powertrain is actuated by cylinder gas pressure. This pressure acts on the piston and drives through its stroke. A piston moves linearly. Its linear movement has to be converted into torque. To accomplish this, a connecting rod links the piston assembly to a throw on the crankshaft. The crankshaft throw is offset from the centerline of the crankshaft. As the piston reciprocates in the cylinder bore, the crankshaft rotates. The drive torque imparted to the crankshaft is delivered to a flywheel. The flywheel is bolted to the crankshaft. One of its roles is to act as coupling so that engine torque can be delivered to the drivetrain components and ultimately to the drive wheels. This chapter begins with a close look at pistons.

PISTON ASSEMBLIES

By definition, a piston is a circular plug that seals the engine cylinder bore and reciprocates within it. The piston is subject to the gas pressure conditions within the cylinder. It can either:

- Impart force: delivers force when traveling upward on its compression stroke
- Receive force: combustion pressures act on it during the powerstroke

The piston assembly consists of the piston, piston rings, and a wrist pin. Piston rings seal the cylinder and lubricate the cylinder walls. The wrist pin is installed through a boss in the piston: It connects the piston to the connecting rod.

Piston Terminology

A diesel engine piston is shown in Figure 6–1 along with the terminology used to describe it. The type of piston shown is a forged steel trunk piston used in many current diesels. However, there are four general categories of pistons used in diesel engines, which are discussed a little later. Following the callouts in Figure 6–1, you should be able to make sense of the description that follows.

Piston Crown

The upper face of the piston is called the **crown**, and it is exposed directly to the cylinder chamber and, therefore, the effects of combustion. Diesel engine pistons absorb up to

20% of the (rejected) heat of the combustion gases. A piston's ability to rapidly dissipate heat is essential, especially when aluminum pistons are used. High-output, turbocharged engines often have piston cooling jets that spray lubricating oil on the underside of the piston to help remove heat. Piston crown temperatures are always high, and the rings assist in cooling by conducting some of this heat to the cylinder walls. The crown geometry has much to do with the gas dynamics (swirl and squish) produced during the compression stroke. This determines the fuel-air mixing characteristics as well as the ignition location within the cylinder. The Mexican hat piston crown design shown in Figure 6–1 is common in low-emissions truck direct-injected (DI) engines. The circumferential edge of the piston around the crown is known as its leading edge. Most modern diesel engines have low clearance volumes. This means that at top dead center (TDC), the piston rises in the cylinder bore to a height that recesses are required in the crown to accommodate cylinder valve protrusion. These recesses are known as **valve pockets**.

Piston Materials

It cannot be emphasized enough how critical it is for a piston to efficiently transfer heat. In some cases, diesel engine combustion temperatures can exceed the melting point of the piston material. This only occurs for fractions of a second, but it means pistons have a role as a sort of heat siphon. Actual piston temperatures depend on the piston material. In current diesel engines, pistons are usually manufactured from aluminum- and iron-based alloys. However, some original equipment manufacturers (OEMs) have experimented with ceramic sections and composites. Some facts:

- Combustion temperatures rise to transient spikes of 2000°C (3630°F).
- Aluminum melts at 660°C (1220°F).
- Cast iron melts at 1540°C (2800°F).

Two basic piston designs are used in current commercial and off-highway diesel engines:

- Trunk-type pistons
- Articulating pistons

Trunk-type Pistons

Trunk-type pistons are single-piece pistons consisting of a crown and integral skirt. There are three categories of trunk-type pistons used in diesel engines today:

- aluminum trunk pistons
- forged steel trunk pistons
- composite steel trunk pistons

Aluminum Trunk Pistons

Until the 1990s most commercial diesel engines used single-piece pistons manufactured from

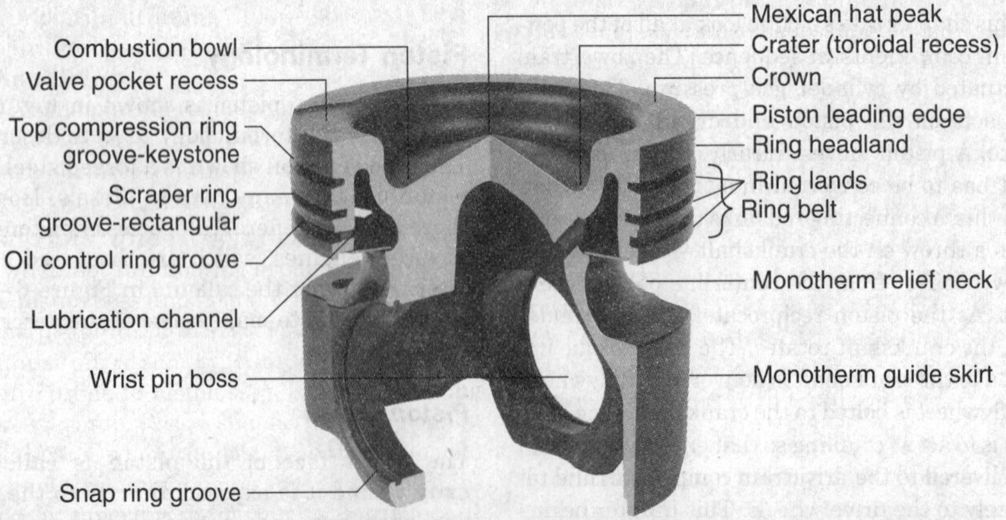

Figure 6–1 *Piston terminology. (Courtesy of MAHLE International GmbH)*

aluminum alloys. The reason for using single-piece pistons in those earlier diesel engines was primarily to minimize piston weight. One of the key aluminum alloying substances was a small percentage of silicon that considerably toughened the aluminum. The low melting temperature of aluminum combined with a lack of toughness when compared with forged steels or cast irons requires most aluminum trunk pistons to use a ring groove insert for at least the top compression ring groove. The insert is usually a **Ni-Resist™ insert**, a nickel bearing, iron alloy with great resistance to high temperatures and wear, but also with a coefficient of heat expansion nearly identical to that of aluminum. The Ni-Resist insert is molecularly bonded to the aluminum piston. Figure 6–2 is an image of an aluminum trunk piston: Note the location of the Ni-Resist insert.

Heat treatment also improves the performance of aluminum as a piston material. In some small-bore diesels manufactured since 2000, hypereutectic (an alloying process) aluminum pistons are used to provide greatly increased wear resistance. Another method of increasing the toughness and high-temperature performance of aluminum (without compromising its primary advantage of light weight) is the ceramic fiber reinforced (CFA) process introduced on smaller bore diesels. The CFA process eliminates the requirement for a Ni-Resist insert because it reinforces the piston at the top of the ring belt, extending up into the top of the crown. This allows placement of the top compression ring close to the leading edge and reduces headland volume (see Figure 6–1). High location of the top compression ring is almost a requirement for low-emissions diesels. Another fiber reinforcement

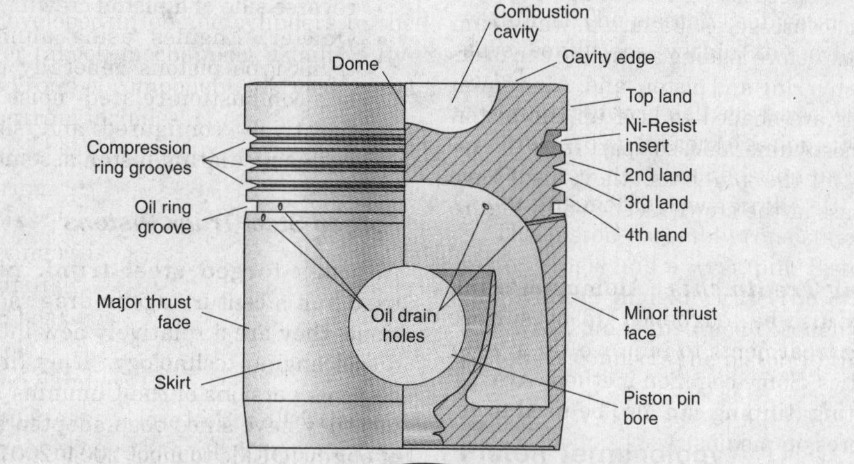

Figure 6–2 *Aluminum trunk-type piston and ring terminology. (Courtesy of Mack Inc.)*

manufacturing practice known as squeeze cast, fiber reinforced (SCFR) is used by some manufacturers to toughen the crown area of the piston. This is an alumina fiber manufacturing process.

Cam Ground Aluminum pistons are light in weight and have a high coefficient of heat transfer, which means they get rid of any heat they are subjected to rapidly. Aluminum also has a high coefficient of heat expansion and contraction, which means that they have to be **cam ground** to be used in diesel engines. A cam ground piston is slightly elliptical (oval) when cold. As the piston heats, the greater mass of material around the **pin boss** expands more than the thinner skirt area between the pin bosses. The idea is that at running temperatures, the piston should expand to a circular shape. The disadvantage of cam ground pistons is that they should be warmed to operating temperature before being subjected to high cylinder pressures. Failure to warm cam ground pistons before loading an engine can overstress the piston ring and piston land. Aluminum trunk pistons are shaped to beef up the piston where it is most vulnerable, so apart from the reinforcement at the pin boss, they also have increased mass at the crown as shown in Figure 6–2.

Toughening Treatments Aluminum trunk pistons may also be toughened up by surface and alloying treatments to improve their wear characteristics. Some common methods are:

- Tin plating (tinning can also be used to repair scores on pistons)
- **Anodizing**
- Chrome plating (usually of the crown)
- Squeeze cast, fiber reinforced (more than just surface treatment, the structural integrity of the piston is increased)

Wrist Pins and Ring Belt Piston wrist pins when not full floating are usually press fit to the piston boss and float on the rod eye. Heating the piston to 95°C (200°F) in boiling water facilitates pin assembly. Aluminum trunk pistons are prone to piston slap, especially during engine warmup. Piston slap is the tilting action of the piston when the piston is thrust loaded by cylinder combustion pressure. It can be minimized by tapering the piston so that the outside diameter at the lower skirt slightly exceeds the outside diameter over the ring belt region. The ring belt region is exposed to more heat and expands more as the piston heats to operating temperatures.

Advantages of Aluminum Trunk Pistons

- Lightweight. This reduces the piston mass and also the inertia forces that the connecting rod and crankshaft have to sustain. The result is that tensile stressing (stretching forces) of the connecting rod and crankshaft are significantly lower, permitting the use of lighter weight components throughout the engine powertrain.
- Cooler piston crown temperatures. The ability of aluminum alloy pistons to rapidly dissipate combustion heat results in lower crown operating temperatures, especially when oil cooling jets are targeted at the piston **undercrown**. The undercrown is the reverse side of a piston crown.
- Quieter. Engines using aluminum alloy trunk-type pistons generally produce less non-combustion-related noise than comparatively configured and sized engines with articulating piston assemblies.

Forged Steel Trunk Pistons

Although **forged steel trunk pistons** were used half a century ago in drag racing applications, they are a relatively new introduction to diesel engine technology. They first appeared on some versions of the Cummins ISX in 2002, but they have since been adopted by a number of engine OEMs to meet 2004/ 2007/ 2010 emissions standards. The forged steel trunk piston design adopted by most diesel engine OEMs originates from piston design specialist Mahle. The brand name used for these pistons now used by most diesel engine OEMs is **Monotherm™** , and it has been industry practice to describe this category of pistons as Monotherm pistons: These are being called *forged steel trunk pistons* in this textbook simply to avoid using the brand name. Figure 6–3 shows two views of a forged steel trunk piston.

Design and Construction The single-piece steel trunk piston has the appearance of an aluminum trunk piston with a large circumferential slot cut away between the pin boss and the

Figure 6–3 *Two views of a forged steel trunk piston (Monotherm). (Courtesy of MAHLE International GmbH)*

ring belt. The skirt is designed to guide the piston over its thrust sides and is recessed across the pin boss transverse: This, plus the fact that a separate skirt does not have to be supported, permits the use of shorter, lower weight wrist pin. It also allows the piston pin boss support area to be increased, making the design well suited to engines producing high cylinder pressures. Mahle claims that its pistons can sustain peak cylinder pressures exceeding 3,500 psi (250 bar).

The problem of piston slap in aluminum trunk pistons has been overcome with the Monotherm design. Monotherm steel alloy pistons expand much less than aluminum trunk pistons when heated from cold to operating temperatures because steel expands less than aluminum: This means they can be manufactured with a much tighter fit to the liner bore. The lubricating oil cooling gallery integrated into a Monotherm piston is closed with cover plates, allowing for higher oil feed and flow-back volumes. The oil feed to the underside of the piston is delivered by a spray cooling jet. In some cases, the connecting rods used with Monotherm pistons are not rifle drilled with a lubrication passage. This means that the piston and wrist pin depend almost entirely on the oil spray cooling jet for lubrication and cooling, emphasizing the importance of precision in setting up the jet.

The Monotherm's micro-alloyed steel construction provides high enough strength to permit much less material to be used. The result is a tough steel piston of about the same weight of the older aluminum trunk pistons. Monotherm pistons also feature a bushingless wrist pin bore. The pin boss is phosphate treated over the pin bearing bore. Figure 6–1 and Figure 6–3 both show forged steel trunk pistons typical to those used in a majority of current, high horsepower highway diesel engines.

Advantages of Forged Steel Trunk Pistons The reasons that so many engine OEMs have adopted forged steel trunk pistons over the past few years are primarily due to the increased cylinder combustion pressures in late model engines, engine longevity, and emissions. Following is a list of these reasons:

- Reduction of headland volume. **Headland volume** is the volume in the cylinder above the top compression ring and below the crown leading edge. This volume tends to be less affected by cylinder turbulence and the effects of cylinder scavenging so it collects dead end gas. Headland volume can be minimized by placing the top compression ring as close as possible to the crown leading edge. Because the tensile strength of forged steel is much greater than that of the aluminum alloy trunk piston and requires no groove insert, the upper ring groove can be located close to the crown leading edge. Monotherm pistons place the top compression ring just slightly under the leading edge of the piston.
- Thermal expansion factors. Thermal expansion is simply how the piston material responds to heat. Forged steel has a much lower coefficient of heat expansion than aluminum, so a cam ground design is not required in steel trunk pistons. Also, due to lower expansion response to heat, piston-to-liner clearances can be lower, reducing the vulnerability of the piston to damage by high cylinder pressures at cold startup.
- Longevity. Micro-alloyed forged steel coated with phosphate provides much longer service life than equivalent aluminum trunk pistons.
- Lightweight. Monotherm pistons are as light in weight as aluminum trunk pistons but have the strength to sustain the increasingly high cylinder pressures of modern diesel engines. Some forged steel trunk pistons are specified to sustain cylinder

pressures exceeding 3,500 psi (250 bar). The light weight also provides the potential for high engine speeds though as yet no engine OEMs are introducing this: Development of higher diesel engine speeds is currently inhibited by emissions controls.

Composite Steel Trunk Pistons

The industry is in the process of introducing **composite steel trunk pistons**. These are a variation of the forged steel trunk-type pistons in that the crown and skirt sections of the piston are manufactured separately: The two sections are then screwed together using a proprietary process. The piston crown section is manufactured from a high-temperature steel then screwed into a steel skirt assembly. Mahle-manufactured versions of these pistons will be known as **Monocomp™** . Figure 6–4 shows a composite steel trunk piston: Study it closely to see where the two sections are joined.

Figure 6–4 *Composite steel trunk piston. (Courtesy of MAHLE International GmbH)*

Articulating Pistons

Articulating pistons were adopted by most engine OEMs during the 1990s and up to 2004 usually in their high-power output models. More recently the use of articulating pistons in diesel engines has dropped as OEMs have opted to use new generation forged steel and composite steel trunk-type pistons. Two types of articulating piston assemblies are used in commercial diesel engines:

- Crosshead pistons. Differentiated by the use of a semifloating wrist pin bolted to the connecting rod.
- Full articulating piston. Use a full floating wrist pin that permits pivot on the connecting rod eye, crown, and skirt assemblies.

Crosshead Pistons

The Detroit Diesel Corporation (DDC) used **crosshead pistons** for many years. The crosshead piston is a two-piece piston assembly consisting of a crown and a skirt linked by a semifloating wrist pin itself bolted directly to the connecting rod. Because this design allowed both the crown and skirt assemblies some degree of independent movement (each could pivot on the wrist pin), the assembly can be said to articulate. The key difference between the crosshead and true articulating piston is that crosshead wrist pins are bolted directly to the connecting rods. In other words, the wrist pins used on crosshead pistons are semifloating rather than full floating.

Full Articulating Piston

The articulating piston consists of a crown usually manufactured of forged steel or cast-iron alloy and a separate skirt, usually (but not always) manufactured of aluminum alloy. A full floating wrist pin permits both the crown and skirt to pivot independently while also linking them to the connecting rod eye. The true articulating piston assembly always uses a free-floating wrist pin and therefore shares bearing surface with both the pin boss and the connecting rod eye. This requires a heavier wrist pin and helps to disadvantage the articulating piston with higher weight. Mahle's **Ferrotherm™** articulating pistons were used by a number of diesel engine OEMs in their medium- and large-bore highway engines. Figure 6–5 shows a Mack Trucks, three-ring articulating piston assembly used in its E-Tech engines: The crown has been raised slightly from the skirt assembly.

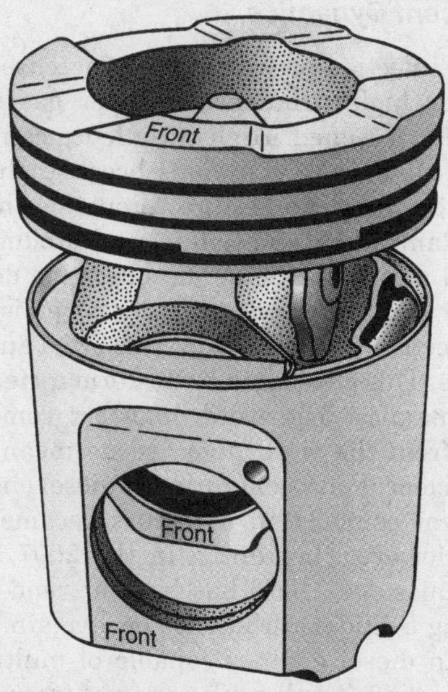

Figure 6–5 *Articulating piston assembly. (Courtesy of Mack Inc.)*

Advantages of Articulating Pistons The crown assembly used in articulating pistons is manufactured from either forged steel or a cast-iron alloy. Both forged steel and cast-iron materials are suitable for the high pressures and temperatures they are directly exposed to. The skirt can be made out of a lighter material, usually an aluminum alloy. Two-piece articulating piston assemblies were widely adopted during the early 1990s despite their heavier weight than the aluminum trunk pistons they replaced because of the following reasons:

- Greater longevity. The industry was progressing to greatly increased engine life and the forged steel crown assembly was tougher and much more durable than aluminum.
- Reduced piston slap. This reduced both noise and wear but also provided better performance during the cold start to engine operating temperature phase.
- More durable crown assembly. Because of the higher combustion pressures, temperatures, and emissions standards required of high-efficiency, low-emissions diesel engines entering the 1990s, using an aluminum alloy crown material was insufficient.
- Reduced headland volume. Another significant reason diesel engine OEMs adopted articulating pistons despite the extra weight penalty was the ability to raise the top piston ring and locate it close to the piston leading edge. The cylinder volume above the top compression ring and below the piston leading edge is less affected by cylinder turbulence and scavenging so it tends to collect dead gas, which is not good when it comes to minimizing emissions. This is known as headland volume and using articulating pistons was the industry's first attempt at minimizing it. The tensile strength of the aluminum alloy trunk piston and its requirement for a Ni-resist insert was insufficient to locate the upper ring groove close enough to the crown leading edge. Some OEMs liked to refer to reduced headland volume pistons as **headland pistons**. They have been proven to reduce smoking and improve fuel economy.

Disadvantages A major disadvantage of articulating piston assemblies over the aluminum trunk design is significantly increased weight, which increases tensional loading on the powertrain. This requires the use of beefed-up engine cylinder block and powertrain components, especially the connecting rods and crankshaft. This weight penalty proved to be a major factor in the demise of articulating pistons in many diesel engines, and thus were replaced by tougher and lighter forged steel, trunk-type pistons.

Piston Thrust Faces

As cylinder gas pressure acts on a piston, there is a tendency for it to cock (pivot off a vertical centerline) in the cylinder bore because it pivots on the wrist pin. This action creates thrust surfaces on either side of the piston. The major thrust face is on the inboard side of the piston as its throw rotates through the power stroke. The minor thrust face is on the outboard side of the piston as its throw rotates through its power stroke. Take a close look at Figure 6–2 in which the piston thrust faces are identified. The major thrust face is sometimes simply called the **thrust face**, whereas the minor thrust face is called the antithrust face. It is important for the diesel technician to identify the thrust faces of a piston for purposes of failure analysis. On a steel trunk-type piston such as the Monotherm, only the thrust faces are skirted with the objective of guiding the piston true in the bore while reducing the weight of the piston.

Combustion Chamber Designs

In DI diesel engines, the physical shape of the piston crown determines what happens to gas in the cylinder on the compression and power strokes. Indirect-injected (IDI) diesel engines were used in a few small-bore and automotive diesel engine applications up until 2001, but none are currently manufactured: IDI is an obsolete technology but the technician can take a look at a typical IDI precombustion chamber in Chapter 19. The commercial vehicle diesel technician is only likely to have encountered IDI technology in the now-obsolete Caterpillar precombustion chambers used with their poppet nozzles.

High Turbulence

DI engines use an open combustion chamber principle. In an open combustion chamber, the injector is usually located in the cylinder head and positioned over the piston crown. The shape of the piston crown therefore defines the type of combustion chamber. Since the mixing of the fuel charge with air takes place within the engine cylinder, the **gas dynamics** (swirl and turbulence) are critical in determining the mixing efficiency and the actual location where ignition occurs. Because the piston is the only moving component in the cylinder after the intake valves close, the geometry (shape) of the piston crown is critical in defining the gas dynamics. Aggressive crown geometry such as more pronounced recesses and protrusions result in more aggressive turbulence.

For many years diesel engine designers attempted to maximize cylinder turbulence but this changed in the early 1990s. A high turbulence or aggressive gas dynamic can be a big help when the atomized droplets of fuel are large, because gas movement in the cylinder can help rip them into smaller droplets. However, today we have the technology to produce much higher injection pressures that result in smaller droplets of injected fuel so high turbulence is no longer a requirement.

Quiescent Dynamics

As fuel injection pressures have increased, the need for high-turbulence cylinder gas movement has lessened. High turbulence can actually disadvantage emissions because clusters of fuel droplets can be thrown outside the primary flame front, ignite late, and produce unwanted afterburn. So as diesel engine designers got more serious about reducing emissions, the objective was to produce less swirl and turbulence. This gave birth to the term **quiescent** gas dynamics. The word *quiescent* comes directly from the word *quiet,* so its meaning is pretty clear. For a generation of diesel engines, quiescent combustion dynamics became common. However, beginning in the 2007 model year emissions, there has been a trend to increasing cylinder turbulence once again, especially in diesel engines capable of multipulse fuel injection. A full explanation of combustion dynamics and multipulse fuel injection appears later in Chapter 15. Most DI engines use one of three basic piston crown designs.

Mexican Hat

The **Mexican hat piston crown** design is by far the most common, and the title perfectly describes its shape. The central area of the piston crown is recessed below the piston leading edge, forming a crater (also known as a *toroidal recess*). Rising from the center of the crater is a cone-shaped protrusion (see Figure 6–1 and Figure 6–2). This aggressiveness of the cone shape is engineered to produce the required amount of turbulence through the compression stroke and gas movement through the power stroke. In most cases, the fuel injector is positioned directly above the center of the cone so that it directs atomized fuel toward the crater where the swirl effect is greatest.

As mentioned before, both the bowl depth and cone height of Mexican hat piston crowns determine how much gas movement is generated. Deep bowl designs produce greater turbulence and are often used with fuel systems with lower peak injection pressures. When engine OEMs describe their engines as having quiescent gas dynamics, it means that they are designed to produce relatively low turbulence and, almost always, higher fuel injection pressures. Another reason diesel engine OEMs frequently use Mexican hat pistons is that diesel fuel droplets can be directed into the crater and be relied on to ignite before contacting the crown material. This provides a low risk of fuel burnout scorching on the piston crown directly below the injector and lengthens service life.

Mann Type (or "M" Type)

The Mann-type piston crown is named after the German company that first designed it. It is usually used with older trunk-type pistons and consists of a spherical recess or bowl located directly under the injector, though not necessarily in the center of the piston crown. Depending on the depth of the recess, the Mann-type combustion chamber generally produces high turbulence but is more vulnerable to localized burnout in the bowl.

Dished

The dished piston crown has a slightly concaved to almost flat design that produces low turbulence when compared with the previous types. You will see this in some current small-bore and older IDI diesel engines.

Piston Cooling

The size of the piston, peak cylinder pressures, and whether the engine is turbocharged all determine what type of piston cooling is required. Within the same engine family, you may find that engines with lower horsepower specifications do not use cooling jets and those with higher horsepower trim do. Because combustion temperatures may peak at values that exceed the melting temperature of the cylinder materials (especially where aluminum is used), it is essential that any heat not converted to usable energy is either exhausted directly or dissipated into the cylinder materials. A percentage of cylinder heat is always transferred through the piston assembly. Three methods are used to cool pistons. Engines may use one or combinations of these piston cooling methods:

1. Shaker—Oil is delivered through the connecting rod to a cell or gallery in the underside of the crown; this oil is distributed by the motion of the piston after which it drains to the crankcase.
2. Circulation—Oil is delivered through the connecting rod rifling, through the wrist pin, and subsequently circulated through a series of grooves machined into the underside of the piston crown. It then drains back into the crankcase.
3. Spray—A stationary jet cylinder block mounted below the cylinder liner and fed by a lubricating oil gallery is directed at the underside of the piston. This oil cools the piston crown and may also lubricate the wrist pin. The jet must be precisely aimed on installation to be effective. This is usually accomplished using a transparent perspex (thermoplastic resin) template that fits over the fire ring groove on the cylinder block deck and an aim rod inserted in the jet orifice; a direction or target window is scribed in the perspex template and the aim rod must be positioned within the window as shown in Figure 6–6. The spray cooling method is used in a majority of current diesel engines.

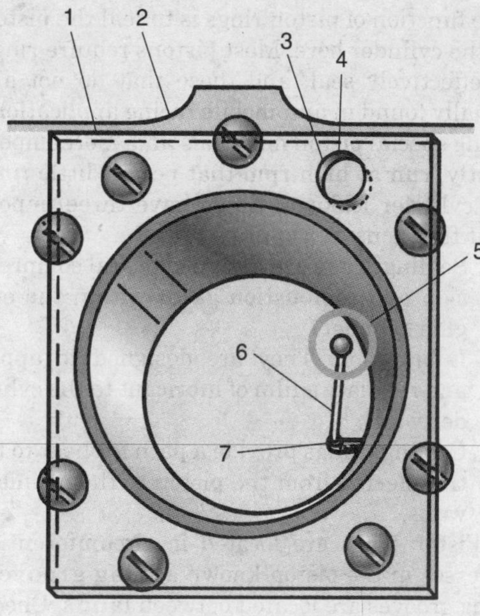

1. Plastic target
2. Cylinder block
3. Alignment hole
4. Cylinder block hole
5. Target area
6. Rod
7. Spray nozzle

Figure 6–6 *Spray nozzle targeting. (Courtesy of Mack Inc.)*

Some Facts

- The rate of heat flow is approximately three times greater in aluminum than in cast iron. Therefore, aluminum will dissipate the heat it is exposed to much more quickly than cast irons and steels.
- Weight of aluminum is 0.097 lb per cubic inch.
- Weight of forged steel or cast iron is 0.285 lb per cubic inch.

Piston Fit Problems

- Excessive piston-to-bore clearance results in the piston knocking against the cylinder wall—this is especially noticeable with aluminum trunk pistons when the engine is cold.
- Too little piston clearance causes piston scoring and scuffing (localized welding)—the film of lube oil on the cylinder wall is scraped off.

PISTON RINGS

The function of piston rings is to seal the piston in the cylinder bore. Most pistons require rings to effectively seal, and those that do not are usually found in automobile racing applications using special piston materials and, more importantly, run at high rpm that permit little *time* for cylinder leakage. Rings have three important functions:

1. Sealing: They are designed to seal compression and combustion gases within the engine cylinder.
2. Lubrication: They are designed to apply and regulate a film of lubricant to the cylinder walls.
3. Cooling: Rings provide a path for heat to be transferred from the piston to the cylinder walls.

Piston rings are located in circumferential recesses in the piston known as **ring grooves**. Ring grooves are located between **lands**. Check out Figure 6–1, paying special attention to the ring area.

Roles of Piston Rings

Piston rings may be broadly categorized as compression and oil control rings. **Compression rings** are responsible for sealing the engine cylinder and play a role in helping to dissipate piston heat to the cylinder walls. The term **scraper ring** is used to describe rings below the top compression ring that play a role in sealing cylinder gas as well as managing the oil film on the cylinder wall. Oil control rings are responsible for lubricating the cylinder walls and also provide a path to dissipate piston heat to the cylinder walls.

Ring Materials

Piston rings are designed with an uninstalled diameter larger than the cylinder bore, so that when they are installed, radial pressure is applied to the cylinder wall. Until recently most piston compression rings were manufactured from cast-iron alloys and tended to be more brittle than today's versions. Today, most piston rings have metal properties more similar to stainless steel than cast iron. This makes them much less likely to fracture and many are highly flexible. The open graphite structure of the cast-iron rings gave them desirable self-lubricating properties but a not-so-desirable shorter service life. In most current diesels these have given way to much tougher and more flexible rings that often show little evidence of wear at engine overhaul. The wall section of the piston in which the set of rings is located is known as the **ring belt**.

Ring Action

The major sealing force of piston rings is high-pressure gas. Piston rings have a small side clearance. The result of this minimal side clearance is that when cylinder pressure acts on the upper sectional area of the ring, three things happen to make it seal:

1. Pressure forces the ring downward into the land.
2. Forcing the ring into the land enables cylinder pressure to get behind the ring.
3. When cylinder pressure gets behind the ring between it and the groove wall, it gets driven outward into the cylinder wall, creating the seal.

The lower right side of Figure 6–7 explains ring action in diagram form. Make sure you understand this. What this tells us is that cylinder

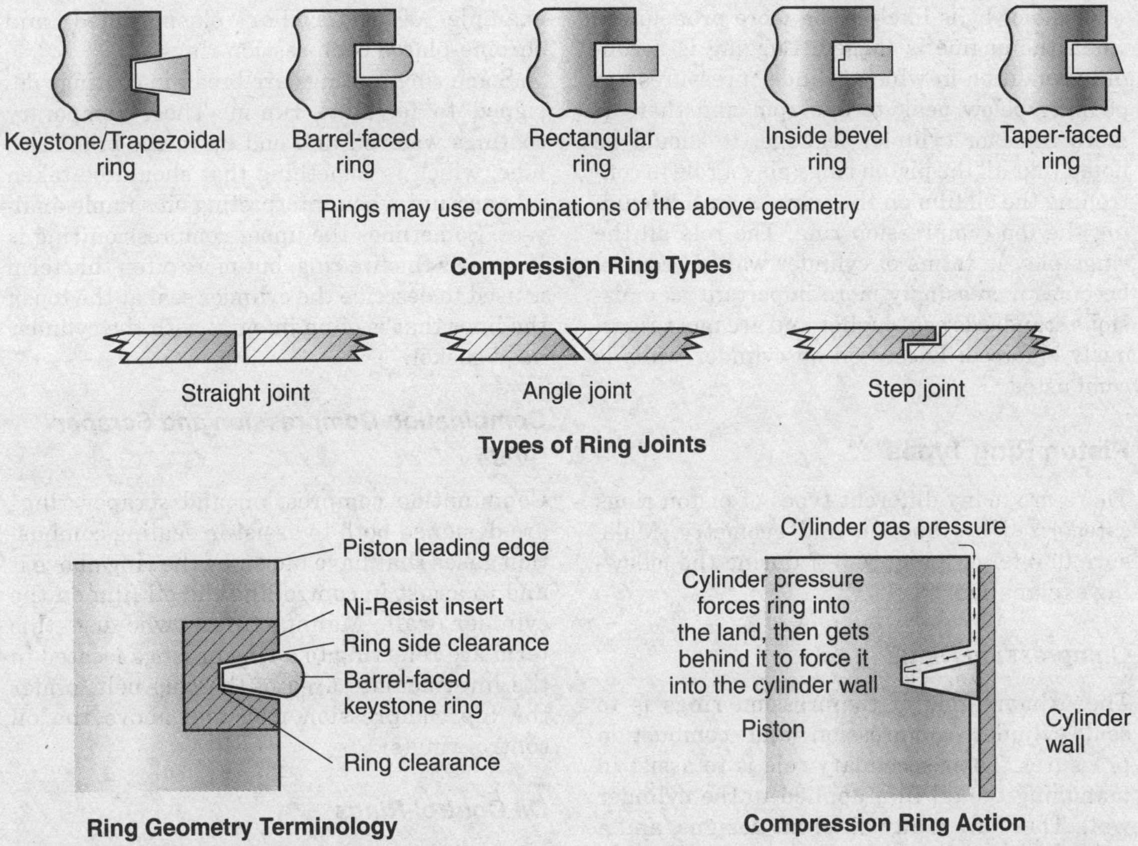

Figure 6–7 *Piston ring geometry and action.*

sealing efficiencies increase with cylinder pressure. The higher the cylinder pressure, the better rings seal the cylinder.

Number of Rings

The number of rings used is determined by the engine manufacturer and the factors are bore size, engine speed, and engine configuration. Time is probably the major factor in determining the number of compression rings. The slower the maximum running speed of the engine, the greater the total number of rings because there is more time for gas blowby to occur. Most current truck and bus medium- and large-bore diesel engines have rated speeds that are in the 2,000 rpm range. Engine OEMs commonly use a three-ring configuration of two compression rings and a single oil control ring. Go back a little in time, and you might have seen a greater number of rings. Four- and five-ring configurations are sometimes still seen but usually in off-shore, off-highway products.

Gas Blowby

The top compression ring gets the greatest sealing assist from cylinder pressures. Gas blowby from the top compression ring passes downward to seal the second compression ring and so on. Gas that blows by all the rings enters the crankcase, so crankcase pressure values are often used as an indication of the overall health of an engine. As a rule of thumb, gas pressure diminishes by around 50% beyond the top compression ring. Because the cylinder is sealed by rings, some cylinder leakage past the ring belt is inevitable, but the limiting factor is time. When an engine is operated at 2,000 rpm, one full stroke of a piston takes place in 15 milliseconds (0.015), so there is quite simply insufficient time for significant cylinder leakage to take place.

Gas blowby is likely to be more pronounced when the engine is lugged. Lugging is a running condition in which cylinder pressures are peaking below peak torque rpm and there is more time for cylinder leakage. It should be noted that all the piston rings play a role in controlling the oil film on the cylinder wall, including the top compression ring. The role all the rings play in terms of cylinder wall oil control becomes increasingly more important as emissions standards get tougher and are more rigorously enforced: Excess oil on cylinder walls is combusted.

Piston Ring Types

There are many different types of piston rings categorized by function and geometry. Make sure to reference Figure 6–7 during the following explanation.

Compression Rings

The primary role of compression rings is to seal cylinder compression and combustion pressures. Their secondary role is to assist in managing the oil film applied to the cylinder wall. There are many different designs, and a variety of materials are used. Malleable and cast irons have generally given way to highly alloyed steels containing molybdenum, silicon, chromium, vanadium, and other alloys. Most current diesel engine piston rings lack the brittleness of the older cast-iron rings. In fact, most can be substantially deformed without fracturing. Compression rings are usually coated by tinning, plasma, and chrome cladding to reduce friction. Figure 6–8 shows an example of Caterpillar plasma-filled and chrome-plated compression rings.

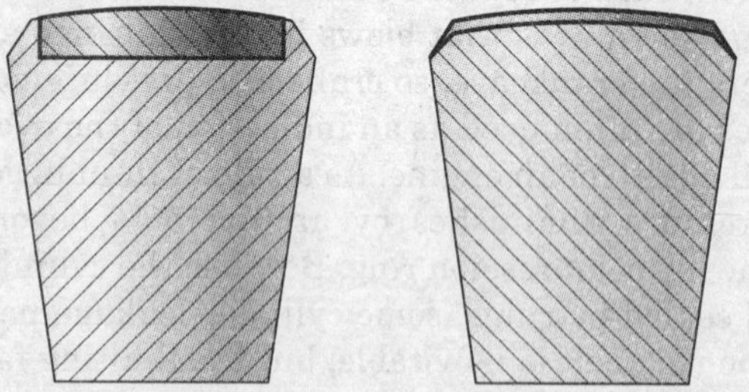

Figure 6–8 *Comparison of plasma-filled ring face and chrome-plated keystone rings. (Reprinted Courtesy of Caterpillar Inc.)*

Some ring coatings are break-in coatings designed to facilitate run-in. These temporary coatings wear off and end up in the crankcase lube, which is something that should be taken into account when interpreting oil sample analyses. Sometimes the upper compression ring is known as the fire ring, but more often this term is used to describe the cylinder seal at the top of the liner that is often integral with the cylinder head gasket.

Combination Compression and Scraper Rings

Combination compression and scraper rings are designed both to assist in sealing combustion gases that have blown by the ring above it and to assist in controlling the oil film on the cylinder wall. Manufacturers who use this term are referring to a ring or rings located in the intermediate area of the ring belt, under the top compression ring and above the oil control ring(s).

Oil Control Rings

Oil control rings are designed to control the oil film on the cylinder wall, which has to be achieved with some precision. Too much oil on the cylinder wall will end up in the combustion chamber, while too little will result in scoring and scuffing of the cylinder wall. The action of first applying and then wiping lubricant from the cylinder wall is also responsible for removing heat from the cylinder and transferring it to the engine lube. Most oil control rings use circumferential scraper rails forced into the cylinder wall by a circumferential expander, usually a coiled spring. They are sometimes known as *conformable* rings, because they will flex to conform to a moderately distorted liner bore.

Piston Ring Geometry

Ring geometry describes the physical shape of the piston rings. Figure 6–7 shows some examples of ring geometry used in diesel engines today. Most piston rings use combinations of the characteristics outlined here. Following are some examples:

Keystone or Trapezoidal Rings

Keystone or **trapezoidal rings** are wedge shaped and fitted to a wedge-shaped ring groove. The design is commonly used for the top compression ring because its shape allows gas pressure exerted on its upper sectional area to easily get behind the ring and act on its inner circumference to load it into the cylinder wall. Keystone rings also inhibit the buildup of carbon deposits due to the angled scraping action caused by the relative motion of the ring within its groove. This makes them less susceptible to sticking. They also function effectively as they wear. Keystone rings have become the ring of choice among diesel engine designers, as shown in the examples of pistons and rings in this chapter.

Rectangular Rings

The sectional shape of the ring is rectangular. Commonly used in the past when the ring material of choice was cast iron, this ring design is loaded evenly (that is, with relatively little twist) into the cylinder wall when subjected to gas pressure, resulting in lower unit sealing pressures but greater longevity. Some pistons that use a keystone top compression ring use a rectangular second or third compression ring as seen in Figure 6–1.

Barrel Faced

The outer face is "barreled" with a radius (rounded), usually with the objective of increasing its service life. There is no sharp edge to bite into the cylinder wall when the ring subjected to gas pressure twists within the groove. Keystone rings are often barrel faced. Figure 6–7 and 6–8 show examples of barrel-faced keystone rings.

Inside Bevel

A peripheral recess is machined into the inner circumference of the ring to facilitate cylinder gas getting behind the ring and causing it to twist within the groove. When a ring twists within its groove the result is usually high unit-sealing pressures; that is, the ring is allowed to bite into the cylinder wall to maximize the seal. A ring with an inside bevel is shown in Figure 6–7.

Taper Faced

The design is much the same as the rectangular ring with the exception that the outer face is angled, giving it a sharp lower edge. Once again, this enables the ring to achieve high unit-sealing pressures; that is, to bite into the cylinder wall when loaded with cylinder gas pressure. A taper faced ring is shown in Figure 6–7.

Channel Section

This design is used exclusively for oil control rings. They usually consist of a channel sectioned or grooved ring with a number of slots to permit oil to be both applied and removed from the cylinder walls. Often an expander ring is used in conjunction with this design to improve its conformability. Conformability refers to a ring's ability to adapt to minor variations in the liner bore. The expander ring is a coiled or trussed spring installed to a groove behind the face rails. See the channel section oil control rings shown in the lower right corner of Figure 6–2.

Ring Joint Geometry

Piston rings must be designed so that when heated to operating temperatures, the ring does not expand so much that the joint edges come into contact: This would buckle the ring. A ring must also be capable of sealing with some efficiency when cold and cylinder pressures are low. Three types of joint design are used:

1. Straight. The split edges of the ring abut. This design has the disadvantage of affording the most potential for gas blowby at the ring joint. It is, however, the most commonly used.
2. Stepped. This design uses an L-shaped step at the joint and affords the least potential for cylinder leakage at the ring joint.
3. Angled. The ring is faced with complementary angles at the abutting joint and seals fairly efficiently at the ring joint.

Figure 6–7 shows some examples of ring joint geometry.

Installing Piston Rings

Rings should be installed to the piston using the correct installation tool. Stretching rings over the piston by hand can crack the plating and cladding materials. Also, try to avoid using generic, multipurpose ring expanders. This is one case where the engine OEM's special tool is usually the best bet: An example of an OEM ring expander is shown in Figure 6–9. *Never* install a piston ring in which the coating appears cracked or chipped. Most rings must be correctly installed and that means they usually have an upside. OEM instructions should be consulted to avoid installing a ring upside down. Most rings are marked, typically by using a dot to indicate the upside.

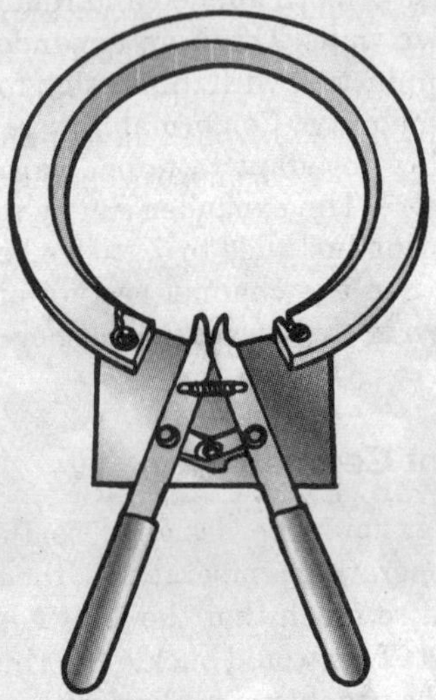

Figure 6–9 *Ring expander. (Courtesy of Mack Inc.)*

Ring End Gap

The ring end gap is checked by installing a new ring into the cylinder bore into the ring belt sweep and measuring with thickness gauges. The specification is usually 0.003″ to 0.004″ per 1 inch of cylinder diameter (0.3–0.4 mm per 100 mm). When checking ring end gap do so by installing a cold ring into a cold liner: The effect is to make the end gap appear large. When installing a new ring into a used cylinder liner/bore, measure ring end gap at the lowest point of the ring belt swept area (tightest section). Remember that, as with most engine specifications, they are established for performance at engine operating temperature and a warm ring will expand to almost close the gap. Never use a ring that measures out of specification. In certain cases where the OEM permits, ring end gap can be adjusted by filing.

Ring Gap Spacing or Stagger

Observe OEM instructions. The gaps are usually offset by dividing the number of rings into 360 degrees, so if there were three rings, the stagger would be 120 degrees offset. However, there are other ways of doing this, and Figure 6–10 and Figure 6–11 demonstrate two different OEM methods: one using a three-ring piston and the other a four-ring piston. It is not recommended that ring gaps be placed directly over the thrust or antithrust faces of the piston.

Ring Side Clearance

Ring side clearance is the installed clearance between the ring and the groove it is fitted to. The dimension is measured using thickness gauges. Figure 6–12 shows the ring side clearance dimension. This dimension must be within specification for the ring to seal properly.

Piston and Cylinder Wall Lubrication

Oil control rings are designed to maintain a precisely managed film of oil on the engine cylinder wall. On the downstroke of the piston, when not loaded by cylinder pressure, lubricating oil is forced into the lower part of the ring groove while the ring is contacting the upper ledge of the land. When the piston changes direction to travel upward, the ring is forced into the lower land of the ring groove, allowing the lubricating oil to pass around the ring to be applied to the liner wall. While the action of simultaneously applying and scraping oil from the cylinder walls ensures that the applied film thickness is minimal, all engines will burn some oil. In the latest generation of low-emissions engines the amount of burned oil is minimal: This is due both to the ring design and properties of CJ-4 lube oil.

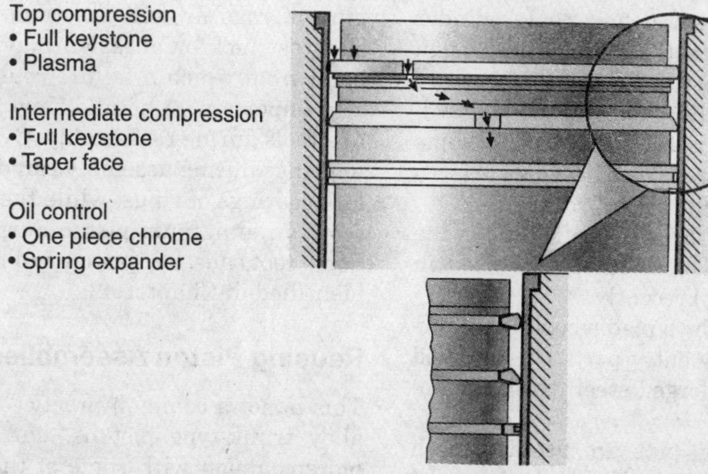

Figure 6–10 *Ring stagger and pressure balance. (Courtesy of Navistar International Corp.)*

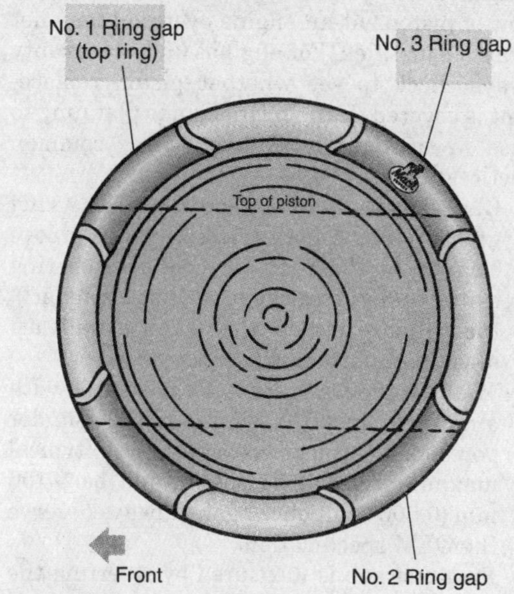

Figure 6–11 *Ring stagger. (Courtesy of Mack Inc.)*

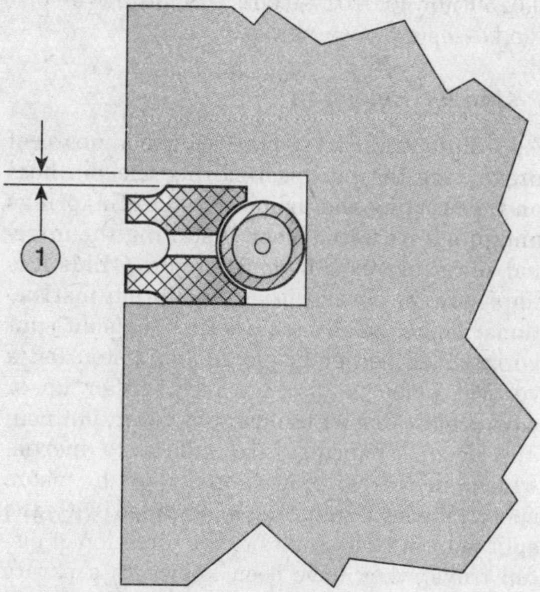

Figure 6–12 *Oil ring side clearance. (Courtesy of Mack Inc.)*

Wrist Pins

The main function of **wrist** or **piston pins** is to connect the piston assembly with the connecting rod eye or small end. In the two-piece, articulating piston assembly, they also link the crown with the skirt, allowing both the crown and skirt to pivot independently. A wrist pin has to be tough enough to transmit loading to (compression stroke) and from (power stroke) the piston assembly while also acting as a pivot point between the connecting rod and piston. The highest expected cylinder pressures determine whether the pin is solid or bored through. Because the weight of the wrist pin adds to the piston total weight, it is designed to be as light as possible while sustaining the loads it is subjected to. This is one of the reasons that forged steel trunk pistons have replaced articulating pistons: The wrist pin can be reduced in length and weight because of the elimination of the separate skirt.

Maximum engine speed and peak cylinder pressures determine actual wrist pin design and material. In most cases, wrist pins are manufactured from mild steel and surface hardened, but middle alloy steels are used in some heavy-duty applications. Their bearing surfaces are lubricated by engine oil in two ways:

- Directed upward through a rifle bore in the connecting rod: This method was used in most engines until recently.
- Sprayed upward by a piston cooling jet targeted at a gallery entry port: This method is used on some forged steel trunk (Monotherm) pistons.

Full-floating piston pins are fitted to both the rod eye and the piston boss with minimal clearance, usually between 0.0025 mm and 0.0250 mm (0.0001″–0.001″). Some newer piston bosses are bushingless.

Piston Pin Retention

All full-floating piston pins require a means of preventing the pin from exiting the pin boss and contacting the cylinder walls. Snaprings and plugs are used. When installing the internal snaprings used by most engine OEMs it is important to observe the installation instructions. These usually require that the split joint be installed perpendicular to the piston (on a vertical plane to piston travel), either up or down depending on the manufacturer, but usually down. Snaprings are subject to inertia, which increases proportionally with piston speed. When a snapring is installed with the split joint at right angles to the direction of piston travel, they have been known to separate from their retention groove in the pin boss, causing an engine failure. Semifloating wrist pins such as those used in crosshead piston assemblies are bolted directly to the connecting rod. In older DDC two-stroke cycle engines, a press fit, sealing cap is used; this component should be vacuum leak tested when assembled. Failure of this cap to seal will result in wrist pin lubricant bleeding to the cylinder walls, and air box pressures may charge the crankcase.

Assembling Piston and Rings

When clamping a piston/connecting rod assembly in a vise, use brass jaws or a generous wrapping of rags around the connecting rod. The slightest nick or abrasion may cause a stress point from which a failure could develop. It is also important to handle rings with the specified tools during reassembly. Overflexing of piston rings during assembly can damage the surface coatings of rings, while the rings ends can score an aluminum piston during installation. Some root causes of piston and ring failures are identified in Chapter 42.

Reusing Piston Assemblies

This is not a common practice with aluminum alloy trunk-type pistons, but it is becoming more common with some of the latest pistons that use forged steels. Attempt to observe OEM recommended practice, but routine replacement of pistons at an engine overhaul may not always be justified. For engines under warranty always check to see whether piston replacement is covered. If the technician is planning to reuse a piston, here are some common practices:

- Clean crystallized carbon out of the ring groove using a correctly sized ring groove cleaner, or if a used top compression ring can be broken (many cannot, use a grinder!), file it square and use this. Then visually assess the condition of the groove.
- The ring groove is correctly measured with a new ring installed square in the cylinder bore using thickness gauges. A typical maximum clearance spec would be 0.150 mm (0.006″) but once again, always observe the OEM specification.
- Ring end gap is measured by inserting the ring by itself into the cylinder bore and measuring gap with thickness gauges. Typically, ring end gaps are 0.075 to 0.100 mm (0.003–0.004″) per 2.5 cm (1″) of cylinder diameter, so for a 5″ cylinder bore end gaps would typically measure between 0.015″ and 0.020″. Remember, check OEM specifications and *always* measure new rings before installation.

Piston Thrust and Antithrust Side Identification

The piston thrust side is that half of the piston divided at the wrist pin pivot on the inboard side of the crank throw during the downstroke. For a typical engine that rotates clockwise viewed from the front, the major thrust side is the right side of the piston observed from the rear of the engine. The opposite side is known as the **antithrust side**. The terms **major thrust side** and **minor thrust side** are also used. Check out Figure 6–2 and make sure this is understood: It could be key to effective troubleshooting later on.

CONNECTING RODS

Connecting rods (also: conn rods) transmit the force developed in the cylinder, acting on the piston to the throw on the crankshaft (Figure 6–13). The end of the conn rod that links to the piston wrist pin is known as either the **rod eye** or the **small end**, while the end that links it to the crankshaft throw is known as the **big end**. Both the rod eye and big end have bearing surfaces. In this way the linear force that acts on the piston and drives it through its stroke can be converted to rotary force or torque by the crank throw, which rotates around the centerline of the crankshaft. Conn rods used with crosshead-type pistons have no rod eye bearing and instead have a saddle that bolts directly to the wrist pin; upper bearing action is therefore provided by the piston pin boss bearing. Most commercial diesel engines use two-piece conn rods: The rod is usually forged in one piece, the big end cap being subsequently cut off, faced, and fastened (bolted) for machining.

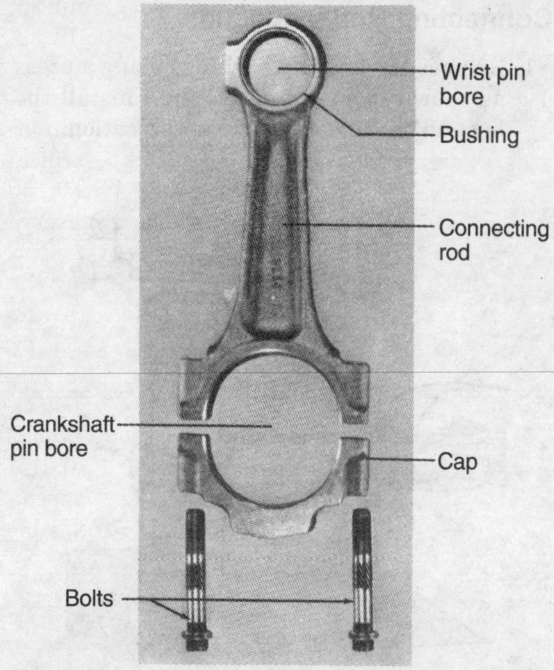

Figure 6–13 *Connecting rod terminology. (Courtesy of Detroit Diesel Corporation)*

Cracked Rods

For many years, **cracked rod** technology has been used in auto racing applications but they have now become common in diesel engines. Some diesel OEMs refer to this as **fractured rod** technology. Cracked rods have a big end that is machined in one piece. After machining, the rod big end is fractured. Depending on the materials used, the fracture process may take

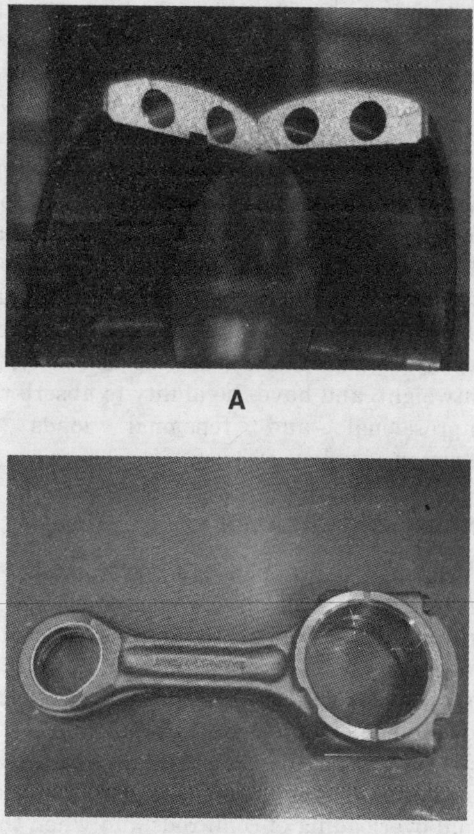

Figure 6–14 *Cracked rod from a Volvo D16 showing the mating faces: A. Disassembled; B. Assembled.*

place at room temperatures or the rod is frozen to subzero temperatures. Fracturing requires a diametric separation through the rod big end. This produces rod and rod cap mating faces that appear rough but form a perfect final-fit alignment. Providing they are assembled properly, cracked rods make the procedure of checking rod sideplay after assembly unnecessary. Most diesel engine OEMs are using cracked rods in at least some of their engines (Figure 6–14).

Connecting Rod Construction

Most rods use an I-beam section design, but round section has also been used. The majority are rifle drilled from big to small ends to carry lubricant from the crank throw up to the wrist pin for purposes of both lubrication and cooling. **Trapezoidal** or **keystone rods** have become commonplace because they reduce bending stresses on the wrist pin by increasing the loaded sectional area, which is ideal for diesels with high cylinder pressures. Keystone rods have a wedge-shaped rod eye or small end. Conn rods are subjected to two types of loading: compressional and tensional. The conn rods shown in Figure 6–16, Figure 6–17, and Figure 6–18 all use a keystone small end. Conn rods have complex metal properties due to the punishment they sustain under normal operation. They are required to have some elasticity, be lightweight, and have the ability to absorb the compressional and tensional loads of the piston.

Compressional Loading

During the compression and power strokes of the cycle, the connecting rod is subjected to **compressional loads**. Another way of saying compression is "squeezing." The extent of compressional loading on a rod can be calculated knowing the cylinder pressure and the piston sectional area values. Commercial diesel engine connecting rods seldom fail due to compressional overloading of conn rods and when they do, it is coincidental with another failure such as hydraulic lock. Hydraulic lock is usually the result of a cylinder head gasket failure that has permitted coolant to leak into the cylinder.

Tensional Loading

Tensional loading is stretching force. At the completion of each stroke, the piston actually stops in the cylinder at either TDC or BDC. This reversal of motion occurs nearly 70 times per second in each connecting rod when an engine is run at 2,000 rpm. The greater the mass of the piston assembly, the greater the inertial forces and therefore the greater the tensile stress the rod and crank throw are subject to. This stress can be extreme in modern engines using heavy articulating piston assemblies. Tensile loading on connecting rods increases with engine rpm and the resulting piston speed increase. When an engine is oversped, the increased tensile loading on the conn rod can result in a tensile failure.

Offset Big End Caps

Many OEMs offset (from horizontal) the mating faces of the two halves of the big end. Doing this ensures that the rod cap fasteners do not have to sustain the full tensile loading of the rod. The conn rod shown in the rod fixture in Figure 6–17 uses an offset big end.

Connecting Rod Inspection

1. Remove the bearing bushings using a press as shown in Figure 6–15, then install the big end cap and torque to specification.

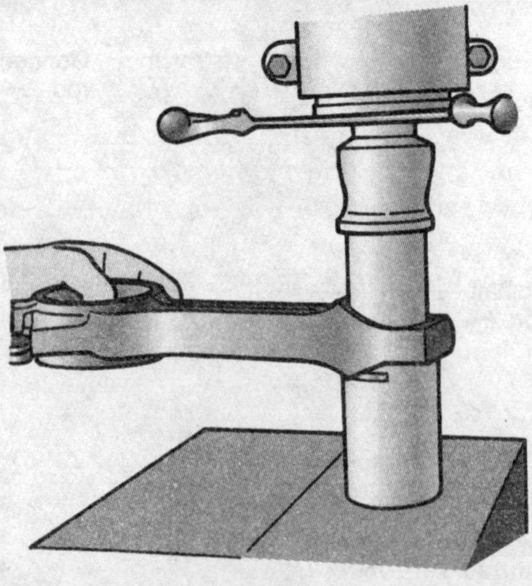

Figure 6–15 *Wrist pin bushing removal. (Courtesy of Mack Inc.)*

2. Measure the big and small end bores with snap or telescoping gauges used in conjunction with an outside micrometer. Big end concentricity is critical and stretch is a result of tensile loading of the rod.
3. Rods should be checked for straightness and twisting in a mechanical rod fixture (Figure 6–16) or electronic rod gauge fixture.
4. The oil passage in the bore should be blown out with compressed shop air and, if necessary, cleaned with a nylon bristle, rifle brush.
5. Finally, the conn rod should be electromagnetic flux tested for cracks. This process involves magnetizing the rod using an electromagnet designed for the purpose and then coating it with a fine magnetic flux, either dry or in a solution. A crack will interrupt the magnetic field, and the flux will concentrate at the flaw. When the magnetic flux is suspended in solution, it is coated in white pigment so that small cracks may be observed using black light (ultraviolet). Any components that have been magnetic flux tested should be demagnetized before being reused.

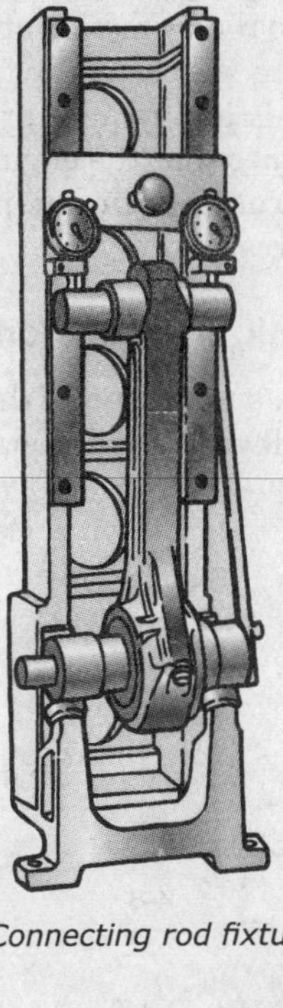

Figure 6–16 *Connecting rod fixture. (Courtesy of Mack Inc.)*

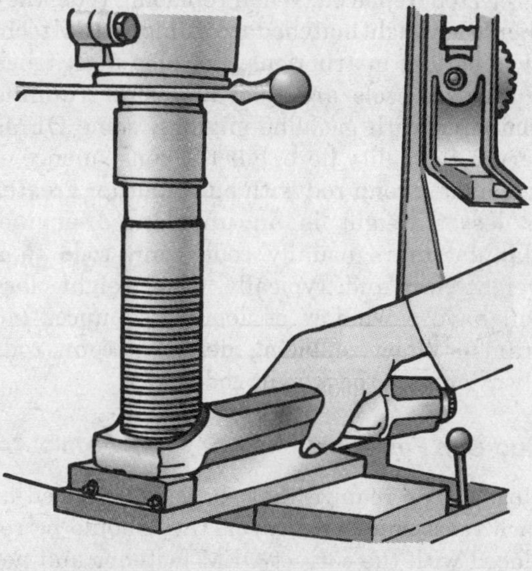

Figure 6–17 *Burnishing a wrist pin bushing. (Courtesy of Mack Inc.)*

6. When conn rod eye bushings are replaced, the newly installed bushings should be sized with a burnishing **broach** as demonstrated in Figure 6–17.

Conn rods should be handled with a great amount of care. When assembling the rod to the piston, a brass jaw vise and light clamping pressure should be used. Slight nicks and scratches on conn rods create stress focal points that can lead to a separation failure. Most diesel engine OEMs suggest that conn rods be electromagnetic flux tested *every* time they are removed from the engine. The cost of magnetic flux examination is small compared to the damage potential of a rod failure. When a conn rod fails in a running engine, the result is often a rod driven through the cylinder block casting. OEMs recommend that connecting rods that fail the inspection procedure be replaced. When reconditioning engines, bear in mind that removing material from a conn rod changes its weight and therefore alters the dynamic balance of the engine.

Replacing Rods

As reconditioning of rods is not widely practiced when rebuilding today's diesel engines, conn rods should be inspected according to the procedures outlined by the manufacturer and if rejected, replaced. When replacing rods, they should be weight matched according to OEM technical service instructions. Because most truck and bus diesels are relatively slow running compared with gasoline engines, some OEMs permit flexibility here, but the consequence of replacing a conn rod with one of either greater or lesser weight is an unbalanced engine. Manufacturers usually code conn rods to a weight class and, typically, each weight class will have a window of about 1.4 ounces (40 grams). When replacing defective conn rods always match the weight codes.

Rod Cap Fasteners

Most OEMs require that these be replaced at each reassembly. Moreover, they should be replaced with the correct OEM fastener and not cross-matched to an SAE-graded bolt. Thread deformation and stretching make these bolts a poor reuse risk, considering the consequences of fastener failure at the rod cap. The actual chances of failure are low but, remember, the consequences of a conn rod failure are severe. If the work is being performed for a customer, recommend that they be replaced and leave the decision to reuse the fasteners to the customer. If you are a betting person, the odds are better when reusing fasteners with an offset mating big end.

Connecting Rod Bearings

Most engine OEMs use a single-piece bushing, press fit to the rod eye, and two-piece friction **bearing shells** at the big end. Rod eye bearings should be removed using an appropriately sized mandrel or driver (see Figure 6–15) and an arbor or hydraulic press. Using a hammer and any type of driver that is not sized to the rod eye bore is not recommended because the chances of damage are high. New one-piece bushings should be installed using a press and mandrel, ensuring that the oil hole is properly aligned, and then sized using a broach or **hone** as shown in Figure 6–17. Split big end bearings should also be installed respecting the oil hole location and the throw journal-to-bearing clearance should be measured. Bearing shells and rod eye bushings should both be installed to a clean dry bore: Remove any packing protective coating from the bearings by washing them in solvent, followed by compressed air drying.

CRANKSHAFTS AND BEARINGS

A crankshaft is a shaft with offset throws or journals to which piston assemblies are connected by means of conn rods. Figure 6–18 shows some typical crank throw configurations but note that in the case of the V-configured engines, OEMs may use different numbering sequences than that shown here. For instance, DDC numbers cylinders sequentially by left and right bank: 1L, 2L, 3L, 4L, 1R, 2R, 3R, 4R. When rotated, the offset crank throws convert the linear, reciprocating movement of the pistons in the cylinder bores into rotary motion at the crankshaft. Crankshafts are supported by friction bearings at main journals and require pressurized lubrication at all times that the engine is run to enable their **hydrodynamic suspension** within the bearing bores.

Hydrodynamic Suspension

Hydrodynamic suspension is the supporting of a rotating shaft on a fluid wedge of constantly

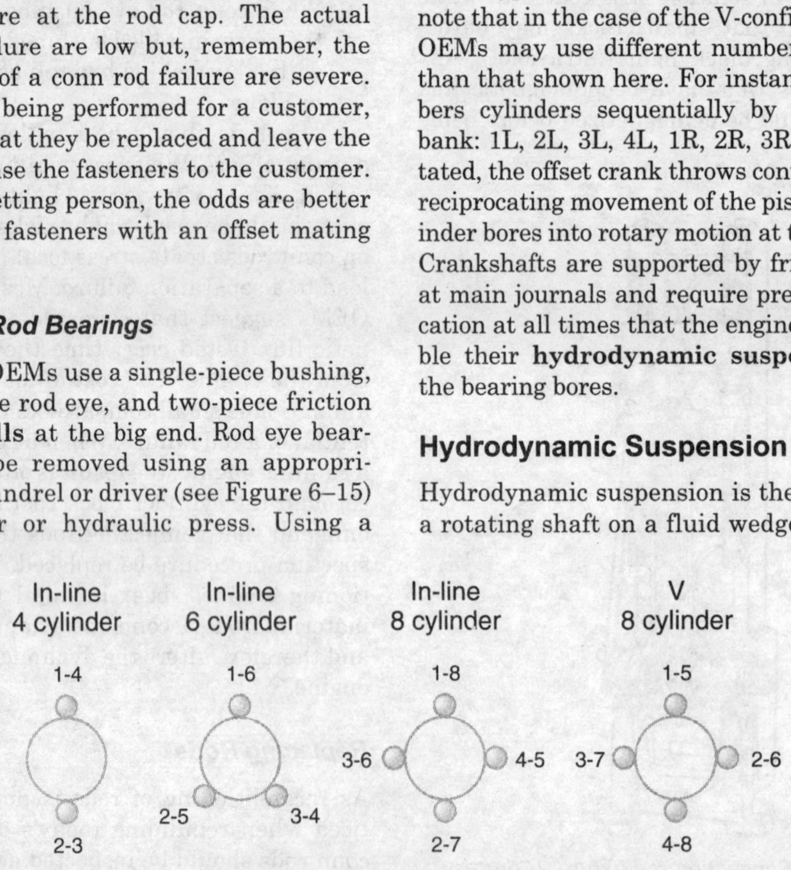

Figure 6–18 *Crank throw configurations.*

changing, pressurized engine oil. In this way, direct shaft main journal-to-bearing bore contact is avoided; an oil film protects the friction bearing shell when the engine is stationary or cold cranked. Friction bearings are also used at the crank throw/conn rod big end. The lubrication required for these bearings is supplied from oil passages routed through the crankshaft and oil holes located at each journal. Offsetting the oil hole angle from the crankshaft axis generally drives a thicker wedge of oil onto which the crankshaft main is rolled as it turns so most OEMs do this. Crankshafts are designed for dynamic balance and use counterweights to oppose the unbalancing forces generated by the pistons. These forces tend to diminish as the number of cylinders in an engine increases, and companion throws (geometrically paired) contribute a counterbalancing effect. Crankshafts are subjected to two types of force: bending forces and torsional forces.

Bending Forces

Bending stress occurs between the supporting main journals between each power stroke. Crankshafts are engineered to withstand the considerable bending forces that result from subjecting a crank throw to the compression and combustion pressures developed in the cylinder. This normal bending stress takes place between the main journals at any time the crank throw between them is loaded by cylinder pressure, and, therefore, it peaks when engine cylinder pressures peak.

Torsional Forces

Torsion is twisting. **Torsional stresses** are the twisting vibrations that a crank is subject to that occur at high speed. Crankshaft torsional vibration occurs because a given crank throw while under compression (that is, driving the piston assembly attached to it upward on the compression stroke) slows to a speed marginally less than average crank speed. Subsequently, the same throw on receiving power stroke forces accelerates to a speed marginally greater than average crank speed. These twisting vibrations or oscillations take place at high frequencies and crankshaft design, materials, and hardening methods must take them into account.

Torsional stresses on a crankshaft tend to peak at crank journal oil holes at the flywheel end of the shaft. These torsional oscillations are amplified when an engine is run at slower speeds with high cylinder pressures because the real time duration between cylinder firing pulses is extended. This type of running (lower speed/high load) is known as **lugging**. However, most current diesel engines are designed to run at up to 30% lower speeds and output 30% more torque than those of a generation ago. These engines produce higher torsional oscillations that are projected through the drivetrain, making the proper matching transmissions and final drive carriers essential.

Crankshaft Construction

Figure 6–19 is a technician's guide to crankshaft terminology. Understanding these terms is critical when measuring and machining crankshafts. Most diesel engine crankshafts are made of steel forgings, but exceptionally special cast-iron alloys have been used. All crankshafts are tempered (heat treated) to provide a tough core with the required flexing characteristics to withstand the bending and torsional punishment they are subject to. Most diesel engine OEMs use proprietary processes to forge crankshafts and maintain some secrecy about the details of the process. An understanding of journal hardening procedures is important as the reconditionability of the crankshaft depends on this.

Journal Surface Hardening Methods

Though OEMs often use proprietary terms to describe the following processes, there are basically three methods of surface hardening main and rod journals:

- Flame hardening. Used on plain carbon and sometimes middle alloy steels. Flame hardening consists of the direct application of heat followed by quenching with oil or water. The result is relatively shallow surface hardening and the actual hardness dependent on the carbon and other alloy content of the steel. Note: Flame is sometimes as used as protection in induction harden-

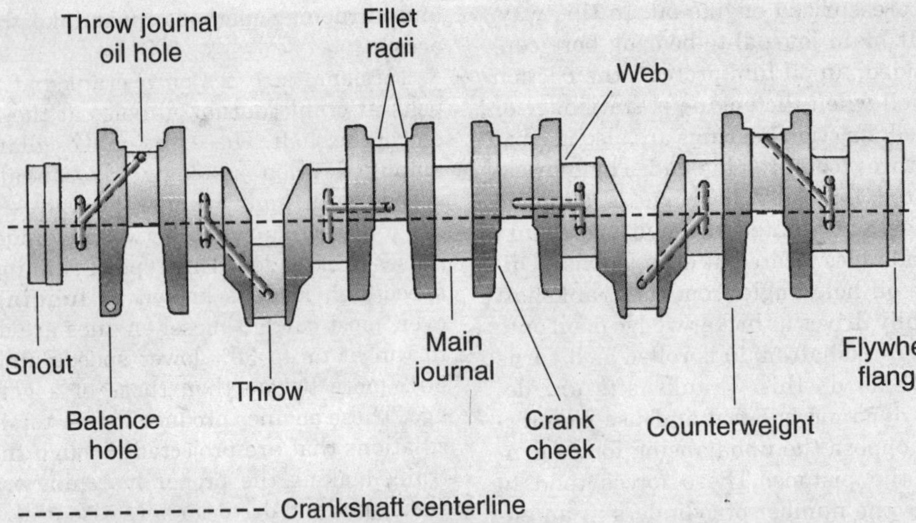

Figure 6–19 *Crankshaft terminology. (Reprinted Courtesy of Caterpillar Inc.)*

ing treatment ovens because it prevents air exposure to the crankshaft during the tempering process.

- Nitriding. Used on alloy steels. Nitriding involves higher temperatures than flame hardening, and the surface hardens to a greater depth—around 0.65 mm (0.025″).
- Induction hardening. Area to be hardened is enclosed by an applicator coil through which alternating current (AC) is pulsed, heating the surface— tempering is achieved by blast air or liquid quenching. This process results in hardening to depths of up to 1.75 mm (0.085″), providing a much wider wear and machinability margin than the preceding methods. Most current diesel crankshafts are surface hardened by this method and sometimes during the process, flame is used in the induction oven to prevent air exposure to the crankshaft when at a cherry-red temperature and therefore vulnerable to oxidation.

Removing Crankshaft from Cylinder Block

The cylinder block should be in an engine stand and in an inverted position. Once the main bearing caps and any other obstructions have been removed, a rigid crankshaft yoke should be fitted to the hoist: The yoke prongs should be fitted with rubber hose (to prevent damage to the journals) and linked to a pair of throws in the center of the crankshaft (for weight balance). Figure 6–20 demonstrates how a yoke connects the crankshaft to a hoist.

Figure 6–20 *Crankshaft removal from cylinder block. (Courtesy of Mack Inc.)*

Torque Twist Control Devices

In an effort to reduce engine weight, engine OEMs have used tougher, lighter alloys in cylinder block manufacture. This makes a cylinder block more susceptible to torque twist flexing when the engine is under extreme loads or there are sudden changes in load. OEMs have attempted to limit cylinder block torque flexing by installing bolster plates across the lower cylinder block flange and transverse bolts known as **buttress screws** into the main caps. When buttress screws are used, the main cap is fastened through a vertical plane in the usual manner—but also transversely by buttress screws. If removing a main cap on a diesel engine after removing the cap screws is difficult, check on the outside of the cylinder block for buttress screws. Buttress screws have to removed before a buttressed main cap can be removed.

Crankshaft Failures

A certain small percentage of crankshaft failures will result from manufacturing and design problems. Design problems are usually revealed shortly after an engine series is introduced and quickly remedied by the OEM because they do not wish to have an engine tagged with a basic engineering fault likely to produce failures. Because research, development, and testing of new engine series are so thorough in today's engines, when crankshaft failures occur, the cause will most likely be attributable to one of the categories listed next.

Bending Failures

Abnormal bending stresses occur when:

- Main bearing bores are misaligned. Any kind of cylinder block irregularity causes abnormal stress over the affected crankshaft area.
- Main bearings fail or become irregularly worn.
- Main caps are broken or loose.
- Standard specification main bearing shells are installed where an oversize is required.
- Flywheel housing is eccentrically (relative to the crankshaft) positioned on the cylinder block. This produces a broken back effect on the drivetrain. The condition usually results from a failure to (dial) indicate a flywheel housing on reinstallation.
- Crankshaft is not properly supported either out of the engine before installation or while replacing main bearings in-chassis. The latter practice is more likely to damage the block line bore but may deform the crankshaft as well.

Bending failures tend to initiate at the main journal fillet and extend through to the throw journal fillet at 90 degrees to the crankshaft axis.

Torsional Failures

Excessive torsional stress can result in fractures occurring from a point in a journal oil hole extending through the fillet at a 45-degree angle or circumferential severing through a fillet. In an inline six-cylinder engine, the #5 and #6 journal oil holes tend to be more vulnerable to torsional failure when the crankshaft is subjected to high torsional loads.

Causes of crankshaft torsional failures are:

- Loose, damaged, or defective vibration damper or flywheel assembly.
- Unbalanced engine-driven components: fan pulleys and couplings, fan assembly, idler components, compressors, and power takeoffs (PTOs).
- Engine overspeed. Even fractional engine overspeeding will subject the crankshaft to torsional stresses that it was not engineered for that may initiate a failure. It should be remembered when performing failure analysis that the event that caused the failure and the actual failure may be separated by a considerable time span.
- Unbalanced cylinder loading. A dead cylinder or fuel injection malfunction of over- or underfueling a cylinder(s) can result in a torsional failure.
- Defective engine mounts. This can produce a "shock load" effect on the powertrain.

Spun Bearing(s)/Bearing Seizure

Spun bearings are lubrication-related failures caused by a lack of oil in one or all of the crank journals. The bearing is subjected to high friction loads and surface welds itself to the affected crank journal. This may result in a spun bearing in which the bearing friction welds itself to the journal and rotates with the journal in the bore or alternatively continues to scuff the journal to destruction. When a crankshaft fractures as a result of bearing seizure, the surface of the journal is destroyed by excessive heat, and it fails because it is unable to sustain the torsional loading. Causes of spun bearings and bearing seizure are:

- Misaligned bearing shell oil hole.
- Improper bearing-to-journal clearance. Excessive clearance will result in excessive bearing oil throwoff, which starves journals farthest from the supply of oil. Insufficient bearing clearance can be caused by overtorquing, use of oversize bearing shells where a standard specification is required, and cylinder block line bore irregularities.
- Sludged lubricating oil causing restrictions in oil passages.
- Contaminated engine oil. Fuel or coolant in lube oil will destroy its lubricity.

Etched Main Bearings

Etched main bearings are caused by the chemical action of contaminated engine lubricant. Chemical contamination of engine oil by fuel, coolant, or sulfur compounds can result in high acidity levels that can corrode all metals. However, the condition is usually first noticed in engine main bearings. It may result from extending oil change intervals beyond that recommended. Etching appears initially as uneven, erosion pock marks, or channels.

Crankshaft Inspection

Most truck and bus engine OEMs recommended that a crankshaft be magnetic flux tested at every out-of-chassis overhaul. Essentially this process requires that the component to be tested be magnetized and then coated with minute iron filings. When AC is pulsed though the shaft, the magnetic lines of force that result will bend into a crack or nick, causing the iron filings to collect in the flaw.

Magnafluxing

Magnafluxing a crankshaft requires that it be magnetized using a circumferential electromagnet. Fine magnetic particles are coated in white pigment and suspended in fluid, allowing them to be sprayed over the crankshaft where they concentrate over any flaw that breaks up the magnetic field force lines. Minute flaws can be identified by inspection with an ultraviolet or black light. Any component that has been magnetized for a magnetic flux examination must be demagnetized before it is reused. It should be noted that most small cracks observed by magnetic flux inspections are harmless. The following cracks require crankshaft replacement:

- Circumferential fillet cracks
- 45-degree cracks extending into critical fillet areas or journal oil holes

Visual Inspection

Evidence of failures in these locations *require* the replacement of the crankshaft; the consequence of not doing so could be total destruction of the engine. Following magnetic flux inspection, the crankshaft thrust surfaces should be checked for wear and roughness. It is usually only necessary to polish or dress these areas using the appropriate crank machining equipment and using the methods described later in this section. Also, the front and rear main seal contact areas should be checked for wear; wear sleeves, interference fit to the seal race, are available for most engines and can be easily installed without specialized tooling. Most crankshafts in current medium- and large-bore highway diesel engines require no special attention during the life of the engine.

Crankshaft Measuring Practices

Specifications listed here are *typical* maximums. The OEM service literature should always be consulted when determining the serviceability of a specific crankshaft. Journals should be measured across axes 90 degrees apart in three places, usually at either end and in the center of the journal and the results matched to OEM specification. Using a measuring chart is the preferred method of organizing the data and retaining it.

- *Journal out of roundness:* 0.025–0.050 mm (0.001″ –0.002″)
- *Journal taper:* 0.0375 mm (0.0015″)
- *Bent crankshaft:* Check in V-blocks with a dial indicator while smoothly rotating by hand—out of service (OOS) specifications differ widely due to differing crankshaft materials and lengths.

Polishing Crankshafts

The objective of polishing a crankshaft is to remove minor main and throw journal scratches, scores, and nicks from a crankshaft that has been measured and gauged to be within the OEM specifications. The use of oversized bearings would not normally result from a polishing procedure because no significant amount of journal surface material should be removed.

The crankshaft should be mounted to the appropriate lathe mandrels with the journal oil holes plugged with flush port caps and rotated at low speed in a crank grinding lathe in the normal engine rotation direction. A long, narrow strip of low-abrasive emery cloth (the type sold in rolls) wetted with diesel fuel or kerosene should be applied for short periods to the journals requiring the polishing and frequent inspection should be performed. When polishing the throw journals, the rotation will be eccentric, so a longer strip of emery cloth should be used.

The crankshaft thrust faces may also have to be similarly dressed. This procedure is a little bit more tricky, and the emery cloth should be wrapped around an oversized, flat wood tongue depressor. Care should be exercised whenever hands have to be placed near rotating machinery. Thrust face scoring may require dressing by grinding, but before performing this procedure check on the availability of oversized thrust washers/bearings.

Reconditioning Crankshafts

While it should be made very clear that most diesel engine OEMs do not approve of the reconditioning of crankshafts, the practice is widespread. In most cases, the reconditioning procedures attempt not to compromise the original surface hardening but where the damage is severe, such as in the example of a spun bearing, journal damage is sustained below the hard surfacing. While journal surfaces can be rehardened, this practice is not so common. The reconditioning methods are simply outlined here and the technician should remember that the industry consensus is that they are bad practice and poor economics in the long term. There are four basic methods of crankshaft reconditioning:

1. Grinding to an undersize dimension. This may require that oversize bearings, not always supplied by the engine OEM, be available.
2. Metallizing surface followed by grinding to the original size.
3. Chroming surface to the original size.
4. Submerged arc welding buildup followed by grinding to the original size.

ROD AND MAIN BEARINGS

There are probably engineers who devote their entire careers to the study and development of **friction bearings** used in diesel engines. It makes sense for the diesel technician to have a basic appreciation of bearing construction (Figure 6–21) and to be able to diagnose characteristic failures. Most diesel engine manufacturers make available excellent bearing failure analysis charts and booklets. These use high-definition color photography and examples of actual failures that make diagnosing failures a cinch. Color Figure C1 shows some examples of classic bearing failures.

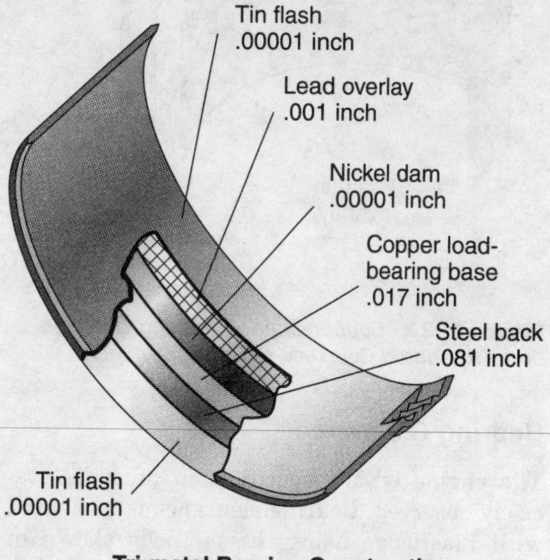

Figure 6–21 *Construction of a typical bearing. (Courtesy of Navistar International Corp.)*

Construction and Design

Two basic designs are used in current applications:

1. Concentric wall—uniform wall thickness

2. Eccentric wall—wall thickness is greater at the crown than the parting faces; also known as deltawall bearings

Materials

Rod and main friction bearings consist of a steel base or backing plate, onto which is layered copper, lead, tin, aluminum, and sintered combinations of metals often with a zinc or tin outer protective coating. Friction bearings are designed to have a degree of embedability. This means that the outer face of the bearing must be soft enough to permit small abrasives to penetrate the outer shell, known as the overlay, to a depth at which they will cause a minimum amount of scoring to the crank journals that rotate in the bearing. Figure 6–22 shows a copper-bonded friction bearing favored by Caterpillar.

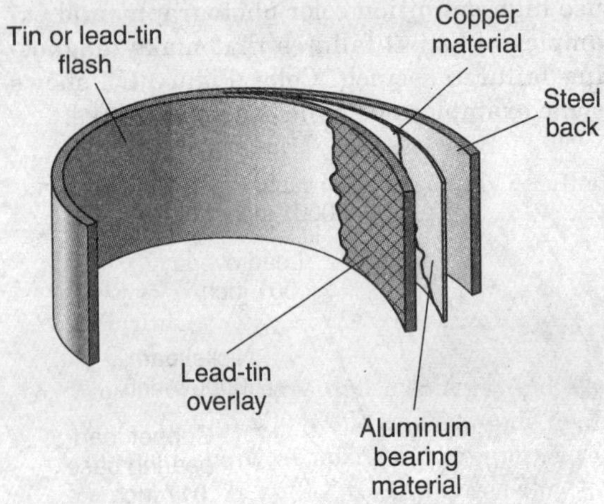

Figure 6–22 *Copper-bonded bearing construction. (Reprinted Courtesy of Caterpillar Inc.)*

Bearing Clearance

The engine OEM's specification must be precisely observed. Bearing clearance is measured with Plastigage using the methods shown in Figure 6–23 and Figure 6–24. Most manufacturers make several oversizes (of bearings) to accommodate a small amount of crankshaft journal wear or machining. It is important not to assume that a new engine will always have standard size bearings. Typical bearing clearances in highway diesel engine applications run from 0.050 mm to 0.100 mm (0.002″ to 0.004″).

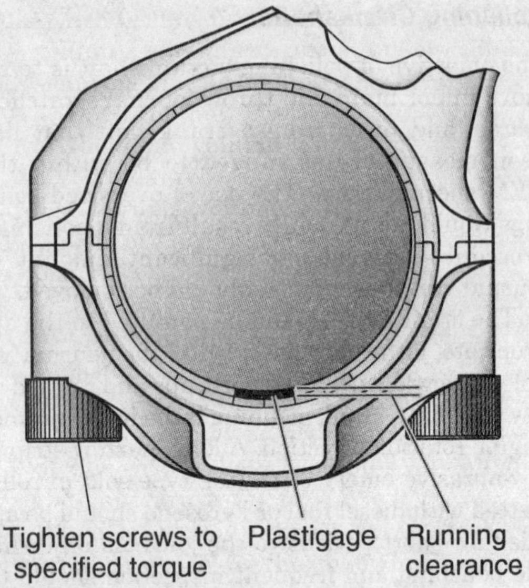

Figure 6–23 *Checking bearing clearance using Plastigage. (Courtesy of Mack Inc.)*

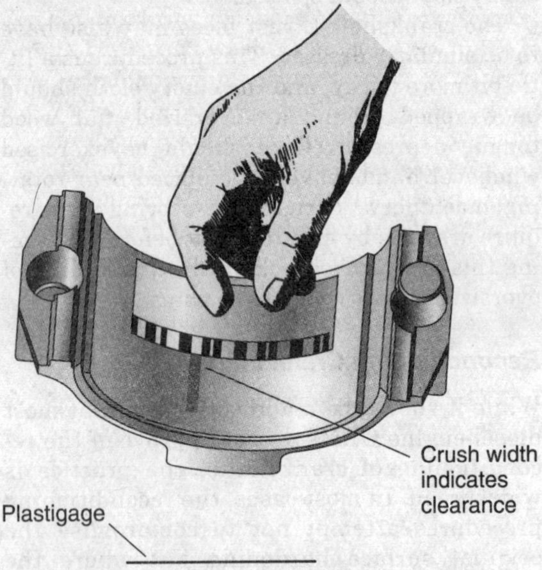

Figure 6–24 *Correct method of locating Plastigage. (Courtesy of Mack Inc.)*

The ability to maintain hydrodynamic suspension of the crankshaft diminishes with increased bearing clearances. When bearing clearance increases above specification, a drop in oil pressure results that may indicate the need for an in-chassis bearing rollover. Increased bearing clearance also increases oil throwoff, and this may result in excessive lube being thrown up onto the cylinder walls.

The technician should not attempt to measure bearing clearance when the engine is in-chassis because the results will have little validity. This is because of the flexibility of crankshafts when not fully supported at each main bearing. The engine should be upside down and level so that there is no weight load acting on the retaining cap of the bearing being measured.

The bearing clearance specifications should be referenced first. Sometimes bearing clearances are expressed in ten thousandths of an inch because they are engineering specs: In this case, round the value to the nearest half thousandths (0.0005″) and work in units of thousandths. For instance, an OEM clearance spec of 0.0027″ can be expressed as 2½ thousandths. Next select the Plastigage color code capable of measuring between the specified parameters. **Plastigage** is malleable plastic thread that easily deforms to conform to whatever clearance space is available when compressed between a bearing and a journal. The crushed width can then be measured against a scale on the Plastigage packaging. The lower the clearance, the wider the Plastigage strip will be flattened.

Using Plastigage™

A short strip of Plastigage should be cut and placed across the center of the bearing in line with the crankshaft. The bearing cap with the bearing shell in place should then be installed and torqued to specification in the incremental steps outlined in the service literature. Do not rotate the engine with the Plastigage in place. Next the bearing cap and shell should be removed and the width of the flattened Plastigage checked against the dimensional gauge on the Plastigage packaging. If clearance is within specifications, carefully remove the Plastigage from the journal before reinstalling the cap and shell assembly.

Selecting the Correct Plastigage Strip

Plastigage is manufactured in four sizes, each color-coded for the range of clearances each is capable of measuring. Table 6–1 identifies each Plastigage color-coded size window in both standard and metric dimensions. While most diesel OEMs still display their specifications using standard measurements, some display in both standard and metric, and one uses metric-only specifications.

Table 6–1 *Plastigage™ Code Identification*

Part Number	*Color*	*Standard Dimension Window*	*Metric Dimension Window*
HPG1	Green	0.001–0.003 inch	0.025–0.76 mm
HPR1	Red	0.002–0.006 inch	0.051–0.152 mm
HPB1	Blue	0.004–0.009 inch	0.102–0.229 mm
HPY1	Yellow	0.009–0.020 inch	0.230–0.510 mm

Crankshaft Endplay

One of the main bearings is usually flanged to define crank endplay; this is known as the **thrust bearing** and it is available in several sizes to accommodate some thrust surface wear as well as some thrust face dressing in the crankshaft. Alternatively, split rings known as thrust washers may be used to control crank endplay. Endplay specifications would typically be in the 0.2 mm to 0.3 mm (0.008″ to 0.012″) range. Use a dial indicator to measure: Use a lever that is not going to damage the crankshaft and force it fore and aft for measurement.

Bearing Retention

Bearings are retained primarily by *crush:* The outside diameter of a pair of uninstalled bearing shells slightly exceeds the bore to which it is installed (Figure 6–25). This creates radial pressure that acts against the bearing halves and provides good heat transfer. The bearing halves are also slightly elliptical to allow the bearing to be held in place during installation. This pro-

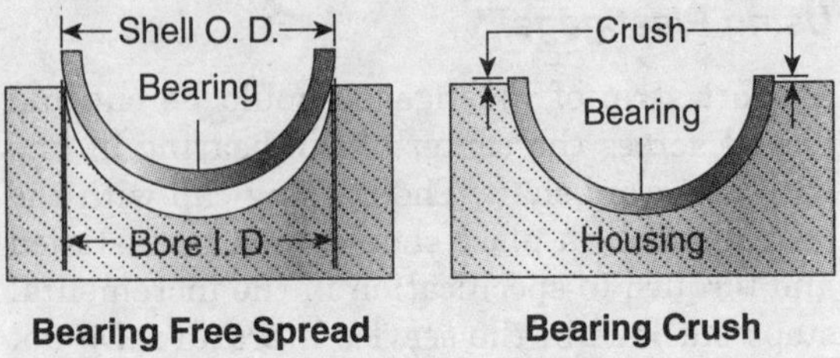

Figure 6–25 *Bearing spread and crush. (Courtesy of Detroit Diesel Corporation)*

Figure 6–26 *Upper and lower bearing shells. (Courtesy of Detroit Diesel Corporation)*

vides what is known as bearing spread (see Figure 6–25). Tangs in bearings (Figure 6–26) are inserted into notches in the bearing bore to minimize longitudinal movement, prevent bearing rotation, and align oil holes.

Bearing Removal and Installation

Service literature should always be consulted and their procedures observed. This operation is very straightforward when performed out of chassis because the crankshaft is removed when installing the cylinder block side bearing shell and the technician is working above the engine with perfect visibility and accessibility for the remainder of the procedure.

Bearing Rollover

For many novice truck technicians, their first experience of engine work will be performing what is known as an in-chassis bearing rollover, a simple procedure but one that can be messy, especially on a hot engine that may drip oil for hours. The procedure is practiced more often than necessary on today's engines because a slight drop in oil pressure is often mistakenly attributed to bearing wear. Many service facilities offer bearing rollover "specials" usually early in the spring season when temperatures begin to warm and coincide with a slight reduction in oil pressure. It is not unusual today when performing a bearing rollover to remove a set of bearings in near perfect condition. This is a testament to the quality of today's bearings (and other factors) compared with those of a generation ago.

In-chassis, main bearings are rolled out using a capped dowel (often "homemade" by machining a bolt) inserted into the journal oil hole. Ensure that the dowel cap does not exceed the thickness of the bearing because if it does, it will destroy the journal. Next, rotate the crank to roll the bearing out, tang side first. Using a screwdriver or similar method will almost certainly result in scratch and score damage to the journal. New bearing halves are installed clean, dry, and with as little handling as possible. If a new bearing is coated with a protective film of grease or light wax, this should be removed using a bristle brush and solvent (do not use a solvent that has any residual oil component such as Varsol™), followed by air drying.

The backing should be installed clean, dry, and preferably without finger contact—trace quantities of lubricant, moisture, or dirt will reduce its ability to transfer heat. The facing may be lightly coated with clean engine lubricant applied by finger. Most engine OEMs recommend avoiding the use of any type of grease, such as white lube (lithium-based grease) or other engine assembly lubes, on the bearing face, because these reduce the ability of the engine lubricant to hydrodynamically support the crankshaft on startup and may, in the crankcase, interact with the additives in the engine oil. After performing any procedure that involves the draining of lubricant from the oil galleries and passages that supply the bearings, it is good practice to prime the engine lubrication circuit using a remote pump before cranking.

Bedplate Main Bearings

Some recent small-bore engines usually of offshore origins have introduced a lower cylinder block bedplate. The bedplate replaces a set of main bearing caps with a single casting plate that bolts to the engine cylinder block. This design can enable a lighter cylinder block to withstand higher resistance to acceleration and deceleration torsionals. It goes without saying that the bedplate design is not conducive to an in-chassis overhaul.

VIBRATION DAMPERS

Vibration damper and harmonic balancer both refer to the same component, but vibration damper is used by most diesel engine OEMs so it is used in this text. A vibration damper is mounted on the free end of the crankshaft, usually at the front of the engine. Its function is to reduce the amplitude of vibration and add to the flywheel's mass in establishing rotary inertia. In other words, its primary function is to reduce crankshaft torsional vibration (there is a full explanation of crankshaft torsional vibration earlier in this chapter under crankshafts).

Vibration Damper Construction

A typical vibration damper consists of a damper drive or housing and inertia ring (Figure 6–27 and Figure 6–28). The housing is coupled to the crankshaft and using springs, rubber, or viscous medium, drives the inertia ring: The objective is to drive the inertia ring at *average* crankshaft speed. Vibration dampers therefore have three main components:

- Drive member: bolted to the crankshaft
- Drive medium: either a fluid (silicone gel) or solid rubber
- Driven member: an inertia ring

There are two types of vibration dampers used in current diesel engines:

- Viscous-drive type
- Solid rubber-drive type

Viscous Drive Operating Principle

Viscous drive-type vibration dampers are more common in truck and bus diesels. The drive member is bolted to the crankshaft and consists of an annular hollow housing. Within the hollow housing, the inertia ring is suspended in and driven by silicone gel. Because the drive member is bolted to the crankshaft snout it is subject to the torsional oscillations occurring at the front of the crankshaft. When the drive member rotates, it rotates the drive medium (silicone gel) and inertia ring with it. Because the inertia ring contains most of the mass (weight) of the vibration damper, it will rotate at *average* crankshaft speed and not be subject to the torsionals occurring at the front of the crankshaft. Because the speeds of the drive member and inertia ring differ, shearing action will take place in the drive medium, which is the silicone gel. This shearing action of the viscous fluid film between the drive housing and the inertia ring effect the damping action.

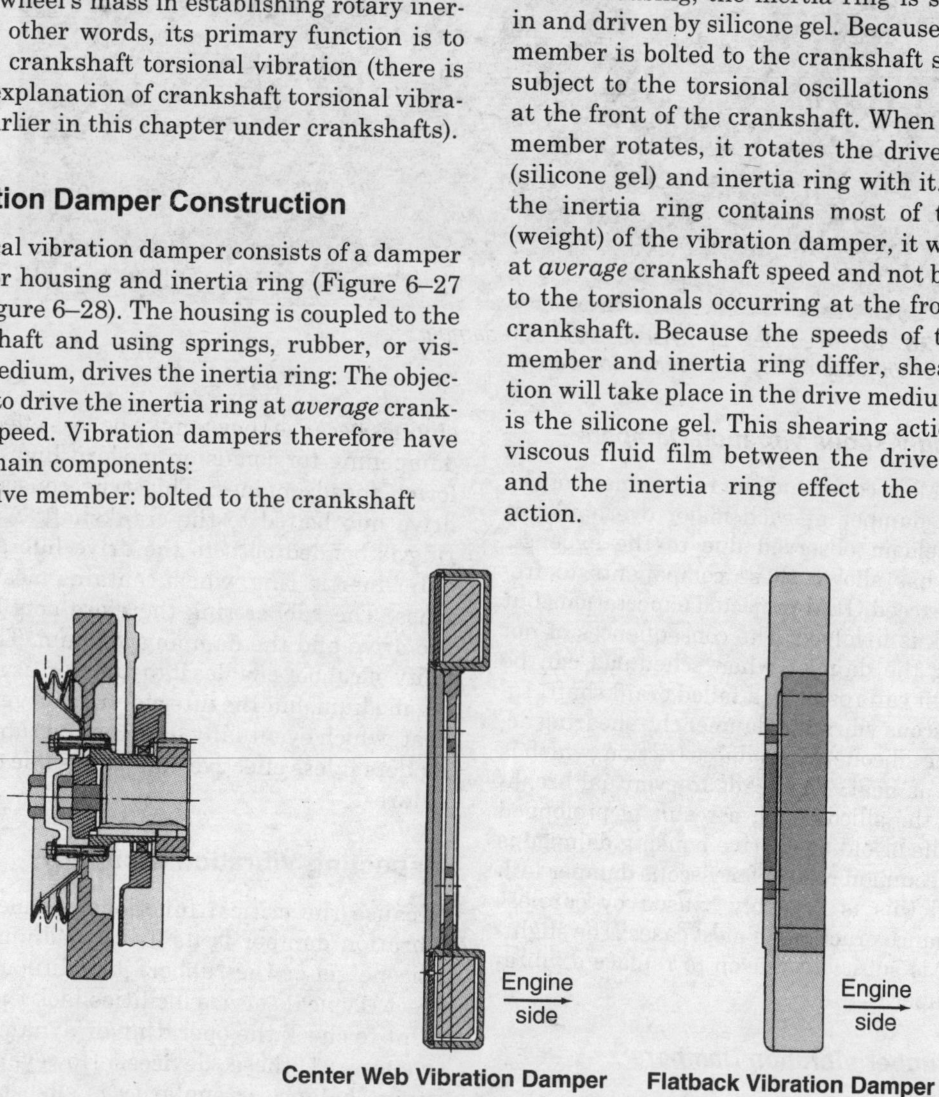

Figure 6–27 *Sectional view of a vibration damper. (Courtesy of Mack Inc.)*

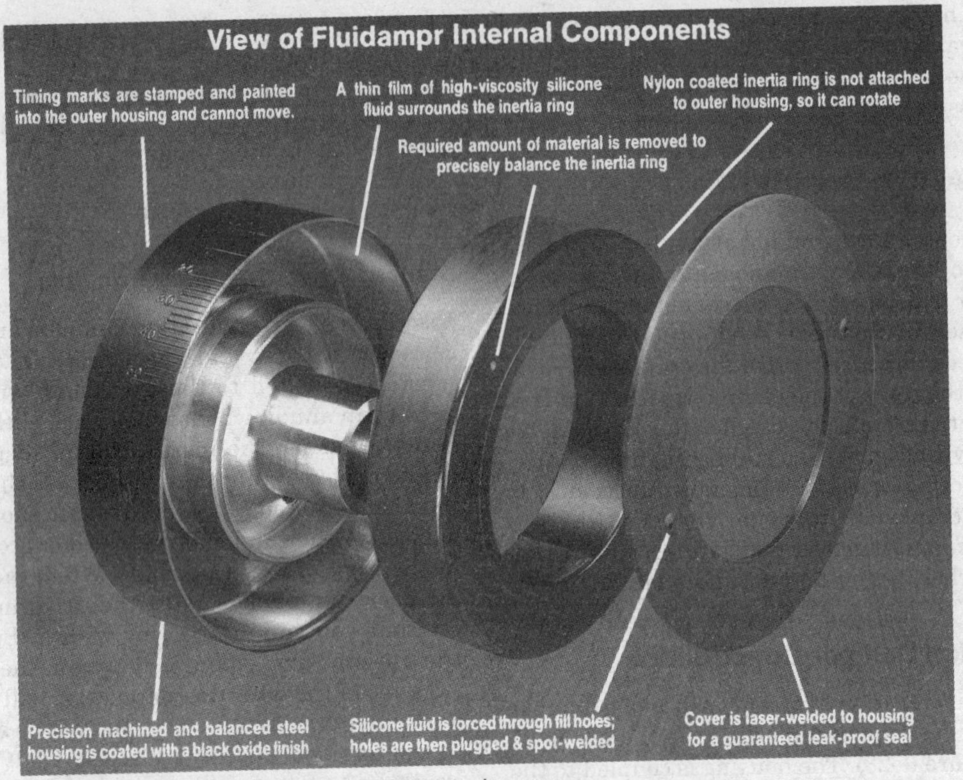

Figure 6–28 *Exploded view of a viscous vibration damper. (Courtesy of Vibratech TVD, A Horschel Company)*

Replacing Viscous Vibration Dampers

Most OEMs recommend the replacement of the vibration damper at each major overhaul but this is seldom observed due to the expense. Practice has shown these components to frequently exceed OEM projected expectations but some risk is involved. The consequences of not replacing the damper when scheduled can be costly as it can result in a failed crankshaft. Inside a viscous vibration damper the shearing action of the silicone gel produces friction, which is released as heat. This leads to eventual breakdown of the silicone gel, a result of prolonged service life or old age. Drive housing damage is another common reason for viscous damper failure, and this is probably caused by careless maintenance practices in most cases: The slightest ding is sufficient reason to replace a vibration damper.

Solid Rubber Vibration Dampers

Solid rubber-type vibration dampers are less often used on today's heavy truck diesel engines because they tend to be less effective at dampening torsionals on modern high-torque, lower speed engines. This type consists of a drive hub bolted to the crankshaft: A rubber ring is bonded both to the drive hub and the outer inertia ring, which contains most of the mass. The rubber ring therefore acts both as the drive and the damping medium. The elasticity of rubber enables it to function as a damping medium, but the internal friction generates heat, which eventually hardens the rubber and renders it less effective and vulnerable to shear failures.

Inspecting Vibration Dampers

Because the critical functional element in a vibration damper is its drive medium, either silicone gel or the rubber, it is difficult to inspect. Typical service facilities lack the equipment to check the operating or dynamic effectiveness of these devices. However, if an engine balance irregularity is suspected and the OEM-projected service life is known to have been exceeded, the damper should be replaced. If the service life has not been exceeded, there are some external checks that the technician may perform to help diagnose the condition:

1. Visually inspect the damper assembly: Any dents or signs of warpage in either viscous or solid rubber drive dampers are reason to reject the component.
2. Using a dial indicator, rotate the engine manually, and check for damper housing radial and axial runout to the OEM specification. This is usually a low tolerance specification (especially on the viscous type), usually 0.005″ (0.127 mm) or less.
3. Check for indications of fluid leakage in the viscous type, initially with the damper in place. Trace evidence of leakage justifies replacement of the damper.
4. If the damper checks out OK using the above tests, remove it from the engine. By hand, shake the damper; any clunking or rattle is reason to replace it.
5. Next, using a gear hotplate or component heating oven, heat the damper to its operating temperature, usually close to the engine operating temperature, around 90°C (180°F)—this may produce evidence of a leak in viscous dampers: The slightest leak is reason to replace the damper.
6. A final strategy is to mount the damper in a lathe using a suitable mandrel. It should be run up through the engine operating rpm range and monitored using a balance sensor and strobe light.

FLYWHEELS

The engine flywheel in the typical commercial diesel engine is normally mounted at the rear of the engine. It has three basic functions:

1. To store kinetic energy (the energy of motion) in the form of inertia and both help smooth out the power pulses in the engine and establish an even crankshaft rotational speed.
2. To provide a mounting for engine output: It is the PTO device to which a clutch or torque converter is bolted.
3. To provide a means of rotating the engine by cranking motor during startup.

As an energy storage device, the flywheel plays a major role in dampening the torsional vibrations that act on the crankshaft during engine operation, and its mass helps rotate the engine between firing pulses. Flywheel mass (weight) depends on a number of factors, such as whether the engine is a two- or four-stroke cycle; the number of engine cylinders; and the engine operating rpm range. Because the number of crank angle degrees between power strokes is half that on a two-stroke cycle compared to a four-stroke cycle of the equivalent number of cylinders, generally less flywheel mass is required.

six-cylinder, two-stroke cycle—frequency of power strokes: 60 crank angle degrees

six-cylinder, four-stroke cycle—frequency of power strokes: 120 crank angle degrees

The flywheel therefore stores potential energy in the form of inertia. Engines designed to be run at consistently high rotational speeds require less flywheel mass, and it should be noted that a heavy flywheel will inhibit rapid response to acceleration demand. While a four-stroke cycle, single-cylinder diesel engine is unlikely to power any modern truck, it is worth taking a look at such an engine (you can still find some old one-cylinder Lister engines on farms) and noting its flywheel mass, a dominant characteristic of its appearance.

Types of Flywheels

The flywheels (and flywheel housings) used on all North American-built trucks are categorized by the SAE by size, shape, and bolt configuration. This allows engines to be easily coupled to different OEM clutch and transmissions. Two basic geometric shapes are used: the *flat face* design and the *pot* design. Figure 6–29 shows a flat face truck flywheel. Most trucks equipped with current medium- and large-bore highway diesel engines use one of two SAE flywheel sizes:

SAE # 4—accommodates a 15½″ clutch assembly

SAE # 5—accommodates a 14″ clutch assembly

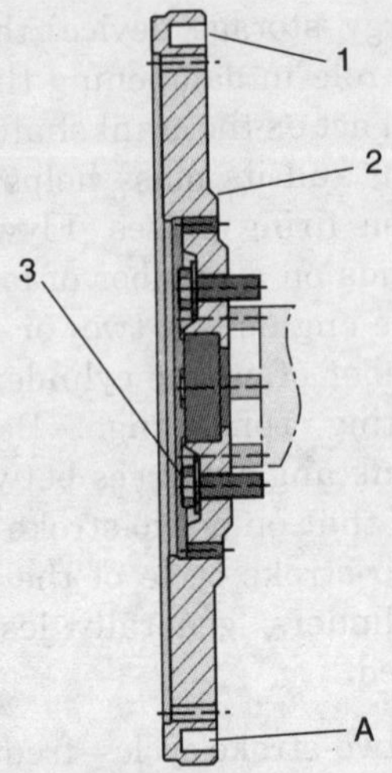

Typical Example

Make reference to Flywheel Runout for correct method of flywheel inspection

(1) Flywheel ring gear. Gear must be assembled against shoulder (A) of flywheel. Maximum temperature of gear before shrinking 315°C (599°F)

(2) Flywheel.

(3) Apply thread sealant to bolts and torque to 300 ± 30 N•m (220 ± 22 lb.-ft.)

Figure 6–29 *Sectional view of a flywheel. (Reprinted Courtesy of Caterpillar Inc.)*

Flywheel Construction

Flywheels used on most truck and bus diesel engines are machined from cast iron or cast steel. Rim velocities are factored by the rotational speed (engine rpm) and the clutch diameter. However, rim stresses seldom generate metallurgical failures in truck and bus diesel engine flywheels due to their relatively low rotational speeds. Stresses tend to peak at the juncture of the hub with the rim, an area that is subject to torsional as well as centrifugal loads, but failures are rare.

Ring Gear Replacement

A ring gear is shrunk fit to the outer periphery of the flywheel. This is the means of transmitting cranking torque to the engine by the starter motor during startup. A worn or defective ring gear is removed from the flywheel by first removing the flywheel from the engine and then using an oxyacetylene torch to partially cut through the ring gear working from the outside on a single tooth. This is in most cases sufficient to expand the ring gear so that the removal can be completed using a hammer and chisel. Care should be taken to avoid heating the flywheel any more than absolutely necessary or damaging the flywheel itself by careless use of the oxyacetylene flame. A hacksaw and chisel may also do the job.

Installing a New Ring Gear

To install a new ring gear, place the flywheel on a flat, level surface, and check that the ring gear seating surface is free from dirt, nicks, and burrs. Ensure that the new ring gear is the correct one and if its teeth are chamfered on one side, that they will face the cranking motor pinion after installation. Next, the ring gear must be expanded using heat so that it can be shrunk to the flywheel. Most OEMs specify a specific heat value because ring gears are heat treated, and overheating will damage the tempering and substantially reduce the hardness. A typical temperature specification would be around 200°C (400°F) but this may be as high as the 315°C (600°F) recommended by Caterpillar (see Figure 6–29).

Because of its size, the only practical method of heating a ring gear for installation is using a rosebud-type (high gas flow), oxyacetylene heating tip. To ensure that the ring gear is heated evenly to the specified temperature and especially to ensure that it is not overheated, the use of temperature-indicating crayon such as Tempilstick is recommended. When the ring gear has been heated evenly to the correct temperature, it will usually drop into position and almost instantly contract to the flywheel. Handle a hot ring gear with blacksmith tongs.

Reconditioning and Inspecting Flywheels

Flywheels are commonly removed from engines for reasons such as clutch damage, leaking rear main seals, or leaking cam plugs, and care should be taken when both inspecting and reinstalling the flywheel and the flywheel housing. Flywheels should be inspected for:

- Face warpage
- Heat checks

- Scoring
- Intermediate drive lug alignment and integrity
- Axial and radial runout using dial indicators, straightedges, and thickness gauges

Damaged flywheel faces may be machined using a flywheel resurfacing lathe to OEM tolerances: Typical maximum machining tolerances range from 0.060″ to 0.090″ (1.50 mm to 2.30 mm). It is important to note that when resurfacing pot-type flywheel faces, the pot face must have the same amount of material ground away as the flywheel face; the consequence of machining only the flywheel clutch face is to have an inoperable clutch.

REVIEW QUESTIONS

1. Which of the following engine conditions would be the likely cause of a tensile failure of a connecting rod?
   a. Engine overspeed
   b. High cylinder pressures
   c. Prolonged lug down
   d. Prolonged engine idling
2. Which of the following is a disadvantage of an articulating piston assembly when comparing it with an aluminum alloy trunk piston?
   a. Piston slap control
   b. Highest possible location of an upper ring groove
   c. Service life
   d. Increased tensile loading on the connecting rods
3. Where would a Ni-Resist insert be located?
   a. Upper ring belt, aluminum trunk piston
   b. Upper ring belt, crosshead piston
   c. Trunk piston, pin boss
   d. Lower ring belt, articulating piston
4. Under which of the following conditions would a piston ring seal more effectively?
   a. High engine temperatures
   b. Low engine temperatures
   c. High cylinder pressures
   d. Low cylinder pressures
5. A crankshaft has a fracture failure that appears to have initiated at a throw journal oil hole. Which of the following would be the more likely cause?
   a. A broken main cap
   b. Excessive cylinder pressure
   c. A failed vibration damper
   d. A spun rod bearing
6. Which of the following journal surface hardening methods would provide the highest machinability margin?
   a. Flame hardening
   b. Nitriding
   c. Shot peening
   d. Induction hardening
7. Rod and main journal bearing clearances are properly measured with:
   a. Tram gauges
   b. Dial indicators
   c. Plastigage
   d. Snap gauges
8. What type of piston is favored by diesel engine OEMs in post-2007 products?
   a. Forged steel trunk
   b. Aluminum trunk
   c. Two-piece articulating
   d. Two-piece crosshead
9. When observing a disassembled engine, the outer crown edges of all six pistons show signs of melting and erosion. This could be caused by:
   a. Advanced timing
   b. A dribbling injector
   c. Retarded timing
   d. Fuel with high sulfur content
10. Which of the following types of pistons use bushingless pin bosses?
    a. Ferrotherm
    b. Monotherm
    c. Crosshead
    d. Aluminum alloy
11. What would be the consequence of a misdirected piston oil cooling jet?
    a. Fuel in lube
    b. Torched piston
    c. Low oil pressure
    d. Overheated engine lube
12. *Technician A* states that keystone rings are most common in current diesel engines. *Technician B* states that keystone connecting rods have a wedge-shaped big end. Who is correct?
    a. Technician A only
    b. Technician B only
    c. Both A and B
    d. Neither A nor B
13. What should be used to install piston rings onto a piston?
    a. The thumb and index finger

b. A ring contractor
c. A ring expander
d. Snapring pliers

14. Which of the following operating modes would be more likely to result in a compressional failure of a connecting rod?
a. Prolonged low idle running
b. Hydraulic lock caused by head gasket failure
c. Prolonged high idle running
d. Overspeeding engine

15. Which of the following would be the more likely outcome of running an engine with a failed vibration damper?
a. Increased fuel consumption
b. Increased cylinder blowby
c. Crankshaft failure
d. Camshaft failure

16. In what area of a Mexican hat combustion chamber are the injected droplets of fuel directed?
a. Center of the cone protrusion
b. Crater
c. Piston leading edge
d. Upper compression ring

17. The upper face of a piston assembly is called the:
a. Skirt
b. Boss
c. Trunk
d. Crown

18. *Technician A* states that a cam ground piston is circular when cold and expands to an elliptical shape at operating temperature. *Technician B* states that the reason the cam ground piston is elliptical when cold is to accommodate the thermal expansion of the greater mass of material at the piston boss as it is raised to operating temperature. Who is correct?
a. Technician A only
b. Technician B only
c. Both A and B
d. Neither A nor B

19. The most commonly used piston crown design by OEMs in today's highway diesel engines is the:
a. Mexican hat
b. Mann type
c. Dished
d. Barrel face

20. Which term is used to describe the cylinder volume between the piston upper compression ring and its leading edge?
a. Dead volume
b. Headland volume
c. Toroidal recess
d. Clearance volume

21. Which of the following is true of a CFA piston?
a. The ring belt reinforcement insert extends up into the leading edge of the piston.
b. It is a type of Ni-Resist insert.
c. It is used only on articulating pistons.
d. The ring belt operates at higher temperatures.

22. When crankshaft thrust surfaces are reground, which of the following would be true?
a. Undersize thrust bearings are required.
b. Oversize thrust bearings are required.
c. The thrust washers must be shimmed.
d. The thrust washers must be machined to a smaller dimension.

23. *Technician A* states that main bearing shells are usually interchangeable and may be installed in either the upper or lower position. *Technician B* states that the main bearing shell with an oil hole must always be installed in the main bearing cap. Who is correct?
a. Technician A only
b. Technician B only
c. Both A and B
d. Neither A nor B

24. Excessive clearance at the crankshaft main bearings could result in which of the following conditions?
a. Aerated oil
b. Fluctuating oil pressures
c. High oil pressure
d. Low oil pressure

25. *Technician A* states that the most commonly used type of vibration damper used on truck diesel engines is the viscous type. *Technician B* states that the inertia ring drive medium on a viscous-type vibration damper is a rubber compound. Who is correct?
a. Technician A only
b. Technician B only
c. Both A and B
d. Neither A nor B

7

Engine Feedback Assembly

INTRODUCTION

The feedback assembly of an engine is the engine's mechanical management apparatus and includes timing and accessory drive gearing, the camshaft, tappets, **valve** and unit injector trains, and fuel pumping mechanisms. Its components are driven by, and often timed to, the engine crankshaft. The drive mechanism for medium- and heavy-duty highway diesel engine feedback assemblies is almost always a gearset. In lighter duty engines pulleys and belts, chains, and sprockets may be used as drives. A typical timing gear set is illustrated in Figure 7–1 and a timing gear cover in Figure 7–2.

TIMING GEARS

Diesel engine timing gears can be located at either the front or rear of the longitudinally mounted engines that power most commercial trucks and the transverse-mounted engines that power most buses. While timing geartrains are more commonly located in the front of diesel engines, Volvo, Mack post-2007, and Detroit Diesel Corporation (DDC) post-2007 DD families of engines are notable exceptions. Timing gears are responsible for driving the camshaft and most of the engine accessories on a diesel-powered truck engine. The geartrain is enclosed in a housing that permits engine oil lubrication of the rotating components. Some larger engines may locate the timing geartrain in the rear or both front and rear. The gear ratios dictate the rotational speed and maintain a fixed location

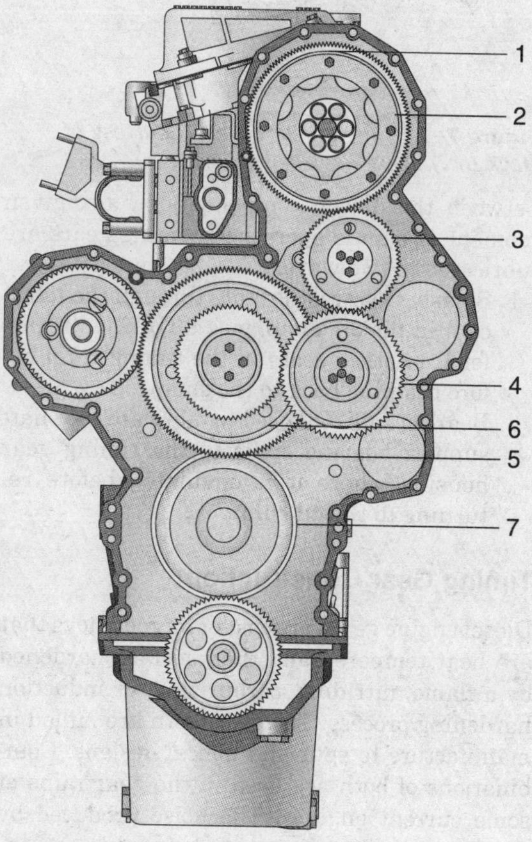

FRONT GEAR TRAIN

1. Timing mark
2. Camshaft gear
3. Adjustable idler gear
4. Idler gear
5. Timing mark
6. Cluster gear
7. Crankshaft gear

Figure 7–1 *Caterpillar 2007 ACERT C13 engine timing geartrain featuring an overhead camshaft. (Reprinted Courtesy of Caterpillar Inc.)*

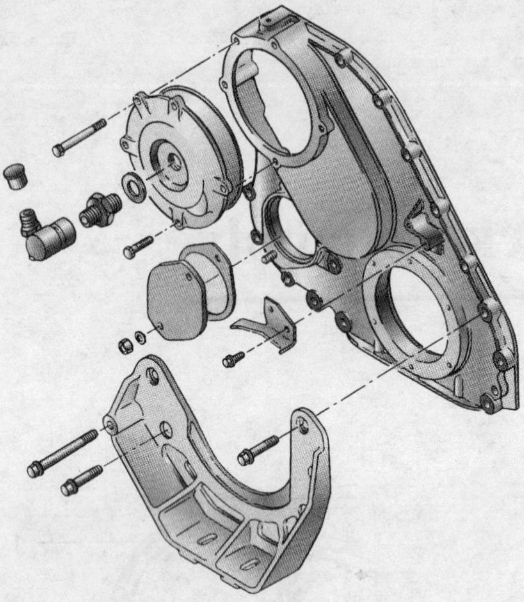

Figure 7–2 *Timing gear cover. (Courtesy of Mack Inc.)*

between the driven components in any given moment of engine operation. Timing gears are lubricated in two ways:

1. Splash: Gear teeth rotate through the lubricant in the oil sump, pick up oil, and transfer it to other gears in the timing train before draining back to the sump.
2. Bearing spill: Oil used to lubricate the shaft support bearing spills to the timing gear housing where it is circulated before returning to the oil sump.

Timing Gear Construction

Diesel engine gears are cast or forged alloys that are heat tempered and then surface hardened by a flame, nitriding, carburizing, or induction hardening process. The gear teeth are milled in manufacture to spur and helical designs: Combinations of both are used in the geartrains of some current engines. The noise produced by the geartrain is a factor and, for this reason, helical cut gears tend to be more common. **Helical gears** have the advantage of increased tooth contact area, lowering unit forces. The **spur gear** design offers the advantage of much lower thrust loads. The gears are commonly press fit to the shafts that they drive and are positioned on the shaft by means of keys and keyways.

Timing Gear Inspection, Removal, and Installation

Visual inspection should be sufficient to determine the condition of timing gears. Indications of cracks, pitting, heat discoloration, or lipping of the gear teeth usually require its replacement. Press fit gears require the use of a mechanical, pneumatic, or hydraulic puller to remove the gear from the shaft. When the shaft and gear can be removed from the engine, a shop air-over-hydraulic press should be used, ensuring that the usual safety precautions are observed and that care is taken to support the components on separation. When a gear has to be separated from a shaft while on the engine, a portable hydraulic press is usually required; while using the press, ensure that it is mounted in such a way that it will not damage either the cylinder block or other gears.

Installing Camshaft Gear

Install the new gear to the shaft by heating to the original equipment manufacturer's (OEM's) specified temperature. To heat the gear to the prescribed temperature use a thermostatically regulated oven or hot plate. If a bearing hot plate is used, ensure that it is large enough to fit the entire surface area of the gear. Typical specified temperatures are around 300°F (150°C). The consequence of overheating a gear will be to destroy the heat treatment/tempering and possibly the surface hardening, resulting in premature failure. A heat-indicating crayon such as **Tempilstick™** may be used to determine the exact temperature. At the required temperature, the gear can be dropped over the shaft and allowed to air cool. Use of a power press to install interference fit gears to shafts should be generally avoided as damage to the shaft may result: A hand-actuated arbor press is less likely to inflict damage. The engine geartrain must be timed according to the OEM procedure to ensure that the relative position of the camshaft and critical accessory drives are correct and balanced. As noted earlier some current engines locate the engine timing geartrain at the rear of the engine. Figure 7–3 shows the rear of engine timing gear rotation on a Volvo D16D engine.

A number of gears are used in the engine geartrain, especially when an OHC has to be driven. If the camshaft is to be rotated in the same direction of rotation as the crankshaft, an idler gear must act as an intermediary between the crankshaft and camshaft gears. There is no requirement that the camshaft be turned in the same direction as the crankshaft as shown in the Volvo D16 example in Figure 7–3. This figure shows the rear of the engine so the actual direction of rotation oriented from the front of the engine is crankshaft CW and camshaft CCW.

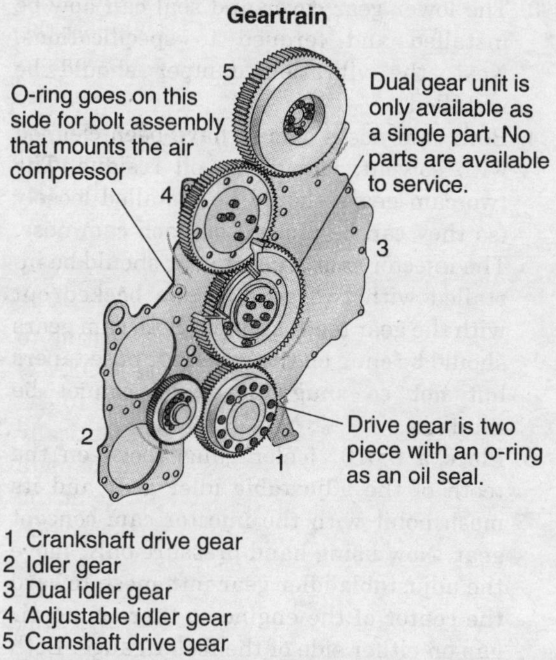

Figure 7–3 *Timing gear rotation on a Volvo D16D engine, which locates the timing gears at the rear of the engine. (Courtesy of Volvo Trucks NA)*

Compound Drive

Chapter 11 addresses turbo-compounding in greater detail; however, it should be mentioned here because turbocharger-sourced drive torque is unloaded to the engine geartrain, providing engine drivetrain-assist. Turbo-compounding has been recently introduced by Detroit Diesel in its DD family of engines and claims it can add up to 50 horsepower (37 kW) when used on a DD15. Although turbo-compounding is not a new technology in diesel engines generally, it is new in an on-highway application. The turbine in the compound turbocharger connects by means of reduction gearing and a viscous drive directly to the engine geartrain: In this way, under conditions of peak turbine gas efficiency, it is able to provide some considerable torque assist.

Timing the Engine Geartrain

The timing procedure is simple but critical and therefore carefully outlined in the OEM technical support literature. As the procedure involves locating the engine powertrain position relative to that of the feedback assembly, the crankshaft should first be located in a specific position usually with the #1 cylinder at top dead center (TDC). A procedure for locating true TDC is outlined in Chapter 13. Gears that are required to be timed to each other are indexed by stamped markings on the teeth that must be intermeshed. Idler gears may have a hunting tooth relationship with the crankshaft and camshaft gears, meaning that after timing the geartrain, the engine may have to be rotated through a large number of revolutions before the timing indexes realign. **Hunting gears** are not uncommon in engine timing geartrains. Engine timing geartrains may be designed with one or more idler gears. Some engines use offset gear-to-shaft locating keys to trim the timing requirements; the procedure for determining the specific offset key required should be precisely observed.

Gear backlash should be checked after the engine geartrain has been assembled and timed. A dial indicator (preferred) or thickness gauges should be used. A typical geartrain backlash specification would be in the region of 200 mm (0.008″), but obviously the OEM specifications should be consulted. A backlash specification higher than the OEM specification is an indication of gear contact face wear and usually requires that a gear or gears be replaced. A lower than specified backlash factor often indicates an assembly or timing problem.

Timing Overhead Camshafts

The procedure for timing overhead camshafts (OHCs) varies and depends on the engine. We have chosen one of the trickier engine timing gearsets, that on a Cummins ISX series engine.

This engine uses a double overhead camshaft arrangement as shown in Figure 7–4. Each camshaft has different functions. The left side camshaft is responsible for actuating the mechanical injectors used on these engines, while the right side camshaft takes care of valve actuation and engine braking. Two **concept gears** are used.

Concept Gears

A concept gear is sectionally divided and uses coaxial springs between the drive hub and the gear to dampen geartrain oscillations. In effect, a concept gear allows a zero lash condition to exist between intermeshing gears regardless of engine temperature and without the risk of gears radially loading each other, which would cause noise and damage. The ISX engine uses two concept gears. The lower concept gear acts as an idler gear and uses seven coaxial springs. The upper concept gear located on the injector camshaft uses four coaxial springs and drives the valve/brake camshaft.

Timing OHC Concept Gears

The procedure used to time an ISX geartrain is briefly outlined in the following sequence.

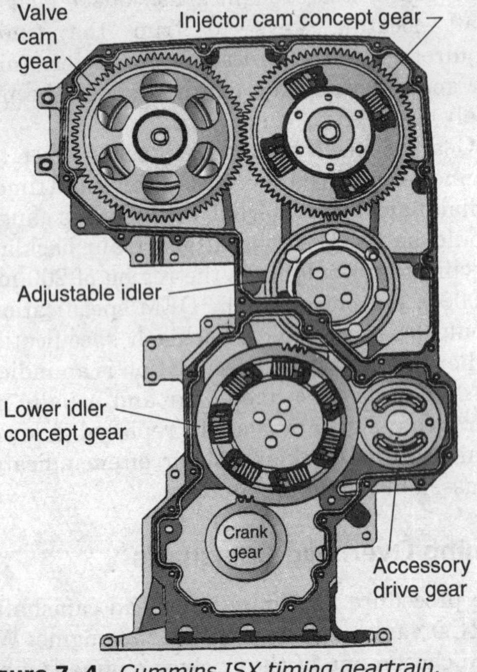

Figure 7–4 *Cummins ISX timing geartrain. (Courtesy of Cummins Inc.)*

Reference Figure 7–4 while going through the procedure:

1. Pin the crankshaft and wedge both of the camshafts: This effectively locks both camshafts and the crankshaft into position. The locking wedges are Cummins special tools and they must be used.
2. First install the lower concept gear, shaft, and retainer plate, making sure that the thrust bearing is behind the gear. Install the gear shaft bolts followed by the gear fasteners and torque to specifications.
3. The lower gear cover and seal can now be installed and torqued to specifications. Next, the vibration damper should be installed.
4. Both cam noses should have been cleaned with solvent, leaving no oil residue. The two cam gears should be installed loosely (so they can be moved) on each cam nose. The injector cam concept gear should be installed with two gear screws backed out with the gear teeth aligned. Both cam gears should be snug on the camshaft nose tapers but not so snug that they cannot be rotated.
5. Place a 0.010″ feeler gauge between the teeth of the adjustable idler gear and its mesh point with the injector cam concept gear. Now using hand pressure only, move the adjustable idler gear into mesh toward the center of the engine so there is no air gap on either side of the feeler gauge: Both cam gears may rotate slightly while performing this step. This sets the lash between the adjustable idler gear and the injector camshaft concept gear. Lash at the lower idler concept gear is set by the spread of the lower idler concept gear teeth, which are constrained by the crank gear.
6. While holding the adjustable idler gear, torque the idler shaft fasteners to specification, and remove the feeler gauge.
7. Next, torque the injector cam concept gear fasteners to specification, followed by the valve cam gear fasteners. Remove the camshaft wedges. Lubricate the engine geartrain with whatever lubricant is normally used in the engine.

CAMSHAFTS

The **camshaft** in most commercial diesel engines is gear driven by the crankshaft through one revolution per complete cycle of the engine. In a four-stroke cycle engine, to complete a full cycle the engine must be turned through two revolutions or 720 degrees, during which the camshaft would turn one revolution. Camshaft speed is therefore one-half engine speed. In a two-stroke cycle engine, the full cycle is completed in a single rotation or 360 degrees; camshaft speed and crankshaft (engine) speed are therefore geared so they are equal.

Cam Profile

The camshaft in a diesel engine actuates the **valvetrains** (the term **train** can describe any components that ride the cam profile and are actuated by it). In many diesel engines, the injection pumping apparatus such as unit pumps and mechanically or electronically actuated unit injectors are also cam actuated. The camshaft is supported at its journals by bushings or bearings that are in most cases pressure lubricated. **Cam geometry** refers to the physical shape of the cams: The term **cam profile** is also used. The profile outside the base circle will actuate the trains riding the cam and convert the rotary movement of the camshaft into reciprocating motion. Overhead camshafts are becoming more common in current truck diesel engines—Cummins ISX, Caterpillar C15, Navistar MaxxForce 13, and the current Volvo and Mack families of engines are examples. Valvetrain timing and unit injection pump or unit injector stroke are determined by cam geometry and the camshaft gear timing to the engine crankshaft. Figure 7–5 shows the sequence of camshaft-actuated events.

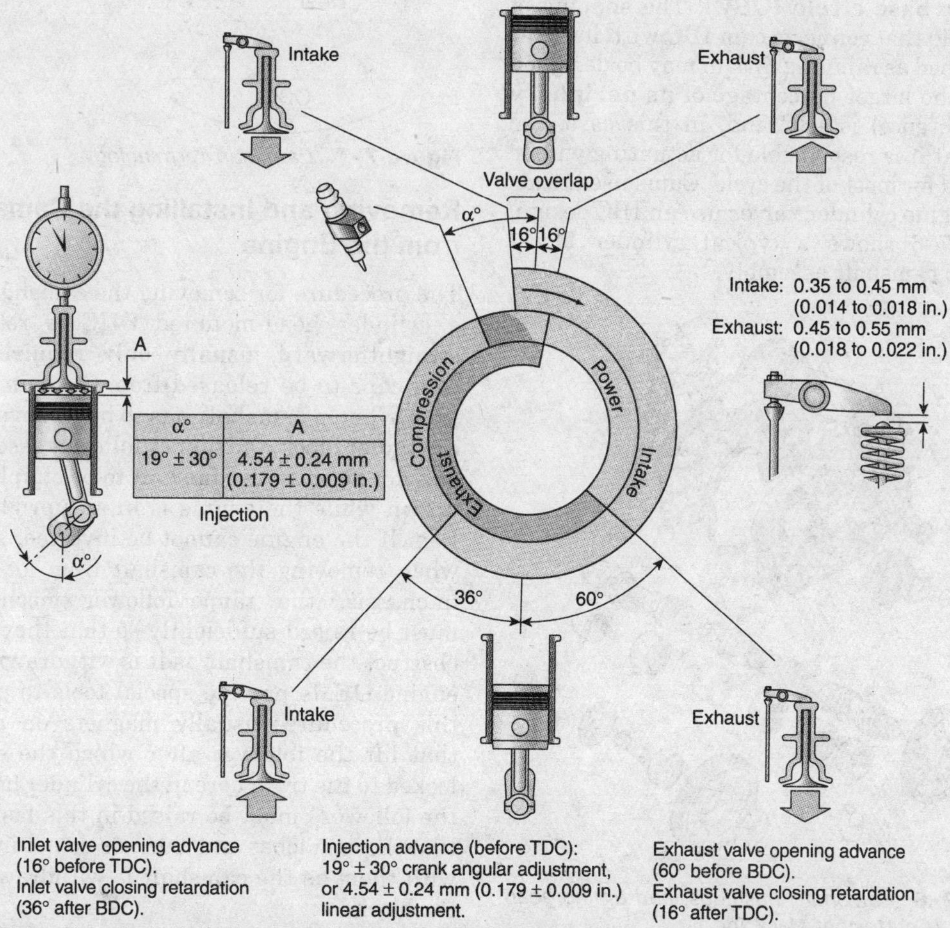

Figure 7–5 *Camshaft timing events. (Courtesy of Mack Inc.)*

Construction and Design

Middle alloy steels are used with hardened journals and cams. Hardening is usually by nitriding or other hard facing processes, followed by finish grinding. Diesel engine camshafts are not usually reconditioned; resurfacing of the journals is possible but most camshaft failures are cam lobe related. Camshafts are supported by bearing journals within a longitudinal bore in the cylinder block or on a pedestal arrangement in the cylinder head in the case of an overhead cam.

Base and Outer Base Circle

Cams are eccentrics machined to convert rotary motion to linear movement. The smallest radius of a cam concentric with the camshaft centerline is known as the **base circle (BC)** or **inner base circle (IBC)**. The largest radial dimension from the camshaft centerline is known as **outer base circle (OBC)**. The shaping of the profile that connects cam IBC with its OBC is described as ramping. A cam may be designed so that the larger percentage of its **periphery** (circumference) is IBC and, in this case, the train that it is responsible for actuating will be unloaded for most of the cycle. Cams used to actuate engine cylinder valves use an IBC design. Figure 7–6 shows a typical cylinder block-mounted camshaft assembly.

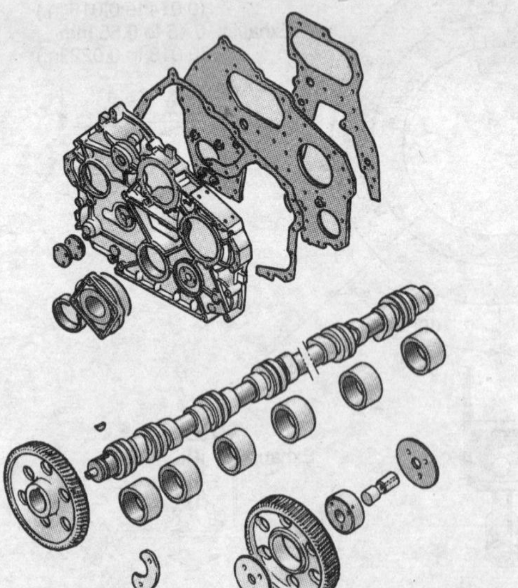

Figure 7–6 *Camshaft, bushings, and timing gear housing. (Courtesy of Mack Inc.)*

Alternatively, cams may be designed so that most of their periphery is OBC, in which case the train it actuates will be loaded for most of the cycle. Cummins TP injectors and some types of electronic unit pump (EUP) fueled engines use mostly OBC cam profiles. Figure 7–7 explains some cam terminology as it applies to a cam whose profile is mostly IBC.

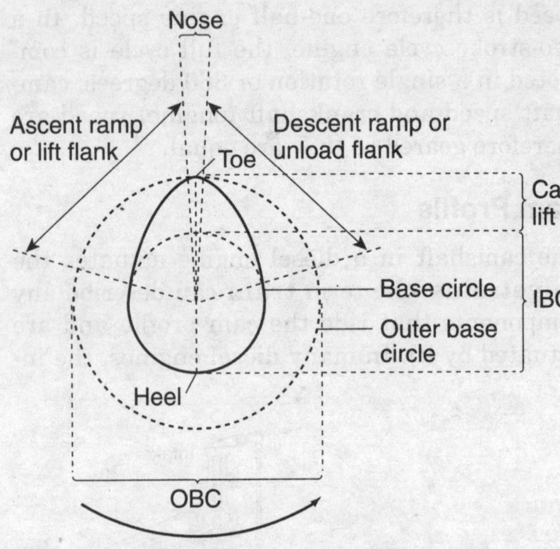

Figure 7–7 *Camshaft terminology.*

Removing and Installing the Camshaft from the Engine

The procedure for removing the camshaft from a cylinder head-mounted OHC is relatively straightforward, usually only requiring the cam caps to be released from the cam pedestals. When a camshaft has to be removed from a cylinder block, the tappet/follower assemblies will not permit the camshaft to be simply withdrawn while the engine is in an upright position. If the engine cannot be inverted, such as when removing the camshaft from an engine in-chassis, the tappet/follower mechanisms must be raised sufficiently so that they do not obstruct the camshaft as it is withdrawn. Most engine OEMs provide special tools to perform this procedure—usually magnets on a shaft that lift the follower after which the shaft is locked to the train bore in the cylinder head. All the followers must be raised in this fashion so that the cam lobes and journals do not interfere with them as the camshaft is withdrawn from the block.

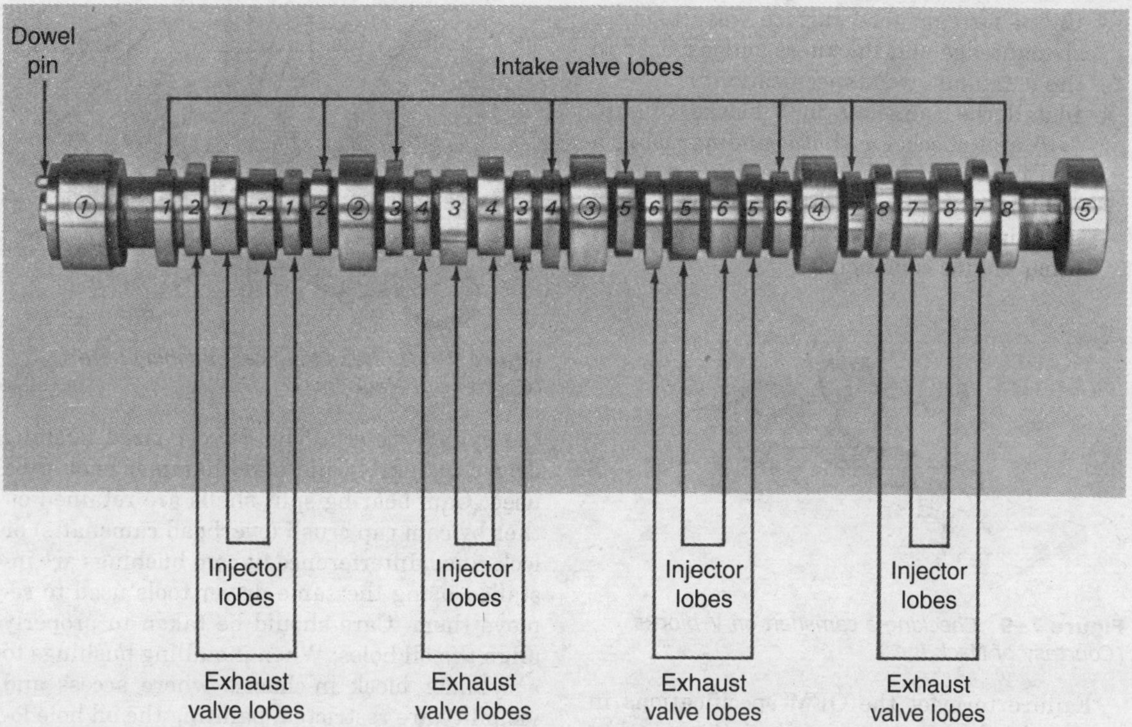

Figure 7–8 *Camshaft configuration used to actuate both valve and injector trains. (Courtesy of Detroit Diesel Corporation)*

When the correct tools are not available, the OEM special tools may by improvised by using coat hanger wire (mechanic's wire is not usually substantial enough) to first hook the follower and then lift it by bending the wire at the top. When withdrawing or installing the camshaft from a cylinder block bore, it is important that it never be forced because the result will almost certainly be damage to the camshaft or its support bearings. In certain engines, powertrain components, such as the crank web and crank throw, protrusions may interfere with the camshaft eccentrics, requiring that the engine be rotated at intervals while the camshaft is being removed and installed. Figure 7–8 shows a camshaft used to actuate both valve and injector trains.

Camshaft Inspection

1. Visual: pitting, scoring, peeling of lobes and scoring, wear, blueing of journals. Check the drive gear keyway for distortion and key retention ability. Any visible deterioration of the hard surfacing on the journal or cam profile indicates a need to replace the camshaft. The inspection can be by touch; a fingernail stroked over a suspected hard surfacing failure can identify hard surfacing failure in its early stages better than the eye.
2. Mike the cam profile heel-to-toe dimension and check to OEM specifications. Mike the base circle dimension and check to specifications. Subtract the IBC dimension from the heel-to-toe dimension to calculate the cam lift dimension. Check to specifications. Cam lift can also be measured using a dial indicator with the camshaft mounted in V-blocks: Zero the dial indicator on the cam IBC and rotate the camshaft to record the lift. Ensure that a dial indicator with sufficient total travel to measure the expected lift is used. When checking the cam lift dimension with the camshaft in-engine using a dial indicator, the measurement may not exactly coincide with the parameters required by the OEM, even when the cam profile is in sound condition, due to the lie of the camshaft in the support journals.

3. Check for cam lobe surface wear using a straightedge and thickness gauges sized to the maximum wear specification.
4. Install the camshaft in V-blocks (Figure 7–9) and check for shaft bending using a dial indicator. It is unusual for camshafts to bend in normal service. Bending a camshaft is usually caused by the failure of another engine component.

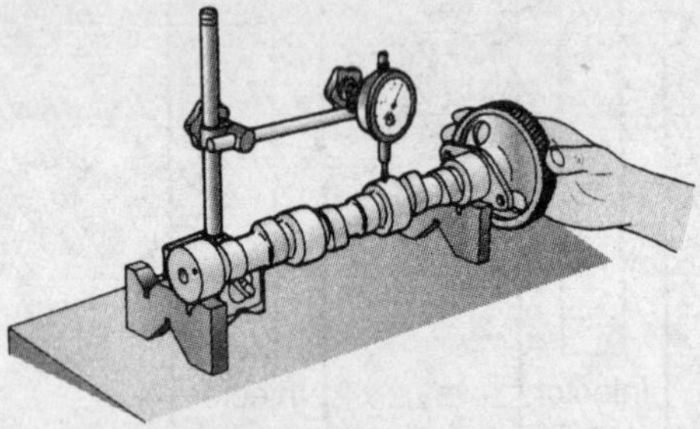

Figure 7–9 *Checking a camshaft on V-blocks. (Courtesy of Mack Inc.)*

Failure to meet the OEM specifications in any of the foregoing categories indicates that the camshaft should be replaced. The technician should be aware that the smallest indication of a hard surfacing failure on the cam profile requires little time to advance to a total failure. Probably the most important inspection is the visual inspection that initiates the listed sequence.

Camshaft Bushings/Bearings

The camshaft is supported by pressure-lubricated, friction bearings at the main journals. The camshaft is subjected to loading whenever a cam is actuating the train that rides its profile. This loading can be considerable, especially when engine compression brake hydraulics and fuel injection pumping apparatus are actuated by the camshaft. Cam bushings are normally routinely replaced during engine overhaul, but if they are to be reused they must be measured with a dial bore gauge (Figure 7–10) or telescoping gauge and micrometer to ensure that they are within the OEM reuse specifications.

Interference fit, cylinder block located, camshaft bushings are removed in sequence, starting usually at the front working to the back of

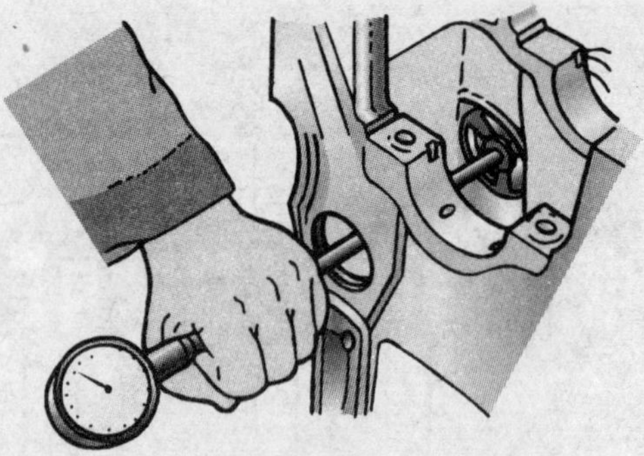

Figure 7–10 *Measuring cam bushing bore. (Courtesy of Mack Inc.)*

the cylinder block. The correct sized bushing driver (mandrel) and slide hammer should be used. Cam bearing split shells are retained either by cam cap crush (overhead camshafts) or lock rings. Interference fit cam bushings are installed using the same driver tools used to remove them. Care should be taken to properly align the oil holes: When installing bushings to a cylinder block in-chassis where access and visibility are restricted, painting the oil hole location on the bushing rim with a shop paintstick or correction white-out may help align the bushing before driving it. Ensure that the correct bushing is driven into each bore.

Bushings and their support bores vary dimensionally, and bushings rarely survive being driven into a bore and then removed. Bearing clearance is the camshaft bushing dimension measured with a dial bore gauge or telescoping gauge and micrometer minus the camshaft journal dimension measured with a micrometer. Prolonged use of certain types of engine compression brake can be especially hard on camshaft bearings.

Camshaft Endplay

Camshaft endplay is defined either by free or captured thrust washers/plates. Thrust loads are not normally excessive unless the camshaft is driven by a helical toothed gear in which instance there is more likely to be wear at the thrust faces. Endplay is best measured with a dial indicator: The camshaft should be gently levered longitudinally rearward then forward and the travel measured and checked to specifications.

Camshaft Position Sensors

Camshaft position sensors (CPS) are used in all current engines. They function to signal shaft speed and position data to the engine control module (ECM). CPS may use either an inductive pulse generator or Hall-effect signaling principle. The operating principles of camshaft position sensors are studied in some detail in Chapter 32. Figure 7–11 shows a typical inductive pulse generator CPS: It signals a frequency value and the irregular tooth at the #1 cylinder indicates TDC.

Camshaft Sensor Location

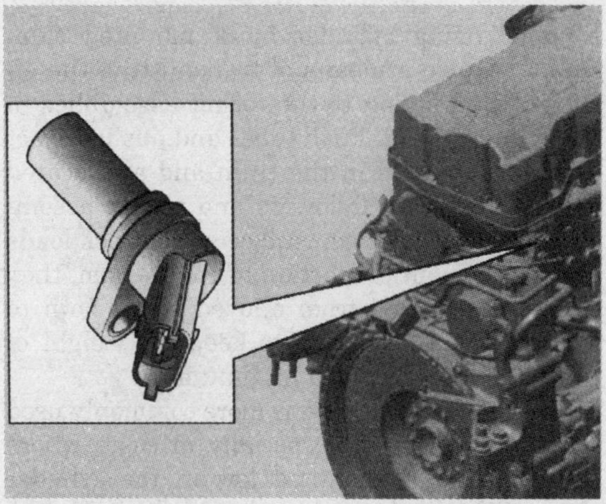

Figure 7–11 *CPS location on a Volvo D16D engine. (Courtesy of Volvo Trucks NA)*

VALVE AND INJECTOR TRAINS

While a camshaft rotates, the trains it actuates move linearly. Valve and injector trains transmit the effects of cam geometry to the cylinder valves and injector assemblies. Followers ride the cam profiles to actuate the trains. The linear movement of the train is converted to rotary motion once again at the rocker. It is its rocking action that opens and closes valves and provides fuel injection pumping force in many engines. Rockers may be used in both cylinder block mounted camshaft engines (shown in Figure 7–12) and those with overhead camshafts.

Followers

"Tappet," "lifter," and "follower" may all mean the same thing depending on which OEM is

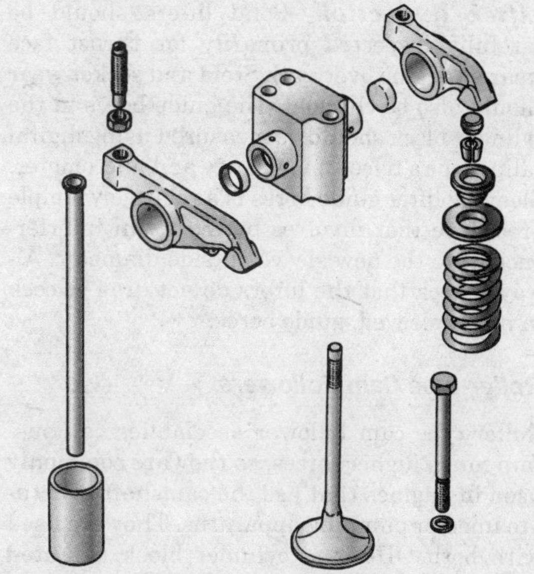

Figure 7–12 *Typical valvetrain assembly. (Courtesy of Mack Inc.)*

using the term. They describe components that usually are positioned to directly ride, or at least be actuated by, a cam profile. The term **tappet** has a broader definition and is sometimes used to describe what is more often referred to as a **rocker** lever. In this text, the term cam **follower** or lifter is used to describe a component that rides the cam profile. The function of cam followers is to reduce friction and evenly distribute the force imparted from the cam's profile to the train it is responsible for actuating. Diesel engines using cylinder block mounted camshafts use two categories of follower, while those using overhead camshafts use either direct-actuated rockers or roller-type cam followers.

Solid Lifters

Lifters are manufactured from cast iron and middle alloy steels and are usually located in guide bores in the cylinder block so they ride the cam profile over which they are positioned. Pushrods or push tubes are fitted to lifter sockets. The critical surface of a solid lifter is the face that directly contacts the cam profile. This face must be durable and may either be chemically hardened, cladded with a toughened alloy, or have a disc of special alloy steel molecularly bonded to the face.

Lifter Inspection Solid lifters should be carefully inspected primarily for thrust face wear at engine overhaul: Stem and socket wear should also be checked. The guide bores in the cylinder block should be measured using digital calipers or a telescoping gauge and micrometer. Sleeving lifter guide bores is a relatively simple procedure that involves boring to an interference fit to the new sleeve outside diameter. Always check that the lifters do not drag or cock in newly sleeved, guide bores.

Roller-type Cam Followers

Roller-type cam follower assemblies can sustain much higher forces, so they are commonly used in engines that use the camshaft to actuate injector pumping apparatus. They are used with both OHC and cylinder block mounted camshafts. Roller-type cam followers usually consist of a roller supported by a pin mounted to a clevis. The clevis can either be cylindrical and mounted in a cylinder block guide bore or be a pivot arm fitted to either the cylinder block or cylinder head.

Roller Cam Follower Inspection In the case of some OHC designs, a roller-type cam follower assembly and the rocker arm are integral.

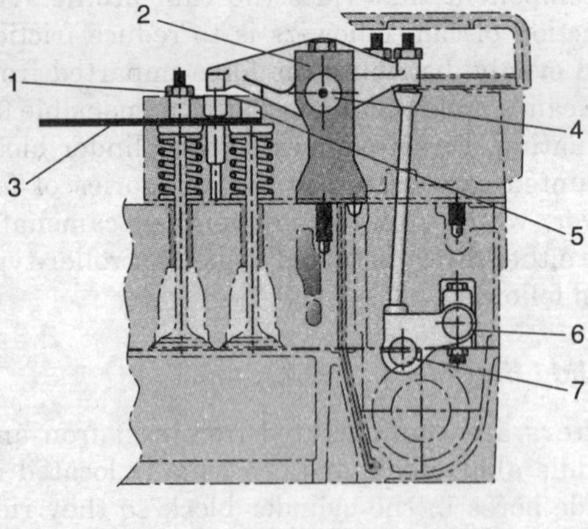

1. Adjusting screw and locknut
2. Rocker pedestal
3. Valve lash—valve stem to rocker pallet clearance
4. Rocker shaft
5. Valve bridge/yoke assembly
6. Cam follower pivot
7. Roller-type cam follower/lifter

Figure 7–13 *Roller-actuated valvetrain. (Reprinted Courtesy of Caterpillar Inc.)*

An example is the overhead camshaft design used to actuate the DDC Series 60 and 50 parallel port valve and EUI configuration. Roller-type cam followers better distribute whatever loading they are subjected to more evenly and outlast solid followers. The roller faces are usually chemically hard surfaced. Roller contact faces should be inspected for pitting, scoring, and indications of hard surfacing disintegration. The roller assembly should also be checked for axial and radial runout. Figure 7–13 shows a Caterpillar valvetrain actuated by a roller-type cam follower.

Pushrods and Tubes

Engines using cylinder block mounted camshafts require a means of transmitting the effects of cam action to the rocker assemblies in the cylinder head. Push tubes and pushrods act as intermediaries in the train and are located between the cam followers and rocker assemblies. Because they are subjected to shock loads that increase in proportion to engine rpm, they are manufactured from alloyed steels both to sustain these loads and to keep the weight of the actuation train to a minimum.

The hollow push tube is more commonly used than solid pushrods, especially in cases where the camshaft is mounted low in the cylinder block, and they are required to be fairly long. A typical push tube is a cylindrical, hollow steel shaft fitted with a solid ball or socket at either end. Balls and sockets form bearing surfaces at the follower and rocker. The bearing contact surfaces are usually lubed by engine oil. This is especially important at the rocker end as the rocker moves through an arc while the push tube moves linearly. Balls and sockets are usually chemically hard surfaced. Push tubes are preferred over solid pushrods because the tubular shape provides nearly as high section modulus (shape characteristic related to rigidity) as a solid cylindrical rod and nearly as much strength with less weight. Pushrods are cylindrical and solid. They are used mostly in applications that permit them to be short and relatively low in weight such as in old Detroit Diesel two-stroke cycle engines.

Inspecting Push Tubes

Both pushrods and push tubes seldom fail

under normal operation but when they fail there are two main causes:

- Inaccurate valve lash or injector adjustments
- Engine overspeeding

Ball and socket wear should first be checked visually. Reject them if there is evidence of hard surface flaking or disintegration. Next check:

1. Ball/socket to tube integrity: Test by dropping onto a concrete floor from a height of 2″. A separating ball will always ring flat. *Never* reuse a push tube with a separating ball or socket.
2. Straightness: Roll the push tube on a known true flat surface such as a (new) toolbox deck. The slightest bend (wobble as it is rolled) is reason to reject a push tube.

Rocker Arms

As indicated earlier, the terms *rocker* and *tappet* are synonymous. Rockers are levers. They transfer camshaft motion to the valves and mechanically actuated injectors. They are used in both in-block camshaft and overhead camshaft configurations. The rocker pivots on a rocker shaft. When a cam **ramps** off its base circle, it acts on the train that rides the cam profile and "rocks" the rocker arm, thereby actuating the components on the opposite side of the rocker arm, either valves or unit injectors. Lubricating oil is normally ducted through the rocker shaft to supply each rocker arm. Figure 7–14 shows an exploded view of the rocker shaft assembly on an engine equipped with a Jacobs-type engine brake (see Chapter 12).

Rocker Ratio

Rocker ratio may be used to amplify the cam lift dimension to increase valve or unit injector plunger travel. This requires that the distance from the centerline of the rocker shaft pivot bore to the pushrod side be less than its distance to the valve or unit injector side. Rocker ratio expresses the mechanical advantage obtained over the actual cam lift dimension. A rocker ratio of 1:2 would convert a cam lift dimension of 1″ into 2″ of valve opening travel. In the case of a rocker with equal distance on either side of the centerline (center of pivot), the rocker ratio would be 1:1, meaning equal cam and valve lift.

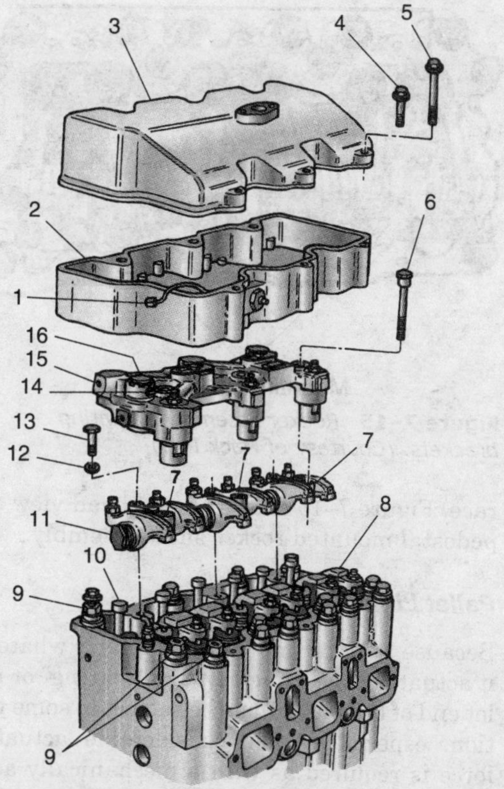

1. Wire connector
2. Riser
3. Valve cover
4. Short capscrew
5. Long capscrew
6. Actuator capscrew
7. Rocket arm shaft bracket
8. Valve yoke
9. Head bolt with threaded hole
10. Pushrod
11. Rocker arm assembly
12. Washer
13. Capscrew
14. Connector
15. Actuator assembly
16. Actuator solenoid

Figure 7–14 *Rocker shaft assembly on a Jacobs brake equipped engine. (Courtesy of Mack Inc.)*

Inspecting Rockers

The rocker arms should be inspected for wear at the push tube end cup or ball socket, the pivot bore (usually a bushing), and the **pallet** end. The pallet end of a rocker is the bearing surface that contacts and actuates the valve stem or valve bridge. If the hard surfacing at either end of a rocker shows signs of deteriorating, the rocker arm should be replaced. The pivot bore bearing/bushing can be replaced if it shows signs of wear as can the adjusting screw ball. The rocker shafts should be checked for straightness and wear at the rocker bearing

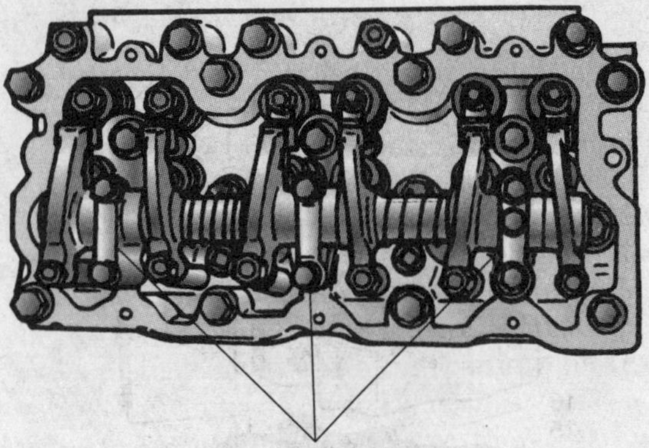

Figure 7–15 *Rocker assembly mounting brackets. (Courtesy of Mack Inc.)*

race. Figure 7–15 shows an overhead view of a pedestal mounted rocker shaft assembly.

Pallet Links

Because a rocker moves radially and whatever it actuates moves linearly, the "bearing" or pallet end of the rocker may be subject to some friction, especially when considerable actuation force is required as with a mechanically actuated fuel injector. It makes sense to use a link between the rocker pallet and the injector tappet. These links, which may clip onto the rocker pallet or be inserted into a bore in the injector tappet, should be inspected for wear at overhaul.

CYLINDER HEAD VALVES

Cylinder head valves provide the means of admitting air into, and discharging exhaust out of, the engine cylinders. Although any movement of cylinder valves in current engines is due to cam profile, the timing of engine valve movement depends on whether the engine is equipped with a compression brake or uses variable valve timing (VVT). With a VVT or engine brake equipped engine, the engine management electronics can manage valve lift and influence timing, sometimes within a wide window. The location of the valves, the lift, and valve timing can influence cylinder gas dynamics. Cylinder gas dynamics refers to the movement of gases in the cylinder during both combustion and exhaust strokes.

Valve Design and Materials

Cylinder head valves are mushroom-shaped, poppet-type valves. The stems are fitted to cylindrical guides in the cylinder head and loaded by a spring or springs to seal to a seat. A disc-shaped, spring retainer locks the spring(s) in position. **Split locks**, also known as **keepers**, are fitted to a peripheral groove at the top of the valve stem to hold the spring retainer in position. Valves can be vulnerable to high cylinder temperatures when they are in an open position; that is, not seated. This is because they rely on transferring heat through seat contact. Thermal failure occurs when valves overheat. Exhaust valves are more vulnerable to thermal failure than intake valves because they open when cylinder temperatures are at their highest. As a consequence the valve head is manufactured using different alloys from the stem: The two sections are then inertia welded. Inertia welding is a type of friction welding that requires spinning the stem at high velocity and pressure until a molecular bond is created. Reference Figure 7–16 while studying this section: It identifies all the critical valve components and terminology.

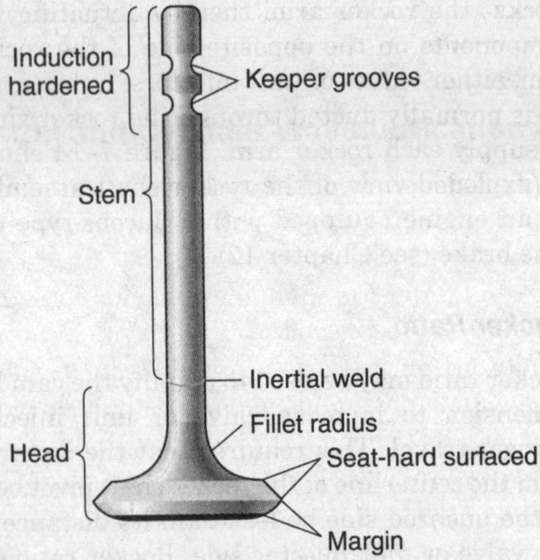

Figure 7–16 *Valve terminology.*

Intake Valves

Intake valves are manufactured from a variety of middle alloy steels. Intake valves do not have to sustain the high temperatures that exhaust valves are exposed to. However, they are actuated at high speeds and must have some degree of flexibility. The flexibility is required so that at high engine speeds when valve closing velocity is at its highest, the valves do not hammer out their seats. While most diesel engine designers use a valve seat angle of 45 degrees due to greater toughness, some choose to use 30-degree cut seats on just the intake valves. This takes advantage of the superior breathing efficiency of 30-degree seat angles over 45-degree seat angles.

Exhaust Valves

Running temperatures of exhaust valves are much higher than intake valves. When the exhaust valve is first opened during the latter portion of the power stroke, flame quench has just taken place and cylinder temperatures are just past their peak. Only about 20% of the heat the valve is subjected to can be dissipated through the stem, so the exhaust valve must sustain high temperatures for the duration of the exhaust process until it seats.

Exhaust valves are manufactured using special ferrous alloys. They may be cladded (coated) at the head and often use chromium, nickel, manganese, tungsten, cobalt, and molybdenum to improve flexibility, toughness, and heat resistance. Exhaust valves are cooled primarily by dissipating heat to the valve seats and, secondarily, by intake air at the valve overlap and through the stem-to-guide contact area. Inertia welding is commonly used to manufacture exhaust valves using separate alloys for the stem and head. In this way, the stem can be made out of an alloy with sufficient material hardness, while the head can be manufactured from an alloy with extreme heat resistance and flexibility. The separate stem and head are then molecularly joined by a friction welding process that requires spinning the stem at high velocity and pressure. Examine Figure 7–16 to see the location where the stem and head are joined on a typical inertia weld valve.

Valve Breathing Efficiency

Given identical lifts, valves machined with 45-degree seats have lower breathing efficiencies than those with 30-degree seats. Valve seats with 45-degree seats are widely used in highway diesel applications simply because they tend to have greater longevity. They can sustain greater dynamic loads, high seating forces, and high temperatures because they have more mass behind the seat. Valves machined with 30-degree seats tend to run hotter than those with 45-degree seats because they have less mass at their periphery. They also have less distortion resistance and lower unit seating force. To get around this problem, some engine OEMs use a 30-degree valve seat angle on their intake valves and a 45-degree seat angle on the exhaust valves.

Valve Operation

Current diesel engine valves rely on maximizing the seat contact area for cooling purposes and, as a consequence, **interference angles** are seldom machined. An interference angle requires that the valve be machined at 0.5 to 1.0 degree more acute angle than the seat. This results in the valve biting into its seat with high unit pressure. Interference angle also breaks up carbon formations on the valve seat but at the cost of reducing the valve-to-seat contact area. This can create a problem when transferring valve heat. A majority of current diesel engines use valve rotators to minimize carbon buildup on the seat and to promote even wear: An interference angle is not machined to valves using valve rotators.

Valve Rotators

Valve rotators use a ratchet principle or a ball and coaxial spring to fractionally rotate the valve each time it is actuated. Valve rotation should be checked after assembly by marking an edge of the stem and guide with a marker pen and then tapping the valve stem with a light nylon hammer a number of times. The valve should visibly rotate each time the stem is struck with the nylon hammer.

Valve Harmonics

Cylinder head valves are seated by a spring or pair of springs. When a pair of springs is used, they are often oppositely wound to help cancel the coincidence of vibration harmonies that may contribute to valve flutter or float. **Valve**

float can occur at higher engine speeds when valve spring force is insufficient (valve operating velocities are highest) and may cause asynchronous (out of time) valve closures. Springs are a critical component of the valvetrain assembly and their importance is increased as engine rpm increases and reduces the real time periods between opening and closing.

Inspecting Valve Springs and Retainers

1. Once the cylinder head has been removed from the cylinder block, the valves may be removed using a valve spring compressor as shown in Figure 7–17. Check the keepers and the valve keeper grooves for wear. Inspect the retainers.

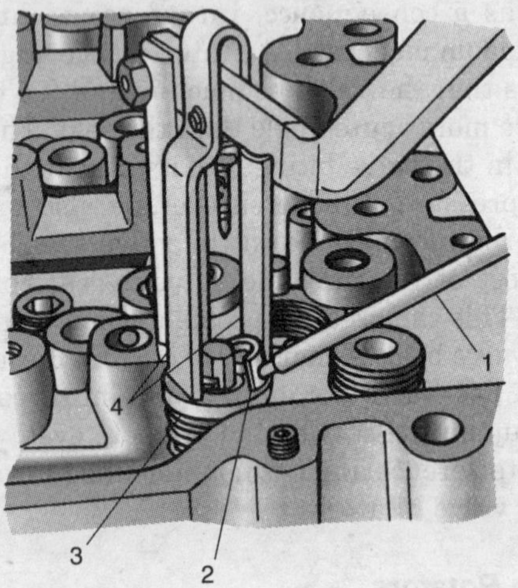

Figure 7–17 *Removing a valve from a cylinder head. (Courtesy of Mack Inc.)*

2. Measure the spring vertical height and compare to specifications. Minor differences do not matter.
3. Test valve spring tension using a tension gauge (Figure 7–18). The valve spring tension should be within 10% of the original tension specification. If not, replace the valve spring.
4. Check the spring(s) for abrasive wear at each end.
5. Check that the spring is not cocked using a straightedge.

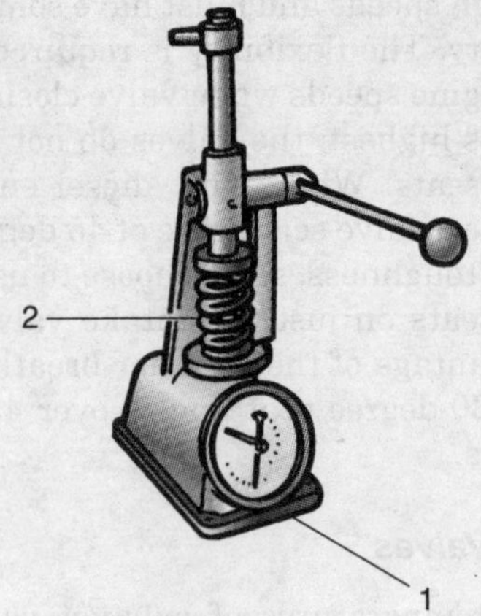

Figure 7–18 *Valve spring tension tester. (Courtesy of Mack Inc.)*

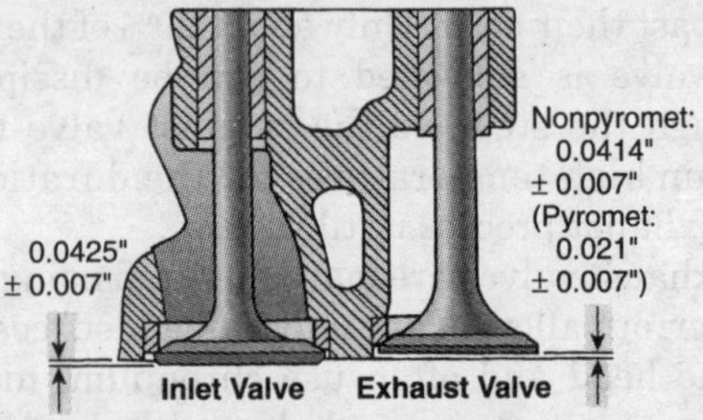

Figure 7–19 *Valve protrusion and recession dimensions. (Courtesy of Mack Inc.)*

6. Check valve spring operation with the valve installed in the cylinder head. Remember that as the cylinder head valve seat material is ground away, the spring operating height is lengthened, reducing spring tension. Valve protrusion and recession must be checked to specification as shown in Figure 7–19 and Figure 7–20.

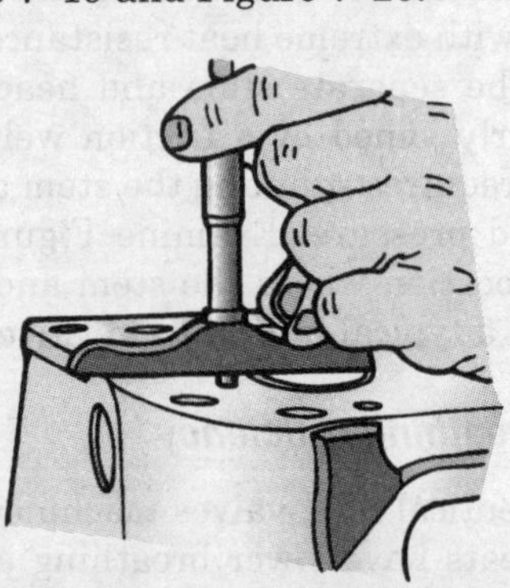

Figure 7–20 *Measuring valve protrusion with a depth micrometer. (Courtesy of Mack Inc.)*

Valve Servicing

The rocker assembly, including engine brake and VVT actuators, should be removed from the cylinder head(s). Use a valve spring compressor (shown previously in Figure 7–17) to remove valves from the cylinder head. The valves should be tagged by location. Valves should first be cleaned on a buffer wheel, taking great care not to remove any metal. A glass bead blaster may be used to remove carbon deposits from the seat face, fillet, and head, but avoid blasting the stem.

Valve inspection should begin with visually checking for dishing, burning, cracks, and pits; evidence of any of these conditions requires that the valve be replaced. Check the valve fillet for cracks and nicks—using a portable magnetic flux crack detector may help here—and again, reject the valve if cracks are evident in this critical area. Next, the valve should be measured. Using a micrometer, mark the valve stem at three points through the valve guide bushing sweep and check to specifications. Valve stem straightness should be checked with a vertical runout indicator.

Valve Margin

Measure the **valve margin** (see Figure 7–16). Valve margin is the dimension between the valve seat and the flat face of the valve head mushroom. This specification is critical when machining valves, and it must exceed the minimum specified value after the grinding process has been completed. A valve margin that is lower than the specification will result in valve failures caused by overheating. The split lock keeper grooves (see Figure 7–16) on the valve stem should also be checked for wear, nicks, and cracks.

Valve Dressing/Grinding

Before servicing a set of valves, the previously outlined checks and measurements should be completed. It is pointless machining valves that have failed in any of the previous categories. First, dress the grinding stone using a diamond dressing tool. Adjust the valve grinder chuck to the specified angle and the carriage stop to limit travel so the stone cannot contact the stem. Run the coolant (soluble machine oil and water solution), and make a single shallow pass. When machining valves, try to make the minimum number of passes to produce a valve seat face surface free of ridging and pitting. The final step should be to check that the valve margin is still within specification.

In cases in which the valve has been loosely adjusted, the stem end may be slightly mushroomed due to hammering from the rocker. Grind a new chamfer, taking care to remove as little material as possible. Too much chamfer will reduce the rocker-to-stem contact area and may damage the rocker pallet. When valves must be replaced, the new valves must be inspected, measured, and sometimes ground using the same procedure as with used valves.

Valve Seat Inserts

Most commercial diesel engines use valve seat inserts rather than integral valve seats machined into the cylinder head. The primary advantage of valve seat inserts is that they can be manufactured from tough, temperature resistant material and then easily replaced when the cylinder head is serviced. Cast-iron alloys toughened with nickel, chromium, and molybdenum may be used, but the use of alloyed steels, some with stellite cladding, is more common. Valve seat inserts are press fit to a machined recess in the cylinder head—sometimes they are staked to position (with a punch) after installation. Since most of the heat of the valve must be transferred from the valve to the seat, it is essential that the contact area of the seat and the cylinder head be maximized.

Valve Seat Removal and Installation

Valve guides must be removed using a removal tool consisting of a sectored collet designed to expand into the valve seat after which it can be either levered out or driven out with a slide hammer. When installing new valve seats, the seat counterbore must be first cleaned out using a low-abrasive emery cloth. The new insert should be inserted into the OEM-specified driver, which has a pilot shaft that fits tight to the valve guide bore. A hammer should be used to drive the insert into its bore until it bottoms. Next check the concentricity of the valve seat with the valve guide bore using a dial gauge; this is a critical specification (Figure 7–21). In most cases the seat will have to be ground, even though new seats are finish-ground. It is important that the valve grinders are serviced before fitting and machining the valve seats so that the new seats are machined to be concentric with the guides that will be put in service.

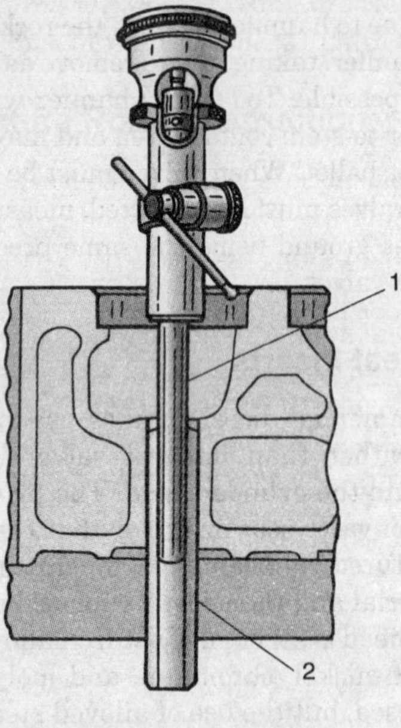

Figure 7–21 *Measuring valve seat concentricity with a concentricity gauge. (Courtesy of Mack Inc.)*

Valve Seat Grinding

Valve seats are ground using a specified grit abrasive stone, dressed to the appropriate angle, with a pilot shaft that fits to the valve guide. The valve head protrusion or recess dimension specifications must be respected when grinding valve seats, and in cases in which the cylinder head deck surface has been machined, undersize valve seat inserts are available from most OEMs. Whenever a cylinder head deck has been machined, the valve stem height, which is dictated by the specific valve seat insert location, must be within specifications to ensure the correct valve spring dynamics.

Valve Lash Adjustment

When valves are properly adjusted, there should be clearance between the pallet end of the rocker arm and the top of the valve stem. Valve lash is required because as the moving parts heat up, they expand, and if clearance were not factored somewhere in the valvetrain, the valves would remain constantly open by the time an engine reached operating temperature. Actual valve lash values depend on factors such as the length of the push tubes and the materials used in valve manufacture. Exhaust valves are subject to more heat and as a consequence expand more than intake valves when heated to operating temperature. Valve lash specifications for exhaust valves are usually greater than the intake valve lash setting specification.

Maladjusted Valves

Loose valve adjustment will retard valve opening and advance valve closing, decreasing the cylinder breathing time. Actuating cam geometry is designed to provide some "forgiveness" to the train at valve opening and valve closure to reduce the shock loading. When valves are set loose, the valvetrain is loaded at a point on the cam ramp beyond the intended point. The same occurs at valve closure when the valve is seated. High valve opening and closing velocities subject the valve and its seat to hammering that can result in cracking, failure at the head-to-stem fillet, and scuffing to the cam and its follower.

Valve Adjustment Procedure

The following steps outline the valve adjustment procedure on a typical four-stroke cycle, inline six-cylinder diesel engine. Valves should always be adjusted using the OEM's specifications and procedure.

In order to set valve lash you should become familiar with the cylinder firing order and companion cylinders on an inline six-cylinder engine.

Six-cylinder engine firing order: 1–5–3–6–2–4
Companion cylinders or cylinder throw pairings:
1–6 at TDC
5–2 at 120 degrees BTDC
3–4 at 120 degrees ATDC

Companion Cylinders

Cylinder throw pairings are called **companion cylinders**. In other words, when #1 piston is at TDC completing its compression stroke, #6 piston (its companion) is also at TDC having just completed its exhaust stroke. If the engine is viewed from overhead with the rocker covers removed, engine position can be identified by observing the valves over a pair of companion cylinders. For instance, when the engine timing indicator (timing marker) indicates that the pistons in cylinders #1 and #6 are approaching TDC and the valves over #6 are both closed (lash is evident), then the point at which the valves over #1 cylinder rock (exhaust closing, intake opening) at valve overlap will indicate that #1 is at TDC having completed its exhaust stroke and #6 at TDC having completed its compression stroke. This method of orienting engine location is commonly used for valve adjustment.

Adjustment

1. Locate the valve lash dimensions. These are often specified on the engine id plate on the rocker housing cover or cylinder block. The lash specification for the exhaust valve(s) is usually (but not always) greater than that for the inlet valve.
2. The valves on current diesel engines should usually be set under static conditions and with the engine coolant 100°F (37°C) or less. Locate the engine timing indicator and the cylinder calibration indexes, 120 degrees apart; depending on the engine, this may be located on a vibration damper, on any pulley driven at engine speed, or on the flywheel.
3. Ensure that the engine is prevented from starting by mechanically or electronically no-fueling the engine. The engine will have to be barred in its normal direction of rotation through two revolutions during the valve-setting procedure requiring the engine to be no-fueled to avoid an unwanted startup.
4. If the engine is equipped with valve bridges or yokes (Figure 7–22) that require adjustment, this procedure should be performed before the valve adjustment. To adjust a valve yoke, back off the rocker arm then loosen the yoke adjusting screw locknut and back off the yoke adjusting screw. Using finger pressure on the rocker arm (or yoke), load the pallet end (opposite to the adjusting screw) of the yoke to contact the valve; next, screw the yoke adjusting screw clockwise until it bottoms on the other valve stem. Turn an additional one flat of a nut (60 degrees) as shown in Figure 7–23, then lock to position by torquing the jam nut.

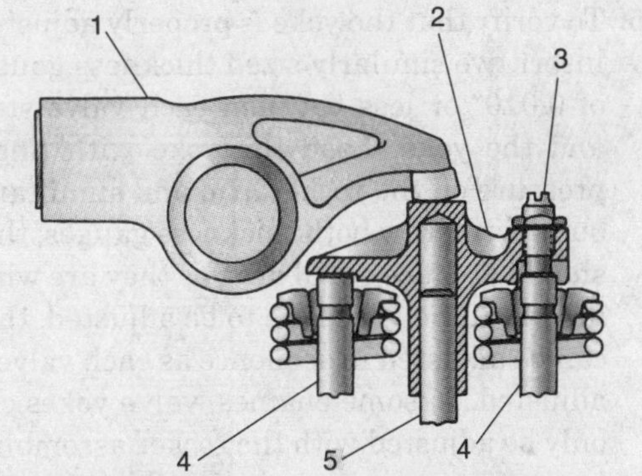

1. Rocker arm
2. Valve yoke
3. Adjusting screw
4. Valve stem
5. Yoke guide pin

Figure 7–22 *Valve bridge/yoke assembly. (Courtesy of Mack Inc.)*

Figure 7–23 *Adjusting a valve bridge/yoke assembly: The procedure on this engine is to bottom the adjusting screw, turn one flat, then torque the jam nut. (Courtesy of Mack Inc.)*

5. To verify that the yoke is properly adjusted, insert two similarly sized thickness gauges of 0.010″ or less between each valve stem and the yoke. Load the yoke with finger pressure on the rocker arm and simultaneously withdraw both thickness gauges; they should produce equal drag as they are withdrawn. If the yokes are to be adjusted, they can be adjusted in sequence as each valve is adjusted. In some engines, valve yokes can only be adjusted with the rocker assemblies removed because it is impossible to access the yoke to verify the adjustment otherwise.
6. If the instructions in the OEM literature indicate that valves must be adjusted in a specific engine location, ensure that this is observed; the cams that actuate the valves may only have a small percentage of IBC. Setting valves requires the lash dimension between the rocker arm and the valve stem on the rocker arm and the valve yoke to be defined. When performing this procedure, the valve adjusting screw jam nut should be backed off and the adjusting screw backed out. Insert the specified size of thickness gauge between the rocker and the valve stem/yoke. Release the thickness gauge. Then turn the adjusting screw clockwise until it bottoms; turn an additional ½ flat of a hex nut (30 degrees). Hold the adjusting screw with a screwdriver and with a wrench torque the jam nut. Now for the first time since inserting the thickness gauge, handle it once again; withdraw the thickness gauge. A light drag indicates that the valve is properly set. If the valve lash setting is either too loose or too tight, repeat the setting procedure. Do *not* set valves too tight. Set all the valves in cylinder firing order sequence, rotating the engine 120 degrees between settings. It is preferable to begin at #1 cylinder and proceed through the engine in firing order sequence. Figure 7–24 shows this procedure: Note how the feeler gauge is not being held during the adjustment.

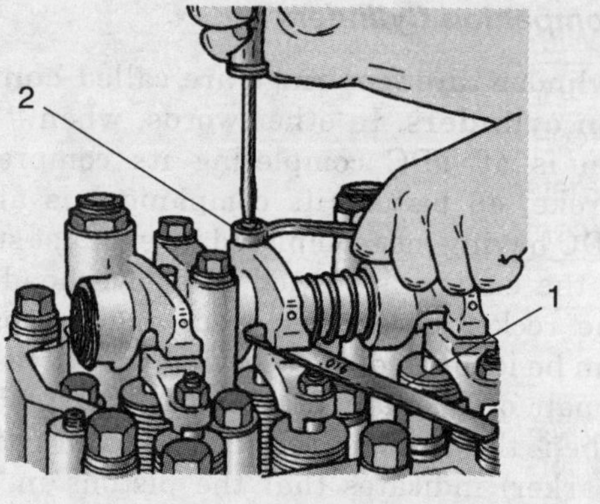

1. Thickness gauge 2. Adjusting screw and nut

Figure 7–24 *Adjusting intake valve lash on a Mack E-Tech engine. (Courtesy of Mack Inc.)*

Shortcuts

When observing experienced technicians performing diesel top end tune-up the student will note that often they shortcut the service literature procedures. When technicians know a specific engine like the back of their hand, they will get a sense of which shortcuts involve zero risk. Until this happens, avoid falling into the trap of thinking that a shortcut on one engine will work on another. The best kind of shortcut is one that the OEM approves of by outlining it in their service literature. For instance, the top end on a Navistar International DT466E can be set using just two engine positions, #1 piston at TDC on compression, followed by #6 piston at TDC on compression as shown in Figure 7–25. Most other engines have similar "shortcuts" and it is fine to use them when approved by the engine OEM.

Barring Engines

When setting valves, you will have to rotate the engine. With older engines using symmetrical cam profiles, you may be able to bunt the engine over using a remote starter switch providing the engine is no-fueled. With today's engines, the technician should use the OEM-approved method in their service literature. Manually barring an engine into overhead setting positions can be made easier if a tool such as that shown in Figure 7–26 is used: This engages with the flywheel ring gear teeth, allowing the engine to be easily rotated with a ½ drive ratchet.

With	Adjust Valves											
Piston 1 at TDC (compression)	INT 1	EXH 2	INT 3			EXH 6	INT 7			EXH 10		
Piston 6 at TDC (compression)				EXH 4	INT 5			EXH 8	INT 9		INT 11	EXH 12

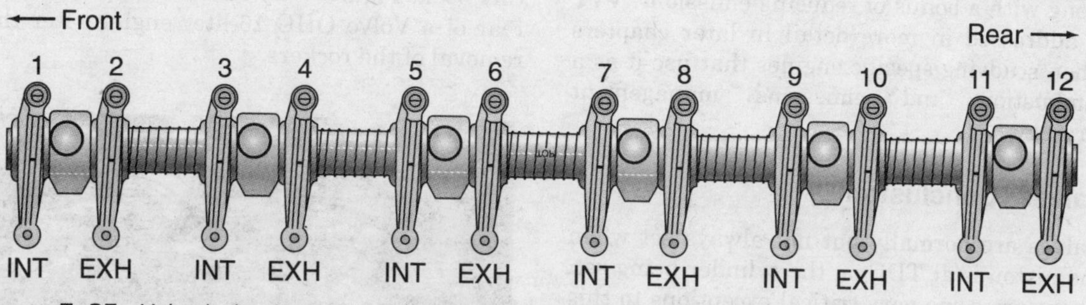

Figure 7–25 *Valve lash adjustment on a Navistar International DT466E. Note that all the valves can be adjusted in just two engine positions. (Courtesy of Navistar International Corp.)*

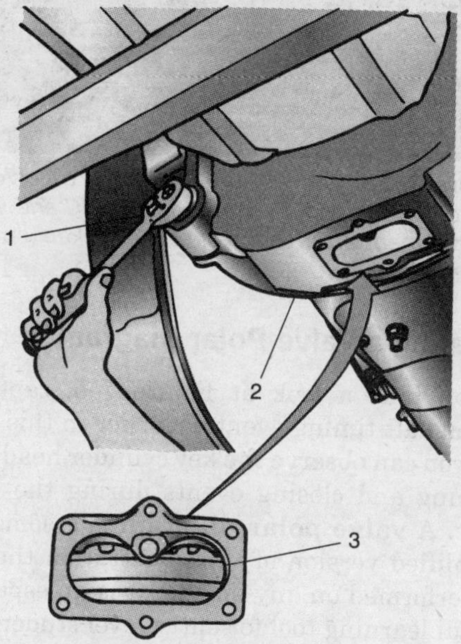

1. Ratchet and engine barring tool
2. Flywheel housing
3. Flywheel and engine position scale

Figure 7–26 *Barring an engine using a toothed barring tool while observing the engine position scale. (Courtesy of Mack Inc.)*

Variable Valve Timing

Variable valve timing (VVT) refers simply to any means used to vary either the opening or closing of cylinder valves. Highway diesel engines have used engine compression brakes for five decades: Engine brake controls were initially hydromechanically controlled and actuated, but since the early 1990s these have been electronically controlled and hydraulically (engine lube) actuated. More recently, VVT of intake valves has been introduced. Figure 7–27 shows the **variable valve actuator (VVA)** assembly used on a Caterpillar C13 engine. Having the ability to manage the moment of intake

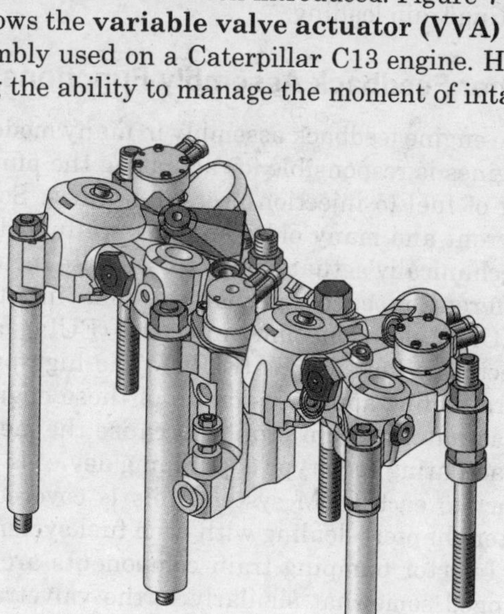

Figure 7–27 *Caterpillar ACERT C13 variable valve actuator (VVA) assembly: The VVA fits over the rocker assembly. (Reprinted Courtesy of Caterpillar Inc.)*

valve closure permits retarding the beginning of the compression stroke, with the net result that the ECM can determine exactly how much intake gas is admitted to the engine cylinder. This enables a large engine to behave like a smaller engine when power output requirement is lower along with a bonus of reducing emissions. VVT is addressed in more detail in later chapters when studying specific engines that use it as a combustion and emissions management strategy.

Valves: Conclusion

Valves are normally but not always set when the piston is at TDC on the cylinder being set. There are some very critical exceptions to this rule that are addressed in later chapters. When the piston is at TDC on the compression stroke, it can usually be assumed that the valves are fully closed. The procedure for setting valves in an engine is sometimes referred to as **overhead adjustment** or tune-up. In most cases, the valve setting procedure is accompanied by procedures such as injector timing and/or lash setting/train loading.

Other Feedback Assembly Functions

The engine feedback assembly in many modern engines is responsible for actuating the pumping of fuel to injection pressure values. Some current and many older engines are fueled by mechanically actuated injectors. These include different types of mechanical unit injectors (MUIs), electronic unit injectors (EUIs), and electronic unit pumps (EUPs). The high-pressure fueling apparatus on all of these engines is actuated by cam profile. Because the means of actuating each type of pumping device is distinct to each OEM system, this is covered in later chapters dealing with each fuel system.

Injector pumping train components are arranged somewhat similarly to the valvetrains and depend largely on the location of the camshaft and the force required to actuate the pump elements, which can be considerable. For instance, in engines with a block-mounted camshaft, the push tubes used in the injector train must be built to sustain much higher compressional loads than those used in the valvetrains. Injector actuation trains usually have to be precisely set because the location of the injector pumping plunger is moved when the train is adjusted. Failure to observe the injector height specification, which is often set with a tram gauge, can affect fuel delivery timing. These critical adjustments are examined later on in this book. Figure 7–28 shows a view from the rear of a Volvo OHC 16-liter engine after the removal of the rockers.

Figure 7–28 *View of the top end of a Volvo 16-liter diesel engine showing the OHC and valve arrangement after removal of the rocker assemblies.*

Creating a Valve Polar Diagram

If you take a look at Figure 7–5, captioned "Camshaft timing events," earlier in this chapter, you can observe the key cylinder head valve opening and closing events during the diesel cycle. A **valve polar diagram** is a somewhat simplified version of this illustration that can be performed on any engine. It is an especially useful learning tool for entry-level students because it requires that the engine be closely observed through an entire effective cycle. The objective is to map the valve opening and closing events by observing the tappets as an engine is barred over. Students can perform this exercise by first mapping the valves only. Then they can advance to include injector actuation train activity on engines equipped with mechanically actuated injection systems such as those using PT, MUI, EUI, EUP, and HPI-TP fuel systems. When performing this exercise, you should try to be as precise as possible. You should end up with a diagram similar to that shown in Figure 7–5 but with specifications for the engine on which you are working.

REVIEW QUESTIONS

1. The location on a cam profile that is exactly opposite the toe is referred to as the:
   a. Nose
   b. Ramp
   c. Heel
   d. Sole
2. On a cam profile described as mostly inner base circle, the profiles between IBC and OBC are known as:
   a. Cam geometry
   b. Ramps
   c. Ridges
   d. Heels
3. Which of the following dimensions would be consistent with a typical engine timing gear backlash setting?
   a. 0.002″
   b. 0.008″
   c. 0.014″
   d. 0.024″
4. In which direction must a camshaft be rotated on a diesel engine with a crankshaft that is rotated clockwise?
   a. Either clockwise or counterclockwise depending on the engine
   b. Clockwise
   c. Counterclockwise
5. Which tool would be required to measure the cam lift of a cylinder block-mounted camshaft in position?
   a. A dial indicator
   b. An outside micrometer
   c. A depth micrometer
   d. Thickness gauges
6. A camshaft gear is usually precisely positioned on the camshaft using a(n):
   a. Interference fit
   b. Key and keyway
   c. Captured thrust washer
   d. Dial indicator
7. When grinding a valve face to remove pitting, the critical specification to monitor during machining would be the:
   a. Shank diameter
   b. Stem
   c. Poppet diameter
   d. Margin
8. Which of the following operating conditions would be more likely to cause a valve float condition?
   a. Engine lug down
   b. Operating in the torque rise profile
   c. Engine overspeed
   d. Operating in the droop curve
9. When grinding a new set of cylinder valve seats, which of the following should be performed before the machining?
   a. Install the valves.
   b. Install the rockers.
   c. Adjust the valve yokes.
   d. Install the valve guides.
10. The reason for using a pair of oppositely wound valve springs is to:
    a. Minimize valve dynamic flutter
    b. Double the closing force of two similarly wound valves
    c. Increase the valve closing velocity
    d. Diminish valve noise
11. *Technician A* states that the backlash in timing gearsets can be measured with a dial indicator. *Technician B* states that if the gear backlash is insufficient, the result can be gear whine. Who is correct?
    a. Technician A only
    b. Technician B only
    c. Both A and B
    d. Neither A nor B
12. *Technician A* states that helical cut gears are often used in engine timing geartrains because they operate more quietly. *Technician B* states that shaft thrust loads are much higher when helical cut gears are used. Who is correct?
    a. Technician A only
    b. Technician B only
    c. Both A and B
    d. Neither A nor B
13. *Technician A* states that when installing an interference-fit gear to a shaft, the gear should not be heated above 400°F. *Technician B* states that if an oxyacetylene rosebud is not available, an interference-fit gear should be pressed to the shaft using a hydraulic press. Who is correct?
    a. Technician A only
    b. Technician B only

c. Both A and B
d. Neither A nor B

14. *Technician A* states that it is good practice to lubricate the inside bores of camshaft bushings to facilitate camshaft installation. *Technician B* states that if the cam bushing oil holes are not properly aligned an immediate camshaft failure could be the result. Who is correct?
    a. Technician A only
    b. Technician B only
    c. Both A and B
    d. Neither A nor B
15. Which of the following methods is used to transfer drive from the crankshaft to the camshaft on most commercial diesel engines?
    a. Belt and pulley
    b. Timing chain and sprocket
    c. Gears
    d. Fluid coupling
16. *Technician A* states that variable valve timing (VVT) is used on some diesel engines to delay the closing of intake valves on the compression stroke. *Technician B* states that VVT is used to open the intake valves early during the power stroke. Who is correct?
    a. Technician A only
    b. Technician B only
    c. Both A and B
    d. Neither A nor B
17. What type of cam geometry is required if the objective is to load the train that rides it for most of the cycle?
    a. Mostly OBC profile
    b. Mostly IBC profile
    c. Symmetrical OBC and IBC profiles
18. *Technician A* states that the camshaft on a two-stroke cycle engine turns through one complete rotation per complete engine revolution. *Technician B* states that the camshaft on a four-stroke cycle diesel engine rotates through one revolution per two engine revolutions. Who is correct?
    a. Technician A only
    b. Technician B only
    c. Both A and B
    d. Neither A nor B
19. *Technician A* states that to check cam lift, the camshaft must be removed from the engine to obtain a valid reading. *Technician B* states that to check for a bent camshaft, the camshaft should be removed from the engine and runout tested in V-blocks. Who is correct?
    a. Technician A only
    b. Technician B only
    c. Both A and B
    d. Neither A nor B
20. *Technician A* states that valves machined with 45-degree seats breathe with higher efficiencies than those machined with 30-degree seats. *Technician B* states that valves machined with 30-degree seats run cooler than those machined with 45-degree seats. Who is correct?
    a. Technician A only
    b. Technician B only
    c. Both A and B
    d. Neither A nor B

8

Engine Housing Components

INTRODUCTION

The engine housing components are those components that enclose the internal engine components. The cylinder block acts as the central frame of the engine: All the other engine components are in some way attached to it, including the remaining engine housing components. The components examined in this chapter are as follows:

1. Engine cylinder block and cylinder liners
2. Cylinder head assemblies
3. Intake and exhaust manifolds
4. Oil pans

ENGINE CYLINDER BLOCK

The engine **cylinder block** is the frame of the engine around which all the other components are assembled in much the same way that subcomponents are assembled around a truck frame. The cylinder block houses the engine cylinders and the engine crankcase. North American highway medium- and large-bore diesel engines all use inline or V-cylinder configurations requiring a single crankshaft and therefore use cylinder blocks cast as a single unit. In all these engines, the cylinder block is manufactured from variations of cast-iron alloys. Although aluminum, composites, and plastics may be used as a material for oil pans, and Caterpillar used an aluminum spacer deck in the original 3176 engine (forerunner of the C10/C11 engine family), because of the forces a diesel engine is subject to, iron-based rather than aluminum-based alloys are used. With the objective of getting weight down and strength up, a majority of recent engine cylinder blocks are alloyed and cast using special processes. Figure 8–1 shows a stripped-down cylinder block used in the MaxxForce 13 engine.

Figure 8–1 *Navistar MaxxForce 13 compacted graphite iron (CGI) cylinder block. (Courtesy of Navistar International Corp.)*

Cylinder Block Forces

The extent to which the cylinder block is alloyed has increased in recent years in the pursuit of lighter, stronger engines that last longer. Typical forces that a cylinder block has to sustain are:

- Torque twist
- Cylinder pressures
- Thermal shock

Torque twist occurs when twist forces from the crankshaft anchor through the cylinder block. A cylinder block can be subject to torque twist effect from either crankshaft input or output. This means that the condition can occur at high cylinder pressures . . . or, oppositely, generated through the drivetrain. Excessively high cylinder pressures, especially in small parent bore engines, can generate failures. And thermal shock occurs when a hot cylinder block is subject to sudden cooling such as from cold water in a wash bay. Torque twist is more likely to be a factor in recent model lightweight cylinder blocks. The following methods are used to control cylinder block torque twist:

- Brace plate: bolted to cylinder block lower flange to add rigidity.
- **Buttress screws:** bolts threaded transversely through cylinder block into main bearing caps.
- Cast oil pan with high section modulus: bolts to the cylinder block crankcase flange and onto the flywheel housing.

Cylinder Block Design and Construction

In most current highway diesel engines, the cylinder block is bored and fitted with cylinder **liners** or **sleeves**, which can be replaced when the engine is overhauled. Those engines that do not use liners or sleeves incorporate integral cylinder bores. An engine cylinder block with integral bores is known as a **parent bore** engine. Until recently parent bore diesel engines were mostly confined to the light-duty sector. More recently heavier duty parent bore engines have appeared on the market in which the cylinder bore surface area has been induction hardened. The Mercedes Benz MB-900 family of engines falls into this category. Figure 8–2 shows an older Detroit Diesel, V-configured cylinder block and identifies some of the terms used in this chapter.

As well as supporting the cylinders, a cylinder block must also support the crankshaft and power takeoff mechanism, usually a flywheel. The crankshaft is underslung from the cylinder block and supported in a cradle of main bearings in what is usually referred to as the **crankcase**, which is located below the cylinder block. The flywheel is usually mounted within a housing attached to the cylinder block.

The engine configuration of choice in medium- and large-bore truck diesel engines has been the inline, six-cylinder engine. Most of the truck diesel engine OEMs have flirted with V configurations in the past, usually V8s with V angles ranging from 90 degrees to a relatively tight 50 degrees. None are in current production for on-highway use. The lower profile of the V-configured engine has made them more successful over the years in transit bus applications but few of these remain on the roads today.

Cylinder blocks may be bored to support the camshaft or camshafts and are cast with coolant passages and a water jacket. Even in a liquid-cooled engine, the cylinder block frame

Figure 8–2 *Cylinder block terminology. (Courtesy of Detroit Diesel Corporation)*

plays a major role in dissipating rejected heat to the atmosphere.

Cylinder Block Materials

Truck and bus cylinder blocks were traditionally manufactured from grey cast irons usually alloyed with at least some silicon to reduce brittleness of the block material. As noted before, most medium- and large-bore highway diesel engines use cylinder liners, meaning that the block base materials are not directly subject to engine cylinder pressures and temperatures. More recently, there has been a trend toward substantially increasing cylinder pressures while also building lighter engines. This has resulted in today's low-weight, high-strength cylinder block castings. Compare the weight-to-horsepower ratings of today's engines with those of the 1980s: Today's engines often produce twice the power for each pound of engine weight—at least until all of the emissions control hardware on the current engine have been weighed. Most diesel engine original equipment manufacturers (OEMs) today use **compacted graphite iron (CGI)** in cylinder block construction because of its ability to withstand much higher forces.

CGI Cylinder Blocks

CGI cylinder blocks (see Figure 8–1) have been a key to reducing engine weight over the past 5 years. Contemporary high horsepower diesel engines today often weigh less than engines of half their horsepower just 10 years ago. Among the benefits of CGI are:

- 70% lighter than an equivalent grey cast iron block
- 200% the fatigue limit of grey
- 40% more rigid than grey cast iron

Cylinder Block Functions

Although there are some differences among manufacturers, a diesel engine cylinder block must perform some or all of the following:

- Incorporate bores for the piston assemblies.
- Incorporate main bearing bores to mount the crankshaft.
- Incorporate coolant passages/water jacket.
- Incorporate lubricant passages/drillings.
- Incorporate mounting locations for other engine components.

A cylinder block may contain longitudinal bore(s) to mount the camshaft(s). Figure 8–2 shows the bores and passages identified above. Figure 8–3 shows the location of expansion or cup plugs in a typical six-cylinder engine block. These are usually fitted with a slight interference and retained by a medium-duty adhesive such as blue Loctite.

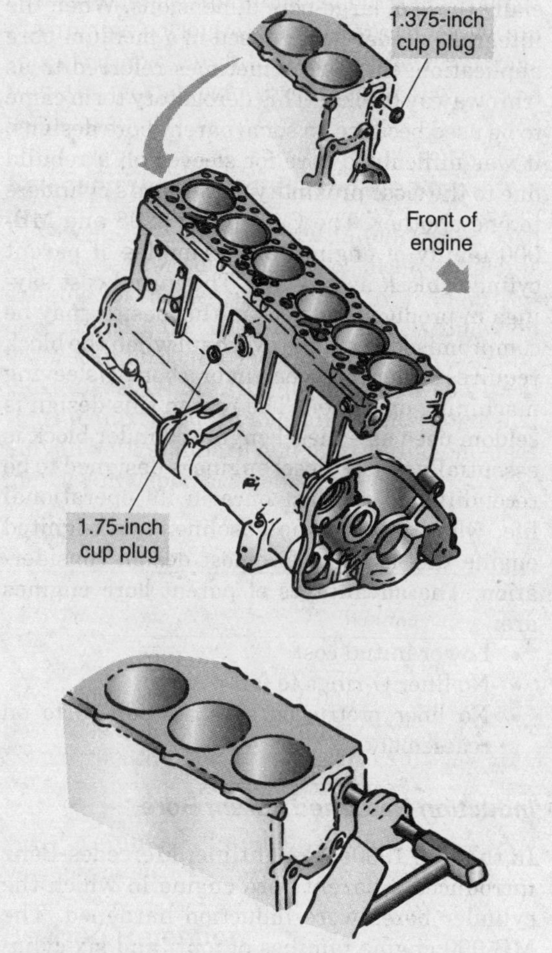

Figure 8–3 *Inline six-cylinder block showing the location of the cup plugs and installation method. (Courtesy of Mack Inc.)*

Categories of Cylinder Block

Cylinder blocks used in commercial diesel engines can be categorized by type:

- Integral cylinder bore or parent bore
- Wet sleeve/liner

- Dry sleeve/liner
- Combination wet/dry liners

Other factors such as two-stroke cycle and air cooled are also reflected in the block design.

Integral Cylinder Bore

Most automobile and small-bore engines use integral or parent bore cylinder blocks. They are less often seen in truck diesel engines, especially those of large-bore dimensions. When the integral cylinder bore is used in a medium-bore application, they are sometimes referred to as "throwaway blocks." This derogatory term came to be used because on some parent bore designs, it was difficult to bore for sleeves on a rebuild due to the close proximity of the bore's cylinders to one another. The Caterpillar 3208 and MB-900 family of engines are examples of parent cylinder block assemblies. The initial cost savings in producing a block of this design may be compromised at engine overhaul when the block requires either replacement or a boring/sleeving machining operation. The reason this design is seldom used as a diesel engine cylinder block is essentially that a diesel engine is designed to be reconditioned at least once in its operational life, whereas with the gasoline, spark-ignited engine this is not a foremost design consideration. The advantages of parent bore engines are:

- Lower initial cost
- No liner O-rings to fail
- No liner protrusion specs to adhere to on reassembly

Induction Hardened Parent Bore

In the late 1990s, Freightliner/Mercedes-Benz introduced a parent bore engine in which the cylinder bores were induction hardened. The MB-900 engine families of four- and six-cylinder engines use a unique, helical-striped, induction-hardening feature in the upper ring belt sweep. This is claimed to give the bores considerably extended service life compared with other parent bore engines. In the event the cylinder bores are worn to oversize, they can be bored and sleeved. Mercedes-Benz recommended that sleeving be performed only by them on these engines. This engine has become very successful in both on- and off-highway applications.

Wet Liner/Sleeve

The cylinder block is designed so that the water jacket is in direct contact with the liner in the block bore; the liner must therefore have a wall thickness sufficient to sustain the engine's peak combustion pressures. **Wet liners** transfer heat efficiently into the coolant and are easily replaced at overhaul. Their main disadvantage is that a seal must be maintained for the life of the liner and that an O-ring failure results in coolant contaminated engine lube.

Cavitation caused by vapor bubble implosion can shorten the life of wet liners, but this is seldom a problem if the coolant chemistry is properly monitored. Despite the fact that a wet liner is constructed from a sizeable mass of cast iron or steel, it is relatively easily flexed. Try installing a telescoping gauge into the bore of a wet liner on the bench and locking it so it grabs in position. Gentle hand pressure is sufficient to flex the liner enough for the telescoping gauge to unseat and drop to the bottom of the liner. Now think about how much flexing could occur when the liner is subject to 3,000 psi (207 bar) of combustion pressure.

How Cavitation Erosion Occurs

Wet liners, when subject to cylinder pressures, which may be as high as 3,000 psi (207 bar), expand outward into the wall of coolant that surrounds them then contract, creating a sort of lower pressure void or "bubble" whose matter is actually boiled-off coolant vapor. This "bubble" almost immediately collapses, causing the wall of coolant to collide on the liner exterior wall. The condition repeats itself at high frequency (17 times per second on each liner at 2,000 rpm) and has been tested to produce pressures of up to 60,000 psi (4137 bar). **Bubble collapse** results in cavitation unless the coolant provides protection in the form of a coating on the exterior of the liner wall. Cavitation can be identified by pitting/erosion that usually appears on the liner outside the thrust faces of the piston.

Liner O-rings

The O-rings used to seal wet liners are made from a variety of rubber-type compounds, and it is important that the OEM installation recommendations be observed. Rubber type means that O-rings may be 100% rubber or be derived entirely from petro-chemicals—or any combination of the two. OEM recommendations can include that O-rings be installed dry or coated in coolant, soap, engine oil, and various other substances. The reason is that some O-rings are designed to react to the coating substance and swell, creating a perfect seal. Wet liners are often alloyed so they possess characteristics metallurgically superior to the block casting, which increase service life. One OEM induction hardens the liner inside bore.

Dry Sleeves

A thinner-walled sleeve than the wet liner, the dry sleeve is installed into the block bore usually with a marginally loose fit and retained by the cylinder head. The dry sleeve does not transfer heat as efficiently as the wet liner, but it is easily replaced and does not present coolant sealing problems.

In older applications, **dry liners** were made of cast-iron material almost identical to that of the engine block; these were installed with a fractional interference to maximize heat transfer. Current dry liners tend to be alloyed for superior toughness, and because a liner's ability to transfer heat is dependent on maximizing its surface contact area in the cylinder block bore, they are manufactured with a slightly greater coefficient of heat expansion than cast iron. They are therefore designed to be installed loose so they expand into the block bore when heated, maximizing the contact surface. If this type of liner is installed with an interference fit, they buckle in use, greatly reducing their service life. Precise measuring of the block bore inside diameter (id), the liner outside diameter (od), and selective fitting eases installation and increases engine longevity. Loose fits of around 0.035 mm (0.0015″) are common.

Combination Wet/Dry Sleeves

The wet/dry sleeve is designed so that the hottest part of the liner at the top is in direct contact with the coolant in the water jacket and the lower portion fits directly to the cylinder block with a fractional loose or interference fit. Consequently, the upper portion in direct contact with the water jacket must have considerably more mass because it has to contain the cylinder combustion pressures. Wet/dry liners are required to seal the water jacket usually with O-rings. They are also known as midstop liners. Figure 8–4 shows a typical wet/dry liner and counterbore shims: The crevice seal (item 1) is also known as a fire ring and is used to seal cylinder gases in the cylinder.

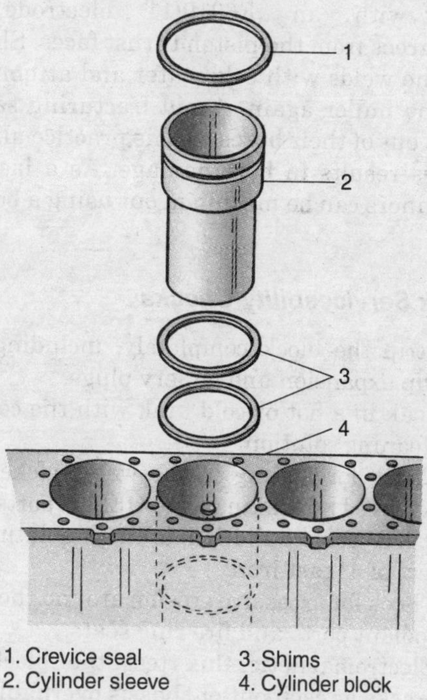

Figure 8–4 *Wet/dry liner components. (Courtesy of Mack Inc.)*

Cylinder Sleeve Removal

Sleeves should be removed with a puller and adaptor plate or shoe. The procedure is obviously more simple on wet liners. Dry sleeves often require the use of mechanical (Figure 8–5), hydraulic, or air-over-hydraulic pullers. When these fail to extract a seized liner, vertical arc weld runs may break a liner free, but great care has to be exercised because the block may become distorted. To use this method of removing a seized liner, first ensure that every critical engine component is moved well

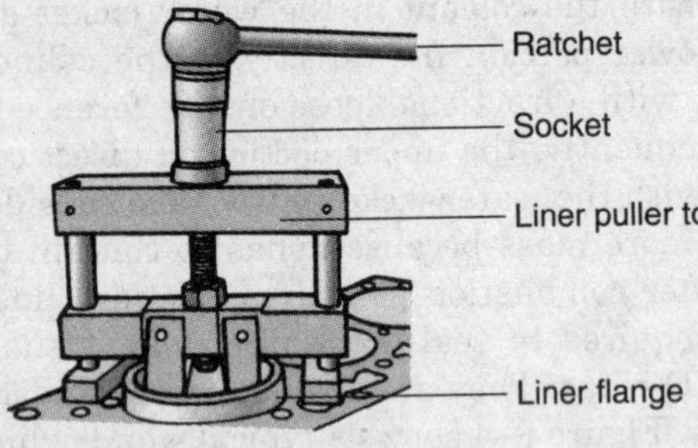

Figure 8–5 *Pulling a liner using a mechanical puller. (Courtesy of Mack Inc.)*

out of the way, then use two vertical-down runs with an E6010/11 electrode at 90 degrees from the piston thrust faces. Shock-cool the welds with cold water and attempt to use the puller again. Avoid fracturing seized liners out of their bores as this practice almost always results in bore damage. As a last resort, liners can be machined out using a boring jig.

Block Serviceability Checks

1. Strip the block completely, including the cup expansion and gallery plugs.
2. Soak in a hot or cold tank with the correct cleaning solution.
3. Check for scaling in the water jacket not removed by soaking. One OEM reports that 0.060″ scale buildup has the insulating effect of 4″ cast iron.
4. Check for excessive erosion around the deck coolant ports and fire ring seats.
5. Electromagnetic flux test the block for cracks at each out-of-chassis overhaul.

Final Inspection and Assembly

1. Check for deck warpage using a straightedge and thickness gauge. A typical maximum specification is approximately 0.004″ but always reference OEM tolerances.
2. Check the main bearing bore and alignment. Check the engine service history to ensure that the engine has not been previously line bored. A master bar is used to check alignment for a specific engine series. Check that the correct bar has been selected for the engine being tested and then clamp to position by torquing down the main caps minus the main bearings. The master bar should rotate in the cylinder block main bearing line bore without binding. Should it bind, the cylinder block should be line bored.
3. Check the cylinder sleeve counterbore (Figure 8–6) for the correct depth and circumference. Counterbore depth should typically not vary by more than 0.001″. Counterbore depth can be subtracted from the sleeve flange dimension to calculate sleeve protrusion. Recut and shim the counterbore using the OEM recommended tools and specifications (Figure 8–7).

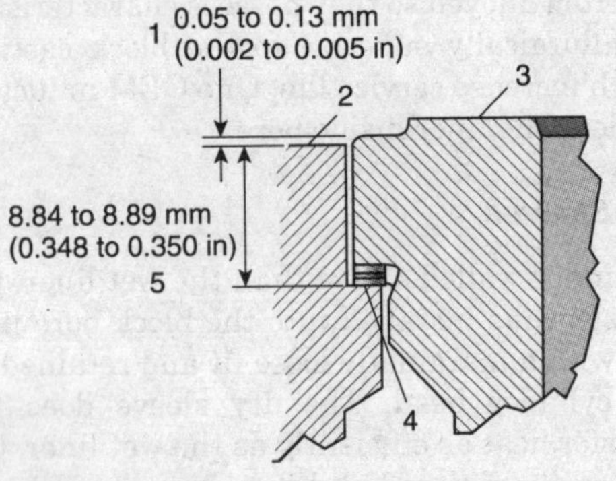

Maximum allowable variation of counter bore between four points—0.025 mm (0.001 in)

1. Sleeve protrusion
2. Crankcase top surface
3. Sleeve
4. Shim to suit
5. Maximum allowable counter bore—10.49 mm (0.413 in)

Figure 8–6 *Installing counterbore shims on a Navistar DT 466E engine. (Courtesy of Navistar International Corp.)*

4. Check the cam bore dimensions and install the cam bushings with the correct cam bushing installation equipment: Use drivers with great care as the bushings may easily be damaged and ensure that the oil holes are lined up before driving each bushing home.
5. Install the gallery and expansion plugs. These are often interference fitted and sealed with silicone, thread sealants, and hydraulic dope.

Liner and Sleeve Reconditioning

Liner and sleeve reconditioning is not a common current practice due to the time limit con-

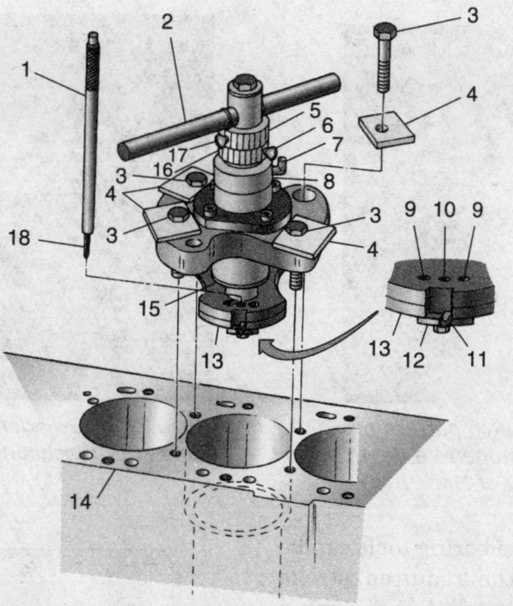

1. Cutter plate holder	7. Oil fill tube	13. Cutter plate
2. T-handle	8. Main housing	14. Cylinder block
3. Capscrews, M16 x 2 x 90	9. Hold-down capscrews	15. Main shaft
4. Special washers	10. Cutter bit adjuster	16. Lower depth-set collar
5. Upper depth-set collar	11. Cutter bit	17. Upper thumbscrew
6. Lower thumbscrew	12. Hold-down cap	18. Hex key wrench

Figure 8–7 *Counterbore cutting tool for use on an engine cylinder block with wet/dry liners. (Courtesy of Mack Inc.)*

straints on the technician and projected service life of reconditioned liners. It is probably good practice to advise customers against reconditioning liners, citing the fact that it will probably cost them more in the long term. OEMs make oversize liners when the block bores are damaged and have to be rebored. Two basic methods of liner reconditioning are used:

Glaze Busting

When checked to be within serviceability specs, the liner should be deglazed. Deglazing involves the least amount of material removal. A power-driven (heavy-duty electric drill with accurate rpm control) flex hone or rigid hone with 200 to 250 grit stones can be used for deglazing. The best type of glaze buster is the flex hone, typically a conical (Christmas tree) or cylindrically shaped shaft with flexible branches of carbon/abrasive balls.

Preserving Crosshatch The objective of glaze busting is to machine away the cylinder ridge above the ring belt travel and reestablish the crosshatch. The drill should be set at 120 to 180 rpm and used in rhythmic reciprocating thrusts: short sequences stopping frequently to inspect the finish produces the best results. A 60- to 70-degree horizontal crossover angle (some OEMs express this as 120- to 130-vertical crosshatch), 15 to 20 micro-inch depth crosshatch should be observed (Figure 8–8). When using a spring-loaded hone to deglaze liners, faster results are produced but there is also more chance of damaging the cylinder liner. Figure 8–9 shows the crosshatch on a Volvo D16 engine after machining.

Honing

Honing is performed with a rigid hone, powered once again at low speeds by either a drill or overhead boring jig. The typical cylinder hone consists of three legs, which are static set to produce a radial load into the liner wall. The abrasive grit rating of the stones determines the aggressiveness of the tool—200- to 250-grit

Figure 8–8 *Cylinder crosshatch pattern produced by a 150- to 250-grit stone to a depth of 15 to 20 micro-inches. (Courtesy of Mack Inc.)*

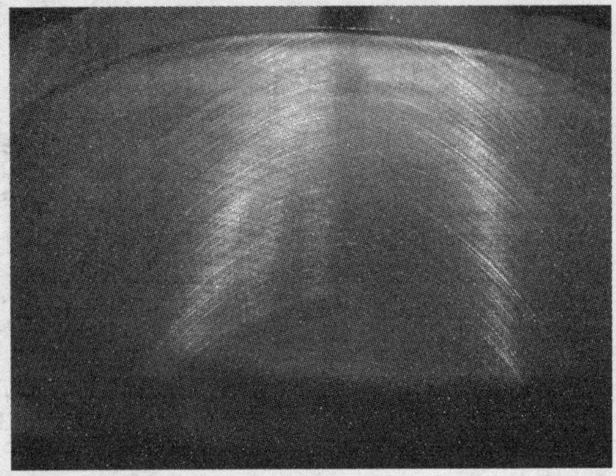

Figure 8–9 *Cylinder crosshatch on a Volvo D16 engine after machining.*

stones are typical. Overhead boring tools can be programmed to produce the required stroke rate for the specified crosshatch, but if using a hand-held tool, remember that a few short strokes with a moderate radial load tend to produce a better crosshatch pattern than many strokes with a light radial load.

Creating Crosshatch Once again, a 60- to 70-degree horizontal crossover angle, 15 to 20 micro-inch crosshatch should be observed. It should be clearly visible by eye as shown in Figure 8–7. The honing process is designed to produce the specified crosshatch geometry so that the liner contact surface retains oil required to enable rings to effectively seal. When installing wet liners, observe the OEM installation procedure. Figure 8–10 shows a typical method of sealing a wet/dry liner in the cylinder bore.

Selective Fitting of Liners

Selective fitting of liners to block bores is good practice whether or not the block bore has been machined. To selective fit a set of dry liners to a block, measure the inside diameter of each block bore across the North-South and East-West faces and grade in order of size.Next, get the set of new liners and measure the outside diameter of each, once again grading in order of size. Ensure that *every* measurement falls within the OEM specifications. Then fit the liner with the largest outside diameter to the block bore with the largest inside diameter and so on, down in sequence.

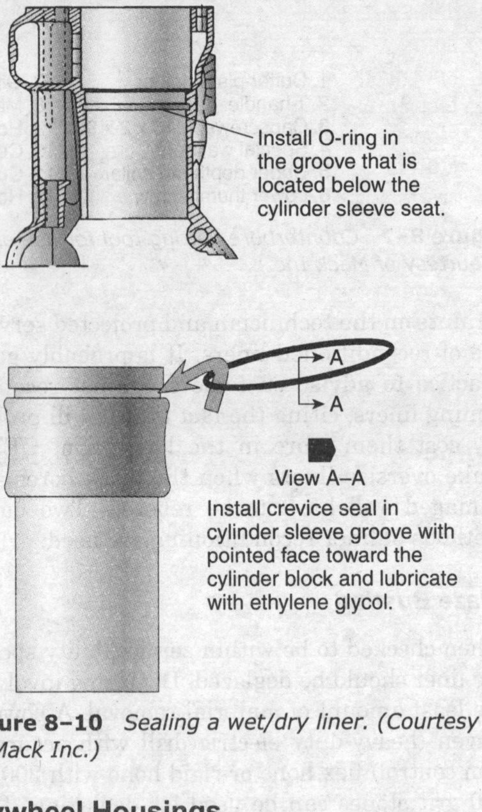

Figure 8–10 *Sealing a wet/dry liner. (Courtesy of Mack Inc.)*

Flywheel Housings

Flywheel housings bolt to the rear of the cylinder block. They form the stationary coupling between the engine and transmission. Flywheel housings used in most medium- and heavy-duty truck drivetrains are categorized by SAE codes. These SAE code designations permit a similarly coded transmission bell housing to couple to the flywheel housing. This is important because it allows transmission OEMs to fit their product to a range of different OEM engines. The flywheel housing encloses the flywheel and clutch pack as it couples the engine and transmission. In some cases the rear engine mounts may be integral with, or attach to, the flywheel housing. Rear engine mounts may also be located on the transmission bell housing. The starter motor is usually flange mounted to the front of the flywheel housing.

Flywheel housings are usually precision located to the engine cylinder block by cylindrical or diamond dowels: Clamping force is provided by fasteners, usually grade 8 bolts. Flywheel housings should be checked for:

- Flange inside diameter radial concentricity. This is an important check; it is covered in some detail in Chapter 13. If it fails to meet specification it produces a broken-back effect, imparting drive torque to the transmission bell housing that can take out clutches, clutch shafts, crankshafts, and transmissions.
- Outer flange face runout. This is the mating face to the transmission bell housing. It seldom fails to meet specification but it should be checked at engine overhaul.

CYLINDER HEADS

Cylinder heads seal the engine cylinders and manage cylinder breathing. In diesel applications they are usually cast-iron assemblies machined with breathing tracts and ports, cooling and lubrication circuit manifolds, fuel manifolds, and injector bores. They may also support rocker assemblies and camshafts when overhead camshaft design is used. There are several configurations used in diesel engines:

1. Multicylinder, single-slab casting
2. Multicylinder, multiple cast units
3. Single cylinder

Cylinder heads usually contain the valve assemblies, breathing ports, injector bores, and coolant and lubricant passages. On older indirect injection (IDI) applications, precombustion chambers were either built in or installed as units in the head. Top and bottom detail views of cylinder heads are shown in Figure 8–11, a sectional view of a cylinder head is shown in Figure 8–12, and a cutaway view is shown in Figure 8–13.

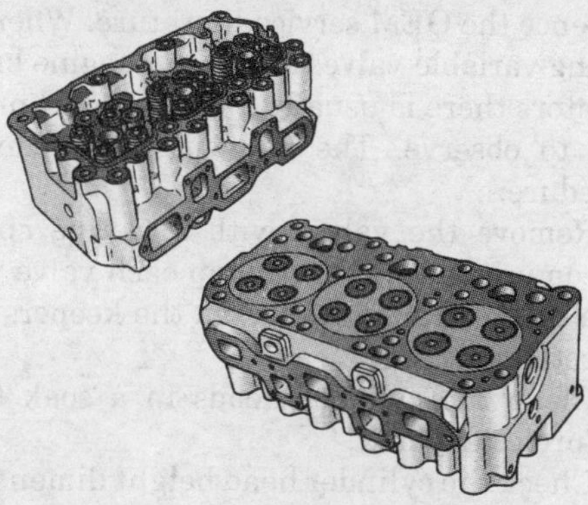

Figure 8–11 *Top and bottom views of a typical cylinder head. (Courtesy of Mack Inc.)*

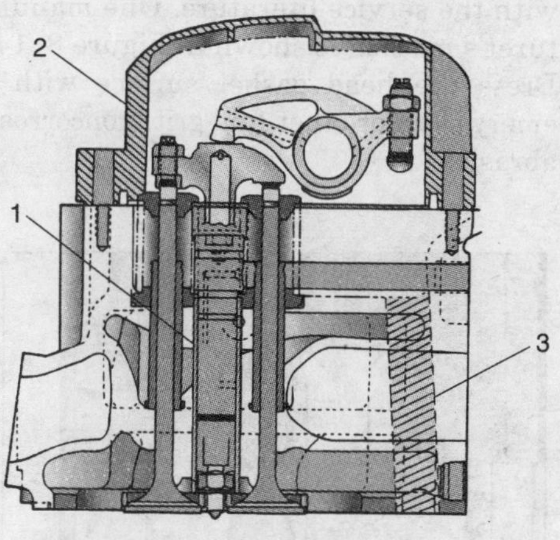

1. Large diameter nozzle sleeve to accommodate 22 mm hydraulic injector nozzle
2. E7 EUP cylinder head cover
3. Pushrod bore angled at four degrees

Figure 8–12 *Sectional view of a Mack E-Tech cylinder head. (Courtesy of Mack Inc.)*

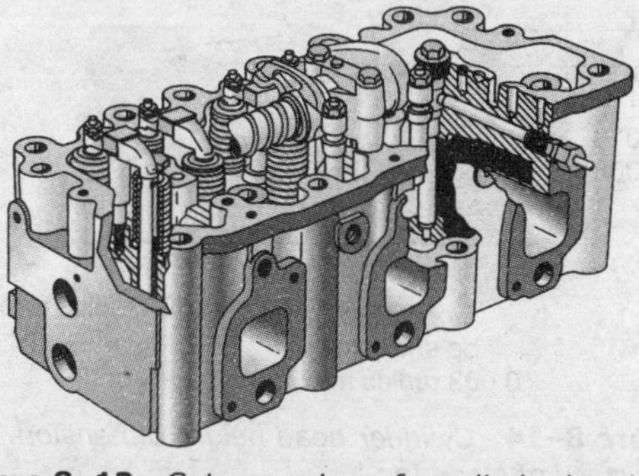

Figure 8–13 *Cutaway view of a cylinder head. (Courtesy of Mack Inc.)*

Cylinder Head Disassembly, Inspection, and Reconditioning

Before beginning any disassembly procedure, reference the OEM service literature. When removing variable valve timing and engine brake actuators there is usually a very specific procedure to observe. The following is a general procedure:

1. Remove the valves with a C-type spring compressor and gently tap each valve with a nylon hammer to loosen the keepers and retainers.
2. Clean the cylinder heads in a soak tank (preferably hot).
3. Check the cylinder head height dimension. Each OEM has its own preference on where to make the measurement so check with the service literature. One manufacturer's method is shown in Figure 8–14.
4. Dress the head gasket surface with an emery cloth or other fine-grit, noncorrosive abrasive.

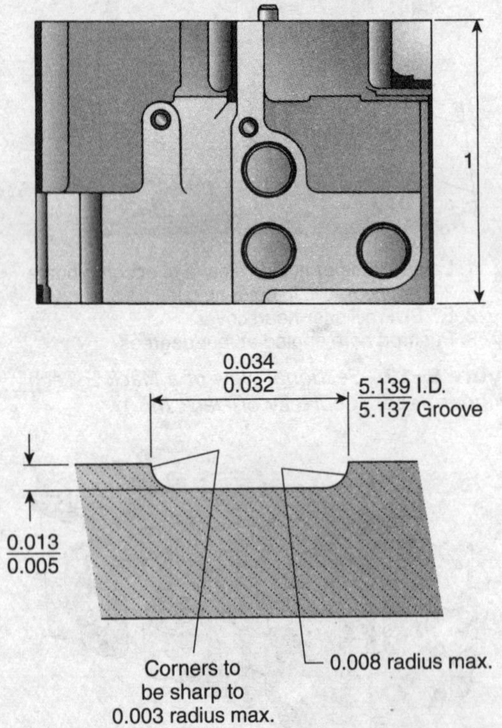

Figure 8–14 *Cylinder head height dimension. (Courtesy of Mack Inc.)*

5. Electromagnetic flux test the head for cracks. Optionally dye penetrant testing can be used, but this is messy and inaccurate.
6. Hydrostatic pressure test: Cap and plug the coolant ports then place the head in a test jig and heat by running hot water through. When hot to the touch, hydrostatically test using shop air at around 100 psi. Areas to watch are valve seats and injector sleeves. When brass injector sleeves are used (usually older engines), perform the hydrostatic test both cold and hot.
7. If the injector sleeves have to be replaced, use the appropriate removal and installation tools and repeat the pressure test after the operation. These are sometimes swaged and sometimes threaded into the injector bore.
8. Check for warpage using a straightedge and an appropriately sized thickness gauge. The specification will vary according to the size of the head.
9. Check valve guide bore to specs with a ball gauge. If in need of replacement, the guide must be pressed or driven out with the correct driver or mandrel. Damage to the guide bore may require reaming to fit an oversize guide. Integral guides can be repaired by machining for guide sleeves or knurling. Installation of new guides can be made easier by freezing (dry ice) or use of press fit lubricant.
10. New guides may require reaming after installation, but some OEMs use cladded guides that should *never* be reamed.
11. Check the valve seats for looseness using a light, ball peen hammer and listening—a loose seat resonates at a much higher pitch. To recondition, select the appropriate mandrel pilot and insert into the guide, then match the valve seat to the mandrel grinding stone. Stellite-faced seats require special grinding stones. Dress the stone to achieve the required seat angle. Interference angles are seldom used in current applications because the valve's ability to transfer heat to the head is compromised. An interference angle is usually not required when valve rotators are used.

New valve seats are installed with an **interference fit** and require the use of the correct driver; after installation the seat is usually knurled or staked in position. Most current diesels use alloy steel seats—cast iron does not require staking of seats. After valve and seat reconditioning, check the valve head height (valve protrusion) with a dial indicator.

12. Valve reconditioning. Clean the valves with a wire wheel or glass bead blaster, then check for stretching, cupping, burning, or pitting. When refacing (dressing) valves, ensure that the valve margin remains within OEM specifications. Check valve seating using Prussian blue (a.k.a. machinist's blue). Lapping is not required if the grinding has been done properly.
13. Check the valve springs for straightness, height, and tension using a right-angle square, tram gauge, and tension gauge.
14. Valve rotators. Positive rotators (rotocoil) can be checked after hand assembly by tapping the valve open with a nylon hammer after assembly to simulate valvetrain action—they should rotate.

Cylinder Head Installation

Most current cylinder head gaskets are integral, meaning the fire rings and grommets are manufactured integrally with the gasket plate. If not, care should be taken to ensure that the **fire rings** (sealing rings at the liner flange) and grommets are properly positioned while the cylinder head is being installed to the block. The head **gasket** must be properly torqued to ensure that its **yield point** is attained for proper sealing. Yield point means that a malleable gasket is crushed to conform to the profile designed to produce optimum sealing. Properly torquing a cylinder head means observing the OEM torque increments and sequencing. An engine cylinder head that is torqued to the correct specification without observing the incremental steps can damage the gasket by deforming it. The objective of torquing cylinder head gaskets is to ensure that the specified clamping force is obtained and the gasket conforms to its engineered yield shape. Cold torquing is required on current engines.

Component Creep

Most head gaskets require no applied sealant and in fact may fail if sealants are used. Head gaskets are designed to seal engine components in a region where both the temperature and pressure are at their highest; they must do this and at that same time accommodate a large amount of component **creep**. Creep is the relative movement of clamped engine components due to different coefficients of heat expansion or different mass. Because a cylinder head has much less material than the cylinder block, it will expand more rapidly as it is heated and contract more rapidly as it cools. While doing this it is clamped to the cylinder block and must maintain a good enough seal to contain cylinder combustion pressures, which can exceed 3,000 psi. The head gasket is the key to ensuring that this seal is effective under all the operating conditions of the engine. Cylinder head fasteners should be lightly lubed with engine oil before installation. Oil should never be poured into the block threads because a hydraulic lock may result. The torque sequence for a Mercedes-Benz MB-906 cylinder head is shown in Figure 8–15.

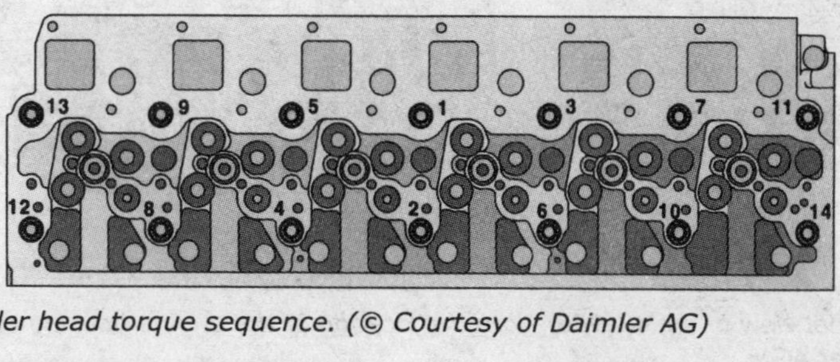

Figure 8–15 *Cylinder head torque sequence. (© Courtesy of Daimler AG)*

Incremental Torque Sequence

On inline multicylinder heads, it is usually required that separate heads be aligned with a straightedge across the intake manifold faces before torquing. Failure to observe torquing increments and sequencing can result in cracked cylinder heads, failed head gaskets, and fire rings that will not seal. Because of the large number of fasteners involved, a click-type torque wrench should be used. Many OEMs use **torque-to-yield** fasteners that require a **template torque** method be used. Torque-to-yield is fully described in Chapter 2 but, simply, this requires setting a torque value first and then turning a set number of degrees beyond that value using a template or protractor. Cylinder head bolts usually should be installed lightly oiled. Excessive quantities of oil should be avoided because the excess can drain into the bolt hole and cause a hydraulic lock.

Rocker Housing Covers

Rocker housing or valve covers seal the upper portion of the engine above the cylinder head. Because valve and injector trains have become more complex over the years, the physical size of rocker housing covers has increased, especially on engines equipped with variable valve timing and engine brakes. Rocker housing covers are the most frequently removed engine component. They have to be removed to access the valves and injectors for purposes of tune-up and injector timing. For this reason, many engine OEMs have adopted the use of multiuse, rubber compound sealing gaskets usually fitted to a captured groove in the rocker housing cover. They also help dampen the top end clatter produced by the rockers. OEM recommendations for installing these gaskets should be observed, especially if the gaskets will be used more than once. Usually a light coating of engine oil is all that should be applied. Try to avoid using aggressive adhesives such as weatherstrip glue because it ruins rubber gaskets, which tend to be much more costly than their fiber predecessors.

INTAKE AND EXHAUST MANIFOLDS

In current diesel engines, the **intake manifold** is required to deliver both air and recirculated exhaust gas (EGR) to the engine cylinders. The intake manifold is bolted to the cylinder head(s) enclosing the intake tracts. Because of turbocharging the runners that extend from the intake plenum can be of unequal lengths without compromising engine breathing. EGR mixing with boost air may take place in the region of the intake manifold plenum.

Intake manifolds can be either wet (coolant ports) or dry. Most current mobile diesel engines use dry intake manifolds. Materials used for intake manifolds are usually aluminum alloy or cast iron, but some OEMs are experimenting with plastics and carbon-based fibers. The gaskets used are usually fiber based and must be able to accommodate a small amount of component creep and the peak boost pressures. Peak manifold boost pressures may be as high as 100 psi on some more recent engines. When

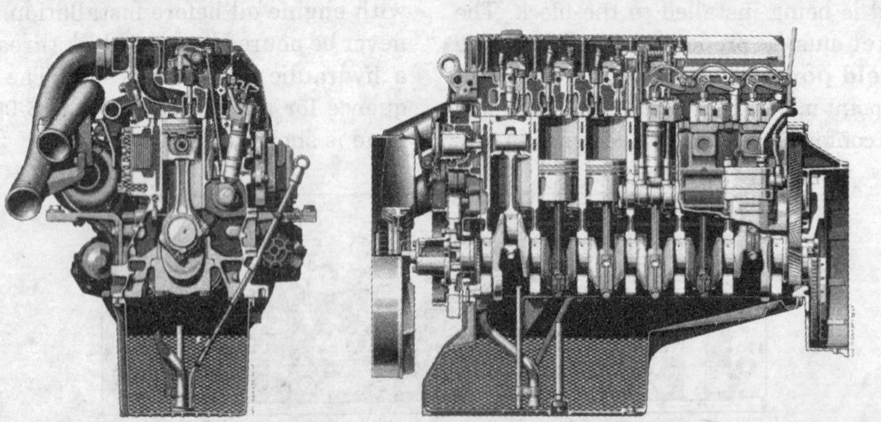

Figure 8–16 *Sectional view of an MB-906: Note the location of the engine housing components. (© Courtesy of Daimler AG)*

a single section cast aluminum manifold is bolted to a multicylinder head configured engine, it is critical that the cylinder heads be aligned with a straightedge before they are torqued down. Figure 8–16 shows a sectional view of an engine: Identify the location of all of the components studied so far in Chapters 6, 7, and 8.

Exhaust Manifold

The function of the **exhaust manifold** is to collect cylinder end gases and deliver them to the turbocharger or in the case of a naturally aspirated engine, directly to the exhaust piping. They are usually manufactured in single or multiple sections of cast iron. The exhaust manifold assembly is bolted to the cylinder head; in engines using multiple heads, it should be aligned before torquing.

Tuned Manifolds

Most current diesel engine exhaust manifolds are described as tuned, meaning that exhaust gas is routed to the turbocharger with minimal flow resistance. The gas dynamic of exhaust manifolds is discussed in more detail in Chapter 11. The term *tuned* is applied to any exhaust system designed with at least some accommodation to optimize exhaust gas flow. If the exhaust manifold and piping is properly designed, as each slug of cylinder exhaust is discharged, it will not "collide" with that from another cylinder but instead be timed to unload into its tailstream.

Minimizing Heat Losses

Because exhaust gas heat can be "recycled" to drive turbochargers, most current exhaust manifolds are designed to minimize heat losses upstream from the turbocharger turbine housing. This may mean that the exhaust tract in the cylinder head and sections of the exhaust manifold are "insulated" usually with stainless steel inserts to reduce heat losses.

Exhaust Manifold Gaskets

Exhaust manifold gaskets in truck/bus engine applications are usually of the embossed steel type, although occasionally fiber gaskets are used. They must be able to sustain a high degree of component creep along with high temperatures. Embossed steel exhaust manifold gaskets are almost always installed dry and should be used once only. In many cases the fastening hardware such as studs, bolts, and nuts is manufactured from highly alloyed steels to accommodate high temperatures and the high thermal expansion/contraction rates.

OIL PANS OR SUMPS

The **sump** is a reservoir usually located at the base of the engine cylinder block enclosing the crankcase. **Oil pans** are manufactured from cast aluminum, stamped mild steel, laminated steels, and various plastics and synthetic fibers. The oil pan acts as the reservoir used to collect the lubrication oil that gravity causes to drain to the crankcase and from which the oil pump pickup can recycle it through the lubrication circuit. Take a look at the oil pan and its pickup tube in Figure 8–16.

Oil Pan Functions

Oil pans can act as a sort of boom box and amplify engine noise, so they are usually designed to minimize noise. Laminated steels are effective at noise dampening and with it provide some desirable toughness to the oil pan. The oil pan also plays a role in dissipating lube oil heat to atmosphere but the effectiveness of this role obviously has a lot to do with the material from which it is manufactured. Aluminum will dissipate heat more effectively than a fiber-reinforced plastic. Also this heat exchanger role of the oil pan can produce problems when running in midwinter or subarctic conditions and cause gelling and sludging. Some off-highway trucks use a deep oil pan design to accommodate a scavenging pump. **Scavenging pumps** are secondary oil pumps designed to ensure that the engine lubrication circuit is provided with adequate oil when the vehicle is operating on steep grades.

Removing an Oil Pan

Oil pans are usually located in the airflow under the frame rails and are vulnerable to damage from objects on the road—from rocks on rough terrain to small animals on the highway. Most highway diesel engines have oil pans that can be removed from the engine while it is in chassis. It is advisable to drain the oil sump before removing it. Oil pans that seal to the engine block using gaskets can be difficult to remove, especially where adhesives have been used to ensure a seal. A 4 lb rubber mallet may make removal easier. Avoid driving screwdrivers between the oil pan and its block mating flange because the result will be damage to the oil pan flange and the cylinder block mating flange.

Oil Pan Inspection

Oil pans that use rubber isolator seals tend not to present removal problems. In fact, sometimes rubber isolator seals can be reused if they are inspected to be in good condition. After removal the oil pan should be cleaned of gasket residues and washed with a pressure washer. Inspect the oil pan mating flanges, check for cracks, and test the drain plug threads. Cast aluminum oil pans that fasten both to the cylinder block and to the flywheel housing are prone to stress cracking in the rear due to cylinder block torque twist. Carefully inspect them before returning to service. When torquing aluminum oil pans always meticulously observe torque sequences and values.

Welding Damaged Oil Pans

Cast aluminum oil pans can be successfully repair welded without distortion using the tungsten inert gas (TIG) process providing the cracks are not too large and the oil saturated area of the crack is ground clean. It is more difficult to execute a lasting repair weld on a stamped steel oil pan due to the distortion that results. Steel oil pans may rust through, in which case they should be replaced rather than welded. Whenever an oil pan has indications of porosity resulting from corrosion, it should always be replaced. The technician should always be aware of the fact that a failure of the oil sump or its drain plug on the road can result in the loss of the engine. It is generally recommended that plastic composite, fiberglass, and carbon fiber oil pans be replaced rather than repaired.

REVIEW QUESTIONS

1. Which of the following cylinder block designs would be the most common in medium- and heavy-duty truck diesel engines?
   a. Eight-cylinder, 90-degree V-configuration
   b. Inline six cylinder
   c. Six-cylinder, 60-degree V-configuration
   d. Inline eight cylinder
2. The main reason for bench pressure testing a diesel engine cylinder head is to:
   a. Test cylinder gas leakage
   b. Check for air leaks
   c. Check for exhaust leaks
   d. Check for coolant leaks
3. When selective fitting a set of dry liners to a cylinder block, which of the following statements should be true?
   a. The liner with the largest od is fitted to the bore with the largest id.
   b. The liners are installed with an interference fit.
   c. The liner with the smallest od is fitted to the bore with the largest id.
   d. The liners are installed with a fractionally loose fit.
4. Which of the following is the recommended method of pulling dry liners from a cylinder block?
   a. The arc welding method
   b. A puller and shoe assembly
   c. Heat and shock cool method
   d. Fracture the liner in position
5. Which of the following tools is recommended for checking cylinder block line bore?
   a. Straightedge and thickness gauges
   b. Dial bore gauge
   c. Dial indicator
   d. Master line bore bar
6. When boiling out a cylinder block, why is it critical that all the scale in the water jacket be removed?
   a. Scale is an effective insulator.
   b. Scale may contaminate the engine lubricant.
   c. Scale can accelerate coolant silicate drop-out.
   d. Scale causes cavitation of wet liners.

7. Which of the following tools should be used to check a cylinder block deck for warpage?
   a. Master bar
   b. Dial indicator
   c. Laser
   d. Straightedge and thickness gauges
8. Which of the following tools should be used to check a cylinder head for warpage?
   a. Master bar
   b. Dial indicator
   c. Laser
   d. Straightedge and thickness gauges
9. *Technician A* states that all the machined surfaces of a cylinder block should be checked to specification with a straightedge and thickness gauges. *Technician B* states that cylinder blocks should be boiled in a soak tank at every major overhaul. Who is correct?
   a. Technician A only
   b. Technician B only
   c. Both A and B
   d. Neither A nor B
10. When cavitation damage is evident on a set of wet liners, which of the following is more likely to be responsible?
    a. Lubrication breakdown
    b. High cylinder pressures
    c. Coolant breakdown
    d. Cold engine operation
11. When cylinder block counterbores are machined, which of the following would be required?
    a. Oversize fire ring
    b. Shims
    c. Undersize liner flange
    d. Cylinder head resurfacing
12. *Technician A* states that it is good practice to magnetic flux test a cylinder block for cracks at each major overhaul. *Technician B* states that before performing a magnetic flux test, the cylinder block should be boiled in a soak tank. Who is correct?
    a. Technician A only
    b. Technician B only
    c. Both A and B
    d. Neither A nor B
13. Most current heavy-duty, diesel engine cylinder blocks are manufactured from:
    a. Cast aluminum alloy
    b. Cast-iron alloys
    c. Composite fibers
    d. Compacted graphite iron (CGI)
14. Once a cylinder head gasket fire ring has been torqued to its yield point, it should:
    a. Not be reused after removal
    b. Be immediately heated to operating temperature
    c. Deform and no longer seal effectively
15. *Technician A* states that all cylinder heads on current engines should be hot torqued after a rebuild procedure. *Technician B* states that cylinder head bolts should be soaked in engine lubricating oil before installation. Who is correct?
    a. Technician A only
    b. Technician B only
    c. Both A and B
    d. Neither A nor B

9

Engine Lubrication Systems

INTRODUCTION

All medium- and large-bore diesel engines currently used on North American highways use a pressure lubrication system to supply the bearings and moving components with engine oil. Engine oil is the medium used in lubrication circuits, and it is especially formulated to fulfill the lubrication and service life requirements of diesel engines. A modern diesel engine with a not-too-ambitious one million miles on it will have had as many power strokes as a human heart beats in five lifetimes. Engine lubricating oil temperatures run somewhat higher than engine coolants, and the actual oil temperature is probably a more reliable indicator of true engine running temperature. The basic components required of a diesel engine lubrication system are shown in Figures 9–1 and 9–2. This chapter discusses:

- Functions of a lubrication circuit.
- Lubricants: usually petroleum-based, liquid media used to reduce friction, hydrodynamically support shafts, help seal pistons, and act as a cooling medium.
- Sump: storage oil space enclosing the crankcase but, exceptionally, a remotely located tank in what is called a **dry sump** system.

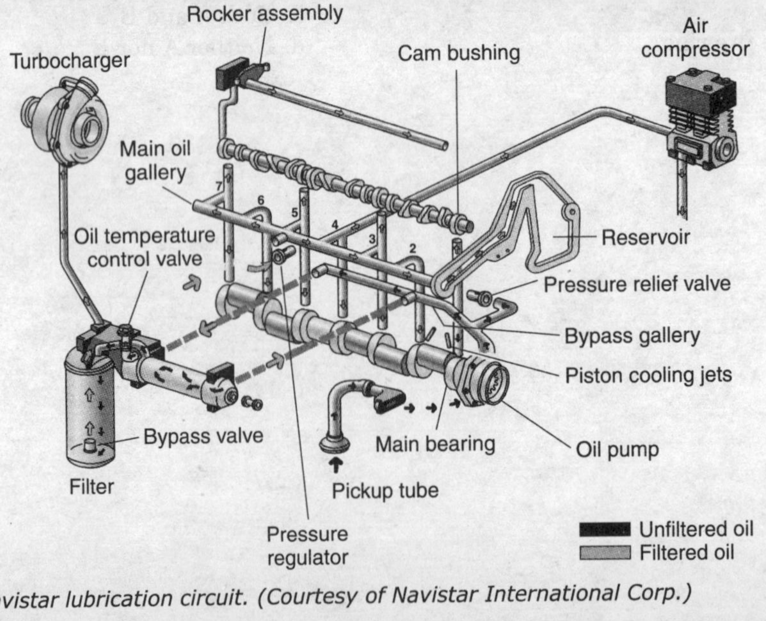

Figure 9–1 *Navistar lubrication circuit. (Courtesy of Navistar International Corp.)*

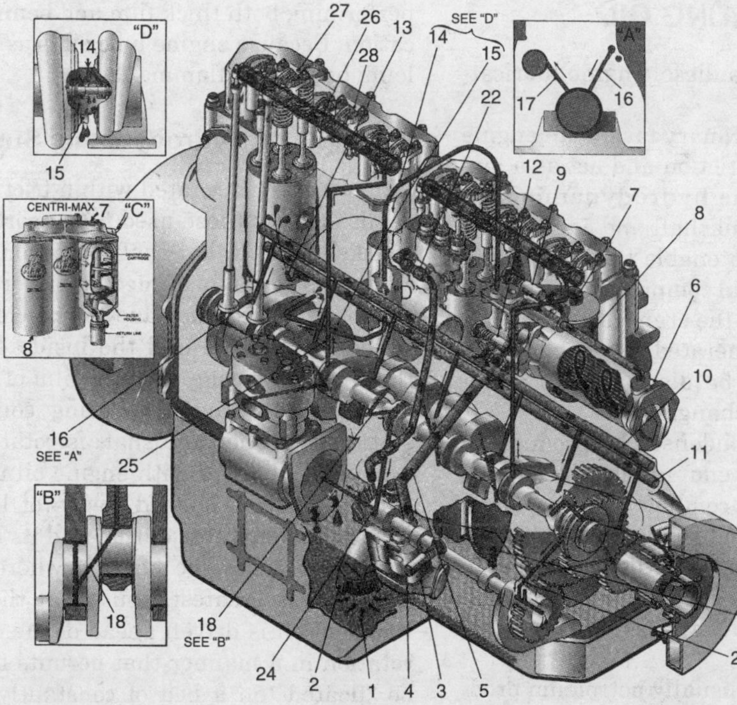

1. Oil pump
2. Oil pump inlet tube
3. Oil pump
4. Oil pressure relief valve (located in pump)
5. Oil pump discharge line to oil cooler
6. Lube oil cooler
7. Oil filter mounting adapter
8. Oil filter(s) (see inset "C")
9. Oil filter bypass valve
10. Main oil gallery
11. Piston oil cooling gallery
12. Oil pressure relief valve (piston cooling gallery)
13. Piston cooling oil spray nozzle
14. Lube feed to turbocharger (external)
15. Turbocharger lube oil return (external)
16. Oil supply from main oil gallery to crankshaft main bearings (see inset "A")
17. Oil flow from main bearings to camshaft bearings (see inset "A")
18. Main bearing and connecting rod bearing lube oil passages (see inset "B")
19. Line to injection pump drive assembly from camshaft bushing number 1
20. Line to auxiliary shaft front bushing from camshaft bushing number 1
21. Oil supply from camshaft bushing number 2 to rocker arm shaft
22. Line to injection pump (external) and governor from camshaft bushing number 4
23. Fuel injection pump lube oil drains
24. Line to auxiliary shaft rear bushing and compressor from camshaft bushing number 4
25. Oil supply from camshaft bushing number 5 to rocker arm shaft
26. Rocker arm shaft
27. Valve rocker arm
28. Oil supply from rocker arm shaft to rocker arm tip

Figure 9–2 *Mack Truck lubrication circuit. (Courtesy of Mack Inc.)*

- Pump: responsible for moving the oil through the lubrication circuit to supply the system oil passages and bearings.
- Filter: commercial diesel engine lubrication systems require multistage filtration systems to remove particulates from the engine oil.
- **Oil cooler**: a heat exchanger that uses engine coolant as its medium. Heat picked up by the engine oil is transferred to the coolant from which it can be dissipated to atmosphere.
- Piston cooling requirements: spray nozzles.

Friction

Friction is a common element in our lives that we simply take for granted. For instance, we can walk up a steep hill without slipping because of high friction between the soles of our feet and the ground surface. We also accept that when that same hill is covered with packed snow, we can ski down it. In the first instance, the coefficient of friction is high, and in the second it is low. Rubber soles have a higher coefficient of friction than skis; packed snow has a lower coefficient of friction than bare earth. Another way of thinking about the coefficient of friction is as a means of rating the aggressiveness of friction surfaces. Lubricants are designed to reduce friction between surfaces that are, or could be, in contact with one another creating wear at the contact points.

ENGINE LUBRICATING OIL

The main functions of a diesel engine lubricating system are:

1. Lubrication. The primary task of an engine oil is to minimize friction and act as a medium to support the **hydrodynamic suspension** of the crankshaft and camshaft.
2. Act as a sealant to enable the piston and ring assembly to seal compression and combustion gases from the crankcase.
3. Coolant. Heat generated by combustion and friction must be dissipated to atmosphere via heat exchangers.
4. Cleaning agent. Condensed by-products of combustion gases end up in the engine crankcase and can combine to form harmful liquids (acids) and particulates (sludge).
5. Maintain an oil film even when subjected to high thrust loads, such as the compressional loading to which a piston/rod assembly is subjected.

Lubricating oils are usually petroleum products that are complex mixtures made up of many different fractions. Like gasoline and diesel fuel, engine lubricant is elementally composed of about 85% carbon and 15% hydrogen. The fractioned compounds that make up an engine oil are refined from petroleum and asphalt bases and subsequently mixed. Some synthetics are also used but the reality is these have not yet proven cost effective as diesel engine lubricants though synthetic oils are used in most other truck chassis systems. The refining processes of petroleum are presented in a little more detail in later in Chapter 16.

The theoretical action of an engine oil is to form a film between moving surfaces so that any friction that results occurs in the oil itself, that is, it is **fluid friction**. Fluid friction generates considerably less heat than dry friction. The lubricating requirements of an engine oil can be classified as **thick film lubrication** and **boundary lubrication** (thin film). Thick film lubrication occurs where mating tolerances between components are wide. Boundary lubrication is required where mating tolerances are narrow, such as when pressure is applied to one of the components: A breakdown of boundary lubrication will result in metal-to-metal contact. Good quality engine oil should be capable of performing both thick film and boundary lubrication. Because engine lube oil is usually petroleum based it is flammable.

Principle of Hydrodynamic Suspension

When a shaft is rotated within friction bearing shells such as those used to support an engine crankshaft, and that shaft is stationary, a crescent-shaped gap is formed on either side of the line of direct contact due to the clearance between the journal and the inside diameter of the friction bearing. A static film of oil (boundary) prevents shaft-to-bearing contact when stationary. When the shaft is rotated and the bearing is charged with engine oil under pressure, a crescent-shaped wedge of lubricant is formed between the journal and its bearing; oil is introduced to the bearing where the shaft clearance is greatest, usually at the top. This wedge of oil is driven ahead of the direction of rotation in a manner that permits the shaft to be "floated" on a bed of constantly changing, pressurized oil. This principle is used in most crankshaft and camshaft dynamics and is known as hydrodynamic suspension.

In hydrodynamic lubrication, maintaining a liquid film between moving surfaces is achieved by pumping lubricant through a circuit. In the case of a rotating shaft and a stationary friction bearing, the shaft acts as a pump to maintain the lubricant film. The result is that the shaft journal floats on a film of oil the thickness of which depends on:

- Oil input: the rate at which oil is delivered to the bearing. This is why there is a problem if oil pressure is low.
- Oil leakage rate: the oil that spills from a bearing during operation. This is why the association between worn engine main bearings and low oil pressure is made.

The thickness of the hydrodynamic wedge therefore depends on the following four factors, any one of which will change that thickness:

1. Load increase: causes oil to be squeezed out of the bearing at a faster rate.
2. Temperature increase: causes oil leakage rate from bearing to increase.
3. Lower viscosity oil: flows with less resistance causing more leakage.

4. Changing shaft speed: reduce speed (remember, the shaft is the "pump") and the film becomes thinner; increase speed and it thickens.

Figure 9–3 shows how hydrodynamic lubrication functions.

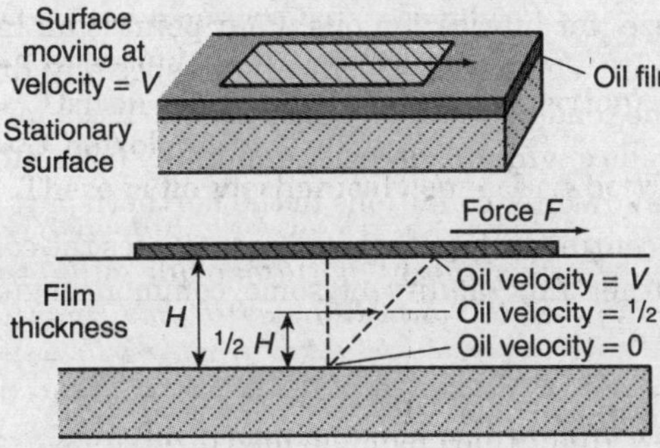

Figure 9–3 *Concept of hydrodynamic viscosity.*

Engine Oil Classification and Terminology

To fully understand the terms and codes used on the label of any engine oil probably requires the learning of a tribologist, that is, an expert in **tribology**, the study of friction, wear, and lubrication. However, every technician should have a rudimentary understanding of the codes and terms used to describe engine oils. The following section introduces some of the basic language of lubricants.

Viscosity

The **viscosity** rating of an oil usually describes its resistance to flow. High viscosity oils have molecules with greater cohesion ability. However, properly defined, viscosity denotes resistance to **shear**. When two moving components are separated by engine oil, the **lamina** (portion of the oil film closest to each metal surface) on each moving component should have the least fluid velocity while the fluid in the center has the greatest fluid velocity. Shear occurs when the lamina fluid velocity is such that it is no longer capable of adhering to the surface of the moving components. When we use the term *friction bearings* to describe crankshaft main bearings, the friction being referred to is the fluid friction of the lube hydrodynamics in supporting the crankshaft.

Viscosity Index

The viscosity index (VI) is a measure of an oil's tendency to thin as temperature increases. Temperature affects viscosity. Viscosity may be considerably reduced with an increase in temperature and, to a smaller extent, as fluid pressure is increased. The greater the VI, the less of an effect temperature will have on the actual viscosity of an oil. In other words, oils that show relatively small viscosity changes with changes in temperature can be said to have high VI.

SAE Numbers and Viscosity

The viscosity of automotive engine oils is graded by the Society of Automotive Engineers (SAE). These gradings and the recommendations that accompany them are listed in Chapter 8. These gradings specify the temperature window within which the engine oil can function to provide adequate shear resistance under boundary lubrication conditions.

Multiviscosity Oils

Multiviscosity engine oils are the lubricating oil of choice of truck and bus diesel engine original equipment manufacturers (OEMs). They have been around for many years but, initially, some manufacturers took a cautious approach before finally recommending them. The main advantage of multiviscosity oils over straight grade oils is that they provide proper lubrication to the engine over a much wider temperature range. In other words, they have a relatively flat viscosity to temperature curve.

Multiviscosity oils are produced by special refining processes and the addition of VI improvers. They possess good cold cranking characteristics and show comparatively small variations of viscosity over their nominal operating range. Synthetic oils, when marketed with an SAE grading, usually greatly exceed the nominal grading window.

Lubricity

Two oils with identical viscosity grades can possess different lubricity. The **lubricity** of an oil properly describes its flow characteristics. Lubricity is also affected by temperature: Hotter oils flow more readily, colder oils less readily.

In comparing two engine oils, the one that has the lowest frictional resistance to flow can be said to possess the greater lubricity. In thick film fluid lubrication, flow friction is determined by the fluid's viscosity, that is, its resistance to shear. In thin film or boundary lubrication, flow friction is determined by the lubricity of the fluid. Sulfur compounds add to the lubricity of diesel engine oils. Synthetic oils tend to have high lubricity with low sulfur content.

Flash Point

Flash point is the temperature at which a flammable liquid gives off enough vapor to ignite momentarily. The **fire point** of the same flammable liquid is usually about 10°C higher and is the temperature at which a flammable liquid gives off sufficient vapor for continuous combustion. The flash point specification has some significance when assessing a diesel engine lubricating oil because a large portion of the cylinder wall is swept by flame every other revolution in a four-stroke cycle engine and every revolution in a two-stroke cycle engine. However, actual cylinder wall temperatures are significantly lower than the temperatures of the combustion gases and the oil is only exposed to them for very short periods of time. Most diesel engine lubricating oils have flash points of 400°F (205°C) or higher.

Pour Point

The temperature at which a lubricant begins to gel or simply ceases to flow is known as the **pour point**. Engine lubricating oils formulated for extreme cold weather operation have pour point additives or depressants that act as "antifreeze" for lubrication oils. Pour point is an important engine oil specification especially due to the tendency of many operators to use a multigrade engine oil viscosity nominally not suited for midwinter conditions in the northern part of the continent. Figure 9–4 demonstrates the cold weather flow ability of some common engine lubes.

Inhibitors

The term **inhibitors**, when applied to an engine lubricant, refers to the additives that protect the oil itself against corrosion, oxidation, and acidity. Their objective is to make the oil less likely to participate in reactions with the contaminants that find their way into the crankcase, such as combustion by-products, moisture, and raw fuel. When lube oil inhibitors fail to work properly (usually due to extended service), the oil lacks protection and begins to degrade.

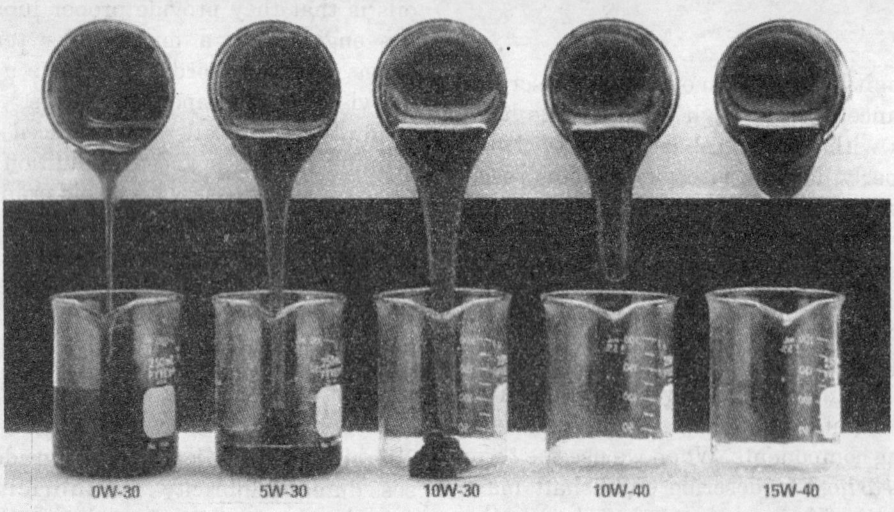

Figure 9–4 *Demonstration of cold weather flow ability of different oil viscosity grades. (Courtesy of Imperial Oil Limited)*

Ash

Ash in lubricant is mineral residue that results from oxide and sulfate incineration. Some ash content in diesel engine lube oils was desirable because it checked the formation of some acids formed. However, in the latest CJ-4 diesel oils (described in detail later in this chapter), the objective is to keep ash levels at a minimum because they can plug emissions control devices. When higher-than-spec-ash levels are reported in oil analyses this is usually caused by high-temperature operation.

Film Strength

Most engine oils possess adequate film strength to prevent seizure and galling of contacting metal surfaces, but those formulated for high-speed, high-output diesel engines usually contain additives to improve film strength and cohesiveness of the oil. Cohesion is the ability of molecules to stick together when pressure is applied. Most synthetic oils have superior cohesiveness to mineral oils. This can be tested by applying a couple of drops of each to a smooth surface: To clean them off the surface, you will find that the mineral oil is easily wiped away while the synthetic oil leaves a waxy film.

Detergents

Detergents are added to engine oils to prevent the formation of deposits on internal engine components. They also help keep soluble oxidation products from becoming insoluble when sludge could be formed. Polymeric detergents and amine compounds are used. Some engine oils require a high detergent load (especially those recommended for newer diesels), while others have almost no detergent.

Dispersants

Dispersants are added to engine oils to help keep insoluble oxidation products in suspension and prevent them from coagulating into sludge and deposits. When sludge and deposits do form in the crankcase, the capability of the dispersants in the engine oil has been exceeded. This is usually an indication that the oil change interval should be reduced because the result is lube oil degradation.

Oil Contamination and Degradation

When oil becomes contaminated, it can lead to complete engine failure. Contaminated engine oil has some characteristic tattletales. These can be quickly identified by visual and odor testing, and confirmed, if required, by oil analysis. The following are some common contaminants:

Fuel

When fuel contaminates engine oil, the oil loses its lubricity and it appears thinner and blacker in color. The condition is usually easy to detect because small amounts of fuel in oil can be recognized by odor. When fuel is found in significant quantities in engine oil, the cylinder head(s) are the likely source.

Coolant

Coolant in the engine lubricant gives it a milky, cloudy appearance when churned into the oil. After settling, the coolant usually collects at the bottom of the oil sump. When the drain plug is removed, the heavier coolant exits first as long as sufficient time has passed for it to settle. When coolant is found in the engine oil, the cylinder head(s), cylinder head gasket, and liners are the probable source. Within the cylinder head, the injector cup seals are usually the culprit.

Oil Aeration

Aerated engine oil, also known as foaming, affects its ability to properly lubricate the engine. Aerated oil can be caused either by the chemical characteristics of the oil itself (such as detergent additives) or by contaminants such as water; surface tension is a factor. The conditions that aerate oil in the crankcase are the rotary action of the crankshaft, sucking of air into the oil pump inlet, and the free fall of oil into the crankcase from the oil pump **relief valve** and cylinder walls. Antifoaming additives inhibit oil aeration. Silicone polymers are used to prevent foaming in diesel engine lubricants.

Cold Sludge

Cold sludge is caused by oil degeneration that occurs during prolonged engine low load operation at low temperatures. Routinely idling diesel engines for prolonged periods at truck stops creates cold sludge. The sludge settles in the engine crankcase and can accelerate engine wear rates. When a diesel engine is operated with its coolant thermostat(s) removed, the result can also be the formation of cold sludge.

API Classifications

The **American Petroleum Institute**, usually known as **API**, classifies all engine oil sold in North America. There are two main classifications designated by the prefix letters *G* and *C*. The G classes of engine oil designate those oils suitable for passenger cars and light trucks. The *G* represents GF or gasoline fueled. The C classes of engine oils designate oils suitable for heavy-duty trucks, buses, and industrial and agricultural equipment. The *C* represents CI or compression-ignited engines. C category and the most recent G category oil classifications are listed and described here because many fleets use engine oils that claim to be suitable for both C and G classifications. In most cases, OEMs have specific requirements for engine lubricants that should be observed because failure to do so could result in higher hydrocarbon (HC) emissions. It should also be noted that using inappropriate oil in current diesel engines can damage emission control hardware.

CA—for Light-Duty Diesel Engine Service. Service typical of diesel engines operated in mild or moderate duty with high-quality fuels. Occasionally has included gasoline engines in mild service. Oils designed for this service were widely used in the late 1940s and 1950s.

CB—for Moderate-Duty Diesel Engine Service. Service typical of diesel engines operated in mild to moderate duty, but with lower quality fuels that require more protection from wear deposits. Occasionally has included gasoline engines in mild service. Oils designed for this service were introduced in 1949.

CC—for Moderate-Duty Diesel and Gasoline Engine Service. Typical of lightly supercharged diesel engines operated in moderate to severe duty and has included certain heavy-duty gasoline engines. Oils designed for this service were introduced in 1961 and used in many trucks, in industrial and construction equipment, and farm tractors. These oils provide protection from high-temperature deposits in lightly supercharged diesels and also from rust, corrosion, and low-temperature deposits in gasoline engines.

CD—for Severe-Duty Diesel Engine Service. Service typical of supercharged diesel engines in high-speed, high-output duty requiring highly effective control of wear and deposits. Oils designed for this service were introduced in 1955 and provide protection from bearing corrosion and from high-temperature deposits in supercharged diesel engines when using fuels of a wide quality range.

CD-II—Service typical of two-stroke cycle diesel engines requiring highly effective control over wear and deposits. Oils designated for this service also meet all performance requirements of API Service Category CD.

CE—Service typical of turbocharged or supercharged heavy-duty diesel engines manufactured since 1983 and operated under both low-speed, high-load and high-speed, high-load conditions. Oils designed for this service may be used when previous API Engine Service Categories for diesel engines are recommended.

CF—May be used in diesel engines exposed to fuels that may have a sulfur content greater than 0.05%. Effective in controlling piston deposits, wear, and bearing corrosion. May be used where API Service Category CD is recommended.

CF-2—Service typical of two-stroke cycle engines that require effective control over cylinder and ringface scuffing and deposits. May be used where API Service Category CD-II is recommended.

CF-4—Service typical of on-highway, heavy-duty truck applications. Designated for multigraded oils and introduced in 1991. Oils meeting this category will also meet API Service Category CE.

CG-4—For use in high-speed, four-stroke cycle diesel engines used on both heavy-duty on-highway and off-highway applications (with less than 0.05%. sulfur fuel). CG-4 oils provide effective control over high-temperature piston deposits, wear, and soot accumulation. Effective in meeting 1994 exhaust emission standards and may also be used where API Service Categories CD, CE, and CF-4 are specified.

CH-4—For use in high-speed, four-stroke cycle diesel engines used in on- and off-highway applications that are fueled with fuels containing less than 0.05% sulfur. Supersedes CD, CE, CF, and CG category oils.

CI-4—Introduced in September 2002 for use in high-speed, highway diesel engines meeting 2004 exhaust emission standards, implemented in October 2002 by EPA agreement with engine OEMs. CI-4 oil is especially formulated for engines using fuels containing less than 0.05% sulfur and maximizes lube oil and engine longevity when exhaust gas recirculation (EGR) devices are used. CI-4 engine oil supersedes CD, CE, CF, CG, and CH category oils. It meets the requirements of heavy-duty engine oil (HDEO) standard PC-9 but should not be used in any post-2007 diesel engines. It should be available up until around 2010 and will probably be used by some operators due to its lower cost than CJ-4.

CJ-4—Formulated for use in 2007 diesel engines equipped with cooled-EGR and diesel particulate filters ([DPFs]; see Chapter 41) but is also backward compatible. The main difference between CJ-4 and CI-4 is a change in the additive package designed to reduce ash generated when the oil is either combusted in the cylinder or routed through the closed crankcase ventilation system. Higher ash loads in the exhaust gas can plug the DPF or poison the catalyst.

The change to CJ-4 requires some changes in the way we interpret diesel engine oil analysis reports (discussed later in this chapter) because sampled **total base number (TBNs)** will increase. TBNs are a measure of used oil acidity. Ash increases the alkalinity of in-service engine oil and helps reduce acids generated by combustion. However, the postcombustion acids were mostly produced by combusting the sulfur component in diesel fuel and the introduction of ultra-low sulfur (ULS) fuel will counter this.

While CJ-4 has backward compatibility, it costs more. This means that some operators may continue to use CI-4 for pre-2007 engines until they are forced to change. Use of other than CJ-4 engine oil post-2007 engines can result in voiding of engine and emission hardware warranty. CJ-4 engine oil will be around for a while. To create a new engine oil category, the **Engine Manufacturers Association (EMA)** must formulate a request to the **American Society for Testing Materials (ASTM)**. From that point, it typically takes around 4 years to formulate the chemistry for the new oil, which typically has to meet the EMA requirements plus have backward compatibility. Figure 9–5 shows the chemical limits required of CJ-4 diesel engine lube.

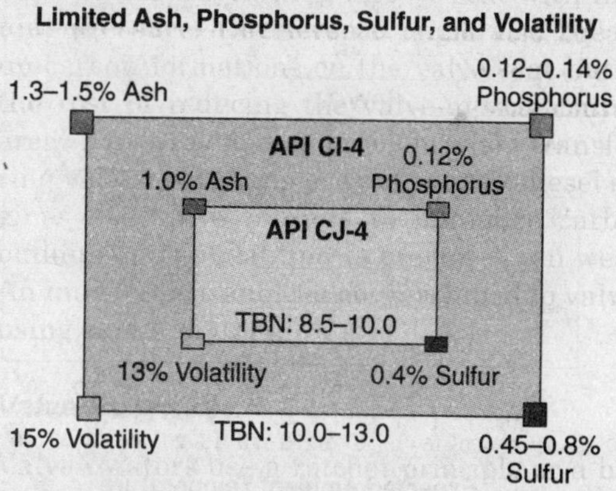

Figure 9–5 *Chemical limits of API CJ-4 engine lube. (Reprinted Courtesy of Caterpillar Inc.)*

Oils for SI Engines

Oils formulated for use in gasoline-fueled engines are categorized in much the same way and they are no longer of much importance to diesel technicians because the days of dual fuel application engine lubes are past. The most recent engine oils formulated for service in gasoline engines areas are:

SH 1993
GF-2 1996
GF-3 2001
GF-4 2004

There are still some engine oils formulated that claim to meet the needs for both SI gasoline-fueled and diesel-fueled engines, a relic from the years when some fleets (such as school bus operators) had a mix of diesel- and gasoline-fueled vehicles. Most diesel engine OEMs recommend that multipurpose, multifuel specified oils be avoided: This is because such oils tend to be loaded with excessive additives in an effort to meet too wide a range of lube specs.

SAE Viscosity Grades

The following list the SAE engine oil grades and the recommended temperature operating ranges. Figure 9–6 shows the temperature range for some common SAE viscosity grades. The *W* denotes a winter grade lubricant.

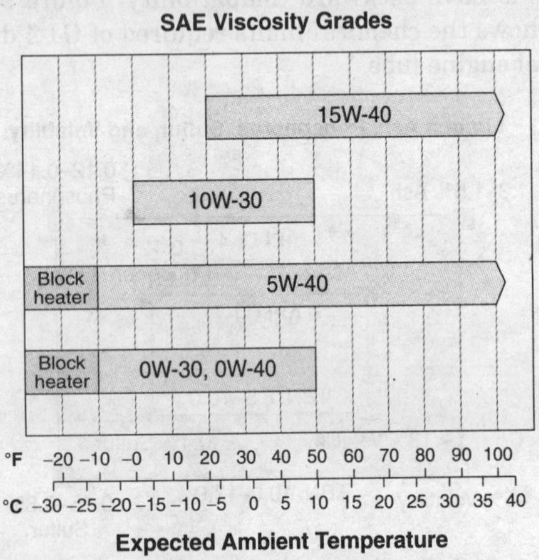

Figure 9–6 *SAE viscosity grades and temperature ranges for 2007 engines. (Courtesy of Navistar International Corp.)*

Multigrade engine oils:

0W-30 Recommended for use in arctic and subarctic winter conditions.

5W-30 Recommended for winter use where temperatures frequently fall below 0°F (–18°C) and seldom exceed 60°F (15°C).

5W-40 Recommended for severe-duty winter use where temperatures frequently fall below 0°F (–18°C).

5W-50 Recommended for severe-duty winter use in arctic and subarctic conditions where temperatures frequently fall below 0°F (–18°C). A synthetic oil viscosity grade.

10W-30 Recommended for winter use where temperatures never fall below 0°F (–18°C).

10W-40 Recommended for severe-duty winter use where temperatures never fall below 0°F (–18°C).

15W-40 Recommended for use in climates where temperatures never fall below 15°F (–9°C). Despite this and diesel engine OEM recommendations that support the use of lighter multigrades in winter conditions, this is by far the most commonly used viscosity grade for truck and bus engines year round, often in climates that have severe winters. A true 15W-40 engine oil will freeze to a grease-like consistency in subzero conditions, making the engine almost impossible to crank; additionally, engine wear is accelerated during the warmup phase of operation. Figure 9–5 is a Navistar schematic showing its recommended viscosity grades by temperatures: Note that 15W-40 lube is not recommended by Navistar for engines operating below 20°F (–6°C).

20W-40 Recommended for use in high-performance engines in climates where temperatures never fall below 20°F (–6°C).

20W-50 Recommended for use in high-performance engines in climates where temperatures never fall below 20°F (–6°C).

Straight grades:

10W Recommended for winter use in climates where temperatures never fall below 0°F (–18°C) and never exceed 60°F (15°C).

20W-20 Recommended for use in climates where temperatures never fall below 20°F (–6°C).

30 Recommended for use in climates where temperatures never fall below 32°F (0°C). This grade of oil is the most widely used truck and bus diesel engine in operations that persist in using straight grade oils.

40 Recommended for severe-duty use in climates where temperatures never fall below 40°F (4°C).

50 Recommended for severe-duty use in climates where temperatures never fall

below 60°F (15°C). Often used to disguise engine problems, especially oil burners and leakage.

Synthetic Oils

Most diesel engine OEMs approve of the use of synthetic lubricants and often recommend them for severe-duty applications such as extreme cold weather operation, providing the oil used meets their specifications. Synthetic lubricants are largely petrochemically sourced, but some plant and coal-sourced additives are also used. Synthesis is the process of creating compounds in laboratory, so while conventional lubricants are obtained from petroleum crude oils by distillation or other refining methods, synthetics are manufactured or "synthesized" in chemical plants by reacting components. Some examples are poly-alpha-olefins, diesters, polyesters, and silicone fluids. Only the high cost of synthetic lubricants is limiting their use in diesel engines. Synthetics have already become the oil of choice in transmissions and final drive carriers, but the frequency of service required of diesel engine lubes have limited their application. That said, most of the indicators suggest that synthetics substantially outperform conventional lubricants, and while OEMs approve of the use of **synthetic oils,** they have not as yet endorsed extended service intervals when they are used.

What Oil Should be Used?

Each engine OEM has specific requirements for the engine oils it wants used in its engines. These requirements are outlined in tests that engine oils are subjected to prior to an approval. Oil refiners attempt to meet as many standards as possible when marketing engine oils for the obvious reason that they can service more potential buyers with a single product. The problem is that oil is like a brew and throwing more additives in does not necessarily improve it. For instance, if your preference is to drink bourbon and water, you probably would not consider it a better drink if some gin, vodka, and cola were added to it. For this reason, a general-purpose, heavy-duty engine oil is not necessarily better than the OEM-labeled oil, although of course it will be both cheaper and more advertised.

Developing a Formula

The best choice for the engine is to use the manufacturer-recommended engine oil. After all, the research and performance profiles have all been performed on the engine using this oil. This may not be the best choice for the pocketbook, and in fairness there are general-purpose diesel engine oils proven to perform well in engines over time. You should also know that engine OEMs usually do not "engineer" their own oils. This development is done by specialist oil research companies, such as Lubrizol (http://www.lubrizol.com), that have been contracted by the OEM. A company such as Lubrizol neither manufactures nor markets oil. The company is paid by an OEM to develop a formula (a "recipe" if you like): After the formula has been developed, the OEM takes it to an oil refiner that then manufactures the product to specification. Developing the right additive package for an engine oil is a science in itself and is usually the result of extensive testing by lubrication engineers known as tribologists.

LUBRICATION SYSTEM COMPONENTS

Engine lubricant must be retained in a reservoir, pumped through the lubrication circuit, filtered, cooled, and have its pressure and temperature monitored. The group of components that performs these tasks is known as the lubricating circuit. The components in a diesel engine lube circuit vary little from one engine to another but OEM service literature should be consulted before servicing and reconditioning any components.

Sump—Oil Pan

The sump is a reservoir usually located at the base of the engine cylinder block enclosing the crankcase (Figure 9–7). Oil pans are manufactured from cast aluminum, stamped mild steel, and various plastics and fibers. The oil pan acts as the reservoir used to collect the lubrication oil, which gravity causes to drain to the crankcase, and from which the oil pump pickup can recycle it through the lubrication circuit. A more complete description of an oil pan is provided in Chapter 8.

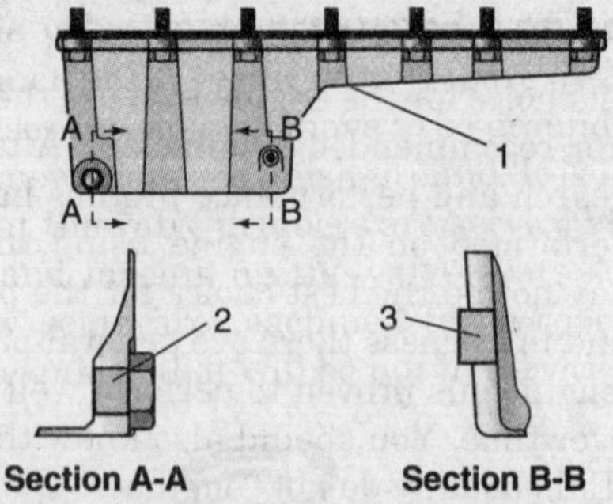

Figure 9–7 *A Caterpillar oil pan. (Reprinted Courtesy of Caterpillar Inc.)*

Dipsticks

The dipstick is a rigid band of hardened steel that is inserted into a round tube to extend into the oil sump. Checking the engine oil level is performed daily by the vehicle operator, so its location is always accessible. In cab-over-engine (COE) chassis, the dipstick must be accessible without raising the cab, so it may be of considerable length. It is crucial that the correct dipstick is used for an engine. When replacing a missing or defective dipstick, replace the engine oil and filters, installing the exact OEM-specified quantity. Run the engine for a couple of minutes then shut down and leave for 10 minutes. Dip the oil sump with the new dipstick and scribe the high-level graduation with an electric pencil. Measure the distance from the high-level to low-level graduations on the old dipstick and then duplicate on the replacement. Remember, the consequences of low or high engine oil levels can be equally serious, so ensure that this operation is performed with precision. Some electronically managed engines have an oil level sensor that signals a low oil level condition to the ECM, which can then initiate whatever failure strategy it is programmed for.

Oil Pump

Engine oil pumps are usually of the positive displacement type and have pumping capacities that greatly exceed the requirements of the engine. They are gear driven and usually located in the crankcase close to the oil they pump, though in some Cummins and Caterpillar applications they are external. Oil pumps are driven either directly or indirectly by the engine geartrain. In cases where the oil pump is located in the crankcase, the drive source is a vertical shaft and pinion engaged with a drive gear on the camshaft. A pickup is located close to but not contacting the base of the oil pan (Figure 9–8). Two basic types of oil pumps are used.

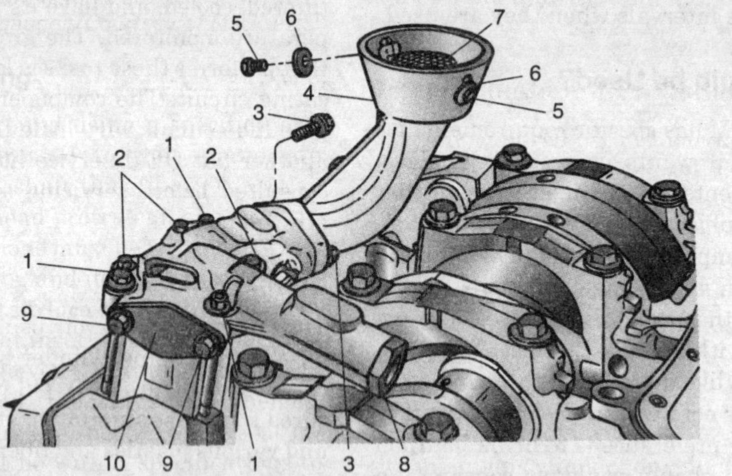

1. Capscrews	4. Sump	7. Screen	10. Plate
2. Capscrews	5. Capscrews	8. Relief valve cap	
3. Capscrews, 12-point	6. Washer	9. Capscrews	

Figure 9–8 *View of an oil pump and pickup on an inverted engine. (Courtesy of Mack Inc.)*

External Gear

External gear pumps consist of two meshed gears, one driving the other within a housing machined with an inlet (suction) port and outlet (charge) port. As the gears rotate, the teeth entrap inlet oil and force it outside between the gear teeth and the gear housing to the outlet port. Where the teeth mesh in the center, a seal is formed that prevents any backflow of oil to the inlet. This is by far the most common oil pump design on current engines. Figure 9–9 demonstrates the operating principle of a typical external gear-type oil pump.

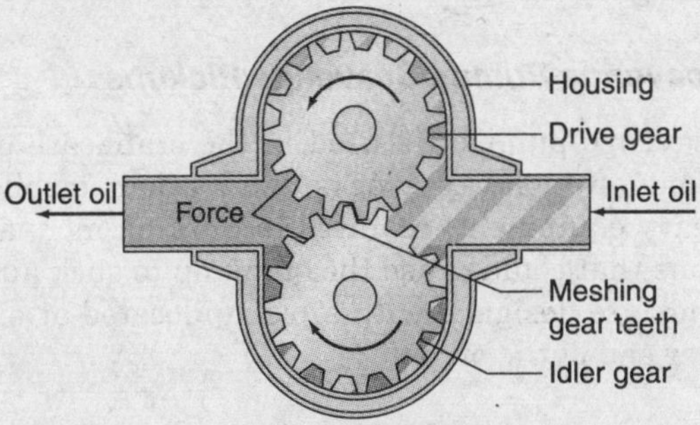

Figure 9–9 *Operating principle of an external gear oil pump.*

Gear Pump Overhaul Gear-type engine oil pumps seldom malfunction. When they show evidence of wear, the underlying reason is usually found in contaminated engine lube. To inspect a gear-type pump, remove the cover plate and, with the gears in the housing, use a thickness gauge to check the gear-to-housing radial clearance as shown in Figure 9–10. If this exceeds the OEM specification, the gears should be replaced.

Next, the gear-to-cover plate clearance (gear member axial clearance) should be measured using a straightedge and thickness gauges as shown in Figure 9–11: Replace the gears if the measurement exceeds specifications.

Before reassembling a gear-type pump, clean thoroughly using solvent and compressed air drying, ensuring that the pickup pipe and inlet screen are not overlooked. Reassemble using engine oil as assembly lubricant and check the drive gear teeth backlash as shown in Figure 9–12.

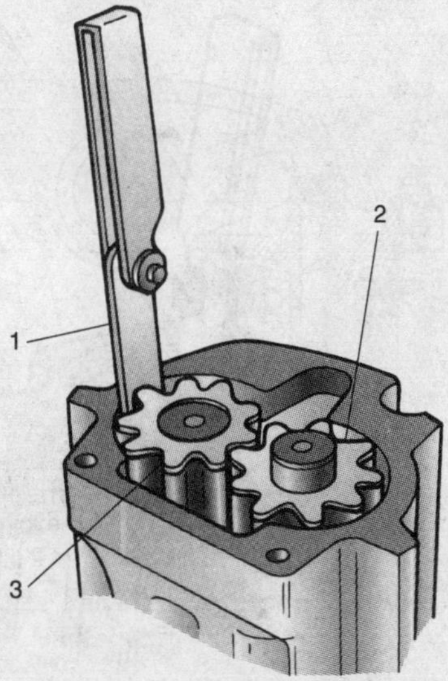

1. Thickness gauge
2. Pump driven gear
3. Pump driven gear

Figure 9–10 *Checking gear side clearance. (Courtesy of Mack Inc.)*

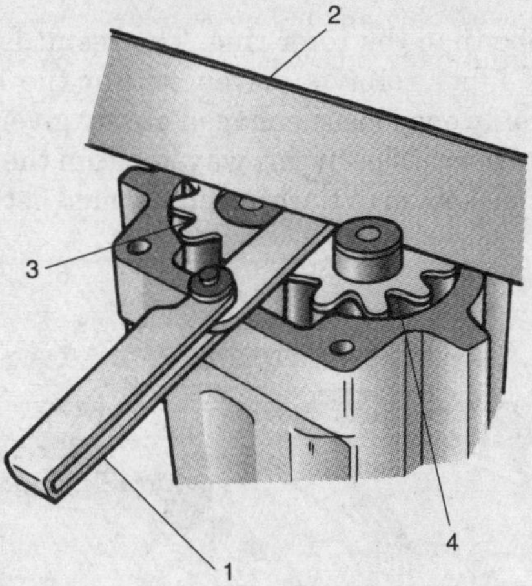

1. Thickness gauge
2. Straightedge
3. Pump driven gear
4. Pump driven gear

Figure 9–11 *Checking gear end clearance. (Courtesy of Mack Inc.)*

Gerotor

Gerotor type oil pumps use an internal crescent gear pumping principle. An internal impeller with external crescent vanes is rotated within an internal crescent gear also known as a rotor ring. The inner rotor or impeller has one

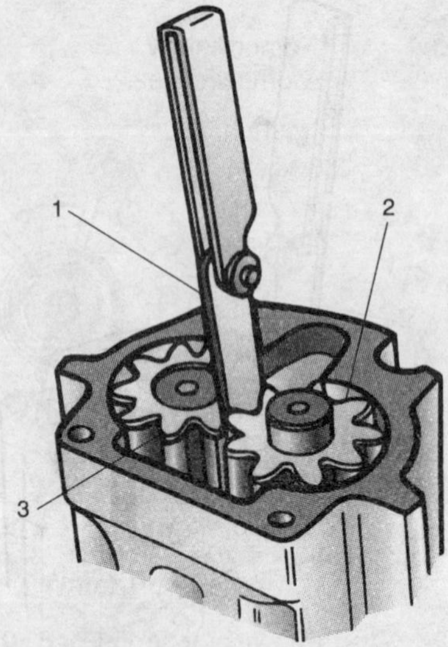

1. Thickness gauge 2. Pump driven gear 3. Pump driven gear

Figure 9–12 *Checking gear backlash. (Courtesy of Mack Inc.)*

less lobe than the rotor ring. The result is that as the inner rotor is driven within the outer rotor, only one lobe is engaged at any given moment of operation. In this way, oil from the inlet port is picked up in the crescent formed between two lobes on the impeller and, as the impeller rotates, forced out through the outlet port as the lead lobe once again engages. The assembly is rotated within the gerotor pump body.

Gerotor pumps tend to wear most between the lobes on the impeller and on the apex of the lobes on the rotor ring. These dimensions should be checked to OEM specifications using a micrometer. The rotor ring-to-body clearance should be checked with a thickness gauge sized to the OEM maximum clearance specification and the axial clearance of the rotor ring and impeller measured with a straightedge and thickness gauges.

Scavenge Pumps/Scavenge Pickups

Scavenge pumps are used in the crankcase of some off-highway trucks (and other mobile heavy equipment) required to work on inclinations that could cause the oil pump to suck air. They are designed with a pickup located at either end of the oil pan.

Pressure-regulating Valves

Pressure-regulating valves are responsible for defining maximum system oil pressure. Most

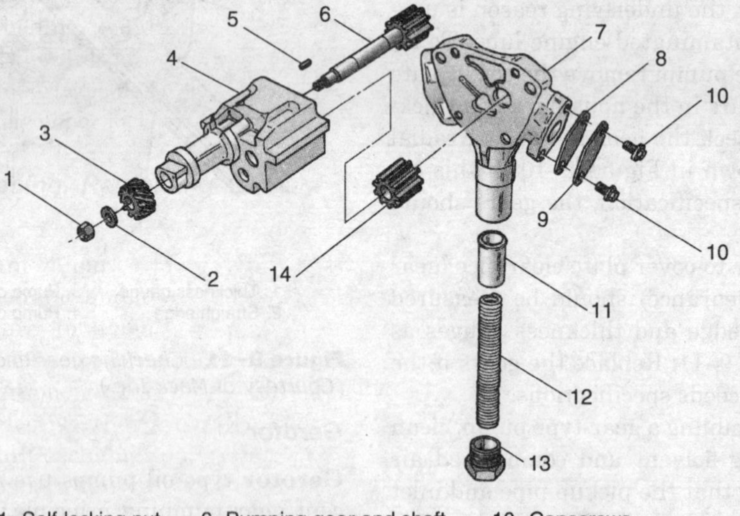

1. Self-locking nut
2. Washer
3. Drive gear
4. Housing
5. Key
6. Pumping gear and shaft assembly
7. Oil pump housing cover
8. Gasket
9. Inlet flange plate
10. Capscrews
11. Gasket
12. Relief valve spring
13. Relief valve cap
14. Idler gear

Figure 9–13 *Exploded view of an oil pump and integral pressure relief valve. (Courtesy of Mack Inc.)*

are adjustable. Typically, an oil pressure-regulating valve consists of valve body with an inlet sealed with a spring-loaded, ball check valve. Other types of poppet valves are also used but the principle is the same. The regulating valve body is plumbed in parallel to the main oil pump discharge line (Figure 9–13). When oil pressure is sufficient to unseat the spring-loaded check ball, it unseals, permitting oil to pass through the valve and spill to the oil sump; this action causes the pressure in the oil pump discharge line to drop. The regulating pressure value is adjusted by setting the spring tension of the spring usually by shims or adjusting screw. Some OEMs use color-coded springs to define the oil pressure values.

Filters

The role of oil filters in the diesel engine lubrication system is to remove and hold contaminants while providing the least amount of flow restriction in the lubrication circuit. Filters use different principles to accomplish this objective. Generally, oil filters can be classified as:

- Full flow: in series between the oil pump and lube circuit
- Bypass: ported off a lube gallery in parallel

The term **positive filtration** is used to describe a filter that operates by forcing all the fluid to be filtered through the filtering medium. Most engine oil filters use a positive filtration principle. It should be noted that filters function at higher efficiencies when the engine oil is at operating temperature. Filters work to clean the engine oil by using the following methods:

Mechanical Straining

Mechanical straining is accomplished by forcing the lubricant through a filtering medium, which if greatly enlarged, would have the appearance of a grid. The sizing of the grid openings defines the size of the particle that may be entrapped by the filtering medium. Most current diesel engine oil filters make use of mechanical straining in conjunction with other principles. Straining media would include rosin-impregnated paper, often pleated to increase the effective surface area, and cotton fibers.

Absorbent Filtration

These filters work by absorbing or sucking up engine contaminants as a sponge would. Effective absorbent filtering media include cotton pulp, mineral wools, wool yarn, and felt. These filters not only absorb coarse particles but may also remove insoluble oxidized particulate, moisture, and acids.

Adsorbent Filtration

Filters adsorb by holding (by adhesion) the molecules of dissolved substances or liquids to the surface of the filtering medium. Adsorbent filtering media include charcoal, Fuller's earth, and chemically treated papers. Because adsorbent filters may act to remove oil additives, they are often used only where low-additive engine oils are specified.

Filter Types and Efficiencies

Most current filters are spin-on disposable cartridges. Older engines may have permanent canisters enclosing a replaceable element; the canister was mounted to the filter pad with a long threaded shaft that extended through the length of the canister. They are seldom seen today as most OEMs made conversion adapters so that disposable spin-on cartridges could replace them. Filters are categorized by the manner in which they are plumbed into the lubrication circuit.

Full Flow

The filter mounting pad is usually plumbed into the lubrication circuit close to the oil pump outlet, and all of the oil exiting the pump is forced through the filter. The filtering medium is usually rosin or otherwise treated paper or cotton fiber. In most cases, these filters employ a mechanical straining principle, so the particles entrapped by the filtration media are those too large to pass through it. All **full flow filters** used on current truck and bus diesel engines use positive filtration and are rated to entrap particulate sized between 25 and 60 microns. **Bypass valves** located on full flow filter mounting pad(s) protect the engine should a filter become plugged. In this event, the oil exiting the oil pump would be routed around the plugged

Figure 9–14 *Volvo VDE-16 filter pad assembly with a pair of full flow filters, a bypass filter, and lube circuit valves.*

filter directly to the lubrication circuit. Figure 9–14 shows a Volvo VDE-16 filter pad assembly to which a pair of full flow filters, a bypass filter, and lube circuit valves are integrated.

Bypass

Bypass filters are used to complement the full flow filters on current highway diesel engines. These are plumbed in parallel in the lubrication circuit, usually by porting them into the main engine oil gallery. They filter more slowly but are rated to entrap particles down to 10 microns in size. Two types are used:

Luberfiner Filter These are large canister-type filters designed to entrap much smaller particles than full flow filters. Luberfiner filters are supplied from any point in the engine lubrication circuit and return the filtered lubricant directly to the oil sump. They may serve an additional role as an oil cooler because they are often mounted in the airflow. Luberfiner filters are large volume filters that substantially increase the amount of engine oil required: This is important to remember when servicing the engine because more oil will be required. A replaceable filter element is installed to a permanent canister; after replacing a filter element, the canister should be purged of air after engine startup.

Centrifugal Filter Centrifugal filtration or centrifuging is used to entrap smaller particles than most full flow filters so they are usually, but not always, of the bypass type. A **centrifugal filter** consists of a canister within which a cylindrical rotor is supported on bearings. Study Figure 9–15, which shows how a Mack Trucks centrifugal filter is arranged in the lube circuit: It is identified as a CentriMax filter on the schematic. The filter is plumbed into the lubrication circuit so that the rotor is charged with engine oil at lube system pressure. It exits the rotor via two thrust jets, which are angled to rotate the assembly at high velocity. The centrifuge forces the engine oil through a stationary cylindrical filtering medium wrapped outside the rotor. The filtering medium is usually a rosin-coated paper element. The filtered oil drains back to the oil sump.

Filter Bypass Valves

Filter mounting pad bypass valves operate in much the same way as the oil pressure-regulating valves, except that their objective is to route the lubricant around a restricted full flow filter to prevent engine damage by oil starvation. When a filter bypass valve is actuated and the check valve unseated, instead of spilling the oil to the crankcase it reroutes the oil directly to the lubrication circuit, effectively shorting out

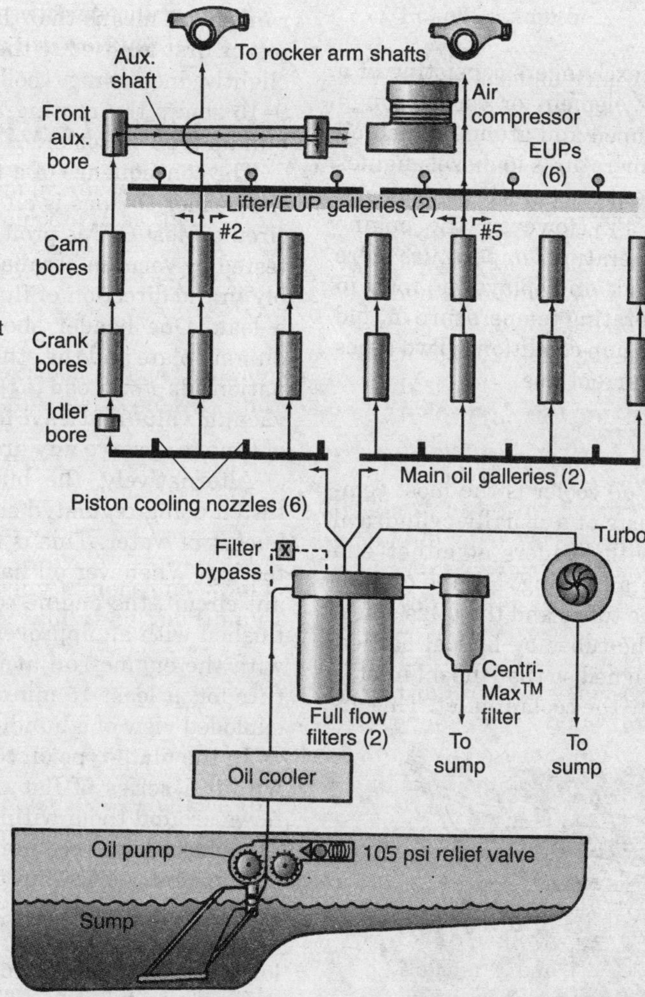

Figure 9–15 *Schematic showing oil flow through a lubrication circuit equipped with a centrifugal filter. (Courtesy of Mack Inc.)*

the filter assembly. So, when a bypass value trips, unfiltered oil is circulated through the lubrication circuit. Note the location of the filter bypass in Figure 9–14 and Figure 9–15.

Replacing Filters

Filters are removed using a band, strap, or socket wrench. Ensure that the filter gasket and seal (if used) are removed with the filter. Precautions should be taken to capture oil that spills when the filter is removed. Disposable filters and elements are loaded with toxins and must be disposed of in accordance with federal and local jurisdiction legislation that applies to used engine oils and filters.

Most OEMs require that new oil filters be primed. If this is not done, in some cases, the lag required to charge the oil filters on startup is sufficient to generate a fault code. Priming an oil filter requires that it be filled with new engine oil on the inlet side of the filter until it is just short of the top of the filter; this will take a little while as the oil must pass through the filtering medium to fill the outlet area inside the element. The sealing gasket should be lightly coated with engine oil. OEMs are usually fairly specific about how much the filter should be tightened and caution against overtightening. In most cases, the filter should be tightened by rotating it one-half to a full turn after the gasket and filter pad mounting face contact.

Oil Coolers

Oil coolers are heat exchangers consisting of a housing and cooling element or core through which coolant is pumped and around which oil is circulated. Oil temperatures in diesel engines run higher than coolant temperatures, typically around 110°C (230°F). However, the engine coolant reaches its operating temperature more rapidly than the oil and plays a role in heating the oil to operating temperature in cold weather startup/warmup conditions. Two types of oil coolers are in current use.

Bundle

The **bundle** type of oil cooler is the most common design. It consists of a usually cylindrical "bundle" of tubes with headers at either end that are enclosed in a housing. Engine coolant is flowed through the tubes and the oil is spiral circulated around the tubes by helical baffles. The assembly is designed so that the oil inlet is at the opposite end to the coolant inlet. This arrangement means that the engine oil at its hottest is first exposed to the coolant at its coolest, slightly increasing cooling efficiency. Figure 9–16 shows the routing flow through a typical bundle-type oil cooler.

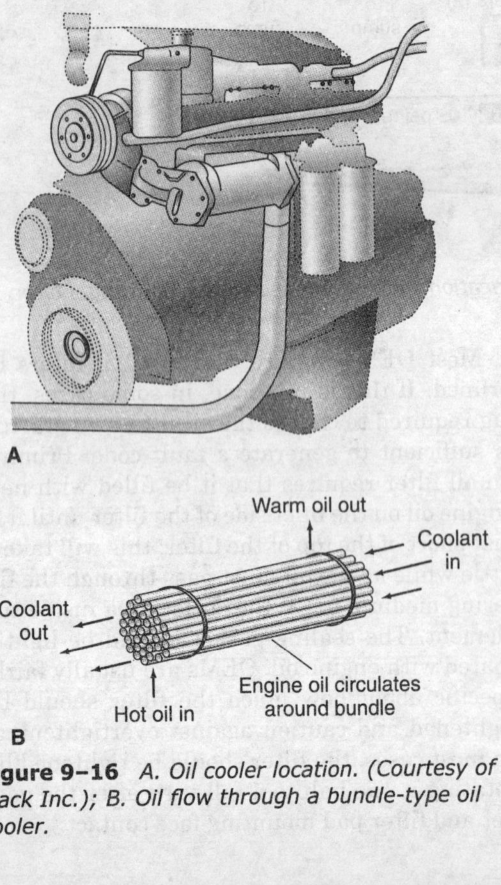

Figure 9–16 *A. Oil cooler location. (Courtesy of Mack Inc.); B. Oil flow through a bundle-type oil cooler.*

The consequence of a failed cooler bundle or the header O-rings is oil charged to the cooling circuit. Most OEMs prefer that bundles be leak tested by vacuum because this tests the assembly in the direction of fluid flow in the event of a leak. One header should be capped with a dummy plate and the other fitted with the evacuation adaptor: Load to the OEM-recommended vacuum value and leave for the required amount of time to observe any dropoff.

Alternatively, the bundle can be pressure tested using regulated shop air pressure and a bucket of water. This is known as reverse-flow testing. Whenever oil has leaked into the coolant circuit, the engine cooling system must be flushed with an approved detergent and water with the engine run at its operating temperature for at least 15 minutes. Figure 9–17 is an exploded view of a bundle-type oil cooler.

In the plate-type oil cooler, the oil circulates within a series of flat plates and the coolant flows around them within a housing assembly. They have lower cooling efficiencies than bundle element coolers, but they are easier to clean and repair. Figure 9–18 shows a typical low-maintenance, plate-type oil cooler. Lower cost, lower maintenance, and more compact design mean that plate-type oil coolers are preferred by some engine OEMs.

Oil Cooling Jets

Piston oil cooling jets have always been a critical component of the engine lubrication circuit. They are plumbed into a cylinder block lubrication gallery and spray engine lube onto the underside of each piston. They have always played a critical role in managing piston temperatures but it could be argued that they have become even more critical in today's engines, many of which have no other means of delivering oil to the pistons. Bushingless piston bosses such as those used in Monotherm pistons and conn rods with no delivery rifling have both added to the already important role played by piston cooling jets.

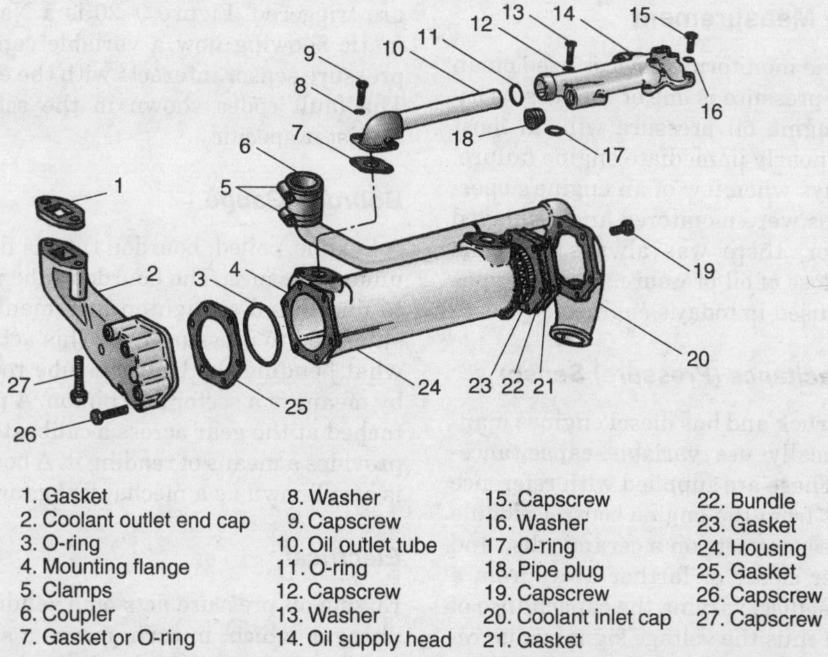

Figure 9–17 *Exploded view of a bundle-type oil cooler. (Courtesy of Mack Inc.)*

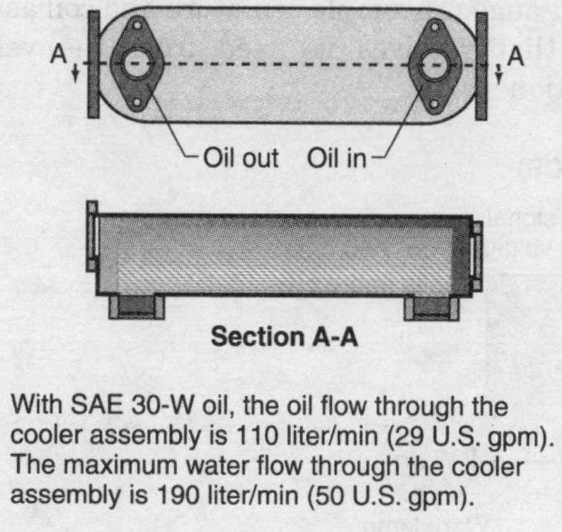

Figure 9–18 *A plate-type oil cooler. (Courtesy of Caterpillar, Inc.)*

Figure 9–19 *Oil cooling nozzle alignment targeting on a MB-4000 engine. (Courtesy of Detroit Diesel Corporation)*

Targeting Piston Cooling Jets

Piston cooling jets usually have to be targeted to a specific area on the underside of the piston. As indicated in Chapter 6, what seems to be a slightly misdirected cooling jet can result in rapid torching of a piston due to overheating. By targeting the cooling jet, oil is delivered to a circulation gallery machined in the underside of the piston: This enables it to route engine lube to those areas of the piston that require lubrication and cooling. The procedure used to target piston cooling jets is covered in Chapter 6. The technician should ensure that the consequences of an improperly targeted cooling jet are understood. Just to make things more confusing, OEMs frequently alter the target window through different engine generations, so it is important to consult OEM service literature before undertaking the procedure. Figure 9–19 shows the target zone on the undercrown of an MB-4000 engine piston: the cooling jet directional target has a way of changing with each successive generation of an engine series so ensure the correct tooling is used.

Oil Pressure Measurement

Of all the engine monitoring devices used on an engine, the oil pressure is one of the most critical. Loss of engine oil pressure will, in most cases, cause a nearly immediate engine failure. Back in the days when few of an engine's operating conditions were monitored and displayed to the operator, there was always a means of signaling a loss of oil pressure. Several types of sensors are used in today's engines.

Variable Capacitance (Pressure) Sensor

Most current truck and bus diesel engines managed electronically use variable capacitance-type sensors. These are supplied with reference voltage output from the engine control module (ECM). Oil pressure acts on a ceramic disc and moves it either closer or farther away from a stationary steel disc, varying the capacitance of the device and thus the voltage signal value returned to the ECM. The ECM is responsible for outputting the signal that activates the dash display or gauge. It should be noted that engine oil pressure usually has to fall to dangerously low levels before programmed failure strategies are triggered. Figure 9–20 is a Navistar schematic showing how a variable capacitance oil pressure sensor interacts with the engine ECM: The fault codes shown in the schematic are Navistar specific.

Bourdon Gauge

A flexible, coiled, bourdon tube is filled with oil under pressure. The bourdon tube will attempt to uncoil and straighten incrementally as it is subjected to pressure rise. This action of somewhat bending the bourdon tube rotates a gear by means of a sector and pinion. A pointer is attached at the gear across a calibrated scale and provides a means of reading it. A bourdon gauge is also known as a mechanical gauge.

Electrical

Engine oil pressure acts on a sending unit diaphragm, which, in turn, moves a sliding wiper arm across a variable resistor that incrementally grounds out a feed from the electric gauge. The gauge is a simple armature and coil assembly that receives its feed from the vehicle ignition switch.

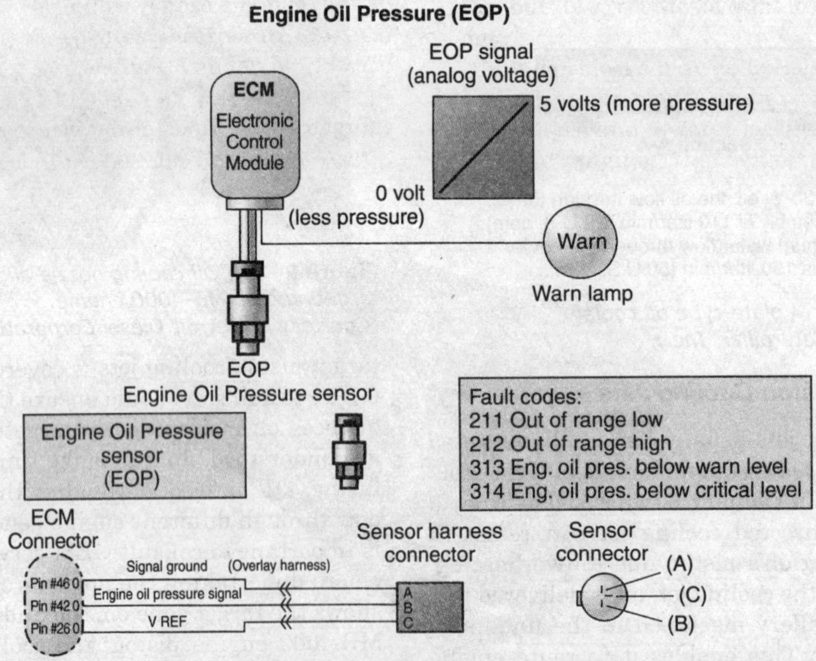

Figure 9–20 *A Navistar oil pressure sensing circuit. (Courtesy of Navistar International Corp.)*

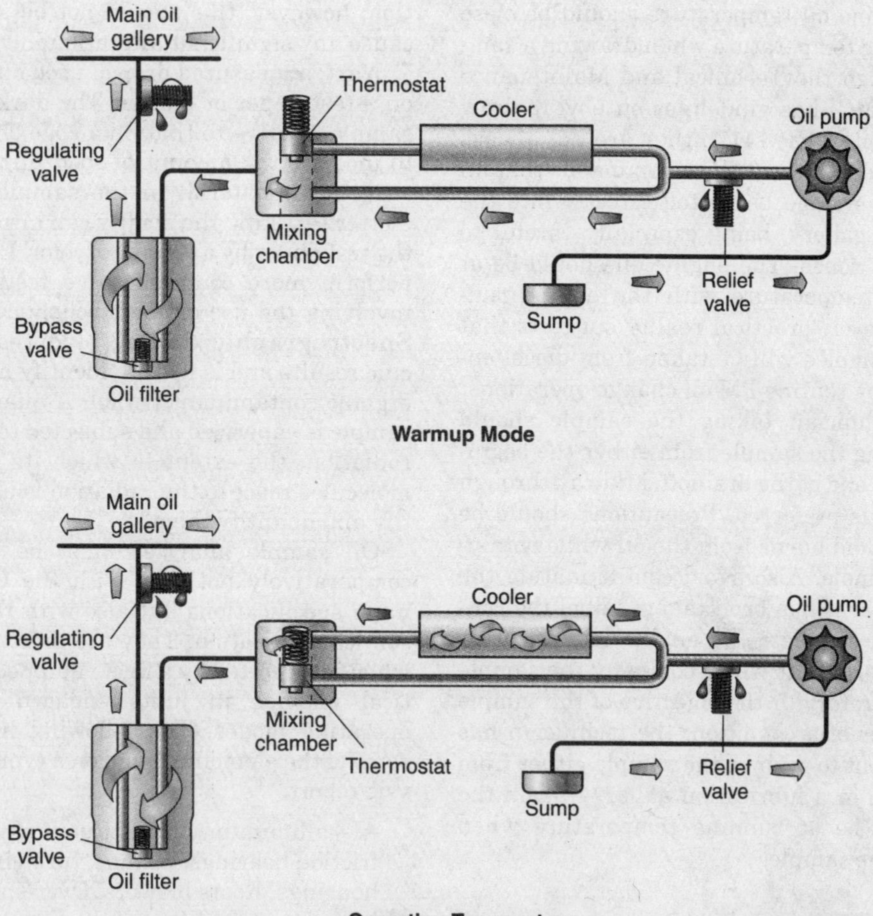

Figure 9–21 *Navistar temperature control circuit. (Courtesy of Navistar International Corp.)*

Oil Temperature Management

As Environmental Protection Agency (EPA) noxious emissions standards become increasingly tougher, there is an ever greater requirement to manage the combustion temperatures to a tight operating window. Engine oil temperatures are one of the most accurate indicators of actual engine temperature. It should also be remembered that engine oil plays a major role as a coolant in engines, especially in managing piston temperatures. In addition, when engine oil is used as a hydraulic medium as it is in variable valve operation, engine brakes, and the hydraulically actuated electronic unit injector (HEUI) fuel system, its performance is to some extent dependent on its operating temperature. On the other end of the scale, the time required to heat engine oil to operating temperature is a factor in the higher emissions output during a warm-up phase. Figure 9–21 shows a Navistar temperature control circuit that enables the engine oil to bypass the oil cooler while the engine is warming up.

INTERPRETING OIL ANALYSES

Oil analysis has become a key component of good preventative maintenance (PM) practice, primarily because it has the potential to save operators money. Performing the analyses at each PM and responding appropriately to the lab reports is key to making them cost-effective. An oil analysis is performed to determine:

- The viscosity
- The presence of coolant
- Dirt contamination
- Abnormal wear contamination

The engine oil temperature should be close to operating temperature when drawing a sample. Although the Technical and Maintenance Council (TMC) has guidelines on how to draw off oil samples (RP 1417), they are not always easy to observe. The TMC recommends that an oil sample syringe be ported directly into the main lube gallery, being especially careful to avoid dead zones: The engine oil should be at operating temperature with the engine running. However, practical reality suggests that most oil samples will be taken from diesel engines during routine PM oil change operations.

The technician taking the sample should avoid taking the sample from either the beginning or the end of the drainoff: Midway through the runoff is preferred. Precautions should be taken to avoid burns from the oil while obtaining the sample. Also avoid contaminating the sample after it has been taken. Label the container and complete the engine identification and data form that will accompany the sample to the laboratory. If the objective of the sample is diagnosis of a condition, the technician has no choice but to syringe the sample either from the oil pan or a lubrication gallery—again the oil should be at running temperature when drawing the sample.

Types of Testing

When used engine oil is analyzed, the objective is to produce a report card on the rate of engine wear, the suitability of the oil change interval to the application, and perhaps head off imminent engine problems before they become major problems. Oil-testing laboratories use primarily blotter and spectrographic testing to produce reports that reference the engine OEM limit specifications. A blotter test kit is available for nonlaboratory testing of engine oil. A **blotter test** is a relatively crude means of testing some of an oil's characteristics. A change in the oil's viscosity can be determined using a flow comparator. A significant increase in the engine oil's viscosity can be caused by prolonged high-temperature operation, aerated lubricant, or coolant contamination. A decrease in viscosity is normally attributable to fuel contamination. It should be noted that all truck diesel engine oil is affected to some extent by fuel contamination; however, this should not be sufficient to cause any significant alteration in viscosity.

Next, a measured drop of used oil is dropped on a test paper or blotter. The darkness of the sample relative to that on a code chart is used to indicate the amount of soot, dirt, and other suspended material in the sample. On some blotter test kits, the acidity can also be read on the test paper by a change of color. Laboratories perform more comprehensive testing usually involving the use of spectrochemical analyses. **Spectrographic testing** produces more specific results and is used to identify metallic and organic contaminants in oil. A quantity of the sample is vaporized and subjected to ultraviolet radiation; the extent to which its constituent molecules react to the radiation is used to identify and quantify them.

Oil sample analyses must be interpreted comparatively not only with the OEM maximum specifications but also with the vehicle's service application. They are most meaningful when interpreted by a fleet equipped with identical engines in units engaged in similar operating modes. The following list includes some of the elements found on a typical oil analysis report.

Al—aluminum. A constituent of many friction bearings, pistons, turbo impeller housings, Roots blowers. Over-spec Al can indicate wear in any of these components.

B—boron. An engine coolant SCA (supplemental cooling additive) and an engine oil additive.

Cr—chromium. A common piston ring coating.

Cu—copper. A constituent of many oil cooler bundles, friction bearings, engine bushings, and thrust washers. Copper levels tend to be higher when an engine is new (100 ppm) and should drop down to lower levels (20 ppm) after 100,000 miles (160,000 km).

Fe—iron. Most of the engine components are iron-based alloys. High Fe readings could be an indication of rapid wear on almost any engine component exposed to the lube oil; engine cylinder and valvetrain components should be suspected first.

Pb—lead. A constituent of most friction bearings. High Pb suggests high load engine operation.

Si—silicon. Silicon dioxide from airborne fine sands and dusts ingested by the engine intake system often indicate malfunctioning air filter or intake ducting leaks. Silicon is a constituent of new engine oil in which it serves as an antifoam agent, so the silicon level reported by the laboratory is corrected to read only externally acquired silicon.

Na—sodium. Found in some lubricants as an additive, so the laboratory reading must be corrected to read acquired sodium. When the source is other than the engine oil, the culprit is usually coolant SCAs.

Sn—tin. A constituent of most friction bearings and occasionally used to coat pistons.

Soot—also known as total solids, soot is a form of fine carbon particle, a combustion by-product that is not readily identifiable in routine lab testing; in fact, some laboratories report total suspended particles as soot level. Fuel soot is present in all engine oil. It is nonabrasive but may increase the oil's viscosity.

LUBRICATING CIRCUIT PROBLEMS

The following are some guidelines for troubleshooting lubrication circuit complaints. Be aware of the sequencing. You would be amazed at how many lubrication circuit problems are root sourced by insufficient oil in the sump.

- When investigating a low oil pressure complaint, first check the oil sump level. Next, check the appearance and odor of the oil.
- Low oil pressure complaints must be verified with a master gauge.
- High sump levels caused simply by excessive oil can aerate the lubricant, causing low pressure and fluctuations or surging.
- After an engine rebuild, it is good practice to pressure prime the engine lubrication system with an external pump.
- Most OEMs recommend priming full flow filters on all turbocharged engines. Oil filters should be primed by filling with new engine oil poured into the inlet side of the oil filter. One OEM (Caterpillar) prefers that oil filters not be primed before installation, due to the risk of dirt contamination caused by careless handling of the filter during priming and improper priming technique. It is not necessary to prime bypass filters, although it may be necessary to purge air from them as the engine is started up after an oil change.
- When taking an oil sample, draw off midway through the sump runoff, never at the beginning or end. If an oil sample is required outside of the normal oil change interval, use a syringe. Oil samples should be taken shortly after an engine has been run to obtain the most reliable results.
- Oil change intervals are usually determined primarily by mileage in linehaul applications in conjunction with OEM recommendations. In fact, mileage rather than engine operating hours is used as the indicator of service intervals in most highway vehicles.

As with *all* troubleshooting on electronic engines, make sure to connect to the chassis data bus: Today this is just part of sound shop practice.

Want to know more? Try these Websites:

http:///www.api.org American Petroleum Institute

http://www.dieselnet.com EPA diesel emissions news

http://www.lubrizol.com Lubrizol home page

Extended Oil Changes

Servicing trucks costs money. When oil change intervals can be extended, it saves money. Engine OEMs, in conjunction with those who research their oil requirements, are continually working on developing engine oils that will sustain longer service life before degradation because their customers are asking for this feature. And until the almost universal adoption of EGR on diesel engines in 2004 they were doing a great job. Rerouting soot particles back into the engine with the intake charge in an EGR mixing chamber has a way of shortening the life of engine oils. The API CI-4 and CJ-4 engine lube categories are formulated to buffer the effect of minute soot particles, but OEM-recommended oil service intervals have not been raised much in recent years. It can be expected that diesel engine OEMs will work toward increasing oil change intervals over the next few years, but nothing dramatic is likely to happen in the near future. The clean gas induction (CGI) system used by Caterpillar ACERT engines produces cleaner combustion, and the key may be in using this technology.

REVIEW QUESTIONS

1. Which of the following SAE multigrade oils would likely be recommended by most highway diesel engine OEMs for North American summertime conditions?
   a. 5W-30
   b. 15W-40
   c. 20W-20
   d. 20W-50
2. Which of the following API classifications would indicate that the oil was formulated for diesel engines meeting 2007 emissions standards?
   a. GF
   b. CC
   c. CI-4
   d. CJ-4
3. Which of the following conditions could result from a high crankcase oil level?
   a. Lube oil aeration
   b. Oil pressure gauge fluctuations
   c. Friction bearing damage
   d. All of the above
4. When investigating a low oil pressure complaint, which of the following should be performed first?
   a. Perform a rod and main bearing rollover.
   b. Install a master gauge to verify the complaints.
   c. Test the oil for fuel contamination.
   d. Check the crankcase oil level.
5. A full flow oil filter has become completely plugged. What is the likely outcome if the engine is running?
   a. A bypass valve diverts the oil around the filter.
   b. Engine lubrication ceases.
   c. The oil pump hydraulically locks.
   d. The engine seizes.
6. Which of the following statements would usually correctly describe the relative temperatures of the coolant and lubricant when the engine is at operating temperature?
   a. Coolant temperatures are higher than engine oil temperatures.
   b. Engine oil temperatures are higher than coolant temperatures.
   c. Coolant and engine oil temperatures should be equal.
7. Which type of oil pump is most commonly used in current highway diesel engines?
   a. External gear
   b. Plunger
   c. Centrifugal
   d. Vane
8. Which of the following is usually the preferred OEM method of testing an oil cooler bundle?
   a. Pressure test using shop air and a bucket of water
   b. Vacuum test
   c. Die penetrant test
9. When interpreting a used engine oil analysis profile, which of the following conditions would be most likely to cause high silicon levels?
   a. Oil cooler bundle disintegration
   b. Plugged air cleaner
   c. Turbocharger bearing failure
   d. Air cleaner perforation
10. Which of the following correctly identifies the source of diesel engine oil?
    a. Vegetable-based oil
    b. Crude petroleum
    c. Asphalttars
11. Engine oil service classifications are standardized by which of the following organizations?
    a. SAE
    b. API
    c. ASTM
    d. EPA
12. *Technician A* states that the operating temperature of an engine oil should be within 5°F of that of the engine coolant operating temperature. *Technician B* states that engine oil temperatures are probably a more accurate means of assessing actual engine operating temperature than the coolant temperature reading. Who is correct?
    a. Technician A only
    b. Technician B only
    c. Both A and B
    d. Neither A nor B

13. An auxiliary oil pump designed to feed oil to the lubrication circuit on a vehicle that operates at extreme working angles is known as a(n):
    a. Gerotor pump
    b. Scavenging pump
    c. Vane pump
    d. Emergency pump
14. Oil that has a milky, clouded appearance is probably contaminated with:
    a. Fuel
    b. Engine coolant
    c. Dust
    d. Air
15. Which of the following statements correctly describes viscosity?
    a. Resistance to flow
    b. Lubricity
    c. Resistance to shear
    d. Breakdown resistance
16. *Technician A* states that an oil cooler operates by having coolant pumped through the bundle tubes and oil circulated around the tubes. *Technician B* states that a leaking oil cooler core must always be replaced. Who is correct?
    a. Technician A only
    b. Technician B only
    c. Both A and B
    d. Neither A nor B
17. *Technician A* states that 15W-40 is the most commonly used truck engine oil. *Technician B* states that 15W-40 oil is not an appropriate all-season oil, because winter temperatures are often much lower than the SAE recommended ambient temperature window for this oil. Who is correct?
    a. Technician A only
    b. Technician B only
    c. Both A and B
    d. Neither A nor B
18. *Technician A* states that the device used to signal oil pressure values to the ECM on most electronically managed engines uses a variable capacitance principle. *Technician B* states that an oil pressure sensor usually requires a V-Ref source from the engine ECM. Who is correct?
    a. Technician A only
    b. Technician B only
    c. Both A and B
    d. Neither A nor B
19. *Technician A* states that an engine equipped with EGR and DPF must use API CJ-4 engine oil. *Technician B* states that engines manufactured before 2007 may use API CI-4 engine oil. Who is correct?
    a. Technician A only
    b. Technician B only
    c. Both A and B
    d. Neither A nor B
20. A diesel engine dash oil pressure gauge displays low and fluctuating oil pressure values when the vehicle is operated. *Technician A* states that this could be caused by low engine oil level. *Technician B* states that this could be caused by high engine oil pressure. Who is correct?
    a. Technician A only
    b. Technician B only
    c. Both A and B
    d. Neither A nor B

10

Engine Cooling Systems

INTRODUCTION

Cooling systems dissipate a percentage of engine **rejected heat**. Rejected heat (see Chapter 5) is that percentage of the potential heat energy of a fuel that the engine is unable to convert into useful **kinetic energy** (the energy of motion). Rejected heat has to be dissipated to atmosphere, either in the exhaust gas or indirectly using the engine cooling system. If an engine is operating at 40% thermal efficiency, 60% of the potential heat energy is rejected. Approximately half of the rejected heat is discharged in the exhaust gas, which leaves the engine cooling system responsible for dissipating the other half to atmosphere.

This task of dissipating rejected heat is complicated by the extremes of the North American climate given that it is necessary to manage a consistent engine-operating temperature at all engine speeds and loads to ensure optimum performance and minimum emissions. Liquid cooling systems are universal in North American truck and bus applications, and only they are addressed in this section. The Deutz engine company of Germany manufactures air-cooled engines in the 200 to 500 BHP power range, but in North America their engines are generally only found in agricultural and mining applications.

Functions of the Cooling System

The functions of a diesel engine liquid cooling system are to:

- Absorb heat from engine components.
- Absorb heat from engine support systems such as exhaust gas recirculation (EGR) and transmission coolers.
- Transfer the absorbed heat by circulating the coolant.
- Supply heat to cab and bunk heaters.
- Dissipate the heat to atmosphere by means of heat exchangers.
- Manage engine-operating temperatures.

The actual coolant temperature at any given moment of operation is one means the ECM uses to determine engine temperature. In fact, in older hydromechanical engines it was the only means used. In today's engines ECMs use both coolant and oil temperature readings to make a determination of *engine* temperature. Note the following:

- Engine coolant warms to operating temperature faster than engine oil.
- Engine oil is regarded as being a better indicator of actual engine temperature.
- When both coolant and lube oil are at operating temperature, engine coolant temperatures are always lower than oil temperatures by a good margin.

Heat Transfer

Combustion heat can be dissipated by the cooling system to atmosphere in three ways:

1. **Conduction**: the transfer of heat through solid matter, such as the transfer of heat through the cast-iron material of a cylinder block.

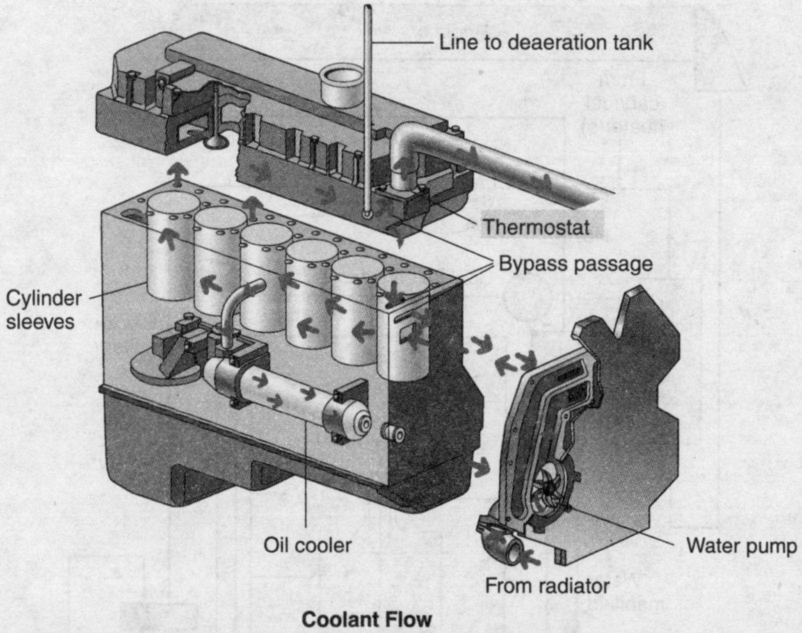

Figure 10–1 *Coolant flow through a Navistar six-cylinder engine with no water manifold: Flow routing is from the front to the back. (Courtesy of Navistar International Corp.)*

2. **Convection**: the transfer of heat by currents of gas or liquids, such as in the movement of ambient air through an engine compartment.
3. **Radiation**: the transfer of heat by means of heat rays not requiring matter, such as a fluid or solid. The turbine housing of a turbocharger radiates a considerable amount of heat.

Antiboil Properties

Cooling systems are sealed and maintained under pressure. By confining a liquid under pressure, its boil point is increased. Most cooling systems are designed to manage coolant temperatures at just below their boil points. The chemistry of the **antifreeze** and its concentration in the coolant will define the actual boil point of a coolant. Most antifreezes double as antiboil agents. When an engine is approaching an overheat condition, the coolant will first boil at the location within the system where the pressure is lowest. In most cases, boiling will occur first at the inlet (suction side) of the system water pump. Figure 10–1 and Figure 10–2 illustrate a couple of diesel engine cooling circuits, the first without and the second with a water manifold: The role of water manifolds is discussed a little later in the chapter.

ENGINE COOLANT

Water-based coolant is the medium used to absorb engine rejected heat, transfer that heat to a **heat exchanger**, and then dissipate it to atmosphere. The coolant is circulated through the engine water jacket to absorb the heat of combustion in the manner shown in Figure 10–3. Engine coolant is a mixture of water, antifreeze, and supplemental coolant additives (SCA). If the only objective of diesel engine coolant was to act as a medium to transfer heat, pure water would accomplish this more efficiently than any currently used antifreeze mixture. However, water possesses inconvenient boil and freeze points, poor lubricating properties, and furthermore promotes oxidation and scaling activity.

Types of Antifreeze

Almost all current truck and bus engines use a mixture of **ethylene glycol (EG)**, **propylene glycol (PG)**, plus water as coolant, or a premix

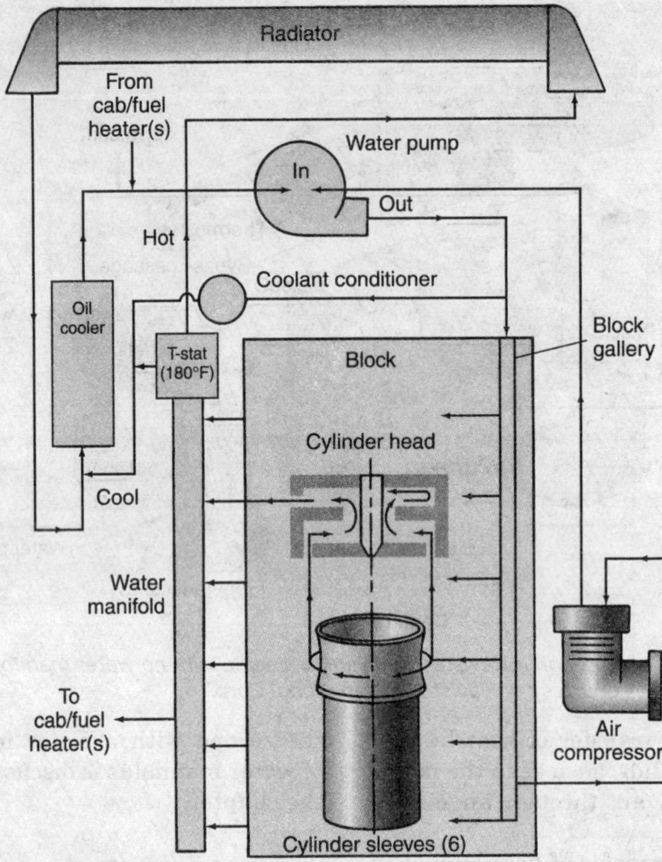

Figure 10–2 *Cooling system flow through an engine using a water manifold that ensures more even cylinder temperatures. (Courtesy of Mack Inc.)*

of carboxylate, EG-based **extended life coolant (ELC)**. Alcohol-based solutions are no longer used as they evaporate at low temperatures. A properly formulated diesel engine coolant is always made up of the correct proportions of water, antifreeze, and SCAs. When the antifreeze is an EG- or PG-based mixture with water, the SCAs require monitoring and routine replenishing. ELC is low maintenance in that the coolant life can be up to 6 years, with only one SCA charge required in that period.

Coolant Expansion and Contraction

Water expands about 9% in volume as it freezes and it can distort or fracture a vessel it is housed even when this container is a cast-iron engine block. Water occupies the least volume when it is in the liquid state and close to its freezing point of 0°C or 32°F. As water is heated from a near freezing point to a near boiling point it expands approximately 3%. A 50/50 mixture of water and EG will expand even more, approximately 4% through the same temperature range. This means that cooling systems must be designed to accommodate the expansion and contraction of the engine coolant while it is in the liquid state. Just as important to its operation is that antifreeze is also antiboil.

Characteristics of Good Antifreeze

The mixture of water, antifreeze, and SCAs that is referred to as engine coolant should perform the following:

1. Corrosion protection. Corrosion inhibitors in both the antifreeze and the SCA package protect the metals, plastics, and rubber compounds in the engine cooling system.

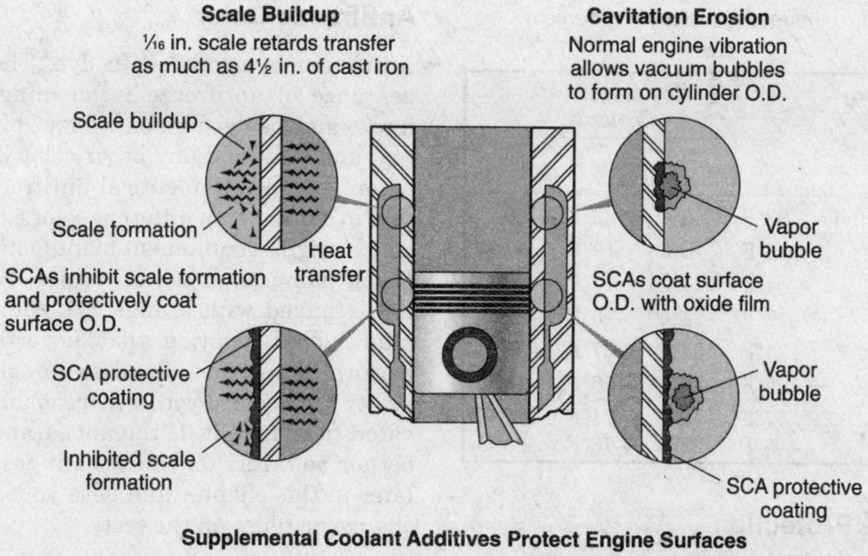

Figure 10–3 *How SCAs help scaling and cavitation problems in wet liners. (Courtesy of Navistar International Corp.)*

2. Freeze protection. The degree of freeze protection of the coolant is directly related to the proportion of antifreeze in the mixture.
3. Antiboil protection. The degree of antiboil protection of the coolant is again directly related to the proportion of antifreeze in the mixture.
4. Antiscale protection. The antifreeze should contain antiscale additives that prevent hard water mineral deposits from adhering to the cooling system heat transfer surfaces.
5. Acidity protection. A pH buffer is used to inhibit the formation of acids in the coolant, which would result in corrosion.
6. Antifoam protection. This prevents aeration of the coolant that could be caused by the action of pumping and flowing it through the cooling circuit.
7. Antidispersant protection. This prevents insoluble matter from coagulating and plugging cooling system passages.

Toxicity of Coolants

Except for the antifreeze and antiboil characteristics of glycol-based coolant mixture, the remaining protection additives deplete with engine operation especially at high temperatures. Because coolant degrades, the protection additives must be evaluated and restored at appropriate maintenance intervals. Both PG and EG are petrochemical products. EG has been used as the standard antifreeze for some time, but the Federal Clean Air Act and Occupational Safety and Health Administration (OSHA) have both come to regard EG as a toxic hazard. PG in its virgin state is said to be less toxic than EG, and for that reason it has gained some acceptance as an antifreeze. However, leaks and spillage of both EG- and PG-based coolants should be regarded as dangerous to mammals (including humans) and plant life. The chemical characteristics of engine coolant are altered in use and the toxicity increases. Although ELC uses an EG base, it is regarded as safer because it requires less handling (it is premixed and does not require periodic testing) in service.

Sources of Glycol

EG is derived from ethylene oxide, which in turn is produced from ethylene—a basic petroleum fraction. PG is derived from propylene oxide, which is produced from propylene—another basic petroleum fraction. When mixed in a solution with water, both PG and EG are described as aqueous, meaning they mix pretty well with it. ELCs are claimed to have lower toxicity but because the antifreeze base is EG, they should be regarded as toxic.

Table 10–1 *Antifreeze Protection by Percent Volume*

Concentration of Antifreeze by Percent Volume	*Freeze Point of Coolant*			
	EG		*PG*	
0% (water only)	32°F	0°C	32°F	0°C
20%	16°F	–0°C	19°F	–7°C
30%	4°F	–16°C	10°F	–12°C
40%	–12°F	–24°C	–6°F	–21°C
50%	–34°F	–37°C	–27°F	–33°C
60%	–62°F	–52°C	–56°F	–49°C
80%	57°F	–49°C	–71°F	–57°C
100%	–5°F	–22°C	–76°F	–60°C

Antifreeze Protection

The temperature at which a glycol-based coolant solution ceases to protect against freezing depends on:

- Whether the antifreeze is EG or PG
- The percentage of antifreeze to water in the solution

Table 10–1 shows how temperature correlates with freeze point in both EG and PG antifreeze solutions. Note that when using EG, increasing the percentage of antifreeze ceases to reduce the freeze point of the coolant once it exceeds 60%. On the other hand, the freeze point of PG is at its maximum at 100% so it is often used as the coolant of choice in subarctic conditions.

Compatibility of EG and PG

PG- and EG-based coolants should *never* be mixed. The mixture in itself will not cause any immediate engine or cooling system problems, but it will be impossible to determine the antifreeze mixture strength with either a refractometer or a **hydrometer**. If a mixture of EG and PG is known to have taken place and the coolant, for whatever reasons, cannot be immediately replaced, use a refractometer with an EG and a PG scale and average the two readings. However, the cooling system should be drained and refilled with either aqueous PG or EG when possible. Only ELC premix should be added to the cooling system, or in conditions of extreme cold, ELC concentrate. ELC is incompatible with PG and EG.

Antifreeze Color

Antifreeze is colored with dye. The color appearance of antifreeze is meaningless. Antifreeze may be dyed green, yellow, blue, orange, red, and pink and any other color in the spectrum. Chemically identical antifreezes can be sold in a number of different colors according to which original equipment manufacturer (OEM) brand name is on the container. When green EG is mixed with orange EG, the result is a mud-colored solution that nevertheless will perform as specified. Because the solution looks pretty gruesome, you will probably be motivated to change it. If the appearance does not bother you, rely on the coolant tests outlined later in this chapter and base your decision to change or reuse on the tests.

Measuring Coolant Mixture Strength

Standard antifreeze hydrometers are calibrated for measuring EG mixtures, and even when measuring these they tend to be inaccurate and require calculation to temperature correct the reading. The Technical and Maintenance Council (TMC) and most OEMs recommend the use of a **refractometer** to test the antifreeze strength of a coolant. A refractometer should have an accuracy variable within 7°F of the actual freeze point of the coolant. A refractometer designed for measuring diesel engine coolant will usually have a PG and EG calibration scale. Ensure that the correct scale is used for the antifreeze being tested. A refractometer measures **refraction** in a liquid: The refractive index of antifreeze increases with an increase in concentration. Refractometer readings are only valid in the context of a specific type of antifreeze.

Supplemental Coolant Additives

Supplemental coolant additives (SCA) are a critical ingredient of the coolant mixture. The actual SCA package recommended by an engine manufacturer will depend on whether wet or dry cylinder liners are used, the materials used in the cooling system components, and the fluid dynamics (high flow/low flow) within the cooling system. Operators may wish to adjust the SCA package to suit a specific operating environment or set of conditions. Abnormally hard water, for instance, requires a greater degree of antiscale protection.

Depending on the manufacturer, SCA may be added to a cooling system in a number of ways. The practice of installing the SCA in the system coolant filter is less commonly used today because it resulted in generally higher levels of additives than required, and an excess of additives can create problems. Most OEMs suggest testing the SCA levels in the coolant then adding SCA such as **Nalcool** or **diesel coolant additive (DCA)** to adjust to the required values. Never dump unmeasured quantities of SCAs into the cooling system at each preventative maintenance (PM).

Cavitation

SCAs are especially important to controlling cavitation on engines that use wet liners. **Cavitation** is caused by vapor bubble collapse, which results in pitting of the exterior of the liner wall across the thrust faces of the piston. When the liner is subjected to the gas pressure of combustion, it first expands then contracts, forming a low-pressure vapor bubble. The vapor bubble almost immediately collapses (implodes), resulting in high-velocity coolant impacting on the liner wall. This has been proved to produce surface pressures up to 60,000 psi, which can cause pitting and erosion. The characteristic will normally appear on the exterior of the liner wall across the thrust faces of the piston.

Liner pitting can be controlled by adding molybdates and nitrite to coolant, which act to form an invisible but very tough, protective oxide film on the liner wall. The vapor bubbles will still form, but the liner is shielded by the oxide film. The bad news is that this film breaks down as the coolant ages.

Scaling

The Cummins Engine Company reported a number of years ago that a scale buildup of 1/16″ (1.5 mm) had the equivalent insulating effect of 4″ (100 mm) of cast iron. Scaling is caused by hard water mineral deposits (especially magnesium and calcium) adhering to the surfaces of the cooling system where temperatures are highest. Scale formation on wet liners is an especially serious condition because it seldom forms in an even layer and acts as an insulator. This limits the liner's ability to transfer heat. The resulting high temperatures may cause buckling and other distortions to the liner. Left unchecked, scale buildup insulates engine components designed to transfer heat, resulting in overheating and subsequent failure as shown in Figure 10–3.

Descaling Engines There are commercially available descalants that may work to remove minor scale buildup followed by a cooling system flush, but most often by the time scale buildup has progressed to the point that it is causing an engine to overheat, the engine must be disassembled and the cylinder block and heads boiled in a soak tank.

Testing SCA Levels

Generally OEMs recommend that the coolant SCA level be tested at each oil change interval. Additionally, whenever there is a substantial loss of coolant and the system has to be replenished, the SCA level should be tested. The test kits usually consist of test strips that must be stored in air-tight containers and that have expiration dates that should be observed. Standard coolant test kits permit the technician to test for the appropriate SCA concentration.

Test Procedure

The coolant sample should be taken from the radiator, not from the recovery or overflow system. The coolant temperature should be between 50°F and 130°F during testing, and the test should be completed within 75 seconds.

1. Remove a strip from the package or bottle without touching the pads. Discard if the nitrite test pad has turned brown.
2. Dip the strip into the coolant for 1 second; remove and shake to remove excess coolant.
3. Wait 45 seconds then match the colors of the test pads to the colors on the chart.

End pad A Tests freezepoint. Low accuracy about 10%.

Middle pad B Tests molybdates. Buffers the formation of acids.

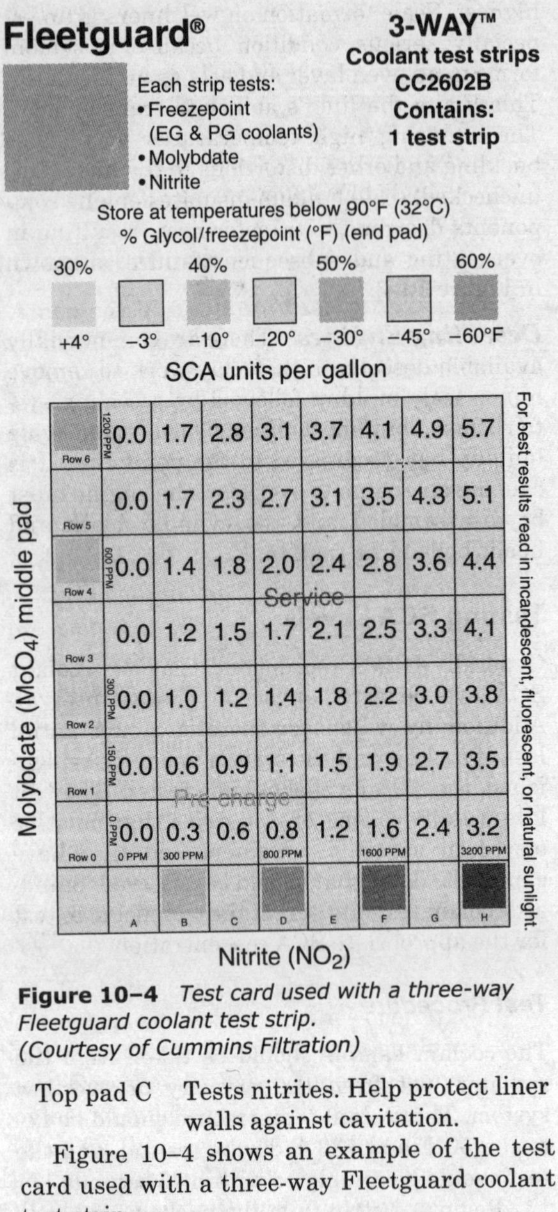

Figure 10–4 *Test card used with a three-way Fleetguard coolant test strip. (Courtesy of Cummins Filtration)*

Top pad C Tests nitrites. Help protect liner walls against cavitation.

Figure 10–4 shows an example of the test card used with a three-way Fleetguard coolant test strip.

Testing Coolant pH

The **pH** level defines the relative acidity or alkalinity of the coolant. Acids may form in engine coolant exposed to combustion gases or, in some cases, when cooling system metals (ferrous- and copper based) degrade. The pH test is a litmus test in which a test strip is first inserted into a sample of the coolant then removed, and the color of the test strip is indexed to a color chart provided with the kit. The optimum pH window is defined by each OEM, but normally falls within 7.5 and 11.0 on the pH scale. A pH of 9.5 is typical. Higher acidity readings (below 8.0 on the pH scale) in tested coolant are indications of corrosion of ferrous and copper metals, coolant exposure to combustion gases, and, in some cases, coolant degradation. Higher-than-normal alkalinity readings indicate aluminum corrosion and possibly that a low silicate antifreeze is being used where a high silicate antifreeze is required.

Testing TDS

Testing for **total dissolved solids (TDS)** requires using a TDS probe, which essentially measures the conductivity of the coolant by conducting a current between two electrodes. Distilled water does *not* conduct electricity. The ability of water to conduct electricity increases with its TDS content. The TDS test is performed by inserting the probe into the top radiator tank. A reading higher than the OEM-specified TDS measured in parts per million (ppm) is an indication the condition of the coolant may be conducive to scale buildup.

Blending Heavy-duty Coolant

Ideally, coolant for use in heavy-duty diesel engines should not be blended in the engine cooling system, but in a container before pouring or pumping it into the engine. Good quality water (known to be not excessively hard and to have no iron content; it should not appear reddish) should always be poured into the container first, followed by the correct proportions of antifreeze and, finally, the SCA package. Mix the solution before adding to the engine cooling system. When adding coolant to compensate for a low coolant level, some OEMs manufacture premixed, heavy-duty coolant. This shifts the responsibility of mixing coolant from the technician because the water, antifreeze, and SCA are premixed in the correct proportions; it also eliminates the problems associated with poor quality water.

Good Quality Water

One of the reasons premixed ELC significantly outlasts EG and PG solutions is that distilled water is used in the solution. There are so many variables associated with tap water that it does not make sense to use anything but distilled water when mixing antifreeze solution. Distilled water is not costly, just not as easily available on the shop floor as tap water. Fleetguard defines good quality water with the following specifications:

- Less that 40 ppm chloride
- Less then 100 ppm sulfate
- Less than 170 ppm calcium/magnesium (hardness)

High Silicate Antifreeze

High silicate concentrations are used in less costly antifreezes to protect aluminum components exposed to the engine coolant. However, most diesel OEMs require that low or no silicate coolant be used in their engines. High silicate and low silicate antifreeze should not be mixed. Premixed ELCs do not use silicates, nitrates, borates, phosphates, and amines to inhibit scaling, but instead use a carboxylate additive.

Extended Life Coolants

Premixed extended life coolants (ELC) use an EG base and promise a service life of up to 600,000 miles (960,000 km) or 6 years with one additive recharge at 300,000 miles (480,000 km) or 3 years. This compares with a typical EG service life of 2 years during which up to 20 recharges of SCA would be required for conventional EGs and PGs. Most ELCs are only available as premixed solution to ensure that the water quality (contaminant level) is at specification. The pricing of premixed ELCs by quantity is generally comparable with EG and PG, but because of reduced cooling system maintenance and extended service life, is fast becoming the coolant of choice of the engine OEMs. No test kits are required to monitor the ELC SCA level. Notable among the advantages claimed for ELCs are:

- Greatly extended service life—6 years or 600,000 miles
- No inhibitor testing required
- Improved water pump life due to much lower TDS content (TDS are often abrasive)
- Reduced hard water scaling
- Improved cavitation protection
- Improved corrosion protection
- Improved heat transfer ability
- No gelling problems: No silicates are used in premixed ELC, which eliminates the problem of silicate dropout responsible for sludging EG and PG
- Improved aluminum corrosion inhibitors
- Better high-temperature performance than EGs and PGs

ELCs may be used in engine cooling systems that have previously used either EG or PG solutions. The EG and PG should be drained from the cooling system, which should then be flushed with clean water. The ELC can then be installed. If a coolant filter is used, replace the existing filter with an SCA-free filter. For midwinter Northern United States and Canadian operation, the ELC premix should be strengthened with concentrate. Check with the ELC manufacturer. Most engine OEMs endorse the use of ELCs in their products.

COOLING SYSTEM COMPONENTS

The components used to store, pump, condition, and manage engine coolant flow and temperature are known as the cooling system components (Figure 10–5). These components vary little from one diesel engine manufacturer to the next but the way in which coolant is routed through these components tends to be specific to each OEM. In Figure 10–1 and Figure 10–2 two quite different diesel engine cooling circuits are shown. For instance, note the location of the oil cooler in each figure. Oil coolers, while integrated into the cooling circuit, were studied in this book as part of the lubrication system in Chapter 9. When servicing and repairing cooling system components, always consult the appropriate service literature.

Radiators

Radiators are heat exchangers. The engine power rating usually dictates the frontal area of a radiator, typically 3″ to 4″ (squared) per BHP unit. The cooling medium is **ram air**, that is, ambient air forced through the radiator core as the truck is driven down the highway. Vehicle

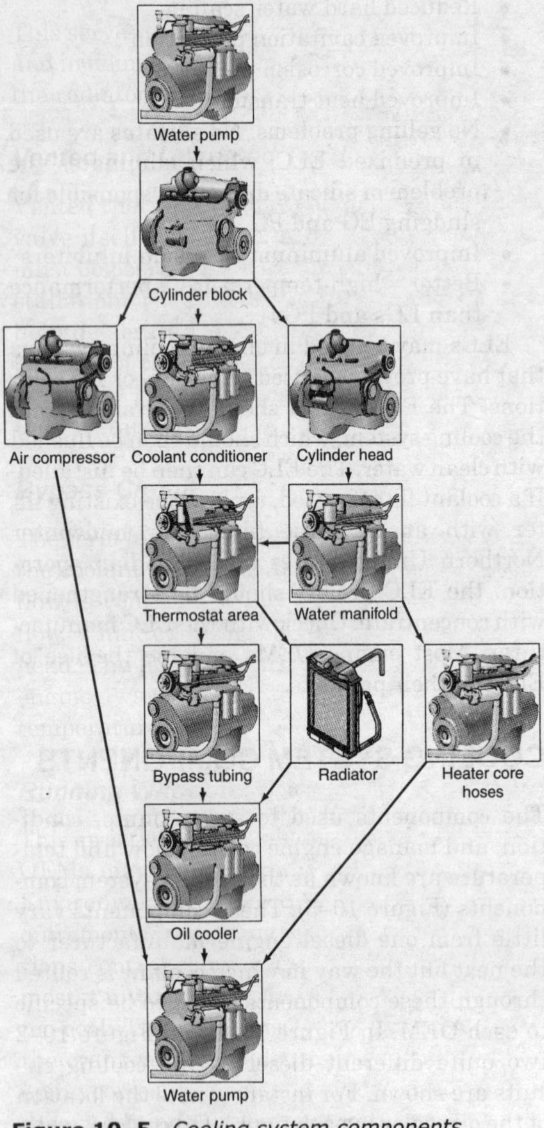

Figure 10–5 *Cooling system components. (Courtesy of Mack Inc.)*

speed and ambient temperatures obviously determine the radiator's efficiency as a heat exchanger. Fan shrouds improve airflow through the radiator core and the efficiency of the fan.

Radiator Materials and Construction

Most truck diesel engine radiators are currently fabricated mainly from copper and brass components, but the use of aluminum and plastics is increasing. Radiators typically consist of bundled rows of round or elliptical tubes through which the coolant is flowed and to which are connected fins that increase the sectional area to which the ram air is exposed. The tubes are usually brass and the fins are copper in truck applications, but OEMs are experimenting with aluminum, which is widely used in automobile radiator construction, due to its lower cost and lighter weight. Copper, brass (an alloy of copper and zinc), and aluminum all have high coefficients of heat transfer and are ideal as base material for radiators. Aluminum is more susceptible to corrosion than copper and brass, both from within (coolant breakdown, poor water quality) and outside (salt, both ambient and road salt). Plastics are being increasingly used in the construction of radiator tanks, replacing metal tanks; plastic tanks are usually crimped onto the main radiator core, enclosing the **headers**.

All radiators are equipped with a drain valve located at the lowest point in the assembly, an inlet and outlet piping, and a filler opening sealed with a radiator cap. Most use a **single pass**, downflow principle, which requires that the cooling tubes run vertically from the top tank to the bottom tank. Radiators are classified by their flow characteristics. The following types may be found in current truck applications:

- **Downflow radiators**: Coolant enters the radiator through the top tank and flows, aided by gravity, to the bottom tank by means of vertical tubes that connect the upper and lower tanks. Downflow radiators are a single pass design, meaning that the coolant is routed from the top tank to the bottom and then exits. The location of a typical downflow radiator is shown in Figure 10–6 and the coolant routing is illustrated in Figure 10–7.
- **Crossflow radiators**: The headers and tanks are positioned on either side of the radiator core. The coolant enters through one of the side tanks and then flows through horizontal tubes to the tank at its opposite side. This design affords a lower profile than the downflow design, and it is used by chassis OEMs using low-nose, aerodynamic designs of trucks. Flow is single pass.
- **Counterflow radiators**: The coolant usually enters through a bottom tank that

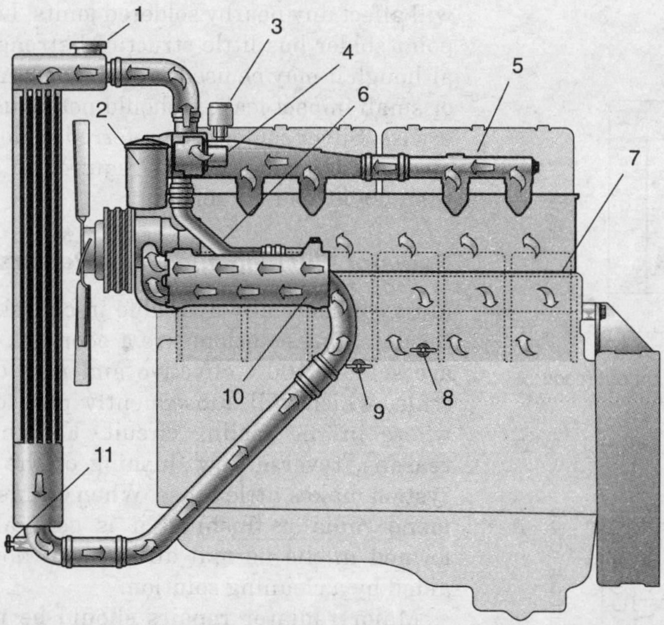

Figure 10–6 *Cooling system flow with a downflow radiator. (Courtesy of Mack Inc.)*

is divided into inlet and outlet sections. The coolant flows vertically upward from the inlet section of the bottom tank to the top tank, then downward to the outlet section of the bottom tank before being returned to the engine cooling circuit. Essentially, the cooling efficiencies of this design are improved mainly because the coolant is retained in the radiator for a longer period. The flow of coolant through the radiator is **double pass**. The design was popular when liquid-cooled, charge-air heat exchangers were common.

Aeration Avoidance

Air in a cooling system can cause many problems. It compromises the coolant's ability to transfer heat, it may promote corrosion, and in extreme instances it can cause a cooling system to become air bound. An air-bound system occurs when air becomes entrapped in the inlet to the coolant pump, and it effectively loses its prime. As a consequence, most cooling systems are designed to limit aeration problems. Some radiators have a divided top tank in which coolant enters the lower section from the cylinder head and the upper section from a standpipe; the two sections are separated by a baffle. This design tends to reduce cooling system aeration problems. Also, vent lines help to de-aerate the cooling system. The radiator top tank stores reserve coolant volume and accommodates the thermal expansion of coolant.

Servicing and Testing Radiators

Radiators in highway trucks tend to be left in the chassis with no maintenance until they fail, either because they leak or fail to adequately cool. An often overlooked PM practice is the external cleanliness of the radiator; buildup of road dirt and summer bugs can severely compromise the radiator's ability to effectively cool. Radiators should be cleaned externally using a low-pressure steamer or regular hose, detergent, and a soft nylon bristle brush. Never use a high-pressure washer because this will almost certainly result in damaging the cooling fins.

Repairing Leaks

Leaks are more often the result of external damage than corrosion failure: They are indicated by white or reddish streaks at the location of the leak. Radiators are commonly pressure tested at around 10% above normal operating pressure, but ensure that the OEM

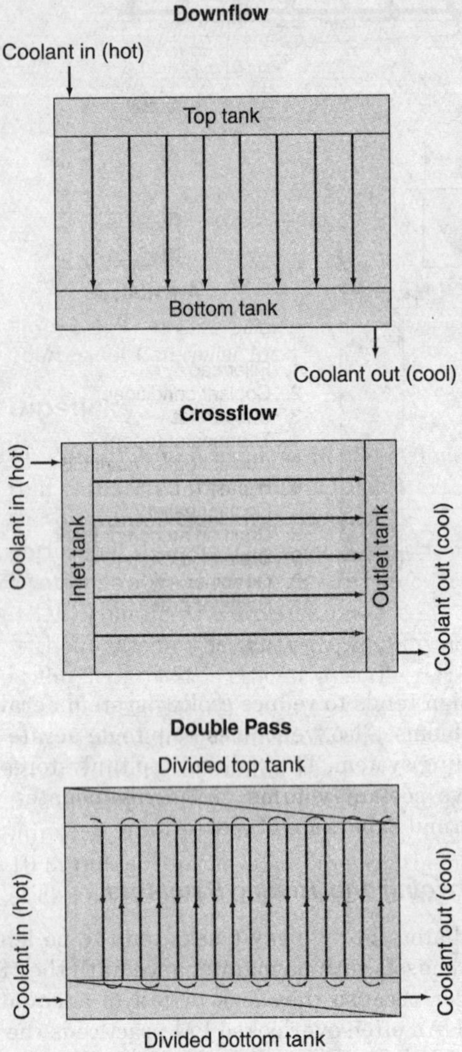

Figure 10–7 *Coolant flow shown on downflow, crossflow, and double pass radiators.*

test specifications are consulted, especially where plastic tanks are used. Pressure testing may help identify the locations of leaks. It is important that radiator leaks be promptly repaired. If the leak has been caused by external damage and the radiator is in otherwise good condition, the radiator may be repaired by shorting out the affected tubes. This usually involves removing the top and bottom tanks and plugging the damaged tube(s) at the headers. If a leaking tube is accessible, a soldering repair may be possible.

Soldering radiators is at best a risky business, so before beginning assess how the heat will affect any nearby soldered joints. Low melt point solder has little structural strength, and although it may be used to seal a hairline crack or small impact leak, it should not be used otherwise. Silver solder is a preferred solder repair medium, but more heat is required to apply it than 50/50 lead-tin solder.

Radiator Flushing and Major Repairs

Most commercially available in-chassis radiator descaling solutions are a poor risk as they are seldom 100% effective and may dislodge scale, which will subsequently plug up elsewhere in the cooling circuit. For the same reasons, reverse flow flushing of the cooling system makes little sense. When OEMs recommend radiator flushing, it is generally performed in the normal direction of flow often aided by a cleaning solution.

Major radiator repairs should be referred to a radiator specialty shop. A scaled radiator falls into the classification of major repair. An ultrasonic bath will remove most scale rapidly and effectively. A properly equipped radiator repair shop will also be able to determine the extent of repairs required and whether recoring is necessary. One of the problems of performing radiator repairs without the proper test equipment is the inability to test the radiator until it is assembled and reinstalled in the chassis. Removing and installing radiators can be a labor-intensive operation in some chassis.

Auxiliary Heat Exchangers

The engine cooling system may be required to provide the cooling media for other heat exchangers plumbed into the cooling system. Other heat exchangers that can be plumbed into the cooling circuit are:

- Aftercoolers and intercoolers
- Tip turbine heat exchangers
- EGR heat exchangers
- Oil coolers
- Clean gas induction (CGI) heat exchangers
- Cab and bunk heaters
- Transmission oil coolers

We will address these heat exchangers in the chapters that deal with the specific systems they address; for instance, oil coolers were discussed in Chapter 9 dealing with engine lubrication systems.

Radiator Cap

Radiators are usually equipped with a pressure cap whose function it is to maintain a fixed operating pressure while the engine is running. This cap is additionally equipped with a vacuum valve to admit surge tank coolant (or air) into the cooling circuit (the upper radiator tank) when the engine is shut down to accommodate coolant thermal contraction. Radiator caps permit pressurization of a sealed cooling system. For each 1 psi (7 kPa) above atmospheric pressure, coolant boil point is raised by 3°F (1.67°C) at sea level; for every 1,000 feet of elevation, the boil point decreases by 1.25°F (0.5°C). System pressures are seldom designed to exceed 25 psi (172 kPa) and, more typically, they range between 7 psi (50 kPa) and 15 psi (100 kPa).

Radiator caps are identified by the pressure required to overcome the cap spring pressure and unseat the seal: When this occurs, the coolant is routed to a surge tank. The surge tank coolant level is always at its highest when the engine is running hottest. As the engine cools, the pressure within the cooling system drops, and when it falls to a "vacuum" value of approximately ¼ psi, the radiator cap vacuum valve is unseated, which allows coolant from the surge tank to be pulled back into the radiator. Table 10–2 shows how much coolant boil point is raised correlated with rad cap trip pressure.

Testing Radiator Caps

Radiator caps can be performance tested using a standard cooling system pressure testing kit. The radiator cap should first be installed to an appropriately sized adapter on the hand pump, then pump to the seal crack value (this should exceed the cap rated value by 1 psi [7 kPa]). Next, release the pressure and once again recharge to the exact rated pressure value of the cap and observe the pump gauge: Pressure drop-off should not exceed 2 psi over 60 seconds.

Water Manifold

Most diesel engines are equipped with a water manifold. OEMs opt to use a water manifold because it ensures an even distribution of coolant through the engine block, and therefore more consistent cylinder temperatures. For instance, if you study the coolant routing shown in Figure 10–1 you will note that coolant flows from the front to the rear of the engine. This results in higher temperatures at the rear of the engine. In the coolant system with a water manifold illustrated in Figure 10–2, note how coolant is routed through the cylinder block to cool the midstop cylinder liners: This ensures more even cylinder block temperatures.

Water Pumps/Coolant Pumps

Water pumps are usually nonpositive, centrifugal pumps driven directly by a gear or by belts. When the engine rotates the coolant pump, an impeller is driven within the housing, creating low pressure at its inlet, which is usually located at or close to the center of the impeller. The impeller vanes throw the coolant outward and centrifugal force accelerates it into the spiraled pump housing and out toward the pump outlet. Because the cooling system pressure at the inlet to the coolant pump is at its lowest (because of low pressure pull of impeller at the inlet), system boiling always occurs at this location first. This will very rapidly accelerate the overheating condition as the pump impeller will be acting on vapor. Coolant pumps are the main reason that engine coolants should have some lubricating properties, because they are vulnerable to

Table 10–2 *System Pressure and Boil Point*

Cooling System Max Pressure	*Increase in Boil Point*	*Coolant Boils at*
7 psi (52 kPa)	21°F (12°C)	233°F (112°C)
10 psi (79 kPa)	30°F (17°C)	242°F (117°C)
15 psi (105 kPa)	40°F (22°C)	252°F (122°C)

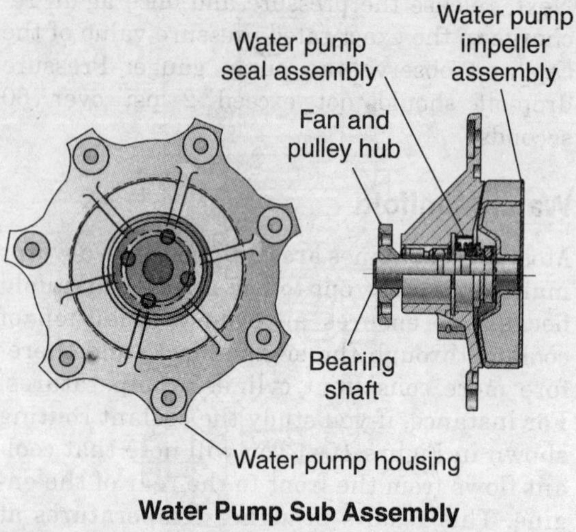

Figure 10–8 *Sectional view of a water pump and housing. (Courtesy of Navistar International Corp.)*

abrasion damage when the coolant TDS levels are high. A sectional view of a Navistar water pump is shown in Figure 10–8.

Coolant pumps fail for the following reasons:

- Overloading of the bearings and seals caused by misalignment or tight drive belts
- High TDS levels in the coolant that erode the impeller
- Mineral scale buildup on the pump housing
- Overheating—boiling usually occurs first at the inlet to the water pump, so a system that is not properly sealed, or hot shutdowns, can cause vapor lock.

Inspecting, Replacing, Rebuilding Coolant Pumps

A defective coolant pump should first be removed from the engine and analyzed for the cause of the failure to avoid a repetition. Although they were commonly rebuilt by the technician a generation ago, this is seldom the case now. When defective, they are replaced as a unit with a rebuilt/exchange unit. Rebuilding of water pumps is usually performed at a rebuild center equipped with the proper equipment by persons who specialize in the process. Although the technician who reconditions one or two coolant pumps a year may not be able to compete with the time of the specialist rebuilder, it is certainly possible to perform the work to the same standard. A slide hammer is usually required to remove the pulley from the impeller drive shaft, and an arbor press is generally preferred over power presses both for disassembly and reassembly. The pump is one of the more simple subcomponents of the engine and essentially consists of a housing, an impeller, an impeller shaft, bearings, and seals. When a pump is rebuilt, inspect the components thoroughly; in many cases, especially where plastic impellers are used, only the housing and shaft are reused. Examine the shaft seal contact surfaces for wear. When gear-driven coolant pumps are reconditioned, pay special attention to the drive gear teeth. The OEM instructions should be observed, and where ceramic seals are used, great care is required to avoid cracking them during installation. A critical specification is the impeller-to-housing clearance, and failure to meet this will reduce pumping efficiency. Figure 10–9 is a sectional view of a Caterpillar water pump with the rebuild specs and tolerances.

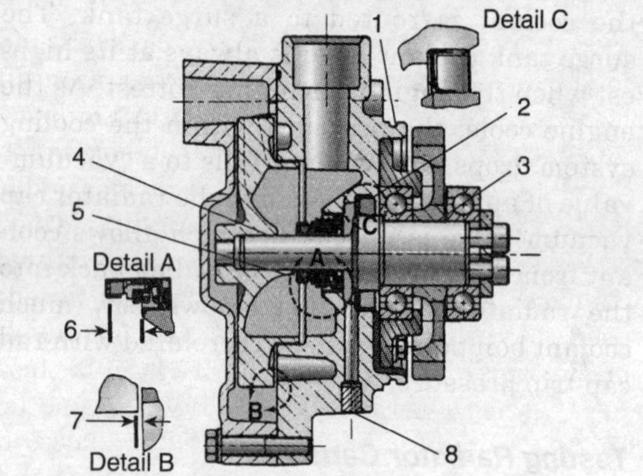

1. Seal groove.
2. Lip type seal.
3. Shaft diameter on drive gear end.
4. Shaft diameter on impeller end.
5. Seal assembly.
6. Clearance from surface in pump housing to top of seal assembly.
7. Clearance between impeller and pump housing.
8. Filter. Install filter flush with pump housing.

Figure 10–9 *Sectional view of a water pump and subcomponents. (Reprinted Courtesy of Caterpillar Inc.)*

FILTERS

Coolant filters are usually of the spin-on cartridge type connected in parallel to coolant flow. Flow through the filter is consistent with that of most other engine filters in that the coolant enters the canister through outer ports and exits through a central single port. Figure 10–10 shows the flow routing through a typical coolant filter. As mentioned earlier in the chapter, SCAs may also be packaged within the OEM coolant filter, so take care to avoid overservicing.

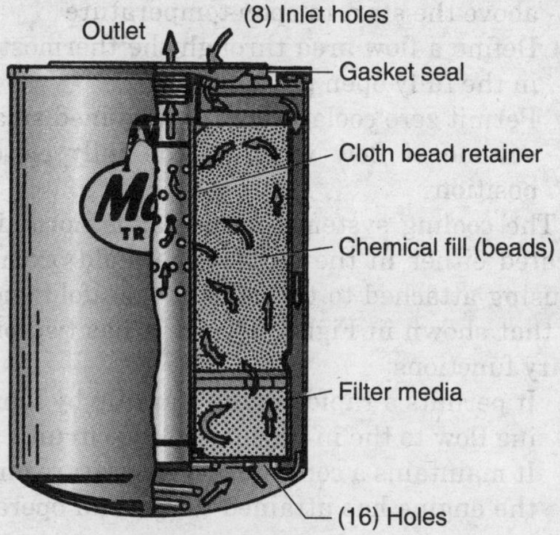

Figure 10–10 *Cutaway view of a coolant filter. (Courtesy of Mack Inc.)*

When coolant filters are required to be changed, check the type of shutoff mechanism used. Most current diesel engines use automatic shutoff valves as shown in Figure 10–11:

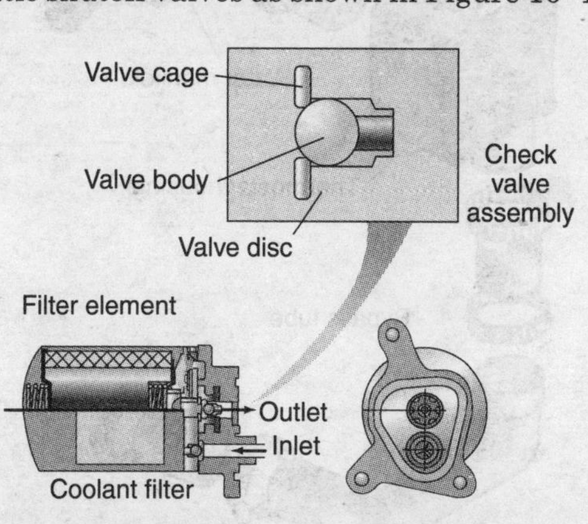

Figure 10–11 *Coolant filter with a mounting pad check valve. (Courtesy of Navistar International Corp.)*

This means that no spillage results when a filter is removed for servicing because a lock-off ball engages. Some engines have manual shutoff valves, which must be turned off before removal. New coolant filters do not require priming. Some coolant filters are equipped with a zinc electrode to neutralize the electrolytic effect of the coolant, although these are more likely to be found in marine applications.

COOLANT MONITORING CIRCUIT

A diesel engine cooling system is monitored in different ways according to the generation of the system. Coolant temperature is often a primary reference to the engine management software when factoring timing and fuel-air ratio parameters on electronically managed engine systems, so the temperature display to the operator is secondary in importance. All current electronically managed engines can be programmed to default to failure strategies, which may include engine shutdown based on the coolant temperature or level readings. Thermistors are almost universally used to sense the temperature of the cooling system, as well as other engine fluid temperatures, including ambient air, boost air, and lubricating oil. The following methods are used to sense coolant temperature:

Thermistors

Thermistors are solid-state, semiconductor devices whose internal resistance varies with temperature change. They are supplied with a specific reference voltage, almost always 5 V-DC and return a signal to the ECM based on temperature. Negative temperature coefficient (NTC) thermistors tend to be more commonly used: The internal resistance in an NTC thermistor decreases as temperature rises. Positive temperature coefficient (PTC) thermistors function oppositely: The internal resistance increases as temperature rises. NTC thermistors are used to read coolant temperature on almost all diesel engines produced for a generation. Because the internal resistance in an NTC thermistor decreases as temperature increases, the signal voltage (returned to the ECM) goes up proportionally with temperature rise. This signal is used by the engine ECM to drive warm-up strategy and help process fuel-air ratio logic. The engine ECM broadcasts engine coolant temperature information to the data bus so that the instrument cluster controller can display this data on a dash digital display or gauge.

Electric

Electric sensors use a bimetal arm in conjunction with a resistor supplied by a modulated electrical signal from the temperature gauge: When the bimetal arm is heated, the greater linear expansion of one of the bimetal strips causes it to bend one way, and as it cools and contracts, bend in the opposite direction. Connected to the bimetal strip is a wiper, which short-circuits the current flow through the resistor to ground, thereby altering the gauge value.

Expansion

An expansion-sensing gauge consists of a tube filled with a liquid that expands as it is heated and in expanding, activates the gauge indicator needle. This gauge tends not to be used much in today's engines but you will likely come across a few examples in older truck engines and many off-highway, heavy-equipment applications.

Coolant Level Indicators

Most current electronically managed engines have radiators equipped with low coolant level warning systems. Most operate using the same principles. The ECM outputs a signal to a probe (or sensor), usually located in the top radiator tank, which grounds through the coolant. When the probe fails to ground through the coolant, a low coolant level warning is generated; the outcome depends on how the ECM has been programmed (this is usually a customer data program option). In most cases, a lag (somewhere around 5 to 12 seconds) is required before the ECM resorts to a programmed failure strategy. This may be simply to alert the operator, ramp down to a default rpm/engine load, or shut down the engine after a suitable warning period. Some radiators are equipped with two probes used to signal low and dangerously low coolant levels.

THERMOSTATS

Thermostats function as a type of automatic valve that senses changes in engine temperature and regulate coolant flow to maintain an optimum engine-operating temperature. To function effectively, a thermostat must:

- Start to open at a specified temperature
- Be fully open at a set number of degrees above the start to open temperature
- Define a flow area through the thermostat in the fully open position
- Permit zero coolant flow or a defined small volume of flow when in the fully closed position

The cooling system thermostat is normally located either in the coolant manifold or in a housing attached to the coolant manifold such as that shown in Figure 10–12. It has two primary functions:

- It permits a rapid engine warmup by limiting flow to the in-engine cooling circuit.
- It maintains a consistent temperature once the engine has attained its normal operat-

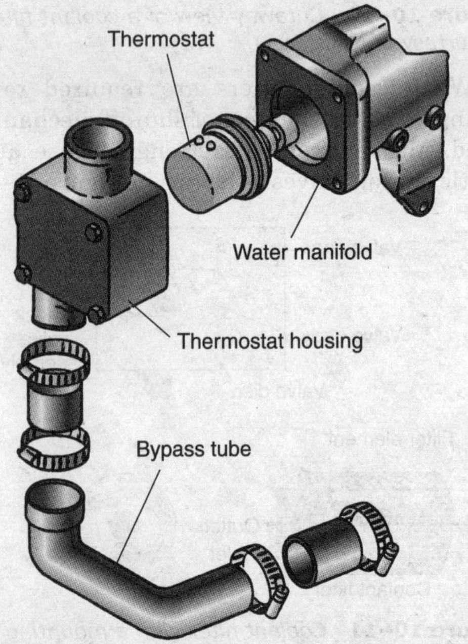

Figure 10–12 *Exploded view of a thermostat housing. (Courtesy of Mack Inc.)*

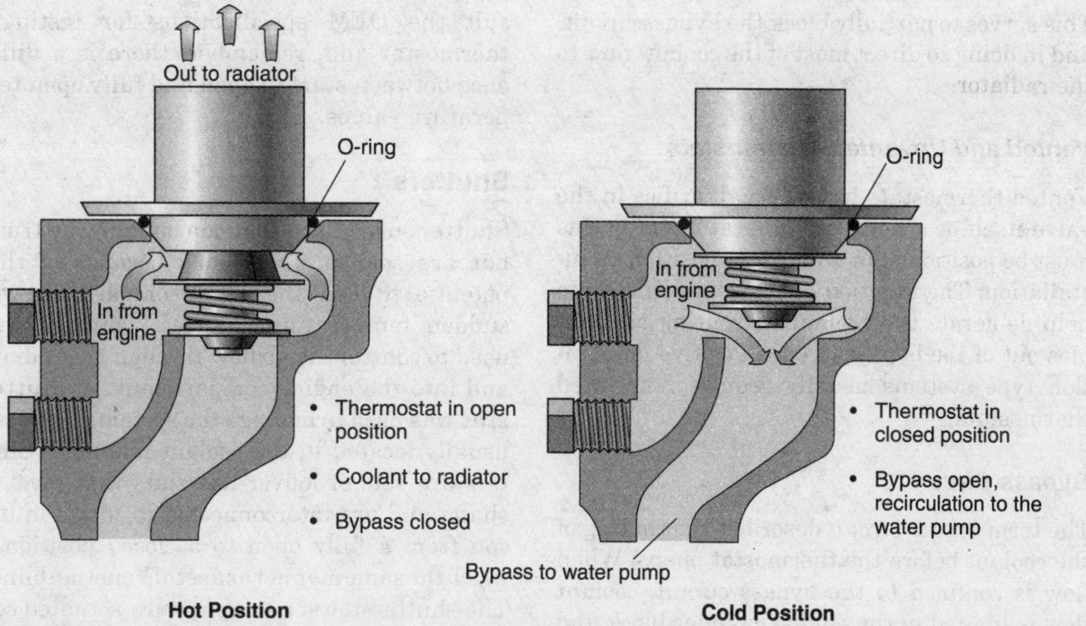

Figure 10–13 *Sectional view showing thermostat operation. (Courtesy of Navistar International Corp.)*

ing temperature by optioning flow to the radiator.

The thermostat therefore functions by optioning flow to either the in-engine circuit or the radiator circuit. It opens and closes based on temperature. As it opens and closes it defines a flow area so, for instance, a partially open thermostat limits the flow area of coolant routed to the radiator.

Operating Principle

Because a thermostat defines the flow area for circulating the coolant, there is often more than one. A heat-sensing element actuates a piston attached to the seal cylinder. When the engine is cold, coolant is routed back to the coolant pump to be recirculated through the engine. When the engine heats to operating temperature, the seal cylinder gates off the passage to the coolant pump and routes the coolant to the radiator circuit. The heat-sensing element consists of a hydrocarbon or wax pellet into which the actuating shaft of the thermostat is immersed. As the hydrocarbon or wax medium expands, the actuating shaft is forced outward in the pellet, opening the thermostat. Thermostats can be full blocking or partial blocking. Figure 10–13 shows a typical engine thermostat in its engine warm and engine cold positions.

Top Bypass Thermostat

The top bypass thermostat simultaneously controls the flow of coolant to the radiator and the bypass circuit. During engine warmup, all of the engine coolant is directed to flow through the bypass circuit. As the temperature rises to operating temperature, the thermostat begins to open and coolant flow is routed to the radiator, increasing incrementally with temperature rise.

Poppet or Choke Thermostats

Poppet-type thermostats control the flow of coolant to the radiator only, with the result that the bypass circuit is open continuously. Flow to the radiator is discharged through the top of the thermostat valve.

Side Bypass or Partial Blocking Thermostat

The side bypass thermostat functions similarly to the poppet type. It has a circular sleeve below the valve that moves with the valve as it opens.

This serves to partially block the bypass circuit, and in doing so direct most of the coolant flow to the radiator.

Vented and Unvented Thermostats

Vented thermostats have a small orifice in the valve itself or a notch in the seat; usually this must be positioned in an upright position on installation. The function of the vent orifice is to help de-aerate the coolant by routing air bubbles out of the bypass circuit. Positive de-aeration type systems usually require nonvented thermostats.

Bypass Circuit

The term *bypass circuit* describes the routing of the coolant before the thermostat opens. When flow is confined to the bypass circuit, coolant flow is limited to the engine cylinder block and head. The flow of bypass coolant permits rapid engine warmup to the required operating temperature.

Running Without a Thermostat

This practice is not recommended and engine OEMs may void warranty. It also violates the Environmental Protection Agency (EPA) requirements regarding tampering with emissions control components. Removing the thermostat invariably results in the engine running too cool. This can cause vaporized moisture in the crankcase to condense and result in developing corrosive acids and sludge in the crankcase. Additionally, low engine running temperatures will increase the emission of hydrocarbons (HC). Conversely, engines that should use top bypass or partial bypass-type thermostats may overheat when the thermostat is removed as most of the coolant will be routed through the bypass circuit with little being routed in the radiator.

Testing Thermostats

Testing a thermostat can be performed using a specialized tool that essentially consists of a tank, a heating element, and an accurate thermometer. Such a test device can be manufactured from an open top electric kettle. Always consult the OEM specifications for testing a thermostat and, remember, there is a difference between start to open and fully open temperature values.

Shutters

Shutters used to be common on highway trucks but are seldom used today because of their potential to *shock* the engine compartment with sudden temperature changes. Shutters were used to control the airflow through the radiator and into the engine compartment. A **shutterstat** was used to manage the system, and it was usually located in the coolant manifold. Shutters are sets of louver-like slats that pivot on shafts and are interconnected to rotate in unison from a fully open to a closed position in much the same manner as a set of Venetian blinds. The shutterstat is a temperature-actuated control mechanism. It receives a feed of system air pressure, which it will allow to pass through until the coolant temperature reaches a predetermined value specified on the shutterstat.

The shutter assembly is mounted on the radiator and is usually spring loaded to the opened position. When air from the shutterstat is delivered to the shutter cylinder, the plunger extends to actuate a lever and close the shutters. When no air is available at the shutter cylinder, the shutters are held open by spring force so we can say that the default position is fully open. During engine warmup, the shutters would normally be held closed once there was sufficient air pressure in the system.

COOLING FANS

OEMs use two basic operating principles of engine compartment cooling, requiring the use of either suction or pusher fans. Suction fans pull outside air into the engine compartment, whereas pusher fans do the opposite and push heated air out. Highway vehicles that receive ram air assistance most commonly use suction fans. In fact, depending on the variables of engine size, radiator design, and vehicle application, ram air is often sufficient to perform cooling 95% of operating time, which is a good thing because driving a fan leaches engine power.

Most engine cooling fans today do not leach significant engine power. Lightweight, variable pitch fan assemblies manufactured from carbon composites and plastics draw as little as 6 BHP at engine rated speed and load. The older, heavy gauge steel fan assemblies with fixed pitch often drew as much 18 BHP at rated speeds. These heavy fan assemblies are still in use today, but not often in highway trucks: Their more likely application is where the cooling system is set up to run the fan constantly such as in slow-moving quarry trucks.

Fan Design

Fan design is important, and fans should draw the least possible amount of engine power: 6 BHP is typical of modern composite, fiberglass, or plastic fans, but twice that may be required to rotate a heavy steel fan. Consequently, many OEMs have adopted lighter fan blade materials in preference to steel or aluminum. Many current designs use flexible pitch fan blades that alter a fan's efficiency proportionally with its driven speed. Flexible pitch blades have variable efficiency. They are aggressive (and efficient) at low rpm but as they are speeded up, the blade pitch flexes and reduces efficiency (and leached engine power). Fan assemblies must be precisely balanced. An out-of-balance fan (a small fragment missing from one blade is sufficient) may unbalance the engine driving it sufficiently to produce a torsional stress failure of the crankshaft. Figure 10–14 shows some examples of fan blades used in some current trucks.

Fan Cycles

Because driving a fan assembly draws engine power, most current truck engines use lightweight, temperature-controlled fans. Fanstat,

Figure 10–14 *Various diesel engine fan blades used in current trucks. (Courtesy of Horton Holding Inc.)*

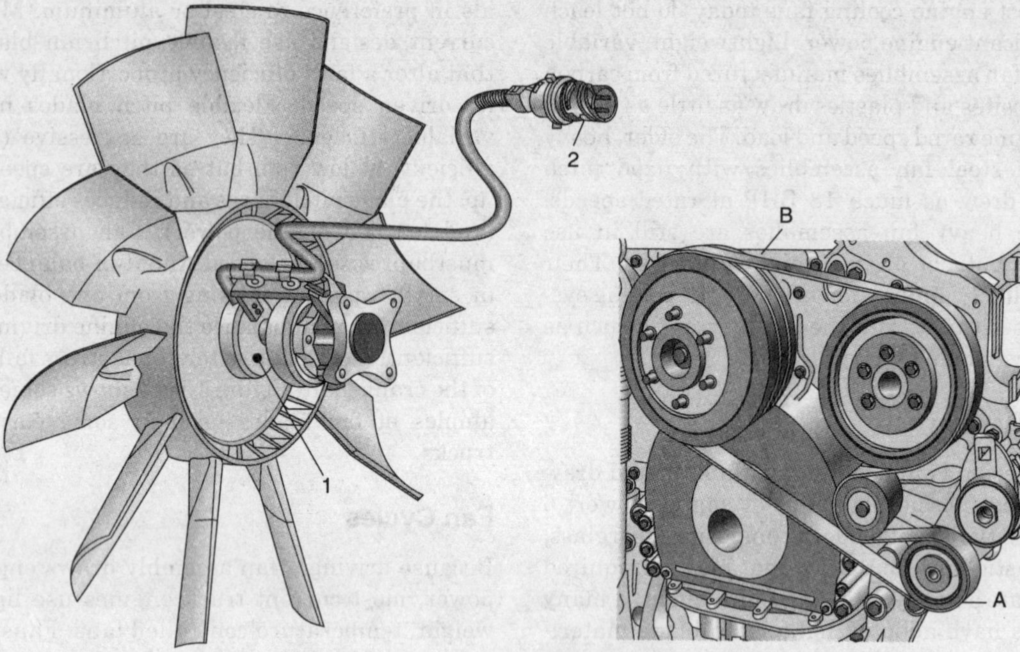

Figure 10–15 *Electrically controlled fan hub and fan hub drive. (Courtesy of Volvo Trucks NA)*

air/electric/oil pressure engaged clutch fans of the on-off type are used as well as thermomodulated, viscous drive fans. The objective is usually to have the fan running as little as possible. In today's electronically controlled diesel engines, fan cycles (on or off) are usually managed by the engine ECM, which can use sensor input from a variety of sources to determine cycling. Figure 10–15 shows a typical ECM-controlled, electrically switched fan assembly.

On-Off Fan Hubs

On-off fan hubs use air, oil pressure, or electrically actuated clutches controlled by a fanstat or the engine management system. Vehicle air-conditioning systems may also control fan hub cycles independently from the ECM using electric-over-pneumatic switching. However, increasingly this type of switching can be handled by the vehicle data bus negotiation; see Chapter 33 on multiplexing. When a **fanstat** is used, it is usually located in the water manifold: It is a temperature-triggered switch that can either lock up or freewheel the fan hub using an electrical signal or chassis system air pressure to control the clutch.

Most on-off fan hubs are spring loaded to the engaged mode: This means that a pneumatic or electrical signal is required to disengage the fan clutch to permit it to freewheel. This means that the default mode is a fully engaged fan drive hub. When an air pressure-engaged, spring-released fan hub is used, the fan hub must be mechanically locked in the event of control circuit failure. On-off types of fan hub leach the least amount of engine power when the coolant temperatures are below the trigger value, which is usually designed to be 90% to 95% of operating time. However, high ambient temperatures and the requirements of the vehicle air-conditioning system may change this ratio.

When engine oil is used as the hub drive coupling medium (some bus applications), the fanstat controls the flow of oil to the hub assembly. At the fanstat nominal value, oil pressure directed to the fan hub acts as a fluid coupling so there is always a percentage of slippage. Some electronically managed engines directly control the fan cycle and use it as a retarding mechanism (in a smart cruise "braking" event) as well as an engine housing cooling aid.

Thermatic Viscous Drive Fan Hubs/ Thermomodulated Fans

Thermatic viscous drive fan hubs are integral units with no external controls. They use silicone fluid as a drive medium between the drive hub and the fan drive plate. The hub assembly consists of three main subcomponents:

- Drive hub (input section)
- Driven fan drive plate (output section)
- Control mechanism

There is no mechanical connection between the drive hub and the driven member. During minimum slip operation, torque is transmitted through the internal friction of the silicone drive fluid in the working chamber that couples the input and output sections. A wiper attached to the driven member continually wipes the fluid, and centrifugal force returns it to a supply chamber where an open valve cycles it back to the working chamber. As ambient temperature drops, a bimetal temperature-sensing strip contracts to close the valve that supplies the silicone drive medium to the working chamber, and the fluid becomes trapped in the supply chamber. An advantage of the viscous drive fan hub is that it is capable of infinitely variable drive efficiencies depending on the temperature of the bimetal strip.

A thermatic fan hub senses engine temperature at the front of the engines so they will be rated at lower specified temperatures than those that directly sense coolant temperature. Because they are capable of (6°C) variable drive efficiencies and never completely lock, they are driven at minimum slippage when the temperature is rising to around 10°F above specified. Once the hub enters minimum slip, it does not drop out until temperature reduces to around 10°F (6°C) below specified. In a thermatic hub spec'd at 150°F (66°C), as underhood temperature rises, the thermatic hub does not achieve minimum slip until approximately 160°F (71°C). Having achieved minimum slip drive, it does not drop out until the underhood temperature decreases to around 140°F (60°C).

Fan Shrouds

Fan shrouds are usually molded fiber devices bolted to the radiator assembly and may partially enclose the fan, providing a small measure of safety if the fan is engaged when the engine is running and the hood is open and exposed to the technician. Shrouds play an important role in shaping the airflow through the engine compartment, and a missing or damaged shroud can result in temperature management problems. In hot weather conditions, fan efficiencies can be dramatically lowered by a defective or missing shroud, so they should be examined at each inspection.

Fan Belts and Pulleys

Fan pulleys use external V or poly-V grooves and internal bearings of the roller, taper roller, and bushing types. The belt tension should be adjusted using a belt tensioner. The consequences of maladjusted belts are:

- Too tight: excessively loads the bearings and shortens bearing and belt life
- Too loose: causes slippage and destroys belts even more rapidly than a too tight adjustment

Belts should be inspected periodically as part of a PM routine. Replace belts when glazed, cracked, or nicked. Replacing belts with early indications of failure costs much less in the long run than the breakdowns that may be caused by belts that fail in service. Figure 10–16 shows a poly-V fan hub drive favored by Caterpillar.

Water Manifolds

The water manifold acts as a sort of main artery for the cooling system. It is usually a cast-iron or aluminum assembly fitted to the cylinder heads and may house the thermostat(s). Figure 10–17 shows a typical water manifold that is divided into two sections coupled by a rubber hose.

COOLING SYSTEM PROBLEMS

Troubleshooting cooling system problems is covered in more detail in Chapter 42. In this chapter we take a brief look at some cooling systems problems and their possible causes. Cooling system performance problems can be grouped into the following categories:

- Overheating
- Overcooling

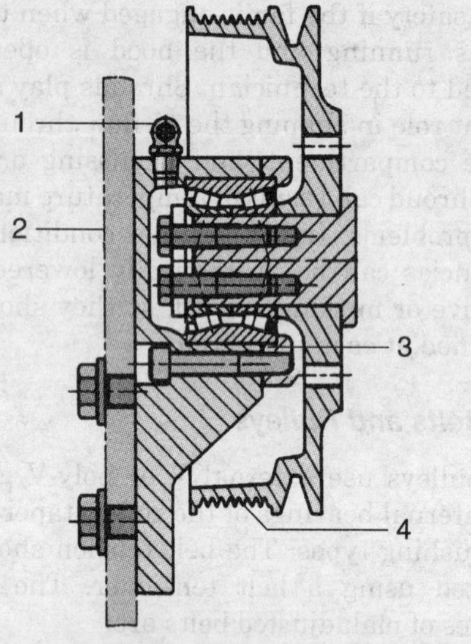

1. Fitting. Fill seal cavity with 2S3230 bearing lubricant.
2. Fan drive mounting plate
3. O-ring seal
4. Poly-V belt pulley

Figure 10–16 *Sectional view of a Poly-V fan drive. (Reprinted Courtesy of Caterpillar Inc.)*

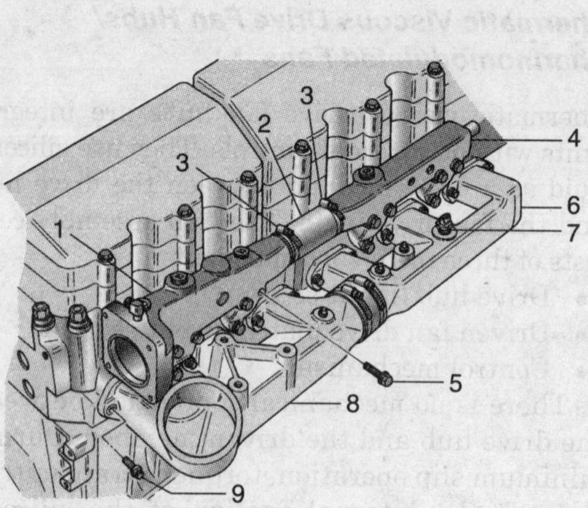

1. Coolant manifold, front section
2. Couple
3. Clamp
4. Coolant manifold, rear section
5. Capscrew
6. Air inlet manifold, rear section
7. Inlet air temperature sensor (sending unit)
8. Air inlet manifold, front section
9. Capscrew, 12-point

Figure 10–17 *Two-section coolant manifold on a six-cylinder engine with a two-cylinder head design: The intake manifold is located below. (Courtesy of Mack Inc.)*

- Loss of coolant
- Defective radiator cap
- Defective thermostat

Leaks

Cooling system leakage is common, and the system should be inspected by the operator daily. Cold leaks may be caused by contraction of mated components at joints, especially hose clamps. Cold leaks often cease to leak at operating temperatures because everything expands. Many fleet operators replace all the coolant hoses after a prescribed in-service period regardless of their external appearance to avoid the costs incurred in a breakdown. Silicone hoses are more expensive than the rubber compound type but they usually have longer service life. Silicone hoses require the use of special clamps, and these are sensitive to overtightening. Take care to torque to the required specification. Pressure testing a cooling system will locate most external cooling system leaks. A typical cooling system pressure-testing kit consists of a hand-actuated pump and gauge assembly calibrated from 0 to 25 psi (170 kPa) plus various adapters for the different types of fill neck and radiator cap. Some are capable of vacuum testing.

Internal leaks can be more difficult to locate. When coolant is present in the engine oil, it appears as a milky sludge before settling to the bottom of the oil pan. It is the first fluid to exit the oil pan when the drain plug is removed. When the leak is sourced to a failed wet liner seal, to locate the engine cylinder, remove the oil pan, allow the oil to drip for a period, then pressurize the cooling system using standard hand pump apparatus and place a sheet of cardboard under the engine. Wet liner O-ring seals may leak either cold or hot and even this method is not surefire.

The appearance of bubbles in the radiator may indicate that combustion gases are leaking into the coolant or that the engine is being run without thermostat(s). There are a couple of ways of determining whether combustion gases are leaking into the cooling circuit, and the best method varies according to the engine being tested. Pressure testing of the cooling system using the method outlined previously is the preferred method. On some engines, if you can disconnect the water pump drive, removing the piping to the upper radiator tank and the thermostat, filling the engine with coolant (water will do), and running the engine can identify cylinder pressure leakage to the cooling system.

Stray Voltage Damage

Stray voltage grounding through engine coolant can result in electrolytic action that can cause considerable engine damage in unbelievably short periods of time. This electrolytic damage ranges from pinholes in heat exchangers to erosion pitting failures of cast-iron liners and cylinder blocks. The incidence of stray voltage damage has become more commonplace due to the increase in electrical and electronic components combined with the introduction of nonconducting components such as plastic radiator tanks. Chassis static voltage buildup can also discharge through engine coolant.

One fleet hauling both tankers and flatbed trailers, using tractors with identical powertrains, reported that engine failures attributed to coolant electrolysis were only occurring in the tractors assigned to flatbeds. Trailer technicians will know that tankers have a dedicated ground circuit (known as *bonded ground*), whereas flatbeds do not. Use the following method to identify stray voltage and eliminate it.

Testing for Stray Voltage

Stray voltage can be alternating current (AC) (alternator diode bridge leakage) or direct current (DC) but is more likely to be the latter. Use a digital multimeter (DMM) on autorange to perform the following tests, checking for DC first.

1. Run the engine and turn on all the vehicle electrical loads.
2. Place the negative DMM probe directly on the battery negative terminal and the other into the coolant at the neck of the radiator without touching metal.
3. Record voltage reading. 0.1 V-DC is OK; the maximum acceptable to most OEMs is 0.3 V-DC. If higher, locate the leakage source. 0.5 V-DC leakage is capable of eating out a cast-iron engine block.
4. Sequentially shut down each electrical component, checking the voltage reading. When the leaking component or circuit has been identified, repair its ground. Attempting to ground the coolant at the radiator will not repair the problem.

COOLING SYSTEM MANAGEMENT

Most truck engine cooling systems are managed with the objective of opening the thermostat first, followed by the shutterstat (if equipped), and finally the fan will engage. You should remember this sequence because this is usually what you are required to remember when tested on the subject, but it is important to remember that this may not necessarily be so. Because the modern fan consumes power, most OEMs go with the premise that it is desirable for it to be driven as little as possible. However, some will elect to open shutters before opening the thermostat to allow the thermostat full control of engine temperature and avoid the sudden changes of engine compartment temperature caused by cycling the shutters. **Thermatic fans** sense underhood or engine compartment temperature. An underhood temperature of 155°F can be generally reckoned to an engine coolant equivalent of 190°F.

Actively Pressurized Cooling Systems

As diesel engines have evolved to adapt for reduced exhaust emissions there has been a significant increase in the amount of heat that the engine cooling system is required to dissipate. Cooled exhaust gas recirculation (C-EGR) heat exchangers in particular have increased the heat unloaded into the cooling system, and in early systems (2004) some engines experienced coolant boiling in the C-EGR heat exchanger. In the past, engine OEMs addressed increased heat dissipation requirements by increasing the core size of the radiator, increasing coolant capacity, and improving under-hood airflow. The limiting factor of this approach was the physical space required by larger radiators and higher energy consumption required by increased coolant volumes. Recently, some engine OEMs have opted to increase diesel engine cooling capacity by using an external means of pressurizing the cooling system and thereby raise the boiling point of the antifreeze mixture. The system is known as **actively pressurized cooling system (APCS)**.

An APCS is managed by the ECM and introduces air under pressure from an external pressurized source. This source is usually the intake manifold on turboboosted engines. This pressurized air from the engine intake manifold is plumbed into a cooling system expansion tank by means of a flow control check valve. The flow control valve is a spring-loaded nonreturn valve located in the connection plumbing for unidirectional flow between the intake manifold and the expansion tank.

Using higher system coolant temperatures requires proper pressurization under all operating conditions in order to control cooling characteristics, ensure adequate coolant flow, prevent cavitation, and especially to prevent coolant boiling and possible overflow. An independent module networked to the chassis data bus is used to manage the APCS system: It is assigned a message identifier MID address (MID 186) on the chassis data bus (MIDs are explained in Chapters 31 and 33). The functions of the APCS controller are to monitor:

- Coolant level
- Coolant pressure
- Coolant temperature
- System malfunctions

The APCS controller manages:

- Pressurization solenoid of the cooling system tank
- Broadcast of system status and faults to the chassis data bus

Because APCS is a networked, computer-controlled system, the specifics of control logic are not covered in this chapter, but later in Chapter 33.

COOLANT HEATERS

With many jurisdictions imposing anti-idling legislation, many more proposing to, there is suddenly some urgency about installing climate control devices on trucks. These function to either keep the cab warm enough or cool enough depending on the season. Some of these devices work independently from the diesel engine heating, ventilating, air conditioning (HVAC) circuit such as Webasto's Blue Cool system, which is air conditioning (A/C) based on the old icebox system. However, most are integrated in some way into the engine HVAC circuit. We are going to take a look at how a typical coolant heater functions. Before beginning we should dispel a truck driver myth, namely, a diesel engine cannot be started in temperatures below 0°F (–17°C). This might have been true with your Dad's diesel engine but it is not today. You can make it true today by ignoring the OEM recommendations for such critical items as the viscosity of the oil you are using.

Diesel-fired Coolant Heater

A diesel-fired coolant heater sources the main fuel system for its supply of fuel. These units are only intended to be operated when the vehicle is parked with the engine off. A typical coolant heater consists of a burner nozzle, combustion chamber, heat exchanger, and coolant circulation pump. System control is electronic and key sensors in the system are a coolant temperature sensor, overheat sensor, and flame sensor. Some use spark ignition but in the unit shown in Figure 10–18 a glow pin is used. Heat regulation can usually be managed at several different levels.

REVIEW QUESTIONS

1. Which type of diesel engine coolant is regarded as potentially the most harmful?
   a. EG
   b. PG
   c. Pure water
   d. ELC
2. Which of the following cooling media would transfer heat most efficiently?
   a. EG
   b. PG
   c. Pure water
   d. ELC
3. Wet liner cavitation is caused by:
   a. Aerated coolant
   b. Combustion gas leakage
   c. Air in the radiator
   d. Vapor bubble collapse

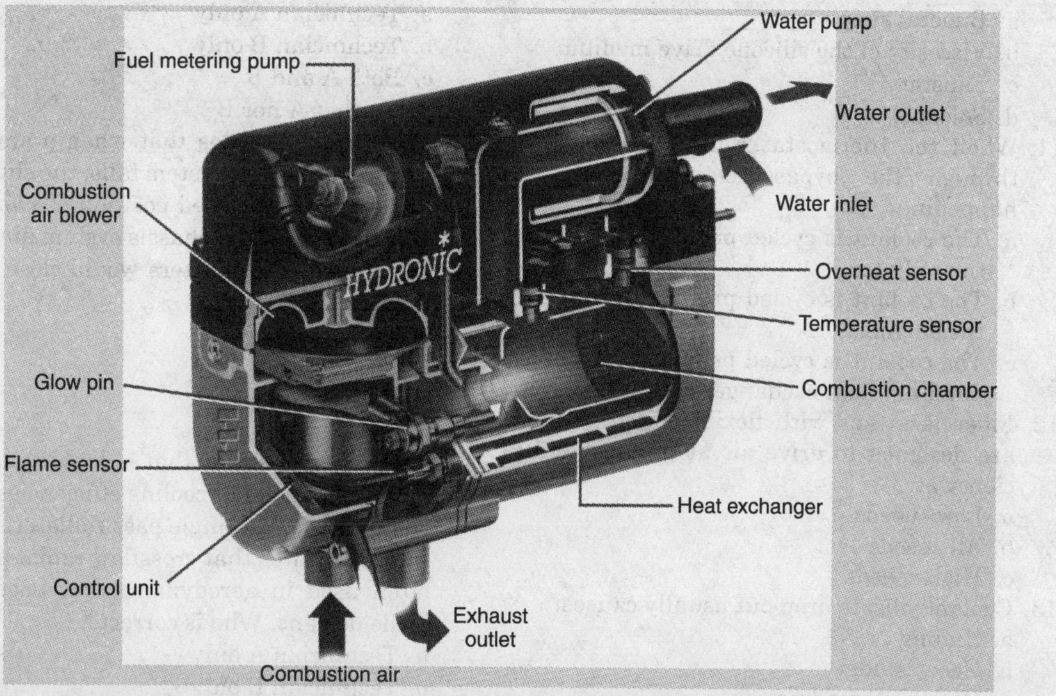

Figure 10–18 *Espar Hydronic coolant heater. (Courtesy of Espar Heater Systems)*

4. The cooling system hoses on an engine collapse when the unit is left parked overnight. Which of the following is the likely cause?
   a. This is normal.
   b. A defective thermostat
   c. Improper coolant
   d. A defective radiator cap
5. If a radiator cap pressure valve fails to seal, which of the following would be the likely outcome?
   a. Coolant boil-off
   b. Cooler operating temperatures
   c. Higher HC emissions
   d. Cavitated cylinder liners
6. Which would be the warmest portion of a typical downflow-type radiator when the engine is at operating temperature?
   a. The top tank
   b. The surge tank
   c. The bottom tank
   d. The center of the core
7. The temperature rating of a thermatic viscous fan hub is a nominal 155°F (68°C). At approximately what equivalent coolant temperature will this produce minimum slip drive as temperatures rise?
   a. 155°F (68°C)
   b. 165°F (74°C)
   c. 190°F (88°C)
   d. 225°F (107°C)
8. In a typical diesel-powered highway truck at operating temperature, which of the following should be true?
   a. Coolant temperatures run cooler than lube oil temperatures.
   b. Coolant temperatures run warmer than lube oil temperatures.
   c. Coolant temperatures should be equal to lube oil temperatures.
9. Which of the following properly describes the operating principle of a typical coolant pump on a highway diesel engine?
   a. Positive displacement
   b. Centrifugal
   c. Constant volume
   d. Gear type
10. Which of the following controls the drive efficiency of a thermatic, viscous drive fan hub?

a. Bimetal strip
b. Viscosity of the silicone drive medium
c. Fanstat
d. Solenoid

11. When the thermostat routes the coolant through the bypass circuit, what is happening?
a. The coolant is cycled primarily through the radiator.
b. The coolant is cycled primarily through the engine.
c. The coolant is cycled primarily through auxiliary heat exchangers.

12. Fiberglass fans with flexible pitch blades are designed to drive air at greatest efficiency at:
a. Low speeds
b. All speeds
c. High speeds

13. Coolant silicate drop-out usually causes:
a. Scaling
b. Gooey sludge
c. High TDS readings

14. In the event of an engine overheating, where is the coolant likely to boil first?
a. At the engine water jacket
b. At the top of the radiator tank
c. At the inlet to the coolant pump
d. At the thermostat housing

15. The correct instrument for checking the degree of antifreeze protection in a heavy-duty diesel engine coolant is a:
a. Hydrometer
b. Refractometer
c. Spectrographic analyzer
d. Color-coded test coupon

16. *Technician A* states that premixed ELC has an approved service life of up to 6 years. *Technician B* states that water should never be added to premixed ELC even when it is distilled. Who is correct?
a. Technician A only
b. Technician B only
c. Both A and B
d. Neither A nor B

17. *Technician A* states that when a shutterstat management system fails, the shutters default to their closed position. *Technician B* states that if the chassis system air pressure failed, the shutters would close. Who is correct?
a. Technician A only
b. Technician B only
c. Both A and B
d. Neither A nor B

18. *Technician A* states that a double pass radiator offers higher cooling efficiencies than a similarly sized single pass radiator. *Technician B* states that crossflow radiators are often used in aerodynamic low hood and grille designs. Who is correct?
a. Technician A only
b. Technician B only
c. Both A and B
d. Neither A nor B

19. Which of the following is true of an APCS?
a. Coolant boil temperature is raised
b. Cooling system pressure is raised by compressed air
c. Helps the cooling system function with higher heat loads
d. All of the above

20. Which of the following strategies is a requirement of starting diesel engines in subzero (below –17°C) temperatures?
a. Use of an auxiliary diesel coolant heater
b. Discharge a can of ether into the engine while cranking
c. Avoid shutting the engine off
d. All of the above

11

Engine Breathing

INTRODUCTION

Over the years, the intake and exhaust systems on commercial diesel engines have become more interconnected because of the universal use of turbochargers and exhaust gas recirculation (EGR) systems, so it makes sense to group them together in one chapter and call it *Engine Breathing*. Turbochargers boost the intake manifold at pressures above atmospheric and they use rejected heat from the exhaust system to do it. EGR systems of one sort or another are also universal on current diesel engines and because these reroute exhaust gas back into to the intake circuit, they play a role in both intake and exhaust systems. In addition, the exhaust system contains an ever more complex aftertreatment canister to control emissions. The objective in this chapter is to identify emissions control hardware: Their operating principles are discussed in Chapter 41.

The manifold-boosted compression ignition (CI) engine cylinder is *charged* with air rather than induced. In a naturally aspirated engine, air or an air-fuel mixture is drawn into the engine cylinder by the below atmospheric pressure created in the engine cylinder on the downstroke of the piston on the first stroke of the four-stroke cycle. This air or air-fuel mixture is therefore *induced* into the cylinder. The correct term to collectively describe the components responsible for delivering breathing air in **naturally aspirated** engines is **induction circuit**. The preferred term for collectively describing the breathing air delivery components in an engine with turbo/manifold boost is air **intake circuit**. For that reason, we use this term exclusively in this chapter.

Gas Flow in Breathing Circuit

Figure 11–1 is an overview of the gas flow of a typical post-2007 engine breathing circuit. We have used a series turbocharged (twin turbo) system because this is common in Caterpillar and Navistar engines. The figure looks complicated and it is. The reality is that this schematic simplifies the circuits shown, which are ECM controlled and monitored by numerous sensors. This chapter focuses primarily on gas flow through the engine breathing circuit. We will take a closer look at sensors and details of emissions controls in later chapters.

Role of Intake System

The function of the air intake system in a diesel engine is to supply a charge of air or air/dead gas mixture to the engine cylinders for combustion, cooling, and scavenging. Diesel engines are designed for lean burn operation. This means that the air charged to the engine cylinder often substantially exceeds that required to combust the fuel. In an earlier generation of diesel engines, the objective was to cram as much air (that is, oxygen) into the engine cylinders. This was great for performance but bad news for emissions. Today, every

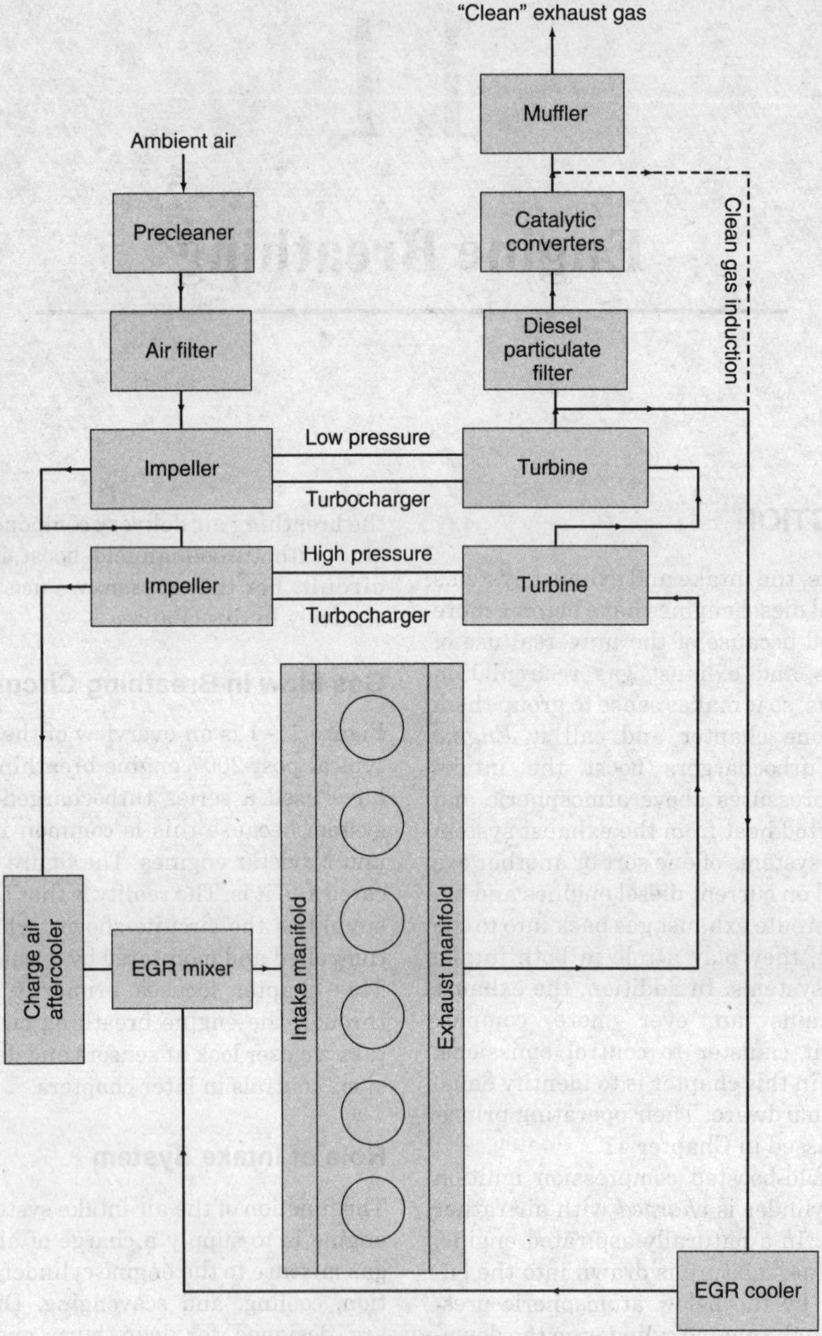

Figure 11–1 *Schematic showing a typical post-2007 engine breathing circuit.*

engine outcome is managed in respect of the emissions it produces. In addition, engine life has increased significantly during the last 20 years and this is in part due to improvements in air intake system components.

Role of the Exhaust System

The function of the exhaust system is to minimize both engine noise and noxious emissions while restricting the exhaust gas flow to a minimum degree. The turbocharger is a component

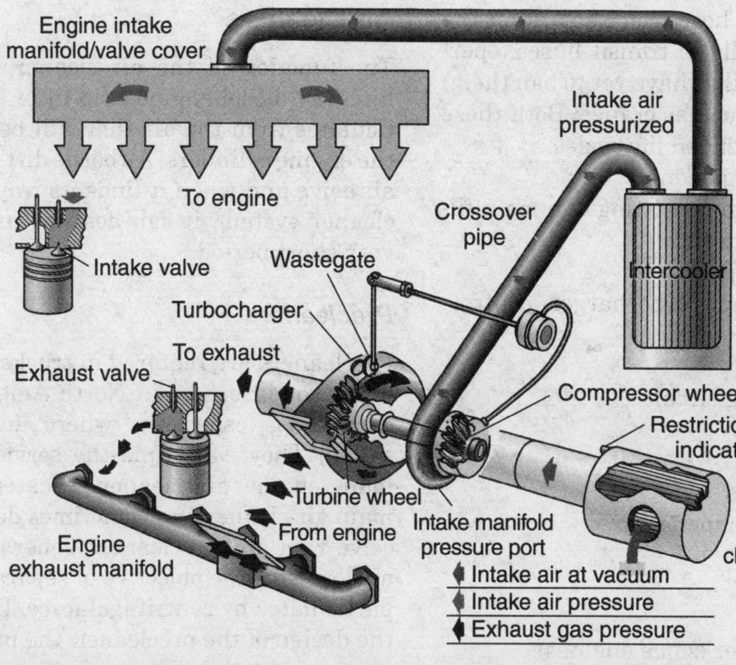

Figure 11–2 *Typical pre-2004 diesel engine breathing circuit. (Courtesy of Navistar International Corp.)*

common to both intake and exhaust systems. Its function is to use the exhaust system to recapture some of the engine **rejected heat** to pressurize the air delivered to the engine cylinders from the intake circuit. All current certified highway diesel engines have complex exhaust gas aftertreatment systems to manage emissions. Diesel engine breathing requires that the gas dynamics of the intake and exhaust systems work to complement each other in managing engine output and minimizing emissions. Because this requires coordinating a wide range of intake and exhaust components, management of the engine breathing circuit is the responsibility of the engine ECM. Figure 11–2 shows the components of a pre-2004 diesel engine breathing circuit.

BREATHING COMPONENTS

We begin by identifying the breathing components used on a current four-stroke cycle, highway-certified diesel engine.

Current Four-stroke Cycle

Bear in mind that older engines may not be equipped with all of the components listed here. EGR became commonplace in 2004 and diesel particulate filters (DPFs) in 2007.

Intake system components:

1. Precleaner
2. Main filter
3. Intake ducting/piping
4. Turbocharger(s) and turbocharger controls
5. Charge air cooling circuit
6. Intake module for mixture management of EGR
7. Intake manifold
8. Valve porting and intake tract design

Exhaust system components:

1. Valve configuration and exhaust tract geometry
2. Exhaust manifold
3. Turbocharger(s) and turbocharger controls
4. EGR system
5. Exhaust piping
6. DPF
7. Pyrometer(s)
8. **Catalytic converter(s)**
9. Muffler

Two-stroke Cycle

Due to emissions regulations, it has not been possible to certify two-stroke cycle diesel engines since 1991. Those that remain on the roads today are usually in transit buses (operating in jurisdictions that have yet to ban them) or costly equipment such as cranes: Both these applications have extended life cycles.

Intake system components:

1. Air cleaner system consisting of a main filter and a precleaner
2. Intake ducting/piping
3. Turbocharger and turbocharger controls (on some applications)
4. Roots blower
5. Charge air cooling circuit
6. Intake manifold
7. Air box
8. Liner intake ports

Exhaust system components:

1. Exhaust valves
2. Turbocharger(s)
3. Muffler
4. Catalytic converter (some engines)
5. Particulate trap (some engines)
6. Exhaust pipingt

AIR INTAKE SYSTEM COMPONENTS

The air intake system components are those responsible for delivering ambient air and dead gas rerouted from the exhaust system to the engine's cylinders. The ambient air is filtered upstream from the turbocharger(s). Turbocharger(s) pressurize intake air well above atmospheric pressure values to increase the oxygen density of the intake charge. In pressurizing the air, the temperature increases and this has a reducing effect on its density. To counter this, most turboboosted engines use some form of heat exchanger to cool the boost air (while maintaining the pressure) before it is directed into the engine cylinders. The term used to describe this heat exchanger is charge air cooler (CAC). When the air exits the CAC it is routed to an engine ECM-controlled EGR mixer. Depending on how the engine is being operated, the boost air is mixed with dead gas routed from the exhaust system. We will follow the intake circuit component by component in this section.

Air Cleaners

The function of the air cleaner system on a highway diesel engine is to filter airborne particulates from the air that will be delivered to the engine cylinders. Airborne dirt can be highly abrasive and when it finds its way past the air cleaner system, it can destroy an engine in a very short period.

Precleaners

Precleaners are required in trucks operating in dusty conditions or in North American winter conditions, especially where highways are salted. They can triple the service life of the main engine air cleaner. Located before the main air cleaner and sometimes designed to receive ram air, precleaners generate a cyclonic airflow with the objective of separating heavier particulates by centrifugal force. Depending on the design of the precleaner, the particulate removed from the airstream can either be discharged or collected in a dust bowl. Trucks operating in conditions in which they are subjected to high levels of larger airborne matter such as ash and grain chaff may use precleaner screens, but these are vulnerable to plugging so they should only be used in conjunction with inlet restriction gauges. In agricultural and mining applications, a precleaner can be connected to the exhaust system so that exhaust backpressure pulls separated particulate into the exhaust gas exit stream: This is not permitted in on-highway applications.

Dry, Positive Filters

Dry, **positive filtration** air cleaners are used in all contemporary North American commercial vehicle diesel engines. Because they use a positive filtration principle, *all* of the air entering the intake system must pass through the filtering media, usually resin-impregnated, pleated paper elements. The filtering medium is surrounded by a perforated steel mesh, which provides the element with a limited amount of structural integrity. If a visible dent is evident in the surrounding mesh, the filter will probably not seal in the canister and should be replaced. Filtering efficiencies are high throughout the speed and load range of the engine, usually better than 99.5% and highest just before replacement is required. Some dry element filters are two stage and may eliminate the need for a precleaner; these induce a vortex flow and use centrifugal force to separate heavier particulate, which is then discharged by an ejector valve.

Servicing Air Filters

Dry paper element filters are designed to last for as long as 12 months in a linehaul application and do not have to be serviced unless the inlet restriction specification exceeds the original equipment manufacturer's (OEM) maximum or a used oil analysis report indicates that dirt may be bypassing the air cleaner. Onboard inlet restriction gauges are not accurate instruments. When a filter mounted inlet restriction gauge indicates that the filter element is plugging before the regular service interval is achieved, check the reading using a water-filled manometer or negative pressure gauge. A **manometer** is a clear tubular column formed in a U shape around a calibration scale marked off in inches: The column is then filled water (usually colored with dye for ease of reading) to a zero point on the calibration scale. When a manometer is connected to a fluid circuit, it produces a reading according to the pull (vacuum circuit) or push pressure acting on the fluid in the column. Never remove an air filter from its canister unnecessarily. Every time a filter is serviced, some dust will be admitted beyond the filter assembly, no matter how much care is taken.

Checking filter restriction with a trouble light is a common practice that finds little favor with OEMs because whether the light is visible or not visible through the filtering medium depends more on the filtering medium materials than the extent it might be plugged up. The only valid reason for inspecting a dry filter with a trouble light is to locate a perforation. OEM specifications should always be consulted, but typical *maximum* inlet restriction specs are:

15″ H_2O vacuum for naturally aspirated (NA) engines

25″ H_2O vacuum for boosted engines

The practice of removing and attempting to clean a dirt-laden, dry filter on the shop floor should be avoided. A dirt-laden filter weighs many times the weight of a new filter and dropping it onto a concrete floor to shake dust free is usually sufficient to damage the filter element (by crumpling the perforated mesh) sufficiently to prevent it from sealing in the housing. Another malpractice that invariably ends up costing more in the long term is reverse compressed air blowing out of filters. Sharp particulate embedded in the filtering media can be dislodged, leaving an enlarged opening through which other larger particles may pass.

Laundering Dry Element Filters

When a dry filter element has become plugged the obviously ideal solution is to replace it. However, dry filter elements are expensive and in some operations, such as on construction sites, mining, and aggregate hauling, these may become restricted in a couple of working days even when precleaners are used. Professional laundering is perhaps a second best but necessary option to replacement. The laundering process usually requires that the filter element be soaked in a detergent solution for a period of time, followed by reverse flushing with low-pressure clean water. The element is next dried with warm air and inspected for perforations in the element and for seal/gasket integrity. OEM testing of professionally laundered filter elements indicates that filtering efficiencies are reduced with each successive laundering.

Inlet Restriction Gauges

Inlet restriction gauges are resettable gauges mounted either on the filter canister or remotely in the vehicle dash. They provide a readout in inches of water vacuum in the same way the manometer does. Although these instruments are not noted for their accuracy, they provide an indication of when filter service is required. See the location of the inlet restriction gauge in Figure 11–2.

Oil Bath Filters—Viscous Impingement Oil Filters

Low operating efficiencies have made these types of nonpositive filters things of the past so you are only going to see them on much older diesel engines and possibly some off-highway engines. They consist of a mesh-filled canister with a sump filled with engine oil. Airflow is cycloned through the canister and wets down a cylindrical mesh with engine oil to which dirt particles attach themselves. This principle works effectively to remove larger particles from the airstream and has higher efficiencies when induced airflow is highest, that is, at rated speed. They possess low filtering efficiencies at low-load, low-speed operation. Oil bath filters are serviced by replacing the sump oil and washing down the mesh with solvent.

ROOTS BLOWERS

The **Roots blower** (also known as a **supercharger**) is a gear-driven, positive displacement air pump used to **scavenge** the engine cylinders in Detroit Diesel Corporation (DDC) two-stroke cycle engines. It produces relatively low-peak pressures compared to those produced by turbochargers. As a scavenging component, the primary function of the Roots blower is to displace end gases from the engine cylinders. DDC two-stroke cycle engines used in on-highway applications use Roots blowers for cylinder scavenging. Some of these engines may also use a turbocharger to produce manifold boost. Under conditions of high engine loads, the turbocharger may be capable of performing cylinder scavenging, allowing the blower to idle. In some nonautomotive applications of DDC two-stroke cycle engines used under continuous high loads, scavenging may be performed exclusively by turbocharger(s), making the blower unnecessary.

Operating Principle

Roots blowers are used as scavenging pumps because of their positive displacement pumping principle. They pump air to an air box, which in turn supplies air to the cylinders by means of intake ports in the cylinder liner. A positive displacement pump will displace a constant slug volume per cycle. This means that the Roots blower will unload a constant slug volume of air per rotation so the volume of air charged to the **air box** is directly dependent on the driven speed. This gives the Roots blower somewhat

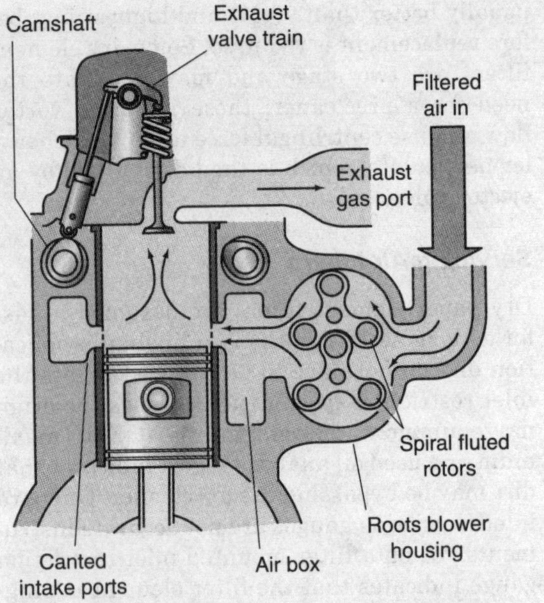

Figure 11–3 *Roots blower aspirated two-stroke cycle diesel engine.*

higher efficiencies (than a turbocharger) as a pump when engine rpm and loads are low. Turbochargers are driven by engine-rejected heat energy and therefore tend to have low efficiency when engine loads are low, plus a lag between power demand and turbo response. Figure 11–3 is a diagram of a Roots blower on an inline engine: Note the gas flow through the engine and key components.

Construction

The DDC Roots blower housing is manufactured from an aluminum alloy casting. Within this housing is a pair of intermeshing, spiral-fluted rotors supported by bearings located at either end of the assembly. One of the rotors receives gear-driven input, and it drives the other. Each rotor has three spiraled lobes timed to intermesh, so one spirals clockwise and the other counterclockwise. The rotor drive may be on either side, depending on the model. The rpm ratio of the blower to that of the engine also varies and depends on factors such as whether a turbocharger and intercooler are used. The fuel pump and water pump are also driven by blower drive shaft, the water pump at one end and the fuel pump at the other end of the assembly.

Lubrication

The DDC Roots blower is supported by pressure-lubricated roller bearings that ensure that the rotors have close clearance in operation but no contact. Lip-type seals prevent engine oil from bypassing the rotor shaft and being pumped into the engine. Evidence of engine oil on the rotor lobes is an indication that the blower shaft seals are failing. This is potentially dangerous. A failed blower shaft seal can result in engine lube being pumped into the intake, causing a runaway. To diagnose Roots blower problems, consult DDC service literature.

TURBOCHARGERS

Turbochargers are used universally in North American medium- and large-bore, highway diesel engines. By definition, a **turbocharger** is an exhaust gas-driven, centrifugal pump that "recycles" some of the rejected heat from the engine's cylinders. Turbochargers may be driven to speeds exceeding 200,000 rpm in certain race car engine applications but maximum speeds are about 30% lower in diesel engines. The turbocharger is used to deliver a pressurized charge of air to the engine's cylinders; in short, it increases the oxygen density in the air charge. In larger, off-highway diesel engines, the exhaust gas-driven **turbine** may be used with a fluid coupling to drive reduction gearing connected to the engine crankshaft; in this instance, the turbocharger assists in driving the crankshaft.

Principles of Operation

A turbocharger is an exhaust gas-driven air pump consisting of a turbine and an **impeller** mounted on a common shaft. The shaft is floated (hydrodynamically suspended) on friction bearings supplied with pressurized lube oil. The turbine wheel is subject to engine exhaust gas energy (heat) and is driven within a turbine housing through which the exhaust is routed. The impeller is enclosed in a separate **compressor housing** and acts on intake system air, pumping it through the charge side of the intake system. The exhaust gas that drives the turbine and the intake air the impeller acts on do not come into contact. Figure 11–4 shows the gas and lubrication flows through a simple turbocharger.

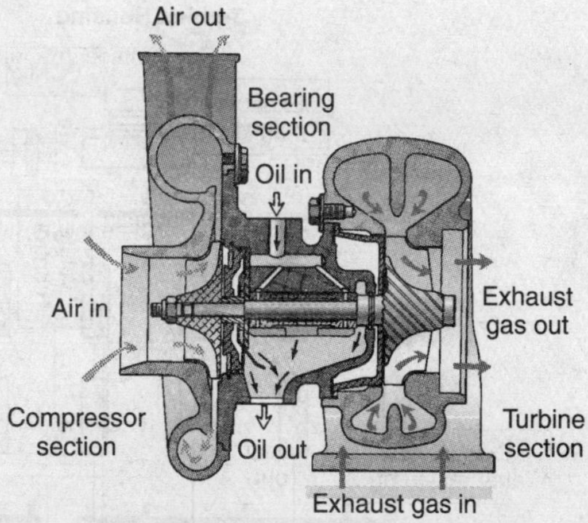

Figure 11–4 *Sectional view of a turbocharger, the lubrication circuit, and gas flow. (Courtesy of Schwitzer Engine Components)*

Compressor Operation

Filtered intake air is pulled into the compressor housing the impeller on the compressor side of the turbine shaft. The turbine drives the impeller on the other side of the turbine shaft so the actual speed of the impeller is determined by what is happening in the turbine housing. As the impeller rotates, the air from the intake system is accelerated to high velocities. High-velocity air flows radially outward to a diffuser. The **diffuser** is a restriction. It is designed to convert the kinetic energy (energy of motion) of the intake air into pressure as it attempts to drive it through the diffuser. The diffuser may be a voluted (snail-shaped) type or a blade type (see Figure 11–4). Blade-type diffusers have higher efficiencies.

Turbine Operation

Exhaust gas is routed to the turbine housing. The greater the engine rejected heat value (this increases somewhat proportionally with engine output), the greater the exhaust gas heat energy. The exhaust gas enters the turbine housing radially and is routed to flow into the **volute**, a snail-shaped, diminishing sectional area. The volute represents a restriction. However, when the exhaust gas exits the volute it then expands, acting on the turbine vanes, after

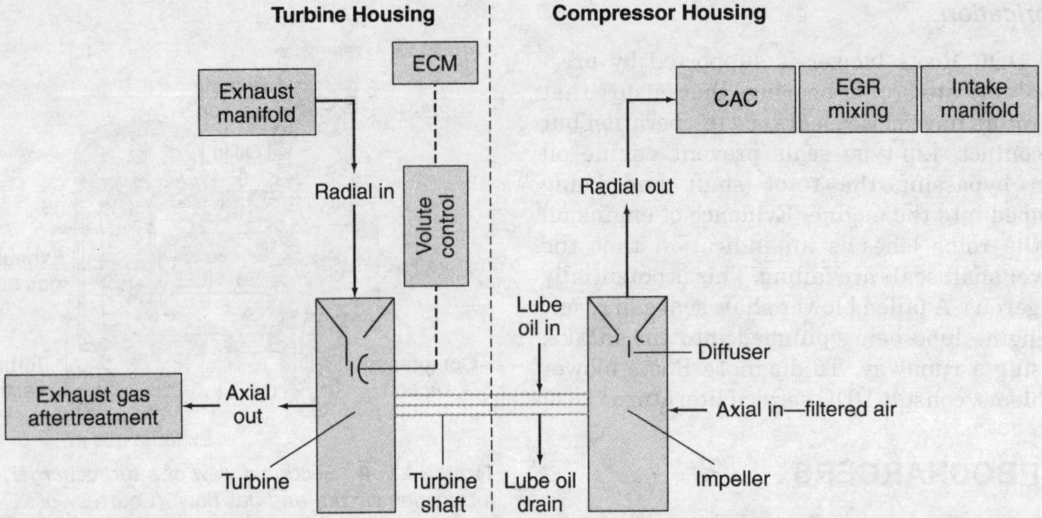

Figure 11–5 *Turbocharger gas flow schematic indicating the roles played by the volute and diffuser.*

which it is routed axially to the exhaust circuit. The extent of gas expansion in the turbine housing is determined by the amount of exhaust gas heat. Under high engine output, engine rejected heat increases so exhaust gas expansion acting on the turbine vanes produces higher turbine speeds. It is important to underline the fact that turbocharger rotational speed is factored primarily by exhaust gas heat and not exhaust gas pressure. The importance of the volute should also be understood. The smaller its size, the greater restriction to flow it represents, but a smaller size volute means that exhaust gas expands more when it exits. The best arrangement is to be able to control the volute flow area, which we will look at next. Figure 11–5 is a turbocharger gas flow schematic indicating the key roles played by the volute and diffuser: Note how gas flow enters and exits the turbine and compressor housings.

Gas Flow Routing

Gas flow into the turbine housing is radial and with an axial outflow (see Figure 11–5). Airflow through the impeller housing is axial inflow and radial outflow. In the simplest type of turbocharger design, the turbine housing intake tract or throat is undivided, meaning that all the cylinders feed gas into a single passage. However, variable geometry turbochargers manipulate gas flows into and out of the turbine housing. This is discussed later in this section.

Types of Turbochargers

First it is important to make a distinction between constant geometry turbochargers and variable geometry turbochargers. The definitions used in this text are as follows:

- Constant geometry (CG): a turbocharger in which all of the exhaust flow is routed through the turbine housing regardless of how the engine is being operated.
- Variable geometry (VG): a turbocharger that uses either external or internal controls to either manipulate the flow areas through the turbine housing or allow some percentage of exhaust gas to bypass the turbine housing.

Just 10 years ago a majority of turbochargers used on highway diesel engines used constant geometry: When we see a constant geometry turbo on a current truck diesel it is usually one of a series-charge pair or in compound turbocharger arrangement. Today, almost all turbochargers use some means of varying (controlling) exhaust flow through the turbine housing.

Constant Geometry Turbos

We begin by describing the operation of a **constant geometry (CG)** turbocharger. These were commonly used on trucks until the mid-1990s but much less so today. CG turbos were designed for optimum performance at a specific rpm and a specific rejected heat load. In other words they were not versatile. This meant that engineers had to select when that moment of optimum turbine efficiency occurred. In most highway engines that used CG turbos, they were usually spec'd to produce peak efficiencies under full load and at peak torque rpm—and when operated outside that spec point, there would be a performance deficiency. CG turbos were simple and that shown in Figure 11–4 is typical.

The reason for spec'ing peak turbine efficiencies at peak torque rather than rated speed provided the CG turbo a self-regulating characteristic, because as engine rpm winds up, the real time available for charging the cylinders with air and injecting fuel diminishes. Operate a CG turbo outside its specified rpm range and the engine torque curve rapidly dropped off. This meant that the engine labored under lug and dropped off rpm rapidly when load was applied at the top end of its rpm range. Fuel economy is also adversely affected.

It should be noted that when we see CG turbos today, it is usually one of the pair of turbos used in series turbocharged engines, or in a compound turbocharger system such as the version described a little later in this section.

Variable Geometry Turbos

A little earlier in this chapter a **variable geometry (VG)** turbocharger was defined as one that uses either external or internal controls to either manipulate the flow areas through the turbine housing or allow some percentage of exhaust gas to bypass the turbine housing. In looking at VG turbos we will divide them as follow:

- Wastegate controlled
- Volute controlled

The objective of all VG turbos is to:

- Allow the turbo to act like a small turbocharger when engine loads are light.
- Allow the turbo to act like a large turbocharger when engine loading is high.

Today's ECM-controlled turbochargers will accomplish both the preceding objectives and every incremental stage in between. The payoff is a much faster response to requests for changes in engine output (reduced turbo lag) and, of course, reduced engine emissions.

Wastegate Controlled VG

Wastegated turbochargers have been around for many years. In operation, they can either allow all of the exhaust gas to be routed through the turbine housing or option a percentage of it to bypass the turbine directly to the exhaust circuit. The *gate* in the wastegate functions as the exhaust gas control door. Two methods are used to control the gate:

- Pneumatic: Wastegate movement depends on manifold pressure.
- Electronic: Wastegate is positioned by the engine ECM.

Pneumatic Wastegate A pneumatically controlled wastegate uses a wastegate actuator, which is usually a canister with an actuator rod to move the gate. The actuator rod connects the wastegate door and the other to a bellows inside the canister. An internal spring in the canister loads the actuator to a (default) closed position. In the closed position, all of the exhaust gas is routed through the turbine housing. On the other side of the bellows, manifold boost acts in opposition to the spring. When manifold boost reaches a prescribed value, it begins to overcome the spring pressure, moving the actuator rod to allow some of the exhaust gas to bypass the turbine housing.

Electronic Wastegate The engine ECM manages electronically controlled wastegates. For instance, some engines use wastegated turbos with dual entry ports. Using the example of a 2001+ Cummins ISX engine, the exhaust manifold was designed so that end gas from the front three cylinders is constantly routed through the turbine housing, and the wastegate managed exhaust flow from the rear three cylinders. This meant that exhaust gas could either be full flowed or incrementally bypass the turbine housing. To achieve this, the engine ECM uses a combination of solenoids (inlet and vent) and manifold boost to precisely manage the pressure acting on wastegate actuator bellows to provide four levels of boost.

Volute-Controlled VG

Volute-controlled VG turbochargers have become commonplace on the diesel engine of today. Race car engines have used variable

Figure 11–6 *Caterpillar C7 variable nozzle turbocharger cutaway. (Reprinted Courtesy of Caterpillar Inc.)*

aperture nozzles to control volute flow area for many years, and recently diesel engines have adapted this technology. The first volute-controlled diesel engine turbochargers appeared in the early 1990s with limited success, but today most diesel engine OEMs are using variable volute in preference to wastegated turbochargers. We are going to model our description of a volute-controlled VG turbocharger on the system used on a post-2007 Caterpillar ACERT C7 engine shown in Figure 11–6.

Variable Nozzle Turbocharger

Figure 11–6 shows a sectioned view of the Caterpillar **variable nozzle (VN)** turbo: Identify the turbine, pin wheel, vanes, and unison ring, cross-referencing this figure with Figure 11–7.

The VN turbocharger manipulates the volute flow area by moving the vane pitch to produce the desired boost. This is accomplished when oil pressure acts on the piston, which is tooth-meshed to the cam gear and crankshaft: This "cranks" or turns the unison ring. The unison ring supports the vane assemblies. Each vane has a spiral slot: When the unison ring rotational position changes, vane pitch angle changes because each vane pivots on wheel pins. Whenever the crankshaft moves the uni-

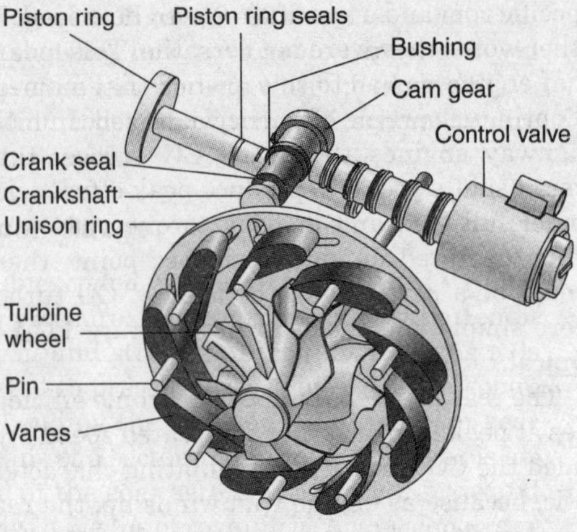

Figure 11–7 *Component operation of a VN turbocharger. (Reprinted Courtesy of Caterpillar Inc.)*

son ring, the vane pitch is altered simultaneously. In this way, the volute (inflow) flow area can be increased or decreased. Volute flow area onto the turbine determines gas efficiency in the turbine housing. In simple terms this defines turbine rotational speed and therefore the boost pressure produced by the turbo compressor.

The VN turbocharger is controlled by the VN control valve. The VN control valve is a proportioning actuator managed by the ECM: It converts the current delivered to the actuator into a specific piston ring position. Maximum boost is produced when the vanes are held in the nearly closed position: They are never fully closed. Boost air is managed by vane position because vane pitch governs how gas acts on the turbine. In Figure 11–8, oil pressure acting on the piston ring actuates the cam and crank assembly, turning the unison ring clockwise. Moving the unison ring clockwise opens the vanes, reducing turbine gas efficiency.

When oil pushes the piston to the left (Figure 11–9), the unison ring turns counterclockwise, moving the vane pitch to the nearly closed position, increasing turbine efficiency to provide maximum boost. You can see this action in Figure 11–10, which follows.

The rotational speed of the VN turbocharger is signaled to the ECM by a shaft speed sensor that uses an inductive pulse generator

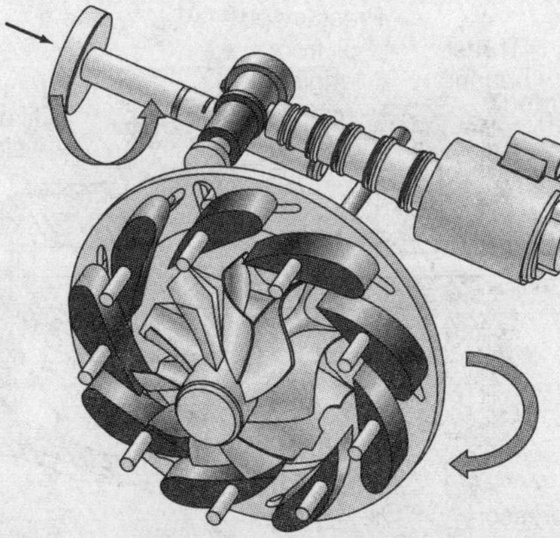

Figure 11–8 *VN turbocharger: opening vane pitch to reduce turbine efficiency. (Reprinted Courtesy of Caterpillar Inc.)*

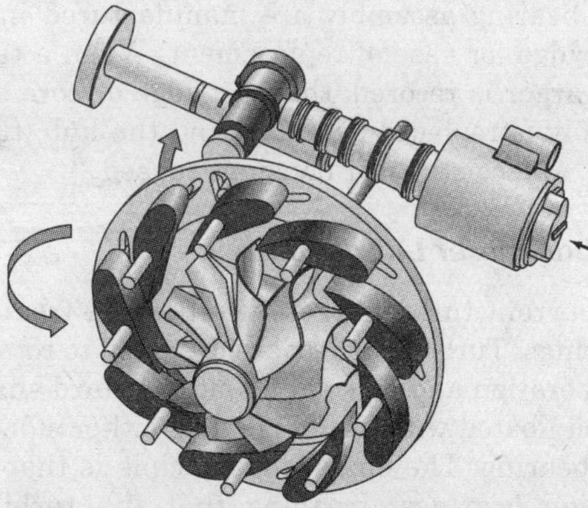

Figure 11–9 *VN turbocharger: closing down vane pitch to produce maximum boost. (Reprinted Courtesy of Caterpillar Inc.)*

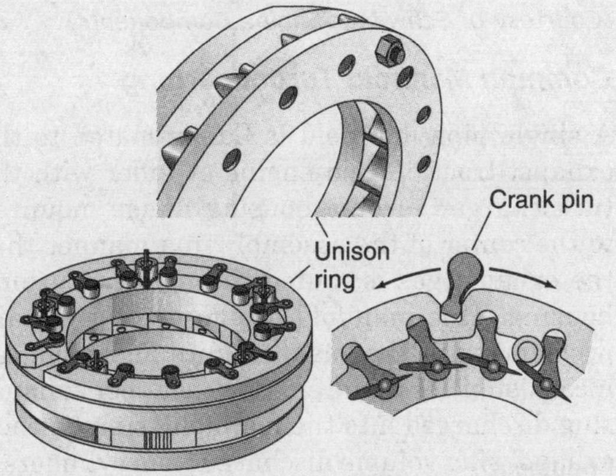

Figure 11–10 *Action of unison ring in controlling vane pitch. (Reprinted Courtesy of Caterpillar Inc.)*

Figure 11–11 *Overhead view of the VG actuator module used on a post-2007 Cummins ISC engine.*

principle: The trigger is a flat on one section of the turbine shaft.

Sliding Ring Volute Turbos

Another way of varying the volute flow area is used by Cummins in some of its Holset turbochargers. The VG actuator moves a sliding ring that determines flow unload area. The objective is identical to that on the Caterpillar VN turbo; it is just achieved in a slightly different way. Figure 11–11 shows an overhead view of the VG actuator module used on a post-2007 Cummins ISC engine, similar to those used on Volvo and Mack engines.

Construction

Turbochargers used in truck diesel engine applications can sometimes wind out at peak rpm of up to 150,000, perform continuously at shaft speeds of 75,000 rpm, and have to sustain temperatures that may exceed 1,400°F (760°C), so the materials used in turbo construction must be able to withstand high temperatures and centrifugal forces. The compressor components are often manufactured from aluminum alloys. Turbine housings must be able to sustain rotor burst. Rotor burst occurs when centrifugal force causes turbine or impeller disintegration. Turbine housings are manufactured from austenitic (a particular crystalline structure of iron and carbon) Ni-Resist^{TM} cast iron, which has high hot strength and resistance to oxidation. The turbines are manufactured from nickel or cobalt steels and ceramics. The turbo shaft is usually an alloy steel to which the turbine is either welded or bolted and the impeller is bolted. Sometimes the turbo shaft, turbine, impeller, and bearing assembly are manufactured in a cartridge for ease of replacement. When a turbocharger is recored, this cartridge or core assembly is replaced as a unit and the hub, turbine, and compressor housings reused.

Turbocharger Lubrication

All current turbochargers use floating friction bearings. Turbo bearings are designed to rotate in operation and do so at about one-third shaft speed floated with a film of oil on either side of the bearing. The operating principle is that of friction bearings, meaning that the turbine shaft is hydrodynamically suspended during operation. Accordingly, the radial play specification is a critical determination of bearing and/or shaft wear. Figure 11–12 shows a method of measuring turbine shaft radial play using a dial indicator. A thrust bearing defines the axial play of the turbine shaft, and gas pressure on both the turbine and compressor housings is sealed by a pair of piston rings. The oil that pressure feeds the bearings spills to an oil drain cavity and then drains to the crankcase by gravity by means of a return hose. The lubricating oil required for the bearings also plays a major role in cooling the turbocharger components. Figure 11–13 shows the oil flow routing through a typical turbocharger.

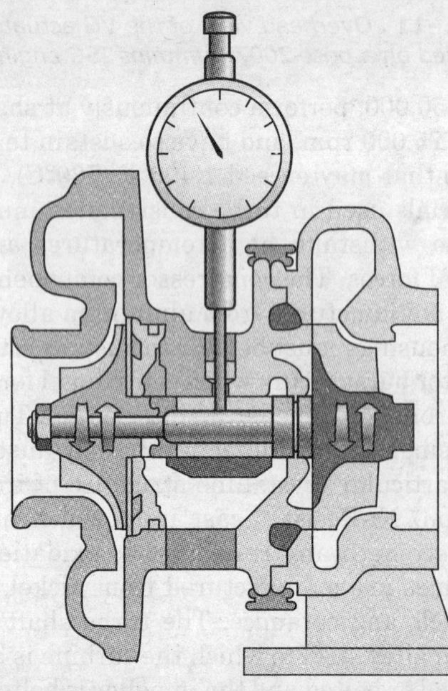

Figure 11–12 *Checking turbine shaft radial play.*

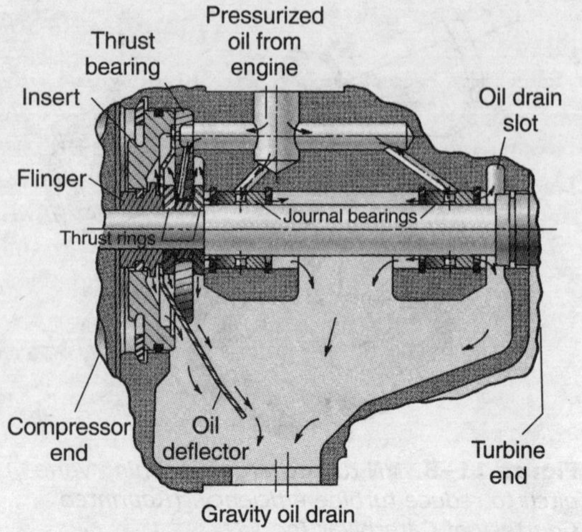

Figure 11–13 *Oil flow through a turbocharger. (Courtesy of Schwitzer Engine Components)*

Common Manifold Turbocharging

A single pipe manifold is flange mated to the exhaust tract of each engine cylinder with the turbocharger turbine housing flange mounted to the center of the assembly in a manner that the exhaust gas is routed through the turbine housing. This manifold design creates certain problems with the gas dynamics and produces inconsistencies in the effective heat per exhaust slug discharged into the manifold; simply, each exhaust slug volume discharge from cylinders 1 and 6 on an inline, six-configured engine must travel considerably farther before they enter the turbine housing tract than those discharged at cylinders 3 and 4 located at the center of the engine.

Pulse/Tuned Exhaust Manifold

OEMs of many current diesel engines often used **pulse/tuned exhaust** manifolds that use geometrically tuned pipes to direct the exhaust gas from each cylinder almost directly into the inlet tract of the turbine housing. This minimizes gas flow interference and exhaust slug pressure variables that reduce the efficiency of common manifold systems. Pulsed manifolds tend to diminish turbo lag (turbocharger response time) and increase low engine load performance.

Twin and Double Flow Turbos In a double flow design, the turbine throat is divided and each half feeds one half of the turbine wheel circumference. Another variation is the twin flow in which two passages are fed from a common throat for the full rotation of the turbine wheel. V-configured engines commonly use either the double or twin flow turbine housing design as shown in Figure 11–14.

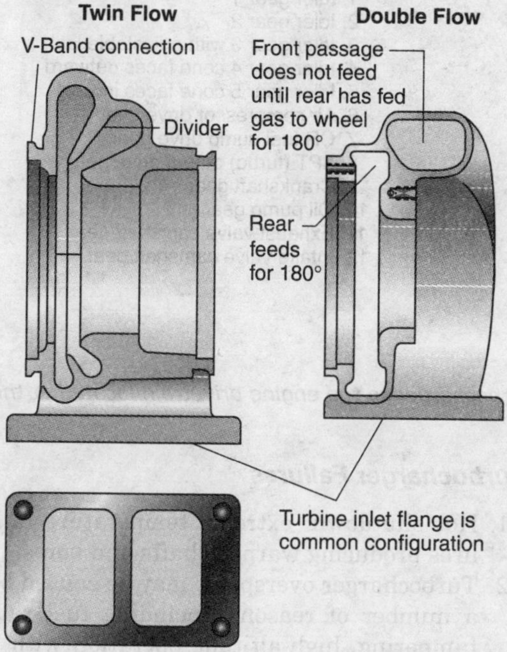

Figure 11–14 *Sectional view of a divided turbine housing. All housing types can have a water passage around the gas passage for cooling. (Courtesy of Schwitzer Engine Components)*

Two-Stage or Series Turbocharging Series turbocharging requires use of two turbochargers in series. The terms *primary* and *secondary turbochargers* (or low-/high-pressure turbochargers) designate the roles played by the two units used. This technology is used by Caterpillar in its electronic unit injector (EUI)-fueled ACERT engine family and by Navistar in MaxxForce 11 and 13 engines. Series turbocharging will become more common as OEMs strive for better management of the combustion process to maximize power and minimize emissions.

Paralleling Paralleling describes the use of multiple turbochargers usually to charge each bank of a V configuration engine. Parallel turbocharger configurations are not used on any current truck diesel engines, but they are used extensively in larger, off-highway applications. Paralleling is often used in railway locomotive applications.

Compound Turbocharging

The term **compound turbocharging** is often incorrectly used to describe two-stage or series turbocharging. In a true compound turbocharger circuit the turbine, in place of driving an impeller, is connected to a reduction gearing coupling with an output shaft indirectly connected to the engine crankshaft. Most turbo compounded systems make use of a viscous drive. In other words, the turbocharger helps directly drive the crankshaft. Compound turbocharging has been used for many years in off-highway diesel engine applications but has more recently been adopted for on-highway engines by DDC in its DD family of engines.

The DD15 engine uses a CG turbocharger. When a CG turbocharger is used in a compounded turbo arrangement, peak gas efficiency is set at the rated speed and load rpm of the engine: This provides maximum torque assist to the engine drivetrain when it is most needed. Figure 11–15 shows the components used in a compound turbocharger circuit. We will base our description of compound turbocharging on the recent DDC system described next.

Figure 11–15 *Components of a compound turbocharger.*

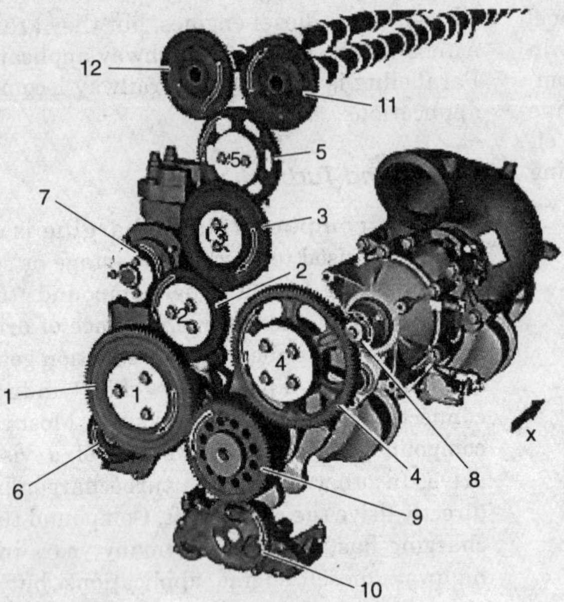

Figure 11–16 *Coupling arrangement of a compound turbocharger to the engine drivetrain located at the rear of the engine.*

Axial Power Turbine DDC refers to its compound turbocharger system as an **axial power turbine (APT)** system. The APT makes use of a CG turbine housing to impart drive torque to the engine geartrain. This is achieved by connecting the exhaust gas-driven turbine to reduction gearing in the APT as shown in Figure 11–16. DDC states that turbo compounding on a DD15 engine can contribute up to 50 horsepower (37 kW) at peak efficiency. The APT reduction gearing connects to the engine geartrain by means of a viscous drive and pinion (see Figure 11–16 #8). The APT pinion is in constant mesh with the APT-driven gear (see Figure 11–16 #7).

Turbocharger Precautions

1. Avoidance of hot shutdown: Allow at least 5 minutes of idling before shutting down the engine after prolonged loading on a dynamometer or road test.
2. Prelubing: Pour oil directly onto the turbine shaft through the oil supply flange when installing a turbocharger before connecting the lube supply line.
3. Some OEMs require that the oil filters be primed when replacing them at service intervals.

Turbocharger Failures

1. Hot shutdown: extreme temperature failures producing warped shafts and bores.
2. Turbocharger overspeed: may be caused by a number of reasons, including fuel rate tampering, high-altitude operation with a defective altitude compensation fuel control, and mismatching of complete turbocharger units.
3. Air intake system leaks: allows dirt to enter the compressor housing, causing the impeller vanes to be eroded wafer thin.
4. Lubrication-related failure: caused by abrasives in the engine oil, improper oil, broken down oil, or restricted oil supply.

CHARGE AIR HEAT EXCHANGERS

The act of compressing the intake air charge by the turbocharger typically produces air temperatures of 300°F (150°C) at a 70°F (21°C) ambient temperature and proportionally more at higher ambient temperatures. The objective of **charge air coolers (CACs)**, whatever their cooling medium, is to cool the air pressurized by the turbocharger as much as possible while maintaining the pressure. As intake air temperature increases, air density diminishes and the oxygen charge in the cylinder is reduced. The result is lower power and higher cylinder temperatures as the proportion of rejected heat increases. A CAC is a **heat exchanger**.

To minimize this loss of power potential, the boosted air charge is cooled using one of several types of heat exchanger. As a rule, a 1°C increase in intake air temperature will produce a 2°C increase in exhaust gas temperature. It should also be noted that because the performance of CACs greatly influences combustion temperatures and the density of the cylinder air charge, they are an integral component in the vehicle's noxious emissions control system. Most North American boosted diesel engines use some type of heat exchanger to cool intake air.

Types of Charge Air Cooler

For purposes of study, we will divide boost air coolers as follows:

- Air-to-air: Cooling medium is ram air.
- Aftercoolers and intercoolers: Cooling medium is engine coolant.
- Tip turbine: Cooling medium is turbine fanned air and engine coolant.

The type of charge cooler used depends on the vehicle application. For instance, air-to-air coolers offer the highest efficiencies but because these rely on a vehicle moving at some speed, they are not going to be effective in many vocational trucks or city buses.

Air-to-Air Heat Exchangers

These have the appearance of a coolant radiator and are often chassis mounted in front of the radiator (see Figure 11–13). As the vehicle moves down the highway, ambient air is forced through the fins and element tubing. **Ram air** is therefore the cooling medium. Cooling efficiencies are highest when the vehicle is moving at higher speeds. They are least efficient when the airflow through the cooler is minimal, and while the engine fan may assist, air-to-air cooling may not be suitable for construction applications. Under optimum conditions, air-to-air heat exchangers have efficiency ratios that better any liquid-cooled heat exchangers. Optimum conditions mean maximum airflow through the element so the vehicle has to be traveling at full speed. Figure 11–17 shows the location and plumbing of a typical CAC.

Figure 11–17 *Charge air cooler plumbing. (Courtesy of Mack Inc.)*

Testing Air-to-Air Coolers

When an engine is operating at close to rated speeds and loads, air enters a CAC at around 300°F (150°C). In cases where an air-to-air heat exchanger is used, the boost air should exit at around 110°F (44°C) when ambient temperature is 75°F (24°C), though there is some variability of this specification depending on the OEM. CACs are designed so that they can leak small amounts of air without any performance drop-off. When testing the air leak-off rate, plug the inlet and outlets of the CAC and pressurize to the OEM-recommended test value. Some examples of OEM-recommended test values are:

Engine OEM	*PSI Start Test Value*	*PSI Drop-off over 15 Sec.*
Caterpillar	30 psi	5 psi
Cummins	30 psi	7 psi
Detroit Diesel	25 psi	5 psi

If the pressure drop-off exceeds specifications, drop the pressure down to 5 psi and hold. Then attempt to locate the leak using a soap and water solution. Figure 11–18 shows the equipment and method Caterpillar recommends to test its CACs.

Dyno Testing and CACs

When chassis dynamometer testing trucks with air-to-air CACs, it is essential to take into ac-

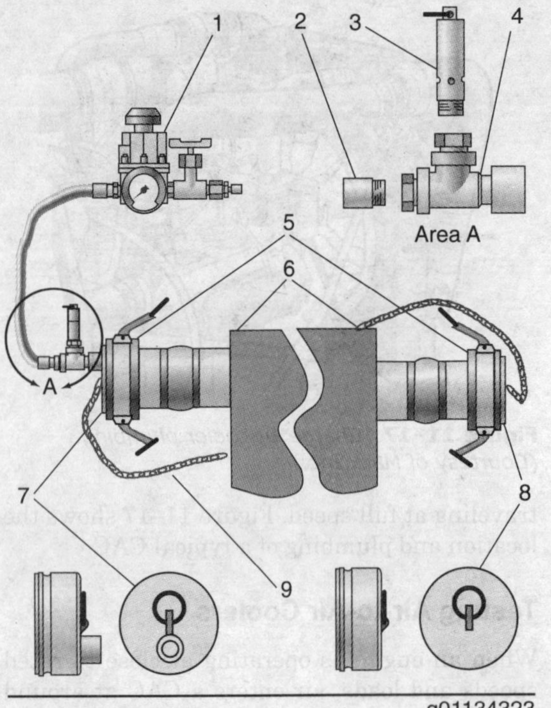

1. Regulator and valve assembly
2. Nipple
3. Relief valve
4. Tee
5. Coupler
6. Aftercooler
7. Dust plug
8. Dust plug
9. Chain

Figure 11–18 *CAC test kit: Charge CAC to 30 psi (200 kPa). Pressure drop-off should not exceed 5 psi (35 kPa) after 5 minutes. (Reprinted Courtesy of Caterpillar Inc.)*

count that airflow through the engine housing drastically reduces cooling efficiency. Using a dyno airflow fan does little compared with the ram air effect that makes an air-to-air CAC so effective. Because of the inefficiency of the CAC under chassis dyno test conditions, reduced peak BHP readings and tailpipe higher NO_x result. During a dyno run cycle, you should test for peak BHP almost immediately after the coolant has reached operating temperature.

Aftercoolers and Intercoolers

These normally refer to heat exchangers that use liquid engine coolant as the cooling medium. The two terms are synonymous and the one used merely indicates the OEM preference. Boosted air is forced through an element containing tubing through which coolant from the engine cooling system is pumped. Cooling efficiencies tend to be lower than with air-to-air heat exchangers due to the heat of the cooling medium; they continue to be used in applications in which low airflow through the engine housing rules out the use of an air-to-air exchanger.

Tip Turbine Heat Exchangers

The **tip turbine** type of heat exchanger uses a combination of ambient air and engine coolant as cooling media. Boosted air is delivered by a main duct to a heat exchanger through which it is forced before entering the engine cylinders. Engine coolant is circulated through tubing within the heat exchanger. However, a small amount of boosted air exiting the impeller housing of the turbocharger is ported off to drive a tip turbine. The tip turbine is an air-driven centrifugal pump whose function it is to force fresh, filtered ambient air through the heat exchanger element. This air does not come into direct contact with the charge air but acts to cool both it and the engine coolant. Tip turbine charge air cooling is popular in construction and stop/start applications in which airflow through the engine housing is insufficient to use air-to-air exchangers. The common application of tip turbines today are off-highway dump trucks and other mining and construction equipment. Figure 11–19 shows the operating principle of a typical tip turbine.

Boost Circuit Troubleshooting

When manifold boost is either too high or too low, engine performance complaints result. Common causes of lower-than-specified manifold boost are:

- Air system restriction upstream or downstream from the turbocharger impeller
- Air leakage downstream from the turbocharger impeller
- Low fuel delivery
- Mismatched turbocharger

Common causes of higher-than-specified manifold boost are as follows:

- Mismatched turbocharger

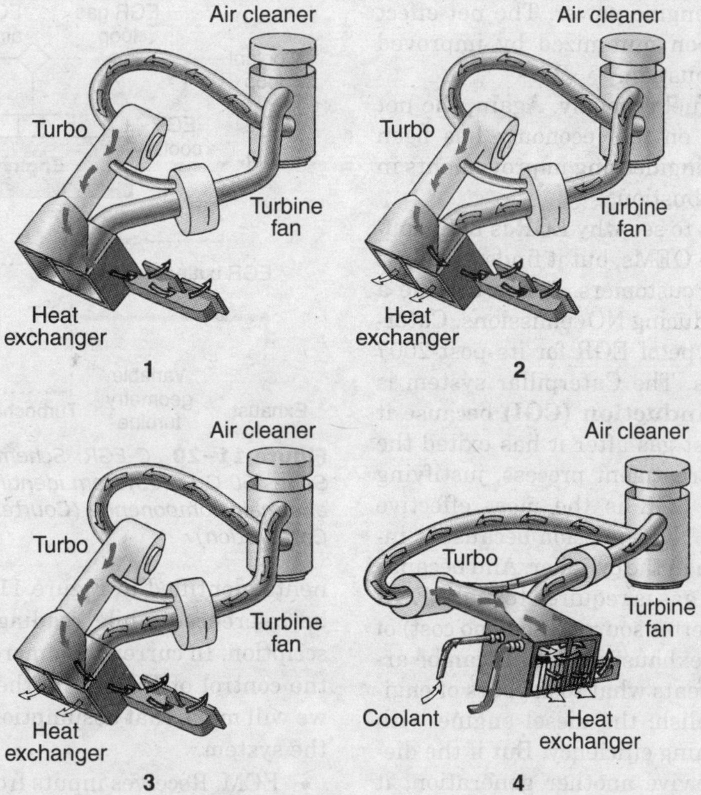

Figure 11–19 *Tip turbine operation. (Courtesy of Mack Inc.)*

- Sticking a unison ring on VG turbos
- Excessively high inlet air temperatures
- Deposits on the turbine volute or nozzle, discharge port, or turbine vanes
- Overfueling caused by the governor or by calibration programming
- Advanced fuel injection timing

EXHAUST GAS RECIRCULATION

Until the introduction of the Environmental Protection Agency (EPA) 2004 emissions standards, truck diesel engine manufacturers had successfully avoided using **exhaust gas recirculation (EGR)** systems. From 2007 onward, all the diesel engine OEMs are using some form of EGR though they may not call it EGR. Because EGR is an emissions control strategy, we will look at it primarily from the point of view of gas flow in this chapter. The role of EGR in the emissions control system is discussed in more depth in Chapter 41.

EGR Operating Principle

Oxides of nitrogen, or NO_x, is produced when engine combustion temperatures are high, which is a condition that occurs during lean burn combustion. Diesels run lean. Although EGR compromises performance output, by 2004 diesel engine manufacturers had no option but to turn to EGR. An EGR system routes some of the exhaust gas back into the intake system. Exhaust gas is "dead" gas. That means it is unreactive. Dead gas is routed back into engine cylinders simply to occupy some space and reduce the extent of lean burn that makes NO_x formation more likely. A better way of saying this is that EGR dilutes the intake charge of oxygen. It is not popular for the following reasons:

- Putting end gas back into the engine cylinder in place of fresh filtered air adds wear-inducing contaminants and increases engine oil acidity, both of which reduce engine longevity.

- EGR reduces engine power. The net effect of this has been minimized by improved control of combustion.
- EGR reduces fuel economy. Again, the net effect of EGR on fuel economy has been minimized by engineering improvements in managing combustion.

It is not difficult to see why EGR is not popular with the engine OEMs, but it finds even less favor among their customers. It has become a key strategy for reducing NO_x emissions. Caterpillar adopted a type of EGR for its post-2007 generation engines. The Caterpillar system is called **clean gas induction (CGI)** because it sources the exhaust gas after it has exited the exhaust gas aftertreatment process, justifying the term "clean." EGR is the most effective means of reducing NO_x emission because it inhibits its formation in the cylinder. And because dead or unreactive gas is required to make EGR work, there is no better source (it has no cost) of dead gas than the exhaust system. It can be argued that EGR defeats what took years of engineering to accomplish: the diesel engine with high engine breathing efficiency. But if the diesel is going to survive another generation, it will do so only by minimizing the emissions it produces.

Cooled EGR

Most diesel engine EGR systems cool the end gas using a heat exchanger before rerouting it back into the engine cylinders. The ECM controls the cycling of the EGR gas. Cooled EGR systems have produced the acronym **C-EGR**. Some OEMs have opted to use internal exhaust gas recirculation or **I-EGR** in engines powering vocational trucks. I-EGR functions by using the valve actuation train to permit some of the combustion end gas to remain in the cylinder. It accomplishes the same objective as C-EGR, that is, it dilutes the cylinder charge of oxygen by simply occupying space with *dead* gas. Figure 11–20 is a schematic of a Detroit Diesel cooled EGR circuit on a Series 60 diesel engine, and Figure 11–21 is the heat exchanger used on a post-2004 Volvo D12D engine.

EGR Components

A typical C-EGR system comprises the components identified in Figure 11–18, which should be referenced while reading through this description. In current commercial diesel engines, the control of EGR is by the engine ECM and we will make that assumption as we go through the system.

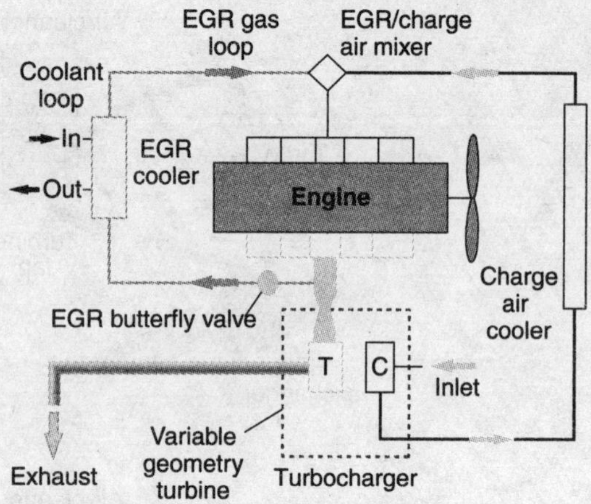

Figure 11–20 *C-EGR: Schematic of a DDC Series 60 C-EGR system identifying flow routing and main components. (Courtesy of Detroit Diesel Corporation)*

- ECM. Receives inputs from system sensors, and output circuit drivers manage the operation of EGR flow gates and mixing.
- Sensors. Key sensors in the EGR circuit are ambient temperature, barometric pressure (altitude), boost pressure, mass airflow (MAF), coolant temperature, and oil temperature.
- Butterfly valve. Manages flow rate from the exhaust circuit to the EGR circuit.
- C-EGR cooler. The C-EGR heat exchanger uses engine coolant to reduce the temperature of EGR gases to be recirculated back to the engine cylinders.
- EGR mixer. Combines exhaust gas with charge air from the intake system to be routed into engine cylinders. ECM controlled. In systems using differential pressure-type MAF sensors, the venturi is usually integrated into the EGR mixer assembly.

The MAF sensors used in the diesel engine EGR breathing system can use either hot-wire or differential pressure operating principles.

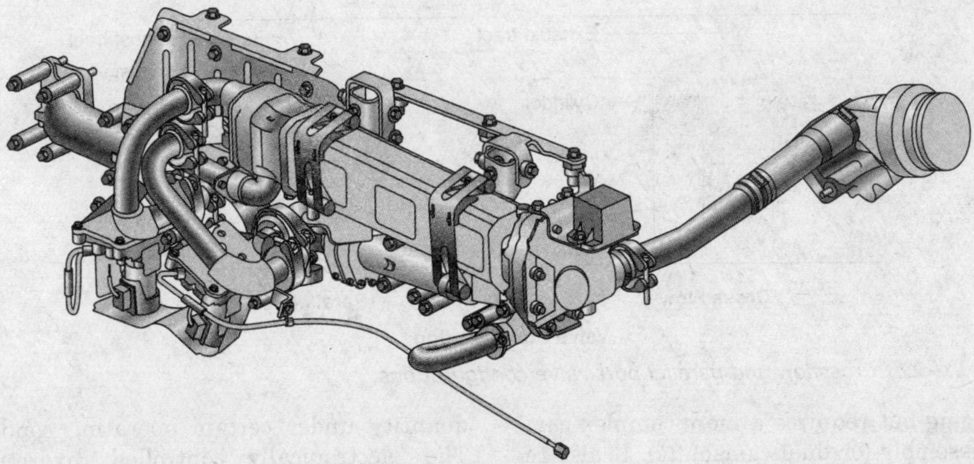

Figure 11–21 *C-EGR heat exchanger used on a Volvo D12D engine. (Courtesy of Volvo Trucks NA)*

Intake Manifold Design

Because of the almost universal use of turbochargers on diesels, intake manifold design is generally less complex than that for naturally aspirated spark-ignited (SI) engines. This usually means that the runners that extend from the plenum are of unequal lengths, and this does not notably compromise engine breathing. A tuned intake manifold is one in which the shape and length of each runner is similar and designed to establish optimum gas dynamics for engine breathing. A single box manifold supplied with boost air and EGR gas is usually all that is required to meet the engine's breathing requirements. Intake manifolds can either be wet (coolant ports) or dry, but dry are used in truck and bus applications. Materials used are aluminum alloys, cast irons, poly-plastics, and carbon-based fibers.

Where a tuned intake manifold is used in a boosted CI engine, it is designed to work with a tuned exhaust manifold and establish ideal gas dynamics for effective scavenging during valve overlap. In full authority, electronically managed engines, a thermistor is located in the intake manifold; this thermistor is often a primary reference for determining fuel-air ratio. Some OEMs use the term "tuned" to describe the mixing of boost air and EGR gas.

VALVE DESIGN AND BREATHING

Cylinder head valves were studied in some detail in Chapter 7 so our purpose here is to look at them from how they influence engine breathing. Most current diesel engines use multivalve configurations consisting of two inlet and two exhaust valves, but other arrangements, such as two inlet and one exhaust, are possible. Two basic breathing configurations are used for the more common four-valve cylinder heads, which are crossflow and parallel port.

Crossflow Configurations

Up until the electronic era, most diesel engines used **crossflow valve configurations**. Crossflow breathing locates both sets of valves transversely, meaning that the flow of intake air to the inboard valve interferes with that from the outboard valve, creating turbulence. This is good if the objective is to maximize gas movement in the cylinder, which it was in older engines. Back in the days when diesel engines produced lower fuel injection pressures and were designed to maximize cylinder turbulence, crossflow four-valve configurations did the job. Figure 11–22 compares crossflow with parallel port valve configurations.

Parallel Port Configurations

Breathing efficiencies can be improved by using **parallel port valve configurations**, which permit each pair of cylinder valves to be responsible for an equal amount of gas flow without crossflow interference. Parallel port valve configuration improves both cylinder charging and

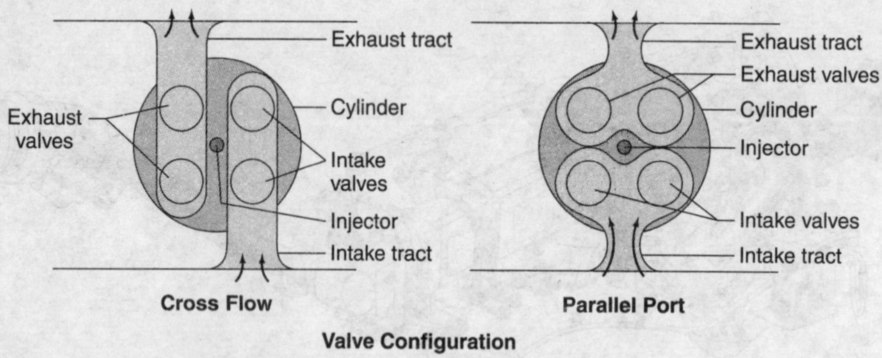

Figure 11–22 *Crossflow and parallel port valve configurations.*

scavenging but requires a more complex camshaft assembly (or dual camshafts). It also reduces cylinder gas velocity and swirl effect. This is desirable in newer engines that use the **quiescent** combustion technology that was explained in Chapter 7.

Valve Seat Angle

The valve seat angle also affects cylinder breathing. Valve seats in diesel engines are usually cut and machined at either 30 degrees or 45 degrees. Gas flow both into and out of the engine cylinder is generally around 20% greater using a 30-degree valve seat angle and the same lift as a 45-degree valve seat angle. However, a 30-degree valve has less mass around the critical seat area so they tend not to last as long. For this reason 45-degree seats tend to be used more in diesel engines due to their higher seating force and distortion resistance. Sometimes both are used. If you have worked on a Volvo VD12 engine you might have observed that the intake valves are machined at a 30-degree seat angle while the (hotter running) exhaust valves are machined at a 45-degree angle.

Variable Valve Timing

One of the strategies Caterpillar uses in its *advanced combustion and emissions reduction technology,* known as **ACERT**, is variable valve timing. In some Caterpillar engines using the ACERT system, intake valve closure can be delayed hydraulically (engine lube is the medium) by the ECM as a NO_x reduction strategy. This innovation allows the ECM to reduce the compression ratio and therefore the actual air quantity under certain operating conditions. The electronically controlled, hydraulically actuated, **variable valve actuator (VVA)** assembly is mounted over the rocker assembly. Its operation can be directly compared with the actuation of engine brake cycles, which open the exhaust valves. Figure 11–23 shows the VVA used on a Caterpillar ACERT C13 engine.

Figure 11–23 *Variable valve actuator (VVA) used on a Caterpillar ACERT C13 engine. (Reprinted Courtesy of Caterpillar Inc.)*

EXHAUST SYSTEM COMPONENTS

Because of the introduction of a complex array of exhaust aftertreatment devices in recent years, we are going to reference Chapter 41 in this textbook for a more detailed study of emissions control equipment. This chapter limits itself to a brief look at emissions controls. Typically, an exhaust system is required to perform the following:

- Assist cylinder scavenging.
- Minimize engine noise.
- Minimize engine noxious gas emission.
- Exhaust heat, noise, and end gases safely to atmosphere.

Exhaust Manifold

The exhaust manifold collects cylinder end gases and delivers them to the turbocharger. The exhaust manifold is usually manufactured in single or multiple sections of cast iron. Most diesel engine exhaust manifolds are "tuned" to deliver to the turbocharger exhaust gas with minimal and balanced flow resistance from each cylinder. The term "tuned" is used to denote any exhaust system designed with at least a little consideration for the exhaust gas flow. If the exhaust manifold and piping is properly designed, as each slug of cylinder exhaust gas is discharged, it will not "collide" with that from another cylinder but instead will be timed to unload into its tailstream. The pulsed manifolds discussed earlier in this section in the context of turbochargers are an example of a tuned exhaust system.

Exhaust backpressure factors in a diesel engine are the turbocharger turbine assembly, the exhaust aftertreatment apparatus, and the engine silencer. Exhaust manifold gaskets in truck/bus engine applications are usually of the embossed steel type, although occasionally fiber gaskets are used; they are usually installed dry. Note that embossed steel exhaust manifold gaskets are designed for one-off usage. When torqued into position, these gaskets are designed to yield (deform) to the two mating surfaces. Attempting to reuse them will almost certainly result in a leak sooner or later.

Tract and Manifold Insulation

One of the roles of the exhaust manifold is to route cylinder rejected heat to the turbine housing of the turbocharger, where it can be "recycled" by producing intake manifold boost. To achieve this at the highest efficiency, some OEMs have reduced heat losses by insulating both the exhaust tract and/or the exhaust manifold. The insulation usually consists of a stainless steel liner.

Pyrometer

Although **pyrometers** are used in emissions control components, especially those built to meet 2007 emissions, they are not often used to signal exhaust gas temperature to the driver in any current highway diesel engines. In the older hydromechanical diesel engines of the 1980s and earlier, a pyrometer located 12-inches (30 cm) downstream from the turbine housing of the turbo was used as a means of alerting the driver to downshift when temperatures rose above a certain threshold. The practice of locating one pyrometer in the exhaust tract on each engine cylinder to indicate cylinder balance is used on larger diesels but seldom on truck and bus engines. A pyrometer is a thermocouple and its electrical principles are explained in Chapter 28. Most diesel particulate filters use several pyrometers to signal temperature conditions to the ECM.

Exhaust Piping

Where mild steel piping, flex piping, and clamps are used, the steel is corrosion protected usually by galvanizing (zinc coating). However, it is becoming more common to use stainless steels, which provide the exhaust system components with at least double the service life. Stainless steel band clamps have been used for many years and are an effective means of sealing flex pipe to straight pipe as they yield to shape at installation. Because they are stretched to yield (deform) at installation they are not designed to be reused.

The function of exhaust piping is to collect the engine exhaust gases unloaded by the turbine housing and route them to the engine aftertreatment canister from which they will discharge to atmosphere clear of the tractor cab structure and the trailer structure. If rain caps are used on vertical exit pipes, they should be mounted transversely, opening away from the chassis. They should never be mounted longitudinally facing either forward or rearward because this can adversely affect exhaust gas flow.

Engine Exhaust Aftertreatment

To meet EPA emissions standards, today's diesel engines have highly complex exhaust gas aftertreatment devices usually housed within a single canister. Emissions control technology is the subject matter of Chapter 41 but at this point it is at least necessary to identify these components in the exhaust circuit. Typical exhaust gas treatment canisters incorporate:

- A muffler or engine silencer
- A DPF
- An oxidation catalytic converter
- A reduction catalytic converter
- Clean air induction piping (routes "dead" gas back to intake)
- Selective catalytic reduction (aqueous urea injection)
- A diffuser (redirects DPF heat away from the trailer/reefer, etc.)

This aftertreatment canister is equipped with many different sensors. Unlike an older exhaust system, the aftertreatment device is usually manufactured by the engine OEM and *must* be replaced with an OEM-approved equivalent. Because they are engine computer controlled and part of the emissions control circuit, we will address them in some detail in Chapter 41 of this textbook. In this chapter, we address only the principles of a muffler or engine silencer on a diesel engine. The design of the aftertreatment canister attempts to minimize restriction in routing "clean" exhaust gas to atmosphere. A more complex, stanchion mounted aftertreatment canister is shown in Figure 11–24: The operation of these assemblies is explained in Chapter 41.

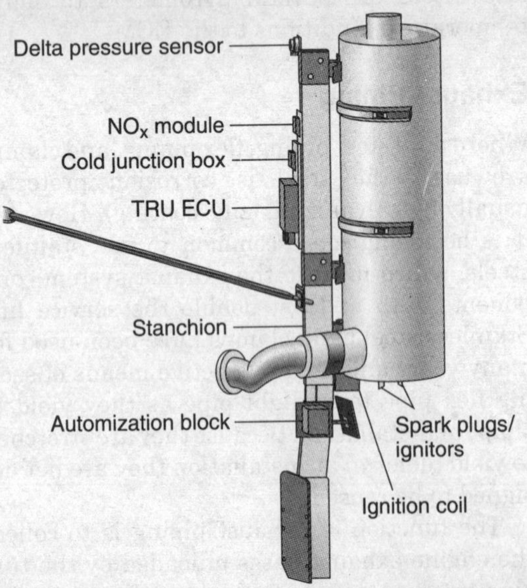

Figure 11–24 *Stanchion mounted, exhaust gas aftertreatment canister. (Courtesy of Mack Inc.)*

Sonic Emissions Control

Sound is an energy form and it is produced in an engine by the firing pulses in engine cylinders. Sound energy from a running engine produces sonic nodes and antinodes at a frequency generally considered to be unpleasant to the human ear. The function of the silencer or muffler is to modify the wavelength or frequency of engine noise to legally acceptable levels. Essentially, this process requires the scrambling of the frequency of the sonic nodes and antinodes produced by engine firing pulsations.

Before the tougher sonic emissions standards of the 1990s, a turbocharger in conjunction with piping design was in some cases able to meet the minimum sonic emissions standards, but this is no longer so. Pressure waves caused by the unloading of each slug of exhaust gas into the piping can have an amplifying effect on engine sound, and the turbocharger is capable of smoothing these pressure waves and enabling a simpler muffler design.

Engine silencers use two basic principles to achieve their objective of dampening sound:

1. **Resonation.** Resonation requires reflecting sound waves back toward the source, thereby multiplying the number of sound emission points. Separate chambers connected by offset pipes as well as baffles are used to achieve the objective of scrambling the frequency of the engine's firing pulses. One of the reasons that the muffler section (usually the last section of the device) of a post-2007 aftertreatment canister looks

more simple is that the presence of diesel particulate filters and catalytic converters have a significant resonation effect.

2. **Sound absorption.** Exhaust gases pass through a perforated pipe enclosed in a canister filled with sound-absorbing material. Sound absorption involves converting sonic energy into heat by friction, and the efficiency of these devices depends on the packing density of the sound-absorbing media and the pipe perforation geometry. The media must be resistant to pressure pulsations and mechanical vibration; mineral and metal wools are used, most often basalt wool. Sound absorption mufflers generally have improved flow resistance characteristics over comparable resonator principle mufflers.

Muffler Volume

Muffler volume required in pre-2007 truck engine systems was typically around five times the total exhaust slug discharge volume per cycle. Note that exhaust slug discharge volume is always somewhat greater than total swept volume in manifold-boosted diesel engines. Since the universal adoption of DPFs and multistage catalytic converters into the exhaust gas aftertreatment canister, there are no easy rules regarding muffler volume. In most cases, it is much less in newer engines due to the resonation effect created by the emissions control devices upstream from the muffler. In simple terms, that means that the mufflers used with post-2007 diesel engines are often more simple than their predecessors.

Diffusers

Because of the amount of heat generated by DPFs during regeneration cycles, it is necessary to fit a heat diffuser on stanchion-mounted, exhaust gas aftertreatment canisters. The heat diffuser simply redirects heat away from a trailing load that may be heat sensitive.

BREATHING CIRCUIT SENSORS

While we are not going to take a look at the operating principles of sensors until Chapter 32, we are going to identify those used in a typical system. Both intake and exhaust circuits are comprehensively monitored in current diesel engines because of the need to manage emissions within statutory limits. The following is a list of sensors that may be found on a diesel engine and, in brackets, the electrical principles that they operate on: Many sensors on today's engines are multifunctional. That means they perform multiple tasks so what appears to be one sensor may read temperature and pressure.

- Ambient temperature (thermistor)
- Barometric pressure (variable capacitance)
- CAC in temperature
- CAC out temperature
- Boost circuit temperature (thermistor)
- Manifold boost pressure (variable capacitance or piezo-resistive)
- MAF (hot-wire or delta pressure differential)
- Turbocharger shaft speed (inductive pulse generator)
- Exhaust gas backpressure (variable capacitance)
- DPF in temperature (pyrometer)
- DPF combustion temperature (pyrometer)
- DPF out temperature (pyrometer)
- NO_x sensor (density)
- Aftertreatment canister mass flow (delta pressure sensor)

REVIEW QUESTIONS

1. The tool of choice to accurately check the inlet restriction of a dry air filter would be a:
   a. Mercury-filled manometer
   b. Trouble light
   c. Water-filled manometer
   d. Restriction gauge
2. A typical maximum specified inlet restriction for an air filter on a turbocharged diesel would be:
   a. 25″ water
   b. 25″ mercury
   c. 25 psi
   d. 25 kPa
3. Which of the following types of filters has the highest filtering efficiencies?
   a. Centrifugal precleaners

b. Oil bath
c. Dry, positive

4. Which of the following should be performed first when checking an engine that produces black smoke under load?
a. Injection timing
b. Plugged particulate trap
c. Fuel chemistry by lab analysis
d. Air filter restriction test

5. Which of the following best describes the function of a C-EGR system on a diesel engine?
a. It increases engine breathing efficiency.
b. It dilutes intake charge with cooled dead gas.
c. It preheats the intake charge to the cylinder.
d. It assists the turbocharger in boosting intake pressure.

6. Which of the following best describes the function of a wastegate in a current turbo-boosted diesel engine?
a. Bleeds down intake boost air when excessively high
b. Options exhaust gas to bypass the turbine
c. Adjusts the volute flow area
d. Holds the exhaust valves open during low boost requirement

7. When series turbocharging is used on an engine, which of the following should be true?
a. A single CG turbocharger is used.
b. Twin turbos are arranged in series.
c. The VG actuator is in series with the EGR mixer.
d. Twin turbos are arranged in parallel.

8. What is the critical flow area managed by a VG turbo using a variable nozzle operating principle?
a. Compressor diffuser
b. Impeller inlet throat
c. Turbine volute
d. EGR gate

9. To produce peak manifold boost from a VN-type turbocharger, which of the following should be true?
a. Vanes are in the nearly closed position.
b. Vanes are in the fully open position.
c. Wastegate is fully open.
d. Wastegate is fully closed.

10. The constant geometry turbocharger used with a high torque rise highway diesel engine is usually designed to produce optimum efficiency at which rpm?
a. Rated
b. Top engine limit (TEL)
c. Peak torque
d. High idle

11. *Technician A* states that an air filter functions at optimum efficiency when it is new and that by the time it requires replacing, it has low filtering efficiencies. *Technician B* states that oil bath filters are seldom used with today's highway diesel engines because they have low filtering efficiencies. Who is correct?
a. Technician A only
b. Technician B only
c. Both A and B
d. Neither A nor B

12. *Technician A* states that a Roots blower is a positive displacement air pump, which volumetrically displaces air proportionally with rpm increase. *Technician B* states that Roots blowers are used to aspirate two-stroke cycle engines and are not found on current highway diesel engines. Who is correct?
a. Technician A only
b. Technician B only
c. Both A and B
d. Neither A nor B

13. *Technician A* states that the term *manifold boost* is used to describe breathing on a turbocharged engine. *Technician B* states that turbine speeds on truck diesel engines can exceed 100,000 rpm. Who is correct?
a. Technician A only
b. Technician B only
c. Both A and B
d. Neither A nor B

14. *Technician A* states that using a light bulb inside an air filter element is a reliable means of determining whether the filter is plugged. *Technician B* states that air filters are best left untouched between service intervals unless the filter restriction gauge indicates an out-of-specification reading. Who is correct?
a. Technician A only
b. Technician B only

c. Both A and B
d. Neither A nor B

15. *Technician A* states that the first item to be checked when an engine produces black smoke should be the air filter inlet restriction gauge. *Technician B* states that whenever an engine produces black smoke the air filter should automatically be changed. Who is correct?
a. Technician A only
b. Technician B only
c. Both A and B
d. Neither A nor B

16. Which of the following methods would identify a hole or tear in an air filter element?
a. Light bulb test
b. Air inlet restriction gauge
c. Manometer test
d. Regulated air pressure test

17. When is a diesel engine turbocharger capable of highest rotational speeds?
a. At peak engine rpm
b. When rejected heat is greatest
c. At peak torque
d. When exhaust manifold pressure is greatest

18. Which of the following components would be located in a diesel exhaust gas after-treatment canister?
a. Diesel particulate filter
b. Catalytic converter(s)
c. Muffler
d. Any or all of the above

19. Which of the following best describes the function of a VG turbocharger?
a. It behaves like a large turbocharger when engine load is high.
b. It behaves like a small turbocharger when engine load is high.
c. It reroutes NO_x back into the engine cylinders.
d. It reroutes EGR gas into the DPF.

20. *Technician A* states that the function of a turbocharger wastegate is to divert exhaust gas around the turbine directly into the exhaust system. *Technician B* states that a wastegated turbocharger can improve turbine efficiencies at lower loads and rpms. Who is correct?
a. Technician A only
b. Technician B only
c. Both A and B
d. Neither A nor B

12

Engine Retarders

INTRODUCTION

Engine compression brakes are widely used on North American truck diesel engines. A large percentage of vehicle braking requires application pressures of 20 psi or less in a typical air brake system. This means that for the majority of vehicle braking requirements, less than a fifth of the air brake system's peak potential is being used. The objective of a supplementary brake system such as an engine compression brake is to relieve the vehicle braking system of some of the light-duty braking. This becomes especially important in instances that require prolonged light-duty braking such as on an extensive downgrade.

Energy Conversion in Air Brakes

A typical truck braking system is air actuated and converts the kinetic energy (energy of motion) of the vehicle to heat energy. This takes place at the brake foundation assembly, and its mechanisms are drums (or rotors), brake shoes (or calipers), and actuating apparatus that converts the potential energy of compressed air into mechanical force. When the brakes are applied, the friction facings on the brake shoes or calipers are loaded into a rotating drum or rotor and resultant resistance or friction retards the movement. The friction produces heat so the brake system must be designed to sustain high temperatures and dissipate braking heat rapidly. The **coefficient of friction** describes the aggressiveness of friction materials. It is desirable to maintain a consistent coefficient of friction in foundation brake components for obvious reasons.

However, coefficient of friction is altered by changes in temperature and the addition of a lubricant such as water. When foundation brake systems are subjected to the most amount of use, they are required to dissipate the most amount of heat to atmosphere. Failure to dissipate this heat can greatly increase foundation brake temperatures, causing both a reduction in the coefficient of friction of the critical friction surfaces at the shoe and drum and an expansion of the drum dimension. When a tractor-trailer combination has to run downhill for prolonged periods, even though brake application pressures are relatively light, excessively high foundation brake temperatures result, causing a reduction in braking efficiency at a time when it is most needed, a condition known as **brake fade**.

Supplementary Brake Systems

A supplementary braking system such as an engine compression brake or driveline retarder can support the main vehicle braking system, enhance peak braking performance, and in cases in which engine braking is engine control module (ECM) managed, reduce some of the driver-actuated braking. The engine compression brake on which most current versions are modeled was invented by Clessie Cummins after he had officially retired from his position as chief executive officer (CEO) of the company that bears his name. The Jacobs Manufacturing Company, perhaps better known for its drill chucks, manufactured engine compression brakes since its introduction of Cummins's first engine brake in the late 1950s to comparatively recent times. Until recently, Jacobs manufactured an engine brake for almost every highway diesel engine and, in doing so, established universal brand identification. Today the term "Jake brake" is used by many drivers to refer to any engine compression brake.

Trucks may also use hydraulic retarders, which may be located anywhere in the driveline, but are usually integral with the engine or transmission. Hydraulic driveline retarders use a torque absorption principle that can be compared to that used in a torque converter. Electric driveline retarders are also currently available and one of them is marketed by Jacobs. The Caterpillar BrakeSaver, which is coupled to the engine is covered in this chapter because it is an engine component rather than a transmission or driveline component.

Retarding Principle

The term **retarder** can be applied to any component or system that effects braking action, that is, retards or slows motion. However, the primary vehicle braking system is seldom referred to as a vehicle retarder system, whereas engine compression brakes and driveline brakes are commonly described as retarders. Compression and driveline brakes do no more than supplement the primary braking system: The retarding powerflow is necessarily confined to those axles directly linked to the drivetrain, that is, through the wheels on the drive axles. The objective of the engine brake system is identical to that of the main vehicle braking system—namely, converting the energy of vehicle motion to heat energy. In the engine compression brake, kinetic energy (that of vehicle motion) is converted first to the potential energy of compressed air and then released into the exhaust system as heat.

Today's Engine Brakes

In recent years as engine control electronics have become more complex, engine manufacturers have been less inclined to entrust the engineering of engine brakes to companies outside of their control. The result of this trend is that most engine original equipment manufacturers (OEMs) today manufacture engine brakes in house. This makes sense. Management of engine brake duty cycles has come a long way from the days when the only way to actuate a retarding cycle was for the driver to toggle a dash switch.

Brake Cycle Controls

With the increased use of electronics in truck chassis systems, the way engine brake cycles are managed has changed considerably over the past few years. In older systems control of the brake retarding cycle was left entirely to the driver, whereas in computer-controlled systems the chassis management electronics are playing an ever-larger role. An obvious use of a computer-controlled engine brake is by the cruise control system to help maintain a consistent road speed.

However, as multiplexing becomes more of a factor on our trucks and we see the emergence of collision warning, electronic lane tracking, and smart braking systems, this opens the door for increased use of engine braking. Engine braking makes sense both from safety and maintenance perspectives. It has become much more than a way of extending foundation brake, friction face service life.

PRINCIPLES OF OPERATION

The term *engine brake* can be used to describe retarder devices that use three distinct operating principles:

1. Internal engine compression brakes
2. External engine compression brakes
3. Hydraulic engine brakes

Internal Engine Compression Brakes

In the four-stroke cycle diesel engine, the piston is required to:

- Perform work (compression stroke)
- Receive work (power stroke)

On its upward travel during the compression stroke, the piston compresses the cylinder air charge. This is immediately followed by the power stroke during which the piston is forced downward through its downstroke by expanding cylinder gases. For an engine to continue turning over, the work received by a piston during the power stroke must exceed the work it has to perform during the compression stroke. Back in Chapter 4 we learned that this net gain is what we describe as mean effective pressure, or MEP. Internal engine brakes use the MEP equation to change the role of the engine.

The **internal compression brake** operates by making the piston perform all of its usual work of compressing the air charge on the compression stroke. Then, just as the power stroke is about to take place, it releases cylinder pressure by opening the exhaust valves. This has the effect of negating (canceling) the power stroke by releasing the compressed cylinder gases to the exhaust system (gas blowdown) somewhere around top dead center (TDC). An internal engine compression brake reverses the normal engine function, converting it from an energy-producing pump to an energy-absorbing compressor.

All engine compression brakes use this principle of operation; the mechanisms used to actuate, time, and control the brakes vary. The braking powerflow of the vehicle begins at the drive axle wheels, extends through the transmissions and drive shafts, through the engine powertrain, and the kinetic energy is converted to pressure; the potential energy of the compressed air is then dumped into the exhaust system. Braking efficiencies are highest when engine rpm is highest.

External Engine Compression Brakes

The operating principle of the **external compression brake** is not that different from the internal compression brake. External engine compression brakes are also known as **exhaust brakes**. They consist of a housing located downstream from the turbine housing discharge in the exhaust system, or they are integrated into the turbine housing of the turbocharger. The principle is simple. Inside the exhaust brake housing is a valve that when actuated, chokes off the exhaust discharge. The result is that the piston on its exhaust stroke is being made to perform compressive work. Restricting exhaust gas flow in this way once again reverses the role of the engine, converting it into an energy-absorbing pump. However, in the internal engine compression brake, the effective pumping stroke is the compression stroke, whereas in the exhaust brake, the effective retarding stroke becomes the exhaust stroke. Therefore, as each succeeding exhaust slug is unloaded into the exhaust manifold, the pressure rises and braking efficiency increases.

Retarding efficiency is to some extent diminished by pressure loss to the intake during valve overlap. Operation of the exhaust brake is managed so that the engine is not fueled during braking and as with the internal engine compression brake, braking efficiencies are highest when the engine rpm is in its higher range. In many current diesel engines, retarding efficiencies are increased by combining internal and external engine compression brakes. This means that both upward strokes of the piston are "braking" strokes. In this way, many current engines are able to absorb (during retarding cycles) nearly as much power as they are capable of producing—at least when engine rpm are close to rated speed.

Hydraulic Engine Brakes

Many types of electric and hydraulic driveline brakes exist, but these are more appropriately dealt with when studying transmission and driveline components. However, the Caterpillar **BrakeSaver™** is a hydraulic retarder that is coupled to the rear of the engine and uses engine lubrication oil as its medium. A rotor is coupled directly to the engine crankshaft, so it rotates at any time the engine is running. The rotor is turned within the BrakeSaver housing, which is coupled to the flywheel housing and to which the transmission is mounted. The engine flywheel is bolted through to the crankshaft, but because of the BrakeSaver, the starter motor must crank the engine by means of a ring gear on a ring gear plate mounted to the crankshaft behind the BrakeSaver rotor. The rotor is therefore driven by the crankshaft between the BrakeSaver housing and a stator.

When the BrakeSaver is actuated, the housing is charged with pressurized engine oil. Vaned pockets on the rotor mean that when the BrakeSaver is charged with oil, the rotor encounters fluid resistance defined by the stator geometry and the rotor rotational speed. Once again, retarding efficiencies are greatest when rotational speed is greatest. The oil that acts as the BrakeSaver hydraulic medium is engine oil, which is supplied from the engine sump by a section of the oil pump dedicated to charging the BrakeSaver.

Control Circuits

Engine brakes typically use electrical control switches. These electrical control switches may use engine lube as the hydraulic medium to control the actuators required by the brake circuit. Because electrical signals control engine brake operation, retarding cycles may occur in three different ways:

- Switched by direct driver command
- Switched by the engine management ECM using internal logic such as in a smart cruise application
- Switched by the engine management ECM using multiplex logic from other modules on the data bus such as in an antirollover strategy.

Electro-hydromechanical

When engine brakes are used in hydromechanically managed engines, the switching of the engine brake is usually electrical but the retarding effect must be actuated either hydraulically or pneumatically. In such engines, a control circuit requires that a series of switches be closed before engine braking can be effected. The first of this series of switches would be the driver control switch, which can be proportional depending on the system. In a proportional engine brake system, typical driver options are:

- Off
- Low: two-cylinder retarding
- Medium: four-cylinder retarding
- High: six-cylinder retarding

Once the dash switch has been closed, the next switch in the series is located at the clutch. The clutch must be fully engaged to operate engine braking, so the clutch switch closes when the clutch pedal is in the fully released position. The third essential switch in the series (though any number of others may be installed) ensures that the accelerator pedal is fully released: In other words, engine fueling is at zero (all engines since 1994) or at least at a minimum in engines before 1994. The objective is to limit the quantity of raw fuel dumped into the exhaust system and to atmosphere. Since 1994 federal regulations have prohibited the discharge of uncombusted fuel to atmosphere.

Electronic Management

Today's engines all use ECM control of engine brake cycles. The driver may still have the option of switching the engine brake on or off, but depending on the system the engine may manage braking cycles in ways that the driver may be entirely unaware of. Because many of today's engine brakes option many different levels of power absorption, they can be used to complement cruise control logic and be made to work in conjunction with data bus-driven strategies such as yaw (directional stability) and antirollover electronics. Although today's engine brakes are electronically managed, the outcomes may be electrically, pneumatically (chassis system air pressure), and hydraulically (engine lube) actuated.

EXHAUST BRAKES

As indicated earlier all external engine compression brakes use the same operating principles. We describe a couple of them here.

Williams Brakes

Williams exhaust brakes are classified as external engine compression brakes (Figure 12–1). Control of the brake is electric over pneumatic. The electrical circuit required to actuate the engine brake consists of three switches, all of which must be closed: a control switch (dash mounted), a clutch switch (clutch must be fully engaged), and an accelerator switch (accelerator must be at zero travel). The sliding gate exhaust brake uses a pneumatically activated gate, which is actuated by chassis system pressure. The air supply to close the gate is controlled by an electrically switched pilot valve. An aperture in the sliding gate permits a minimal flow through the brake gate during engine braking. The butterfly valve version operates similarly (Figure 12–2 and Figure 12–3).

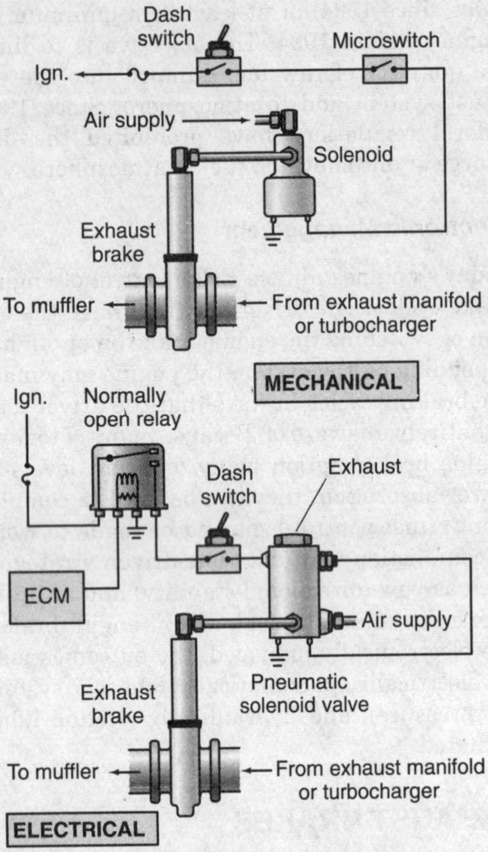

Figure 12–1 *Control and actuation circuits on Williams mechanical and electronic exhaust brakes.*

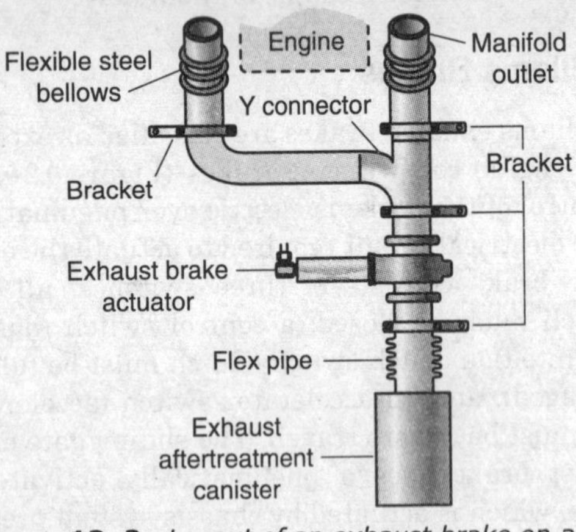

Figure 12–2 *Layout of an exhaust brake on a V-configured engine.*

Exhaust Pressure Governor

Volvo Trucks have used what they call an **exhaust pressure governor (EPG)** on their

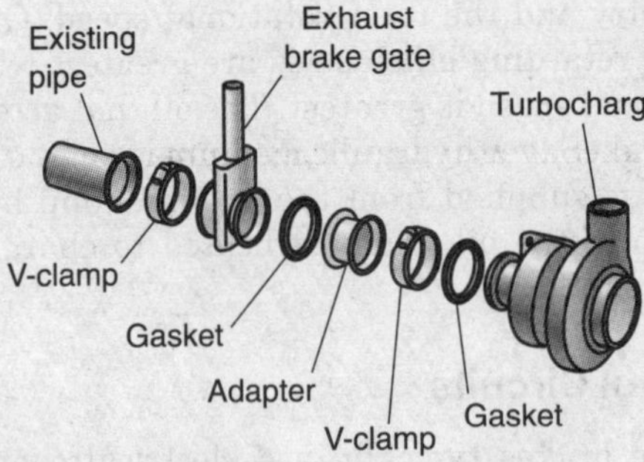

Figure 12–3 *Exhaust brake components.*

engines. These use the principles of exhaust engine brakes we have described earlier, specifically the choking down on exhaust flow downstream from the turbocharger. The EPG doubles as an engine warm-up device because of the high backpressure created by choking of the exhaust flow during actuation. The Volvo EPG is integrated downstream from the turbine housing of the turbocharger as shown in Figure 12–4.

Figure 12–4 *Volvo's exhaust pressure governor (EPG) variation of an exhaust brake.*

INTERNAL COMPRESSION BRAKES

In approaching the operation of internal engine compression brakes we will take a look at an electrical-over-hydraulic engine brake, followed by a couple of more recent examples of electronically managed versions. The power absorption equation is the same in all internal engine compression brakes:

Drive axle tires to *chassis drivetrain* to *engine crankshaft* to *piston* (on compression stroke).

Jacobs Brakes

People in the trucking industry use the term **Jake brake** to refer to any internal engine compression brake because for a number of years, this manufacturer dominated the marketplace by designing its engine brakes for most of the major OEMs. However, there were other manufacturers of similar devices but that did not stop them from being referred to as Jake brakes.

Operating Principle

As an internal engine compression brake, a **Jacobs brake** operates by having the piston perform its usual work of compressing the cylinder air charge on the compression stroke and then negating the power stroke by opening the exhaust valves somewhere around TDC and releasing the cylinder compressed air to the exhaust. Jacobs engine brakes use electric switching of a hydraulic actuating circuit, so they adapt readily to electronic management, and have been used as an integral component in "smart" cruise control systems.

Jacobs brakes hydraulically actuate the opening of the exhaust valve(s) in the engine cylinder performing the braking. The manner in which this hydraulic circuit is timed and actuated varies with the engine; however, the mechanical force required to actuate the hydraulic circuit is typically provided by the engine camshaft. This means that the timing of the exhaust valve opening is also governed by cam profile, either the injector actuating cam profile or that on an adjacent exhaust valve train.

An example of an engine brake actuated by movement from an adjacent exhaust valve is shown in Figure 12–5. When you study this illustration, note that the image shows two different rocker arms. In this circuit, once oil is trapped in the engine brake hydraulic circuit,

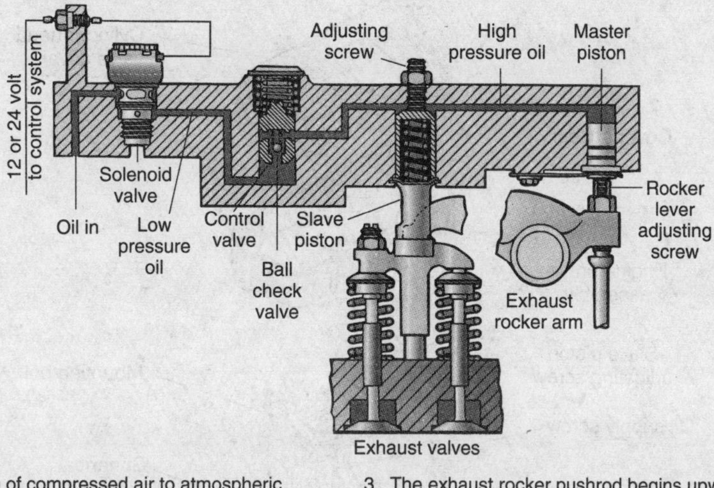

The blowdown of compressed air to atmospheric pressure prevents the return of energy to the engine piston on the expansion stroke, the effect being a net energy loss since the work done in compressing the cylinder charge is not returned during the expansion process.

Exhaust blowdown of the braking cylinder is accomplished by utilizing the pushrod motion of an exhaust valve of another cylinder during its normal exhaust cycle as follows:

1. Energizing the solenoid valve permits engine lube oil to flow under pressure through the control valve to both the master piston and the slave piston.
2. Oil pressure causes the master piston to move down and come to rest on the corresponding exhaust rocker arm adjusting screw.
3. The exhaust rocker pushrod begins upward travel (as in normal exhaust cycle), forcing the master piston upward and directing high pressure oil to the slave piston of the braking cylinder. The ball check in the control valve traps high pressure oil in the master/slave piston circuit.
4. The slave piston (under the influence of the high pressure oil) moves down, momentarily opening the exhaust valves while the engine piston is near its top dead center position, releasing compressed cylinder air to the exhaust manifold.
5. Compressed air escapes to the atmosphere, completing a compression braking cycle.

The level of engine braking is controlled by using the solenoid to turn each housing ON or OFF. The above figure shows the relationships between master pistons, slave pistons, and control valves within the housing.

Figure 12–5 *Jacobs compression brake schematic (on Caterpillar or Mack) and operating description. (Courtesy of Jacobs Vehicle Systems)*

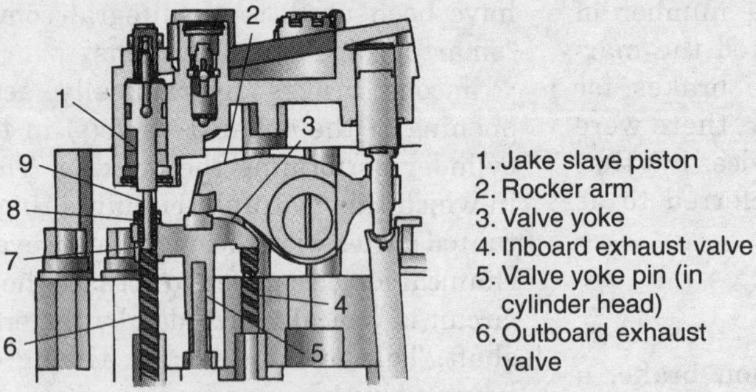

Figure 12–6 *Jacobs brake on a Mack E-Tech engine. (Courtesy of Mack Inc.)*

the master piston is forced inboard by exhaust rocker upward (opening) movement that creates pressure rise. Pressure in the engine brake actuation circuit acts on the sectional area of the slave piston, which in turn acts on the exhaust valve bridge on the cylinder being braked. The slave piston thereby forces the valve bridge downward, opening both exhaust valves, dumping cylinder pressure created by the piston compression stroke.

Figure 12–6 and Figure 12–7 show, respectively, sectional and exploded views of a typical

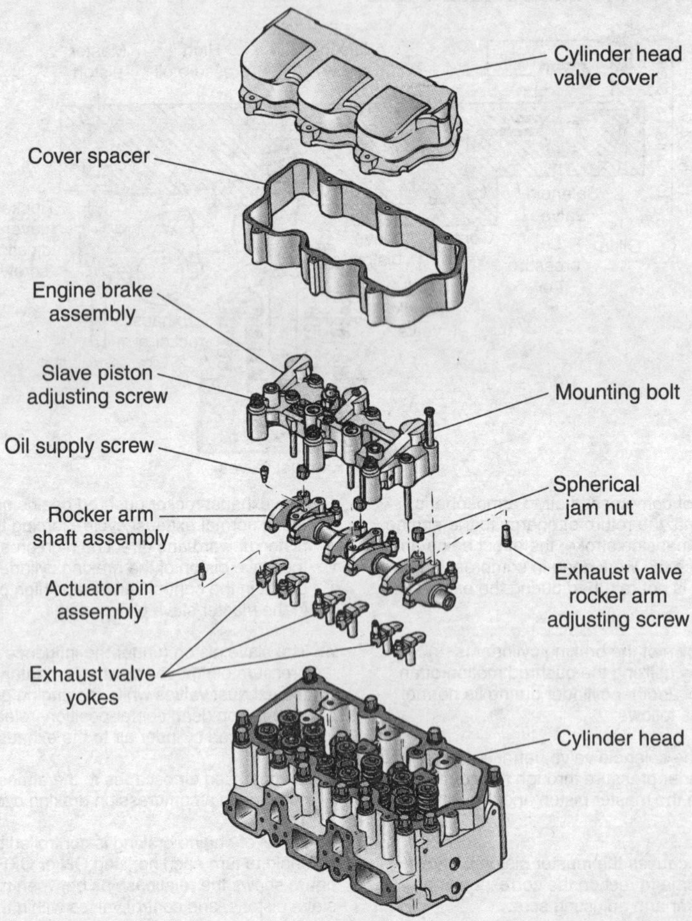

Figure 12–7 *Exploded view of Jacobs compression brake components. (Courtesy of Mack Inc.)*

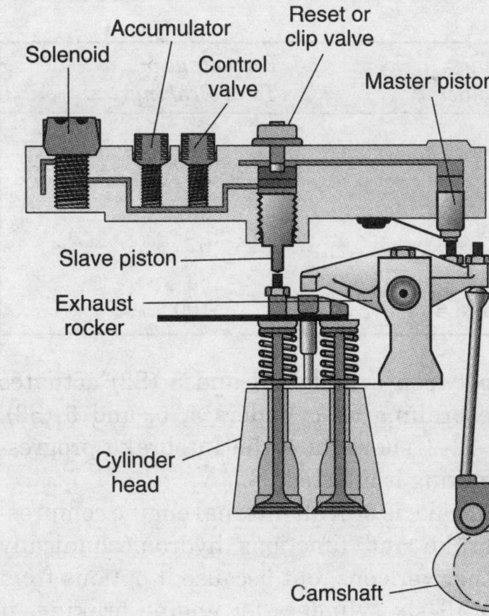

Figure 12–8 *Sectional view of Jacobs engine brake components.*

internal engine compression brake hardware. Remember that all internal engine compression brakes use the same general principles despite minor differences in the hardware. Figure 12–8 shows the setup of a Jacobs brake used on Mack Trucks E-Tech engines up to 2007.

Caterpillar Compression Brake

Caterpillar uses its in-house design of engine compression brakes on its C13 and C15 family of electronic unit injector (EUI)-fueled engines. We will reference Figure 12–9 in outlining the operating principles of this device. The Cat compression brake sources engine oil to use as the hydraulic actuation medium from around the studs of the rocker shaft pedestals. The actuator valve (see Figure 12–9 #5) controls oil flow in the compression brake housing. When the control valve is switched by the ECM, oil at lube system pressure passes from the actua-

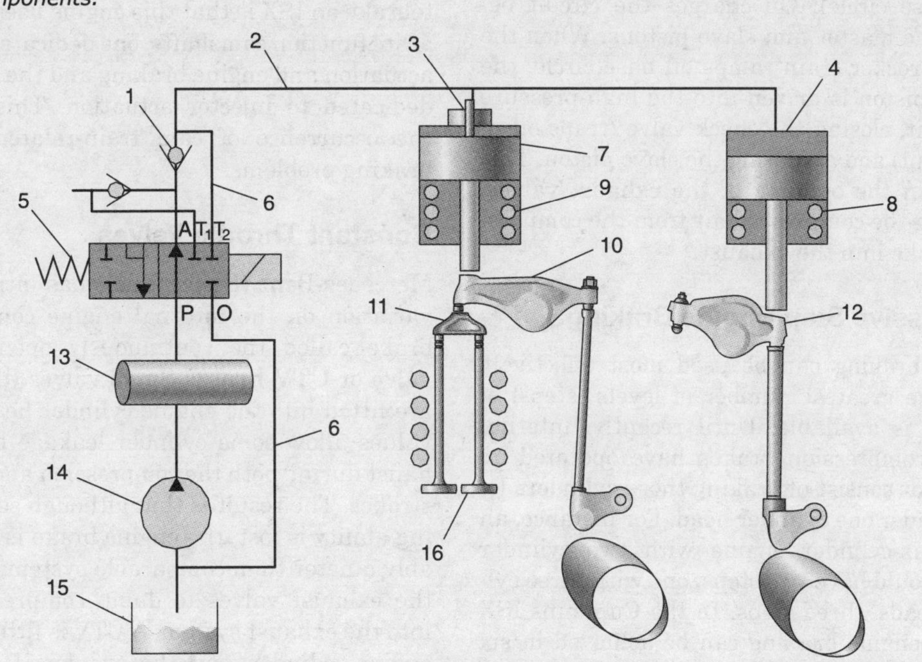

1. Check valve
2. High pressure oil passage
3. Slave piston adjustment screw
4. Master piston
5. Actuator valve
6. Oil drain passage
7. Slave piston
8. Master piston spring
9. Spring for the slave piston
10. Exhaust rocker arm
11. Exhaust bridge
12. Fuel injector rocker arm
13. Rocker arm shaft oil passage
14. Engine oil pump
15. Engine oil pan
16. Exhaust valve

A. Actuation port
T1. Drain port
T2. Drain port
P. Actuator spool for supply port

Figure 12–9 *Operation of a Caterpillar C13 internal engine compression brake. (Reprinted Courtesy of Caterpillar Inc.)*

Table 12–1 *ISX Intebrake Stepped Braking Options*

No. of Braked Cylinders	*Solenoids Energized*	*Actual Braked Cylinder Number*	*Percentage of Total Braking*
1	S1	1	17
2	S2	2 and 3	33
3	S3	4, 5, and 6	50
4	S1 and S3	1, 4, 5, and 6	67
5	S2 and S3	2, 3, 4, 5, and 6	83
6	S1, S2, and S3	1, 2, 3, 4, 5, and 6	100

tor spool supply port (12–9 P) to the actuation port (12–9 A). The resulting oil flow opens the check valve (12–9 #1) and passes into the high-pressure oil passage (12–9 #2): This routes oil to both the slave (12–9 #7) and master pistons (12–9 #4).

As oil pressure overcomes the spring (12–9 #8), the master piston moves toward the injector rocker arm, which is in its unactuated position (base circle): Oil charges the circuit between the master and slave pistons. When the injector rocker train ramps off base circle, the master piston is driven into the high-pressure oil circuit, closing the check valve (traps oil in the circuit) and actuating the slave piston. This results in the opening of the exhaust valves, dumping the compressed air from the compression stroke into the exhaust.

Progressive Step Engine Braking

Engine braking can be used most effectively when the greatest number of levels (steps) of braking is available. Until recently, internal engine compression brakes have operated on steps that consist of braking those cylinders located under one cylinder head. For instance, an inline six-cylinder engine with two cylinder heads would have two steps, one with three cylinder heads, three steps. In the Cummins ISX engine, engine braking can be achieved in six steps, meaning that under a braking cycle, one cylinder progressing to all six can be used. Cummins calls this **Intebrake** and it is achieved using engine oil as the hydraulic medium and three solenoids.

When one cylinder is selected for braking, only the solenoid between cylinders 1 and 2 (S1) is energized. The solenoid between 2 and 3 actuates, braking for cylinders 2 and 3. The solenoid between cylinders 4 and 5 (S2) actuates engine braking on cylinders 4, 5, and 6 (S3). Table 12–1 shows how the Intebrake progressive braking is effected.

Intebrake is still an internal engine compression brake and functions hydromechanically like other versions, but because it options from one to all six cylinders for engine braking, it will probably be copied. A real advantage of Intebrake on ISX is that this engine uses two separate function camshafts, one dedicated to valve actuation and engine braking and the other one dedicated to injector actuation. This reduces the occurrence of cam train-related engine-braking problems.

Constant Throttle Valves

Mercedes-Benz (Freightliner) has introduced a variation on the internal engine compression brake called the continuously open throttle valve or **CTV brake**. Small valves (the CTVs) are fitted into the engine cylinder head. These values allow some cylinder leakage to the exhaust during both the compression and exhaust strokes. The result is that although some braking ability is lost, the engine brake is considerably quieter than comparable systems that use the exhaust valves to dump compression gas into the exhaust system. A CTV is fitted to each engine cylinder and the system is managed electrohydraulically using engine oil as the medium. When activated, the CTVs remain open throughout the braking cycle. Actuation of the CTV brake requires that the accelerator pedal be at zero travel and the clutch be fully engaged. The system is designed to cease retarding when engine rpm drops below 1,100 rpm (factory preset), but this value can be reprogrammed up or

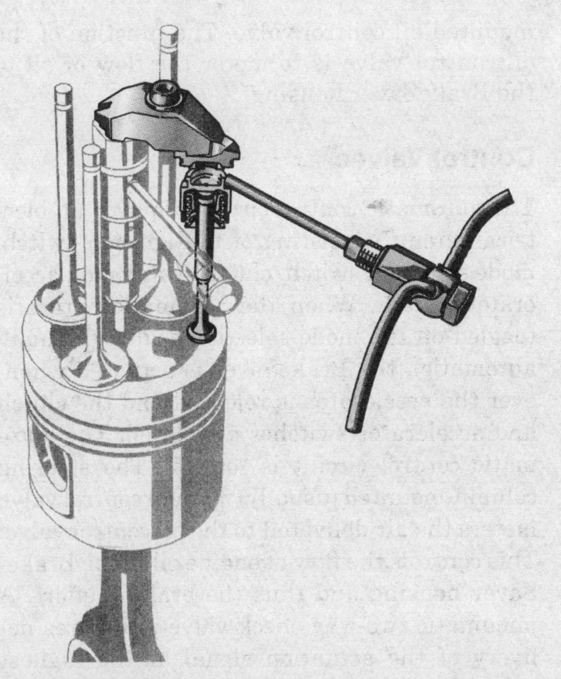

Figure 12–10 *Location of a CTV on the engine. (© Courtesy of Daimler AG)*

down depending on application. Figure 12–10 shows an MB-906 CTV.

When the CTV circuit is actuated on the compression stroke, the CTV permits cylinder compressed air to bleed through the valve into the exhaust circuit. By the time the piston moves away from TDC at the completion of the compression stroke, most of the compression charge has been dumped into the exhaust circuit. Figure 12–11 will help guide you through the following explanation. After the electrical actuation of the solenoid slide valve (4) by the control module, the 3/2-way valve spool (5) opens, allowing engine oil to be directed into the cylinder head to each CTV holding each open. Unintentional closure (fluttering) of the CTV by cylinder backpressure is prevented by a pressure holding valve (6). To take the system out of braking cycle, the solenoid slide valve is de-energized, closing the oil supply at the 3/2 valve spool routing the oil back to the sump. This causes oil pressure in the pressure chambers above the CTV pistons (1.1) to collapse, and the CTVs close to seal the cylinder by spring force. The normal engine cycle resumes at this point.

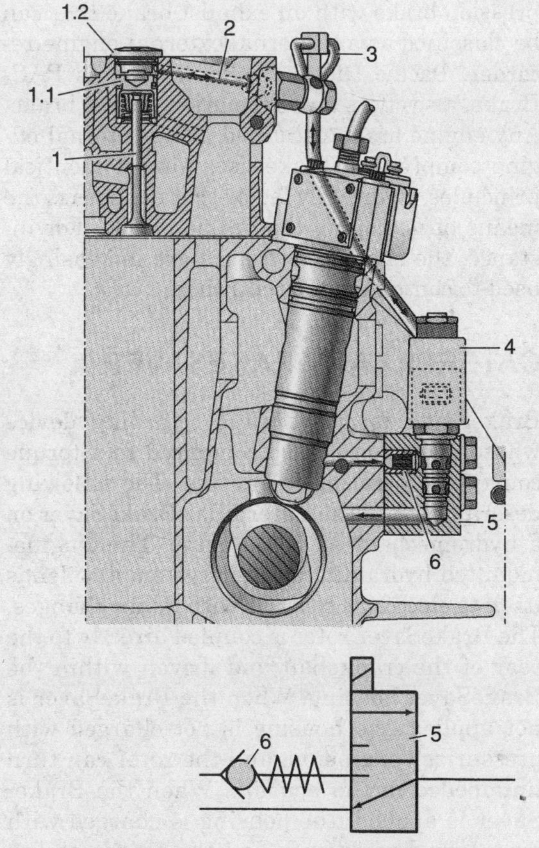

1. Constantly-open throttle valve
1.1. Constantly-open throttle piston
1.2. Constantly-open throttle valve cover
2. Oil port
3. Oil line
4. Solenoid slide valve
5. 3/2-way valve
6. Pressure holding valve (0.2 bar)

Figure 12–11 *Components in a CTV engine brake on an EUP-fueled engine. (© Courtesy of Daimler AG)*

The constant throttle valve retarder is internal engine compression braking with a difference. By not using the engine camshaft to manage the engine braking, camshaft problems associated with engine braking are eliminated. The system is also notably quiet in operation.

Other Internal Retarders

Although they may be referred to as Jake brakes, there are other internal engine compression brakes available on diesel engines marketed in North America, many of them manufactured offshore. Volvo and Mack use a two-stage engine brake on their current engine families: This combines an internal engine compression brake with an exhaust brake that can be classified as an internal/external engine retarder. Pacific Diesel Brake, known as PAC-Brake, also offers an aftermarket engine brake. Any engine brake described as an internal engine compression brake uses almost identical principles, so the only factor that changes is the means of managing the braking cycle. For instance, the engine electronics are increasingly used to control engine retarding.

CATERPILLAR BRAKESAVER

BrakeSaver is a hydraulic retarding device whose operation can be compared to a torque converter operating in reverse. The following description is of the Caterpillar BrakeSaver on a hydromechanical 3406 engine. This engine-mounted hydraulic retarder system also lends itself to electronic controls with minor changes. The BrakeSaver rotor is coupled directly to the rear of the crankshaft and driven within the BrakeSaver housing. When the BrakeSaver is not applied, the housing is not charged with pressurized engine oil, and the rotor can turn unimpeded in the housing. When the BrakeSaver is applied, the housing is charged with pressurized engine oil, and the rotor's ability to turn is modulated by fluid resistance and the pocket or vane geometry, producing the retarding effect. The BrakeSaver offers modulated retarding; the control circuit is electrically or manually controlled and pneumatically actuated. Chassis system air pressure is reduced by a pressure-reducing valve, which drops the pressure to a maximum value of 50 psi (345 kPa): This feeds the manual control valve and the solenoid control valve. Both the automatic solenoid valve and the manual control valve use air to control the flywheel-mounted oil control valve. The function of the oil control valve is to meter the flow of oil to the BrakeSaver housing.

Control Valves

The automatic control circuit requires an electrical circuit consisting of the ignition switch, mode selection switch, clutch switch, and accelerator switch. When the automatic option is toggled on the mode selection switch (manual/automatic), the BrakeSaver is actuated whenever the accelerator is released and the clutch and accelerator switches are closed. The automatic control circuit is "on/off." The steering column-mounted (usually) manual control valve meters the air delivered to the oil control valve. This controls the flow of engine oil to the BrakeSaver housing and thus the braking effort. A pneumatic two-way check valve prioritizes delivery of the actuation signal at the highest pressure to the oil control valve.

Oil Flow

Engine oil is the hydraulic medium used by the BrakeSaver. It is supplied to the oil control valve by the engine oil pump, which is designed with a front and a rear section. The front section feeds the engine lubrication circuit while the rear section feeds exclusively the BrakeSaver oil control valve. BrakeSaver-equipped engines require larger capacity oil sumps. When pneumatically actuated by either the automatic or manual control valve, the oil control valve routes oil to the BrakeSaver housing at a high rate of flow and at a pressure value of around 70 psi (480 kPa). Because of the pumping action within the BrakeSaver, pressures at the outlet increase 50% over the charge pressure. When the

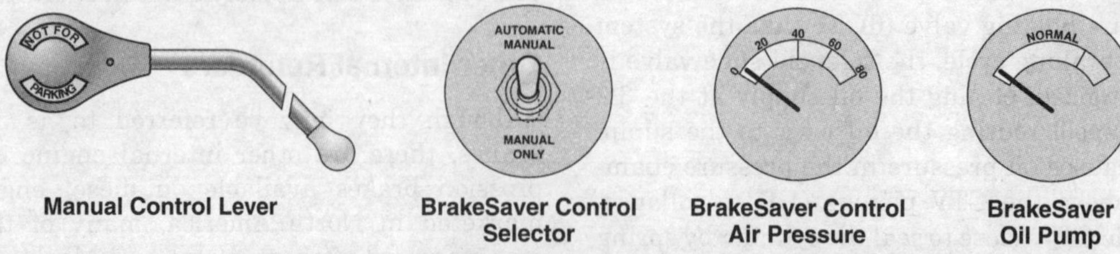

Figure 12–12 *Caterpillar BrakeSaver controls. (Reprinted Courtesy of Caterpillar Inc.)*

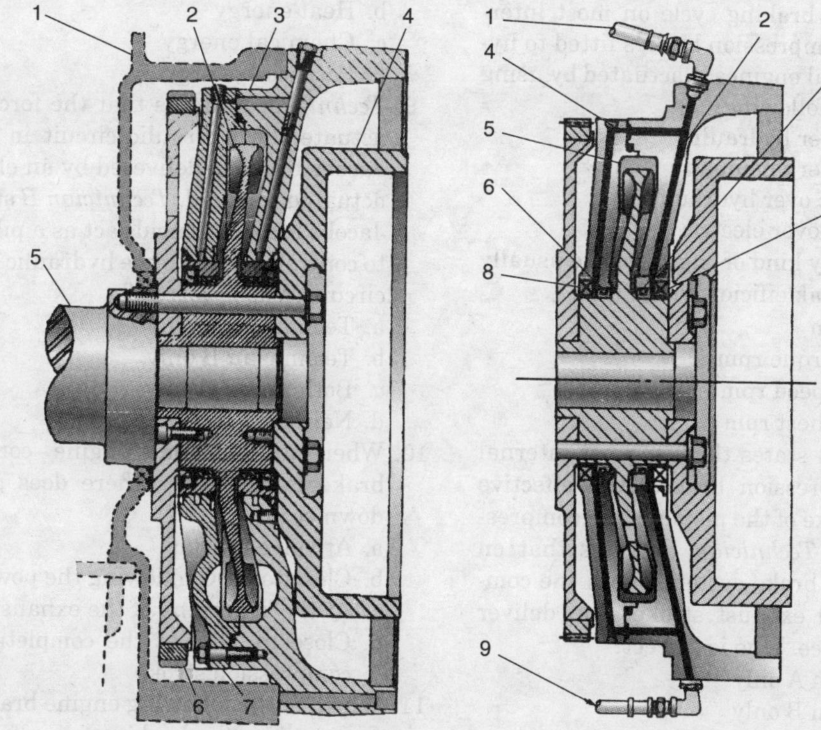

BRAKESAVER COMPONENTS

1. Flywheel housing
2. Rotor
3. BrakeSaver housing
4. Flywheel
5. Crankshaft flange
6. Ring gear plate
7. Stator

1. Oil line
2. Orifice
3. Piston-type ring seal
4. Orifice
5. Chamber
6. Piston ring seal
7. Lip-type seal
8. Lip-type seal
9. Oil line

Figure 12–13 *Sectional view of the Caterpillar BrakeSaver. (Reprinted Courtesy of Caterpillar Inc.)*

BrakeSaver is not actuated, a small amount of oil is pumped through the housing to lubricate the seals. Oil exiting the housing is routed directly to the oil cooler (Figure 12–12 and Figure 12–13).

REVIEW QUESTIONS

1. Which type of engine brake uses the compression stroke of the piston as its retarding stroke?
   a. Internal compression brake
   b. External compression brake
   c. Hydraulic retarder
   d. Volvo EPG
2. Which type of engine brake uses the exhaust stroke of the piston as its retarding stroke?
   a. Internal compression brake
   b. External compression brake
   c. Hydraulic retarder
3. *Technician A* states that the Cummins ISX Intebrake is capable of optioning engine braking on just one engine cylinder. *Technician B* states that the Cummins ISX Intebrake is an external engine compression brake. Who is correct?
   a. Technician A only
   b. Technician B only
   c. Both A and B
   d. Neither A nor B

4. The effective braking cycle on most internal engine compression brakes fitted to hydromechanical engines is actuated by using which of the following?
   a. Electric over hydraulic
   b. Electric over pneumatic
   c. Pneumatic over hydraulic
   d. Hydraulic over electric
5. When will any kind of engine brake usually operate at peak efficiency?
   a. At idle rpm
   b. At peak torque rpm
   c. At rated speed rpm
   d. At the highest rpm
6. *Technician A* states that on most internal engine compression brakes, the effective braking stroke of the piston is the compression stroke. *Technician B* states that on some engine brake systems, both the compression and exhaust strokes can deliver retarding force. Who is correct?
   a. Technician A only
   b. Technician B only
   c. Both A and B
   d. Neither A nor B
7. The operating principle of an internal engine compression brake is to convert the engine into an:
   a. Energy-absorbing compressor
   b. Energy-releasing machine
   c. Inertia-absorbing device
   d. Inertia-producing device
8. The energy of vehicle motion is ultimately converted to what energy form when an engine compression brake is used?
   a. Kinetic energy
   b. Heat energy
   c. Chemical energy
   d. Potential energy
9. *Technician A* states that the force used to actuate the hydraulic circuit in a typical Jacobs brake is delivered by an electrically actuated solenoid. *Technician B* states that Jacobs brake solenoids act as a pilot switch to control oil flow to the hydraulic actuation circuit. Who is correct?
   a. Technician A only
   b. Technician B only
   c. Both A and B
   d. Neither A nor B
10. When an internal engine compression brake is actuated, where does gas blowdown occur?
    a. At valve overlap
    b. Close to BDC following the power stroke
    c. At the beginning of the exhaust stroke
    d. Close to TDC at the completion of the compression stroke
11. Which of the following engine brakes offers progressive step braking using from one to all six cylinders?
    a. Cummins Intebrake
    b. Mercedes-Benz CTV
    c. Mack Trucks Jacobs on E-Tech
    d. Caterpillar BrakeSaver
12. When a Caterpillar BrakeSaver is used, which of the following is also true?
    a. External exhaust brake principles apply.
    b. A larger capacity engine oil sump is used.
    c. It must be controlled by the vehicle ECM.
    d. Retarding efficiency can never be modulated.

13

Engine Removal, Disassembly, Cleaning, Inspection, and Reassembly Guidelines

INTRODUCTION

Most of the procedures and terms used in this chapter have been used elsewhere in this textbook. The objective of this chapter is to put some of the content together in sequences typical of disassembly, cleaning, inspection, and reassembly of a truck diesel engine. It should never be used as a replacement for original equipment manufacturer (OEM) literature, and remember that every engine and, for that matter, every truck chassis, presents its own small and larger distinct problems that can only be addressed using product-specific guidelines.

Service Literature

In this textbook we have used the term **service literature** to refer to what used to be known as a service manual. Service literature today is seldom available in hard format. Most manufacturers provide service literature using their online service information systems (SIS) (covered in Chapter 31) because information can easily be corrected, updated, and enhanced by tagging it with the latest technical service bulletins (TSBs). In this chapter the term *service literature* is used to include the following:

- Hard (paper) service manuals
- Data disk (CD, DVD) service literature
- Online service information systems
- Technical service literature

REMOVAL OF AN ENGINE FROM A VEHICLE

Details of this procedure obviously vary from one vehicle to another and depend to large extent on the specific engine and whether it is hydromechanically or electronically managed. The following is intended as a rough set of guidelines and the sequencing of the steps required to remove an engine from a highway tractor chassis.

Getting Ready

1. It is good practice before beginning to pressure wash or steam clean the engine and engine compartment to remove road grime, grease, and oil. This will provide a clean work area that will enable more precise inspection and promote a more professional work environment.
2. Park the vehicle on a clean, level surface and ensure that the engine compartment can be accessed by the hoist required to lift the engine. Engage the vehicle parking brakes, block wheels, and ensure that there is sufficient bench space or a mobile steel cart on which removed components can be placed. Ensure that some adhesive labels and masking tape are at hand so that electrical wiring and fluid hoses can be identified. Failure to do this can considerably lengthen the reinstallation time. If the

truck is of the **cab-over-engine (COE)** design, ensure that the cab lift hydraulics operate properly and that the positional locks are in place. With the COE chassis, the cab must usually be elevated to its extreme travel position, so check that the mechanical stops are in place and that the extended lift rams are not supporting the weight of the cab for the duration of the engine job.

3. If the engine is electronically managed, download the engine/chassis identification data and the customer data programmable options using a personal computer (PC) (and OEM software) or ProLink and retain on a disk or printout. Disconnect the main battery leads at the battery terminals.
4. Remove the radiator cap. To avoid injury when removing a radiator cap, rotate it counterclockwise (CCW) to the first stop, but do not depress. This will allow any residual pressure to bleed from the cooling system. After the pressure has equalized, press the cap downward and continue turning to remove. Open the cooling system drain cocks usually located on the lower radiator tank and somewhere on the engine block to drain the coolant. If the coolant is to be reused, store in a sealed container: Do not leave in drainage tubs exposed to the shop atmosphere. If the coolant is to be replaced (and it really should be at an engine overhaul), ensure that the used coolant is disposed of in an environmentally safe manner consistent with federal and local jurisdiction regulations.
5. Shut off the air supply to any engine pneumatic controls such as shutterstats, fan controls, puff limiters, air-controlled exhaust brakes, and any other pneumatic apparatus in the engine compartment that may have to be removed or disconnected.
6. Remove the oil pan plug and drain the engine oil. Remove the oil, fuel, and coolant filters and dispose of them and the engine oil in an environmentally safe manner consistent with federal and local jurisdiction regulations.

Removing the Engine from the Chassis

1. Disconnect the piping/ducting to charge air heat exchangers such as aftercoolers, air-to-air coolers, and tip turbine assemblies, labeling any plumbing that could be a problem during reassembly. Remove the intake ducting, capping the turbocharger intake and exhaust porting. Ensure that all intake ducts, air hoses, and the air filter assembly are capped to prevent contaminants from entering the intake system.
2. On vehicles with **air conditioning (A/C)**, it is often possible to remove the engine without losing the refrigerant charge. Remove the condenser from the heat exchanger cluster at the front of the engine housing and fully support it so that it is not hanging by its hoses. Remove the A/C compressor, again without disconnecting its plumbing, and fully support it somewhere where it is not going to hinder further work in the engine compartment. If the refrigerant must be discharged, connect the appropriate recovery station and evacuate the system, plugging any open hoses.
3. Usually the radiator should be removed. In most cases the radiator tie rods obstruct access to the engine, so at minimum these should be removed. To remove the radiator, when possible, leave the air-to-air type, charge air coolers attached to the rad assembly and attach chains to the radiator tie-rod upper brackets. Support the assembly using a shop crane (chain hoist or cherry picker). Next disconnect the upper and lower radiator hoses. Inspect the hoses for cracks and general overall condition to note whether replacement is required: This should be done on removal of the engine from the chassis so that quotes to customers are reliable and parts can be ordered when necessary. Remove the radiator lower mounting support bolts and rubber insulators. Lift the radiator/air-to-air charge air cooler assembly from the chassis using the shop crane. Lift slowly and carefully to avoid damaging any components.
4. When the assembly is out of chassis, separate the two heat exchangers. Carefully place the radiator upright on a flat surface. Use extreme care when handling heat exchanger cores—they are easily damaged. Remove the fan shroud, brackets, and other hardware if the radiator is to be recored.

5. Remove the fan and fan hub assembly.
6. Clearly mark all the electrical leads before disconnection. Do the same with fuel lines, all linkages, oil lines, and water hoses. Cap all disconnected fuel lines to keep out dust.
7. Locate the rear engine mounts. These are either on the engine flywheel housing or on the transmission bell housing. When the rear engine mounts are located on the flywheel housing, the transmission must be fully supported. Position a mechanical transmission jack or blocks under the transmission. In some vehicles it may be necessary to completely separate the transmission from the engine *before* attempting to remove the engine.
8. Loosen and remove the capscrews that hold the transmission bell housing to the engine flywheel housing.
9. Obtain or fabricate a suitable **spreader bar** and lifting chain so that the engine can be lifted without lift chain contact to critical engine components. Fit engine lifting eyes to the OEM recommended location on the engine. Move into position the hoist to be used to lift the engine from the chassis. Attach lifting chain hooks to the engine lifting eyes.
10. After the hoist is chained to the engine lift eyes, apply a slight load to the lifting chain.
11. Remove the engine mounting bolts. Perform a thorough visual inspection to ensure that everything that has to be uncoupled is and that nothing is obstructing the planned removal path of the engine.
12. Separate the engine from the transmission, taking care not to force anything or overload the transmission input shaft. Use the hoist to remove the engine from the chassis. This may be a two-person operation, depending on the chassis. If unfamiliar with the chassis, ensure that someone is around to assist, even if it means just having an extra pair of eyes.
13. Remove external accessories as required: alternator, starter, power steering pump, oil, coolant and fuel filter mounting pads, and anything else on the block that might obstruct the engine overhaul mount stand. Bolt the engine into the overhaul stand and do not release the weight from the hoist until the engine stand mounting plate bolts are tight.
14. Remove the hoist chain hooks and then the lifting eyes from the engine unless they were originally on the engine.
15. Remove the turbocharger oil supply and return lines. Remove the turbocharger and assess its condition.
16. Remove the crankcase breather assembly if fitted.

ENGINE DISASSEMBLY

In most cases the engine will be removed from the location of the chassis before it is disassembled. If an engine is to be rebuilt in a general area of the shop floor, it is essential to ensure that no work, such as welding, be undertaken in close proximity to the engine work.

1. For electronic engines using injectors or pump units with programmable fuel flow codes, it is critical that you tag each unit by cylinder number if you intend to reuse them. On engines that use integral or unit high-pressure injection pumps, remove the injection lines and any electrical wiring by separating the connectors. On engines using integral hydraulic injectors accessible without removing the valve covers, remove the injectors, capping the leakoff and feed ports. On some engines the high-pressure pipe is connected to a cylindrical injector through a port in the rocker housing. The two nuts that locate and seal this connector pipe must be backed off and the pipe removed before any attempt is made to remove the injector.
2. Remove the exhaust and intake manifolds plus the exhaust gas recirculation (EGR) assembly. Carefully observe the location of sensors and actuators, tagging each for location. Be especially cautious when disassembling EGR components: The sensors and actuators located in the EGR heat exchanger(s), venturi, and mixing chamber(s) are sensitive and easily damaged.
3. Remove the water manifold assembly, complete with the thermostat housing when

possible. Sometime the EGR heat exchanger is integrated into the water manifold and it may be easier to remove both as a unit.

4. Separate the vibration damper using great care, especially with viscous-type units. Thoroughly inspect the vibration damper to assess its serviceability. Next remove the crank pulley and crankshaft hub assembly. Usually this requires the removal of the hub retaining capscrew(s) and then using a universal T-puller (Figure 13–1).
5. Remove the engine timing gear cover. If the timing gear cover has thrust buttons fitted, back these off before removing. On engines where the timing indicator is fastened by the timing gear cover screws, remove them first and ensure that they are not damaged.
6. Remove the oil cooler assembly. With bundle-type oil coolers, test the core using either the pressure or vacuum methods outlined in Chapter 9 and then seal using plastic wrap.
7. Disconnect any plumbing to the water pump not already removed, then remove the water pump itself. It has become almost standard practice to replace water pumps with a rebuilt exchange unit when reconditioning engines today. But if you are reusing the pump visually check it for evidence of weeping at the ceramic seal, and the seal itself for cracks.
8. Remove the clutch pack assembly from the flywheel and assess its serviceability.
9. Remove the flywheel assembly using a suitable hoist.
10. Remove the flywheel housing and any attached components, checking for sensors and connecting wires. Check for eccentricity at the locating dowel holes.
11. Remove all accessory drive components not already removed, such as power steering pumps, air compressors, and so forth. Remove accessory drive pulleys using a suitable puller (see Figure 13–1).
12. Remove the oil filler and dipstick tubes.
13. Remove the mechanical tachometer drive, if fitted.
14. If the engine is fitted with electronic unit pumps (EUPs), remove them and store them as per OEM recommendations. If the engine has an inline injection pump, remove any electrical wiring from the rack actuator housing (electronically managed, port-helix metering), transducer module, or governor housing. Disconnect the throttle arm and fuel stop arm from hydromechanical pumps. Check that all the plumbing is disconnected from both sides of the pump and remove any support brackets: Loosen the pump mounting capscrews, and, using a smooth motion, pull the injection pump away from its mounting flange. Pressure-time (PT), common rail (CR), and Cummins Accumulator Pump System (CAPS) pumps can be removed in a similar fashion. The various types of transfer pumps used with the electronic unit injector (EUI) systems are relatively straightforward to remove, but the OEM literature should be consulted whenever hydromechanical or electronic fuel pumping apparatus is removed from the engine. When performing a complete engine overhaul, it may be part of the procedure to recondi-

Figure 13–1 *Removing a pulley using a T-puller. (Reprinted Courtesy of Caterpillar Inc.)*

tion or replace with rebuilt/exchange fuel injection components, so check with the OEM and shop work order instructions.

15. Remove the rocker housing covers.
16. Remove the rocker shaft assemblies including valve bridges where used. On overhead camshaft engines, remove the camshaft assembly.
17. Remove the injectors, using the appropriate puller. When cylindrical hydraulic injectors are used, make sure that all the plumbing is clear of the injector seat before attempting to remove them. Where EUIs, electrohydraulic injectors (EHIs), EUPs, and hydraulically actuated electronic unit injectors (HEUIs) are used, ensure that all electrical and hydraulic connections are removed: This is especially important where transverse cylinder head, high-pressure pipes are used.
18. Check the injector nozzle gasket: If it is not on a removed injector, remove from the injector bore with a magnet or small tapered rod. Do not leave in the cylinder head because this could cause problems on reassembly.
19. Remove the push tubes or rods where fitted. It is good practice to inspect each for serviceability on disassembly. Some OEMs recommend that the push tubes be tagged for position in the engine.
20. Remove fuel manifolds (internal charge and return lines) where fitted, jumper pipes, and crossover pipes.
21. Remove all the cylinder head capscrews and washers. Remove the cylinder head(s).
22. Remove the cylinder head gasket and the sealing grommets and fire rings if not integral with the cylinder head gasket. Inspect the cylinder head gasket for signs of failure, paying special attention to the fire rings.
23. Remove the oil pan capscrews and the oil pan.
24. Remove the oil pump, which may be located in the crankcase or outside fastened to the engine cylinder block.
25. Before removing each piston and connecting rod assemblies, scrape carbon deposits (ring ridge) from the upper inside wall of each cylinder liner using a flexible knife.
26. Plan to remove the pistons on an inline six-cylinder engine in companion pairs; that is, 1 and 6, 2 and 5, 3 and 4.
27. First, remove the connecting rod bearing capscrews and separate the cap from the rod by tapping with a light-duty nylon hammer.
28. Remove each connecting rod and piston assembly carefully, guiding the rod bottom end clear of the piston cooling oil nozzle and avoiding contact with the liner wall. Arrange the piston assemblies sequentially on a bench, making note of any unusual characteristics.
29. Remove the cylinder block located piston cooling nozzles. Oil will be trapped in the gallery that supplies the piston cooling jets, so be prepared to capture it. If the engine is rotated to an angle on the stand, remove either 1 or 6 first and allow the oil in the gallery to drain.
30. Disassemble the engine timing geartrain. Remove accessory drive gears, idler gears, and an in-block camshaft gear and camshaft as an assembly. Rotating the engine upside down will ease the removal of an in-block camshaft because this will drop the cam follower/lifter assemblies out of the way so they do not interfere with the cam lobes and journals as the shaft is withdrawn.
31. Remove the cam follower/tappets/valve lifter assemblies. If the camshaft is to be reused, inspect and tag individual lifters for reassembly to position. Where cam follower housings are used, tag for position. In engines where the cam follower housing shimming defines engine timing, it is advisable to measure and record the shim (gasket thickness) with a micrometer even when the engine is to be retimed on reassembly.
32. Disassemble any remaining gears in the engine geartrain and remove any auxiliary or balance shafts remaining in the engine cylinder block.
33. Remove the crankshaft rear seal housing assembly.
34. Remove all the camshaft bushings using the correct sized driver and slide hammer.

35. With the engine still inverted, remove the main bearing cap capscrews, the cap brackets, and the main bearing caps by gently tapping with a nylon hammer. Tag the bearings by location for inspection and failure analysis purposes.
36. Fit a crank-lifting yoke to the crank throws on a pair of companion cylinders (preferably 2 and 5 on an inline six): The yoke should have rubber conduit installed over the throw hooks to prevent damage to the throw journals. Lift the crankshaft free from the cylinder block using a hoist.
37. Remove the cylinder liners using a suitable puller as shown in Figure 13–2. Wet liners are retained by O-rings and do not usually require a great amount of force to separate them from the block. Dry liners and combination wet/dry liners may require considerable force, and hydraulic and air over hydraulic pullers with the appropriate adapter plates are usually required. In cases where even hydraulic pullers fail to move the liner, a single vertical-down welding bead using a 3.2 mm (1/8″) deep penetration electrode (such as E6010 or E6011) followed by the immediate application of cold water will often work. Whenever such a practice has to be resorted to, it should be recognized that a mistake may destroy the cylinder block. Always ensure that the welding is performed well separated from any engine components. And always use a fast drag, vertical-down technique to avoid overheating the liner.
38. Remove all expansion (frost) press fit plugs and threaded oil passage plugs from the cylinder block before cleaning.

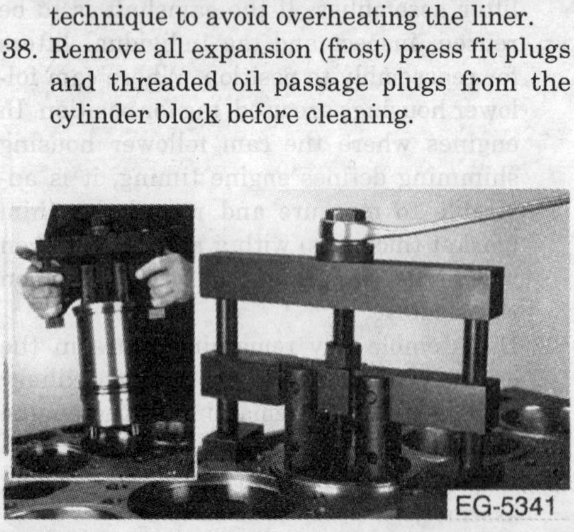

Figure 13–2 *Removing a cylinder sleeve. (Courtesy of Navistar International Corp.)*

CLEANING AND INSPECTING COMPONENTS

The cleaning and inspection of an engine is a critical stage in the engine reconditioning procedure. If the reason for the engine rebuild is a failure, ensure that the cause is identified before any attempt at reassembly is made.

1. Use a gasket scraper to remove all the gasket material and heavy dirt from the cylinder block. Install the cylinder block into a preferably heated, **soak tank** using a heavy-duty alkaline soak cleaner for a period of 1 to 2 hours. One OEM reports that 1/16″ of coolant scale has the insulating properties of 4″ of cast iron. It is therefore important not to skip this procedure.
2. Remove the cylinder block from the cleaning or soak tank. Thoroughly flush the cylinder block using a shop **high-pressure washer**, ensuring that all particles and sludge produced by the block boiling are removed.
3. Visually inspect the cylinder block, checking all the coolant passages and ensuring they are clean and unobstructed. Ream or drill out if necessary to dislodge any deposits.
4. Check to see that there are no casting fins or residues that might obstruct coolant flow by removing any casting irregularities with a prybar.
5. Run a cylindrical wire brush in all the oil passages to ensure they are unobstructed.
6. Flush the oil passages with air and solvent.

Magnetic Flux Test

It makes sense to **magnetic flux test** engine cylinder blocks, crankshafts, and all connecting rods at every major engine overhaul. These processes are neither expensive nor time consuming and as the consequences of a single warranted engine failure out of 20 overhauls will demolish the profits of the other 19, it is shortsighted to overlook it. Remember, this expense is passed on to the customer, and if it is to be passed over, it is the customer who should make the decision: Get a refusal in writing. The usual drawback is not having the equipment on site, but in most cases, the equipment is accessible and machine shops usually pick up and deliver.

ENGINE REASSEMBLY GUIDELINES

Although engines may be disassembled with little or no reference to service literature (not recommended practice), they should be assembled precisely according to the sequencing in the OEM workshop manual. Maybe the only excuse for not observing this practice could be made by the technician who continually overhauls the same engine series and who makes a point of studying every OEM technical service bulletin (TSB). Even for the experienced engine overhaul technician, using service literature as a guide is simply good practice. For the novice engine overhauler, the service literature should guide every move: It is important that short cuts are never experimented with. As every experienced diesel technician will acknowledge, there are a few short cuts to some procedures, but these must involve a zero risk factor, and that requires an intimate knowledge of the engine that only comes with experience.

To provide a general guide to an engine assembly procedure would simply reverse the general guidelines for disassembly, and the point is engines cannot be successfully rebuilt according to general guidelines but should be assembled in the precise sequence outlined by the OEM. However, the following lists some general precautions that the engine rebuilder may encounter when assembling engines.

O-Rings on Wet Liners

Use the lubricant recommended by the manufacturer when installing full wet and mid-stop, wet/dry liners. The lubricant recommended may be antifreeze, engine oil, silastic, or nothing at all. Using an inappropriate lubricant may chemically interact with the O-ring material and cause it to swell, contract, or disintegrate, generating a premature failure.

Use of Anaerobic Sealants

The use of **anaerobic sealants** (silicone, silastic/RTV, and others) has become popular in recent years. Only use where specified, and do not overuse as it may plug up apertures.

Cylinder Block Inspection

The flowchart illustrated in Figure 13–3 is an excellent guideline for evaluating the serviceability of an engine cylinder. Although it relates specifically to a recent model Detroit Diesel Corporation (DDC) Series 60, most of the steps apply to any diesel engine. Make sure you understand exactly what is required in each step.

Test the Cylinder Block Line Bore

This check is performed most easily using a **master bar**, which is clamped into the block line bore without the bearings in place. When available, a master bar can be obtained from the OEM specialty tool supplier. The test is used to determine whether the cylinder block line bore must be machined. After torquing the main bearing caps, the master bar should rotate by hand: Binding indicates that line boring is required because the cylinder block has become distorted.

Cylinder Block Fretting and Warpage

This should be checked to specification using a straightedge and thickness gauges. Today's high brake power, lightweight engines make this a critical check because they are more subject to torque twist.

Sleeve Height Protrusion Specification

This is a critical specification. The protrusion variance of the liners under one cylinder head determines how evenly each liner is clamped to the cylinder block. A liner protrusion error can damage a cylinder head and/or cause an immediate head gasket failure. To determine liner protrusion, the liner should be installed using the proper tooling as shown in Figure 13–4: A dial sled gauge can then be used to measure the protrusion.

Machining Counterbore

A liner protrusion or counterbore irregularity will require recutting the counterbore and

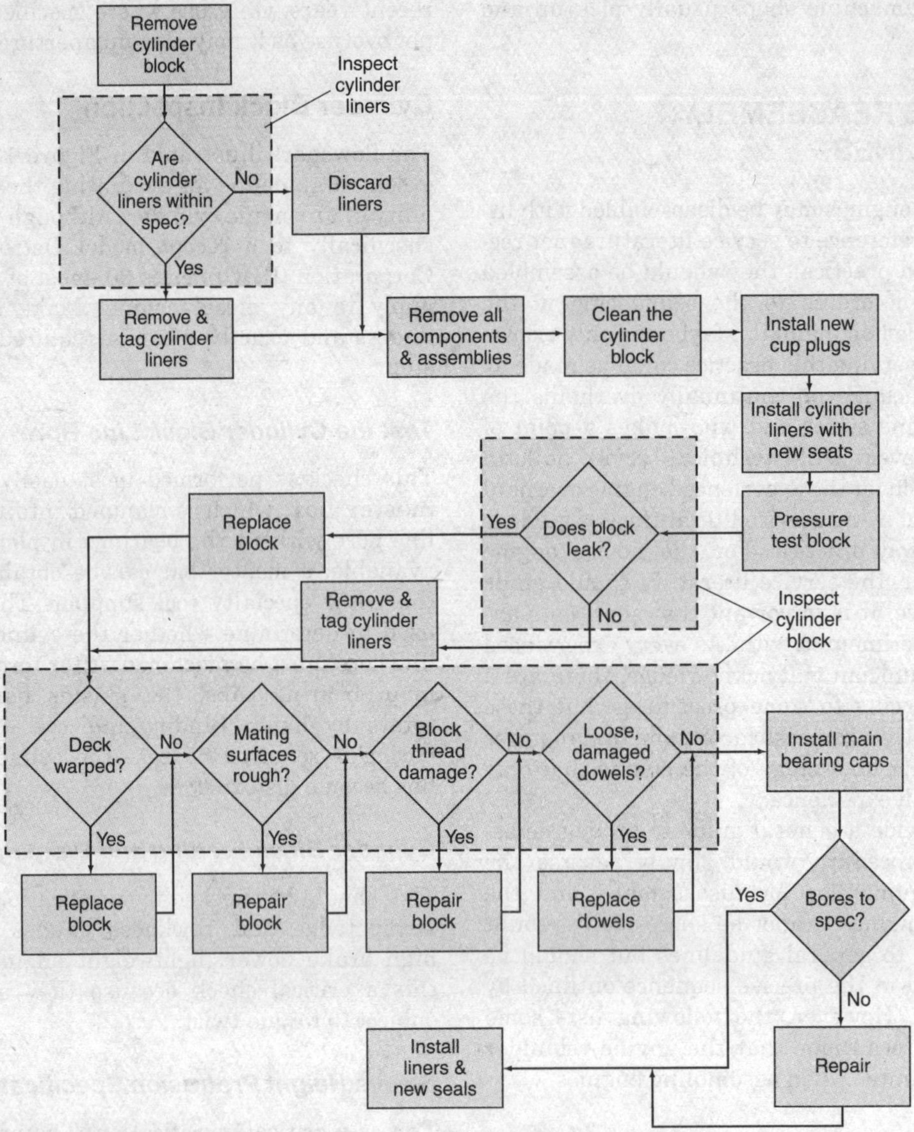

Figure 13–3 *Flowchart for repair or replacement of a cylinder block. (Courtesy of Detroit Diesel Corporation)*

shimming to set specified protrusion. The general procedure for recutting a liner counterbore is shown in Figure 13–5 but always check OEM service literature. Some diesel engines use angled or stepped counterbores that require specialized tooling.

Inspect Crankshaft

Crankshafts should be inspected using the procedures outlined in Chapter 6. Some OEMs emphasize the importance of magnetic flux testing crankshafts at each major overhaul. Others do not. This reflects the fact that crankshaft manufacturing technology since the early 1990s has developed to the point that very few crankshaft failures occur. Figure 13–6 shows a flowchart produced by Detroit Diesel, which emphasizes a thorough visual and measurement assessment of crankshaft serviceability but does not mandate magnetic flux inspection.

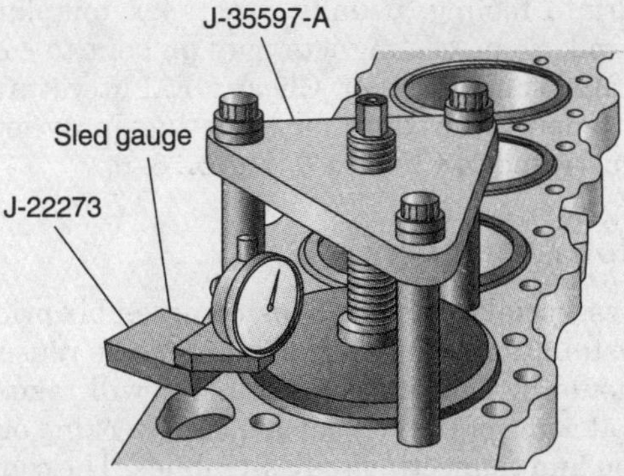

Figure 13–4 *Measuring liner protrusion. (Courtesy of Navistar International Corp.)*

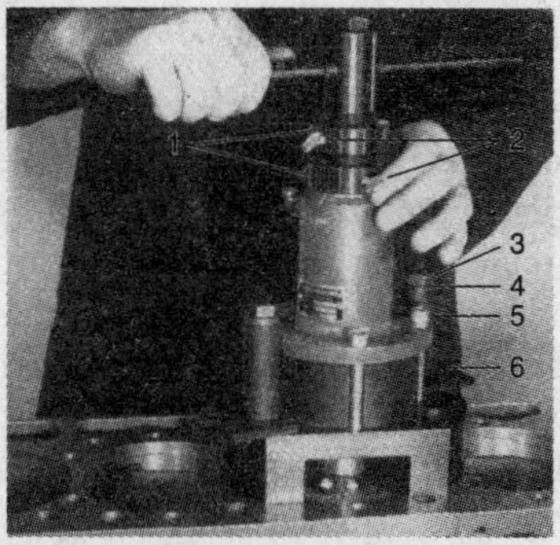

1. Turn knuckles
2. Locking screws
3. Mounting bolts (2)
4. Spacers (2)
5. Washers (2)
6. Locking plates (2)

Figure 13–5 *Machining a cylinder block counterbore. (Courtesy of Navistar International Corp.)*

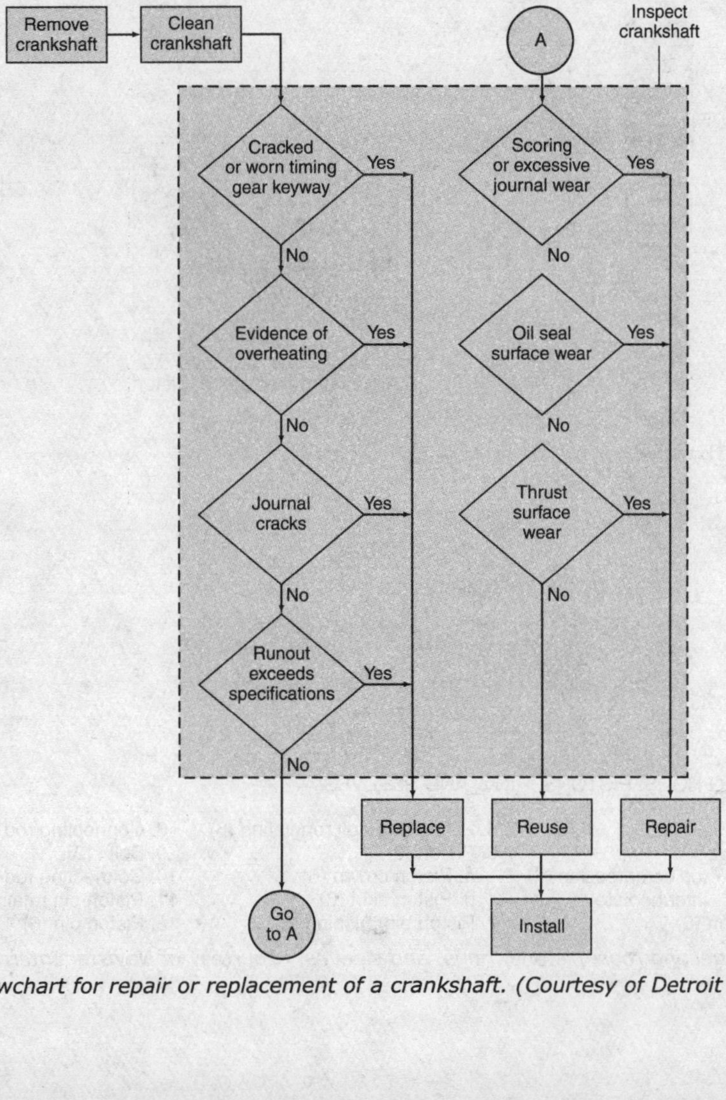

Figure 13–6 *Flowchart for repair or replacement of a crankshaft. (Courtesy of Detroit Diesel Corporation)*

Checking Main Bearing Clearance

This procedure is designed to be performed with the engine upside down using Plastigage™ ; in other words, when the engine is being overhauled outside of the chassis. The specification will differ and have less relevance when performed with the engine in-chassis, primarily because crankshafts flex when not fully supported by their mains. Identify the OEM specification first, then select the appropriate Plastigage color code. When making the measurement, ensure that the main bearing cap is torqued to specification. The procedure for using Plastigage is illustrated in Figure 6–23 and Figure 6–24 in Chapter 6.

Piston Cooling Jets

Always check the directional aim using the appropriate tooling, usually a perspex template and rod. Some cooling jets may be bent to correct aim, some may not: Check OEM literature. The general procedure for targeting spray jets is illustrated in Chapter 6, Figure 6–6.

Piston Assembly

It is essential to inspect a piston assembly prior to installation. When the piston is to be reused it should be carefully checked. We will take a look at some of the critical steps in checking out piston and connecting rod assemblies. The components of a piston assembly are shown in Figure 13–7. Note that what is often referred to as

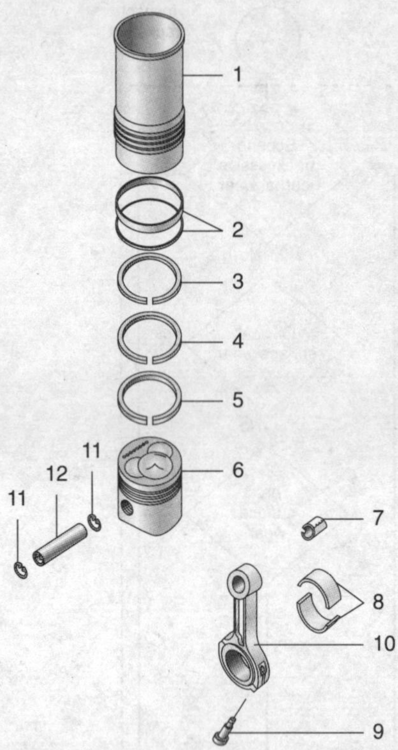

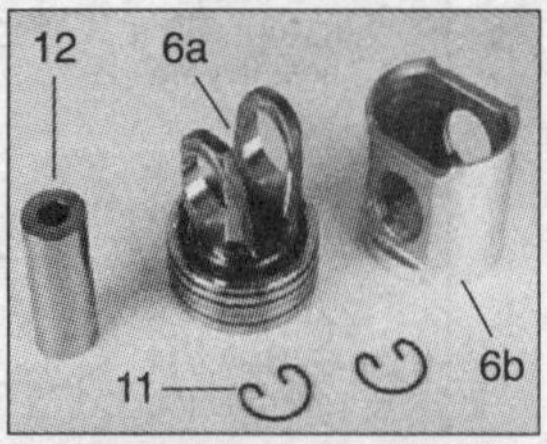

Articulated (two piece) piston

CONNECTING RODS, PISTONS, RING, AND SLEEVES

1. Sleeve (6)
2. Sleeve seals
3. Piston ring - top compression (6)
4. Piston ring - intermediate compression (6)
5. Piston ring - oil regulating (6)
6. Piston (6)
   a. Piston crown (6)
   b. Piston skirt (6)
7. Piston pin bushing (6)
8. Connecting rod bearing (12)
9. Bolt (12)
10. Connecting rod and cap (6)
11. Piston pin retainer (12)
12. Piston pin (6)

Figure 13–7 *Connecting rods, pistons, rings, and sleeves. (Courtesy of Navistar International Corp.)*

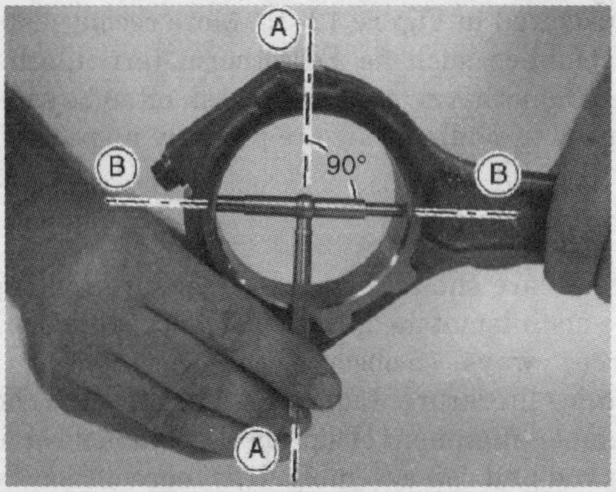

Figure 13–8 *Measuring for out-of-round big end. (Courtesy of Navistar International Corp.)*

a *piston kit* includes the liner in some OEMs' terminology. Figure 13–8 shows the procedure used to measure the big end of a connecting rod: Note that the rod cap is assembled and torqued into position.

Measuring Connecting Rods

The procedure for verifying the serviceability of conn rods is addressed in Chapter 6 and you should be aware of the importance some OEMs place on magnetic flux testing of conn rods at each major overhaul. The flowchart shown in Figure 13–9 is a guide to assessing conn rod serviceability.

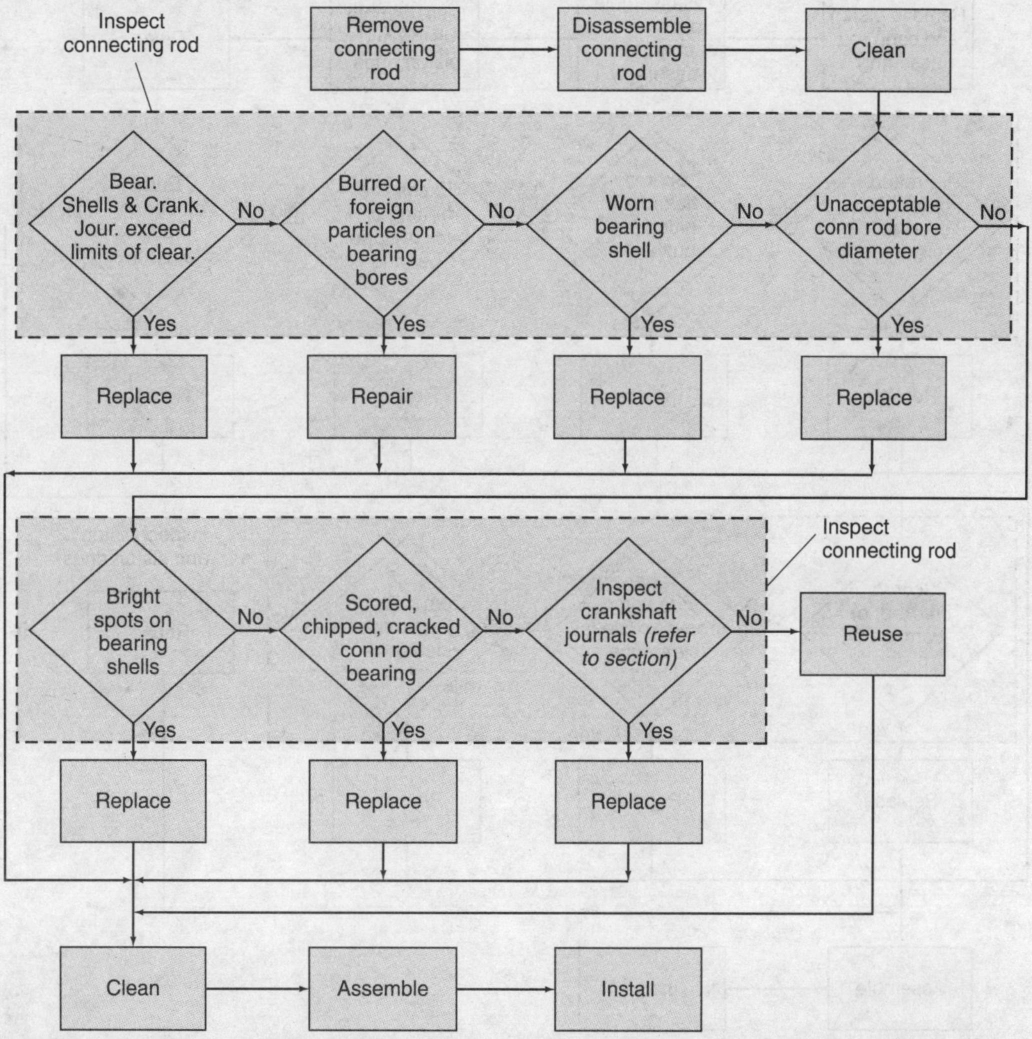

Figure 13–9 *Flowchart for repair or replacement of a connecting rod. (Courtesy of Detroit Diesel Corporation)*

Connecting Rod Weight Classifications

Connecting rod weights are coded by manufacturers. When replacing connecting rods, it may be possible to either increase or decrease the weight class by one increment, but not more. The consequence of an overweight or underweight rod is to unbalance the engine and generate a premature failure. Always check OEM service literature.

Assessing Pistons and Rings

Pistons and rings should be assessed for serviceability using a flowchart such as the one illustrated in Figure 13–10. More recent piston assemblies such as Ferrotherm (articulating) and Monotherm (steel trunk) can often be safely reused providing they are properly inspected.

Rod Assembly

Great care should be taken to ensure rod caps are not mismatched. Each OEM does this in different ways so observe the instructions in service literature. Figure 13–11 shows the code number alignment of a properly matched Navistar rod and cap assembly. It is more difficult to mismatch a cracked rod assembly but you can

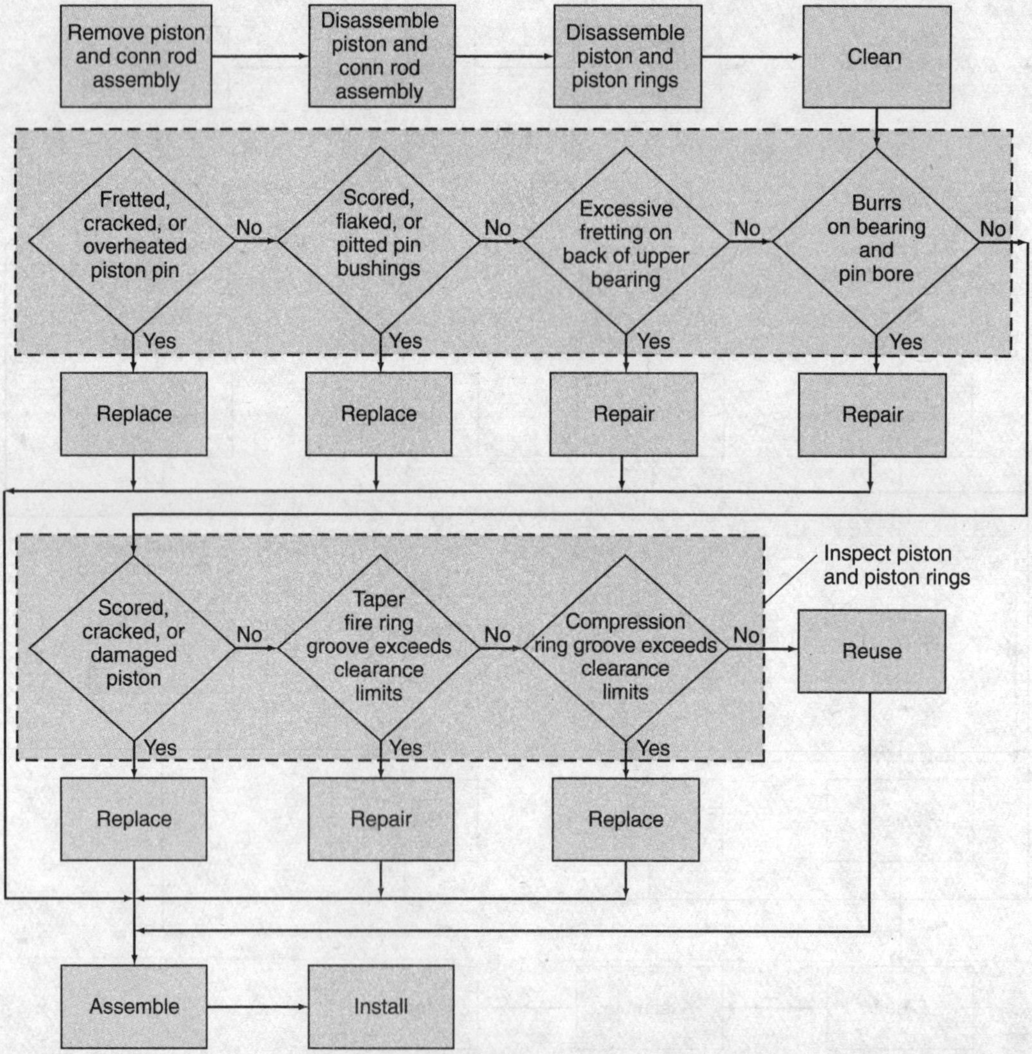

Figure 13–10 *Flowchart for repair or replacement of piston and piston rings. (Courtesy of Detroit Diesel Corporation)*

1. Numbers on connecting rod.
2. Numbers on connecting rod cap.

Figure 13–11 *Match and align numbers on a connecting rod and cap. (Courtesy of Navistar International Corp.)*

bet it has been done: Figure 13–12 shows a mismatched cracked rod assembly.

Ring End Gap

A procedure often overlooked by experienced diesel technicians probably due to the reliability of machining and packaging today (that is, the nominal dimension is usually correct), checking the ring end gap specification takes little time and should be done. Insert the ring into the cylinder bore and measure the gap with thickness gauges.

Piston Wrist Pin Retainer Snaprings

These should be installed flat side out and with the gap either up or down (varies with OEM, but usually down) and never crossways to the direction of piston travel. Believe it or not, the inertia forces on acceleration or deceleration acting on the snapring have been known to unseat them.

Ring Stagger

Ring stagger is crucially important when reassembling engines. Most engine OEMs prefer that no end gap is placed across either thrust face of the piston. Figure 13–13 shows DDC's preferred method for installing and staggering ring end gap on its recent model Series 60 engines using Monotherm pistons.

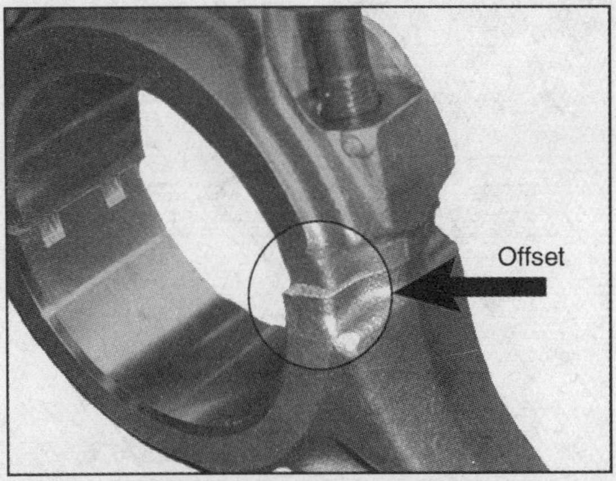

Figure 13–12 *Incorrect rod cap assembly on a cracked/fractured rod assembly. (Courtesy of Navistar International Corp.)*

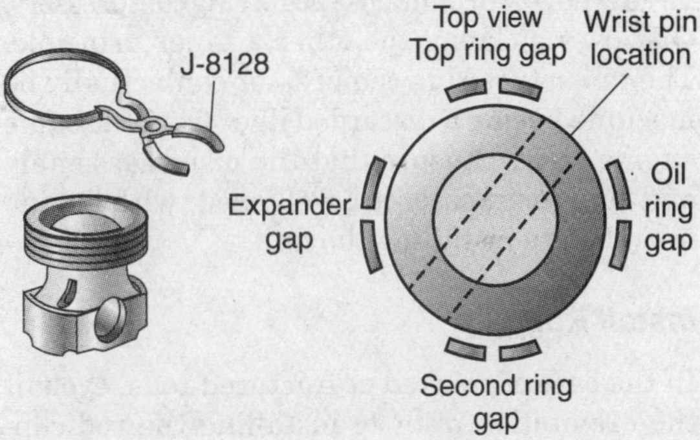

Figure 13–13 *DDC Monotherm piston ring stagger. (Courtesy of Detroit Diesel Corporation)*

Installing the Piston Assembly

Ensure that both the piston and connecting rod are properly oriented, remembering that OEMs usually do this by stamping the components with an arrow pointing toward the front. Use of multipurpose band-type ring compressors should be avoided when installing piston assemblies into the cylinder block because more often than not they damage rings. Figure 13–14 and Figure

Figure 13–14 *Clamp-type ring compressor. (Courtesy of Navistar)*

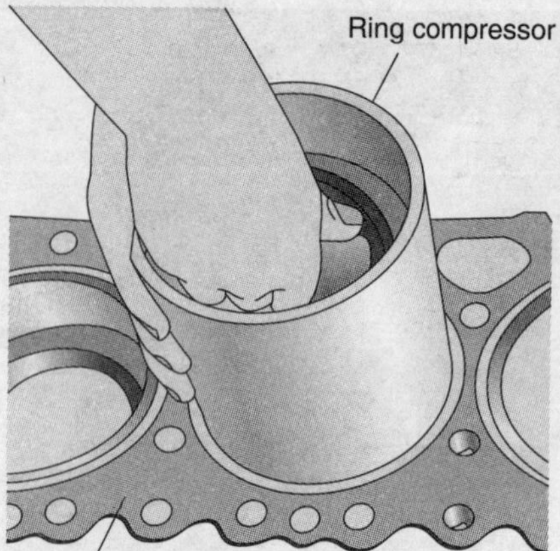

Figure 13–15 *Taper-type ring compressor. (Courtesy of Detroit Diesel Corporation)*

13–15 show two common types of ring compressors. One uses a clamp, the other a taper principle. The taper-type ring compressor can actually be machined using a discarded liner from the same engine series. Ensure that the piston assembly and ring compressor are well oiled (with engine lube) before installing the unit.

Install Rod Cap

In the case of cracked or fractured rods, eyeball the orientation prior to installing the rod cap: Although cracked rod caps will only fit when correctly orientated, it is easy enough to damage the mating faces by attempting to fit them

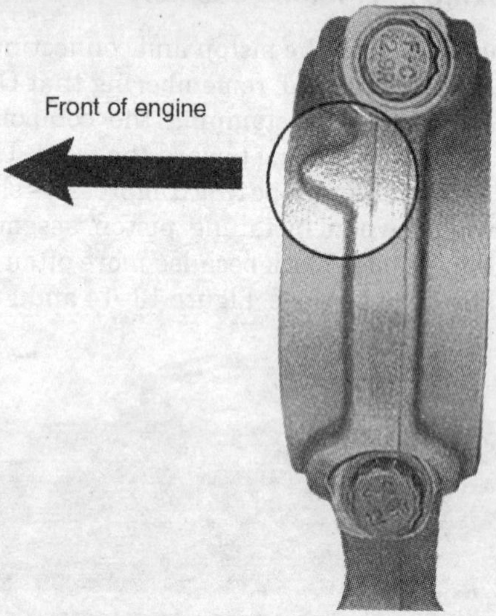

Figure 13–16 *Correct orientation. (Courtesy of Navistar International Corp.)*

improperly. In the case of stepped or clean face machined rod caps, make sure to fit the clamp as indicated by the directional arrow or number on the cap. Figure 13–16 shows the directional arrow on a typical rod cap.

Rod Side Play

This is a critical specification on any rods with machined big end faces. Slightly cocked rod caps can damage crankshafts and promote rapid bearing failure. The use of a pair of equal-sized thickness gauges close to the side play specification value inserted on one side of the cap may assist with alignment problems while torquing the rod cap. Cracked rods used on some engines make this check unnecessary. All correctly assembled rods must have side-play.

Buttress Screws on Main Bearing Caps

Some engines use **buttress** screws on some of the main bearing caps to add rigidity to the block assembly. These are usually installed after the main bolts have been torqued: Do not omit them. Cylinder blocks for the high-horsepower engines of today are made from castings of higher strength and lower weights, enabling them to flex more than those of a generation ago, which is the reason the buttress screws are now engineered into some engines.

Timing Geartrain Assembly

The procedure for setting up and timing the engine geartrain varies considerably from engine to engine. It may be a simple procedure but in the example of some of today's overhead camshaft (OHC) engines such as the Cummins ISX with its double OHCs, some specialized tooling (and training) is required. Some of the procedures you may encounter are listed next.

Camshaft Inspection

There is a detailed account of camshaft inspection in Chapter 7. The camshaft should first be visually inspected, and if this appears OK, it should be mounted in V-blocks as shown in Figure 8–9. The journals should be miked and the profiles indicated. Pay special attention to areas of the cam profile that sustain the most abuse. This would include peak cam lift on symmetrical cam profiles. For camshafts that actuate open nozzle injectors, pay special attention to the end of the inner base circle (IBC) to outer base circle (OBC) ramp: This bottoms the plunger train components into the injector cup to establish mechanical crush and can be especially tough on the cam profile.

Heat Shrinking Gears to Shafts

A bearing hotplate may be used to heat interference-fit gears to the OEM-specified temperature. A kitchen toaster oven (dedicated to the purpose of heating engine components) has an accurate thermostat and works better: The gear is more evenly heated. **Tempilstick** crayon may be used to check the temperature of a heated component and is required if you are using the hotplate method. Do not use an oxyacetylene torch to heat components because the steel may carburize. Gears should not be heated for periods exceeding 45 minutes.

Timing the Geartrain

The OEM procedure must be adhered to. Some guidelines are provided in Chapter 7 of this book but it should be emphasized that this is no substitute for OEM service literature. Figure 13–17 shows the location of the timing gear marks on a Caterpillar C15 engine. This is an overhead camshaft engine with a front timing geartrain. The timing indices used on this engine are Vs and circles.

Rear Cam Bushings

When performing an in-chassis engine overhaul, when the cam bushings are removed, it should be noted that the rear cam bushing in some engines may not be removed without unseating the rear cam plug. Because the replacement of the rear cam plug involves the removal of the transmission/clutch assembly and flywheel housing—procedures that are unlikely to have been factored into the estimate—it is advisable to leave the rear cam bushing in place. Record the fact that it was not replaced on the hardcopy of the work order.

Cylinder Head Servicing

Because OEMs have adopted single slab cylinder heads manufactured from lighter (and stronger) base metal, a cylinder head must be measured and pressure tested before returning to it to service. Some of the required checks are outlined here.

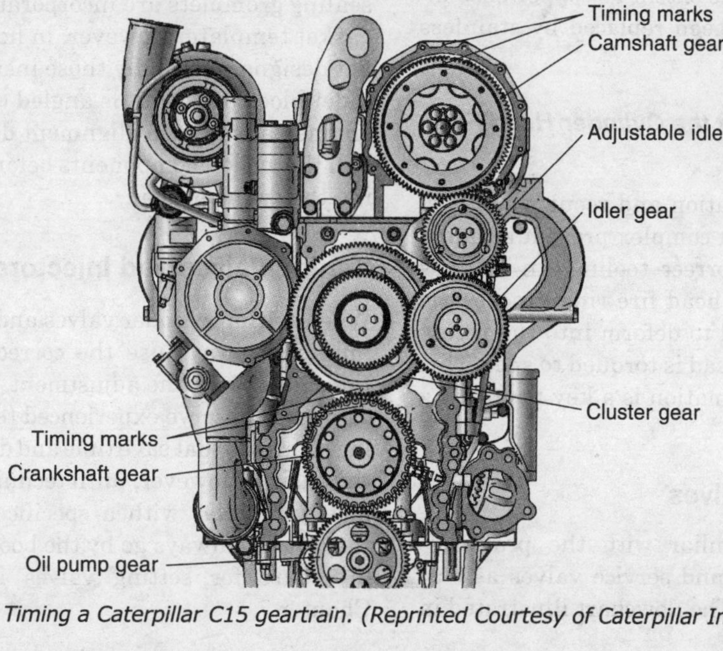

Figure 13–17 *Timing a Caterpillar C15 geartrain. (Reprinted Courtesy of Caterpillar Inc.)*

Cylinder Head Warpage

Check the cylinder for head warpage with a straightedge and thickness gauges. This test is critical in today's low-weight, high-brake power engines, especially those using single slab castings on six-cylinder engines. Check the OEM maximum warpage spec. Select a feeler gauge equivalent to the maximum spec, and while positioning the straightedge corner to corner on the inverted cylinder head, attempt to insert the gauge under the straightedge. If it can be inserted this fails the cylinder head and it should be reconditioned.

Pressure Testing Cylinder Heads

It is preferable to run hot water through cylinder heads for about 10 minutes to heat them before hydrostatically testing them. Hot water and shop air regulated at 100 psi should locate any leakage problems. Consult with an experienced technician to check whether that engine series has a track record of cylinder head problems and where leakage problems occur. The number one source of cylinder head leaks is the injector sleeve with most engine series: Because the injector sleeve is in direct contact with the cylinder head water jacket, coolant leakage into the engine cylinder can occur. The problem has been rectified to some extent as copper base injector sleeves have been replaced by stainless steel (SS) sleeves.

Measure and Recut the Cylinder Head Fire Ring Groove

Measure to specification and recut when necessary. This is not a complex procedure but it does require the correct tooling. In engines that use a cylinder head fire ring groove, the fire ring is designed to deform into the recess when the cylinder head is torqued to spec. This fire ring yield deformation is a key to creating a good seal.

Cylinder Head Valves

You should be familiar with the procedure required to inspect and service valves as outlined in Chapter 7. The flowchart illustrated in Figure 13–18 sets some guidelines for assessing cylinder valve serviceability.

Valve Margin

When dressing valves remember that the valve margin specification is critical, and if it cannot be met the valve should be replaced. Reference Chapter 7 if you are uncertain what the valve margin spec represents.

Valve Interference Angle

Valves are seldom designed with an interference angle in today's diesel engines because this compromises the valve's seating contact surface area. Additionally, most diesel engines use valve rotators and when they are used, an interference angle is never machined into the valve.

Cylinder Head Alignment

When installing multiple heads on an engine block deck, align the heads before torquing using a straightedge. Perform this even when the cylinder block has cylinder head alignment dowels. This enables improved sealing and lower stress loads on the manifolds.

Most current cylinder head gaskets are of the integral design; that is, the fire rings and sealing grommets are incorporated in the head gasket template. However, in nonintegral gasket designs, especially those installed on a cylinder block deck that is angled (such as on a V engine), always use alignment dowels and eyeball the gasket components before finally decking the cylinder head.

Setting Valves and Injectors

When setting cylinder valves and cam-actuated injectors, always use the correct engine locations to perform the adjustment. Novice technicians may observe experienced technicians taking shortcuts that save time and do not endanger the engine. However, until technicians are completely familiar with a specific engine series, they should always go by the book. The general procedure for setting valves is outlined in Chapter 7.

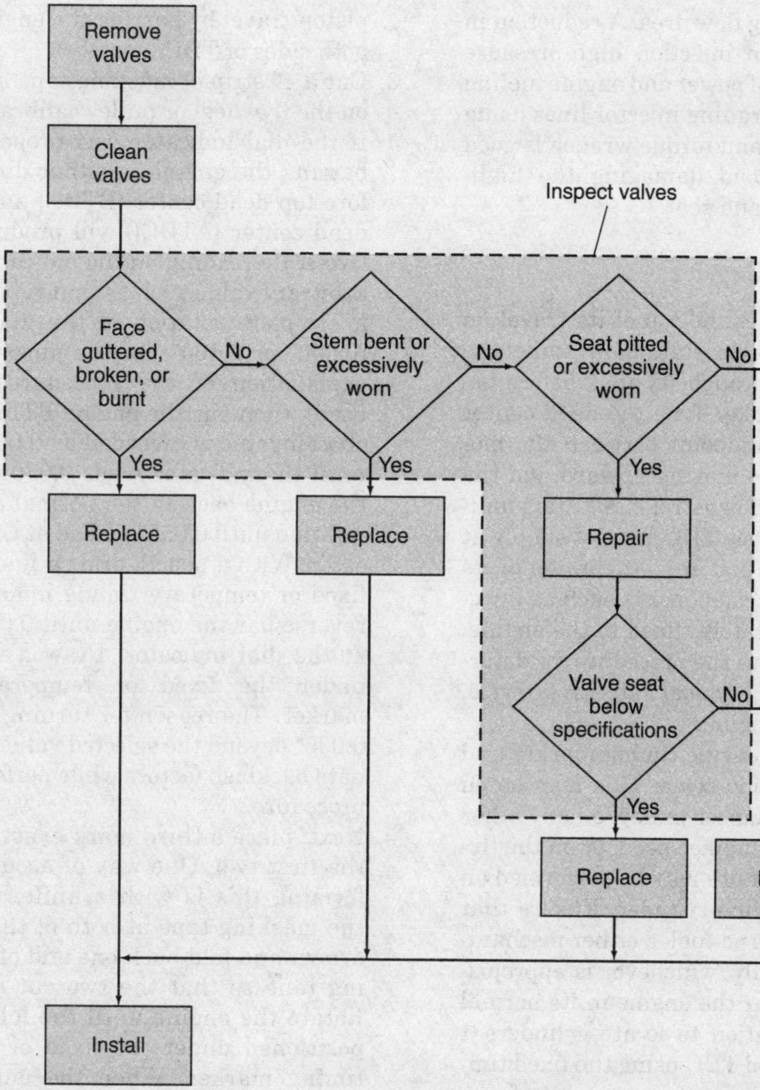

Figure 13–18 *Flowchart for repair or replacement of cylinder head valves. (Courtesy of Detroit Diesel Corporation)*

Injector Installation

When installing hydraulic, EHIs, MUIs, EUIs, and HEUIs, always use the OEM recommended procedure. Lubricating the O-rings with the correct medium is critical. Generally, the use of antiseize compound on cylindrical injectors with O-ring seals is never approved and where the technician may believe he is making the next removal of that set of injectors easier, their removal may come a lot sooner than anticipated. Use the correct lubricant medium for the O-rings and accept that occasionally injectors do seize in their bores. Note that when injectors are removed and replaced either or both of the following may be required:

- Mechanically actuated injectors may have to be timed (height adjusted).
- Calibration codes reprogrammed to ECM

These procedures are covered in later sections of this textbook in chapters addressing each specific OEM system.

Torque High-Pressure Injection Pipe Nuts

Overtorquing high-pressure injection pipe nuts at the injector and the injection pump can ridge the nipple seat and, more significantly, collapse the aperture, reducing flow area. A reduction in flow area through an injection high-pressure pipe may create lack of power and engine fueling balance problems. Torquing injector lines using a line wrench socket and torque wrench is good practice and will avoid damaging the high-pressure pipe nipple and seat.

Verifying True TDC

When a piston reaches the top of its travel on its upstroke, it remains stationary while the crank throw turns through its apex before beginning the downstroke. True top dead center (TDC) is the exact midpoint between the moment the piston stops moving upward and the moment it begins its downstroke. A timing indicator must be set at true TDC and not simply at any point where the piston is at the top of its travel to ensure that components such as injection pumps are accurately timed to the engine. The following list gives the procedure for determining true TDC for a typical four-stroke cycle, inline, six-cylinder engine.

1. Locate the fixed engine timing marker and rotating calibration scale: This may be on the harmonic balancer pulley, on any other pulley driven at engine speed, or on the flywheel. The procedure may be performed on either 1 or 6 engine cylinder. Ensure that the fuel system is no-fueled either mechanically or electrically, whichever is appropriate. Manually bar the engine in its normal direction or rotation to locate cylinders 1 and 6 at indicated TDC using the fixed timing indicator or by making one out of mechanic's wire and clamping it close to where the fixed timing indicator is to be positioned. The flywheel (using a gear and ratchet barring tool) or crank hub (using a barring fixture) should be used to rotate the engine.
2. If the cylinder heads are installed, remove the 1 or 6 injector and install in its place a dial indicator fitted with an extension probe long enough to contact the piston crown. In cases where the cylinder heads are removed from the engine, the dial indicator may be positioned on the cylinder block with the probe contacting the piston crown. Next, zero the indicator at the highest point of piston travel by barring the engine slightly both sides of TDC.
3. Cut a 4″ strip of masking tape and place it on the flywheel or pulley calibration scale.
4. If the dial indicator was properly zeroed, barring the engine in either direction (before top dead center [BTDC] and after top dead center [ATDC]) will produce a negative scale reading on the indicator. Pick an arbitrary value such as that represented by a complete rotation of the dial indicator (0.050″ or 0.100″; for the purposes of this explanation of the procedure, 0.050″ is used), then bar the engine BTDC until the chosen value is exceeded by 0.030″, that is, until the indicator reads 0.080″. Next, bar the engine back in the normal direction of rotation until 0.050″ is read at the dial indicator. With a pencil, draw a line under the fixed or temporary timing indicator. Then reverse bar the engine until 0.050″ is read at the dial indicator. Draw a second line under the fixed or temporary timing marker. The reason for turning the engine 0.030″ beyond the selected value is to eliminate backlash factors while performing this procedure.
5. Next, place a third mark exactly between the first two. One way of accurately performing this is with a knife. Neatly cut the masking tape at both of the two lines drawn and fold back one end of the masking tape so that the two cut ends align. Rotate the engine until the fold crease is positioned under the fixed or temporary timing marker. When the engine is rotated back to the exact midpoint between the two lines, it will be positioned at true TDC.
6. With the engine located at true TDC, the tape can be removed and the fixed timing marker adjusted to the TDC point on the engine calibration scale.

Flywheel Housing Concentricity

Never take any risks with a flywheel housing. They are regarded as indestructible by many technicians but can cause more problems than you may think. Flywheel housings are subjected to high torque shock loads and minor misalignment problems can cause a range of problems. Always observe the OEM measurements when reinstalling flywheels *any* time they are removed.

Diamond Dowels

Diamond dowels work to retain the alignment of mated components much better than cylindrical dowels. However, all the critical alignment checks should still be performed. For instance, if a flywheel housing has been removed, the housing should be radially indicated (dial indicator) for its concentricity to the crankshaft even when diamond dowels are used. If they are used, there is much less chance of an eccentric measurement (Figure 13–19).

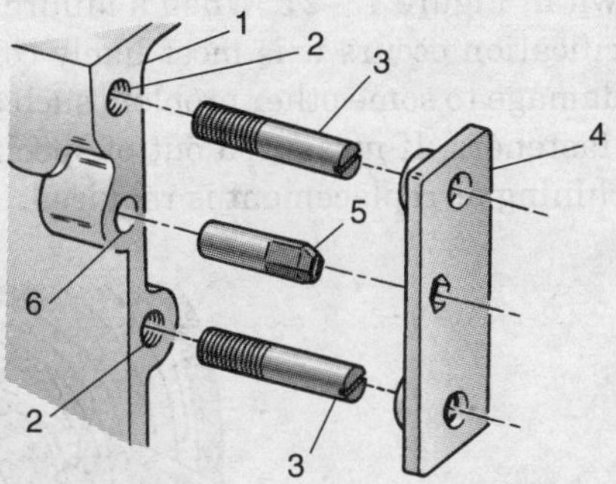

1. Engine cylinder block
2. Flywheel housing bolt holes
3. Guide stud
4. Diamond dowel alignment/installation fixture
5. Diamond dowel
6. Diamond dowel interference fit bore

Figure 13–19 *Diamond dowel installation. (Courtesy of Mack Inc.)*

Indicating a Flywheel Housing

Any time a flywheel is removed from the cylinder block, the housing inner flange face concentricity with the crankshaft should be checked using a dial indicator. The maximum tolerance for the crankshaft-to-flywheel housing eccentricity is low, typically around 0.012″ (0.3 mm) or less total indicated runout (TIR), and the consequences of installing a flywheel housing that exceeds the allowable specification are severe. When this occurs the drive axis is broken, which can result in clutch, engine mount, transmission, and engine failures. The flywheel housing to crankshaft concentricity is preserved by interference fit cylindrical or diamond dowels. These may be relied on to properly realign the flywheel housing on the cylinder block at each reinstallation, but the specification is so critical, it should be checked. The following procedure outlines a method of checking **flywheel housing concentricity**, and a couple of typical strategies for correcting an out-of-specification condition.

1. Ensure that the engine is properly supported. Indicating a flywheel housing may be performed with the engine in-chassis or out; obviously a procedure such as this is made rather easier if the engine is out of chassis. Locate the flywheel inner flange face-to-crankshaft concentricity TIR tolerance in the OEM technical literature. Either mechanically or electrically ensure that the engine fuel system is no-fueled.
2. Mount the flywheel housing to the engine cylinder block using the dowels to align the assembly and snug the flywheel fasteners at about half of the OEM-specified torque value.
3. Using fabricators chalk, mark the flywheel flange face with indices in the following rotational positions: NE, NW, SW, SE, or N W S E (shown in Figure 13–20).
4. Next, thoroughly clean the inside face of the flywheel housing with emery cloth and fix a magnetic base dial indicator to any position on the crankshaft, setting the probe to contact the flywheel housing inside face as demonstrated in Figure 13–20. Using an engine barring tool, rotate the engine in its normal direction of rotation until the indicator probe is positioned at any one of the chalk indexes. Now set the indicator at zero.
5. Bar the engine through a full rotation, recording the indicator reading at each of the indexes. The indicator should once again read zero when the revolution is complete. If the readings were as follows: NE:0, NW: minus 0.003″, SW: minus 0.005″, SE: plus 0.004:, the TIR would be the highest negative value (0.005″) added to the highest positive value (0.004″), giving a reading of 0.009″. If the OEM TIR maximum specification were 0.012″, this reading would be within it.

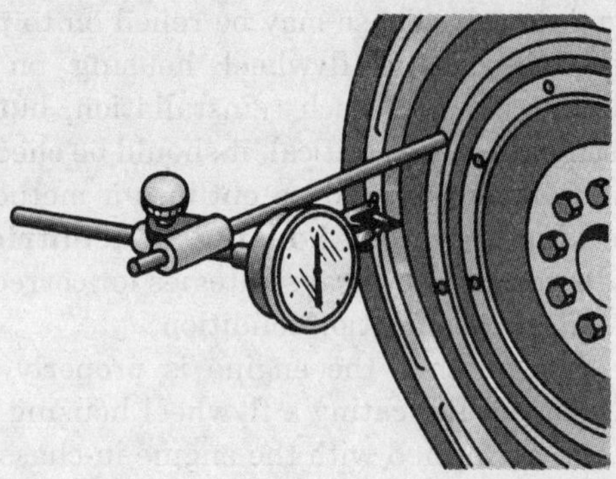

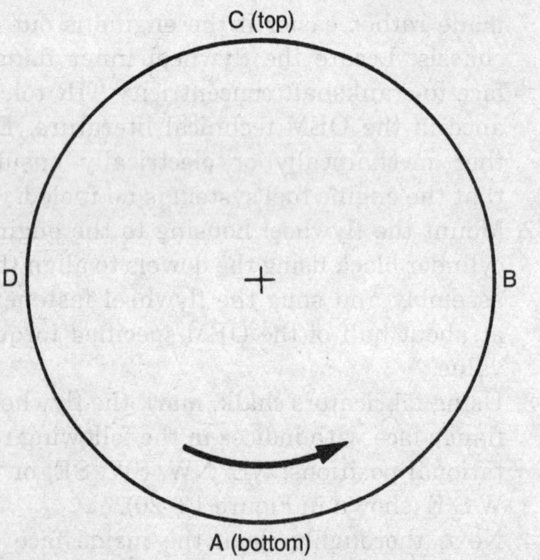

Figure 13–20 *Caterpillar recommended method for checking flywheel housing radial concentricity. (Reprinted Courtesy of Caterpillar Inc.)*

Correcting Flywheel Housing Radial Runout

If the flywheel housing to crankshaft concentricity is outside of the specification, it must be reset. The following outlines this procedure for an engine that aligns the flywheel housing using cylindrical locating dowels. Begin by removing the dowels. Check the availability of oversize locating dowels with the parts department. Loosen the flywheel mounting capscrews until they are barely snug, permitting the flywheel housing to be moved fractionally when struck with a rubber mallet. Perform the flywheel housing concentricity sequence as in the foregoing list. Adjust the position of the flywheel housing by tapping with the rubber mallet on the basis of the readings of the dial indicator. Repeat the procedure until the readings meet the required specification. When they do, fully torque the flywheel housing capscrews to specification. Next, the locating dowel holes in the cylinder block and flywheel housing will have to be reamed to an oversize specification factoring in the interference fit requirement. Drive the oversize dowels into the reamed holes.

Checking Flywheel Housing Axial Runout

Flywheel housing axial runout failures are rare but you should still perform the check. Check the flywheel housing mating face runout as shown in Figure 13–21. When a failure to meet specification occurs it is most likely coincidental damage to some other problem such as missing fasteners. If measured out of specification, machining or replacement is required.

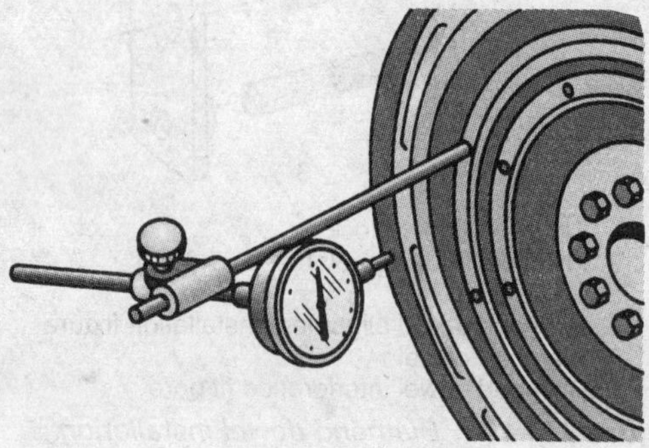

Figure 13–21 *Checking flywheel housing axial concentricity. (Reprinted Courtesy of Caterpillar Inc.)*

Reinstall Engine to Chassis

If an adequate amount of care was taken in removing the engine from the chassis, the reinstallation should be straightforward. Usually this takes less time than the removal because the components are clean and the hoses and connectors are tagged. Never hesitate to obtain assistance, especially in snub nose conventional chassis with an engine compartment that extends into the cab. The overhaul procedure should be completed by running the break-in test profiles covered in Chapter 14.

REVIEW EXERCISES

The contents of this chapter are entirely practical. Simply, it is the application of some of the theory dealt with in previous chapters. At this point it is probably essential to take down and reassemble a diesel engine and put to practice some of the theory covered thus far.

The following are some tasks to prepare you for the experience.

1. Identify the engine to be worked on and obtain the required OEM service literature.
2. Review the disassembly procedure of any diesel engine according to the OEM service literature. Ensure that something to write on and a pencil are available so you can record any observations, especially those procedures that are not common to all engines.
3. Review the OEM recommended cleaning and testing procedure for all the engine components, including the cylinder heads, cylinder block, connecting rods, and crankshaft. List all those procedures requiring the use of specialized equipment.
4. Review the OEM recommended reassembly procedure for the same engine, once again noting those procedures that are distinct to that particular engine.
5. Review the OEM requirements for initial startup of the engine following an engine overhaul. Make a list of each step in the procedure. Take special notice of the recommended break-in procedure.

14

Diesel Engine Run-In and Performance Testing

INTRODUCTION

Running-in and testing a rebuilt diesel engine is key to ensuring that it performs properly. The procedures outlined here are designed to complement those produced by the manufacturers of engine test apparatus and the engine original equipment manufacturers (OEMs) and not to replace them. In the larger truck facility, the chassis dynamometer (dyno) is commonly used as a tool for loading an engine with the objective of diagnosing malfunctions, and it may not always be necessary to connect all of the instrumentation with which the **dyno** is equipped. Also, most electronically managed engines are capable of replicating the dynamometer data. It is strongly recommended, however, that all the available dynamometer instrumentation is used for every dyno test regardless of what data is generated by the vehicle management system. This will verify the engine control module (ECM) data and help develop a dynamometer performance database, which can be referenced in future testing of similar engine/chassis configurations.

When using the road test for engine break-in and testing, the technician should be always aware that the primary responsibility is to safely operate the vehicle on the road. The road test is definitely the least preferred method of engine testing, but nevertheless one that many repair and service operations must use if they cannot access a dynamometer. It makes sense to develop a set of guidelines to be observed by technicians when road testing at a service operation to ensure some consistency of test conditions; this can be useful as a comparative performance reference. If the data generated by dynamometer and road testing is recorded to a database, it will not be long before sufficient data is retained to provide an invaluable reference for diagnosing engine conditions.

The terms used in this chapter concerning power terminology are covered in an earlier chapter of this textbook. The terms and principles concerning engine governing, electronic management systems, and diagnostic instruments are covered in the later sections. In other words, to fully understand all of the concepts introduced in this chapter, you will have to reference other sections of the textbook.

REBUILT ENGINE RUN-IN PROCEDURE

When OEMs talk about **engine longevity**, they are referring to the projected operational service life of the engine from manufacture to first overhaul and thereafter, between engine overhauls. Actual engine longevity is determined by a large number of factors including the type of chassis it powers, the physical conditions under which it is operated, preventative maintenance practices of the operator, and the skills of the driver. There is no doubt that in recent years, expected engine life of electronically managed, highway diesels has more than doubled, sometimes tripled.

Engine Longevity

Engine longevity can be measured either in engine hours or by highway miles completed. In most truck operations, highway miles are used to evaluate engine life. This can be misleading as it would not take into account factors such as prolonged idling, power takeoff (PTO) operation, and the conditions of the operating environment. For instance, a truck engine used for short pickup and delivery runs in a large city will have a much tougher life than one used in a long distance, linehaul application. The same would apply in the comparison between the city bus and the intercity, highway coach. The type of service a vehicle is subjected to is always a factor.

Linehaul Miles

Today it is not unreasonable to expect a service life of 3,000,000 *linehaul* miles or around 5,000,000 km from a current truck engine. Within this life span, one major out-of-chassis and two in-chassis overhauls would be expected at appropriate intervals. **Linehaul** refers to terminal-to-terminal, long-distance highway travel, which is usually regarded as the least punishing operating condition for every chassis system. The same engine in a truck hauling aggregates out of a quarry and running them short distances might be lucky to achieve a sixth of that mileage and that would require conscientious **preventative maintenance (PM)** and a knowledgeable operator. Diesel engines are not designed to be idled. The toughest job a linehaul diesel has to routinely perform is to idle while its driver eats lunch or sleeps. With many states and provinces adopting tough new anti-idling legislation for reasons of emissions and fuel economy, a major beneficiary will be engine longevity.

Run-in

A critical factor in the expected operational life of a rebuilt engine is the initial **run-in** following the overhaul. After a complete overhaul or major repair job involving the installation of piston rings, pistons, cylinder liners/sleeves, or bearings, the engine must be "run-in" before release for service. The procedure for run-in varies depending on the method used and that largely depends on the equipment available. Engine dynamometer, chassis dynamometer, or highway run-in methods may be used. There is no doubt that the preferred diesel engine run-in method is the engine dynamometer, but time constraints make this method impractical in many operations. The chassis dynamometer is probably the most commonly used method of engine break-in, but dynamometer equipment is costly and only larger truck service facilities can justify the expense. The least preferred method is the highway run-in. Many rebuilders have no choice but to use this method of engine break-in, but even when the operator is knowledgeable it is difficult to fulfill the proper requirements of the break-in procedure on a highway road test with little control over load and limited monitoring instrumentation.

Preparing an Engine for the First Startup

Regardless of the method to be used for engine run-in, the engine must be properly prepared before starting for the first time. The following list may be used as a general guide:

1. Lubrication System: The lubricating oil film on the rotating parts and bearings of an overhauled engine is usually insufficient for proper lubrication when the engine is started up for the first time after an overhaul. Install new oil filters. Fill a pressure **prelubricator** (usually an electrically or pneumatically actuated external oil pump) with the OEM-recommended oil and connect the supply line to the main oil gallery. Prime the engine lubrication system with sufficient oil. There are various points on the engine into which the pressure line may be tapped, but if no other is apparent, the oil gauge line may be disconnected and the pressure tank applied at that point. Remove the oil level dipstick, and check crankcase level. Add oil, if necessary, to bring it to the full mark on the dipstick. *Do not overfill!* When using a prelubricator it is

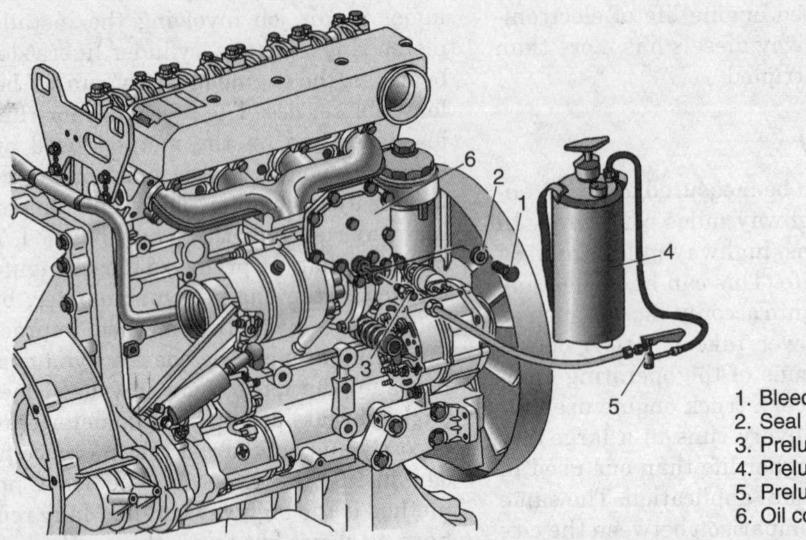

Figure 14–1 *Prelubricator set up to prime an MB-900. (© Courtesy of Daimler AG)*

unnecessary to prime the oil filters. Some OEMs prefer that the oil filters be never primed due to the risk of contaminating the oil during the procedure. When priming filters, pour the oil into the inlet side (outer annulus) of the filter and never into the outlet side (center) (Figure 14–1).

2. Turbocharger: Disconnect the turbocharger oil inlet line and pour approximately half a liter (1 pint) of clean engine oil into the turbo, ensuring that the bearings are lubricated for the startup. Reconnect the oil line.
3. Air Intake System: Check the integrity of the air intake system, checking all the hose clamps, support clamps, piping, charge air cleaner, and the air cleaner element(s). Always replace the air cleaner element after an engine overhaul.
4. Cooling System: Fit a new coolant filter and, if required, separate conditioning additives. Fill the cooling system with the recommended coolant mixture. Ensure that all or at least most air is purged from the cooling system; remove a plug from the water manifold during filling to allow air to escape.
5. Fuel System: Install new fuel filters, priming them as required, with the correct grade of filtered fuel. Next, prime the fuel system by actuating a hand pump or external priming pump. Avoid priming a fuel system by charging the fuel tanks with compressed air; the practice can be particularly dangerous when ambient temperatures are high.
6. Electrical System: Ensure that the batteries hold a proper level of charge. This is especially critical with some electronically managed engines in which the ECM requires a specific minimum operating voltage.
7. Initial Startup: Crank the engine with no-fuel for 15 seconds; check for leaks. Next, start the engine, run for 2 minutes, and then shut down. Check for leaks. Do not run for a period longer than 2 minutes.

Loaded Engine Run-in

Regardless of the run-in method, the technician performing the procedure must be observant throughout to detect and act on any problems that may develop. Instrumentation displaying engine and chassis functions must be monitored constantly and all readings recorded. If a computerized dynamometer with a printout capability is used, also record the cab instrument display and data from the chassis bus. Compare the downloaded data where possible. During dynamometer or road testing on electronically managed systems, use an electronic service tool (EST) to snapshot data as required.

If an engine develops any abnormal running characteristics during run-in, it should be immediately shut down and the problem investigated and corrected before continuing the run-in procedure. The technician should use all his or her senses during the entire startup/break-in process. The following should act as a guideline for conditions that may develop and require immediate investigation:

- Any unusual noises such as knocking, scraping, squealing, or combustion knock.
- Any vibration in the engine or elsewhere in the drivetrain.
- Any significant drop in engine oil pressure. Check the OEM parameters for oil pressure values—they vary widely.
- A rise in coolant temperature that exceeds the maximum specified by the OEM, typically around 200°F (95°C) (maximum values vary—check OEM specifications).
- A rise in engine oil temperature that exceeds the maximum specified by the OEM, typically 240°F (115°C) (maximum values vary—check OEM specifications).
- An exhaust temperature that exceeds maximum limit for the specific engine being tested, as measured by a chassis pyrometer or a dynamometer pyrometer. If the latter is used, ensure that it is positioned in the correct location for the specification reading (12″ downstream from the turbine exhaust outlet is typical), and the probe is not contacting any metal components.
- Any oil, coolant, fuel, manifold boost, or air intake system leaks.
- Any nonengine chassis system malfunction. When using a chassis dynamometer or road test to make engine break-in easier, never forget that while the engine may be the focus of the testing, every chassis system is getting a thorough workout.

Dynamometer Test Objectives

The **dynamometer** is primarily designed to measure power, a term fully defined in an earlier chapter. Essentially, the dynamometer is an instrument that applies turning resistance to the torque output (twisting effort) of another machine and accurately measures the applied resistance. Power is the *rate* of accomplishing work. When power is tested on a dynamometer, its factors are torque and time. The **torque** output of an engine is accurately measured by the dynamometer and factored with time (rpm) to calculate its power. Most dynamometers take care of the math and usually display power in units of **brake horsepower (BHP)** or kilowatts (kW). For a full explanation of power characteristics and terminology, refer to Chapter 5.

When you use a dynamometer to break-in an engine, the test is used to fully load the engine rather than simply measure its power output. There are two general categories of dynamometers used in the diesel service facilities, and they are defined by the method used to apply a resistance to the turning effort of the engine or chassis being tested.

Electromotive Dyno

An electromotive dynamometer is basically an electric motor turned in reverse. The engine is coupled to the dynamometer armature and rotates it: As current is switched to flow through the induction coils of the electromotive dynamometer, resistance to the turning effort of the engine increases. The more current flowed through the dynamometer coils, the stronger the electromagnetic field produced and the greater the amount of torque required by the engine.

Hydraulic Dyno

The second category of dynamometer is hydraulic. The hydraulic medium used is usually water but other types of hydraulic media are also used. The critical component in the hydraulic dynamometer is a load cell or multiple load cells. These use a principle similar to that used by a hydraulic driveline retarder such as the Caterpillar brake saver: Water is flowed through the cell and acts on an impeller. Inlet and outlet valves are used to control the flow of the hydraulic medium into and out of the load cell and define the torque required to rotate the impeller.

Dyno Data

All dynamometers measure torque. Most also measure rotational speed. When both torque and rpm are known, BPH can be calculated using the equations introduced in Chapter 5:

$$BHP = \frac{T \times rpm}{5252}$$

T = Torque expressed in lb-ft.

or

$$kW = \frac{T \times rpm}{9429}$$

T = Torque expressed in N·m.

$$Torque = \frac{BHP \times 5252}{rpm} \text{ or } T = \frac{kW \times 9429}{rpm}$$

When performing chassis dynamometer testing of diesel engines whether for purposes of diagnosis or to perform an engine run-in routine, all of the gauges and instrumentation should be connected to the engine being tested. Loading an engine down on a dynamometer is the ultimate performance test for an engine, and it should be as thoroughly monitored as possible. It takes a little extra time to connect all the required gauges and instruments, but they will assist to ensure accurate diagnoses and complete reports. Electronic printouts do not look professional when half the data categories are left blank.

CHASSIS DYNAMOMETER TESTING

Because of the time required to mount and dismount an engine to an engine dynamometer test bed, the chassis dynamometer is used for most truck engine diagnosis and testing. **Chassis dynamometers** use a roller or rollers to receive input. The truck chassis therefore must be located so that its drive axle(s) are properly aligned and contacting the rollers, and then be chained into position. Determine that the dynamometer is in calibration before starting the test. When double roller chassis dynamometers are used, periodic testing by running a single drive axle chassis on each roller separately will provide an indication of the condition of each absorption unit and the display instrumentation. The following outlines the procedure to be followed when running a truck chassis on a dynamometer.

1. Check all the chassis fluid levels including engine oil, fuel, coolant, transmission, and drive axle carrier. Check the tire pressure and the wheel lug integrity. Inspect suspension components, driveline, and brake adjustment, ensuring the chassis rolls freely with brakes released.
2. If the dynamometer is of the double roller type, measure the truck chassis drive axle spread and set the rollers on the dynamometer to the required spread (Figure 14–2 and Figure 14–3). Then engage the roller brake to lock the roller(s) to enable the truck to be driven onto the dynamometer test bed. Next, drive the truck onto the dynamometer roller(s), release the brake, and align by running up the "road speed" to around 10 mph (16 km/h) in one of the low range gears for 15 seconds or so, then shut down the engine. Chock the front wheels using four wheel wedges and install the safety chains; when performing the latter,

Figure 14–2 *Double roller, chassis dynamometer. (Courtesy of SuperFlow Technologies Group)*

Figure 14–3 *Front axle tie-downs to dynamometer test bed. (Courtesy of SuperFlow Technologies Group)*

be sure to securely fasten the safety chains. Generally, single roller dynamometers require the truck to be fastened down with more chain slack than double roller versions. Check on the procedure outlined by the dynamometer manufacturer.

3. All chassis operating on a dynamometer must be provided with adequate cooling air. This is almost impossible to achieve in the modern, sound insulated dynamometer room. A large fan can assist, but it does not come close to inducing the airflow transfer rate of a tractor driving down the highway at 60 mph (100 km/h). To make matters worse, air-to-air charge air coolers have generally replaced coolant medium heat exchangers, and while these boast higher efficiencies in highway operation, they have lower operating efficiencies when ambient air transfer rates are low, such as operation in a mining pit or an enclosed chassis dynamometer. Fit the exhaust gas extractor pipes to the truck exhaust pipes.
4. Connect all the dynamometer gauges (these are usually of superior quality and accuracy) to the vehicle dash instruments, and, when possible, additionally monitor the engine running data using an EST and the appropriate software. Do not omit connecting monitoring gauges with the belief that it will save time. In the event of a malfunction, critical diagnostic data may be unavailable. Typical dynamometer gauges include: crankcase pressure, air inlet pressure, manifold boost pressure, fuel subsystem charging pressure, coolant temperature, coolant pressure, oil temperature, oil pressure, ambient air temperature (engine compartment), exhaust temperature (pyrometer), and exhaust back pressure. Dynamometer data displays are shown in Figure 14–4 and Figure 14–5.
5. If the engine is hydromechanically managed, ensure that the accelerator linkage is properly set. This should be tested by having someone fully depress the accelerator pedal with the engine stopped while observing the throttle arm/fuel control lever. Normally, a visible travel breakover is required, which is typically ±0.125″ (3 mm). With electronically managed engines, ensure that the throttle position sensor (TPS) calibration programming is correct. Figure 14–6 shows the monitoring and display circuits used in a typical chassis dynamometer.
6. On certain vehicles with front axle/rear axle drives, the transfer case to front axle drive will have to be disabled. In some cases, this might involve the removal of the propeller shaft. Check with the OEM technical service literature on the appropriate test method.

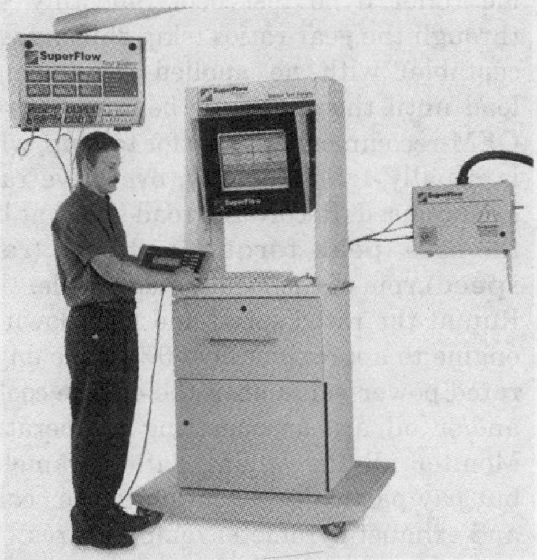

Figure 14–4 *Data display console used on a truck chassis dynamometer system. (Courtesy of SuperFlow Technologies Group)*

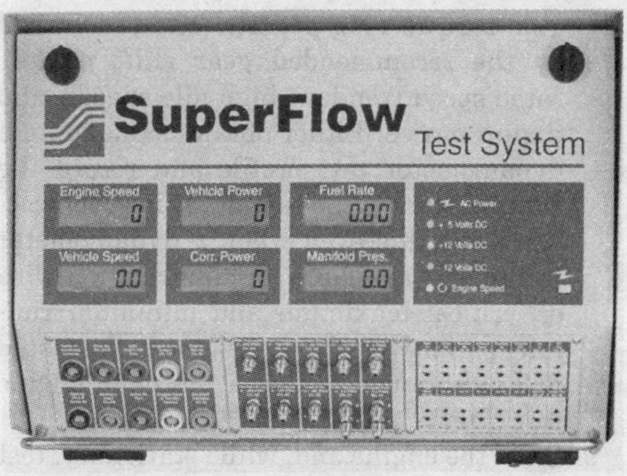

Figure 14–5 *Data display console. (Courtesy of SuperFlow Technologies Group)*

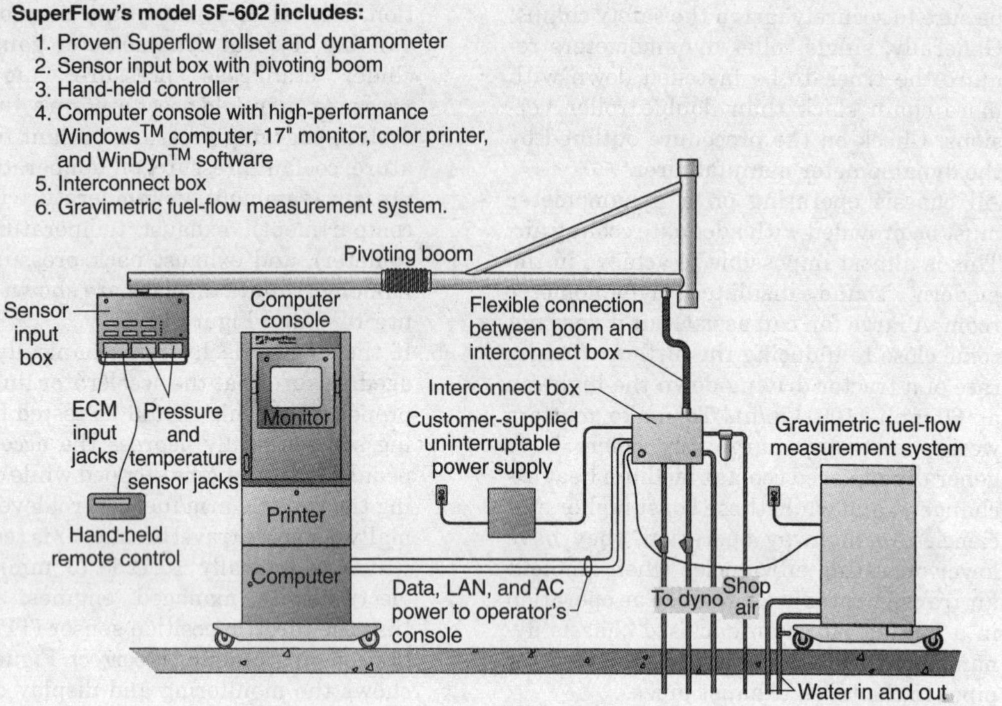

Figure 14–6 *Chassis dynamometer monitoring and display circuits. (Courtesy of SuperFlow Technologies Group)*

7. Remove any stones lodged in the tire grooves. Switch on the exhaust extraction system and the ambient air transfer system. The fumes produced by burning tire rubber while dyno testing are harmful and should not be inhaled. Wear hearing protection.
8. Obtain the OEM test data for dynamometer testing and take particular note of the engine **torque rise profile** (often identified by the recommended gear shift points), rated speed rpm low, high idle speeds, and **droop curve**. Either obtain or make up a dynamometer test profile and record the running data while testing. It is good practice to do this even when a dynamometer printout is available. Familiarization with all critical maximum and minimum running parameters for the engine being tested will help in conducting a thorough engine test (Figure 14–7).
9. Start the engine and, while stationary, test idle and high idle engine speed. On a hydromechanically governed engine, if there is any irregularity with either parameter, cease the test and adjust the governor/set engine overhead as necessary before resuming. On a tandem drive axle unit, be sure to engage the interaxle differential lockout, especially when testing on a double roller dyno test bed. Smoothly shift through the gear ratios (skip shifting is acceptable) with no applied dynamometer load until the chassis is being run in the OEM-recommended ratio for testing, which is usually 1:1. Generally, overdrive ratios are not used. Record the road speed at both the base (**peak torque**) and peak (**rated speed**) rpm of the torque rise profile.
10. Run at the rated speed and load down the engine to approximately 50% of the engine rated power value until the engine coolant and/or oil are at operating temperature. Monitor all the engine data parameters, but pay particular attention to the coolant and exhaust pyrometer temperatures.
11. Never exceed a maximum limit but remember that "maximum" means you are within a specification: Dynamometers are designed to explore the maximum an engine is capa-

ENGINE TEST REPORT

Date: ______	Unit Number: ______
Repair Order Number: ______	Model Number: ______
PROM I.D.: ______	Max. N/L RPM: ______
Rated F/L RPM: ______	
Idle RPM: ______	

A. PRESTART			
1. PRIME LUBE OIL SYSTEM	2. PRIME FUEL OIL SYSTEM	3. FILL COOLING SYSTEM	
B. START-UP AND IDLE FOR 30 SECONDS			
START______ STOP______	OIL PRESSURE______	WATER TEMPERATURE______	
C. WARM-UP — 5 MINUTES	START ______	STOP ______	
RPM MAX. SPEED	LOAD 50%	OIL PRESSURE	WATER TEMPERATURE
1. LUBE OIL LEAKS	2. FUEL OIL LEAKS	3. COOLANT LEAKS	4. LOOSE BOLTS
D. RUN-IN — 5 MINUTES	START ______	STOP ______	
RPM MAX. SPEED	LOAD 75%	OIL PRESSURE	WATER TEMPERATURE
E. FINAL RUN-IN — 20 MINUTES	START ______	STOP ______	
RPM MAX. SPEED	LOAD 100%	CRANKCASE PRESSURE AT F/L	EXHAUST BACK PRESSURE AT F/L
LUBE OIL PRESS. AT F/L	LUBE OIL TEMP. AT F/L	FUEL OIL TEMP. AT F/L	FUEL OIL PRESSURE AT F/L
WATER TEMP. AT F/L	TURBO BOOST PRESS. AT F/L	LUBE OIL PRESSURE AT IDLE	IDLE RPM
REMARKS:			
OK ______ Reject ______	Dynamometer Operator ______	Date ______	

Figure 14–7 *Typical dynamometer test profile recommended by DDC for engine testing. (Courtesy of Detroit Diesel Corporation)*

ble of. However, when excessively high temperatures are observed and the dynamometer ambient temperature is not suspected to be the cause, check the boost air temperature. Install a calibrated temperature gauge downstream from the charge air cooling core, preferably at the intake manifold. Temperatures should generally not exceed 150°F (65°C), but check OEM specifications because exhaust gas recirculation (EGR) mixing variables can result in unusual conditions.

12. When the engine is at operating temperature and the displayed data indicates that its systems are properly functional, the dynamometer test sequence may proceed. Full load rated power or "governed speed" is normally tested first. Running the engine rpm held at the rated speed (not higher than rated speed), load down the roller(s) until the accelerator pedal is at full travel, and at the point at which the engine is unable to sustain the rated speed rpm, record the brake power value. Ensure that the air compressor is unloaded, the air conditioning (A/C) system is switched off, and that the engine fan is disengaged so that the results of the test are as accurate as possible. Check twice. Allow power readings to stabilize before recording. Next, perform the same test at 100 rpm below rated speed. Use the dyno loading to drop the engine

rpm in 100 rpm increments, recording the engine power descending through the torque rise profile to its base rpm. This may be (but does not have to be) the "peak torque" rpm.

13. If the objective of the dynamometer test is to run-in a rebuilt engine rather than to diagnose engine performance, the engine should be run at full power at 100 rpm increments through the torque rise profile for a period not less than 20 minutes. This should be sufficient to properly seat the rings.
14. The technician undertaking the dynamometer test should constantly monitor the engine data displays for abnormal readings, especially those indicating temperature values. It is not unusual for engines using air-to-air charge air coolers to approach overheat values when performing full load chassis dynamometer testing due to low airflow factors.

Test Condition Variability

Chassis dynamometer test results vary from dynamometer to dynamometer. It is up to each service facility to establish acceptable at-the-wheel horsepower ranges on the dynamometer they use for testing. Generally, BHP output at the wheels should approximate 80% to 85% of the OEM engine brake power specification for tandem drive axle chassis and 85% to 90% of the OEM brake power specification for single drive axle chassis. The variables in the vehicle drivetrain are always a factor. Variations from one dynamometer to another can be due to variables such as the altitude of the location, wheel roller diameter, ambient temperature, and humidity. The type and condition of the tires on a vehicle will cause traction variations, and small roller dynamometers generally produce more tire slippage than dynamometers equipped with larger diameter rollers. Single, large diameter roller dynamometers used on tandem drive trucks produce less slippage and generally more accurate results than multiple roller types. The specific values obtained from chassis dyno tests are most useful for comparisons to data obtained from testing similarly configured chassis on the same dynamometer. See Photo Sequence 1.

ENGINE DYNAMOMETER TESTING

The **engine dynamometer** method of checking engine power is more accurate than the chassis dyno method because it removes all the variables involved when transmitting power through a transmission, drive shafts, drive axle carriers, and wheel assemblies. This method is ideal for engine run-in because it affords the test technician perfect observation of the engine throughout the test sequence, something not possible when the engine is in-chassis.

Two types of engine dynamometer are used: the test bed type, which is a permanent fixture usually isolated in an insulated, soundproof room often with a separate observation room, or the portable unit, which is coupled directly to the engine to be tested. The disadvantage of engine dynamometer testing is that it adds considerable time to an engine overhaul to fit and subsequently remove the engine from the dynamometer test bed, and test conditions are far removed from the more realistic chassis dyno test.

In this respect, the portable unit usually requires less time to fit to the engine, but these can seldom be safely used in a large truck garage due to the noise levels emitted and the potential for danger when operating an engine under high loads in a room not dedicated to the purpose.

Before using an engine dynamometer for run-in or testing, the technician should become familiar with the procedure outlined previously for checking power using a chassis dynamometer. The first step is to obtain the OEM dynamometer test data, connect and properly phase and align the drive shaft, connect the appropriate gauges, plumb in a cooling tower, and check all the engine fluid levels.

1. With the engine no-fueled, crank until an oil pressure value is observed on the oil gauge. Program the engine to run in <dyno test> mode. Then start the engine.
2. Check the engine idle and high idle rpm, and on hydromechanical engines ensure that the dyno fuel control linkages produce full travel plus the required breakover factor. On electronic engines ensure that the TPS is properly calibrated and programmed to the engine ECM.

Photo Sequence 1

Chapter 14 Chassis Dynamometer Test Procedure

P1-1 The dyno we are using is a Mustang electromotive tandem roller test bed with Powerdyne electronics and a J1939 data bus connection. Begin by backing the truck onto the dynamometer rollers, ensuring that it is perfectly aligned.

P1-2 After spinning up the wheels to relieve any suspension stresses, chain the truck to the test bed in the manner shown. The chain should be wrapped around the rear cross-member and not contact any pneumatic hoses or electrical wires.

P1-3 It is essential that the chain have some slack. The amount of slack shown here is correct. An excessively tight chain can result in tire overheating and suspension damage.

P1-4 Either clamp the front wheels or insert a pair of wheel chocks on both sides of each front wheel. For this single-drive axle truck, we have opted to drive the rear rollers for the test so front wheel chocks are used.

P1-5 Front wheel chocks in position. These should be securely jammed under the wheels. Their function is to check forward and rearward thrust forces.

P1-6 Next, establish the handshake connection between the truck chassis data bus and the dynamometer electronics. Connect the 9-pin J1939 connector to the chassis data link. The chassis data link is located somewhere on the left side of the steering column. Boot the PC.

3. Start the engine, and run at around 1,000 rpm with no load for 5 minutes to check for oil pressure, noises, vibration, leaks, and drive shaft runout. Shut down the engine, and check and correct oil and coolant levels.
4. Run the engine at a midpoint rpm within the torque rise profile, and load the engine at around 50% of its rated power value. Operate the engine under this half-load condition until the coolant and oil temperatures are each in the operating temperature window.
5. Set the accelerator/throttle to produce the rated speed rpm and smoothly load the engine until it is unable to sustain the rated speed: Determine the peak power setting and then test at descending increments of 100 rpm engine power through the torque rise profile.

Photo Sequence 1

Chapter 14 Chassis Dynamometer Test Procedure (continued)

P1-7 This is the dynamometer console. After booting the PC, select Powerdyne menu from the GUI.

P1-8 Once into the Powerdyne software, you can customize the view according to which data you would like displayed during the test run. Because this system will interact with the J1939 data bus electronics, any data off the chassis data bus can be displayed.

P1-9 With no load applied to the dyno rollers, accelerate the truck until it can be shifted into 1:1 ratio: Accelerate to rated engine speed (rpm). To assist in airflow during the test, a fan is positioned directly in front of the nose of the truck. Note the data on the digital dash display and compare with data on the computer monitor.

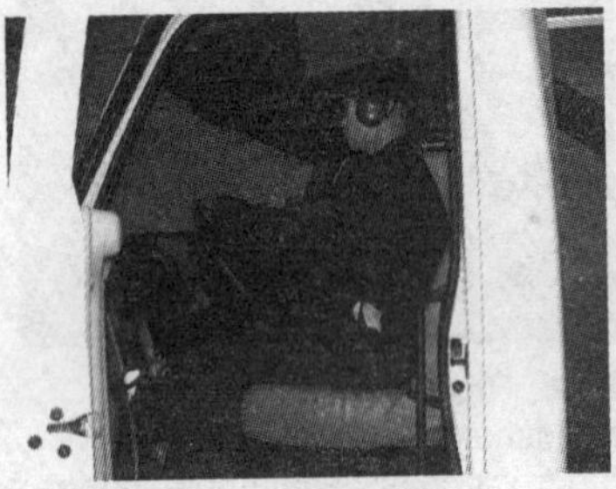

P1-10 When the truck is running at rated engine rpm with no load applied to the dyno rollers, begin to apply load in gentle and gradual increments. At the point coolant temperature exceeds 160 F, you can apply full load to the rollers. Test torque rise power in 100 rpm increments.

P1-11 During the test sequence, visually monitor all the dash gauges in the truck while maintaining close observation of the dynamometer display console. Abort the test if anything irregular is observed. Pay close attention to the tires during the test. A typical test will cull data through the torque rise rpm: Snapshots at rated and peak torque speeds should be downloaded.

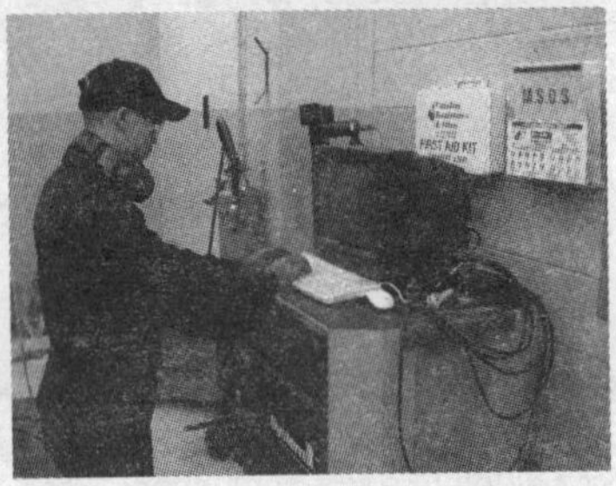

P1-12 At the conclusion of the test, use the PC electronics to analyze the data from the test. This data can be imported into spreadsheet and text files for further analysis.

6. Monitor and record all the engine running data during the test sequence, including: crankcase pressure, air inlet pressure, manifold boost pressure, fuel subsystem charging pressure, coolant temperature, coolant pressure, oil temperature, oil pressure, exhaust temperature (pyrometer), and exhaust backpressure.
7. Run under conditions of peak loading for a minimum period of 20 minutes when the dyno test is being performed to run-in the engine.

Photo Sequence 2 demonstrates the procedure of how to run a diesel engine on an engine test bed.

ROAD TESTING

Before performing a road test, the technician should ensure that the qualifications for oper-

Photo Sequence 2

Chapter 14 Engine Dynamometer Test Procedure

PS2–01 *The engine to be tested should be mounted in an appropriate cradle on the test bed deck. Plumb in the cooling tower (located on the left).*

PS2–02 *It is good practice to connect all the dynamometer test and monitoring instrumentation to the engine for a test sequence.*

PS2–03 *The engine mounted from the dynamometer test bed viewed from the rear.*

PS2–04 *The engine monitoring and test instrumentation may have to be read from within the test cell ..... or from an .....*

PS2–05 *..... adjacent observation room.*

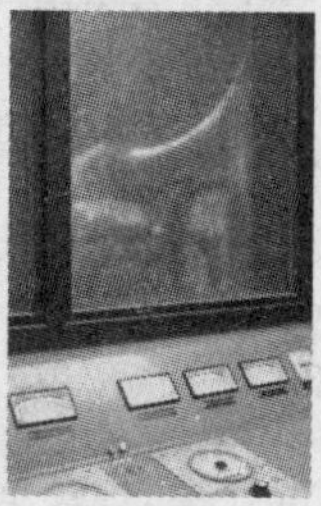

PS2–06 *The view of the dynamometer observation room panel display and the observation window. The engine should be run through the OEM test sequence and the results downloaded on a hard printout or to a diskette. Use the chassis serial number as the file ID.*

PS2–07 *Following the test sequence, the engine should be allowed to idle for at least 5 minutes before shutting down. Ensure that the engine has cooled completely before removing the cooling tower plumbing and the test instrumentation.*

ating a heavy-duty truck or bus on the highway required by the state or provincial jurisdiction are met. It should be emphasized that although the engine may be the chassis system being tested, all the vehicle systems are subjected to a performance workout and it is any vehicle operator's responsibility to ensure that they are both safe and functional. When diagnosing an engine malfunction, the road test usually takes second place to a chassis dynamometer test. And when the objective is the run-in of a rebuilt engine, the road test is the least preferred method after engine and chassis dyno loading, simply because there are too many variables and unknown factors that may arise during a road test.

When running-in an engine, it should be loaded close to its maximum for at least some of the procedure and, therefore, the truck should be coupled to a preferably loaded trailer. Using the vehicle braking system to achieve this can seriously overheat the foundation brakes.

When road testing an electronically managed engine, the operator should be thoroughly familiar with the characteristics of the management system and its programming. Before the road test, scan the ECM for active and historic codes, download any tattletales, and the customer and proprietary data programming that might impact on the diagnosis. During the road test, connect an EST loaded with the appropriate software to the ATA J1708/J1939 connector to scan and snapshot, when required, vehicle running data. If a digital dash display is installed in the truck, determine what data may be displayed during the test that would be most useful in assessing the engine condition.

The technician should never forget while road testing that the primary responsibility is to safely operate the vehicle and the distractions of the dash and electronic displays used to diagnose must not interfere with that responsibility. The following is intended to act as a guide to road test performance whether the objective is engine run-in or malfunction diagnosis:

1. Perform a pretrip inspection or circle test. Most state and provincial jurisdictions require this test to be performed by the driver of any heavy-duty highway vehicle before operating it on a road. This test at minimum would require that the following be checked and corrected before running the vehicle on a highway: vehicle lighting, tire inflation, tire condition, brake adjustment, air system pressure buildup time and values, audible air leaks, mudflap condition, suspension condition, fifth wheel or pintle hook integrity, and load securement.
2. If the objective of the road test is to run-in a newly rebuilt engine, perform the sequence outlined earlier in this chapter entitled "Preparing the Engine for the First Startup." If the objective is to diagnose a malfunction, the engine fluid levels should be checked.
3. Next, the technician should become familiar with the setup of the primary drivetrain components, making a mental note of the engine rated speed, torque rise profile, and any customer-programmed data such as progressive shifting, road speed limit, and governor type that could be a factor while performing the road test. Note the recommended gear shift points (often identified on the visor or elsewhere in the cab) when mechanical engine/transmissions are to be tested.
4. Start the engine and check the idle and no-load high-idle speeds, correcting the values as required. Warm the engine up with the vehicle stationary but avoid idling for longer than 5 minutes, especially where the engine has been newly rebuilt. The test procedure outlined for chassis dyno testing should be reviewed because a road test attempts to replicate some of this. Road traffic largely dictates how thoroughly a road test can be performed, so try to select a highway where traffic is known to be light.
5. Once on the highway, bring the vehicle up to speed and test the performance through the torque rise profile. It really helps if the operator is not forced to maintain a consistent road speed, but this factor will be determined by the amount of road traffic. A road test of 30 minutes after the specified engine-operating temperature has been reached is usually sufficient to break in a rebuilt engine (seat the rings). When diagnosing engine malfunctions, the road test duration will depend on the data revealed during the procedure.

REVIEW EXERCISES

1. Obtain an OEM engine power and torque performance chart. Using calculation, confirm that the nominal BHP value is correct at three points in the torque rise profile.
2. Use a college dynamometer or visit a local service facility with the objective of observing the procedure for mounting an engine to a dynamometer test bed.
3. Visit a facility with a chassis dynamometer and observe the procedure for harnessing a truck to the test bed.
4. Access the data bus of a current truck using the appropriate EST and find out what is required to electronically prep the vehicle for a chassis dyno test.
5. Using an OEM dynamometer performance test profile for either a chassis or engine dynamometer, record the preparation and test sequence.

REVIEW QUESTIONS

1. *Technician A* states that a truck rebuilt diesel engine run-in procedure usually requires that the engine be operated at medium to low loads for the first 1,000 miles (1,600 km) of operation. *Technician B* argues that it is usually essential to run a rebuilt diesel engine at full load for a short period after a rebuild to seat the piston rings. Who is correct?
   a. Technician A only
   b. Technician B only
   c. Both A and B
   d. Neither A nor B
2. *Technician A* states that an engine dynamometer is likely to produce a brake power rating closer to the OEM-specified power than a chassis dynamometer test. *Technician B* states that the chassis dynamometer produces brake power readings at the wheels that are always lower than OEM-rated power. Who is correct?
   a. Technician A only
   b. Technician B only
   c. Both A and B
   d. Neither A nor B
3. If all three options were available and time is not a consideration, which of the following options would be the preferred method of running-in a rebuilt engine?
   a. Engine dynamometer
   b. Chassis dynamometer
   c. Road test
4. *Technician A* states that when performing dynamometer testing of an engine, the altitude of the test location must be considered when analyzing the power. *Technician B* states that ambient temperature variations have a greater effect on brake power produced than altitude. Who is correct?
   a. Technician A only
   b. Technician B only
   c. Both A and B
   d. Neither A nor B
5. Which of the following electronic controllers would have to be disabled before performing a chassis dynamometer test?
   a. Engine ECM
   b. Transmission ECM
   c. Automatic traction control (ATC)
   d. Collision warning system (CWS)

PART 2

Hydromechanical Diesel and Alternate Fuel Systems

15

Chemistry and Combustion

INTRODUCTION

It is important for any mechanical technician to have a fundamental understanding of chemistry for purposes of explaining fuel composition and combustion dynamics in engines, and for developing an ability to work with electricity and electronics. Some knowledge of basic chemistry can be an especially useful diagnostic tool for the technician specializing in engines and fuel systems. This chapter briefly addresses basic chemistry, atomic structure, combustion dynamics, and the language of fuel technology—all topics that have individually been the subject of many textbooks. Our approach will be to look at the basics. Technicians are encouraged to explore this subject matter to a greater depth using other specialized texts.

Chemistry is the science that seeks to understand the composition of **matter** (physical substance in general), the **elements** (any substance that cannot be resolved into simpler substances), the **compounds** (substances containing two or more elements combined in definite proportions and held together by chemical force) they form, and the reactions they undergo. It is one of the oldest sciences, and it evolved from the observations of thinkers in times well before the modern age. The chemist would observe natural phenomena, hypothesize (attempt to reason) its causes, then confirm the **hypothesis** (supposition) by experiment. When it seemed that a hypothesis could be conclusively proven by documented observation of cause and effect, a law was established.

Of course, the history of science is generally littered with hypotheses that became irrelevant as knowledge progressed, so a theory of physics or chemistry today may not necessarily be so tomorrow. And the opposite can occur. Something that seemed insignificant in the past can become the foundation of a core technology as in the case of Hero's development of the reaction turbine aeolipile (see Chapter 3) 2,000 years ago. Not only was the reaction turbine the first heat engine, it was the basis for both steam and jet turbine engines, which are technologies that had to wait 1,600 and 2,000 years, respectively, after the initial invention.

BASIC CHEMISTRY

The building blocks of all matter are **atoms**. All atoms are electrical. Electrical charge is a component of all atomic matter. This section examines some key definitions required to develop an understanding of atomic theory. Some of this information is revisited in Chapter 28 on electrical fundamentals.

Elements

An element is any one of more than 100 substances, most naturally occurring, that cannot be chemically resolved into simpler substances. They consist of minute particles known as *atoms,* which, for purposes of study, are considered to be indivisible. An atom is the smallest particle of an element that can take part in a chemical reaction. The atoms of any one

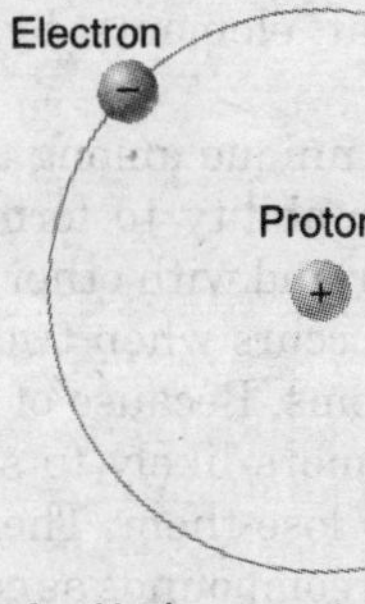

Figure 15–1 *Hydrogen atom.*

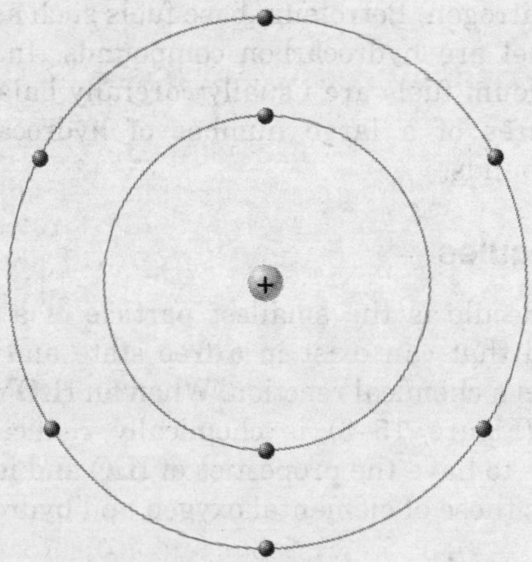

Figure 15–2 *Oxygen atom.*

element are exactly alike and possess identical **mass** (quantity of matter a substance contains: weight). The name of an element is always a single word such as hydrogen or oxygen (Figure 15–1 and Figure 15–2).

All elements have a short form symbol usually derived from the Latin name for the substance. Sometimes the symbol is obvious because the Latin word for an element is similar to the English word (much of the English language is derived from Latin), such as that for the element oxygen, which is O. However, the symbol for the element iron is Fe, derived from the Latin word for iron, which is ferrum. Atomic number identifies the number of **protons** in an atom of the element. Table 15–1 shows the symbols and atomic number of some common elements that you are more likely to come across if you work with vehicle technology.

Mixtures

A **mixture** is composed of two or more elements and/or compounds, all of which retain their own characteristics and identity. Air is an example of a mixture; it is composed of 23% oxygen and 76% nitrogen by mass (plus about 1% of inert gases). These percentages change to 78% nitrogen and 21% oxygen by volume so we can immediately determine that oxygen is slightly heavier than nitrogen. This fact explains why there is less oxygen on top of Mount McKinley than at sea level.

Both oxygen and nitrogen retain their own identity and can take part in reactions independently of each other when they are in this mixture we call air. The oxygen in air, for instance, is commonly involved in reactions that do not involve nitrogen. The properties of a mixture depend on the substances in it; the

Table 15–1 *Symbols and Atomic Number of Common Elements*

Some Common Metallic Elements	*Atomic Number*	*Some Common Nonmetallic Elements*	*Atomic Number*
Iron—Fe	26	Hydrogen—H	1
Sodium—NA	11	Carbon—C	6
Magnesium—Mg	13	Helium—HE	2
Aluminum—Al	13	Sulfur—S	16
Nickel—N	28	Silicon—SI	14
Rhodium—Rh	35	Selenium—Se	34
Silver—Ag	47	Oxygen—O	8
Zinc—Zn	30	Nitrogen—N	7
Gold—Au	79	Argon—Ar	18
Platinum—Pt	78	Radon—Rn	86

compound water (H_2O) is a pure substance with identifiable characteristics that are dissimilar from any other pure substance. However, salt water is a mixture of salt and H_2O; it boils at a higher temperature and freezes at a lower temperature than pure H_2O.

Chemical Bonding

Chemical bondings are interactions that account for the association of atoms into molecules, ions, and crystals. When atoms approach one another, their nuclei and **electrons** interact and distribute themselves in such a way that their combined energy is lower than it would be in an alternative arrangement. Whenever the total energy of a group of atoms is lower than the sum of energies of its component atoms, they are chemically bonded; bonding energy accounts for the overall lowering of energy.

Valency

The number of bonds an atom can form is called its valency (or valence number). The valency of an atom is simply the number of unpaired electrons in its valence shell. The valence shell is the outermost shell of electrons. An H_2O molecule consists of an oxygen atom with a valency of two combined with two hydrogen atoms, which have a valency of one. An electrovalent or ionic (**ion**: an atom with either an excess or deficiency on electrons) bond occurs when an electron is transferred from one neutral atom to another—the resultant charge results in the atoms being held together by electrostatic attraction.

Compounds

A compound is a substance composed of two or more elements combined in definite proportions and held together with chemical force. A compound is composed of identical molecules made up of atoms of two or more elements. Most commonly occurring materials are mixtures of different chemical compounds. Pure compounds can usually be obtained by physical separation processes such as filtration and distillation. Compounds themselves can be broken down into their constituent elements by chemical reactions.

Carbon atoms are unique among the chemical elements in their ability to form covalent bonds with each other and with other elements. **Covalent bonding** occurs when two electrons are shared by two atoms. Because of its atomic structure, carbon is more likely to share electrons than to gain or lose them. There are ten times as many carbon compounds as compounds of all other elements combined; common carbon compounds are formed with hydrogen, oxygen, and nitrogen. Petroleum base fuels such as diesel fuel are hydrocarbon compounds. In fact, petroleum fuels are usually carefully balanced mixtures of a large number of hydrocarbon compounds.

Molecules

A molecule is the smallest particle of a compound that can exist in a free state and take part in a chemical reaction. When an H_2O molecule (Figure 15–3) is chemically reduced, it ceases to have the properties of H_2O and it possesses those of elemental oxygen and hydrogen.

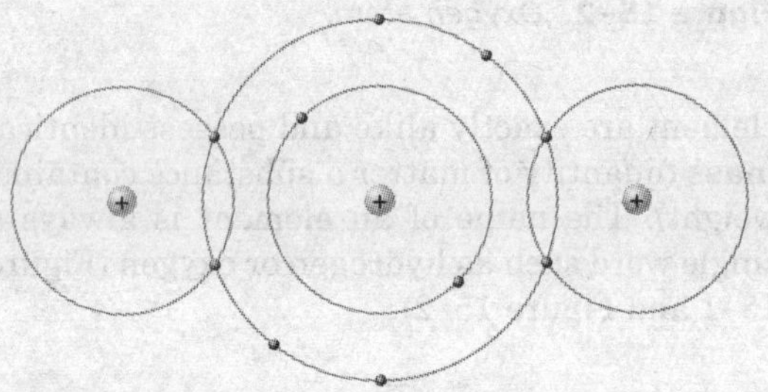

Figure 15–3 *Water molecule: Observe the shared electrons.*

The oxygen found at sea level in the earth's atmosphere combines to form **diatomic** molecules, that is, two oxygen atoms to a molecule (Figure 15–4). Diatomic oxygen is expressed as O_2.

Atomic Structure

Atomic theory seeks to explain both the composition of matter and the laws that apply to chemical combination. It was conceived in 1801 by a chemist, John Dalton, who confirmed

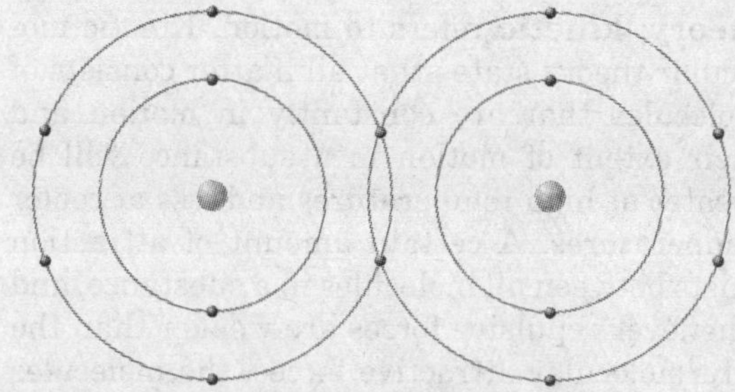

Figure 15–4 *Oxygen molecule: Oxygen in the lower atmosphere is mostly diatomic.*

theorizing by some great thinkers in the era around 400 years B.C., suggesting that matter could not be subdivided indefinitely without altering the properties of that substance. Dalton's atomic theory suggested:

1. Each element is made up of minute, indivisible particles called atoms.
2. Atoms cannot be created or destroyed.
3. Every atom that a specific element is composed of is identical.
4. Atoms of different elements are different.
5. Atoms of one element may combine with other elements to form compounds.

Dalton's atomic theory developed into our current notions of the structure of an atom by the scientists Thomson, Rutherford, and Chadwick in the early part of the twentieth century. These scientists identified the following subatomic particles:

1. The electron—symbol e–. Discovered by J. J. Thompson in 1897. Electrons carry a negative charge and orbit in the shells around the atom's nucleus. An electron has 1/1837 of the mass of a proton.
2. The proton—symbol p^+. Discovered by Ernest Rutherford in 1911. Protons carry a positive charge and are located in the atom's nucleus. A proton has 1,837 times the mass of an electron.
3. The **neutron**—symbol n°. Discovered by James Chadwick in 1932. Neutrons are electrically neutral and are located in the atom's nucleus. A neutron has slightly more mass than a proton.

In 1911, Ernest Rutherford described the nuclear model of the atom and asserted:

1. An atom consists mostly of empty space.
2. Each atom has a minute, extremely dense nucleus.
3. The nucleus is surrounded by a large volume of nearly empty space.
4. Most of the mass and *all* of the positive charge is located in the nucleus.
5. The nearly empty space surrounding the nucleus is sparsely occupied by electrons that possess a fraction of the mass but a negative charge that balances the positive charge of the protons.

Balanced Atoms

A hydrogen atom (see Figure 15–1) consists of a single proton in the nucleus and one electron in its orbital shell; the electron orbits the nucleus in much the same manner Earth orbits our sun. The electrical force attracting the electron toward the positive charge of the nucleus is balanced by the mechanical force acting outward on the rotating electron; this ensures that the electron remains in its orbital shell and is not drawn into the nucleus. All atoms are electrical. An electrically **balanced atom** is one in which there is an equal number of electrons and protons (Figure 15–5 and Figure 15–6). While electrons can move from atom to atom in many substances, protons do not, and they are held in the nucleus of the atom. An atom with either an excess or a deficit of electrons is known

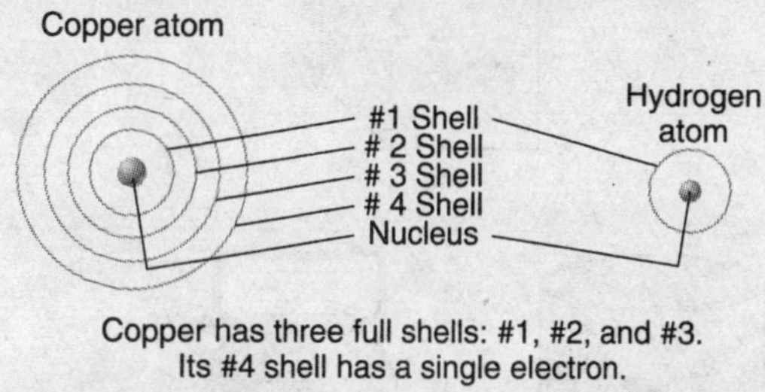

Figure 15–5 *Copper atom: A copper atom has four shells, three of which are full. (Courtesy of Daimler Trucks North America)*

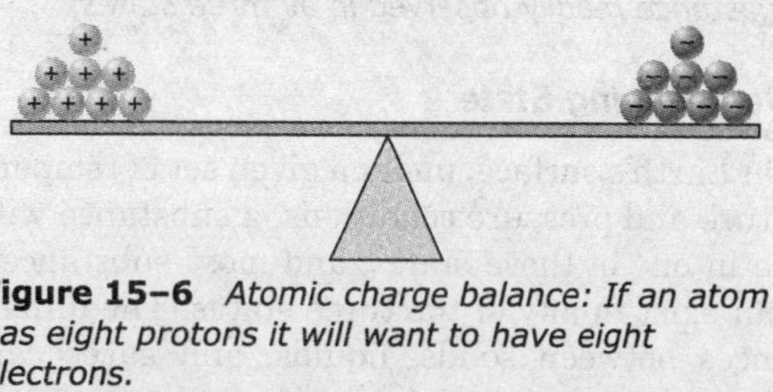

Figure 15–6 *Atomic charge balance: If an atom has eight protons it will want to have eight electrons.*

as an ion. An ion is an atom in an electrically unbalanced state.

Current atomic theory suggests that, in much the same way protons and neutrons make up atomic nuclei, these particles are themselves made up of quarks. Quarks are therefore subatomic particles that are theorized to be the fundamental constituents of all matter. However, the study of quarks probably goes beyond what today's technician needs to know of atomic structure.

States of Matter

In physics, matter is defined as anything that has mass and occupies space. Matter can be generally classified into one of three states or phases within the earth's atmosphere: solid, liquid, or gas. Water is the only substance that is familiar in all three states, which we describe as ice, water, and steam (Figure 15–7). A solid does not readily change in size or shape when subjected to physical forces. Like a solid, a liquid has a definable volume but it has **fluidity** and will change its shape if poured from one container to another. A gas generally expands to fill the volume of the vessel in which it is contained.

Figure 15–7 *States of matter: H_2O is the only substance readily observed in all three states.*

Determining State

On Earth's surface, under a given set of temperature and pressure conditions, a substance will be in one of these states; and most substances can exist in any of the three states. The differences between solids, liquids, and gases can be explained in terms of **kinetic molecular theory**; **kinetic** refers to motion. Kinetic molecular theory states that all matter consists of molecules that are constantly in motion and their extent of motion in a substance will be greater at high temperatures and less at cooler temperatures. A certain amount of attraction exists between all molecules in a substance, and whenever repulsive forces are weaker than the intermolecular attractive forces, the molecules stick together. However, as temperature increases, so does the molecular motion.

In a solid, intermolecular attractive forces overcome the disruptive thermal (heat) energy of the molecules and they are bound together in an orderly arrangement called a crystal. The molecules in a crystal vibrate due to their thermal energy and, again, the amount they vibrate is determined by temperature or the amount of retained heat energy.

As a solid is subjected to heat, its molecules vibrate with increasing energy until they are capable of overcoming the intermolecular attractive forces. At this point the substance melts or changes to the liquid state and acquires fluidity. A change of state works in reverse; the freeze point of a substance is the same as its melt point.

Changes of State

Kinetic molecular theory also explains the change of state from that of a liquid to a gas known as **vaporization**. As heat is applied to a liquid, some molecules acquire sufficient thermal energy to overcome the intermolecular attraction and break free from the surface of the liquid; these molecules are converted to a gaseous state. If heat continues to be applied to the liquid, more molecules will move away from the liquid until the substances' boil point is achieved and all the molecules escape the liquid state. When vaporization is reversed, it is known as **condensation**. The distance between molecules of a gas is large in comparison with that of a liquid because the intermolecular forces are weak. A gas molecule will move until it either strikes another gas molecule or the walls of the vessel that contains it. The effect of many gas molecules striking a container wall is known as pressure.

It is possible for a substance to pass directly from the solid to the gaseous state in a process known as **sublimation**. Dry ice (CO_2) sublimates at atmospheric pressure and temperature. Snow can also be sublimated.

Plasma and Other States of Matter

At very high temperatures, atoms may collide with each other with a force that results in some electrons being jolted free of the nuclei. A mixture containing these positive and negative ions (ion is an atom that has lost or gained one or more electrons) is not defined as a gas but as plasma. Scientists commonly refer to a plasma state as being the fourth state of matter; in fact, most of the known universe is in this state. In the universe, matter may be found in forms that cannot be grouped into these previously described states of matter. Dying stars are theorized to collapse into forms in which matter is so dense that their gravitational force pulls in all matter and radiation. Because no light can escape from a black hole little is known of them.

States of Matter—Conclusion

As far as the technician is concerned, a basic understanding of what we know as the three states of matter should suffice. A practical example of some of the preceding information can be observed in what happens in a typical injection pulse. Fuel directly injected to a diesel engine cylinder is atomized. Atomized fuel is in the liquid state. This fuel, when exposed to the heat of compression in the engine cylinder, vaporizes; it must change state before it can be combusted. If liquid fuel contacts the piston crown, it remains condensed longer and can result in a failure because more heat energy is required to vaporize it, retarding its ignition. Gases condensing in the exhaust gas stream will be observed as white smoke. Particulate or minute solids in the exhaust gas stream will be observed as black smoke. In diagnosing these and many other conditions, the technician will benefit from a basic understanding of chemistry.

Chemical Reactions

There are many classes of chemical reactions, most of which are not of significant interest to the student of diesel technology. However, combustion is an oxidation reaction. The reactant in the engine cylinder is whatever oxygen is present at the time of ignition. In a chemical reaction, the ingredients present undergo significant chemical alteration. In the engine cylinder the ingredients of the chemical reaction are the fuel (hydrocarbon molecules) and the reactant (oxygen in the cylinder). What results from the oxidation reaction in the engine cylinder are compounds with fundamentally altered states. In other words, what goes into an engine cylinder is quite different from what comes out. We will attempt to explain some of the reasons for this in the following section, beginning with learning a little more about elements and compounds.

PROPERTIES OF SOME COMMON ELEMENTS

Each element has a special identity and set of characteristics that make it unique in terms of both its behavior and appearance. Some of the characteristics of common elements are outlined below.

Hydrogen (H)

Hydrogen is a colorless, odorless, tasteless flammable gas—the simplest member of the family of chemical elements with a nucleus consisting of one proton that is orbited by one electron. Hydrogen atoms are reactive and combine in pairs, forming diatomic (combinations of two atoms to a molecule) molecules: H_2. Hydrogen is the most abundant element in the known universe, accounting for about 75% of matter, although only around 1% of all matter on Earth. It is present in all animal and vegetable substances in the form of compounds, most often carbon. It also constitutes 11% of the mass of water. Combustion of hydrogen with oxygen produces temperatures in the region of 2,600°C (4,700°F).

Atomic No.: 1
Melting point: –259.2°C (–435°F)
Boiling point: –252.8°C (–423°F)

Carbon (C)

Carbon is a nonmetallic element that exists in a number of forms and combines to form compounds more readily than any other element. About 0.2% of the earth's crust is composed of carbon. Elemental carbon exists in three forms: diamonds, graphite, and carbon black, which includes charcoal and coal. Pure diamond is the hardest substance known. It is a poor conductor of both electricity and heat, and it is used in industrial applications as a cutting and drilling medium. Graphite has many opposite characteristics: Notably, it is a conductor of heat and electricity and commonly used as a lubricant. When iron is alloyed with small quantities of carbon, *steel* is formed.

Atomic No.: 6
Melting point: 3,550°C (6,420°F)
Boiling point: 4,827°C (8,721°F)

Oxygen (O)

A colorless, odorless, tasteless gas, oxygen is the most plentiful element in the earth's crust. Oxygen is present in air (23% by mass), water (86%), and in the earth's crust (47%). Almost all atmospheric oxygen is the result of photosynthesis. During respiration, animals and some plants process atmospheric oxygen to form carbon dioxide; by photosynthesis, green plants assimilate carbon dioxide in the presence of sunlight to produce oxygen. Most of the gaseous oxygen found in the atmosphere consists of molecules of two atoms known as diatomic oxygen (O_2).

Monatomic (O) and **triatomic** (O_3) oxygen (ozone) are more predominant in the upper strata of the atmosphere where ozone filters the sun's ultraviolet radiation. Commercial oxygen is produced by fractional distillation of liquid air. Oxygen has a valence of 2 and forms a large range of covalently bonded compounds. Molten oxygen forms a blue liquid, and solid oxygen is attracted by a magnet.

Atomic No.: 8
Melting point: –218°C (–361°F)
Boiling point: –183°C (–297°F)

Nitrogen (N)

A colorless, odorless, tasteless gas, nitrogen is the most plentiful element in the earth's atmosphere; the mixture *air* contains approximately 76% nitrogen by mass. Nitrogen is a constituent of all living matter. Chemically, nitrogen is fairly inert at ambient temperatures but during combustion it will combine to form various oxides, especially in high-temperature, lean burn conditions. The automotive industry identifies those various oxides collectively as NO_x and they are subject to Environmental Protection Agency (EPA) standards and regulation as noxious emission.

Atomic No.: 7
Melting point: –210°C (–346°F)
Boiling point: –196°C (–320°F)

Sulfur (S)

A nonmetallic element, sulfur is also one of the more reactive elements. Pure sulfur is a tasteless, odorless, brittle solid that is a pale yellow in color and a poor conductor of electricity. Sulfur is perhaps important to the diesel technician mostly because it is a constituent of petroleum. Sulfur appears prominently in the residual oil fractions (those left over after distillation and other refining practices) of a diesel fuel that provide the fuel with its lubricating properties.

Atomic No.: 16
Melting point: 113°C (235°F)
Boiling point: 445°C (832°F)

Sulfur in Diesel Fuel

Current United States and Canadian statutory requirements of on-highway diesel fuels permit no more than 0.0015% sulfur content. This fuel is known as ultra-low sulfur (ULS) fuel. ULS will be progressively introduced from 2007 to 2012 to off-highway diesel users. This means that sulfurous emissions from diesel engines have been all but eliminated. Sulfur is oxidized to sulfur dioxide (SO_2) in the combustion process. SO_2 is a colorless and poisonous gas regarded by the EPA as a noxious emission and one that readily forms acidic compounds such as H_2SO_4 (sulfuric acid). Sulfur is also a constituent of lubricating oils.

Iron (Fe)

A metallic element that makes up about 5% of the earth's crust, Fe is the second most common metal after aluminum. It combines with many other elements to form hundreds of minerals. Fe almost always contains small amounts of carbon, which modifies its properties. When alloyed with small amounts of carbon, steel is produced—the most widely used of all metals. Elements other than carbon can also be added to Fe in the alloying process to modify the characteristics of steel. Common alloying metals are nickel, chromium, molybdenum, tungsten, vanadium, titanium, and manganese. Fe is extensively used in vehicle technology, mostly as steel.

Atomic No.: 26
Melting point: 1,535°C (2,795°F)
Boiling point: 3,000°C (5,432°F)

Aluminum (Al)

A lightweight, silver-white colored metal that makes up about 8% of the earth's crust, Al is therefore the most common metal. Al never occurs in the metallic form naturally, but its compounds are present in most rocks, vegetation, and animals. Modern commercially produced Al is obtained from bauxite and isolated by electrolysis. Pure Al tends to be soft and weak, but when alloyed with small amounts of silicon and Fe, it becomes harder and tougher. It is an excellent conductor of heat and electricity. Like Fe it is extensively used in vehicle technology.

Atomic No.: 13
Melting point: 660°C (1,220°F)
Boiling point: 2,467°C (4,473°F)

PROPERTIES OF SOME MIXTURES AND COMPOUNDS

The way in which mixtures and compounds behave when they undergo chemical reactions can be explained by the constituent elements of which they are composed. The following section describes some of the mixtures and compounds that occur in or result from combustion reactions in an internal combustion engine. Many of the by-products of combustion described here are referred to in more detail in Chapter 41.

Air

Air is a mixture of oxygen and nitrogen. This mixture contains a group of gases with nearly constant concentrations and another group of gases that are variable. Atmospheric gases of more or less consistent concentration by volume are:

Nitrogen (N_2)	78.084%
Oxygen (O_2)	20.946%
Argon (Ar)	0.934%
Neon (Ne)	0.0018%
Helium (He)	0.000524%
Methane (CH_4)	0.0002%
Krypton (Kr)	0.000114%
Hydrogen (H_2)	0.00005%
Nitrous oxide (N_2O)	0.00005%
Zenon (Xe)	0.0000087%

Of gases present in variable concentrations, H_2O vapor, ozone, and carbon dioxide are of principal importance and are critical to the maintenance of life on the planet. H_2O vapor is the source for all forms of precipitation and is also both an absorber and an emitter of infrared radiation. Carbon dioxide is also an absorber and an emitter of infrared radiation and is critical to the process of photosynthesis. Ozone (triatomic oxygen, O_3) is found principally in the upper strata of the atmosphere 10 to 50 kilometers (6 to 30 miles) above the earth's surface. It shields Earth from most cosmic radiation waveforms. Photochemical reactions between nitrogen oxides (NO_x) and hydrocarbons (HC) can produce ozone in quantities large enough to cause respiratory problems in animals, including humans. Concentrations of these gases in air are:

Water vapor (H_2O) 0–7%
Carbon dioxide (CO_2) 0.01–0.1%
Ozone (O_3) 0.01%

Carbon Dioxide (CO_2)

CO_2 is a compound. It is a colorless gas with a sharp odor and a sour taste. It is an oxidized form of carbon produced in a number of ways, most notably in respiration by animals, as a by-product of the combustion of carbon-containing substances (such as a HC fuel); in the fermentation of vegetable matter; and by plants in the photosynthesis process. The presence of CO_2 in the atmosphere permits the retention of radiant energy (primarily from our sun) received by the earth. CO_2 is commonly used as refrigerant and in fire extinguishers. It is also, by mass, the primary product of the combustion of HC fuels. However, CO_2 is not classified or regulated as a noxious emission, although the excess of it in our atmosphere contributes to global warming. In its solid form it is known as dry ice.

Water

Water is a compound. It can be readily observed in its three physical states: ice, water, and steam. It is vital to sustenance of all life forms. In its liquid state it is colorless, tasteless, and odorless. An oxygen atom has six electrons in its valence shell; when bonded to two hydrogen atoms, it shares an electron from each hydrogen atom and fills its valence shell. Like CO_2, it is a by-product of the combustion of a HC fuel (the result of oxidizing hydrogen), so it is emitted from any internal combustion engine in the form of steam.

Melting point: 0°C (32°F)
Boiling point: 100°C (212°F)

Carbon Monoxide (CO)

Carbon monoxide is a compound of carbon and oxygen. It is a highly toxic, colorless, odorless, and flammable gas. Concentrations of 0.3% can be lethal within 30 minutes.

CO is produced in the stoichiometric or slightly fuel rich combustion zone. It is associated with incomplete combustion in spark-ignited (SI) engines but tends not to be an important factor as in compression ignition (CI) engines due to the considerable excess air factors in diesel engines. CO's toxicity results from its absorption by red blood cells in preference to oxygen. CO poisoning initially produces headaches, dizziness, nausea, and culminates in respiratory failure.

Melting point: –192°C (–314°F)
Boiling point: –199°C (–326°F)

Sulfur Oxides (SO_x)

SO_2 emission is the primary product of oxidizing sulfur impurities in HC fuels. Sulfur tends to be present in the residual oil fractions constituent in diesel fuel, and legislation limiting the sulfur content in on-highway diesel fuel to 0.0015% (from 0.335% in 1994) has diminished the diesel engines' responsibility for producing this emission. SO_2 is a *compound* of sulfur, is odorless and nonflammable, and is a potential source of H_2SO_4. SO_2 smog tends to be more of a problem in geographic areas where burning of solid fuels and heavy oils is not regulated. This type of smog is especially worsened by dampness and high concentrations of suspended particulate.

Oxides of Nitrogen (NO_x)

By mass, nitrogen represents the largest ingredient in the engine cylinder during the combustion process in any engine aspirated with air. At ambient temperatures and pressures, nitrogen is relatively inert, that is, unlikely to become involved in chemical reactions. However, subjected to engine cylinder heat and pressure, nitrogen may become involved in the oxidation process, producing compounds of nitrogen known collectively as NO_x:

1. Nitric oxide (NO)—a colorless, odorless gas that in the presence of oxygen will convert to NO_2
2. Nitrogen dioxide (NO_2)—a reddish orange gas that has corrosive and toxic properties
3. Nitrous oxide (N_2O)—a colorless, odorless gas better known as laughing gas

NO_x emission is generally the result of a number of fuel/engine parameters, the most important of which are actual reaction temperatures and the excess air factor. Generally but not consistently, high-combustion temperatures tend to produce higher NO_x emission. NO_x is a major contributor to photochemical smog because it reacts with HC to form O_3 molecules, or ozone.

Unburned Hydrocarbons

Unburned hydrocarbons (UHC) consist of any emitted unburned fuel fractions and represent the range of fractions in the fuel. They include paraffins, olefins, and aromatics (see Chapter 16). The least volatile elements of a fuel are more likely to result in UHC emissions. UHCs are classified as potentially harmful emissions, so the exhaust gas aftertreatment systems we have seen introduced on more recent diesel engines have focused on combusting in the exhaust system what was not properly burned in the engine cylinder.

Partially Burned Hydrocarbons

A result of low-temperature combustion, partially burned hydrocarbons (PHCs) are substances (aldehydes, ketones, and carboxylic acids) resulting from quenching (extinguishing the flame front) before a molecule is completely combusted. One way in which this can happen is when an engine is overspeeded by overfueling, resulting in insufficient time for the combustion reaction to be taken to completion.

Particulate Matter

Any liquid or solid matter emitted from an exhaust stack that can be detected in light extinction test apparatus, such as a smoke opacimeter, is classified as **particulate matter (PM)** for purposes of emissions certification and enforcement. However, the term is more appropriately applied to emitted ash and carbon soots that are in the solid state. Diesel engines, even when operating under conditions of optimum efficiency, can emit carbon soot particulate sized 1 micron (μ) or less in small quantities. Microscopic particulates are considered to be as harmful to human respiratory systems as larger particulates. This is a contributing reason for the adoption of diesel particulate filters (DPFs) in all post-2007 highway diesel engines.

COMBUSTION

Despite the fact that human society has used fire in one form or another since prehistoric times, no person was able to explain it in appropriate chemical terms until Antoine Lavoisier in 1783. Lavoisier proved that combustion, the rusting of metals, and the breathing of animals all involved the combining of oxygen with other chemicals. It is from him that we began to use the term *oxidation*.

When a fuel is heated to its ignition temperature in the presence of oxygen, a chemical reaction takes place in which the heat energy contained in the fuel is liberated, resulting in a large volume of hot gases. In the oxidation reaction that results from igniting a fuel charge in an engine cylinder, oxygen molecules attack HC molecules, and if they succeed in penetrating their central nuclei (that is, completely combust/oxidize the fuel), they produce the compounds H_2O (in the form of steam) and CO_2 (carbon dioxide). Despite much research concerning the chemical processes that take place during combustion in an engine cylinder, knowledge of the thermodynamic process is largely limited to initial and end states and the intermediate pressure rise profile because it occurs within a minute time frame and at high temperatures.

Combustion with Ambient Air

Combustion calculations are simplified when the reaction ingredients are limited to a HC fuel and pure oxygen. But combustion in an engine cylinder uses the oxygen available in the ambient air mixture, so proportionally the largest ingredient of the reaction is always the element nitrogen. Ideally, the elemental nitrogen aspirated by the engine should remain inert and be exhausted, unaffected by the oxidation of the fuel. However, under some circumstances the nitrogen is oxidized, producing a number of oxides (known collectively as NO_x) that are identified as a noxious emission. The tendency to involve nitrogen in the oxidation reaction taking place in the cylinder generally increases as combustion temperatures increase and the fuel concentration in the mixture decreases, that is, under lean burn conditions.

Combustion in an Engine Cylinder

Compression pressures in diesel engines range generally from 27 bar/400 psi (2,750 kPa) to 48 bar/700 psi (4,862 kPa), but in most current **direct injection (DI)** engines, actual pressures

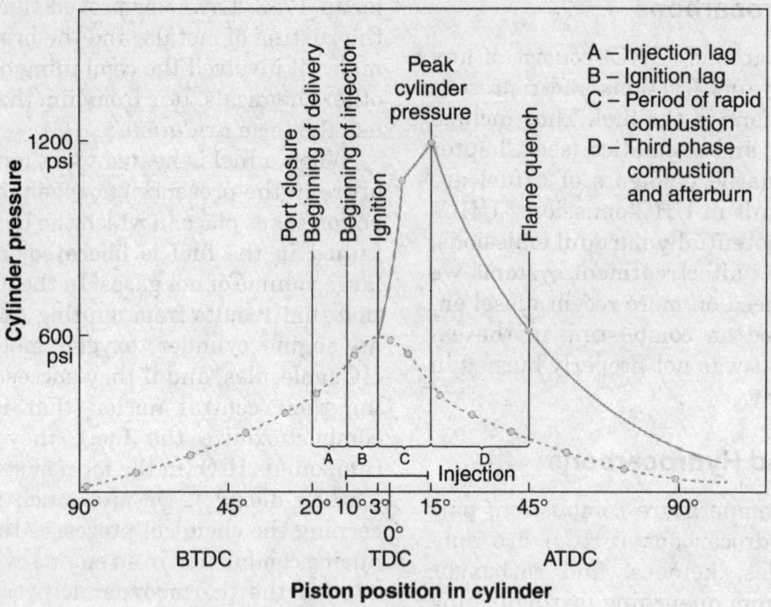

Figure 15–8 *Pressure volume curve in a diesel engine.*

typically would be close to the middle of that range. This produces cylinder temperatures at the end of the compression stroke approximating 500°C, which is around twice the minimum required to ignite the fuel. Peak combustion pressures of between two and five times the compression pressure values result, depending on the actual engine and how it is being fueled. Peak cylinder pressures are managed (by the fuel system) to occur somewhere between 10 to 20 degrees after top dead center (ATDC) (Figure 15–8) to optimize the relationship between cylinder pressure and the mechanical advantage of the crank throw vector angle (see Chapter 5). Some of the most recent diesel engines are capable of introducing fuel into the engine cylinder in multiple short shots of fuel. Because the quantity and duration of each shot of fuel into the engine cylinder is managed by computer, the resulting cylinder pressures closely complement the crank throw angle and leverage this produces. The result is a much smoother application of torque to the engine drivetrain components.

Cylinder Gas Dynamics

Piston crown design, valve configuration, manifold boost, and the engine breathing manifolds all play a role in determining the **cylinder gas dynamics**, that is, how injected fuel is dispersed, mixed, and subsequently combusted in the cylinder. The intent is to create cyclonic turbulence (swirl) in the cylinder as the piston is driven upward through its compression stroke (squish), and current piston design theory for diesel engines suggests that the Mexican hat crown best enables this. The way in which cylinder gas behaves under compression and through the expansion or power stroke governs both the engine's performance efficiency and noxious emissions. The way in which cylinder gas behaves during breathing (intake and exhaust strokes) determines how effectively combustion end gases are expelled and a new air charge is induced.

Because commercial diesel engines are almost always turboboosted, volumetric efficiencies exceed 100% by a wide margin in most performance operating phases. In fact, it would be true to say that all diesel engines are designed to operate with an excess air factor. The actual excess air percentage is usually highest when the injected fuel quantity is lowest (engine loads are lightest). The excess air factor may exceed 600% at low idle and drop to values that only slightly exceed the stoichiometric requirement when operating in the upper portion of the torque rise profile at full power loads. The use of exhaust gas recirculation (EGR) in all current engines has reduced excess air factors when they are more likely to contribute to NO_x formation.

Stoichiometry

The term **stoichiometric** is derived from the Greek school of philosophy (called the Stoics), which, among other things, preached the avoidance of excess. In chemistry this has come to mean the relationship between the relative quantities of substances involved in a reaction. In engine technology, managing a stoichiometric burn ratio means controlling fueling so that the air in the engine cylinder is precisely that required to completely oxidize the fuel. No more, no less. A more comprehensive definition of stoichiometric is: the actual ratio of the reactants in any reaction (not necessarily just a combustion reaction) to the exact ratios required to complete the reaction.

The stoichiometric ratio or **lambda** (λ) factor is therefore dependent on the actual chemical composition of the fuel to be burned: When this is known, the stoichiometric air-to-fuel ratio may be calculated. Many SI automobile engines are electronically managed to run at stoichiometric air-fuel ratios. This is not a requirement of the diesel engine, although it should be noted that the ignition location in the cylinder will occur where the mixture proportions are close to stoichiometric.

$>$ = greater than

$<$ = smaller than

$$\lambda = \frac{\text{Actual air supplied}}{\text{Stoichiometric requirement}}$$

$\lambda > 1$ lean burn

$\lambda < 1$ rich burn

$\lambda = 1$ stoichiometric AFR

Calculating Air-Fuel Ratio

Gasolines and diesel fuels are refined from crude oil petroleum. Petroleum on a mass (weight) basis consists of the following elements:

Carbon 84–87%

Hydrogen 11–15%

Sulfur 0–2%

So on a mass basis, petroleum products, whether they are gasoline, diesel fuel, or engine lubricating oil, are not going to possess widely differing stoichiometric requirements in combustion reactions. That is, the theoretical **air-fuel ratio (AFR)** will not change that much when different petroleum products are combusted. This means that diesel fuel, gasoline, or, for that matter, engine oil all have equivalent stoichiometric ratios. As soon as the exact chemical composition of the fuel is known, the stoichiometric air requirement to combust it can be calculated using the following data. To oxidize:

1 kg carbon (C) requires 2.66 kg of oxygen (O)

1 kg hydrogen (H) requires 8.0 kg of oxygen (O)

1 kg sulfur (S) requires 1.0 kg of oxygen (O)

Air is a mixture composed of 23% oxygen and 76% nitrogen by mass. Therefore, 1 kg of air would contain 0.23 kg of oxygen. So:

$$\frac{1}{0.23} = 4.35$$

Therefore, 1 kg of O would be contained in 4.35 kg of air.

Example:

1. A hypothetical diesel fuel contains by mass 86% carbon, 13% hydrogen, and 1% sulfur. What *mass* of air would be required to completely oxidize 1 kg of this fuel?

$$\text{O required} = \underbrace{(2.66 \times .86)}_{\text{carbon}} + \underbrace{(8.00 \times .13)}_{\text{hydrogen}} + \underbrace{(1.00 \times .01)}_{\text{sulfur}}$$

$$\text{Air required} = 4.35\ \{(2.66 \times .86) + (8.00 \times .13) + (1.00 \times .01)\}$$

$$\text{Air required} = 4.35\ (2.29 + 1.04 + .01)$$

$$\text{Air required} = 4.35 \times 3.34$$

$$\text{Air required} = 14.5\ \text{kg}$$

$$\therefore \text{AFR} = 14.5{:}1$$

2. A hypothetical gasoline contains by mass 86% carbon and 14% hydrogen. Calculate the stoichiometric ratio for this fuel.

$$\text{Air required} = 4.35\ \{(2.66 \times .86) + (8.0 \times .14)\}$$

$$\text{Air required} = 4.35\ (2.29 + 1.12)$$

$$\text{Air required} = 4.35 \times 3.41$$

$$\text{Air required} = 14.8\ \text{kg}$$

$$\therefore \text{AFR} = 14.8{:}1$$

Some Facts

In combustion:

1 lb carbon combines with oxygen and releases 14,540 Btu.

1 lb hydrogen combines with oxygen and releases 62,000 Btu.

AFR is normally expressed as a ratio of masses. AFR by mass of diesel fuels and gasolines falls within a fairly narrow window ranging from 14.5:1 up to about 15.2:1, depending on the exact chemical constituents of the fuel. AFR by volume ranges from around 10,000:1 up to around 14,000:1.

THE ACTUAL COMBUSTION CYCLE IN A DIESEL ENGINE

Ricardo was the first diesel engine designer to focus on emissions. During World War II, a column of diesel-powered tanks could be seen by the enemy 50 miles away by the cloud of smoke produced. His research helped reduce visible smoke from diesel engines but he also produced detailed analyses of diesel combustion that we still reference today. The following briefly outlines Harry Ricardo's description of the diesel engine combustion cycle, which we have updated a little where necessary:

Ignition Delay or Ignition Lag

Ignition delay or ignition lag occurs between the *events* of the start of injection (injector nozzle opening) and the moment ignition occurs. Although some chemical reaction begins the moment the first liquid fuel droplet enters the cylinder, the moment of ignition is normally considered to occur either when a visible flame or measurable pressure rise occurs.

In an SI, gasoline-fueled engine, the moment of ignition is controlled by a spark to the engine cylinder in which the fuel is already vaporized and mixed. Things are not so precise in a diesel engine. Injecting fuel merely begins a sequence of events we know as ignition delay. The duration of the ignition delay is defined by the evaporation rate of the fuel, itself factored by the ignition quality of the fuel and the actual temperature in the cylinder, which rises during injection and combustion.

In a DI diesel engine, the term *AFR* cannot be used to compare the conditions in the cylinder with those evident in the premixed charge present in an SI engine because as long as the fuel is not completely evaporated, the complete range of fuel-air ratios must be present from zero (no fuel) to infinity (no air within the fuel droplets). Ignition will occur where local AFR is most favorable. That is going to be where the mixture conditions are somewhere close to the stoichiometric requirement. In modern diesel engines, this location is engineered by the design of the combustion chamber. For obvious reasons the landscape of the piston crown toward which the injection pulse is directed is the major influencing factor. However, the variability of the fuel chemistry and actual cylinder temperature influence the rate at which fuel evaporates and therefore the specific location of ignition in the cylinder.

Period of Rapid Combustion

In this phase the fuel that evaporated and mixed during the ignition delay period is burned, so the rate and duration of *rapid combustion* are closely associated with the length of the delay period. Generally, as ignition delay is prolonged (for whatever reasons), the rate and resultant pressure rise increase in the second phase. In modern diesel engines with electronically managed variable timing, ignition timing can be controlled so that the period of rapid burning produces peak cylinder pressure at an ideal crank angle under all operating conditions from idle speed/load up to rated speed/load. This may be assisted by using the multipulse injection potential of more recent diesel fuel systems. However, in many older engines with static injection timing (fuel delivery timing could neither be advanced or retarded), this period of high cylinder pressure could occur out of phase with the mechanical dynamics of the crank throw angle.

Third Phase of Combustion

The third phase of the combustion cycle begins at the moment of peak cylinder pressure (wherever that happens to occur) and ends when combustion is measurably complete; that is, the available fuel has been oxidized. Under conditions of an extended fuel delivery pulse (high engine loads), some portion of the fuel will be injected into the cylinder during this third phase, so the burn rate will be influenced by the rate of injection as well as the mixing rate. Generally, engines are designed with cylinder gas dynamics to enable rapid mixing of fuel and air during the third phase so that the combustion process is completed as early as possible.

Afterburn Phase

Afterburn in the diesel combustion cycle is a period in which any unburned fuel in the cylinder may find oxygen and burn. Most modern engines incorporate this phase in the third phase of combustion because with improvements in cylinder gas dynamics and computer-controlled injection timing, it should not be a significant factor in the present-day diesel engine.

Dosing Injection

In some of the latest multipulse diesel injection systems capable of up to seven injection events per cycle, the final shot of fuel into the engine cylinder is often not intended to be combusted in the cylinder. The **dosing** shot of fuel is injected with the intention it is discharged into the exhaust system as raw fuel to be combusted in exhaust gas aftertreatment systems. Original equipment manufacturers (OEMs) are calling this final shot of fuel dosing or afterburn injection.

Detonation

Detonation describes the phenomenon that the diesel technician describes as "**diesel knock**" and a driver of any car knows as "ping." In the diesel engine it is identified by sound caused by intense pressure rise that vibrates the cylinder walls.

Following the ignition of the fuel charge, as the flame **propagates** (spreads) through the combustion chamber, the portion of the charge farthest from the primary flame front is subject to both radiated heat and compression caused by the gas expansion of the primary flame front. The heat and compression may cause unburned portions of the charge to ignite before the arrival of the primary flame front. This multiple flame front condition causes an abnormally high rate of combustion and the resultant pressure rise in the cylinder known as diesel knock.

When the terms *combustion* and *explosion* are used to describe oxidation reactions in an engine cylinder, the physical results of the reaction are the same; however, the term *explosion* is used to describe an oxidation reaction that takes place at higher speeds. Detonation is therefore correctly described as combustion occurring at explosive rates. Its causes relate generally to the fuel chemistry, prolonged ignition lag, and advanced injection timing but, generally, detonation is seldom a significant problem in contemporary electronically managed diesel engines using North American fuels. It was, and still is, a problem in diesel engines fueled by hydromechanical injection, and the condition can be observed when starting a cold engine during winter conditions.

REVIEW QUESTIONS

1. Which of the following statements correctly describes air?
   a. A mixture of nitrogen, oxygen, and inert gases
   b. A compound of nitrogen, oxygen, and inert gases
   c. A molecule of nitrogen, oxygen, and inert gases
   d. An element of nitrogen, oxygen, and inert gases

2. A cubic centimeter of a typical diesel fuel would require approximately what volume of air at sea level to completely oxidize it?
   a. 12 cc
   b. 15 cc
   c. 1,000 cc
   d. 12,000 cc
3. The smallest particle of a compound that can exist in the free state and retain the chemical identity of the compound is known as a(n):
   a. Quark
   b. Nucleus
   c. Atom
   d. Molecule
4. The theory that explains the atomic differences in the three states of matter is known as:
   a. Covalent bonding
   b. Kinetic molecular theory
   c. Atomic theory
   d. Plasma theory
5. The most abundant element found in the known universe is:
   a. Oxygen
   b. Carbon
   c. Hydrogen
   d. Nitrogen
6. The most abundant element found in the earth's crust is:
   a. Oxygen
   b. Carbon
   c. Hydrogen
   d. Nitrogen
7. Oxygen has a valence of:
   a. 0
   b. 2
   c. 4
   d. 8
8. Proportionately, the largest ingredient in an in-engine cylinder, combustion reaction is:
   a. HC fuel
   b. Oxygen
   c. Water
   d. Nitrogen
9. The term used to describe a combustion reaction in which the exact proportions of fuel and air are present is:
   a. Stoichiometric
   b. Perfect burn
   c. Afterburn
   d. Gas blow-down
10. Combustion in an engine cylinder that takes place with more than one flame front is correctly known as:
    a. Stoichiometric
    b. Normal
    c. Detonation
    d. Afterburn
11. Calculate the stoichiometric requirement of a fuel composed of 84% carbon, 15% hydrogen, and 1% sulfur when combusted in pure oxygen.
12. Calculate the stoichiometric requirement of a fuel composed of 84% carbon, 15% hydrogen, and 1% sulfur when combusted in ambient air.
13. How many protons would a balanced atom of oxygen contain?
    a. 1
    b. 2
    c. 8
    d. 16
14. How many electrons would a balanced atom of carbon contain?
    a. 2
    b. 6
    c. 12
    d. 24
15. The atomic number of an element indicates the number of ____________ in an atom of the element.
    a. Electrons
    b. Neutrons
    c. Protons
    d. Quarks
16. *Technician A* states that the stoichiometric requirement of a gasoline combusted in air at sea level must be precisely 14.7:1. *Technician B* states that the stoichiometric requirement of a diesel fuel ranges between 14.5:1 and 15.1:1. Who is correct?
    a. Technician A only
    b. Technician B only
    c. Both A and B
    d. Neither A nor B
17. *Technician A* states that when diesel fuel is combusted one of the products of combustion is water. *Technician B* states that carbon dioxide results from burning diesel fuel. Who is correct?

a. Technician A only
b. Technician B only
c. Both A and B
d. Neither A nor B

18. *Technician A* states that most electronically managed diesel engines are run at precise stoichiometric ratios of air and fuel after the engine is at operating temperature. *Technician B* states that most diesel engines are managed to run at rich fuel-to-air ratios. Who is correct?
a. Technician A only
b. Technician B only
c. Both A and B
d. Neither A nor B

19. *Technician A* states that hydrogen is an inert element. *Technician B* states that carbon can be classified as a reactive element. Who is correct?
a. Technician A only
b. Technician B only
c. Both A and B
d. Neither A nor B

20. *Technician A* states that the end products of both an explosion and a combustion reaction are similar. *Technician B* states that an explosion is an oxidation reaction that occurs in much less time than what would be described as combustion in an engine cylinder. Who is correct?
a. Technician A only
b. Technician B only
c. Both A and B
d. Neither A nor B

16

Diesel Fuel Characteristics

INTRODUCTION

In a couple of earlier chapters we have suggested that diesel engines will happily burn just about any fuel introduced into the engine cylinders and this is true. For sure an engine oil can fuel an engine to a runaway destruction in just seconds. The key to modern highway diesel fuels is to optimize performance over an ever-increasing engine life, which is generally measured in highway miles. Only fuels meeting the highest standards are able to achieve this. In this chapter we examine current petroleum-based diesel fuels.

Almost all current diesel engines are engineered to run on petroleum-based diesel fuel and although we are seeing some increase in the use of biodiesel, this is mainly confined to low percentage cuts into petroleum diesel. What is described as diesel fuel today is likely to change a decade from now. Mercedes-Benz has recently introduced a direct-injected, gasoline-fueled engine that options a combination of spark and compression ignition depending on how the engine is run at any given moment. Its advantage is to produce better thermal efficiency and that means less carbon dioxide (CO_2). While considerable attention has been devoted to hybrid technology in North America, at the time of writing this has not resulted in significant sales, especially in the auto market. While in North America we see diesel power as being primarily confined to commercial vehicles, in Europe the diesel engine has captured around 50% of the automobile market. A betting person would probably be smart to back the advance of diesel over hybrid technology at least in the near future.

FUEL TERMINOLOGY

Before studying some basic fuel chemistry and what happens in the cylinder combustion process, it helps to understand some key terms. This chapter uses the building blocks of the combustion chemistry introduced in the previous chapter and uses some of those terms that apply to the characteristics of diesel fuel. It is probably important, if not essential, that the diesel technician have a fundamental understanding of fuel chemistry and combustion when it comes to troubleshooting engine malfunctions at a higher level. You do not have to memorize most of the terms defined here, but it is important to be familiar enough with them so you can reference them in the future.

AFR—air-to-fuel ratio. The actual ratio of air to fuel in a combustion reaction. Distinct from stoichiometric AFR, which describes the proportions of the reactants in a combustion reaction required to complete the reaction. Normally expressed by mass (weight).

Ash—Diesel fuels normally contain a certain quantity of suspended solids or soluble metallic compounds such as sodium and vanadium. The fuel's ash content can affect injector, fuel pump, and any engine components subjected to high temperatures such as piston rings, exhaust valves, and turbochargers.

ASTM—American Society for Testing and Materials. Organization that classifies diesel fuel (and other fuels) to a standard.

ASTM #1D Fuel—fuel recommended for use in high-speed, on-highway diesel engines required to operate under variable load and variable speeds. Minimum cetane number (CN) must be above 40. In theory, the ideal fuel for highway truck and bus diesel engines but in practice, it is not as often used as 2D fuel because it has less heat energy by weight, making it less economical.

ASTM #2D Fuel—fuel recommended for use in high-speed, on-highway diesel engines required to operate under constant loads and speeds. Like 1D fuel, the minimum CN is required to be above 40. Widely used in highway truck operations because it produces better fuel economy than 1D fuel due to its higher calorific value albeit at the expense of slightly inferior performance.

Boil point—the temperature at which a liquid vaporizes. When applied to liquid hydrocarbon (HC) fuels, it becomes a measure of volatility.

Calorific value—heat energy. The potential heat energy of a fuel is measured in British thermal units (Btus), joules, or calories.

Catalyst—a substance that enables a chemical reaction without itself undergoing any change.

Cetane number (CN)—a measure of the ignition quality of a diesel fuel defined in some detail later in this section.

Cloud point—the temperature at which the normal paraffins in a fuel become less soluble and begin to precipitate as wax crystals. When these become large enough to make the fuel appear cloudy, this is termed the "cloud point." Cloud point exceeds the pour point by 3°C (5°F) to 15°C (25°F) in HC fuels.

Compressed natural gas (CNG)—See *Natural gas.*

Crude oil—raw petroleum. It consists of a mixture of many kinds of HC compounds of differing molecular weights and small quantities of organic compounds such as sulfur. Crude oil is distilled and cracked in the refining process to produce residual oils, distillates, and fractions, which are blended to manufacture fuels, oils, and tars.

Diesel fuel—term used to describe distillate petroleum compounds and fractions formulated for use in on-highway compression ignition (CI) engines. They are generally composed of fractions from the paraffin (the most volatile range used in a diesel fuel), naphthalene, and aromatic (the least volatile range found in a diesel fuel) series of crude oil fractions and are graded by the ASTM. Highway-use diesel fuel has a distillation range between 150°C (300°F) and 290°C (550°F) and specific gravity values ranging from 0.78 to 0.86. (See further references to diesel fuel in this text.)

Distillate—term sometimes used to describe diesel fuel formulated for on-highway use. The term diesel fuel is preferred.

Fire point—the temperature at which a liquid HC fuel evaporates sufficient flammable vapor to burn continuously in air. Fire point generally exceeds flashpoint by about 10°C in HC fuels.

Flame front—the forward boundary of the reacting zone in-cylinder combustion. Usually luminous depending on the fuel used.

Flame propagation—the way in which a fuel combusts inside the engine cylinder as determined by the manner the flame front spreads; dependent on cylinder gas dynamics, the actual AFR, temperature, and fuel chemistry.

Flash point—the temperature at which a liquid HC fuel evaporates sufficient flammable vapor to momentarily ignite when a flame is brought near its surface.

Fractions—a portion of a mixture separated by distillation or a cracking procedure such as hydrocracking or catalytic cracking. Most fuels are carefully balanced brews of combustible petroleum fractions with a range of volatility. Each fraction possesses distinct characteristics.

Fuel-air ratio—commonly used to describe the ratio of fuel to air in diesel combustion: "fuel" appears before "air" in the ratio because it is the control variable.

Gasoline—the group of liquid petroleum fuels blended for use in spark ignited (SI) engines. The actual composition of gasolines varies according to the crude oil source, refining processes, and blend requirements. Typically they have a volatility range that extends from 35°C (90°F) to 210°C (400°F) and a specific gravity that ranges from 0.70 to 0.78.

Kerosene—made up of heavier fractions than gasoline, kerosene is widely used in heating oil and jet fuel. Kerosene typically has a distillation range between 150°C (300°F) and 270° (510°F) and specific gravity values ranging from 0.78 to 0.85.

Liquefied natural gas (LPG)—See Natural gas.

Microorganism growth—airborne bacteria and fungi commonly enter vehicle and storage tanks through their venting systems. When water is present in the bottom of the tank, bacteria may reside in it and feed off the fuel HCs. The metabolic waste from such microorganisms is acidic and may corrode fuel injection components. Fungal growth can plug fuel filters when pumped through the fuel subsystem. It is good practice to fill truck onboard tanks before parking the vehicle to minimize water condensation problems; water draining tanks daily (at the sump tap) also helps. However, when the problem is sourced at the storage tank, chemical treatment of the fuel with biocides is required.

Natural gas—the gaseous product of petroleum either suspended above liquid crude oil or dissolved in it, in which case it becomes the first product to be separated in the distillation process. Natural gas is composed primarily of methane CH_4 with a lesser amount of ethane C_2H_6, propane C_3H_8, and butane C_4H_{10}. Propane and butane are extracted from natural gas and stored as liquids under pressure. Both are used as an automotive fuel and are usually known as liquefied petroleum gas or LPG.

Octane rating—a measure of the antiknock quality of a fuel, usually a gasoline. Knock in an actual engine depends on complex combustion reaction phenomena and on the engine design. However, as a method of classifying gasoline, the ASTM has standardized two methods: the Motor method (MON) and the Research method (RON). Both relate the antiknock performance of a test gasoline to an actual fuel. Two primary reference fuels are used: iso-octane with "ideal" antiknock characteristics and assigned an octane number of 100 and heptane with poor antiknock characteristics and assigned an octane number of 0. A mixture of these two primary reference HC fuels is then used as the means for grading the actual performance of a gasoline. Therefore, a mixture of 90% iso-octane and 10% heptane would theoretically have equivalent antiknock characteristics as a gasoline sold at the pumps as "90 octane."

The actual test conditions used by Motor and Research methods differ. The Motor method testing operates at higher speed and inlet mixture temperatures than the Research method. The Research method is generally the better indicator of fuel antiknock quality for engines operating at full throttle and low speeds, whereas the Motor method is the better indicator at full throttle and high speeds. Federal regulations in the United States and Canada require the posting of the average of Research Octane Number (RON) and Motor Octane Number (MON) at dispensing pumps. This is expressed (R + M)/2.

A gasoline's tendency to knock can be decreased by the addition of antiknock additives, which are metalo-organic compounds that precipitate in the burn and slow the combustion rate. Tetraethyl lead (TEL) and tetramethyl lead (TML) were used until legislated out of use. Currently potassium and oxygenated compounds are used as octane boosters. Manganese compounds are still used in some Canadian gasolines to increase octane number, but these are not permitted in the United States.

Finally, when the octane rating of a gasoline is greater than 100, it is based on the milliliters of TEL required to be added to the reference fuel iso-octane to produce the improved rating. While the sale of poisonous tetraethyl is generally prohibited in Western countries, it is still used in research and development. Unleaded fuels are available at the pumps with (R + M)/2 of between 85 and 93, generally depending on geographic region. Higher ratings are used for performance and specialty applications.

Oxidation stability—the products of oxidizing stored diesel fuel can result in deposits, filter plugging, and lacquering of fuel injection equipment. Antioxidants in the fuel inhibit the condition.

Photochemical smog—results from the photochemical reaction of HCs and oxides of nitrogen (NO_x) with sunlight in the lower atmosphere. It manifests itself as a brownish haze and the ozone that results may cause reduced visibility, plant damage, eye irritation, and respiratory distress. A photochemical reaction occurs when a physical substance absorbs visible, infrared, or ultraviolet radiation.

In the case of photochemical smog the radiation source is the sun; its ingredients are HC and NO_x from combustion processes and subjecting it to still air and sunlight will cause the required molecular decomposition, producing ozone. Photochemical smog is primarily a problem in urban metropolis with ample sunshine and low air movement. Of these, the metropolis of greater Los Angeles, California, is most notable—having an urban population of 17.7 million inhabitants (wikipedia.com) effectively walled-in on its east side by the San Gabriel mountains, limiting air movement that is generally inclined to move from west to east. Chapter 41 gives a full account of photochemical smog.

Pour point—as fuel temperature drops below the cloud point, paraffin wax crystals increase in size and the pour point generally denotes the lowest temperature at which the fuel can be pumped. Pour point is generally 3°C (5°F) to 15°C (25°F) below cloud point. Pour point depressant additives are required for extreme cold weather operation in 1D and 2D on-highway fuels. Pour point depressants have no effect on cloud point.

Specific gravity—the specific gravity of a liquid is the weight of a volume of the liquid compared to the weight of the same volume of water. The specific gravity of a petroleum-based fuel is a direct measure of its heating value (calorific value).

At 15°C (60°F) a diesel fuel measured to a specific gravity of 0.85 releases 139.50 Btu per gallon when combusted.

Stoichiometric ratio—the stoichiometric ratio is an expression of the exact ratio of the reactants required for a chemical reaction to take place. In the internal combustion engine, the specific stoichiometric ratio of a fuel at sea level depends on the chemistry of the fuel and not on the conditions of combustion. It is a ratio of masses.

Sulfur content—on recommendation of the American Petroleum Institute (API), the Environmental Protection Agency (EPA) effected a maximum sulfur content of 0.0015% by weight effective in October 2006 in the United States and Canada. This fuel is known as ultra-low sulfur (ULS) fuel. ULS replaced low-sulfur (LS) fuel, which was specified with a maximum sulfur load of 0.05%.

Viscosity—a measure of a liquid's resistance to shear, a value that generally decreases as temperature increases. It also influences its resistance to flow and therefore its fluidity. The fluidity of a liquid is graded by seconds Saybolt universal (SSU)—a measure of its ability to flow through a defined flow area against time at 100°F (39°C). The SSU rating is used for comparing the viscosity of diesel fuels; 1D is typically 34.4 SSU, 2D typically 40 SSU, and 4D 125 SSU (preheating required). Viscosity and volatility in diesel fuels are closely associated. Fuel viscosity directly affects fuel injection pump service life.

Volatility—the tendency of a liquid to vaporize. The volatility rating of a fuel is more important in SI engines as it determines the vapor-to-air ratio at the time of ignition. The volatility rating of diesel fuel is critical in summer operation when its higher fractions tend to boil off, reducing CN.

PETROLEUM

The word *petroleum* is derived from the Latin words "petra," meaning rock, and "oleum," meaning oil. Petroleum is broadly used to describe HC fossil fuels found in the upper strata of the earth's crust, ranging from solid tars through crude oil to natural gas. Somewhere around 70% of the energy consumed in North

America and 40% worldwide is derived from petroleum products.

Most of this fuel is extracted from crude oil and the natural gases that are often contained in the proximity of petroleum deposits or boils off from them at surface pressures and temperatures. Crude oils are usually black, but many reflect a yellow or greenish tint. They range in density from light, very fluid liquids of high volatility to viscous tars of lower volatility.

Formation of Petroleum

The origin of the carbon and hydrogen that are the elemental components of petroleum is in the organic materials that made up the primordial (existing at the beginning) Earth. These elemental components passed through an organic phase, mostly single-celled plants and algae and mostly located in aquatic environments. Many such simple organisms were known to have been abundant 570 million years ago; rapid burial of these organisms preserved them in sedimentary rock and enabled a series of biological, physical, and chemical changes that permitted them to evolve to what we know as petroleum. Put simply, this evolution requires heat, pressure, and eons of time. Petroleum is therefore a **fossil fuel**, an unrenewable energy source with limited reserves. It has never been duplicated in a commercial laboratory.

Crude Oil

Crude oils are loosely classified by their content, which indicates exactly what can be extracted from them in refining processes. There are three types:

1. Asphalt-based crudes: These are usually black in color. Generally, the higher the fluidity of a crude oil, the better its quality, meaning that a larger percentage of lighter (by mass) fractions can be extracted from it. In fact, the market value of a crude oil is often determined by the number of gasoline-suitable fractions that can be refined from it.
2. Paraffin-based crudes: These often reflect a greenish tint. When refined, they produce paraffin waxes and lubricating oils. They tend to be marketed at lower values than asphalt-based crudes.
3. Mixed-base crudes: These combine the characteristics of asphalt-based and paraffin-based crudes. Accordingly, they are usually priced in between.

The area of the earth's crust in which a crude oil is formed plays a role in defining its properties. A minimum temperature of 50°C (120°F) is required to enable the natural formation of crude petroleum. Oils formed at lower temperatures tend to be heavier and located in the upper strata of the **oil window**. The oil window is the area of the earth's crust in which petroleum oils can be formed and extends from a depth of 1,500 meters (5,000 ft.) up to 6,500 meters (20,000 ft.). Oils formed in the deeper portion of the oil window at temperatures up to 175°C (350°F) tend to be lighter and commercially more valuable. Petroleum located at depths below the oil window is usually in the form of natural gas.

Oil may also be obtained from oil shales and oil sands. In these, organic matter has been reduced to kerogen, an intermediate stage between the organic matter and oil. Oil shale and oil sands originate from the fine-grained sand deposits below ancient lakes and oceans. The kerogens in oil shales and sands can be processed to produce around 100 liters (27 gallons) per ton when heated to 350°C (660°F). Both Canada and the United States have large reserves of oil shale and sands and as world petroleum prices increase, the economics of mining these improves. The downside is that significant amounts of energy are required to process them to petroleum oils. After processing, essentially cleaning them, they can be refined.

Refining Petroleum

The refining of crude petroleum begins with two processes: vaporization and condensation. A crude oil is made up of many different HC compounds, each with different characteristics. Each has distinct boil and condensation temperatures, permitting relatively easy separation. These are known as fractions.

Boiled-off vapors are passed through a distilling column. The distilling column is a vertical cylinder containing a series of stacked vertical trays. Rising vapors bubble upward through holes in the trays and through any liquid that has already condensed. As these vapors rise they cool, condensing on the trays. Condensed liquid is permitted to spill from the trays to external containers. Each batch of liquid drawn off from the distillation column is called a "cut" or "fraction."

The fractions that are drawn off at high temperatures are called heavy fractions and those at low temperatures, light fractions. Straight run distillation therefore grades the fractions extracted from a crude petroleum by cut points or boil temperature. The gas taken off the top of the distilling column (or tower) is natural gas and in descending order of boil temperature, gasoline, naphtha, kerosene, light gas oil, and heavy gas oil. The liquid remaining at the bottom is known as residuum. After removal from the column, the cuts are purified (Figure 16–1).

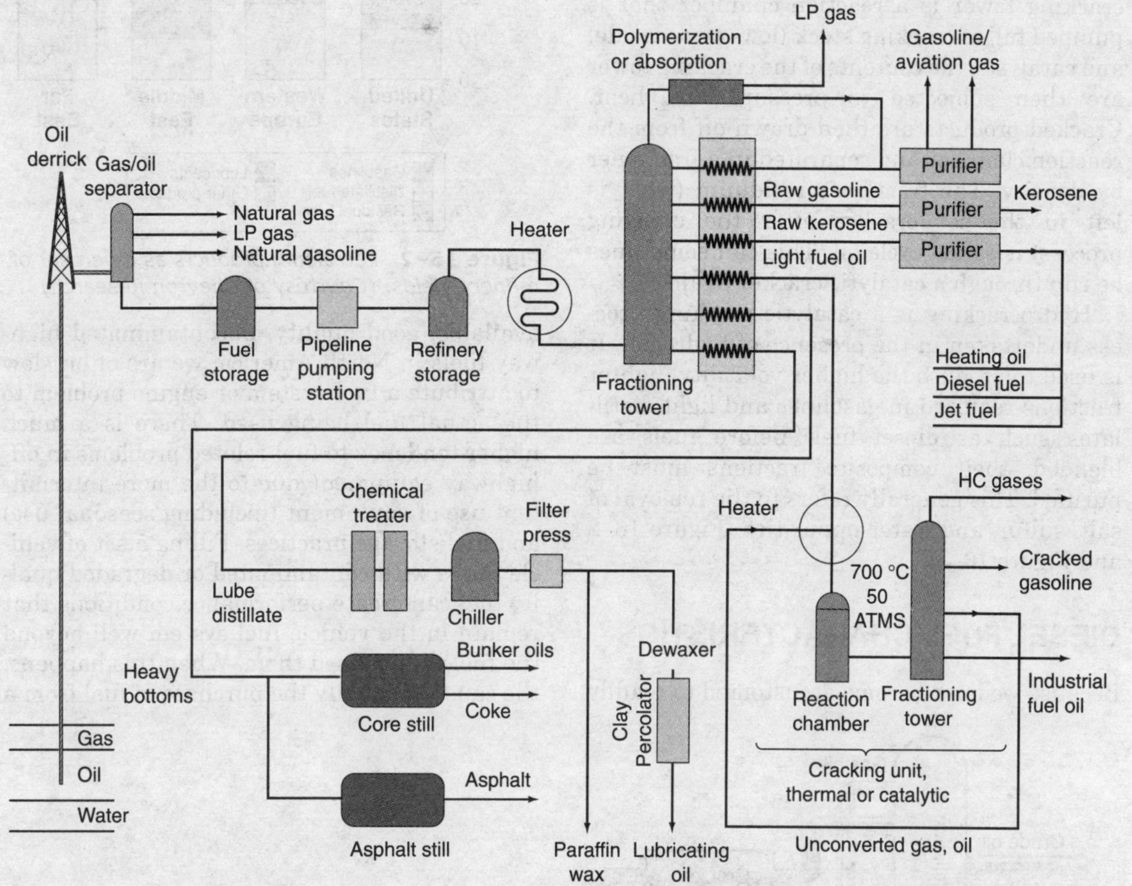

Figure 16–1 *Petroleum refining processes.*

Thermal and Catalytic Cracking

Most highway fuels are a "brew" of fractions carefully blended by the refiners. Racing gasolines are highly complex petroleum brews made from formulae that are sometimes carefully kept secret from competitors. The gasolines and diesel fuel purchased at the pumps are a blend of fractions with a range of volatility designed to produce balanced combustion. The higher or lighter fractions predominate in gasolines, and while a diesel fuel possesses some lighter fractions it also contains heavier fractions with higher lubricity and calorific value than those found in a gasoline. The lighter fractions are more highly valued than the heavier ones.

A problem inherent in the distilling of crude petroleum is that it produces too few of the lighter, more in-demand fractions. Cracking describes processes by which heavier fuel oils may be chemically modified by dividing its heavy molecules into smaller light molecules. Two types of cracking are used. Thermal cracking involves subjecting crude petroleum to high temperatures and pressures in a cylindrical tower. Catalytic cracking is a more efficient method of accomplishing the same objective. A cracking tower is a reaction chamber that is pumped full of cracking stock (low-grade crude) and catalysts; the contents of the cracking tower are then subjected to pressure and heat. Cracked products are then drawn off from the reaction chamber and separated in a fractioner by density. The fractioner residuum (what is left in the cat cracker after the cracking process) is called cycle oil. This can sometimes be run through a catalytic cracker again.

Hydrocracking is a catalytic cracking process undertaken in the presence of hydrogen. It is used to produce the higher volatility, lighter fractions required in gasolines and light distillates such as diesel fuel. Before fuels are blended, their composite fractions must be purified. This generally refers to the removal of salt, sulfur, and water impurities (Figure 16–2 and Figure 16–3).

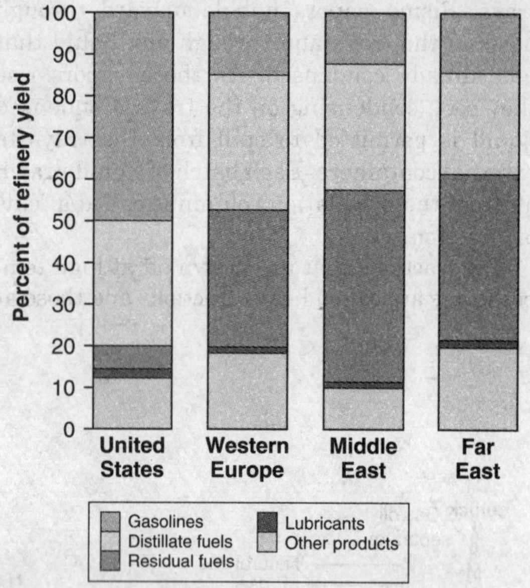

Figure 16–2 *Petroleum products as a percent of refinery yields. (Courtesy of Chevron Research)*

DIESEL FUEL CHARACTERISTICS

Because we have become accustomed to readily available, good quality, uncontaminated highway fuels in North America, we are often slow to attribute a fuel system or engine problem to the actual fuel being used. There is a much higher tendency to fuel-related problems in off-highway equipment due to the more intermittent use of equipment (including seasonal use) and fuel storage practices. Filling a set of vehicle tanks with contaminated or degraded quality fuel can create performance conditions that remain in the vehicle fuel system well beyond the fuel that caused them. When this happens, the cause is usually the purchase of fuel from a

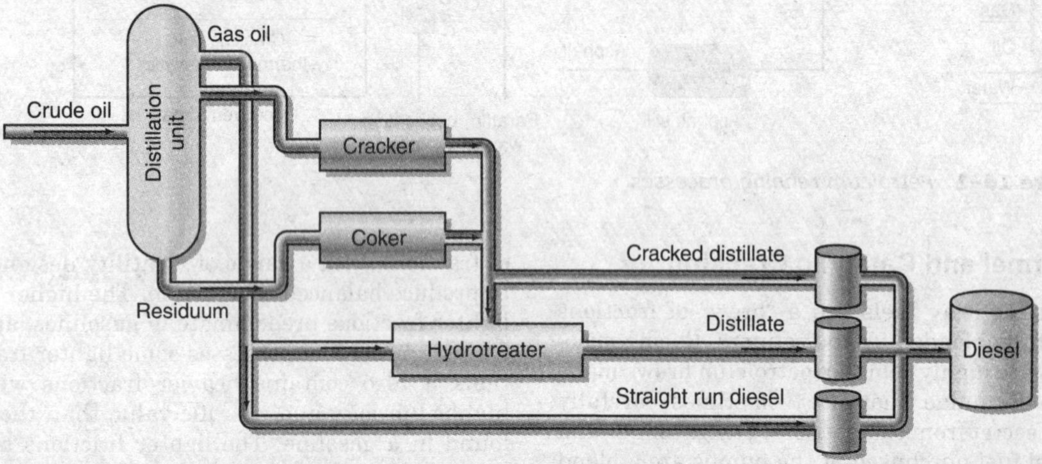

Figure 16–3 *Diesel fuel blending. (Courtesy of Chevron Research)*

source in which the fuel is retained for prolonged periods in storage tanks (most fuel degrades in contact with air). For this reason, many original equipment manufacturers (OEMs) recommend bulk fuel filtration procedures prior to refueling.

Fuel filtration is especially important before testing an engine on a dynamometer or running fuel economy profiles. Figure 16–4 shows a bulk fuel filtration station. As North American trucking activity extends into Central and South America and fuel is purchased in those countries, the chances of fuel contamination-related problems are going to increase. You should also be aware of some of the potential problems of using biodiesel cuts. There are a number of advantages to using biodiesel and biodiesel cuts, but they have a short shelf life compared with diesel fuel. When biodiesel degrades it can create difficult to diagnose fuel problems. We will take a closer look at biodiesel in Chapter 26.

Cetane Number

Cetane number (CN) is a measure of the ignition quality of a diesel fuel. CN by ASTM definition is the percentage by volume of a test fuel consisting of cetane (ideal ignition quality, CN 100) mixed with heptamethylnonane (poor ignition quality, CN 0) required to match the fuel to be classified. A mixture of 45% cetane with 55% heptamethylnonane would have a CN of 45. ASTM 1D and 2D diesel fuels for on-highway use must have a minimum CN of 40 in North America. The CN directly determines the ignition delay phase of the diesel combustion cycle and it is not uncommon for diesel fuel refiners to adjust CN seasonally. However, increasing CN generally reduces fuel density and, therefore, the fuel mileage. As the CN of a diesel fuel increases, its ignition temperature decreases (Figure 16–5).

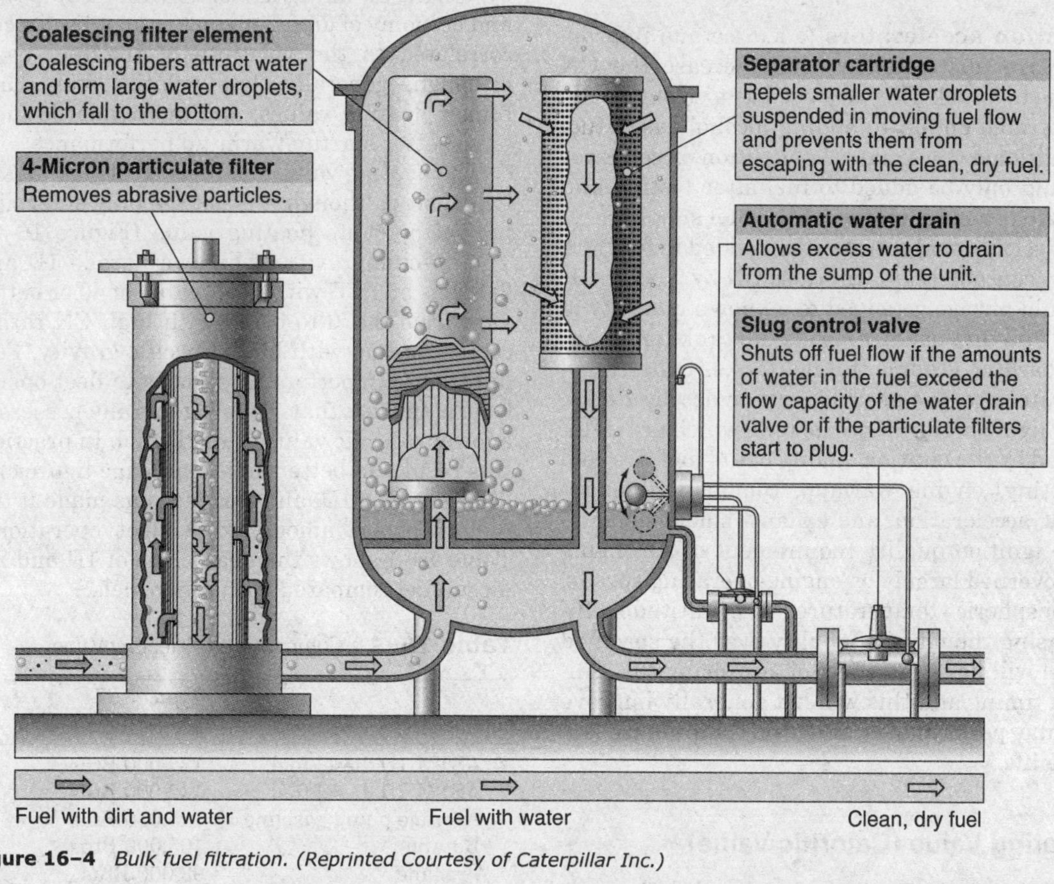

Figure 16–4 *Bulk fuel filtration. (Reprinted Courtesy of Caterpillar Inc.)*

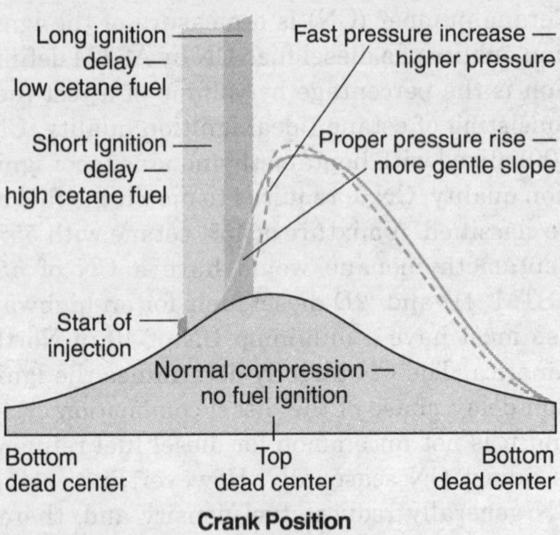

Figure 16–5 *Cetane number versus ignition delay. (Courtesy of Chevron Research)*

Ignition Accelerators

Ignition accelerators (a.k.a. cetane improvers) are fuel additives that increase the CN value of a fuel. They are used as a scientifically preferable option to adding alcohol-based fuel conditioner to tanks. An ignition accelerator should only be added to fuel after testing and on the recommendation of the fuel supplier.

Cyclohexanol nitrate when added to fuel at a 0.2% concentration will raise CN by 7 points. It should always be added to a known quantity of fuel and that usually means full tanks on either onboard or ground storage tanks. Other alkyl nitrates are also used as ignition accelerators.

Engine performance factors, which are influenced by the ignition quality of the fuel, are cold starting, engine warmup, combustion roughness, acceleration, and exhaust smoke density. The ignition quality requirement of an engine is governed largely by engine-operating speeds, atmospheric temperature, and altitude. Increasing the ignition quality over the specified level will have the effect of advancing the ignition timing and this will not generally improve engine performance and can compromise engine life.

Heating Value (Calorific Value)

An important property of a diesel fuel is the **heat energy** it releases during combustion. This value is used to reckon the thermal efficiency of an engine's ability to produce power. The gross heat of combustion at a constant volume is determined by a **bomb calorimeter test**, which measures the amount of heat released by burning a known quantity of fuel outlined by ASTM standard D240. Heating values may be expressed as:

J/kg	joules per kilogram
J/L	joules per liter
cal/g	calories per gram
cal/L	calories per liter
Btu/lb	Btu per pound
Btu/gal.	Btu per gallon

Both volume and weight bases have significance to the engine designer and equipment user because diesel fuel is normally purchased by volume while fuel consumption is expressed by weight as in: kg/kW or lb/BHP. The power and economy of diesel engines cannot be directly correlated to the volatility of the fuel used, although, generally, less volatile fuels have higher heating values, and more volatile fuels have better startup/warm-up performance.

The heating value of a fuel directly relates to its density although it is not the only determining factor of its heating value (Figure 16–6). The specific gravity of highway grade 1D and 2D fuels at 15°C with a CN value of 40 or better vary from 0.870 to 0.780 with high CN rating corresponding with lower specific gravity. Perhaps more important, especially to fleet operators, is the fact that 1D fuel generally possesses a lower calorific value than 2D fuel. In practice, this results in better fuel-to-mileage figures for 2D fuels over 1D, and this fact has made it the fuel of choice among truck fleet operations. Table 16–1 shows the heat value of 1D and 2D diesel fuel compared with other fuels.

Table 16–1 *Comparisons of Heating Values*

Fuel	*Typical BTUs per U.S. Gallon*
ASTM 1D diesel fuel	137,000 Btus
ASTM 2D diesel fuel	142,000 Btus
Average pump gasoline	125,000 Btus
Butane	103,000 Btus
Propane	93,000 Btus

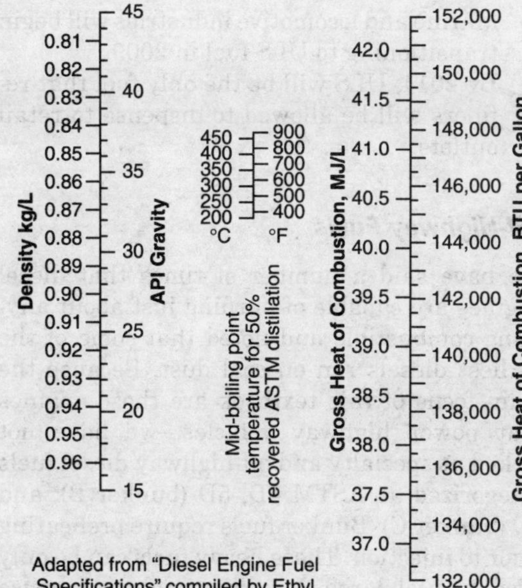

Adapted from "Diesel Engine Fuel Specifications" compiled by Ethyl Corporation, August 1967

Figure 16–6 *Heating value relationships. (Courtesy of Chevron Research)*

API Gravity

American Petroleum Institute gravity, usually known as **API gravity**, is a measure of how the weight of a petroleum liquid compares with the weight of water. For instance, if API gravity is specified as greater than 10, the petroleum-based liquid is lighter than water and would therefore float on water. If the API gravity specification was less than 10, however, it is heavier than water and would sink. API gravity is used to compare the relative densities of petroleum liquids and is important if you are doing performance testing on engine dynamometers.

When a diesel fuel is specified in terms of API gravity it is usually referred to as being in degrees. So for purposes of specifying diesel fuels with API gravity, a hydrometer graduated in degrees is used. For the full range of petroleum products, API gravity would run between 10 and 70 API gravity degrees: But for diesel fuels, most fall within a window of between API 25 degrees and API 48 degrees. Table 16–2 shows a comparison between API gravity, specific gravity, and the density of diesel fuels.

Understanding a little about API gravity can help when attempting to troubleshoot persistent low-power and marginal emissions test failures. With the increase in the cost of diesel fuel in recent years, there has been some reduction in fuel quality. This reduction in fuel quality can cause problems for diesel engines that are operated in environments where emissions are critical, such as in underground operations.

Table 16–2 *Specific Gravity and Density of Fuels*

Gravity in Degrees API at 15°C (60°F)	*Specific Gravity at 15°C (60°F)*	*Density in Pounds per U.S. Gallon*
25	0.9042	7.529
28	0.8870	7.387
31	0.8708	7.250
34	0.8550	7.120
37	0.8398	6.993
40	0.8250	6.870
44	0.8062	6.713
46	0.7972	6.637
48	0.7882	6.562

Ultra-low Sulfur Fuels

Sulfur is present in most crude petroleum and is more prominent in the heavier residual fractions from the refining process. When combusted, sulfur in diesel fuel is oxidized to form sulfur dioxide (SO_2), which combines with water to form sulfuric acid (H_2SO_4). SO_2 produced from the combustion of diesel fuel is recognized as environmentally hazardous and the sulfur content of diesel fuels has progressively been legislated to lower levels in recent years.

Today, **ultra-low sulfur (ULS)** fuel is classified as diesel fuel containing 0.0015% sulfur or less. This is equivalent to 15 parts per million (ppm). By EPA accord, 80% of 1D and 2D diesel fuel currently sold in North America at the pumps as clear fuel has to be of ULS grade. The other 20% of the fuel marketed at the pumps will be required to meet the former **low-sulfur (LS)** fuel maximum sulfur rating of 0.05%. LS fuel was used between 1993 and the nationwide introduction of ULS fuel in October 2006. Up until October 2006, some jurisdictions required the use of a ULS fuel that could contain up to 0.005% sulfur: This fuel ceased to be available after October 2006.

When LS fuel was introduced in 1993, it replaced diesel fuel that could contain up to 0.300% sulfur. This much older standard of diesel fuel possessed better lubricity than LS fuel. At the transition point to the lower LS fuel, some failures resulted in certain older hydromechanical fuel injection pumps. Because of this, some engine and fuel system manufacturers recommended the use of additives that enhanced the lubricity of the fuel in certain older diesel engines. When using a lubricity-enhancing additive, ensure that only the OEM-recommended product is added to the tank in the correct proportions. The addition of lubricity enhancers should be undertaken only when the vehicle tanks are full; that is, when the quantity of fuel to be treated is known exactly so the recommended proportions can be adhered to.

ULS Introduction Schedule

At its introduction in October 2006, ULS applied only to highway fuels and the 80%/20% proportions was a federal requirement. Some jurisdictions such as California adopted a tougher standard from the beginning but the objective is to work toward 100% usage of ULS by 2014. The schedule is as follows:

- ULS must be available in 80% of highway fuel retail outlets from October 2006.
- Some jurisdictions such as California and Canada require ULS usage in 100% of highway fuel retail outlets from October 2006.
- ULS will become mandatory in all highway and off-highway fuel retail outlets from December 2010.
- Marine and locomotive industries will begin transitioning to ULS fuel in 2009.
- By 2014, ULS will be the only fuel that refiners will be allowed to dispense to retail outlets.

Off-Highway Fuels

We have said a number of times that diesel engines are capable of burning just about anything combustible and noted that some of the earliest diesels ran on coal dust. Because the main focus of this textbook are those engines that power highway vehicles, we have not looked at specialty and off-highway diesel fuels categorized as ASTM 4D, 5D (bunker B), and 6D (bunker C): Bunker fuels require preheating prior to injection. These heavy fuels can be only used to fuel large, low rpm engines because they require longer to vaporize (even when preheated) and longer to properly combust. Fuels used in subarctic regions (Alaska and Northern Canada) are chemically similar to ASTM 1D diesel except they are cut with glycol and other antigel compounds similar to jet fuel.

FUEL STORAGE, FUEL DETERIORATION, AND PERFORMANCE

The fuel chemistry of both gasolines and diesel fuels is adjusted seasonally by the refiner/supplier largely because what is required of a fuel varies with temperature. To further complicate matters, a truck hauling a load of fruit from Miami, Florida, to Edmonton, Alberta, can be subject to temperature extremes in excess of 30°C (85°F) to –17°C (0°F) within the span of a single trip. Generally, fuel purchased directly from the refiner/supplier presents fewer problems than fuel that is bulk stored in reservoirs by operators.

Whenever fuel is drawn out of a tank it is replaced by air. Replacement air always contains some percentage of moisture and in areas of high humidity, this can be extreme. When temperatures cool, vaporized moisture in the tank condenses to water, which because of its greater weight than the fuel, settles at the bottom of the tank. Most vehicle and storage tanks contain some water, and draining the tanks periodically (just the water trap sump of the tank: Ensure that fuel is settled before attempting to do this) helps to prevent excessive water buildup. Both storage reservoirs and vehicle fuel tanks locate the pickup tube in a position so that trace quantities of water present in the bottom of the tank are not drawn out.

Water in Fuel

Water presents two main problems while it is actually in a fuel tank. However, it presents a host of others, should it actually be drawn out and pumped through a fuel system. Some trucks carry fuel in a pair of cylindrical tanks mounted at either side of the chassis that have a single pickup tube located in one of the tanks; the tanks are connected by a crossover pipe at the lowest point. If an appreciable quantity of water were present in such an arrangement, that water would tend to collect in the crossover pipe, which becomes vulnerable to freeze-up given its location in the airflow. The common preventative and antidote for this condition is to add methyl hydrate (alcohol) to the tank. While this will prevent water freeze-up, the water and alcohol form a solution that can be absorbed by the fuel and then be pumped through the fuel system.

Microorganism Growth

The other problem associated with water in tanks is microorganism growth. Airborne source bacteria reside in the tank sump water and feed off the HCs in the fuel, causing a certain amount of degradation of the fuel. More serious, their metabolic waste is acidic and has a corrosive effect on fuel system components besides being capable of plugging up water separators and filters. This type of a problem would be more likely in an emergency power, genset engine and its fuel system that only runs a couple of times a year than in mobile diesel engine applications that are likely to run at least daily. Fungi may also develop in fuel tanks and can plug fuel filters and damage in-tank, transfer pumps.

Influence of Climate

Hot weather presents other problems. The year-round average CN value of 1D and 2D ASTM fuels sold in North America is around 47. The legal minimum for both fuels is 40. It makes sense that fuel supplied in the northern states and Canada in winter has higher mean CN values than that sold in the same geographic areas in the summer. However, when fuel is stored and exposed to high temperatures, its more volatile fractions are boiled off, effectively lowering the CN value at a time of year when the supplier has already lowered the CN because the seasonal requirement has dropped.

Cloud Point and Pour Point

Cloud point and pour point are important diesel fuel characteristics in frigid temperature conditions. The cloud point value indicates the temperature at which filters can become plugged, especially the secondary filters in some current applications designed to entrap particulate sized between 2μ and 4μ. At the fuel's pour point temperature, it is on the verge of gelling. Pour point is the lowest temperature a fuel can be pumped through the fuel system. The viscosity of the fuel is temperature related. High-viscosity fuel can result in excessive pump resistance and filter damage. Low-viscosity fuel may result in excessive wear in fuel injection apparatus and account for leaks. As temperature-conditioned ASTM 2D fuel is, in practice, the fuel used by a majority of operators year-round, fuel heaters are sometimes used to help compensate for deficiencies in cloud point, pour point, and viscosity ratings.

Generally, onboard fuel tanks and stationary reservoirs should be maintained as full as possible to minimize the quantity of moisture-laden air the fuel is exposed to. It is good practice for fleets to require refilling of vehicle tanks at the point of returning from a journey and before overnight parking.

Fuel Conditioners

There are no Society of Automotive Engineers (SAE)/ASTM standards that apply to diesel fuel conditioners. While most engine OEMs disapprove of their use, they are commonly sold in their dealerships, a measure of the profitability of the product rather than worth. Diesel fuel conditioners are a vague mixture of cetane improvers, cleansing additives, and pour point depressants in an alcohol base, usually priced well beyond the cost of their ingredients. In a freeze-up, reality dictates that the addition of methyl hydrate is the fastest way out of the problem, and in those circumstances the negatives of putting water in solution with the alcohol are understandably overlooked. However, the addition of methyl hydrate to the tank under these circumstances is arguably

Classification	*Type*	*Function*
Contaminant Control		
Biocides	Boron compounds, ethers of ethylene glycol, quaternary amine compounds	Inhibit growth of bacteria and fungi—prevent filter clogging
Demulsifiers and Dehazers	Surface-active material that increase water/oil separation	Improve separation of water and prevent haze
Rust and Corrosion Inhibitors	Organic acids, amines and amine phosphates	Prevent rust and corrosion in fuel systems, pipelines, and storage facilities
Fuel Stability		
Metal Detectors	Chelating agents	Inhibit gum formation
Oxidation Inhibitors	Alkyl amines	Minimize oxidation, gum, and precipitate formation
Dispersants	Polymeric amine surfactants	Prevent agglomeration and disperses residue
Engine Performance		
Detergents	Polyglycols, polyether amines	Prevent injector deposits Increase injector life
Dispersants	Basic nitrogen polymeric amine surfactants	Peptize injector deposits Increase filter life
Cetane Improvers	Alkyl nitrates	Increase cetane number
Smoke Suppressants	Overbased barium compounds	Minimize exhaust smoke
Fuel Handling		
Pour Point Depressants	Polymeric compounds	Reduce pour point and improve low-temperature fluidity properties
Could Point Depressants	Polymeric compounds	Reduce cloud point and improve low temperature filterability
De-icers	Low molecular weight alcohols	Reduce freezing point of small amounts of water to prevent fuel line plugging
Antifoam	Silicone and nonsilicone surfactant	Minimizes the formation of fuel foam

Figure 16–7 *Diesel additive classification. (Courtesy of Chevron Research)*

preferable to pouring in the contents of a "fuel conditioner" can in that it specifically addresses the problem.

Similarly, when the objective is to improve the CN value of the fuel, then adding a cetane improver in a precisely measured quantity is a more scientific and less costly strategy. Most diesel fuels are carefully balanced by the refiner and the use of additives should be avoided whenever possible. Under circumstances such as the effect of reduced lubricity of modern 0.0015% ULS fuel (compared with pre-LS fuel) on some fuel systems, the OEM will recommend an additive on certain older engines to specifically address the lubricity problem. Aftermarket fuel conditioners generally create more problems than they solve but the reality is that there are occasions when they have to be used. The technician should recognize the problems that can be caused by fuel conditioners and use them only when there is no other option. Diesel fuel conditioner should always be used according to the manufacturer's recommendations (Figure 16–7).

Costs of Operating a Truck

The major expense in truck-operating costs is fuel. Nothing else comes close to it. A minor upsurge in the price of diesel fuel reverberates through the trucking industry and significantly impact the operating profitability. Operators on tight margins go out of business. The increase in fuel prices since 2002 has shifted business away from brokers and small operations into the major fleets. The major fleets can manage the logistics required to make money on low margins so they have been the big winners—and will continue to be as fuel prices rise. Sadly, this is resulting in the gradual disappearance of independent truckers, who have been a part of our culture for over 50 years. Figure 16–8 shows a pie chart that represents the operating costs of a truck in 2005 with fuel costs representing 75%.

The emerging use of biodiesel is examined later in this book, but we can summarize this chapter with a reminder that a diesel engine can run on nonpetroleum-based fuels, including one that flows as sap in trees. Figure 16–9 is a photograph taken at the Audubon Museum of the Americas in New Orleans.

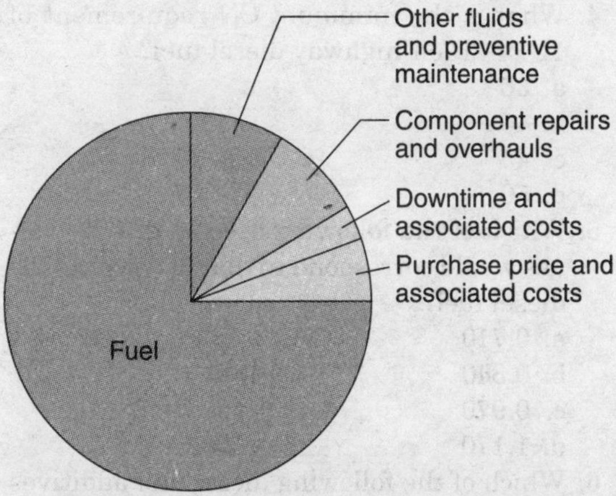

Figure 16–8 *Costs of operating a truck.*

Figure 16–9 *Tree sap as fuel?*

REVIEW QUESTIONS

1. As the octane rating value of a gasoline increases, what can be said of its burn rate?
   a. It is not affected.
   b. It decreases.
   c. It increases.
2. As the CN value of a diesel fuel increases, what happens to its ignition temperature?
   a. It is not affected.
   b. It decreases.
   c. It increases.
3. The temperature at which a fuel evaporates sufficient flammable vapor to momentarily ignite when a flame is brought close to its surface is known as:
   a. Flash point
   b. Fire point
   c. Boil point

4. What is the minimum CN requirement of 1D or 2D on-highway diesel fuel?
   a. 30
   b. 40
   c. 45
   d. 50
5. Which of the following specific gravity values would correspond to that of a typical 2D diesel fuel?
   a. 0.710
   b. 0.840
   c. 0.970
   d. 1.170
6. Which of the following diesel fuel additives is a commonly used cetane improver?
   a. Isopropyl alcohol
   b. Kerogen
   c. Kerosene
   d. Cyclohexanol nitrate
7. The temperature at which a diesel fuel begins to form paraffin wax crystals is known as:
   a. Cloud point
   b. Pour point
   c. Flash point
8. The lowest temperature at which a diesel fuel can be pumped through a fuel system is known as:
   a. Cloud point
   b. Pour point
   c. Gel point
9. If an additive were required to prevent freeze-up of a fuel crossover pipe, which of the following would be the preferred option?
   a. Add 8 oz. of fuel conditioner to the tanks.
   b. Add 8 oz. of methyl hydrate to the tanks.
   c. Add 8 oz. of gasoline to the tanks.
   d. Add 8 oz. of cetane improver to the tanks.
10. Current diesel fuel formulated for on-highway use and classified as ULS is required to have what maximum sulfur content?
    a. 0.0015%
    b. 0.005%
    c. 0.300%
    d. 0.500%
11. Technician A states that 2D diesel fuel will, in most cases, produce slightly better fuel mileage from a highway truck engine than a 1D fuel. Technician B states that 2D usually has more heat energy than a 1D fuel. Who is correct?
    a. Technician A only
    b. Technician B only
    c. Both A and B
    d. Neither A nor B
12. Technician A states that all 1D and 2D fuels must be sold with a CN number of exactly 40. Technician B states that refiners and fuel suppliers seasonally adjust ignition quality of fuels. Who is correct?
    a. Technician A only
    b. Technician B only
    c. Both A and B
    d. Neither A nor B
13. Which of the following best describes the term octane rating when applied to a fuel?
    a. Volatility
    b. Burn rate
    c. Antiknock resistance
    d. Ignition temperature
14. Which of the following components would be more likely to fail if LS fuel were used instead of ULS in a highway diesel engine manufactured after 2007?
    a. Exhaust gas recirculation (EGR) circuit
    b. NO_x reduction catalyst
    c. Diesel particulate filter (DPF)
    d. Crankcase ventilation system (CVS)
15. Technician A states that a catalyst is a substance that enhances or enables a chemical reaction without itself undergoing any change. Technician B states that the laws of thermodynamics dictate that a catalytic reaction must involve a change in the state of the catalyst. Who is correct?
    a. Technician A only
    b. Technician B only
    c. Both A and B
    d. Neither A nor B
16. Technician A states that denser fractions of crude petroleums tend to command higher prices than lighter fractions. Technician B states that diesel fuel tends to be more volatile than gasoline. Who is correct?
    a. Technician A only
    b. Technician B only
    c. Both A and B
    d. Neither A nor B

17

Fuel Subsystems

INTRODUCTION

The diesel **fuel subsystem** on a commercial vehicle is best defined as the group of components responsible for fuel storage and its transfer to the injection pumping apparatus. While injection pumping mechanisms can differ greatly from manufacturer to manufacturer, the fuel subsystems that supply them tend to have much in common. Many of the problems that

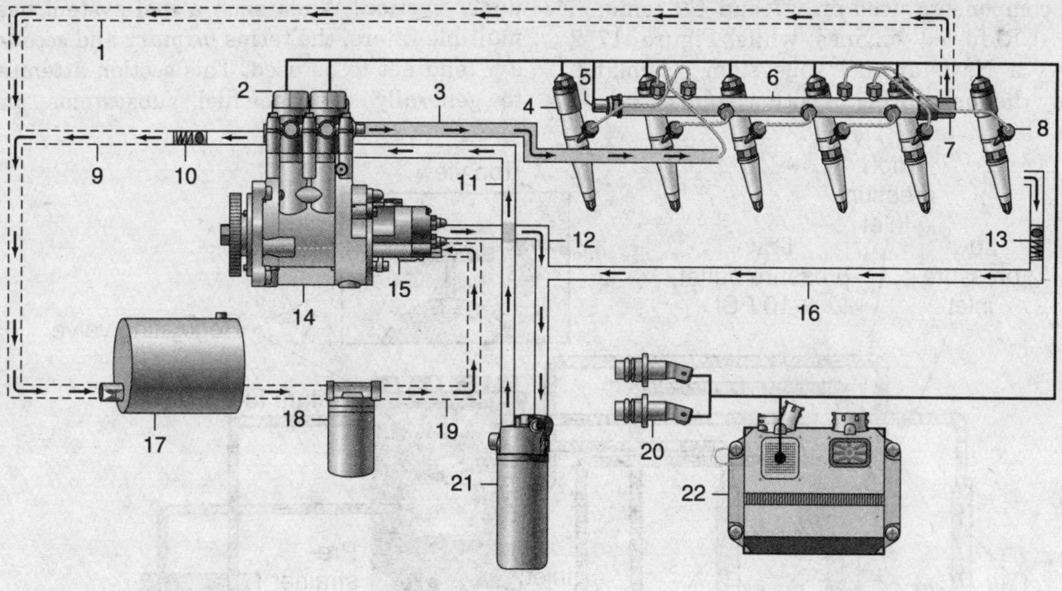

FUEL SYSTEM DIAGRAM

1. Return line from pressure relief valve to tank
2. Solenoid for the fuel pump
3. High pressure supply line
4. Electrohydraulic injector (EHI)
5. Fuel pressure sensor
6. Fuel rail
7. Pressure relief valve for the fuel rail
8. Quill tube
9. Return line from fuel pump to tank
10. Pressure relief valve for the fuel pump
11. Fuel line from secondary filter to fuel pump
12. Fuel line from transfer pump to fuel filter
13. Pressure regulator for the drain line
14. Fuel pump
15. Transfer pump
16. Fuel line for the drain back to filter
17. Fuel tank
18. Primary fuel filter
19. Fuel line from the primary filter to the transfer pump
20. Speed/timing sensor
21. Secondary fuel filter
22. Engine control module (ECM)

Figure 17–1 *Fuel system components and routing used on C7 and C9 CR-fueled engines. (Reprinted Courtesy of Caterpillar Inc.)*

challenge the diesel technician at the novice level focus on the fuel subsystem. The principal components described in this chapter are often dismissed as being straightforward and exempted from study.

However, a thorough knowledge of how these components interact and how they affect the performance of the fuel injection apparatus is essential. Although the basic components used in a fuel subsystem have not altered much as we have progressed through different generations of emissions controls, the way in which fuel subsystem is monitored has changed. A generation ago, the fuel subsystem had one monitoring sensor called a sending unit. Today, depending on the original equipment manufacturer (OEM), system pressure and temperature are monitored at different locations in the circuit. **Fuel tanks**, **fuel filters**, **water separators**, **transfer pumps**, **fuel heaters**, and their interconnecting plumbing are examined in this chapter. Figure 17–1 shows the main fuel system components used on C7 and C9 common rail (CR)-fueled engines while Figure 17–2 shows a Navistar fuel subsystem schematic. Study the fuel routing in both systems.

FUEL SUBSYSTEM OBJECTIVES

If you take a look at the fuel systems shown in Figure 17–1 and Figure 17–2, you will observe a clear divide between the low-pressure side and the charge side of the fuel subsystem. The fuel transfer pump divides the two sections. Most fuel subsystems are of this type and the terms **suction circuit** (low-pressure side) and **charging circuit** (high-pressure side—do not confuse with injection pressures, which are much higher) are commonly used to describe each. A full explanation of the hydraulics of how fuel is moved through the fuel subsystem appears later in this chapter under the heading Fuel Charging/Transfer Pumps.

A **primary filter** is most often located on the suction side of the transfer pump, while the **secondary filter** is located on its charge side. However, there are some fuel systems, notably some manufactured by Cummins, in which all movement of fuel through the fuel subsystem is under suction. When such a fuel system uses multiple filters, the terms *primary* and *secondary* tend not to be used. This section attempts to generally describe fuel subsystems and

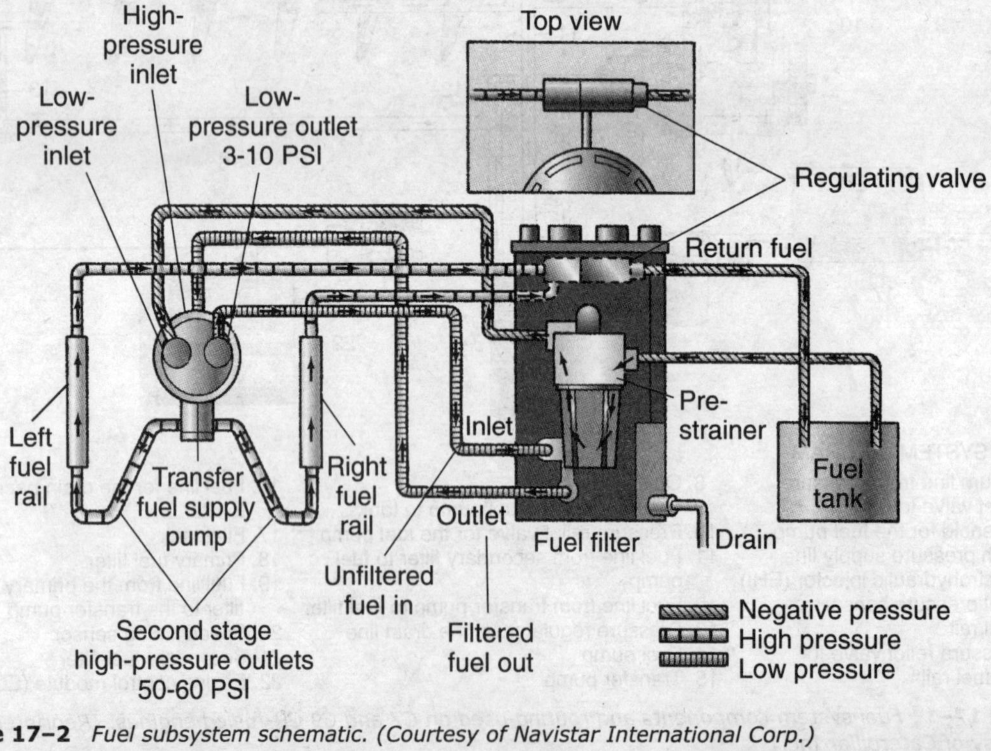

Figure 17–2 *Fuel subsystem schematic. (Courtesy of Navistar International Corp.)*

troubleshooting methods. The characteristics of particular fuel subsystems are described in sections dealing with proprietary systems. We can summarize the objectives of the fuel subsystem as follows:

- Stores fuel in tanks until required
- Removes moisture from the fuel
- Filters fuel to remove abrasive particulates
- Delivers fuel to the injection components at the proper temperature

FUEL TANKS

Fuel is stored on commercial vehicles in fuel tanks. In most highway trucks, fuel tanks are mounted to the frame. It is common to use a pair of fuel tanks. This balances the considerable weight of onboard fuel as it is consumed. A typical fuel tank arrangement is shown in Figure 17–3. Many diesel fuel management systems are designed to pump much greater quantities of fuel through the system than that required for actually fueling the engine. This excess fuel factor varies from a minimal amount to values exceeding 60% of pumped fuel. The excess fuel is used to:

- Lubricate high-pressure injection components
- Cool high-pressure injection components (especially those exposed to extreme temperatures)
- Cool electronic components such as engine control modules (ECMs) and injector drivers

As a cooling medium, the fuel transfers heat from engine components to the fuel tank. This provides the fuel tank(s) with a role as heat exchanger.

Fuel Tank Recommendations

Engine and fuel injection system manufacturers tend not to manufacture most of the fuel subsystem components but often make specific recommendations to the chassis OEM, which may or may not be observed. The fuel tank is always a chassis OEM-supplied component. Figure 17–3 shows a typical truck fuel tank. Essentially, a vehicle fuel tank will function most effectively as a heat exchanger if the following is true:

1. Located in the airflow. Truck fuel tanks tend to be mounted in cradle brackets bolted to the outboard side of laddered frame rails such as those shown in Figure 17–3. This ensures fairly good airflow around the tank into which heat removed

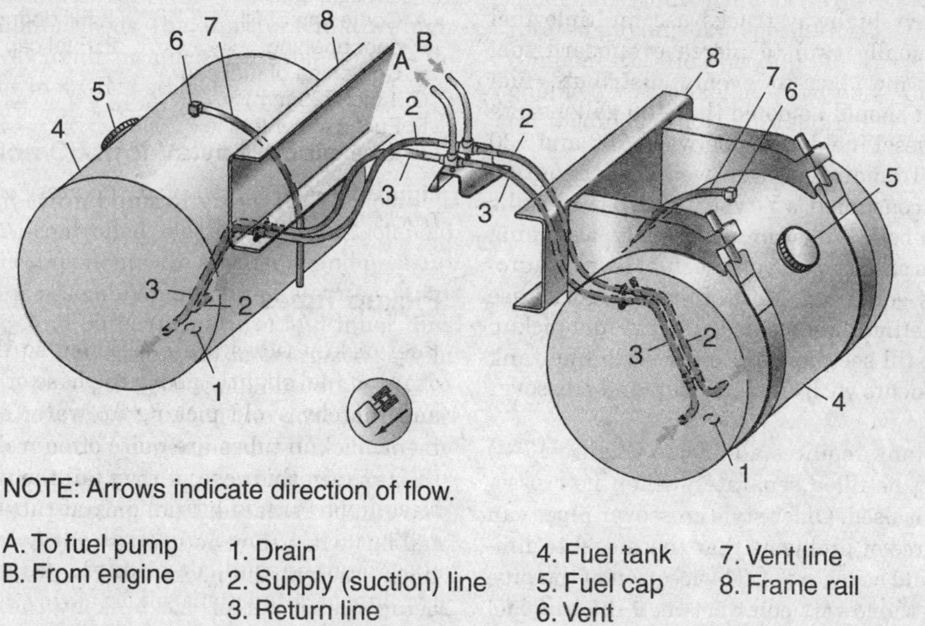

Figure 17–3 *Dual fuel tank arrangement that eliminates the fuel crossover pipe. (Courtesy of Daimler Trucks North America)*

from the cylinder head by the fuel can be dissipated.

2. Cylindrically shaped. A cylindrically shaped vessel (see Figure 17–3) helps maximize the surface area of the tank exposed to the airflow. This shape additionally has higher **section modulus** (relates the shape of a vessel or beam to rigidity) than a rectangular-shaped vessel, permitting thinner wall thickness. That means an overall lighter fuel tank.
3. Aluminum construction. The coefficient of heat transfer of aluminum is high, enabling heat to be transferred efficiently from the fuel to atmosphere.
4. Aluminum is less susceptible than steel to water corrosion at the base of the tank and is much lighter.
5. Maintained 25% full or better. In a fuel system that circulates fuel through the system at a high rate such as the Cummins HPI-TP system, the fuel can heat up to temperatures where its lubricity is compromised when the tanks are near empty. It is good practice in such systems to maintain the tank level at better than 25% full.

Dual Tanks

Most heavy highway trucks use multiple fuel tanks (usually two) to increase onboard fuel capacity and also to evenly distribute fuel weight. It should be noted that 100 gallons of a typical diesel fuel weighs between 700 and 730 pounds. To maintain even weight distribution as fuel is consumed, a Y-type **pickup tube** pulls fuel from both tanks simultaneously; assuming the tanks are of equal volume, fuel level is automatically equalized. Most of today's trucks use a Y (sometimes known as a T-type) fuel pickup but you still see plenty of older style fuel tank arrangements with single pickup and crossover pipes.

Each tank requires a fuel cap (Figure 17–4) and must be filled separately when no crossover pipe is used. Older style crossover pipes can be a source of problems that the diesel technician should be aware of. The location of the pipe slung at the lowest point between the two fuel tanks makes them vulnerable to damage from road debris and animals; the angle iron bracket that supports the crossover line provides little more than minimal protection. The crossover is also exposed to the airflow under the truck and in the middle of winter, water present in the line will freeze. When crossover lines freeze up, alcohol (methyl hydrate) has to be added to the fuel tanks. Most current trucks use dual fuel tank designs that eliminate the crossover pipe such as that shown in Figure 17–3; in this configuration, the pickup lines draw in a parallel arrangement from each tank. This arrangement is known as a Y- or T-type pickup depending on who manufactures the device.

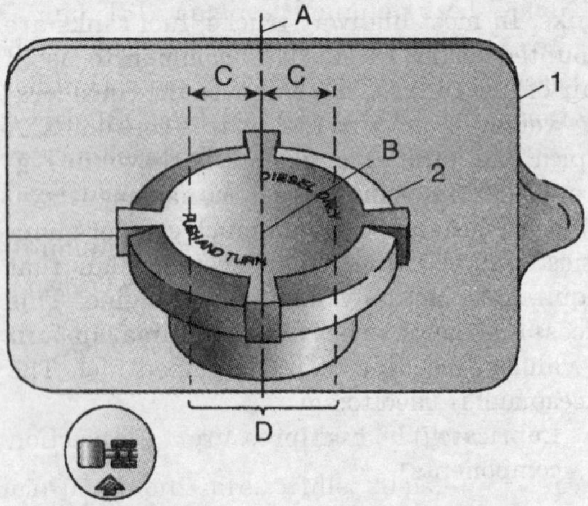

A. Centerline of fill door opening
B. Centerline of fuel cap
C. 1 inch (25 mm)
D. Fuel cap centerline to be within this area.

1. Fill door opening
2. Fuel cap

Figure 17–4 *Fuel tank cap. (Courtesy of Daimler Trucks North America)*

Pickup Tubes

Fuel pickup tubes are positioned so that they draw on fuel slightly above the base of the tank and thereby avoid picking up water and sediment. Pickup tubes are quite often welded into the tank; in this case, if they fail the tank may have to be replaced. Fuel pickup tubes seldom fail but when they do it is usually by metal fatigue crack at the neck; this results in no fuel being drawn out of the tank by the transfer pump whenever the fuel level is below the location of the crack.

Fuel Tank Sending Units

Most commercial truck fuel subsystems use remote (from the tank) fuel transfer pumps and not assemblies that incorporate the **sending unit** and a transfer pump. Therefore, the fuel sending unit is an integral assembly flange fitted to the tank. It consists of a float and arm connected to a variable resistor whose function it is to control current flow to a cab gauge proportionally with fuel tank level.

Most older dual fuel tank truck chassis in which the tanks were connected underneath by a **crossover** pipe (a crossover pipe connects a pair of cylindrical fuel tanks, mounted on the outside of opposing frame rails) often used a single sending unit located in one of the tanks. The tank with the sending unit was that opposite the tank with the pickup tube. It is generally preferable to locate a sending unit in each tank and provide a dash gauge for each, providing the operator with some advance warning of a crossover pipe restriction.

Testing Sending Units

Fuel sending unit problems can be diagnosed by disconnecting the terminals and using a digital multimeter (DMM) in resistance mode. When the float arm is moved through its arc the readings observed should change as the arm angle changes.

Diagnostic Procedure for a 240-Ohm Fuel Sending Unit

1. Remove the wire lead from the sensor terminal.
2. Connect the ohmmeter across the sensor terminal and mounting flange (ground).
3. Visually estimate the amount of fuel in the tank.
4. Correlate the visual fuel level estimate with the following values:

Full tank	About 20 to 50 ohms
Half tank	About 80 to 120 ohms
Empty tank	About 220 to 260 ohms

5. Values that differ widely from those listed here indicate a defective sending unit.

Diagnostic Procedure for a 90-Ohm Fuel Sending Unit

1. Remove the wire lead from the sensor terminal.
2. Connect the ohmmeter across the sensor terminal and mounting flange (ground).
3. Visually estimate the amount of fuel in the tank.
4. Correlate the visual fuel level estimate with the following values:

Full tank	About 86 to 94 ohms
Half tank	About 40 to 50 ohms
Empty tank	About 0 to 4 ohms

5. Values that differ widely from those listed here indicate a defective sending unit.

Venting

Currently, most jurisdictions in North America permit **venting** of diesel fuel tanks to atmosphere. Venting permits gas movement both in and out of a fuel tank. As fuel is pumped out of onboard tanks to fuel the engine, it is replaced by ambient air drawn in through a breather. Gas movement is reversed on refueling or if the fuel becomes heated. In hot weather conditions, some of the lighter fuel fractions boil off coincidentally, causing a slight reduction in the fuel cetane number (CN) value. However, this boil-off seldom occurs at a rate sufficient to significantly compromise the fuel except in instances in which a tank of fuel is retained for a prolonged period in extreme heat.

Boiled-off fuel fractions from diesel fuel have so far not been considered a noxious hydrocarbon (HC) emission worthy of legislated controls. Fuel tank vents or breathers should be routinely inspected for restrictions and should be protected from ice buildup. A plugged fuel tank vent or breather will rapidly shut down an engine, creating a suction side **inlet restriction** value the transfer pump will not be capable of overcoming.

Breather Filters

As fuel is drawn out of a fuel tank, ambient air is drawn in. This ambient air contains whatever dust and dirt contaminants happen to be suspended in the air. One OEM has determined that over 60% of the dirt trapped by a 4-micron secondary filter has passed through unfiltered fuel cap vents or breathers. As a consequence, the OEM recommends the use of a high-efficiency fuel tank breather similar to that shown in Figure 17–5.

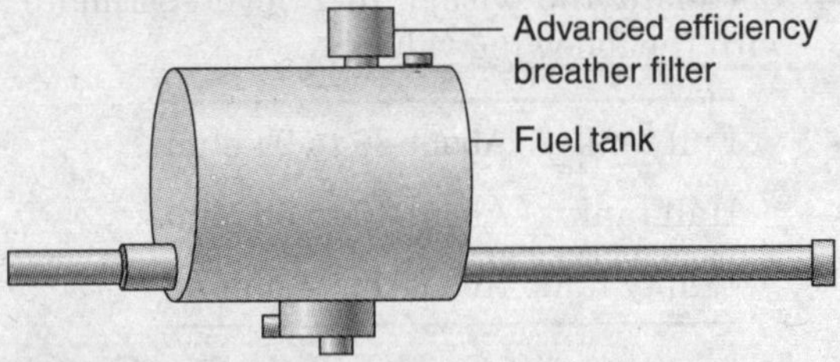

Figure 17–5 *Fuel tank breather filter recommended by Caterpillar. (Reprinted Courtesy of Caterpillar Inc.)*

WIF Checking

Diesel fuel tanks should be periodically checked for water in fuel (WIF). To check for the presence of water in fuel tanks, first allow the fuel tanks to settle then insert a probe (a clean aluminum welding rod) lightly coated with water detection paste through the fill neck until it bottoms in the base of the tank; withdraw the rod and examine the water detection paste for a change in color. This test will give some idea of the quantity of water in the tank by indicating the height on the probe where the color has changed. Trace quantities (just the tip of the probe changes color) in fuel tanks are not unusual and will not necessarily present any problems.

Fuel Tank Repair

Generally, OEMs recommend that a defective fuel tank be replaced rather than repaired. However, corroded and punctured fuel tanks are going to be repaired regardless of OEM recommendation so the following suggestions are offered, not to endorse fuel tank repair practice but to promote some safety awareness. It is essential to recognize the explosion hazard represented by diesel fuel vapors.

If a metal fuel tank must be welded, all fittings and plugs should be removed and the tank high pressure steamed until the temperature exceeds 200°F (95°C) throughout or for a minimum of 1 hour; it should also be remembered that in some jurisdictions, the practice of steaming fuel storage vessels to atmosphere is prohibited. Fuel tanks sectioned with internal baffles must be directly subjected to the steam within each section. After steaming, the tank should be filled with nitrogen gas before repair welding and/or weld patching. When a gasoline tank is to be repair welded, it should be steamed then evacuated with a vacuum pump to a moderate vacuum before filling with nitrogen gas.

The practice of welding full fuel tanks in position, while arguably safer than welding on an empty tank filled with fuel vapors, should never be undertaken. A single mistake or metallurgical defect in the tank wall would result in the death of the welder and any other person in the vicinity.

FUEL FILTERS

Diesel fuel injection equipment is manufactured with minute clearances and impurities in fuel, which if not removed by the fuel subsystem, can cause premature failures. Most dirt found in fuel is a result of conditions in stationary fuel storage tanks, refueling practices, and improper fuel filter priming techniques by service technicians. The function of a fuel filter is to entrap particulate (fine sediment) in the diesel fuel, and while some current secondary filters filter to the extent that water in its free state will not pass through the filtering media, a water separator is often used to remove water (H_2O). All diesel fuel systems require clean fuel, and the function of the filters in a fuel system is to ensure that the fuel is as clean as possible before it is delivered to the injection pumping components.

A typical fuel subsystem with a primary circuit and a secondary circuit in most cases uses a two-filter arrangement, one in each circuit. Two basic types of filter are used: the currently more common spin-on, disposable **cartridge** type and the **canister** and disposable element type. Spin-on filters are obviously easier to service and are the filter design of choice by most manufacturers. Figure 17–6 shows the flow routing in some typical filters.

Fuel Filter
Two-stage box-type filter

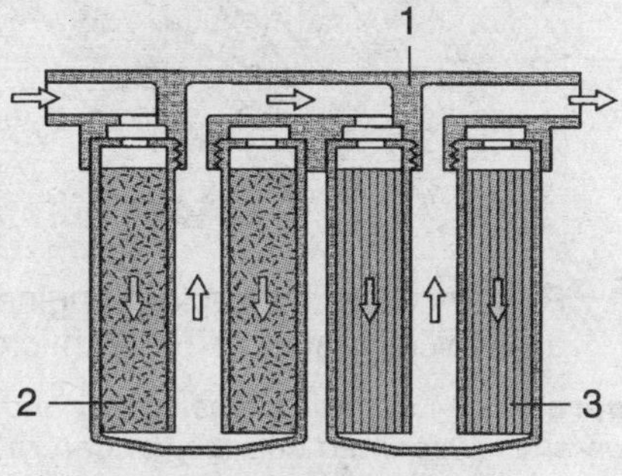

Multistage Filter
With spiral V-form filter element

1. Filter cover with mounting
2. Coarse filter
3. Fine filter

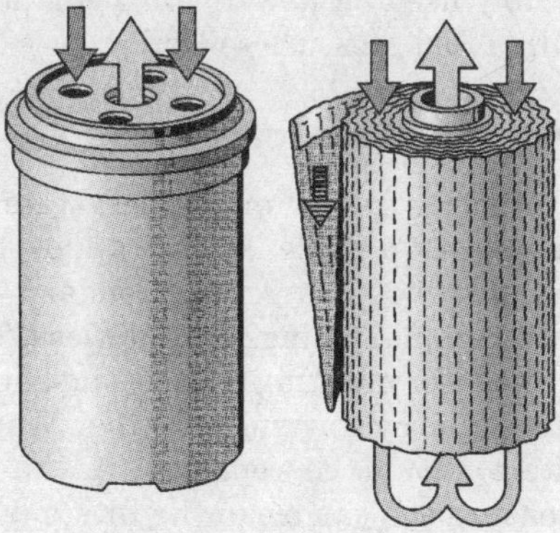

A **B**

Fuel Filter

A. Easy-charge filter
B. Spiral V-form filter element

Figure 17–6 *Types of fuel filters. (Courtesy of Robert Bosch LLC)*

Defining Clean Fuel

Fuel filters are rated by their ability to entrap particulates. They function on a sort of a fishnet principle whereby particles smaller than a certain size pass through the mesh—and those too large get entrapped. The unit we measure particle size by is the **micron**. One micron (1 μ—the Greek letter mu) is equal to one millionth of a meter. So a typical secondary fuel filter rated at 5 μ is designed to entrap most solids of that size that attempt to pass through it. You should note that the human eye is unable to see a particle smaller than 40 μ in size. There is an International Standards Organization (ISO) standard used to identify the particles. Figure 17–7 shows the sectional size of a human hair in relation to some of the dimensions we reference when talking about clean fuel.

The ISO fuel standard relates to the number of fuel particles suspended in fuel in three critical sizes: those of 4 μ, 6 μ, and 14 μ, respectively. Caterpillar's definition of clean fuel is ISO standard 18/16/13, and this refers to the counts of particles of those sizes and in that sequence as shown in Figure 17–7. Caterpillar also states that the maximum H_2O content in its fuel should not exceed 1,000 parts per million or 0.01%. So what they are saying is that provided fuel cleanliness is 18/16/13 or better, then the engine filtration system can perform the job it needs to do in cleaning the fuel. If fuel contaminants are worse than specified, then the filters are likely to plug. All fuel filters are rated by their ability to trap particles of a specified size. A filter with a specified rating of 2 microns at 99% efficiency means that it will entrap particles that are 2 microns and larger, 99% of the time.

Primary Filters

Primary filters represent the first filtration stage in a typical two-stage filtering fuel subsystem. Primary filters are therefore usually under lower than atmospheric pressure in operation, plumbed in series between the fuel tank and the fuel transfer pump. They are designed to entrap particles larger than 10–30 μ depending on the fuel system. They achieve this using media ranging from cotton threaded

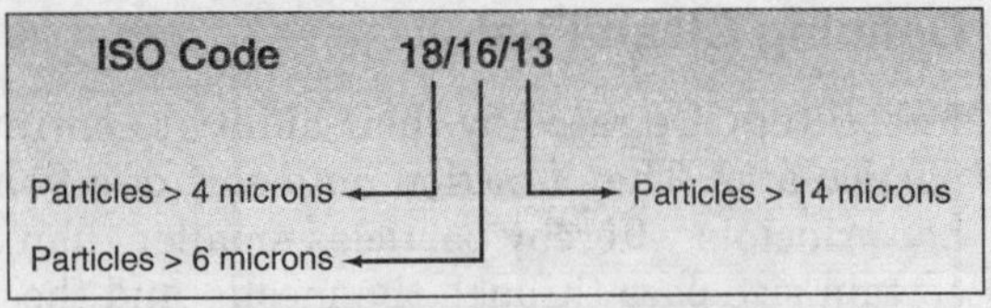

Range numb	Micron	Actual particle count range (per ml)
18	4+	1,300 - 2,500
16	6+	320 - 640
13	14+	40 - 80

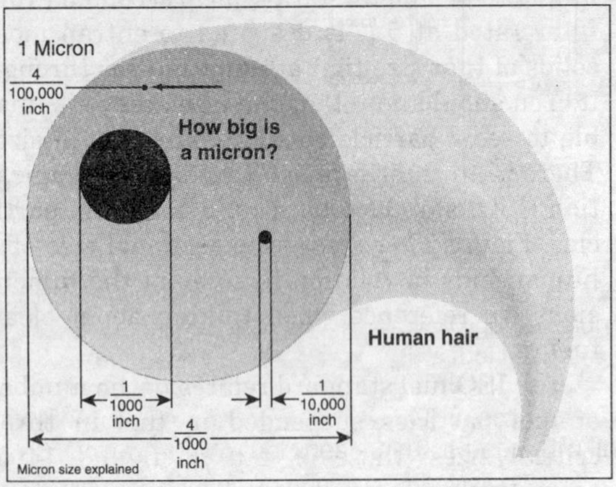

Figure 17–7 *ISO fuel code interpretation and definition of a micron. (Reprinted Courtesy of Caterpillar Inc.)*

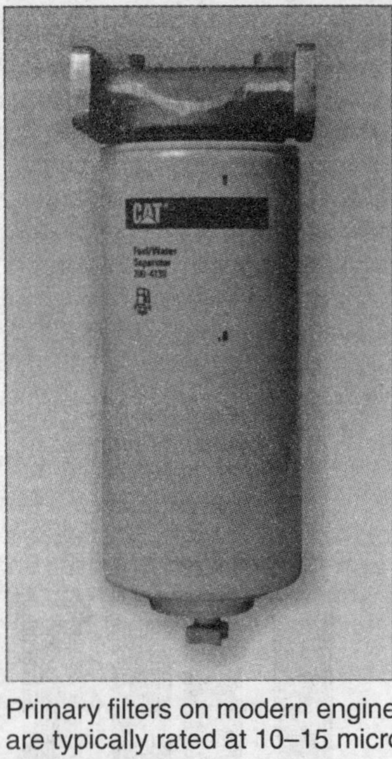

Figure 17–8 *Spin-on primary filter and water separator. (Reprinted Courtesy of Caterpillar Inc.)*

fibers, synthetic fiber threads, and resin-impregnated paper. Figure 17–8 shows a typical spin-on type of primary filter.

Secondary Filters

Secondary filters represent the second filtration stage in two-stage filtering. In a typical fuel subsystem, the secondary filter is charged by the transfer pump, and this enables use of more restrictive filtering media. The secondary filter would therefore normally be located in series between the transfer or **charging pump** (the pump responsible for pulling fuel from the fuel tank and charging the fuel injection components) and the fuel injection apparatus. In some diesel fuel subsystems using two-stage filtering, the primary and secondary filters may be both located on the same circuit, usually the charge circuit. In such cases, both filters are mounted on the same base pad with the primary filter feeding the secondary filter in series. Such an arrangement is more likely to be found in off-highway applications of diesel engines. Current secondary filters may be specified to entrap particulate sized as small as 1μ, but filtering efficiencies of 2–5 μ are more common. Secondary filters use a variety of media, including chemically treated pleated papers and cotton fibers. Figure 17–9 shows the secondary filter pad assembly used on newer Volvo engines equipped with a fuel pressure sensor.

Water and Secondary Filters

H_2O in its free or emulsified state (terms defined just a little later) cannot be pumped through most current secondary fuel filters with nominal ratings of 5 μ or less. This results in the filter plugging on H_2O and shutting down the engine by starving it for fuel. Water-plugged filters should be replaced. As an emergency measure you can clean the filters using methyl hydrate or other pure alcohol, then reprime with fuel.

Single Circuit Subsystems

In a fuel subsystem that is entirely under suction, such as some Cummins systems, the terms

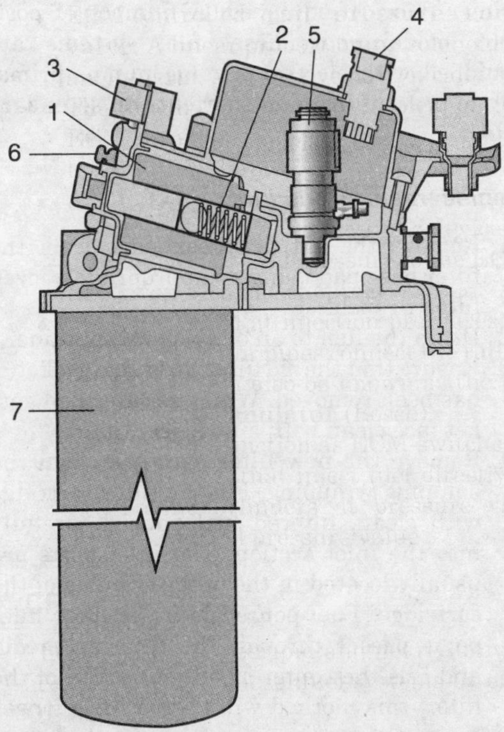

1. Electric pump
2. Lock-off valve
3. Air vent
4. Terminals
5. Fuel pressure sensor
6. Fuel drain fitting
7. 5 micron secondary filter

Figure 17–9 *Volvo secondary filter pad assembly. (Courtesy of Volvo Trucks NA)*

primary and *secondary* are not used to describe multiple filters when fitted to the circuit. Because every filtering device used in the fuel subsystem is held at a lower-than-atmospheric pressure (i.e., under "suction"), the inlet restriction specification is critical and, if exceeded, would result in a loss of power caused by fuel starvation.

Servicing Filters

Most fuel filters are routinely changed on preventative maintenance schedules that are governed by highway miles, engine hours, or calendar months. They are seldom *tested* to determine serviceability. When filters are tested, it is usually to determine if they are restricted (plugged) to the extent they are reducing engine power by causing fuel starvation.

Primary filters or a filter under suction should be tested for inlet restriction with a negative pressure gauge calibrated in inches of mercury (Hg) restriction. The negative pressure specifications are based on using an **Hg manometer**. A manometer is a clear tubular column formed in a U shape around a calibration scale marked off in inches: The column is then filled with either Hg or H_2O (usually colored with dye for ease of reading) to a zero point on the calibration scale. When a manometer is connected to a fluid circuit, it produces a reading according to the pull (vacuum circuit) or push pressure acting on the fluid in the column. The negative pressure gauge produces the same readings without the danger potential associated with Hg. Actual inlet restriction values vary considerably with the fuel system, and specifications should always be referenced.

The Hg calibrated, negative pressure gauge should be connected into the circuit between the filter mounting pad and the transfer pump. Transfer pumps are usually **positive displacement** (unload a constant slug volume of fluid per cycle), so they pump more fuel proportionally with rpm increase; this means valid test results can be obtained without loading the engine. When testing circuit restriction on a fuel subsystem that is entirely under suction, the specifications are likely to be fairly exacting. Circuit restriction specification readings on some Cummins systems should be 4–6″ Hg with a new filter but 7″ Hg is the maximum specification; if this value is exceeded, it will result in fuel starvation.

Secondary filters are usually charged by the transfer pump. Testing **charging pressure** (the pressure downstream from the charging/transfer pump) is normally performed with an accurate, fluid-filled pressure gauge (Figure 17–10) plumbed in series between the transfer pump and injection pump apparatus; it is not generally used as a method of determining the serviceability of a secondary filter. Secondary filters tend to be changed by preventative maintenance schedule rather than by testing or when they plug on H_2O or midwinter fuel waxing and shut down an engine. In summary:

- Primary filters are tested for inlet restriction measured in inches of Hg.
- Secondary filters are restriction tested using a pressure gauge in psi.
- The pressure downstream from the transfer pump is known as charging pressure.

Figure 17–10 *Fluid-filled pressure gauge used to measure charging pressure. (Reprinted Courtesy of Caterpillar Inc.)*

Procedure for Servicing Spin-on Filter Cartridges

Much dirt is placed in diesel fuel systems by technicians using improper service techniques. Most diesel service technicians realize that sets of replacement filters should be primed, that is, filled with fuel before installation, but few concern themselves about the source of the fuel. Filters should be primed with filtered fuel. Shops performing regular engine service should have a reservoir of clean fuel; any process that requires the technician to remove fuel from vehicle tanks will probably result in it becoming contaminated at least to some extent, however much care is exercised. The container used to transport the fuel from the tank to the filter should be cleaned immediately before it is filled with fuel. Paint filters (the paper cone-shaped type) can be used to filter fuel. The inlet and outlet sections of the filter cartridge should be identified.

The filter being primed should be filled only through the inlet ports usually located in the outer annulus (ring) of the cartridge and never directly into the outlet port, which is usually located at the center. Most manufacturers prefer that only the primary filters be primed before installation during servicing. After the primary filter(s) has been primed and installed, the secondary filter should be installed dry and primed with a hand primer pump or inline electric primer pump if equipped. A number of post-2007 model year diesel fuel systems are equipped with electric priming pumps primarily to prevent priming shortcuts on secondary filters.

Replacement Procedure

1. Remove the old filter cartridge from the filter base pad using an appropriately sized filter wrench.
2. Drain the fuel to an oil disposal container.
3. Ensure that the old filter cartridge gasket(s) has been removed. Wipe the filter pad gasket face clean with a lint-free wiper.
4. Remove the new filter cartridge from the shipping wrapping. Fill the filter cartridge with clean, filtered fuel poured carefully into the inlet section. The inlet ports are usually located in the outer annulus of the cartridge. Fuel poured into the filter inlet ports passes through the filtering media and fills the center or outlet section of the filter; this method will take a little longer because it requires some time for the fuel to seep through the filtration medium.
5. The fuel oil itself should provide the gasket and/or O-ring and mounting threads with adequate lubricant; it is not necessary or good practice to use grease or white lube on filter gaskets.
6. Screw the filter cartridge **clockwise (CW**: right-hand threads are used) on to the mounting pad; after the gasket contacts the pad face, a further rotation of cartridge is usually required. In most cases, hand tightening is sufficient but each filter manufacturer has its own specific recommendations on the tightening procedure and these should be referenced.

Water Separators

Most current diesel engine-powered highway vehicles have fuel subsystems with fairly sophisticated H_2O removal devices. H_2O appears in diesel fuel in three forms:

- Free state
- Emulsified
- Semiabsorbed

H_2O in its free state appears in large globules and, because of its greater weight than diesel fuel, readily collects in puddles at the bottom of fuel tanks or storage containers.

H_2O emulsified in fuel appears in small droplets; because these droplets are minutely sized, they may be suspended for some time in the fuel before gravity takes them to the bottom of the fuel tank. When free state H_2O collects in the base of fuel tank, 3 miles (5 km) driving on a rough class B road is enough to **emulsify** it (finely dispersing it into the fuel), making it more of a problem. Semiabsorbed H_2O is usually H_2O in solution with alcohol, which is a direct result of the methyl hydrate (type of alcohol added to fuel tanks as deicer or in fuel conditioner) added to fuel tanks to prevent winter freeze-up. H_2O that is semiabsorbed in diesel fuel is in its most dangerous form because it may emulsify in the fuel injection system where it can seriously damage components.

Why Water Damages Fuel Systems

Generally, H_2O damages fuel systems for three reasons. H_2O possesses lower lubricity than diesel fuel, has a tendency to promote corrosion, and its different physical properties affect the pumping dynamics. Diesel fuels are compressible at approximately 0.5% per 1,000 psi; H_2O is less compressible at approximately 0.35 per 1,000 psi. Modern fuel injection pumping apparatus is engineered to pump diesel fuel at very high pressures, and if H_2O with its lower lubricity and compressibility is pumped through the system, the resultant pressure rise can cause structural failures, especially at the sac/nozzle area of the fuel injectors. When you see a modern fuel injector with its tip blown off, the cause can often be traced to a water-in-fuel condition.

H_2O Separator Operating Principle

H_2O separators have been used in diesel fuel systems for many years. Often these were crude devices that used gravity to separate the heavier H_2O from the fuel. However, over the past two decades as both injection pumping pressures have steadily risen and the consumer's expectation of engine longevity has greatly increased, H_2O separators have developed accordingly. Often a H_2O separator will combine a primary filter and H_2O separating mechanism into a single canister. Many of these combination primary filter/H_2O separators are manufactured by aftermarket suppliers such as Racor, CR, Davco, Dahl, and others. These use a variety of means to separate and remove H_2O in free and emulsified states; they will not remove H_2O from fuel in its semiabsorbed state.

H_2O separators use combinations of several principles to separate and remove H_2O from fuel. The first is gravity. H_2O in its free state or emulsified H_2O that has been **coalesced** (where small droplets come together to combine into larger droplets) into large droplets will, because of its heavier weight, be pulled by gravity to the bottom of a reservoir or sump. Some H_2O separators use a **centrifuge** to help separate both larger globules of H_2O and emulsified H_2O from fuel; the centrifuge subjects fuel passing through it to centrifugal force, throwing the heavier H_2O to the sump walls where gravity can pull it into the sump drain. A centrifuge acts to separate particulate from the fuel in the same manner. Fuel directed through a fine resin-coated, pleated paper medium passes through the medium with greater ease than H_2O. H_2O entrapped by the filtering medium can collect and coalesce in large enough droplets to permit gravity to pull it down into the sump drain. In many cases, aftermarket fuel filter/H_2O separators are designed to replace the fuel system OEM's primary filter; in others this unit may work in conjunction with the primary filter. Figure 17–11 shows an assortment of combination filter/H_2O separators used on diesel engines.

When installing an aftermarket filter/H_2O separator unit on the suction side of a fuel subsystem, it is good practice to locate the manufacturer's maximum restriction specification and test that it is not being exceeded. An Hg manometer or negative pressure gauge is the appropriate test tool. In cases in which the entire fuel subsystem is under suction, the consequences of exceeding the restriction specification are generally more severe, the result being fuel starvation to the engine. Servicing H_2O separator units is a simple process but one that should be undertaken with a certain amount of care because it is easy to contaminate the fuel in the separator canister either by priming it with unfiltered fuel or by permitting dirt to

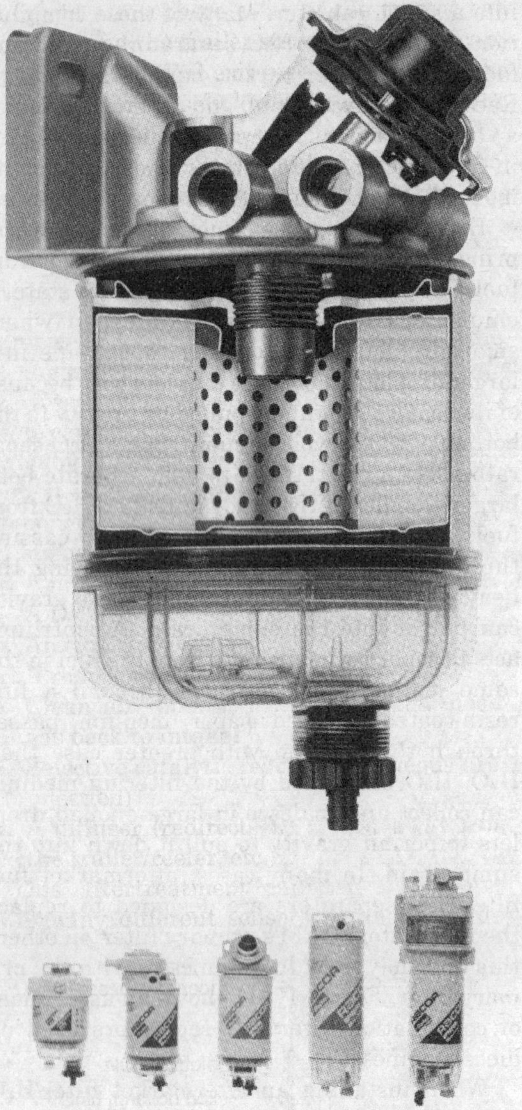

Figure 17–11 *An assortment of water separators used on diesel engines. (Courtesy of Parker Hannefin, Racor Division)*

enter when the canister lid is removed. Most aftermarket H_2O separators have a clear sump through which it is easy to observe the presence of H_2O.

All H_2O separators are equipped with a drain valve. The drain valve may be operated manually or electrically. The purpose of this valve is to siphon H_2O from the sump. H_2O should be routinely removed from the sump using the drain valve. The filter elements used in combination primary filter/H_2O separator units should be replaced in most instances with the other engine and fuel filters at each full service. However, some manufacturers claim that their filter elements have an in-service life that may exceed the oil change interval by two or more times. Whenever an H_2O separator is fully drained, it should be primed before attempting to start the engine.

Fuel Heaters

In recent years it is more common to find trucks equipped with fuel heaters. In fuel systems in which fuel is flowed through the injection system circuitry at a rate much higher than that required for fueling the engine, constant filtering of fuel removes some of the wax and therefore some of its lubricity even when the appropriate seasonal pour point depressants are present.

Pour point depressants tend not to have too much effect on the cloud point of a fuel, which is its first stage of waxing. This is a condition to which ASTM 2D fuels are more prone than 1D. However, there is some debate about the use of fuel heaters, and the fuel system/engine manufacturer should always be consulted when fitting such a device. One engine manufacturer warns that its warranty is voided if electric element-type fuel heaters are used in its system.

There are two types of fuel preheaters in current use:

1. Electric element type. An electric heating element uses battery current to heat fuel in the subsystem. This type offers a number of advantages, most notable of which is that the heater can be energized before startup so that cranking fuel is warmed up. Electric element fuel heaters may be thermostatically managed so that fuel is only heated as much as required and not to a point that compromises some of its lubricating properties.
2. Engine coolant heat exchanger type. This type of fuel heater consists of a housing within which coolant is circulated in a bundle (heat exchanger core) and over which the fuel is passed. A disadvantage of this type is that the engine cooling system must be at operating temperature before the fuel can be heated.

Fuel heaters exist that use both electric heating elements and coolant medium heat exchangers and, furthermore, manage the fuel temperature. Optimally, fuel temperature should be managed not to exceed 90°F (32°C). Once fuel exceeds this temperature its lubricating properties start to diminish and the result is reduced service life of fuel injection components. Figure 17–12 shows a Detroit Diesel Fuel Pro assembly: This combines a filter, H_2O separator, and thermostatically controlled heating element.

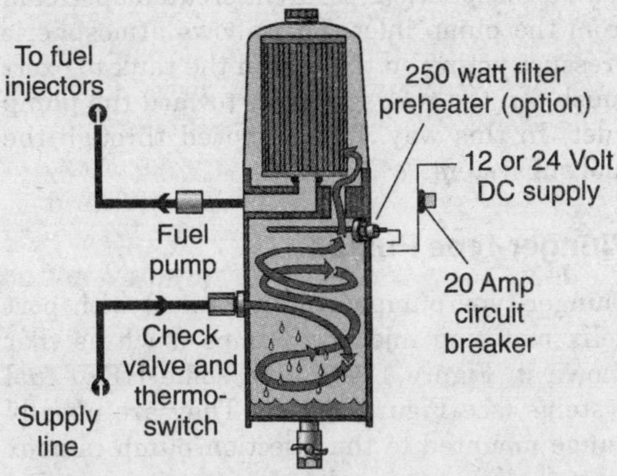

Figure 17–12 *Detroit Diesel Fuel Pro assembly that combines a filter, water separator, and thermostatically controlled heating element. (Courtesy of Detroit Diesel Corporation)*

Water-in-fuel Sensors

Most current systems use a **water-in-fuel (WIF) sensor** to alert the data bus of H_2O contamination of fuel. A WIF sensor can be built into a replaceable filter cartridge or be integrated into a combination filter/H_2O separator assembly. The sensor uses a couple of probes and a 12-volt supply. Because H_2O has different electrical resistivity (known as dielectric properties) to fuel, a return signal is output from the sensor when the electrical path across the probes acts through H_2O rather than fuel. At this point the WIF broadcasts a service alert. Note that sometimes a WIF can produce a service alert immediately after draining the H_2O sump: The reason is that water-resident bacteria can coat the probes after draining and trigger a false signal. Figure 17–13 shows a typical WIF sensor and its circuit used in a Volvo fuel subsystem.

FUEL CHARGING/TRANSFER PUMPS

Fuel charging or transfer pumps are positive displacement pumps driven directly or indirectly by the engine. A positive displacement pump displaces the same volume of fluid per cycle and, therefore, fuel quantity pumped increases proportionally with rotational speed. Because of this, if a positive displacement pump unloads to a defined flow area, pressure rise is proportional with rpm increase. On most current diesel fuel systems two types of transfer pump are used:

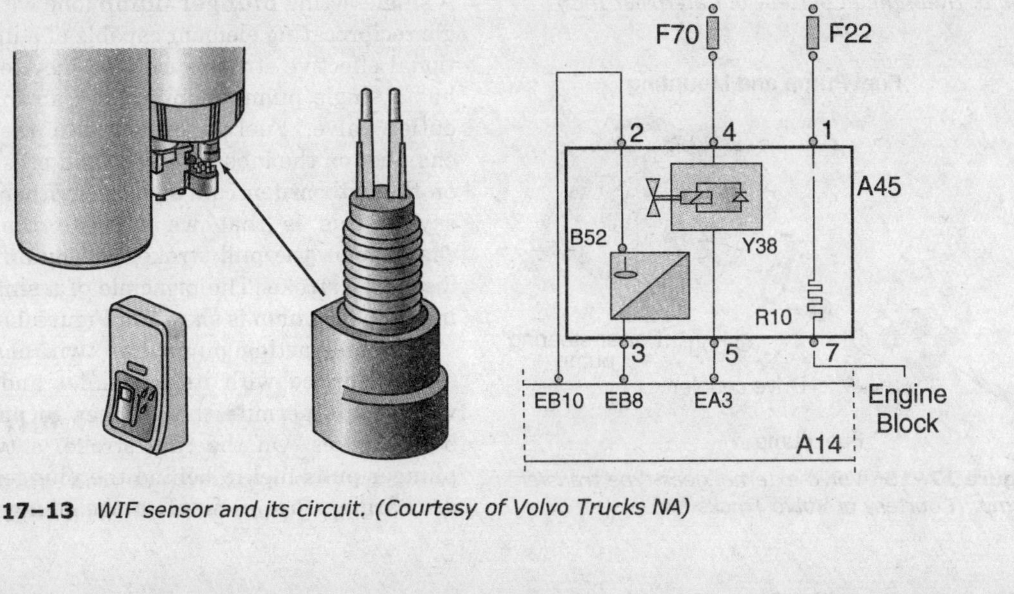

Figure 17–13 *WIF sensor and its circuit. (Courtesy of Volvo Trucks NA)*

- Plunger-type pumps
- Gear-type pumps

Plunge -type pumps are less commonly used today. Figure 17–14 shows a typical plunger-type pump used by Caterpillar on its hydraulically actuated electronic unit injector (HEUI) fueled engines: The reciprocating pumping action is created by a cam and the assembly is mounted behind the HEUI swash plate (oil) pump. Figure 17–15 shows a Volvo external gear-type pump and its mounting location behind the power steering pump. Gear pumps are the commonly used fuel transfer pump today and they share common operating principles with other external gear pumps such as oil pumps.

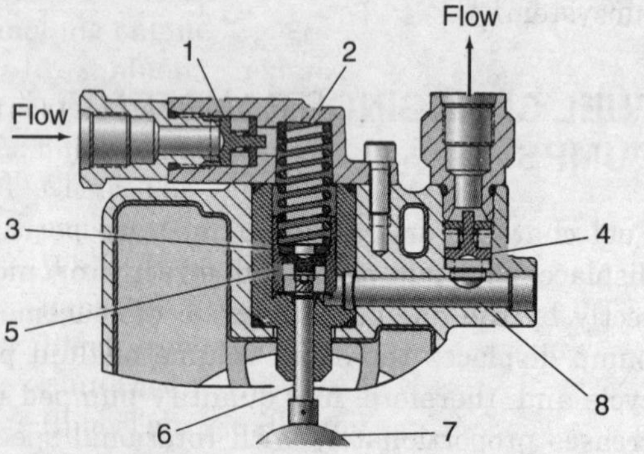

FUEL TRANSFER PUMP
1. Inlet check valve
2. Spring
3. Piston assembly
4. Outlet
5. Piston check valve
6. Tappet assembly
7. Cam
8. Passage

Figure 17–14 *Caterpillar plunger-type transfer pump. (Reprinted Courtesy of Caterpillar Inc.)*

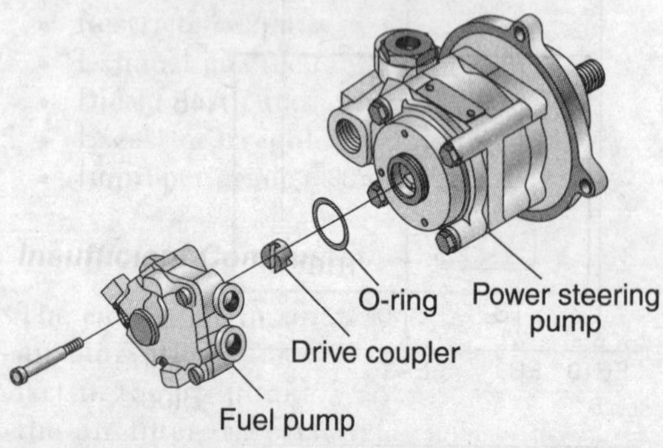

Figure 17–15 *Volvo external gear-type transfer pump. (Courtesy of Volvo Trucks NA)*

Pumping Principle

In describing pump operation in the fuel subsystem, in common with most truck OEMs we use the terms *suction circuit* and *charge circuit*. Fuel movement through a fuel subsystem is created by a positive displacement pump, sometimes known as a **prime mover**, though some would prefer this term be used only for the engine that drives the pump. The way a transfer pump works is by creating flow that forces fuel out its discharge circuit: In doing this, lower-than-atmospheric pressure is created upstream from the pump inlet. This allows atmospheric pressure acting on the fuel in the tank to exert "push" on the fuel, forcing it toward the pump inlet. In this way, fuel is moved through the fuel subsystem.

Plunger-type Pumps

Plunger-type pumps are often used with port helix metering injection pumps (such as that shown in Figure 17–16) and some HEUI fuel systems (see Figure 17–14). They are usually flange mounted to the injection pump cambox (port helix) or swash plate pump housing (HEUI), and driven by a dedicated cam or eccentric on the pump camshaft. Single-acting and double-acting plungers may be used, with the latter type specified in higher output engines requiring more fuel.

Single and Double Acting

A single-acting **plunger pump** (one with a single reciprocating element capable of a unidirectional effective stroke such as a bicycle pump) has a single pump chamber and an inlet and outlet valve. Fuel is drawn into the pump chamber on the inboard stroke and pressurized on the outboard or cam stroke. Another way of saying this is that we aspirate the pump chamber on the pull stroke, and pump on the discharge stroke. The principle of a single-acting plunger pump is shown in Figure 17–17.

A double-acting pump has twin chambers each equipped with its own inlet and outlet valve. This permits the plunger to pump on both strokes. On the cam stroke, a two-way plunger pulls fuel in behind the plunger while discharging fuel in front of the plunger. The

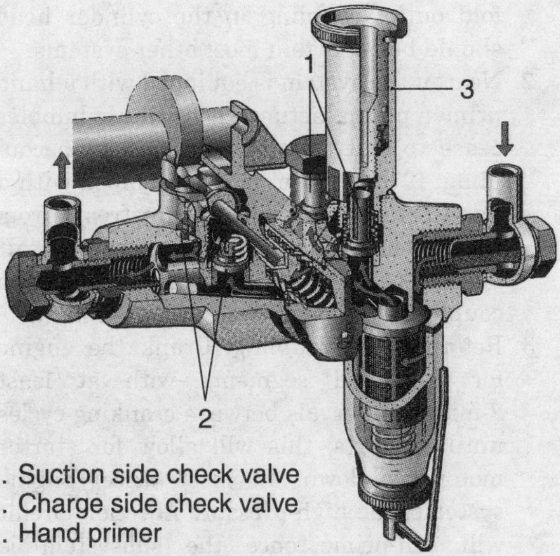

Figure 17–16 *Bosch charging pump with integral hand primer and primary filter. (Courtesy of Robert Bosch LLC)*

reverse occurs as the plunger is pulled back on its retraction stroke. The principle of a double-acting plunger pump is shown in Figure 17–17.

Gear-type Pumps

Gear pumps are also commonly used as transfer pumps, especially on most electronically managed engines. These are normally driven from an engine accessory drive and are located wherever convenient. Gear pumps usually have an integral relief valve that defines the system charging pressure. Fuel injection systems designed to be charged at pressure values higher than typical tend to use gear-type transfer pumps instead of cam-actuated plunger pumps. In instances in which a gear pump feeds an injection system with no main filter in series, a filter mesh is sometimes incorporated to protect injection pumping apparatus; when gear pumps are used, there is a small chance that gear teeth cuttings can be discharged into the system. A majority of the full authority, electronic management fuel systems use gear-type transfer pumps. They can also produce the much higher flow and pressures required by DPF dosing systems used on post-2007 diesel engines.

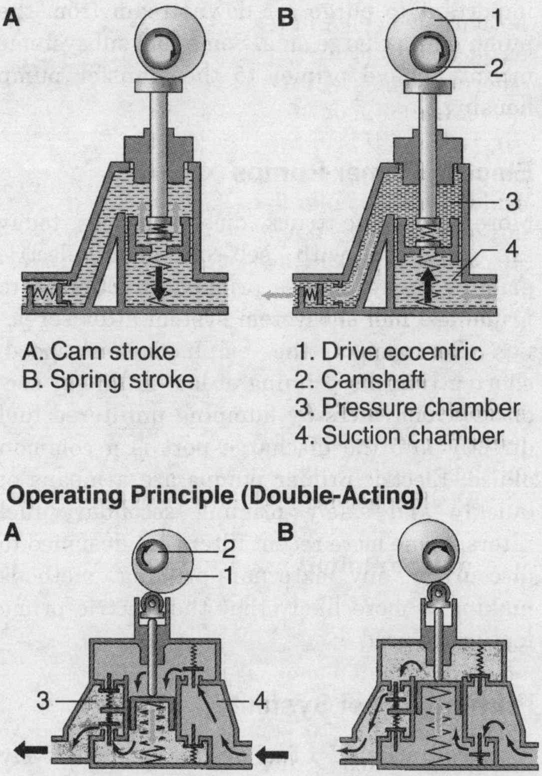

Figure 17–17 *Action of single- and double-acting plunger pumps. (Courtesy of Robert Bosch LLC)*

Hand Primer Pumps

A hand primer pump may be a permanent fixture to a fuel subsystem located on the fuel transfer pump body or a filter mounting pad (see Figure 17–9 and Figure 17–16). A hand primer pump can be a useful addition to the technician's tool kit, and it can be fitted to a fuel subsystem when priming is required. The function of a hand primer pump is to prime the fuel system whenever prime is lost. Typically, they consist of a hand-actuated plunger and use a single-acting pumping principle. On the outward stroke, the plunger exerts suction on the inlet side, drawing in a charge of fuel to the pump chamber; on the downward stroke, the inlet valve closes and fuel is discharged to the outlet. When using a hand primer pump, it is important to purge air downstream from the pump on its charge side. Some fuel subsystems mount a hand primer to the transfer pump housing.

Electric Primer Pumps

More and more truck diesel engines today are equipped with self-contained, electric priming pumps whose primary function is to prime the fuel subsystem system after servicing. Because of the emphasis on speed, shortcutting the priming of last-in-line or secondary fuel filters by dumping unfiltered fuel directly into the discharge port is a common abuse. Electric primer pumps are a means of quickly and safely priming secondary fuel filters. Some more recent filters are designed to discourage any external priming methods making it more likely that the electric prime feature be used.

Priming a Fuel System

Although priming a fuel system is a relatively simple procedure, it is usually advisable to consult the appropriate service literature. Most OEMs prefer that the technician avoid pressurizing air tanks with regulated air pressure to prime a diesel fuel system. It should be remembered that diesel fuel contains volatile fractions and the act of pressurizing a fuel tank with air pressure will vaporize some fuel. Air exiting an air nozzle creates friction and the potential for ignition. Avoid this practice in extreme hot weather conditions.

Recommended Procedure

When a vehicle runs out of fuel and it is determined that the fuel subsystem requires priming, remove the filters and fill with filtered fuel.

1. Locate a bleed point in the system—often on an inline injection pump system, this will be at the exit of the charging gallery—and crack open the coupling. A fuel manifold outlet coupling in the cylinder head should be opened in most other systems.
2. Next, if the system is equipped with a hand primer pump, actuate it until air bubbles cease to exit from the cracked open coupling. If the system is not equipped with a hand primer pump, fit one upstream from the transfer pump and actuate until air bubbles cease to exit from the cracked open coupling.
3. Retorque the coupling. Crank the engine for 30-second segments with at least 2-minute intervals between cranking cycles until it starts; this will allow for starter motor cool-down. In most diesel engine systems, the high-pressure injection circuit will self-prime once the subsystem is primed.

Refueling

It is good practice to refuel tanks on returning from a trip; filling the tanks with fuel displaces the air present in the tank that replaced the fuel as it was consumed. When tanks are left in a near empty condition for any length of time (overnight is long enough), the moisture in the air condenses and contaminates the fuel.

COMPLETE FUEL CIRCUIT

Now that we have taken a look at all of the individual components that make up a fuel subsystem, we should take a brief look at how they interact as a system. Figure 17–18 shows how a typical post-2007 fuel subsystem supplies fuel to an electronic unit injector (EUI) fuel circuit. See if you can identify the sensors used in the circuit: The electrical principles of these sensors are not covered until Chapter 30, but you can see how they are used to signal status conditions of the fuel subsystem. Some examples of sensors used in diesel fuel subsystems are:

- Sending unit
- Fuel temperature
- WIF
- Fuel pressure

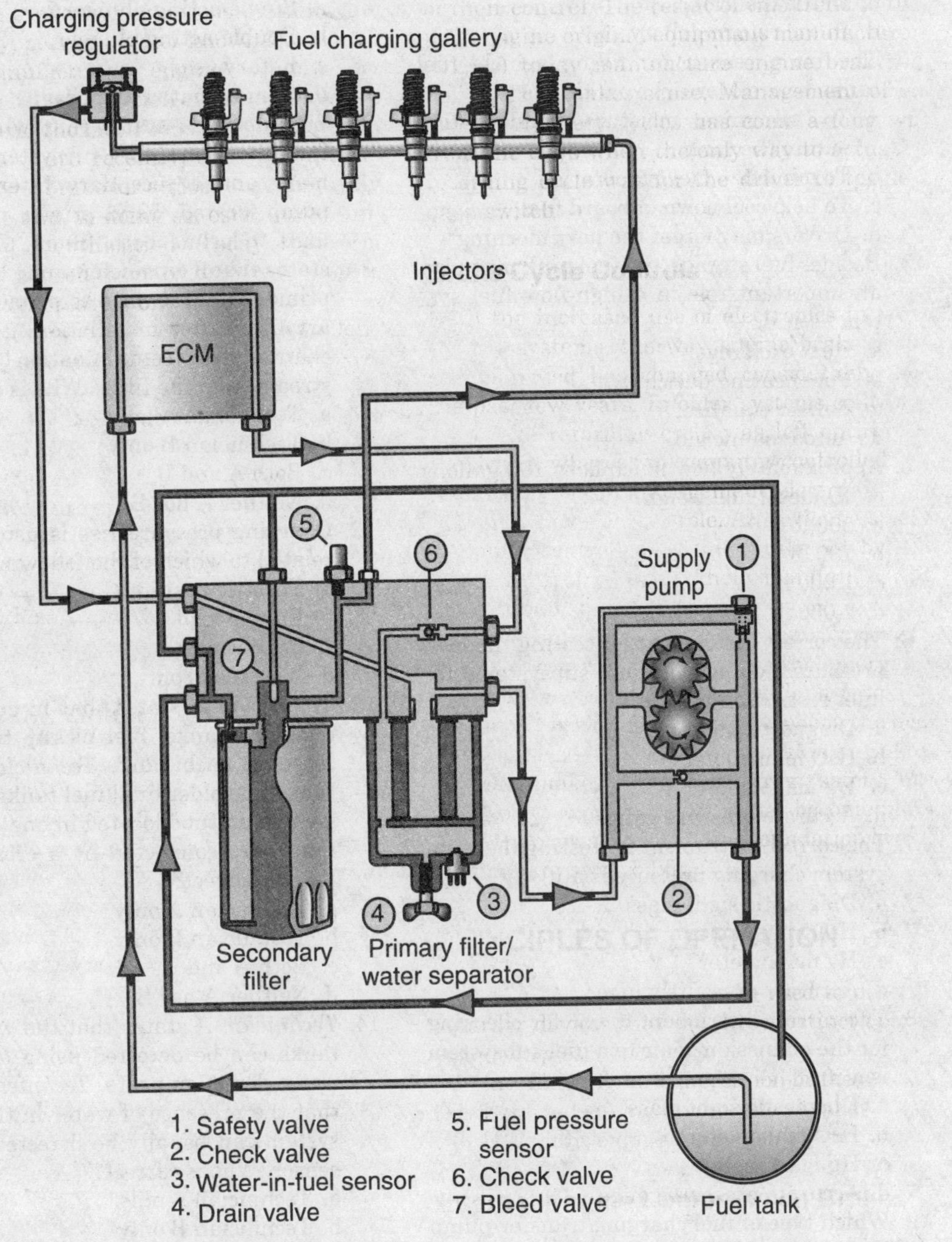

Figure 17–18 *Fuel subsystem supplying EUIs.*

REVIEW QUESTIONS

1. On the typical truck diesel fuel subsystem, which of the following is usually subject to the lowest pressure in the circuit?
   a. Fuel heater
   b. Primary filter
   c. Secondary filter
   d. Charging gallery
2. A secondary filter is located:
   a. Upstream from the transfer pump
   b. On the charge side of the transfer pump

c. In the fuel rail
d. In the return gallery

3. The main reason for filling vehicle fuel tanks before overnight parking is:
   a. To minimize moisture condensation within the tank
   b. To minimize fuel evaporation
   c. To help cool down onboard fuel
   d. Drivers may forget the next morning
4. Besides fuel storage, the fuel tank may play an important role in a high-flow fuel system as a(n):
   a. Heat exchanger
   b. Fuel heating device
   c. Ballast equalizer
   d. Aerodynamic aid
5. Approximately how much does 100 gallons of 2D diesel fuel weigh?
   a. 500 lb
   b. 730 lb
   c. 1,000 lb
   d. 2,000 lb
6. The correct instrument for testing the low-pressure side of most fuel subsystems for inlet restriction is a(n):
   a. Diagnostic sight glass
   b. H_2O manometer
   c. Hg manometer
   d. Accurate pressure gauge
7. The correct instrument for testing fuel subsystem charging pressure is a(n):
   a. Diagnostic sight glass
   b. H_2O manometer
   c. Hg manometer
   d. Accurate pressure gauge
8. The correct instrument to use for checking for the admission of air to a fuel subsystem is a(n):
   a. Diagnostic sight glass
   b. H_2O manometer
   c. Hg manometer
   d. Accurate pressure gauge
9. Which type of fuel charging/transfer pump is more commonly used by electronically managed engine/fuel systems?
   a. Plunger pumps
   b. Centrifugal pumps
   c. Diaphragm pumps
   d. Gear pumps
10. What type of pumping principle is used by a typical hand primer pump?
    a. Single-acting plunger
    b. Double-acting plunger
    c. Rotary gear
    d. Cam-actuated diaphragm
11. *Technician A* states that when servicing fuel filters on a fuel subsystem with a primary and a secondary filter, a transfer pump located, hand primer pump can be used to prime both filters. *Technician B* states that if a transfer pump located, hand primer pump is used to prime the secondary filter after installation, it reduces the chances of dirt contamination beyond downstream from the filter. Who is correct?
    a. Technician A only
    b. Technician B only
    c. Both A and B
    d. Neither A nor B
12. Charging pressure rise is usually directly related to which of the following?
    a. Throttle position
    b. Engine load
    c. Peak power
    d. Increased rpm
13. *Technician A* states that in current truck dual fuel tanks, fuel pickup tubes are located in each tank. *Technician B* states that some older dual fuel tanks used a single pickup tube located in one of the tanks that were connected by a crossover pipe. Who is correct?
    a. Technician A only
    b. Technician B only
    c. Both A and B
    d. Neither A nor B
14. *Technician A* states that the water-in-fuel tanks can be detected using a probe and water detection paste. *Technician B* states that the presence of water in the fuel subsystem can usually be detected by a WIF sensor. Who is correct?
    a. Technician A only
    b. Technician B only
    c. Both A and B
    d. Neither A nor B
15. *Technician A* states that some water separator units use a centrifuge to separate water from fuel. *Technician B* states that because fuel is heavier than water, fuel can be easily separated from the water because it always settles under it. Who is correct?

a. Technician A only
b. Technician B only
c. Both A and B
d. Neither A nor B

16. *Technician A* states that gear pumps are of the positive displacement type. *Technician B* states that plunger pumps are positive displacement. Who is correct?
a. Technician A only
b. Technician B only
c. Both A and B
d. Neither A nor B

17. If the fuel tank vent valve on a diesel-powered vehicle is plugged, which of the following would be the likely outcome when running the engine?
a. The tank would explode.
b. The engine would run away due to excess fuel.
c. The tank would implode.
d. The engine would shut down due to fuel starvation.

18. *Technician A* states that diesel engine fuel tank caps are fitted with a two-way valve that permits both fuel vapor seepage and air admission to the tank. *Technician B* states that most truck diesel engines use a vane-type transfer pump located in the fuel tank to pump fuel to the transfer pump. Who is correct?
a. Technician A only
b. Technician B only
c. Both A and B
d. Neither A nor B

19. *Technician A* states that diesel fuel can combine with air to form potentially explosive mixtures. *Technician B* states that diesel fuel CN is likely to degrade more quickly during the winter than in the heat of summer. Who is correct?
a. Technician A only
b. Technician B only
c. Both A and B
d. Neither A nor B

20. At refueling on the completion of a journey, fuel is steaming in the tanks. *Technician A* states that this condition is a result of normal operation in some fuel systems. *Technician B* states that such a condition can be a result of running the tanks low on fuel in a high-flow fuel system. Who is correct?
a. Technician A only
b. Technician B only
c. Both A and B
d. Neither A nor B

18

Overview of Diesel Fuel Injection Basics

OVERVIEW OF DIESEL FUEL INJECTION PRINCIPLES

Before beginning the study of **hydromechanical engine management** (term used to describe a diesel engine that is managed without a computer) and **electronic management** (refers to any system that is managed by computer) of the diesel engine, a thorough understanding of engine components and principles of operation is required. Fuel systems manage diesel engines. In any **compression ignition (CI)** engine, a power stroke will only take place if the fuel system is accurately phased to the engine and performs the following four objectives:

1. Timing

Fuel delivery **timing** is critical during all engine operating phases. Typically, fuel is injected into the engine cylinder slightly before the piston completes its compression stroke. While older engines tended to have hard (nonvariable) timing values, current engines have variable timing features that can be managed mechanically, hydraulically, or electronically. Variable timing is required in today's engines to produce optimum performance and minimal noxious emissions. **Indirect injection (IDI)** diesel engines belong to an earlier generation of diesel engines. They are seldom seen today in truck applications. IDI diesels require much advanced injection timing compared to **direct injection (DI)** engines for two reasons:

- Larger droplet sizes emitted from the low-tech injectors used
- Much higher cylinder clearance volumes required by IDI

The larger droplets emitted from IDI engines rely on turbulence created in prechambers to help rip them down to a smaller size suitable for combustion in the short amount of time available.

Hydromechanical DI engines with high-pressure injection (and therefore smaller emitted droplet sizes than IDI engines) were usually base injection timed at around 20 degrees **before top dead center (BTDC)**, and most had variable timing capability. Today we use computers to control injection timing. Most current diesel engine use **multipulse injection** timing. This means the injection pulse is broken up into several separate injection events during one cycle. Multipulse injection may consist of two pulses that we describe as:

- Pilot pulse
- Main injection

The objective of pilot injection is to deliver a short initial pulse of fuel into the engine cylinder. Following the pilot pulse, the engine control module (ECM) calculates how much time is required (based on input sensor signaled conditions) to ignite it. At the point of igniting the pilot pulse of fuel in the engine cylinder, fuel injection resumes with the delivery of the main pulse.

Multipulse injection today can mean breaking an injection pulse into up to seven events per injection cycle. Timing of each successive injection pulse is managed by the engine ECM. Computer-timed, multipulse injection is used to make the best use of heat energy released by the fuel along with minimizing emissions and smoothing the application of force to the piston.

In a general sense the timing of fuel injection also influences tailpipe emissions. Modern diesel fuel systems are required to precisely control injection timing to contain exhaust emissions within legal limits, and a strategy that best minimizes one controlled emission may have a counterproductive effect on another. In terms of injection timing and emissions, it can be said that the following is true:

- Advanced timing means more time to completely burn injected fuel so it has a reducing effect on hydrocarbon (HC) emissions.
- Advanced timing produces higher cylinder temperatures and tends to increase the chances of oxidizing nitrogen to produce harmful oxides of nitrogen (NO_x) compounds.
- Retarded timing results in lower cylinder temperatures and therefore reduces the tendency to produce NO_x.
- Retarded timing reduces the time available to completely burn injected fuel and can result in higher HC and soot at the tailpipe.

We will take a much more detailed approach to emissions in Chapter 41 of this book and at how different engines manage timing strategy in the various chapters on engine management.

2. Pressurizing and Atomization

The fuel system must be capable of pressurizing the fuel sufficiently to prepare the fuel charge for combustion. Fuel exiting diesel fuel injectors is atomized. When high-pressure fuel is forced through a restricted flow area, it breaks up into minute droplets: This is known as **atomization**. The restricted flow area is located in injectors. Fuel exits from injectors by means of an **orifice** (single hole) or **orifii** (multiple holes). Atomized fuel is in liquid state. The degree of atomization is defined by the size of the atomized fuel droplets. The size of fuel droplets emitted from an injector is determined by:

- The pressure of the fuel supplied to the nozzle orifii
- The flow area or sizing of the nozzle orifii

The means used to pressurize the fuel to the required injection pressures varies with the type of fuel system. In most cases the fuel injection pumping apparatus is actuated mechanically by pumping elements. The source of the pump drive may be the engine timing geartrain or main engine camshaft. Current fuel systems are capable of producing up to 35,000 psi (2,450 bar) peak pressures and in most cases are required to produce these kinds of pressures independently from engine rpm. Pressure is always the control variable when it comes to determining droplet size.

In order to meet stringent emissions controls and optimize fuel economy, the ECM managing diesel engines is required to control the droplet size emitted from injectors. The ECM achieves this by managing injection pressures. Fuel droplets burn from the outside in (Figure 18–1), meaning that:

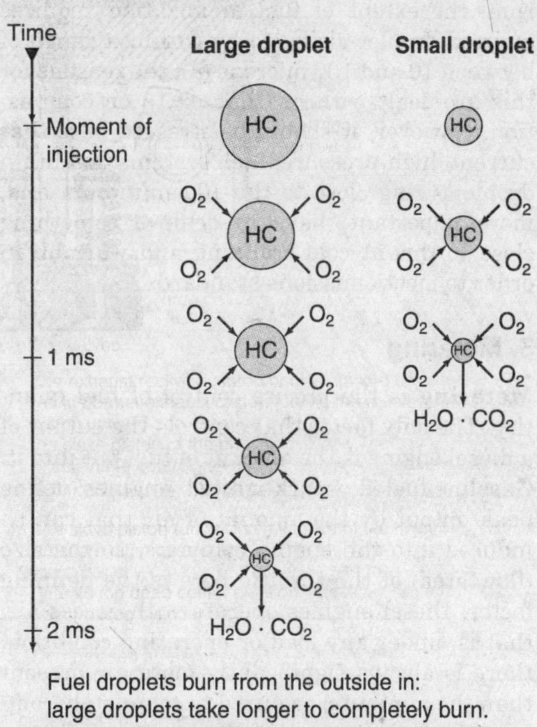

Figure 18–1 *Fuel droplets burn from the outside in: Large droplets take longer to burn completely so droplet size influences combustion rate.*

- Larger fuel droplets require more time to completely burn: this means that larger fuel droplets have to be injected earlier in the cycle if they are going to burn properly during the power stroke.
- Smaller fuel droplets respond more rapidly to whatever heat is in the engine cylinder and therefore take less time to completely burn.

Older diesel engines had injector nozzles with hard opening and closing values. This meant that if the nozzle opening pressure (NOP) of an injector was specified at 5,000 psi (350 bar), at the beginning of the injection pulse fuel droplets were always the same size regardless of how the engine was being run. These older diesel fuel systems had fuel systems designed to increase pressure during the injection pulse: This was generally desirable because as fuel injection extended into the cycle, the amount of real time available to complete the burn was reduced.

In diesel engines of the past couple of decades with wind-out speeds not exceeding 2,000 rpm, the extent of fuel atomization requires emitted droplet sizing to be within a range of between 10 and 100 microns (μ); the reasons for this are dealt with in Chapter 15 on combustion. However, it should be stressed here that current high-pressure fuel systems maintain droplet sizing close to the 10μ minimum and, more important, have to achieve something close to this at cold cranking and warmup in order to meet emissions standards.

3. Metering

Metering is the precise control of fuel quantity. The only factor that controls the output of a diesel engine is the amount of fuel put into it. Gasoline-fueled, spark-ignited engines define peak output by the amount of air that can be induced into the engine cylinders, so the size (flow area) of the throttle bore is the defining factor. Diesel engines operate with excess air, that is, under any load or operating condition, there is always significantly more air present than the minimum required to completely combust the fuel. Because of this excess air factor, most mobile diesel engines given unlimited fuel will accelerate at a rate of up to 1,000 rpm per second—to inevitable self-destruction.

All diesel engines precisely meter fuel into the engine's cylinders. Metering in the engines of today is achieved by several distinct means, the subject matter of most of the latter chapters in this book. When technicians refer to the metering of fuel into diesel engine cylinders two methods are used:

- **Injection rate** is the fuel quantity injected per **crank angle** degree. This term was primarily used to describe metering in older hydromechanical diesel engines. It was one of the trim factors when calibrating older injection pumps. Having said that, at least one current truck diesel engine original equipment manufacturer (OEM) has software that analyzes injection rate.
- *Duty cycle* is the term more commonly used today to describe metering in electronically controlled diesels. It is displayed as pulse width (PW) on electronic service tools (ESTs). PW is usually measured in milliseconds.

4. Distribution

In multicylinder engines:

a. The fuel system must be phased (sequenced) to deliver the fuel to each engine cylinder at the correct time and in the correct firing order. Correct fuel system phasing is required to balance the engine output. If in a given engine-operating range the ECM computes that ignition should occur at 3 degrees BTDC, then injected fuel must be delivered sufficiently in advance of that in each engine cylinder in the engine so that ignition really does occur at exactly 3 degrees BTDC.
b. The fuel must be dispensed to the correct area of the combustion chamber so that the fuel droplets combust at the appropriate time (spray dispersion). The position of the fuel injector is critical in ensuring that the injected fuel droplets are directed to the correct area of the combustion chamber in DI engines. For instance, if an injector was seated on double sealing washers (a not uncommon occurrence), the nozzle would be effectively too high in the cylinder; the result would be to inject fuel above the combustion bowl on a

Mexican hat piston crown, often causing it to condense on the piston headland and cylinder walls. Fuel exiting each nozzle orifice should be propelled toward the recess around the peak of the Mexican hat in the piston and vaporized before it actually comes into contact with any hardware. Fuel that contacts the relatively cooler hardware in the cylinder before reaching its ignition point may condense. Fuel that condenses on the piston boils off during combustion and results in erosion (pitting).

UNITS OF PRESSURE AND LINEAR MEASUREMENT

Although the tendency in North America is to speak and think using English (standard) unit values of measurement, most diesel engines and fuel systems engineered during the past two decades have used the metric system. In fact, most North American-built diesel fuel injection equipment has always used the metric system. As our manufacturers become increasingly multinational, the need to understand and work with the metric system becomes increasingly important. All this means is that the diesel technician today should be prepared to readily convert values from standard to metric systems and vice versa.

While a domestic diesel engine manufacturer engineers every system in an engine using nothing but the metric system, that same OEM then sets about converting all the specifications into standard values for publication in the technical service literature. This is changing. Some U.S. manufacturers are beginning to produce specifications exclusively in the metric system, which is a trend that could make our lives somewhat easier. That said, if you want to work on engines at this moment in time you should have some familiarity with both the metric and standard systems.

Manufacturers of older fuel injection testing apparatus often expressed pressure values in units of atmospheric pressure known as *atms.* This had the advantage of being easily converted either to the metric or standard systems. In Europe, units of *bar* (105 Newtons per square meter) are used to express units of atmosphere: One unit of bar is not precisely equivalent to one unit of atm. Most fuel injection test equipment is calibrated in atms and bar, but because of the international influence on our technology, we tend to use the unit *bar* more often than *atms* today. Consequently, in this book we commonly use bar to express metric pressure values. The following are some units of measurement used in this and following chapters that deal with fuel management:

1 **atm** = 14.7 **psi** = 101.3 **kPa** (kilopascals) = 1.013 bar

200 **atms** = 2,940 psi = 20.26 **MPa** (megapascals)

1 **micron** = millionth of a meter

1 **millimeter** = 0.0394 inches

1 **centimeter** = 0.394 inches

1 **meter** = 39.4 inches

Engine Management Objectives

The fuel system manages *engine* output. The overall objective of any engine whether it powers a lawnmower, a ship, or a highway truck is to transmit the force developed in its cylinders as smoothly and evenly as possible to the power takeoff device, which is usually a flywheel. To achieve this objective, the fuel system must manage fueling to complement the geometry of the engine powertrain components or, most specifically, that of the crankshaft.

Throw Leverage

First, it is important to understand the mechanical relationship of the **crank throw** (the crankshaft connecting rod journal) with the crankshaft axis (centerline); it is a **lever**. The greater the offset of a crank throw, the greater leverage it can exert. High-compression diesel engines use crank throws with greater offset than their gasoline-fueled counterparts so they can achieve higher compression ratios and apply greater leverage through the power stroke. When a piston is at **top dead center (TDC)**, the crank throw-to-**crank axis** (center point of crank main journal) angle is zero, so there is zero leverage. In this position no matter how much force is applied to the piston by cylinder pressure, none can be transmitted to the engine powertrain because of the zero leverage. However, the piston only has to move slightly beyond TDC and some throw leverage is achieved. As the piston continues to be driven downward through its stroke, the crank throw-to-connecting rod angle increases; this means that the leverage also increases. Given an exactly even amount of pressure applied to the piston on its downward travel, this would mean that the torque applied to the crankshaft would progressively increase consistent with increase in leverage.

Cylinder Pressure

Remembering that the objective is to *smoothly* and *evenly* transmit force developed in the engine cylinders to the engine flywheel, ideally the fuel system should be managed so that cylinder pressures peak when the crank throw angle offers low **mechanical advantage** (but not zero, which would be true when the throw and piston are at TDC) and progressively diminish as the crank throw is driven to a 90-degree angle with the connecting rod in which position mechanical advantage (leverage) peaks. With this in mind, most spark-ignited and diesel cycle engines are engineered to attempt to produce a peak cylinder pressure in the 10- to 20-degree **after top dead center (ATDC)** range at any phase of load or speed operation. Figure 18–2 shows how managing cylinder pressure to synchronize with crank throw leverage can achieve even and smooth application of torque.

Synchronizing Pressure with Leverage

To get this relationship between cylinder pressure (controlled by engine fuel system) and crank throw leverage (hard value dependent on specific engine rotational position) right, today's engines use computerized management. The ability of older hydromechanical fuel systems to achieve this was limited at best. It was based on strategies such as advancing either the ignition timing (SI engines) or the injection timing (CI engines) as engine speed increased and the real time available for combustion diminished. In fact, if the only objective was to get the absolute best fuel economy without worrying about tailpipe emissions, today's ECM-managed diesels could produce far better combustion efficiency than they are actually capable of.

Real Time and Metering

A modern engine management system has to manage engine fueling and the resulting

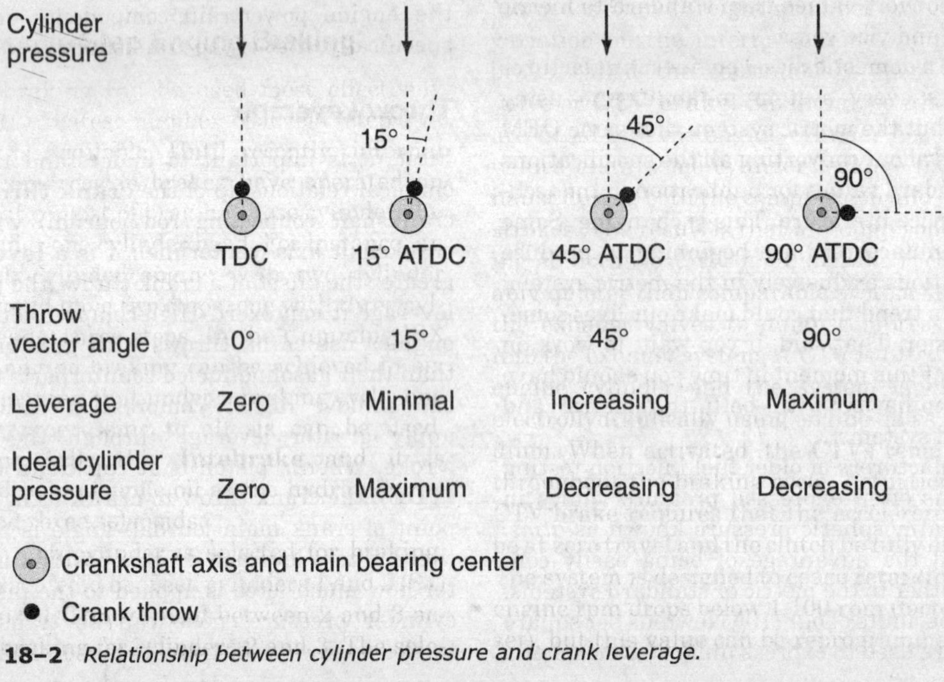

Figure 18–2 *Relationship between cylinder pressure and crank leverage.*

cylinder pressures within time dimensions so small that they are difficult to envisage. Think of this. A complete engine effective cycle is 720 degrees or two engine revolutions. At its maximum, a diesel fueling pulse is spread over less than 40 crank angle degrees of the 720 degrees. A four-stroke cycle diesel engine rotating at 2,000 rpm produces approximately 17 power strokes in each of its cylinders every second. It should come as no surprise that the pumping, metering, and injection activity required to manage the engine has to measured in milli- and microseconds.

REVIEW QUESTIONS

1. The precise control of injected fuel quantity is known as:
   a. Timing
   b. Metering
   c. Atomization
   d. Spray dispersion
2. Which two factors define the emitted droplet sizings from a nozzle assembly?
   a. Injection pressure and orifice size
   b. Load and rpm
   c. Accelerator position and governor
   d. Cylinder pressure and crank angle
3. Given unlimited fuel a diesel engine may accelerate at a rate up to:
   a. 60 rpm per second
   b. 120 rpm per second
   c. 120 rpm per minute
   d. 1,000 rpm per second
4. If a four-stroke cycle engine is run at 1,000 rpm, approximately how many power strokes per second will occur in any one of its cylinders?
   a. 1,000
   b. 8
   c. 17
   d. 32
5. In what position through the power stroke is crank throw leverage at its maximum?
   a. TDC
   b. 10 degrees BTDC rod-to-throw angle
   c. 10–20 degrees ATDC rod-to-throw angle
   d. 90-degree ATDC rod-to-throw angle
6. An ideal fuel system manages fueling to produce peak cylinder pressure at which location on the power stroke to produce optimum performance on any given engine cycle?
   a. TDC
   b. 15 degrees ATDC
   c. 90 degrees ATDC
   d. BDC
7. The specification for an injector nozzle opening pressure (NOP) is 350 bar. Convert this to a standard U.S. value.
   a. 35 psi
   b. 350 psi
   c. 3,500 psi
   d. 5,000 psi
8. When the term *pilot injection* is used to describe the initial shot of fuel in a multipulse injection system, which of the following best describes its function?
   a. It lubricates the cylinder walls.
   b. It doses the diesel particulate filter.
   c. It initiates the flame front for successive pulse(s).
   d. It cools the piston crown.
9. In which engine position does the crank throw provide the least amount of leverage?
   a. TDC
   b. 15 degrees ATDC
   c. 90 degrees ATDC
   d. 120 degrees BTDC
10. Which of the following diesel fuel injection terms is used to describe the amount of fuel injected per crank angle degree?
    a. Metering
    b. Injection rate
    c. Timing
    d. Phasing
11. *Technician A* states that an advantage of variable fuel delivery timing is better control over the noxious exhaust gas emissions. *Technician B* states that another reason for variable fuel delivery timing mechanisms on engines is to produce good fuel economy. Who is correct?
    a. Technician A only
    b. Technician B only
    c. Both A and B
    d. Neither A nor B
12. *Technician A* states that the term *metering* refers to the control of ignition timing in a diesel engine fuel system. *Technician B*

states that the term *injection rate* refers to the speed of the injection pulse in real time values. Who is correct?

a. Technician A only
b. Technician B only
c. Both A and B
d. Neither A nor B

13. *Technician A* states that the degree of atomization is controlled by managing injection pressure. *Technician B* states that atomized droplets are in a gaseous state. Who is correct?
a. Technician A only
b. Technician B only
c. Both A and B
d. Neither A nor B
14. What units are used to express *duty cycle* in electronically managed diesel engines?
a. Crank angle degrees
b. Microns
c. Milliseconds
d. Microseconds
15. What acronym is commonly used to display *duty cycle* on electronic service tools (ESTs)?
a. PW
b. PWM
c. DC
d. IRT
16. A post-2007 Caterpillar C15 produces peak injection pressures exceeding 34,000 psi. Convert this to units of metric bar and round to the nearest of the following values.
a. 1,847 bar
b. 2,378 bar
c. 2,672 bar
d. 3,048 bar
17. *Technician A* states that the greater the offset of a crank throw in relation to the crankshaft centerline, the greater the potential leverage. *Technician B* states that diesel engines almost always have greater throw offsets than gasoline-fueled engines. Who is correct?
a. Technician A only
b. Technician B only
c. Both A and B
d. Neither A nor B
18. *Technician A* states that some multipulse diesel fuel injection systems can break up an injection pulse into up to seven separate injection events in one cycle. *Technician B* states that multipulse diesel fuel injection can be hard on powertrain components because it produces spiked cylinder pressures. Who is correct?
a. Technician A only
b. Technician B only
c. Both A and B
d. Neither A nor B
19. When attempting to cold start a diesel engine in subzero (lower than –17°C) conditions, which of the following would provide the most help?
a. Larger than normal fuel droplet sizing
b. Smaller than normal fuel droplet sizing
c. Substantial advance to injection timing
d. Minimum fuel quantity
20. *Technician A* states that an injector seated on doubled seal washers causes fuel to be injected above the combustion bowl in an engine that uses Mexican hat piston crowns. *Technician B* states that an injector nozzle located either too high or too low in the cylinder of a direct-injected engine can cause the fuel to condense on the cylinder hardware before ignition occurs. Who is correct?
a. Technician A only
b. Technician B only
c. Both A and B
d. Neither A nor B

19

Injector Nozzles

INTRODUCTION

Diesel fuel is injected into an engine cylinder using an injector. When the diesel technician refers to an injector, the nozzle holder assembly is generally being described. A nozzle holder assembly may be a single function device we know as an integral injector assembly or a subcomponent of a multifunction assembly such as mechanical unit injectors (MUIs), electronic unit injectors (EUIs), and hydraulically actuated, electronic unit injectors (HEUIs). In a modern high-pressure fuel injection system, the velocity of fuel exiting a fuel injector can exceed the speed of sound. In describing nozzle assemblies and their operating principles in this chapter, we will reference:

- Integral injector assemblies: single function devices used to atomize fuel and may define the hydraulic opening pressure. They are usually connected to the metering and high-pressure pump by a line.
- Subcomponent nozzles: a subassembly incorporated into MUIs, EUIs, and HEUIs that are responsible for performing the metering, timing, and pumping functions of diesel fuel injection.

In general terms, we can describe injector nozzles as devices with the following functions:

- Valves that open and close to begin or end fuel injection
- Define **nozzle opening pressure (NOP)** (hydraulic nozzles only)
- Atomize fuel to the correct size for combustion

In this chapter we reference both **hydraulic injectors** (one original equipment manufacturer [OEM] refers to these as **mechanical injectors**) and **electrohydraulic injectors (EHIs)**. Until the 2007 model year, almost all the nozzles used in highway diesel engines were hydraulic nozzles. However, the reverse is true with the post-2007 model year, following which all but one diesel fuel system uses variations of EHI nozzles. Most off-highway diesel engines followed suit in 2008.

Functions

Injectors are classified by nozzle design. Nozzles are simple hydraulic or electrically switched valves whose functions are to open and close at the correct moment to atomize fuel as it exits the nozzle assembly, thereby injecting fuel into the engine cylinders. There are four basic types of injector nozzles, two of which are effectively obsolete in medium- and heavy-duty truck and bus engine applications. However, both the obsolete types are briefly described in this chapter for two reasons:

- Some older equipment with these nozzle types are still in service.
- They are still included in college curriculum and certification testing.

Older poppet and pintle nozzles are best suited to indirect-injected engine applications and therefore are not used in any current North American medium- and heavy-duty engines. Our main focus will be on:

- **Multi-orifii nozzles** used in most **direct-injection (DI)** diesel engines using

inline, port helix metering injection pumps, and as an integral subcomponent of mechanically and electronically controlled unit injection systems.
- Electrohydraulic injector (EHI) nozzles used in common rail (CR) and EUI fuel systems.

All of the injector nozzles covered in this chapter are closed-nozzle systems. Open nozzle injectors exist but they have very different operating principles; they are covered in later sections of this textbook.

NOP in Hydraulic Injectors

An injector nozzle is a hydraulically or electrically actuated switch. All older hydraulic injector nozzles were specified with a hard value opening pressure. This opening pressure was defined by the hydraulic pressure required to switch the device and open it: It could be adjusted by a technician but only when out of the engine. This trigger pressure required to open a nozzle valve was known as either nozzle opening pressure (NOP) or by the term **popping pressure**. In all hydraulic injector nozzles, the NOP value was defined by the mechanical spring tension of an internal nozzle spring. This spring tension loaded the nozzle valve onto its seat: It therefore defined specifically how much hydraulic pressure was required to unseat (and open) the valve. Most hydraulic injectors incorporated a means of adjusting the injector spring tension so that the NOP value could be set to a precise specification. The means of adjusting the spring tension that defined NOP was either internal shims or an adjusting screw and jamnut. The NOP value is always one of the first performance specifications to be evaluated when performing a **pop test**. Pop testing of injector nozzles is performed on a bench test fixture also known as a pop tester.

EHI Opening and Closing

In more recently introduced electrohydraulic injectors, NOP is a *soft* value. We use the term *soft value* because both opening and closing of the injector valve is controlled by the engine control module (ECM) managing the engine. This gives EHI injectors the advantage of opening at whatever pressure the ECM calculates is appropriate for combustion. Almost all post-2007 are using some form of EHI nozzles or injectors, so in this chapter we will take a close look at the advantages they provide. One term we will not use when describing EHI operation is NOP because it is not a fixed value.

POPPET NOZZLES

Poppet nozzles were used by Caterpillar in the past on **indirect injection (IDI)** engines using a precombustion chamber. They are perhaps the most simple of the hydraulic injector nozzles and as such were the least costly to manufacture. Poppet nozzles are not easily reconditioned; the technician is normally required to bench test the nozzle for the correct NOP value and observation of spray pattern and either reject or accept it for continued service.

Poppet nozzles use an outward or forward opening valve principle. This means that all the fuel pumped to the injector ultimately ends up in the engine cylinder, eliminating the necessity for the leakoff lines (return circuit) used with most other nozzle types. NOP values were low (compared to today's values), generally ranging between 500 and 1,800 psi (35 and 125 bar). For example, the Caterpillar poppet nozzles referenced in this section had NOPs of around 800 psi (55 bar).

Poppet nozzles operate as hydraulic switches to define the NOP and atomize the fuel. The nozzle spring loads the poppet valve onto its seat. Hydraulic line pressure (from the injection pump) acts on the sectional area of the seated upper portion of the poppet valve, and when the injection pump delivers pressure rise sufficient to unseat the poppet valve, fuel passes around the poppet to exit from the nozzle's single orifice.

When a precombustion chamber design is used, the engine cylinder clearance volume is generally too large to generate compression temperatures in cold weather sufficiently high to ignite the fuel charge (Figure 19–1). Glow plugs were often designed into IDI engine systems as a cold weather starting aid. It should be remembered by technicians who come across older precombustion chamber engines that the use of ether should be avoided whenever a glow plug circuit is energized. The result of ether usage in IDI engines is possible explosion damage to the prechamber.

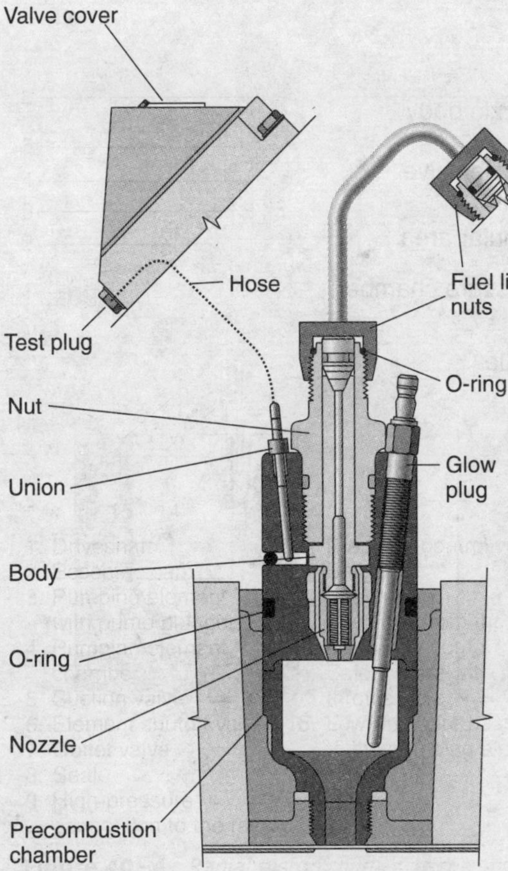

Figure 19–1 *Poppet injector and prechamber assembly.*

PINTLE NOZZLES

Like poppet nozzles, **pintle nozzles** are used mainly in IDI engine applications and seldom in truck/bus-sized compression ignition (CI) engines of the recent past. You can find them in older automobile diesel engines manufactured by GM, Ford, and VW. The pintle nozzle body is designed with a single exit orifice. The valve pintle protrudes through this single orifice even when in the fully open position. This means that the fuel exiting the nozzle is forced around the pintle, producing a conical (cone-shaped) spray pattern. The valve body and pintle valve were lapped together in manufacture often to a tolerance of around 2 μ (μ: **micron** or one millionth of a meter), which does not permit the components to be interchanged.

Pintle nozzles generally achieve better **atomization** (reduction of a liquid to minute droplets) than poppet nozzles and generally much greater service life as they are easily reconditioned. Nozzle opening pressure parameters range between 100 and 150 bar (1,470 and 2,200 psi) with peak system pressures seldom exceeding 500 bar (7,350 psi).

Action

Fuel is delivered to the pintle injector from the fuel injection pump by means of a high-pressure pipe. The nozzle valve is held in the closed position mechanically by spring pressure either acting directly on the pintle valve or transmitted by means of a **spindle** (shaft that transmits spring force to the nozzle valve). Fuel is ducted through the injector assembly to the pressure chamber within the nozzle; this permits the line pressure to act on the sectional area of the pintle valve exposed to the pressure chamber. Whenever the hydraulic pressure generated by the injection pump acting on the sectional area of the pintle valve exposed to the pressure chamber is sufficient to overcome the spring pressure loading the valve on its seat, it retracts, permitting fuel to flow past the seat and exit through the nozzle orifice. Fuel will continue to flow through the nozzle orifice for as long as hydraulic pressure exceeds the mechanical force represented by the injector spring. The length of a fueling pulse is determined by the fuel injection pump and when line pressure collapses, the spring reseats the nozzle valve, ending injection.

The pintle nozzle uses an inward opening valve principle requiring the use of **leakoff lines/pipes** to return fuel to the tank. The leakage takes place at the pintle valve-to-nozzle body clearance and tends to increase as these matched components age. The shaping of the spray pattern emitted from a pintle nozzle depends on the pintle design. Throttling nozzles are designed to emit less fuel at the beginning of the injection pulse and may be identified by their conically shaped pintle valves. Figure 19–2 shows the operation of a typical pintle nozzle closed, at NOP, and fully open. Figure 19–3 compares the external appearance of a pintle nozzle with the multi-orifii nozzles we will study next.

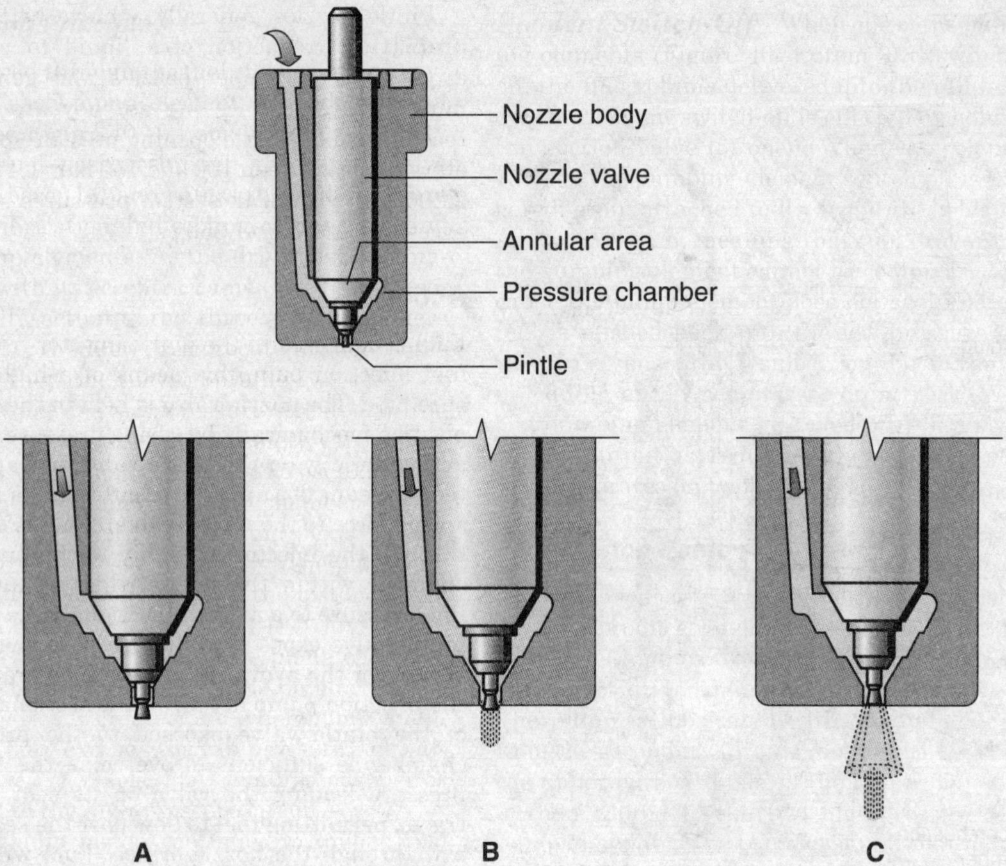

Figure 19–2 *Pintle nozzle operation: Terminology. A. Nozzle seated; B. Opening; C. Open. (Courtesy of Robert Bosch LLC)*

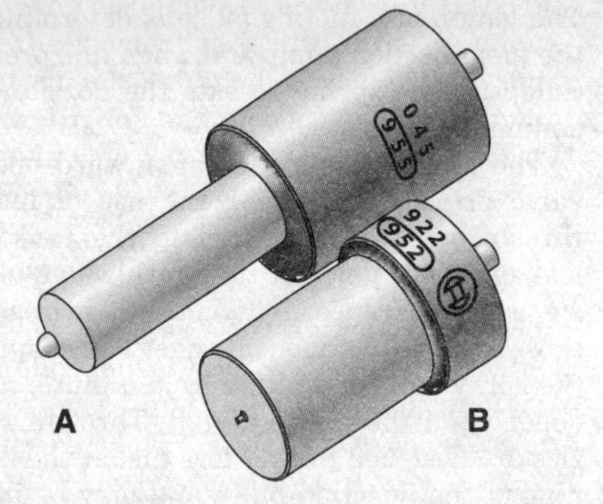

Figure 19–3 *A. Multi-orifii nozzle; B. Pintle nozzle.*

Nozzle Valve Motion Sensor

Some Bosch hydraulic injectors with pintle nozzles are equipped with **nozzle valve motion sensors (NVMS)**. The NVMS is integrated into the injector body and consists of a coil and pressure pin. Pressure pin movement through the coil winding changes the coil reluctance, inducing a signal voltage proportional to the valve's speed of movement. The NVMS enables the ECM to determine the precise moment of start of injection, which helps build the algorithms required for injection timing and exhaust gas recirculation (EGR) management. It is designed to detect and signal nozzle valve movement (at NOP) because the pressure pin moves with the nozzle valve. Examples of NVMS pintle injectors equipped with NVMS are shown in Chapter 34 (see Figure 34–2 and Figure 34–3) dealing with electronically controlled Bosch VE injection pumps.

Electromechanical Pintle

The gasoline fuel injectors (GFI) used in automobile engines use an electromechanical principle. So do some of the dosing injectors we use in diesel engines equipped with diesel particulate filters (DPFs). The function of a dosing injector is to deliver fuel to the DPF so it can effect a regeneration combustion event. The pintle-style nozzle valve in a typical diesel dosing injector is integral with an armature: When the solenoid is energized, the dosing injector opens in exactly the same way a GFI injector opens. When opened electrically, the fuel is forced around the pintle to exit through a single orifice: This produces a conical spray pulse. Diesel dosing pressures are usually around 250 psi (17 bar).

ORIFICE NOZZLES

As mentioned in the introduction to this chapter, until recently most current truck and bus diesel engines use closed hydraulic injector nozzles. Until 2007, an overwhelming majority of injector nozzles could be classified as multi-orifii. Multi-orifii nozzles appear in truck engines in traditional, pump-line-nozzle configurations, but in addition they were an integral subcomponent in EUIs and HEUIs. They were also used with electronic unit pump (EUP) fueled engines. **Orifice nozzles** may have a single orifice or several orifii, almost always the latter. Multi-orifii and the closely related EHI nozzles are required in most high-speed DI diesel engines because only these produce the necessary degree of atomization: droplets sized between 10 μ and 100 μ. Figure 19–4 shows a sectional view of a typical multi-orifii injector. It should be stressed that when EUIs or HEUIs use hydraulic nozzles as a subassembly, the operating principles are identical to those of the injector shown in Figure 19–4.

Droplet Sizing

The emitted droplet sizing from an injector nozzle depends on two factors:

- Flow area (size of nozzle orifii)
- Pressure (managed by injection pumping apparatus)

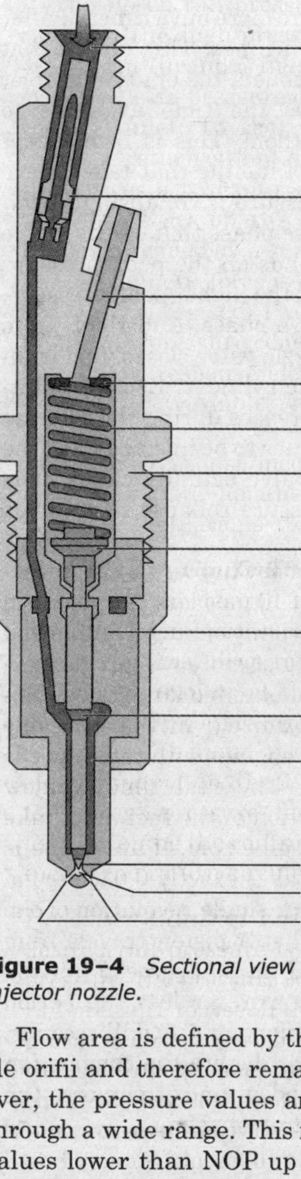

Figure 19–4 *Sectional view of a multi-orifii injector nozzle.*

Flow area is defined by the sizing of the nozzle orifii and therefore remains constant. However, the pressure values are designed to vary through a wide range. This range extends from values lower than NOP up to the **peak pressure** (the highest pressure attainable in a fuel injection system) value. Because the flow area remains constant, as pressure increases, the droplet sizing decreases. The more prolonged the injection pulse (that is, the longer the duration of the effective pump stroke), the higher the circuit pressure at the injection pump port opening (end of delivery) and, therefore, the smaller the atomized droplet sizing.

The reduction in droplet sizing that occurs as the fueling pulse is extended is generally favorable for the complete combustion of the fuel. Fuel droplets burn from the outside inward.

Larger droplets require more burn time; smaller droplets, shorter burn times. This means that as a fuel pulse is extended, the emitted droplet size reduces as does the time available to completely combust them. This is true except for the short period of fueling that takes place while the system pressure is collapsing to *end* injection. This collapse phase that occurs at the end of injection depends on the peak pressure attained in that cycle: the higher peak pressure, the longer the collapse phase. And right up to the point that the nozzle valve closes, fuel is injected into the engine cylinder. Because emitted droplet sizing increases during the collapse phase and fuel continues to be injected until the moment the nozzle valve actually closes, it is desirable for the pressure collapse to occur as rapidly as possible.

You might recall from Chapter 18 that atomized droplet sizings of 10 μ or less can result in erratic ignition or no ignition at all. Atomized droplet sizings of 100 μ or more will produce extended ignition lag and are too large to be completely combusted (oxidized) in the available time window of diesel engines with rated speeds of 1,600–2,600 rpm. The real time window within which combustion of the fuel must take place is defined by engine rpm. If an engine is run at 2,000 rpm, it takes exactly 0.030 second or 30 milliseconds for a single revolution of the engine; this means that a piston travels from top dead center (TDC) to 90 degrees after top dead center (ATDC), more or less the period available for combustion, in 7.5 milliseconds. Top end engine speeds dictate the combustion or burn duration window and, therefore, the maximum atomized droplet sizing.

Back Leakage

Orifice nozzles use an inward opening valve principle and that requires leakoff lines. The leakage is that occurring at the nozzle valve-to-nozzle body clearance and measured on the injector test bench as **back leakage**. Where orifice nozzles are used in diesel engines the following range of values is typical:

- NOP (opening pressures) range from 1,800 to 5,500 psi (120 to 400 bar).
- Peak system injection pressures range from two to ten times the NOP value.

Multi-orifii nozzles used in older hydromechanical engines typically have NOPs in the region of 200 bar or 3,000 psi. The multi-orifii nozzles used in EUIs, EUPs, and HEUIs generally have NOP values close to 5,000 psi (350 bar).

Action

The multi-orifii nozzle is usually dowel positioned in the injector assembly to ensure that the spray pattern is directed to a specific location of the engine cylinder. In most DI engines, the correct area is the toroidal recess (crater) around the Mexican hat peak in the piston (see Figure 6–1 in Chapter 6). Fuel from the injection pump mechanism is delivered to the nozzle holder and then ducted to an annular recess in the upper nozzle valve body. Either single or multiple fuel ducts extend from the annular recess to the pressure chamber, meaning that the pressure here will be the same as that in the high-pressure pipe.

The nozzle valve is loaded to a closed position on its seat, either directly by the injector spring or indirectly by spring pressure relayed to the nozzle valve by means of a spindle. This spring tension is adjustable by either a screw or by shims and will set the NOP value. When line pressure is driven upward by the injection pump, pressure increases in the nozzle pressure chamber. When this pressure is sufficient to overcome the mechanical force of the injector spring, the nozzle valve retracts, permitting the fuel to flow past the seat into the **sac** and exit the nozzle orifii. At the center of the nozzle valve seat, a single duct connects to the nozzle sac, which is a spherical chamber into which the nozzle orifii are machined. The sac hydraulically balances the fuel exiting the orifii, so that droplets start to exit from each orifice at approximately the same moment at the beginning of the fuel pulse.

The injection pulse continues for as long as the nozzle valve remains open. Depending on the specific means used to generate injection pressures, *maximum* fueling pulses can extend from 25 to 50 engine crank angle degrees. Minimum fueling pulses may be as short as 5 engine crank angle degrees. A fueling or injection pulse ends when the pressure is collapsed by the metering/pumping element and the pressure in the nozzle assembly drops to a value below NOP. Control of the injection pressure always occurs outside the nozzle assembly.

Nozzle Differential Ratio

Nozzle differential ratio describes the geometric relationship between the sectional area of the **nozzle seat** and that of the pressure chamber or valve shank. Nozzle valves are opened by hydraulic pressure acting on the sectional area of the valve subject to the pressure chamber; when this overcomes the spring pressure that loads them on their seat, they retract. However, the instant the nozzle valve unseats, hydraulic pressure is permitted to act over the whole sectional area of the nozzle valve. The whole sectional area of the nozzle valve is the sectional area of both the seat and that of the pressure chamber combined (Figure 19–5).

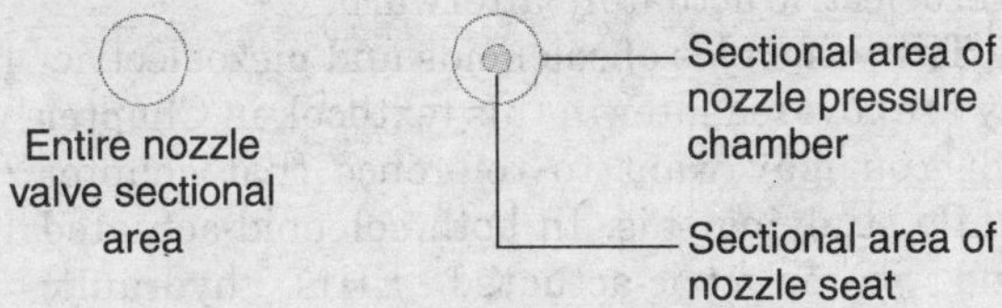

Figure 19–5 *Nozzle differential ratio: sectional areas of a nozzle valve.*

When pressure rise begins in the injection line circuit before the opening of the nozzle valve, the hydraulic circuit being acted on is closed (it is sealed at the nozzle seat). The instant the NOP value is achieved, the nozzle valve unseats and opens the circuit. This must result in a drop in line pressure. However, because at the moment the nozzle valve opens the sectional area over which the hydraulic pressure acts is increased by that of the seat sectional area, less pressure is now required to hold the nozzle open. Nozzle differential ratio must be sufficient to prevent nozzle closure when this pressure drop occurs. After NOP, fuel passing around the seat fills the sac and pressure rise resumes because of the restriction represented by the minute sizing of the nozzle orifii. Nozzle differential ratio means that nozzle closure at the end of injection occurs at a value somewhat below the NOP value. It also helps define the specific **residual line pressure** (pressure that dead volume fuel is retained at in the high-pressure pipe) value.

The bottom line is that more pressure is required to unseat a nozzle valve than that required to hold it off its seat due to nozzle differential ratio. This means that any hydraulic injector (including those used in conjunction with EUIs, EUPs, and HEUIs) will have a nozzle closing pressure somewhere around 20% lower than its specified NOP.

Nozzle Holders/Injectors

Injectors are simply mounting devices for hydraulic nozzles. They come in many different shapes and sizes for a variety of reasons; for instance, long stem nozzles are easier to cool. Older hydraulic injectors tended to be of the high spring design. **High spring injectors** located the spring in the upper portion of the holder and relayed the spring tension to the nozzle value by means of a spindle. Spring tension was altered directly by an adjusting screw and setting the NOP value on the pop tester was a quick and easy procedure. A disadvantage of the high spring design was at NOP, the nozzle valve slammed open until it was mechanically prevented from further inboard travel. Because of its high opening velocity, the spindle was driven into the spring, which rebounded sufficiently to hammer the nozzle valve from its open position back into the seat, interrupting the injection pulse.

More recent hydraulic injector nozzles have tended to use a low spring design that eliminates the spindle and thus reduces the mass of moving parts; this has minimized rebound interference of the injection pulse. However, **low spring injectors** generally use shims acting on the spring to define the NOP value, which extends the time required to set the NOP value on the test bench.

Pencil-type Injector Nozzles

Pencil injector nozzles are a type of multi-orifii injector nozzle and, in fact, share common operating principles with a couple of small exceptions. They are seldom found in any current truck engine applications. Caterpillar used them in some of its older hydromechanical engines. The pencil nozzle assembly is cylindrical and has the approximate appearance of a pencil. Within the nozzle body, the nozzle valve extends through nearly the full height of the injector assembly, and the injector spring acts directly on top of the nozzle valve shaft. Adjustment is by means of an adjusting screw and locknut located at the top of the assembly. Some pencil nozzles have no leakoff lines; fuel that bleeds by the nozzle valve during injection is accumulated in a chamber above the valve, and pressure equalization with line pressure occurs after nozzle closure. Pencil nozzles are most often damaged in the process of removal and service. When making any adjustments, use the recommended mounting fixture and observe the specified torque procedure.

VCO Nozzles

Valve closes orifice (VCO) nozzles have eliminated the sac. The function of the sac in the nozzle is to provide balanced fuel dispersal, which is especially important in keeping the ignition lag time consistent. However, at the completion of the injection pulse, the volume of fuel in the sac was essentially wasted fuel that added to hydrocarbon (HC) emission. At the instant of nozzle closure, the sac and the nozzle orifii would contain fuel that would be vaporized due to the heat of combustion but at best, only partially combusted. Current injector nozzle designs have either substantially reduced the sac volume or eliminated it entirely. A true VCO nozzle has orifii that extend directly from the seat. VCO nozzles are most often used on electronically managed injection systems that use high NOP values, which can to some extent compensate for the compromising of the balanced fuel dispersal offered by nozzle sacs. Figure 19–6 shows a VCO nozzle design.

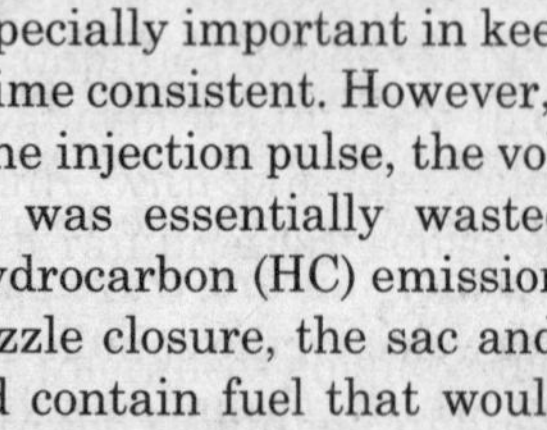

Figure 19–6 *Valve closes orifice (VCO) nozzle. (Courtesy of Robert Bosch LLC)*

ELECTROHYDRAULIC INJECTORS

Electrohydraulic injectors (EHIs) were introduced in the late 1990s and used on the first generation of CR, diesel fuel injection systems such as those used in Mack Trucks light-duty engines, Cummins ISB, and the GM/Isuzu family of engines. The first generation of EHIs were similar to multi-orifii hydraulic injectors in most ways but were controlled electrically by solenoids rather than hydraulically. More recently, **piezoelectric actuators** have been introduced as the need for more precise management of combustion within the engine cylinder has increased. The operating principles of solenoid-actuated and piezo-actuated injectors are almost identical. The **piezo injectors** (most OEMs use this term) respond faster to control inputs and require less current than solenoid-actuated versions. We will take a look at the general operating principles first then examine piezoelectric actuators afterward.

The principles of solenoids and piezoelectricity are covered later in this textbook in Chapter 28: You may want to reference that chapter while studying this. In both solenoid-actuated and piezoelectric-actuated EHIs, hydraulic pressure in the rail is used to effect the opening of the nozzle when the unit is switched. The EHI (Figure 19–7) can be subdivided as follows:

- Nozzle assembly
- Hydraulic servo system
- Actuator valve (either a solenoid or piezoelectric actuator)

EHI Operation

Referencing Figure 19–7, fuel at rail pressure is supplied to the high-pressure connection (4), to the nozzle through the fuel duct (10), and to the control chamber (8) through the feed orifice (7). The control chamber is connected to the fuel return (1) via a bleed orifice (6) that is opened by the actuator valve. With the bleed orifice closed, hydraulic force acts on the valve control plunger (9), exceeding that at the nozzle-needle pressure chamber (located between the shank and the needle of the nozzle valve). Although the hydraulic pressure values acting on the top of the nozzle valve and that in the lower pressure chamber are identical, the sectional area at the

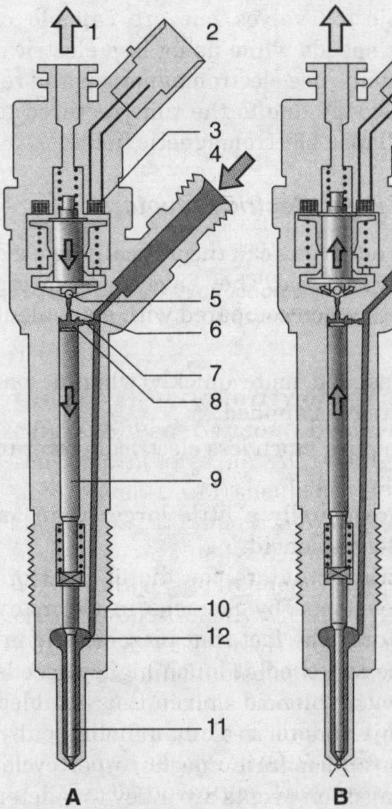

INJECTOR (schematic)
A. Injector closed (at-rest status)
B. Injector opened (injection)
1. Fuel return
2. Electrical connection
3. Triggering element (solenoid valve)
4. Fuel inlet (high pressure) from the rail
5. Valve ball
6. Bleed orifice
7. Feed orifice
8. Valve control chamber upper pressure field
9. Valve control plunger
10. Feed passage to the nozzle
11. Nozzle needle
12. Lower pressure field

Figure 19–7 *Electro-hydraulic injector. (Courtesy of Robert Bosch LLC)*

top of the nozzle valve is greater. As a result, the nozzle needle is loaded into its seated position, meaning that the injector is closed.

When an ECM signal triggers the actuator valve, the bleed orifice opens. This immediately drops the control-chamber pressure and, as a result, the hydraulic pressure acting on the top of the nozzle valve (11) also drops. When hydraulic force acting on top of the nozzle valve drops below the force on the nozzle-needle pressure shoulder, the nozzle valve retracts and allows fuel to pass around the seat to be injected through orifii into the combustion chamber. The hydraulic assist and amplification factor are required in this system because the forces necessary for rapid nozzle valve opening cannot be directly generated by the actuator valve alone. Fuel used as hydraulic media to open the nozzle valve is in addition to the injected fuel quantity, so this excess fuel is routed back to the tank. In addition to this fuel, some leak-by fuel losses occur at the nozzle valve-to-body clearance and the valve-plunger guide clearance.

EHI Operating Phases

When the engine is stopped, all injector nozzles are closed, meaning that their nozzle valves are loaded onto their seats by spring pressure. In a running engine, injector operation takes place in three phases.

Injector Closed In the at-rest state, the actuator valve is not energized and, therefore, the nozzle valve is loaded onto its seat by the injector spring combined with hydraulic pressure (from the rail) acting on the sectional area of the valve control plunger. With the bleed orifice closed, the solenoid valve spring forces the armature ball check onto the bleed-orifice seat. Rail pressure builds in the injector control chamber, but identical pressure will be present in the nozzle pressure chamber. Given equal pressure acting on the larger sectional area of the nozzle control plunger (which is mechanically connected to the nozzle valve) and in the nozzle pressure chamber, this pressure and the force of the nozzle spring combine to load the nozzle valve on its seat, holding the injector closed.

Nozzle Opening When the actuator valve is energized by the ECM injector driver, it is actuated at spiked voltage value typically around 100 V-DC: Current draw varies depending on whether a solenoid or piezoelectric actuator is used. Force exerted by the actuator valve overcomes that of the valve spring, and the bleed orifice opens. As the bleed orifice opens, fuel flows from the valve-control chamber into the cavity above it and out to the return circuit. This collapses the hydraulic pressure acting on the valve control plunger that was helping to hold the nozzle valve closed. Now, pressure in the valve-control chamber (upper pressure field) is much lower than that in the nozzle pressure chamber (lower pressure field) below it that is maintained at the rail pressure. The result is the force that was holding the nozzle valve closed collapses and the nozzle valve opens, beginning the injection pulse.

The nozzle needle opening velocity is determined by the difference in the flow rate through the bleed and feed orifices. When the actuator control plunger contacts its upper stop, it is cushioned by fuel generated by flow between the bleed and feed orifices. When the injector nozzle valve has fully opened, fuel is injected into the combustion chamber at a pressure very close to that in the fuel rail.

Nozzle Closing When the actuator valve is de-energized by the ECM injector driver, its spring forces the valve downward and the check ball closes the bleed orifice. The closing of the bleed orifice creates pressure buildup in the upper control chamber via the input from the feed orifice. This pressure should be the same as that in the rail and now it exerts an increased force on the nozzle valve control plunger through its end face. This force, combined with that of the nozzle spring, exceeds the hydraulic force acting on the nozzle valve sectional area and the nozzle valve closes, ending injection. Nozzle valve closing velocity is determined by the flow through the feed orifice. The injection pulse ceases the instant the nozzle valve seats.

Piezoelectric Injectors

Piezoelectric actuators are based on reversibility of piezoelectric effect. For a full explanation of piezoelectricity, reference Chapter 28; we only briefly describe the principles here. When certain crystals are subjected to pressure, a voltage is produced in a process known as **piezoelectricity**. These same crystals are capable of reversing the process, so when we subject them to a voltage, the crystal bonds change almost instantly, resulting in either expansion or contraction. This change of shape produces a powerful force that makes it possible for them to be used as actuators. Wafered layers of piezo crystals are stacked within the actuator body. As such, they function much like solenoids in producing a mechanical response to an electrical signal. They can therefore be used for actuating injector valves but are capable of much higher speeds while using less electrical energy. Solenoids use electromagnetism and respond more slowly due to the time required to build and collapse electromagnetic fields.

Features of Piezoelectric Actuators

Piezoelectric actuators can theoretically be used to replace any solenoid. They have the following characteristics when compared with equivalent solenoids:

- They respond more quickly when a control voltage is applied.
- They require much less electrical power to operate.
- They are usually a little larger than an equivalent solenoid.

Piezoelectric actuators for diesel injectors were first developed by Siemens in Germany for CR injectors. The fact that piezoelectric injectors can be switched at much higher speeds (compared with solenoid valves) has enabled multipulse injectors to break up a fueling pulse into up to seven separate injections per cycle. Multipulse injection events are a key to quieter diesel engines that emit less pollutants.

Diesel engine fuel systems are now developing smaller-sized piezoelectric actuators that are integrated directly into the injector valve shaft. This is allowing current CR systems to produce rail pressures up to 26,000 psi, and rail pressures up to 35,000 psi will be possible by 2010, near the equivalent of diesel solenoid-actuated EUI systems.

Piezoelectric Injector Operation

A piezo actuator is constructed of several hundred piezo crystals in stacked wafers. These piezo crystal wafer stacks have the ability not only to operate nozzle needles against the extremely high diesel fuel injection pressures, but they do so quickly and repetitively. The high frequency at which the injectors can be switched makes it possible to reduce the intervals between each injection pulse during a single power stroke. This capability can divide the total fuel injected in one cycle into up to seven separate injection events. Piezoelectric actuator technology has been a key to improving fuel economy, emissions, and noise levels in the latest generation of diesel engines. It allows the ECM precise control over combustion and cylinder pressure.

Nozzle Hole Geometry

CR injectors use both sac-chamber nozzles and what are generally referred to as VCO nozzles. Both sac and VCO nozzles were covered a little earlier in this chapter but VCO nozzles tend to be required in most current engine families. Current EHI injectors use a 4 mm to 6 mm nozzle valve diameter.

Both sac-hole and seat-hole nozzles have input edges of each orifice rounded (radiused) by hydroerosive (HE) machining. HE machining helps prevent edge wear caused by the abrasive particles in the fuel and carbon coking that can reduce flow and disrupt the spray geometry. For reasons of structural strength, the nozzle tip is conically shaped.

Summary of EHIs

Because of their soft opening and closing pressures and their capability of providing high injection pressures at low engine rpm, the use of CR fuel systems and EHIs, especially those with piezoelectric actuators, have become commonplace. The advantages of EHIs include:

- They can be integrated as a subcomponent of an EUI to provide it with soft (computer-controlled) opening and closing values.
- They are designed to open and close precisely when the ECM electrically switches them so they are not limited by factors such as cam profile.
- They produce injection pressure values that remain relatively consistent through a fueling pulse.
- They close almost instantly, eliminating the collapse phase that disadvantages comparable hydraulic injectors.
- They may use either solenoid or piezoelectric actuators. Piezoelectric actuators respond much more rapidly to control signals and are used in more recent fuel systems requiring multipulse injection.

NOZZLE TESTING AND RECONDITIONING

Injector nozzles are probably subjected to more abuse than any other engine component. They are exposed to temperature peaks of 2,550°F (1,400°C) outside and to pressures exceeding 30,000 psi (2,000 atms) internally. Manufacturers' service intervals are seldom respected and maintenance often is overlooked until a problem occurs. In today's world, the diesel technician is seldom required to fully recondition nozzle assemblies due to the high cost of labor and service tooling versus the relatively low cost of nozzle replacement. However, because most manufacturers can determine injector efficiency by electronically running multicylinder cutout tests (these can accurately measure cylinder balance), it is usually possible to identify problems before they can create serious engine damage.

Consult OEM service literature before attempting to service injectors. In cases where hydraulic injector nozzles are a subcomponent of EUIs and HEUIs, OEMs require that a problem injector be identified and replaced rather than reconditioned. However, where integral injectors are used in pump-line-nozzle (PLN) and in EUP fuel systems, most manufacturers include the injectors in preventative maintenance schedules. Servicing this type of nozzle requires a simple bench test fixture (or pop tester) and ultrasonic bath. Servicing a set of injectors is a safe and easy procedure, but some precautions are required.

Removal of Injectors from the Cylinder Head

The OEM-recommended method of removing an injector from a cylinder head must be adhered to and failure to do so can result in failed injectors. Before attempting to remove injectors, thoroughly clean the surrounding area, remembering that any dirt around the injector bore can end up in the engine cylinder below it. It is also good practice to drain high-pressure lines when removing the pipe fittings. Thin-bodied, pencil-type nozzles are vulnerable to any side load force and the correct pullers should always be used.

Additionally, whenever pencil-type nozzles are removed from a cylinder head, they must be tested in a test fixture before reinstallation because they are so easily damaged during the removal process.

Many injector assemblies are flanged, in which case the hold-down fasteners and clamps should first be removed and the injector levered out using an injector heel bar. The injector heel bar is 8″ to 12″ (19–30 cm) in length to prevent the application of excessive force to the injector flange. Where cylindrical injectors are used, a slide hammer and puller nut that fits to the high-pressure inlet of the injector should be used to pull the injector. With certain cylindrical hydraulic injectors, the high-pressure delivery pipe fits to a recess in the injector through the cylinder head; the high-pressure pipe must be backed away from the cylinder head before attempting to remove the injector, or both the injector and the seating nipple on the high-pressure pipe will be damaged.

When removing injectors from cylinder heads, make a practice of removing the injector nozzle washer at the same time; these have an inside diameter (id) less than the cylinder injector aperture id, so providing the piston is not at TDC, they can usually be removed either by inserting an O-ring pick or, in cases of difficult-to-remove washers, the tapered end of the injector heel bar and jamming it into the washer. It is not good practice to reuse injector washers. Both steel and copper washers harden in service. The copper washers can be annealed by heating to the point they shimmer green, then quenched with water. Steel washers are a poor reuse risk and no attempt should be made to anneal them.

When the injector has been removed, use plastic caps to seal the cylinder head injector bores, the injector inlet, and high-pressure pipe nipples. Ensure that a set of injectors is marked by cylinder number and properly protected in an injector tray; wrapping the injectors in shop rags will do if no injector tray is available.

Seized Injectors

Occasionally, an injector may seize in a cylinder head and defy any normal means used to remove it. Once the injector becomes damaged (for instance, the line threads that fit the injector to the puller are destroyed), the technician has no option but to remove the cylinder head. Never risk damaging a cylinder head attempting to remove an injector. With the cylinder head removed, a seized injector can usually be removed with a punch and hammer.

Testing

OEM specifications and procedures should be consulted before placing the nozzle to be tested in the nozzle test fixture. A typical procedure for testing nozzle assemblies is as follows:

Assessment Procedure

Perform the following tests in sequence. At the first failure, cease following the sequence and proceed to the next section.

1. Clean the injector externally with a brass wire brush (Figure 19–8).
2. Locate the manufacturer's test specifications.
3. Mount the injector in the bench test fixture. Build pressure slowly using the pump arm, watching for external leakage.
4. Bench test the NOP value and record. Use three discharge pulses and record the average value. Sticking or a variation in NOP value that exceeds 10 bar (150 psi) fails the nozzle.
5. Test **forward leakage** by charging to 10 bar (150 psi) below the NOP value and holding the gauge pressure at that value while observing the nozzle. Any leakage evident at the tip orifii fails the nozzle.
6. Check back leakage by observing pressure drop from a value 10 bar (150 psi) below NOP. Pressure drop values should typically be in the range of 50–70 bar (700–1,000 psi) over a 10-second test period. Pressure drop that is less than OEM specification indicates too little valve-to-body clearance (possibly caused by valve-to-body mismatch). A rapid pressure drop exceeding the OEM specification indicates excessive nozzle valve-to-body clearance, which is a condition usually caused by wear.
7. Once again, actuate the bench fixture pump arm and observe the nozzle spray

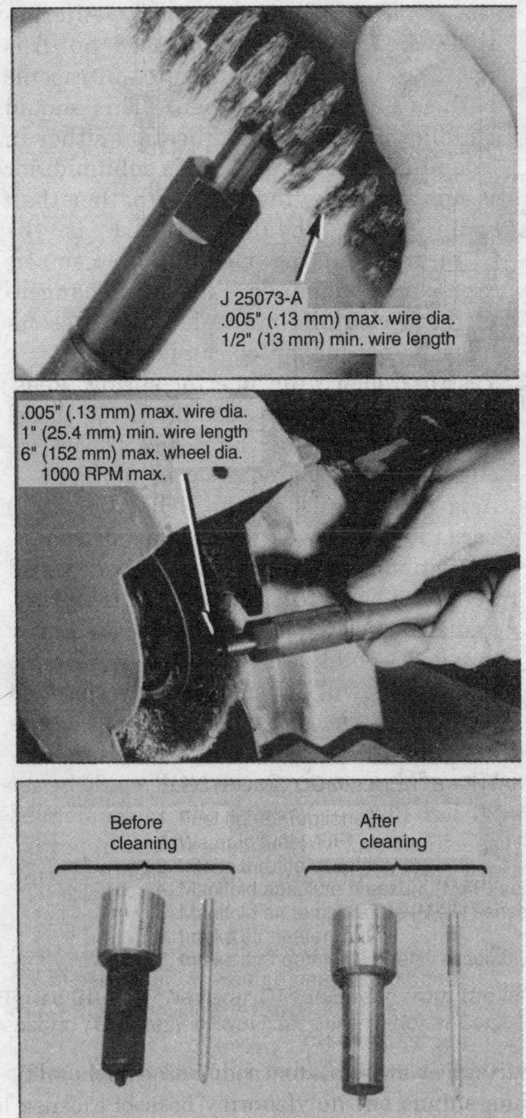

Figure 19–8 *Cleaning the exterior of the nozzle assembly. Top: Brass wire brush; Middle: Brass wire wheel; Bottom: Before and after. (Courtesy of Mack Inc.)*

pattern, checking for orifii irregularity. Ignore nozzle **chatter** (rapid pulsing of the nozzle valve); this can be regarded as a test bench phenomenon due to *slow* rate of pressure rise and inability to drive pressure much above NOP. In some modern injectors, the intensity of chatter noise is reduced due to the use of double seats.

Reconditioning

The function of the assessment procedure is to determine whether the nozzle should be reconditioned or placed back in service. Reconditioning in the service shop of today seldom means regrinding of the valve nozzle seat or reaming nozzle orifii (Figure 19–9) due to high labor and tooling costs in comparison with the cost of nozzle replacement. The following simple procedure probably summarizes what most operations interpret as reconditioning:

1. A nozzle that fails the inspection procedure should first be removed from the injector. Disassemble the injector and remove the nozzle, separating the nozzle valve from the nozzle body. Ensure that nozzle valves remain matched with the nozzle bodies they are removed from because they are not interchangeable.

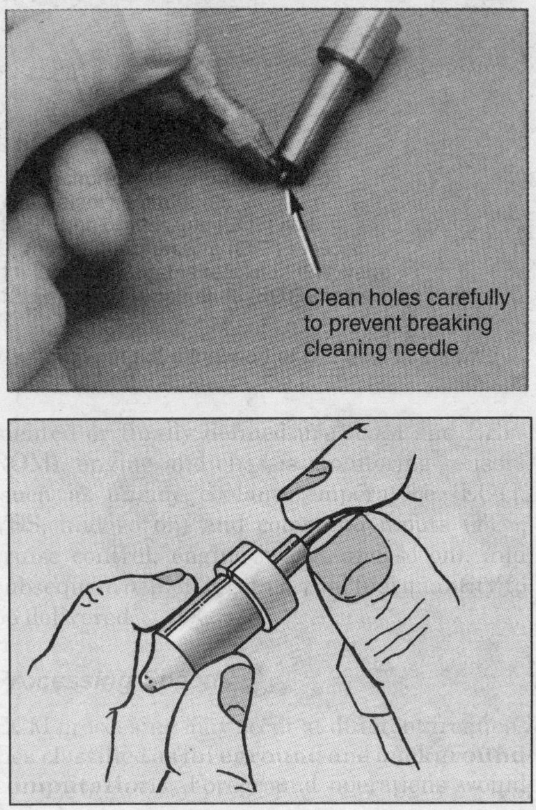

Figure 19–9 *Top: Cleaning nozzle orifii; Bottom: Performing valve-to-nozzle slide test. (Courtesy of Mack Inc.)*

2. Prepare the nozzle valve and body for ultrasonic cleaning (Figure 19–10). An ultrasonic bath filled with clean soapy water should be used. In other words, change the cleaning solution regularly. Alternatively diesel test oil can be used, but it is not as effective. A 5-minute, ultrasonic bath is generally sufficient. This should be followed by dehumidification either in a dehumidifier or an oven. If a dehumidifier is not available, fuel test oil rather than soapy water should be used as the ultrasonic soaking medium. Remember, lapped components are not interchangeable so if cleaning multiple nozzle assemblies, do not mismatch components.

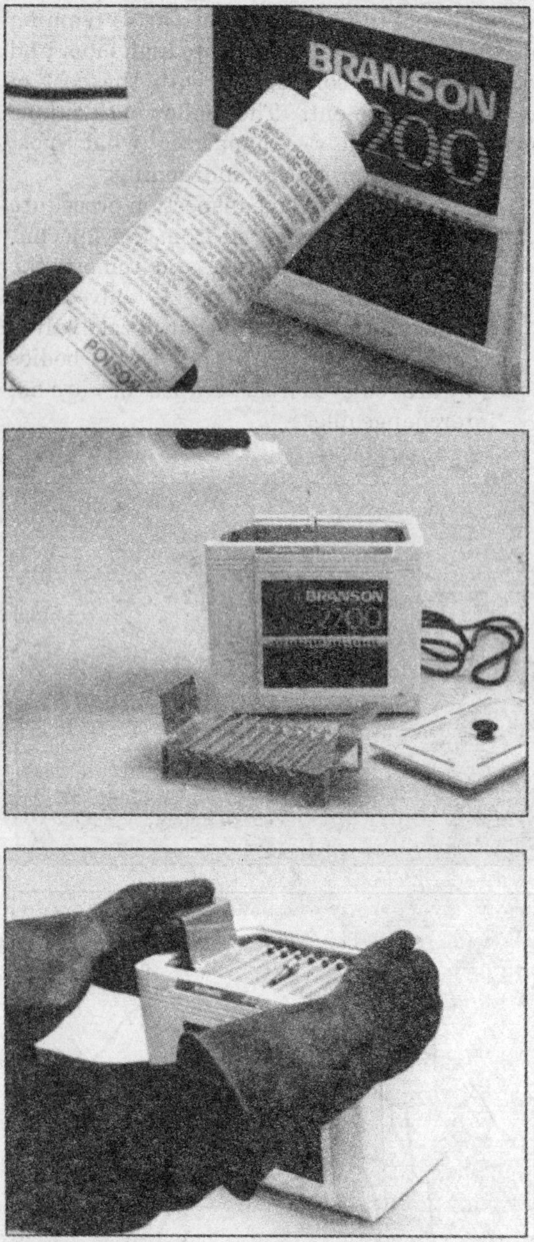

Figure 19–10 *Ultrasonic bath procedure. Top: Follow instructions; Middle: Fill the tank with solution; Bottom: Lower the tray with the nozzle assemblies into the tank. (Courtesy of Mack Inc.)*

3. Test or replace the injector spring. Reassemble the injector assembly using diesel fuel as an assembly lubricant. Never use engine oil or any grease-based lubricant.
4. As a general rule, set the NOP value at 10 bar (150 psi) above the recommended specification if no parts have been replaced. Add another 5 bar (75 psi) if using a new spring, and a further 5 bar (75 psi) if using a new nozzle assembly. After 100 engine hours, the nozzle will function at the specified NOP value. In other words, when reassembling an injector with a new nozzle assembly and a new spring, the NOP would be set 20 bar (300 psi) above the recommended specification. Always consult the appropriate service literature for the NOP setting procedure.
    - Screw type: Back off the locknut, then turn the adjusting screw clockwise (CW) to increase or counterclockwise (CCW) to decrease NOP.
    - Shim type: Add or subtract shims acting on the injector spring.
5. Check forward leakage.
6. Check back leakage.
7. Check back leakage pipe restriction.
8. Observe spray pattern and check for nozzle orifii irregularity, preferably using a template.

A critical function of the nozzle washer (spacer) is to ensure optimum droplet penetration into the engine cylinder. It is not a mere joint gasket. Spacer thickness is critical and determines the nozzle protrusion dimension. Use of an incorrectly sized washer, two washers, or no washer can result in rapid engine failures.

Reinstallation of Injectors

The injector bore should be cleaned before installation using a bore reamer if necessary to remove carbon deposits. Blow out the injector bore using an air nozzle. Where injector sleeves are used, service/replace as per OEM instructions. Most diesel engine OEMs prefer that injectors be installed dry; lubricants such as Never-Seize™ can hinder the ability of the injector to transfer heat to the cylinder head. It is good practice to turn the engine over and pump fuel through the high-pressure pipe before connecting it to the injector; this will purge the pipe of air and possible debris that may have intruded into the line while it was disassembled.

Always torque the nut on the high-pressure pipe with a line socket; failure to do this can result in damage to both the line nut and the nipple and seat.

A Reason Not to Recondition Nozzles

Earlier we said that it is not good practice to recondition hydraulic injector nozzles because even with the best shop practice procedures and equipment, reconditioned nozzles tend to have much shorter service life than new nozzles—and may also produce poor performance. Here is one reason why. When carbon residues collect on nozzle orifii, they initially have a soot-like appearance. However, once subjected to the inevitable high-temperature/cool-off cycles of an engine cylinder, the carbon begins to crystallize, and in the process of doing so, leaches iron from the nozzle orifice. This process is known as coking. When a nozzle with established carbon deposits is ultrasonically cleaned, crystallized carbon is vibrated free, taking with it some of the base material at the nozzle orifice. This alters the orifice geometry and therefore the injection spray pattern.

NOZZLE TROUBLESHOOTING

While there are some general guidelines to injector nozzle troubleshooting, the nature of problems observed depends on which of the four general types of nozzle is used with a system. Table 19–1 outlines some typical engine symptoms that result from certain types of nozzle malfunctions.

High-pressure Pipes

In most hydromechanical fuel injection systems and some electronically managed systems, fuel is delivered from the pumping elements in the fuel injection pump to the nozzle holders by high-pressure pipes. These are subject to very high-pressure values and are manufactured from alloyed steels with substantial wall thickness. Pressure wave reflection caused by pressure fluctuations and high-speed switching must also be accommodated. Failures are caused by pipe erosion, **cavitation** (vapor bubble collapse), and technician abuse-related problems such as overtorquing and bending.

High-pressure pipes should never be repair welded or braised when worn through due to the extreme pressures they are required to contain. Weld or braising repairs can penetrate into the worn-away area and reduce the flow area within the high-pressure pipe. They may also completely plug the pipe, and when a pump is capable of generating pressures exceeding 30,000 psi, something has to let go. Perhaps the most important reason that high-pressure pipes should never be repair welded is that if the weld is unseated by the pipe pressure in operation, it can be propelled with substantial force and have the potential to injure anyone in the vicinity.

In multicylinder engines, it is critical that the high-pressure pipes be of equal length and identical bores. Small variations in length or bore size can substantially alter injection timing and unbalance engine performance. High-pressure pipe nuts at both the injector and injection pump ends should always be torqued to specification. When attempting to locate a cylinder misfire, technicians sequentially crack the high-pressure pipe nuts in a procedure known as "shorting out injectors." In practice, where no torque wrench is used, the result is usually overtorquing—this has the effect of crushing the sealing nipple of the high-pressure pipe and reducing flow through the line. The nipple sealing seats on high-pressure pipes should be inspected for ridging and aperture (bore) size at each removal. It should also be noted that this condition can result in low power and/or unbalanced fuel delivery.

TABLE 19–1 *Nozzle-Related Engine Performance Problems*

Engine Symptom	*Nozzle Malfunction*	*Possible Nozzle-related Causes*
• Smoking • Higher fuel consumption • White smoke at startup • Slightly higher power output	• NOP is slightly below specification: usually within 70% of OEM spec • Enlarged nozzle orifii	• Fatigued nozzle valve spring • Erosion of nozzle orifices caused by wear or ultrasonic reconditioning
• Misfire • Smoking • Overfueling	• NOP more than 70% below OEM-specified value • Severely eroded nozzle orifii	• Fatigued injector spring • Severely enlarged nozzle orifii caused by erosion, carbon leeching, or physical damage
• Misfire • Light smoking • Erosion of center piston crown	• Nozzle valve not seating • Injector dribble	• Cylinder gas leaking into the nozzle • Carbon nozzle seat and tip
• Severe smoking • Misfire at all loads and speeds • Fuel in the crankcase • Combustion gases in the fuel system • Overfueling	• Nozzle stuck open	• Broken injector spring • Coked (carboned) nozzle valve • Seized nozzle valve • Short-circuited nozzle armature (electrohydraulic nozzles)
• Slightly low power • Intermittent engine miss	• High nozzle back leakage caused by excessive clearance of nozzle valve shank to nozzle body	• Abrasive wear of nozzle valve shank caused by fuel contamination • Mismatched nozzle valve to nozzle body after reassembly
• Erratic engine surge below requested rpm • Misfire	• Nozzle seizure, higher NOPs, nonpositive nozzle closure	• Almost always caused by mismatch of nozzle valve to nozzle body during reassembly
• Low power • High fuel consumption • Rough operation • White smoke at startup	• Plugged nozzle orifice/orifii	• High sulfur fuel • Contaminants in fuel such as ATF or other oils • Prolonged cold operation
• Low power • Black smoke emission • Poor pickup	• In EUI, EUP, and HEUI systems, internal leakage in the control cartridge	• Nozzle functions OK: Replace the control cartridge or the EUI, EUP, or HEUI as per OEM
• Engine misfire • Piston damage • White smoke at startup • Black smoke under load	• Nozzle tip blown away	• Usually caused by water ingression into high-pressure injection circuit

REVIEW QUESTIONS

1. Which type of injection nozzle would be found on most DI truck diesel engines before 2007?
   a. Poppet
   b. Pintle
   c. Throttling
   d. Multi-orifii
2. When injector back leakage is bench checked, which of the following is being tested?
   a. NOP
   b. Nozzle seat seal
   c. Valve-to-body clearance
   d. Injector spring
3. When nozzle forward leakage is bench checked, which of the following is being tested?

a. NOP
b. Nozzle seat seal
c. Valve-to-body clearance
d. Injector spring

4. Setting the injector NOP value is accomplished by:
a. Adding or subtracting shims
b. An adjusting screw
c. Both a and b
d. Neither a nor b

5. An injector spring that is fatigued would likely produce which of the following conditions?
a. Higher NOP
b. Higher NOP and retarded injection timing
c. Nozzle chatter
d. Lower NOP and advanced injection timing

6. Replacing a single high-pressure injection pipe on a multicylinder engine by one of greater length would likely have what effect on injection timing in the affected cylinder?
a. Advance
b. Retard
c. None
d. Decrease the fuel pulse width

7. *Technician A* states that nozzle chatter while bench testing an injector is an indication of a fatigued injector spring. *Technician B* states that a high injector back leakage rate is an indicator of excessive nozzle valve-to-nozzle body clearance. Who is correct?
a. Technician A only
b. Technician B only
c. Both A and B
d. Neither A nor B

8. What is the distinguishing feature of a VCO nozzle?
a. The sac is eliminated.
b. High spring location
c. Low spring location
d. Used only in IDI engines

9. Which of the following injector types would be more likely to require a spindle?
a. High spring
b. Low spring
c. Poppet
d. VCO

10. Which type of injector nozzle would be incorporated into post-2007 EUIs?
a. Pintle
b. Poppet
c. Multi-orifii
d. Electrohydraulic

11. *Technician A* states that hydraulic injector nozzles are designed so that the pressure required to open the nozzle valve is greater than the pressure required to hold it open. *Technician B* states that after NOP, the nozzle line pressure steadily decreases until the nozzle closes at the end of the injection pulse. Who is correct?
a. Technician A only
b. Technician B only
c. Both A and B
d. Neither A nor B

12. Nozzle differential ratio is a ratio of:
a. Nozzle seat sectional area and total nozzle sectional area
b. The high-pressure pipe sectional area and the orifii flow area
c. The nozzle pressure chamber sectional area and the high-pressure pipe sectional area
d. The ratio of mechanical spring force and hydraulic pressure

13. Which of the following factors does most to prevent secondary injections?
a. Pressure wave reflection
b. Nozzle differential ratio
c. The nozzle sac
d. The pressure chamber sectional area

14. Which of the following is another way of stating the NOP specification?
a. Not operating properly
b. Residual line pressure
c. Peak pressure
d. Popping pressure

15. Most diesel engine OEMs recommend that injectors be installed into the cylinder head injector bore:
a. Dry
b. Coated with engine oil
c. Coated with Never-Seize
d. Coated with lubriplate

16. *Technician A* states that when a hydraulic multi-orifii injector is used in an electronically managed fuel system, nozzle opening

and closing is switched hydraulically. *Technician B* says that the opening and closing of an electrohydraulic injector is controlled the ECM. Who is correct?

a. Technician A only
b. Technician B only
c. Both A and B
d. Neither A nor B

17. Excessive fuel returned into the leakoff lines from a set of hydraulic injector nozzles is an indication of:
a. Nozzle seat leakage
b. Wear in the nozzle body-to-nozzle valve fit
c. Wear in the nozzle seat
d. A weak injector spring

18. *Technician A* states that EHI injector nozzles can use solenoid actuators. *Technician* B states that EHIs can use piezoelectric actuators. Who is correct?
a. Technician A only
b. Technician B only
c. Both A and B
d. Neither A nor B

19. *Technician A* states that the injector nozzles used on pre-2007 EUIs and HEUIs use the same operating principles as those on any hydraulic pump-line-nozzle system. *Technician B* states that hydraulic injector nozzles use a soft value NOP. Who is correct?
a. Technician A only
b. Technician B only
c. Both A and B
d. Neither A nor B

20. Which of the following types of injectors would be best suited to multipulse injection?
a. Hydraulic poppet
b. Hydraulic multi-orifii
c. EHI with solenoid actuator
d. EHI with piezoelectric actuator

20

Port-Helix Metering Injection Pumps

INTRODUCTION

The first high-pressure liquid fuel injection to a high-speed diesel combustion chamber was developed in 1927 by Robert Bosch. This evolved into the fueling apparatus used by Caterpillar, Mack, Navistar, Deere, and other diesel engine manufacturers into the 1990s. The principles of pumping and metering a fuel charge have changed little since Bosch's first 1927 high-pressure injection pump. However, the port-helix (helix: scroll shaped) metering injection pump, emissions certified for the 1990s, will, in most cases, be managed electronically rather than hydromechanically, and the hard (hard: not changeable) delivery timing window defined solely by the geometry of actuating cam profile and helix geometry has given way to variable timing devices.

Electronic controls permitted the technology of the inline, port-helix metering pump to live on into the electronic age at least for a short while. In most cases, when the truck technician uses the term **pump-line-nozzle (PLN)**, an inline port-helix metering injection pump is being referred to. Today when you see a port-helix metering injection pump it is either in a much older truck or in older off-highway diesels. But the inability of PLN fueling to meet current emissions standards is only one part of the story. Port-helix metering pumps neither produce the fuel economy nor the precise control over combustion demanded by today's diesel engines.

This chapter describes the port-helix metering pumps engineered and manufactured by Bosch (Figure 20–1 and Figure 20–2), Delphi-Lucas, and Caterpillar and used on Caterpillar, Mack, Cummins, Navistar, and other older highway and off-highway diesel engine applications.

TECHNICAL DESCRIPTION

The typical port-helix metering pump used to fuel a diesel engine is inline configured and flange mounted to an engine accessory drive to be driven through one complete rotation (360 degrees) per complete engine cycle (720 degrees). The internal pump components are housed in a frame constructed of cast aluminum, cast iron, or forged steel. The engine crankshaft drives the injection pump by means of timed reduction gearing. The gear-driven pump drive plate is connected to the injection pump **camshaft** (the fuel pump shaft fitted with eccentrics designed to actuate the pump elements), so rotating the pump drive plate rotates the pump camshaft. The camshaft is supported by main bearings and rotates within the injection pump **cambox**. The cambox is the lower portion of the injection pump that houses the camshaft, **tappets**, and an integral oil sump (Figure 20–3 and Figure 20–4).

The camshaft is designed to have one cam profile dedicated to each engine cylinder. Riding each cam profile is a tappet assembly driving a pump element consisting of a **plunger** and a **barrel**. The barrel is stationary and drilled with two ports in its upper portion, which are exposed to the fuel charging gallery.

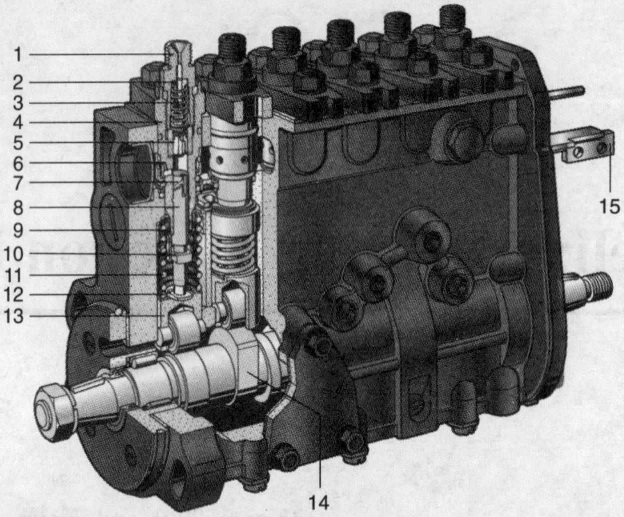

1. Delivery valve
2. Filler piece
3. Delivery valve spring
4. Pump barrel
5. Delivery valve
6. Inlet and spill port
7. Helix
8. Pump plunger
9. Control sleeve
10. Plunger control arm
11. Plunger return spring
12. Spring seat
13. Roller tappet
14. Actuating cam
15. Control rod

Figure 20–1 *Inline, port-helix metering injection pump components. (Courtesy of Robert Bosch LLC)*

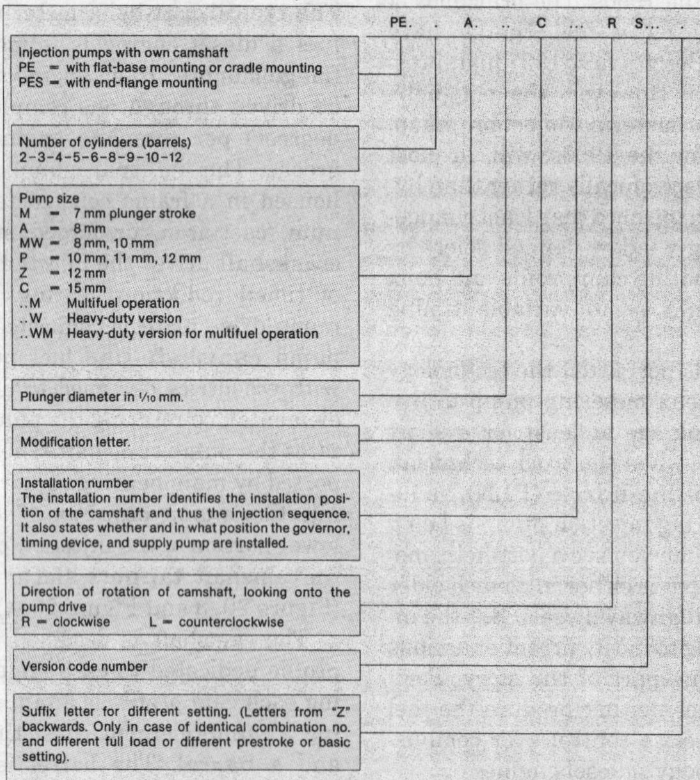

Figure 20–2 *Interpreting a Bosch injection pump serial number. (Courtesy of Robert Bosch LLC)*

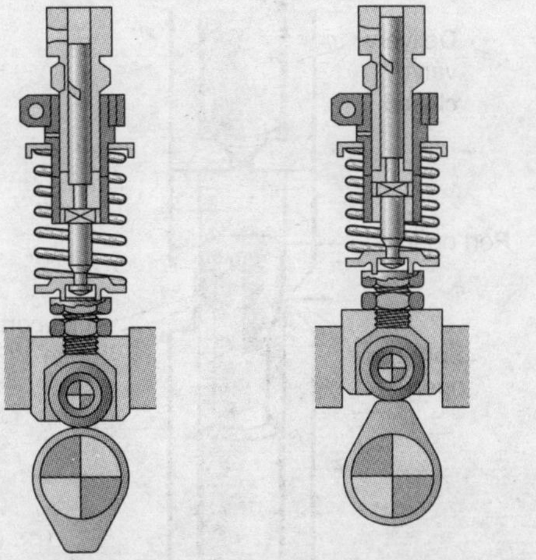

Figure 20–3 *Actuating a port-helix type pump element. (Courtesy of Robert Bosch LLC)*

Fuel Routing

The fuel gallery is charged with low-pressure fuel, typically between 1 and 5 **bar** (15 and 75 psi). The metric unit bar is not precisely equivalent to one unit of atmosphere, which is the standard unit known as **atm**. This charging pressure fuel flows into and through the barrel ports when they are not obstructed by the plunger. The plunger reciprocates within the barrel; it is loaded by spring pressure to ride its actuating cam profile. Therefore, actual plunger stroke is constant. Plunger-to-barrel tolerance is close, the components being lapped in manufacture to tolerance of 2 μ to 4 μ. Fuel quantity to be delivered in each stroke is controlled by managing plunger **effective stroke** (see Figure 20–5, Figure 20–6, Figure 20–7, and Figure 20–8). The plunger is milled with a vertical slot or cross and center drillings and helical recesses. The function of the vertical slot or cross and center drillings is to maintain a constant conduit between the pumping chamber above the plunger and the helical recesses. In other words, whatever pressures exist in the pumping chamber must also exist in the helical recess. Initially, only plunger designs with a **lower helix** or **helices** are discussed.

Effective Stroke

Effective stroke describes the delivery stroke. The delivery stroke begins when the plunger is

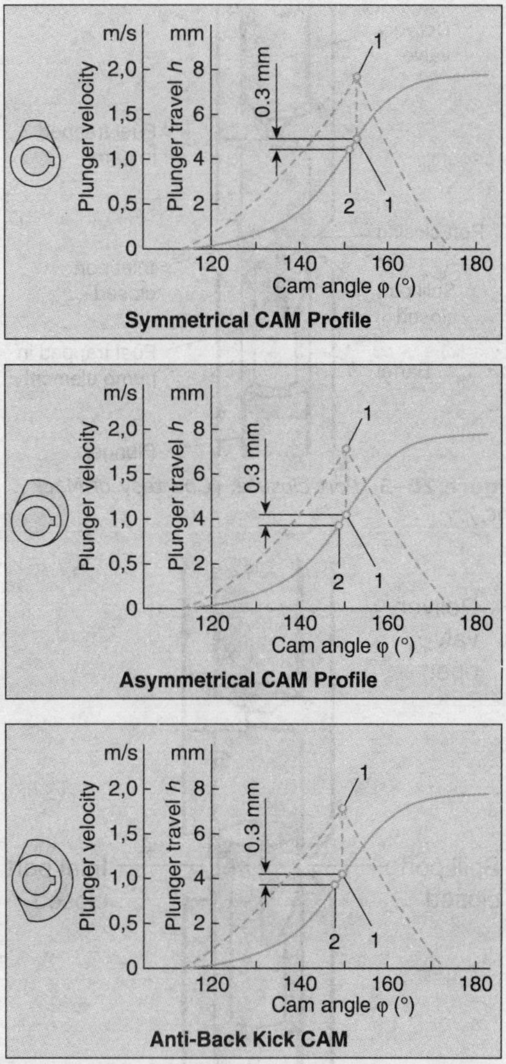

Figure 20–4 *Actuating cam geometry. (Courtesy of Robert Bosch LLC)*

forced upward by cam profile and the **plunger leading edge** (uppermost part of the plunger) traps off the spill port(s). As the plunger rises through its stroke in the barrel after trapping off the spill port, rapid pressure rise occurs, creating the required injection pressures. The precise moment that begins effective stroke is known as **port closure** (Figure 20–5). It is of critical importance to the diesel technician because its precise setting is used to control ignition timing. As pressure rises (Figure 21-6) in the pump chamber, it acts first on a **delivery valve**; next on the fuel confined in the

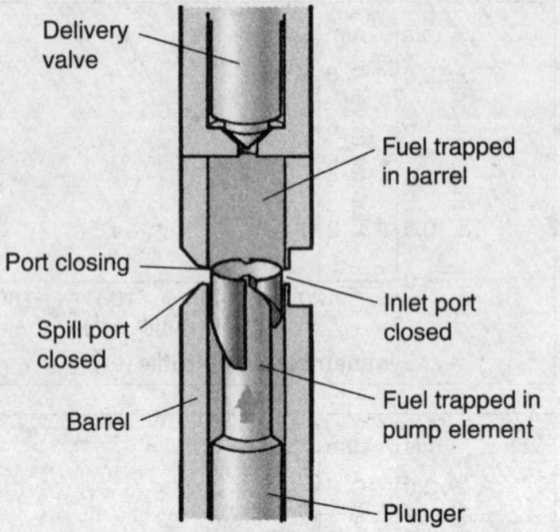

Figure 20–5 *Port closure. (Courtesy of Mack Inc.)*

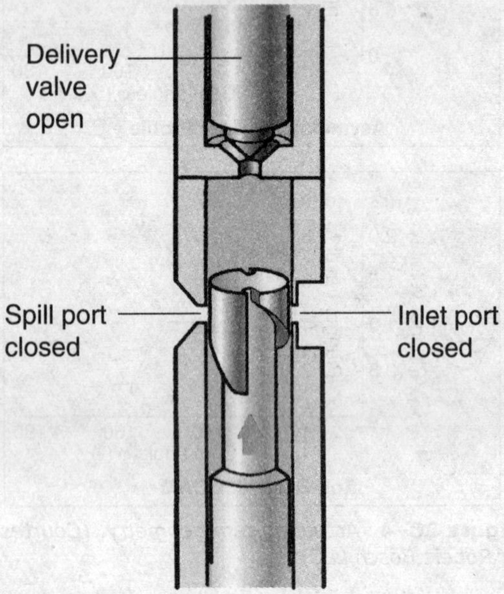

Figure 20–6 *Effective stroke. (Courtesy of Mack Inc.)*

high-pressure pipe, transmitting fuel to the injector nozzle; and, finally, delivering a fuel pulse to the engine cylinder. Effective stroke ends at **port opening** (Figure 20–7). This is the precise moment that the upward travel of the plunger exposes the helical recess(es) to the spill port. High-pressure fuel is spilled back to the charging gallery, causing a rapid collapse of pressure in the pump chamber, line, and nozzle.

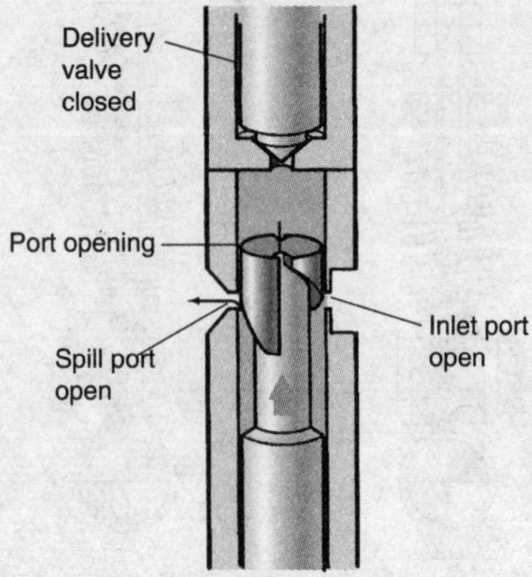

Figure 20–7 *Port opening. (Courtesy of Mack Inc.)*

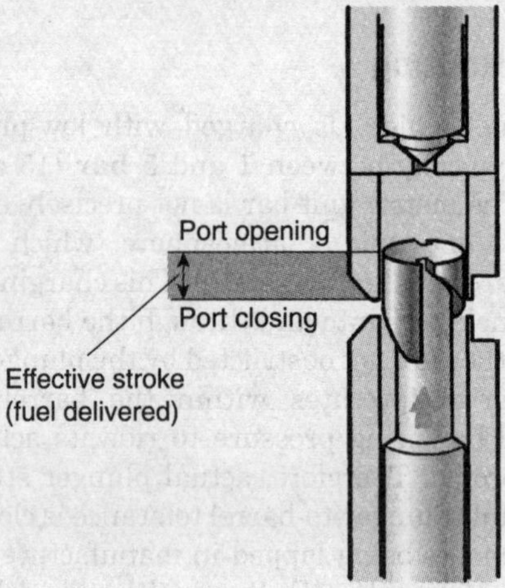

Figure 20–8 *Plunger travel through the effective stroke. (Courtesy of Mack Inc.)*

Collapse Phase

The injection pulse ceases when there is no longer sufficient pressure to hold the delivery and nozzle valves open. Port opening always occurs while the plunger is moving in an upward direction, that is, not at plunger top dead center (TDC) or beyond. This is required because the pressure in a port-helix pump element is designed to rise through the delivery stroke, thereby producing smaller atomized droplets from the injector toward the end of effective stroke; however, at the point of port opening, regardless of the length of the effective stroke, pump pressure should collapse as rapidly as possible and minimize the larger droplets emitted from the injector as pump pressure falls to a value below nozzle opening pressure (NOP).

Control of Effective Stroke

The length of plunger effective stroke (Figure 20–8) depends on where the plunger helix **registers** (vertically aligns) with the spill port. **Control sleeves** lugged to the plunger permit the plunger to be rotated while reciprocating. Rotating the plunger in the bore of the barrel changes the location of register of the spill port with the helix. Therefore, plunger effective stroke depends entirely on the rotational position of the plunger. In multiple cylinder engines, the plungers must be synchronized to move in unison to ensure balanced fueling at any given engine load.

The control sleeves are tooth meshed to a governor **control rack**, which when moved linearly, rotates the plungers in unison (Figure 20–9). This is important. It means that in any linear position of the rack, all of the plungers will have identical points of register with their spill points, resulting in identical pump effective strokes. The consequence of not doing this would be to unbalance the fueling of the engine, that is, deliver different quantities of fuel to each cylinder (Figure 20–10).

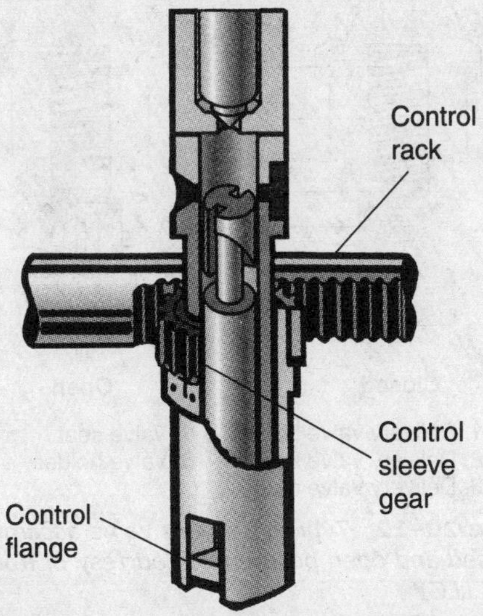

Figure 20–9 *Control rack and sleeve gear. (Courtesy of Mack Inc.)*

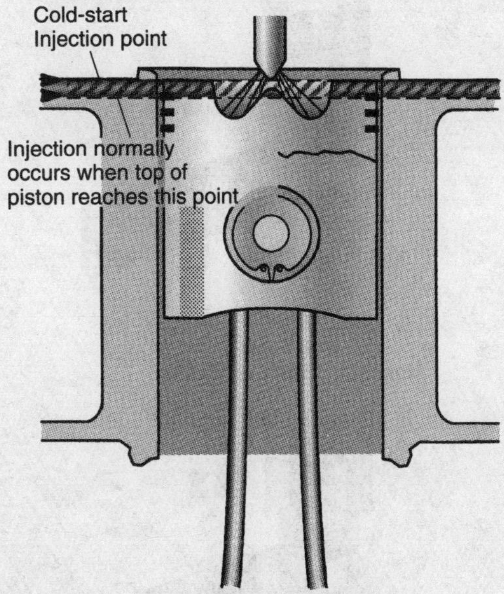

Figure 20–10 *Cold-start, retarded injection timing. (Courtesy of Mack Inc.)*

No-fuel

Engine shutdown is achieved by moving the control rack to the no-fuel position. The rotational position of the plungers is now such that the vertical slot will be in register with the spill port for the entirety of plunger travel; the plunger will merely displace fuel as it travels upward, with no pumping action possible. In other words, as the plunger is driven into the pump chamber, the fuel in the chamber will be squeezed back down the vertical slot to exit through the spill port and return to the charging gallery (Figure 20–11).

Most port-helix metering injection pumps use delivery valves to reduce the amount of work required of each pump element per cycle. Delivery valves function to isolate the high-pressure circuit that extends from the injection pump chamber to the seat of the nozzle valve. Fuel retained in the **high-pressure pipes** (pipes that connect the injection pump elements with hydraulic injectors) between pumping pulses is known as **dead volume fuel.**

Residual Line Pressure

Dead volume fuel is retained at a pressure necessarily somewhat below the NOP value; the

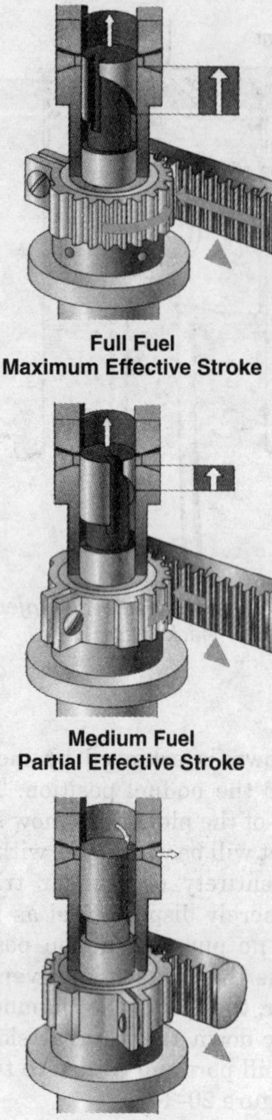

Figure 20–11 *Rack position and relationship to fuel delivery quantity. (Courtesy of Robert Bosch LLC)*

high-speed hydraulic switching that occurs in the high-pressure circuit creates pressure wave reflections, which may have the effect of spiking (causing surges) line pressures. To ensure that these pressure spikes do not exceed the NOP value and cause secondary injections, the dead volume fuel is retained at about ²/₃ of the NOP value; this is known as **residual line pressure**.

Role of Delivery Valve

The delivery valve is loaded into its closed position on its seat by a spring and by the residual line pressure. If, for whatever reason, the residual line pressure value were zero, hydraulic pressure of around 20 bar (300 psi) would have to be developed in the pump element to overcome the mechanical force of the spring. This mechanical force is therefore compounded when the residual line pressure acts on the sectional area represented by the delivery valve and establishes the pressure value that must be developed in the pump chamber before it is unseated.

The delivery valve **flutes** provide a means of guiding the valve in its bore while permitting hydraulic access between the **retraction collar** or **retraction piston** (both terms are used) and the pump chamber (plunger and barrel assembly). The retraction collar seals the pump chamber from the dead volume fuel, which is retained at a higher pressure value. Consequently, when the delivery valve is first unseated, it is driven upward in its bore by rising pressure in the pump chamber and acts as a plunger driving inward onto the fuel retained in the high-pressure pipe (Figure 20–12 and Figure 20–13).

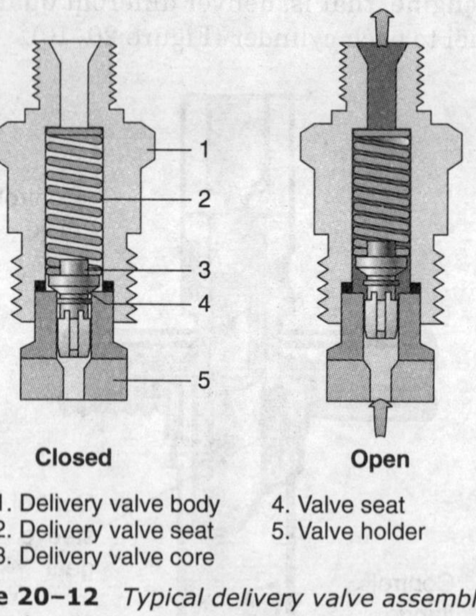

Figure 20–12 *Typical delivery valve assemblies in closed and open positions. (Courtesy of Robert Bosch LLC)*

The moment the retraction collar clears the delivery valve seat, fuel in the injection pump chamber and that in the high-pressure pipe

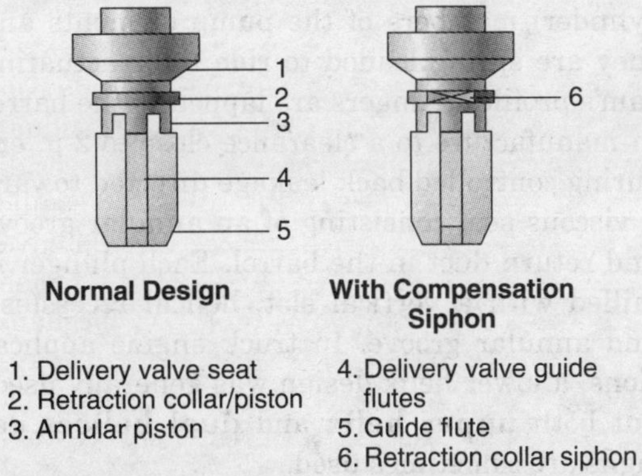

Figure 20–13 *Delivery valve core terminology. (Courtesy of Robert Bosch LLC)*

unite and the injection pump plunger is driven into a volume of fuel that extends from the plunger to the nozzle valve seat. Rising pressure subsequently unseats the injector nozzle valve (NOP) and forces atomized fuel into the engine cylinder.

At port opening, the effective pump stroke ends, beginning a rapid pressure collapse as fuel spills from the barrel spill port. When there is insufficient pressure in the pump chamber to hold the nozzle valve in its open position, spring pressure overcomes hydraulic pressure and it seats, sealing the nozzle end of the high-pressure pipe. Almost simultaneously, the delivery valve begins to retract. The instant the retraction collar passes the delivery valve seat, it hydraulically seals the pump end of the high-pressure circuit. However, after sealing the pump end of the high-pressure circuit, the delivery valve must travel farther before it seats, increasing the volume available for dead volume fuel storage. This causes a drop in line pressure and defines the residual line pressure value.

In short, the volume available for fuel storage in the high-pressure pipe is increased by the swept volume of the retraction collar. This volume of fuel is known as dead volume fuel and is retained at residual line pressure. Retraction collar swept volume is matched to the length of the high-pressure pipe to achieve a precise residual line pressure value so that pipe length should not be altered.

Most delivery valves are of the constant volume type described here. However, some fuel systems have a constant pressure delivery valve that consists of a forward delivery valve in the forward flow direction and a pressure holding valve (spring and ball) in the reverse flow direction. This rather more complex valve design helps minimize the effects of pressure wave reflection, which causes wear and cavitation.

INJECTION PUMP COMPONENTS

The following is a list of subcomponents that, assembled, will form a typical port-helix metering injection pump.

Pump Housing

The pump housing is the frame that encases all the injection pump components and is a cast aluminum, cast-iron, or forged steel enclosure. The pump housing is usually flange mounted by bolts to the engine cylinder block to be driven by an accessory drive on the engine geartrain. In some offshore applications of inline, port-helix metering injection pumps, the pump assembly is cradle mounted on its base, in which case it is driven by an external shaft from the timing geartrain.

Cam Box

The cam box is the lower portion of the pump housing incorporating the lubricating oil sump and main mounting bores for the pump camshaft. Camshaft main bearings are usually pressure lubed by engine oil supplied from the engine crankcase, and the cam box sump level is determined by the positioning of a return port. In older injection pumps, the pump oil was isolated from the main engine lubricant, and the oil was subject to periodic checks and servicing.

Camshaft

The camshaft is designed with a cam profile for each engine cylinder and supported by main bearings at the base of the pump housing. It is driven at one-half engine rotational speed in a four-stroke cycle engine by the pump drive plate, which is itself either coupled directly to the **pump drive gear** or to a variable timing device. Camshaft actuating profiles are usually symmetrical (that is, geometrically similar on both sides of the toe) and mostly inner base circle (IBC: the smallest radial dimension of an eccentric) in profile, though asymmetrical (the geometry of each cam ramp or flank differs) and mostly outer base circle (OBC: the largest radial dimension of an eccentric) designs are used. For a full explanation of cam geometry, reference Chapter 7.

Tappets

The tappets are arranged to ride the cam profile and convert the rotary motion of the camshaft to the reciprocating action required of the plunger. A **retraction spring** is integral with the tappet assembly. This is required to load the tappet and plunger assembly to ride the cam profile, and it is necessarily large enough to overcome the low pressure (vacuum) established in the pump chamber on the plunger return stroke. This low pressure can be considerable when plunger effective strokes are long, but it enables a rapid recharge of the pump chamber with fuel from the charging gallery. The time dimension within which the pump element must be recharged decreases proportionately with pump rpm increase.

Barrel

The barrel is the stationary member of the pump element located in the pump housing so its upper portion is exposed to the charging gallery. This upper portion of the barrel is drilled with diametrically opposed ports known as inlet and spill ports that permit through flow of fuel to the barrel chamber to be charged. Because it contains the spill ports, both its height and rotational position in relation to the plunger are critical. Barrels are often manufactured with upper flanges so that their relative heights can be adjusted by means of shims, and fastener slots permit axial movement for purposes of **calibration** and **phasing** (the procedures of calibration and phasing are covered in more detail later in this chapter).

Plunger

Plungers are the reciprocating (something that reciprocates, moves backward and forward such as in the action of a piston in an engine cylinder) members of the pump elements and they are spring loaded to ride their actuating cam's profile. Plungers are lapped to the barrel in manufacture to a clearance close to 2 μ, ensuring controlled back leakage directed toward a viscous seal consisting of an annular groove and return duct in the barrel. Each plunger is milled with a vertical slot, helical recess(es), and annular groove. In truck engine applications, a lower helix design was generally used, but both **upper helix** and **dual helices** designs are sometimes used.

Plunger Geometry

The positioning and shape of the **helices** (plural of **helix**) on a plunger are often described as the **plunger geometry**. Plunger geometry describes the physical shape of the **metering recesses** machined into the plunger, and this defines the injection timing characteristics. The function of the vertical slot is to ensure a constant hydraulic connection between the pump chamber above the plunger and the plunger helical recess(es).

Timing and Plunger Geometry

A plunger with a lower helix will have a constant beginning, variable ending of delivery timing characteristic, whereas an upper helix design will be of the variable beginning, constant ending type. When port helix metering injection pumps were used in truck applications, lower helix plunger geometry was used: An external means of varying port closure timing was used in these pumps. **Double helix** designs have both an upper and a lower helix and a variable beginning and variable ending of delivery; this geometric design tends not to be often used in highway diesel engines. Another perhaps more common plunger design is the dual helices design—identically shaped helices are machined into the plunger, diametrically opposite each other. Plungers with diametrically opposed helices were used in more recent high-pressure injection pumps to provide hydraulic balance to the pump element; specifically, it prevents the side loading of the plunger into the barrel wall that may occur at high-pressure spill-off.

A further feature of some plungers is a start retard notch. Start retard notches are milled recesses in the leading edge of plungers with lower helix geometry. The start retard notch is usually on the opposite side of the vertical slot from the helix and in a position that would correlate close to a full-fuel effective stroke; the governor of the injection pump is designed to permit the start retard notch to register with the spill port only at cranking speeds (under 300 rpm) and with the accelerator fully depressed.

The objective of the start retard notch on a lower helix design plunger is to retard the injection pulse until there is a maximum amount of heat in the engine cylinder, usually when the piston is close to TDC. The instant the engine exceeds 300 rpm, it becomes no longer possible for the start retard notch to register with the spill port. Figure 20–10 illustrates the delay produced when the start retard notch is positioned in register with spill port(s).

Rack and Control Sleeves

The rack and control sleeves permit the plungers in a multicylinder engine to be rotated in unison to ensure balanced fuel delivery to each cylinder. Plungers must therefore be timed either directly or indirectly to the control rack. The rack is a toothed rod that extends into the governor or **rack actuator** housing. The rack teeth mesh with teeth on plunger control sleeves, which are either lugged or clamped to the plunger. It must be possible to rotate the plungers while they reciprocate to permit changes in fuel requirements while the engine is running. Linear movement of the rack will rotate the plungers in unison, alter the point of register of the helices with their respective spill ports, and thereby control engine fueling. Timing of the plungers to the control rack is the means used to adjust the effective stroke in individual pump elements in a procedure known as calibration.

This timing procedure is effected either directly, by adjusting the relationship of the plunger with the rack, or indirectly, by adjusting the rotational position of the barrel and therefore its relationship with the plunger geometry. The means of calibrating an injection pump depends on the make and model.

Delivery Valves

Delivery valves isolate the high-pressure circuit that extends from the injection pump chamber to the seat of the injector nozzle valve. They act as somewhat like check valves. Because they seal before they seat, they permit line pressure to drop to a residual value well below the NOP value, and this helps prevent secondary injections. The delivery valve is machined with a seat, retraction collar, and flutes to guide it in the bore of the delivery valve body (see Figure 20–13).

Charging Pumps

The terms **charging pump** and **transfer pump** are used interchangeably, depending on the original equipment manufacturer (OEM). The charging pump is responsible for all fuel movement in the fuel subsystem. In truck applications using port-helix metering injection, the charging pump is normally a plunger pump flange mounted to the fuel injection pump and actuated by a dedicated eccentric on the injection pump camshaft. The charging or transfer pump is responsible for producing **charging pressure**.

Charging pressures range from 1 to 5 bar (15 to 75 psi) depending on the system. The role of charging/transfer pumps is dealt with in greater detail in the context of the fuel subsystem, the subject of Chapter 17.

Governor or Rack Actuator Housing

Either a governor or rack actuator housing must be incorporated to a port-helix metering injection pump. This acts as the control mechanism for managing fueling. Because both the subject of governors and the rack actuators used in electronically managed versions of port-helix metering injection pumps are dealt with in detail elsewhere in this textbook, only a brief description is provided here.

The only factor that determines the output of a diesel engine is the amount of fuel metered into its cylinders. Given unlimited fuel, a diesel engine is capable of accelerating at a rate of 1,000 rpm per second until self-destruction occurs. To prevent such an "engine runaway" and to provide the engine with a measure of protection from abuse to ensure that it meets the manufacturer's expectations of longevity, fuel quantity delivered to the engine cylinders must be precisely managed by a governor under all operating conditions. The fuel control mechanism on a port-helix metering injection pump is the rack. Therefore, the governor or rack actuator controls engine fueling by precise positioning of the rack.

Hydromechanical governors are not so often used to manage medium- and heavy-duty, on-highway diesel engines of today, but when they are they consist of a self-contained housing mounted to the rear of the injection pump. In a simple mechanical governor, the vehicle accelerator linkage connects to a throttle arm or governor input lever located on the side of the governor housing. Enclosed within the governor housing, a weight carrier is mounted to the rear of the camshaft. Flyweights are mounted in the weight carrier, which is loaded inboard by spring force.

Governor Operation

As the camshaft rotates, centrifugal force acting on the flyweights thrusts them outward. The faster the rotational speed, the more centrifugal force is generated. Countering the centrifugal force generated by the flyweights is an accumulation of spring forces. The amount of spring force is usually variable and is increased as the accelerator arm is moved. A thrust collar acts as an intermediary (that is, something in between two other components) between the mechanical force of the spring(s) and the centrifugal force generated by the flyweights. The fuel control mechanism or rack is connected to the thrust collar. Spring force acting on the thrust collar attempts to increase engine fueling. Centrifugal force acts in opposition to the spring force and attempts to diminish engine fueling.

Maximum engine speed is defined by ensuring that at a set speed, the centrifugal force generated by the flyweights will overcome any amount of spring force the governor can counter it with. Even the simplest mechanical governor suitable for managing a highway truck engine will define idle speed, rated speed, and the droop curve (graduated engine defueling as engine speed rises from rated speed to the high-idle speed), leaving intermediate speed selection to the operator. A simple governor may also provide torque rise fuel enhancement and an **aneroid** (light pressure sensing device) to sense manifold boost and limit fueling until manifold pressure achieves a predetermined value.

Two main categories of hydromechanical governor exist. The first is the LS or limiting speed governor, which defines idle and high-idle engine speeds, leaving the intermediate speed ranges to the operator. The second category is the variable speed governor in which throttle arm position defines an engine rpm value; the governor then attempts to maintain this speed while engine loading varies. Hydromechanical governors are not capable of managing today's high-power engines that must achieve statutory and customer requirements of low-emissions operation and high fuel economy. They have been seldom used since 1991. Governors are looked at in some detail in Chapter 25.

Rack Actuator Operation

When the port-helix metering injection pump is managed by a computer, engine governing depends on how the electronic control module (ECM) is programmed. The governor housing attached to the rear of the injection pump is replaced by a rack actuator housing consisting of switched ECM output devices and sensors. The housing contains a rack actuator, either a linear magnet (proportional solenoid) controlled directly by the ECM or an electric-over-hydraulic device (engine oil acts as the hydraulic medium) also ECM controlled. The rack actuator sets rack position on ECM command and nothing more.

A rack position sensor reports the exact rack position to the ECM. Additionally, the rack actuator housing may house other sensors to report rotational speed (pulse wheel fitted to the rear of the camshaft), engine position, and timing data to the ECM. Rack actuator housings are not governors. Where a rack actuator housing is fitted to an inline, port-helix metering injection pump, the engine governing functions are undertaken by the software programmed into the fuel management ECM.

Electronic governing offers infinitely more control over engine fueling, and it allows OEMs to meet a higher level of statutory noxious emissions requirements and achieve better fuel economy; but it additionally can be easily programmed/reprogrammed with customer data to tailor the engine for varying engine and chassis applications. This subject matter is dealt with in detail elsewhere in this text in Chapter 35 relating to partial authority, electronic engine management.

Lubrication

The lower portion of the port-helix metering injection pump is lubricated by engine oil. In older versions, the injection pump lubricating circuit was often isolated from that of the engine and the lube level would be checked and replenished through a dipstick located in the cambox. Contemporary injection pumps tend to be plumbed into the main engine lubricating circuit. The camshaft main bearings are often pressure lubed directly from an engine oil gallery, and the remainder of the cambox components are splash lubricated from the oil held in the sump.

The upper portion of the injection pump is lubricated by the diesel fuel being pumped through the charging gallery, so the lubricity of the diesel fuel is critical. Compromising the lubricating qualities of the fuel may cause premature failure and/or leakage and some manufacturers have reported such problems following the introduction of low-sulfur, low-lubricity fuels mandated by Environmental Protection Agency (EPA) standards. It is critical for the operation of both the fuel injection pump and the engine it manages that the fuel used to lubricate the upper portion of the pump never come into contact with the engine oil in the cambox. Plunger-to-barrel clearance is lapped in manufacture to a minute tolerance and fuel that leaks by the recessed metering areas bleeds to an annular belt in the barrel.

A duct connects the annular belt in the barrel to the charging gallery, permitting bleed-by fuel to be routed there. This is known as a viscous seal. Trace leakage of a viscous seal can rapidly cause engine oil contamination and lead to lubricant breakdown. Viscous seals fail due to plunger side loading (single helix design), prolonged usage wear, and fuel contaminants—especially water. Fuel that bleeds by the metering recesses serves to guide the plunger true in the barrel bore, minimizing metal-to-metal contact. When studying the results of engine oil analysis, it should be remembered that failure of the injection pump's viscous sealing ability is probably the least likely source of a fuel-in-engine oil condition.

Injection Rate and Fuel Rate

Injection rate is defined as fuel quantity injected per crank angle degree; it is therefore determined by the geometry of the pump element actuating cam profile. Most port-helix metering injection pumps use a symmetrical cam profile whose periphery (outer surface) is mostly IBC; this design holds the plunger at the bottom of its travel for most of the cycle. Occasionally asymmetrical cams and cams whose periphery is mostly OBC are used.

The shaping of the cam contour affects not only injection rate but also pump chamber breathing and cooling. A cam profile that is mostly OBC is sometimes known as a back-kick cam, which will prevent the engine from being started in reverse rotation.

Fuel rate is a term usually used to define the actual rate of delivering fuel to an engine cylinder, so it is factored by both cam profile and engine rpm.

Timing Advance and Variable Timing Mechanisms

Older port-helix metering injection pumps were usually directly driven by reduction gearing from the engine camshaft gear. Such a system would dictate that the **static timing** value (port closure) occur at the same number of crank angle degrees before top dead center (BTDC) regardless of engine load or speed when lower helix geometry is used. Fuel economy and noxious emission considerations led to the development of first mechanical advance mechanisms and, more recently, electronically managed, variable timing.

Mechanical Variable Timing

Mechanical timing advance mechanisms (Figure 20–14) are actuated using a set of flyweights and eccentrics to advance the drive angle of the pump camshaft in relation to the pump drive gear, using a spiral gear on a shaft. In other words, the position of the fuel injection pump relative to that of the engine is advanced in direct proportion to the centrifugal force generated by the weight carrier. Port closure was therefore advanced as engine rpm increased. The extent of the advance offered by mechanical advance mechanisms can be as little as 3 degrees crank angle and seldom more than 10 degrees crank angle.

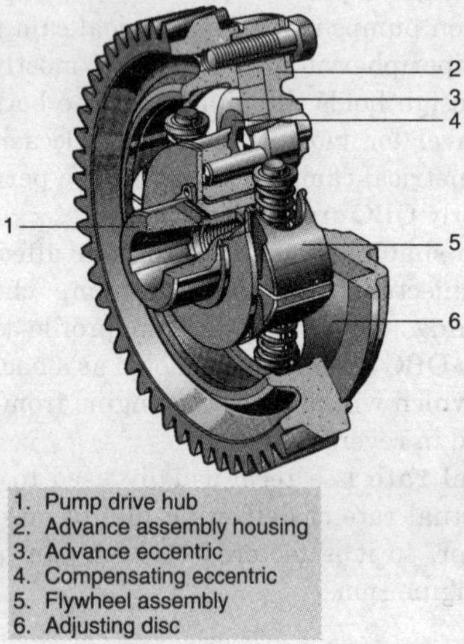

Figure 20–14 *Mechanical timing advance mechanism. (Courtesy of Robert Bosch LLC)*

Electronic Variable Timing

Electronically managed variable timing devices such as those used by Mack and Caterpillar partial authority PLN management employ a variable timing coupling mechanism between the pump drive gear and the pump camshaft. This intermediary was designed to establish a variable timing window of up to 20 degrees crank angle, which is managed by the engine electronics.

The static timing value (that is, port closure) specified is usually the most retarded parameter in the variable timing window. For example, an injection pump timed at 7 degrees BTDC with a 20-degree variable timing mechanism would permit the ECM to select any port closure value between 7 degrees BTDC and 27 degrees BTDC. Injection pump static timing (port closure) specs of as little as 4 degrees BTDC are used to limit combustion heat and, therefore, oxides of nitrogen (NO_x) emission; substantially retarded injection timing does not generally enhance either performance or fuel economy. In fact, the tradeoff of retarded injection timing is an increase of hydrocarbon (HC) emission but all emissions legislation is about keeping the exhaust noxious emissions within a window of acceptability. Both Mack and Caterpillar manage their variable timing devices in different ways. The Mack system is discussed later in Chapter 35.

GOVERNOR TRIM DEVICES

Governor trim devices were used primarily on hydromechanical PLN pumps. Their objective was to modulate fueling as either ambient pressure (altitude) or manifold boost pressure changed. Their objective was essentially to prevent overfueling the engine when running or ambient conditions resulted in lower oxygen content in the cylinder. Two basic types are described here:

Aneroids

By definition, an aneroid is a low-pressure sensing device. In application, it is used on a turbocharged diesel engine to measure manifold boost and limit fueling until the boost pressure achieves a predetermined value. Such devices are known variously as puff limiters, turboboost sensors, AFC valves, and smoke limiters, and all seek to accomplish the same objective. They typically consist of a manifold containing a diaphragm; boost air is piped from the intake manifold to act on the diaphragm. Such devices were used on most PLN **boosted engines** (turbocharged engines).

When an aneroid was used on an inline, port-helix metering pump, it was usually a mechanism consisting of a manifold, spring, and control rod. The manifold is fitted with a port, and a steel line connects it directly to the engine intake manifold. In this way, boost pressure is delivered to the aneroid manifold where it will act directly on the diaphragm within. Attached to the diaphragm is a linkage connected either directly or indirectly to the fuel control mechanism (rack); a spring loads the diaphragm to a closed position, which limits fueling by preventing the rack from moving into the full-fuel position.

When manifold boost acting on the diaphragm is sufficient to overcome the spring pressure, it acts on the linkage, permitting the rack full travel and thus maximum fueling. Such systems were easily and commonly shorted out by operators in the mistaken belief that aneroids reduce engine power. The emission of a "puff" of smoke from the exhaust stack at each gear shift point is an indication that the aneroid-/boost-sensing mechanism has been tampered with. As aneroids evolved they were designed to be more tamperproof. However, when older diesel engines with PLN fueling fail roadside emissions tests, it is a safe bet that the engine aneroid system has been tampered with. Figure 20–15 and Figure 20–16 illustrate the operation of a typical aneroid under conditions of low and high manifold boost.

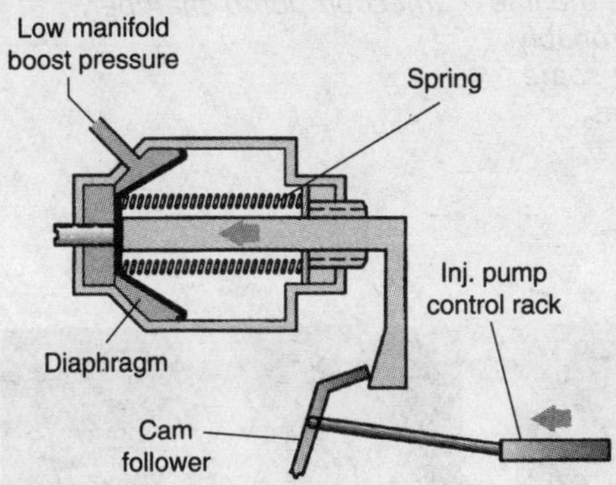

Figure 20–15 *Aneroid operation at low boost. (Courtesy of Mack Inc.)*

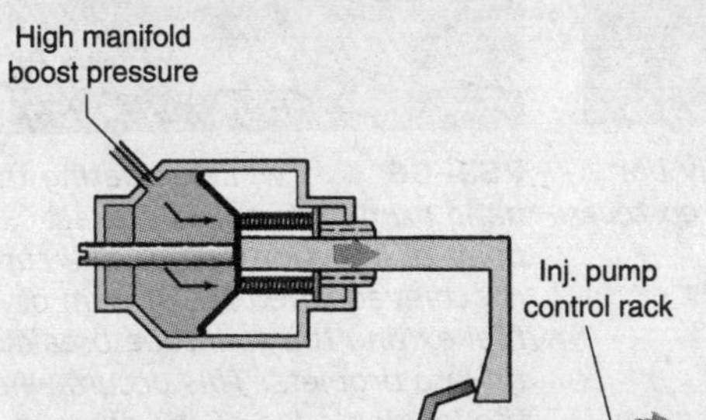

Figure 20–16 *Aneroid operation at high boost. (Courtesy of Mack Inc.)*

Altitude Compensator

An **altitude compensator** device contains a **barometric capsule** that measures barometric pressure and on this basis downrates engine power at higher altitudes to prevent overfueling. They are required when running hydromechanical PLN-fueled engines at higher altitudes because the oxygen density in the air charge decreases with an increase in altitude and, unless the fuel system responds to this, it effectively overfuels the engine. The critical altitude at which some measure of injected fuel quantity deration is required is 1,000 ft., but in older engines this may not actually occur until 3,000 ft. in altitude. The result of an altitude compensator malfunction is engine smoking under load at altitude.

TIMING INJECTION PUMPS TO AN ENGINE

Port-helix metering injection pumps are timed to the engine they manage by phasing port closure on the #1 cylinder (usually in North American-engineered engines; check the specification—for a variety of reasons, the pump may be required to be timed to the #6 cylinder) to a specific engine position. All injection pumps must be accurately timed to the engine they will fuel.

This usually means the phasing of pump port closure to a specific number of degrees BTDC on the cylinder to be timed to a specification that seldom can be outside of 1 degree crank angle and may have to be within ¼ degree crank angle. Methods used to time the injection pump to the engine vary by OEM. Actual spill timing of the pump to the engine is a procedure that has become obsolete on highway diesel engines; when you are required to spill time an engine, it is most likely to be an off-highway, usually small-bore engine. However, most would argue that it continues to be a procedure that diesel technicians should understand because it can be used in the absence of OEM tooling.

Spill Timing Procedure

Before beginning the **spill timing** procedure, check the engine (and OEM chassis) manual for positioning of the fuel control lever, stop fuel lever, brake valve, and gear shift lever position. See Photo Sequence 3.

1. Check the injection pump data plate for port closure value. Manually bar the engine in the direction of rotation to position the piston within the #1 cylinder on its compression stroke. Locate the engine calibration scale, usually to be found on the front pulley, vibration damper, or flywheel. Position the engine roughly 20 degrees before the port closure specification.
2. Remove the high-pressure pipe from the delivery valve on the injection pump #1 cylinder. Unscrew the delivery valve body and

Photo Sequence 3

Chapter 20 Spill Timing for an Inline, Port-Helix Metering Injection Pump

PS3–01 *Check the pump port closure specification on the injection pump ID plate, noting which engine cylinder the pump is timed to. Locate the engine in the correct position for the spill timing procedure by observing the valve rocking action on the companion cylinder to the one being timed. Position the engine to ± 20 before the port closure specification value.*

PS3–02 *The injection pump in the photograph is timed on its #1 cylinder. Remove the hydraulic nut on the #1 high-pressure pipe and move the line out of the way; it will probably be necessary to remove some line clamps to enable this.*

PS3–03 *Remove the delivery valve body and the delivery valve core and spring. Ensure that no dirt gets into the exposed injection pump chamber.*

PS3–04 *Insert a spill tube (an old injection high-pressure pipe, cut and shaped to a gooseneck) into the delivery valve body in place of the delivery valve. Actuate the hand primer pump to ensure that fuel exits from the spill tube. Capture the fuel exiting the spill tube in a container.*

PS3–05 *Next, manually bar the engine slowly in the correct direction of rotation . . .*

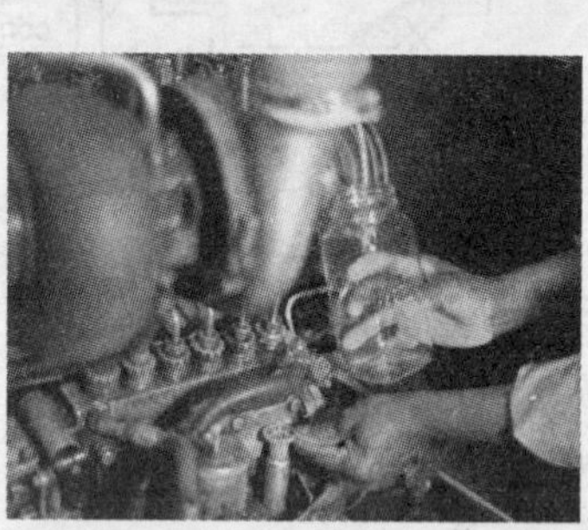

PS3–06 *. . . while actuating the hand primer pump. The exact point of injection pump, spill cutoff is achieved when the stream of fuel exiting the spill tube breaks up into droplets. This occurs when the leading edge of the plunger rises in the barrel bore to begin to cut off the spill port.*

Photo Sequence 3

Chapter 20 Spill Timing for an Inline, Port-Helix Metering Injection Pump (continued)

PS3–07 *Spill cutoff. Be sure that when the hand primer pump is actuated, droplets of fuel still exit the spill tube. If no fuel exits the spill tube, the engine has been barred past the point of injection pump port closure. If this is so, back engine up at least 20 degrees before the port closure specification and repeat the previous steps.*

PS3–08 *After locating the port closure on the pump, check the engine calibration plate. On this engine, the calibration plate and pointer are located on the vibration damper: in some engines the calibration plate is on the flywheel. If the engine location reading is within 1 crank degree of the port closure specification, assume the engine to be properly timed. If not . . .*

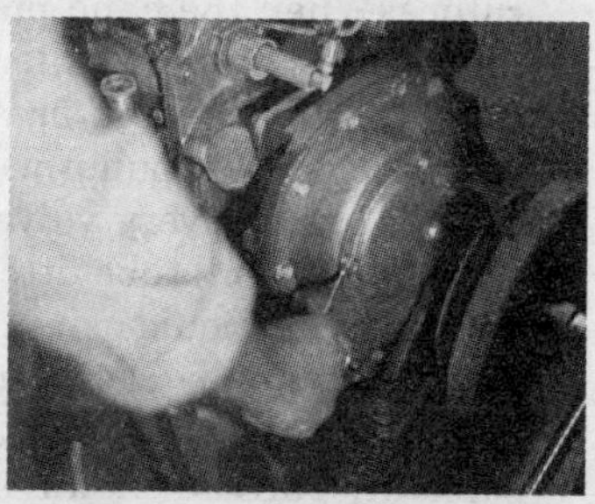

PS3–09 *. . . remove the pump drive gear cover plate located directly in front of the injection pump.*

PS3–10 *Loosen the pump drive gear to pump drive plate fasteners: this should be loose enough to permit the engine to be rotated independently of the gear pump.*

PS3–11 *Carefully bar the engine to the correct port closure location on the engine calibration plate: the objective is to rotate the engine while the injection pump remains stationary in its location port closure at #1 pump element.*

PS3–12 *With engine in the correct location, torque the pump drive gear to pump drive plate fasteners. Then repeat the sequence outlined in steps 4 through 8 to confirm that the injection pump is properly timed to the engine.*

PS3–13 *When the port closure specification has been confirmed, remove the spill tube and carefully reassemble the injection pump delivery valve, reconnecting the high-pressure pipe.*

remove the delivery valve core and spring. Replace the delivery valve body and install a spill tube. A discarded high-pressure pipe neatly cut and shaped to a goose neck will suffice. When the hand primer pump is actuated, the charging gallery will be pressurized. The amount of pressure created by the hand primer pump will be insufficient to open the delivery valves, so it will exit through the spill tube fitted to the #1 cylinder. The fuel should exit in a steady stream, and it should be captured in a container held under the spill tube. Next, slowly and smoothly bar the engine in its direction of rotation, observing the stream of fuel exiting the spill tube.

When the plunger leading edge rises to trap off the spill port, the steady stream of fuel exiting the spill tube will break up first into droplets and then cease as the plunger passes the spill port.

The objective of this step is to locate the pump precisely at port closure—this means that the flow area at the spill port should exist but be minimal; two to six drops per 10 seconds should be within the specification window. Ensure that the pump has not been barred past port closure cutting the fuel off altogether as it is impossible to determine how much beyond port closure the plunger has traveled.

3. Next, check the engine calibration scale. The specification is typically required to be within 1 degree crank angle of the port closure (PC) specification. If this is so, the pump can be assumed to be correctly timed to the engine. If it is not, proceed as follows.
4. Remove the accessory drive cover plate. Loosen the fasteners that couple the pump drive gear to the pump drive plate. Bar the engine to the correct PC specification position. By uncoupling the pump drive plate from the pump drive gear, it is hoped that the pump will remain stationary at PC on the #1 cylinder while the engine is barred independently. When the engine is in the correct position, torque the fasteners that couple the pump drive gear to the pump drive plate. Back the engine up roughly 20 degrees before the PC specification and then repeat steps 1 through 3.

Every diesel technician should be acquainted with the foregoing procedure even though he or she may seldom practice it. Spill timing can also be performed using compressed air as the test medium; the setup procedure is the same as just listed, but regulated compressed air is ported into the charging gallery. Instead of a spill tube, a flexible hose is connected to the pump element used for the spill timing procedure and immersed in a water-filled glass jar; when air is supplied to the charging gallery, it will pass through the barrel spill ports and exit through the flexible hose, producing bubbles in the glass jar. As the pump is rotated toward the PC location and the plunger leading edge starts to trap off the spill port, the stream of bubbles will turn into smaller bubbles produced less frequently. Using air to spill time injection pumps is probably the least recommended method. The air supply should always be equipped with an air dryer/filter assembly; and even when it is so fitted, the danger of moisture and other air-suspended contamination does not warrant the risk of using this method. OEMs prefer alternative methods of static timing of port-helix metering injection to be used that do not require the removal and disassembly of the delivery valve. Following are some of the alternatives.

1. Timing pin. This is probably the simplest method and least likely to present problems. It is also the least accurate. The engine is located to a specific position by inserting a timing pinion or bolt usually in the cam gear but sometimes in the flywheel. Similarly, the injection pump is pinned by a timing tool to a specific location. The injection pump is always removed and installed with the timing tools in position. It goes without saying that the timing tools must be removed before attempting to rotate or start the engine.
2. High-pressure pump. This involves connecting a high-pressure timing pump into the circuit. This portable electric pump charges the charging gallery at a pressure in excess of the 20 bar required to crack the delivery valves, which causes them all to open; consequently the injector leak-off circuit and pump gallery return must be plugged off. Next, the high-pressure pipe on the #1 cylinder is removed at the pump, and a spill pipe discharging into the portable pump

sump is fitted. The procedure then replicates that used to spill time the injection pump—the engine must be correctly located well before the PC specification (to eliminate engine gear backlash variables) and then barred to spill cutoff.

If the settings are out of specification, they are rectified by altering the coupling location of the pump drive gear to the pump drive plate.

3. Electronic. This method tends to be used with later generation hydromechanical and electronically managed, port-helix metering injection pumps. The static timing value is checked with an electronic timing tool that senses the positioning of a raised notch located on the pump camshaft. The procedure is outlined in some detail using a specific injection pump and engine in Chapter 35.

Injection pump-to-engine timing is critical. It should be checked each time a pump is reinstalled to the engine and as a step in performance complaint troubleshooting. Pump timing may be checked using a diesel engine timing light. This consists of a transducer that clamps to the high-pressure pipe and signals a strobe light when it senses the line pressure surge that occurs when the delivery valve opens. The test is not an accurate one and the values read on the timing light should not be confused with the pump manufacturer's PC specification because the pressure rise pulse used to trigger the light occurs after PC. The test has some validity as a comparative test (using data obtained from other engines in the same series, which are timed at the same PC value) and is a good means of verifying the operation of timing advance mechanisms.

Removing a Port-Helix Metering Injection Pump from an Engine

Probably close to half the port-helix metering injection pumps that are removed from engines have nothing wrong with them. This is due to poor field diagnostics caused by a low level of understanding of the operating principles of injection pumps combined with inaccurate interpretation of service literature. Another factor is a tendency of field technicians to black-box fuel pump technology as the responsibility of the pump room technician and, as a result, package up problems for someone else to repair.

The correct location to diagnose most fuel injection pump problems is with the pump on the truck engine. If the OEM-recommended on-board tests are performed before removing the fuel injection pump, then if the pump has to be removed from the engine, the job of the pump room technician has been made easier. When it has been determined that the injection pump is responsible for a fuel problem and it must be removed, the procedure should be as follows:

1. Power wash the injection pump and surrounding area of the engine.
2. If the fuel pump and engine are equipped with locking/timing pins, position the engine to install them.
3. Remove the fuel supply and return lines. Remove the lubricating oil supply and return lines.
4. Remove the accelerator and fuel shutoff linkages (if equipped) or the electronic connector terminals.
5. Disconnect the high-pressure pipes from the delivery valves. Ensure that the high-pressure pipes can be moved away from the delivery valves without forcing or bending them; this usually means releasing insulating and support clamps. Cap both the high-pressure pipe nipples and the delivery valves.
6. Unbolt the fasteners at the pump mounting flange. Depending on the pump, the pump may have to be removed separately from the variable timing/advance timing mechanism. Some pumps may require that the pump drive gear be separated from the pump drive plate or the variable timing device from the front of the pump drive gear, requiring the removal of the timing gear cover plate. Most pumps should slide back easily after this but care should be taken not to support the weight of the injection pump on its drive gear.

Reinstallation

Essentially, the foregoing procedure is reversed but the PC timing should be confirmed using the OEM-recommended method (outlined earlier in this section). When installing the pump, ensure that it is installed with the pump drive gear/plate correctly registered with the engine timing gear. Resistance can be an indication that the pump is being installed a tooth out of register and contacting a dowel or key; never force an injection pump into position. When resistance is encountered, remove the pump and check both the pump position and its drive mechanism for problems.

PHASING AND CALIBRATION

Calibration and phasing of port-helix metering injection pumps is the responsibility of the manufacturer-trained pump technician; these operations require the use of specialized tooling, a comparator bench, and the manufacturer's technical data. However, the diesel technician should be capable of defining the procedures.

Phasing

Phasing the injection pump sets the phase angle between the individual pump elements, essentially to ensure that PC on each occurs at a precise spacing. On a six-cylinder engine this would be exactly 60 degrees. Phasing the injection pump ensures balanced, multicylinder engine timing. It is checked on the comparator bench using a degree wheel (protractor located on the comparator drive turret) and adjusted by varying either the barrel height or shimming the plunger tappets. The critical specification is the *lift to PC*, which is a measure of the positioning of the plunger in relation to cam lift. Depending on the type of pump used, either the tappet height or the barrel height is adjusted to ensure that PC in each pump element is phased to occur at a precise interval.

A typical method of phasing a port-helix pump is outlined in the following procedure. A dial indicator assembly fitted with a spill tube is mounted in place of the delivery valve on the cylinder to be phased; when a hand primer pump is actuated and the pump rotated, fuel will be cut off by the leading edge of the plunger at PC. The dial indicator probe rides on top of the pump plunger and thus its position relative to the movement of PC can be determined. An adjustment would require the raising or lowering of either the barrel height (by means of split shims) or the raising or lowering of the tappet height (by means of phase angle shims or an adjusting screw).

For a pump required to fuel a six-cylinder engine, PC in each of the six cylinders should be phased to be exactly 60 degrees apart. Using the degree wheel as a reference, PC in each cylinder would be set in engine firing sequence at the required interval. Phasing can be also checked/ adjusted using the spill method in conjunction with the degree wheel on the comparator bench turret. It should be noted that primary phasing is the responsibility of the camshaft geometry, and a failing cam profile will result in a phasing problem.

Calibration

Calibrating the injection pump dynamically balances fuel delivery quantity output from the individual pumping elements. The injection pump is mounted to a comparator bench with sufficient power to handle the torque required to drive the pump when it is delivering full-fuel loads. The output from each pump element is measured in calibrated **vials** (graduates/burets) and often displayed on an LCD monitor. In a typical test profile, fuel output is recorded through a sequence of specified rotational speeds and simulated fuel demands.

Fuel quantity output in a pump element is adjusted by one of two methods. In certain fuel pumps, the barrels are manufactured with upper flanges that are located by means of eccentric stud holes; these permit some rotation of the barrels in their mounting bores. This rotation turns the spill ports relative to the plunger geometry and therefore alters plunger effective stroke. Another method used to balance fueling in port-helix pump elements is the use of control sleeve lock collars to clamp the plungers. When the lock collar is loosened, the plunger can be rotated within and retorqued to adjust the delivered fuel quantity.

Bench Testing

Comparator bench testing is a specialty. It requires manufacturer training and a properly equipped fuel injection pump repair facility to develop the techniques required to test fuel injection pumps. It also requires the manufacturer's service literature and specifications. The term *comparator* is used to describe the test bench because the performance output values of the pump being tested are *compared* directly with the manufacturer's specifications. A typical comparator bench procedure requires a test sequence that includes:

1. Phasing
2. Full-fuel quantity calibration
3. Peak torque rpm fuel quantity calibration
4. Droop calibration
5. High-idle rpm calibration
6. Idle speed calibration
7. Cranking fuel calibration
8. Start retard fueling
9. Aneroid operation and adjustment
10. Supply pump charging pressure

CRITICAL SYSTEM PRESSURE VALUES

The following is a review of some of the system values in a typical port-helix metering injection pump system. Much of the language of fuel injection systems stems from this original high-pressure injection pump system and is used in many of the systems that have evolved from it.

Charging Pressure

Charging pressure is generated by the transfer pump. It is the pressure in the injection pump charging gallery usually between 1 and 5 bar (15 and 75 psi), and it varies according to the system.

Delivery Valve Crack Pressure

When there is no residual line pressure in the high-pressure pipe, pressure equivalent to about 300 psi (20 bar) is required to crack the delivery valve that is held closed only by the delivery valve spring. This plus the residual line pressure is required to crack the valve when the engine is running.

Residual Line Pressure

Residual line pressure is the pressure the dead volume fuel is retained at in the high-pressure pipe. It is usually about two-thirds of the NOP value. Most fuel pumps attempt to hold this value static during operation, but in some injection pumps, the delivery valve seat has a compensation siphon (see Figure 20–13). The siphon port permits priming through the high-pressure pipe to the injector (unlike other systems) and also limits the time high pressure can be retained in the injection pipe.

Nozzle Opening Pressure

Also known as popping pressure, NOP is the pressure at which the nozzle valve unseats in a hydraulic injector. NOP parameters range between 2,200 and 5,100 psi (150 and 375 bar) in PLN multi-orifii nozzles. NOP represents the beginning of injection.

Peak Pressure

Peak pressure is the highest pressure a system can generate. In most systems, the pressure increases as the injection pulse is prolonged, so peak pressure will be attained only at full fuel. Peak pressure parameters range between 2 and 10 times NOP values.

DELIVERY, INJECTION, AND COMBUSTION

The function of any port-helix metering injection pump is to manage the fueling of an engine. This section attempts to match the critical events in the injection pump with the critical events in the engine; in other words, to describe how the fueling pulse effects combustion in the engine cylinders. Figure 20–17 demonstrates the relationship between what is occurring in the fuel injection pump, the injector, and the engine cylinder in real time. This graphic can actually be used to demonstrate fueling in most types of diesel fuel injection system.

Injection Lag

Injection lag is the time measured in crank angle degrees between PC in the injection pump

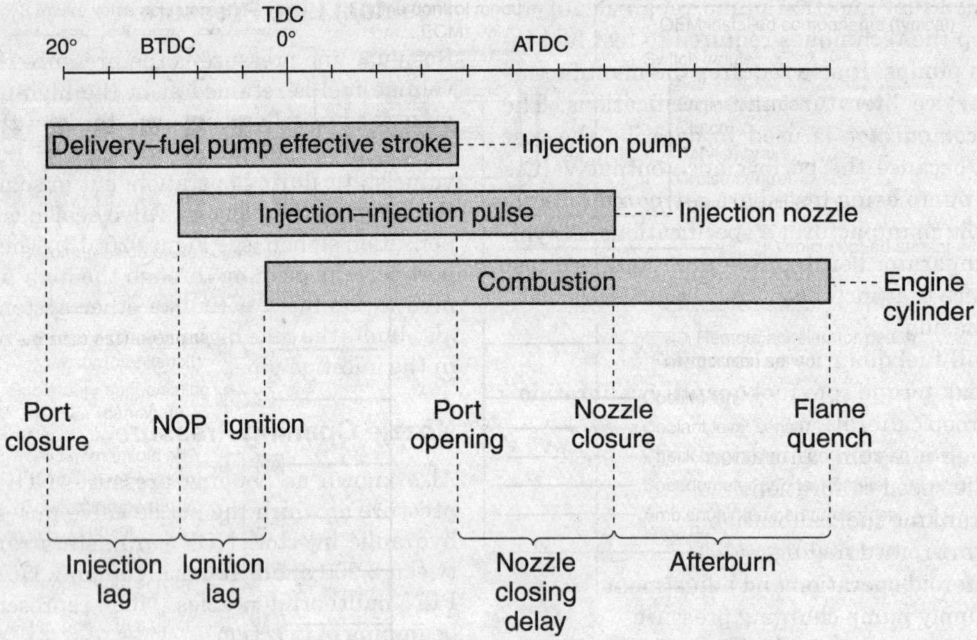

Figure 20–17 *The phasing of the events of delivery, injection, and combustion.*

and NOP, which is accounted for primarily by the time required to raise pump chamber pressure first to the residual line pressure value, then to the NOP value. Secondary causes of injection lag are:

1. The elasticity of the high-pressure pipe.
2. The compressibility of the fuel. This is approximately 0.5% per 1,000 psi (67 bar) but it must be accounted for when the system is engineered.

Injection lag can be regarded by the diesel technician as a constant.

Ignition Lag

Ignition lag is the time period between the events of NOP and ignition of the fuel charge. It is usually measured in crank angle degrees at any given engine rpm and load. Liquid droplets exiting the nozzle orifii must be vaporized and ignited. This time period is variable and depends on the ignition quality of the fuel (cetane number [CN] value) and the actual compression temperature. Because it is variable, so will be the amount of fuel in the cylinder at ignition. In cold weather startup conditions, this can be excessive and can result in a detonation condition known as diesel knock.

Combustion

The duration of combustion depends on the length of the injection pulse (that is, the total quantity of fuel injected to the cylinder and the point in crank angle degrees when injection ceases) and the engine rpm. The rate of injection is optimumly designed so that cylinder pressure and crank angle leverage are synchronized to deliver power to the flywheel as smoothly as possible.

Nozzle Closure Lag

Nozzle closure lag is the time period between the end of injection pump delivery and actual nozzle closure. This varies, increasing as the injection pressure pulse increases.

Afterburn

Afterburn is the normal combustion of fuel in the engine cylinder after injection nozzle closure. Its duration depends on the length of the injection pulse, the actual quantity of fuel in the cylinder, and many other factors including droplet sizing just before nozzle closure. Afterburn duration is managed to be as short as possible in later versions of hydromechanical, port-helix injection pumps by delivering higher peak pressures and using smaller injector nozzle orifii sizing, the result of which are smaller droplets of fuel that oxidize more quickly.

REVIEW QUESTIONS

1. The beginning of the injection pump effective stroke is known as:
   a. Port closure
   b. Port opening
   c. Injection lag
   d. Afterburn
2. The ending of injection pump effective stroke is known as:
   a. Port closure
   b. Port opening
   c. NOP
   d. Flame quench
3. The device used to rotate the injection pump plungers in unison is known as a:
   a. Barrel
   b. Tappet
   c. Rack
   d. Control sleeve
4. Charging pressures in a port-helix metering injection pump are created by a(n):
   a. Supply pump.
   b. Accumulator
   c. Vane pump
   d. Centrifugal pump
5. What component replaces the governor housing at the rear of a port-helix metering injection pump if it is managed by a computer?
   a. Electronic governor
   b. Rack actuator housing
   c. ECM
   d. Module
6. Which of the following defines injection rate?
   a. Engine speed
   b. Fuel demand
   c. Cam profile geometry
   d. NOP
7. Static timing can also be referred to as:
   a. Port closure timing
   b. NOP timing
   c. Point of ignition
   d. Completion of injection
8. Which of the following components would limit engine fueling at high altitudes?
   a. Barometric capsule
   b. Aneroid
   c. Governor
   d. Variable timing
9. A port-helix metering injection pump managing a direct-injected diesel engine would typically be port closure timed to what position on the engine?
   a. 50 degrees BTDC
   b. 50 degrees ATDC
   c. 20 degrees BTDC
   d. 20 degrees ATDC
10. A common cause of diesel knock is:
    a. Prolonged injection lag
    b. Afterburn
    c. Prolonged ignition lag
    d. Preignition
11. Technician A states that when static timing a port-helix metering pump to an engine, the pump #1 element must always be timed to the #1 engine cylinder. Technician B states that the spill timing method is seldom recommended by OEMs to time later-model port-helix metering pumps to engines. Who is correct?
    a. Technician A only
    b. Technician B only
    c. Both A and B
    d. Neither A nor B
12. When a port-helix metering injection pump that requires to be locked into position by a timing pin is to be installed to an engine, what else must also be done?
    a. The pump should be high-pressure timed.
    b. The engine should be locked into position by a timing bolt.
    c. Timing should be checked with a pulse-actuated timing light.
    d. Spill timing.
13. At what speed is a port-helix metering injection pump driven on a four-stroke cycle engine in relation to engine speed?
    a. 1/4 engine speed
    b. 1/2 engine speed
    c. Engine speed
    d. 2 times engine speed

14. 360 degrees of pump rotation is equal to how many degrees of engine rotation?
    a. 120 degrees
    b. 180 degrees
    c. 360 degrees
    d. 720 degrees
15. Which type of cam geometry is used to actuate the high-pressure pump elements in most inline port-helix metering injection pumps?
    a. Symmetrical of mostly base circle design
    b. Asymmetrical of mostly outer base circle design
    c. Antikick back
    d. Double lobe
16. Technician A states that the phasing of an inline port-helix metering pump involves the correct spacing of port closure in each pump element through its effective cycle and is usually performed with a degree wheel (protractor) on the comparator bench drive turret. Technician B states that calibration of an inline port-helix metering injection pump can be performed on the engine with the appropriate governor adjusting tools. Who is correct?
    a. Technician A only
    b. Technician B only
    c. Both A and B
    d. Neither A nor B
17. What is the function of the flutes machined into a typical delivery valve core?
    a. Guide the valve in the body
    b. Permit the valve to seal before it seats
    c. Define the residual line pressure
    d. Define valve core travel in the valve body
18. What is the function of the retraction collar/piston on the delivery valve core?
    a. Guide the valve in the body
    b. Permit the valve to seal before it seats
    c. Limit pressure wave reflection
    d. Limit back leakage to the pump chamber
19. Which of the following does most to define the actual residual line pressure value in a running diesel engine using a port-helix metering PLN fuel system?
    a. The swept volume of the retraction collar/piston
    b. The mechanical force of the delivery valve spring
    c. The dead volume fuel
    d. The length of the high-pressure pipe
20. An electronically managed port-helix metering injection pump has ECM-managed variable timing with a 20-degree window statically timed at 8 degrees BTDC. What is the timing window within which the ECM can select port closure?
    a. 28 degrees BTDC to 8 degrees BTDC
    b. 8 degrees BTDC to 12 degrees ATDC
    c. 8 degrees BTDC to 28 degrees ATDC
    d. 12 degrees ATDC to 8 degrees BTDC

21

Detroit Diesel Mechanical Unit Injection

INTRODUCTION

Detroit Diesel Corp. (DDC) hydromechanical, two-stroke cycle engines have not been Environmental Protection Agency (EPA) certifiable for use on North American highways since 1991. However, they were the bus and coach engine of choice for nearly 50 years, and due to the tendency of transit corporations to recondition components many times over, these engines will be with us for a little while yet with their numbers diminishing each year. DDC two-stroke cycle engines belong to a past era when considerations such as emissions and fuel economy were not a factor. The most likely venue to see one of these engines today is in a truck or tractor pull where the 12V–71 version of the engine lives on.

DDC two-stroke cycle engines enjoyed some popularity as a truck engine; this popularity peaked in the 1970s (I6–71, 8V–71, and 12V–71) and declined through the 1980s (8V–92). DDC was owned by General Motors until 1987 when the division was purchased by Penske Corporation; in 2001, DaimlerCorporation assumed ownership.

Features

In a **mechanical unit injector**-fueled engine, each engine cylinder has its own **unit injector**, which is essentially a rocker-actuated pumping and metering element and hydraulic injector nozzle combined in one unit. Each mechanical unit injector, or **MUI**, has fuel delivered to it at charging pressure, which is variable between 30 and 70 psi (200 kPa and 470 kPa) depending on engine speed.

Charging pressure is generated by a positive displacement **gear pump**, driven by the **Roots blower** drive shaft, and responsible for all movement of fuel through the fuel subsystem. Engine output is controlled by a mechanical governor (in most truck and bus applications) that regulates fuel quantity by controlling the unit injector fuel racks.

DDC mechanical unit injection system engines were engineered using the English measurement system, and most of the technical specifications are listed by DDC in standard-only values. As a consequence, standard system values are used where specifications are referred to in this chapter.

ENGINE IDENTIFICATION

DDC uses the alphanumeric codes shown in Figure 21–1 to identify its two-stroke cycle engines. To perform a tune up on a two-stroke diesel engine the technician should be familiar with these engine identification codes and those used to identify the governor.

It is essential to identify the governor used on a specific DDC two-stroke diesel engine before beginning a tune up because the procedure varies with each. The following lists governor acronyms used in highway applications:

LS—Limiting speed
VS—Variable speed

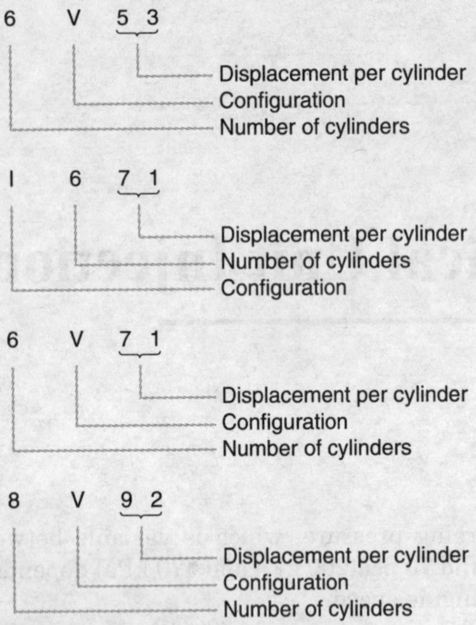

Figure 21–1 *DDC engine identification.*

DW—Double weight
SW—Single weight
TT—Torque tailored

FUEL SYSTEM COMPONENTS

The layout and most of the fuel system components are almost identical for 53, 71, and 92 series engines. The components listed in the fuel subsystem are for typical applications; some differences may occur in specific chassis requirements of the fuel subsystem such as in bus, coach, vocational, and truck applications due to such factors as distance of engine from fuel tanks and controls.

Fuel Filters

Most DDC two-stroke cycle engines with MUIs use a typical diesel fuel filter configuration consisting of two filters, one of which is under suction and the other under charging pressure produced by the gear pump.

Primary Filters

The primary filter is located in series between the fuel tank and the engine-mounted gear pump; this filter is under suction. Older applications used nondisposable canisters, within which was a disposable element. This element was surrounded by a cloth enclosure and the flow of fuel was from the outside of the element to the inside, with filtered fuel exiting the center of the assembly. Most newer engines use disposable spin-on cartridges. Older engines may be retrofit with filter mounting pads that use the spin-on cartridges for ease of servicing and higher filtering efficiencies. The filter mounting pads are fitted with an inlet check valve to prevent fuel siphoning (drain back) to the tank.

Secondary Filters

The secondary filter is located in series between the gear pump and the inlet port in the cylinder head fuel manifold. Secondary filters on most current engines are disposable, spin-on filter cartridges, although some of the older canister and element filters can still be seen.

Current spin-on filter cartridges entrap particulate down to 1 micron and inhibit water from passing through the filtering media to the extent that they will plug and starve the engine for fuel.

Gear Pump

The transfer pump is an external gear pump. It is an engine-driven, positive displacement pump responsible for charging the unit injectors with pressures between 30 and 70 psi (200 and 470 kPa). It is driven by the Roots blower drive shaft.

Fuel Lines and Fuel Manifold

Hydraulic hose is used to connect the fuel tank with the filters and pump and deliver fuel to the fuel manifolds. In 53, 71, and 92 series, the fuel manifolds are drillings within the cylinder head(s) that deliver fuel to the unit injectors at charging pressure and return fuel back to the tank(s). A restriction fitting is located at the exit port of each return fuel manifold. This defines the flow area that establishes the charging pressure window. A relief valve in the transfer pump prevents the charging pressure from exceeding 70 psi (470 kPa).

Jumper Pipes

Jumper pipes connect the fuel manifolds in the head with the unit injectors; one charges the unit injector, the other returns fuel to the return manifold. Fuel is cycled through the system whenever the engine is running; in other words, all of the fuel delivered to the MUIs is not used for fueling the engine. Fuel cycled through the MUI circuitry is used for cooling and lubricating the internal components of the assembly.

DDC MECHANICAL UNIT INJECTORS

A mechanical unit injector combines a complete pumping, metering element and a hydraulic injector nozzle in a single, cam-actuated unit. There are two types differentiated by nozzle valve design: **crown valve nozzle** and **needle valve nozzle**. However, the crown valve type with NOP values of 800 psi can be considered obsolete and is not discussed here. The needle valve type was introduced in the early 1970s and is almost universal in any DDC mechanical engines likely to be around today. Figure 21–2 shows a cutaway of a V-configured DDC MUI-fueled engine, and Figure 21–3 shows a sectional view of an MUI: Note the plunger recess geometry.

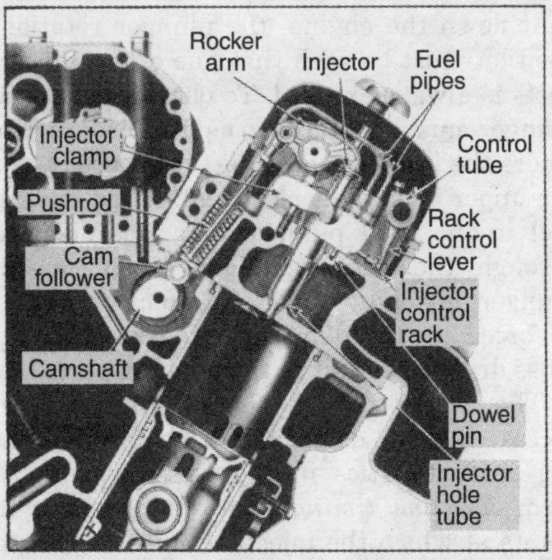

Figure 21–2 *DDC engine cutaway showing the location of the MUI. (Courtesy of Detroit Diesel Corporation)*

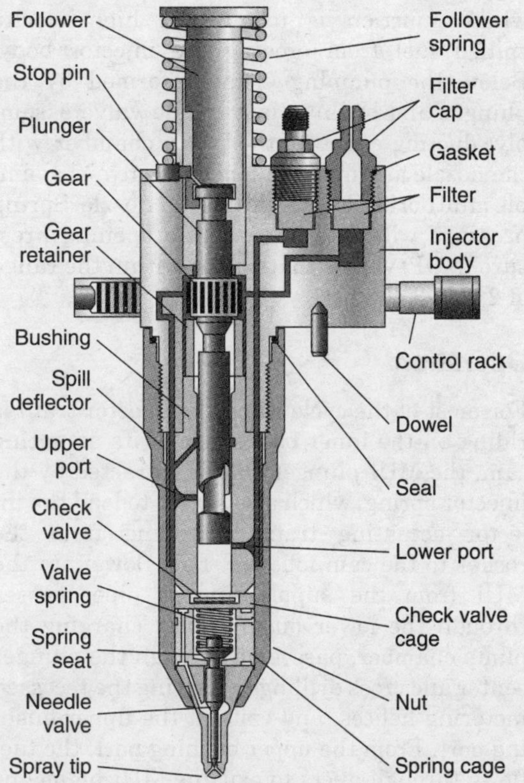

Figure 21–3 *Sectional view of a DDC mechanical unit injector.*

Design

The **plunger** and **bushing** within the unit injector can be compared to the pumping element in an inline, port-helix metering pump. The bushing is stationary and machined with upper and lower ports, 180 degrees offset. The plunger reciprocates within the bushing. It is actuated by the engine camshaft by a cam profile that is mostly inner base circle and an injector train consisting of a **follower**, rocker, and pushrod. The plunger is milled (a machining process) with helical **metering recesses** and is center and cross drilled.

The unit injector follower is lug connected to the plunger and the injector follower spring serves to load the plunger to ride its actuating cam profile. A gear positioned over the upper portion of the plunger is tooth-meshed with the **control rack**, permitting the plunger to be rotated within the bushing when the rack is moved linearly. Surrounding the bushing is a **spill deflector**; this is a stellite alloy sleeve whose function is to prevent high-velocity spilled fuel from eroding the injector body. Below the pumping element formed by the plunger and bushing is a needle valve assembly; ducting connects the pump chamber with the nozzle assembly. The needle valve is a simple multi-orifii, hydraulic injector nozzle. Spring pressure will determine nozzle opening pressure (NOP) values that will be within the range of 2,200–3,400 psi.

Operation

For most of the cycle when the injector train is riding on the inner base circle of its actuating cam, the MUI plunger will be retracted by the injector spring, which also serves to load the injector actuating train, extending from the rocker to the cam follower. Fuel flows into the MUI from the supply jumper pipe, passes through the lower bushing port charging the pump chamber, passes up through the plunger center and cross drillings, charging the recessed metering helices, and exits at the upper bushing port. From the upper bushing port, the fuel flows through ducts to exit the MUI by means of return jumper pipes.

Actual plunger stroke is defined by cam profile geometry and therefore will not vary. At each rotation of the camshaft, each MUI plunger is driven through a single stroke that begins when the cam ramps off its base circle toward peak lift. Cam peak lift represents the maximum point of downward travel of the plunger into the bushing pump chamber. After peak lift, the cam ramps back toward cam base circle and the injector spring lifts the plunger back into its retracted position.

Effective Stroke

The term **effective stroke** is used to describe that portion of the plunger stroke where fuel is actually being pumped. The actual fuel quantity injected to the engine's cylinders fueling is therefore the result of the MUI effective stroke. The length of the effective stroke depends on the rotational position of the plunger, which is controlled by the MUI rack. Specifically, it depends on where the recessed metering helices register with the upper and lower ports machined into the bushing.

For an effective stroke to occur, both bushing ports must be closed for some part of the downward stroke of the plunger. When the injector cam rotates off base circle, the cam ramp loads the injector train and the rocker arm drives the plunger through its downward stroke within the bushing. As the plunger begins to move, its leading edge first closes the lower bushing port. As the plunger descends through its stroke, fuel in the pump chamber is displaced, passing through the plunger center and cross drillings and exiting the upper bushing port. Fuel continues to be displaced by the plunger until the helical edge of the plunger metering recess closes off the upper bushing port; because the lower bushing port has already been closed, the fuel is trapped in the pump chamber, and the effective stroke begins.

Effective stroke or pumping stroke will continue until the lower edge of the metering recess is exposed to the lower bushing port; at this point, the high-pressure fuel in the pump chamber spills, exiting the pump chamber through the now exposed lower bushing port. Figure 21–4 shows how MUI effective stroke takes place.

The length of the effective stroke depends on the rotational position of the plunger and the points at which the plunger metering recesses register with the upper and lower bushing ports. Because an effective stroke can only take place when both bushing ports are closed off, to shut down the engine, the plunger rotational position must be such that one of the bushing ports is always exposed. To obtain no-fuel, the plunger must be rotated to a position of register where the lower bushing port is exposed before the upper bushing is closed off. At no-fuel, the fuel in the MUI pump chamber is displaced through the entire plunger stroke; as the plunger descends into the pump chamber, fuel is forced up through the plunger center and cross drillings to exit at the bushing ports.

The geometry (shape) of the metering recesses/helices determines the injection timing characteristics in the unit injector. The term *injection timing* is used to denote the points at which the injection pulse begins and ends; these points are usually specified in engine crank angle degrees. Depending on application, these may be:

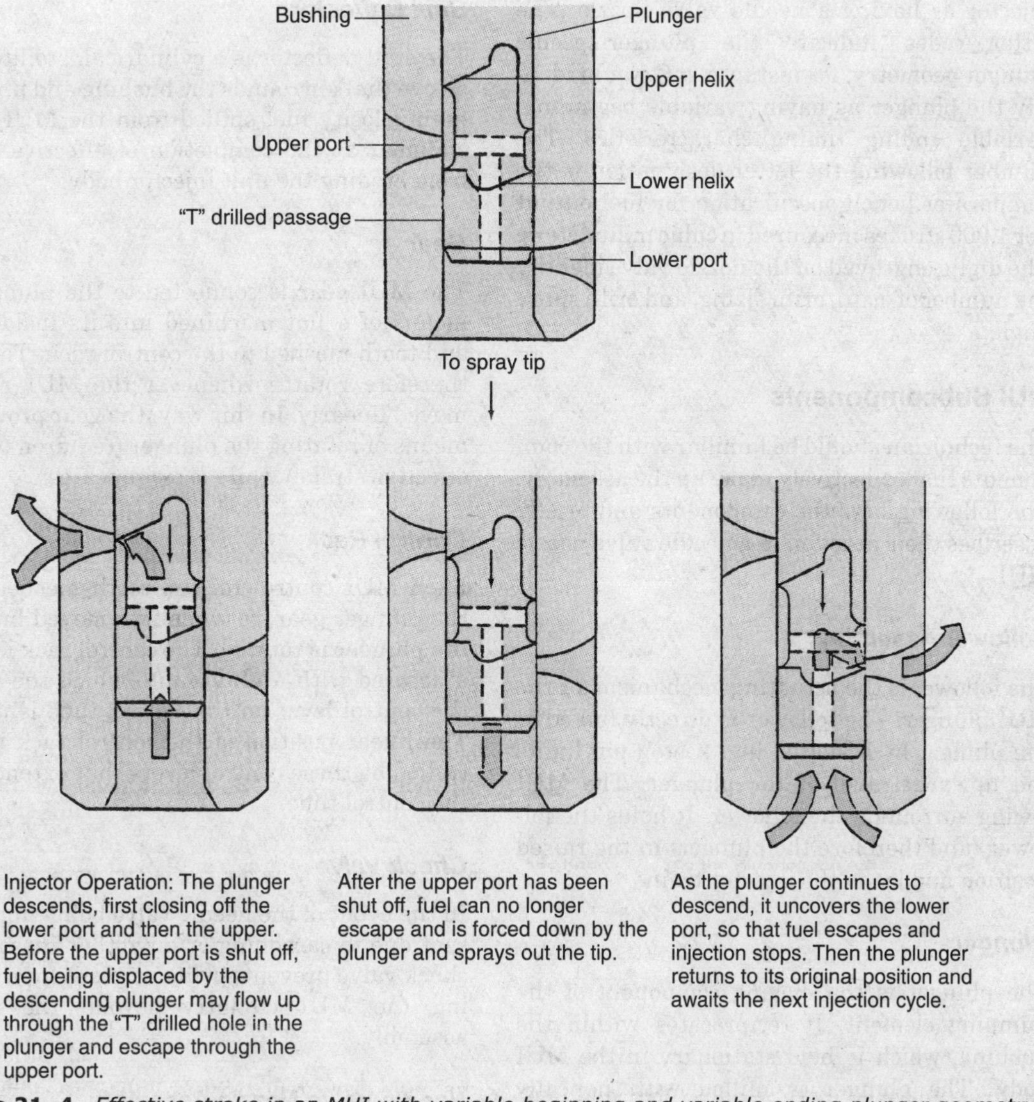

Injector Operation: The plunger descends, first closing off the lower port and then the upper. Before the upper port is shut off, fuel being displaced by the descending plunger may flow up through the "T" drilled hole in the plunger and escape through the upper port.

After the upper port has been shut off, fuel can no longer escape and is forced down by the plunger and sprays out the tip.

As the plunger continues to descend, it uncovers the lower port, so that fuel escapes and injection stops. Then the plunger returns to its original position and awaits the next injection cycle.

Figure 21–4 *Effective stroke in an MUI with variable beginning and variable ending plunger geometry. (Courtesy of Detroit Diesel Corporation)*

- Variable beginning, variable ending of injection pulse
- Constant beginning, variable ending of injection pulse
- Variable beginning, constant ending of injection pulse

Unit Injector Identification

A circular identification tag pressed into the unit injector body identifies the class number. Unit injectors with no line under the manufacturer name are generally those with the obsolete crown valve nozzles. A needle valve nozzle would have this line under the manufacturer's name:

Manufacturer ID	GM	Reliabilt	DDC	DDC
Code	N60	B60	N70	C70
Nozzle type	Needle	Crown	Needle	Needle

In the first example in the table, the line under the manufacturer name identifies the injector as having a needle valve nozzle. The letter codes indicate the plunger-specific plunger geometry; for instance, a C would identify the plunger as having variable beginning, variable ending timing characteristics. The number following the letter designation is the comparator bench specification for fuel output per 1,000 strokes measured in cubic millimeters. The digits engraved on the nozzle valve identify the number of orifii, orifii sizing, and orifii spray angle.

MUI Subcomponents

The technician should be familiar with the components that collectively make up the assembly. The following lists the components and briefly describes their function in a needle valve nozzle MUI.

Follower Assembly

The follower is the actuating mechanism for the MUI plunger. The follower is directly linked to the plunger by a slotted lug. A stop pin limits the upward travel of the plunger. The MUI spring surrounds the follower. It holds the follower (and therefore the plunger) in the raised position and loads the injector train.

Plunger

The plunger is the moving component of the pumping element. It reciprocates within the bushing, which is held stationary in the MUI body. The plunger is milled with helically shaped metering recesses, for purposes of varying effective stroke, and center and cross drilled, to provide a hydraulic connection between the pumping chamber below the plunger and the metering recesses. Plungers are lapped to bushings in manufacture. Lapped components are not interchangeable.

Bushing

The MUI bushing is a cylindrical housing within which the plunger reciprocates and with which it forms the pump element. The bushings are drilled with diametrically opposed upper and lower ports and are positioned in the MUI by dowels.

Spill Deflector

The spill deflector is a cylindrical, stellite alloy sleeve that surrounds the bushing and prevents high-velocity fuel spilled from the MUI pump chamber at the completion of effective stroke from eroding the unit injector body.

Gear

The MUI gear is connected to the plunger by means of a flat machined into its inside bore and tooth meshed to the control rack. The gear therefore rotates whenever the MUI rack is moved linearly. In this way, the gear provides a means of rotating the plunger (required to alter effective stroke) while it reciprocates.

Control Rack

Each MUI control rack is tooth meshed with the plunger gear, so when it is moved linearly, the plunger is rotated. The control rack is manufactured with a **clevis** into which the foot of the control lever on the control tube is placed. The linear position of the control rack is controlled by these control levers that extend from the control tube.

Check Valve

In the event of the needle valve being unable to seat due to carboning or any other reason, the check valve prevents cylinder gases from entering the MUI circuitry beyond the nozzle assembly.

Needle Valve Nozzle Assembly

The DDC needle valve assembly uses all the same operating principles as any other multi-orifii nozzle assembly (this subject matter is handled in detail in Chapter 20). As with any hydraulic injector nozzle, an MUI needle valve nozzle is held closed and seated by the nozzle valve spring. Fuel pressures developed in the MUI pump chamber below the plunger are ducted to act on the pressure chamber sectional area of the nozzle valve. When that pressure is sufficient to overcome the nozzle valve spring tension, the needle valve retracts, allowing high-pressure fuel from the MUI pump chamber to pass around its seat to a sac and exit through nozzle orifii, beginning the injection pulse. When the pump element effective stroke ends, fuel starts to spill from the pump chamber, exiting the lower bushing port. The moment that fuel pressure acting on the sectional area of the needle valve is insufficient to hold the valve spring retracted, it closes, ending the injection pulse.

Injector-to-Governor Linkages

The output of the mechanical unit injectors is governor controlled by means of a set of linkages that mechanically connect the governor with the MUI. Governor **control rods** extend from the **governor differential lever** (twin arm lever that pivots on a fulcrum) in the governor housing and connect to a control tube lever by means of a clevis and pin; linear movement of the governor control rods will by this means be converted into rotary movement of the control tube, which runs lengthwise through the upper cylinder head. Extending from the control tube are the **rack levers**. When the control tube is rotated by the control rods, the rack levers rotate with it. The feet of the rack levers connect to the MUI fuel racks by means of a clevis, so once again, rotary movement is converted to linear movement to position the racks. The mechanical relationship between the MUI rack and the rack lever is adjusted by an adjusting screw (or pair of screws in older engines) on the rack lever to control tube clamp.

Therefore, the rack levers link the individual MUIs with the control tube. When the control tube is rotated by the governor control rods, the rack levers extending from the control tube move the unit injector racks linearly. A critical procedure in DDC engine tune up is ensuring that the unit injectors are balanced; that is, that the point of register of the helices with the bushing ports in each unit injector in the engine is identical. This is achieved by correctly setting the control tube adjusting screw (newer engines) or adjusting screws (older engines). The result of balancing is equal fueling of each cylinder.

GOVERNOR

A simple mechanical governor is used in most truck and bus applications. Governor components consist of a set of engine-driven flyweights, a thrust collar, **bell crank**, spring set, and differential lever. As in most mechanical governors, centrifugal force exacted by the flyweights attempts to diminish engine fueling while spring force moderated by speed control lever position attempts to increase engine fueling. The thrust collar acts as an intermediary between the spring force defined by precise adjustment and moderated by the speed control lever position and the centrifugal force that directly correlates to engine speed. Extending from the governor housing are the control rods that link the MUI fuel control linkages with the governor assembly.

Both limiting speed and variable speed governors are used. The governor type should always be identified prior to engine tune up as the procedure varies with governor type.

DETROIT DIESEL TUNE UP

Although it is not an objective of this textbook to address specific workshop procedure, the DDC tune up is one instance in which the workshop text might benefit from some enhancement. The following is intended to be used as a guide to DDC tune up of two-stroke-cycle engines and it should always be used in conjunction with DDC workshop texts, specifications, and service bulletins. Initially, the engine and its governor type should be identified.

DWLS—double weight, limiting speed (mobile equipment)

SWLS—single weight, limiting speed (mobile equipment)

SWVS—single weight, variable speed (industrial, marine)

DWDRG—double weight, dual range governor (mobile/highway)

The tune-up procedure does not require a readjustment of every setting—it is a procedural method of checking for possible change in the settings.

The following description is loosely based on V–92 or 71 series tune-up procedure, and it should be remembered that the procedures will differ somewhat from engine to engine.

GENERAL TUNE-UP GUIDELINES

DDC prefers that all tune-up adjustments be made with the engine at 160–185°F coolant temperature. As many of these adjustments have to be made with the engine stopped, it may have to be run between adjustments to maintain the required temperature. Ensure before every startup, that when the stop lever is engaged, the injector racks move to the no-fuel position. The consequence of not doing this could result in engine overspeed or even a runaway.

If the engine classification has a **TT** suffix, it is a **tailored torque**, highway application fuel squeezer engine with a Belleville spring retainer governor. The spring retainer nut must be backed out until there is approximately 0.060″ clearance between the washers and the retainer nut (Figure 21–5).

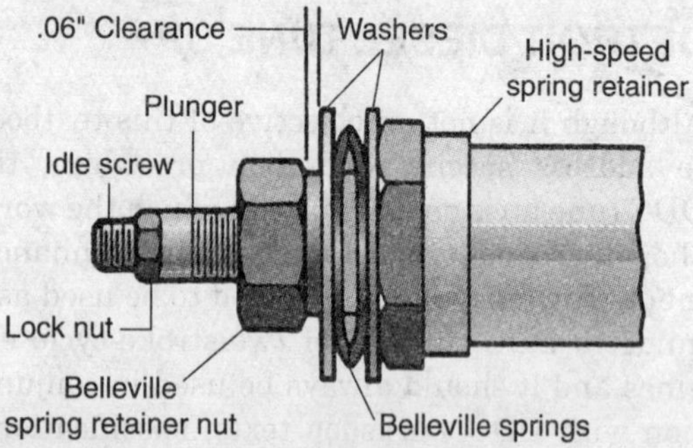

Figure 21–5 *Belleville washer location. (Courtesy of Detroit Diesel Corporation)*

Adjust Exhaust Valve Bridges

Adjusting exhaust **valve bridges** procedure is required when a cylinder head is removed/overhauled/replaced or if there is a reason to suspect that it should be checked. Remove the bridge assembly from the cylinder head (essential: Attempting to do this on the engine can bend the bridge guide stem) and mount it in a soft-jawed vise; loosen the locknut on the adjusting screw. Remove the bridge from the vise and install on the bridge guide; turn the adjusting screw clockwise (CW) until it contacts the valve stem then rotate an extra 1/8 of a turn. Hold with locknut finger tight, remove the bridge, and install once again in a soft-jawed vise; hold the adjusting screw stationary and torque the locknut to specification (25 lb-ft.). Reinstall the bridge saddle on the bridge guide. To check, cut two pieces of 0.0015″ shimstock (or thickness gauges) so they fit under the bridge saddle. Load the bridge onto the guide with finger pressure and check for even drag as the two pieces of shimstock are simultaneously withdrawn from either side of the bridge. Ensure that all the valve bridges are properly positioned when the rocker assembly is reinstalled. Figure 21–6 shows a valve bridge adjustment.

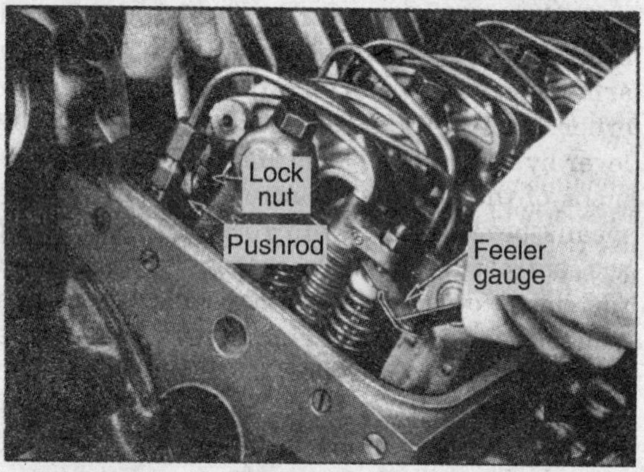

Figure 21–6 *Adjusting valve bridges. (Courtesy of Detroit Diesel Corporation)*

Valve Lash Adjustment

Unless there is substantial reason to suspect the valve bridge adjustment such as excessive valve clatter, this adjustment can be omitted from the routine tune up. Valve lash in a DDC two-stroke cycle engine can be set/checked in one engine rotation. Check the engine cylinder firing sequence. Engine rotation is always determined by observing the engine from the front. In a V configuration engine, the cylinder banks are oriented by observing the engine from the rear. Following is the cylinder firing sequence for an 8-V-configured engine.

CW = RH–IL 3R 3L 4R 4L 2R 2L 1R
CCW = LH–IL IR 2L 2R 4L 4R 3L 3R

It is not necessary to align timing marks to set the valves on these engines; observing the action of the injector actuating train will ensure that the valves are closed. Precise positioning of the engine on the timing indices is not required.

A remote starter switch can be used to bunt the engine to each valve set position.

Bunt the engine using an auxiliary switch and the cranking motor to the required position for adjusting each set of valves. Adjustment should be performed when the injector rocker is fully actuated. Check data in the workshop manual for the lash specifications. DDC prefers the use of go-no-go thickness gauges. Remember also that cold lash settings differ from hot lash settings. Figure 21–7 shows a valve lash adjustment.

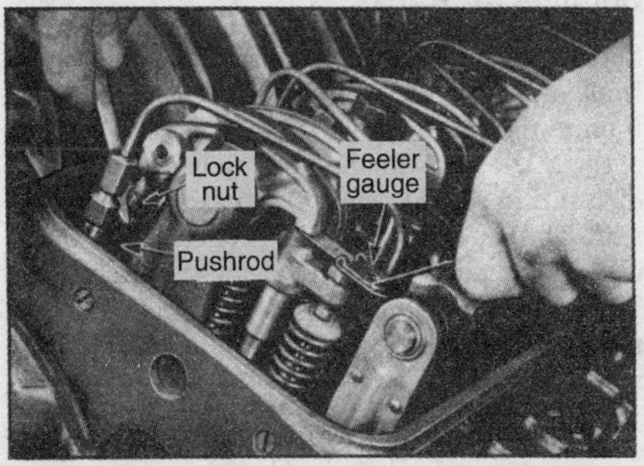

Figure 21–7 *Adjusting valves on a four-valve cylinder head. (Courtesy of Detroit Diesel Corporation)*

Fuel Injector Timing

The injector **timing dimension** specification can be found stamped on the valve/rocker cover plate. A timing pin with the correct dimension stamped on its flange should be selected. The engine may be rotated in the same manner as when performing the valve adjustment, and there is no reason why both adjustments cannot be set simultaneously, observing correct engine position and sequencing. Set the timing dimension when the valves in the cylinder are fully depressed. Insert the dowel in the timing gauge into the injector timing hole located in the upper flange of the MUI. Smear a drop of engine oil on the injector follower flange and then adjust the injector pushrod so that when the timing tool is rotated, it should gently wipe the oil film from the follower.

The injector-actuating train consists of a solid pushrod that connects to the rocker by a pivot pin and clevis; the pushrod connects to the clevis by screw threads and a locknut. An adjustment is therefore performed by backing off the locknut and either reducing or lengthening the pushrod total length to set the specified plunger travel. All the injectors are set in the same way; that is, setting actual plunger travel by adjusting the MUI actuating train dimension. Figure 21–8 shows the timing procedure used on an MUI.

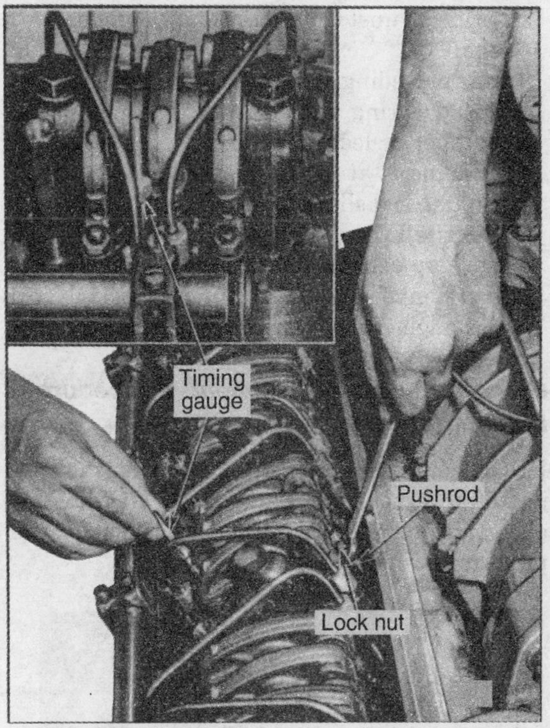

Figure 21–8 *Fuel injector timing. (Courtesy of Detroit Diesel Corporation)*

Governor Gap Adjustment

Governor gap adjustment is a critical step to engine performance and if not properly set, the result will be a lack of power and fueling flat spots. The procedure varies somewhat according to engine series and governor type so check DDC service literature.

Governor Gap Adjustment Procedure—Engine Stationary

1. Remove the governor cover and back out the **buffer screw** and the **starting aid screw**, if equipped.

2. Next, bar the engine to position the governor weights in a horizontal position, then insert a **governor weight fork** (the common practice of using a screwdriver to spread the weights is not recommended) down between the inner lobe of either one of the low-speed governor weights (the larger weights) and the weight riser shaft. Use the governor weight fork to force the weights outward to their fully extended position against the stop on the weight carrier shaft.
3. The preceding step will have forced the low-speed spring cap against the high-speed plunger. Select the correct thickness gauge and insert it at the governor gap; there should be a slight drag. Adjust if required.
4. Reinstall the governor cover. If the governor gap setting was checked cold, the adjustment should once again be checked with the engine at operating temperature.

Governor Gap Adjustment Procedure—Engine Running

DDC has technical service bulletins on the subject of governor gap adjustments with the engine running (Figure 21–9) but typically the procedure would be as follows:

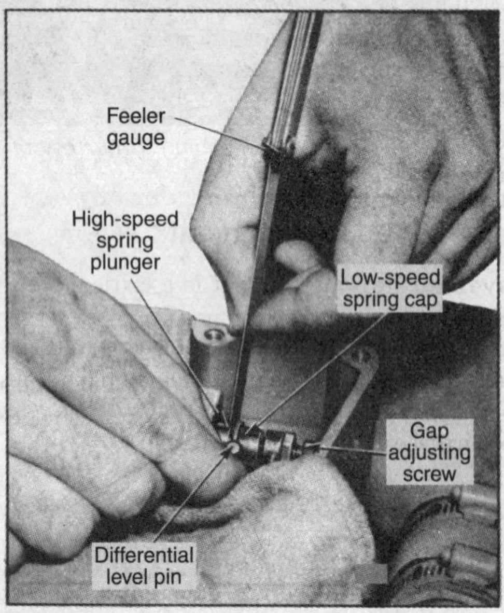

Figure 21–9 *Governor gap adjustment. (Courtesy of Detroit Diesel Corporation)*

1. Nonfuel squeezer engines: The engine should be run between 800 and 1,000 rpm and the governor gap should be 0.0015″. When using shimstock cut with mechanics shears or scissors, and ensure that the edge has not been distorted to a much greater-than-specified dimension.
2. Fuel squeezer (TT) engines: The engine should be run between 1,100 and 1,300 rpm and the governor gap should be 0.002″ to 0.004″.

Set Injector Racks

Setting injector racks is a critical adjustment that must be precisely executed to balance fueling to each engine cylinder. Erratic setting of the MUI racks during tune up is a main cause of hunting, a rhythmic fluctuation in engine rpm. The exact linear position of each MUI rack is determined by its control lever adjustment. The governor's location on the engine determines which injector rack should be set first; this is always that closest to the governor.

1. Disconnect all linkages connected to the governor speed control lever.
2. Back out the idle speed adjusting screw until there is no tension on the low-speed spring (LS governors). When backed out approximately 1/2″, the low-speed spring will have little or no tension, permitting the closing of the low-speed gap without risk of bending the fuel rods or damaging the yield link spring mechanism (used with **throttle delay** mechanisms). Failure to back out the idle speed adjusting screw will result in inaccurate fuel rack settings.
3. The throttle delay mechanism should have been removed when setting the governor gap; ensure that it remains off. Also, back out the buffer screw approximately 5/8″ as it was before setting the governor gap.
4. On TT engines, the Belleville spring retainer nut should have 0.060″ clearance, and on engines with a starting aid screw, this should be left backed out.
5. Loosen the adjusting screw (inner and outer adjusting screws on older, two-screw rack lever adjusters) of each MUI rack control lever on the control tube. Ensure that the foot of the rack lever is not binding into the

MUI **rack clevis**. More recent model engines use a single adjusting screw and locknut with a torsion spring instead of the two-screw rack adjusters to prevent jamming a whole bank into the full-fuel position in the event of a single MUI rack seizure.

6. On V-configuration engines, remove the clevis pin from the right bank fuel rod connection to the control tube, leaving the left bank control tube connected to its fuel rod.
7. On engines with VS mechanical governors, the speed control lever should be pinned into the full-fuel position while setting the injector racks, and the stop lever should be held in the run position. On LS mechanical governors, load the speed control lever into its full-fuel position with light finger pressure.
8. Next, with both LS- and VS-governed engines, observing the previous step, turn CW the inner adjusting screw (two-screw type) or adjusting screw (one-screw type) on the injector closest to the governor until the rack control clevis can be observed to kick up; at this point, a slight increase in effort required to turn the screwdriver should be detected. On the single-adjustment screw type, turn the screw CW on further 1/8 turn, then lock in position with the locknut. On the older two-screw adjustment rack levers, turn the inner screw a further 1/8 turn CW, then screw in the outer screw until it bottoms. Next, alternately tighten each of the screws by turning 1/8 turn CW, finally torquing to specification (24–36 lb-in.).
9. Next, on LS governor engines, move the speed control lever through its travel arc or, likewise, the stop lever on VS engines through its arc; if any drag is detected, the rack should be readjusted.
10. On all engines, the first MUI rack to be adjusted is the one closest to the governor; this becomes the master rack once adjusted according to the procedure outlined in the previous step. The remaining rack control levers are adjusted as follows:
    - Inline engines: Remove the fuel rod clevis pin from the control tube actuating arm, then hold the injector control racks in the full-fuel position using the control tube. Adjust each MUI rack lever in sequence moving away from the master rack down the cylinder bank. Check to ensure that each rack clevis has equal spring back when tested with the screwdriver. Never readjust the master rack once this is properly set.
    - V-configuration engines: Remove the clevis pin from the fuel rod at the left bank injector control lever. Reinstall the clevis pin in the fuel rod in the right bank, reconnecting it with the governor. Adjust the #1 right bank cylinder injector rack exactly as outlined for the left bank #1 cylinder. To verify that both #1 left and #1 right racks are balanced, connect the left bank fuel rod again, then move the speed control lever (LS governor) to maximum speed or the stop lever (VS governor) to the run position. Check for drag or binding at the clevis pin connecting each bank. Next, check the spring back at the rack clevis on both the #1 left and #1 right cylinders; this should be identical. If not, readjust the #1 right cylinder. This must be correct before proceeding.

Now remove the fuel rod clevis pins and readjust each bank of injector racks sequentially, moving away from the "master" on each bank. Figure 21–10 shows an injector rack adjustment on a two-screw rack lever.

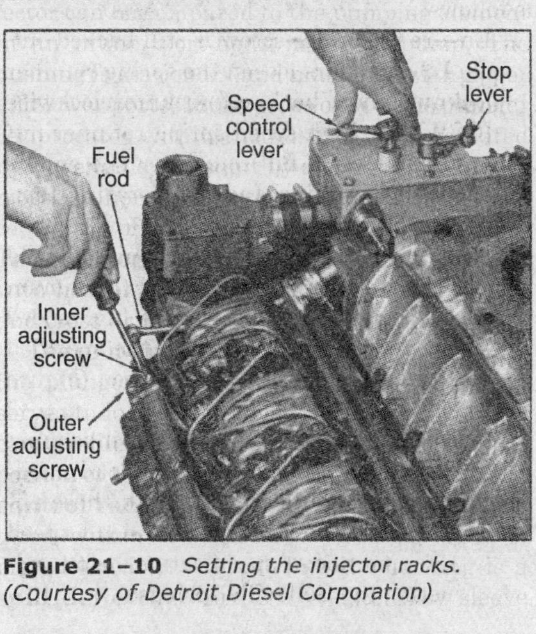

Figure 21–10 *Setting the injector racks. (Courtesy of Detroit Diesel Corporation)*

11. When all the injector racks have been set, install the fuel rod clevis pins. Check for drag on each clevis pin and once more re-check spring back at rack clevises (Figure 21–11).

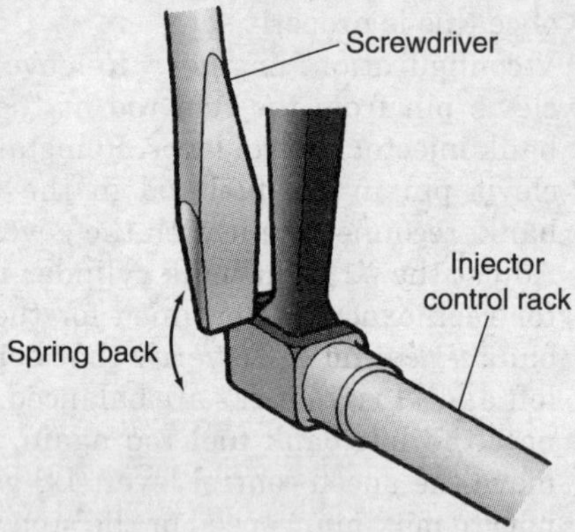

Figure 21–11 *Checking rack clevis spring back. (Courtesy of Detroit Diesel Corporation)*

12. Finally, secure clevis pins with cotter pins and screw the idle adjust screw until it projects 3/16″ from the locknut; this will allow the engine to start.

Adjust Maximum No-load (High-Idle) Engine Speed (LS Governors)

The specification should be stamped on the engine option plate on the rocker cover. Proceed as follows:

1. Ensure the buffer screw is still backed out. On LS governors loosen the spring retainer locknut, then back off (counterclockwise [CCW]) the high-speed spring retainer nut approximately 5 full rotations. The engine should be running and at operating temperature with no load on it.
2. Next, using an accurate tachometer, hold the speed control lever in the maximum speed position then rotate the high-speed spring retainer nut CW until the specified maximum no-load rpm is achieved (Figure 21–12).
   - TT engines: The Belleville springs must be readjusted following a no-load maximum speed adjustment. With these engines, set the maximum no-load speed 150 rpm above specification before adjusting the Belleville springs.

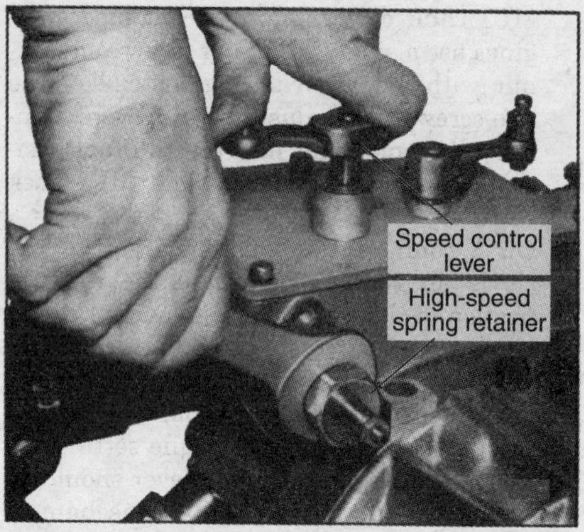

Figure 21–12 *Adjusting the maximum no-load speed. (Courtesy of Detroit Diesel Corporation)*

   - VS governor: Shims are used on the bell crank end of the governor spring to set the maximum no-load speed (spring tension is set by shims rather than a nut). Consult the workshop manual for specifications. This method is also used on some 53 series with LS governors.

Adjust Idle Speed

The engine should still be at operating temperature and the buffer screw backed out. Check the idle speed specification. Loosen the locknut on the idle screw and rotate either CW or CCW to set idle speed 15 rpm below the recommended value. It may be necessary to use the buffer screw to reduce engine rpm roll/hunting while performing this operation. However, once set, lock the idle screw with the locknut and back out the buffer screw (Figure 21–13).

Belleville Spring Adjustment (TT Engines Only)

This is a very critical adjustment and fractional inaccuracies can result in greatly reduced power output. Reference Figure 21–5.

1. Idle drop method—The engine tune up should have been performed up to this stage. Use an accurate tachometer for rpm readings. Disconnect the accelerator linkages. The engine should be at running temperature. Refer to DDC service literature to

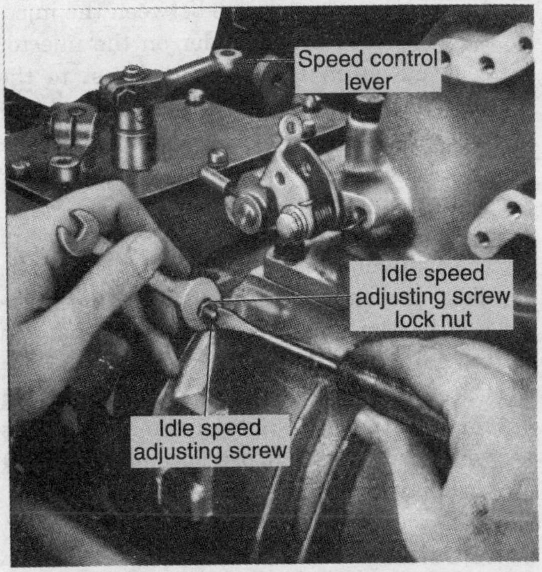

Figure 21–13 *Adjusting idle speed. (Courtesy of Detroit Diesel Corporation)*

obtain the idle drop specifications, peak brake horsepower, and rated engine speed values. Set the initial idle speed by using the idle adjust screw. Then with the speed control lever in the idle position, turn the Belleville spring retainer nut CW until the specified idle drop value is obtained exactly. This must be performed with precision, and it should be noted that a difference of 1 rpm off the specified idle drop value can result in a 2 to 3 BHP power loss. Then, lock the Belleville spring retainer nut. Next, lower the idle speed to the required specification using the idle adjust screw.

2. Chassis dynamometer method—This adjustment can be made on a chassis dynamometer, but it is not recommended due to the variables of dynamometer calibration, chassis driveline efficiency, air density, tire slippage, and the fuel cetane number (CN). It does set the peak brake horsepower (BHP) value, however, and may be preferred by some operators. Consult the engine specifications, and remember that engine power at the flywheel and engine power at the wheels are entirely different.

Adjust Buffer Screw

Engine idle speed should have been set at a value below specification in a previous step (15 rpm) and the engine should be at operating temperature. Turn the buffer screw CW so that it just begins to contact the governor differential lever; this should have the effect of fractionally increasing engine speed and leveling off engine rpm roll. Use an accurate tachometer and ensure that the idle speed does not exceed the required specification. Next, check the high-idle speed and if this has increased by more than 25 rpm, back off the buffer screw. Lock up the buffer screw when rpm values are set to specification. Figure 21–14 shows a buffer screw adjustment.

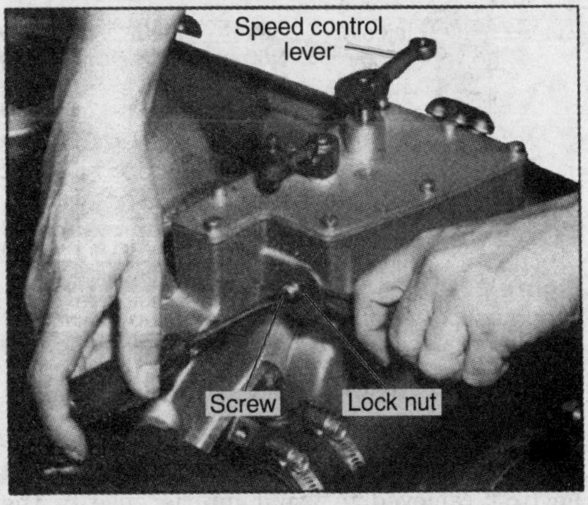

Figure 21–14 *Adjusting the buffer screw. (Courtesy of Detroit Diesel Corporation)*

Adjust Starting Aid Screw (DWLS-Turbo Boosted Engines)

The starting aid screw is designed to minimize fuel injected during cranking to reduce exhaust smoke. It is necessary because the turbocharger impeller restricts airflow to the cylinders until the engine fires, and exhaust heat and flow act to rotate the turbine. Adjust the external (internal on some governors) adjusting screw on the governor as follows:

1. With the engine not running, the stop lever in the run position, and the speed control lever in the idle position, select the correct gauge as per DDC technical literature. The setting can be measured at any convenient cylinder location.
2. When the starting aid screw is properly adjusted, the gauge should have an end clearance of 1/64″. Lock up the starting aid screw as shown in Figure 21–15.

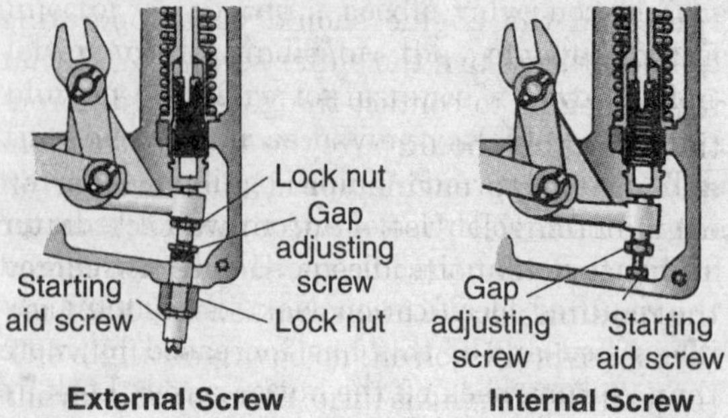

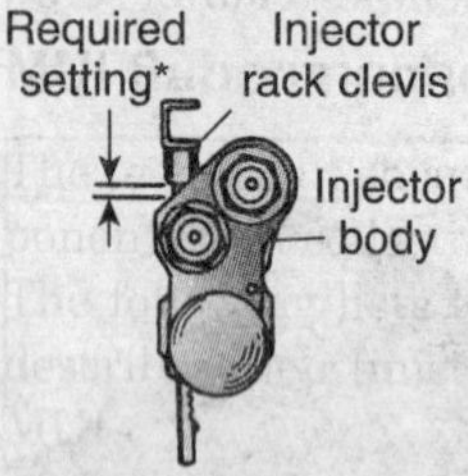

Injector	Gage Setting*	Tool Number
S80	.345"	J 24889
N60	.345"	J 24889
N65	.385"	J 24882
N70	.385"	J 24882
N75	.385"	J 24882
N80	.385"	J 24882
N90	.454"	J 23190
C65	.385"	J 24882
C80	.385"	J 24882

Figure 21–15 *Adjusting the starting aid screw. (Courtesy of Detroit Diesel Corporation)*

Adjust Throttle Delay Cylinder

If the tune up is being performed on an engine equipped with a throttle delay device, it must be first removed to avoid interference in the governor or injector control linkage. Used on engines with and without turbos, its purpose is to retard movement of fuel control tube and injector racks on acceleration. It is a crude smoke-limiting device that functions by preventing the control tube from snapping the MUI racks immediately to full fuel when requested by the operator. It also prevents drivers from over-revving the engine while upshifting. Any time the injector racks are moved toward the no-fuel position, free movement of the throttle delay piston is ensured by air drawn into the throttle delay cylinder. However, movement of the piston toward the rack to no-fuel position allows oil from the reservoir to enter and displace the air.

Sudden fuel demand such as that required for acceleration causes the racks to move toward full-fuel position, but the throttle delay piston must first displace the oil in the throttle delay cylinder through an orifice. This momentarily retards the inboard travel of the MUI racks toward the full-fuel position.

1. Insert the correct gauge between the injector body and the shoulder on the injector rack clevis on the injector closest to the throttle delay cylinder. Move the speed control lever to full-fuel and hold in that position, causing the injector rack to move against the gauge.
2. Insert DDC pin gauge (J25558:- 0.072") in the cylinder oil fill hole. Rotate the throttle delay lever until further movement is limited by the piston coming up against the pin gauge.
3. Carefully hold the throttle delay piston in this position and tighten the U-bolt nuts while exerting light pressure on the throttle delay lever in its direction of actuation.

Conclusion

Adjusting the throttle delay cylinder (when equipped) completes the DDC two-stroke cycle tune up. It is important to check that when accelerator linkage is reconnected to the speed control lever, it moves to the maximum speed position when the accelerator pedal is fully depressed.

A DDC tune up sets all the critical fuel system adjustments. The first time the technician undertakes DDC tune-up procedures, it is essential that the appropriate DDC service literature and specifications be consulted and that the novice technician is guided by one with some DDC tune-up experience. Although the chances of a runaway are rare, the technician should be aware of how to trip the blower air gates (these cut intake air to the engine) and take the precaution of having a CO_2 fire extinguisher close by.

REVIEW QUESTIONS

1. Which of the following correctly describes the type of fuel transfer pump used on DDC highway, two-stroke cycle engines?
   a. Plunger
   b. Diaphragm
   c. Vane
   d. Gear

2. Charging pressure values to DDC mechanical unit injectors would approximate which of the following?
   a. 17 psi
   b. 70 psi
   c. 800 psi
   d. 2,300 psi
3. Which of the following correctly describes the component(s) that feed and return fuel to DDC mechanical unit injectors?
   a. High-pressure pipes
   b. Annular charge galleries
   c. Jumper pipes
   d. Rail pipes
4. Which of the following is most responsible for developing injection pressures in a DDC mechanical unit injector?
   a. Charge pump
   b. Rack actuator
   c. Control tube
   d. Injector cam
5. Which of the following will determine the amount of fuel injected per crank angle degree in a DDC mechanical unit injection system?
   a. Rack position only
   b. Engine rpm
   c. Cam profile only
   d. Rack position and cam profile
6. Which of the following is *not* a function of a DDC mechanical unit injector?
   a. Converts charging to injection pressures
   b. Meters fuel quantity
   c. Atomizes fuel
   d. Determines injection rate
7. Effective stroke in a DDC mechanical unit injector occurs when:
   a. Both upper and lower bushing ports are exposed.
   b. Both upper and lower bushing ports are covered.
   c. The plunger is on its downstroke.
   d. The plunger is on its upstroke.
8. When considering the injection timing characteristics of a DDC mechanical unit injector, which of the following is true?
   a. Variable beginning, constant ending
   b. Constant beginning, variable ending.
   c. Variable beginning, variable ending
   d. All of the above
9. What would be the total swept volume of a DDC 8V–92?
   a. 8 liters
   b. 736 cc
   c. 736 cu. in.
   d. 460 BHP
10. Which of the following conditions would cause a hunting engine roll on a two-stroke cycle DDC engine?
    a. Weak high-idle spring
    b. Belleville spring adjustment
    c. Injectors out of balance
    d. Seized control tube
11. NOP parameters in DDC mechanical unit injectors with needle valve nozzles would be:
    a. 600–850 psi
    b. 2,200–3,400 psi
    c. 10,000–12,000 psi
    d. Up to 25,000 psi
12. *Technician A* states that hunting is a condition that may result from the erratic setting of MUI racks during tune up. *Technician B* states that engine hunting can always be removed by adjusting the buffer screw CW until the rpm smooths out. Who is correct?
    a. Technician A only
    b. Technician B only
    c. Both A and B
    d. Neither A nor B
13. *Technician A* states that the exhaust valves on DDC two-stoke cycle engines are set with identical lash to the intake valves. *Technician B* states that the adjusting screw on DDC valve bridges is used to set valve lash. Who is correct?
    a. Technician A only
    b. Technician B only
    c. Both A and B
    d. Neither A nor B
14. If a single MUI is seized in the full-fuel position in a DDC two-stroke cycle engine, which of the following devices would do most to prevent an engine runaway?
    a. Two-screw type, rack lever adjuster
    b. Single-screw type, rack lever adjuster

c. Throttle delay cylinder
d. A limiting speed governor

15. *Technician A* states that the length of a DDC MUI effective plunger stroke is determined by the rotational position of the plunger in the bushing. *Technician B* states that plunger rotational position is determined by the linear position of the rack. Who is correct?
a. Technician A only
b. Technician B only
c. Both A and B
d. Neither A nor B

22

Caterpillar Mechanical Unit Injection

INTRODUCTION

Truck and bus technicians familiar with Detroit Diesel Corporation (DDC) **mechanical unit injector (MUI)** systems will require little introduction to the Caterpillar MUI system. However, there are some critical differences and Caterpillar technical literature must be referenced when performing its version of the tune-up sequence. Caterpillar MUIs can be found on 3114, 3116, and 3126 engines. Most technicians will see the 3126 version of this engine with Caterpillar's HEUI fuel system, a computer-controlled, hydraulically actuated unit injector system; this is covered later on in this text.

The 3116 was available exclusively to General Motors Corporation (Figure 22–1). To interpret the numeric engine series code, the first digit (3xxx) means Caterpillar, the next two digits denote the displacement volume per cylinder (x12x = 1.2 liters) and the fourth digit (xxx6) denotes the number of cylinders. The system described in this chapter relates to MUI-fueled small-bore highway diesel engines manufactured by Caterpillar, but it has plenty in common with Caterpillar large-bore off-highway engines using MUI fueling.

FUEL SUBSYSTEM COMPONENTS

Fuel from the fuel tank is pulled from the fuel tank by a plunger-type transfer pump; a primary fuel filter may be fitted between the fuel

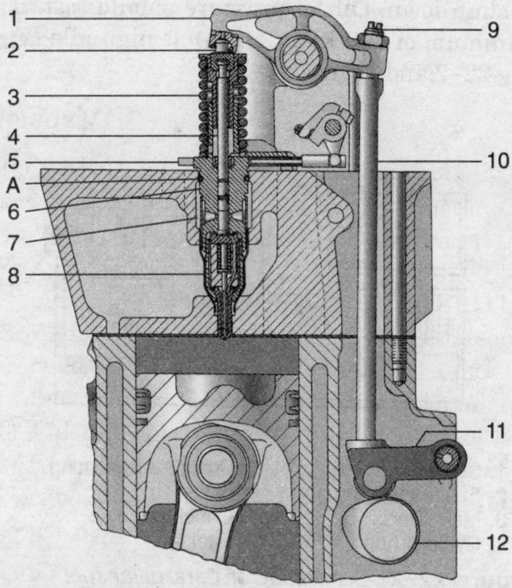

FUEL INJECTION SYSTEM

1. Rocker arm
2. Follower
3. Tappet spring
4. Plunger
5. Rack
6. Barrel
7. Fuel gallery
8. Sleeve
9. Setscrew
10. Pushrod
11. Lifter
12. Cam

A. O-ring

Figure 22–1 *Cutaway view of a Caterpillar MUI engine. (Reprinted Courtesy of Caterpillar Inc.)*

tank and the transfer pump. The transfer pump is integral with the governor housing.

The transfer pump charges a secondary or main fuel filter and the unit injector assemblies. The charge side of the transfer pump is pressure protected by an outlet check valve and a pressure relief valve. When equipped, the hand primer pump is located on the secondary filter mounting pad; its function is to purge air from the fuel subsystem responsible for charging the unit injectors. Fuel is delivered from the fuel subsystem to drilled passages in the cylinder heads that intersect annular galleries around the cylindrical bore of the unit injectors.

The fuel transfer pump is a cam-actuated plunger pump located in the front housing of the governor. Its actuating cam is located on the governor drive shaft and it provides the pump plunger with the required reciprocating action working in conjunction with a retraction spring. Inlet and outlet check valves aspirate the pump chamber; these close when the engine is shut down. Outlet pressure should test at a minimum of 200 kPa (29 psi) at high idle (Figure 22–2 and Figure 22–3).

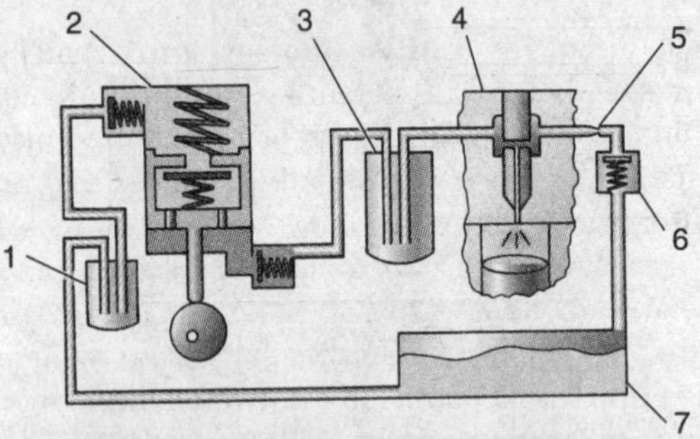

1. Screen
2. Fuel transfer pump
3. Main filter
4. Cylinder head
5. Pressure regulating orifice
6. Check valve
7. Fuel tank

Figure 22–2 *Schematic of Caterpillar fuel subsystem layout used with MUI-fueled engines. (Reprinted Courtesy of Caterpillar Inc.)*

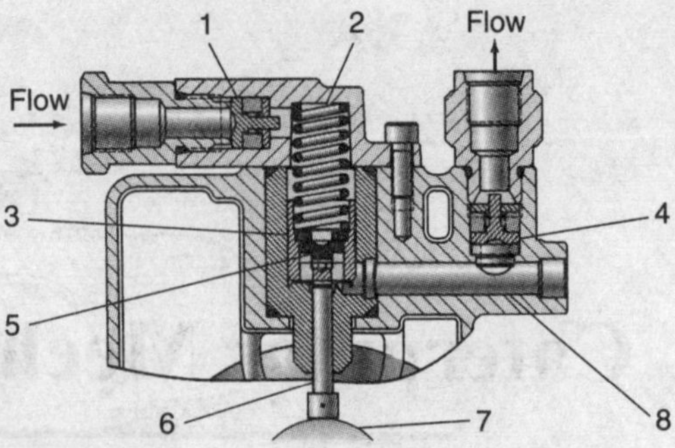

1. Inlet check valve
2. Spring
3. Piston assembly
4. Outlet check valve
5. Piston check valve
6. Tappet assembly
7. Cam
8. Passage

Figure 22–3 *Caterpillar plunger-type fuel transfer pump. (Reprinted Courtesy of Caterpillar Inc.)*

PRINCIPLES OF OPERATION OF CATERPILLAR MUIs

The MUIs are responsible for metering, pumping to injection pressure values, and atomizing fuel directly to the engine cylinders. The hydraulic principles of the Caterpillar MUI relate closely to those of the inline port-helix metering pump and a hydraulic, multi-orifii injector nozzle. In an MUI (Figure 22–4), the pump is actuated by a dedicated cam profile on the engine camshaft, the high-pressure pipe is eliminated, and the hydraulic injector nozzle is integral with the assembly. The MUI is installed within

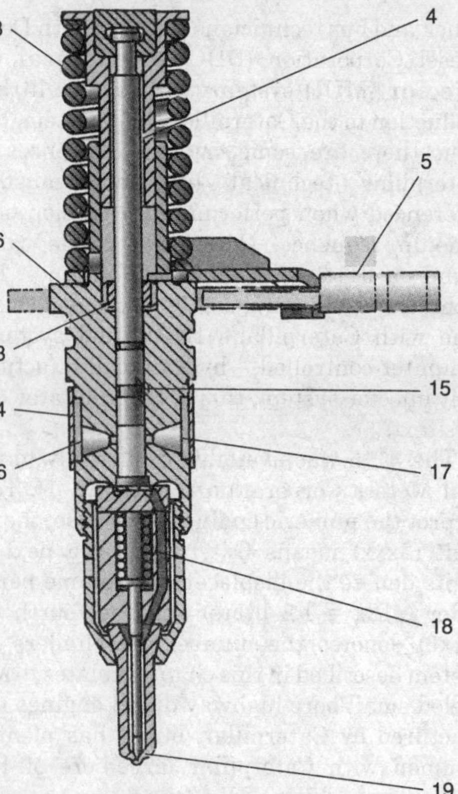

FUEL INJECTOR PUMP (Unit Injector)

3. Tappet spring
4. Plunger
5. Rack
6. Barrel
13. Gear
14. Sleeve filter
15. Helix
16. Lower port
17. Upper port
18. Spring
19. Check (needle valve)

Figure 22–4 *Caterpillar MUI. (Reprinted Courtesy of Caterpillar Inc.)*

a sleeve press-fit to the cylinder head; this is directly exposed to engine coolant passages. The sleeve should only be serviced with Caterpillar special tools.

The injector cam profile actuates a train consisting of a roller-type lifter, pushrod, and rocker arm. The rocker arm acts on the tappet assembly of the MUI using a "floating button" as an intermediary; this prevents side loading of the tappet by the rocker arm.

The plunger is linked to the tappet and reciprocates within a stationary barrel. It is milled with a helix that can be classified as a lower helix design, resulting in a constant beginning, variable ending of fuel delivery timing. With this design of helix, the beginning of the effective pumping stroke is determined by the angular location of the cam. The barrel is machined with an upper port and a lower port diametrically offset. Metering of fuel is accomplished by rotating the plunger in the barrel bore and altering the point of register of the spill ports with the helix. The plunger is rotated by means of a gear that is tooth meshed to the rack. The gear collars the plunger in a manner that permits plunger rotation while it reciprocates.

Effective Stroke

Plunger actual stroke is determined entirely by cam profile and, therefore, remains constant. Control of fuel quantity depends on the rotational position of the plunger, specifically the point of register of the helix, with the upper barrel port. When the actuating cam ramps off its base circle, the plunger descends within the barrel and first, its leading edge will close off the upper barrel port. As the plunger continues to descend, it displaces fuel below it in the pump chamber, forcing it out of the lower barrel port. Effective stroke begins when the lower barrel port is closed by the plunger leading edge. Pressure rise occurs in the pump chamber below the plunger. Connected to the pump chamber by a duct is the pressure chamber sectional area of the needle valve in the hydraulic injector nozzle at the base of the MUI.

At nozzle opening pressure (NOP), sufficient pressure rise has occurred to overcome the spring pressure that loads the needle valve on its seat, causing it to retract. Fuel passes around the needle valve seat and is forced out through the nozzle orifii. Effective pumping stroke will continue as long as both the upper and lower barrel ports are closed off by the plunger. It ends at the moment the upper barrel port is exposed by the recessed helix milled into the plunger, permitting fuel to be spilled to the return circuit. The resultant pressure collapse enables the injector nozzle spring to close and seat the needle valve, ending the injection pulse. Plunger descent continues until the injector train is fully extended by the cam nose. As the injector cam ramps toward its base circle the plunger retracts, creating low pressure, enabling rapid recharging of the pump chamber.

Fuel Rack Control

The governor is mechanically connected to the rack shaft assembly by means of a **link and lever assembly** (Figure 22–5). When the governor demands fuel, the injector rack must be moved linearly away from the injector body, rotating the plunger counterclockwise (CCW). The governor output shaft, therefore, moves to rotate the shaft, causing the rack levers to move the injector racks simultaneously outboard. To

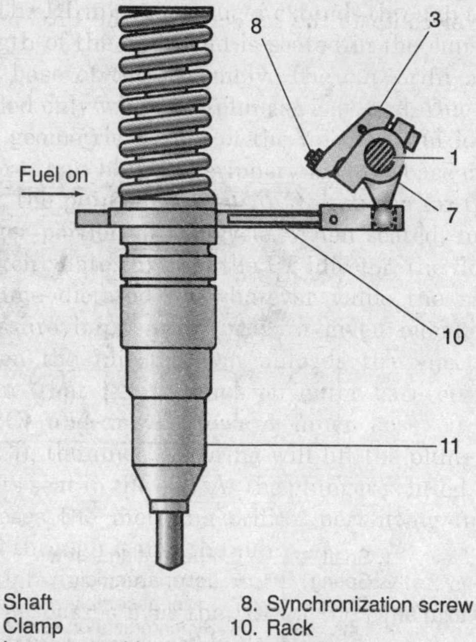

1. Shaft
3. Clamp
7. Lever assembly
8. Synchronization screw
10. Rack
11. Injector

Figure 22–5 *MUI fuel-on position. (Reprinted Courtesy of Caterpillar Inc.)*

no-fuel the engine, the injector racks must be forced inboard, causing the plungers to move clockwise (CW). A torsion spring at each rack lever assembly allows the rack control linkage to return to a no-fuel position should one injector seize in a fuel-on position.

Power setting of the #1 cylinder injector is made with the fuel setting screw on the rack lever assembly. As the fuel setting screw is turned, shaft position relative to the governor link is altered. Adjusting screws on each clamp and lever assembly permit synchronization of the remainder of the unit injectors to that on the #1 cylinder. Synchronization of the unit injector racks is essential to balance engine fueling (Figure 22–6 and Figure 22–7).

Fuel Ratio Control

Turbocharged engines use a **fuel ratio control (FRC)** to limit engine smoke during conditions of low boost. Essentially, the FRC limits fueling to the cylinders until there is sufficient air in them to combust it. Boost air is ported from the engine intake manifold to the FRC inlet port. At low boost, when the governor output shaft moves to the fuel-on position attempting to draw the unit injector racks outboard, the limit lever contacts the set screw on the FRC lever.

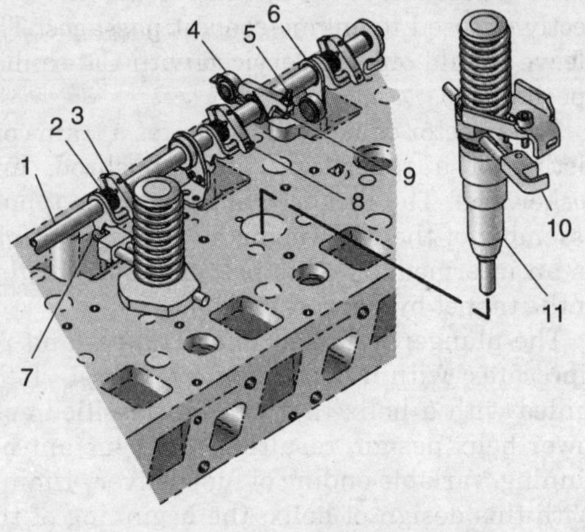

1. Shaft
2. Spring
3. Clamp
4. Link
5. Fuel setting screw
6. Lever assembly
7. Lever assembly
8. Synchronization screw
9. Clamp assembly
10. Rack
11. Injector

Figure 22–6 *MUI location and rack control assembly. (Reprinted Courtesy of Caterpillar Inc.)*

As manifold boost increases, it acts against the FRC diaphragm and overcomes the FRC spring pressure; this permits the retainer shaft to move outward, enabling the FRC lever and the limit lever to rotate CW and the rack shaft to rotate toward full-fuel.

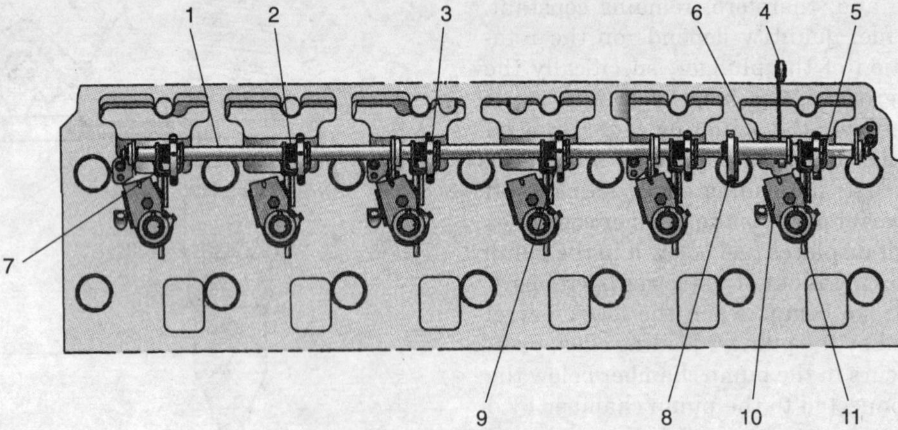

1. Shaft
2. Spring
3. Clamp
4. Link
5. Fuel setting screw
6. Lever assembly
7. Lever assembly
8. Synchronization screw
9. Clamp assembly
10. Rack
11. Injector

Figure 22–7 *Overhead view of the MUIs and the rack control linkages. (Reprinted Courtesy of Caterpillar Inc.)*

Fuel Shutoff Solenoid

The fuel shutoff solenoid used in off-highway applications is an **energized-to-run (ETR) latching solenoid** (latching solenoid: a pilot valve that stays in the energized position even when not electrically energized and remains so until the circuit it controls is shut down). When the engine is used in trucks, the shutoff solenoid does not latch. A spring-loaded plunger inside the solenoid acts on a lever assembly located in the front housing of the governor. This lever assembly forces the governor output shaft to the fuel-off position when the solenoid armature is released at engine shutdown.

At startup the solenoid armature is energized to latch in the run position, permitting the governor output shaft to move to the fuel-on position. Engines with latching solenoids can be manually shut off by depressing the override button. This solenoid cannot be manually latched in the run position for starting (Figure 22–8 and Figure 22–9).

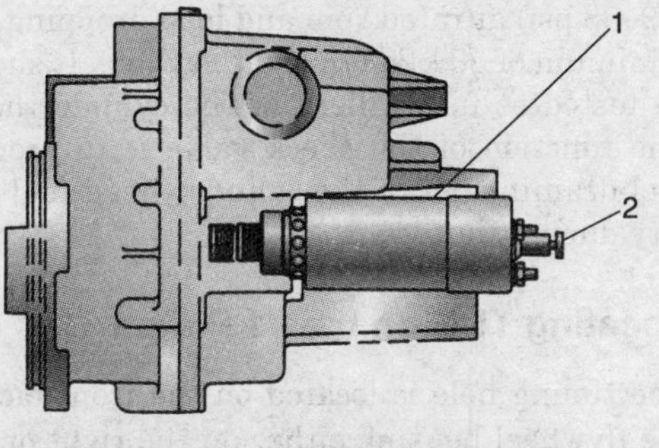

Figure 22–8 *Location of fuel shutoff solenoid. (Reprinted Courtesy of Caterpillar Inc.)*

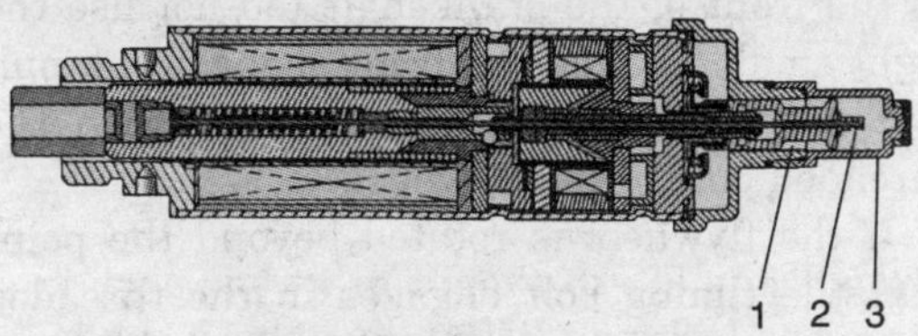

Figure 22–9 *The ETR latching solenoid. (Reprinted Courtesy of Caterpillar Inc.)*

GOVERNORS

Six different governor types are used on 3116, 3114, and 3126 engines. Each is identified sequentially by Roman numerals. Governors for Caterpillar MUI-fueled engines are covered only briefly in this section. Proprietary literature (SEN6454) and test equipment (IU7326 Governor Test Stand) are required to service and adjust these governors. The governor is gear driven by the engine camshaft, rotating a weight carrier and flyweight assembly.

When the flyweights are rotated, centrifugal force exacted will be proportional to rotational speed and thus to engine rpm. Opposing the outward thrust of the flyweights is spring force, which uses a riser (thrust collar) as an intermediary. Fuel request from the operator is delivered to the governor by means of a control lever. This input is relayed to the governor output shaft, but the governor will moderate engine fueling when rated load or a lug condition is attained. The only on-engine adjustments to the governor are the final low- and high-idle settings (Figure 22–10).

MUI ENGINE TESTING AND ADJUSTMENTS

Caterpillar service literature should be consulted when performing all engine and testing procedures. Technical service bulletins should also be referenced.

Testing Injectors

Caterpillar recommends a preliminary test for MUIs using an **infrared thermometer**. When performing this test, the engine should be operated at *low idle* and the temperature at the exhaust manifold ports measured. Low temperature at one of the ports can indicate no-fuel from an injector, while excessively high temperature is evidence of overfueling.

The difference between cylinders should not be greater than 70°C (158°F).

With the valve cover removed and the engine idling, each unit injector may be tested individually by moving its rack to the fuel-on position; this should immediately result in combustion knock in the cylinder being tested. If this com-

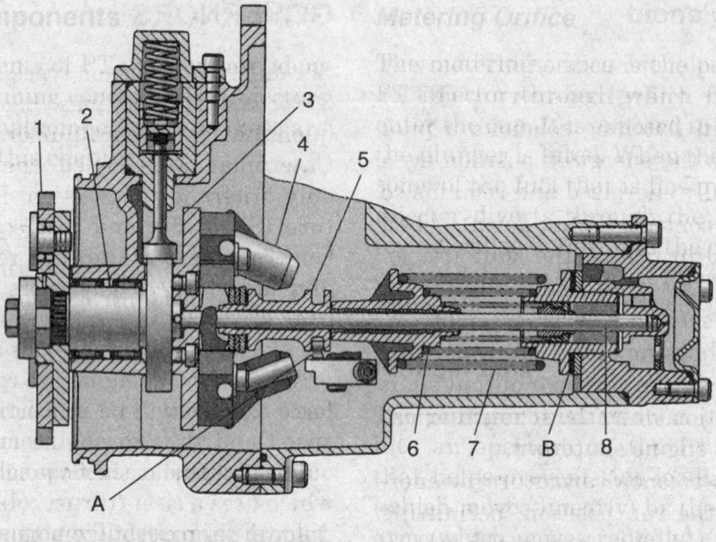

1. Governor drive gear
2. Shaft
3. Flyweight carrier
4. Flyweights
5. Riser
6. Low idle spring
7. High idle spring
8. Shaft
A. Pin
B. Shims

Figure 22–10 *Sectional view of one of the six governors used on Caterpillar MUI-fueled engines. (Reprinted Courtesy of Caterpillar Inc.)*

bustion knock does not result, there may be a problem with the injector assembly, the fuel supply to it, or the seal between the injector and sleeve.

Out-of-Engine Injector Inspection

Out-of-engine injection inspection requires the use of a Caterpillar test fixture. It should be performed by a Caterpillar dealership only.

Governor Adjustment

The only on-engine adjustments of the governor are the final low- and high-idle settings. Both settings should be trimmed to specification on a running engine by loosening the locknut and adjusting the screw against throttle arm stops.

Fuel Pressure Test

A fuel pressure test measures the charging pressure to the engine cylinder head. To check the filter, the pressure gauge should be installed at the taps in the filter mounting pad. A restricted filter causes lower pressure at the outlet tap than at the inlet tap. Charging pressure to the cylinder head should be 200 to 400 kPa (29–58 psi) at rated rpm and load dropping to a minimum of 50 kPa (7 psi) at low idle. It should be tested at fuel gallery in the cylinder head. The function of the check valve is to prevent fuel draining out of the cylinder head fuel gallery during shutdown.

Locating TDC on the #1 Piston

The timing hole is located on the front face of the flywheel housing either on the right or left side depending on engine application. The timing hole plug should be removed and **timing bolt** 8T0292 inserted in the timing hole. Next, bar the engine CCW using the four large bolts at the front of the crankshaft (do not use the eight small bolts on the front of the crank pulley) until the timing bolt engages with the threaded hole in the flywheel.

If the flywheel is rotated beyond the point that the timing bolt engages in the threaded hole, rotate the flywheel CW approximately 30 degrees, then CCW to the timing point; this is important to ensure that gear backlash is not a factor when determining the exact top dead center (TDC) position. Insert the timing bolt into the threaded hole. The valve cover should be removed and the rocker arms on the #1 cylinder should both show lash, indicating that the cylinder valves are closed. If so, then the #1 piston is at TDC on the compression stroke. If the rocker arms on #1 evidence no lash, the timing bolt must be removed, the engine rotated 360 degrees, and the timing bolt reinstalled.

Fuel Timing, Valve Setting, and Injector Synchronization

3116/3126 crankshaft positions for fuel timing and valve setting:

3116/3126	*Valve Lash*
Intake valves	0.38 + 0.08 mm (0.015″ + 0.003″)
Exhaust valves	0.64 + 0.08 mm (0.025″ + 0.003″)

The firing order is 1 5 3 6 2 4.
With the #1 piston at TDC on compression stroke:

Adjust injectors: 3–5–6
Adjust intake valves: 1–2–4
Adjust exhaust valves: 1–3–5
Rotate the engine 360 degrees.
With the #1 piston at TDC on exhaust stroke:

Adjust injectors: 1–2–4
Adjust intake valves: 3–5–6
Adjust exhaust valves: 2–4–6

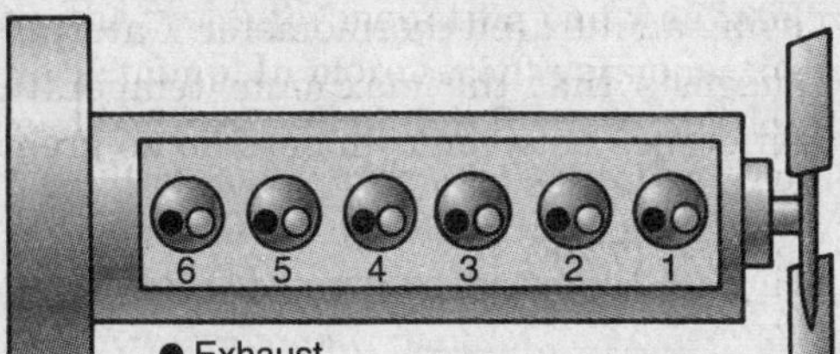

Figure 22–11 *The valve locations on 3116 and 3126 engines. (Reprinted Courtesy of Caterpillar Inc.)*

Figure 22–11 shows the intake/exhaust value locations.

Injector Synchronization

Injector synchronization is the setting of all the control racks on a set of unit injectors on the engine so that each delivers the same amount of fuel to each cylinder. This is performed by setting each injector rack to the same position using a 3.5 mm calibration block while the control linkage is in a fixed position called **synchronizing position**. Synchronizing position is when the injector at #1 cylinder is at fuel shutoff. Because the #1 injector is the reference point for the other injectors, no synchronizing adjustment is made to the #1 injector rack. An injector should always be synchronized whenever it has been removed and reinstalled or replaced. When the #1 injector has been removed and reinstalled or replaced, all the injectors must be synchronized. Figure 22–12 shows the synchronizing tool.

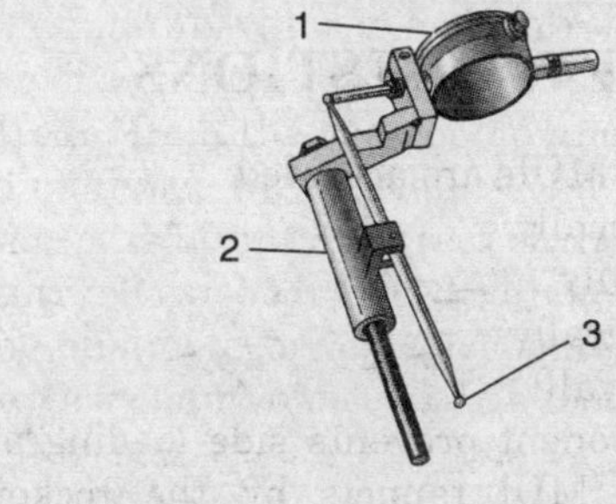

1. 6V6106 dial indicator
2. 1U6679 indicator group
3. Lever (part of 1U6679 indicator group)

Figure 22–12 *Fuel synchronization tool used to set Caterpillar MUIs. (Reprinted Courtesy of Caterpillar Inc.)*

Fuel Timing

The fuel timing dimension is a critical tune-up adjustment that sets the unit injector tappet dimension using the injector actuating train. Essentially, this process positions the plunger at a specific location in the barrel while the injector cam is on base circle; this "times" the activity in the unit injector with that in the engine and ensures that the beginning of injection occurs at a consistent crank angle in each cylinder. Figure 22–13 shows the fuel timing dimension (D), which is set using a timing pin.

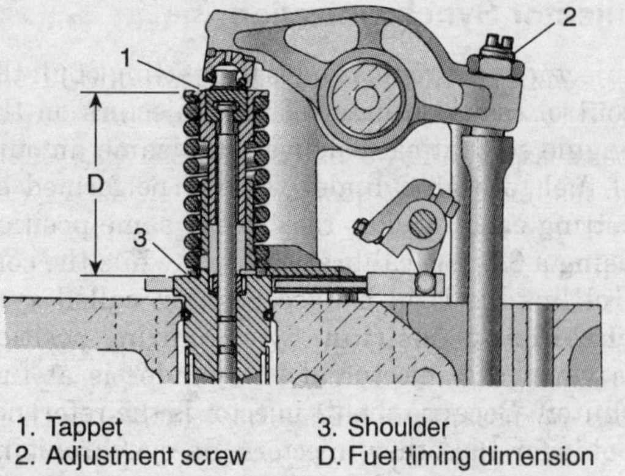

Figure 22–13 *The fuel timing dimension setting. (Reprinted Courtesy of Caterpillar Inc.)*

REVIEW QUESTIONS

1. Caterpillar MUIs are actuated:
   a. Hydraulically
   b. Electrically
   c. Electronically
   d. Mechanically
2. What component prevents side loading of Caterpillar MUI tappets by the rocker arm?
   a. Injector link
   b. Floating button
   c. Tappet spring
   d. Toothed collar
3. The Caterpillar MUI actuating train uses what type of lifter?
   a. Roller
   b. Solid
   c. Hydraulic
   d. Pneumatic
4. What component in a Caterpillar MUI-fueled engine dictates the length of the actual plunger stroke?
   a. Injector cam
   b. Rack
   c. Governor
   d. Crankshaft rpm
5. Altering a Caterpillar MUI effective stroke is accomplished by:
   a. Rotating the plunger in the barrel
   b. Rotating the barrel within the MUI body
   c. Varying plunger stroke
   d. Varying cam lift
6. In the event of an MUI seizure in a fuel-on position, what component permits the remaining MUIs to return to a no-fuel position?
   a. Governor spring
   b. Torsion spring
   c. MUI tappet spring
   d. Hydraulic charging pressure
7. What event occurs at the precise moment that begins effective stroke?
   a. The lower barrel port is covered.
   b. The upper barrel port is covered.
   c. The plunger begins its downstroke.
   d. The plunger completes its downstroke.
8. Synchronizing MUIs is accomplished by:
   a. Setting the MUI timing dimension
   b. Adjusting screws on the clamp and lever assemblies
   c. Adjusting screws on the injector train rockers
   d. Setting the governor link
9. What component is used on Caterpillar MUI systems to limit fueling under conditions of low manifold boost?
   a. ETR solenoid
   b. Aneroid
   c. FRC device
   d. AFC valve
10. When testing for a malfunctioning MUI using an infrared thermometer, Caterpillar suggests that the maximum temperature difference between cylinders be no greater than:
    a. 25°C (77°F)
    b. 70°C (158°F)
    c. 100°C (212°F)
    d. 150°C (302°F)

23

Cummins PT System

INTRODUCTION

Until the aftermath of World War II, diesel engine manufacturers had been generally content to use port-helix metering fuel injection systems on medium- and large-bore diesel engines. There were a few exceptions but these tended not to survive. However, with port-helix metering injection, a disproportionate percentage of the total engine cost must be allocated to the fuel management system and engine designers began to explore alternative systems. The Cummins engine company introduced its **pressure-time (PT) fuel system** in the early 1950s and by 1970, PT-managed engines were the North American market sales leaders in the 200–500 BHP range. By 1980 Cummins PT engines, which were manufactured in facilities in the United States, Britain, and Germany and supported by casting and subassembly plants in half a dozen other countries, outsold all others worldwide in the 200 BHP and larger market. This success continued until 1994 North American Environmental Protection Agency (EPA) noxious emissions legislation rendered the PT-managed engine ineligible for on-highway certification.

Unlike some of the other hydromechanical fuel injection systems, the PT system did not readily lend itself to electronic management, and Cummins developed its own full authority electronic management systems to replace the PT **common rail system**. A common rail fuel system is one in which a fuel pump supplies injectors from a common pipe or accumulator. Despite the diminished numbers of PT-fueled engines on our highways, the PT fuel system endured in off-highway applications and continues to be addressed in college curriculum and certification testing. Learning the PT fuel system should not represent wasted time because it evolved into a couple of electronically managed systems we continue to use today. PT fuel systems continued to be used on off-highway engines up until around 2004. The layout of a typical PT-managed engine is shown in Figure 23–1.

PT SYSTEM THEORY

The power output of any diesel engine is determined by the amount of fuel metered into its cylinders. The Cummins PT system manages fueling using a set of principles quite distinct from those used in port-helix metering systems. The sequence of the two letters in the acronym PT is important because Cummins also uses the acronym TP, representing time/pressure; in each case, the first letter of the acronym denotes the *control variable*. In the case of the PT system, that would be the *pressure* factor. In the case of the TP fuel system studied later in this book, the control variable is *time*.

A simple hydraulic equation constructed to determine a volume of flow requires the following data:

1. The fluid pressure
2. The flow time
3. The flow area

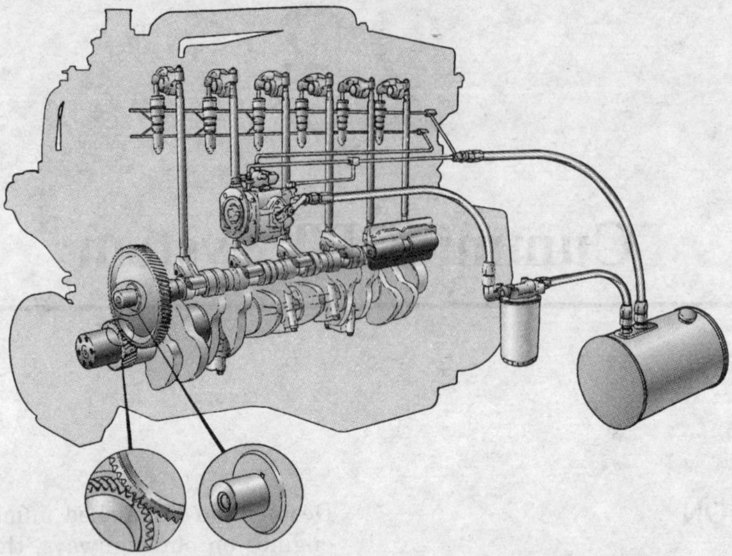

Figure 23–1 *Layout of a Cummins PT fuel system. (Courtesy of Cummins Inc.)*

These three factors produce the product of the equation: the volume of flow. Changing any of the values in the preceding hydraulic equation has to change its outcome, which is a specific volume of flow that we might better refer to as metered fuel quantity. In the Cummins PT system, the critical **flow areas** (flow area: the most restricted sectional area of a hydraulic circuit under flow) are designed to remain constant, so the variables in the hydraulic equations are:

1. *Flow time*—variable, but dictated by engine speed, which has little relationship with fuel quantity required. The time available for metering diminishes inversely with rpm increase.
2. *Fluid pressure*—variable and precisely managed by the system to control engine output.

Flow Control

The critical component in terms of managing the fueling is the PT pump; this is *not* an injection pump. It is best regarded as a flow control device. The PT pump feeds all the injectors by means of a common rail; that is, a single outlet pipe from the PT pump that supplies a rail gallery in the cylinder heads. The injectors are cylindrical and receive and return fuel through exterior annuli separated by O-rings. As the PT injectors are mounted in parallel and the only way in which the rail fuel can pass into the cylinder head return gallery is by passing through the PT injectors, their inlet orifii will collectively identify the flow area the PT pump unloads to. This flow area remains constant (they are accurately sized to the application on the test bench) so it can be said that the PT pump directly manages the rail pressure. Rail pressure values are used to manage engine output through the entire rpm and load ranges of the engine; they must, therefore, be widely variable. Typically they range from 8 psi to 210 psi.

Our approach to the PT system is handle it component by component. However, it is important to bear in mind the system objectives while doing this. Remember that the PT fuel system does what any other fuel system attempts to achieve: precisely manage the fuel quantity metered per cycle. This actual quantity of fuel-to-metered per cycle depends on the following five factors:

Rail Pressure

The PT pump is a flow control device responsible for managing rail pressure. Actual rail pressure is factored by the quantity of fuel unloaded into the single rail pipe that supplies all of the PT injectors and the flow area as defined collectively by the PT injectors.

Balance Orifice Sizing

All of the fuel unloaded by the PT pump to the rail is routed through the PT injectors mounted in parallel. The **balance orifice** is the inlet orifice of each PT injector; each balance orifice is accurately sized because collectively (that means all six PT injectors in an I-6 engine), they define the flow area used to factor the actual rail pressure. Although the term *balance orifice* is used in this text, Cummins also uses the term **calibrating orifice** to describe this precisely sized inlet to the PT injector.

Engine Speed

Engine speed defines the real time available for metering and pumping in the PT injector. It also determines the PT gear pump (a subcomponent of the PT pump) pressure; this increases proportionally with a rise in rpm given a constant flow area and accordingly influences rail pressure values.

Metering Orifice Sizing

The PT injector is essentially a pumping device responsible for converting rail pressure values to injection pressure values. The larger percentage of fuel entering the PT injector circulates (for purposes of lubing and cooling) and exits to the return gallery in the cylinder head, from which it is routed back to the fuel tank. The percentage of fuel to be used for fueling the engine is that forced through the **metering orifice** within the PT injector in the time period factored by the cam geometry (the PT injector pumping stroke is actuated by cam profile) and engine rpm.

Injector Timing

The pumping action of the PT injector is provided by a dedicated cam profile (located between the valve cams in the cylinder) and injector train, consisting of a cam follower, push tube, and rocker assembly. Timing adjustments are critical, and marginal inaccuracies can cause both fueling problems and physical damage to the injector and injector train.

Common Rail

The PT fuel system uses a common rail fueling principle. The pump is a flow control device that manages rail pressure. It feeds PT injectors positioned in parallel extending from the common rail. Fuel exiting the rail must pass through the PT injectors where a portion is metered for injection and the remainder routed to the return manifold. Fuel is flowed through a circuit in the PT injectors at any time the engine is running. The fuel used for metering is diverted from this circuit within the PT injector.

THE PT INJECTOR

The PT injector is a rocker-actuated pumping element responsible for converting rail pressures to injection pressures. It also defines a couple of critical flow areas essential for metering. In mobile diesel engine applications, it is cylindrically shaped and has two exterior annuli separated by O-rings. When installed in the injector bore in the cylinder head, the lower annular area in which the balance orifice is located is subjected to rail pressure. The balance orifice is the injector inlet orifice. In this way fuel enters the PT injector, circulates within it, and exits through an orifice in the upper exterior annulus to be returned to the fuel tank.

The PT injector plunger extends through the length of the body and is seated in the cup at the base of the assembly. The cup orifii are sealed only when the plunger is seated. Due to the geometric design of the injector cam lobe (more than half its periphery is outer base circle), the plunger is loaded into the cup for the larger portion of the cycle. When seated, fuel will circulate through the PT injector, the flow volume dictated by whatever value the rail pressure happens to be at a given moment. When the injector cam unloads the injector train (that is, it ramps off outer base circle [OBC] and moves toward inner base circle [IBC]), the injector spring will lift the plunger off its seat in the cup. As the plunger is lifted, it exposes the metering orifice, permitting fuel flow through it into the cup.

Metering continues until the injector cam profile passes over the IBC of the cam profile onto the ramp toward OBC. This ramp drives the injector train (follower, push tube, and rocker) and, therefore, drives the plunger downward into the cup, first purging air from the cup and next, acting on whatever amount of fuel was metered to pump it through the cup orifii. This action creates the required injection pressures as the fuel metered into the cup must be forced out through the cup orifii directly into the engine cylinder. The sequence ends with the plunger being loaded into the cup with full mechanical crush (considerable force: equivalent to 380 lb of linear force), this being required first, to seal the PT injector from the engine cylinder, and, second, to enable effective heat transfer from the cup to the plunger. The cup orifii are sized to produce an average emitted droplet sizing of 50 μ when the engine is run at rated speed and loads, which Cummins considers ideal for PT engines.

Open Nozzle Concept

PT injectors use an open nozzle design. This means that during metering, air from the engine cylinder (under compression) can siphon into the cup. As engine compression pressures exceed rail pressure by a considerable margin, this could cause fuel flow reversal. To prevent this from occurring, a check ball is located at the metering orifice, which is designed to engage any time the pressure in the cup exceeds the metering pressure. Although this interferes with the metering process, it tends only to present a problem at low engine speeds. The minute sizing of the cup orifii means that at higher speeds, there is insufficient time during metering for cylinder pressures to siphon to the injector cup. The check ball also engages whenever a full charge of fuel is metered into the cup. In this instance, the plunger acts directly on the fuel metered into the cup before it has closed off the metering orifice. Once again, the check ball (the term *lock-off* ball is also used) engages whenever cup pressure exceeds metering pressure regardless of the reason.

Typically, only about 35% of the fuel exiting the PT pump rail pipe actually fuels the engine. The remainder is used for lubrication and, very importantly, taking heat away from the PT injectors. This heat is substantial and the fuel subsystem should be designed to dissipate it; cylindrical aluminum tanks placed in the airflow and maintained at better than 25% full help this process. The specific percentage of the rail volume used to fuel the engine is known as the fuel rate.

Cam profile dictates the fuel discharge rate. Cam profile, engine speed, and the sizing of the cup orifii together determine the actual droplet sizing. The ideal 50 μ would be obtained at rated speed and load. In injector comparator testing, fuel rate, rather than total flow, is monitored through the injector. It would be incorrect to describe PT injectors as unit injectors because they do not manage metering. Their function is to create the injection pressure and to atomize the fuel. The two flow areas they define are critical for determining engine fueling. For instance, if a set of low flow balance orifii were installed in a set of injectors, rail pressure values would go up; that is, the flow area the PT pump unloads to would be reduced so the pressure in the rail increases. It is important to remember that all of the rail fuel is forced to pass through the PT injectors; there is no other connection between the cylinder head rail manifold and the return manifold. Figure 23–2 illustrates an older PT injector and Figure 23–3 details the metering and injection cycles on PT-fueled engines.

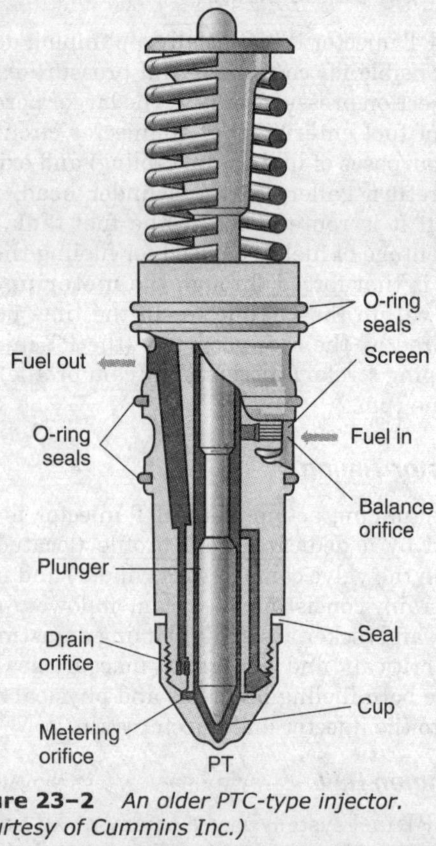

Figure 23–2 *An older PTC-type injector. (Courtesy of Cummins Inc.)*

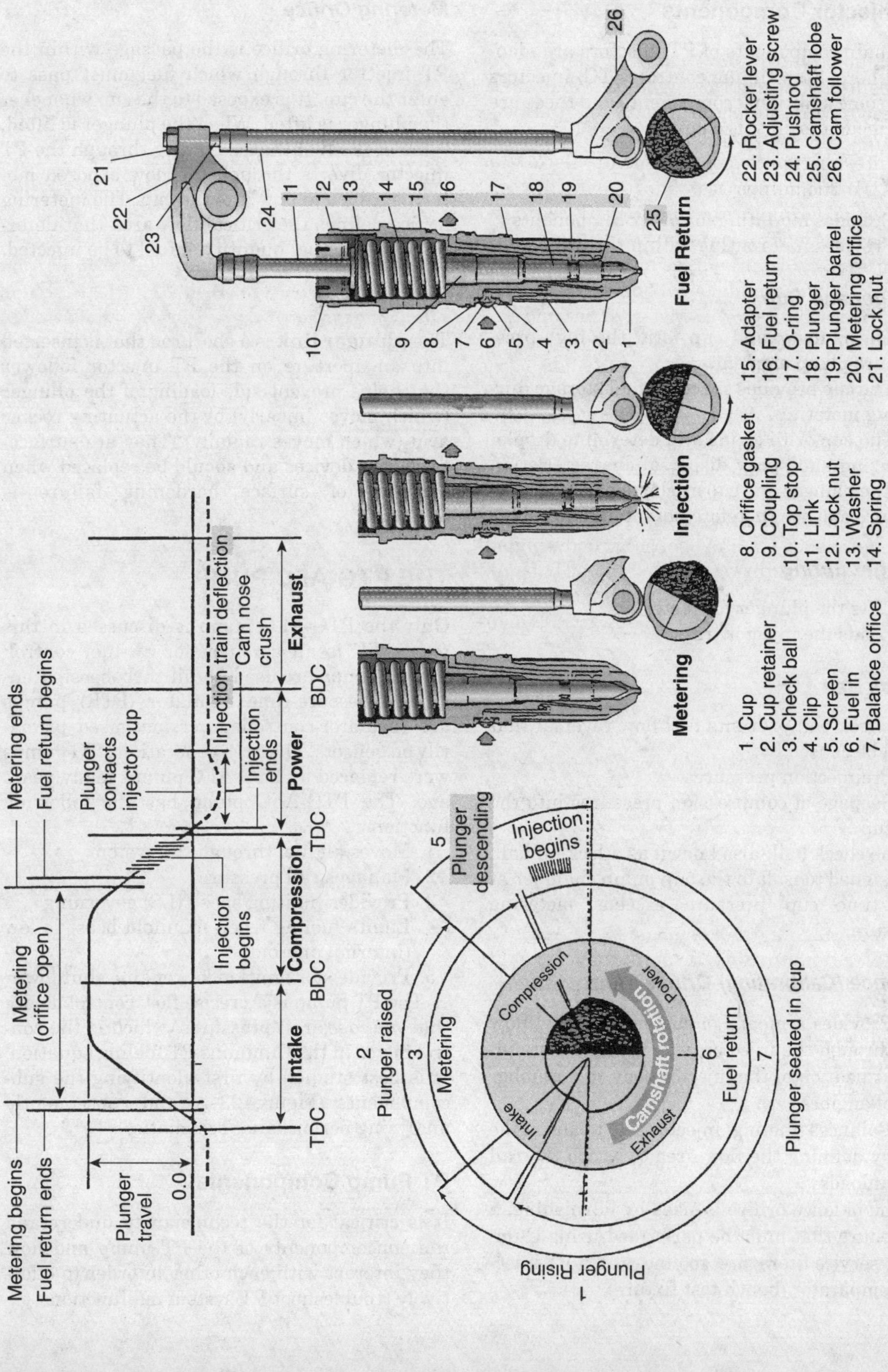

Figure 23–3 *The metering and injection cycle of a PT injector. (Courtesy of Cummins Inc.)*

PT Injector Components

The main components of PT injectors are identified here. Step timing control (STC) injectors have some additional components and these are described later in this chapter.

Body

a. Provides mounting for other components
b. Provides fuel routing within the body

Plunger and Cup

a. The plunger and cup form the high-pressure pump element.
b. The cup provides space for fuel storage during metering.
c. The cup orifii sizing will determine droplet sizing and spray dispersal characteristics. The plunger and cup are lapped in manufacture and are therefore not interchangeable.

Plunger Spring

a. Lifts the plunger for metering
b. Loads the injector train

Check Ball

The check ball prevents fuel flow reversal that could be caused by:

- Preinjection pressures
- Seepage of compression pressures into the cup

The check ball, also known as a lock-off ball, is designed to isolate the cup pump chamber at any time cup pressure exceeds metering pressure.

Balance (Calibrating) Orifice Plug

a. Provides a means of customizing fuel flow through the injector, which permits standardization of the injector body to a number of engines
b. Balances (fueling) injector sets to an engine by defining the flow area to which the rail unloads

The balance orifice is sized by burnishing, a procedure that must be performed using Cummins service literature, tooling, and a PT injector comparator (bench test fixture).

Metering Orifice

The metering orifice is the passage within the PT injector through which fuel must pass to enter the cup. It is exposed to the cup whenever the plunger is lifted. When the plunger is lifted, some of the fuel that is flowing through the PT injector diverts through the now exposed metering orifice to flow into the cup. The metering orifice defines the critical flow area that determines the actual quantity of fuel to be injected.

Plunger Link

The **plunger link** is a short rod that is inserted into an aperture on the PT injector follower that helps prevent side-loading of the plunger (which moves linearly) by the actuating rocker arm (which moves radially). They are surface-hardened devices and should be replaced when evidence of surface hardening failure is observed.

THE PTG-AFC PUMP

Only the PTG-AFC pump is discussed in this text. AFC is an acronym for air-fuel control: These pumps are used on all turboboosted engines. Pressure-time regulator (PTR) pumps are regulator-controlled versions used primarily on gensets. The first-generation PTG pumps were replaced by PTG-AFC pumps many years ago. The PTG-AFC pump has the following functions:

1. Moves all fuel through the system
2. Manages rail pressure
3. Provides limiting speed (LS) governing
4. Limits fueling when manifold boost is low (internal aneroid)
5. Provides for ignition key engine shutdown

The PT pump is a precise flow control device that *manages* rail pressures, which is the control factor in the Cummins PT fueling equation. It is best studied by first identifying the subcomponents (Figure 23–4) and subsequently analyzing performance schematics.

PT Pump Components

It is critical for the technician to understand the subcomponents of the PT pump and how they interact with each other in order to effectively troubleshoot PT system malfunctions.

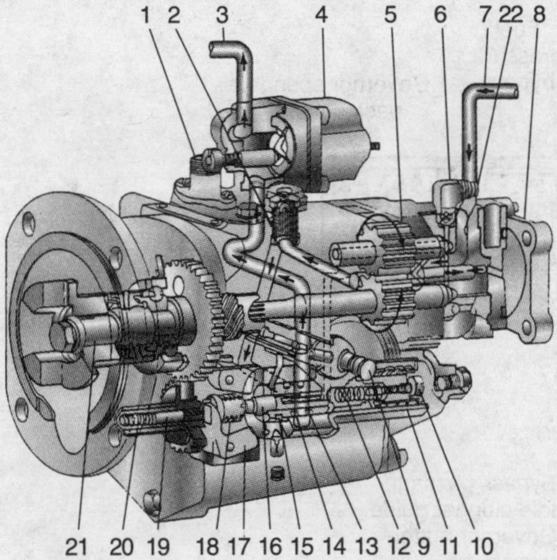

1. Tachometer drive shaft
2. Filter screen
3. Supply to injector rail
4. Shutdown solenoid
5. Gear pump
6. Check valve elbow
7. Fuel from tank
8. Pulsation damper
9. Throttle shaft
10. Idle adjust screw
11. High-speed spring
12. Idle spring
13. Gear pump pressure
14. Fuel manifold pressure
15. Idle pressure
16. Governor plunger
17. Governor weights
18. Torque spring
19. Weight assist plunger
20. Weight assist spring
21. Main shaft
22. Bleed line

Figure 23–4 *PT pump components. (Courtesy of Cummins Inc.)*

Gear Pump

The gear pump is located at the rear of the PT pump and is encased within a cast-iron housing. It is responsible for all movement of fuel in the system. Therefore, the entire fuel subsystem is under suction. In early PT pump versions, the fuel filter mounting pad was integral with the pump. In more recent versions, the filter pad is remote mounted. Remote positioning of the filter mounting pad facilitates troubleshooting the admission of air into the fuel subsystem (a not uncommon problem) by permitting the insertion of a diagnostic sight glass in series between the gear pump and the filter pad. The gear pump is positive displacement; that is, through each cycle, it unloads a constant slug volume into the outlet. Because it unloads to a defined flow area, pressure rise will be proportional to its rotational speed. It is driven at engine rpm. The pressure values produced by the pump always exceed rail pressure.

Pulsation Damper

The gear pump produces pressure in pulses, that is, there is a surge each time a slug of fuel is unloaded by the gear teeth into the ducting supplying the governor assembly. To smooth these pressure pulses, a **pulsation damper** consisting of a steel disc supported on either side by a pair of O-rings within an aluminum housing is mounted directly onto the gear pump housing. A fractured pulsation damper disc can produce erratic/fluctuating rail pressures, the result of which is engine rpm surging.

Internal Filter

The internal filter is a cylindrical core consisting of a mesh and a magnet. It is located downstream from the gear pump and its purpose is to minimize the possibility of a gear tooth cutting passing into the rest of the PT pump circuitry. It can be replaced without disassembling the pump but is not considered part of routine service maintenance and, in fact, should be left untouched.

Governor Assembly

The governor assembly (Figure 23–5) is driven at approximately two times the PT pump-driven speed, allowing smaller weights to be used. The governor uses all the principles of a typical mechanical governor we describe in more detail in Chapter 25. In any mechanical governor, centrifugal force generated by rotating flyweights attempts to diminish engine fueling at speeds beyond governor break (rated speed), while spring force attempts to increase engine fueling. At engine high-idle speed, the centrifugal force generated by the flyweights *must* be capable of overcoming any of the spring forces that are applied against it to prevent engine overspeed.

At any engine speed, the position of the **governor plunger** within the **governor barrel** is determined by the balance between flyweight force acting on one end of the plunger and spring force acting on its opposite end. The governor plunger is machined with a recessed annulus, which is cross and center drilled; it is driven by a radial lug that rests between the flyweight feet and therefore rotates within the stationary governor barrel. As speed increases, the plunger is forced down the governor barrel.

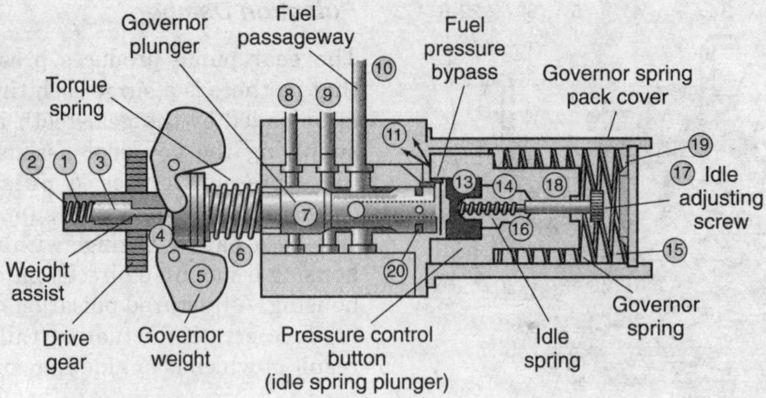

Figure 23–5 *PTG governor. (Courtesy of Cummins Inc.)*

Fuel from the gear pump enters the governor barrel through the supply passage at any time the engine is running.

The supply passage is always exposed to the recessed annular area of the governor plunger. The cross and center drillings hydraulically connect the plunger annulus with the **governor button**, which is spring loaded to contact the inboard end of the plunger. The sizing of the recess in the governor button will determine how much fuel pressure is required to overcome the spring pressure applied to the opposite side of the governor button and separate the button from the plunger to spill fuel back into the bypass passage for recirculation through the PT pump circuitry. The smaller the surface area of the governor button recess, the more pressure is required in the governor barrel before button separation and spill occurs. A governor button with a small recess has the effect of increasing supply pressure downstream of the governor assembly.

Fuel is spilled to the bypass passage continually; the amount spilled will determine the actual pressure in the plunger annular recess. Fuel may also exit the plunger recess through the idle passage and the main passage. At idle speeds, both the idle and main passages are exposed to the plunger annulus; as the speed increases and centrifugal force drives the governor plunger inboard, the idle passage ceases to register with the plunger recessed annulus. Through torque rise, the torque control spring becomes a factor; this spring helps define torque rise rpm. Beyond rated speed, the governor plunger has been driven inboard sufficiently that the flow to the main passage begins to diminish, reducing the flow potential to the rail.

As engine speed progresses through the droop curve (graduated fuel deration that occurs as the engine rpm increases from the rated to the high-idle speeds), the flow area to the main passage continues to diminish, meaning that most of the supply fuel passes through the plunger cross and center drillings, acts on the governor button, and spills to the bypass passage. At the end of the droop curve or high idle, transverse drillings in the governor plunger, called overspeed dump holes, register directly with the bypass passage and most of the fuel entering the governor barrel is recirculated to the gear pump intake.

The tension of the governor spring assemblies determines how much centrifugal force (this correlates to an rpm value) is required to drive the plunger to a position where the main passage is taken out of register with the plunger annular recess. Governor spring tension is set with shims and defines the high-idle speed. The sizing of the governor button recess moderates the actual supply pressure value and affects engine fueling throughout the speed range of the engine. The governor button is loaded against the inboard end of the plunger by the idle spring. Final adjusting of this spring tension is performed dynamically (engine running) and sets the engine idle speed.

Throttle Shaft Assembly

The throttle shaft assembly receives fuel from two passages in the governor assembly and supplies fuel to the air-fuel control (AFC) circuit and the no-air set screw. It is actuated by a throttle lever (accelerator lever) with a spring breakover; the throttle lever is clamped to the throttle shaft. The throttle shaft is set to move through an arc of 28 degrees ± 1 degree. Moving the throttle shaft will set a flow area from the main passage to the AFC circuit by setting the extent of register of the throttle shaft fuel orifice with the main passage. When the throttle lever travel is measured with a protractor and exceeds the required 28 degrees ± 1 degree, the complaint is usually that of low power, and it is an indication that the throttle arm stops have been tampered with.

When the throttle shaft is in the idle position, a small quantity of fuel flows through the throttle shaft fuel orifice; this fuel is defined as throttle leakage. The idle passage feeds an eccentric in the throttle assembly and bypasses the throttle fuel orifice. As the throttle shaft is rotated, the throttle shaft fuel passage is brought into full register with the main passage. Flow area through the throttle fuel orifice is set by a fuel adjusting screw within the throttle shaft bore; it is used to adjust rail pressure on the pump calibration stand.

No-Air Set Screw

The no-air set screw defines the maximum flow area from the throttle assembly to the rail *before* a predetermined manifold boost value has been achieved. It should only be adjusted on the PT pump calibration fixture.

AFC Circuit

The **air-fuel control (AFC) circuit** is an acceleration smoke control mechanism that replaced the external aneroid device used in earlier PT pump models. It is a requirement of all turboboosted engines that fuel delivery be modulated to the actual amount of air charged into the engine cylinders. In a Cummins PT engine, intake manifold pressure is piped to the AFC manifold on the PT pump, where it acts on an AFC diaphragm to which the AFC plunger is attached. A spring acting on the AFC plunger loads it into its closed position. When manifold boost reaches a predetermined value, usually 15 psi, it acts on the diaphragm to overcome spring tension to drive the AFC plunger inboard, exposing the AFC ducting and increasing the fuel flow potential to the rail. Unlike most aneroid devices that are on-off, the PTG-AFC increases flow to the rail in proportion to manifold boost increase. The AFC circuit is usually fully open by the time 25 psi of manifold boost is attained, regardless of peak boost on the engine, which is usually around 35 psi.

Shutdown Solenoid

The **shutdown solenoid** is an energized-to-run (ETR) device that permits the engine to be shut down using the ignition key; it acts by gating the rail. When the solenoid is not energized, a spring loads a disc into a closed position that gates the supply passage from the AFC ducting to the rail pipe. When the solenoid is energized the disc is pulled into the solenoid, permitting flow. A manual override is incorporated in the device in the event of a chassis electrical failure. When turned clockwise (CW), the shutdown solenoid override displaces the shutdown disc by forcing it back against the spring, thus opening the rail circuit.

PT PUMP CIRCUIT SCHEMATICS

Probably the best way to understand the operation of a PT pump is to interpret the following sequence of schematics that show how every internal pump component responds to any given set of running conditions. The technician who is serious about accurately troubleshooting Cummins PT systems should etch these schematics into memory where they will be invaluable.

Idle Speed

(See Figure 23–6.)

1. Fuel is pulled through the fuel subsystem by the gear pump.
2. Pressure waves caused by the gear pump are smoothed by the pulsation damper.
3. Fuel is forced through an internal filter consisting of a mesh and a magnet.
4. Fuel enters the governor assembly through the supply port in the governor barrel. The governor weights generate little centrifugal force at idle speed, and, therefore, the governor plunger (which rotates within the governor barrel) is in a position in which both the idle and main ports are in register with the recessed annular area of the plunger—fuel at supply pressure will exit through the idle and main ports. Because the governor plunger is center and cross drilled, supply pressure will also act on the governor button, separating it, and recirculate fuel to the bypass circuit.
5. Fuel from the governor assembly arrives at the throttle assembly from the idle and main passages. However, the throttle is in the idle fuel position, which locates the throttle fuel orifice out of register with the main port; a small quantity of fuel, known as throttle leakage, is permitted to pass through the throttle fuel orifice. Fuel from the idle passage is permitted to bypass the throttle shaft.
6. Fuel from the throttle assembly is ducted to the AFC circuitry. However, at idle speed there is little or no manifold boost to act on the AFC diaphragm; accordingly the AFC plunger is in the closed position. The flow area defined by the no-air set screw setting is the total flow area feeding the rail pipe.
7. Fuel passes around the no-air set screw and exits the PT pump through the solenoid to which the rail pipe is connected.

Normal Engine Operation—Operation in Torque Rise RPM

(See Figure 23–7.)

1–3. As in the previous schematic, but pressure within the defined flow area represented by the passage feeding the governor barrel will be higher as the positive displacement gear pump is being rotated at a higher speed.

4. Fuel enters the governor assembly through the governor barrel supply port and circulates in the annular recess of the governor plunger. However, due to the higher rpm and greater centrifugal force generated by the flyweights, the governor plunger has been driven inboard into the barrel sufficiently to have taken the idle passage out of register with the plunger recessed annulus. Supply fuel exits the governor barrel through the main passage and also acts on the governor button, spilling fuel to the bypass for recirculation. This action moderates supply pressure. The torque spring is now a factor; this helps resist further inboard plunger travel toward a diminished fuel position.
5. Fuel from the governor is ducted to the throttle assembly by the main passage. As the engine is running somewhere within

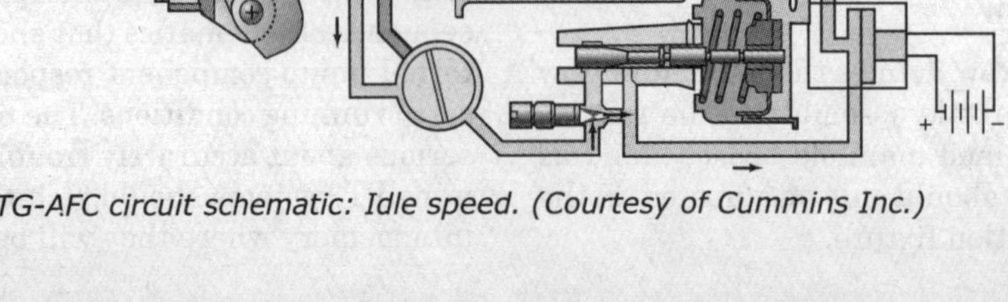

Figure 23–6 *PTG-AFC circuit schematic: Idle speed. (Courtesy of Cummins Inc.)*

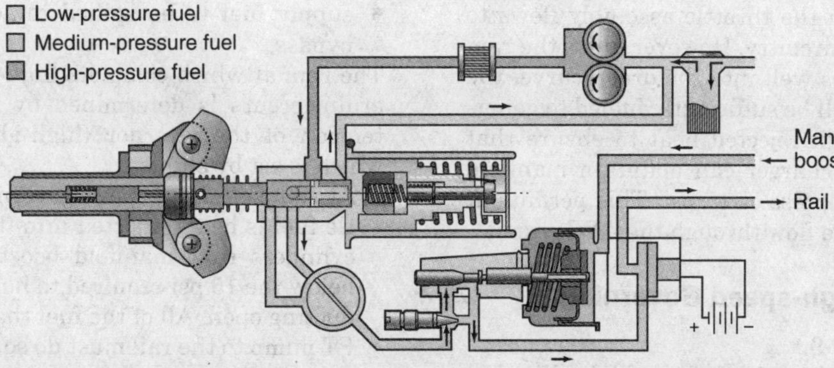

Figure 23–7 *PTG-AFC circuit schematic: Normal operation. (Courtesy of Cummins Inc.)*

the torque rise rpm, the throttle shaft is positioned so there is some degree of register of the fuel orifice with the main passage, defining a flow area.

6. Fuel flows from the throttle assembly to the AFC circuitry. As before, it flows up to and around the no-air set screw, but now there is sufficient manifold boost acting on the AFC diaphragm to overcome the AFC spring and drive the AFC plunger inboard sufficiently to permit flow through the AFC circuit, increasing flow area to the rail.
7. Fuel from the AFC circuitry passes through the shutdown solenoid and out to the rail pipe.

High-speed Governing

(See Figure 23–8.)

High-speed governing begins at governor break, that is, as the engine rpm enters the droop curve. High-speed governing is a common running condition for Cummins PT engines in which the accelerator is held in the full-travel position by the operator.

1–3. Supply pressure is higher than in previous schematics due to higher rpm but otherwise the operating conditions are the same.

4. Fuel enters the governor assembly through the governor barrel supply port—but now the centrifugal force exacted by the flyweights has driven the governor plunger inboard to the extent that the main passage flow area is diminished. More fuel is being spilled by passing through the center and cross drillings acting on the governor button to enter the bypass circuit.
5. For the engine to be run in this condition, the accelerator would necessarily be fully depressed, allowing the throttle shaft fuel orifice full register with the main passage.

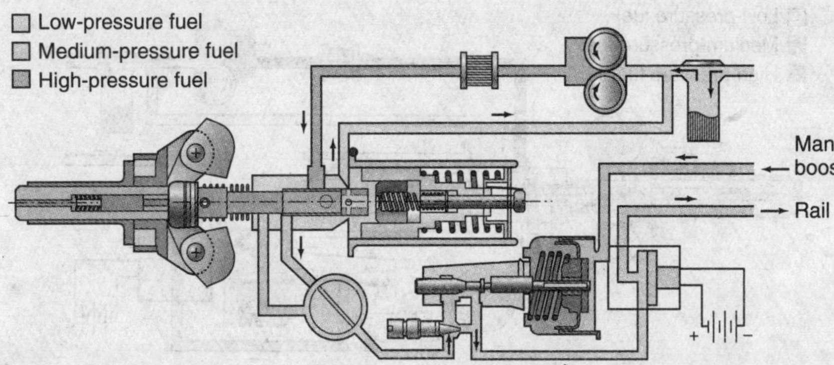

Figure 23–8 *PTG-AFC circuit schematic: High-speed governing. (Courtesy of Cummins Inc.)*

6. Fuel from the throttle assembly flows to the AFC circuitry. However, until the rpm penetrates well into the droop curve, the engine will be sufficiently fueled to generate enough rejected heat to ensure that the turbocharger can maintain manifold boost in excess of 15 psi. This permits at least some flow through the AFC circuit.

Complete High-speed Governing

(See Figure 23–9.)

Engine rpm has risen above high idle; that is, the engine is being run at an rpm beyond the droop curve. In most cases, this means that the engine is being driven by the drivetrain of the vehicle—a condition that would occur when running on a prolonged downhill. The schematic shows what is occurring in the PT pump circuitry when the engine speed exceeds the high-idle speed.

1–3. Supply pressure being dependent on rpm is at its highest.

4. Fuel enters the governor barrel through the supply port in the governor barrel. In this running condition, the governor flyweights are fully extended and have driven the governor plunger inboard against the governor spring (high-speed spring) to the extent that:
   - The main passage has been entirely or almost entirely taken out of register with the recessed plunger annulus.
   - The plunger has been driven inboard to the extent that a transverse drilling through the plunger has extended beyond the governor barrel, allowing most of the supply fuel to be spilled directly to the bypass.

The rpm at which complete high-speed governing occurs is determined by the spring tension of the governor (high-idle) spring, which is set by shims.

5. At complete high-speed governing, so little fuel is being injected into the engine's cylinders that manifold boost will drop below the 15 psi required to hold the AFC ducting open. All of the fuel that exits the PT pump to the rail must do so by passing around the no-air set screw.

PT PUMP TESTING AND ADJUSTMENTS

Cummins PT pumps are calibrated on a comparator bench—a procedure that should be performed using the Cummins service literature and specifications. This procedure is the specialty of the diesel pump room specialist and is not within the scope of this textbook. The following outlined procedures are those that the truck diesel technician would use to set up and diagnose problems occurring in a PT fuel system.

Dynamic Idle Adjust Tool

After the PT pump has been calibrated on the comparator bench or if an exchange pump has been installed on an engine, the final adjustment to idle speed should be set to specification with the engine running. Install an idle adjust tool. This tool can be made as shown in Figure 23–10.

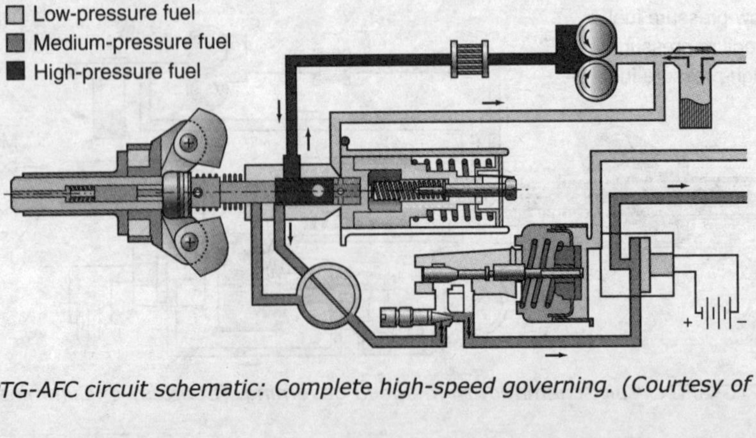

Figure 23–9 *PTG-AFC circuit schematic: Complete high-speed governing. (Courtesy of Cummins Inc.)*

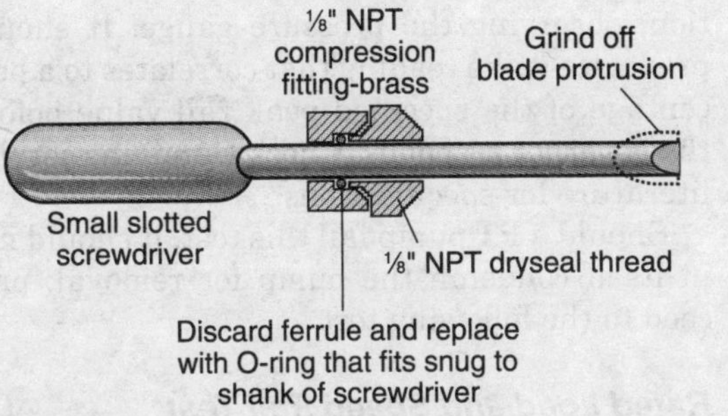

Figure 23–10 *Idle adjust tool.*

Dynamic-Idle Adjustment

This should be set whenever a PTG-AFC is replaced.

The idle adjust tool should be installed to the access plug on the governor spring assembly cover. The engine should then be started and run until there is no performance evidence of air in the PT pump circuit, which is usually about 5 minutes. The idle adjust screw is then turned CW to increase the idle speed or counterclockwise (CCW) to decrease it; this procedure sets the spring tension that acts on the governor button, and the higher this is, the less easily it spills from the governor barrel assembly (Figure 23–11). Always consult Cummins specifications before setting the idle speed; this setting will impact on engine fueling overall when maladjusted. The idle speed specification is stamped on the control parts list (CPL) plate.

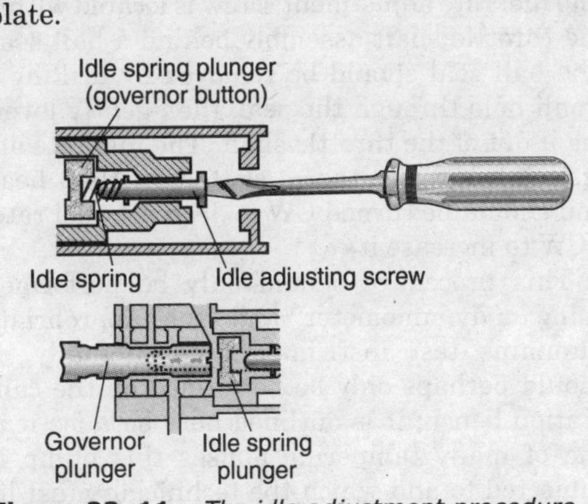

Figure 23–11 *The idle adjustment procedure. (Courtesy of Cummins Inc.)*

High-Idle Adjustment

On PTG-AFC pumps, the high-idle spring tension is set by shims. To adjust the high-idle speed, the governor spring cover should be removed. Next remove the snapring, spring assembly retainer, and shim pack. When adding or subtracting shims, 0.001″ shim thickness is approximately equivalent to 2 rpm.

Fuel Subsystem Inlet Restriction

The entire fuel subsystem is under suction in the Cummins PT system so ensuring the inlet restriction values are within specification is critical. These values should be tested with a mercury (Hg) manometer or pressure gauge reading Hg (inches of mercury).

Specifications:
With a new filter: 4–6 in.-Hg
Typical in-service specification: 7 in.-Hg
Replace filter when specification exceeds: 8 in.-Hg

When inlet restriction readings exceed specification after the filter has been replaced, any component between the PT pump and the pickup tube inlet in the tank may be responsible. Start by checking filter gaskets and remember that internal failure of the fuel hoses can cause the problem.

Admission of Air to the Fuel Subsystem

This condition is characterized by rough engine running overall and an erratic rpm roll at all no-load speeds. Check using a diagnostic sight glass installed in series between the gear pump and the filter mounting pad. In some older applications, the filter mounting pad assembly was integral with the pump, so the sight glass would have to be located upstream from the filter.

After installation the engine should be run for about 5 minutes to purge the air that was admitted when installing the sight glass. To analyze what is observed through the sight glass, the engine should be run at a high-idle speed; with good light focused on the sight glass, the fuel should appear transparent. A cloudy appearance is an indicator that air is being pulled into the fuel system; only in cases in which a large quantity of air is being admitted to the suction side of the fuel system will actual bubbles be observed.

Checking Throttle Arm Linkages

The PT throttle is set to travel 28 degrees ± 1 degree on the comparator bench. The travel stops should not be adjusted on-engine. The technician investigating a low-power complaint should first check the accelerator linkage by having someone fully depress the accelerator in the cab; a visible breakover should be observed at the throttle arm (1/8 inch). If not, adjust the accelerator linkage to obtain the required throttle breakover, ensuring that it returns to the idle position after each adjustment.

Should this first test perform to specification, remove the accelerator linkage from the throttle arm and use a protractor to check that the throttle arm moves through the required 28-degree arc. Throttle lever travel should be set on the PT pump comparator bench. Tampering with the throttle stop settings almost always results in reducing flow at the throttle shaft bore at full travel. Should an out-of-specification throttle arm travel reading be observed, the pump should be recalibrated.

Testing PT Rail Pressure

The following tests are used by the technician to diagnose PT system malfunctions arising from incorrect rail pressure values. When performing these tests, Cummins service literature should be consulted, especially when using the chassis dynamometer as a diagnostic tool.

Snap Rail Test

The snap rail test is a fast means of checking that the rail pressure is at least approximately achieving specification. It is outlined in some of the older Cummins service literature with the caution that it is inclined to produce erratic results. The test works by "surprising" the governor into producing a surge of rail pressure. It has validity because it is a means of roughly checking rail pressure without the use of a dynamometer. Locate the specification for peak rail pressure on the engine to be tested. Insert a fluid-filled pressure gauge calibrated to read up to 250 psi in the rail pipe plug in the shutdown solenoid assembly. Start the engine and idle. Check the base rail pressure value. Next, "snap" the throttle arm to the full-speed position, observing the pressure gauge. It should produce a spike reading that correlates to a percentage of the specified peak rail value before the governor responds. Check Cummins service literature for specifications.

Should a PT pump fail this test, it should not in itself condemn the pump for removal; proceed to the following test.

Rated Load and Speed Rail Test

This test should be performed on a dynamometer, preferably in conjunction with full monitoring instrumentation hookup. A fluid-filled pressure gauge capable of reading to at least 250 psi should be fitted to the rail pipe plug in the shutdown solenoid and positioned so it can be read while running the dynamometer. Manifold boost should also be monitored while performing this test. Incrementally load the engine at 100 rpm steps through the torque rise profile (see Chapter 15), recording the rail pressure values. Peak rail pressure should read to within 2% (about 4 psi) of specification.

Fuel Rate Adjustment

This tests the fuel quantity pumped through the system that is actually used for fueling the engine. It is measured to specification by the technician at rated speed and load on an engine or chassis dynamometer in pounds per hour. The fuel rate adjustment screw is located within the throttle shaft assembly behind a ball seal. The ball seal should be removed by drilling a small hole through the seal, then gently levering it out of the throttle shaft. The fuel rate adjusting screw is either a slotted or Allen head and should be turned CW to decrease fuel rate, CCW to increase it.

This procedure should only be performed using a dynamometer and the appropriate Cummins test instrumentation. Actually, it should perhaps only be performed on the calibration bench; it is outlined here because it is one of many tampering abuses this pump is subjected to and which the technician must be aware of when troubleshooting the system. It should be especially noted that backing the fuel rate screw too far out of the throttle shaft (at least 0.250″ depth into the shaft bore is required) can result in the screw being forced out of the shaft.

Adjusting the No-Air Set Screw

This setting defines the maximum flow area from the pump circuitry to the rail when there is insufficient manifold boost to have activated the AFC circuit. The no-air set screw is located behind the throttle cover plate, directly above the throttle shaft. The adjustment of the no-air set screw should be confined to the comparator bench but again, it is yet another setting that is commonly tampered with, so the technician should be aware of it. The checking procedure seeks to determine a specific (within a range) rail pressure for a loaded engine at a set engine speed with the AFC signal (that is, manifold boost) pipe plugged. Fit a liquid-filled pressure gauge capable of reading up to 250 psi to the rail plug in the shutdown solenoid assembly. The pump code and engine specifications should be referenced for the specification window permitted. If the no-air rail pressure value is out of specification it can be adjusted by backing off the no-air set screw locknut. Rotating the screw CW will lower rail pressure; CCW will increase rail pressure.

Testing Peak Manifold Boost

Cummins PT engines use fixed geometry turbochargers so testing manifold boost is a relatively easy procedure. A dynamometer MUST be used. Attempting to test manifold boost on a road test will produce generally meaningless data. Fit a good-quality, liquid-filled pressure gauge (calibrated to 50 psi) to an intake manifold port and position so that it can be read while operating the dynamometer. Reference Cummins service literature for specifications. Peak manifold boost will be observed at rated speed and load.

PT Pump Installation

The PTG pump is flange mounted to an accessory drive rotated at engine rpm; it is not timed to the engine. When removing the pump from the engine, all the plumbing ports should be sealed with dust caps. When reinstalling the pump, the flange gasket should be replaced and the drive coupling spider inspected for serviceability. The pump is often installed to the rear of the compressor housing in truck applications.

1. Install the coupling spider to the drive lugs, place a new flange gasket in position, and torque the four mounting bolts.
2. Connect the fuel and AFC plumbing. Connect the shutdown solenoid wiring. Connect the tachometer cable when fitted.
3. Remove the priming plug (a standard pipe plug located at the upper deck of the PT pump) and fill the pump housing with filtered fuel. Replace and prime the fuel filter.
4. Attach the accelerator linkage. Adjust to ensure the throttle arm moves through the required 28-degree arc and produces a breakover of 1/8 inch. Set the accelerator pedal stop.
5. Start the engine. There will usually be evidence of air in the system for about 5 minutes but generally it will self-purge.

PT Injector Installation and Adjustment

The engine firing order is 1-5-3-6-2-4.

Short cuts should never be used when performing Cummins PT overhead adjustment. It should also be noted that when the A, B, and C (1-6, 5-2, and 3-4 on older engines) calibration indices on the accessory drive pulley are aligned with the calibration notch on the engine block, no pair of pistons is at top dead center (TDC). See Table 23–1. Always consult Cummins technical data when setting the overhead adjustments. The following is a brief description of the methods used.

Table 23–1

Engine Position	*Piston Location*					
A or 1–6	1–6	90 degrees ATDC	2–5	30 degrees BTDC	3–4	150 degrees BTDC
B or 2–5	2–5	90 degrees ATDC	3–4	30 degrees BTDC	1–6	150 degrees BTDC
C or 3–4	3–4	90 degrees ATDC	1–6	30 degrees BTDC	2–5	150 degrees BTDC

Pretorque

Injector cam lobe must be on OBC (outer base circle: the largest radial dimension of a cam profile), that is, *not* as shown in Table 23–2. This method sets the actual loading of the plunger into the cup, a linear force of around 380 lb. This is done by torquing the injector train adjusting screw on its actuating rocker to a specific value. The torque applied to the adjusting screw varies between 45 lb-ft. and 90 lb-ft. depending on the rocker housing casting material (aluminum/cast iron) and whether the adjustment is performed hot or cold.

Table 23–2 *Top end set positions—rotate the engine CW.*

Engine Position	*Set Injector*	*Set Valves*
A or 1–6 (#5 on compression)	3	5
B or 2–5 (#3 on compression)	6	3
C or 3–4 (#6 on compression)	2	6
A or 1–6 (#2 on compression)	4	2
B or 2–5 (#4 on compression)	1	4
C or 3–4 (#1 on compression)	5	1

Dial Indicator

The injectors should be set with the engine location as shown in Table 23–2. That is, each injector is set when its actuating cam is at a specific location of its base circle. This adjustment requires that plunger free travel be set to a precise dimension. A dial indicator is located in position to ride on the plunger flange, and the rocker is actuated using a rocker actuating claw. The travel of the plunger is then set by adjusting the rocker adjusting screw to an exact value; 0.170″ ± 0.001″ is typical for N series, but again, always consult the OEM specifications. Setting PT injectors by this method should ensure that the required loading of the plunger into the cup on OBC is achieved and the method is preferred by Cummins.

Top Stop

Top stop injectors are adjusted as shown in Table 23–2; in other words, as with the dial indicator method, they are set on a specific location of IBC (inner base circle or the smallest radial dimension of the cam profile). However, these are actuated by a train that Cummins wants set with zero lash. Zero lash is identified by Cummins as having a slight load on the injector actuating train. The setting is achieved by torquing the train adjusting screw to 5–6 lb-in. It is good practice to initially overtorque the specification by about 50%, back off, then retorque to specification. Correctly set, the setting can be checked by attempting to rotate the injector link—it should rotate but with some difficulty.

STC Injector Setting

Step timing control (STC) injectors can be set using an OBC method. It is not straightforward, however, and Cummins Service Bulletin #3666006–00 of May 1991 should be consulted.

Locating Cylinder Misfire

This process is the equivalent of shorting out hydraulic injectors by cracking the union nuts on high-pressure pipes. A tool called a rocker claw or rocker lever is used to perform this test on a PT-fueled engine. This tool has an arced yoke that fits under the rocker shaft and with the help of a ¾″ or 19 mm wrench levers the injector rocker into its fully actuated position. This action holds the PT injector plunger seated in the cup, preventing it from metering and pumping. An idling engine may be tested for cylinder misfire using this test; each cylinder should be cut out (the term *shorted* is often used) in sequence using the rocker claw. If rpm drops each time a PT injector is cut out, that injector can be presumed to be functioning properly. If the engine rpm does not drop when an injector is shorted out, that injector should be removed and tested.

Engine Timing

Actuation of the fuel system pumping pulse by the engine timing geartrain must be precisely phased in any diesel engine. This process essentially times the events that manage the fuel system pumping pulse to a specific engine position. Cummins achieves this, using two methods depending on the engine series.

Offset Camshaft Key Timing

The **offset camshaft key timing** (the camshaft gear is set to a specific axial position on the camshaft by the offset of the locating key) is used on L, M, and K series (Figure 23–12), whereas **cam follower housing timing** is used on N series.

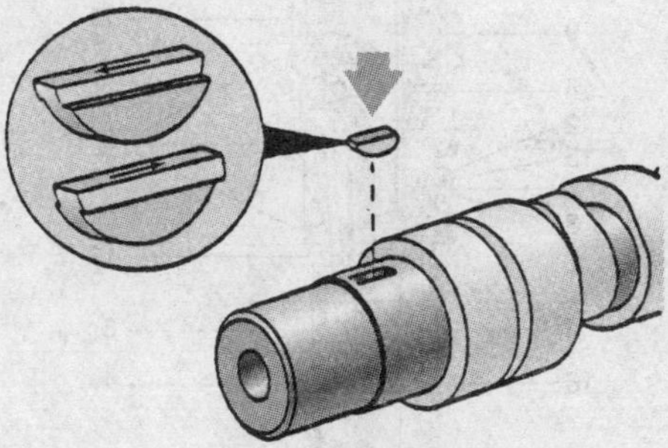

Figure 23–12 *The cam timing procedure used on some PT engines showing the offset camshaft key method. (Courtesy of Cummins Inc.)*

Cam Follower Housing Timing

To perform cam follower housing setting on N-series, PT engines, a Cummins engine timing fixture is used. The timing fixture consists of a bracket that mounts to the cylinder head (Figure 23–13). The timing fixture is fitted with two dial indicators and probe assemblies. One probe is positioned so that it can be inserted through the injector bore in the cylinder head to contact the piston crown, and the second probe is positioned to ride on the injector push tube. In this way, one probe is set to read engine position while the other reads camshaft position.

An adjustment is made by either adding or subtracting gaskets that seal the cam follower housing and cylinder block. Seen from the front of the engine, the camshaft rotates CCW (viewed from the front); the camshaft and cam follower assemblies are positioned on the left side of the engine (viewed from the rear). The result is that if shims are added to the cam follower housing, the cam follower rollers (which actuate the injector train) will register with any portion of the cam geometry slightly earlier. Similarly, if shims are removed from the cam follower housing, the cam follower rollers will move farther inboard and away from the camshaft, registering with any portion of the cam geometry fractionally later. Therefore, in N-series engines,

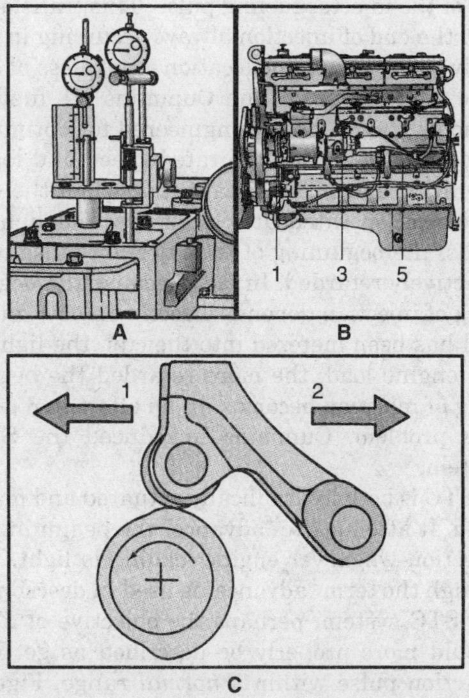

Figure 23–13 *A. Injection timing tool in position: The tool determines injector pushrod travel in relation to piston travel; B. Location of the cam follower housing and reference cylinders; C. 1. Advance 2. Retard. (Courtesy of Cummins Inc.)*

shimming up the cam follower housing advances the injector train actuation timing, whereas subtracting shims retards timing.

The procedure is outlined in Cummins technical service literature and is straightforward, but the method must be meticulously adhered to. The procedure should be undertaken at each major overhaul: Measuring and replicating the shim (gasket) thickness on disassembly is not considered sufficiently accurate; however, it is a good starting point when beginning the timing procedure. The offset camshaft gear achieves the same objective; it phases fuel injection actuation with a precise engine position. Several offset keys are available and their function is to alter the axial relationship between the camshaft gear and the camshaft.

STEP TIMING CONTROL

A characteristic of PT, common rail injection is to have a variable beginning and constant ending of the injector fueling pulse. This translates into the end of injection always occurring in exactly the same engine location regardless of engine load or speed. The Cummins PT fueling window was originally engineered for optimum performance around the rated speed and load-operating zone. This meant that when the engine was operated at low speeds and lighter loads, the beginning of the injection pulse was effectively retarded. In fact, because the beginning of injection depends directly on how much fuel has been metered into the cup, the lighter the engine load, the more retarded the beginning of injection becomes. In an effort to rectify this problem, Cummins introduced the STC system.

STC is both hydraulically actuated and managed. It attempts to "advance" the beginning of injection whenever engine loading is light. Although the term "advance" is used in describing the STC system, perhaps the objective of STC should more properly be described as getting injection pulse within a *normal* range. Figure 23–14 shows the components used in an STC injector.

STC hardware consists of PTD top-stop style injectors fitted with tappets and an STC control valve, plus an interconnecting hydraulic circuit (Figure 23–15). The STC tappet is located at the top of the PT injector; when this tappet is charged with engine oil, it effectively lengthens the PT plunger. The STC actuating cam profile has greater lift than non-STC injector cams.

As a result, when the injector train is actuated with the STC tappet charged, the plunger both begins its downstroke earlier then bottoms into the cup before peak cam lift is attained. When the plunger bottoms into the cup, the STC actuating oil is trapped in the tappet so the oil pressure increases until a load check ball is unseated. This occurs at between 1,100 and 1,500 psi, spilling the oil from the tappets into the rocker housing from which it is returned to the sump.

Whenever the STC tappets are charged with engine oil by the STC control valve, the timing operates in advanced mode. This occurs at light engine loads. When the STC control valve sets normal timing mode, the STC tappets no longer charge with oil, so the first portion of the injector cam flank travel into the PT injector is re-

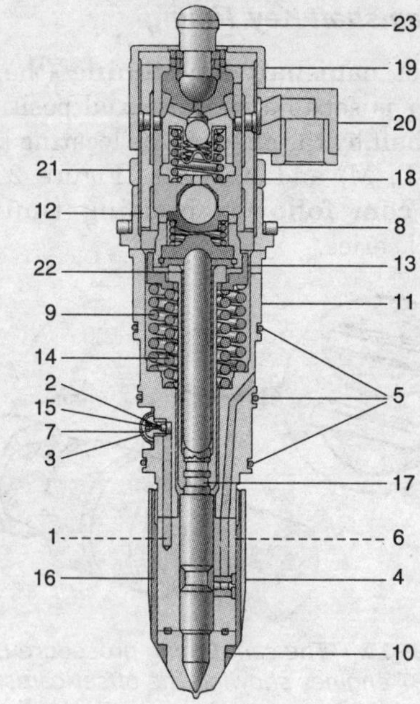

1. Check ball
2. Injector gasket
3. Screen retainer
4. Cup retainer
5. O-ring seal
6. Roll pin
7. Filter screen
8. Washer
9. Plunger spring
10. Injector cup
11. Injector adapter
12. Stop screw
13. Spring retainer
14. Plunger spring
15. Metering orifice plug
16. Barrel
17. Plunger
18. STC tappet
19. Tappet stop cap
20. Tappet top stop
21. Lock nut
22. Lock nut
23. Injector link

Figure 23–14 *STC injector components. (Courtesy of Cummins Inc.)*

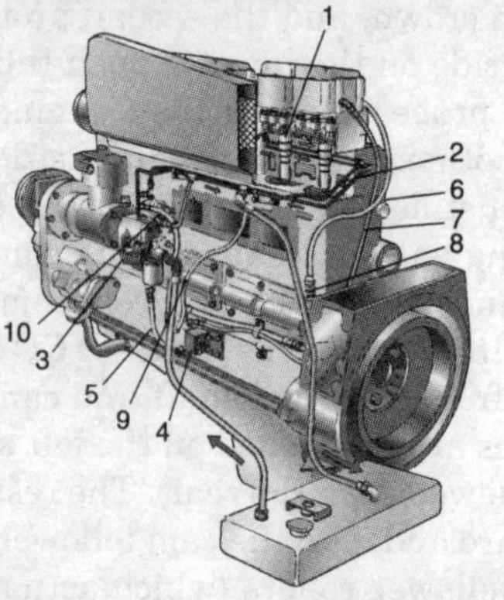

1. Injector
2. Injector drain line
3. Fuel pump
4. STC control valve
5. STC oil supply line
6. STC oil outlet line
7. C-brake sensing line/oil drain line
8. Check valve
9. STC fuel drain line
10. STC fuel pressure sensing line

Figure 23–15 *STC circuit layout. (Courtesy of Cummins Inc.)*

sponsible for collapsing the STC tappet and the PT plunger is then actuated mechanically rather than hydraulically.

The STC control valve monitors actual rail pressure and uses predetermined values to differentiate between normal and advanced modes. Most commonly, the critical value is 53 psi. In other words, when rail pressure is 53 psi or less, the STC control valve will operate in advance mode and charge the STC tappets. As rail pressure rises beyond the critical 53 psi parameter, the STC control valve switches to normal operation. Normal operation continues until rail pressure decreases to 25 psi; at this point, the system switches to advanced mode once again.

Direct fuel feed (DFF) injectors are a later version of the STC injector. DFF injectors are designed to prevent metering when the engine is in a motoring condition (throttle arm at zero travel, chassis momentum driving engine). This reduces noxious emission, cup orifii carboning, and, of course, wasted fuel.

PACE AND PACER

PACE and **PACER** are Cummins PT system, partial authority computer-controlled systems. A **partial authority** engine management system is one in which a hydromechanical fuel system is adapted to computer controls; Chapter 30 on computerized engine management should be referenced to understand the operating principles of electronic control module (ECM)-managed engines. A brief account is provided here.

System Objectives

The PACE/PACER management system adapts the PT system for some electronic controls but leaves most of the key components of the system operating as if there were no computer controls. The primary feature of PACE/PACER is road speed governing. In fact, PACE and PACER are not true acronyms but instead refer to the road speed governing feature. A major disadvantage of the hydromechanical PT system in truck applications was the tendency of drivers to operate the engine in the governor droop curve rpm. That is, drive the vehicle with foot to the floor, resulting in an engine speed somewhere above rated and below high idle. When an engine is run in the droop curve it will usually run at lower-than-optimum efficiency and produce higher levels of noxious emissions. The PACE/ PACER essentially governs the operator out of the droop curve.

The PT fuel system remains largely intact. PACE/ PACER is mastered by a **PT control module (PTCM)** supplied with inputs from a transmission tailshaft-mounted vehicle speed sensor (VSS), a brake switch sensor, clutch switch sensor, and engine position sensor.

The PTCM processes input data and controls fueling by means of an electronic fuel control valve assembly mounted on top of the PT pump. This unit is equipped with a throttle bypass valve positioned in parallel with the PT throttle shaft, which is capable of routing fuel alternately (bypassing the rail) when energized. Its function is solely to limit fueling; the PTCM can thereby define a maximum road speed or a set power takeoff (PTO) mode speed.

The PACE/PACER system (Figure 23–16) is a very early example of electronic management of a highway truck diesel engine. The foregoing description is included here to emphasize that the standard PT system is left largely intact. In short, PACE/PACER is a hydromechanically controlled PT fuel system with electronic governing of road speed, cruise control, and PTO operation. Should the PACE/PACER system fail, the PT system continues to operate but at slightly reduced power.

REVIEW QUESTIONS

1. The PT injector is actuated by:
   a. Spring pressure
   b. Hydraulic pressure
   c. Mechanical pressure
   d. Electrical pressure
2. Which of the following represents the critical flow area when factoring PT rail pressures?
   a. Collective restriction of the balance orifii in the rail
   b. Collective restriction of the metering orifii in the rail
   c. Outlet orifice in the PT pump solenoid
   d. PT pump housing pressurizing valve

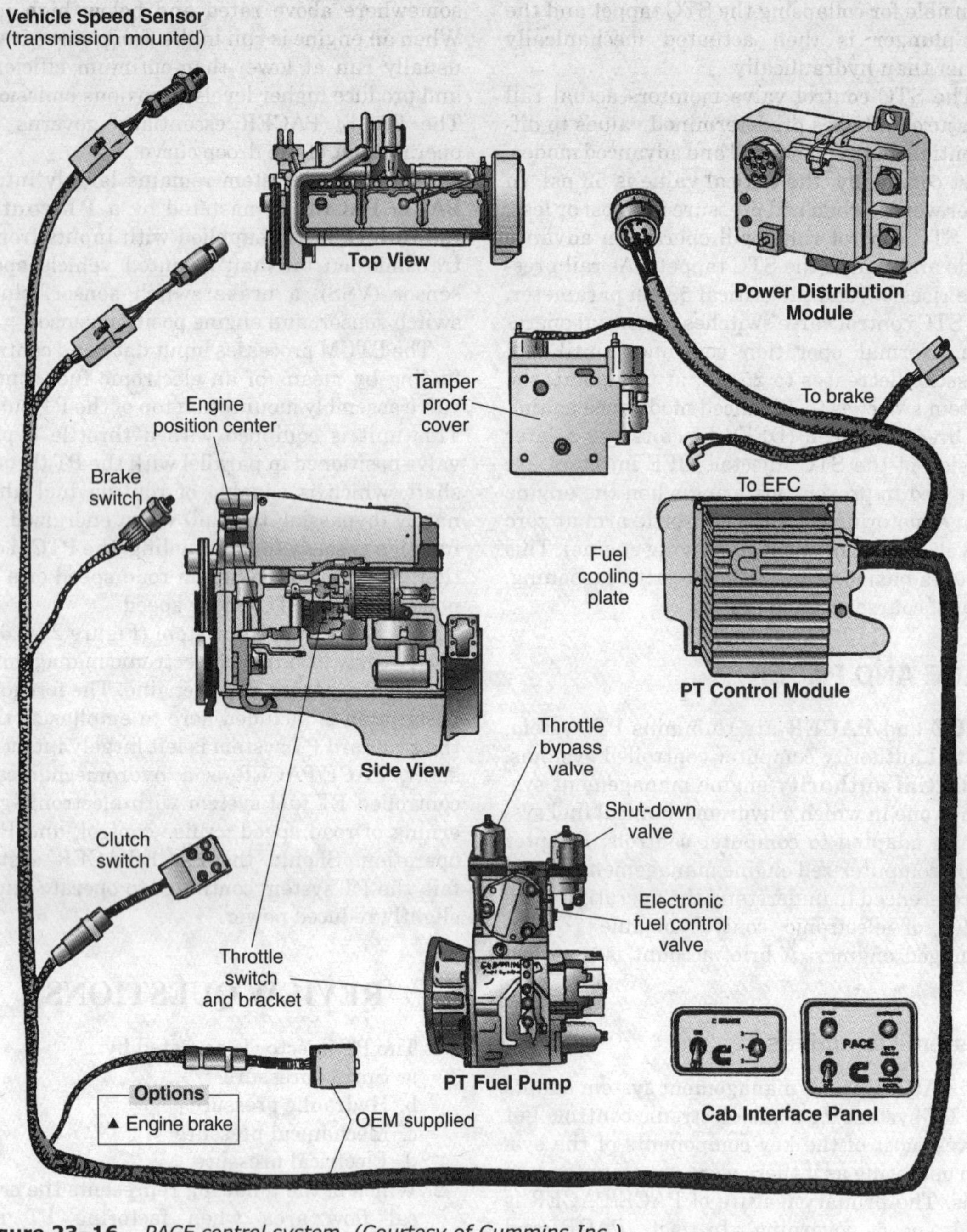

Figure 23–16 *PACE control system. (Courtesy of Cummins Inc.)*

3. When considering the actual quantity of fuel delivered to the cylinders of a PT-fueled engine, which of the following represents the critical flow area?
   a. Balance orifice
   b. Metering orifice
   c. No-air screw
   d. AFC valve
4. Which of the following represents the correct maximum fuel inlet restriction specification for a PTG pump?
   a. 8 inches H_2O

b. 8 inches Hg
c. 8 psi
d. 15 psi

5. Removing shims/gaskets from the cam follower housing assembly would have what effect on PT injector timing?
a. Advance
b. Retard
c. No difference

6. When setting an injector using the pre-torque method, which of the following should be true?
a. Injector cam profile is on outer base circle.
b. The piston is at TDC completing its compression stroke.
c. The injector cam profile is on inner base circle.

7. Which would be the correct amount of travel of the throttle arm on a PTG pump?
a. 15 degrees
b. 28 degrees
c. 90 degrees
d. 170 degrees

8. When using the dial indicator method of setting PT injectors, which of the following should be true?
a. Injector cam profile is on outer base circle.
b. Injector cam profile is on inner base circle.
c. The piston is at TDC completing its compression stroke.

9. When setting 0 lash (top-stop) D injectors, which of the following would result in a correct adjustment?
a. Click-type torque wrench set at 0–1 lb-in.
b. Dial-type torque wrench indicating 0–1 lb-in.
c. Dial-type torque wrench indicating 6 lb-in.
d. Click-type torque wrench set at 72 lb-in.

10. After diagnosing fluctuating rail pressure values on a PTG-AFC-fueled engine, which of the following components would most likely be at fault?
a. Solenoid
b. Throttle arm
c. AFC diaphragm
d. Pulsation damper

11. When checking for air in a PTG pump, which would be the location of choice for an inline diagnostic sight glass?
a. In the rail
b. Between the gear pump and filter pad
c. Between the filter pad and separator
d. Between the separator and tank

12. When considering how a PT pump is coupled to an engine, which of the following would be true?
a. Pump timed to engine, driven at crank speed
b. Pump not timed to engine, driven at crank speed
c. Pump timed to engine, driven at cam speed
d. Pump not timed to engine, driven at two times crank speed

13. The PTG governor weight carrier is driven at:
a. Camshaft speed
b. Crankshaft speed
c. More than two times crank speed
d. Pump-driven speed

14. Which of the following would increase the high-idle value in a PTG pump?
a. Subtract shims acting on the maximum speed spring.
b. Add shims acting on the maximum speed spring.
c. Turn CW the maximum speed spring adjusting screw.
d. Turn CCW the maximum speed spring adjusting screw.

15. Which of the following would increase the idle speed value on a running Cummins PT engine?
a. Add shims acting on the idle spring pack.
b. Subtract springs acting on the idle spring pack.
c. Turn CW the idle spring adjust screw.
d. Turn CCW the idle spring adjust screw.

16. Approximately what average percentage of fuel exiting the PTG-AFC pump actually fuels the engine?
a. 10%
b. 35%
c. 65%
d. 100%

17. If a set of lower flow injectors were installed in a Cummins engine, rail pressures would:
    a. Increase
    b. Decrease
    c. Not be affected
18. Under which running condition would engine compression pressures be most likely to interfere with the metering process in a PT-fueled engine?
    a. Operation in the droop curve
    b. Rated speed
    c. Peak torque
    d. Idle speed
19. *Technician A* states that the AFC circuit helps minimize smoking while the engine is being accelerated. *Technician B* states that the AFC circuit can help reduce overfueling when running the engine at high altitudes. Who is correct?
    a. Technician A only
    b. Technician B only
    c. Both A and B
    d. Neither A nor B
20. *Technician A* states that when the engine calibration index C aligns with the timing notch, the companion pistons #3 and #4 are at TDC. *Technician B* states that when the engine calibration index B aligns with the timing notch with #3 piston on its compression stroke, the PT injector on cylinder #3 can be adjusted. Who is correct?
    a. Technician A only
    b. Technician B only
    c. Both A and B
    d. Neither A nor B

24

Rotary Distributor Pumps

INTRODUCTION

In the days before electronic engine management systems, a significant disadvantage of port-helix metering injection pumps was upfront cost due to its requirement for close tolerance manufacture and a pump/metering element for each cylinder on the engine. This cost factor became increasingly noticeable in small-bore engines in which the fuel system alone (pump/lines/nozzles) represented up to 35% of the total market cost of the engine. This fact gave manufacturers plenty of incentive to look for options to Robert Bosch's 1927 fuel pump. Over the years, many alternate designs of diesel fuel pumping apparatus have been used to manage diesel engines, but few have lasted. Many of these alternate designs used rotary distributor injection principles, the objectives of which were to reduce complexity, cost, and the number of pumping elements. Among the advantages offered by most rotary distributor injection pumps are:

- Lower upfront cost
- Simple disassembly and reassembly (for purposes of cleaning) not requiring costly calibration equipment
- Hard value phasing of injection pulse not influenced by wear
- Balanced cylinder fueling

Rotary distributor injection pumps have a significant initial cost savings advantage over equivalent port-helix metering pumps, but they also have the disadvantage of overall reduced longevity. Over the years, rotary distributor pumps have used a number of diferent operating principles, but only a couple of designs have survived Today, rotary distributor pumps are mostly associated with small-bore diesel engines of the 1990s and earlier. It has been many years since they have been used on American built truck engines in the medium- to large-bore range.

In theory, larger bore engines could be fueled by rotary distributor pumps, but this was seldom the case in North America where the market requires good longevity and minimal downtime, both of which are disadvantages of rotary distributor pumps. Because rotary distributor pumps are simple in construction, they have been popular in geographic regions where fuel quality is less reliable and necessitates frequent pump disassembly and cleanup. Many distributor pumps can be disassembled for cleaning without requiring complex comparator bench adjustments following reassembly.

Roosamaster

The first practical design for a rotary distributor pump was engineered by Vernon Roosa, a diesel mechanic employed by the City of New York, where he maintained and repaired diesel-driven generators. Roosa patented his design in 1941, but it took him some time to generate any interest in his invention. He tried to interest both American Bosch (now Ambac) and CAV (now Delphi/Lucas CAV) in his patent. Following the conclusion of World War II, a machine shop best known for the manufacture of aircraft engine components for Pratt and Whitney, the Hartford Machine and Screw Company, decided to test out the Roosa design and brought its inventor to Hartford to assist in the research and development. At the same time, CAV in the United Kingdom started to work with the Vernon Roosa blueprints, resulting in the simultaneous introduction of the Roosa pump (manufactured by Hartford Machine and Screw Company) and a CAV version of the pump in the year 1952. The CAV DPA version pump became the biggest selling fuel injection pump ever manufactured worldwide, and it is still used today in areas of the world in which noxious emissions and fuel economy are subject to less regulation than in North America.

Although the Roosa design for a rotary distributor injection pump was not the first, it was easily the most successful and, accordingly, it generated a succession of copycat designs. Only one of these, the Bosch VE, lasted into the 1990s and is used primarily on small automotive diesel engines. The attraction of the rotary distributor injection pump was low cost, simplicity, and easy serviceability. The Hartford Machine and Screw Company was absorbed by the Stanadyne Corporation but the Roosa injection pump continued to be known as a RoosaMaster. Currently, the two versions of the Roosa pump (Stanadyne and Delphi/ Lucas/ CAV) and the Bosch design are still occasionally seen with partial authority management electronics.

In this chapter, two distinct types of rotary distributor pumps are addressed. It is important that you do not confuse them because they use very different operating principles. We begin by looking at the Roosa pump, which is generically described as an **inlet-metering**, opposed-plunger, distributor injection pump, then examine Bosch hydromechanical distributor pumps. Truck technicians should be aware that, while they are unlikely to come across either of these injection pumps in current highway trucks, in rural areas where the truck shop does all the diesel repair work in the vicinity (including all the agricultural equipment), having some knowledge of these pumps can be invaluable.

OPPOSED-PLUNGER, INLET-METERING INJECTION PUMPS

In terms of operating principles, the Roosa injection pump and its close relative, the CAV DPA are nearly identical. Although the DPA significantly outsold Stanadyne/Roosa pumps worldwide, this was not true in North America, where both GM and Ford used the Stanadyne pump. For purposes of describing the operation of the pump, the DB 2 version of the Roosa pump is primarily referenced, but it should be remembered that there are a number of slightly different versions within the Stanadyne Roosa and Delphi/Lucas/ CAV families of opposed-plunger pump.

Roosa DB2

When the Roosa DB2 pump was first introduced, it was marketed as a RoosaMaster pump. Older technicians might often use this term today. The DB2 injection pump is an opposed-plunger, inlet-metering, distributor-type, diesel injection pump. It was designed for low-cost production and simplicity. A typical DB2 pump has a total of around 100 component parts and only 4 main rotating members. There are no spring-loaded components, none are lap-finished in manufacture, and there are no ball bearings or gears. The pump has a single pumping chamber in which two **opposed plungers** are actuated by an **internal cam ring**. Because these injection pumps are used primarily with small-bore, low-cost diesel engines, low initial cost is a feature of the pump.

The geometry of the **hydraulic head** determines the distribution of fuel between cylinders, and because fuel flow can be preset, lengthy periods on the fuel pump test bench are avoided. DB2 pumps are self-lubricated and contain roughly the same number of component parts regardless of the number of cylinders served. They may be mounted in any position on the engine.

Main Pump Components

The operation of an inlet-metering, opposed-plunger distributor pump is not complicated. First, it is necessary to be able to identify the major components and circuits of the pump. A cutaway version of the DB2 version of the Roosa pump, which we use as our primary reference for study, is shown in Figure 24–1.

Using Figure 24–1, note that the drive shaft directly engages with the **distributor rotor** in the hydraulic head. In other words, the distributor rotor rotates within the stationary hydraulic head. It is difficult to see in Figure 24–1, but the drive end of the DB2 rotor incorporates two pumping plungers that oppose each other in a single chamber. These plungers are actuated by the inlet-metering circuit internal cam ring. (See Figure 24–3 for a closer look at the pump chamber components.)

Overview of Operating Principles

The DB2 injection pump is driven through one complete rotation per complete engine cycle. This means that it is rotated at camshaft speed, that is, to turn the pump through one full 360-degree rotation, the engine has to be turned through two complete rotations, or 720 degrees. The pump chamber is formed between the pump plungers, and they are actuated toward each other simultaneously by an internal cam ring. The plungers ride on rollers and shoes that are carried in slots at the drive end of the rotor. The number of cam lobes normally equals the number of engine cylinders and in turn determines how many effective strokes will occur during each pump cycle.

Pump Components

A fuel transfer pump is located at the rear of the rotor. The pump is a positive displacement vane type so it will displace fuel in proportion to rotational speed. The assembly is enclosed in the end cap, which also houses the fuel inlet strainer and the transfer pump pressure regulator. The face of the regulator assembly is compressed against the liner and distributor rotor

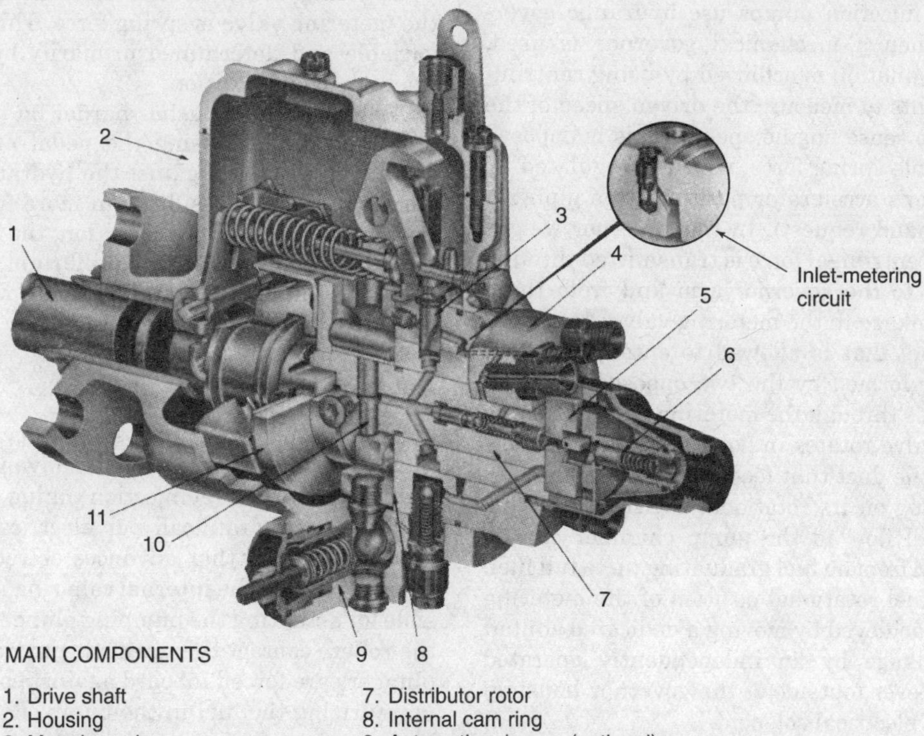

MAIN COMPONENTS

1. Drive shaft
2. Housing
3. Metering valve
4. Hydraulic head assembly
5. Transfer pump blades
6. Pressure regulator assembly
7. Distributor rotor
8. Internal cam ring
9. Automatic advance (optional)
10. Pumping plungers
11. Governor

Figure 24–1 *Cutaway of a Stanadyne DB2 injection pump. (Courtesy of Stanadyne Corporation)* and forms an end seal for the transfer pump. The injection pump is designed so that end thrust acts against the face of the transfer pump pressure regulator. The distributor rotor incorporates two charging ports and a single axial bore with one discharge port. This discharge port aligns with stationary distribution ports in the hydraulic head dedicated to each engine cylinder. As the distributor rotor rotates, it is brought in and out of register with each discharge port once per pump cycle. The hydraulic head contains the bore within which the distributor rotor rotates, the metering valve bore, charging ports, and the hydraulic head discharge fittings. High-pressure injection pipes connect these discharge fittings to hydraulic fuel injectors. In most cases, the hydraulic fuel injectors use pintle-type nozzles.

Governing and Metering

The DB2 pump uses a mechanical governor. Other versions of inlet-metering, opposed-plunger injection pumps use hydraulic governors. When a mechanical governor is used, speed regulation is achieved by using centrifugal weights to measure the driven speed of the pump (to sense engine speed). This is opposed by variable spring force, which is regulated by the driver's accelerator position (this inputs a fuel demand request). In the governor weight carrier, centrifugal force is transmitted through a sleeve to the governor arm and from there uses a linkage to the metering valve.

All fuel that is allowed to enter the pump chamber formed by the two opposed plungers must pass through the metering valve. The metering valve rotates in bore and defines a flow area to the duct that feeds the pump chamber. Depending on its rotational register with this duct, fuel flow to the pump chamber can be managed from no fuel graduating up to full fuel. The no-fuel rotational position of the metering valve is achieved by moving a dedicated shutoff solid linkage by an independently operated shutoff lever (outside of the governor housing) or by an electrical solenoid.

Hydraulic Governing

Many versions of the Delphi Lucas CAV pumps use hydraulic governing. Hydraulic governing simplifies the injection pump and reduces its size. In hydraulically governed versions of this pump, the metering valve moves longitudinally in its bore instead of rotationally.

Because the fuel transfer pump is a positive displacement vane type of pump, it displaces fuel proportionately to engine rpm. If this fuel is unloaded into a defined (and unchanging) flow area, it makes sense that engine fuel pressure correlates with rpm.

Metering Valve Operation The metering valve is cross and center drilled and milled with a graduated metering recess. This graduated metering recess can be brought into and out of register with the metering duct that directs fuel to the pump chamber: Its longitudinal position is determined by the forces that act above and below it. Below the metering valve, fuel pressure acts against the sectional area of the valve, attempting to force it upward, tending to reduce fuel delivery. Fully upward, no fuel is metered to the pump chamber and shutdown occurs.

Opposing the hydraulic force acting below the metering valve is spring force. This force is variable and determined primarily by the accelerator pedal position.

As the operator pushes harder on the accelerator pedal, increasing the pedal angle, the spring force acting against the hydraulic pressure is increased, resulting in more fuel delivery. So, in any running condition, the inlet metering valve establishes equilibrium between the spring and the hydraulic forces that position it to define a fuel quantity.

Advance Mechanism

Most inlet-metering, opposed-plunger pumps are equipped with an automatic advance device that can advance fuel injection timing. This can be either a hydraulically or electrically actuated mechanism that advances or retards the pumping cycle. The internal cam ring is responsible for actuating the pumping plungers. When the rollers contact the internal cam ramps, the plungers are forced inboard against each other, pressurizing the fuel in the pump chambers in between them. If the cam ring is rotated toward the direction of rotor rotation, an advance is achieved. An external lug protrudes from the cam ring into the advance chamber below it, permitting a window of advance.

In the case of hydraulic advance, spring force acts on one side of the cam ring lug, positioning it in the most retarded location. An advance piston is located on the other side of the lug, and fuel pump pressure from the vane pump is allowed to act on this piston. Because a vane pump is positive displacement, when it unloads to a defined flow area, pressure rise will be proportional to pump rotational speed. When this fuel pressure acts on the advance piston, it overcomes the spring pressure, opposing it to move the cam ring to advance fuel delivery timing. The result is that any advance achieved will be entirely speed sensitive. The same thing can be accomplished electrically in later versions of the pump by using a linear proportioning solenoid controlled by the engine electronic control module (ECM).

Fuel Flow

Now that you have some understanding of the pump operation, we can take a look at exactly how the fuel is routed through the pump during its cycle. It makes sense to use both the text and the figures to understand how this is achieved. Reference Figure 24–2, which is a fuel flow schematic of the DB2 pump.

Movement of fuel through the fuel subsystem is the responsibility of the transfer pump, which is the vane pump integral with the injection pump. Fuel is pulled from the main fuel tank through a filter or filters into the pump inlet. At the pump inlet, fuel is pulled through the inlet filter screen by the transfer pump. Some fuel is bypassed through the pressure regulator and routed back to the suction side.

Fuel under transfer pump pressure flows through the center of the transfer pump rotor shaft past the rotor retainers into an annular groove on the rotor. It next passes through a connecting passage in the head to the advance cylinder, up through a radial passage, and then through a duct to the metering valve. The rotational position of the metering valve, which is directly controlled by the governor, regulates fuel flow to the radial charging passage, which incorporates the hydraulic head charging ports.

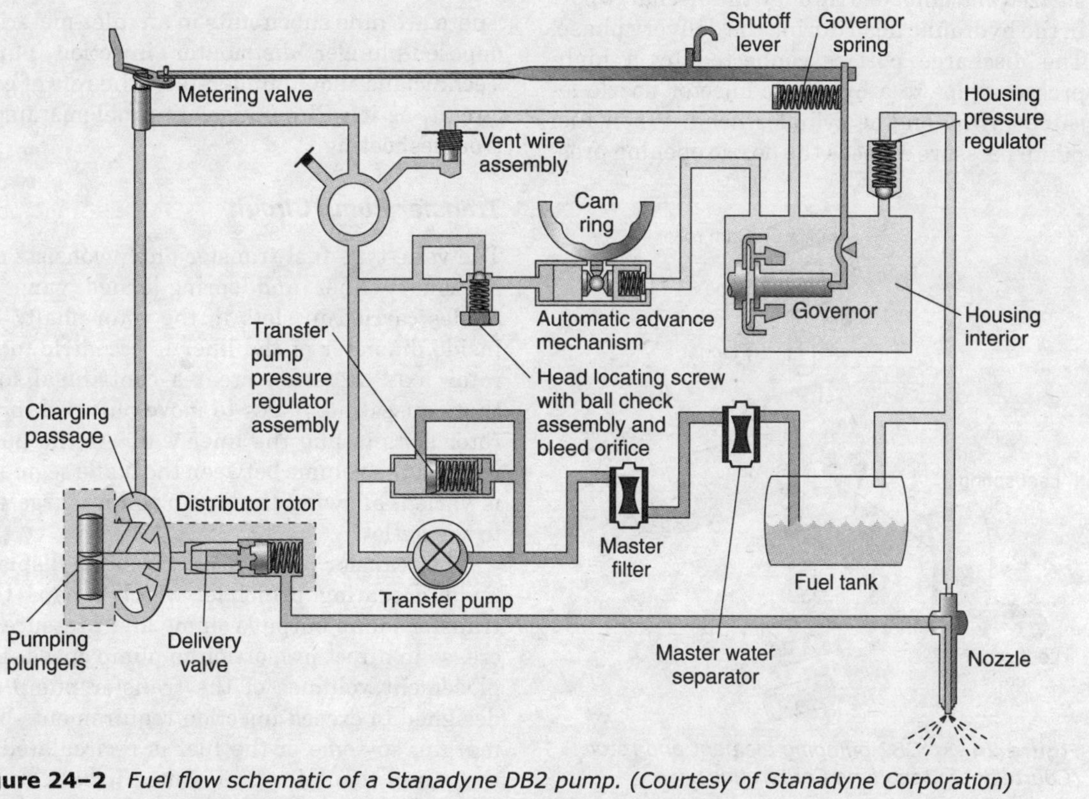

Figure 24–2 *Fuel flow schematic of a Stanadyne DB2 pump. (Courtesy of Stanadyne Corporation)*

As the rotor revolves, the two rotor inlet passages register with charging ports in the hydraulic head, allowing fuel to flow into the pumping chamber. Fuel flowing into the pump chamber spreads the pumping plungers outward. The extent to which they spread outward depends entirely on the fuel quantity metered into the pump chamber: This fact tends to mean injection timing is advanced proportionate with load. With further rotation, the inlet passages move out of register and complete the metering phase. Next, the discharge port of the rotor is brought into register with one of the hydraulic head distribution outlets that connect the pump with an individual fuel injector. During the register of the discharge port with the rotor, the pumping phase takes place. The plunger-actuating rollers contact the lobes of the internal cam ring, forcing the opposed plungers inboard into the pump chamber.

The pumping element is shown in Figure 24–3. The higher the fuel load in the pump chamber, the more advanced the timing of this moment that begins delivery. Fuel trapped in the pump chamber between the plungers is pressurized and unloaded through the discharge port in the hydraulic head during the delivery phase. The discharge port is connected by a high-pressure pipe to a hydraulic injector nozzle located in the engine cylinder head. When fuel pump pressure exceeds the nozzle opening pressure (NOP), the injector nozzle valve unseats and fuel is injected. Most applications using the Roosa pump use indirect injection and pintle-type nozzle assemblies.

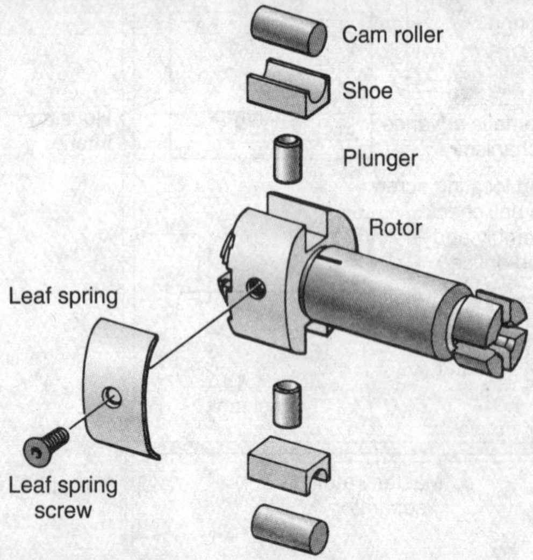

Figure 24–3 *DB2 pumping element and rotor. (Courtesy of Stanadyne Corporation)*

Inlet-metering, opposed-plunger, distributor injection pumps are self-lubricating and self-priming. The lubricity of diesel fuel accomplishes the lubrication of the internal components of the pump. As fuel at transfer pump pressure reaches the charging ports, slots on the rotor shank allow fuel and any entrapped air to flow into the pump housing cavity. An air vent passage in the hydraulic head connects the outlet side of the transfer pump with the pump housing. This permits air and some fuel to bleed back to the fuel tank by means of a return line. Bypass fuel fills the pump housing, lubricates the internal components, acts as coolant, and purges the housing of small air bubbles. The pump is therefore designed to operate with the housing charged with fuel. During operation there should be no air within the pump.

Inlet-metering Distributor Pump Circuits

There are nine subcircuits in an inlet-metering, opposed-plunger distributor injection pump. Technicians should understand the role of each circuit, as it will help isolate problems during troubleshooting.

Transfer Pump Circuit

The vane-type fuel transfer pump consists of a stationary liner and spring-loaded vanes or blades carried in slots in the rotor shaft. The inside diameter of the liner is eccentric to the rotor axis. Rotation creates centrifugal force that causes the blades to move outward in the rotor slots to hug the liner wall. As the pump rotates, the volume between the blade segments is varied, allowing the pump to discharge fuel to the outlet.

The transfer pump uses a positive displacement operating principle, which means that transfer pump output volume and pressure increase in direct proportion to pump speed. Displacement volumes of the transfer pump are designed to exceed injection requirements by a margin, so some of the fuel is recirculated by the pump regulator that routes it back to the

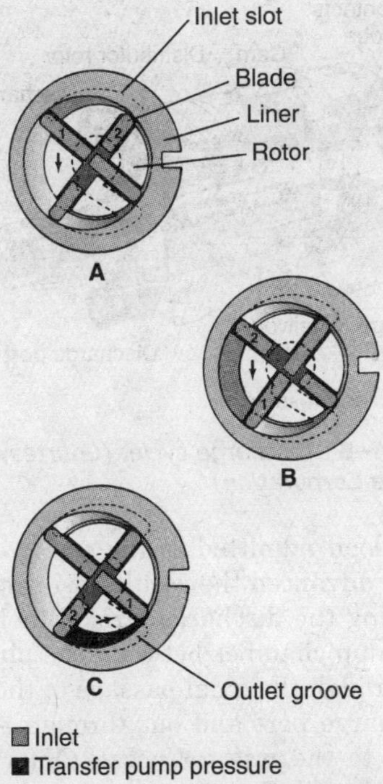

Figure 24–4 *Roosa vane-type transfer pump. (Courtesy of Stanadyne Corporation)*

inlet side of the transfer pump. Figure 24–4 shows the vane-type transfer pump.

Figure 24–4 illustrates the pumping principle. Radial movement causes a volume increase in the quadrant between blades 1 and 2 (Figure 24–4A). In this position, the quadrant is in register with a kidney-shaped inlet slot in the top portion of the regulator assembly. The increasing volume causes fuel to be pulled through the inlet fitting and filter screen into the transfer pump liner. Volume between the two blades continues to increase until blade 2 passes out of register with the regulator slot. At this point, the rotor is in a position in which there is little outward movement of blades 1 and 2 and the volume is not changing (Figure 24–4B). The slug of fuel between the blades is now carried to the bottom of the transfer pump liner.

As blade 1 passes the edge of the kidney-shaped groove in the lower region of the regulator assembly (Figure 24–4C), the eccentric liner wall compresses blades 1 and 2 in an inward direction (Figure 24–4A). This action reduces the volume available for fuel storage but not the quantity, so the fuel is pressurized. Pressurized fuel is unloaded through a groove in the regulator assembly and directed into a channel on the rotor leading to the hydraulic head. As the rotor continues with its rotation, volume between blades continues to decrease, pressurizing the fuel in the quadrant, until blade 2 passes the groove in the regulator assembly.

Regulator Assembly Circuit

The regulator prevents excessive vane pump pressures in the event of an engine overspeed condition. Fuel from the discharge side of the transfer pump forces the regulator piston against the regulator spring. As pump speed and therefore flow increases, the regulator spring is further compressed until the leading edge of the regulator piston starts to expose the pressure-regulating slot. Because fuel pressure acting on the regulator piston is opposed by the regulator spring, delivery pressure of the transfer pump is controlled by:

- Regulator spring tension
- Flow area defined by the regulating slot

Metering Circuit

Inlet metering refers to the fact that only fuel permitted to pass through the metering valve enters the pump chamber. The fuel delivered to the pump chamber during each pump cycle is the metered fuel quantity and defines the injection pulse duration. The metering valve rotates in the bore. It receives fuel from the charging passage. Because it is milled with a scrolled metering recess, it directs this fuel to the opposed plunger injection pump through a charging passage. In other words, the rotational position of the metering valve controlled by the governor defines the flow area to the duct that feeds the pump chamber. This flow area must accommodate all the fuel requirements of the engine from no fuel to full fuel. The no-fuel rotational position of the metering valve is achieved by moving a dedicated shutoff linkage or by an electrical solenoid.

Charging Circuit

The charging circuit determines the exact quantity of fuel that enters the pumping chamber. As the rotor revolves (Figure 24–5), the two

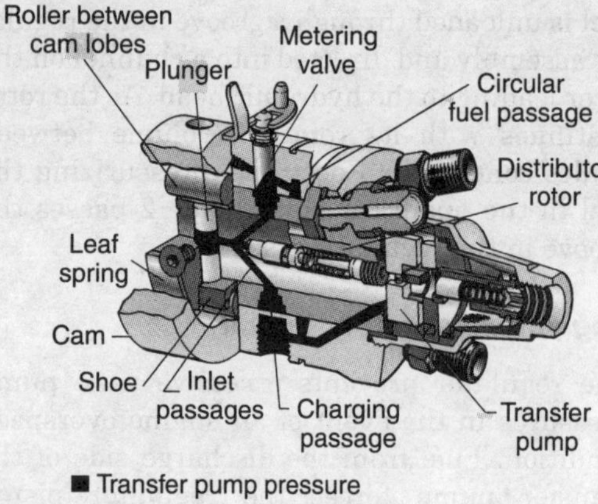

Figure 24–5 *Charging cycle. (Courtesy of Stanadyne Corporation)*

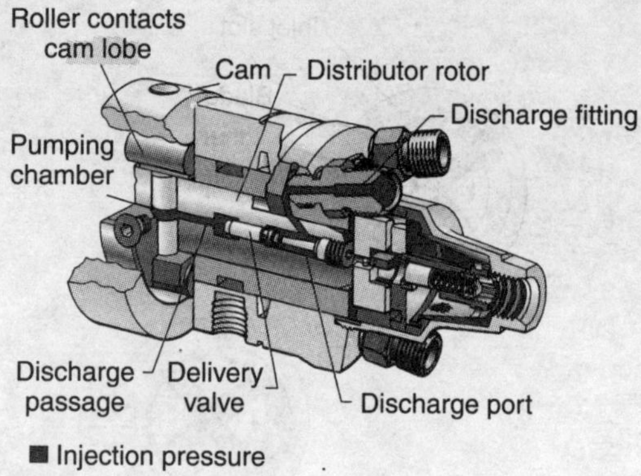

Figure 24–6 *Discharge cycle. (Courtesy of Stanadyne Corporation)*

inlet passages in the rotor register with charging passage ports. Fuel from the transfer pump, which is controlled by the opening of the metering valve, flows into the pumping chamber thereby forcing the plungers apart.

The plungers move outward only to the extent required to accommodate the fuel required for injection on the following stroke. If only a small quantity of fuel is charged to the pumping chamber, for instance that required to idle the engine, the plungers spread only a short distance. When engine fuel load is high, the plungers are forced outward further to accommodate the higher volume of fuel. Maximum plunger travel is limited by a leaf spring that contacts the edge of the roller shoes. Only when the engine is operating at full load will the plungers move to the most outward position.

Discharge Circuit

As the cycle continues, the inlet passages are moved out of register with the charging ports. Next, the rotor discharge port is brought into register with one of the outlets in the hydraulic head that will direct fuel to an injector. The plunger-actuating rollers then contact the cam profiles in the internal cam ring, driving the shoes against the plungers and forcing them inboard into the pump chamber. This action begins the high-pressure pumping phase called the discharge cycle shown in Figure 24–6.

The beginning of injection varies according to the required fuel load. Assuming that the internal cam ring remains stationary, the higher the fuel load admitted to the pump chamber, the more advanced the resultant injection timing. During the discharge stroke, fuel trapped in the pump chamber between the plungers is forced through the axial passage of the rotor to the discharge port and out through the injection line to the injector nozzle. Fuel delivery continues until the plunger-actuating rollers are ramped beyond the highest point of the cam lobe and are allowed to move outward. Pressure in the axial passage is allowed to collapse, dropping line pressure and allowing the injector nozzle to close. Nozzle closure ends delivery.

Delivery Valve Circuit

Delivery valves are an option on most inlet-metering, opposed-plunger distributor pumps. The delivery valve accelerates injection line pressure drop after injection to a value approximately two-thirds of the specified nozzle closing pressure. This reduction in line pressure permits the nozzle valve to seat abruptly and minimizes the fuel injected during the collapse phase of injection. When larger droplets of fuel are injected to the engine cylinder during the collapse phase at the end of the injection pulse, there is insufficient time available to properly combust them, which causes higher hydrocarbon (HC) emissions. Delivery valves are found on most highway applications.

The delivery valve is located in a bore in the center of the distributor rotor. This type of delivery valve requires no seat and uses just a mechanical stop to limit travel. Sealing is accomplished by the close tolerance fit between the valve and bore. When injection begins, fuel pressure moves the delivery valve slightly out of its bore, adding its swept volume displacement to the spring chamber. Because at this moment the discharge port is already exposed to a hydraulic head outlet, the retraction volume and plunger displacement volume are discharged under high pressure to the nozzle.

Delivery ends when pressure on the plunger side of the delivery valve drops as the cam rollers ramp over the high point of the cam profile. Next, the rotor discharge port is closed off completely and a residual line pressure is sealed in the high-pressure injection pipe. The delivery valve seals only while the discharge port is open: Once the port is closed, residual line pressure is maintained by the close tolerance fit of the hydraulic head and rotor.

Fuel Return Circuit

Fuel under transfer pump pressure is discharged into a vent passage in the hydraulic head. Flow through the passage is restricted by a vent wire assembly to prevent excessive return fuel that could cause undue pressure loss. The actual amount of return fuel is controlled by the size of wire used in the vent bore assembly. The smaller the wire, the greater the return flow and vice versa. Vent wires are available in several size options to meet differing specifications. The vent wire assembly can be accessed by removing the governor cover. The vent passage is located behind the metering valve bore and connects with a short vertical passage containing the vent wire assembly, leading to the governor housing.

Any air entering the transfer pump is routed to the vent passage as shown. Both air and fuel then flow from the housing to return to the fuel tank via the return line. Housing pressure is maintained by a spring-loaded ball check return fitting in the governor cover of the pump.

Mechanical Governor

The governor used can be classified as a variable speed governor, also known as an all-speed governor. This type of governor functions to maintain desired engine speed (requested rpm) and will do so within a certain window as load changes.

Speed sensing is performed by the flyweights as shown in Figure 24–1 and indicated by arrow 11. Movement of the flyweights acting against the governor thrust sleeve rotates the metering valve in its bore by means of the governor arm and linkage hook. Rotation of the metering valve varies the register of the metering valve scroll to the passage from the transfer pump, thereby controlling how much fuel is metered to the pump element. Centrifugal force, which directly correlates with rotational speed, forces the flyweights outward, moving the governor thrust sleeve against the governor arm, actuating the linkages that rotate the metering valve. Force of the flyweights acting on the governor arm is balanced by the governor spring force. This spring force is variable and controlled by the manually positioned throttle lever. The throttle lever connects to the vehicle accelerator linkage.

Load Decrease A variable speed governor defines a specific rpm value. In the event of a load reduction on the engine, engine speed would tend to increase. This would in turn increase the centrifugal force produced by the flyweights, which would act to rotate the metering valve clockwise to reduce engine fueling.

Load Increase When the load on the engine is increased, engine speed initially tends to drop. As engine speed reduces, centrifugal force generated by the weights drops, permitting the spring forces that oppose it to rotate the metering valve in the counterclockwise direction, increasing fuel metered to the pumping element. Engine speed at any point in the operating range of the engine is dependent on the combination of forces that act on the governor thrust lever. As with any mechanical governor, centrifugal force acting on the thrust lever will result in speed reduction (less fuel) and spring forces acting on the thrust lever will attempt to increase speed (more fuel).

Governor Operation Summary A light idle spring is provided for more sensitive regulation when flyweight centrifugal force is low, such as when the engine is operating close to idle speeds. The limits of throttle travel are set by adjusting screws to define low-idle and high-idle speeds. A light tension spring on the linkage assembly takes up slack in the linkage joints and permits the shutoff mechanism to close the metering valve without having to overcome the governor spring force, which means that only a little force is required to rotate the metering valve to the closed position to shut the engine down.

The function of any variable speed governor is to attempt to hold engine rpm at a consistent value (based on a given throttle position) and let the governor adjust fueling to hold that rpm constant as engine load varies. Positioning the throttle lever simply defines a request for an rpm value, not a fuel quantity, which would be the case in a limiting speed governor.

Advance Circuit

Opposed-plunger, inlet-metering distributor pumps permit the use of a simple, direct-acting hydraulic mechanism powered by fuel pressure from the transfer pump to rotate the internal cam within the pump housing. This changes the phasing of injection pump stroke with that of the engine that drives the injection pump and varies delivery timing. The advance mechanism defaults to the most retarded position when there is no or low pressure developed by the vane-type transfer pump. As vane pump pressure increases, fuel pressure acting on the advance piston overcomes the advance spring pressure and rotates the internal cam ring to advance the start of fuel delivery. Total movement of the cam is limited by the piston length. A trimmer screw is provided to adjust the advance spring tension, which essentially determines how much fuel pressure is required (that is, the specific rpm) to begin the advance. It can be incorporated at either side of the advance mechanism and may be adjusted on the test bench while running. Because the extent of advance depends on vane pump pressure, it is speed sensitive.

Figure 24–7 shows the advance circuit components and operating principle.

Advancing fuel injection timing compensates for inherent injection lag and greatly improves high-speed engine performance. Beginning the injection pulse earlier when the engine is operating at higher speeds ensures that peak cylinder combustion pressures are developed when the piston is ideally positioned in its downstroke to optimize torque transfer to the crankshaft. Advancing the beginning of the injection pulse results in the plungers' completing their pumping stroke earlier.

Opposed-plunger, Inlet-metering Pump Summary

The worldwide success of Roosa's injection pump design is a testament to its low initial cost, ease of maintenance, and ability to manage a range of engines. Though not seen in any-

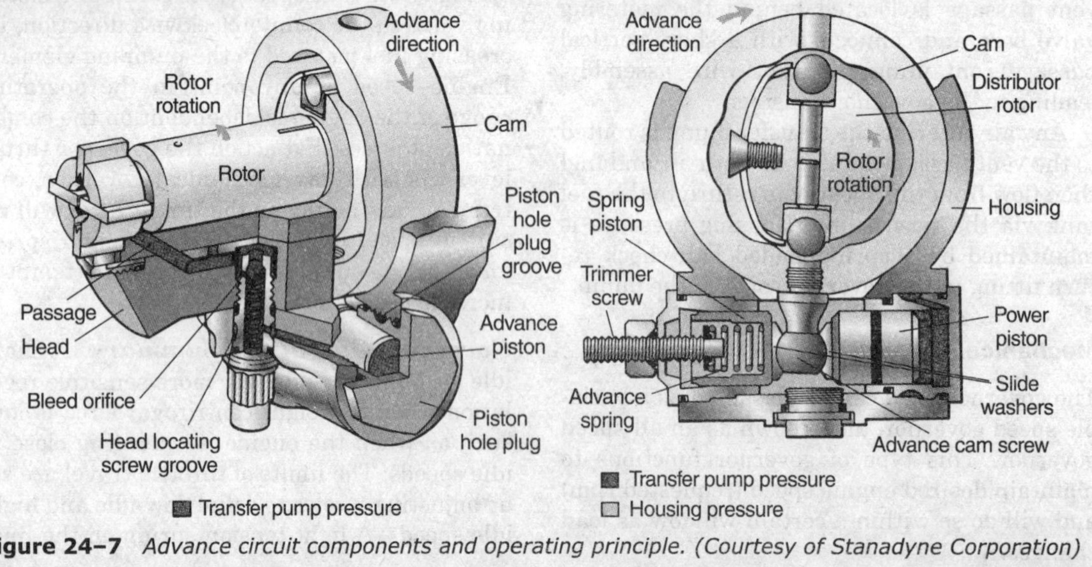

Figure 24–7 *Advance circuit components and operating principle. (Courtesy of Stanadyne Corporation)*

thing but light-duty trucks, opposed-plunger, inlet-metering, rotary distributor pumps are found in many small diesel engine applications. They are especially prominent in agricultural equipment. The design lends itself to some limited enhancement by using partial authority computer controls, usually governing (ECM control of metering valve) and advance functions: In these, the critical components of the fuel pump are retained. However, with today's expectations for minimum emissions, better fuel economy, and higher longevity, this category of fuel injection pump has some limitations and can be regarded as a disappearing technology, at least in North America.

SLEEVE-METERING, SINGLE PLUNGER DISTRIBUTOR PUMPS

Bosch hydromechanical, **sleeve-metering** distributor injection pumps use a single plunger pumping element. In describing this technology, the popular VE pump is used as our primary reference. The VE pump has been used in a number of off-shore small-bore engines, including Volkswagen, and was an option on some early versions of the Cummins B series engine. These pumps manage fueling for engines with up to six cylinders. A VE pump has four primary circuits:

1. Fuel-supply pump
2. High-pressure pump
3. Governor
4. Variable timing

The VE distributor pump has been used in passenger cars, commercial vehicles, agricultural tractors, and stationary engines.

Subassemblies

The VE distributor pump uses only one pump cylinder and a single plunger. It is designed to fuel multicylinder engines. The general layout of a typical VE fuel system is shown in Figure 24–8.

All movement through the fuel subsystem is the responsibility of a vane-type transfer pump integral with the VE pump assembly. Once fuel from the vehicle tank enters the VE pump, fuel is pressurized to injection pressures and routed to the engine cylinders by means of high-pressure pipes. Injection pressures are created by a single plunger-type pump. Fuel delivered by the pump plunger is routed by a distributor groove to the outlet ports, which connect to hydraulic injectors located at each engine cylinder. The VE distributor pump housing contains the following subcircuits:

- High-pressure (injection) pump with distributor
- Mechanical (flyweight) governor
- Hydraulic timing device
- Vane-type, fuel supply pump
- Shutoff device
- Engine-specific add-on modules

Figure 24–9 shows a cutaway view of the subcircuits used on a typical VE pump. Add-on modules, some of which are described later in the chapter, allow the pump to be adapted to the requirements of specific diesel engines.

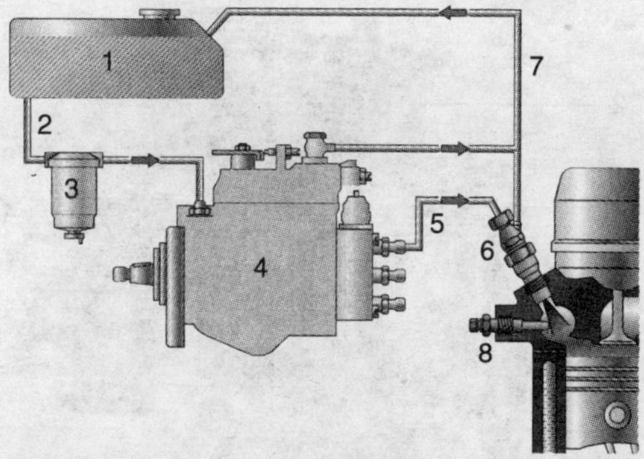

Figure 24–8 *VE distributor pump fuel system layout. (Courtesy of Robert Bosch LLC)*

Design and Construction

The pump drive shaft is supported by bearings in the pump housing and drives the vane-type fuel supply pump. A roller ring is located inside the pump at the end of the drive shaft although it is not connected to it. Figure 24–10 shows the arrangement of the pump drive shaft and components in the front section of the pump. A ro-

1. Pressure-control valve
2. Governor assembly
3. Overflow restriction
4. Distributor head with high-pressure pump
5. Vane-type fuel supply pump
6. Timing device
7. Cam plate
8. Electromagnetic shutoff valve

Figure 24–9 *Location of pump subassemblies. (Courtesy of Robert Bosch LLC)*

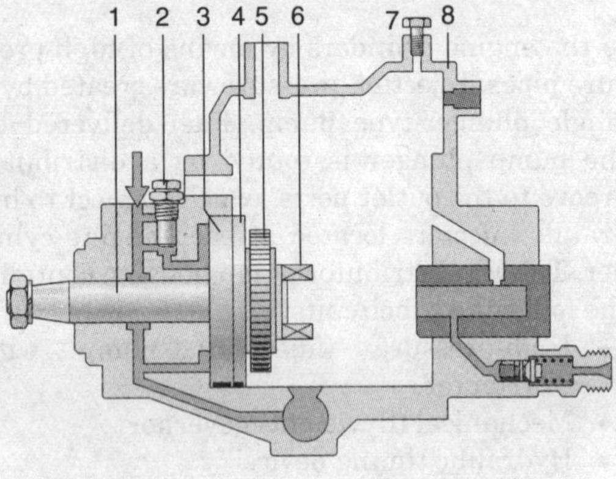

1. Drive shaft
2. Pressure control valve
3. Eccentric ring
4. Support ring
5. Governor drive
6. Drive shaft dogs
7. Overflow restriction
8. Pump housing

Figure 24–10 *Interaction of fuel supply pump, pressure control valve, and overflow restriction valve. (Courtesy of Robert Bosch LLC)*

tating-reciprocating movement is imparted to the distributor plunger by means of a **cam plate** driven by the input shaft that rides on the rollers of the roller ring. The plunger moves inside the distributor head, which is itself bolted to the pump housing. Located in the distributor head are the electric fuel shutdown, screw plug with vent screw, and the delivery valves. If the distributor pump is also equipped with a mechanical fuel shutdown, it is mounted to the governor cover.

The governor assembly, which includes flyweights and the control sleeve, is driven by the drive shaft. The governor linkage, made up of control, starting, and tensioning levers, pivots in the housing. The governor shifts the position of the control sleeve on the pump plunger and, in this way, defines plunger effective stroke. Located above the governor mechanism is the governor spring that connects with the external control lever by means of the control-lever shaft, itself held in bearings in the governor cover.

The control sleeve is used to control pump output. The governor assembly is located at the top of the pump, and it contains the full-load adjusting screw, the overflow restriction or the overflow valve, and the engine speed adjusting screw. The variable timing device is located under the pump assembly and it functions to advance pump timing based on fuel pressure developed by the internal vane pump.

Pump Drive

The sleeve-metering, distributor injection pump is direct driven by the engine it manages and it must be driven through one complete rotation per full engine cycle. This means that on a four-stroke cycle engine, the pump is driven through one complete revolution per two crankshaft revolutions, in other words at camshaft speed. In common with most other injection pumps, sleeve-metering injection pumps must be precisely timed to the engine it manages.

The injection pump can be driven by toothed timing belts, a pinion, gear wheel, or chain. The direction of rotation can be either clockwise or counterclockwise, depending on the engine manufacturer requirements. The fuel delivery outlets in the **distributor head** are supplied with fuel in rotational geometric sequence and each is identified with a letter, beginning with A and following through with B, C, D, and so on, up to the total number of engine cylinders. This is done to avoid confusion with engine cylinder numbering. VE distributor pumps will fuel an engine with up to six cylinders.

Fuel Subsystem

The fuel subsystem of a sleeve-metering distributor injection pump is also known as the low-pressure circuit. It consists of a fuel tank, fuel lines, fuel filter, vane-type fuel supply pump, pressure control valve, and overflow restriction. The vane-type supply pump is responsible for all movement of fuel in the fuel subsystem. It pulls fuel from the fuel tank and routes it through a filter before it enters the injection pump. The vane pump is positive displacement in operating principle, so the volume of fuel it pumps is directly related to rotational speed. A pressure control valve ensures that injection pump internal pressure is managed as a function of vane pump speed. This valve sets a defined internal pressure for any given speed, meaning that pump internal pressure rises directly in proportion to engine speed. In operation, some fuel flows through the pressure-regulating valve and is routed back to the suction side of the vane pump: Some fuel also flows through the overflow restriction located at the top of the governor housing, and this fuel is routed back to the fuel tank. By flowing excess fuel through the injection pump, fuel is used for cooling and venting the injection pump housing.

In some applications, an overflow valve is fitted instead of the overflow restriction. The interaction between the supply pump, pressure control valve and overflow restriction valve is shown in Figure 24–10. In some applications, the vane pump exerts insufficient suction to pull fuel through the fuel subsystem, and in these applications, a presupply pump is required. This pump is usually located in or close to the fuel tank.

Supply Pump Operation

The prime mover of the fuel subsystem, the vane-type transfer pump is located on the injection pump drive shaft. The pump impeller assembly is concentric with the shaft and lugged to it by means of a Woodruff key. The impeller assembly rotates within an eccentric liner. When the drive shaft rotates, centrifugal force throws the four vanes in the impeller outward against the wall of the eccentric liner. Fuel enters the impeller assembly through an inlet passage and a kidney-shaped recess in the pump housing and charges the cavity formed between the vanes and the liner wall. As the pump rotates, fuel between adjacent vanes is forced into the upper or outlet, kidney-shaped recess and from there, directed to the injection pump circuitry. Some discharge fuel is also directed to the pressure control valve, a spring-loaded spool valve, which options fuel to the return circuit when pressure exceeds a specified value.

Developing Injection Pressures

Fuel injection pressures are produced by the high-pressure stage of injection pump assembly. High-pressure fuel is then routed to delivery valves and from there to injection nozzles located at each engine cylinder by means of high-pressure pipes.

Distributor Plunger Drive

Rotary movement of the drive shaft is transferred to the **distributor plunger** by a coupling unit as shown in Figure 24–11. Here, drive lugs or dogs on the cam plate engage with the recesses in a yoke, located between the end of the drive shaft and the cam plate. The cam plate is loaded onto the roller ring by a spring, so when it rotates, the cam lobes riding on the ring rollers convert the rotational movement of the drive shaft into a rotating-reciprocating movement of the cam plate.

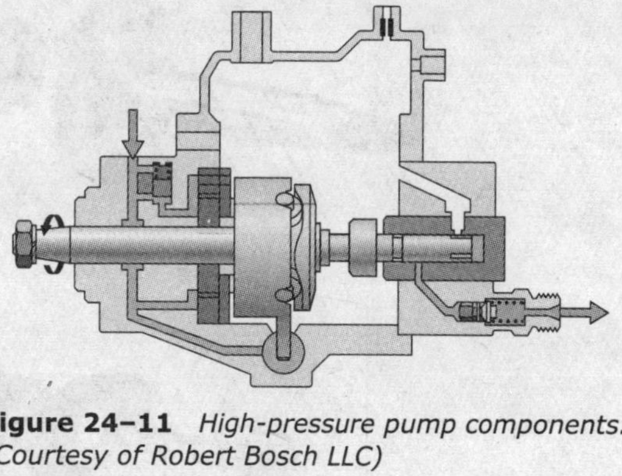

Figure 24–11 *High-pressure pump components. (Courtesy of Robert Bosch LLC)*

The distributor plunger is locked into position relative to the cam plate by a pin. The distributor plunger is actuated upward through its stroke by cams on the cam plate while a pair of symmetric return springs forces it back downward. Because the plunger is actuated by cam profile and loaded to ride that profile by springs, its actual stroke does not vary.

The plunger return springs contact the distributor head at one end, and at the other, act on the plunger by a link element. These springs also have a dampening effect and can prevent the cam plate jumping off the rollers during sudden speed change. Return spring length must be carefully matched so that the plunger is not side-loaded in the pump bore. (See the lower portion of Figure 24–12.) An exploded view of the high-pressure pumping components is shown in Figure 24–12.

Cam Plates

The cam plate and its cam contour help define fuel injection pressure values and injection duration, along with pump-driven speed that determines plunger actuation velocity. Because of the different requirements of each type of engine, fuel injection factors produced by an injection pump are distinct to each engine. That means a specific cam plate profile is required for each engine type and technicians should remember that cam plates are generally not interchangeable because of the engine-specific machining of each.

Distributor Head Assembly

The distributor plunger, the distributor head bushing, and the control collar are each precisely fitted by lapping into the distributor head. These components are required to seal at injection pressure values. Some internal leakage losses do occur and serve to lubricate the plunger. In common with other components that are lap finished in manufacture, the distributor head should always be replaced as a complete assembly.

Metering

Metering and the development of injection pressures take place in four distinct phases shown in Figure 24–13. In a four-cylinder engine, the distributor plunger has to rotate through 90 degrees for a complete pumping stroke to occur. A complete pump stroke means that the plunger has to be stroked from bottom dead center (BDC) to top dead center (TDC) and back again. In the case of a six-cylinder engine, the plunger has to complete a pumping stroke in 60 degrees of pump rotation.

Generates the high pressure and distributes the fuel to the respective fuel injector.

1. Yoke
2. Roller ring
3. Cam plate
4. Distributor plunger foot
5. Distributor plunger
6. Link element
7. Control collar
8. Distributor head flange
9. Delivery valve holder
10. Plunger return spring

4–8. Distributor head

Figure 24–12 *Exploded view of the high-pressure pump components. (Courtesy of Robert Bosch LLC)*

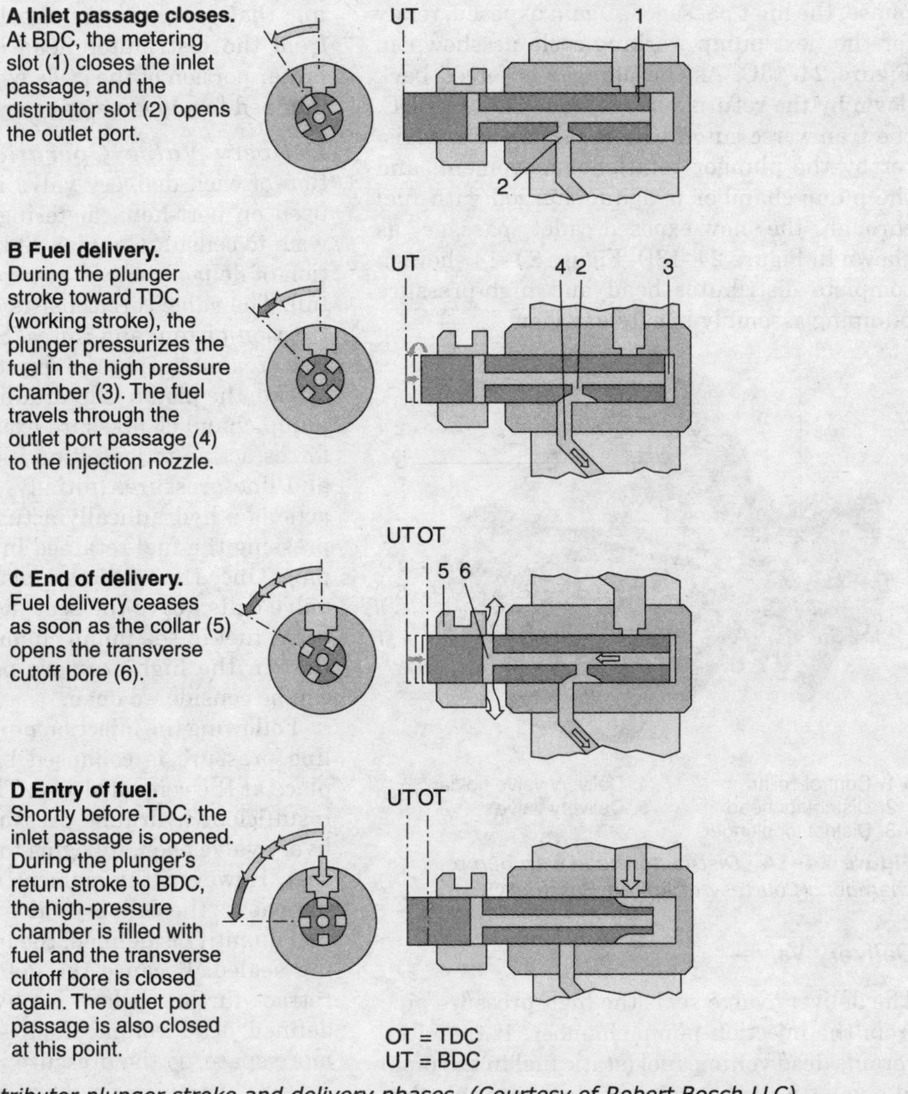

Figure 24–13 *Distributor plunger stroke and delivery phases. (Courtesy of Robert Bosch LLC)*

Use Figure 24–13 and correlate the text with the four phases of a complete pumping stroke. As the distributor plunger is forced down from TDC to BDC, fuel flows through the open inlet passage and passes into the pumping chamber located above the plunger. At BDC, plunger rotational movement takes the plunger out of register with the inlet passage and exposes the distributor slot for one of the outlet ports as shown in Figure 24–13A. The plunger now reverses direction and is driven upward to begin the working stroke. Pressure rise is created in the pump chamber above the plunger and when sufficient, opens the delivery valve and forces fuel through the high-pressure pipe to the injector nozzle. The working (delivery) stroke is shown in Figure 24–13B. The delivery stroke is complete when the plunger transverse cutoff bore (cross drilling) protrudes beyond the metering sleeve, collapsing the pressure. When collapse is initiated, pressure drops in the high-pressure line and pump chamber until there is no longer sufficient pressure to hold the nozzle valve open. Injection ceases the instant the nozzle valve closes.

As the plunger continues to move upward to TDC, fuel spills backward through the cutoff bore to the pump housing. During this collapse phase, the inlet passage is again exposed, ready for the next pump working cycle as shown in Figure 24–13C. As the plunger is forced back down by the return springs from TDC to BDC, the transverse cutoff bore is taken out of register by the plunger rotational movement, and the pump chamber is again charged with fuel through the now-exposed inlet passage as shown in Figure 24–13D. Figure 24–14 shows a complete distributor head and high-pressure pumping assembly in cutaway view.

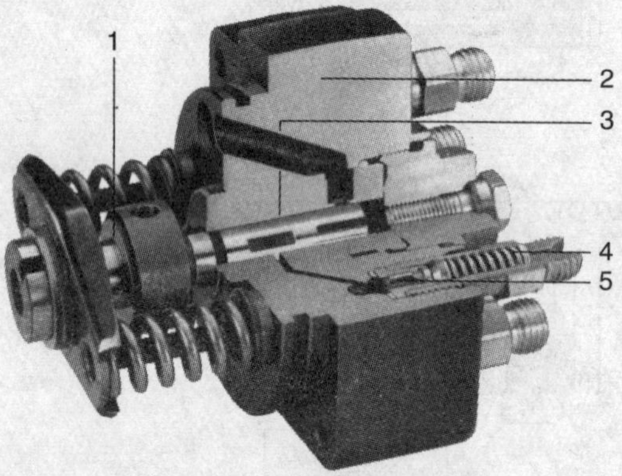

1. Control collar
2. Distributor head
3. Distributor plunger
4. Delivery valve holder
5. Delivery valve

Figure 24–14 *Distributor head with pump chamber. (Courtesy of Robert Bosch LLC)*

Delivery Valve

The delivery valve seals the high-pressure pipe from the injection pump chamber. It therefore retains dead volume fuel (static fuel in the pipe) at a value well above that in the pump chamber but comfortably below that required to open the injector nozzle. This static pressure is known as residual line pressure, and it ensures precise closure of the injector nozzle at the end of injection. It also ensures that a stable pressure is maintained in the high-pressure pipe between injection pulses, regardless of injected fuel quantity.

One delivery valve is used per engine cylinder. The delivery valve is a spring-loaded plunger. It is opened by delivery pressure developed in the injection pump chamber and closed by a return spring. Between injection pulses, the delivery valve remains closed. Its function can be most simply described by stating that it separates the high-pressure pipe from the distributor head outlet port for the larger portion of the cycle when no fuel is being pumped to a given engine cylinder.

Delivery Valve Operation The construction of each delivery valve is identical to that used on port-helix metering pumps: You may wish to consult Chapter 20 in which the operation of delivery valves is covered in greater detail. The valve core consists of a stem, seat, retraction piston, and flutes. Subject to pressure rise created during the injection pump delivery stroke, the piston is hydraulically opened when pump chamber pressure exceeds the combined forces acting to load it on its seat: spring force and line pressure. Initially, the delivery valve acts as a hydraulically actuated plunger, compressing the fuel retained in the high-pressure pipe. Once the retraction piston of the delivery valve exits above the seat in the delivery valve body, fuel in the pump chamber is united with that in the high-pressure pipe and the valve can be considered open.

Following the injection pump delivery stroke, line pressure is collapsed by the spill taking place at the control sleeve: The instant there is insufficient hydraulic pressure to hold the delivery valve open, it begins to retract toward its seat. However, the moment the retraction piston passes the delivery valve seat, dead volume fuel quantity is defined, the injector nozzle having sealed. Because the delivery valve has to retract further before it physically seats, the defined dead volume fuel is given a fraction more space, so the pressure is further reduced. This reduction in pressure establishes the residual line pressure value, usually about two-thirds of that required to open the injector nozzle.

Delivery Valve with Return-Flow Restriction Pressure drop-off to a precise value in the high-pressure pipes is desirable at the end of injection. However, the high-speed, high-pressure switching that the high-pressure pipe is subject to creates pressure waves that are reflected between the delivery valve and injector nozzle seat. This pressure wave reflection causes local spiking of line pressure and may cause undesirable nozzle opening known as secondary injection or vacuum phases in the high-pressure pipe, causing cavitation.

Using a delivery valve with a restriction bore that is only a factor in the direction of return (spill) fuel flow can minimize pressure wave reflection. The return-flow restriction consists of a valve plate and a pressure spring arranged so that the restriction is only effective in the return direction when it dampens pressure spikes and vacuum phases.

Constant-Pressure Valve

Another means of dealing with the problems associated with pressure wave reflection (pressure spikes and vacuum phases) is to use a constant-pressure-type delivery valve. These are usually found on high-speed engines using direct-injection (DI) engines. Constant-pressure valves relieve the high-pressure pipe pressure by means of a single-acting, nonreturn valve set to a specific pressure. The specific pressure would define the residual line pressure. A diagram of a constant-pressure delivery valve is shown in Figure 24–15.

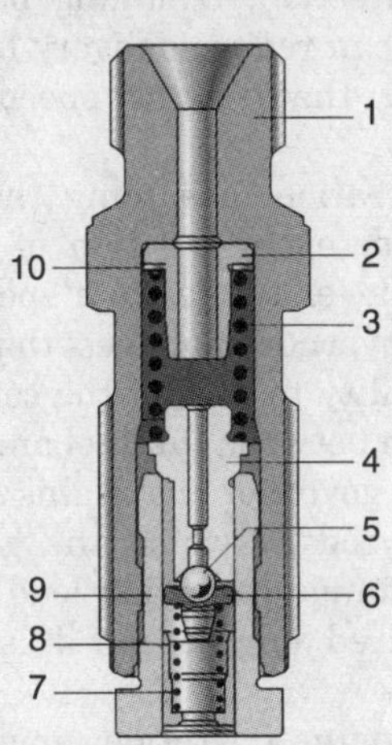

1. Delivery valve holder
2. Filler piece with spring locator
3. Delivery valve spring
4. Delivery valve plunger
5. Constant pressure valve
6. Spring seat
7. Valve spring (constant pressure valve)
8. Setting sleeve
9. Valve holder
10. Shims

Figure 24–15 *Constant-pressure delivery valve. (Courtesy of Robert Bosch LLC)*

Governing

Bosch sleeve-metering, rotary distributor pumps are available with both variable speed and limiting speed governing options. These are usually described by Bosch using the British terms:

- Variable speed governor = all speed governor
- Limiting speed governor = min-max governor

A full description of governor principles of operation is provided in Chapter 25 so it is not repeated here. Instead, we look at how the governor functions to control the positioning of the sleeve-metering collar under various operating conditions. In addressing the governors used on Bosch sleeve-metering distributor pumps, we reference a variable speed governor. Remember, regardless of governor type, the position of the sleeve-metering collar defines the injection pump effective stroke, so engine output is entirely dependent on this.

Governor Design

The governor assembly is attached to the governor drive shaft located in the governor housing. It consists of a flyweight carrier, **thrust collar**, and tensioning lever. When the flyweights rotate, they are forced outward due to centrifugal force. This radial outward movement is converted to axial movement of the thrust collar or what Bosch describes as a sliding sleeve. Thrust collar travel is allowed to act on the governor lever assembly, which is made up of a start lever, tensioning lever, and adjusting lever.

As in any mechanical governor, spring forces defined at the governor lever assembly act in opposition to the centrifugal force produced by the flyweights. Centrifugal force tends to reduce fueling/rpm, while spring force tends to increase fueling/rpm. Therefore, the interaction of spring forces and centrifugal force acting on the thrust collar (sliding sleeve) defines the positioning of the governor lever assembly. The governor lever assembly controls the position of the control sleeve or collar. As we learned earlier, the position of the control sleeve determines the plunger effective stroke that defines the quantity of fuel to be delivered. The governor assembly on a VE distributor pump is shown in Figure 24–16.

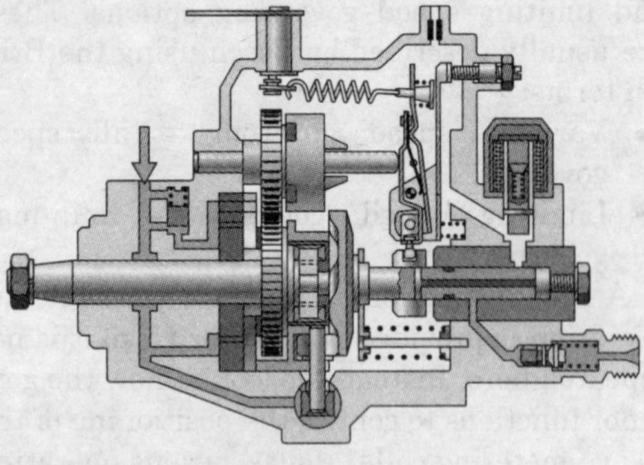

Figure 24–16 *VE governor assembly. (Courtesy of Robert Bosch LLC)*

Startup Fueling

When the engine is stationary, the flyweights and thrust collar are in their initial position. This results in the start lever being forced into the start position by the starting spring, moving the control sleeve on the distributor plunger to its start fuel position. When the engine is cranked, the distributor plunger travels through a complete working stroke before the cutoff cross-drilling is exposed to end delivery. The result is a full-fuel delivery pulse.

Low-Idle Operation

Once the engine is running with the accelerator pedal released, the engine speed control lever shifts to the idle position. This positions the control sleeve to provide a short plunger effective pumping stroke and represents the lowest fuel delivery condition of a running engine. Idle speed can be adjusted independent of the accelerator pedal setting and can be increased or decreased if temperature or load conditions require it.

Operation under Load

During actual operation, the driver requests the required engine speed by accelerator pedal angle. If higher engine speeds are required, the driver pushes harder on the pedal, increasing the angle and, therefore, the governor spring force. If a lower engine speed is required, the driver reduces the pedal angle. At any engine speed above idle, the start and idle springs are compressed completely, so they do not influence governing. Governed speed becomes the responsibility of the governor spring.

Under load, the driver sets the accelerator pedal at a specific position, increasing the pedal angle. If a higher engine speed is required, the pedal angle has to be further increased. As a result, the governor spring is tensioned, increasing the spring force available to counter the centrifugal force produced by the flyweights. This acts through the thrust collar governor levers to shift the control sleeve toward the full-fuel direction, increasing the injection pump effective stroke. As a result injected fuel quantity increases and engine speed rises.

The control collar remains in the full-fuel position until equilibrium is established once again between the centrifugal force generated by the flyweights (now greater) and the governor spring forces that oppose it. Should engine speed continue to increase, the flyweights extend further, resulting in governor thrust collar movement that forces the control sleeve toward the no-fuel direction, trimming back fueling. The governor can reduce delivery fueling to no fuel, ensuring that engine speed limitation takes place.

During operation, assuming the engine is not overloaded, every position of the engine speed control lever relates to a specific engine speed: The governor manages that speed by having the ability to control the control sleeve in any position between full fuel and no fuel. In this way, the governor maintains the desired speed. The speed at which the governor responds to a change in engine load in order to maintain desired engine speed is known as *droop*.

If during engine operation, engine load increases to the extent that when the control sleeve is in full-fuel position engine speed continues to drop, the engine can be assumed to be overloaded and the driver has no option but to downshift.

Governor Break

During the downhill operation of a vehicle, the engine is driven by vehicle momentum and engine speed tends to increase. This results in the governor flyweights moving outward, causing the governor thrust collar to press against the tensioning and start levers. Both levers react by changing position and pushing the control sleeve toward the no-fuel position until a reduced fuel equilibrium is established in the governor that corresponds to the new load/speed condition. At engine overspeed, the governor can always no-fuel the engine. With a variable-speed governor, any control sleeve position can be set by the governor in order to maintain desired speed.

Variable Timing Device

A hydraulically actuated timing device is located under the main pump housing at right angles to the pump longitudinal axis as shown in Figure 24–17. The variable timing device is a speed-sensitive advance mechanism that defaults to the most retarded delivery position in the absence of hydraulic (fuel) pressure.

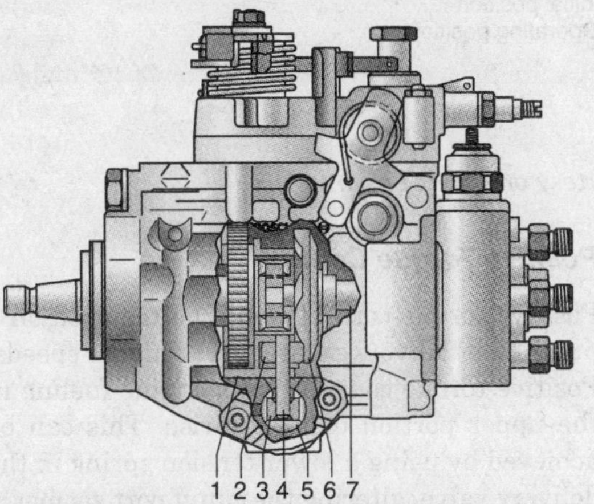

1. Roller ring
2. Roller ring rollers
3. Sliding block
4. Pin
5. Timing-device piston
6. Cam plate
7. Distributor plunger

Figure 24–17 *VE variable timing device. (Courtesy of Robert Bosch LLC)*

The timing device housing is closed with a cover on either side. A passage is located in one end of the timing device that allows fuel from the vane pump to enter. This fuel is allowed to act on the sectional area of the advance piston. On the opposite side of the piston, spring force opposes the hydraulic pressure of the fuel. The piston is connected to the roller ring by means of a sliding block and pin, enabling piston linear movement to be converted to rotational movement of the roller ring.

Timing Device Operation

The timing device is held in its initial or default position by the timing device spring as shown in Figure 24–18A. When the engine is started, the pressure control valve regulates fuel pressure so that it is exactly proportional to engine speed. As a result, this engine-speed-dependent fuel pressure is applied to the end of the timing device piston in opposition to the spring force acting on the other side of it.

At a predetermined vane pump fuel pressure, the advance piston overcomes the spring preload and shifts the sliding block and the pin that engages with the roller ring. The roller ring is rotated, moving its relative position with the cam plate, which results in the rollers lifting the rotating cam plate earlier. This action means that the actuation of the injection plunger is advanced. The maximum advance angle achievable in Bosch sleeve-metering distributor injection pumps is limited by timing piston linear movement and is usually 24 crank angle degrees. Figure 24–18B shows the maximum timing angle location of the advance piston.

Add-on Modules and Shutdown Devices

Bosch sleeve-metering, distributor injection pumps are available with a variety of add-on modules and shutdown devices. Because of the modular construction of the pump, these supplementary devices can be added to optimize engine torque profile, power output, fuel economy, and exhaust gas composition. A brief description of add-on modules and how they impact on engine operation follows. Figure 24–19 is a schematic that shows how these add-on modules interact with the basic distributor pump.

Torque Control

Torque control relates to how fuel delivery is managed with respect to engine speed and the engine load requirement characteristic. Generally, engine fuel requirement increases when a

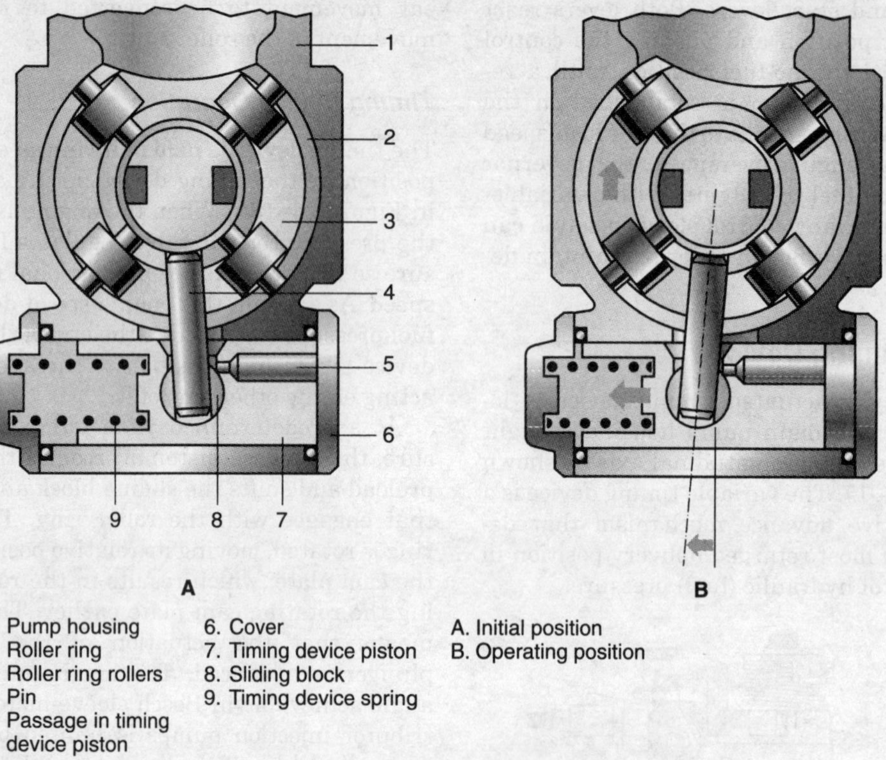

Figure 24–18 *Timing device operating principle. (Courtesy of Robert Bosch LLC)*

request for higher engine speed is made, that is, the driver pushes his foot harder on the accelerator pedal. Fueling should then level off as actual engine speed approximates desired engine speed.

Note that the same position of the fuel control sleeve will result in slightly more fuel delivery at higher engine speeds than at lower rpm due to the throttling effect that occurs at the distributor plunger cutoff port. This means that if the fuel trim settings were set to produce maximum torque at low engine speeds, the engine would be overfueling at high engine speeds, resulting in smoking and possible engine overheat.

Conversely, if the fuel trim settings were set to produce optimum performance at rated speed, the engine would not be able to develop sufficient power at lower-than-rated speeds. Getting fueling quantities optimized throughout the engine-operating range is known as torque control.

Positive Torque Control

Positive torque control would be required on a pump that delivers excess fuel at higher speeds. Positive torque control limits engine fueling in the upper portion of torque rise. This can be achieved by using a lower tension spring in the delivery valve, altering the cutoff port geometry producing a throttling effect, or using additional torque control springs in the governor.

Negative Torque Control

Negative torque control would be required in an engine that overfueled in the lower speed ranges but produced satisfactory performance at speeds closer to the rated speed of the engine. Negative torque control limits fueling in the lower portion of torque rise speed. It achieves this by governor spring pack trim or hydraulically, using pump housing fuel pressure.

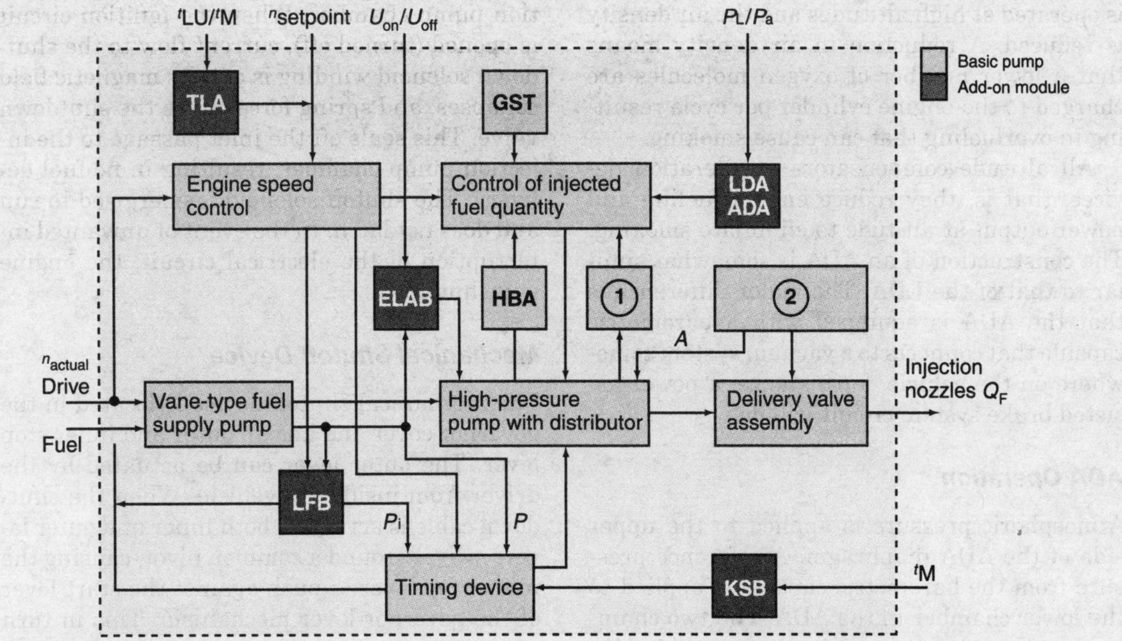

LDA Manifold pressure compensator.
Controls the delivery quantity as a function of the change-air pressure.

HBA Hydraulically controlled torque control.
Controls the delivery quantity as a function of the engine speed (not for pressure-charged engines with LDA).

LFB Load-dependent start of delivery.
Adaptation of pump delivery to load. For reduction of noise and exhaust gas emissions.

ADA Altitude pressure compensator.
Controls the delivery quantity as a function of atmospheric pressure.

KSB Cold start accelerator.
Improves cold start behavior by changing the start of delivery.

GST Graded (or variable) start quantity.
Prevents excessive start quantity during warm start.

TLA Temperature-controlled idle speed increase.
Improves engine warmup and smooth running when the engine is cold.

ELAB Electrical shutoff device.

A Cutoff port, n_{actual} Actual engine speed (controlled variable), $n_{setpoint}$ Desired engine speed (reference variable). Q_F Delivery quarterly, t_M Engine temperature, t_{LU} Ambient air temperature, P_L Change-air pressure, p_i Pump interior pressure.

①Full-load torque control with governor lever assembly.

②Hydraulic full-load torque control.

Figure 24–19 *VE add-on modules. (Courtesy of Robert Bosch LLC)*

Aneroid

The manifold pressure compensator (known by Bosch as an LDA, a German acronym) is a simple aneroid device. As such, it reacts to the manifold boost pressure generated by the turbocharger and essentially limits fueling until there is sufficient air in the engine cylinder to properly combust it.

Charge air is ported directly to the LDA assembly. The LDA is divided into two separate airtight chambers divided by a diaphragm. Manifold boost is applied to one side of the diaphragm while spring force is applied to the other. The diaphragm is connected to the LDA sliding pin, which has a taper in the form of a control cone. This is contacted by a guide pin through which levers act on the full-fuel stop setting for the control sleeve.

At lower engine loads, there is insufficient manifold boost developed to effect movement on the diaphragm so the spring remains in control, with the result that the LDA limits maximum fueling. When engine load is increased and a predetermined manifold boost pressure is achieved, the LDA spring pressure is overcome, allowing the control sleeve to permit a longer pump effective stroke.

Altitude-pressure Compensator

The objective of an altitude-pressure compensator (known by Bosch as ADA, a German acronym) is to limit engine fueling when the vehicle is operated at high altitudes and the air density is reduced. A reduction in air density means that a lower number of oxygen molecules are charged to the engine cylinder per cycle resulting in overfueling that can cause smoking.

All altitude compensators are deration devices, that is, they reduce engine fueling and power output at altitude to eliminate smoking. The construction of an ADA is somewhat similar to that of the LDA. The major difference is that the ADA is equipped with a barometric capsule that connects to a vacuum system somewhere on the vehicle, for instance, a power-assisted brake system circuit will do.

ADA Operation

Atmospheric pressure is applied to the upper side of the ADA diaphragm. A reference pressure from the barometric capsule is applied to the lower chamber in the ADA. The two chambers are separated by a diaphragm. Should a drop in atmospheric pressure occur such as would be experienced by driving a vehicle up a mountain, the barometric capsule (constant pressure) would be at higher pressure than that on the opposing side of the diaphragm. This pressure would cause the diaphragm to move the sliding bolt vertically away from the lower fuel stop and reduce engine fueling.

Engine Shutoff

Diesel engines are shut down by no-fueling them. Most Bosch sleeve-metering distributor pumps are shut down using a solenoid-operated shutoff (ELAB). A few of these pumps are equipped with a mechanical shutoff device but you will probably only see them in off-highway applications.

Electrical Shutoff Device

Electrical shutoff is desirable because the solenoid can be energized by the vehicle ignition circuit. On the Bosch distributor pump, the fuel shutoff solenoid (acronym ELAB is German) is installed in the distributor head. When the engine is running, the solenoid is energized. When energized, the solenoid shutoff valve is positioned open, allowing fuel to pass into the injection pump chamber. When the ignition circuit is opened (turned off), current flow to the shutdown solenoid winding is cut, its magnetic field collapses, and spring force closes the shutdown valve. This seals off the inlet passage to the injection pump chamber, resulting in no-fuel delivery. The shutoff solenoid is energized to run and does not latch. In the event of unwanted interruption of the electrical circuit, the engine will shut down.

Mechanical Shutoff Device

The mechanical shutoff device is located in the governor cover and has an outer and inner stop lever. The outer lever can be actuated by the driver from inside the vehicle. When the shutdown cable is actuated, both inner and outer levers swivel around a common pivot, causing the inner stop lever to push against the start lever of the governor-lever mechanism. This in turn moves the control sleeve to the no-fuel position; therefore the distributor plungers cutoff port remains open throughout plunger stroke and no fuel can be pressurized.

REVIEW QUESTIONS

In answering these questions, remember that two different pump systems are studied in this chapter: The first ten questions (1–10) address inlet-metering, opposed-plunger distributor pumps; the next ten (11–20) address the Bosch sleeve-metering distributor pump.

1. If a four-stroke cycle diesel engine fueled by an inlet-metering, opposed-plunger fuel pump is rotated through a complete cycle, how many times is the pump rotated?
   a. One-half revolution
   b. One revolution
   c. Two revolutions
   d. Four revolutions
2. What actuates the plungers of an inlet-metering, opposed-plunger fuel injection pump to create fuel injection pressures?
   a. Metered fuel quantity
   b. Charging fuel pressure
   c. External cam profile
   d. Internal cam ring

3. In an inlet-metering, opposed-plunger fuel pump, what causes the outward movement of the opposed plungers in the pump chamber during metering?
   a. Centrifugal force
   b. Metered fuel quantity
   c. Spring force
   d. Pump housing backpressure
4. How is an inlet-metering, opposed-plunger fuel pump lubricated?
   a. With a dedicated supply of engine lubricating oil
   b. With pressurized engine lubrication oil
   c. With fuel from the vane pump
   d. Prelubed on assembly
5. *Technician A* states that when a mechanical governor is used on an inlet-metering, opposed-plunger fuel pump, metering valve rotational position determines how much fuel is metered to the pump element. *Technician B* states that the metering valve determines the pressure value developed by the vane-type transfer pump. Who is correct?
   a. Technician A only
   b. Technician B only
   c. Both A and B
   d. Neither A nor B
6. When an inlet-metering, opposed-plunger fuel pump fuels an automotive eight-cylinder diesel engine, how many delivery valves would be used?
   a. One
   b. Two
   c. Four
   d. Eight
7. *Technician A* states that on an inlet-metering, opposed-plunger fuel pump, plunger actual stroke is constant regardless of fuel requirement. *Technician B* states that in the same pump, plunger effective stroke is determined by vane pump pressure. Who is correct?
   a. Technician A only
   b. Technician B only
   c. Both A and B
   d. Neither A nor B
8. How is the timing advance mechanism on a hydromechanical, inlet-metering, opposed-plunger fuel pump actuated?
   a. Hydraulically by the vane pump fuel pressure
   b. Hydraulically by engine oil pressure
   c. Pneumatically by manifold boost
   d. By manifold vacuum
9. What limits the maximum fuel metered to the pump chamber in an inlet-metering, opposed-plunger fuel pump?
   a. Vane pump pressure
   b. Metering valve flow area
   c. Leaf spring stops
   d. Roller outboard travel
10. *Technician A* states that pintle-type injector nozzles are usually used with an inlet-metering, opposed-plunger fuel pump system. *Technician B* states that this type of fuel system is more likely used on indirect-injected diesel engines rather than direct-injected engines. Who is correct?
    a. Technician A only
    b. Technician B only
    c. Both A and B
    d. Neither A nor B
11. When a Bosch sleeve-metering, rotary distributor pump is used to fuel a four-stroke cycle engine, at what speed is the fuel pump driven?
    a. Engine camshaft speed
    b. Engine crankshaft speed
    c. Two times engine speed
    d. Four times engine speed
12. *Technician A* states that Bosch sleeve-metering, rotary injection pumps can be used to fuel high-speed diesel engines of up to eight engine cylinders. *Technician B* states that when this type of pump is used on a V-configured engine, two distributor heads are used. Who is correct?
    a. Technician A only
    b. Technician B only
    c. Both A and B
    d. Neither A nor B
13. When a Bosch sleeve-metering, rotary distributor pump is used on a six-cylinder, four-stroke cycle diesel engine, how many delivery valves would be required?
    a. One
    b. Two
    c. Four
    d. Six

14. *Technician A* states that the pumping plunger used on a Bosch sleeve-metering, rotary distributor pump reciprocates. *Technician B* states that the pumping plunger used in this fuel system rotates. Who is correct?
    a. Technician A only
    b. Technician B only
    c. Both A and B
    d. Neither A nor B
15. What determines plunger stroke in a Bosch sleeve-metering, rotary distributor pump?
    a. Position of the sleeve-metering collar
    b. Cam geometry
    c. Rotational speed of the pump
    d. Metered fuel quantity
16. *Technician A* states that the cam geometry used on a Bosch sleeve-metering, rotary distributor pump is specific to the engine that it fuels. Referring to the same pump, *Technician B* states that the cam ring rotates at engine camshaft speed. Who is correct?
    a. Technician A only
    b. Technician B only
    c. Both A and B
    d. Neither A nor B
17. How is timing advance achieved in a Bosch hydromechanical, sleeve-metering, rotary distributor pump?
    a. Centrifugal force generated by flyweights
    b. Hydraulic pressure generated by the vane pump
    c. Spring force relayed by the governor
    d. Mechanical force from the accelerator linkage
18. *Technician A* states that a variable-speed governor on a Bosch sleeve-metering, rotary distributor pump functions by defining a specific fuel quantity that corresponds to throttle lever angle. *Technician B* states that a variable-speed governor will attempt to maintain a specific rpm value as engine load changes. Who is correct?
    a. Technician A only
    b. Technician B only
    c. Both A and B
    d. Neither A nor B
19. *Technician A* states that most Bosch sleeve-metering, rotary distributor pumps use a shutoff solenoid known as an ELAB. *Technician B* states that an ELAB does not latch so in the event of a loss of electrical power, the fuel system will shut the engine down. Who is correct?
    a. Technician A only
    b. Technician B only
    c. Both A and B
    d. Neither A nor B
20. Which of the following could address the problem of a Bosch sleeve-metering, rotary distributor pump fueled engine that functioned well at peak torque but was overfueling and smoking when run at rated speed?
    a. Positive torque control
    b. Negative torque control
    c. Trim back maximum speed stop screw
    d. Replace the aneroid

25

Governors

INTRODUCTION

The output of any diesel engine is limited only by the quantity of fuel injected into its cylinders. When we talk of a gasoline-fueled engine, there is some emphasis on managing fueling at a stoichiometric ratio. A stoichiometric ratio is one in which a cylinder is provided with the exact quantity of both air and fuel required to complete the combustion reaction: no more and no less.

The output of a gasoline-fueled engine is limited by the volume of *air* that can pass through the throttle body. A diesel engine is aspirated with excess air. This excess air is usually considerably above the stoichiometric requirement, especially under conditions of high turboboost. Given unlimited fuel, diesel engines will accelerate at rates of up to 1,000 rpm per second, often to self-destruction. For this reason a governor is required.

Electronic Governing

The primary function of a diesel engine governor is to sense engine speed and limit fueling when the engine is run at its specified maximum speed. Until about 20 years ago, most mobile diesel engines used **hydromechanical governing**. Today **electronic governors** are used on all highway, and most off-highway, diesel engines. Electronic governors are built into the engine management software programmed into ECM memory. That said it does not hurt to know something about hydromechanical governors for two reasons:

- The objectives of hydromechanical and electronic governors are identical, but there is no doubt that computer-controlled governors accomplish these objectives more effectively.
- All of the language we use to describe electronic governing was borrowed from the previous generation of hydromechanical governors.

An electronic governor manages engine output based on command inputs from the vehicle operator and the range of sensor signaling information sent to the engine and chassis electronic control modules (ECM). The governor software (in the engine ECM) then maps a fueling profile. The fueling profile is the output commands that control such things as:

- Fuel quantity
- Injection timing
- Pilot and multipulse injection events

The ECM in a truck diesel engine is responsible for both the "thinking" (computations) and switching (actuation) required of a computer. We spend some time explaining vehicle computers later in Chapter 30 and we should make it clear here that to properly understand today's highway diesel engine governors, you have to understand the basics of vehicle computer controls. When we talk about an ECM's role in governing engine output, we should identify two distinct phases:

- The "processing" (thinking) phase. Engine OEMs call this the governing **algorithm**. Another way of saying this is the **governing map**. Based on its software instructions, the ECM analyzes all of the inputs such as engine speed and throttle position and computes a fueling "plan." This fueling plan then has to be communicated to the part of the ECM responsible for switching.

- The switching phase. After the ECM has processed the governing algorithm, this has to be converted into action by the ECM switching apparatus. In an ECM the components responsible for converting "thinking" into outputs are known as drivers. The drivers responsible for controlling the fuel injectors are known as injector drivers. Injector drivers control the duty cycle of fuel injectors.

Today's ECMs may be programmed with comprehensive fuel maps, engine malfunction strategy, and a wide variety of customer data options, but most of the language used to describe engine governing conditions originated from James Watt's 1786 centrifugal governor. James Watt (see Chapter 3) invented the **mechanical governor** because he wanted to regulate the speed of his steam engines. To regulate the speed of an engine you have to have a way of knowing its exact speed. We call this speed sensing.

James Watt accomplished speed sensing by driving a set of flyweights in a carrier at a speed proportional to engine speed; the flyweights pivoted in the carrier and were loaded into their most retracted position by a spring. As the carrier was rotated, the centrifugal force exacted from the flyweights would act against the applied spring force; the spring tension could then be set so that at a specified maximum speed, centrifugal force would overcome the spring force to act directly on a fuel control mechanism to limit fueling. Most truck and bus governors classified as "hydromechanical" use variations of Watt's governor.

WHAT GOVERNORS DO

A governor determines how an engine is fueled. At minimum it limits fueling at the highest intended engine rpm to prevent overspeed. In addition it is also required to define the lowest no-load rpm of the engine to enable the engine to idle without any input from the driver's accelerator pedal. A governor must also provide a means of no-fueling the engine to shut it down. Also most modern governors have many other features such as the ability to provide extra fuel over the intended operating rpm window: We call this normal engine operating rpm.

Torque Rise

The "normal engine operating rpm" is the range of the rpm spectrum where the engine original equipment manufacturer (OEM) calculates that the engine should be operated within to produce the best combination of power and fuel economy. As truck technicians, we call this **torque rise** rpm. Torque rise is identified to the truck driver by a sticker on the visor identifying the upper and lower shift rpm. In simple terms, this tells the driver the engine rpm at which he should be making upshifts and downshifts. A good driver will always observe these shift points. These shift points are broadcast to the chassis data bus so when an automatic or automated transmission is used, shifting is phased to the appropriate engine rpm.

Figure 25–1 shows the fuel map used on a typical highway diesel engine: This engine has its fuel system set up to produce peak torque at 1,400 rpm and rated power at 1,800 rpm. The torque rise profile on this engine is therefore between 1,400 and 1,800 rpm. This can be expressed as percentage torque rise. Because the torque rise duration is 400 rpm, when we divide 400 by 1,800 the result can be rounded to 22%. Back in the days when we did not care too much about fuel consumption, it was not unusual to find *high torque rise* engines with 50% torque rise. One way of producing this would be to trim

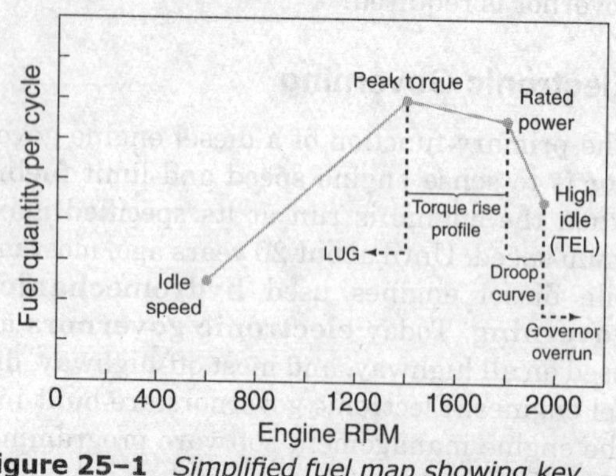

Figure 25–1 *Simplified fuel map showing key governor points and how they correlate with rpm and fuel quantity.*

the fuel system so that rated speed and power was achieved at 2,000 rpm and peak torque at 1,000 rpm. Drivers liked high torque rise engines because when one was coupled with a manually shifted transmission, it would require a lot less shifting.

Rated Speed and Load

One important role of the governor is to define the specific engine rpm at which peak power is achieved. This is usually expressed as horsepower in North America but with the growing influence of European and Asian technology on our trucking industry, we hear kilowatts used to express power more and more frequently. When an OEM specifies diesel engine power (see Chapter 5) it is always done so correlated with the rpm at which that power is achieved. So peak horsepower is expressed as **rated power**, but more often engine OEMs use the term **rated speed**. If the peak power in an engine is 600 horsepower at 1,800 rpm, this could also be referred to as the rated speed.

Peak Torque

If you read Chapter 5, you will know that **peak torque** in a diesel engine is produced whenever cylinder pressures peak. Peak torque rpm is important in understanding governor operation because in most cases it represents the lowest rpm within the "normal engine-operating range." This means that it is represents the lowest speed the driver wants to run an engine before he makes a downshift. In the simplified fuel map in Figure 25–1, peak torque represents the base (lowest rpm) of the torque rise profile. Assuming the driver has his foot to the floor, as engine rpm is driven down by load, if a downshift fails to occur at the peak torque rpm, the engine drops into the lug rpm. Lugging a diesel engine can result in a number of problems that may include high cylinder pressures, dangerously high driveline torsionals, fuel wastage, and exhaust smoking.

Idle and High Idle

Governors also have to define idle speeds. Whenever we use the term *idle* in reference to a diesel engine, it means there is no load on the engine. There are a couple of key idle rpm. The first is usually just called **idle speed**. In Figure 25–1, the idle speed is 600 rpm. On a highway truck this is the speed that a diesel engine runs without any input from the accelerator pedal. You can also refer to it as low idle. Next, the term **high-idle speed** is important. High idle is the fastest speed that the engine should run under no load: Caterpillar refers to this speed as **top engine limit (TEL)**. In Figure 25–1, high idle is 1,950 rpm. On hydromechanical engines governed for mobile applications, high-idle speeds exceed rated speeds by something between 5% and 20%.

Droop Curve

Rated speed is the highest rpm that a diesel engine should be operated and most ECM-managed engines will only allow engine rpm to run above rated under exceptional circumstances. However, all diesel engines powering mobile equipment are top end governed with **droop curve**. The droop curve rpm range runs between the rated speed rpm and the high idle. Its objective is to prevent abrupt changes in fueling when the engine rpm exceeds rated speed.

Another way of defining the role of governor droop curve is to describe it as the graduated fuel deration designed to occur above high idle. To abruptly cut off fueling a highway diesel engine the instant the rpm exceeded rated speed could be potentially dangerous in terms of vehicle handling on the highway. Droop curve fueling ensures that fueling adjustments at speeds exceeding rated are gentle. Droop curve is only a requirement of highway diesel engines. Droop curve is expressed as a percentage by diesel engine OEMs. The percentage is typically between 5% and 20%.

Droop curve on hydromechanical engines can be calculated by determining the high-idle speed then dividing it by the specified rated speed. Using the data in Figure 25–1, where the engine has a specified rated speed of 1,800 rpm and high-idle speed of 1,950 rpm, the droop curve extends over 150 rpm. Then 1800 divided by 150 produces a rounded result of 8%—a typical droop curve percentage found on hydromechanical engines.

Lug

An engine is put into a lug mode when the rpm drops below the lowest rpm in the intended engine-operating range. This is known as **lugging** and it occurs when engine load drops the rpm below the base rpm of torque rise. Using the data as it appears on the graph in Figure 25–1, if the rpm dropped below 1,400 during a shift, this would put the engine into lug. The consequences of lugging an engine today tend to be less noticeable due to innovations such as variable geometry turbochargers, soft fuel injection timing, and multipulse fuel injectors, but a generation ago an engine could labor and produce dangerously high cylinder pressures attempting to pull out of a severe lug condition.

Startup Strategy

A generation ago cold-start strategy meant providing excess, usually absolute full fuel, during cranking. Some systems would also accommodate some timing adjustment during startup, usually retard, so that cylinder heat was maximized at the beginning of injection. The combination of full fuel and retarded timing resulted in some pretty dense smoke emission during the start/warm up phase. Each diesel engine had some differences required of the operator to put the governor in cold-start mode, usually simply that of flooring the accelerator and holding it there until the engine coughed into motion. If the conditions were cold enough, the driver often had to nurse the engine with the accelerator for 5 minutes, dumping excess fuel into the engine cylinders.

Today cold-start strategy is precisely managed by the governor software, and input from the operator such as flooring the accelerator pedal is ignored. Furthermore, many engines (and transmissions) run in inhibited mode until the temperature conditions measured by input sensors clear the vehicle out of the inhibit mode. Startup strategy in current diesel engines is ECM managed with the objectives of minimizing fueling and tailpipe smoking and the potential to damage the engine or other drivetrain components.

GOVERNOR TYPES

Governors are classified by type, that is, by operating principles. The oldest type, the mechanical governor, was most common on hydromechanical engines until 1990. Every governor managing a truck or bus engine has the benefit of an operator or driver who regulates engine output according to how that driver interprets the requirements of the trip being undertaken. The accelerator therefore represents an essential input to the governor. The governor must also know exactly how fast the engine is rotating in any given mode of operation; so another essential input to the governor is a means of precisely determining engine rpm. In vehicle engine governors, these two inputs would be regarded as essential.

Mechanical Governors

Speed sensing is by means of rotating flyweights driven at a speed proportional to engine rpm; in a mechanical governor, centrifugal force generated by the rotating flyweights (Figure 25–2, #6) acts directly on the fuel control mechanism

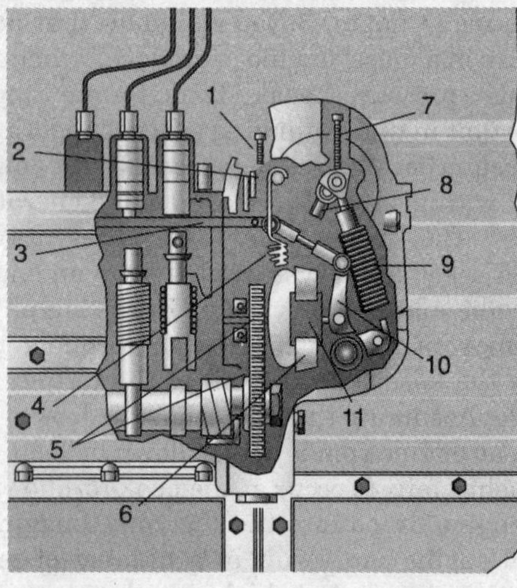

1. Low-idle adjusting screw
2. Spring-loaded stop assembly
3. Fuel rack
4. Shutoff spring
5. Gears
6. Governor weight
7. High-idle adjusting screw
8. Low-idle stop lever
9. Governor spring
10. Terminal lever
11. Thrust bearing and sliding sleeve

Figure 25–2 *Cutaway view of a simple mechanical governor. (Reprinted Courtesy of Caterpillar Inc.)*

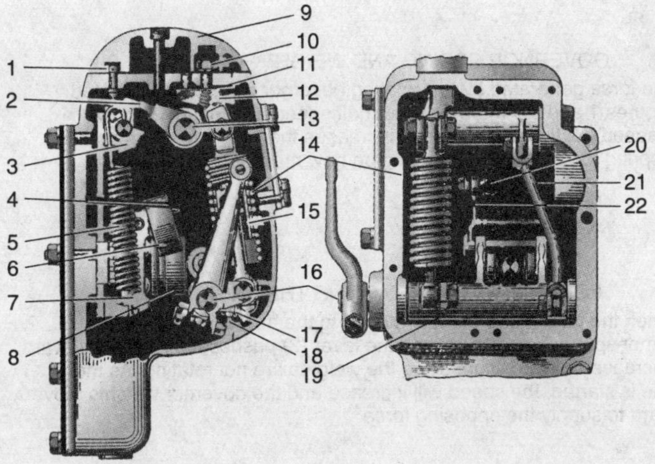

1. Low-idle adjusting screw
2. Low-idle stop lever
3. Stop
4. Small governor weight
5. Shutoff spring
6. Large governor weight
7. Spindle assembly
8. Thrust bearing and sliding sleeve
9. Top cover
10. Capscrew
11. Shims
12. Anchor
13. Stop lever shaft
14. Governor spring
15. Small spring
16. Terminal lever shaft
17. Throttle shaft control lever
18. Setscrew (earlier engines)
19. Terminal lever
20. Link
21. Rod
22. Lever

Figure 25–3 *Sectional and overhead views of a simple mechanical governor and components. (Reprinted Courtesy of Caterpillar Inc.)*

(see Figure 25–2, #s10 and 11). The **governor spring** tension is set (Figure 25–3, #14) to oppose the centrifugal force and will define the TEL) or high-idle speed (the highest speed at which an engine is designed to run). The **thrust collar** (a.k.a. thrust washer or thrust sleeve) acts as an intermediary between the spring force and centrifugal force and is connected to the fuel control mechanism. Governor spring tension is usually designed to be variable and increases with accelerator pedal travel.

A simple way of looking at mechanical governor operation is to see the thrust collar to which the fuel control mechanism is attached as being subject to two opposing forces. On one side, centrifugal force (produced by the flyweights) is attempting to reduce rpm by trimming back fueling. Opposing the centrifugal force is spring force. Spring force is variable and increased when the accelerator pedal angle increases. By opposing the centrifugal force, spring force attempts to increase rpm by requesting more fuel. Spring force is generally allowed to have its own way until governed speed is achieved. At governed speed, centrifugal force has to be able to overcome any amount of applied spring force if the governor is to properly do its job.

Figure 25–4 shows how a mechanical governor responds through a range of speeds beginning at engine stationary and increasing to rated speed and load.

Hydraulic Governor—Servo Type

In a **hydraulic**, **servo-type governor**, speed sensing is accomplished once again by rotating flyweights driven at a speed proportional to engine speed, and opposing this force is the governor spring. However, the intermediary between the centrifugal and spring forces is a control valve that feeds oil (usually pressurized oil from the engine lubricating circuit) to a servo, which moves the fuel control mechanism, so the actual force used to move the fuel control mechanism is hydraulic. This design permits the use of smaller flyweights and more sensitive response to speed and load fluctuations. Caterpillar used servo-type hydraulic governors to manage some of its older hydromechanical engines.

Hydraulic Governor—Nonservo

In a **hydraulic nonservo governor**, speed sensing is accomplished by using a positive displacement engine-driven transfer pump that unloads to a defined (and constant) flow area, so fuel pressure rises proportionally with engine speed. This type of governor is used on inlet metering, distributor pump fuel systems, and speed regulation is achieved by gating or diverting fuel from the inlet metering apparatus at a specified fuel pressure. They are therefore not used on any current medium- and

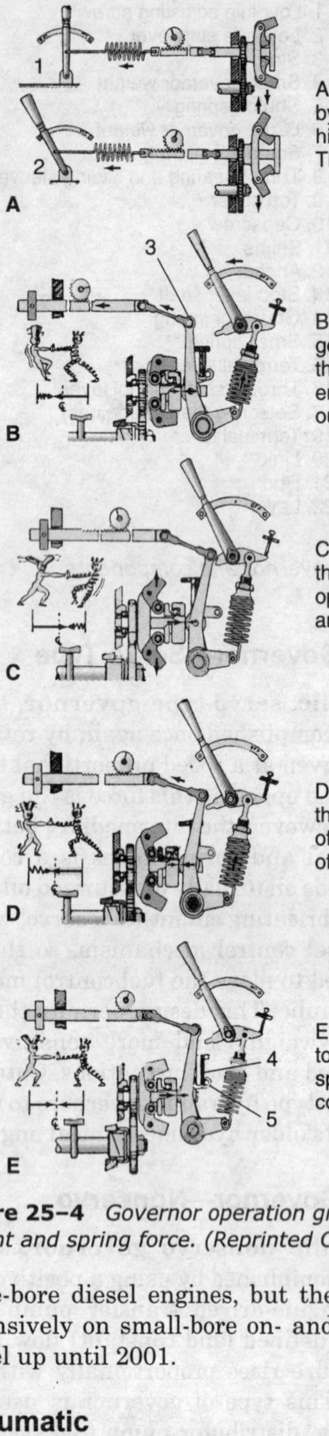

Figure 25–4 *Governor operation graphics: Note the position of the rack and the figures representing weight and spring force. (Reprinted Courtesy of Caterpillar Inc.)*

large-bore diesel engines, but they were used extensively on small-bore on- and off-highway diesel up until 2001.

Pneumatic

Pneumatic governors are used on naturally aspirated diesel engines and therefore seldom in truck and bus applications. Fueling is factored by manifold vacuum, which is regulated by a **throttle** valve controlled by an operator. Pneumatic governors are common in older marine and agricultural applications. The chances of seeing a diesel engine governed by a pneumatic governor today are remote: Your best chance is to check out older boats in a marina.

Electronic Governing

Electronic governing is used to govern current

diesel engines. Engine speed sensing is usually by means of a inductive pulse generator, which sends its signal to an ECM. The ECM first "processes" the sensor information, then plots a fueling profile, and then finally effects it by switching actuators, or fuel injectors. Today's commercial vehicles are drive-by-wire, meaning that there is no direct mechanical connection between the accelerator pedal and the fuel control mechanism. The accelerator pedal is just one of a number of sensor inputs to the ECM, so pushing it to the floor makes a "request" for full fuel that may not be fulfilled: Just try it during engine startup.

Electronic governing is used by all of today's diesel engine electronic management systems. It goes by names such as Detroit Diesel/Mercedes-Benz DDEC, Cummins CELECT and IS (Interact System), Caterpillar ADEM, Mack Trucks V-MAC, Navistar Diamond Plus, and Volvo VECTRO. Each of these systems is handled in some detail elsewhere in this textbook.

GOVERNOR CLASSIFICATION

Any governor, hydromechanical or electronic, is classified by how it manages engine performance. In the days of hydromechanical governors, it was impossible to change governor classifications without significant changes to hardware and complex adjustments. But that has changed with the introduction of computer controls. Because the nature of electronic engine governing is determined by programming, in most cases the classification field can be altered in a matter of minutes using the appropriate electronic service tool (EST). An explanation of the governor classifications found on contemporary commercial vehicle diesel engines fuel management systems follows.

Limiting Speed (LS)

A **limiting speed governor (LSG)** sets the engine idle speed, defines the high-idle speed, and permits fueling between those values to be controlled by an operator (driver). LSGs are the most common in commercial vehicle applications. One of their advantages is to make the diesel engine respond to accelerator input in much the same manner as a spark-ignited (SI), gasoline-fueled engine responds to throttle control. While the gas engine responds to an increase in accelerator pedal angle by providing more airflow though the throttle body, the LSG diesel governor does so by providing more injected fuel. A governor classified as LS will in most cases provide excess startup fuel, define a **torque rise profile**, define droop curve, and be capable of no-fueling the engine for shutdown. A mechanical LSG is sometimes known as an **automotive governor** or a **min-max governor**.

Driving with LS Governing

Driving a vehicle with LS governing is easy because the accelerator pedal is designed to respond to the foot in much the same way as the accelerator in a car. If a truck was being driven on a flat profile highway, the accelerator pedal angle would help define a specific fuel quantity injected to the engine cylinders. Because of this, when the vehicle encountered an uphill gradient, more engine power and, therefore, more fuel would be required. If the driver wished to maintain the same road speed traveling up the gradient, the accelerator pedal angle would have to be increased so that more fuel could be delivered to the engine. The opposite would be true when negotiating a downhill gradient.

- Accelerator pedal angle in an LSG-managed truck diesel engine defines a fuel quantity. If the driver wants either more or less fuel, the pedal angle must be either increased or decreased.

Variable Speed (VS)

A **variable speed governor (VSG)** (a.k.a. *all-speed*) sets engine idle speed and defines high idle and *any* speed in the intermediate range depending on accelerator pedal position. A given amount of accelerator pedal travel corresponds to an engine rotational speed; as engine loading either increases or decreases, the governor manages fueling to attempt to maintain that engine speed. All VSGs, whether hydromechanical or electronic, are designed to respond to a change in engine load without any rpm variation. If an rpm variation occurs, this is known as **droop**. Hydromechanical VSGs were common in many Mack Trucks and Caterpillar applications and others where **power takeoff** (**PTO**: where the engine is used to drive auxiliary equipment) management was a consideration.

From the driver perspective, the VS governor takes a little getting used to. Most of today's electronic management systems can be toggled (as customer data programming) to either LS or VS mode. A hydromechanical governor classified as VS usually provides excess startup fuel, defines a torque rise profile, defines droop curve, and is capable of no-fueling the engine for shutdown.

Driving with VS Governing

Driving a vehicle with VS governing is often the preference of more experienced truck drivers because it ultimately requires less effort. Accelerator pedal angle is essentially a "request" for a specific engine rpm, so within the capabilities of the fuel system this request will be effected. Therefore, if a truck was being driven on a flat profile highway, the driver-controlled accelerator pedal angle defines a specific engine rpm. When the vehicle encounters an uphill gradient, more engine power is required to maintain the set rpm: The reverse is true with a downhill gradient. With VS governing, the driver requests a set engine rpm by simply maintaining the same pedal angle. While the truck travels uphill more injected fuel is required to maintain the set rpm, and within the limitations of the fuel system this is provided with VS governing.

The same would be true when negotiating a downhill gradient. Once again if the driver wants to maintain the same road speed traveling downhill, he is not required to change his accelerator pedal angle. The fueling trimback is accomplished automatically because the pedal angle correlates to a requested engine rpm, not a fuel quantity. Use a gentle foot when changing accelerator pedal angle in any VS-governed truck.

- Accelerator pedal angle in a VS-managed truck diesel engine requests a specific engine rpm: Within the limitations of the fuel system, the governor attempts to maintain that rpm as engine load conditions change.

Isochronous

An **isochronous governor** is only required when driving a generator. In this application the engine must respond instantly to load changes with zero droop (*no* rpm fluctuation when engine load changes) or the (electrical) frequency will alter. However, the term is being used to describe an option in diesel engine electronic management systems. In this instance, isochronous governing mode would be used to manage PTO fueling while stationary, and one OEM uses the term to describe engine fueling at an electronically managed, default (when critical input signals are lost) rpm.

Governor Trim Options

Governor trim options vary widely on hydromechanical diesel engines depending on the type of governor and the OEM. Trim refers to any adjustable value. If you increase the high-idle trim, you are adjusting the high-idle rpm. When performing any governor trim adjustments, consult the OEM service literature and make sure you understand the exact consequences of the adjustment, one of which could be that you are doing something illegal. Figure 25–5 shows some trim adjustments on an older Caterpillar governor.

Governor Auxiliary Devices

Hydromechanical governors functioned on engines that lacked the comprehensive monitoring circuit we take for granted on today's engines. Conditions such as manifold boost and ambient pressure had to be factored into engine governing without the benefit of a computer. Although a couple of these devices are discussed elsewhere in this book, here are some trim devices.

Aneroids

In a general sense, an aneroid is used to measure pressure values. As a truck governor auxiliary device, an aneroid measures manifold (turbo) boost and attempts to limit fueling until there is sufficient (boost) air in the cylinder to properly combust it. Excessive smoking would occur if an aneroid failed to do its job. Aneroids go by different names depending on the engine OEM. Here are some examples:

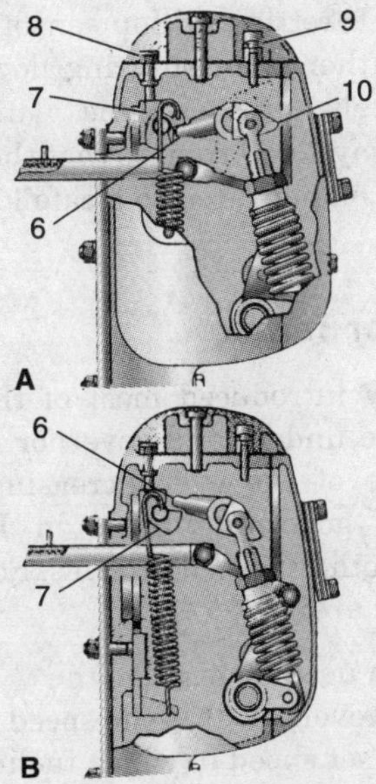

HIGH-AND LOW-IDLE ADJUSTMENTS

A. When the low-idle stop lever (6) is in contact with the spring-loaded stop (7) as illustrated, the engine operates at low idle. The screw assembly (8) controls the low-idle setting. When the lever (10) contacts the high-idle screw (9), which is also adjustable, the engine operates at the high-idle speed.

GOVERNOR SHUTOFF POSITION

B. When the governor control is in the shutoff position, it causes the stop lever (6) to pivot the spring-loaded stop (7), and it will hold the fuel rack in the "fuel off" position.

Figure 25–5 *Governor adjustments and shutoff position. (Reprinted Courtesy of Caterpillar Inc.)*

Bosch	LED (on Mack/IH/Deere)
Caterpillar	Fuel-air ratio control (FARC) valve
Cummins	Air-fuel control (AFC)
Mack Trucks	Puff limiter

Altitude Compensators

An altitude compensator used on a hydromechanically governed diesel engine measures ambient pressure using a barometric capsule (similar to that used in a household barometer) and limits engine fueling as the oxygen density in air decreases with an increase in altitude. They begin to become a factor at an altitude of 1,000 ft. (300 m): Essentially they derate the engine to prevent overfueling.

Shutdown Solenoids

Some hydromechanical engines use shutdown solenoids as an additional engine no-fuel device. Examples are the Cummins PT energized-to-run (ETR) solenoid and the solenoid used on the Caterpillar new scroll family of port-helix metering injection pumps. Some are ETR and others latch. A latching solenoid requires switched current to change its status whether shutting down or turning on. Figure 25–6 shows an image of a typical shutdown solenoid: This example acts directly on the governor thrust collar.

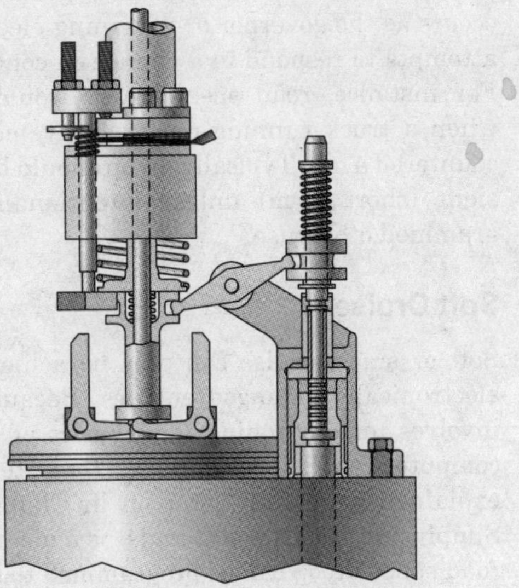

Figure 25–6 *Shutdown solenoid. (Reprinted Courtesy of Caterpillar Inc.)*

Road Speed Governing

All of today's drive-by-wire commercial vehicles offer a range of road speed governing options. This section briefly describes some of them, but for a more detailed understanding you should read ahead to those chapters that deal with the specific OEM engine management systems.

Road Speed

Road speed governing (RSG) is usually, but not always, the maximum programmed road speed permitted by the vehicle electronics. Some vehicle electronic management systems will permit the programming of maximum cruise speed at a higher value than "maximum" road speed to encourage drivers to use cruise control. The reason for this is that most truck cruise control systems have been proven to produce significantly better fuel economy than an "average" driver is capable of achieving.

Droop

The term *droop* means a drift outside a programmed value. It can be applied to an engine rpm or to a road speed value. Try not to confuse this term with droop curve, which was discussed a little earlier in this chapter. Typically droop occurs as the governor or governing electronics attempts to respond to a change in conditions. For instance, road speed droop would occur when a truck running on level pavement encountered a hill. Typically droop should be transient (short term) unless intentionally programmed otherwise.

Soft Cruise

Soft or smart cruise can only be achieved on electronically managed engines. Because this involves understanding some basics of vehicle computer-controlled chassis, this term is explained more fully later on in Chapter 30. Simply put, smart or soft cruise is a means that road speed droop can be programmed to benefit fuel economy: Instead of attempting to maintain road speed at a programmed "hard" value, the electronics permit some drift, both higher and lower, than the set mph. Soft cruise may also use other vehicles systems such as the engine brake to manage the up-and-down droop of programmed cruise speed.

Default Governing

The term *default mode* is used to describe how the engine manages itself in the event of the loss of a key input sensor. The best example would be loss of throttle position sensor (TPS) input signal. Rather than shutting down the engine completely, the engine runs at a default speed, say 1,000 rpm, so that the vehicle can be safely maneuvered to a safe location on the highway.

Other Governor Terms

We have already introduced most of the key terms required to understand governor operation but because of the ever-increasing programmability of governing fields in ECMs, some OEMs use other governor terms. Some of the more popular are:

Deadband: term used to describe the sensitivity of a governor. It is the speed window around set speed in which fueling correction is made by the governor.

Governor cutoff: speed at which the governor cuts off fueling.

Hunting: rhythmic change in engine speed often caused by unbalanced fuel delivery in multicylinder engines.

Motoring: running an engine at zero throttle, with chassis momentum driving engine.

Overrun: the inability of a governor to keep the engine speed below the high-idle speed when it is rapidly accelerated.

Overspeed: any speed above high idle.

Road speed governing (RSG): any governor system in which engine fueling is moderated by a predetermined road speed value.

Sensitivity: ability to respond to maintain a set rpm without rpm fluctuation as load changes.

Speed drift: in which engine speed rises above or below the set speed, often in surges; differentiated from hunting by the fact that it is not rhythmic.

Stability: ability to maintain a set rpm.

Underrun: a governor's inability to maintain the engine low-idle speed when rpm is quickly dropped.

WOT: wide-open throttle. A term used mainly on SI automotive engines and less often on diesels; it means *full-fuel request*.

REVIEW QUESTIONS

1. Rhythmic fluctuation in engine rpm is known as:
   a. Governor overrun
   b. Governor underrun
   c. Dieseling
   d. Hunting
2. An engine run at a speed exceeding rated, but less than high-idle, speed would be operating:
   a. In the torque rise profile
   b. At peak torque
   c. In the droop curve
   d. In an overrun condition
3. Which type of hydromechanical governor would most commonly be used to manage a highway truck or bus diesel engine?
   a. Mechanical, limiting speed
   b. Hydraulic, limiting speed
   c. Mechanical, variable speed
   d. Hydraulic, variable speed
4. Speed sensing in most truck diesel engine mechanical governors is performed by:
   a. A pulse wheel
   b. Matching centrifugal force to spring force
   c. Precise determination of fuel pressure
5. A mechanical, limiting speed governor has a fatigued governor spring. What would be the more likely consequence?
   a. Erratic low-idle speed
   b. Hunting
   c. Lower high-idle speed
   d. Engine runaway
6. Under which of the following conditions would a properly functioning mechanical governor on a truck or bus diesel engine *not* be capable of controlling engine rpm?
   a. Loaded, running downhill
   b. Unloaded, running level
   c. Loaded, running level
   d. Unloaded, running uphill
7. Which of the following governors would sometimes be known as an *automotive governor*?
   a. Hydraulic, servo type
   b. Hydraulic, nonservo
   c. Mechanical LS
   d. Mechanical VS
8. When an engine is rapidly accelerated and momentarily exceeds the nominal high-idle speed, the condition is known as:
   a. Runout
   b. Governor cutoff
   c. Speed drift
   d. Overrun
9. *Technician A* states that governor rated speed and governor high-idle speed should occur at the same rpm in a truck engine hydromechanical governor. *Technician B* states that governor droop curve is typically between 5% and 20% in truck hydromechanical engines. Who is correct?
   a. Technician A only
   b. Technician B only
   c. Both A and B
   d. Neither A nor B
10. *Technician A* states that a limiting speed governor is designed to produce accelerator response close to that of an automobile throttle. *Technician B* states that a variable speed governor is capable of increasing and decreasing fuel quantity per cycle without moving the accelerator pedal. Who is correct?
    a. Technician A only
    b. Technician B only
    c. Both A and B
    d. Neither A nor B
11. *Technician A* states that a variable speed governor provides zero droop performance. *Technician B* states that a variable speed governor usually provides faster response to a load change than an isochronous governor. Who is correct?
    a. Technician A only
    b. Technician B only
    c. Both A and B
    d. Neither A nor B
12. *Technician A* states that an isochronous governor could be used on a vehicle engine providing it had a fluid clutch. *Technician B* states that an isochronous governor is normally used to manage a genset. Who is correct?
    a. Technician A only
    b. Technician B only
    c. Both A and B
    d. Neither A nor B

13. *Technician A* states that a high-speed diesel engine given unlimited fuel will accelerate to self-destruction. *Technician B* states that an engine fueled on engine oil can also run away to self-destruction. Who is correct?
    a. Technician A only
    b. Technician B only
    c. Both A and B
    d. Neither A nor B
14. *Technician A* states that most computer-managed engines still use hydromechanical governors. *Technician B* states that electronic governors can be programmed to function in either LS or VS modes. Who is correct?
    a. Technician A only
    b. Technician B only
    c. Both A and B
    d. Neither A nor B
15. *Technician A* states that when a governor high-idle speed is increased, an increase in brake horsepower is a more likely result than a decrease. *Technician B* states that drivers are more likely to complain about a lack of power on an engine with a short torque rise profile than one with high torque rise. Who is correct?
    a. Technician A only
    b. Technician B only
    c. Both A and B
    d. Neither A nor B

26

Biodiesel, Alternate Fuels, and Hybrid Drives

INTRODUCTION

In the first edition of this book in a section on alternate fuels, it was suggested that diesel fuel might not be around in the foreseeable future as a highway truck fuel. We could say that again, but it is just as unlikely now as it was then that diesel fuel could be replaced as the primary fuel used by commercial vehicles in anything less than 10 years after the innovation of a viable technology. Currently, there are no alternate fuel technologies on the horizon that could compete with diesel fuel. This fact means that changes will focus on the fuel itself and the way diesel engines combust it. However, it is worthwhile to look at potential challengers to diesel fuel, some of which are actually used and others that are just talked about. Most of the alternate fuel technologies are marginal at best and do not pose any immediate threat to the diesel engine. In order to accurately assess alternate fuel technologies as they are outlined in this chapter, the student might also reference Chapter 41, which addresses diesel emission in some detail.

The most unlikely scenario is that some revolutionary power plant technology will replace the diesel engine in the immediate future. The diesel engine has been around for a hundred or so years and is supported by a massive engineering, production, fueling, and repair infrastructure that would take years to replace. Although the use of diesel fuel in some jurisdictions is statutorily limited, usually within the limits of some especially polluted city, the consequence has been to put the alternate fuel (compressed natural gas [CNG]/propane/alcohol based) under the microscope, and the emissions and performance results have not been as good as expected.

And despite all of the recent talk about fuel cells, these have generated a lot more talk than electricity, it is unlikely that the fuel cell will ever be a viable technology to replace the diesel engines in our commercial trucks. It is more likely to be developed as a transit bus power plant. In the next decade or so, we are likely to see a continued effort led by the Environmental Protection Agency (EPA) and California Air Resources Board (CARB) to improve the chemical characteristics of diesel fuel in an effort to reduce or eliminate its more toxic combustion by-products.

For two decades, transit corporations have used gaseous and alcohol-based fuels in jurisdictions that limit the use of diesel fuel. The downside has been decreased engine life, poor fuel economy, and higher maintenance costs. These disadvantages arise partly because few engines used in this way have been designed from scratch as an "alternate" fuel engine. Instead, an existing commercial diesel engine has been modified to combust the alternate fuel. In such engines fueled by **alcohol**-based fuels, **liquefied petroleum gas (LPG)**, or **compressed natural gas (CNG)**, the fuel charge will not auto-ignite at diesel compression temperatures, so a means of igniting the fuel must be provided. The result is that either a spark ignition system has to be incorporated or a small shot of diesel fuel has to be injected as an ignition pilot. When an alcohol-based fuel is used in pumping hardware designed for diesel fuel, the low lubricity of the alcohol necessitates frequent overhauls of expensive fuel pumping and injection apparatus.

In this chapter, we look at some friendlier diesel fuels, at least from the environmental perspective, and at some of the technologies that are, or may be, able to replace the diesel engine. We also examine fuel cells, because, after all, everybody is talking about them and they are finding their way into some intracity transit buses, and they could even be used in a limited way on trucks, say, as an **auxiliary power unit (APU)**.

Auxiliary Power Units

An APU is used on a truck to provide heating, cooling, and electrical power when the engine is not running. Most current APUs are fueled using diesel fuel: This is contentious in itself. While some might argue that a diesel-fueled APU is preferable to running the main vehicle engine, others would object, including CARB (see Chapter 41), that it is just another source of harmful environmental emissions. The EPA is currently debating whether to mandate diesel particulate filters (DPFs) (see Chapter 41) for these as yet unregulated small power plants. A couple of facts emerge. APUs are required on vehicles today because of increasingly aggressive anti-idling legislation. And technologies to manage more environmentally friendly APUs represent much less challenge than those presented by the motive power plant of the truck chassis.

BIODIESEL

We begin with a fuel, **biodiesel**, that actually has a growing track record of success as a diesel fuel, largely because it combusts in the diesel engine without any significant changes having to be made to the engine. Pure biodiesel known as B100 has some real limitations because it is best suited to operate in warm climates, but it is cleaner burning than petroleum-based fuels. However, when B100 is mixed with standard diesel fuel at concentrations ranging from B20 (20% biodiesel/80% petroleum diesel) down to B2 (2% biodiesel/98% petroleum diesel), it improves its temperature-operating range while somewhat lowering harmful emissions depending on the mixture cut.

One of the factors that has increased biodiesel usage in the recent past has been the introduction of B5 and B2 fuel. While cynics might argue that this is more about optics than substance, a response might be that if every fleet changed to B2 fuel this would represent at least a 2% reduction in the dependence on imported petroleum products—with the additional bonus of at least a small decrease in noxious emissions. Operators who use B2 or B5 fuel get to put a sticker on the truck, advertising how environmentally conscientious they are. This factor, added to which some jurisdictions throw in a tax break, has led most major consumers of diesel fuel to at least consider using B fuels. Using low concentration cuts of biodiesel such as B2 and B5 avoids many of the performance problems experienced by users of higher percentage biodiesel cuts.

What is Biodiesel

In the past, biodiesel conjured images of hippies concocting fuels from McDonald's waste cooking fat—an operation that could take place in a barn or perhaps a kitchen. The definition of biodiesel is a little hazy because the term can be applied to a number of homemade fuels as well as those refined under more stringent guidelines. However, we will say here that any biodiesel that finds its way into a truck fuel system should conform to American Society of Testing and Materials (ASTM) standard D6751.

ASTM Definition

ASTM defines biodiesel as a domestic, renewable fuel for diesel engines derived from natural oils such as soybean and other vegetable oils. It is *not* raw vegetable oil: A biodiesel must be refined (processed) so that all natural glycerin is removed from the oil. The term *biodiesel* can refer to 100% vegetable-based fuel oil known as B100—or a range of blends (known as *cuts*) of biodiesel and petroleum fuel oil. For instance, B20 is petroleum-based diesel fuel blended with 20% ASTM D6751 biodiesel. Despite some reports to the contrary, using spent frying oil from fast food outlets will not be a likely large scale source of biodiesel due to the variables (read "contamination") of the base oil and the cost of refining this to a reliably performing highway fuel. In describing biodiesel in this chapter, we use ASTM standard D6751 and the U.S. Department of Energy (DOE) research rather than referencing some of the cruder do-it-yourself technologies.

Biodiesel Chemistry

The chemical composition of biodiesel is given in Table 26–1. In the table, biodiesel manufactured to ASTM D6751 is compared with a highway diesel fuel ASTM 975 manufactured to 2002 standards.

Biodiesel Characteristics

The advantage of biodiesel over petroleum-based diesel fuel is its biodegradability. However, this biodegradability is also the major disadvantage of the fuel because it results in rapid degradation for two reasons:

- Hygroscopic: Biodiesels readily attract water from atmosphere, giving it a much shorter "shelf life" than petroleum-based diesel.
- Vulnerability to microbial contamination: The problems petroleum-based diesel fuels are subject to (see Chapter 16) can be multiplied by ten times.

A chemist would describe biodiesel as a straight chain fatty acid, and as such it is a culinary delicacy for microorganisms. Given that biodiesel is also hygroscopic, it provides an ideal environment for microbes that tend to reside in the water and consume the hydrocarbon fractions of the fuel. We can summarize the features of B100 as follows:

- No aromatics or nitrogen (these can form toxins during combustion)
- Nontoxic and biodegradable
- Zero sulfur content (if ASTM D6751 standard)
- 10% less heat energy than ASTM #2 diesel fuel
- Fractionally less heat energy than ASTM #1 diesel fuel

Combusting Biodiesel

Biodiesel burns more cleanly than petroleum-based diesel fuel. Most of the harmful emissions produced by combusting petroleum diesel fuel are significantly reduced with the exception of oxides of nitrogen (NO_x). NO_x emission is slightly increased over diesel fuel because biodiesel is an oxygenate fuel; that is, like alcohol, there is an oxygen component in the fuel. Perhaps the biggest disadvantage of biodiesel is its inability to flow at colder temperatures. This

Table 26–1 *Biodiesel Compared with Highway Diesel Fuel*

Fuel Property	*Diesel (ASTM average 2D)*	*Biodiesel (B-100)*
Specific gravity	0.85	0.88
Caloric value (heat energy)	140,000 Btu/gal	118,000 Btu/gal
Density (lb per gal at 15°C)	7.079	7.328
Water percentage of volume	0.05	0.05
Carbon (by weight)	87%	77%
Hydrogen (by weight)	13%	12%
Oxygen (by weight)	0%	13%
Sulfur (by weight)	0.05%	0 to 0.0025%
Flash point	60–80°C	100–170°C
Cetane number	40 to 55	48 to 60
Pour point	–35 to –15°C	–15 to 16°C

Source: Data from U.S. DOE, Feb. 2002.

means that as a pure fuel, its primary markets are going to be in warmer climates. When biodiesel is cut by petroleum diesel it becomes a more viable fuel in cooler climates. There have been some government initiatives to popularize **B20**. Biodiesel B20 is able to retain some of the advantages of biodiesel **B100** (100% vegetable base) and is probably more practical as a future commercial vehicle fuel. B20 and B100 are compared in Table 26–2.

Table 26–2 *Biodiesel Emissions Compared with EPA-Certified Diesel Fuel*

Emission	*B100*	*B20*
Carbon monoxide	–43%	–12%
Hydrocarbons	–56%	–11%
Nitrogen oxides	+6%	+1.2%
Air toxics	–80 to –90%	–20%

Source: Data from U.S. DOE, Feb. 2002.

Costs and Availability

Because reduced-cut biodiesels such as B5 and B2 have been recently introduced, their use has increased. This increased usage has been aided by steep increases in the cost of petroleum. Combine this effort with EPA efforts to make highway fuel cleaner (therefore more expensive!), reducing the cost differential, and the future for biodiesel may be bright. For the time being, the DOE suggests that it is best to avoid any biodiesel that does not meet the ASTM D6751 standard.

Color and Odor

Biodiesel and biodiesel cuts can meet ASTM D6751 and D975 while having a variety of colors and odors. Examples of some acceptable colors are orange, red, yellow, brown, and green. These fuels can also have a range of odors, so fuels can seldom be condemned on the basis of odor alone. However, fuel that has a distinct "septic tank" odor is a common indicator of bacterially degraded fuel.

Biocides

Biocides can be added to biodiesel to kill bacteria in the same way antibiotics seek out and destroy bacteria in the human body, so you might ask why biocides are not added to all biodiesel fuel stocks. This is not done for the same reason that antibiotics are not routinely taken by humans. Bacteria have an uncanny way of developing a resistance to any substances humans develop to kill them. When a biodiesel fuel comes under attack by bacteria, shocking the system by dosing the fuel with biocides may rescue the fuel before it becomes chemically depleted to the point that it can no longer be used.

Power and Fuel Economy

Using B100 in place of #2 (ASTM standard 975) diesel fuel will reduce power by about 10%. Put simply, there is less heat energy in biodiesel; therefore, more of it has to be used to accomplish the work equivalent to a similar volume of petroleum fuel. For every gallon of diesel fuel consumed, approximately 1.1 gallons of B100 will have to be consumed. The difference is obviously much less noticeable when B20 and lower concentration cuts are used.

Low-Temperature Operability

Even when biodiesels meet the ASTM 6751 standard, they have some unreliability when it comes to defining the cloud point of the fuel. In Chapter 16 we described cloud point of a fuel, which is the temperature at which it begins to form wax crystals large enough to start plugging a fuel filter. This unreliability has given biodiesel some bad press in colder parts of North America. Other factors that impact on low-temperature operability are:

- Filter location on chassis
- Low porosity (high efficiency) filters
- Absence of a block heater
- Prolonged shutdown periods

Biodiesel Conclusions

The use of biodiesel will increase because there are enough very good reasons to justify its use. Not the least of these reasons is a 2007 EPA requirement that by 2012, the equivalent of 7.5 billion gallons of agricultural-source fuels must be blended into vehicle fuel sold in the United States: Not clear is what percentage of this requirement will apply to biodiesel blended to diesel fuel (the remainder will be ethanol cut into gasoline). One major problem at present is that the ASTM has structured standards for B100 (ASTM 6751) and ASTM #1 and #2 petroleum diesel fuels (ASTM D975); there is no standard for any of the reduced concentration cuts. It should be emphasized that any biodiesel cut (B20, B5, B2) is only as good as the B100 it was sourced from. When experiencing problems with biodiesel blends there is currently no standard that the fuel supplier can be held accountable to: This makes it almost impossible to attribute blame for a bad fuel. In most cases, when a biodiesel cut fails to perform, the fuel becomes useless and should be recycled. Attempting to rectify the problem by adding kerosene may sometimes work but this practice results in a nonapproved fuel.

ELECTRICITY

Most of the automobile and some of the bus OEMs are producing an increasing number of electrically powered and hybrid or dual-powered vehicles. Electric motors have powered vehicles such as subway trains and streetcars since the early years of the last century. When electricity can directly power drive motors on vehicles they are capable of thermal efficiencies of up to 60%. However, in the absence of a rail or other conduit to provide a source of direct electrical energy to a vehicle, when electrical power is used, it has to be produced on the vehicle. This is a significant limiting factor in the use of electrically powered vehicles. Electrically powered vehicles can be categorized as follows.

Battery Powered

The mass and the space occupied by the battery packs are just the beginning of the problems of battery driven, electrically powered vehicles. The challenge to industry is to be able to build batteries of much larger storage capacity combined with much lower weights. Nevertheless, cars, vans, buses, and light trucks powered by nickel metal hybrid (NiMH) battery packs are being manufactured in limited numbers, mostly for intracity use. A few years back, the state of California pushed the electric vehicle concept, but this has mostly gone by the wayside, though some of the vehicle-charging stations remain. At present most battery-powered vehicles are used in applications such as golf carts and forklift trucks, but this could change in a hurry as industry looks at improved electrical energy storage systems. Some cities are experimenting with all electric transit buses but to this point they have produced poor results due to a high incidence of on-the-road breakdowns.

Evolution of Electrical Storage Devices

Developments in NiMH and other high-energy density electrical storage systems are hastening the day when we commonly accept battery-powered commercial vehicles. Recent developments in **ultracapacitors** that function like a pair of capacitors in series can be used for short-term storage of electrical potential. They are currently used in conjunction with battery storage systems in some electrically powered vehicles but their use is expected to expand. Other battery technologies undergoing vehicle testing are cobalt-based, lithium-ion (Li-ion), lithium polymer (Li-Po), and lithium metal polymer (LiMePo).

Electric Motor, Gas Turbine Driven

The thrust toward higher efficiency power plants has been driven by the Europeans who have the most motivation for increasing fuel efficiency, namely, astronomically high fuel prices. They are engineering and manufacturing long-range, light vehicles that use a super efficient gas turbine fueled with kerosene to drive an electric motor. Combusting kerosene has a tendency to produce high NO_x so the key to expanding this technology is in using small, highly efficient turbines.

Electric Motor, Hard Wire Powered

Electric motors have been used for decades to power trolley buses and streetcars, these power plants have obvious limitations and will probably never be a viable truck fueling technology. Electrically powering large commercial vehicles is not generally as environmentally friendly as some would like to believe, depending on the means used to generate the electricity. Fossil fuels and nuclear reactors are used to produce electricity in many parts of North America and, even if the source is hydroelectric, some would argue that there is an ecological cost.

Fuel Cells

Fuel cells generate electricity and we focus on how they operate later in this chapter. They have been appearing on our roads in limited numbers as city bus power plants, and some OEMs option a fuel cell APU on certain trucks. Fuel cells can produce thermal efficiencies of up to 50%.

Parallel Hybrid, Diesel Electric

Dual power plant or **parallel hybrid (PH)** vehicles overcome some of the disadvantages, and exploit the advantages, of the electric motor as a vehicle engine. This is achieved by using a traditional internal combustion engine as a generator to power an electric motor, and the simplest possible PH drive arrangement is shown in Figure 26–1. We usually describe these as **hybrid electric vehicles (HEVs)**. Several design variations are available in the marketplace, and some large cities have invested heavily in parallel hybrid city buses led by New York City's Metropolitan Transportation Authority (NYC-MTA) and Toronto's Transit Corporation (TTC), which between them now run over 1,000 PH buses. The success of the New York and Toronto ventures into parallel hybrid, diesel electric technology will almost

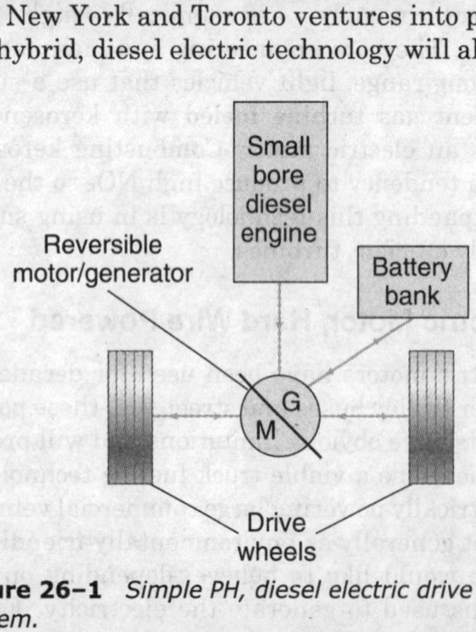

Figure 26–1 *Simple PH, diesel electric drive system.*

certainly spread to other major cities, many of which are currently undertaking HEV trials. Our focus in this section is on this particular HEV system because it has already achieved a successful track record.

How It Works

A parallel hybrid drive system uses two sources of power to move a vehicle. In the case of the HEV buses we want to reference, the two power sources are a diesel engine and a battery bank. The diesel engine/generator unit in an HEV is coupled to provide two outputs:

- Generate electrical energy to maintain charge to the batteries
- Deliver continuously variable drive to the drive axle wheels

The objective is to allow the diesel engine to run at optimum efficiency. In practice, as an HEV bus accelerates from a standstill, the batteries power a high-torque electric drive motor off the line. As the bus accelerates to its cruise speed, drive is provided by a blending of mechanical and electric motor sourced torque. Once at the cruise speed, the light-duty diesel engine may take over as the drive source but also continues to generate charge to the batteries. In other words, there are three power delivery paths:

- Pure electrical path: batteries power electric motors
- Torque blending: diesel engine and electric motor blended torque
- Pure mechanical path: diesel engine direct drive

HEV Powertrain

At the moment of writing, most HEV buses use a powertrain consisting of a Cummins ISB diesel engine typically rated at 275 BHP to provide mechanical drive or generate electrical power supplied to three-phase AC electric motors. On its own, the Cummins ISB 275 engine (shown in Figure 26–2) would have insufficient power and torque to drive a full-size city bus, but in a PH arrangement it becomes an ideal power source. The key to making the HEV powertrain work as a package is the **blended torque transmission**.

Figure 26–2 *Cummins ISB 6.7 liter engine: common power plant used in HEV drivetrains. (Courtesy of Cummins Inc.)*

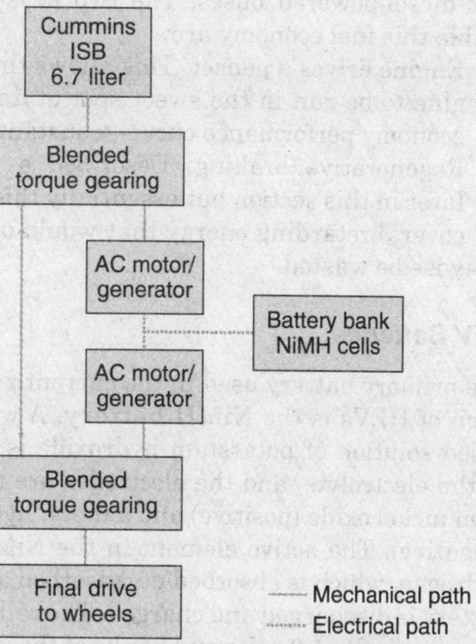

Figure 26–3 *Mechanical and electrical powerflows in a transit bus PH drivetrain.*

Blended Torque Transmission In describing HEV transmissions we are going to assume that the reader fully understands the operation of a planetary gearset (see Chapter 17 of Heavy Duty Truck Systems 4E). The HEV transmission uses a two-mode, split compound principle that provides continuously variable torque (CVT) and speed. In appearance, it is not unlike a typical heavy-duty automatic transmission. To understand the blended torque feature of the HEV transmission we have to first change some notions we have about the "laws of planetary gearsets," namely, that to produce an output, you have to hold one of the three members (sun, carrier, or ring gears) and deliver input torque to another member, so that the third can become the output. Hybrid drive requires two input sources to the planetaries, and that one of those "inputs," namely, the electric motor/generator, can also double in the role of an "output." Figure 26–3 shows the arrangement of powertrain components of a transit bus PH drive system along with the mechanical and electrical powerflows.

Blended torque takes this concept one stage further. Instead of holding one of the three planetary members, why not "partially" hold it. Doing this permits the gearset to produce an output from two input sources, each contributing at different levels of torque. It is this feature that provides the infinitely variable torque output that we know as blended torque. We can summarize the blended torque output principle in the following way:

- The two input devices rotate at different shaft speeds and generate a complementary output based on the input torque of each.

HEV Performance Bus drivers will tell you that HEV buses produce better off-the-line acceleration than any conventional diesel-powered bus they have driven. This makes them ideal people movers on busy city streets with a high frequency of stop-start events. The puff of black smoke and engine roar at each takeoff is eliminated, providing benefits to those not riding the bus in addition to a much smoother ride for the driver and passengers. The Cummins ISB engines used in current applications are equipped with a diesel particulate filter and two-stage catalytic converter. Testing with the MTA in New York City has shown a 50% reduction in NO_x, 98% less CO, and better than 200% improved fuel economy than equivalent diesel-powered buses. The two keys that enable this fuel economy are:

- Engine drives a genset: This allows the engine to be run in the sweet spot of its fuel economy performance curve—constantly.
- Regenerative braking: Described a little later in this section but essentially this "recovers" retarding energy that would otherwise be wasted.

HEV Batteries

The primary battery used in the current generation of HEVs is the **NiMH battery**. A water-based solution of potassium hydroxide is used as the electrolyte, and the electrodes are made from nickel oxide (positive) and a metal hydride (negative). The active element in the NiMH is hydrogen, which is absorbed/de-adsorbed as the battery is discharged and charged. The cell voltage is typically 1.2 volts and banks of these batteries are used in HEV buses. NiMH batteries have good energy density, sustain deep cycling, and are environmentally friendly in that they contain no lead, mercury, or cadmium. Even more important, they are one-third the weight and half the size of equivalent lead acid batteries. They are disadvantaged by having a high self-discharge rate but this is generally not a problem in a transit vehicle that is used daily. Figure 26–4 shows the location of all of the mechanical and electrical powertrain components on a transit bus.

Regenerative Braking

A bonus of a diesel electric hybrid drive is achieved with **regenerative braking**. When the drive electric motor magnetic field is reversed, it applies retarding torque and generates electricity used to charge the batteries. Regenerative braking significantly contributes to the energy equation in a city bus application featuring constant stop-starts, because somewhere around 40% of the power required to accelerate a bus to a cruise speed comes from the energy "recovered" from stopping it.

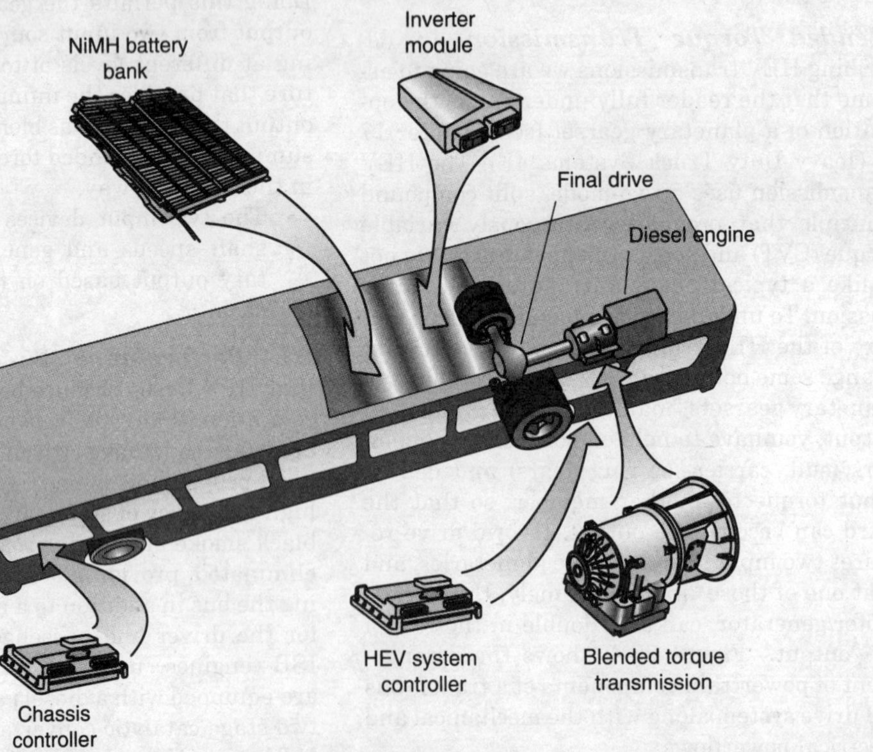

Figure 26–4 *Location of HEV powertrain components on a transit bus.*

Drive Motor/Generators

The drive motors used are synchronous alternating current (AC) electric motors that must be precisely phased. The drivetrain powerflow when operating under electric drive alone is engine-to-generator-to-battery-to-electric motor. Because batteries are direct current (DC) storage devices and the motors used are AC motor/generators, the electrical characteristics to drive or effect regenerative braking must undergo cycles of rectification and inversion. These are managed by the system controller.

Inverters The inverters used in HEV buses tend to be their most vulnerable component because they produce large amounts of heat, which has to be dissipated or the unit will fail. A typical inverter module is a bulky, heavy unit weighing up to 75 kg (165 lb). They are rated at a nominal 600 V-DC with 150 kW continuous three-phase AC output. The units are oil cooled.

System Controller The system controller used on commercial HEVs is networked to the J1939 data bus (see Chapter 33) and masters the operation of all system hardware. Operation of system controllers is the subject of Chapter 30.

HEV Summary

Diesel electric drive is not a new. It has been used for many decades on railway locomotives. The HEV system we have discussed here simply takes this one stage further to exploit the best features of the diesel engine mechanical and electric motor drive options. Many experts believe that this will become the drive system of choice for city transit over the next decade. It is also has plenty of potential for certain types of vocational truck used in stop-start applications.

HYDRAULIC HYBRIDS

Parallel hybrid hydraulic (PHH) drive systems began with an Eaton Corporation collaboration with the United States Army aimed at reducing fuel consumption in military ground vehicles: The program was known as **hydraulic hybrid, advanced materials and multifuel engine research (HAMMER)**. With the military trials in progress, Eaton developed a version for nonmilitary purposes known as **hydraulic launch assist (HLA)** intended to power small intracity vehicles such as courier vans and shuttle buses. In addition, a **series hybrid hydraulic (SHH)** drive, which is a hydraulic vehicle propulsion system, is being used in similar applications.

- PHH: Hydraulic circuit "assists" a conventional drivetrain.
- SHH: Hydraulic circuit replaces the conventional drivetrain.

HLA Operating Principles

The Eaton HLA is a PHH drive system. It functions by using a conventional diesel-driven powertrain assisted by a hydraulic system. The hydraulic system consists of the following:

- Reversible hydraulic piston pump/motor
- Hydraulic pump/motor clutch coupling to the conventional drivetrain
- High-pressure accumulator
- Reservoir and circuit plumbing

In the first available commercial versions of HLA, a Ford (Navistar International) 6-liter PowerStroke engine rated at 235 horsepower was used to master the powertrain. Much lower fuel consumption is achieved by regenerative braking and an accumulator and by managing the powertrain so that the engine shuts down, eliminating idling. When the vehicle brake pedal is actuated, deceleration energy converts the pump/motor into its "pump" role, so that it charges hydraulic fluid from the reservoir into a high-pressure accumulator. This "potential energy" is then stored in the nitrogen gas-loaded high-pressure accumulator.

During acceleration, the role of the pump/motor is reversed so that the stored energy in the high-pressure accumulator is used as the driving force of the hydraulic motor: This allows the pump/motor to apply torque to the drivetrain by means of the HLA clutch. The nitrogen gas-loaded accumulator has a capacity of 10 liters (2.6 gallons) and is capable of storing hydraulic pressures of up to 345 bar (5,000 psi). Figure 26–5 is a schematic that shows the powerflow arrangement used in a PHH system equipped with an accumulator.

A major advantage of the Eaton HLA system is greatly improved power density enabled by

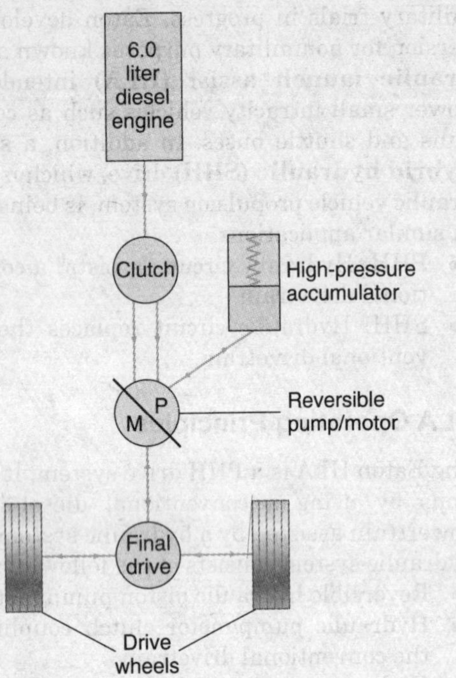

Figure 26–5 *Parallel hydraulic hybrid (PHH) system capable of mechanical or hydraulic drive-assist and regenerative braking.*

the hydraulic accumulator. A Navistar 6.0 PowerStroke diesel typically produces 235 BHP and around 440 lb-ft. of torque. HLA boosts the conventional drivetrain output to around 300 BHP and nearly 1,000 lb-ft. of torque.

SHH Operating Principles

In the SHH drive system, hydraulics are used as the vehicle propulsion system. In both the United States Army and commercial vehicle versions of SHH, the engine is a Navistar 6.0-liter PowerStroke. However, in SHH the conventional drivetrain consisting of a transmission, drive shaft, and final drive carrier is dispensed with. The diesel engine transfers drive torque to a reversible pump/motor that can both power the vehicle and recycle regenerative braking effort using an accumulator. SHH is said to achieve up to a 70% improvement in fuel economy. This can be achieved because:

- The diesel engine power plant can be run at a constant, optimum speed during all phases of operation.
- Regenerative braking energy is unloaded into an accumulator where it can be stored until required.

A stated disadvantage of the SHH system over a PHH system is that it lacks redundancy. However, so does any other type of series drive system; therefore, this is really not that significant, especially in view of the potential advantages of the technology. Series hydraulic propulsion has been used by the off-highway equipment segment for decades. SHH promises:

- Significantly greater engine longevity
- Reduced maintenance costs due to savings in brake serving and powertrain components
- 60 to 70% lower fuel costs
- Lower emissions, including 40% reduction in carbon dioxide
- Quieter vehicle noise and reduced operator fatigue

Figure 26–6 shows an SHH drive system that features hydraulic propulsion and regenerative braking.

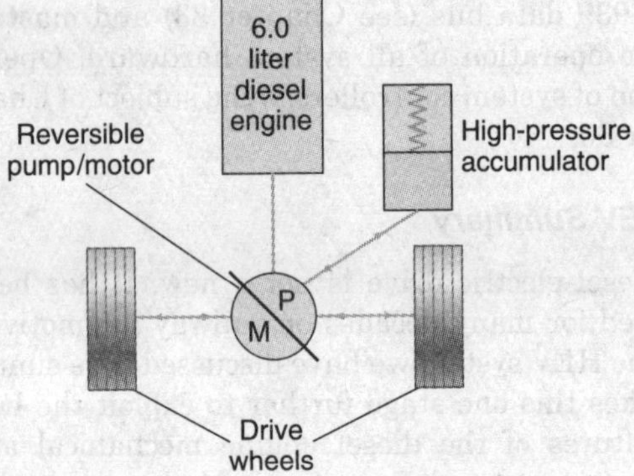

Figure 26–6 *Series hydraulic hybrid (SHH) drive featuring hydraulic propulsion and regenerative braking.*

FUEL CELLS

A fuel cell is an electrochemical device in which the energy of a chemical reaction is converted into electricity. It does nothing more than produce electricity. If a vehicle is to be powered by a fuel cell, the fuel cell simply produces electricity, and electric motors are used to drive the vehicle. Fuel cells are not new, having been invented by Sir William Grove in 1839. After their invention, not a whole lot was done with fuel cells until the National Aeronautics and Space Administration (NASA) used them to provide some of the electrical requirements of its spacecraft in the 1960s. Figure 26–7 shows the power equation of a fuel cell.

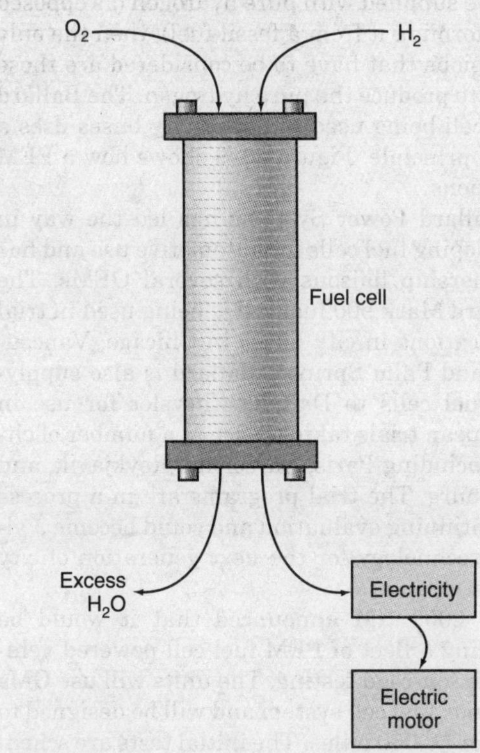

Figure 26–7 *Fuel cell power equation.*

Fuel cells are not "free" energy as some have described them; they need **hydrogen** to function, which creates problems. Commercial hydrogen is usually produced electrolytically from water, which is a process that requires large amounts of electricity. Further, safe storage of highly explosive hydrogen on a vehicle is not easy. The option is to use fuel reforming to extract the hydrogen from hydrocarbon (HC) fuels such as gasoline, natural gas, or alcohols. The fuel cells themselves do not discharge like a battery; they will run as long as supplied with hydrogen fuel. The claim that fuel cells are zero-emission energy producers is only true when the fuel provided is pure hydrogen.

Reforming Hydrogen

Most current fuel cells used in small vehicles require an HC fuel such as gasoline as the hydrogen source, using a reforming process. Reforming is the separating of H_2 from the HC fraction molecule in the gasoline or other HC fuel. When a fuel cell is fueled with reformed hydrogen, the results are not as environmentally friendly as some might claim because the carbon by-product of the process has to be considered.

So although hydrogen is the most abundant element in the known universe, to produce it on Earth either takes lots of electricity (the production of which usually causes noxious emissions) or it has to be reformed from a fossil fuel, the greater percentage of which is carbon. These factors are likely to make the fuel cell a very limited mobile technology in the immediate future.

How a Fuel Cell Works

A fuel cell functions on the principle of the thermodynamic reversibility of the electrolysis of water. For many years we have produced hydrogen and oxygen by the electrolysis of water, and, as mentioned before, it does take a lot of electricity. When supplying hydrogen and oxygen to the two electrodes of an electrolysis cell, a potential difference is created, and electric current begins to flow. Join several cells together and you create a multicell fuel cell. The electrochemical reaction is:

$$2H_2 + O_2 \text{ produces } 2H_2O + \text{electricity}$$

Figure 26–8 demonstrates the operating principle of a fuel cell: Hydrogen and oxygen are electrochemically reacted to produce water and electricity.

Types of Fuel Cells

There are a number of different types of fuel cells. We look briefly at some of them and in more detail at one currently being tested in city buses that may perhaps have a future as a short-range, small truck power plant. Each type of fuel cell is distinguished by the electrolyte material used.

Alkaline Fuel Cells

Alkaline fuel cells have been used by NASA in the Apollo and space shuttle programs. To function they must be provided with pure hydrogen

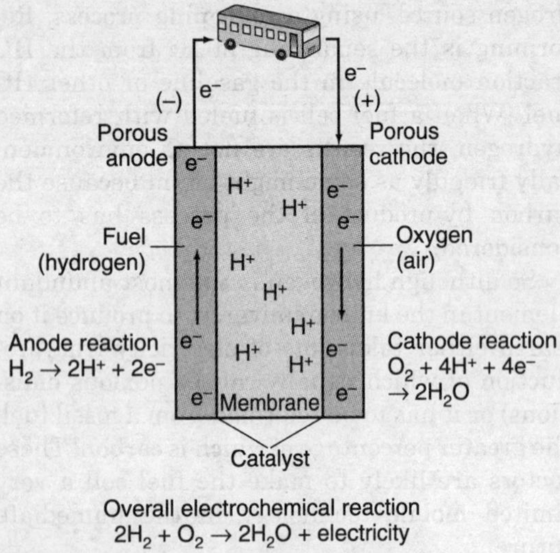

Figure 26–8 *Fuel cell chemical reaction.*

and pure oxygen, substances that are stored and readily available on a spacecraft, but difficult to store on a truck or automobile.

Phosphoric Acid Fuel Cells

Phosphoric acid fuel cells are suitable only in stationary power generation applications due to the corrosive nature of the liquid electrolyte and high-operating temperatures, usually in the range of 200°C (392°F). They have been used as electrical generators in large buildings, such as hospitals and hotels, with a need for a power source independent from the grid.

Solid Oxide Fuel Cells

Solid oxide fuel cells operate at the highest temperatures, between 700° (1,292°F) and 1,000°C (1,832°F), but because of this, they will tolerate relatively impure fuels such as the HC gas obtained from the gasification of coal. These fuel cells have a relatively simple design and are thought to be most suitable for large stationary power generators. It is unlikely that there would be any mobile vehicle application in the near future.

Proton Exchange Membrane Fuel Cells

Proton exchange membrane (PEM) fuel cells have created some excitement in the automotive industry because they are the most viable mobile fuel cell technology. The PEM fuel cell has been used as a primary power source in some vehicles, and Freightliner chose this technology for its optional APU unit. A PEM fuel cell uses a solid polymer membrane (a thin plastic film) as the electrolyte. If the fuel cell can be supplied with pure hydrogen (as opposed to reforming it from a fossil fuel), then the only emissions that have to be considered are those used to produce the pure hydrogen. The Ballard fuel cell being used to power city buses uses a PEM principle. Figure 26–9 shows how a PEM functions.

Ballard Power Systems has led the way in developing fuel cells for automotive use and has partnership liaisons with several OEMs. The Ballard Mark 900 fuel cell is being used in trial applications in city buses in Chicago, Vancouver, and Palm Springs. Ballard is also supplying fuel cells to Daimler-Chrysler for use in European trials taking place in a number of cities, including Paris, Barcelona, Reykjavik, and Hamburg. The trial programs are in a process of continuing evaluation and could become a viable technology for the next generation of city buses.

In 2007 GM announced that it would be building a fleet of PEM fuel cell-powered vehicles for on-road testing. The units will use GMs Equinox fuel cell system and will be designed to run for 50,000 miles. The initial tests are scheduled to take place in the Los Angeles, New York City, and Washington, D.C., urban areas.

FUEL ALTERNATIVES FOR ENGINES

Generally, where engines and fueling apparatus have been successfully adapted for fueling with any of the alternatives to diesel fuel, their use is confined to short-haul and intracity transit applications. The thrust toward maximizing engine efficiency and finding alternatives to diesel fuel and gasoline has taken a different path in North America than in Japan and Western Europe. The objective of lower reliance on imported fossil fuels has driven development in Japan and Western Europe due to the much higher cost of fossil fuels in those geographic areas. It should be stated that a large portion of this high cost is a result of government taxation of fuels.

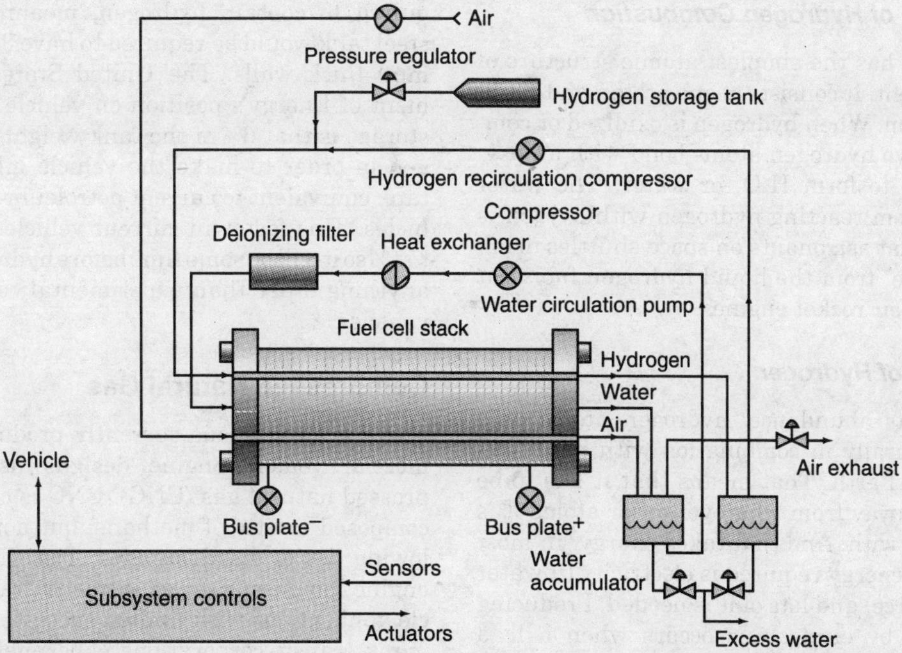

Figure 26–9 *PEM fuel cell principle.*

Development of low-emissions engines has been the primary objective of North American OEMs, which are driven by clean air legislation. We now look at the characteristics of some of the possible alternatives to the gasoline and diesel fuel we use today. Table 26–3 shows the energy density of some alternative fuels: Note that heat energy by volume in diesel fuel exceeds that in the alternatives by a considerable margin.

Table 26–3 *Comparisons of Heat Energy in Different Fuels*

Fuel	*Typical BTUS per U.S. Gallon*
ASTM diesel fuel average	140,000 Btus
Average pump gasoline	125,000 Btus
CNG	75,000 Btus
Alcohol (methanol M-100)	76,000 Btus
Propane	93,000 Btus

Hydrogen

Hydrogen could be the fuel of the future. It is the most abundant element in the universe and it oxidizes (combusts) producing no noxious emissions. There are no current commercial vehicle engine designs in production, but most of the major manufacturers have experimented with using hydrogen as a fuel. If you read the earlier section in this chapter on fuel cells, you know that the uses of hydrogen are not limited to burning it; it also can be used to produce electricity using fuel cells.

A rocket engine runs on hydrogen fuel reacted with oxygen. When you observe a rocket launch, the clouds of white smoke you see emitted are composed of vaporized water or steam. Combusted hydrogen combines with oxygen to form water (H_2O), so if the oxygen is pure, the result is a zero-toxic emissions engine. However, if hydrogen is reacted with the oxygen in air, then NO_x can result from the combustion reaction. Because of its explosiveness and difficulty in the storage/refueling process, hydrogen is unlikely to be combusted in mobile engines in the near future. For instance, a hydrogen tank similar in size to one 100-gallon truck diesel fuel tank would provide the truck with only enough fuel for a single 100-mile trip. If hydrogen is ever to be adopted as commercial vehicle fuel, it will more likely be to fuel advanced versions of the fuel cells described in the previous section.

Products of Hydrogen Combustion

Hydrogen has the simplest atomic structure of any element. It consists of one proton and a single electron. When hydrogen is oxidized or combusted, two hydrogen atoms bond with an oxygen atom to form H_2O, or water. The water formed from reacting hydrogen with oxygen is so pure that astronauts on space shuttles drink the "waste" from the liquid hydrogen fuel that powers their rocket engines.

Sources of Hydrogen

Despite its abundance, hydrogen atoms only exist naturally in combination with other elements on Earth. That means that it has to be stripped away from whatever other atoms it is combined with. And that takes energy. In most cases the energy required is electricity (if water is the source) and lots of it is needed. Producing hydrogen by electrolysis occurs when a lead acid battery is overcharged.

Advantages

High heating value, three times that of diesel fuel
Clean burning—oxidizes to H_2O (water)
Lightweight
Abundant

Disadvantages

Expensive to produce
Low boiling point –423°F (–252°C)
Difficult to store safely on a vehicle
Highly explosive

Current Uses of Hydrogen-Fueled Engines

Some experimentation has been undertaken by auto OEMs using hydrogen to fuel adapted versions of their gasoline-fueled vehicles. In the commercial vehicle arena, the Ford Motor Company has recently begun producing hydrogen-fueled V-10 engines installed in shuttle buses. The first of these entered service in Florida during 2007 and are currently under evaluation.

Hydrogen Storage

While hydrogen is in some ways an ideal fuel, one of the biggest challenges is how to store it on a vehicle. A combination of high pressure and low temperature (boils at –252°C) is required to contain hydrogen, meaning that a steel tank would be required to have 3-inch (750 mm) thick walls. The United States Department of Energy's position on vehicle hydrogen storage is that 6% of the tank weight be hydrogen in order to make the vehicle mileage per tank equivalent to current petroleum-fueled vehicles. This factor in current vehicles is closer to 1% so it will be some time before hydrogen fuels anything other than experimental commercial vehicles.

Compressed Natural Gas

There are numerous currently produced, commercial vehicle engine designs using compressed natural gas (CNG). CNG is chemically composed mostly of methane, but note the following list of disadvantages. The CNG-fueled engine has been used in intracity transit vehicle applications with limited success. As a fuel, many transit corporations experimented with CNG in the early 1990s, but most did not like the results. These trials continue in transit operations but it is unlikely that CNG will replace diesel fuel as the fuel of the twenty-first century. Often low-pressure, plumbed city gas is compressed into vehicle tanks, but this procedure is usually slow and can take a number of hours.

A number of established diesel engines such as the Cummins L10, Mack E7, and Detroit Diesel Corporation (DDC) Series 50 have CNG counterparts, but these tend to be most used where a local jurisdiction has banned the use of diesel fuel. Part of the problem with using CNG as a fuel for commercial engines is that no engines have been designed specifically for CNG. In most cases, diesel engines have been adapted to burn it. CNG does not autoignite as does diesel fuel, so a means of igniting the fuel charge has to be provided. This characteristic means that either a spark ignition (SI) system or diesel pilot must be used. In most cases, an SI system is used.

Advantages

Nontoxic
Abundant with good distribution infrastructure
Technology already exists for use as a vehicle fuel

High octane (permits a high-compression ratio)
Relatively clean burning—less HC, CO, and NO_x

Disadvantages
Low calorific value compared with diesel fuel
Bulky storage tanks of large volume
Reduced peak power performance
Refueling can be a problem—very slow
Range limitations
Electronic or diesel pilot ignition system required

Liquefied Petroleum Gas

LPG (chemically composed mostly of methane) or **propane** is a by-product of the process of extracting crude petroleum from the ground. As with CNG, there are numerous existing proven designs, but none of them have prevailed beyond some limited, usually intracity, light-duty, commercial vehicle designs. Propane does not autoignite under compression pressures so, as with CNG, when used to fuel an engine designed as a diesel, a means of igniting the fuel charge must be provided. Ignition can be by spark or pilot pulse.

Advantages
Equals fuel weight-to-mileage range of gasoline
Technology already exists for use as a vehicle fuel
High octane
Clean burning—emits less HC, CO, and NO_x
Increased engine longevity

Disadvantages
Heavier than diesel fuel by 20%
Must be stored at –260°F (–162°C) or less to remain liquid
Lower heating value than diesel fuel
Reduced peak power performance
Electronic or diesel pilot ignition system required

Alcohol-based Fuels (Methanol/Ethanol)

There are some current alcohol-fueled engines used in commercial vehicles, mostly intracity buses. Engine longevity can be drastically reduced when alcohol fuels are used in existing diesel engines adapted for alcohol fueling because of the low lubricity of alcohol. Its use in commercial vehicles has been confined to short-haul operations within cities where local jurisdictions have banned or discouraged the use of diesel fuel.

Advantages
Produced from natural gas, coal, or vegetation (garbage)
Adapts readily to current liquid fuel delivery technologies
High octane rating—up to 110
Used in current bus diesel applications
Reduced HC emissions
Technology exists for use in diesel engines

Disadvantages
Half the calorific value of diesel fuel
Low lubricity (a disadvantage for use in current diesel engines)
Low cetane number (CN)—requires spark or piloted ignition
Poor cold weather performance (does not readily vaporize)
Higher acidity promotes corrosion and lube oil breakdown

REVIEW QUESTIONS

1. Which of the following is a correct description of a B5 fuel?
   a. 95% biodiesel cut with 5% petroleum-based diesel
   b. 95% petroleum-based diesel cut with 5% biodiesel
   c. Biodiesel rated with a CN of 50
   d. Petroleum-based diesel with an ignition temperature of 500°C.
2. What would a B100 fuel be composed of?
   a. 100% petroleum-based diesel fuel
   b. 100% alcohol-based stock
   c. Even mixture of petroleum- and vegetable-based stock
   d. 100% vegetable-based stock
3. Which of the following makes pure biodiesel (B100) less suitable for use in cooler geographic locations?
   a. Pour point
   b. Cetane number

c. Sulfur content
d. Particulate emissions

4. Which of the following noxious emissions tends to increase when the combustion of pure biodiesel (B100) is compared with the combustion of petroleum-based diesel fuel?
a. Sulfur oxides
b. Particulate matter
c. Hydrocarbons
d. Nitrogen oxides

5. What is the fuel required by a fuel cell?
a. Nitrogen only
b. Hydrogen and oxygen
c. Hydrogen only
d. Water

6. What results from the electrochemical reaction in a fuel cell?
a. Electricity only
b. Water only
c. Hydrogen and oxygen
d. Electricity and water

7. Which type of fuel cell is currently being used in trials in city buses?
a. Proton membrane exchange
b. Alkaline
c. Solid oxide
d. Phosphoric acid

8. Which of the following has the most amount of heat energy by mass (weight)?
a. Propane
b. Compressed natural gas
c. Hydrogen
d. Diesel fuel

9. When propane or compressed natural gas is used as the only fuel in an internal combustion engine designed for diesel fuel, what must also be provided?
a. Particulate trap
b. Means of igniting the fuel
c. Cooled EGR
d. Distributor

10. *Technician A* states that a disadvantage of alcohol as an internal combustion engine fuel is its low octane rating. *Technician B* states that an advantage of LPG over gasoline as an internal combustion engine fuel is its high calorific value. Who is correct?
a. Technician A only
b. Technician B only
c. Both A and B
d. Neither A nor B

11. When a regenerative braking cycle is engaged in an HEV (diesel electric) hybrid drive system, which of the following should be true?
a. The diesel engine turns in reverse.
b. The diesel engine compression retarder is actuated.
c. The AC motor-generator reverses.
d. The DC motor magnetic field is reversed.

12. Which of the following fuels has the most heat energy (calorific value/Btus by mass)?
a. Alcohol
b. Compressed natural gas
c. Propane
d. Diesel fuel

13. When hydrogen is burned using pure oxygen, what end gas results from the combustion process?
a. Hydrogen
b. Oxygen
c. Water
d. Carbon dioxide

14. Which of the following fuels is often a byproduct of crude petroleum?
a. Biodiesel
b. Propane
c. Alcohol
d. Hydrogen

15. Which of the following elements is the most abundant in the known universe?
a. Iron
b. Oxygen
c. Nitrogen
d. Hydrogen

16. *Technician A* states that a disadvantage of a parallel hydraulic hybrid (PHH) drive system is that a medium- to large-bore diesel engine is required. *Technician B* states that a PHH system can only impart hydraulic drive torque so that in the event of a leak, the vehicle would be crippled. Who is correct?
a. Technician A only
b. Technician B only
c. Both A and B
d. Neither A nor B.

17. Which type of batteries tends to be used in the current HEV transit buses?
a. Lead acid
b. Nickel cadmium (Ni-cad)

c. Alkaline
d. Nickel metal hybrid (NiMH)

18. When a blended torque transmission is used on an HEV transit bus, which of the following features are possible?
    a. Input torque from more than one source
    b. Continuously variable output ratios
    c. Infinitely variable output speeds
    d. All of the above

19. Which of the following is true of a motor/generator used in an HEV transit bus?
    a. It produces three-phase AC power.
    b. It requires the use of an inverter.
    c. Output must be rectified to DC to charge battery banks.
    d. All of the above.

20. What type of hydraulic accumulator is used in an Eaton PHH and an SHH drive system?
    a. Mechanically loaded with a maxi-spring rated at 3,000 psi (207 bar)
    b. Nitrogen gas loaded at 5,000 psi (345 bar)
    c. Hydraulically loaded at 29,000 psi (2000 bar)
    d. Any of the above, depending on application

27

Failure Analysis

INTRODUCTION

Failure analysis is all about problem sourcing and solving. It is about getting to the root cause of a failure and ensuring that it does not repeat itself. For instance, if an 18-gauge electrical wire is overloaded with current due to a short circuit, we know that it will likely overheat tripping a circuit breaker and possibly melt the insulation on the wire. If the repair plan was to simply replace the damaged wire with a thicker 12-gauge wire, it would be just a matter of time before the failure recurred because the source of the original problem was not repaired. The job of the technician is to determine the **root cause** of any failure. The root cause is the source of a failure in our example, which is a short circuit. The technician's task should be to source the cause of the high current draw, not simply replace the most evident failed component in hopes that the failure would not recur. The molten insulation on the wire in this example is known as **coincidental damage**: It is also referred to as **collateral damage**.

Failure analysis training requires learning a structured approach to problem solving. Technicians must learn to distinguish between causes and results. They must learn how to analyze evidence. To develop skillful failure analysis techniques, technicians must learn how to:

- Find and focus on facts
- Eliminate collateral effects of the root cause
- Analyze those facts to produce findings
- Communicate the findings to a report (work order)
- Develop a repair strategy that prevents a recurrence

Failure analysis in itself is a process and should take place in five steps:

1. Observe and organize the evidence of the failure.
2. Identify the facts that have contributed to the failure.
3. Determine the causes of the failure.
4. Analyze the facts and causes of the failure.
5. Communicate the root causes of the failure.

Caterpillar defines the term *failure analysis* as "a thoughtful review of product and environmental facts which leads to identification of root causes of product problems." Its definition emphasizes *facts* and *root causes*. Both are keys to developing an effective failure analysis process.

The root cause of a failure is the specific condition that initiated a failure. Root cause has to be the focal point of failure analysis. The challenge for the technician is to eliminate collateral damage evidence while zeroing in to identify the root cause of a failure.

Failure Analysis Training

Developing failure analysis skills is not something learned overnight because experience and exposure to failed systems and components are required. However, while experience is a component of developing failure analysis skills, so is training. Technicians should also be aware of the fact that advanced failure analysis can often require lab analysis and equipment not normally available in the typical service facility. Broadly speaking, failure analysis training falls into two areas:

- The technician must first learn an elementary understanding of some of the basic principles of metallurgy, fracture, wear, and visual examination of failed components.
- Next, the technician must learn structured methods for conducting failure analysis procedures and learn to eliminate collateral damage factors. Poor failure analysis results in a repair strategy that focuses on collateral damage because the root cause has not been identified, leading to a recurrence.

STRUCTURED FAILURE ANALYSIS

The technician who has had many years experience of diesel engine technology has also probably developed an inventory of failure analysis profiles that make identifying root causes of a problem easy. With such a person, all it may take is a quick glance at a failed component. For those who do not have years of hands-on knowledge, it is a safer bet to use a more structured approach. A structured method for conducting failure analysis can be organized into eight steps.

1. State the problem clearly.
2. Check repair history and service records.
3. Observe and record the facts.
4. Eliminate collateral damage.
5. Analyze the facts.
6. Identify possible and probable root cause(s).
7. Communicate the conclusions (work order/warranty/customer).
8. Perform the repairs required.

Skipping a step, or performing steps out of sequence, is likely to result in a failure to identify the root cause of a problem. This means that the resulting repair strategy will be flawed because collateral damage will be repaired while the root cause is overlooked. This results in comebacks, unnecessarily high invoices, policy adjustments to work orders, downtime, and dissatisfied customers.

Failure analysis is necessarily a process that relies on facts to determine events and zero-in on the root cause of system and component failures. Facts have to be located in the material evidence observed on failed components. The technician is then tasked with interpreting the facts to identify the events that produced the facts and this must include an examination of the vehicle repair history. Finally, all this data should be organized in a logical sequence so that the root cause can be identified.

PRINCIPLES OF FAILURE ANALYSIS

As indicated a little earlier, in order to better understand how components fail on engines, the technician has to understand some basic principles. These basic principles can be divided into the following four areas:

- Metallurgy: the science of metals
- Wear: identifying normal and abnormal wear conditions
- Fractures: primary indicator of the manner and speed of a failure
- Visual examination: ability to correlate visual evidence to a type of failure

Metallurgy

Metallurgy is the science of metals. It deals with properties, structure, and behavior of metals. A diesel or truck technician who aspires to develop failure analysis expertise must have a basic understanding of metallurgical principles. Vehicles are constructed from a range of alloys. An **alloy** is a mixture of two or more elements, at least one of which is a metal. In vehicle technology, we are primarily concerned with **ferrous metals**. Ferrous metals are composed primarily of iron. For instance, **steel** is an alloy of iron and a small quantity of carbon. There are many different types of steel. Because there are many different elements that can mix with iron, many types of steel can be manufactured with characteristics that suit a variety of purposes.

Metal Structure

All metals are composed of grains. The grain is produced by crystals that form from molten metal as it solidifies. For instance, in cast irons and steel, the grain boundaries almost always contain some minor impurities such as refractory brick or trapped slag left over from the casting and production processes. Grain impurities are known as inclusions (just as we describe them when entrapped in a weld). Most inclusions that result from a manufacturing process are small enough not to significantly weaken the metal. But a larger inclusion in the wrong place, such as a high-stress focal point, could result in a failure. This is known as a metallurgical failure. In performing failure analysis, the technician should always keep in mind that the root causes of a small percentage of failures are metallurgical in origin. When warranty is a factor, a metallurgical sourced failure would be an original equipment manufacturer (OEM) responsibility.

Iron and Steel

Iron and steel are made into useful forms by casting and shaping processes. Casting requires that the base metal be reduced to a completely molten state, then set in a form. Both iron and steel can be cast. In vehicle technology, we use a variety of irons and steels. In production, steel can be cast or formed by one of more of the following shaping processes:

1. Rolling
2. Forging
3. Extruding
4. Drawing

Steel that is reshaped by one of the preceding processes is referred to as wrought. The term *wrought* means that something has been worked into a specific shape. Cast and wrought steels form two general categories of metal with quite different sets of properties.

Rolling Rolling is the most common steel working process. Rolling steel produces wrought shapes such as plates, sheets, slabs, and bars. Rolling requires steel to be heated to about 1,200°C (2,200°F). Heating the steel to this temperature makes it plastic, that is, susceptible to yield when pressure is applied. When high pressure is applied to heated steel using large rollers it can be formed or squeezed into a desired shape. Often rolling takes place in multiple stages. For instance, rolled steel slabs are often further reworked by extruding, forging, drawing, or a variety of other metal forming processes.

When steel is subjected to rolling, the grains in the metal reform. This means that the grain boundaries take on a new shape without tearing apart. The new grain structure of rolled steel aligns in the rolling direction to form new grain flow lines. This has the desirable effect of making a steel stronger in the direction of the grain flow. This *grain flow* in production formed metals causes different material properties in different directions. Grain flow also affects the appearance of a fracture surface when observing fracture faces. Sometimes a fracture follows the grain flow while at other times it will run against it. Learning to identify grain flow in fracture faces can tell us a lot about the root cause of the fracture.

Forging Forging is a common manufacturing process. A century ago, forging was something performed by a blacksmith: At its simplest, forging required heating a metal to a specific temperature, then shaping it using percussive force delivered by a hammer. Today it is a manufacturing process that is more scientific. Metals are heated to precise temperatures and the applied force is effected hydraulically. Drop-forging is used to form components ranging from connecting rods to crankshafts.

Extruding Extruding literally means squeezing. In metal forming, extrusion requires forcing a semimolten metal through a die. This process both shapes the metal and gives it some desirable surface characteristics.

Drawing Drawing is another forming process in which heated steel is pulled through an aperture. This provides the steel with a specific shape and sectional area. Drawing is commonly used to produce individual strands of copper, aluminum, and steel wire.

Heat Treatment

Heat treatment is a common method used to redefine the structure and properties of a range of metals. The term *tempering* is used to describe a range of heat treatment processes. Heat treatment of metals predates accurately recorded history and we know that it was first used by the ancient Greeks 3,000 years ago. Heat treatment processes require controlled heating and cooling to alter the characteristics of a metal and produce the desired material properties. A metal can be heated in several ways, but in modern steel production, the most common method is to use a furnace. The manner in which heat is applied to a metal and the manner in which it is cooled are both important in determining the properties of a metal. For instance, rapid cooling of heated steel known as quenching produces the hardest characteristics, sometimes even making the material **brittle** (susceptible to fracture). A slower cooling process to heated steel generally produces a less hard steel but provides higher plasticity.

In truck and engine technology, heat treatment processes are usually designed to make steels tougher and harder. But the opposite may be true. For instance, to anneal copper (make it softer), the metal is first heated then rapidly cooled by quenching in water. Heat treatments change the structure of a metal, especially when the metal has been alloyed with carbon and other alloying elements such as nickel, chromium, titanium, and molybdenum. If metal that has been tempered in manufacture is subjected to temperatures that exceed the tempering temperature, the effects of the controlled tempering process can be entirely lost. This may result in a metal that is either much harder or much softer than desirable.

Stress Raisers

When vehicle components are designed it is necessary to calculate the applied loads, determine where loads will be applied to the component, and determine how the geometry (shape/section modulus) of the component will respond to applied loads. Components respond to applied loads by setting up internal stresses. A geometric change in a component caused by an applied load can result in a **stress raiser** located in the deformed area. A stress raiser is a feature in either the shape or composition of a component that causes a localized increase in stress. The greater the change in geometric shape, the more aggressive the stress raiser effect. This can initiate metallurgical failure and result in a crack.

Apertures (holes), keyways, threads, shoulders, and other irregular shapes can all be stress raisers. Where known stress raisers occur in a component, they can be compensated for by design, material selection, heat treatment, or forming techniques. Stress raisers due to metallurgical problems, manufacturing process abnormalities, or abuse can result in component failures when there is an insufficient safety factor built into the design.

Examples of Stress Raisers There are many causes of potential stress raisers that designers must take into account. They can be generally divided into the following categories with the first three relating to design and production defects:

Geometry

- Fillets: should be radiused rather than sharp-angled
- Holes: chamfering at top and radiused at bottom reduces localized stresses
- Keyways: should be chamfered at edges

Metallurgical Defects

- Inclusions: usually a manufacturing (casting) or production (welding) defect
- Voids: manufacturing defect
- Cracks: can be embedded into castings while cooling from liquid to solid state in the manufacturing process

Production Defects

- Forging laps
- Quench cracks
- Weld crystallization
- Grinder cracks

Abuse

- Overloading
- Scratches
- Nicks
- Gouges

Root Causes of Failures

The root causes for most failures can usually be sourced to the application, operation, and maintenance areas. In other words, a large percentage of failures can be traced to the way in which equipment is operated and maintained. After this the next most probable root cause area is installation. Installation malpractice is traceable to the technician and caused by either lack of training or carelessness. The least likely root cause sources are deficiencies in manufacturing and design. The good news is that failures due to design or manufacturing shortcomings tend to be the easiest to diagnose. Failures generated by the other root cause areas are often harder to diagnose because it is more difficult to establish the facts connected to the failure.

Collecting Evidence

When investigating an equipment failure the first step should be to identify all of the systems affected, then collect and isolate the damaged components and fragments. This first step should include visually damaged components and those components that interact with them. Using the example of a connecting rod failure, the evidence should include the failed conn rod, plus the piston assembly, wrist pin and bushing, rod cap, friction bearings, rod fasteners, and crankshaft throw journal. This evidence should be laid out in a systematic manner so that the clues in sourcing the root cause of the failure can be identified.

Collecting Facts

Failure analysis facts come from four sources, the first three of which are the responsibility of the technician:

1. The operator's report
2. The failed component(s)
3. Vehicle and engine service history
4. OEM failure analysis lab

Having stated this, most failure analysis problems should be solved by properly undertaking steps 1 through 3. In a small number of cases, microstructure facts using source 4 are required to determine root causes: This requires the use of lab instruments and specialized expertise, so the problem exceeds the capabilities of the typical service facility. In reality, source 4 type problems are only investigated by the manufacturer of the failed component. It is in the interests of the manufacturer to do this to avoid costly recurrences that could damage the product credibility and increase warranty claims.

Conducting an Examination

Because visual examination is such an important component of failure analysis it is important that the technician optimizes the inspection environment. Good lighting is essential. Before cleaning, failed components should be visually inspected and record anything unusual. Use a magnifying glass when examining fractures. Next, carefully clean the components in a way that is guaranteed not to damage surfaces. The use of mild solvents, soft bristle brushes, and air drying will clean components without destroying evidence. Take special care when cleaning components with soft surfaces such as engine friction bearings. Gently rinse in solvent then allow to air dry. Key indicators of a failure can too easily be destroyed by aggressive cleaning so the use of glass bead blasters, grit blasters, wire brushes, or grinders should be avoided.

During any failure analysis procedure, try to avoid resorting to preconceived notions about what might have caused the failure. For instance, if a truck driver complained that his engine produced a loud knocking noise before seizing, the diagnostic technician might easily conclude that low crankcase oil level producing insufficient lubrication could have been the cause. And it could have. But a thorough technician should treat every failure analysis as a challenge worthy of a CSI ("Crime Scene Investigation" TV show) plot. Doing this will allow more facts to be gathered and they might suggest that the oil pump gears failed, shutting down the lube circuit—or any of half a dozen other conditions. Using preconceived ideas based on past experience of similar failures can often lead the technician in the wrong direction and result in coming to faulty conclusions. A good failure analysis technician has to have a little of Sherlock Holmes's blood.

As the facts of a failure are collected, the technician should carefully identify and sort through the evidence. This means regarding the wear and fracture facts of the failure as road signs along a highway that will lead to the root cause of the problem—and establish a means of avoiding a recurrence.

Wear

Wear may occur over a period of time and often be regarded as expected. However, there are many different types of wear and some can act as key tattletales in the search for a root cause of a failure. Specific environments cause different types of wear. While all components age in operation and produce wear, the interest of the failure analysis technician obviously falls into recognizing abnormal types of wear.

There are many types of wear, but whether the wear is normal or abnormal can usually be categorized into one of the following seven areas:

1. Abrasive wear
2. Adhesive wear
3. Corrosion
4. Erosion
5. Cavitation erosion
6. Contact stress fatigue
7. Fretting

Each type of wear has its own characteristics that stem from operation in a specific environment. Recognizing specific wear characteristics helps technicians to identify the environment that caused the wear.

Abrasive Wear

Abrasive wear accounts for a majority of wear failures. It may occur in both lubricated and nonlubricated components. When particulate grit comes into contact with metal surfaces, it begins to wear it down: For instance, take a look at the dump box in any aggregate hauler. In lubricated components, any time hard particles larger than the lubricant film thickness get between two moving surfaces, wear will result. Surface faces are first scored then this produces secondary debris that accelerates the condition. Hard surfaces such as bearings do not score easily when exposed to particulate, but more heat is generated so wear rate depends on how effectively the lubricant acts as coolant. Given adequate lubrication, the heat generated is held in check and can slow down the abrasive wear effect.

Adhesive Wear

Adhesive wear tends to produce accelerated failures. It occurs when two moving surfaces contact each other with insufficient lubrication. An example would be when the hydrodynamic suspension of a rotating shaft in a friction bearing fails. The resulting contact produces friction and enough heat to approach melting point: This creates localized welding or an adhesive effect.

Adhesive wear in engines can usually be attributed to a lubrication problem. This type of wear failure is often initiated in extreme cold conditions, especially midwinter cold start in applications where an inappropriate engine lube is the culprit. Flow-ability is an important characteristic for good lubrication. At low temperatures, viscosity increases significantly (see Chapter 9), and at –30°F (–35°C), the commonly used 15W-40 engine oil has the constituency of axle grease; that is, it has no flow-ability. A single cold-start event under these conditions using this engine oil can destroy an engine. Figure 27–1 shows examples of both the early stages and advanced stages of adhesive wear.

Start of adhesive wear. Advanced adhesive wear. Damage from large soot particles.

Figure 27–1 *Examples of adhesive wear and soot particles on friction bearings. (Reprinted Courtesy of Caterpillar Inc.)*

Corrosion

Corrosion is chemical deterioration of a metal surface. Unrefined ore is oxidized metal. During metal processing, ores are refined to produce pure metals that tend to be less stable. Pure metals often revert back to the more stable oxidized states and will do so when exposed to electrolytic substances. An electrolyte is a liquid solution (like that in a lead acid battery) that enables electrolytic interactions on metals such as steel or aluminum.

Most corrosion on engine components is oxidation related, but nonoxidation corrosion can also occur when metals are exposed to acids and alkalines within engine fluids. A severe type of corrosion can result when engine coolant deteriorates and becomes acidic: The role of the engine coolant becomes that of electrolyte, and metal surfaces become electrodes, resulting in rapid failures.

Erosion

Erosion occurs when abrasive particles suspended in fast-moving fluids contact surrounding surfaces, resulting in fine impact and abrasive damage. An eroded surface usually has a rough, matt appearance. Erosive wear can occur anywhere in an engine where fluids move at some velocity, but it will be especially noticeable when dust enters the intake circuit. Observing filter change intervals is one of the best strategies for controlling erosive wear.

Cavitation

Cavitation erosion occurs when gas bubbles in liquids collapse against metal surfaces, creating a percussive effect. Under some circumstances liquids can suspend dissolved gases, which form bubbles in low-pressure areas. Also, the pulsing action of wet liners against the wall of coolant creates vapor bubbles as the liner pulses inward. Bubble implosion can generate forces exceeding 60,000 psi (414 MPa) capable of first pitting and ultimately eroding through a wet liner. Cavitation usually occurs on the outside of wet and wet/dry liners in line with the piston thrust faces. It results when the cavitation protection of an antifreeze mixture depletes or in high-pressure fuel pumping mechanisms when charging pressures fall below spec.

Contact Stress Fatigue

Contact stress fatigue occurs when two surfaces slide, roll, or are dragged against each other. This results in high stress, surface face removal, and fatigue cracks in one or both surfaces. Unintended contact surface movement can occur if applied stresses become excessive, or if a component is underspecified and cannot withstand intended working loads. Cyclic movement of a pair of surfaces exceeding design limits can result in cracks, surface pitting, and spalling: This condition is known as contact stress fatigue.

Fretting

Fretting corrosion occurs when two parts that fit tightly, usually under significant clamping force, are allowed to move slightly against each other. This can result in microwelding and producing patterned surface irregularities. Left unchecked, fretting can result in small pieces separating from each surface. Sometimes these small pieces corrode and form reddish-brown oxides on either or both clamped surfaces. When replacing worn components, their mating surfaces should be carefully checked for fretting because the failure could recur.

Most tightly clamped components in an engine have to be provided some ability to move independently. For instance, a cylinder head has much less mass than a cylinder block so as an engine heats up, the cylinder head will expand more quickly than the cylinder block. It will also contract more quickly when the engine is shut down and cools. The same would apply to many other mating surfaces represented by manifolds. This differential in expansion and contraction is known as **creep**. Therefore, a key function of a gasket is to permit thermal creep in mated components while effectively maintaining a seal.

Loads

When components fail, each failure produces a pattern of tattletales that the failure analysis technician learns to identify. When we took a look at wear, each type of wear correlates with a specific operating environment. In the same way, when components fracture, the type of fracture correlates with the specific way the load was applied to the failed component. In this section, we will review some of the basics of fracture mechanics.

Components generally fracture when they either contain material defects or they are subjected to excessive loads. Loads can be generally divided into three types:

- *Impact loads* are sudden and excessive.
- *Overloads* are also excessive but applied over a period of time.
- *Fatigue* or *cyclic loads* are loads that are applied repeatedly over a period of time: They usually involve much lower applied loads than impact or overloads.

The direction or plane on which a load is applied to a component will have an effect on the nature and direction of a fracture. Also significant are the metallurgical properties of the material under load. A brittle steel (one that cannot sustain any plastic deformation) will produce a different fracture characteristic from one that is **ductile** (flexible or malleable). If you understand some of the metallurgical basics of the component that fractured and you can correlate the appearance of the fracture characteristics to the type of load that caused it, you should then be able to make an accurate assessment of root cause.

Types of Load

Generally, loads can be divided into the following categories:

- Bending
- Compressional
- Tensional
- Torsional

Bending Loads When a component is bent, it is deformed in such a way that one side can be said to be convex in shape and the other, concave. *Bending* loads place the convex surface under tension (stretch) and the concave surface under compression (squeeze). A fracture initiates on the tension side and runs through toward the compression side of the component. The actual appearance of the fracture surface depends on the type of load (impact, overload, or cyclic) and the physical properties of the material. A bending failure of a crankshaft could take place in the event of a main bearing cap separation.

Compressional Loads When something is compressed it is squeezed by opposing forces. *Compressional* loads produce compressive stress in the direction of the applied load and tensile loads perpendicular to the applied load. When compressive loads are applied to ductile materials, depending on the shape of the material, the result may be to cause a bulge or other plastic deformation. Any resulting deformation will usually be perpendicular to the applied load or in the cases of thinner components, buckling.

Components manufactured from brittle materials are more likely to fracture in a direction parallel to the applied load when subjected to compressional loads, depending on the physical shape of the material. Fractures result because compression loading produces tensile stress in the direction perpendicular to the applied load. This often initiates with multiple minor cracks that eventually converge, resulting in shatter. A connecting rod could be subjected to a compressional overload in the event of hydraulic lock in an engine cylinder resulting from head gasket failure and coolant leakage into the cylinder.

Tensional Loads *Tensional* loads are the result of stretching force applied to a component. Stretching force produces tensile stress in the direction of loading and compressive stress perpendicular to the direction of loading. Tensional loads cause materials to elongate in the direction of the applied load. In the case of ductile materials, the cross section perpendicular to the applied load may decrease in size. This characteristic can be seen in an overtorqued bolt that produces a characteristic known as "necking." More brittle materials such as cast irons may initially elongate (slightly) but then fracture perpendicularly. In most cases, brittle materials that fail due to tensional overloads produce little observable plastic deformation prior to fracture. Tensional failures of connecting rods may occur during an engine overspeed event.

Torsional Loads Torsion is twisting force. Failures resulting from *torsional* loading tend to be more complex. In a shaft that conducts torque, shear stresses are generated both across and through the length of the shaft. This results in tensile and compressive stresses at 45-degree angles along the shaft. Shear stresses tend to predominate in shafts that have a ductile characteristic producing both transverse (across the shaft) and longitudinal fractures. When torsion is applied to shafts manufactured from more brittle materials, tensile stresses tend to predominate in brittle materials, so when a fracture occurs it is in a spiraling, 45-degree direction. Cyclic loads applied to either brittle or ductile shaft materials can produce fractures in any of the three directions, making them more difficult to diagnose. Torsional failures occur in shafts when they are subjected to more twisting force than they were engineered to sustain.

Fractures

When components fracture, the cause can usually be categorized as:

- Brittle fracture
- Ductile fracture
- Fatigue fracture

Brittle fractures are the result of a shock load or high force impact. Brittle fractures take place rapidly producing crystalline or grainy fracture face surfaces. In most cases, little or no plastic deformation occurs during the fracture event. Depending on the specific material fractured, there may or may not be chevrons on the fracture face surface. Fracture surfaces can range from bright and sparkly to dull in appearance, this depending once again on the material.

Ductile fractures are often the result of a component overload. Ductile fractures occur quickly, but not as fast as brittle fractures. In most cases, they produce rough, dark surfaces, with some evidence of plastic deformation (bending, twisting, or stretching) prior to the failure. The plastic deformation that may accompany ductile fractures often results in shear lips at the outside edge of the fracture.

Fatigue fractures are typically the result of cyclic overloading. Fatigue fractures are the result of much lower loads than those characteristic of brittle or ductile fractures. They are also less catastrophic, and because they occur much more slowly, they produce lighter, smoother fracture face surfaces. Fatigue fractures can often leave **beach marks** (ripples) as a tattletale.

COMPONENT FAILURES

Identifying the root cause of a component failure is always made much easier when the disassembly is organized and methodical. It helps to label and tag components on removal from the engine. Clean and inspect every component as it is removed from the engine. When inspected components are to be reused on reassembly, they should be sealed in polyplastic paper. Make a point of not beginning an engine reassembly until the exact cause of the failure has been determined. This section deals with some common failure modes of engine components, but it should be remembered that it is by no means comprehensive.

Pistons

When disassembling an engine, numerically tag pistons for location on removal, even when the primary cause of the failure has been identified and you plan to replace the pistons. Most OEM warranty processing requires that pistons be tagged. You should have a basic knowledge of the operating principles of diesel engine piston assemblies as outlined in Chapter 6. The pistons used in diesel engines have undergone some significant changes in recent years: For instance, Caterpillar's popular 3406/C15 engine has progressed through aluminum trunk, steel/aluminum articulating, and single-piece, forged steel trunk pistons within a period of a dozen years. Some typical piston failure conditions correlated with possible causes are identified in Table 27–1. In order to maximize piston life, the piston must be properly assembled and then properly installed. Once in operation, piston life depends both on how the engine is operated and the performance of the engine support circuits that remove combustion heat from the pistons:

- Air intake circuit
- Lube circuit
- Cooling system
- Exhaust circuit

Heat is the biggest enemy of pistons. This is what OEMs like to call **thermal failures**. A performance breakdown in any of the systems de-

Table 27–1 *Piston Failure Chart*

Condition	*Causes/Results*
Excessive combustion pressure	Caused by overfueling and ether abuse (Figure 27–2 and Figure 27–3).
Advanced injection timing	Outer edge of the piston melts—erosion of the top ring land above the Ni-Resist insert. The erosion is caused by fuel dispersal/condensation outside the combustion bowl (Figure 27–4 and Figure 27–5).
Retarded ignition/detonation	Failure occurs near the center of the piston crown, producing pitting erosion. In severe cases, the pitting can rapidly eat through a piston. The pitting is caused by fuel condensation on the piston crown before ignition.
Piston **scuffing** (localized welding)	Produced by metal-to-metal contact. On thrust sides only—caused by high cylinder or boost pressures (overfueling, severe lug). Scuffing overall—caused by poor cylinder wall lubrication, contaminated lube oil, overheating, or insufficient ring end gap.
Carbon buildup	Overfueling, retarded injection timing, inappropriate fuel, injector nozzle failure.
Crown **pitting** (pocked erosion)	Retarded injection timing, injector nozzle failure/dribbling, moisture from intake air condensing on the piston crown.
Lacquering	High sulfur content in fuel.
Torched piston	Advanced timing or extreme overheating. Single piston torching: usually caused by a failure to dissipate heat resulting from a misaimed cooling jet. Figure 27–6, Figure 27–7, and Figure 27–8 show some examples of piston failure due to extreme overheating.

signed to take heat away from pistons can produce a failure. Some of the tattletales of piston thermal failures are:

- Piston crown carboning
- Ring carboning
- Discoloration of the underside of the piston crown
- Erosion of the wall around the toroidal crater of a Mexican hat crown
- Melting of the piston leading edge
- Blackening and/or lacquering of the piston crown

When attempting to identify the root cause of piston thermal failures, make sure you examine all the pistons. For instance, a thermal failure limited to one piston would indicate a heat dissipation problem from just one engine cylinder, so you should focus on that. A misdirected oil cooling jet (this should be aimed at a specific location on the underside of the piston crown) can cause a "torched" piston. A single torched piston in an engine is often due to assembly carelessness such as contact between the rod big-end and oil cooling jet.

Piston Scuffing

Scuffing is adhesive wear. Another way of saying this is localized welding. Piston scuffing begins in one area but usually spreads, making it difficult to identify the source. Crown scuffing occurs due to high cylinder temperatures that cause the crown to expand relatively more than the skirt. It is much less common in modern diesels using articulating or drop-forged steel pistons so when you do see this type of failure, it will be more likely in aluminum trunk pistons. Some causes of piston crown scuffing are:

- Injector dribble
- Inadequate undercrown cooling
- Restricted exhaust system or other breathing circuit problem
- Overfueling
- High air intake temperatures
- Prolonged lug operation

Skirt scuffing occurs more commonly than crown scuffing and, once again, is more likely to occur in aluminum trunk-type pistons. When an aluminum trunk piston skirt is overheated, overexpansion usually occurs where the mass is

Figure 27–2 *Piston failure due to high cylinder pressure. (Courtesy of Mack Inc.)*

Figure 27–5 *Early indication of an advanced timing failure of a piston. (Courtesy of Mack Inc.)*

Figure 27–3 *Ring lands blown out due to high cylinder pressure. (Courtesy of Mack Inc.)*

Figure 27–6 *Piston torched due to advanced timing or excessive overheating. (Courtesy of Mack Inc.)*

Figure 27–4 *Major piston failure due to advanced timing. (Courtesy of Mack Inc.)*

Figure 27–7 *Complete crown meltdown caused by excessive cylinder temperatures. (Courtesy of Mack Inc.)*

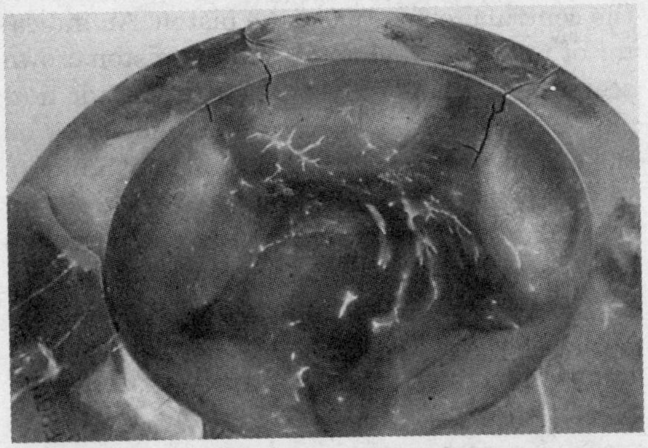

Figure 27–8 *Piston failure due to overheating, causing cavity edge cracks. (Courtesy of Mack Inc.)*

highest: This means that the "corners" of the piston bosses expand relatively more, creating a "four-corner" vertical scuff profile. As the scuffing continues so does the heat buildup and may result in engine seizure or disintegration of the piston. Skirt scuffing is caused by:

- Insufficient cooling
- Lack of lubrication
- Distorted or fractured cylinder liner

Cold Start

When diesel engines are cold started, the pistons and rings are cold and ring end gaps are wider than at operating temperatures. In addition, only residual oil (boundary lubrication) is present to act as lube. The consequence is that rings have to sustain cylinder pressures while being supported by a lower sectional area until the ring land warms to its operating temperature. For this reason, it is important to run a diesel engine under no load long enough on cold starts to enable the pistons to warm to operating temperature. This is especially important on engines using aluminum trunk-type pistons.

Fast Track Analysis of Piston Failures

The following is a summary of typical piston failures and some pointers toward their root causes. Always remember that every failure has a certain uniqueness to it so there are no absolutes when it comes to accurate identification of root cause.

Torched Pistons Torching of pistons occurs when either one piston or a set of pistons have been overheated to such an extent that meltdown has occurred. The condition was obviously more common in engines using aluminum trunk-type pistons. Diagnosing a torched piston in isolation from the engine from which it was removed may be a game of guesswork rather than sound failure analysis practice. However, the cause of the condition must be unmistakably diagnosed before the engine is reassembled, or a recurrence of the failure will result. If only one piston is affected, the diagnosis may be simpler, but it is always important to remember that piston torching may be related to lubrication, cooling system, or fuel injection causes.

Skirt Scoring The usual causes of skirt scoring are overheating, overfueling, insufficient piston clearance, lack of lubrication, and fuel injector nozzle problems.

Cracked Skirt The causes of a cracked piston skirt are excessive use of ether, excessive piston-to-bore clearance, and foreign objects in the cylinder.

Uneven Skirt Wear The common causes of uneven skirt wear are contaminated lube oil (either with abrasives or fuel/coolant), abrasives in air charge (dusting), and too little piston-to-bore clearance. The latter can be caused by aftermarket pistons: This can measure to specification when cold but expand to a greater extent when at operating temperature.

Broken Ring Lands Broken ring lands are caused by excessive use of ether, excessive piston-to-bore clearance, and foreign objects in cylinder. Engines subjected to high cylinder pressures before warming to operating temperatures can also produce ring land fractures.

Worn Piston Pin Bosses The causes of worn piston (wrist) pin bosses are old age (normal wear), contaminated lube oil, and insufficient lubrication. When a pin boss is starved of lubrication, the root cause is in the lube delivery system so be sure to check this out.

Burned or Eroded Center Crown When the center of a piston crown is burned or eroded, the usual causes are plugged injector nozzles, injector dribble, cold loading of engine, retarded injection timing, and water/coolant leakage into cylinder.

Fuel Injection/Ignition Timing Related Failures of Pistons

With the almost universal adoption of computer controls on diesel engines today, most of the problems associated with fuel injection timing have disappeared. It is fair to say that most fuel injection timing problems were the result of abuse, either intentional or because of a lack of training. Although there are ways of manipulating fuel injection timing in modern diesel engines, most OEM audit trails can pinpoint the time of any tampering abuse so it is not common.

Advanced Injection Timing Advanced engine timing causes excessive combustion pressures and higher than specified temperatures. It may result in torching or blowing out the lower ring lands in severe cases. In less severe cases, erosion and burning of the top ring land or headland area result, which are evidenced by pitting. The latter condition is more common and is often the result of abuse by technicians attempting to increase engine power and fuel economy. It often does but at a heavy cost because of a significant reduction in engine longevity. Advancing injection timing results in high NO_x output. Total failure resulting from advancing timing usually occurs in one cylinder with evidence appearing in the remaining cylinders.

Retarded Injection Timing Retarded engine timing will cause excessive cylinder temperatures and burning/erosion damage through the central crown area of the piston. An indicator of retarded injection timing is piston crown scorching immediately under the injector nozzle. This is caused by droplets condensing on the piston crown before vaporizing and ignition. In days when engine timing was a technician responsibility, this condition occurred much less frequently than advanced injection timing because it was seldom intentional. It was usually the result of technician error or component failure though it can result from actuating component wear. Examples of actuating component wear are cam profile and camshaft bearing degradation.

Piston Rings

Piston rings provide plenty of tattletales when it comes to determining the root cause of an engine failure. Most engine overhauls are due to normal wear of piston rings over the life of the engine. As piston rings wear, their ability to effectively manage the oil film on the cylinder wall and seal becomes reduced as illustrated in Figure 27–9. Here are some of them:

Cold Stuck/Gummed Rings

Cold stuck rings are a condition in which the piston ring seizes in the ring groove when the engine is cold but frees and starts to seal the cylinder when the engine is at operating temperature. Cold stuck or gummed rings are caused by carbon/sludge buildup in the ring groove. The condition can be caused by a defective thermostat, operating for short periods in cold weather, and constant stop/start operation. It results in high crankcase pressure during engine warmup and fuel contaminated engine lubricating oil.

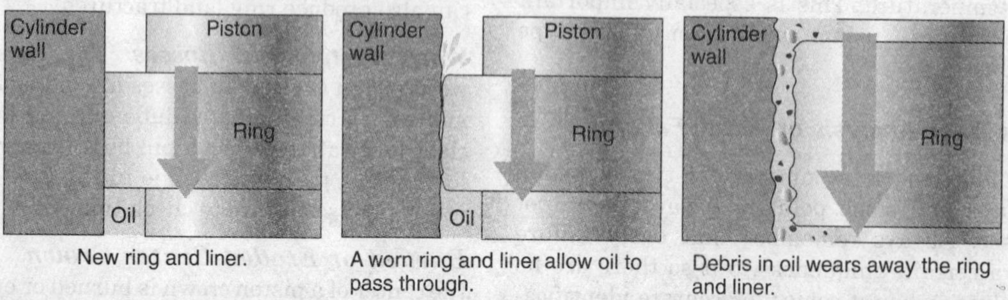

Figure 27–9 *How wear affects the ability of rings to effectively form a seal with the liner. (Reprinted Courtesy of Caterpillar Inc.)*

Hot Stuck Rings

Hot stuck rings are caused by crystallized carbon buildup resulting from the use of contaminated or improper fuel or the failure of oil control rings. The ring is usually captured stationary in the groove and the contact face blackened by cylinder blowby gas.

Scuffed Rings

Scuffed rings are usually related to an improper ring-to-cylinder bore fit such as would occur with an out-of-specification ring end gap, piston-to-bore clearance, or ring-to-groove side clearance. The condition may also be caused by cylinder wall lubrication failure. Because scuffing is a localized weld condition, created by frictional heat resulting from metal-to-metal contact, the result is often the seizure of the piston in the liner bore.

Eroded/Glazed Rings

Eroded or glazed rings are usually a result of poorly filtered or unfiltered intake air to the engine cylinders. Diesel engines used in construction and aggregate hauling applications often ingest enough ultra-small particles of abrasive silica sand through a properly functioning air cleaner system to glaze and erode rings and liners and shorten engine life. However, whenever this condition is identified, always test the engine intake system for leaks.

Broken Rings

Broken rings are usually caused by improper installation or crystallized carbon buildup in a sector of the ring groove. Broken compression rings are not often observed in current engines because flexible centrifugally cast or special alloy steels are used. More brittle cast irons are still used as oil control rings and these are vulnerable to fracture on installation if not handled correctly. A ring fracture that occurs 180 degrees from the end gap is classic indicator of improper installation (overstretch).

Plugged Oil Control Rings

Plugging of oil control rings is often caused by low engine operating temperatures or broken down, sludged engine lubricant. The result of this condition is usually an engine that burns engine oil, because the oil applied to the liner walls by the piston and bearing throwoff is not controlled. The condition causes coincidental problems such as compression ring gumming or sticking.

Carefully remove the piston rings and tag them for location on the piston and on the engine in sequence as they are removed on disassembly of the engine. Some common causes of piston ring failures are shown in Table 27–2.

Cylinder Head Valves

You should understand the operating principles and construction of cylinder head valves to the extent they are examined in Chapter 7. Cylinder head valves may be of a one-piece, two-piece, or multiple-piece construction depending on engine manufacturer. For instance, some manufacturers will use a single-piece construction on their intake valves and two-piece construction on their exhaust valves: Because one way in which valves may fail is at the weld joints on multiple-piece construction valves, it helps to know something about how the valve was manufactured.

Table 27–2 *Piston Ring Failures*

Condition	*Causes/Results*
Land failures	Excessive manifold boost, ether abuse, overfueling
Sticking/stuck rings	Lacquering and **gumming** caused by high sulfur or contaminated fuel, overfueling, or prolonged lugging. May result in engines fueled with fuel cut with used engine oil
Broken rings	Overfueling, ring erosion followed by fracture caused by dirt, improper installation (fracture 180 degrees from the end gap)
Cylinder scuffing	Insufficient ring end gap
Seized rings	High-operating temperatures, overfueling, contaminated fuel

Valve Terminology

The technician should be familiar with cylinder head valve terminology so that the location of a failure can be identified. Cross-reference the following parts of a valve with Figure 7–16, which shows a typical, inertia-welded, two-piece construction valve:

- Stem
- Head
- Keeper grooves
- Inertial weld
- Fillet radius
- Hard surfacing

Valve Construction

Valves are required to sustain high temperatures (especially exhaust valves) and rapidly disperse heat. They must also be capable of flexing during closing while maintaining a positive seal. These required characteristics determine what materials are used in valve manufacture. One-piece valves tend to be manufactured from low alloy, medium carbon steels that are tempered to permit some flexing at the lower stem and fillet, and are either induction or hard-surfaced at the keeper groove section of the stem and head. Two-piece valves use a low alloy, medium carbon steel stem that is joined by a friction weld to a heat resistant, high alloy steel head assembly. The head assembly is often so highly alloyed that it may not attract a magnet. In this type of cylinder valve, the two pieces are joined by an inertia weld in the stem, usually just above the fillet. In addition, some multiple-piece valves use a friction-welded, stellite head plate manufactured from a high alloy steel.

Hard Surfacing Hard surfacing of cylinder head valves is performed either chemically or by an arc welding procedure. In cases where an arc welding process is used, a high alloy (nickel, chrome, molybdenum) steel is applied to the seat circumference of the valve head.

Friction Welding Friction welding, sometimes known as inertial welding, is used to join a low alloy valve stem to a high alloy valve head. The process requires that the stem be rotated by a spun-up turret (establishing a fixed amount of energy) that is then forced against a held-stationary valve head. The energy of the rotating stem is released as frictional heat, resulting in melting and the welding of the two sections. The location of a friction weld in a two-piece valve can be easily identified because the low alloy upper stem attracts a magnet, and the high alloy lower stem and head either does not, or barely, attracts a magnet. If a valve were to fail by separating at the friction weld, this would usually be considered a manufacturing defect. It is not a common failure but it is also not unheard of.

Valve Loading

Cylinder head valves are subjected to three types of loads:

1. Spring forces
2. Valve closing forces
3. Cylinder pressures

Spring Forces When a cylinder valve is in its static closed position, it is subjected to a light tensile load between the valve seat and the keeper grooves. As a valve is opened, a compressive load is applied between the keeper grooves and the stem end. Neither of these normal light loads is sufficient to cause fatigue failures.

Valve Closing Forces As a valve closes, it slams back into its seat, coming to an abrupt halt, shock loading the valve between the seat and keeper grooves. Valves are designed to yield (to a small extent) to accommodate this type of tensile shock loading. However, the tensile shock loading increases with engine rpm and can be further increased by valve maladjustment. Loose valve adjustments can result in higher valve closing speeds because seat contact occurs before the valvetrain has ridden the valve transition geometry of the actuating cam profile. Additionally, engine overspeed increases both the frequency and velocity of the valves as they seat, and this tensile overloading can result in valve failures. When an engine "drops a valve," in more cases than not this can be related to an engine overspeed event.

Stresses concentrate anywhere there is a physical irregularity. This means that cracks often initiate in valve fillet radii and the keeper grooves in the stem when subject to tensile overloads. A majority of catastrophic valve failures can be attributed to their being subjected to excessive closing forces.

Cylinder Pressure During recent years engine cylinder combustion pressures have greatly increased in diesel engines. Some engines produce peak cylinder pressures of 3,000 psi. Because the head of a cylinder valve is subject to cylinder pressure, the head of a cylinder valve is designed to flex. This flexing action places the underside of the valve head under compression and the top side under tension. When an engine is operated under abnormally high-combustion pressures, fatigue cracks usually initiate on the top side of the valve head close to the fillet. Valve failures due to excessive cylinder pressures are comparatively rare.

Reusing Cylinder Valves

During an engine overhaul, cylinder valves can normally be reused. They should be checked for:

- Fatigue cracks
- Thermal cracks
- Pitting
- Carbon coking
- Dishing distortion

Loose Valve Adjustment

Apart from producing a lack of power, loose valve adjustment results in cyclic overloading of cylinder head valves and the valvetrain. This occurs because the transition geometry on the camshaft lobe that governs how a valve is accelerated on opening and decelerated on closing is eliminated. This results in cyclic:

- Compressive overloads on the valve during opening
- Tensile overloads on the valve as it seats on closing between the valve seat and the keeper grooves
- Shear overloads on the keeper groove lands

Figure 27–10 illustrates the action of a properly set cylinder valve versus one set loose: Excessive lash tends to increase the impact force as the valve seats on closing.

Tight Valve Adjustment

Valves cold set with too little lash or set when hot to a cold engine specification can also be damaged because when the engine is at operating temperature, the valve may not seat with the required amount of force. Valves, especially exhaust valves, depend on being properly seated to dissipate the heat they are subjected to. When cylinder head valves overheat, they thermally distort and fail rapidly.

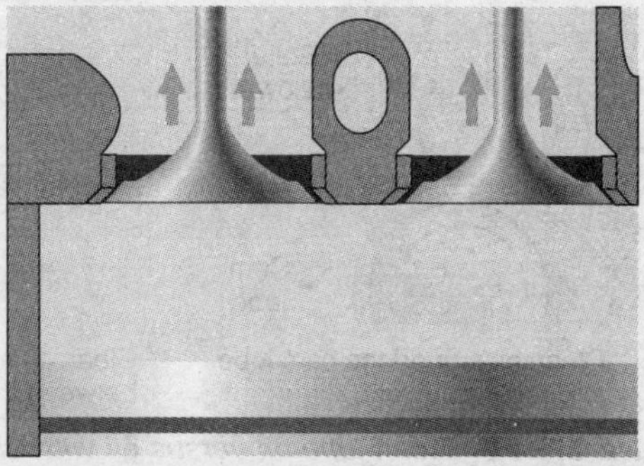

Valvetrain operation with valve lash properly adjusted.

Excessive lash causes increased closing impact.

Figure 27–10 *Closing forces of a properly set valve versus a loose set valve. (Reprinted Courtesy of Caterpillar Inc.)*

Camshafts

There are two primary reasons for premature camshaft failure:

- Poor lube quality
- Excessive valvetrain lash

When poor lube quality is the cause of a camshaft failure the cause may be high contaminant levels (soot load) or low viscosity. High soot loads suspended in lube oil results in abrasive wear to both cam lobes and the followers that ride the cam profile. Abrasive wear of camshafts is illustrated in Figure 27–11, and the consequences of low viscosity oil providing poor boundary lubrication is shown in Figure 27–12.

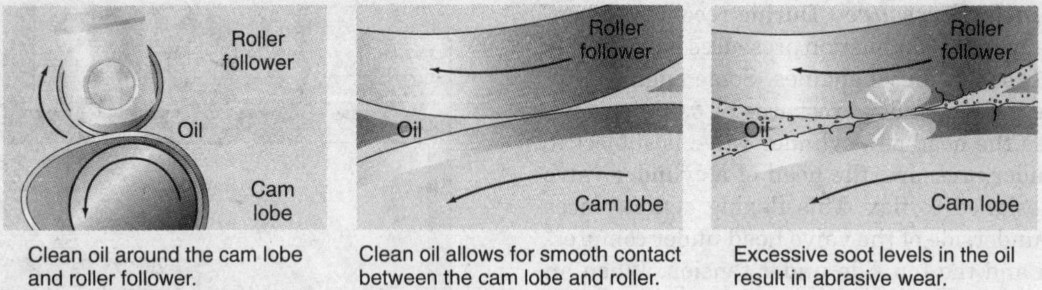

Figure 27–11 *Clean lube oil versus oil with elevated soot levels on cam contact surfaces. (Reprinted Courtesy of Caterpillar Inc.)*

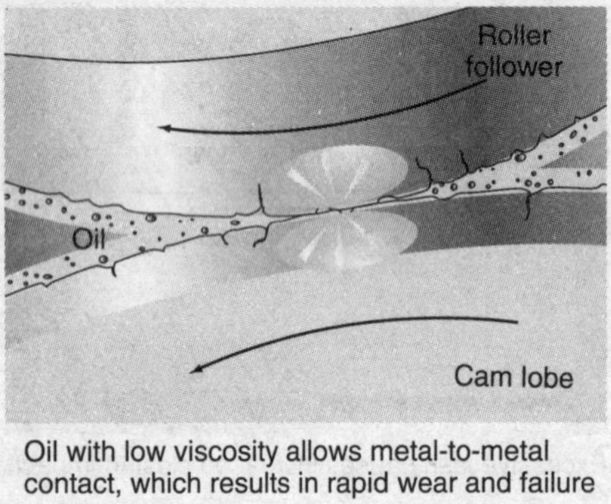

Figure 27–12 *Oil with low velocity provides inadequate boundary lubrication. (Reprinted Courtesy of Caterpillar Inc.)*

Excessive valve lash results in a hammering effect on the cam profile. On actuation, the valvetrain is accelerated to the extent it loses contact with the valve lobe: Almost immediately, the valve spring drives the follower back down onto the cam profile, pounding out the subsurface of the cam profile below the hard surfacing. Figure 27–13 shows the effect of excessive valvetrain lash on a camshaft.

Liners

The inside diameter of most liners is surface hardened and honed to provide a horizontal crosshatch of 60 to 70 degrees. This provides a surface that can retain oil to lube and seal the ring contact faces while wearing well. Having the specified crosshatch is essential to prevent combustion gases from blowing by the rings while limiting oil consumption. Anything that compromises the liner contact surfaces will result in cylinder gas blowby and oil burnoff during combustion.

When attempting to determine the root cause of an engine failure, ensure that the liners are removed, numbered by cylinder using a paint stick or crayon, and paired with their pistons. Do the same with the rings. Clean up the components using a noncorrosive solvent. The objective is not to remove any of the indicators that may have caused the failure so never use bead or grit blasting, wire brushes, or aggressive cleaning solutions.

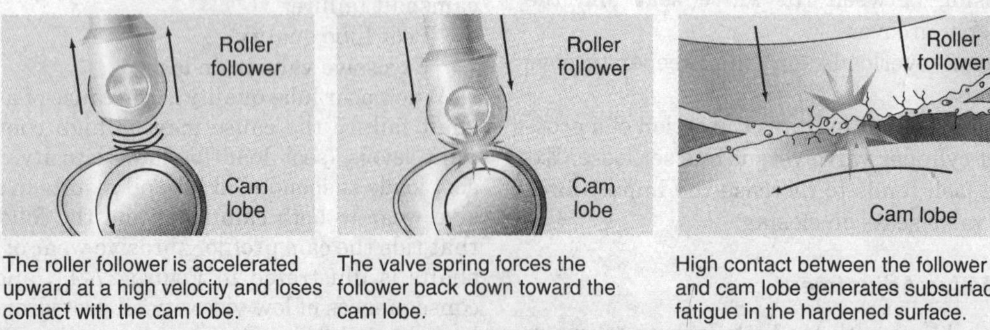

Figure 27–13 *Effect of excessive valve lash on cam profile. (Reprinted Courtesy of Caterpillar Inc.)*

Scoring

A tight fit of a dry sleeve od to its cylinder bore can cause sleeve distortion. A modern dry sleeve is usually designed to expand into the cylinder block bore as it heats up so if fitted with an interference fit it can cause buckling. A tight fit of the piston assembly to the sleeve id can also result in scoring.

Scuffing

Scuffing results when the sleeve-to-bore fit is too loose in dry liners, resulting in poor heat transfer. Sleeve temperatures are higher than usual and discoloration often results.

Etching

This has become a problem in some post-2004 exhaust gas recirculation (EGR) engines. Soot routed back into the intake from the EGR circuit can create etching at the top of the ring belt sweep. Soot, while in itself abrasive, also combines with moisture and fuel to form acidic residues that can etch a groove in the liner.

Running Cool

Running an engine at lower than specified operating temperature may result in condensing combustion end gases, and this can produce acids. Although most diesel engine oils specified by the engine OEM are designed to neutralize these acids, inferior oils or prolonged cold operation can deplete the additives, resulting in liner pitting or etching. When the duration of this type of corrosive attack is prolonged, graphite (carbon) flakes start to separate from the liner cast iron. The result is an engine that burns oil.

Prolonged Low-Load Operation

Despite the fact that idling a diesel engine is now illegal in most jurisdictions, truck drivers can be pretty inventive when it comes to defeating idle-shutdown programming and not too many states aggressively enforce anti-idling regulations as yet. Diesel engine piston rings and liners are engineered to operate with high surface contact loads. Prolonged periods of idling create frictional heat that erodes the liner crosshatch, producing a glazed finish. Once the liner ring belt surface area becomes glazed, cylinder blowby increases and the engine begins to burn oil. Most OEMs agree that prolonged idling creates oil-burning diesel engines. The good news is that engine control module (ECM) audit trails can identify the incidence and duration of unnecessary idling, and this may be used by fleets to crack down on this type of abuse.

Exterior Scaling

Exterior scale on wet liners can be caused by using depleted coolant mixtures, improper coolants, or water in place of coolant mixtures. Any type of scale formation on a wet liner has a considerable insulating effect, resulting in higher cylinder temperatures. If a water-only coolant was used in a diesel engine, the result can be rapid formation of rust on the outside of the liner. The liner then overheats, causing heat discoloration and heat checks on the inside diameter of the liner. The quality and mineral load in water that is mixed with diesel engine antifreeze/coolants is also a factor of liner exterior wall scaling. You can avoid many problems by mixing antifreeze with pure distilled water—or better still, use long-life, premixed diesel engine coolants.

Liner Cavitation

Cavitation caused by vapor bubble implosion can shorten the life of wet liners but this should not be a problem if the coolant chemistry is properly monitored. A full explanation of cavitation erosion is provided in Chapter 8. Wet liners, when subject to cylinder pressures, which in today's engines may be as high as 3,000 psi (24 MPa), expand outward into the wall of coolant that surrounds them then contract, creating a vapor bubble. This "bubble" almost immediately collapses, causing the wall of coolant to collide onto the liner exterior wall. When this condition repeats itself at high frequency (17 times per second at 2,000 rpm), cyclic pressures of up to 60,000 psi (414 MPa) eat away at the liner wall. Bubble collapse erosion only occurs when diesel engine coolant is depleted or contaminated with high-mineral content in the water.

Wet liner cavitation can be identified by pitting/erosion that usually appears on the liner outside the thrust faces of the piston. If left, the erosion can actually perforate the liner.

When liners show evidence of cavitation erosion, the coolant quality and service frequency should be examined. The practice of reusing a cavitated liner by rotating it 90 degrees on reinstallation is probably not cost-effective. Persistent cavitation problems on wet liners can usually be cured by using extended life coolant (ELC).

Liner O-Ring Failure

Check seating areas thoroughly and also check that correct sealing medium has been used. The sealing medium used on the O-ring during installation depends on the O-ring material. This can be dry (nothing), antifreeze (ethylene glycol [EG], propylene glycol [PG], etc.), soap solution, OEM sealants, and so forth. It is important to remember that what may be referred to as a rubber can actually be one of a number of different rubber, silicone, and petrochemical compounds; therefore, the OEM instructions should be observed. When performing failure analysis on failed O-rings, check for chemically degraded O-ring material caused by inappropriate sealing compounds.

Coolant in Combustion Chamber

This condition is evidenced by white smoke emission and the characteristic bittersweet odor of vaporized coolant: A cracked wet liner flange can leak coolant into the combustion chamber.

Wet liners should be installed by hand and subsequently checked with a plug gauge. This is machined to within 0.0015″ (0.038 mm) of the liner id and ensures that it is not distorted. Liner flange protrusion of a set of liners clamped under one head should normally be within 0.001″ (0.025 mm) of each other; check OEM specification.

Installation Malpractice

Diesel engine dry sleeves and wet liners are routinely replaced at a scheduled engine overhaul without much inspection. Correct installation procedures should be observed to ensure maximum engine life. It is common practice to install current sleeves 0.0005″ (0.0127 mm) to 0.0015″ (0.038 mm) loose; this means that in some cases, the sleeve can be pressed into its cylinder bore entirely by hand. Interference-fit sleeves are seldom used in current diesel engines, and machining an interference can result in premature failure because the sleeve will buckle. Selective fitting of sleeves to bores should be practiced; this requires measuring the od of each new sleeve and each bore id, then getting the largest sleeve to the largest bore, down to the smallest sleeve to the smallest bore.

Counterbore depth should be inside the specification window and usually within 0.001″ of each other under one cylinder head. Recutting counterbores and shimming are common practices, but these procedures usually require special tools to properly perform them. When honing cylinder dry sleeve bores in the engine block, ensure that the machining residues of the procedure are completely removed. This is especially important when undertaking the procedure in chassis with the crankshaft in place.

Friction Bearings

Despite some OEMs' opinions to the contrary, it does not make economic sense to reuse connecting rod and main bearings. When bearings are removed from an engine, whether at overhaul, bearing rollover, or other engine problem, they should be numbered and examined with a view to determining the cause of failure. Many failures are lube oil related, especially contaminated engine oil. When engine lube becomes contaminated, the bearings become especially vulnerable as shown in Figure 27–14. Improper installation practices can also shorten a bearing's life. An example of a color bearing failure analysis guide is shown in Figure C1. Table 27–3 lists some friction bearing failures and identifies some causes and outcomes of those failures.

Figure 27–15 shows a broad range of typical engine bearing failures.

B-Life

Projections of engine longevity often refer to **B-life** ratings. The term *B-life* actually originated

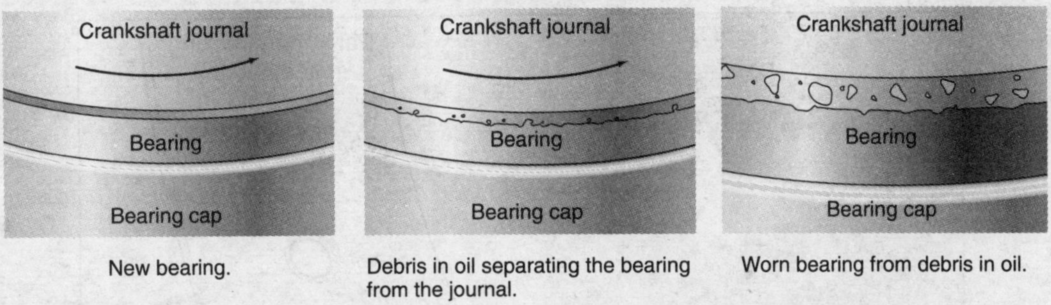

Figure 27–14 *Consequences of contaminated engine lube on friction bearings. (Reprinted Courtesy of Caterpillar Inc.)*

Table 27–3 *Friction Bearing Failures*

Condition	*Causes/Results*
Etching	Chemical contamination of lube—usually fuel, coolant, or sulfur.
Glazing	Broken down lube or improper lube.
Lacquering	Orange/brown film on bearing face caused by poor quality lube or high sulfur content in lube.
Pitting	Particulate/abrasives in lube oil.
Spalling (localized welding)	High cylinder pressures caused by high manifold boost, overfueling, lugging, or detonation—less commonly a result of excessively high lube oil temperatures.
Arcing	Incorrect chassis welding practices.
Low oil pressure	Associated with bearing wear and may be grounds for a bearing rollover.
Bottom end failures	Spalling enlarges the bearing id and the resultant hammering increases tensional loading on the throw/conn rod assembly.
Insufficient bearing crush	Results in poor heat dissipation and a failure characterized by heat discoloration.

from engineers' analyses of ball bearing life: It was a statistically based study of how long a given percentage of bearings would last in terms of operational hours. Today we use B-life as a means of reckoning costs over the life of engines and other powertrain components. The *B* refers to "bearing" and the number that follows is a percentage. Most B-ratings in the trucking industry relate to linehaul miles (truck vocational and off-highway engines use engine hours).

Table 27–4 shows some Navistar data that relates to two of its older engine series. The B10 mileage indicates the specific mileage achieved before 10% of the engines in the model family failed. Similarly, the B50 rating indicates that 50% of the engines had failed at the next mileage value. This type of data today tends to be pretty accurate due to the audit trails recorded by engine ECMs and downloaded for OEM analysis.

Table 27–4 *B-Life Ratings of Two Engines*

Engine Model	*B10*	*B50*
DT 466E	300,000 miles	450,000 miles
T 444E	200,000 miles	350,000 miles

In Table 27–4, the T 444E and DT 466 are small- to medium-bore engines used in applications such as school buses and pickup and delivery vehicles. Both these applications tend to be tough on diesel engines because they involve a lot of stop/start running. Most fleet managers will say that the B10 rating is the better predictor of how long an engine will reliably last.

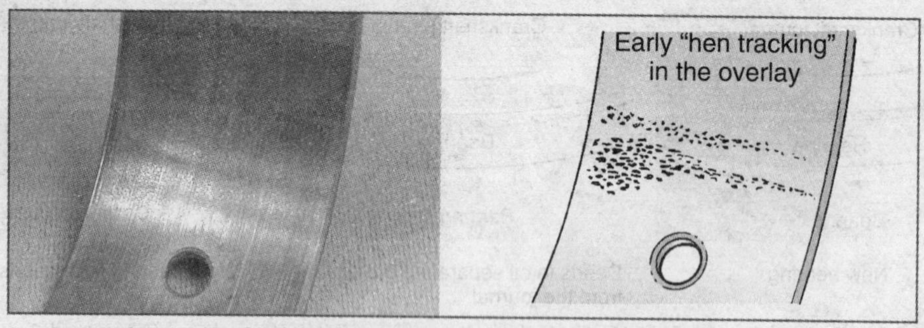

A. Overlay fatigue, normal wear: early stages

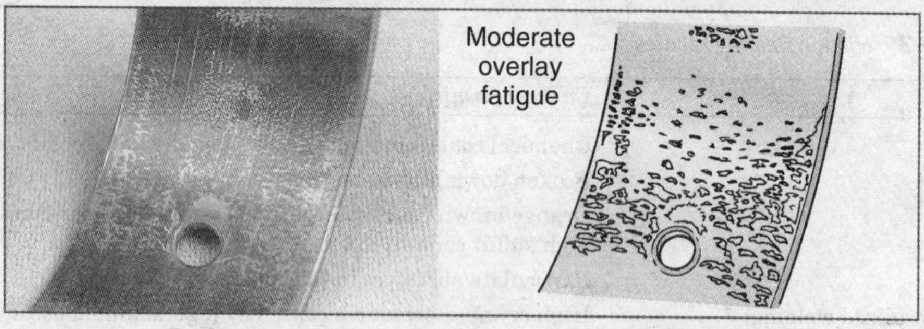

B. Overlay fatigue, normal wear: intermediate stage

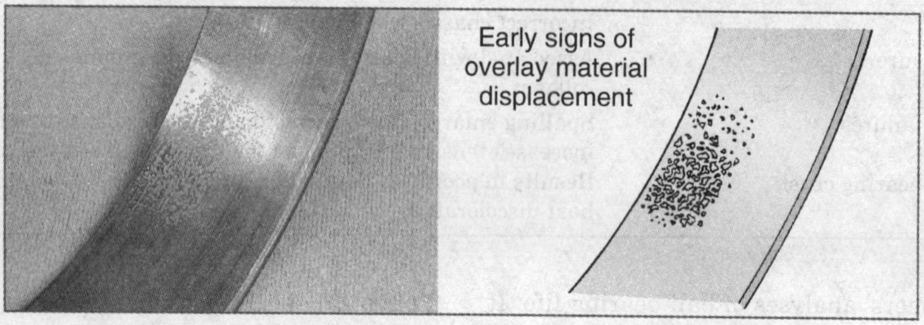

C. Orange peel appearance, normal wear: early stages

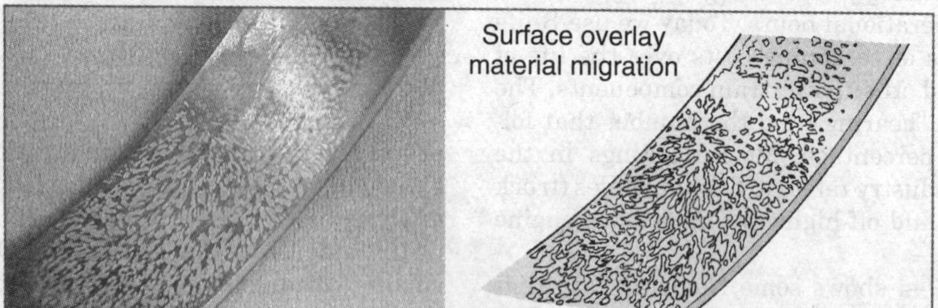

D. Orange peel appearance, normal wear, high mileage condition, intermediate stage

Figure 27–15 *Engine bearing distress analysis. (Courtesy of Federal Mogul Corporation)*

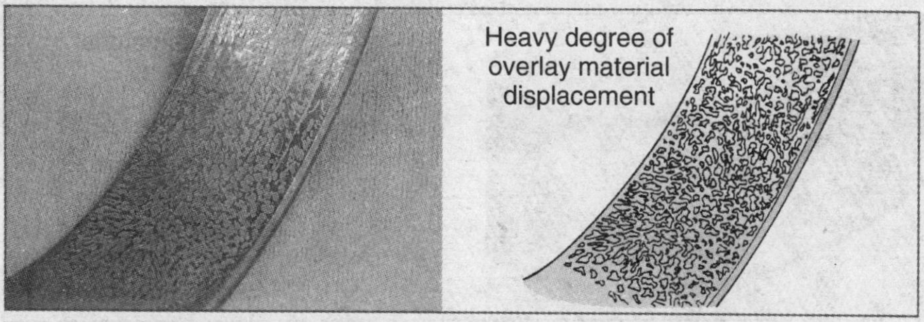

E. Orange peel appearance: advanced suction erosion condition caused by high mileage

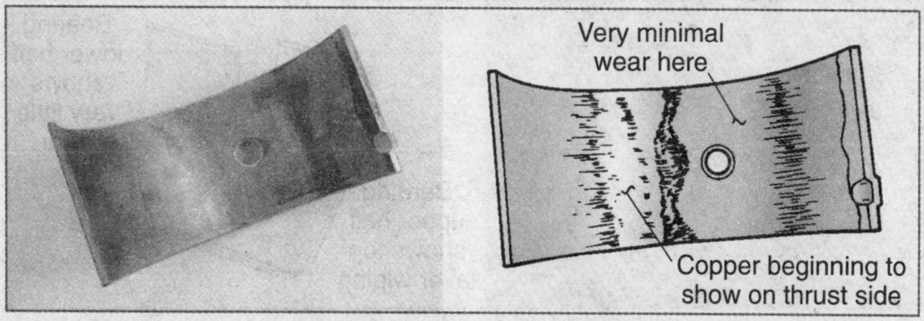

F. High mileage, normal wear

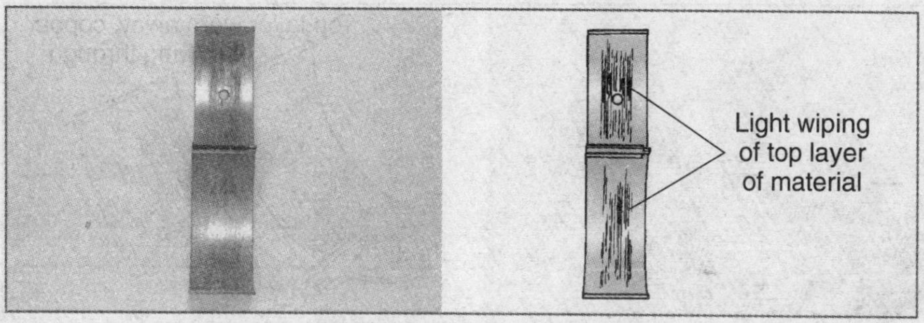

G. Lack of lube: early stages

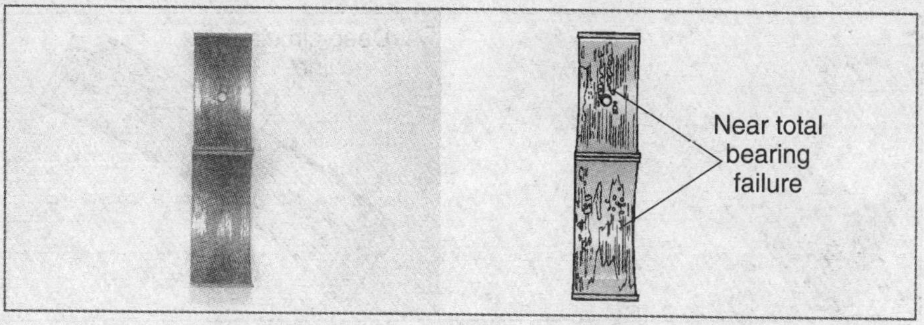

H. Lack of lube: advanced stage

Figure 27–15 *(Continued)*

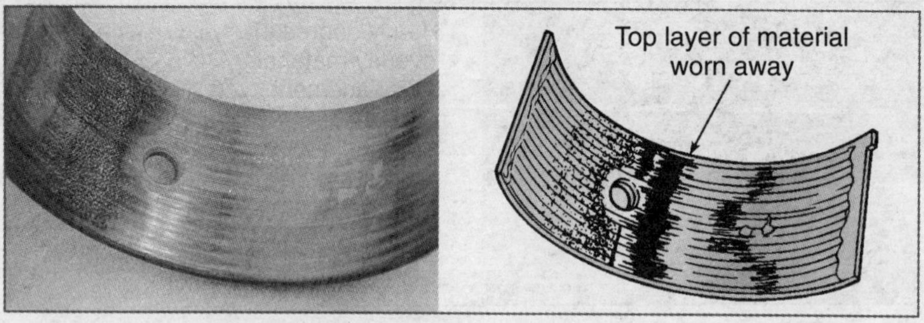

I. Coolant contaminated lube

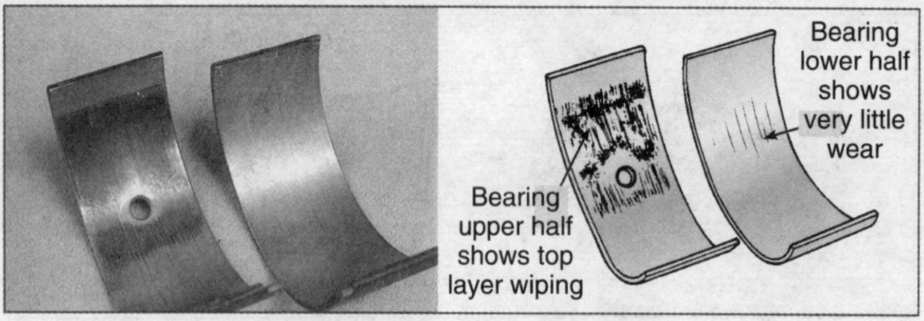

J. Fuel contaminated lube: early stages

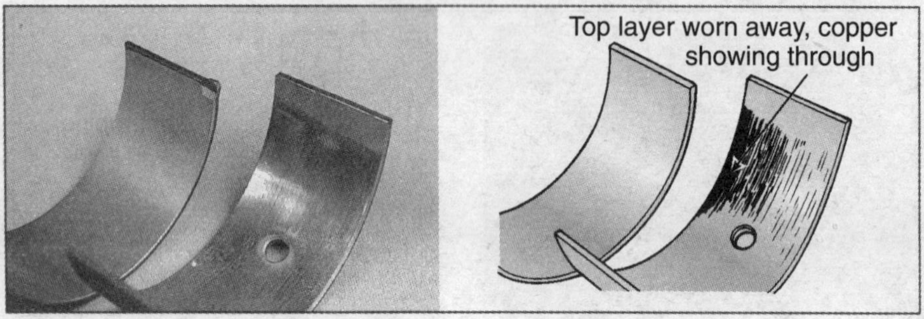

K. Fuel contaminated lube: advanced stage

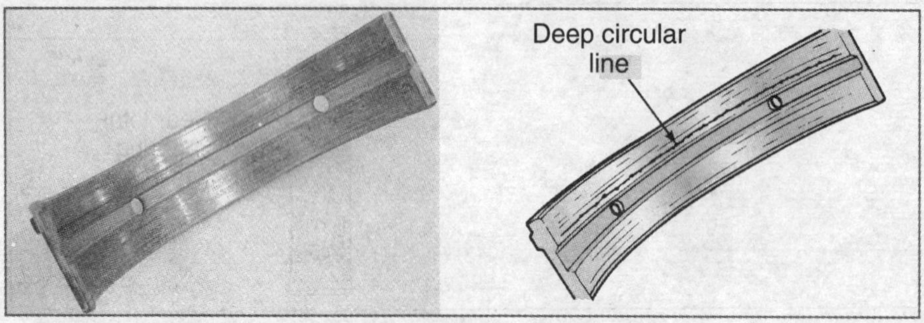

L. Metal particles in lube: early stages

Figure 27–15 *(Continued)*

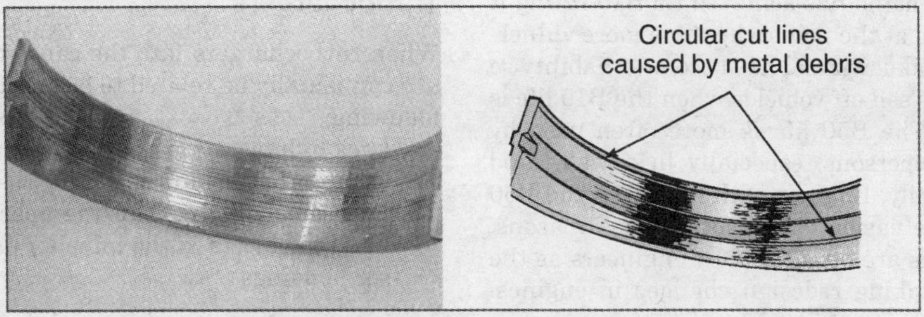

M. Metal particles in lube: advanced stage

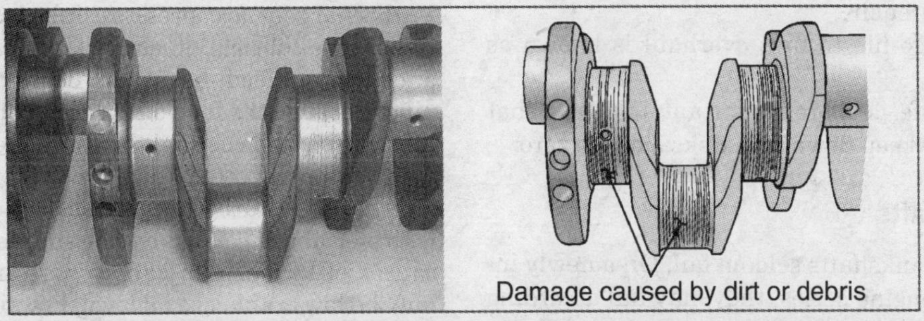

N. Crankshaft journals damaged by dirt or metal debris in lube

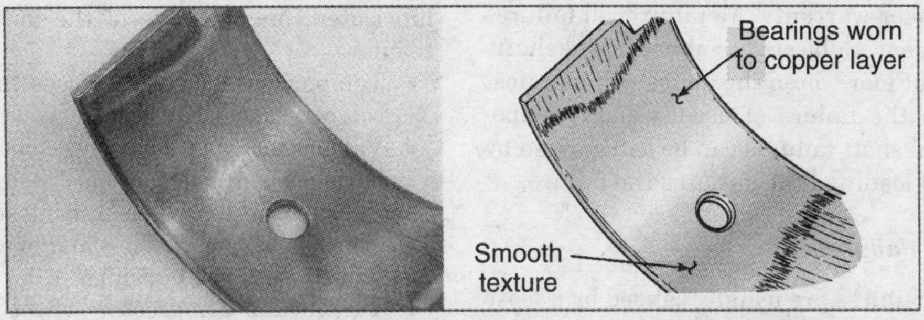

O. Airborne dirt in oil: abrasive action wear to a smooth texture

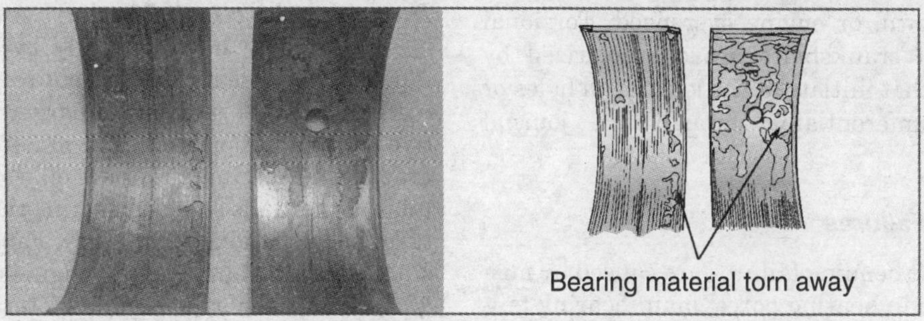

P. Hot short rod bearing failure caused by high shear loads and extreme temperatures

Figure 27–15 *(Continued)*

When an engine has achieved its B10 rating it has arrived at the point when it is more vulnerable to breakdown and has become a liability so some fleets sell off vehicles when the B10 life is achieved. The B50 life is more often used by OEM salespersons, especially in a well-rated engine family. Engines with high B10 and B50 ratings are easier to sell for obvious reasons. B50 ratings are often used by engineers as the basis of making redesign changes in engines. We can summarize B ratings by saying:

- Bx life is a prediction of engine life before major repair.
- Average life to first overhaul is known as B50.
- B10 life is more important to fleets that like to keep downtime risks close to zero.

Crankshafts

Modern crankshafts seldom fail. On a newly introduced engine series, they may fail according to a pattern. Always report crankshaft failures to the manufacturer whether or not the equipment is under warranty. Metallurgical failures are infrequent. Subjected to abuse, crankshafts will fail, but more often the abuse will manifest itself with the failure of a subsidiary component. Crankshaft failures can be categorized by the type of loading that initiates the failure.

Torsional Failures

Torsional failures are usually caused by a loose or defective crank vibration damper or flywheel, out-of-balance engine-driven components, engine lugdown, or engine overspeed. Torsional failures of crankshafts are characterized by fractures that initiate at the journal oil holes or run circumferentially through the journal fillets.

Bending Failures

Crankshaft bending failures are caused by misaligned main bearing bores, main bearing failure, or broken/loose main bearing caps. This type of failure is characterized by cracks that initiate at the main journal fillet and extend at a 45-degree angle through the throw journal fillet.

Turbochargers

When turbochargers fail, the cause of the failure can usually be related to one or more of the following:

- Lack of lubrication
- Contaminated lube oil
- Excessively high exhaust temperatures
- Foreign objects in the impeller or compressor housings

Lack of Lubrication

Turbochargers are pressure lubed directly by the engine lubrication circuit, in most cases by a dedicated feed hose and dedicated return pipe. Because the lube oil is responsible for both lubrication and cooling, the consequences of oil deprivation are often almost immediate. A common cause of turbocharger lubrication failures is driver abuse caused by hot shutdown. Other causes of lack of lubrication in a turbocharger may be low crankcase oil level, low oil pressure, inappropriate oil, or excessively high oil temperatures. When a turbocharger is not properly lubricated, one or more of the following may result:

- Tempering of metal surfaces indicated by coloration, usually blue
- Varnishing of lube oil onto bearing surfaces
- Failure of the turbine housing seal
- Contact of the turbine and/or compressor wheels with the housings
- Turbine or compressor wheel separation from the turbo shaft

Inappropriate Lubrication A turbo shaft is supported hydrodynamically by the engine lube and because it can wind out revolutions as high as 140,000 rpm, turbochargers are sensitive to inappropriate, contaminated, or degraded lube oil. Inappropriate oil would include lube oil that falls short of meeting the manufacturer specifications or viscosity. For instance, a single cold-start event in an engine using 15W-40 lube oil in northern, midwinter conditions can trigger a range of engine problems, including damage to the turbocharger. The use of low-cost bulk lube oils is widespread among fleets: Because these oils often lack additives that define the hydrodynamic and boundary throwoff properties required by the OEM specification, when turbine shafts are wound out at their highest speeds, centrifugal force throws oil from the turbine shaft, resulting in accelerated wear and bearing failures.

When oil becomes contaminated with abrasives or liquid contaminants such as fuel or coolant, the hydrodynamics of the turboshaft are compromised: The result can wear out the shaft and destroy seals and bearings. Another common cause of turbocharger lubrication failures is degraded lube oil. Not observing the OEM-recommended oil change intervals results in engine lube supporting a higher load of blowby residues and reduction in the ability of the additives to function effectively to maintain the hydrodynamic and boundary lubrication requirements of the turbo.

Hot Shutdown When an engine is shut down immediately after a high-load run, the oil feed to the turbocharger immediately ceases. Lubrication oil to the turbocharger supports the turbine shaft hydrodynamically but also plays a major role in cooling. When a hot shutdown occurs, the turbine shaft is usually rotating, so initially lube oil is thrown off the shaft by centrifugal force. Next, high shaft temperatures boil off oil and bake residual oil to the shaft, causing varnishing. When the engine is next started, the turbo remains dry until lube oil reaches it from the lube circuit: This dry-start condition, accompanied by the abrasive effect of the varnished oil, scores the shaft and rapidly wears the softer bushings.

Because it is such a common cause of turbocharger failure, the tattletales associated with hot shutdown should be identified. Hot shutdowns often result in varnishing of the turbine side bearing: When the cooked varnishing is peeled away, abrasive scoring of the shaft is often evident. A key to identifying hot shutdown abuse are quench spots or rings on the turbine shaft close to the turbine housing: This is caused when drops of oil fall on a severely overheated shaft, causing quenching discoloration. The abrasive effect of hot shutdowns impacts on the bushings, wearing them rapidly to the point that both the turbine and compressor wheels contact their housings.

External Debris Failures

When considering "external" debris damage to a turbocharger, the gas flow through the assembly obviously has to be accounted for to source the material. Debris entering the impeller housing is sourced upstream from the compressor or "cold" side, while debris entering the turbine housing or "hot" side has been discharged through the engine cylinder or cylinder head. External debris entering a turbocharger impeller housing has usually bypassed the air intake system components upstream from the housing, meaning that the failure begins there. When intake ducting, hose clamps, or the engine air filter fail, in more cases than not, the damage that results is gradual rather than catastrophic in nature. Dusting of an engine occurs when an air filter develops a small puncture or tear, an intake air hose clamp loosens, or a small hole develops in ducting or hoses upstream from the impeller housing. The fine dust ingested by an engine in this location may be extremely abrasive. Its effect on the turbocharger impeller can be to wear its vanes or blades paper thin, resulting in loss of power and possible fracture.

When external debris enters a turbine housing, the turbocharger suffers collateral damage caused by an internal engine condition such as a dropped valve, separated valve seat, broken ring(s), or injector damage. Unlike most cold-side originated turbocharger failures, external debris originating from within the engine cylinder tends to cause immediate, catastrophic failures. When investigating external debris turbocharger failures, it is important to describe the appearance of debris so that it can be sourced and/or referenced to others for analysis:

- Size
- Shape: uniform, nonuniform
- Material: magnetic, nonmagnetic
- Color: may indicate temperature at which the failure occurred

Some typical indicators of foreign debris in a turbocharger are:

- Bent vanes
- Broken vanes
- Bent turbine shaft
- Wheel separation

Note that damage to the impeller wheel is more likely to occur at its inside diameter. Similarly, damage to the turbine wheel is more likely to occur at its outside diameter. This is common sense really and results from the gas flow through the turbocharger.

High Exhaust Temperatures

Turbocharger failures due to excessively high exhaust temperatures are less common today simply because with computer-controlled engines it has become more difficult to either overfuel or advance injection timing. High exhaust temperatures heat the turbine housing and this heat can penetrate the turbo core. The result can be oxidation and distortion of the core components, including the seals and bearings. High engine exhaust temperatures cause:

- Carburized, degraded engine oil
- Oxidation, surface erosion
- Scaling and coking (carbon crystallization) on components
- Tempering of components (indicated by surface coloring blue/purple/yellow)
- Turbine seal failure
- Worn bearings
- Turbine and impeller wheel housing contact
- Wheel separation from the turbine shaft

EGR-related Problems

EGR failures are commonly sourced to the EGR heat exchanger, the valve that controls EGR mixture concentricity, or the Delta pressure sensor circuit. Symptoms of EGR circuit failure relate to soot and sludge buildup in the venturi, the control gate, or in the intake manifold. When there is evidence of excessive soot or sludge in the intake manifold, troubleshooting the EGR circuit should be undertaken beginning with checking any logged codes. Most OEMs comprehensively monitor EGR circuit performance, so in almost all cases either fault codes or an audit trail should provide some clues to source the problem. Investigating EGR failures should take into account EGR valve malfunctions and EGR heat exchanger internal failure. Some indicators for troubleshooting EGR problems are provided in Chapter 42, but this should always be undertaken guided by the OEM service literature when working on actual engines.

Diesel Particulate Filter (DPF)-Related Problems

DPF failures are often characterized by black smoke emission but when attempting to source the problem, you will be required to check out all the circuits upstream from the DPF. As in the case of the EGR circuit, all OEMs comprehensively monitor the DPF and its regeneration history so you must begin any troubleshooting procedure by connecting to the chassis data bus. When checking the cause of black smoke emission from an EGR- and DPF-equipped diesel engine you should use the OEM electronic service tool (EST) and service information systems (SIS) to check out the following:

- Check for any diagnostic trouble codes that relate to the engine breathing circuit.
- Inspect the DPF for external dents and cracks (impact damage).
- Remove the DPF and check for the presence of oil or fuel. If either is present, this should direct you to source the fuel or oil ingression into the circuit (e.g., fuel injection, turbocharger, etc). If not,
- Check out the venturi for the EGR or clean gas induction (CGI) system. Check out the EGR or CGI control valve and the EGR or CGI heat exchanger.

Injector-related Problems

We are going to take a brief look at injector-related failure problems. To do this we are going to focus on commonly used, electronic unit injector (EUI) assemblies despite the fact these are not studied in detail until the third section of this book. Before beginning, it should be pointed out that the self-diagnostic capability of most of today's electronically controlled diesel engines identifies most injector failures. Most injector failures tend to produce no external evidence that is visible to the naked eye.

Internal Leakage

The number one cause of EUI failure is internal leakage. Internal leakage is caused by a failure of the poppet valve to seal in the control cartridge (see Chapters 36 and 37). An EUI poppet valve has an angled seat that when closed traps fuel in the EUI pumping circuit. Because this valve opens and closes with some force up to 19 times per second at rated engine speed, any abrasives in the fuel can create wear. When an EUI poppet valve fails to seal, the result is low power sometimes accompanied by fueling imbalance and visible smoke emission. In most engines manufactured after 2004 (and some before), performing an engine cylinder balance test (or multicylinder cutout test) can identify internal leakage before it begins to seriously affect performance. Contaminated fuel is the usual cause of internal leakage in an EUI but it may also be caused by cavitation erosion, which is described in the next section.

Fuel Supply Pressure

The fuel subsystem must charge injection pumping chambers (on EUIs, hydraulically actuated electronic unit injectors [HEUIs], electronic unit pumps [EUPs], and pump-line-nozzle [PLN] systems) at their specified pressure, or cavitation damage can result. After the completion of the effective stroke in all types of high-pressure injection pump, the pumping plunger is retracted by a powerful spring. As the plunger lifts to aspirate the pump chamber, supply fuel at charging pressure must flow into the pump chamber rapidly to avoid creating a vacuum. If fuel subsystem charging pressures are lower than specification due to restricted filters, vacuum bubbles may be formed: As system pressure recovers (almost instantly afterward) this fuel pressure acts on vacuum bubbles, causing them to implode. The result is what we know as cavitation erosion. Vacuum bubble implosion can eat away the surface material of the plunger and barrel and also damage sealing surfaces. Cavitation erosion usually requires the replacement of the injector pumping mechanism. Figure 27–16 shows how cavitation can damage an EUI.

Seized Injector Failure

Seized injectors are usually easy to identify because the follower usually binds well below the stop tram as shown in Figure 27–17. The usual cause is contaminated fuel, usually with water.

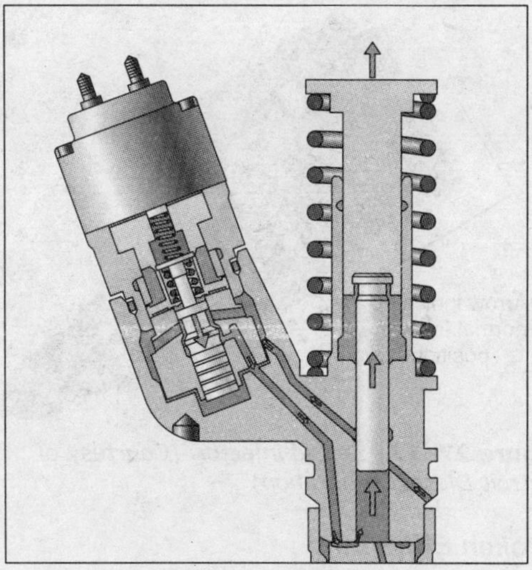

On most EUI injectors, the tappet and tappet spring lift the plunger and pull supply back into the plunger cavity.

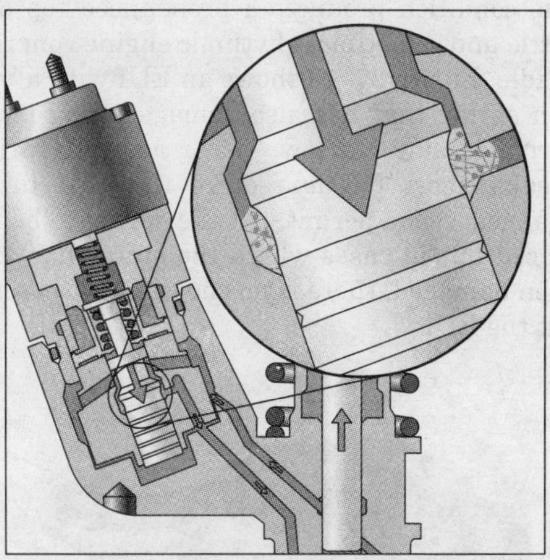

Low fuel supply pressure from plugged fuel filters can cause cavitation damage to injector poppet valve during injector fill.

Figure 27–16 *Low fuel subsystem charging pressure can result in injector cavitation damage due to delayed pump chamber fill. (Reprinted Courtesy of Caterpillar Inc.)*

If this occurs, check the fuel tanks for water and investigate to determine whether excessive methyl hydrate (alcohol) is being dumped into the vehicle fuel tanks. If an EUI seizes when it is nearly new, it may indicate an OEM quality control problem. In all cases, when an EUI seizes, change it: Do not attempt to unseize and reuse it.

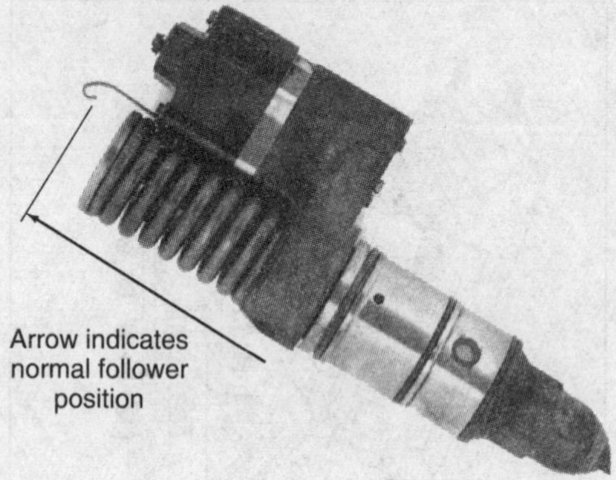

Figure 27–17 *Seized injector. (Courtesy of Detroit Diesel Corporation)*

Broken EUI Spring

When an injector retraction spring breaks, it may also damage the control cartridge. The breakage is usually the result of metal fatigue. The condition produces a pronounced top end rattle and sometimes rhythmic engine rpm role at idle. Figure 27–18 shows an EUI with a broken spring that has also damaged the control cartridge stator screw. When an EUI spring breaks, most OEMs require the EUI to be changed: Some permit just the spring to be replaced but in cases where the stator has also been damaged, there is no choice but to change out the EUI.

Figure 27–18 *Broken injector follower spring and damaged stator. (Courtesy of Detroit Diesel Corporation)*

Loose or Missing EUI Stop Plate Screw

EUI stop plate integrity should be checked when you have a fuel-in-oil complaint that has no other obvious cause. The stop plate should be checked for looseness along with the stop plate screws. The actual stop plate seal is an O-ring so even in cases where the stop plate screw torque appears OK, a rolled or pinched O-ring may be the cause. Figure 27–19 shows the location of the stop plate and its three screws on a DDC EUI.

Figure 27–19 *Loose stop plate screw failure. (Courtesy of Detroit Diesel Corporation)*

Cracked EUI Body

At minimum, a cracked injector body will result in a fuel-in-oil problem: It is usually accompanied by cylinder misfire. The cause is usually fatigue caused by prolonged high-load operation or plain old age. It is mostly commonly seen in EUIs that have been in service for a decade of operation. The crack may be difficult to identify with the naked eye. Figure 27–20 shows an EUI with a cracked body.

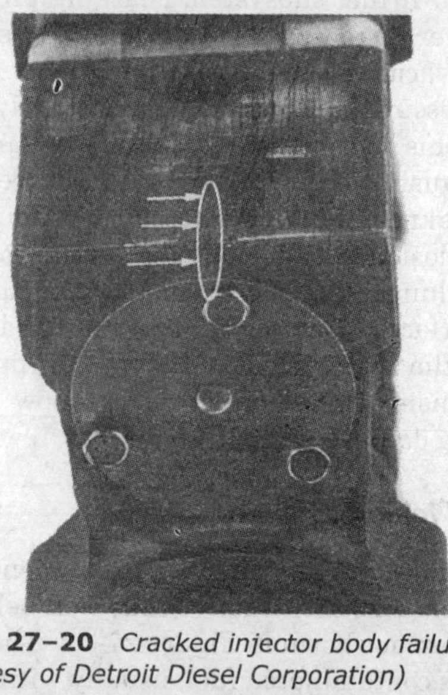

Figure 27–20 *Cracked injector body failure. (Courtesy of Detroit Diesel Corporation)*

Broken EUI Terminal

The usual cause of broken EUI terminals is overtorquing during installation. It is a common problem that you get to know about quickly if you have caused it because it usually produces a next day "come-back." Although one OEM permits control cartridge replacement, usually you have no choice but to replace the EUI, this time observing the torque specification. Figure 27–21 shows a sheared EUI terminal.

Figure 27–21 *Solenoid terminal screw failure. (Courtesy of Detroit Diesel Corporation)*

Blown Nozzle Tip

The symptoms of a blown EUI nozzle tip are misfire, low power, and engine smoking. The more common cause is water in fuel so when this occurs, check the tanks. To get water pumped through a modern fuel system as far as the injectors usually requires substantial abuse of methyl hydrate (alcohol): Reference Chapter 17 if you are not familiar with the consequences of methyl hydrate/fuel conditioner dosing in diesel fuel. The other cause of blown nozzle tips is physical damage caused by dropping the EUI prior to installation or a dropped cylinder valve. Figure 27–22 shows an EUI with the nozzle tip blown away.

Figure 27–22 *Blown nozzle spray tip failure. (Courtesy of Detroit Diesel Corporation)*

COMMON ABUSES OF ENGINES AND FUEL SYSTEMS

The following are some common abuses of engines and fuel systems that the diagnostic technician should always keep in mind when performing failure analysis.

- Cold-startup procedure: excessive use of ether, excessive engine speeds when oil temperatures are low, excessive engine loading when engine temperature is low.
- Excessive engine idling: This can glaze cylinder walls or break down engine lube (rings do not seal well when cylinder pressures are lower); prolonged idling at low temperatures may cause fuel contamination of engine lube.
- Overfueling: causes high cylinder pressures (which can reduce engine life or cause outright mechanical failure), increases noxious emissions, reduces engine longevity, and costs money through wasted fuel.
- Use of inappropriate engine oil: Using 15W-40 engine oil during winter in northern United States, Canadian, and subarctic conditions damages engines and on the way to doing that, can make cold-start cranking extremely difficult.
- Engine overspeed: can cause valve float, which may result in dropping a valve, and increases tensional loading on connecting rods and crank throws, which may result in complete engine failure.

- Advancing fuel injection timing: A common abuse, the objective being to improve engine performance; the consequences of this practice vary from greatly reducing engine life at best, to rapid mechanical engine failure.
- Fuel additive abuses: These include adding to fuel moth balls, gasoline, kerosene, alcohols, aftermarket fuel conditioners, engine oil, and ATF.
- Fuel heaters are not generally recommended by OEMs as the fuel's lubricity is compromised by heating it. The preferred method of running an engine through arctic, subarctic, Canadian, and northern United States midwinter conditions is to use #1D fuel, which is temperature conditioned with cloud point depressants and glycol additives (arctic fuel) by the refiner.
- Hot shutdown: can damage turbocharged engines especially. The turbine bearings are vulnerable because the turbine shaft is still rotating after the oil flow has ceased. Additionally, the heat from the hot turbine exhaust cooks the oil at the bearings, rendering them baked dry for the next startup.

Review Questions

1. An aluminum trunk-type piston has failed with a hole burned through the center of the crown. *Technician A* states that this might be due to a retarded timing condition. *Technician B* states that such a failure could occur caused by a dribbling injector nozzle. Who is correct?
   a. Technician A only
   b. Technician B only
   c. Both A and B
   d. Neither A nor B
2. Which of the following conditions could cause carbon buildup on pistons?
   a. Overfueling
   b. Injector nozzle failure
   c. Retarded injection timing
   d. All of the above
3. Which of the following conditions is the more likely result of advanced timing in a diesel engine?
   a. Erosion near the center of the piston crown
   b. Erosion pitting above the top compression ring land
   c. Overall piston scuffing
   d. Fractured rings
4. Which of the following operating modes would be most likely to result in a tensile conn rod failure?
   a. Overfueling
   b. Lugdown
   c. Ether abuse
   d. An overspeeding engine
5. Etched main bearings could be caused by all of the following conditions *except:*
   a. Fuel in lube
   b. Coolant in lube
   c. Sulfur in lube
   d. Particulate in lube oil
6. Evidence of cavitation erosion on a set of wet liners would most likely be caused by:
   a. High cylinder pressures
   b. Incorrect coolant chemistry
   c. Piston spalling
   d. Liner O-ring failure
7. Which of the following might result in glazed cylinder sleeves?
   a. High cylinder pressures
   b. Prolonged engine idling
   c. Ether usage
   d. Engine lugging
8. When observing a disassembled engine, the outer edges of all the pistons have evidence of melting and erosion. This could be likely caused by:
   a. Advanced timing
   b. A dribbling injector
   c. Retarded timing
   d. High sulfur content fuel
9. When examining a set of pistons from a disassembled engine, scuffing is observed on the thrust sides of all of the pistons. Which of the following would be the more likely cause?
   a. High cylinder pressures
   b. Overheating
   c. Contaminated engine lube
   d. High sulfur content fuel
10. Which of the following components when worn may cause lower oil pressures?
    a. Main bearings

b. Valve guides
c. Oil control rings
d. All of the above

11. What is the most common cause of EUI failure?
    a. Cavitation
    b. Internal leakage
    c. Metal fatigue
    d. Fried stator
12. You have determined that a set of EUIs has failed due to cavitation. Which of the following would be the more likely *root cause* of the failure?
    a. Restricted secondary filter
    b. Prolonged high-load operation
    c. Incorrect injector timing setting
    d. Double gasketing the EUI seat
13. *Technician A* states that B-life is used as a means of reckoning costs over the life of an engine. *Technician B* states that the letter B in the term *B-life* references predicted bearing longevity. Who is correct?
    a. Technician A only
    b. Technician B only
    c. Both A and B
    d. Neither A nor B
14. Which of the following is the correct technical description of what we call *scuffing*?
    a. Scraping
    b. Adhesive failure
    c. Thermal failure
    d. Catastrophic
15. *Technician A* states that B10 life is used by some fleets to determine how long they should keep trucks before dealing them. *Technician B* states that Bx life is a prediction of engine life before major repair. Who is correct?
    a. Technician A only
    b. Technician B only
    c. Both A and B
    d. Neither A nor B

PART 3

Electronics, Engine Management, and Emissions

28

Review of Electrical and Electronics Fundamentals

INTRODUCTION

This chapter reviews some of the basic electrical principles required to understand how an electronically managed engine functions. As it encompasses a range of principles that have been the subject of many textbooks, the approach here is that of reviewing the material, so you may find that you need to use other texts. There is no doubt that the modern truck technician should have a thorough grasp of the basics of electricity and electronics, which has become the single most important subject area. The days when many truck technicians could avoid working on an electrical circuit through an entire career are long past.

Learning Electricity

Mechanical technicians learn best by a hands-on approach. This fact tends to make electricity more difficult to learn because you cannot see or touch it. Some learning programs insist on beginning with the study of the science of electricity before progressing to its more practical aspects. Other programs teach practical electricity, the need-to-know material first, and introduce its theoretical aspects later. Both methods work. The approach used in this chapter is to begin by introducing some of the science of electricity and then progress toward applied electricity. You can skip content within the chapter or use it in any sequence you wish.

Most of the contents of this chapter are used throughout the remainder of the book. You can perform many repairs on truck electrical and computer control systems without understanding the building blocks of electrical theory, but when you encounter problems that fall outside the scope of troubleshooting software or flow charts, if you have good electrical instincts (you can *think* electrically), you are more likely to get to the root of that problem. So whichever way you study electricity, make sure you do not underestimate its importance. This chapter is key to understanding later chapters on computer technology. Try to learn to love electricity, even if you initially find it a real turnoff.

In this chapter, the theoretical aspects of electricity beginning with atomic structure are dealt with first, followed by a look at electrical circuits and components. Toward the end, some of the concepts behind information processing are introduced. Theoretical electricity begins with the study of atomic structure. If you find this confusing, skip forward and get back to it later when you feel more comfortable with some of the practical aspects of electricity.

Brief History

All matter has some electrical properties, but it is only comparatively recently in history that humans have been able to make it work for them. The first type of electricity to be identified was **static electricity** by the Greeks more than 2,000 years ago when they observed that amber rubbed with fur would attract lightweight objects such as feathers. The Greek word for "amber" (a translucent, yellowish resin, derived from fossilized tree sap) is **electron** from which the word "electricity" is derived. Probably the next significant step forward occurred toward the end of the sixteenth century when the English physicist William Gilbert (1544–1603) made the connection between electricity and magnetism. Shortly after, Benjamin Franklin (1706–1790) proved the electrical nature of thunderstorms in his famous kite experiment, established the terms *positive* and *negative*, and formulated the conventional theory of current flow in a circuit. From this point forward, progress accelerated. In 1767, Joseph Priestly established that electrical charges attract with a force inversely proportional to distance, and in 1800, Alessandro Volta invented the first battery.

Michael Faraday (1791–1867) opened the doors of the science we now know as **electromagnetism** when he published his law of induction that simply states that a magnetic field induces an electromotive force in a moving conductor. Thomas Edison (1847–1931) invented the incandescent lamp in 1879, but perhaps even more importantly built the first central power station and distribution system in New York City in 1881. This station provided a means of introducing electrical power into industry and the home. The discovery of the electron by J. J. Thomson (1856–1940) in 1897 introduced the science of electronics and quickly resulted in the invention of the diode (1904) and triode (1907).

Definitions

Electricity is a form of energy. This energy form can be observed in charged particles such as electrons or protons either statically, as accumulated charge, or dynamically, as current flow in a circuit. **Electronics** is a branch of electricity that addresses the behavior of flows of electrons through solids, liquids, gases, and across vacuums. However, in automotive technology, the terms *electronics* and *electronic engine management* are generally used to describe systems that are managed by computers.

ATOMIC STRUCTURE AND ELECTRON MOVEMENT

An **atom** is the smallest particle of a chemical element that can take part in a chemical reaction. An atom is usually made up of **protons**, **neutrons**, and electrons. Protons and neutrons form the center or **nucleus** of each atom while the electrons orbit the nucleus in a manner similar to that of planets orbiting the sun in our solar system. The simplest atom is hydrogen. It has a single electron orbiting a nucleus consisting of a single proton: It has no neutron. A representation of the structure of a hydrogen atom is shown in Figure 28–1.

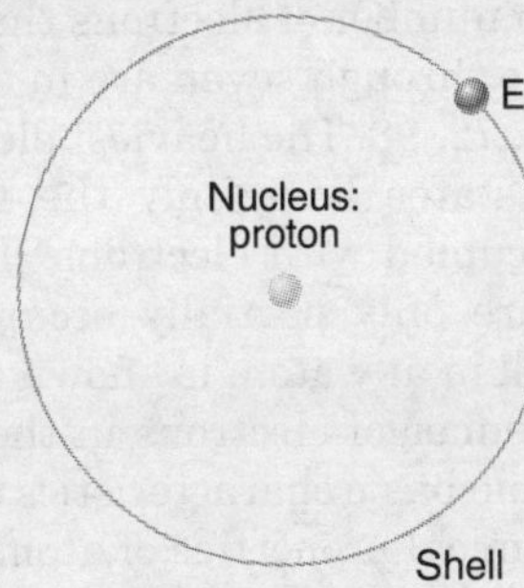

Figure 28–1 *Hydrogen atom.*

Electrons and Protons

Electrons orbit the nucleus of an atom in a concentric ring known as a shell. Electrons have a negative electrical charge. Protons and neutrons form the nucleus of all atoms around which the electrons orbit. Protons have a positive electrical charge, whereas neutrons have no electrical charge. The nucleus of an atom comprises 99.9% of its mass. The number of protons in the nucleus is the atomic number of any given element: The sum of the neutrons and protons is the atomic mass number. In a balanced or electrically neutral atom, the nucleus is surrounded by as many electrons as there are protons.

All electrons are alike. All protons are alike. The number of protons in the nucleus of an atom identifies it as a specific element. Electrons have 5/10,000 (0.0005) of the mass of a proton. Under normal conditions electrons are bound to the positively charged nuclei of atoms by the attraction between opposite electrical charges.

Unbalanced Atoms

Any atom may possess more or fewer electrons than protons. Such an atom would be described as negatively (an excess of electrons) or positively (a net deficit of electrons) charged and known as an **ion**. The concentric orbital **shells** (a shell is an orbital path) of an atom proceed outward from the nucleus of an atom. The electrons in the shells closest to the nucleus of an atom are held most tightly: Those in the outermost shell are held more loosely.

Shell Occupation

The most complex atoms may have seven shells. The maximum number of electrons that can occupy shells one through seven are in sequence: 2, 8, 18, 32, 50, 72, 98. The heaviest elements in their normal states have only the first four shells fully occupied with electrons: The outer three shells are only partially occupied. The outermost shell in any atom is known as its **valence**. The number of electrons in the valence will dictate some basic characteristics of an element. The chemical properties of atoms are defined by how the shells are occupied with electrons.

Charge Differential

An atom of the element helium whose atomic number is 2 has a full inner shell. An atom of the element neon with an atomic number of 10 has both a full first and a second shell (2 and 8): Its second shell is its valence (Figure 28–2). Other more complex atoms that have eight electrons in their outermost shell, even though this shell might not be full, will resemble neon in terms of their chemical inertness. Remember that an ion is any atom with either a surplus or deficit of electrons. Free electrons can rest on a surface or travel through matter (or a vacuum) at close to the speed of light. Electrons resting on a surface cause it to be negatively charged. Because the electrons are not moving, that surface is described as having a negative static electrical charge. The extent of the charge is measured in **voltage** or **charge differential**. A stream of moving electrons is known as an electrical current. For instance, if a group of positive ions passes in close proximity to electrons resting on a surface, they will attract the electrons by causing them to fill the "holes" left by the missing electrons in the positive ions. Current flow is measured in **amperes**: 1 ampere equals 6.28×10^{18} electrons (a **coulomb**) passing a given point per second.

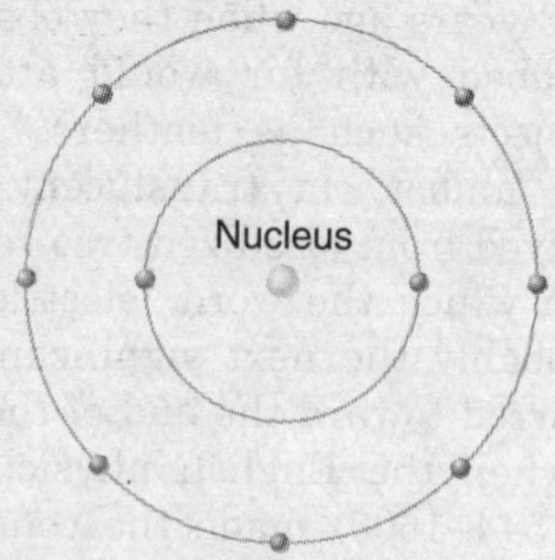

Figure 28–2 *Atomic structure of a neon atom: The outer shell is full.*

Factors of Charge Imbalance

A number of factors such as friction, heat, light, and chemical reactions can "steal" electrons from a surface, and when this occurs the surface becomes positively charged. If the positive ions remain at rest, the surface will have a positive static electrical charge differential. Every time a person walks across a carpet, electrons are "stolen" from the carpet surface and this has an electrifying effect (electrification) on both the substance from which electrons are stolen (the carpet) and the moving body that performs the theft. When the moving body has accumulated a sufficient charge differential (measured in voltage), the excess electrons will be discharged through an arc and balance the charge.

Electrification results in both attractive and repulsive forces. In electricity, like charges repel and unlike charges attract. When a plastic comb is run through hair, electrons are stolen by the comb, giving it a negative charge. The comb may subsequently attract small pieces of paper as shown in Figure 28–3. The experiment always works better on a dry day because electrons can travel more easily through humid air and any accumulated charge will dissipate rapidly. Two balloons rubbed on a woolen fiber will both acquire a negative charge and therefore tend to repel each other. An atom is held together because of the electrical tendency of unlike charges attracting each other.

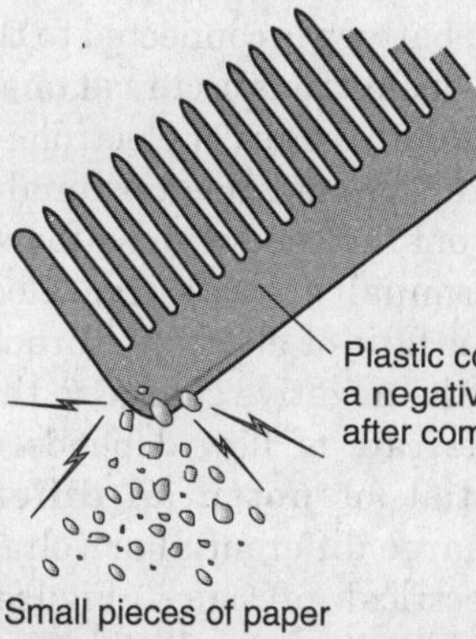

Figure 28–3 *Unlike charges attract.*

Positively charged protons hold the negatively charged electrons in their orbital shells and because like electrical charges repel each other, the electrons do not collide. All matter is composed of atoms. Electrical charge is a component of all atoms. When an atom is balanced, that is, the number of protons matches the number of electrons (Figure 28–4), the atom can be described as being in an electrically neutral state. So all matter is electrical in essence. The phenomena we describe as electricity concerns the behavior of atoms that, for whatever reason, have become unbalanced or electrified.

CONDUCTORS AND INSULATORS

The ease with which an electron moves from one atom to another determines the conductivity of the material. **Conductance** is the ability of a material to carry electric current. To produce current flow, electrons must move from atom to atom as shown in Figure 28–5. Materials that readily permit this flow of electrons from atom to atom are classified as **conductors**. A conductor is generally a metallic element that contains fewer than four electrons in its outer shell. Examples of conductors include copper, aluminum, gold, silver, iron, and platinum.

Materials that inhibit or perhaps prevent a flow of electrons are classified as **insulators**. An insulator is a nonmetallic substance that contains more than four electrons in the outer shell. Examples of insulators include glass, mica, rubber, and plastic.

Semiconductors are a group of materials that cannot be classified either as conductors or insulators: They have exactly four electrons in their outer shell. Silicon is an example of a semiconductor.

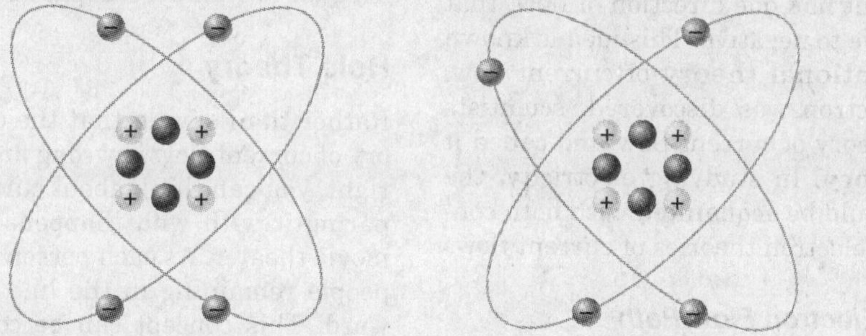

Figure 28–4 *Balanced atom: The number of protons and electrons are equal.*

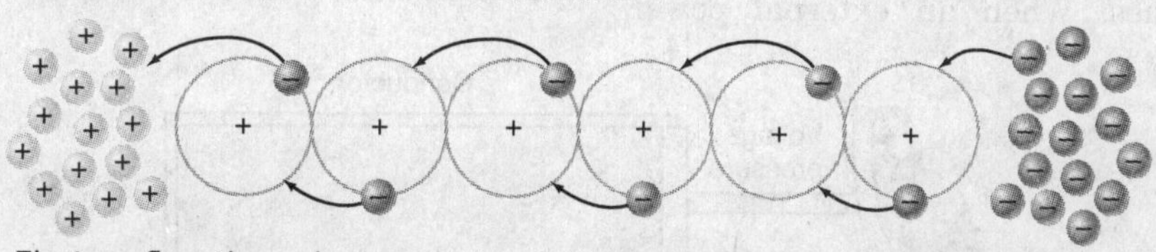

Figure 28–5 *Electron flow through a conductor.*

Conductivity in a Circuit

Conductive metals even when in an electrically neutral state contain vast numbers of moving electrons that move at random from atom to atom. When a battery is placed at either end of a conductor, such as copper wire, and a complete circuit is formed, electrons are pumped from the more negative terminal to the more positive until either the charge differential ceases to exist or the circuit is opened. The number of electrons does not change.

CURRENT FLOW

Current flow only occurs when there is a path and a difference in electrical potential: This difference is known as charge differential and is measured in voltage. Charge differential exists when the electrical source has a "shortage" (deficit) of electrons and therefore is positively charged. Because electrons are negatively charged and unlike charges attract, electrons flow toward the positive source.

Conventional and Electron Theory of Current Flow

Initially it was thought that current flow in an electrical circuit had one direction of flow, that is, from positive to negative. This idea is known as the **conventional theory** of current flow. When the electron was discovered, scientists revised the theory of current flow and called it **electron theory**. In studying electricity, the technician should be acquainted with both conventional and electron theories of current flow.

Creating an Electron Flow Path

A conductor, such as a piece of copper wire, contains billions of neutral atoms whose electrons move randomly from atom to atom, vibrating at high frequencies. When an external power source such as a battery is connected to the conductor, a deficit of electrons occurs at one end of the conductor and an excess of electrons occurs at the other end: The negative terminal repels free electrons from the conductor's atoms while the positive terminal attracts free electrons. This results in a flow of electrons through the conductor from the negative charge to the positive charge. The rate of flow depends on the charge differential (or **potential difference/** voltage). The charge differential or voltage is a measure of electrical pressure. Figure 28–6 shows the effect of voltage on electron flow. The role of a battery, for instance, is to act as a sort of electron pump. In a closed electrical circuit, electrons move through a conductor, producing a displacement effect close to the speed of light.

The physical dimensions of a conductor are also a factor. The larger the sectional area (measured by wire gauge size), the more atoms there are over a given sectional area, therefore the more free electrons. This means that as wire cross-section size increases, so does the ability to flow more electrical current through the wire. The rate of electron flow is called **current** and it is measured in amperes. A current flow of 6.25 billion billion electrons per second is equal to 1 ampere as shown in Figure 28–7.

Hole Theory

Rather than saying that the conventional theory of current flow is wrong and electron flow is right, you can think about current flow by comparing it with what happens in a lineup at a movie theater. As each person buys a ticket, the people remaining in the line move a step forward. This concept can be compared to what happens to each electron in an energized electrical circuit —it moves forward. However, as the line moves forward, the gap moves backward. You can compare this gap to the hole left

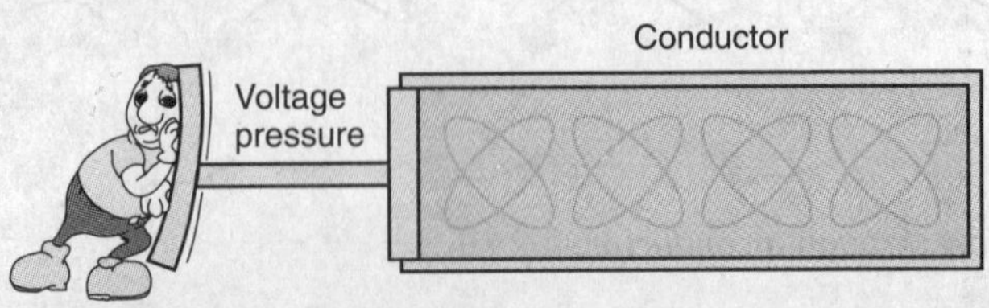

Figure 28–6 *Voltage or charge differential is the pressure that causes electrons to move.*

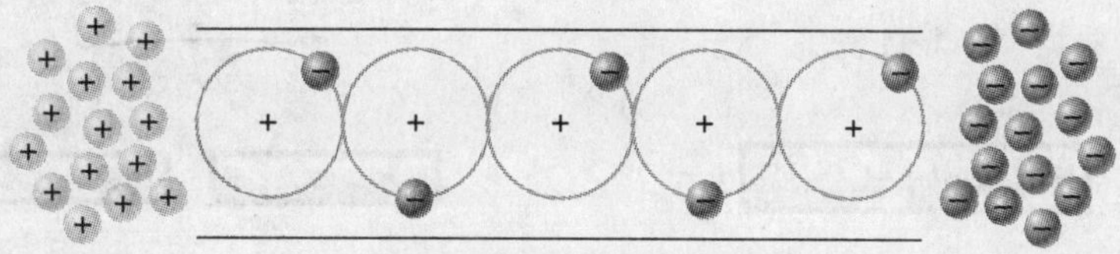

Figure 28–7 *The rate of electron flow is measured in amperes.*

by an electron in the electrical circuit as it moves forward. There is nothing different in what is happening in the movie theater line, just whether you think of its movement from the point of view of a person in it or the gap.

Terminals

The terminal from which electrons exit in an electrical device is known as the **anode** or positive terminal. The terminal through which electrons enter an electrical component is known as the **cathode** or negative terminal. The suffix *~ode* is taken from the term *electrode.* It might help you to think of an electrode as a sort of gate in an electrical circuit.

MAGNETISM

The phenomenon of magnetism was first observed in lodestone (magnetite) and the way ferrous (iron-based) metals reacted to it. When a bar of lodestone was suspended by string, the same end would always rotate to point toward the earth's North Pole. The molecular theory of magnetism tends to be the most widely accepted: In most materials, the magnetic poles of the composite molecules are arranged randomly so there is no magnetic force. In certain metals such as iron, nickel, and cobalt, the molecules can be aligned so that their north or N poles all point in one direction and their south or S poles point in the opposite direction. In lodestone, the molecules align themselves naturally. Some materials have good magnetic retention, which means that when they are magnetized they retain their molecular alignment. Other materials are only capable of maintaining their molecular alignment when positioned within a magnetic field; when the field is removed, the molecules disarrange themselves randomly and the substance's magnetic properties are lost.

All magnetism is essentially electromagnetism in that it results from the kinetic energy of electrons. Whenever an electric current is flowed through a conductor, a magnetic field is created. When a bar-shaped permanent magnet is cut in two, each piece assumes the magnetic properties of the parent magnet with individual north and south poles. Figure 28–8 shows some magnetic fields and how separate magnetic fields react when in proximity with one another.

The term **reluctance** describes resistance to the movement of magnetic lines of force. Reluctance can be reduced by using permeable (susceptible to penetration) materials within magnetic fields. The permeability of matter is rated by giving a rating of 1 for air, which is generally considered to be a poor conductor of magnetic lines of force. In contrast, iron is ascribed a permeability factor of 2,000, and certain ferrous alloys may have values exceeding 50,000.

The force field existing in the space around a magnet can be demonstrated when a piece of cardboard is placed over a magnet and iron filings are sprinkled on top of the cardboard. The pattern produced as the filings arrange themselves on the cardboard is referred to as flux lines. Flux lines are directional and exit from the magnet's north pole and enter through the south pole. Flux lines do not cross each other in a permanent magnet. Flux lines facing the same direction attract, while flux lines facing opposite directions tend to repel.

The flux density (concentration) determines the magnetic force. A powerful magnetic field exhibits a dense flux field, whereas a weak magnetic field exhibits a low-density flux field. Flux density is always greatest at the poles of a magnet.

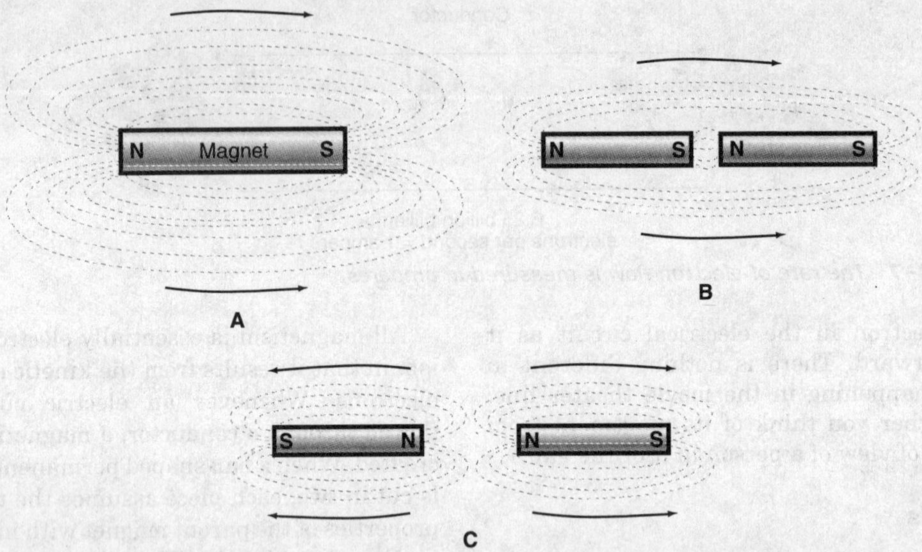

Figure 28–8 *Magnetic principles: A. All magnets have poles. B. Unlike poles attract each other.*

Atomic Structure and Magnetism

In an atom, all of the electrons in their orbital shells also spin on their own axes in much the same way the planets orbit the sun. They rotate axially, each producing magnetic fields. Because of their axial rotation, each electron can be regarded as a minute permanent magnet. In most atoms, pairs of electrons spinning in opposite directions produce magnetic fields that cancel each other out. An atom of iron has 26 electrons, 22 of which are paired. In the second from the outermost shell, four of the eight electrons are not paired, meaning that they rotate in the same direction and do not cancel each other out. This fact accounts for the magnetic character of the metal iron. A large number of vehicle electrical components use electromagnetic principles in some way or other, and magnetic field strength is the reason that **coils** and other electromagnetic devices contain iron cores.

ELECTROMAGNETISM

Electrical current flow through a conductor such as copper wire creates a magnetic field surrounding the wire. This effect can be observed by passing a copper wire through which current is flowing, which is lengthwise over a compass needle: The needle will deflect from its north-south orientation when this occurs. Any magnetic field created by electrical current flow is known as electromagnetism. Study of the behavior of electromagnetic fields has proved the following:

- Magnetic lines of force do not move when current flow through a conductor remains constant. When current flowing through the conductor increases, the magnetic lines of force will extend farther away from the conductor.
- The intensity and strength of magnetic lines of force increase proportionally with an increase in current flow through a conductor: Similarly they diminish proportionally with a decrease in current flow through the conductor.
- The right-hand rule is used to denote the direction of the magnetic lines of force: The right hand should enclose the wire with the thumb pointing in the direction of conventional current flow (positive to negative), and the fingertips will then point in the direction of the magnetic lines of force as shown in Figure 28–9 and Figure 28–10.

Using Electromagnetism

When wire is coiled and electric current flows through it, the magnetic field that would be produced if the wire were straight combines to form a larger magnetic field identified by north

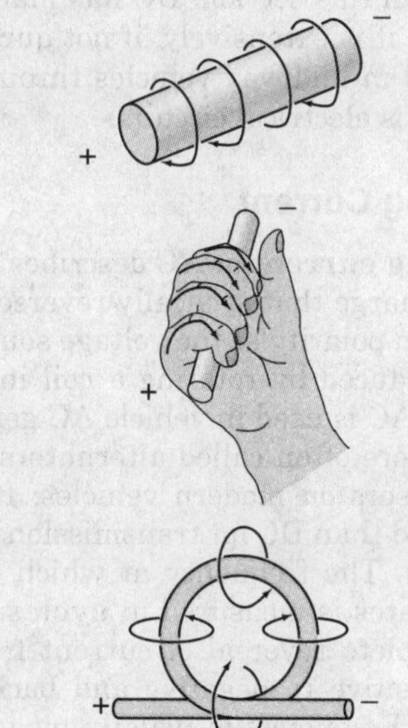

Figure 28–9 *Electromagnetic field characteristics.*

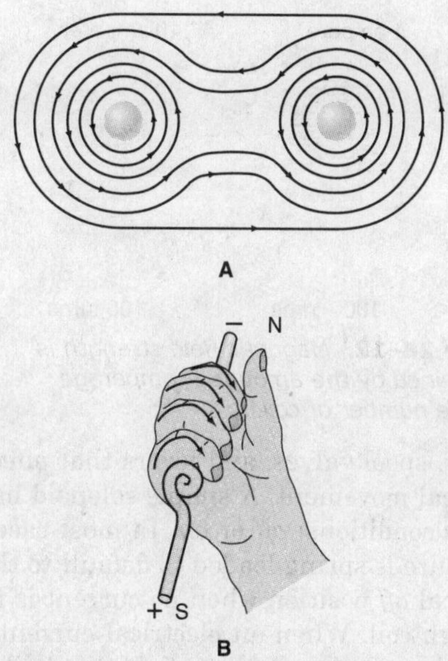

Figure 28–10 *A. Magnetic lines of force join together and attract each other. B. The right-hand rule.*

and south poles. This effect can be amplified by placing an iron core through the center of the coil (Figure 28–11), which reduces the reluctance of the magnetic field. The polarity of the electromagnet created can be determined by the right-hand rule for coils: The coiled wire should be held with the fingers pointed in the direction of conventional current flow (positive to negative) and the thumb will point to the north pole of the coil (check out Figure 28–10

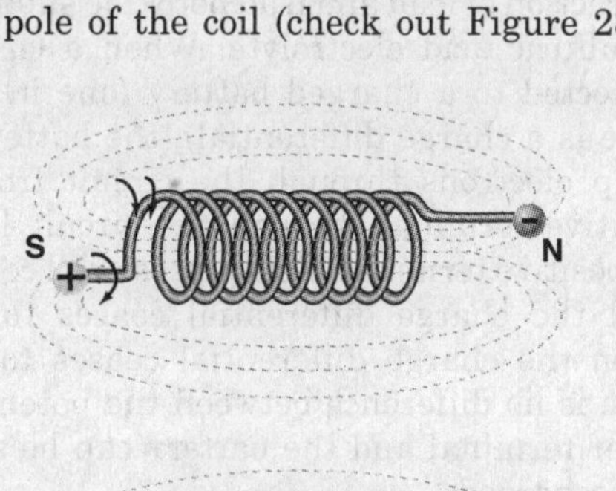

Figure 28–11 *Magnetic field characteristics of a coiled conductor.*

again). Electromagnetic field force is often described as magnetomotive force (mmf). The mmf is determined by two factors:

1. The amount of current flowing through the conductor
2. The number of turns of wire in a coil

Magnetomotive force is measured in **ampere-turns (At)**. Ampere-turn factors are the number of windings (complete turns of a wire conductor) and the quantity of current flowed (measured in amperes). Referencing Figure 28–12, if a coil with 100 windings has 1 ampere of current flowed through it, the result will be a magnetic field strength rated at 100 At. An identical magnetic field strength rating could be produced by a coil with 10 windings with a current flow of 10 amperes. The actual field strength would have to factor in reluctance. In other words, the actual field strength of both the foregoing coils would be increased if the coil windings were to be wrapped around an iron core.

Solenoids

A simple solenoid consists of a coil and an armature. The coil is usually stationary and the armature moves within it. The armature can be integral with devices such as hydraulic poppet

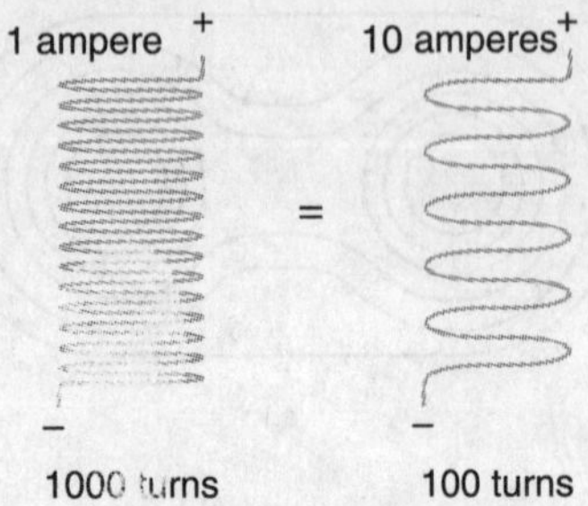

Figure 28–12 *Magnetic field strength is determined by the amount of amperage and the number of coils.*

valves, spool valves, and levers that effect mechanical movement. A simple solenoid has two status conditions: *off* or *on*. In most cases, the armature is spring-loaded to default to the mechanical *off* position when no current is flowed through coil. When an electrical current energizes the coil, the magnetic field that builds mechanically moves the armature. This is the principle that is used to engage a starter motor or move the poppet valve in a solenoid-actuated injector.

ELECTRICAL CURRENT CHARACTERISTICS AND SOURCES OF ELECTRICITY

Electrical current is classified as either direct current (DC) or alternating current (AC). The technician should understand the basic characteristics of both because both are used in vehicles. Although most of the chassis electrical circuit uses DC, an alternator and an inductive pulse generator both produce AC. In fact, an alternator produces AC internally and rectifies it to DC externally.

Direct Current

Electrical current can be made to flow in two directions through a conductor. If the current flows in one direction it is known as **direct current**, usually abbreviated as **DC**. The direction of flow can be expressed with respect to either the electron or conventional theories of current flow, but if the current characteristic is DC, then it does not alternate. The current flow may be steady or have a pulse characteristic. DC can be produced in a number of ways outlined later in this section. DC has many applications and it is extensively, if not quite exclusively, used in highway vehicles through most of the chassis electrical circuits.

Alternating Current

Alternating current or **AC** describes a flow of electrical charge that cyclically reverses due to a reversal in polarity at the voltage source. It is usually produced by rotating a coil in a magnetic field. AC is used in vehicle AC generators (perhaps more often called alternators) and in certain sensors on modern vehicles: It is also better suited than DC for transmission through power lines. The frequency at which the current alternates is measured in **cycles**. A cycle is one complete reversal of current from zero through positive to negative and back to the zero point. Frequency is usually measured in cycles per second.

Sources of Electricity

There are a number of ways of producing electricity and modern trucks use most of them in some way or another.

Chemical

Batteries are a means of producing DC from a chemical reaction. In the lead acid battery, a potential difference is created by the chemical interaction of lead and lead peroxide submerged in sulfuric acid electrolyte. When a circuit is connected to a charged battery (one in which there is a charge differential), the battery will pump electrons through the circuit from the negative terminal through the circuit load to the positive terminal. This process will continue until the charge differential ceases to exist. When the charge differential ceases to exist, there is no difference between the potential at either terminal and the battery can be said to be discharged.

Types of Battery Batteries used in highway commercial vehicles whose motive power source is the diesel engine are standard lead acid or its gel cell relative:

- Standard lead acid battery. Suitable for providing short bursts of peak currents for cranking but otherwise are not deeply discharged.

- Gel cell (recombinant) lead acid battery. Adapted for deep cycle applications such as for providing chassis power when the engine is not running.

NiMH Batteries At this moment in time, the primary battery used in hybrid electric vehicles (HEV) is the **NiMH battery**. An aqueous solution of potassium hydroxide is used as the electrolyte and the electrodes are made from nickel oxide (positive) and a metal hydride (negative). The active element in the NiMH is hydrogen, which is absorbed/de-adsorbed as the battery is discharged and charged. The cell voltage is 1.2 volts and banks of these batteries are used in HEV buses.

Lithium-Based Batteries Lithium is the highest electrochemical potential of any current batteries, producing a per-cell potential of between 2 and 5 volts. Their storage and power density features make them attractive for HEV applications where they are in trial development. The common categories are cobalt based, lithium-ion (Li-on) and lithium-ion polymer (Li-po).

Static Electricity

The term *static electricity* is somewhat misleading because it implies that it is unmoving. Perhaps it is more accurately expressed as frictional electricity because it results from the contact of two surfaces. Chemical bonds are formed when any surfaces contact, and if the atoms on one surface tend to hold electrons more tightly, the result is "theft" of electrons. Such contact produces a charge imbalance by pulling electrons of one surface from that of the other. As electrons are pulled away from a surface, the result is an excess of electrons in one surface (result = negative charge) and a deficit in the other (result = positive charge). The extent of the charge differential is measured in voltage. While the surfaces with opposite charges remain separate, the charge differential will exist. When the two polarities of charge are united, the charge imbalance will be canceled.

Static electricity is an everyday phenomenon as described in examples earlier in this chapter. It usually involves voltages of more than 1,000 V and perhaps rising to as much as 50,000 V. A fuel tanker trailer towed by a highway tractor steals electrons from the air as it is hauled down the highway (as does any moving vehicle) and can accumulate a significant and potentially dangerous charge differential. This charge differential, which can be as high as 40,000 V, must be neutralized by grounding before any attempt is made to load or unload fuel that could otherwise be ignited by an electrostatic arc.

Electromagnetic Induction

Current flow can be created in any conductor that is moved through a magnetic field or alternatively by a mobile magnetic field and a stationary conductor. The voltage induced increases both with speed of movement and the number of conductors, so densely wound conductors tend to produce higher voltage values. Generators, alternators, cranking motors, and shaft speed sensors all use the principle of electromagnetic induction. Figure 28–13 shows what happens when a conductor is moved through a magnetic field.

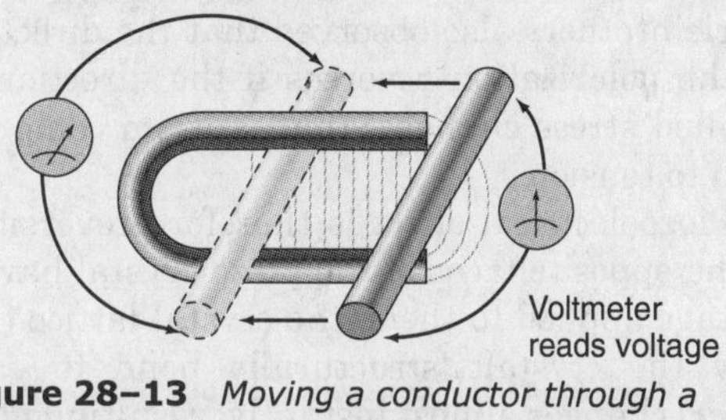

Figure 28–13 *Moving a conductor through a magnetic field induces an electrical charge differential.*

Thermoelectric

Electron flow can be created by applying heat to the connection point of two dissimilar metals in a device known as a **thermocouple**. This principle is used in the **pyrometers** used to monitor exhaust gas temperature in recently introduced diesel particulate filters (DPFs): In addition, pyrometers were often used as a driver downshift alert device in diesel engines a generation ago. A thermocouple consists of two dissimilar metals (iron and constantin, a coppertin alloy) joined at the "hot" end and connected to a sensitive voltmeter at the gauge end. As temperature increases at the hot end, the reading will increase at the display gauge, which functions like millivoltmeter calibrated to display Celsius or Fahrenheit.

Photoelectric

When light contacts certain materials such as selenium and cesium, electron flow is stimulated. Photoelectric cells are used as sensors that can control headlight beams and automatic daylight/night mirrors.

Piezoelectricity

Piezoelectricity is a word taken from the Greek words meaning "pressure electricity." It was discovered in 1880 by the brothers Pierre and Jacques Curie. They observed that some crystals, notably quartz (in their original experiments), become electrified when subjected to direct pressure, the potential difference increasing with pressure increase. When mechanical force is applied across the opposite faces of certain nonconducting crystals a voltage is generated. The actual voltage value that results is proportional to the applied stress. The Curie brothers also observed that the direction of the polarization reverses if the direction of applied stress changes, that is, from compression to tension.

Piezoelectrical-effect is therefore reversible. If the opposite faces of the piezo crystal have a voltage applied to them, the crystal lattice (the way the crystals structurally bond to each other) changes almost instantly, resulting in either expansion or contraction. This change of shape produces a powerful force that can be used for actuating a valve in much the same way as a solenoid but at much higher speeds. Solenoids use electromagnetism and respond more slowly due to the time required to build and collapse electromagnetic fields.

Uses of Piezoelectricity Piezoelectricity can be used as both sensors and actuators in vehicle electronics. Piezoelectric sensors are used as detonation sensors used on spark-ignited (SI) electronically controlled engines. Reverse piezoelectric-effect is becoming common in diesel engine actuators replacing solenoids as fuel injector switches. The reason for adopting piezoelectric actuators is much faster response times along with lower electrical power draw.

ELECTRICAL CIRCUITS AND OHM'S LAW

The German physicist Georg Ohm (1787–1854) proved the relationship between electrical potential (pressure or voltage), electrical current flow (measured in amperes), and the **resistance** to the current flow. In order to understand the behavior of electricity in an electrical circuit, it is necessary to understand the terminology used to describe its characteristics. When introducing electricity, comparisons are often made between electrical circuits and hydraulic circuits: These analogies are used in the following explanations.

Voltage

Voltage can be described in a number of different ways. Charge differential, electrical pressure, and potential difference are all correct ways of describing voltage. Using a hydraulic analogy, voltage in an electrical circuit can be compared to fluid water pressure. Similar to pressure in the hydraulic circuit, voltage may be present in an electrical circuit without any current flow. You measure voltage using a digital multimeter (DMM) on the V-DC or V-AC settings. You can measure voltage in both a closed (energized) and an open electrical circuit.

Resistance

Resistance is the opposition current flow. The resistance to the flow of electrons through a circuit is measured in **ohms**, the symbol for which is Ω. The resistance to the flow of free electrons through a conductor results from the innumerable collisions that occur and, generally, the greater the sectional area of the conductor (wire gauge size), the less resistance to current flow, simply because there are more available paths for free electrons.

Once again, using the hydraulic analogy, the resistance to fluid flow through a circuit would be defined by the pipe internal diameter or flow area. In an electrical circuit, resistance generally increases with temperature because of collisions between free electrons and vibrating atoms. As the temperature of a conductor increases, the tendency of the atoms to vibrate also increases and so does the incidence of colliding free electrons. The categories of resistance we see in vehicle electrical circuits are:

- Load—could mean lights, motors, anything that is powered electrically
- Corrosion—eats away from outside in reducing the effective surface contact area
- Wiring—size and material should provide as low resistance as possible to the electron flow

Resistance in an electrical circuit is measured with a DMM set in ohms: It can only be measured when the circuit is not electrically energized.

∞ = infinite resistance, out of limit, open circuit

OL = infinite resistance, out of limit, open circuit

0 Ω = zero or minimum circuit resistance, complete or **closed circuit**

There can be both voltage and resistance in an open electrical circuit. For current to flow in a circuit, both voltage (charge differential) and a physical circuit (resistance) must first exist: When such a circuit is closed, a path for current flow is provided.

Variable Resistance

We can use variable resistance devices in a number of ways to monitor parameters such as temperature and position status in modern engines. These will be examined in greater detail in later chapters of this book but some examples are:

- Thermistors: Designed to precisely vary resistance as temperature rises or falls. This means that when a constant value reference voltage is supplied to the thermistor, a signal can be output that exactly correlates with a temperature value.
- Rheostats: These are two-wire, mechanically variable resistors. A mechanical wiper into which the return signal path is integrated moves over a resistor. Its actual position determines how much of the electrical path has to pass through the resistor.
- Potentiometers: These are three-wire, mechanically variable voltage dividers. A reference voltage is supplied to the potentiometer. The mechanical position of a wiper on a resistor determines how the voltage is divided between the signal and ground paths.

Current

Current flows when the conditions of voltage and a closed circuit are met. Current is specifically the flow of electrons measured in amperes. One ampere is equal to 6.28×10^{18} electrons passing a given point in an electrical circuit in 1 second—or 1 coulomb. If the comparison with a hydraulic circuit is used once again, amps can be compared to gallons per minute (gpm). If current flow is to be measured in an electric circuit, the circuit must be electrically active or closed, that is, actively flowing current.

Closed Circuit

A complete electrical circuit is an arrangement that permits electrical current to flow. At its simplest this arrangement would require a power source, a load, and a means of connecting supply and return paths to the power source. A circuit is described as closed when current is flowing and open when it is not. Figure 28–14 shows the simplest type of electrical circuit.

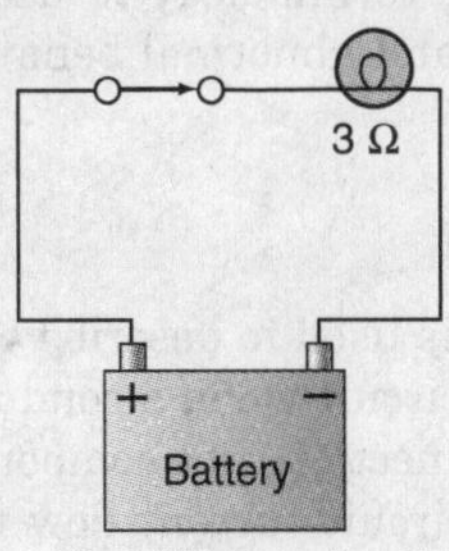

Figure 28–14 *Simplified series light circuit with 3-ohm resistance.*

Series Circuits

A **series circuit** may have several components such as switches, resistors, and lamps, but they are connected so that there is only one path for current flow thorough the circuit, such as that in Figure 28–14. In this figure, if the element in the light bulb were to fail, the circuit would open and no current would flow.

Parallel Circuits

A **parallel circuit** is one with multiple paths for current flow, meaning that the components in the circuit are connected so that current can flow through a component without having first flowed through other components in the circuit. Figure 28–15 shows an example of a parallel circuit using three paths for current flow, each through a separate load.

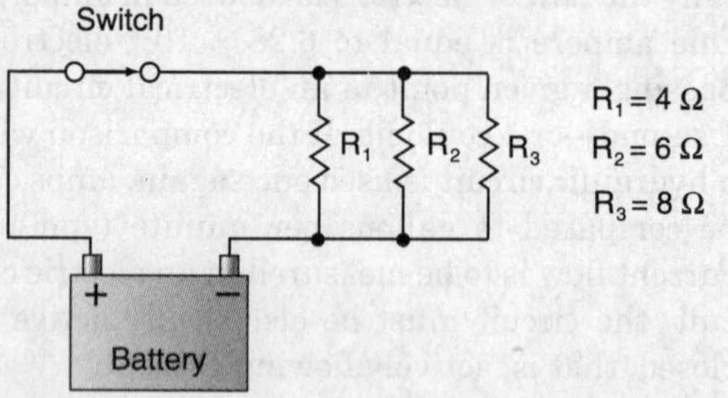

Figure 28–15 *A parallel circuit with different resistances in each branch.*

Series-Parallel Circuit

Many circuits are constructed using the principles of both series and parallel circuits and are known as series-parallel circuits.

Electrical Circuit Terminology

The following terminology is used to describe both normal and abnormal behavior in electrical circuits.

Short Circuit

Short circuit is used to describe what occurs in an electrical circuit when a conductor is placed across the connections of a component and some or all of the circuit current flow takes a shortcut. Short circuits are generally undesirable and can quickly overheat electrical circuits. Electricity will generally choose to flow through the shortest possible path in order to complete a circuit.

Open Circuit

The term **open circuit** describes any electrical circuit through which there is no current flow. A switch is used in electrical circuits to intentionally open them. An electrical circuit may also be opened unintentionally, and this might occur when a fuse fails, a wire breaks, or connections corrode.

Ground

The term *ground* represents the point of a circuit with the lowest voltage potential. In vehicles, ground or chassis ground is integral in the electrical circuit as it provides the return electrical path for all components in the circuit. Technicians used to working on vehicles always use the term *ground* to mean chassis ground. To power a light bulb in a vehicle, a single wire can supply one terminal on the bulb with a feed from the positive terminal of the battery and the return path is completed by connecting the other bulb terminal directly to the chassis ground.

Ohm's Law

Ohm's law tells us that an electrical pressure of 1 V is required to move 1 A of current through a resistance of 1 Ω.

I = intensity = current in amperes

E = EMF (electromotive force) = pressure in voltage

R = resistance = resistance to current flow in ohms

$$E = I \times R \qquad R = \frac{E}{I} \qquad I = \frac{E}{R}$$

Ohm's law can also be expressed in the units of measurement used in the following formulas: See Figure 28–16. Better still, memorize it.

$$V = A \times \Omega \qquad \Omega = \frac{V}{A} \qquad A = \frac{V}{\Omega}$$

Ohm's Law Applied to Series Circuits

In a series circuit, all of the current flows through all of the resistances in that circuit, so the sum of the resistances in the circuit would define the total circuit resistance. If a series circuit were to be constructed with 1-ohm and 2-ohm resistances (Figure 28–17), total circuit resistance (Rt) would be calculated as follows:

$$Rt = R_1 + R_2$$

$$Rt = 1\,\Omega + 2\,\Omega$$

$$Rt = 3\,\Omega$$

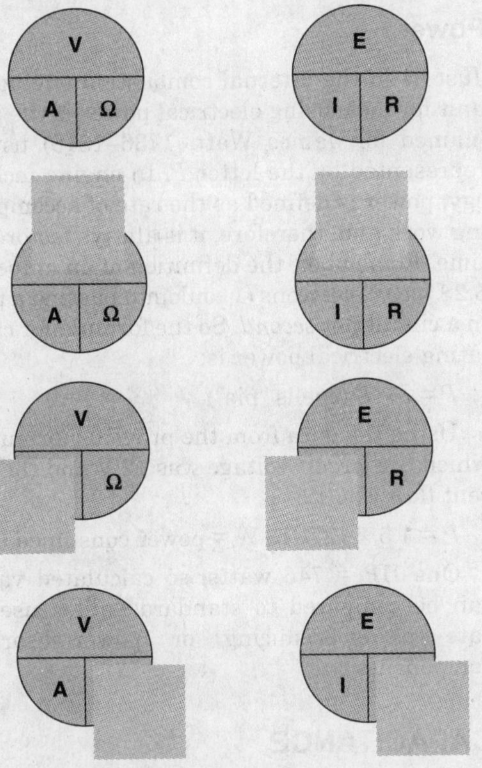

Figure 28–16 *Ohm's law formula.*

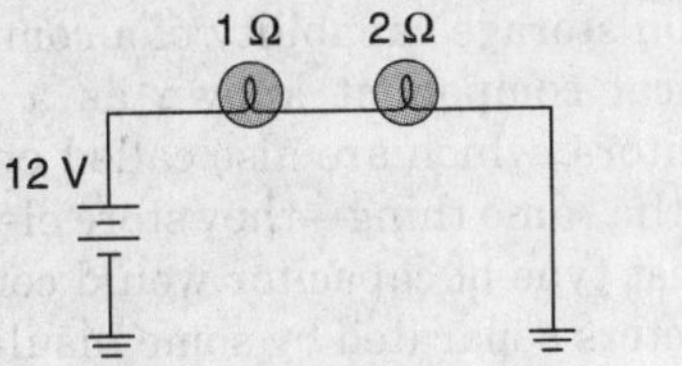

Figure 28–17 *Series circuit calculation.*

If the power source in the foregoing series circuit is a 12 V battery, then when the circuit is closed, current flow can be calculated using Ohm's law as follows:

$$I = \frac{E}{Rt}$$

$$I = \frac{12\ V}{3\Omega}$$

$$I = 4 \text{ amps}$$

Ohm's Law Applied to Parallel Circuits

According to Kirchhoff's law of current (see the next section), current that is flowed through a parallel circuit divides into each path in the circuit. When the current flow in each path is added, the total current will equal the current flow leaving the power source. When calculating the current flow in parallel circuits, each current flow path must be treated as a series circuit or the total resistance of the circuit must be calculated before calculating total current. When performing calculation on a parallel circuit, it should be remembered that more current will always flow through the path with the least resistance. If a parallel circuit is constructed with 2 Ω and 6 Ω resistors in parallel fed by a 12 V power source (Figure 28–18), total current can be calculated by treating each current flow path separately as follows:

$$I_1 = \frac{12\ V}{2\Omega} = 6 \text{ amps}$$

$$I_2 = \frac{12V}{6\ \Omega} = 2 \text{ amps}$$

$$I_T = 8 \text{ amps}$$

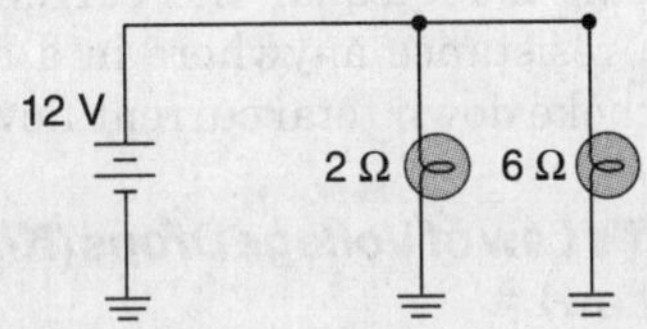

Figure 28–18 *Parallel circuit calculation.*

An alternative method would be to calculate *Rt* using the following formula:

$$Rt = \frac{R_1 \times R_2}{R_1 + R_2}$$

$$Rt = \frac{6\ \Omega \times 2\ \Omega}{6\ \Omega + 2\ \Omega} = \frac{12\ \Omega}{8} = 1.5\ \Omega$$

then

$$I = \frac{E}{R}$$

$$I = \frac{12\ V}{1.5\ \Omega} = 8 \text{ amps}$$

Reciprocal Formula The inverse or reciprocal formula can be used to calculate resistance in a parallel circuit with more than two paths for current flow. Each resistance value is divided into 1 and expressed as follows:

$$\frac{1}{Rt} = \frac{1}{R_1} + \frac{1}{R_2} + \frac{1}{R_3} + \frac{1}{R_4} \quad \ldots \text{etc}$$

Bear in mind when calculating resistance in parallel circuits that total circuit resistance always has to be less than the resistance in any individual path.

Most programs of study emphasize circuit calculations basically to get students of technology to "think" electrically. For that reason you should remind yourself that the information presented in this section is presented more as a review guideline than a basic introduction. Many vehicle electrical circuits are of the series-parallel type, that is, they combine the characteristics of both the series circuit and the parallel circuit. When performing circuit analysis and calculation, it helps to visualize the circuit in terms of paths for current flow.

Kirchhoff's Law of Current (Kirchhoff's First Law)

Kirchhoff's law of current states that the current flowing into a junction or point in an electrical circuit must equal the current flowing out. High resistance anywhere in a closed circuit will choke down total current flow.

Kirchhoff's Law of Voltage Drops (Kirchhoff's Second Law)

Kirchhoff's law of **voltage drops** states that voltage will drop in exact proportion to the resistance and that the sum of the voltage drops must equal the voltage applied to the circuit. Measuring or calculating voltage drop is frequently performed by technicians troubleshooting electrical circuits. If a series circuit is constructed with a 12 V power source and 2 Ω and 6 Ω resistors, the voltage drop across each resistor when the circuit is closed and subject to a current flow of 1.5 A can be calculated:

Voltage drop for $R_1\Omega$:

$E_1 = I \times R_1$

$E_1 = 1.5\ \text{A} \times 2\ \Omega$

$E_1 = 3\ \text{V}$

Voltage drop for R_2:

$E_2 = I\,\Omega\,R$

$E_2 = 1.5\ \text{A} \times 6\ \Omega$

$E_2 = 9\ \text{V}$

Total voltage drop through the circuit should equal that of the sum of the two calculations:

$E_1 + E_2 =$ source voltage

$3\ \text{V} + 9\ \text{V} = 12\ \text{V}$

Power

Just as in the internal combustion engine, the unit for measuring electrical power is the **watt** (named for James Watt, 1736–1819) usually represented by the letter *P*. In engine technology, power is defined as the rate of accomplishing work and, therefore, it is always factored by time. Remember, the definition of an ampere is 6.28×10^{18} electrons (1 coulomb) passing a point in a circuit per *second*. So the formula for calculating electrical power is:

$P = I \times E$ (spells "pie")

Using the data from the previous formula in which the circuit voltage was 12 V and the current flow was 1.5 A

$P = 1.5 \times 12 = 18\ \text{W} =$ power consumed

One HP = 746 watts, so calculated values can be compared to standard values used to rate power-producing or power-absorbing components.

CAPACITANCE

The term **capacitance** is used to describe the electron storage capability of a commonly used electrical component known as a **capacitor**. Capacitors, which are also called **condensers**, all do the same thing—they store electrons. The simplest type of capacitor would consist of two conductors separated by some insulating material called **dielectric**. The conductor plates could be aluminum and the dielectric may be mica (silicate mineral): The greater the dielectric properties of the material, the greater the resistance to voltage leakage. When a capacitor is connected to an electrical power source, it is capable of storing electrons from that power source. When the capacitor's charge storage capability is reached, it ceases to accept electrons from the power source. The charge is retained in the capacitor until the plates are connected to a lower voltage electrical circuit. At this point, the stored electrons are discharged from the capacitor into the lower potential (voltage) electrical circuit. Figure 28–19 shows the operating principle of a capacitor.

As a capacitor is electrified, for every electron removed from one plate, one is loaded onto the other plate. The number of electrons in a

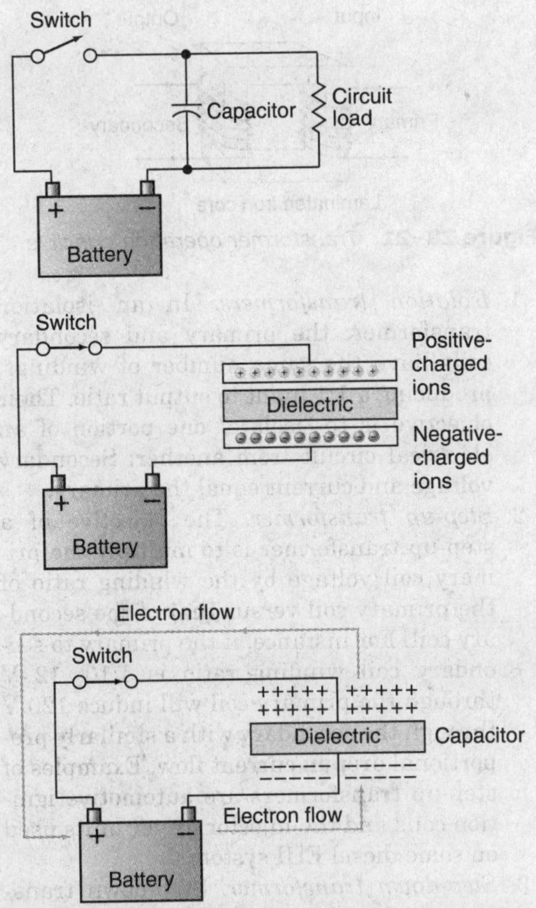

Figure 28–19 *Operating principle of a capacitor.*

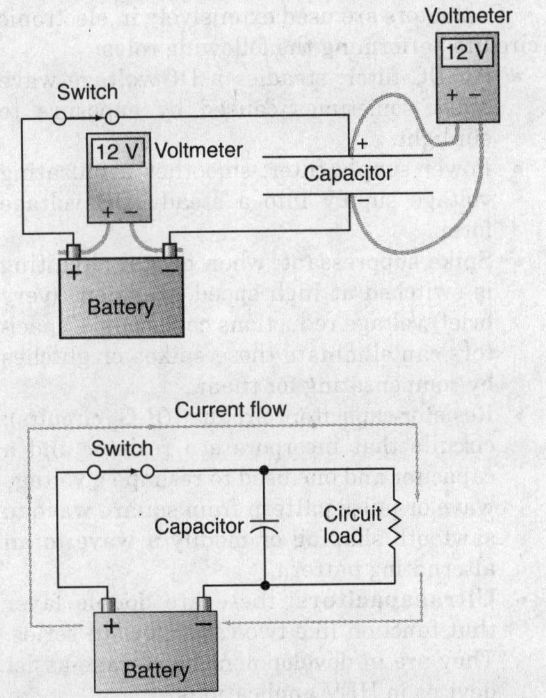

Figure 28–20 *Current flow with a fully charged capacitor.*

capacitor is identical when it is in both the electrified and neutral states. What changes is the location of the electrons. The electrons in a fully charged capacitor (Figure 28–20) will in time "leak" through the dielectric until both conductor plates have an equal charge: At this point, the capacitor would be described as being in a discharged condition. The ability to store electrons is known as capacitance, and this is measured in **farads** (named after Michael Faraday [1791–1867], the discoverer of the principle). One farad is the ability to store 6.28×10^{18} electrons at a 1-volt charge differential. Most capacitors have much less capacitance so they are rated in picofarads (trillionths of a farads) and microfarads (millionths of a farad).

1 farad = 1 F

1 microfarad = 1 µF = 0.000001F

1 picofarad = 1 µF – 0.000000000001F

Types of Capacitors

Many capacitors used in electric and electronic circuits are fixed value capacitors: These are coded by capacitance and voltage rating. Some capacitors have a variable capacity. These may have a combination of fixed and moving conductor plates, and the capacitance is varied by a shaft that rotates the moving plates. In some variable capacitance-type capacitors, the dielectric may be air. One truck engine manufacturer used a variable capacitance-type sensor to signal throttle position in its first generation of electronically managed engines. Electrolytic-type capacitors tend to have much higher capacitance ratings than nonelectrolytic types, and they are polarized so they must be connected accordingly in a circuit. The dielectric in this type of capacitor is the oxide formed on the aluminum conductor plate.

When working on electrical circuits, it should be noted that capacitors can retain a charge for a considerable time after the circuit current flow has ceased. Accidental discharge can damage circuit components and personal injury through electrical shock.

Capacitors are used extensively in electronic circuits performing the following roles:

- AC-DC filter: steadies a DC voltage wave offset sometimes caused by exposure to sunlight.
- Power supply filter: smoothes a pulsating voltage supply into a steady DC voltage form.
- Spike suppressant: when digital circuiting is switched at high speed, transient (very brief) voltage reductions can occur. Capacitors can eliminate these spikes or glitches by compensating for them.
- Resistor-capacitor circuits (R-C circuits): circuits that incorporate a resistor and a capacitor and one used to reshape a voltage wave or pulse pattern from square wave to sawtooth shaping or modify a wave to an alternating pattern.
- **Ultracapacitors**: these are double layer that function like two capacitors in series. They are in development as storage-assist devices in HEV applications.

COILS AND TRANSFORMERS

When electrons are moved through a conductor, an electromagnetic field is created surrounding the conductor. When the conductor is a wire wound into a coil, the electromagnetic field created is stronger. **Coils** are the basis of electric solenoids and motors. They are used in electronic circuits to shape voltage waves because they tend to resist rapid fluctuations in current flow. Like capacitors, coils can be used in electrical circuits to reshape voltage waves. Also, the energy in the electromagnetic field surrounding a coil can be induced to any nearby conductor. If the nearby conductor happens to be a second coil, then a current flow can be induced in it. The principle of a transformer is essentially that of passing current through a primary coil and inducing current flow in a secondary or output coil as shown in Figure 28–21. Variations on this principle would be coils constructed with a movable core, which permits their inductance to be varied, thereby altering output.

Transformers of various different types are used in vehicle electrical circuits, but they all fall into three general categories:

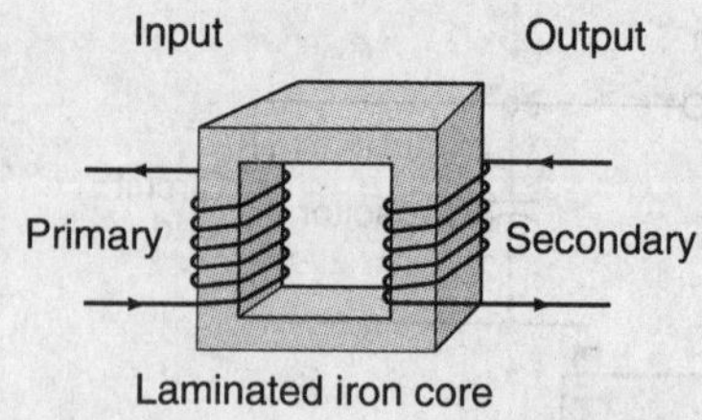

Figure 28–21 *Transformer operating principle.*

1. *Isolation transformers.* In an isolation transformer, the primary and secondary coils have the same number of windings, producing a 1:1 input to output ratio. Their objective is to "isolate" one portion of an electrical circuit from another: Secondary voltage and current equal the primary.
2. *Step-up transformer.* The objective of a step-up transformer is to multiply the primary coil voltage by the winding ratio of the primary coil versus that of the secondary coil. For instance, if the primary to secondary coil winding ratio is 1:10, 12 V through the primary coil will induce 120 V through the secondary with a similarly proportional drop in current flow. Examples of step-up transformers are automotive ignition coils and the injector driver units used on some diesel EUI systems.
3. *Step-down transformer.* Step-down transformers function oppositely from step-up transformers. The primary coil winding ratio exceeds that of the secondary coil, resulting in a diminished output voltage and increased output current.

SEMICONDUCTORS

Semiconductors are a group of materials with exactly four electrons in their outer shell. As such, they cannot be classified either as insulators or conductors. Silicon and germanium are commonly used semiconductor materials in the manufacture of electronics components. Both silicon and germanium have four electrons in their outer shells but would "prefer" to have eight: This means that silicon and germanium atoms readily unite in clusters called crystals, sharing electrons in their outer shells. Silicon and germanium can be grown into large crystals by applying heat to melting point followed by a period of cooling. Pure semiconductors are of little use in electronics components. For semiconductors to be useful they must be doped, that is, have small quantities of impurities added to them. The doping agents are usually phosphorus and boron. The doping intensity defines the electrical behavior of the crystal. After doping, semiconductor crystals may be sliced into thin sections known as wafers.

The type of doping agent used to produce semiconductor crystals defines the electrical properties of the crystals produced. A boron atom has three electrons in its outer shell. The outer shell is known as the valence and an atom with three electrons in its outer shell is known as trivalent. A boron atom in a crystallized cluster of semiconductor atoms produces an outer shell with seven electrons instead of eight. This "vacant" electron opening is known as a hole. The hole makes it possible for an electron from a nearby atom to fall into the hole. In other words, the holes can move, permitting a flow of electrons. Semiconductor crystals doped with boron (or other elemental atoms with three electrons in the outer shell—trivalent) forms a P-type semiconductor.

A phosphorus atom has five electrons in its outer shell. It is pentavalent. In the bonding between the semiconductors and the doping material, there is room for only eight electrons in the center shell. Even when the material is in an electrically neutral state, the extra electron can move through the crystal. When a semiconductor crystal is manufactured using a doping material with five electrons in the outer shell (pentavalent) it forms an N-type semiconductor. Figure 28–22 shows a germanium atom in a crystallized cluster with shared electrons.

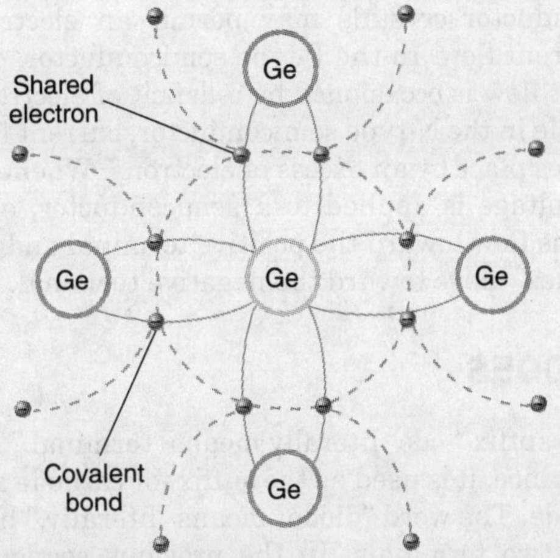

Figure 28–22 *Structure of a germanium atom with four electrons in the outer shell, shown in a crystallized cluster with shared electrons.*

The doping of a semiconductor crystal will always define its electrical characteristics. In Figure 28–23, germanium is used once again to show how the element forms a P-type semiconductor when doped with arsenic and an N-type semiconductor when doped with boron. Semiconductor materials are classified as P type and N type. In different ways, P-type and N-type sem-

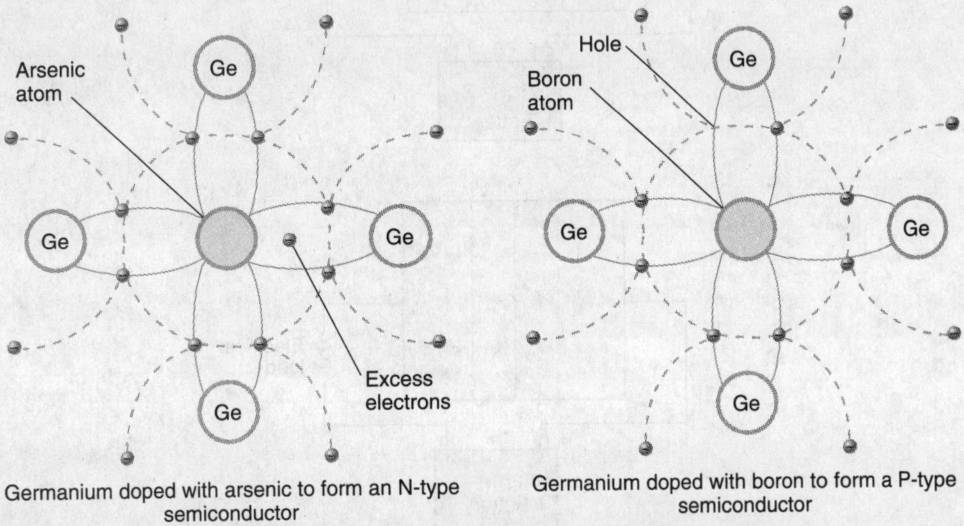

Figure 28–23 *Doped germanium semiconductor crystals of the N and P types.*

iconductor crystals may permit an electrical current flow. In the P-type semiconductor, current flow is occasioned by a deficit of electrons, while in the N-type semiconductor, current flow takes place by an excess of electrons. Whenever a voltage is applied to a semiconductor, electrons flow toward the positive terminal and the "holes" move toward the negative terminal.

DIODES

The suffix "-ode" literally means "terminal." For instance, it is used as the suffix for cathode and anode. The word "diode" means "literally," having two terminals. In the previous section it was explained how both P-type and N-type semiconductor crystals can conduct electricity: The actual resistance of each type is determined by either the proportion of holes or surplus of electrons. When a chip is manufactured using both P- and N-type semiconductors, electrons will flow in only one direction. The diode is used in electronic circuitry as a sort of one-way check valve that will conduct electricity in one direction (forward bias) and block it in the other (reverse bias). Figure 28–24 shows how a diode operates in forward and reverse bias modes.

The positive terminal (+) is called the anode and the negative terminal (–) the cathode. As an electrical one-way check valve, diodes will permit current flow only when correctly polarized. Diodes are used in AC generators (alter-

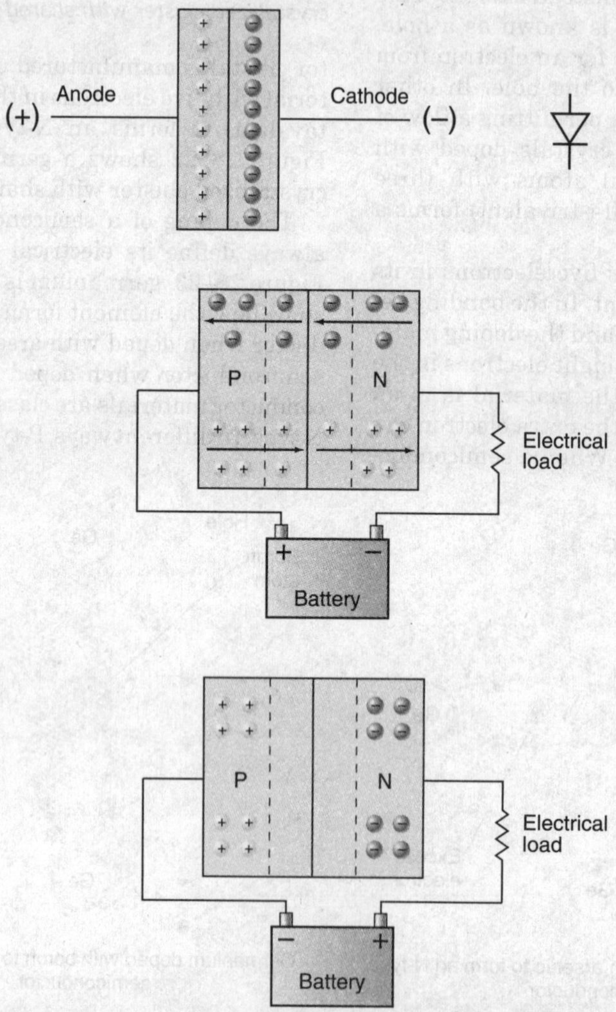

Figure 28–24 *Diode operation and diodes in forward and reverse bias.*

nators) to produce a DC characteristic from AC. They are also used extensively in electronic circuits.

Diodes may be destroyed when subjected to voltage or current values that exceed their rated capacity. Excessive reverse current may cause a diode to conduct in the wrong direction and excessive heat can melt the semiconductor material. Diode voltage and current specifications are written as follows:

V_F = forward voltage I_F = forward current
V_R = reverse voltage I_R = reverse current

Numerous types of diodes play a variety of roles in electronic circuits. The following list gives examples of some of the more common types:

- Small signal. Small signal diodes are used to transform low current AC to DC (rectification), perform logic data flows, and absorb voltage spikes.
- Power rectifier. Power rectifier diodes function in the same manner as small signal diodes but are designed to permit much greater current flow. They are used in multiples and often mounted on a heat sink to dissipate excess heat caused by high current flow.
- Zener. A **zener diode** functions as a voltage sensitive switch. Named after its inventor (Clarence Zener, who invented it in 1934), the zener diode is designed to block reverse bias current but only up to a specific voltage value. When this reverse breakdown voltage is attained, it conducts the reverse bias current flow without damage to the semiconductor material. Zener diodes are manufactured from heavily doped semiconductor crystals. They are used in electronic voltage regulators and in other automotive electronic circuitry. Zener diodes are rated by breakdown voltage (V_Z) and this can range from 2 V to 200 V. A schematic of a zener diode is shown in Figure 28–25.
- Light-emitting diode. All diodes emit electromagnetic radiation when forward biased, but diodes manufactured from some semiconductors (notably gallium arsenide phosphide) emit it at much higher levels. **Light-emitting diodes (LEDs)** may be constructed to produce a variety of colors and are commonly used for digital data displays. For instance a digital display with seven linear LED bars arranged in a bridged rectangle could display any single numeric digit by energizing or not energizing each of the seven LEDs. LED arrangements constructed to display alpha characters are only slightly more complex. LEDs convert electrical current directly into light (photons) and therefore are highly efficient as there are no heat losses. Figure 28–26A shows an LED and Figure 28–26B shows how one is represented on a schematic.

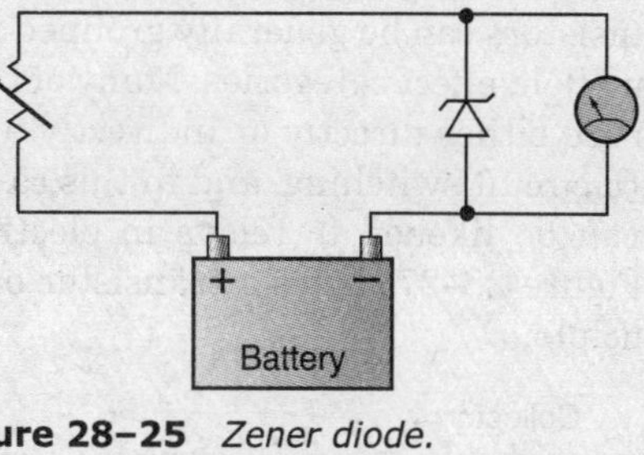

Figure 28–25 *Zener diode.*

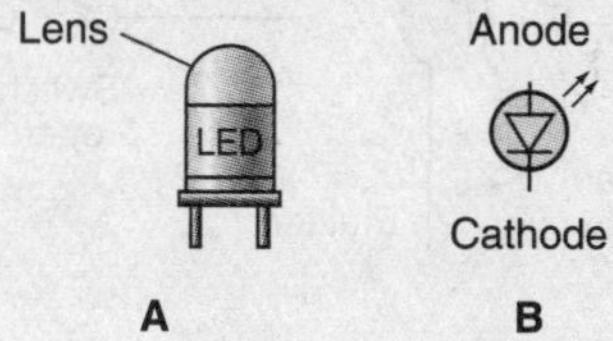

Figure 28–26 *Light-emitting diode (LED).*

- Photo diode. All diodes produce some electrical response when subjected to light. A photo diode is designed to detect light and therefore has a clear window through which light can enter. Silicon is the usual semiconductor medium used in photo diodes.

Diodes operate in either forward or reverse bias.

- Forward bias: A positive voltage is applied to the P-type material and a negative voltage to the N-type material, permitting current flow.
- Reverse bias: A positive voltage is applied to the N-type material and a negative voltage to the P-type material, meaning that current flow is blocked.

TRANSISTORS

Transistors are three-terminal semiconductor chips that are used extensively in electronic circuits. Transistors can be generally grouped into bipolar and field effect categories. Many of their functions are either directly or indirectly associated with circuit switching, and in this capacity they can be likened to relays in electrical circuits. Figure 28–27 shows a transistor-operating principle.

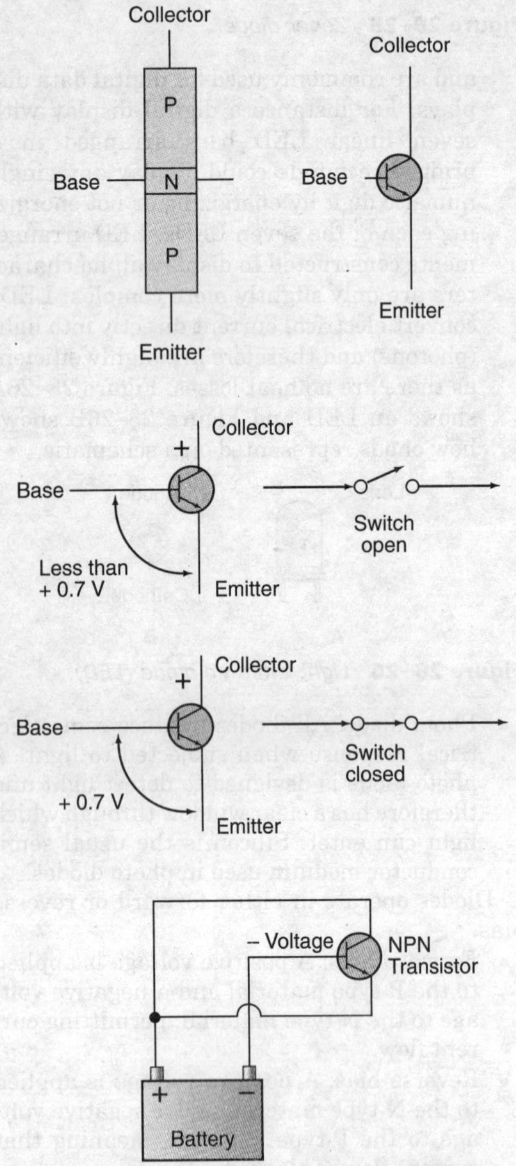

Figure 28–27 *Transistor-operating principle.*

Bipolar Transistors

A bipolar transistor functions as a sort of switched diode with three terminals known as collector, emitter, and base, each connected to semiconductor media in what is known as a silicon sandwich. The sandwich may either be NPN or PNP. The middle of the sandwich always acts as a *gate* capable of controlling the current flow through all three layers: The base is fairly thin and it has comparatively fewer doping atoms than that of the semiconductor material on either side of it. Bipolar transistors are designed so that a small emitter-base current will "ungate" the transistor and permit a larger emitter-collector current flow. Figure 28–28 shows the action of a PNP transistor.

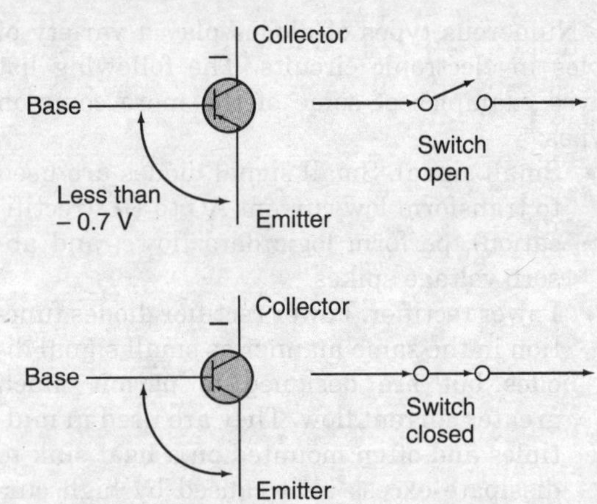

Figure 28–28 *Action of a PNP transistor.*

Facts about Bipolar Transistor Operation

- The base-emitter junctions will not conduct until the forward bias voltage exceeds ±0.6 V.
- Excessive current flow through a transistor will cause it to overheat or fail.
- Excessive voltage can destroy the semiconductor crystal media.

Small Signal Switching Transistors

Small signal transistors are used to amplify signals. They may be designed to fully gate current flow in their off position and others may both amplify and switch. These are used in many vehicle electronically controlled circuits and are often the ones you get to test and build into circuits in class projects.

Power Transistors

Used in power supply circuits, power transistors may conduct high current loads and may be mounted on heat sinks to enable them to dissipate heat. They are sometimes known as drivers because they serve as the final or output switch in an electronic circuit used to control a component such as a solenoid or pilot switch. The injector drivers used in many current diesel engine management systems use power transistors.

Field Effect Transistors

Field effect transistors or **FETs** are more commonly used than bipolar transistors largely because they are cheaper to manufacture. They may be divided into junction type and metal-oxide semiconductors. The FET has three terminals: source, drain, and gate. The source supplies the electrons like the emitter in a bipolar transistor. The drain collects the current so it can be compared to the collector in the bipolar transistor. The gate creates the electrostatic field that "switches" the FET and permits electron flow from the source to the drain so it can be compared to the gate in a bipolar transistor.

Both junction-type and metal-oxide FETs are controlled by a very small input or gate voltage. They are used in many smart switch multiplexing circuits used today in trucks. An FET functions like a relay with no moving parts. It is connected into a circuit that uses the FET like a switch. Figure 28–29 shows the operating principle of an FET.

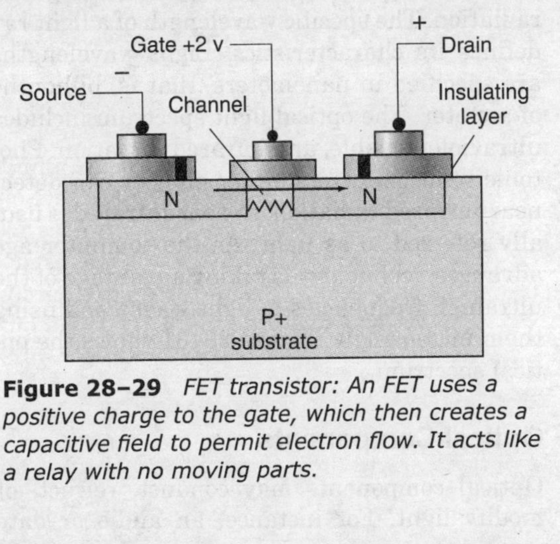

Figure 28–29 *FET transistor: An FET uses a positive charge to the gate, which then creates a capacitive field to permit electron flow. It acts like a relay with no moving parts.*

Facts about FETs

- Gate-channel resistance is very high so the device has almost no effect on external components connected to the gate.
- Almost no current flows in the gate circuit because the gate-channel resistance is so high. The gate and channel form a "diode," and so long as the input signal reverse biases this diode, the gate will show high resistance.

Thyristor

Thyristors are three-terminal, solid-state switches. A small current flow through one of the terminals will switch the thyristors on and permit a larger current flow between the other two terminals. Thyristors are switches, so they are either in an on or off condition. They are classified by whether they switch AC or DC current. Some thyristors are designed with two terminals only: They will conduct current when a specific trigger or breakdown voltage is achieved, functioning somewhat like a diode.

Silicon-Controlled Rectifiers

Silicon-controlled rectifiers (SCRs) are similar to a bipolar transistor with a fourth semiconductor layer added as shown in the upper section of Figure 28–30. They are used to switch DC. When the anode of an SCR is made more positive than the cathode, the outer two PN junctions become forward biased: The middle PN junction is reverse biased and will block current flow. However, a small gate current will forward bias the middle PN junction, enabling a large current to flow through the thyristor. SCRs will remain *on* even when the gate current is removed: The *on* condition will remain until the anode-cathode circuit is opened or reversed biased. SCRs are used for switching circuits in vehicle electronic and ignition systems.

Darlington Pairs

A **Darlington pair** (named after the inventor) consists of a pair of transistors wired so that the emitter of one supplies the base signal to a second, through which a large current flows.

The objective once again is to use a very small current to switch a much larger current. This type of application is known as **amplification**. Darlington pairs are used extensively in vehicle computer control systems. Figure 28–30 shows a Darlington pair relationship.

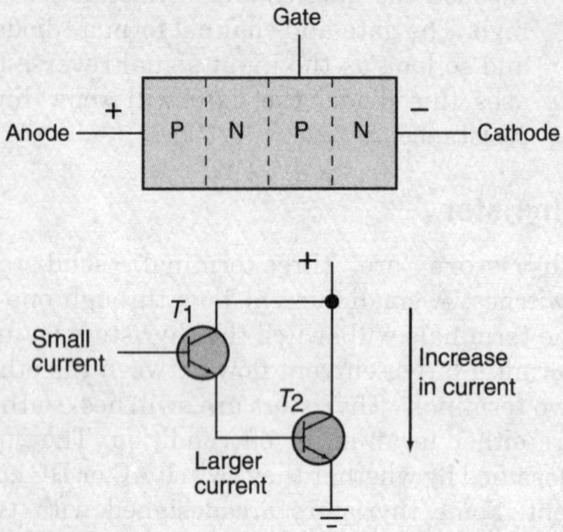

Figure 28–30 *A forward direction SCR and Darlington pair relationship.*

Summary of Transistor Operation

A transistor has three operating states:

- Cutoff: Reverse bias voltage is applied to the base terminal. In cutoff state, no current will flow through the semiconductor.
- Conduction: Bias voltage between the base and emitter has increased and switched the transistor to its on state. In its on state, the transistor is conducting, permitting electron flow. Output current is proportional to base current.
- Saturation: When collector-to-base voltage drops off to a near-zero value by voltage drop across the collector, the transistor enters saturation state.

TESTING SEMICONDUCTORS

Diodes and transistors are normally tested using a DMM. The semiconductor being tested should be isolated from the circuit. Diodes should produce a high-resistance reading with the DMM test leads in the reverse bias and low resistance when the leads are switched. Low-resistance readings both ways indicate a shorted diode. High resistance both ways indicates an open diode. Similarly you can set the DMM to the diode test mode.

A functional transistor should test:

- Continuity between the emitter and base
- Continuity between the base and the collector when tested one way and high resistance
- When DMM test leads are reversed
- High resistance in either direction when tested across the emitter and collector terminals Testing semiconductors is discussed in Chapter 31 in more detail with diagrams.

PHOTONIC DEVICES

Photonic semiconductors emit and detect light or photons. Photons are produced when certain electrons excited to a higher-than-normal energy level return to a more normal level. Photons behave like waves and the distance between the wave nodes and antinodes (wave crests and valleys) is known as **wavelength**. Electrons excited to higher energy levels emit photons with shorter wavelengths than electrons excited to lower levels. Photons are not necessarily visible and it is perhaps important to note that they may only truly be described as light when they are visible. Phototransistors are used in conjunction with an LED and slotted beam interrupter to accurately signal shaft/rotor speeds in vehicle electronics.

The Optical Spectrum

All visible light is classified as electromagnetic radiation. The specific wavelength of a light ray defines its characteristics. Light wavelengths are specified in nanometers, that is, billionths of a meter. The optical light spectrum includes ultraviolet, visible, and infrared radiation. Photonic semiconductors either emit or can detect near-infrared radiation, so near-infrared is usually referred to as light. As the computer age advances, technology is taking advantage of the ultrahigh frequencies of light waves and using them increasingly. Figure 28–31 shows the optical spectrum.

Optical Components

Optical components may conduct, refract, or modify light. For instance, an audio or data

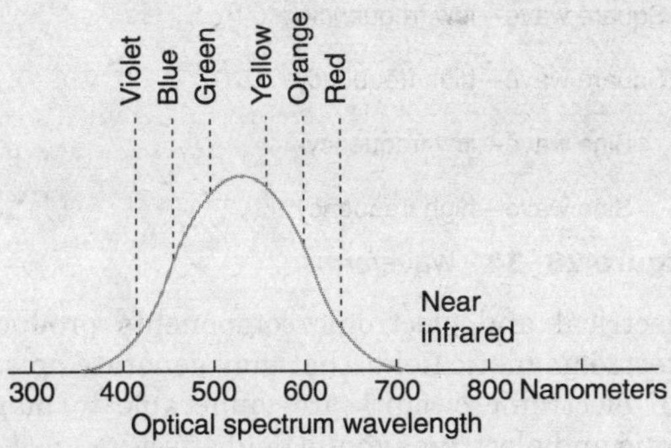

Figure 28–31 *The optical spectrum.*

compact disk (CD) retains data that has to be read optically, usually by a laser. Some of the more common optical components are:

- Filters. Filters transmit only a narrow band of the spectrum and block the remainder.
- Reflectors. In much the same way a mirror functions, reflectors bend back a light beam, or at least most of it.
- Beam Splitters. Beam splitters transmit some of the optical wavelength and reflect back the rest of it.
- Lenses. Lenses bend light waves. They are often used in conjunction with semiconductor light sources and detectors. They are often used to collect and focus light onto a detector.
- Optical Fibers. Increasingly used to transmit digital data by pulsing light, optical fibers are thin, flexible strands of glass or plastic that conduct light: The light travels through a core surrounded by conduit or cladding. The use of fiber optics is growing rapidly, and it is only a matter of time before they become extensively used in vehicle technology.
- Electromagnetic Spectrum. The electromagnetic spectrum is shown in Figure 28–32 so that you have an idea of the relative frequencies of the wavelengths. The higher the frequency, the higher the intensity of data that can be relayed (reason for shift to fiber optics), and we use frequency shift analysis in technologies such as the Doppler radar used in truck collision warning systems.
- Solar Cells. A solar cell consists of a PN or NP silicon semiconductor junction built onto contact plates. A single silicon solar cell may generate up to 0.5 V in ideal light condition (bright sunlight), but output values are usually lower. Like battery cells, solar cells are normally arranged in series groups, in which case the output voltage would be the sum of cell voltages, or in parallel, where the output current would be the sum of the cell currents. They are used as battery chargers on vehicles.

USING ELECTRONIC SIGNALS

Simple electronic circuits can be designed to transmit relatively complex data by switching subcircuit components on or off, making use of the binary voltage or no-voltage features that are the essence of computer transactions. The idea is that a series of input signals work to produce a defined output signal. Signal processing includes filtering, amplification, and rectification performed by the devices presented earlier in this chapter. Digital circuits handle pulsed waveforms that can do things like add, sub-

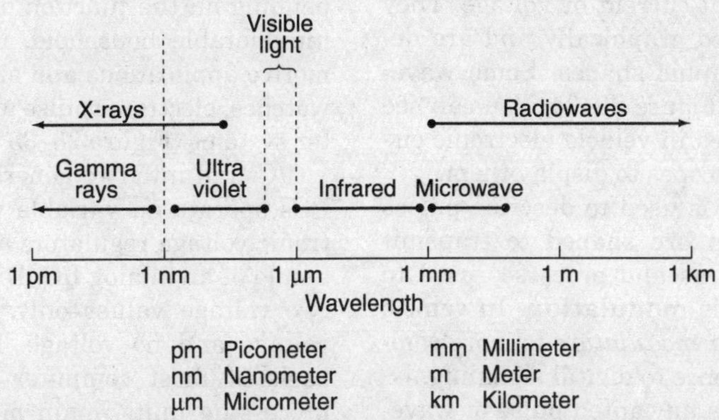

Figure 28–32 *The electromagnetic spectrum.*

tract, and multiply and compare and perform logic processing. An example of a pulsed waveform is **pulse width modulation (PWM)**, which is used frequently in truck digital electronics both for input signaling (Caterpillar TPS) and to drive electronic unit injector (EUI) control cartridges or proportioning solenoids such as the injection pressure regulator (IPR) used on hydraulically actuated electronic unit injector (HEUI) fuel systems. If you want to take a look at a graphical representation of a PWM signal you get the best image using a PWM-actuated proportioning solenoid.

Gates are used to route signals. If a circuit consisting of a power source, light bulb, and switch is constructed, the switch can be used to "pulse" the *on/off* time of the bulb. If you know Morse code, think of what can be expressed using the two Morse signals, a dot and a dash. This pulsing can be coded into many types of data such as alpha or numeric values. Pulses are controlled immediate variations in current flow in a circuit: Ideally, the increase or decrease in current would be instantaneous. If this were to be so, the pulse could be represented graphically as in Figure 28–33.

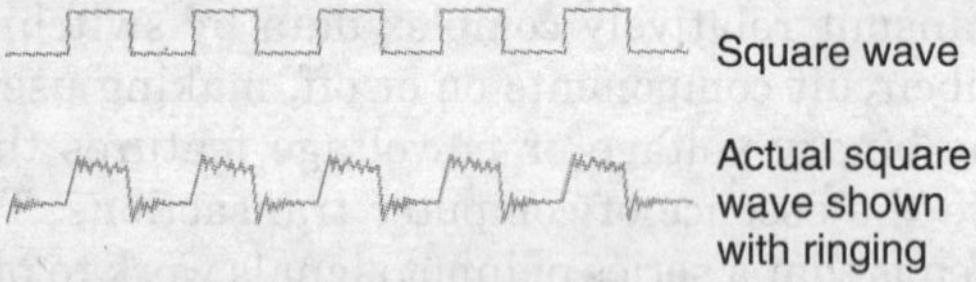

Figure 28–33 *Square sine waves.*

However, true pulse shaping results in graduated rise when the circuit is switched to the *on* state and graduated fall when the circuit is switched to the *off* state. Waves are rhythmic fluctuations in circuit current or voltage: They are often represented graphically and are described by their graphic shapes. Some waveforms are shown in Figure 28–34. You can see these waveform types in vehicle electronic circuits when you use scopes to display them.

The term *signals* is used to describe pulses and waveforms that are shaped to transmit data. The mechanisms and processes used to shape data signals is **modulation**. In vehicle electronics, the term *modulation* is more commonly used in reference to digital signaling.

Electronic noise is unwanted pulse or waveform interference that can scramble signals. All

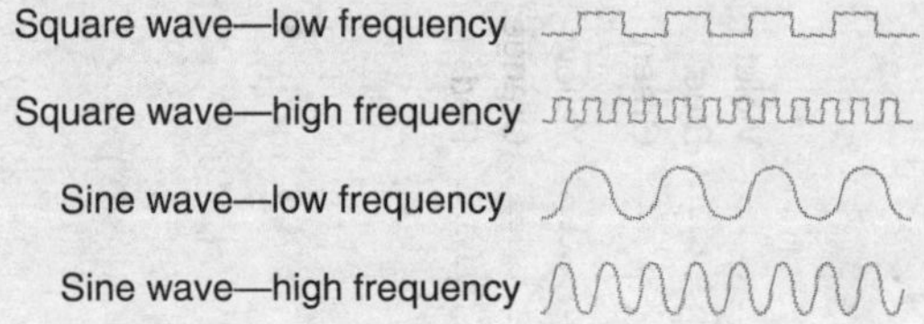

Figure 28–34 *Waveforms.*

electrical and electronic components produce electromagnetic fields that may generate noise. All electronic circuits are vulnerable to magnetic and electromagnetic field effect.

Computer Basics

Computers process information (data). They do this using simple but long-winded methods of saying yes and no (presence of voltage versus no voltage) thousands, sometimes millions, of times per second to express simple messages, using the means described in the previous section. But even though they use this long-winded process, the transactions occur at incredible speeds, that is, at very high frequencies. In this chapter, we begin to look at how computers function, a subject that is continued in Chapters 29, 30, and 33. We also look at some of the hardware and data packaging used in computer technology.

Integrated Circuits

Integrated circuits or **I/Cs** are the hardware, the building blocks of computers. They consist of resistors, diodes, and transistors arranged in a circuit on a *chip* of silicon. The number of electronic components that comprise the I/C varies from a handful to hundreds of thousands, depending on the function of the chip. I/Cs have innumerable household, industrial, and automotive applications and are the basis of digital watches, electronic pulse wipers, and all computer systems. Figure 28–35 shows a typical I/C.

I/Cs fall into two general categories. Analog I/Cs operate on variable voltage values: Electronic voltage regulators are a good vehicle example of an analog I/C. Digital I/Cs operate on two voltage values only, usually presence of voltage and no voltage. Digital I/Cs are the basis of most computer hardware including processing units, main memory, and data retention chips. I/C chips can be fused into a

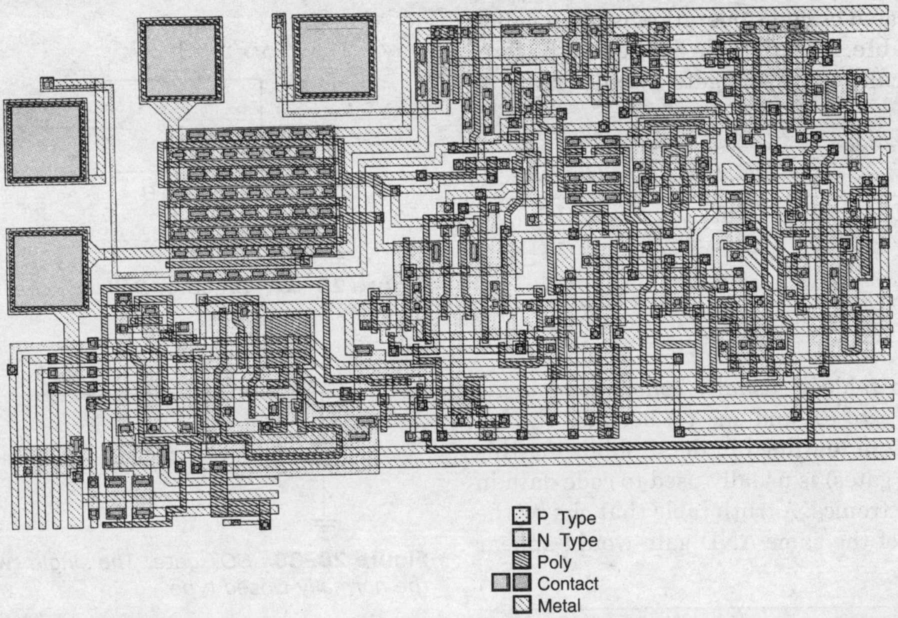

Figure 28–35 *Integrated circuit.*

motherboard (main circuit) or socketed: The latter has the advantage of removal and replacement. A common chip package used in computer and vehicle engine control modules (ECMs) is the dual in-line package (DIP), a rectangular plastic enclosed I/C with usually 14 to 16 pins arranged evenly on either side. DIPs may be fused (not removable) or socketed to the motherboard.

Gates and Truth Tables

In outlining the operation of transistors earlier in this chapter, the importance of **gates** was emphasized. Gates are the electronically controlled switching mechanisms that manage the operating mode of a transistor. Digital I/Cs are constructed by using thousands of gates. In most areas of electronics, gates can be either open or closed: In other words in-between states do not exist. The terms used to describe the state of a gate are *on* or *off*. In a circuit, these states are identified as presence of voltage or no voltage. By saying yes or no through a number of channels, we can actually say quite a bit, as shown by what we can get a single byte of information to represent (shown a little later in this chapter).

The best way to learn to understand the operation of gates in a digital circuit is to observe the operation of some electromechanical switches in some simple electrical circuits such as the ones described next, but because these switches change state and therefore, outcome, on the simple basis of voltage *on*, voltage *off*, they adapt ideally to electronic digital circuits.

AND Gates

In Figure 28–36, a power source is used to supply a light bulb in a series circuit: In the circuit, there are two push-button switches that are in the normally open state. In such a circuit, the light bulb will only illuminate when both switches are closed. This kind of switch is known as an **AND gate**.

The operation of the foregoing circuit can be summarized by looking at the circuit and coming to some logical conclusions: A table that as-

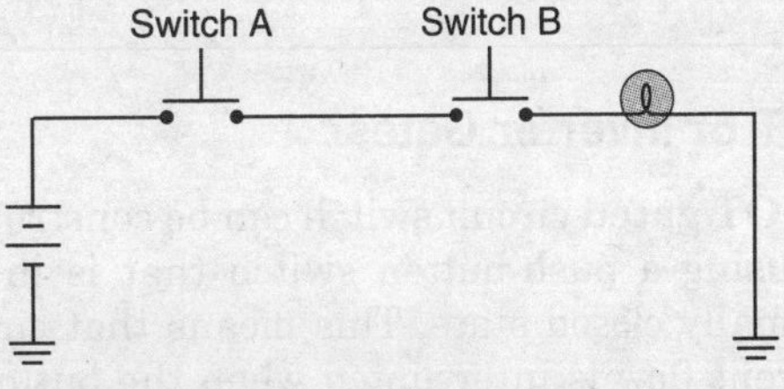

Figure 28–36 *AND gate: Both switches are of the normally open type.*

sesses a gated circuit's operation is often called a **truth table**. A truth table applied to the foregoing circuit, an AND gate, would read as follows:

Switch A	*Switch B*	*Outcome*
Off	Off	Off
Off	On	Off
On	Off	Off
On	On	On

A truth table is usually constructed using the digits *zero* (0) and one (1) because the **binary system** (outlined in detail following this section on gates) is usually used to code data in digital electronics. A truth table that charts the outcomes of the same AND gate would appear as follows:

Switch A	*Switch B*	*Outcome*
0	0	0
0	1	0
1	0	0
1	1	1

OR Gates

If another circuit is constructed using two normally open electromechanical switches but this time they are arranged in parallel, the result would be an **OR gate**. A schematic of an OR gate circuit is shown in Figure 28–37. A truth table constructed to represent the possible outcomes of this circuit would appear as follows:

Switch A	*Switch B*	*Outcome*
0	0	0
1	0	1
0	1	1
1	1	1

NOT or Inverter Gates

A **NOT gated** circuit switch can be constructed by using a push-button switch that is in the normally closed state. This means that circuit current flow is interrupted when the button is pushed. In the series circuit shown in Figure 28–38 current will flow until the switch opens

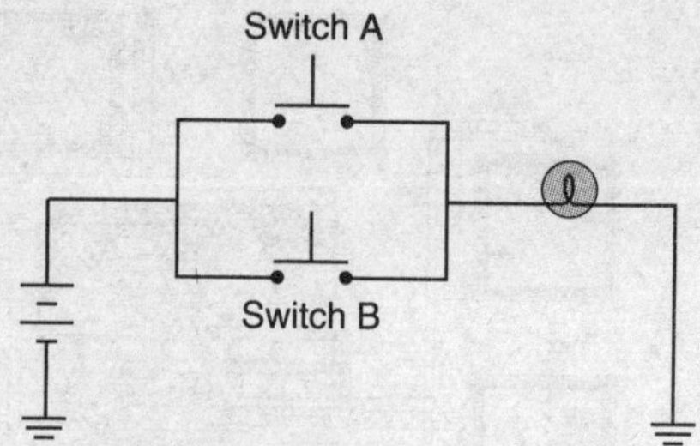

Figure 28–37 *OR gate: Both switches are of the normally open type.*

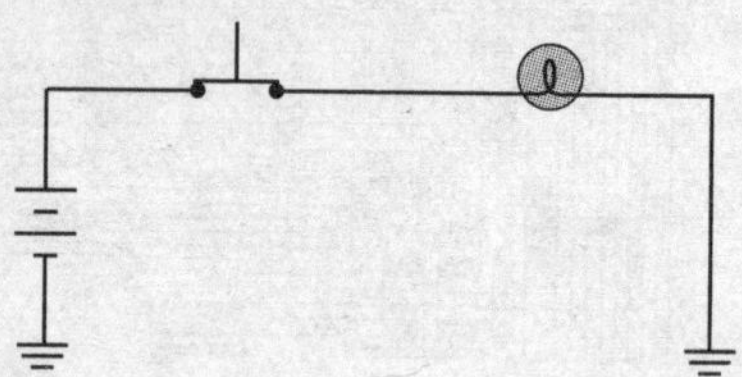

Figure 28–38 *NOT gate: The single switch is of the normally closed type.*

the circuit. A truth table constructed to graphically represent the outcomes of the NOT gated circuit shown in Figure 28–38 would look like the following:

IN	*OUT*
0	1
1	0

While the examples presented here used electromechanical switches for ease of understanding, in digital electronics, circuit switching is performed electronically using diode and transistor gates. AND, OR, and NOT gates are three commonly used means of producing an outcome that depends on the switching status of components in the gate circuit. If you understood how transistors function earlier in this chapter, you should now begin to understand how they can function as the switches in gate circuits.

Gates, Truth Tables, and Basic Data Processing

Figure 28–39 shows some simple two-input logic gates and the outcomes that can be produced from them. Logic gates are the basis of **data processing**.

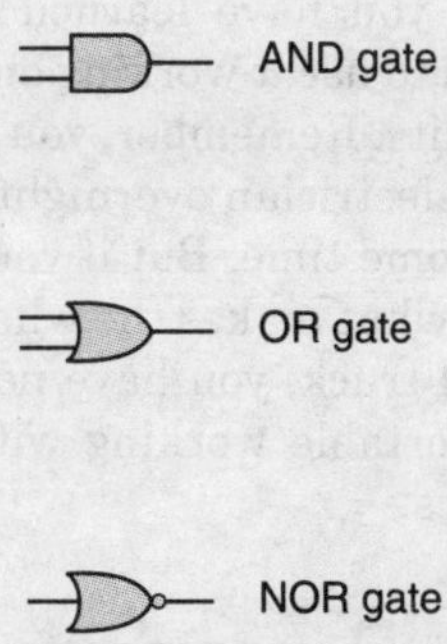

A	B	Out
0	0	0
0	1	0
1	0	0
1	1	1

A	B	Out
0	0	0
0	1	1
1	0	1
1	1	1

A	B	Out
0	0	1
0	1	0
1	0	0
1	1	0

Figure 28–39 *Logic gates and truth tables showing the switching outcomes.*

Gates are normally used in complex networks in which they are connected by buses to form a logic circuit. Bits of data can be moved through the logic circuit or data highway to produce outcomes that evolve from inputs and retained memory. By massing hundreds of thousands of logic circuits, the processing of information becomes possible. A computer works simply by processing input data and generating logical outcomes based on the input data.

Binary System Basics

The binary system is an arithmetic numeric system using two digits: It therefore has a base number of 2. The base number of a numeric system is the number of digits used in it. The decimal system is therefore a ten base number system. The binary numeric system is often used in computer electronics as it directly correlates to the on or off states of switches and circuits. In computer electronics, the binary system is the primary means of coding data using the digits 0 and 1 to represent alpha, numeric, and any other data. Numeric data is normally represented using the decimal system: Again, this is simple coding. When we see the digit 3 it is a representation of a quantitative value that most of us have been accustomed to decoding since early childhood. If decimal system values were coded into binary values, they could look like this:

Decimal digit	*Binary digit*
0	0
1	1
2	10
3	11
4	100
5	101
6	110
7	111
8	1000
9	1001
10	1010

In digital electronics a **bit** is the smallest piece of data that a computer can manipulate: It is simply the ability to represent one of two values: either *on* (1) or *off* (0). When groups of bits are arranged in patterns they are collectively described as follows:

4 bits = nibble

8 bits = **byte**

Most data is referred to quantitatively as bytes. Computer systems are capable of processing and retaining vast quantities of data, so in most cases millions (megabytes) and billions (gigabytes) of bytes are quantified. A byte is 8 bits of data, so it has the ability to represent up to 256 data possibilities. If a byte were to be used to code numeric data, it could appear as follows:

Decimal digit	*Binary coded digit*
0	0000 0000
1	0000 0001
2	0000 0010
3	0000 0011
4	0000 0100
5	0000 0101
6	0000 0110
7	0000 0111
8	0000 1000

and so on.

A number of methods are used to code data. If you are familiar with some computer basics, you may be acquainted with some American Standard Code for Information Interchange (ASCII) codes. This coding system has its own distinct method of coding values and would not be compatible with other coding systems without some kind of translation. Because on-off states can so easily be used to represent data, most digital computers and communications use this technology: It also is used in optical data processing, retention, and communications. Digital signals may be transmitted in series, one bit at a time, which tends to be slower, or in parallel, which is much faster. If the numbers 0 through 3 had to be transmitted through a serial link, they would be signaled sequentially as shown in Figure 28–40.

Decimal number	Binary number
0	0 0 0 0
1	0 0 0 1
2	0 0 1 0
3	0 0 1 1

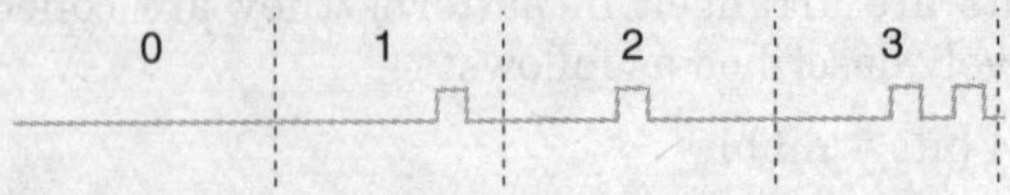

Figure 28–40 *Sequential switching of binary coded numbers: serial link.*

If the same numbers were to be transmitted through a parallel link, they would be outputted simultaneously as shown in Figure 28–41.

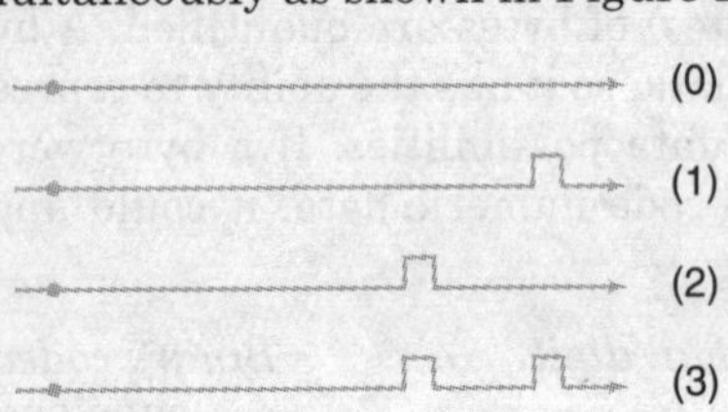

Figure 28–41 *Parallel link: Transmission time is greatly reduced.*

Conclusion

If you have read through this chapter, you will see that electricity has some complex aspects. However, it may comfort you to know that there are only three types of faults that can occur in an electrical circuit:

- Opens
- Shorts
- High resistance

When you understand the operating principles of the electrical circuits you are working on, it makes finding then classifying electrical circuit faults that much easier. Most of the remainder of this textbook requires some understanding of the electrical and electronic basics introduced in this chapter. Chapter 32 uses some of the theory you have learned in this chapter and gets you to use it working on actual truck electrical circuits. Remember, you are not going to become an electrician overnight; it is a process that takes some time. But if you really want to understand what makes the wheels roll on any recent model truck, you have no choice but to become comfortable working with electricity and electronics.

REVIEW QUESTIONS

1. A material described as an insulator would have how many electrons in its outer shell?
   a. Less than 4
   b. 4
   c. More than 4
2. Which of the following is a measure of electrical pressure?
   a. Amperes
   b. Ohms
   c. Voltage
   d. Watts
3. Which of the following units of measurement quantitatively expresses electron flow in a circuit factored with time?
   a. Coulombs
   b. Watts
   c. Farads
4. How many electrons does the element silicon have in its outer shell?
   a. 2
   b. 4
   c. 6
   d. 8
5. Who originated the branch of electricity generally described as electromagnetism?
   a. Franklin
   b. Gilbert
   c. Thomson
   d. Faraday
6. Which of the following elements could be described as being electrically inert?
   a. Oxygen
   b. Neon
   c. Carbon
   d. Iron

7. Which of the following is a measure of charge differential?
   a. Voltage
   b. Wattage
   c. Amperage
   d. Ohms
8. An element classified as a semiconductor would have how many electrons in its outer shell?
   a. Less than 4
   b. 4
   c. More than 4
   d. 8
9. Which of the following describes resistance to movement of magnetic lines of force?
   a. Reluctance
   b. Inductance
   c. Counterelectromotive force
   d. Capacitance
10. Use Ohm's law to calculate the current flow in a series circuit with a 12 V power source and a total circuit resistance of 6 ohms.
11. Calculate the power consumed in a circuit through which 3 amperes are flowed at a potential difference of 24 V.
12. A farad is a measure of:
    a. Inductance
    b. Reluctance
    c. Charge differential
    d. Capacitance
13. The term *pulse width modulation* refers to:
    a. Waveforms shaped to transmit data
    b. Unwanted voltage spikes
    c. Electronic noise
14. To form a P-type semiconductor crystal, the doping agent would be required to have how many electrons in its outer shell?
    a. 3
    b. 4
    c. 5
    d. 8
15. To form an N-type semiconductor crystal, the doping agent would be required to have how many electrons in its outer shell?
    a. 3
    b. 4
    c. 5
    d. 8
16. The positive terminal of a diode is correctly called a(an):
    a. Electrode
    b. Cathode
    c. Anode
    d. Emitter
17. Which of the following terms best describes the role of a typical transistor in an electronic circuit?
    a. Check valve
    b. Relay
    c. Rectifier
    d. Filter
18. When testing the operation of a typical transistor, which of the following should be true?
    a. High resistance across the emitter and base terminals
    b. Continuity across the emitter and collector terminals
    c. Continuity across the base and emitter terminals
19. What would be the outcome in an OR gate if one of two switches in the circuit was closed?
    a. Off
    b. On
20. How many different data codes could be represented by a byte?
    a. 2
    b. 8
    c. 64
    d. 256
21. A potentiometer is a:
    a. Two-terminal, mechanically variable resistor
    b. Three-terminal, mechanically variable resistor
    c. Two-terminal, thermally variable resistor
    d. Three-terminal, thermally variable resistor
22. Which of the following devices can be used to block AC and pass DC?
    a. Rheostat
    b. Inductor
    c. Capacitor
    d. Thermistor
23. Which of the following components is typically used in a half-wave rectifier?
    a. Rheostat
    b. Transformer
    c. Diode
    d. Zener diode

24. An N-type semiconductor crystal is doped with:
    a. Trivalent atoms
    b. Pentavalent atoms
    c. Carbon atoms
    d. Germanium atoms
25. A pulse width modulated (PWM) signal is:
    a. Analog
    b. Digital
    c. Either analog or digital
    d. Neither analog or digital
26. Which one of the following can store energy in the form of an electric charge?
    a. Thermocouple
    b. Induction coil
    c. Potentiometer
    d. Capacitor
27. Which of the following can store energy in the form of an electromagnetic charge?
    a. Thermocouple
    b. Induction coil
    c. Potentiometer
    d. Capacitor
28. In which of the following devices is polarity always important?
    a. Coil
    b. Capacitor
    c. Resistor
    d. Diode
29. Which of the following is a three-terminal device?
    a. Zener diode
    b. Transistor
    c. Capacitor
    d. Diode
30. Which of the following is a two-terminal device?
    a. Potentiometer
    b. Thermocouple pyrometer
    c. NPN transistor
    d. PNP transistor

29

Computers in the Trucking Industry

INTRODUCTION

Today's vehicle technicians are required to have a working knowledge of **personal computers (PCs)** as well as the computers that manage vehicle systems. Truck diesel engine and chassis original equipment manufacturers (OEMs) use the PC as their primary diagnostic electronic service tool (EST). Since the early 1980s, PCs have filed and organized the business activity of the trucking industry, both in fleet and service/repair operations. From the early 1990s, the PC has played an ever-increasing role on the shop floor where it has been used to read, diagnose, guide sequential troubleshooting, and reprogram vehicle onboard computers. Often the PC is networked to a **mainframe** (a high-volume data processing hub, often the anchor for Intranet systems) for purposes of logging repairs or reprogramming proprietary **data** (information) to onboard vehicle computer systems. By definition, a computer is a device used for information processing, arithmetic calculation, and information storage.

There is no doubt that apprentice truck technicians with basic computer skills learn the technology of vehicle electronic management systems much more quickly than those without those skills. Technicians with 20 years experience before the electronics revolution tend to adapt to the changes and challenges less readily than those less set in their ways. This fact has opened up many opportunities for younger technicians to advance at an accelerated rate in the truck and bus repair and maintenance industries. Most students entering a program of study in a college today have grown up with the computer and have probably been using them since the earliest grades.

The objective of this chapter is to introduce computer technology in a general sense. A little later in this book we will take a look at vehicle computer technology and you will note that many of the concepts introduced here are common to all computer systems.

This chapter introduces some of the communications technology that diesel technicians should be familiar with. More truly it describes the communications technology any person living on planet Earth should be aware of to survive in the workplace today. Along with **cybernetics** (the science of computer control systems), today's trucking industry has widely embraced the full spectrum of data communications. It is used to download engine control module (ECM) system management programming from a **data hub** (focal point of a communications system sometimes known as a *network star*) using a **modem** (telecommunications device that translates digital signals to analog for transmission) and network cards in conjunction with personal computer (PC) systems. It is used to geographically track fleets of vehicles using a combination of global positioning and telecom satellites.

A majority of today's computers, whether they be based in a phone, personal digital assistant (PDA), a portable notebook PC, or desktop PC, are networked via the **Internet** (global multimedia communications network) and various in-house **Intranet** (an internal network using Internet protocols) systems. To perform a simple shop procedure, such as removing and replacing an injector today, almost always requires the technician to use telecommunications technology. Telecommunications systems enable devices to talk to other devices, people to communicate with devices, and people to network with other people. The communications technology introduced in this section will become faster and more powerful with each passing year. And as this happens the diesel technician will be required to understand it as it evolves. Like it or not, you have chosen to work in an industry that embraces the leading edge of technology.

A BRIEF HISTORY OF COMPUTERS

The first computer is generally believed to be a simple counting and calculation tool called an abacus. The abacus is thought to have had its origins in Babylonia some 2,500 years ago. It survives as a calculating instrument in remote parts of East Asia and the Middle East and as a child's toy in the West. An expert practitioner can compete with electronic calculation in both speed and accuracy. The abacus would be classified as an **analog** computer. An analog device operates on data represented by numbers or voltage values, which it processes to output a physical analogy of a mathematical problem to be solved. A slide rule, much used until the widespread acceptance of the electronic multitasking calculator in the 1960s, is another example of an analog computer. In computer terminology, **multitasking** describes the ability to handle multiple processing paths simultaneously.

After the abacus, the next significant events in the development of the computer took place in Europe during the seventeenth century. In 1620, Edmund Gunter, an English mathematician, invented the slide rule, and in 1642, the French scientist Blaise Pascal introduced the first mechanical digital calculating machine. These significant innovations, plus the ideas of many scientists and mathematicians who followed, paved the road to the development of the modern computer.

Most modern computers are **digital**. The digital computer works with data in **discrete** form, that is, expressed or coded as digits of binary codes. Binary coding is explained in Chapter 28. Any digital computer processes data by counting, listing, and rearranging bits of data in accordance with detailed program instructions. The results of **processing** are then translated into **outputs** that may be displayed on a screen, printed onto paper, or used to actuate other **peripherals**. Peripherals are input and output devices that are part of the computer system but not integral with main housing.

The invention of the first electronic digital computer is attributed to Presper Eckert and John Mauchly of the University of Pennsylvania in 1946. By the 1960s the computer had developed into a business tool used for data storage and arithmetic processing. The computers of the 1960s were physically vast, power-hungry machines full of vacuum tubes that required considerable maintenance.

The modern computer evolved from these beginnings. Its widespread use as a business tool began in the 1980s and by the 1990s, a PC could be found in a majority of homes. The typical home computer system of the twenty-first century has more computing power than that on board *Apollo 11* (the first manned moon landing craft). Today, computers have a role in driving almost every aspect of our lives. Figure 29–1 shows different types of computers.

COMPUTER HARDWARE

A computer system is made up of a number of pieces of **hardware**. Computer hardware components are physical; they can be touched, unplugged, removed, repaired, and replaced. Figure 29–2 and Figure 29–3 show different computer system components.

The following components would be typical:

Computer System Housing

The computer/system housing is a **desktop** case or tower constructed of plastic or steel. Or it can be a portable computer that we call a **lap-**

1. Notebook Laptop

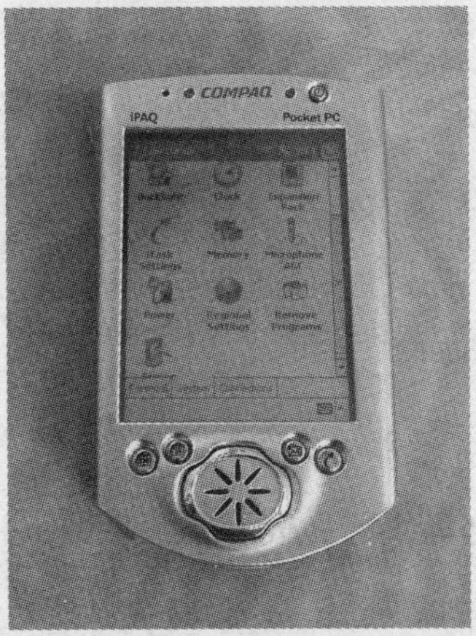

2. Personal Digital Assistant (PDA)

3. Tablet PC

4. Desktop PC

Figure 29–1 *Types of computers.*

top or **notebook computer**. Typically, the system housing contains:

Motherboard

The motherboard is the main circuit board or deck to which the internal chips, cards, and other components are attached either directly by sockets or indirectly by sockets and bus cables.

Central Processing Unit

The **central processing unit (CPU)**, also known as the **microprocessor** unit (MPU), is the "brain" of the computer system. CPUs are usually rated by processing speed. A CPU is usually attached to the motherboard by a multipin connector. The functions of the CPU are as follows:

- Timing and logic control
- Arithmetic and logic operations
- Fetching data and program instructions from memory
- **Decoding** (translating) instructions
- Responding to control signals from input and output devices (I/O)

Microprocessors contain a complex of many electronic circuits photo-infused on a silicon chip. The CPU individual subcircuits are referred to as blocks and collectively as the architecture of the microprocessor. Though recent CPUs have not been reduced in size, their power and speed has risen exponentially. Figure 29–4 shows a typical CPU.

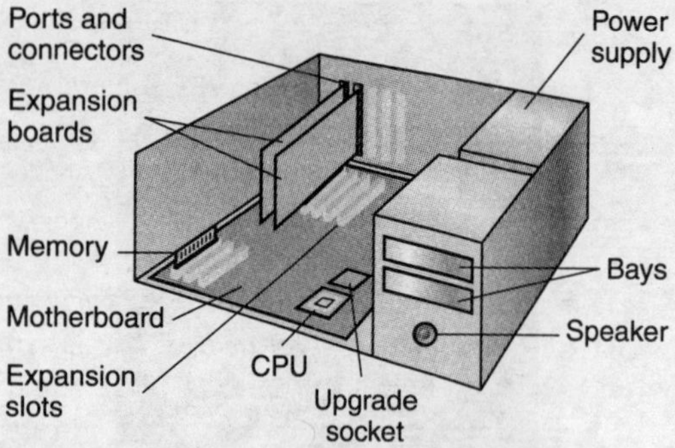

Figure 29–2 *PC system unit components.*

Figure 29–4 *CPU size matched with a bumblebee: The CPU is the brain of the computer processing cycle.*

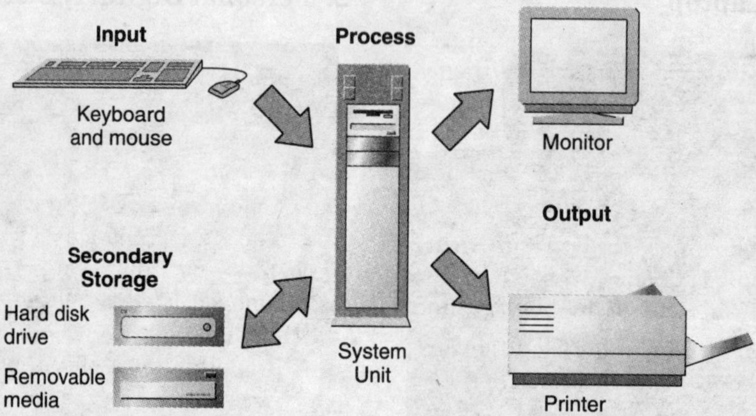

Figure 29–3 *A computer system consists of input devices, the system unit, output devices, and secondary storage devices.*

Word Size CPUs are often described by word size such as 64- or 128-bit processor. Word size describes the number of bits that can be processed simultaneously. The larger the word size, the faster the processor. As each year goes by, computer word size grows. Not all computers need a vast word size. In vehicle computer technology, we use many 8-bit CPUs to drive simple chassis circuits. They are inexpensive and reliable. More important, they are all that is required.

Clock Speed Clock speed is a rating of computer processing speed measured in megahertz (MHz) or millions of cycles per second. Today it is probably only important in PCs because faster clock speeds are marketed as being desirable. Immensely high clock speeds are really only a requirement for virtual games that require instantaneous response from input to output. An example would be driving a racecar on a virtual track with steering wheel, accelerator, and brakes as inputs.

System Internal Buses

Buses are simply connection and data transfer devices. They may be wires, plates, or integrated into motherboards or other circuit boards. System internal buses link the main modules in a typical computer system as follows:

- Control bus—bidirectional. It transmits and receives operational commands to all parts of the computer system.
- Address bus—unidirectional. It connects the CPU to the memory banks and I/O ports. The CPU selects the address or location in memory to which data is targeted.

- Data bus—bidirectional. It acts as a data conduit to and from the memory banks, the CPU, and the I/O ports.

Input Devices

Two input devices that are found on almost every PC are the keyboard and **mouse**. The keyboard (Figure 29–5) is required to input alpha and numeric data into **software** programs and a mouse is required to navigate the **graphical user interface (GUI)** by moving a **cursor** displayed on the **monitor** (display screen). GUIs are standard in today's computers whether they use Mac or a Windows environment. Figure 29–6 shows a typical GUI. *Software* is the term used to describe the digital programs that structure a computer processing cycle to produce outcomes.

Figure 29–5 *A typical PC keyboard.*

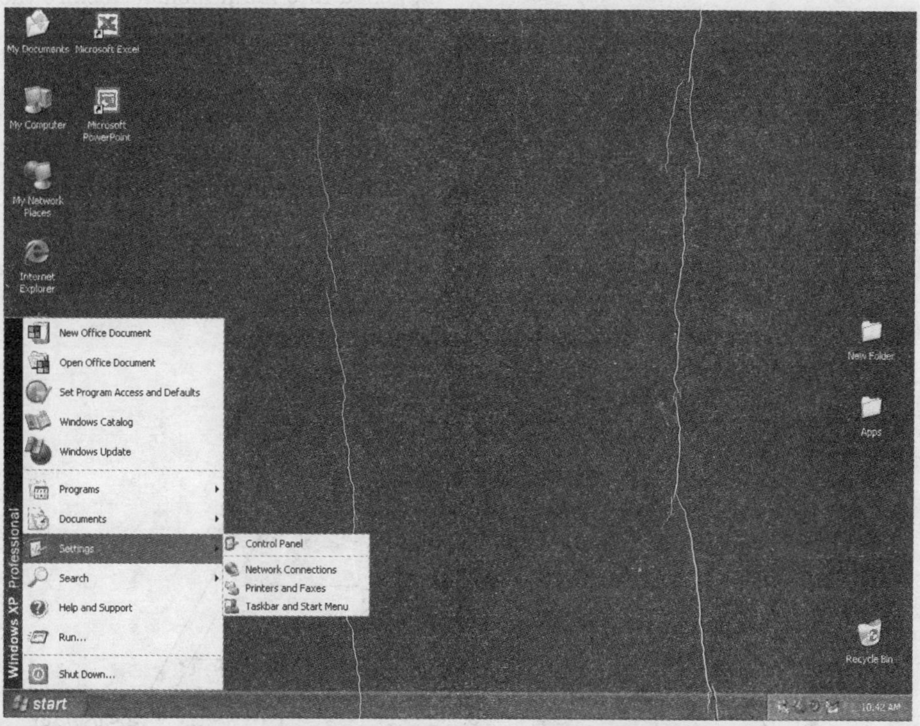

Figure 29–6 *A graphical user interface such as MS Windows makes the computer easier to use: The icons represent different processing and software selection options.*

Output Devices

Required output devices on a PC are a monitor and speaker system. Older **cathode ray tube (CRT)** monitors have given way to **liquid crystal display (LCD)** type monitors in most of today's computer systems. Most desktop computer systems probably also have a printer.

USB Ports

A **universal serial bus (USB)** has become the standard means of connecting peripheral computer devices and portable memory storage. USBs are also used as a means of electronically connecting devices such as cameras, televisions, camcorders, and other devices.

Combination Input and Output Devices

Many computer devices that we know as either an input or an output device actually perform as both. For instance, a keyboard is primarily an input device but most will have some output functions by incorporating LCD status lights. A touch screen monitor acts as a display device but acts also as an input device as it responds to fingertip commands. **Modems** (digital/analog communication translators) and network cards necessarily perform both as input and output devices as does a printer/scanner/fax machine.

Memory Categories

The term *memory* is used to describe the storage of digital data generally in computer terminology. Memory or data can be retained using a number of different means determined by factors such as whether the information must be accessed at high speeds by the processor and whether it needs to be retained more or less permanently, or dumped when the work session is completed. Figure 29–7 shows access speeds of different data retention media. This section categorizes memory by general type:

Random-Access Memory (RAM)

Random-access memory (RAM) is often referred to as "main memory" or "primary storage." It is retained electronically in semiconductor chips. RAM data includes the computer **operating system (OS)** software (e.g., Windows), which must be loaded into main memory each time the computer is booted. The OS is the core set of rules and regulations (we call these protocols) that coordinate computer operations. In a computer system the OS needs to be launched into main memory before the computer can actually do anything useful. Once the OS system is electronically logged into RAM, the **program application software** can be run by executing commands to log it into RAM.

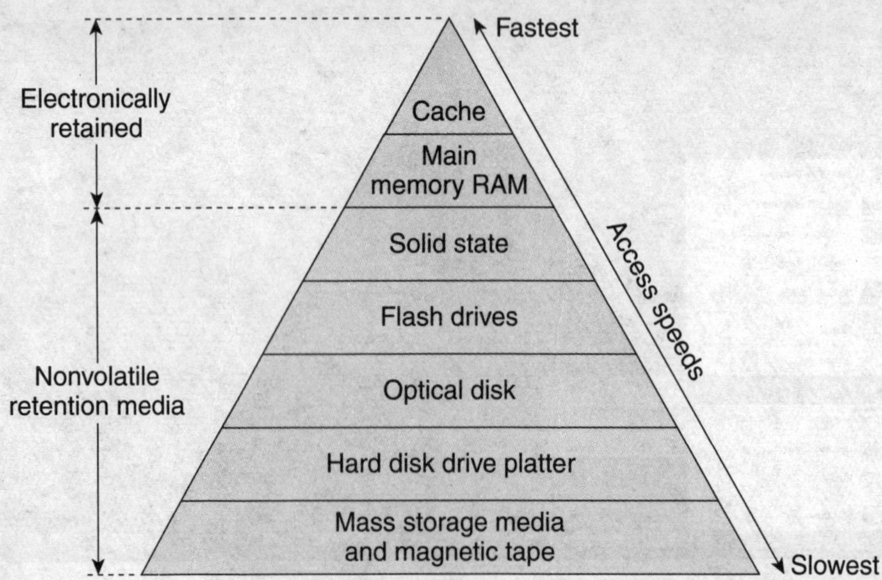

Figure 29–7 *Access speeds of data retention media.*

Program application software is the software required to actually make the computer do something: It is explained a little later in this chapter.

RAM data is volatile. Because it is electronically retained, it is lost when the electrical circuit supplying it is opened. Some vehicles have onboard computers with RAM capability described as both volatile RAM and nonvolatile RAM. Memory that is described as being **volatile memory** is lost when power is interrupted. Vehicle volatile RAM can be likened to main memory in a PC in that it is switched on and off by the vehicle ignition circuit. Some vehicles, especially those with first generation computerized system management electronics, had a nonvolatile RAM memory component; this data was electronically retained but supplied by a circuit directly from the vehicle battery after the ignition circuit was opened. Fault codes and failure mode strategy were written to nonvolatile RAM and "kept alive" until either the battery was disconnected or the engine control module (ECM) reset.

In the PC processing cycle, **cache** is used to retain frequently consulted program instructions or data. Cache is high access speed RAM. As such it is also volatile. When the CPU requires data, it checks cache first, enabling faster execution than retrieval from (slower) main memory.

Read-Only Memory

Read-only memory (ROM) chips retain data, protocols (rules and regulation), and instructions that once written cannot be changed. In other words, the information can be read and used but never altered (although it can be destroyed or corrupted). The **boot-up** procedure in a PC system is used repeatedly and does not change so it makes sense that it is written to ROM. Truck and other vehicle computers (ECMs) have major portions of the total amount of required management data written to ROM. A specific diesel engine series may be used to power a truck, bus, generator, boat, or a grader; in each, the master program for engine management would be common and written to ROM. This data in itself may not be sufficient to run the engine without additional program data. However, it would contain all the common running characteristics required to run that engine in each of the very different applications. ROM describes nonvolatile data retention media; it is usually retained magnetically in current systems.

Programmable Read-Only Memory

Programmable read-only memory (PROM) is written to a chip or set of chips (card) and socketed to the PC or ECM **motherboard** (main circuit board). In a vehicle engine management system, PROM is often used to define application. For instance, the diesel engine designed for use in many different applications and manufactured with an ECM ROM common to all those applications would require function-specific running data before it would actually manage an engine. The data contained in the PROM chip or card would define the actual run parameters of the engine.

Continuing to use the example of an engine ECM, PROM data would qualify that contained in ROM to a specific system or chassis application. In some early vehicle computer-controlled systems, the only means of altering any program was to physically remove and replace PROM chips. This is not the case today. In most cases, PROM is soft programmed into an ECM memory address that also contains the next memory category, electronically erasable programmable read-only memory (EEPROM).

Electronically Erasable Programmable Read-Only Memory

EEPROM is used in most current truck engine and chassis management ECMs. It can be written to by the onboard software, receive customer and proprietary data programming, record driving and performance analyses, and retain codes manifestly and covertly. Data programmed to EEPROM would include customer data, such as a road speed limit, cruise parameters, transmission ratios, final drive ratios, tire rolling radii, and failure mode strategy. This type of data could require alteration a number of times during the vehicle or engine life.

EEPROM is also used to log data such as the fuel map profiles and covert documentation of failure/abuse condition. This type of data on chassis computer systems can often only be read or rewritten directly by the OEM mainframe via networking. A couple of truck OEMs describe their combined PROM and EEPROM capability as a "personality module." The personality module can be physically separate from the ECM, a removable card that bolts to the ECM housing, or completely integrated into the ECM housing. The term *personality module* is an excellent description of the actual role played by this type of data in managing the system.

Flash Memory

Flash memory is another type of nonvolatile computer memory that can be electronically erased and reprogrammed. We use this technology in portable data **storage media** such as memory cards, and USB flash drives (jump drives and thumb drives). These are used for backing up data, segmented file storage, and transfer of data between computers. It differs from EEPROM in the way it is erased and programmed in blocks on the card.

Flash memory is usually less costly than EEPROM and in recent years has taken over as the primary means of portable data storage and data transfer. If you have ever used a flash memory jump drive, you will have noticed that read and access times are much faster than other types of portable data storage media. Small size USB jump drives can hold large volumes of data while also being somewhat, but not totally, immune to physical shock and temperature change: This accounts for their popularity as data storage on digital cameras and camcorders and other battery-powered devices. We will discuss USB jump drives in more detail a little later in this section.

Secondary Data Storage Media

Secondary data storage or auxiliary storage devices are means, such as CDs, DVDs, USB flash drives, or portable hard drives, used for writing, storing, and reading data. The earliest home PCs used cassette tapes on which data was magnetically encoded in just the same way as an audio cassette tape. A cassette tape stores data sequentially. Sequential data storage works fine to retain an album of music where the objective is to listen to the notes and tracks in their originally recorded sequence. But computers need to retrieve data randomly and at high speeds. So in today's computers, cassette tapes are only used for large volume data backup, limited by the speed at which magnetic tape can be driven through the read heads.

Cassette tapes were superseded by 5¼ -inch floppy diskettes and those by 3½-inch disks. The 3½-inch diskette was encased in rigid plastic and only the internal circular disk was "floppy"; this circular disc was nylon plastic coated with metal oxide. The diskette was encoded with data magnetically in the same way an audiotape is encoded and read by a rotating drive mechanism and read heads. They are still used today but it is difficult to find a contemporary PC with an integral diskette drive.

Hard Disks

Hard disks retain data magnetically. They consist of multiple rigid platters coated with metal oxide that permit data to be magnetically written to the surface. On a hard disk assembly, the read-write heads, the drive mechanism, and data retention platters are incased in an airtight, vacuum-sealed case. Hard disks are usually a permanent fixture within a PC housing, although portable hard drives, which are as little as 4 inches across and can fit in a pocket, are available now. We can also use them in devices such as camcorders and external PC back-ups. They rotate at high speeds (7,200 rpm or more) to receive and unload data (Figure 29–8).

Basic Input/Output System

The act of turning a computer on is called bootup. In the boot sequence, the CPU looks to **basic input/output (BIOS)** for startup instructions. BIOS instructions are loaded onto a chip. When the CPU looks to the BIOS chip as the computer is switched on, it receives sequential instructions on how to **interface** (interact) between the disk-operating system and the system hardware.

Optical Data Storage

Optical disks such as **CDs** and **DVDs** are used to store data. A laser is used to burn

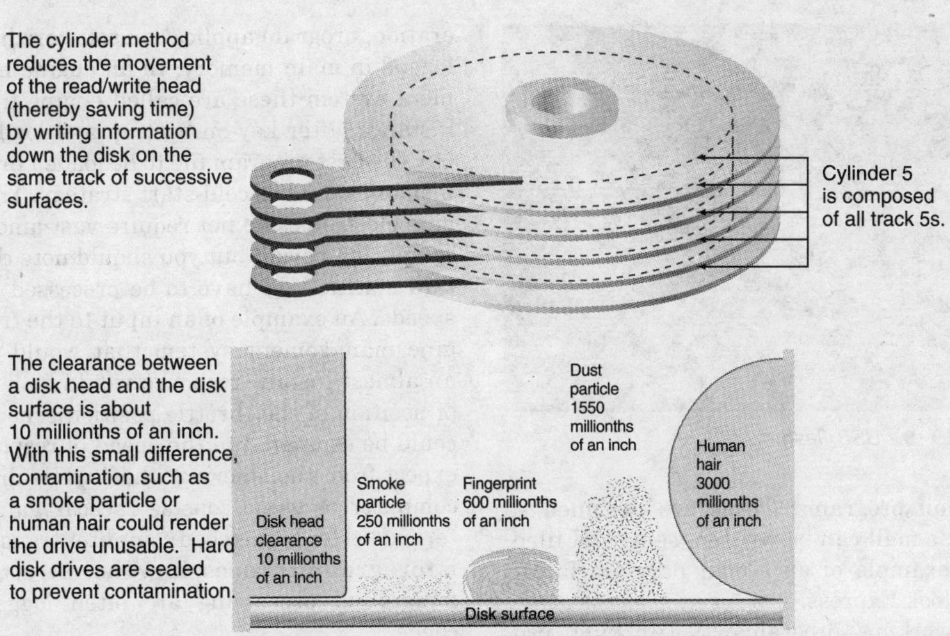

Figure 29–8 *Hard disk drive. (Courtesy of Course Technology Cengage Learning)*

microscopic holes on the surface of a rigid plastic disk. The disk can then be scanned by a low-power laser that reflects light off the encoded disk surface to read digital data. CDs and DVDs can hold large volumes of digital data and may be read-only or rewritable. Both these types of optical data retention media are being widely used in vehicle technology industries. Because of the large amount of data that a CD or DVD can retain, many computer programs are now sold in this format. Optical disks have the additional advantage that the data written to them cannot be damaged by magnetic force.

USB Flash Drives

USB flash drives are memory data storage devices integrated with a USB interface. We mentioned these briefly earlier in this chapter when we were discussing flash memory. USB flash drives are small, portable, very lightweight, and rewritable by flashing. Data capacity of USB flash drives is increasing exponentially and as it does, the cost goes down.

One of the primary advantages of the USB flash drive over other portable data storage devices is their reliability due to an absence of moving parts. A USB flash drive consists of a small printed circuit board encased in a either plastic or metal casing. They tend to withstand some abuse and are commonly carried about in a pocket or in a cluster of keys. Only the USB connector is truly vulnerable and this should be covered by a protective cap when not inserted into a USB port.

Data is accessed on a USB flash drive by connecting it to a computer either indirectly by plugging it into a USB host controller or directly into an integral USB port. Flash drives can be considered inactive when not connected to a computer USB hub. They draw whatever electrical power they require from the USB port. Some high-speed flash drives driven on the USB 2.0 standard require more power than that supplied at a USB hub so they have to be connected either directly into a host controller (the ports found on the PC) or a self-powered hub. Figure 29–9 shows a typical USB flash drive.

Program Application Software

Examples of program application software include programs such as:

- Word processing programs: This type of program is designed to make the computer behave like a typewriter. Examples of word processing software are Microsoft Word or Corel WordPerfect.

Figure 29–9 *USB flash drive.*

- E-mail programs: These are designed so that e-mail can be written, sent, and filed. An example of an e-mail program is MS Outlook Express.
- Networking programs: Networking programs are designed to allow people to navigate the Internet. Examples of networking programs are Mozilla Firefox and MS Internet Explorer.
- Spreadsheet programs: Spreadsheets are designed to handle numeric data. Examples of spreadsheet programs are MS-Excel and Lotus.

Operating Systems

You are probably somewhat familiar with both OS and program application software as they apply to PCs, but we also can use these terms to describe the software architecture on vehicle computer systems. For instance, any vehicle computer system requires an OS: This OS is OEM specific. It is also important that a vehicle computer OS loads into RAM much faster than its equivalent on your home computer: A truck driver is likely to get impatient if forced to wait 2 minutes before boot-up permits the starting of the vehicle engine.

Boot Sequence

After loading the OS, to get the truck computer system to actually do things like start and manage the engine through its various phases of operation, program application software has to be logged in main memory. In an engine management system these are called engine files. For instance, after key-on boot-up and loading the OS, the first program file to be loaded into main memory would be cold-start strategy. Truck engine electronics do not require vast amounts of computing power, but you should note that certain instructions have to be processed at high speeds. An example of an input to the truck engine management system that would require an almost instant response would be a change of position of the throttle position sensor. This could be compared to the speed of response we expect from the steering wheel while driving a computer or game console racecar game: The term used for high-speed computer transactions is **foreground processing**. The instructions for foreground processing are often logged into cache.

ELECTRONICALLY REPRESENTED DATA

In a digital computer, binary digits are used to represent data. The digits 0 and 1 are the only numbers used to represent billions of data possibilities. These digits represent the on (1) or off (0) status we talked about in Chapter 28. Each on or off digital value is called a bit. The word *bit* is derived from the blending of the words *binary digit*.

Bits and Bytes

A **byte** is composed of 8 bits. A byte is capable of representing 256 data possibilities. ASCII is a commonly used secondary data storage code. ASCII is an 8-bit storage code used on PC systems. Observe how the following alpha, numeric, and symbolic values are coded in ASCII.

Quantitative Values

The following values are used to represent data quantity and time values when we describe computer operation. Computer memory in bytes and computer speed is expressed in fractions of a second. Here are some examples:

SYMBOL	ASCII
0	01100000
1	01100001
2	01100010
3	01100011
4	01100100
5	01100101
6	01100110
7	01100111
8	01101000
9	01101001
A	01000001
B	01000010
C	01000011
D	01000100
E	01000101
F	01000110
G	01000111
H	01001000
I	01001001
J	01001010
K	01001011
L	01001100
M	01001101

Quantity

One kilobyte = 1,000 or 10^3 or 1 thousand bytes

One megabyte = 1,000,000 or 10^6 or 1 million bytes

One gigabyte = 10^9 or 1,000,000,000 or 1 billion bytes

One terabyte = 10^{12} or 1 trillion bytes

One exabyte = 10^{18} bytes

Time

One pico-second = 1 trillionth or 10^{-12} second

One nano-second = 1 billionth or 10^{-9} second

One micro-second = 1 millionth or 10^{-6} second

One milli-second = 1 thousandth or 10^{-3} second

SUMMARY OF COMPUTER OPERATION AND THE PROCESSING CYCLE

The system unit in a typical computer consists of a housing containing a motherboard, CPU, electronic and magnetic data retention media, data readers, coprocessors, connection slots and ports, and a power supply (see Figure 29–2). The motherboard is a circuit board; most of the system unit components are either permanently connected or socketed (allowing the option for replacement or upgrade) to the motherboard. In a typical PC, the CPU is contained in a single integrated circuit called a *microprocessor*. The CPU contains the control/command unit and the arithmetic/logic unit (ALU).

Four-part Processing Cycle

The control/command unit in the CPU operates by sequencing a four-part processing cycle below and illustrated in Figure 29–10.

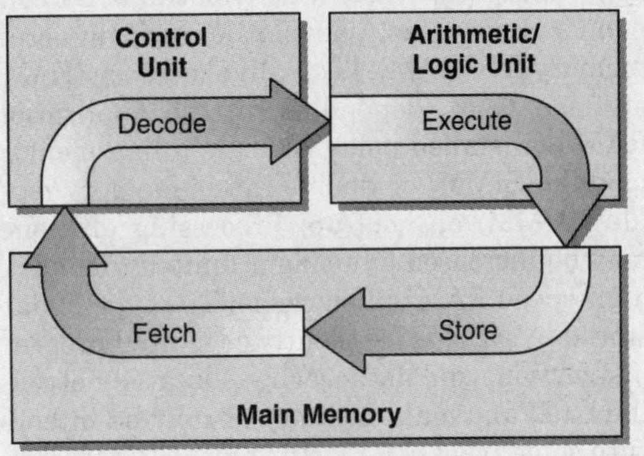

Figure 29–10 *The processing cycle.*

The four-part processing cycle incorporates the following phases:

1. Fetching
2. Decoding
3. Executing
4. Storing

The control unit uses a **system clock** to synchronize all computer operations; the system clock generates electronic pulses at a fixed rate specified in megahertz (MHz). The CPU ALU incorporates the electronic circuitry required to perform arithmetic and logic data manipulation. Arithmetic operations include addition, subtraction, multiplication, and division. Logical operations consist primarily of data comparison and sequential strategizing. The CPU contains registers, which are memory storage locations that may temporarily store certain types of data. A critical factor of computer processing speed is word size: This indicates the number of bits that a CPU can process simultaneously.

Memory capacity generally refers to data quantitatively. This memory can be retained either electronically or stored in a more permanent manner magnetically or optically. RAM refers to main memory or primary storage and retains data electronically. Electronically retained data is volatile; when the computer is switched off it is dumped. After booting up a computer, RAM is loaded with the following data:

1. The OS
2. The application program instructions
3. The data currently being processed

Memory

Some computers have a nonvolatile RAM component known as nonvolatile random-access memory (NV-RAM), keep-alive memory (KAM), or flash RAM, depending on the application. RAM is installed using a single inline memory module (SIMM) or double inline memory module (DIMM) on boot-up. Processing efficiency may be increased by using a limited amount of high-speed RAM called cache. The type of data stored in cache is frequently consulted program instructions and data; cache is located between the CPU and main memory or may be integral with some types of CPU. In vehicle ECMs, cache is used to enable high-speed processing reactions to such events as a change in accelerator pedal angle.

ROM describes chips that retain magnetically encoded data designed to be permanent. ROM chips are often fused to a motherboard and are not removable. PROM chips also retain data magnetically; however, these are often designed for removal and replacement. Electronically erasable PROM (EEPROM) is data retention that can be read, written to, and erased, either by the computer system it is integral with, or from an external source such as a network or portable data retention media. Flash memory devices can also be read, written to, and erased and their use today is becoming more common.

Memory Access Speeds

The speed of memory is usually measured in nanoseconds or fractions of nanoseconds. Registers in the CPU chip have access times of much less than a single nanosecond, and these speeds increase as each new generation of CPUs is introduced. ROM memory loaded onto chips has slower access times that range between 10 and 250 nanoseconds. Slower still is accessing data on a typical PC hard disk drive, which takes as long as 10 to 20 milliseconds. But remember, short periods of time have a way of rapidly dating time specifications, so what is fast today will be considered slow 5 years from now. Computer processing efficiency may be increased by using a coprocessor, a chip, or card that functions to perform specific tasks.

Power Supply

The computer power supply depends on the computer's functions. A PC typically receives an electrical supply at 110 to 120 volts alternating current (AC) and transforms this to an operating value ranging from 3 to 12 volts direct current (DC). A vehicle ECM is usually supplied with 12 volts DC and transforms this to a lower voltage value, usually 5 volts or less within the system. Operating voltage values tend to be lower in more recent computers, the objective being to reduce heat. PCs normally incorporate a cooling fan to remove heat from system components. Diesel-powered vehicle ECMs are sometimes mounted on a heat sink or heat exchanger through which fuel is routed on its way to the injection pumping apparatus. This helps cool the processor and some hardware drivers that may generate heat.

Connecting Peripherals and Memory

Expansion slots are used in PCs to connect in devices that increase computing power or capability. Ports are used to connect peripherals such as printers, modems, and keyboards. Parallel ports are capable of transmitting data at faster rates than serial ports so they function at higher speeds. For the most part, the USB interface (see below) has replaced both the serial port and parallel port. Most modern computers are connected to devices through a USB connection; some are no longer equipped with serial or parallel port connections.

USB is a serial bus standard to interface devices. It was originally designed for PCs, but it has become commonplace on handheld devices such as portable memory devices, video game consoles, PDAs, and portable media players. Most of the EST cables we use to access truck data buses now connect to the EST device by means of a USB port. They are used because they are simple, cheap, and universal. The USB was devised to help retire serial and parallel ports on PCs because these were not standardized and required an expensive range of device drivers. A bay is simply a location within the system unit used to store hardware such as disk and CD drives. Two or more bays stacked vertically can be called a *cage*.

NETWORKING

A **network** in its simplest form would be a "**handshake**" connection between two computers as shown in Figure 29–11. You may use something like the hardwired system shown in Figure 29–11 if you have a couple of computers at home, though you are more likely to use wireless connectivity if security is not an issue. *Handshake* is the term used to describe the establishing of a communications link between one computer and another—or a network of others. A network refers to a series of interconnected computers, usually PCs, for purposes of sharing data, software, and hard resources. A large office with several computer stations may network them to permit data exchange and perhaps have a single printer to serve all the stations. The means of connecting the computers in a network varies, but if it extends beyond a single building, it usually involves using telecommunications hardware and software.

A modem is a device used for converting the digital output of the computer to the analog signals required for transmission by the phone system. The speed at which a modem or network card can transmit data is specified by its **baud rate**. Baud rate indicates the number of times per second 1 bit can be transmitted. The baud rate specification is critical when considering the purchase of any equipment that must be networked. Truck engine original equipment manufacturers (OEMs) always specify a minimum baud rate for networking computers with their mainframe systems and today usually recommend high-speed (large bandwidth) connections.

A **star network** is one that is constructed around a central computer with multiple terminals or **nodes**. *Node* is a term used to describe each user station. The central computer may be a mainframe but it does not have to be. Often a star network will use **dumb nodes**, that is, stations that consist of a screen and keyboard but no processing or data storage capability independent of the central computer.

THE INTERNET

The Internet is a network of networks. It now connects most of the countries in the world and is said to be used by nearly two billion people worldwide, a number that grows daily. Internet penetration in the United States is 69%, the highest in the world: But other countries are catching up—fast. It is very appropriately referred to as the World Wide Web or WWW. The

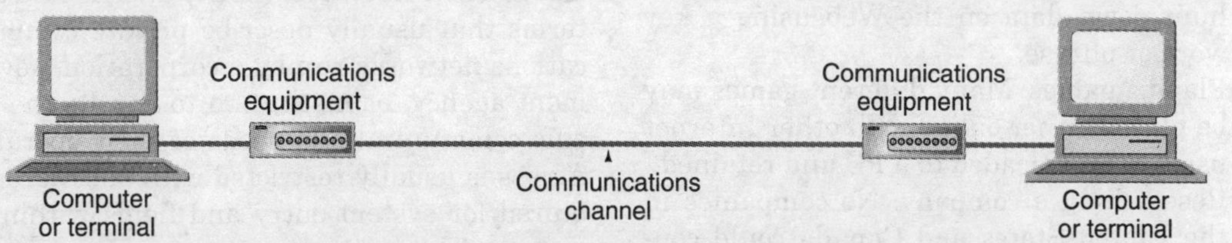

Figure 29–11 *Hard-wired model of a communications system between two computers.*

following list outlines some of the ways we use the Internet system:

1. Sending e-mail (electronic mail). This is a highly efficient means of messaging users served by the Internet. It has the advantage of nearly immediate transfer and is made easier without the protocols (and delay) of hard mail. Users are required to have an e-mail address and minimal computer skills: Some keyboarding ability is desirable.
2. File transfer. Files may be transferred from the Internet to a user PC system using a graphical user interface (GUI) and dragging the file title from the Internet menu to the user menu. Most files are exchanged with minimal security, but in cases where higher levels of security are required, this can be provided using the appropriate software.
3. Newsgroups. Internet users may belong to a newsgroup, which is essentially a discussion group made up of Internet users with common interests. Messages on a newsgroup are titled and may consist of a single phrase or be pages long. A good example would be the International Automotive Technicians Network (iatn.net). Message titles may be browsed and only those of interest downloaded for reading. People with some time on their hands can blog the world with their opinions.
4. Searching databases. Most government agencies and educational institutions make archives of information available on the Internet. Good search software makes accessing and sorting information, especially detailed, subject-specific data, easy, which is certainly much easier than attempting to locate it by phoning bureaucrats or searching a public library with its limited "hard" resources. A variety of **search engines** are used on the Internet. Search engines will hunt down data on the Web using a key word or phrase.
5. Playing games. Many different games may be played either online with other Internet users or downloaded to a PC and retained.
6. Researching a company. No companies in the United States and Canada could conduct business today without having some kind Web profile. The type of information you can access varies with the company. Some companies conduct business solely on the Internet.
7. Shopping. Thousands of products and services are advertised on the Internet and may be purchased using a credit card number. Although most are probably reputable, "buyer beware" should be exercised at all times.
8. Finding a job. More and more companies list job vacancies on the Internet. There are search engines dedicated exclusively to job search on the Internet. An example would be monster.com.
9. Receiving online news. Magazines, newspapers, and other news services all make information available on the Web and, depending on the service it may be up to date. Sports scores throughout the world are an example of data that is almost constantly updated. The scores are usually accompanied by stories that can be read or downloaded for printing. Major newspapers would lose status immediately if they did not maintain a Web version. Astoundingly, newspaper sales have not been that badly affected by Web news.
10. Public relations (PR). Almost every company, government agency, and organization in North America and Western Europe realizes the importance of the Internet as a communications tool. Connect to the White House, for instance, and the user may select anything from a pictorial tour to messages from the president and vice president. A presence on the Internet is probably essential for any major company transacting business in North America.

LANs AND INTRANET SYSTEMS

Local area network (LAN) and Intranet are terms that usually describe private communications networks run by a corporation, government agency, or institution to handle the specific communications needs of their operation. Access is usually restricted with passwords required for system entry and fields within the system. The protocols used are Internet derived, and while general access from the Internet may be to some extent blocked, access to it is usually enabled. A LAN could be housed in a single room and consist of two or three nodes or it may be spread over the country and incorporate thousands of nodes. A LAN may use wireless connectivity or for security reasons (and reliability), retain a hardwired setup as shown in Figure 29–12.

THE TELECOMMUNICATIONS SYSTEM

Telephone systems are responsible for both the hard (wiring/switching/decoding) and soft (microwave/radio waves, and so forth) linkages that enable networking. Of the telecommunications systems in use in the world today, the telephone system is probably the most used. The telephone system is continually being expanded, largely to handle the immense growth in the data communications traffic over the past decade. Figure 29–13 should give you an idea of some of the ways we use a telephone system today.

How a Phone Works

A telephone consists of a transmitter, a receiver, and a dial or push-button mechanism. The tele-

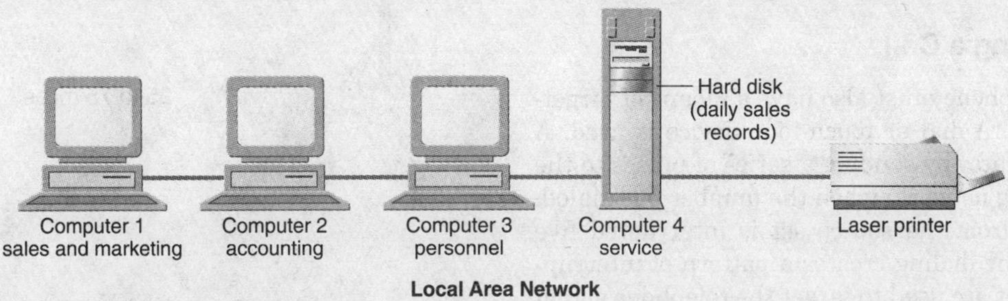

Figure 29–12 *A LAN consists of multiple PCs connected to one another for purposes of sharing hardware and information.*

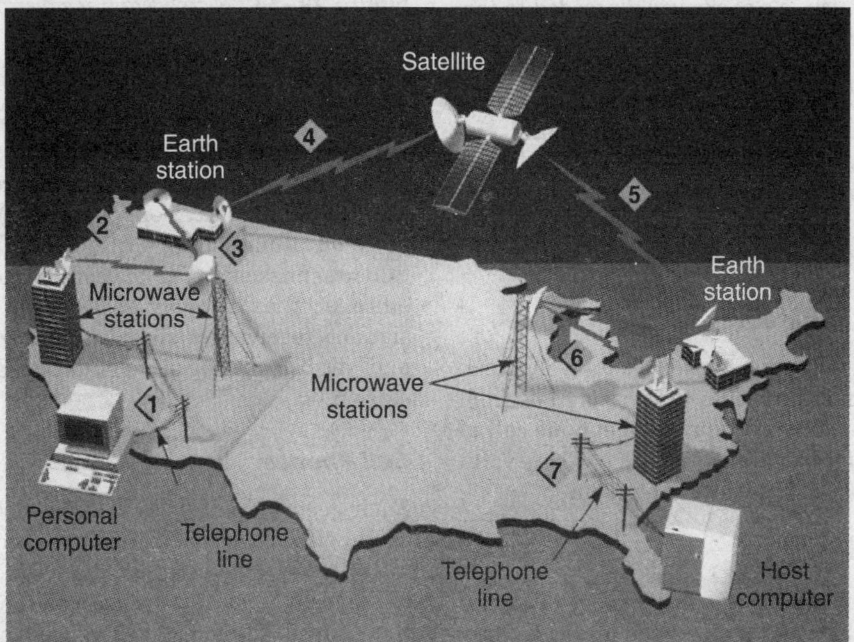

Figure 29–13 *The use of telephone lines, microwave transmission, and a communications satellite allows a PC to communicate with a large host computer, star network, or data hub. (Courtesy of Course Technology Cengage Learning)*

phone transmits voice or sound waves by having them act to vibrate a thin metal diaphragm backed by carbon granules through which current flows; when a sound wave acts to pulse the diaphragm inward, the carbon granules pack closely, permitting higher current flow. Therefore, the current flow depends on sound waves. This fluctuating current flows to a receiver. The receiver earpiece consists of an annular armature located between a permanent magnet and a coil; attached to the armature is a plastic diaphragm. When fluctuating current flows into the coil, it becomes an electromagnet. This attracts the armature, causing it to vibrate and create waves that duplicate speech.

Targeting a Call

The telephone must also have a means of targeting calls. A dial or touch-tone device is used. A dial operates by sending a series of pulses to the switching network; when the number 5 is dialed, current from the source set is interrupted five times. The dialing creates a pattern of interruptions that are used to target the telephone signal to the correct location. With a touch-tone device, each numeric button produces a different *tone* or frequency, which is electronically decoded to target the telephone call.

Most current telephone switching is performed digitally. Incoming signals from a telephone set are first converted into digital signals, which are patterns of on-off pulses. These pulses are then processed by computer to search the shortest (and least overloaded) path to connect the target telephone.

Transmission Media

Early telephone systems used exclusively hardwired transmission media to carry signals from one point to another. Today, a telephone call to a foreign country can involve the use of several different types of transmission media.

Copper Wire

Most local calls are carried over pairs of copper wires. Coaxial cable is also used and can carry signals of higher frequency and allow more transmissions on a single line using multiple coaxial conductors.

Microwave

Microwave is the modulation of high-frequency radio signals. These radio signals are transmitted between antennae located on towers that are within sight of one another. Microwaves travel only in straight lines and do not refract from the ionosphere layer of the atmosphere.

Because they only travel short distances, multiple relays must be used. Microwaves have frequencies of many gigahertz, permitting large numbers of simultaneous transmissions on each frequency band. A typical microwave transaction is illustrated in Figure 29–14.

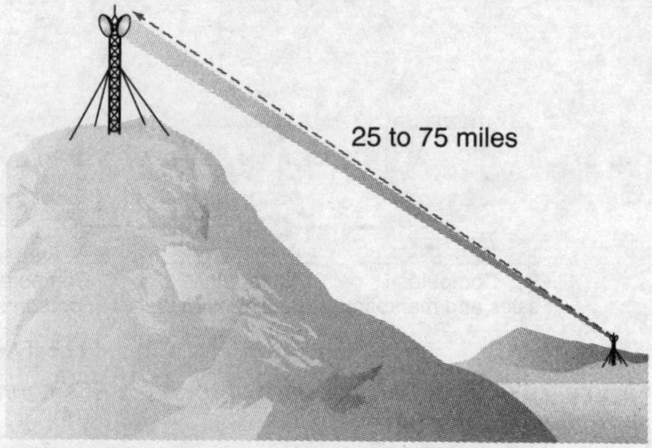

Figure 29–14 *Microwave transmission.*

Satellites

Telecommunications satellites parked in **geosynchronous orbit** (see later on in this section for a full explanation of geosynchronous orbit and telecommunications satellites) handle a large but diminishing portion of transcontinental transmissions. The telecommunications satellite acts as a relay station for microwave signals, receiving a signal from a ground base and retransmitting it to another Earth-based station.

Cell Phones

Each geographic region is divided into "cells," each with its own transmitter-receiver. The cells are arranged so that the frequencies used in adjacent cell locations are different, although the same frequencies are reused many times over throughout the whole system. If a call is made from a cell phone in a car traveling over the boundary of one cell to another, the phone automatically switches frequencies, permitting a continuous phone conversation (Figure 29–15).

Wireless Data Connectivity

A large number of truck stops in the United States and Canada have understood the importance of wireless connectivity systems to the trucking industry both in terms of serving mobile and stationary rigs. **Wireless fidelity**, almost always known as just **WiFi**, has become a key communications technology over the past few years. As far as the trucking industry is concerned, WiFi services adhering to TMC RP2402(T) standards should be used. WiFi client devices in trucks can be divided into:

- Driver devices: These include laptop computers, fixed cab PCs, and PDAs. Driver devices are used for both personal and fleet functions.
- Vehicle devices: These include a range of devices used to broadcast data bus data off-vehicle. For instance, truck chassis J1939 data backbone communications can be monitored at a central data hub. Many city transit buses use a separate data bus in addition to J1939 (they interface) to drive such things as door, air conditioning (A/C), buzzers, lighting, video, and passenger security. By providing wireless connectivity, vehicle and passenger security can be increased.

A WiFi footprint is the geographical area over which client devices can network: This is usually known as a **hotspot**. When a client device enters a WiFi hotspot, more than one WiFi service may be provided. In cases of vehicle client devices, the selection of WiFi provider may be automatic, whereas with driver client devices, the user may have to select the provider. WiFi is more commonly used in trucking than Bluetooth because it functions on higher power, better security, and is capable of high volumes of data transfer. Figure 29–16 illustrates how wireless connectivity can be made to work for the trucking industry, providing for personal communications, truck data bus communications, and trailer tracking.

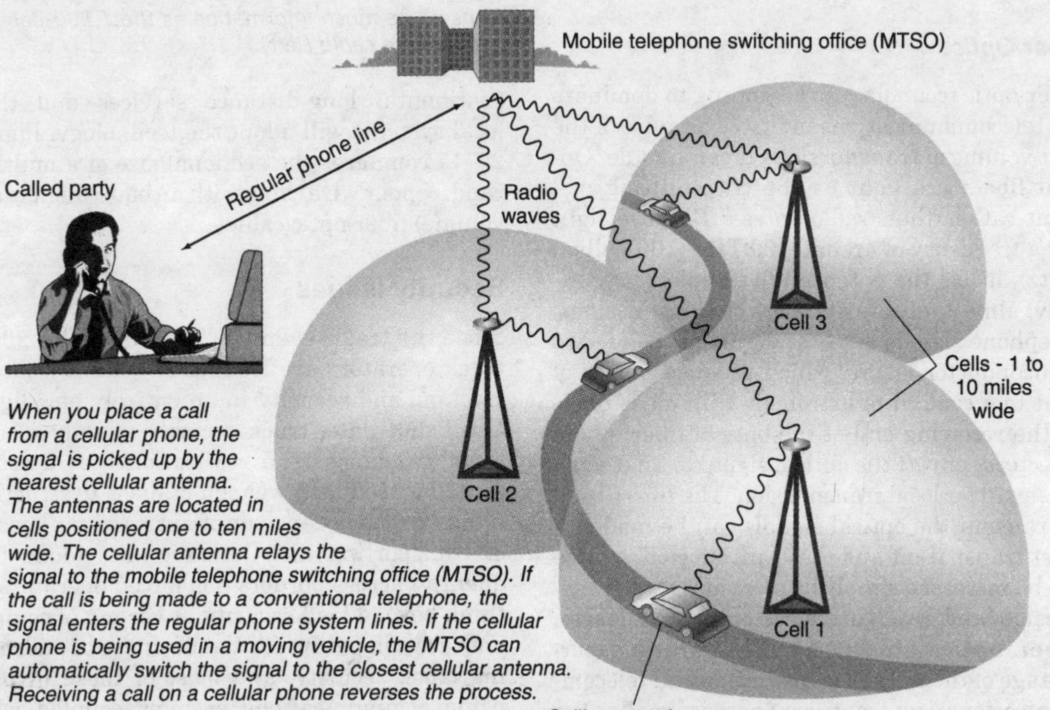

Figure 29–15 *Cellular phone operation. (Courtesy of Course Technology Cengage Learning)*

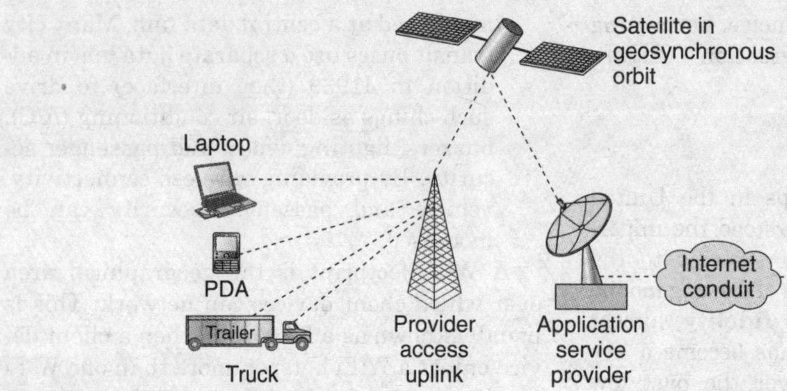

Figure 29–16 *How the trucking industry puts wireless connectivity to work, enabling such things as personal communications, truck data bus communications, and trailer tracking.*

Bluetooth **Bluetooth** is also a wireless network but it is better suited to applications where lower volumes of data have to be transferred. It is used primarily for phones, headsets, and PDAs and suits situations in which two client devices are in close proximity. Bluetooth uses the same radio frequencies as WiFi but is regarded as an "open" network: This makes it more accessible and less subject to permissions.

Fiber Optics

Fiber-optic technology is beginning to dominate the telecommunications industry because of the vast volume of transmissions it can handle. Optical fibers use light as the transmitting medium rather than radio waves. Because light has a frequency of around 100 THz (100 trillion hertz), it has the potential to handle (theoretically) almost infinite numbers of transmissions. Telephone signals are first digitized then fed to semiconductor lasers, which produce pulses of light to a conduit of extremely thin glass fiber. At the receiving end of the optical fiber, photo detectors convert the optical signal to an electrical signal for local transmission. The fibers used to transmit the optical signals can be made almost transparent and the transmission volume can be increased simply by upgrading the transmission and receiving electronics and lasers. **Fiber optics** already handle a growing percentage of the world's transcontinental telecommunications—up-front expense is really the only thing hindering its rate of growth. In time, continental long-distance services and then local systems will adopt the technology. Figure 29–17 compares the sectional size of a multistrand, copper wire cable with an equivalent (data volume) fiber-optic cable.

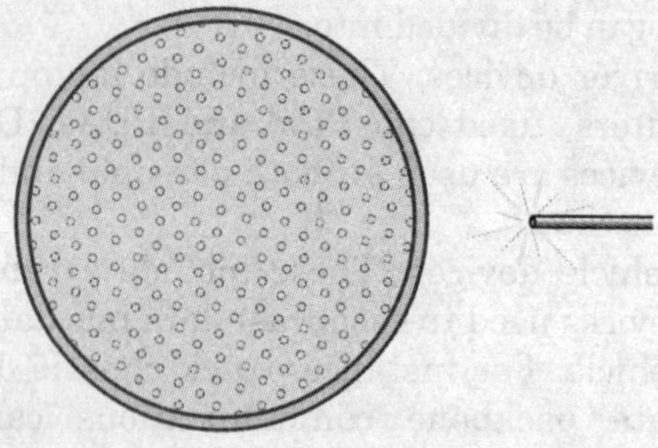

Figure 29–17 *The fiber-optic cable (right) can transmit as much information as the 1,500-pair copper wire cable (left).*

Security Issues

Since the tragic events of September 11, 2001, truck operators are looking at vehicle location tracking and security in an entirely new light. Until that date, truck security usually meant theft avoidance. The realization that a truck could be used as a weapon of mass destruction (it has already been done offshore) or as a means of transporting illegal immigrants (done frequently) across borders has forced operators to think beyond load security. A variety of companies are competing for the business of addressing truck security, and most of these involve using communications systems coupled with fast response strategies.

Companies such as Qualcomm, Corp Ten, and Siemens are marketing systems designed to comprehensively enable driver and vehicle communications. These systems range from simple wireless panic buttons up to powerful broadband off-vehicle networking. This in conjunction with sophisticated vehicle tracking technology is being used by large fleets that can afford telecommunications networking. The major operators have been using asset protection systems of one kind or another for a generation now because they cannot afford not to have it. When a vehicle deviates from a preprogrammed, dispatch route, most fleets like to know about the event immediately.

Satellite Communications Technology

A satellite is by definition a small body that orbits a larger astronomical object; by this definition, the earth's moon is a natural satellite. The first artificial satellite was launched into orbit in 1957 by the Soviet Union. It was called *Sputnik I.* Today, thousands of satellites orbit the earth; all but one are artificial, that is, manufactured on Earth. These satellites orbit the planet at altitudes of 100 miles (160 kilometers) or higher and are used for weather study, spying (military), navigation, and communications.

Geosynchronous Orbit

A communications satellite is often placed in geosynchronous orbit. The concept of geosynchronous orbit was conceived by a science fiction writer by the name of Arthur C. Clarke, whom you might have heard of because he wrote *2001: A Space Odyssey.* Anyway, to the irritation of many true scientists of the day, Clarke came up with the concept of geosynchronous orbit in 1945 and 15 years later, the telecom satellite was born. Clarke figured that the time that it takes for a satellite to orbit the earth once correlates exactly to its altitude. Accordingly, a satellite parked exactly 22,300 miles (35,900 km) above the earth's surface on the equator will be positioned at a fixed point above the earth's surface throughout the rotation. Such a satellite is said to be in geostationary orbit. If its altitude were to increase slightly, its orbit would cause it to fall behind the fixed point—if it descended below the geosynchronous altitude, the reverse would occur. Figure 29–18 demonstrates the concept of the Clarke orbit.

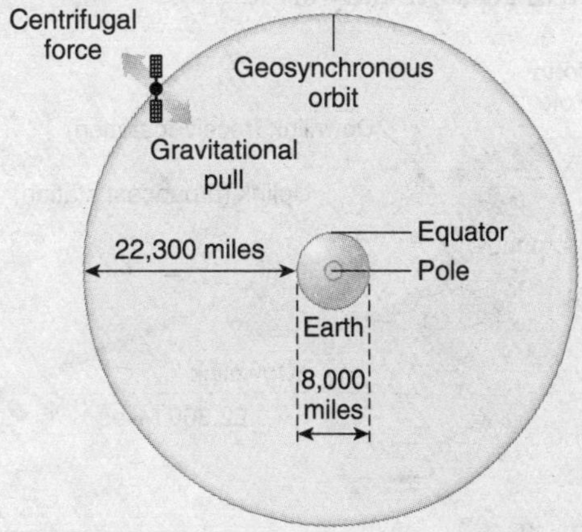

Figure 29–18 *The Clarke orbit: concept of geosynchronous orbit conceived by science fiction writer Arthur C. Clarke (2001: A Space Odyssey) in 1945.*

The first telecom satellite was launched by the United States in 1961; it was named *Telstar,* and it became the subject of a popular hit musical recording of the day. To maintain a geosynchronous orbit, satellites must have onboard rocket engines and a fuel supply, usually hydrazine, to adjust the orbit periodically. The power supply required to run the onboard electronics of a communications satellite is obtained from solar cells or nuclear-fueled thermoelectric generators. Onboard computers control the satellite's operation and orbit and manage the filtering, storing, and relaying of signals from and to ground stations.

Uplinks and Downlinks

To avoid signal interference, telecom satellites in geosynchronous orbit are parked 2 degrees apart from each other on the earth's equator. Therefore, there is limited parking space. The life of such a satellite is often governed by the fuel supply required to run the rockets that maintain the park position. The satellite receives ground signals from **uplinks** (ground-to-satellite communications hardware and software) on Earth that fall within its footprint. An uplink is any Earth-based station (including mobile units) capable of relaying signals to a telecom satellite. The satellite's footprint is the geographic area on Earth over which it can receive and send signals. Figure 29–19 shows how uplinks and downlinks function in relation to a satellite in geostationary orbit.

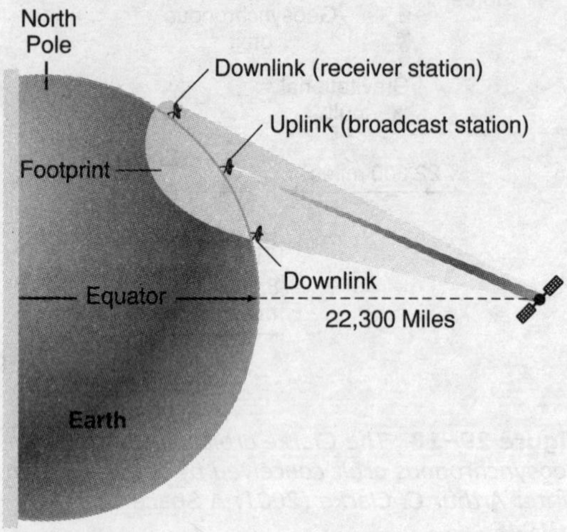

Figure 29–19 *Operating principle of a communications satellite in geosynchronous orbit.*

There are 32 telecom satellites parked with footprints directed on the American continents. Typically, a satellite receives a signal from an uplink, filters it, and rebroadcasts it to a **downlink**. A downlink is an Earth-based station equipped to receive signals from a telecom satellite. In such a way, a radio signal may be relayed through several satellites and ground stations. Until recently, television networks were the biggest users of telecom satellites. For instance, a baseball game that was televised in Seattle could be watched live in Montreal; the signal was relayed through a number of telecom satellites, uplinks, and downlinks. Today many of the communications transactions that were relayed using telecom satellites are undertaken by our fiber-optic networks, especially those by big-money network TV. This fact has now made using telecom satellites much less costly for anyone else who might have a use for it. Telecommunications technology has been used by the trucking industry since the late 1980s, but its use has grown significantly since 2002. Satellite communications technology was first used mainly by major fleets, but now most OEMs offer a range of telecom options.

How Trucks Use Satellite Communications

Until fiber-optic networks took over, the television networks and cable network providers were the major consumers of telecommunications satellite air time, making air time most expensive during prime time television hours. Because a comprehensive fiber-optic network is now in place, television networks and telephone service providers rely less on satellite communications, meaning that the cost of air time for satellite communications has dropped. This has opened the door and made satellite communications technology affordable for the trucking industry as well as other potential users.

Satellite communications technology tracks a vehicle by placing an uplink, sometimes called a **transponder**, on the mirror brackets or cab roof of a truck. The uplink beams a signal to a telecom satellite, which may then relay it across the continent, if necessary, targeting a downlink at a data hub. Any data that can be transmitted digitally can be sent in this manner. Figure 29–20 shows a dish-shaped Earth station designed for telecom transactions.

Figure 29–20 *Earth stations use large dish-shaped antennae to communicate with satellites.*

Vehicle Navigation

Vehicle navigation systems vary in complexity (and cost) but a comprehensive system consists of:

- Geographic positioning
- Target destination

- Route mapping, including real time updates
- Route guidance, including real time updates
- Map display

The detail that the navigation system offers in each of the foregoing categories depends on the stored data or telecommunications capability the system provider offers. Details such as real time traffic reports and detour negotiation tends to be pretty expensive because of the cost of collating data, but maybe it is not so expensive when compared with the cost of a lost truck driver navigating Manhattan in the rush hour. **Global positioning satellites (GPS)** are the key to vehicle navigation technology. In taking a look at GPS communications, remember that this is a one-way communications technology. To achieve two-way communications, GPS must be additionally supported by satellite communications or other wireless technologies such as WiFi or Bluetooth.

Trilateration

To understand how GPS functions, you have to understand a simple geometric principle called **trilateration**. Suppose you are completely lost somewhere in America, you pull into a truck stop to ask someone where you are, and the answer is that he is not sure but he does know for sure that he is exactly 490 miles from Cleveland. At this point you can get a map of the United States, find Cleveland, and draw a circle with a radius on a scale that equals 490 miles. You ask the next trucker, and you get the answer that he does not know exactly, but he has come directly from Houston and has traveled 710 miles. Now you can go back to the map and draw a circle with a radius representing 710 miles around Houston. At this point you still cannot say exactly where you are until you approach a third trucker and ask him the same question. He informs you that he is exactly 520 miles from Jacksonville. Again, you draw a circle with a radius representing 520 miles around Jacksonville. Now if you look at your map (Figure 29–21), you will see that the three circles you have drawn all intersect at one point and you can determine your exact position.

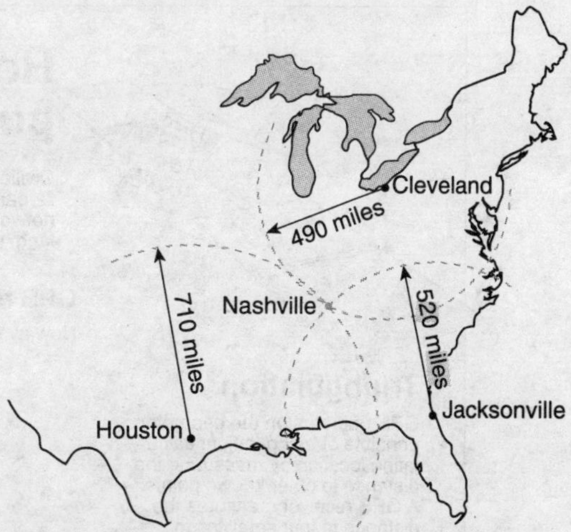

Figure 29–21 *Position location by trilateration.*

GPS Location

If the world were flat, trilateration could be used to determine position. For the principle to work in three-dimensional space, you have to use spheres instead of circles, so in order to pinpoint exact position, four GPS satellites are required. If the GPS receiver can only locate three satellites, then its software creates a *virtual* sphere, which is a dummy sphere that allows it to use the coordinates established by the three satellites it has located. This means that your position can be calculated but not your altitude.

Measuring Distance GPS satellites broadcast radio signals that a GPS receiver can detect. The receiver then calculates how long the signal has taken to travel from the satellite to the receiver. For a GPS system to function, 24 satellites are required so that there is a minimum of four on the horizon at any time or location during the day. The signals received tend to be weak, so this technology works better driving through the countryside than it does in the middle of a large city. Figure 29–22 shows how trilateration is put to use: Note that the term **triangulation** is used in place of trilateration; the two words are synonomous.

Mapping The raw positional data produced by the GPS receiver unit on your car or truck can next be connected to the geographic infor-

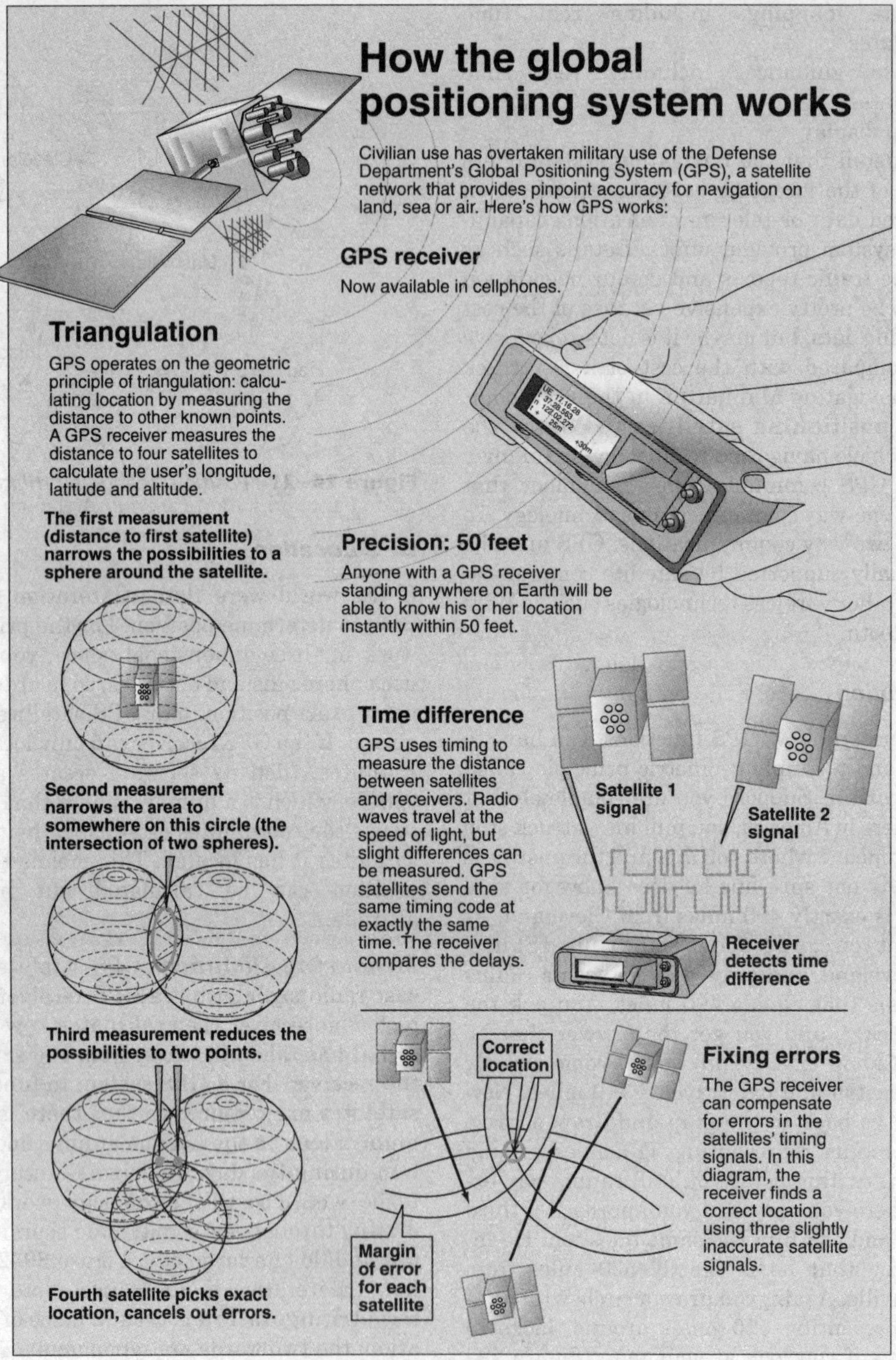

Figure 29–22 *How GPS works. (Knight-Ridder Tribune graphic by Jenny Anderson, San Jose Mercury, May 22, 1996)*

mation retained in the memory of the GPS ECM. In a top-of-the-line system, the navigational display may produce color map readouts and provide detail on street addresses. The retained data may be on compact disk—read-only memory (CD-ROM), but the system works best with a telecom link found on some trucks that can provide live data on such things as traffic congestion and road construction.

Navigational Aids to GPS Depending on the sophistication of the system, the effectiveness of a GPS unit can be improved by inputting data from the chassis electronics (road speed) and dedicated sensors such as a gyrometer to signal change of direction data. A top-of-the-line system uses Doppler effect (frequency shift analysis) to constantly monitor directional and vehicle speed data.

Vehicle Control Center (VCC)

Keeping track of vehicle communications in modern trucks (and agricultural, off-road, and mining equipment) is complex and requires plenty of computing and networking capability. For this reason, there are a number of different vehicle-based communications management systems available. We use the generic term **vehicle control center (VCC)** to describe these comprehensive networking systems. VCC telematics, satellite communications, and navigation systems enable external communications and because each is networked to the truck chassis data bus (CAN 2.0/J1939 discussed in Chapter 33), permit any ECU with a bus address to both broadcast and receive data. The systems can accommodate both secure and open off-board communications and can be configured for GPS, WiFi, Bluetooth, two-way satellite, and proprietary communications systems.

A key feature of VCC systems is the driver display. This can be configured in many different ways according to how the vehicle owner wants the vehicle operated and may include features such as touch screen to minimally distract the operator. One OEM describes its VCC system as the "nerve center" of truck operations. VCC can be programmed for semicovert management operations. For instance, if the navigation system detects a forthcoming hill, VCC messages the chassis data bus to command an automated or automatic transmission to downshift into an optimum gear just ahead of the climb. Perhaps the most cost-effective benefit of a VCC system is predictive remote diagnosis that allows a fleet data hub to be alerted the moment a potential problem is detected. This minimizes downtime and usually avoids costly on-the-road breakdowns.

IRIS

A popular, low-cost method of transferring information from a truck to the land station is to use infrared. Detroit Diesel's infrared information system (IRIS) works in conjunction with DDEC, ProDriver, and ProManager software and can upload and download. IRIS requires a transceiver to be located on the truck and another at a stationary location such as a refueling island. A green dash light alerts the driver when an IRIS has made a connection: The system can be used by dispatch for logging outgoing and incoming vehicles along with other data bus upload and downloads. Figure 29–23 shows the horizontal and vertical window alignment required for infrared data exchange.

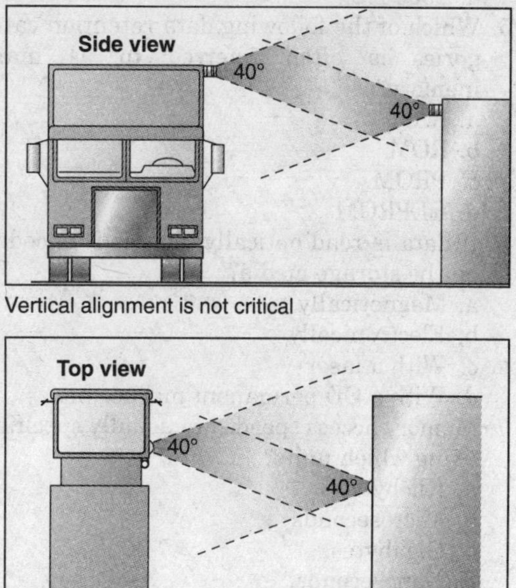

Figure 29–23 *IRIS horizontal and vertical window alignment required for data exchange. (Courtesy of Detroit Diesel Corporation)*

REVIEW QUESTIONS

1. In which of the following is data retained electronically?
   a. RAM
   b. ROM
   c. PROM
   d. EEPROM
2. Which of the following components is sometimes referred to as the *brain* of a computer system?
   a. CRT
   b. VDT
   c. CPU
   d. CD-ROM
3. Which scientist is credited with having invented the first mechanical digital computer?
   a. Hero
   b. Gunter
   c. Pascal
   d. Watt
4. What year was the world's first electronic digital computer introduced?
   a. 1960
   b. 1946
   c. 1642
   d. 2500 B.C.
5. Which of the following data retention categories is often referred to as main memory?
   a. RAM
   b. ROM
   c. PROM
   d. EEPROM
6. If data is read optically, how is it encoded on the storage media?
   a. Magnetically
   b. Electronically
   c. With a laser
   d. With a CD permanent marker pen
7. Memory access speeds are usually specified using which units?
   a. Kilobytes
   b. Microseconds
   c. Gigabytes
   d. Nanoseconds
8. How many bits are there in a byte?
   a. 8
   b. 64
   c. 256
   d. 1,024
9. How many different data possibilities could be contained in a single byte?
   a. 8
   b. 64
   c. 256
   d. 1,024
10. The main circuit board in a computer is often referred to as a:
    a. Motherboard
    b. ROM card
    c. CPU chip
    d. BIOS
11. The word used in computer terminology to describe communication rules and regulations is:
    a. Acronyms
    b. Buffers
    c. Hypermedia
    d. Protocols
12. *Technician A* states that the keyboard on a PC station is an *input* to the computer. *Technician B* states that a touch screen could be described as both an *input* and *output* device. Who is correct?
    a. Technician A only
    b. Technician B only
    c. Both A and B
    d. Neither A nor B
13. Which of the following PC components would not be classified as a peripheral?
    a. Modem
    b. Printer
    c. Scanner
    d. CPU
14. *Technician A* states that RAM data is retained electronically. *Technician B* states that RAM is the main factor in defining a computer's processing speed. Who is correct?
    a. Technician A only
    b. Technician B only
    c. Both A and B
    d. Neither A nor B
15. *Technician A* states that digital CDs are magnetically encoded with data, and close contact with a magnetic field can corrupt the data. *Technician B* states that exposing a hard disk drive to a powerful magnetic field can corrupt its data. Who is correct?
    a. Technician A only
    b. Technician B only
    c. Both A and B
    d. Neither A nor B

16. Which of the following is best described as *a network of networks?*
a. The Internet
b. Intranet
c. E-mail
d. LAN
17. Which of the following components can be used to condition the output of a computer for transmission through the telephone system?
a. CPU
b. Transponder
c. CRT
d. Modem
18. What is the transmission medium used by the telephone companies for local telephone service?
a. Microwave
b. Copper wire
c. Radio waves
d. Telecommunications satellites
19. In what format is data transmitted using fiber-optic technology?
a. Analog
b. Pulses of voltage
c. Digital
d. Nodes and antinodes
20. Which altitude from the earth's surface would describe a geosynchronous orbit?
a. 160 miles (257 km)
b. 22,300 miles (35,888 km)
c. 35,900 miles (57,775 km)
d. 56,700 miles (91,250 km)
21. The first telecommunications satellite was called:
a. *Sputnik I*
b. *Apollo XI*
c. *Telstar*
d. *Discovery*
22. The Earth-based station responsible for broadcasting microwave signals to a telecommunications satellite is called a(n):
a. Modem
b. Uplink
c. Downlink
d. Footprint
23. The uplink/downlink component used on trucks equipped with GPS technology is often called a(n):
a. Aerial
b. Transponder
c. Modem
d. Footprint
24. *Technician A* states that a telecommunications satellite orbits the earth at a distance of 35,900 miles. *Technician B* states that all telecommunications satellites use fuel to maintain a fixed orbit on the earth's equator. Who is correct?
a. Technician A only
b. Technician B only
c. Both A and B
d. Neither A nor B
25. *Technician A* states that telecom satellites in geosynchronous orbit are required to park at about 2 degrees apart from one another. *Technician B* states that the geographic zone over which a telecom satellite's signals are effective is known as its footprint. Who is correct?
a. Technician A only
b. Technician B only
c. Both A and B
d. Neither A nor B
26. *Technician A* states that the hardwired telephone system can only handle analog transactions. *Technician B* states that most telecommunications systems in North America rely on hardwired mediums for at least some of the transmission. Who is correct?
a. Technician A only
b. Technician B only
c. Both A and B
d. Neither A nor B
27. *Technician A* states that a LAN is sometimes known as a network of networks. *Technician B* states that the Internet is owned by the Microsoft Corporation. Who is correct?
a. Technician A only
b. Technician B only
c. Both A and B
d. Neither A nor B
28. What is a transponder?
a. An uplink/downlink device
b. A satellite dish
c. A footprint
d. A CB antenna
29. Which of the following means the same thing as network star?
a. The Internet

b. The Intranet
c. Remote node
d. Data hub

30. Which of the following broadcast mediums relies on modulating high-frequency radio waves?
a. Hardwired telephone system
b. Microwave
c. Fiber optics

31. The software tool capable of finding information on the Internet and on databases is known as a(n):
a. Search engine
b. Operating system
c. Mouse
d. Scanner

32. *Technician A* states that microwaves travel in geometrically straight lines. *Technician B* states that the point-to-point communications distance of microwaves is indefinite and a signal should be able to travel halfway around the world. Who is correct?
a. Technician A only
b. Technician B only
c. Both A and B
d. Neither A nor B

33. Which of the following mediums should be able to transmit the most data given a similar radial diameter of each?
a. Aluminum wire
b. Copper wire
c. Coaxial cable
d. Fiber-optic cable

34. What is the minimum number of GPS satellites a vehicle GPS system must locate to identify both vehicle position and altitude?
a. One
b. Two
c. Three
d. Four

35. To calculate a specific position on a map using trilateration geometry, what is the minimum number of coordinates required?
a. One
b. Two
c. Three
d. Four

30

Vehicle Computer Systems

INTRODUCTION

Most highway truck and, increasingly, bus and off-highway equipment, use computers to manage engine and other onboard system functions. They are cybernetically controlled. **Cybernetics** is the science of computer control systems. The technician is required to troubleshoot system problems and reprogram **parameters** (values), in most cases using software loaded to personal computers (PCs), laptops, or handheld units. Onboard vehicle computers are referred to as **engine/electronic control modules (ECM)** or **electronic/engine control units (ECU)**. ECM is the acronym used in this text except when discussing a product whose original equipment manufacturer (OEM) chooses to use the acronym ECU.

Engine Controller

An ECM is a modular housing containing a **microprocessor**, data retention media, and, usually, an output or switching apparatus. The ECM is the engine controller. It is usually but not always mounted somewhere on the engine. Other locations for mounting an ECM are the firewall or inside the vehicle cab. Although in the early days of vehicle computers the only system that was electronically managed was the engine, today most chassis systems are computer controlled. To enable the various chassis controllers to exchange data, each controller is assigned an address on a chassis data bus. The data bus is the information exchange backbone of the chassis. We use the term **multiplexing** to refer to a communications on a vehicle data bus.

We will look at multiplexing in more detail a little later in Chapter 33. Technicians are required to have a basic understanding of both vehicle and personal computers to perform even the most simple chassis functions. This chapter introduces the essentials of electronic management of vehicle systems. The terms introduced in Chapter 29 are used extensively in this chapter so if you do not understand a concept here, you may wish to refer back.

Data Processing

A vehicle computer may be simple, such as that required to manage a pulse wiper circuit, or complex, such as that required to manage a diesel engine and its emissions controls. However, whether simple or complex, all chassis computers manage an information-processing cycle composed of three distinct stages:

1. Data input
2. Data processing
3. Outputs

Figure 30–1 shows a schematic layout of a more complex management system that is used on post-2007 Caterpillar ACERT C13 and C15 engines. This circuit is required to drive twin actuator injectors, variable valve timing, and

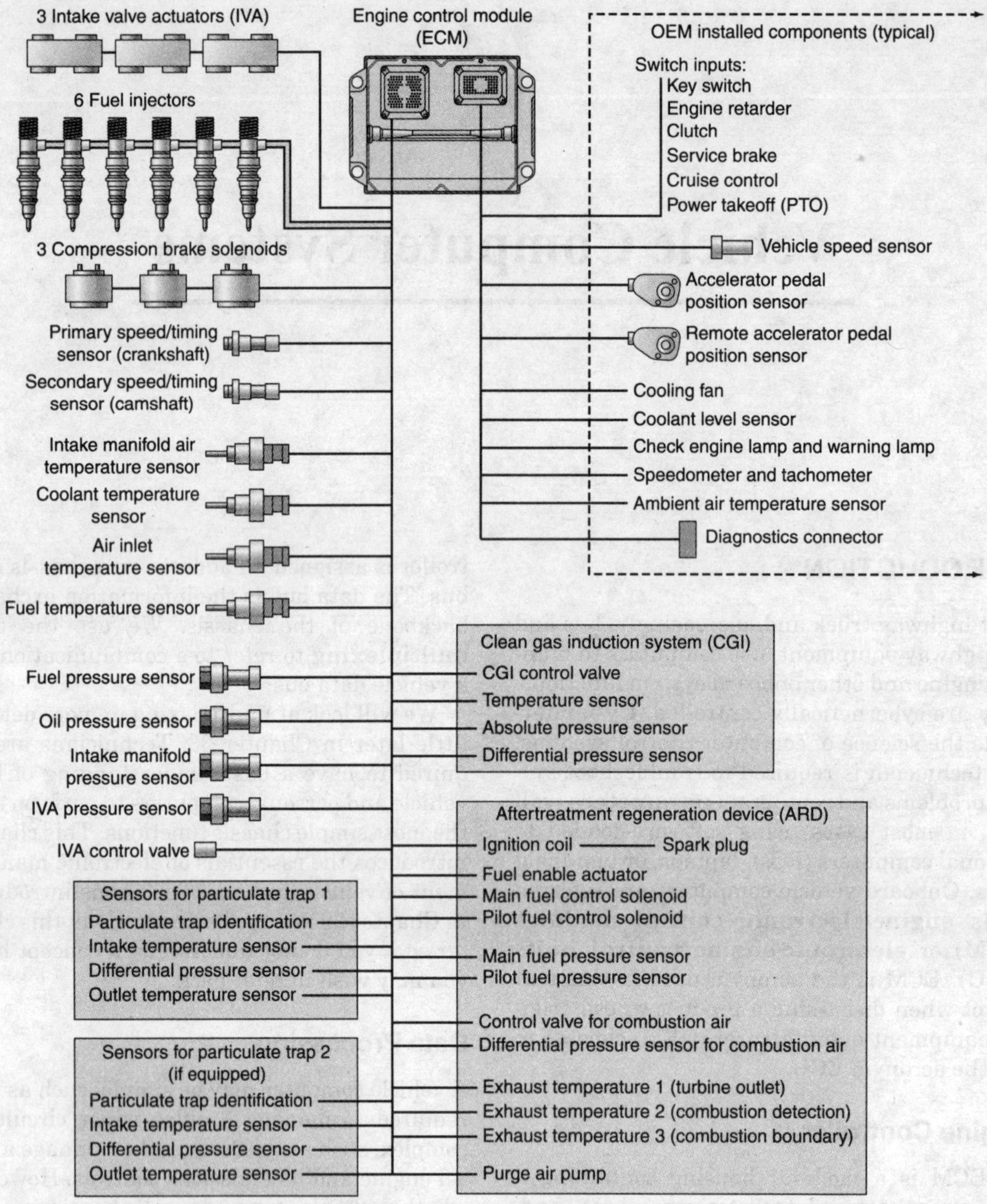

Figure 30–1 *Schematic of a post-2007 Caterpillar ACERT system used to manage C13 and C15 engines. (Reprinted Courtesy of Caterpillar Inc.)*

an assortment of aftertreatment devices, including a diesel particulate filter.

Data Input

Data is simply raw information. Most of the data to be inputted to a truck diesel engine ECM comes from monitoring sensors, such as rpm signals, and command sensors, such as the throttle position sensor. This data may be in analog or digital formats. If the input data is in analog format, such as voltage values, it has to be digitized before it can be processed by the ECM. Figure 30–2 shows some different types of input circuit devices: In this figure the switch status should be open in order to validate the pair of sensors above it.

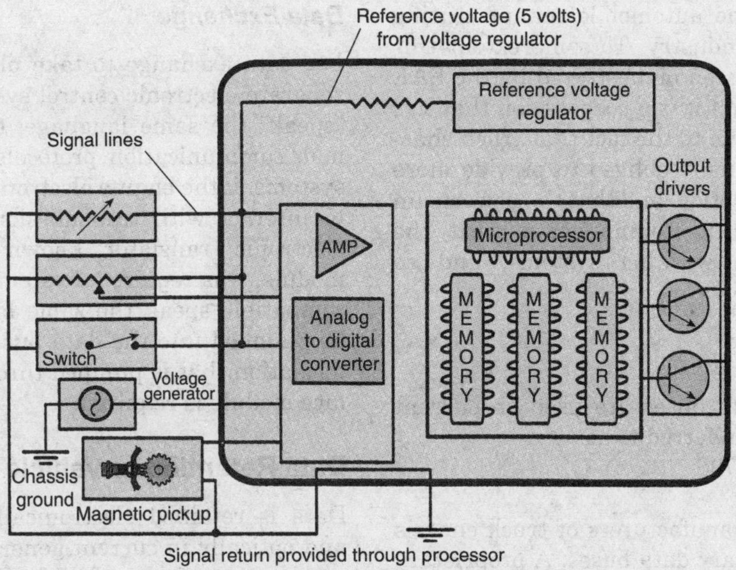

Figure 30–2 *Types of ECM input signals. (Courtesy of Navistar International Corp.)*

Data Processing

A **central processing unit (CPU)** contains a control unit that executes program instructions and an arithmetic logic unit (ALU) to perform numeric calculations and logic processing such as comparing data. Random-access memory **(RAM)** is data that is electronically retained in the ECM—this data can be manipulated by the CPU because it can be accessed at high speed. RAM is often referred to as main memory. The ECM can also call up more permanent types of data that is retained magnetically or optically. Read-only memory **(ROM)**, programmable read-only memory **(PROM)**, and electronically erasable, programmable read-only memory (EEPROM) data can be transferred to RAM by the CPU for processing. Because RAM data is electronically retained, it is lost when its power-up circuit is opened. Figure 30–3 is a simplified schematic of an ECM showing some of its basic functions: The output drivers shown in the schematic are transistors that convert processing logic into action such as switching injectors on and off.

Outputs

The results of processing operations must be converted to action by switching units and **actuators**. In most (but not all) truck/bus management ECMs, the switching units are integral with the ECM. In an ECM managing a full authority diesel engine, the **injector driver** unit would be one of the primary outputs to be switched. ECM commands would be converted to an electrical signal that would determine the pulse width or duty cycle of an electronic fuel injector (EUI) and define its effective pumping stroke. Note the output drivers in Figure 30–2 and Figure 30–3: It helps if you can develop a mental picture of what happens in an ECM rather than simply regarding it as a black box.

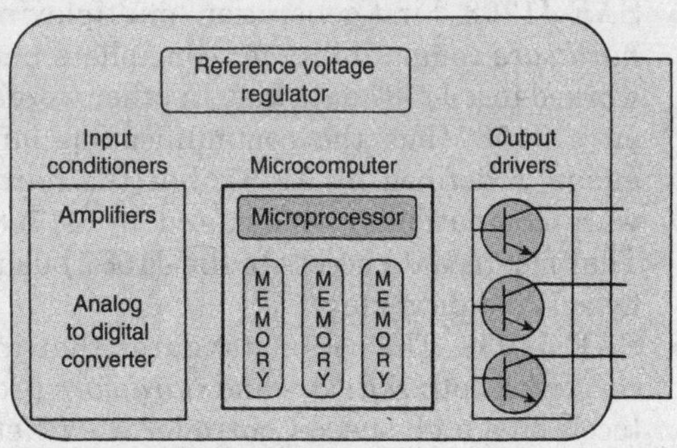

Figure 30–3 *ECM functions. (Courtesy of Navistar International Corp.)*

SAE Hardware and Software Protocols

Among the OEMs of truck and bus engines in the United States, there has been a generally higher degree of cooperation in establishing shared electronics hardware and software protocols than in the automobile manufacturing segment of the industry. To some extent this cooperation has been orchestrated by the SAE and ATA-TMC, but it is a cooperation that has been necessary due to the fact that truck chassis OEMs are usually obliged to provide more than one engine option, and that the powertrain electronics have to communicate with the brakes, collision warning systems, and so forth.

J-Standards

Two primary data buses are used on current trucks. They are referred to as:

- SAE J1587/1708
- SAE J1939

In addition, manufacturers of truck chassis may use proprietary data buses. A proprietary data bus is usually designed to communicate with just one other controller. The other controller is likely networked to one of the two primary data buses.

SAE J standards (surface vehicle recommended practice) dictate the hardware and software protocols that govern multiplexing data exchange between modules. These protocols are covered by the following J standards:

- **SAE J1587**. First generation, multiplexing data exchange *protocols* used in data exchange between controllers networked on a data bus.
- **SAE J1708**. First generation, multiplexing *hardware* compatibility for controllers networked to a J1587 data bus. In other words, on a J1587 bus the communication language is defined by J1587, but the hardware compatibility is dictated by J1708. That means data access to the J1587 bus is by a J1708 data plug.
- **SAE J1939**. The set of standards that incorporate both *software* and *hardware* protocols on a high-speed Controller Area Network generation 2.0 (CAN 2.0) data bus—the current serial communication bus. We will take a close look at this in Chapter 33. When studying vehicle controllers, we often describe a controller by stating which of the two data buses it is compatible with. The light-duty/automotive equivalent of J1939 is J1850: Both use CAN 2.0 architecture.

Data Exchange

For data exchange to take place between two separate electronic control systems, they must "speak" the same language, that is, use common communication protocols. In many early systems, if the engine electronics were required to interact with transmission electronics, an electronic translator, known as an interface module, was required. Systems that are J1939 compatible speak the same language and can be plumbed into the data bus to share any information that is pumped through it. No interface module is required.

Data Retention In Vehicle ECMs

Data is retained electronically, magnetically, and optically in current generation truck, bus, and heavy equipment ECMs. Optically retained data in laser-read systems, once only common in home computing, is now being used in vehicle computers. We categorize data in vehicle ECMs as follows:

Random-Access Memory

The amount of data that may be retained in RAM is a primary factor in quantifying the computing power of a system. It is also known as *main memory* because the CPU can only manipulate data when it is retained electronically. At startup, RAM is electronically loaded with the vehicle management operating system instructions and all necessary running data retained in other data categories (ROM/PROM/EEPROM) so that the CPU can access it at high speed.

Volatility RAM data is electronically retained, which means that it is volatile. Another way of saying this is that RAM data storage is temporary. It requires power-up. If the circuit that supplies the RAM chip(s) is opened, the data in RAM is lost. Most truck ECMs use only fully volatile RAM; in other words, when the ignition circuit is opened, all RAM data is dumped.

Data such as coolant temperature, oil pressure, and boost pressure are also logged into RAM and, providing the values are within a threshold window defined in ROM/PROM/EEPROM, they only have significance at that specific moment; they can therefore be safely discarded when the ignition key is opened. This category of input data is continually monitored by a process known as **sampling**.

NV-RAM A second category of RAM is used in some truck and older automobile systems, usually ones that have no EEPROM capability. This is nonvolatile RAM (NV-RAM) (or KAM—keep-alive memory) in which data is retained until either the battery is disconnected or the ECM is reset, usually by depressing a computer reset button that temporarily opens the circuit. Like RAM, NV-RAM requires power-up to retain its memory. Often this is delivered outside the ignition circuit. For instance, when you disconnect a vehicle battery in some vehicles, the preset station selections are recorded in NV-RAM so these have to be reestablished. Codes and failure **strategy** (action sequence) are the type of things that could be written to NV-RAM in systems that use it.

Read-Only Memory

ROM data is magnetically or optically retained and designed not to be overwritten. It is "permanent" though it can be corrupted (rare), and may be susceptible to damage when exposed to powerful magnetic fields if it is magnetically retained. Low-level radiation, such as that encountered routinely driving on a highway from police radar and high-tension electrical wiring, will not affect any current ECMs. A majority of the total data retained in the ECM is logged in ROM. In an engine ECM, the master program for the system management is loaded into ROM. Production standardization is permitted by constructing ROM architecture so that it is written with common requisite data for a number of different systems. For example, identical ROM chips can be manufactured to run a group of different engines in a series—but to actually make an engine run in a specific chassis application, the ROM data would require further qualification from data loaded into PROM and EEPROM.

Open vs. Closed Loop ROM contains all the protocols (rules and regulations) to master engine (or the system) management, including all the **threshold values** (limits). The term **hard** is often used to describe **parameters**/values that are interpreted by the ECM in rigid terms. For instance, the temperature at which the ECM is programmed to identify an engine overheat condition is always a fixed value: At this specific value, a code and failure strategy are effected. Engine overspeed is another example of a hard parameter. In these examples, the processing loop is closed. The result is either OK or Not-OK.

A **soft parameter**/value is one that would be interpreted by the ECM with a cushion around the value. Soft parameters are those that the processing cycle may use the input of many sensors to compute the output algorithm. When an engine or system is run under open loop conditions, and most current diesel engines are, the processing path to produce outcomes is sometimes known as a **fuzzy logic** path. In a truck you have a *hard cruise* option in which road speed is held at a set value. Or you can option *soft cruise* (a.k.a. *smart cruise*). In soft cruise mode, the objective might be to achieve the best fuel economy, so the ECM permits some latitude (both up and down) around the set road speed.

Programmable ROM

Programmable ROM (PROM) is magnetically retained data—usually a chip, set of chips, or card socketed into the ECM motherboard. PROM can sometimes be removed and replaced. PROM function is to qualify ROM to a specific chassis application. In the earliest truck engine management systems, programming options, such as idle shutdown time, could only be altered by replacing the PROM chip; customer programmable options are written to EEPROM in current systems where they can be easily altered by flash programming. Some OEMs describe the PROM chip as a "personality module," which is an appropriate description of its actual function of trimming or fine tuning the ROM data to a specific application.

Electronically Erasable PROM

The **electronically erasable, programmable read-only memory (EEPROM)** data category contains customer data programmable options and proprietary data that can be altered and modified using a variety of tools ranging from a generic reader programmer to a mainframe computer. It is usually magnetically retained in current systems, but this will probably change in the near future. EEPROM provides the ECM with a write-to-self capability. This allows it to log fault codes, audit trails, tattletales, and failure strategies in a way that is more permanent than NV-RAM. Other examples of the type of data programmed to EEPROM are tire rolling radius, governor options (LS or VS), cruise control limits, road speed limit, and many others. EEPROM data can be rewritten on an as-needed basis. Usually only the owner password is required to access the EEPROM and make any required changes.

Flash Memory

The term *flash programming* is commonly used to refer to both EEPROM and true flash memory. **Flash memory** is magnetically retained, nonvolatile, solid-state memory. We use it extensively for portable digital storage (camera memory) and increasingly in vehicle computers. It is faster to log and erase than EEPROM because memory segments are batched in large blocks of bytes rather than byte by byte. The term has become important because OEMs are using it to describe any type of programmable onboard memory. In this textbook, the term *EEPROM* is used to describe both EEPROM and flash memory.

Proprietary Data Proprietary data logged into EEPROM is more complex both in character and methods of altering. An example of proprietary EEPROM data is the fuel map in an engine ECM. The procedure here normally requires accessing a centrally located mainframe computer (usually at the OEM headquarters location) via the Internet, **downloading** the appropriate files to a PC, and finally **reprogramming** (altering or rewriting the original data) the ECM.

INPUT CIRCUIT

ECM inputs can be divided into sensor inputs and switched inputs. They are covered in this chapter in some detail because there is a high degree of input component commonality among the OEMs. When we deal with specific OEM management systems we will assume that these input circuit components are understood. Inputs to the ECM can be divided into:

- Monitoring
- Command

Sensors and switches are used as input circuit components. A temperature signal to the ECM is an example of a monitoring input. The throttle position sensor is an example of a command input.

Sensors

Anything that signals input data to a computer system can be described as a sensor. Sensors may be simple switches that an operator toggles open or closed to ground a reference voltage, modulate a reference voltage, or be powered up either by V-Ref or require power-up outside of the V-Ref circuit.

Sensors Using a Reference Voltage to Produce an Analog or Digital Signal

These sensors receive a constant reference voltage (usually 5 V) and the ECM compares the return signals with values logged in its data retention banks. There are four types:

1. Thermistors. **Thermistors** precisely measure temperature. There are two types of thermistor defined by whether resistance increases or decreases when temperature rise occurs:
   - Negative temperature coefficient (NTC): Temperature goes up, resistance goes down.
   - Positive temperature coefficient (PTC): Temperature goes up, resistance goes up.

   The ECM receives temperature data from thermistors in the form of analog voltage values. The coolant temperature sensor, ambient temperature sensor, and oil temperature sensor are usually thermistors. Almost without exception, NTC-type temperature sensors are used on truck diesel engine systems, meaning that as temperature increases, resistance decreases, and therefore signal voltage (to the ECM) increases. Temperature sensors are critical inputs to the ECM when factoring injection timing and injection fuel quantity data. Figure 30–4 is a schematic of a thermistor.

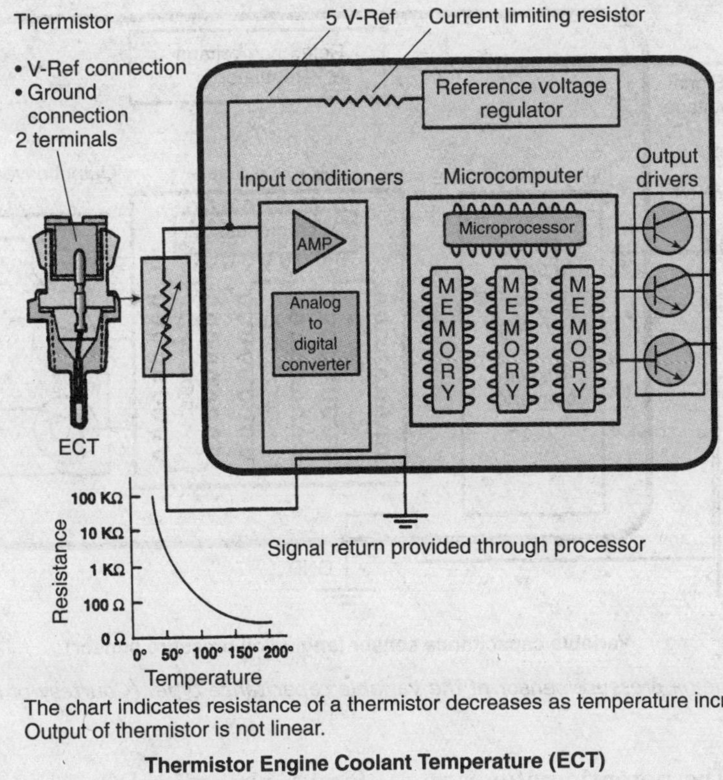

Figure 30–4 *An engine coolant temperature sensor of the NTC thermistor type. (Courtesy of Navistar International Corp.)*

Check Functionality of an NTC Thermistor

Pin 1 = V-Ref (5 V-DC)

Pin 2 = GND

Pin 3 = Signal

Using a DMM on V-DC the following voltage readings indicate a functional sensor.

Pin 1 to pin 2 = 5 V-DC

Pin 2 to pin 3 = 0.2–0.5 V-DC to 4.5–4.8 V-DC proportional with temperature rise

2. Variable capacitance (pressure) sensor. These are three-wire sensors that are supplied with reference voltage and usually designed to measure pressure or linear position values. The medium whose pressure is to be measured acts on a ceramic disc and moves it either closer or farther away from a steel disc, varying the capacitance of the device and thus the voltage value returned (signaled) to the ECM.

   Variable capacitance-type sensors can be used for oil pressure sensing, manifold actual pressure (MAP) sensing (turboboost pressure), barometric pressure sensing (BARO), and fuel pressure sensing. In addition, they can be used in pairs to measure mass airflow (these are described a little later in this chapter). Figure 30–5 shows an engine oil pressure sensor that uses a variable capacitance electrical operating principle. A variable capacitance principle has been used in the past by one OEM (Caterpillar) in a type of throttle position sensor (TPS) it phased out in the mid-1990s.

Check Functionality of a Variable Capacitance Sensor

Pin 1 = V-Ref (5 V-DC)

Pin 2 = GND

Pin 3 = Signal

Using a DMM on V-DC the following voltage readings indicate a functional sensor.

Pin 1 to pin 2 = 5 V-DC

Pin 2 to pin 3 = 0.2–0.5 V-DC to 4.5–4.8 V-DC proportional with pressure rise

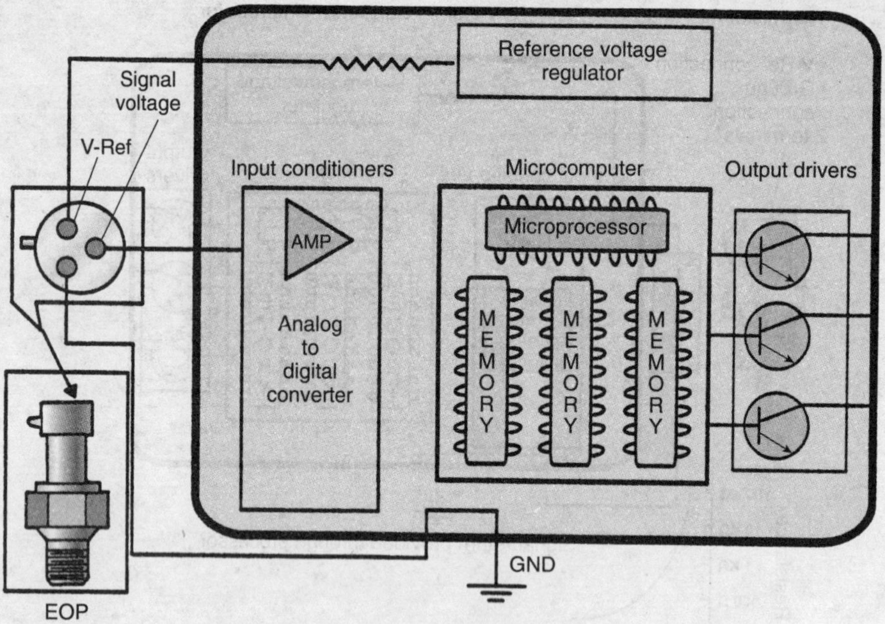

Figure 30–5 *An engine oil pressure sensor of the variable capacitance type. (Courtesy of Navistar International Corp.)*

3. Potentiometers. The **potentiometer** is a three-wire (V-Ref, ground, and signal) variable resistor. Once again, these receive a reference voltage and output a signal proportional to the motion of a mechanical device. Potentiometers are voltage dividers. The moving mechanical device moves a contact wiper over a variable resistor. As the wiper is moved over the variable resistor, the resistance path is altered, with the supply voltage being divided between the signal (sent to the ECM) and ground. TPSs are a typical application for a potentiometer. Figure 30–6 shows a TPS that uses a potentiometer principle of operation.

Check Functionality of a Generic TPS Note that Caterpillar TPSs are supplied with a power-up voltage of 8 V-DC to output a digital signal and should not be tested in this manner.

Pin 1 = V-Ref (5 V-DC)
Pin 2 = GND
Pin 3 = Signal

Using a DMM on V-DC the following voltage readings indicate a functional sensor.

Pin 1 to pin 2 = 5 V-DC

Pin 2 to pin 3 = 0.2–0.5 V-DC to 4.5–4.8 V-DC proportional with pedal angle

4. Piezoresistive pressure sensor. This type of pressure sensor is often used to measure manifold pressure, especially where greater accuracy is required. In trucks it is used as a manifold boost sensor and it is sometimes referred to as a Wheatstone bridge sensor. A doped silicon chip is formed in diaphragm shape so that it measures 250 microns around the outside and reduces to about 25 microns at the center: This means that it forms a sort of drumskin that is 10 times thinner at its center. This permits the diaphragm to flex at that center when subjected to pressure. A set of sensing resistors is formed around the edge of a vacuum chamber over which the diaphragm is stretched. When subjected to pressure, the diaphragm is deflected, causing the resistance of the sensing resistors to change in proportion to the increase in pressure.

Wheatstone Bridge An electrical signal proportional to pressure is produced by con-

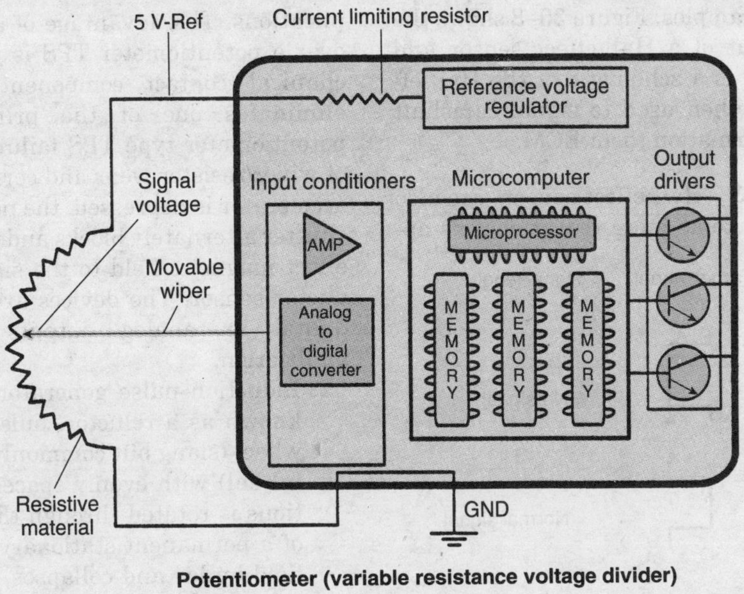

Figure 30–6 *An accelerator position sensor of the potentiometer type. (Courtesy of Navistar International Corp.)*

necting the sensing resistors into a Wheatstone bridge circuit in which V-Ref is used to supply a constant DC voltage value across the bridge. When no pressure acts on the silicon diaphragm, all the sensing resistance will be equal and the bridge can be said to be balanced. When pressure causes the silicon diaphragm to deflect, resistance across the sensing resistors increases, unbalancing the bridge and creating a net voltage differential that can be relayed to the ECM as a signal. Figure 30–7 shows the operating principle of a typical piezoresistive sensor.

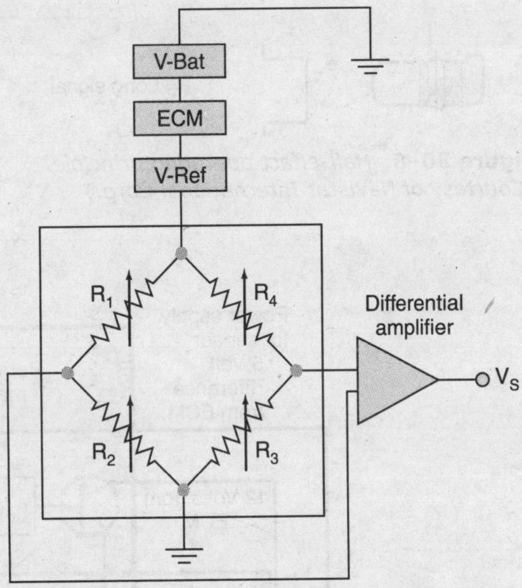

Figure 30–7 *Piezoresistive sensor: Wheatstone bridge circuit, output signal is represented by Vs*

Signal-Generating Sensors

1. Hall-effect sensors. Hall-effect sensors generate a digital signal as timing windows or vanes on a rotating disc pass through a magnetic field. They can be used for both linear and rotary position signaling. In a rotary Hall-effect sensor, the rotating disc is known as a **pulse wheel** or **tone wheel.** Because these are terms that are used to describe the rotating members of induction pulse generators, care should be taken to avoid confusion. The frequency and width of the signal provides the ECM with speed and position data. The disc incorporates a narrow window or vane for relaying position data. The Hall-effect sensor outputs a digital signal by blocking a magnetic field with a vane from a semiconductor sensor.

   Hall-effect sensors are used to input engine position data for purposes of event timing computations such as the beginning and duration of the pulse width. Camshaft position sensor (CPS), timing reference sensor (TRS), and engine position sensors

(EPS) are examples. Figure 30–8 shows the signal output of a Hall-effect sensor and Figure 30–9 is a schematic of the Hall-effect sensor when used to signal camshaft position information to an ECM.

Hall-effect TPS Hall-effect sensors are also used as TPSs in some more recent chassis applications. The advantage of a Hall-effect TPS over a potentiometer TPS is that it lacks mechanical contact components. This feature eliminates one of the primary causes of potentiometer type TPS failures, namely, that of wiper/resistor wear and corrosion. When the accelerator is depressed, the noncontact sliding shutter alternately blocks and exposes the Hall-effect magnetic field to the signaling semiconductor sensor. The devices are configurable to digital or analog outputs depending on application.

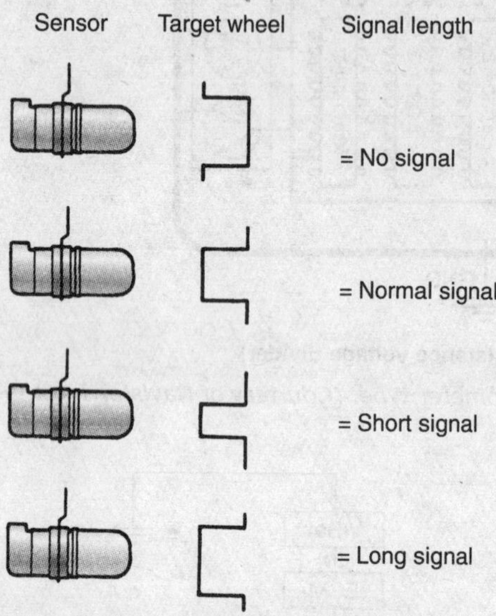

Figure 30–8 *Hall-effect operating principle. (Courtesy of Navistar International Corp.)*

2. Induction pulse generator. A toothed disc known as a reluctor, pulse wheel, or tone wheel (slang but commonly used: **chopper wheel**) with evenly spaced teeth or serrations is rotated through the magnetic field of a permanent stationary magnet. As the field builds and collapses, alternating current (AC) voltage pulses are generated and relayed to the ECM. Reluctor-type sensors are used variously on modern truck chassis in applications such as antilock brake system (ABS) wheel speed sensors, vehicle speed sensors (VSS) located in the transmission tailshaft, and engine shaft speed sensors. Figure 30–10 shows the operating principle of an induction pulse generator.

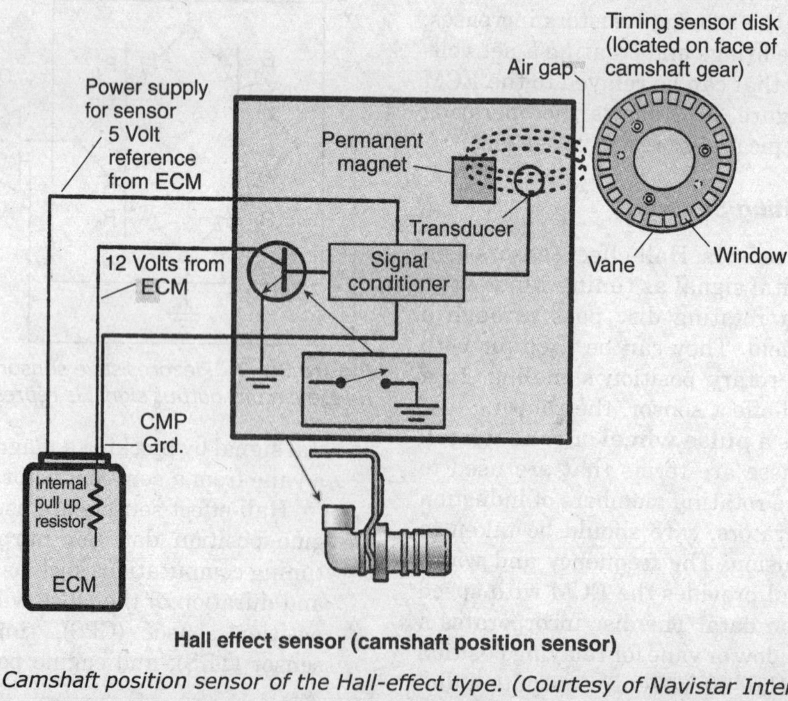

Figure 30–9 *Camshaft position sensor of the Hall-effect type. (Courtesy of Navistar International Corp.)*

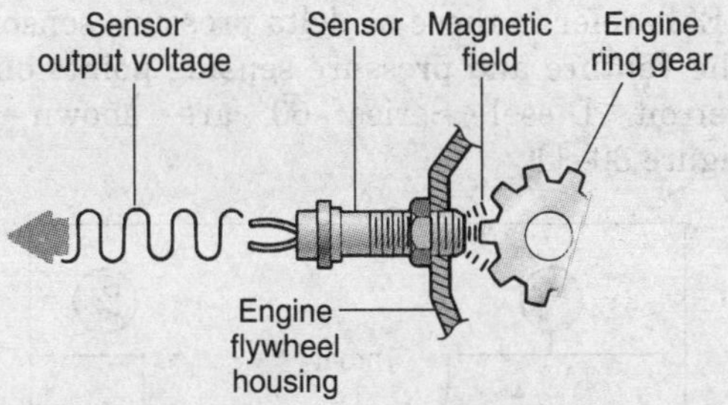

Figure 30–10 *Induction pulse generator shaft speed sensor. (Courtesy of Navistar International Corp.)*

Check Functionality of a Two-Terminal Shaft Speed Sensor Because this type of sensor generates an AC voltage, the digital multimeter (DMM) should be set to V-AC. When the shaft is stationary, zero volts should be read across the terminals. When the shaft is rotated, V-AC voltage and frequency readings should increase proportionally with rpm.

3. Galvanic sensors. Although primarily used on gasoline-fueled automotive engines, diesel technicians should have rudimentary understanding of lambda (λ) sensors (also known as oxygen (O_2) sensor or exhaust gas sensor) operation. They are used on natural gas (NG) adapted diesel engines and downstream from reduction converters (in the exhaust canister) in some post-2007 engines. A lambda sensor was used as the closed loop driver on many spark-ignited (SI) engines; that is, it is the primary input reference the ECM used to monitor air-fuel ratio (AFR). Until the introduction of gas density-type O_2 sensors, galvanic-type sensors were used in most engines required to be fueled at close to stoichiometric ratios.

   The most common lambda sensor is a galvanic device that produces an electric current by chemical action. The lambda sensor is usually located in the exhaust piping exposed to the engine exhaust gas on one side and ambient air on the other. When an O_2 sensor is used on a diesel engine, it plays a different role and is primarily a monitoring device for oxides of nitrogen (NO_x): The ECM calculates NO_x based on the O_2 present in the exhaust compared with the mass of air that entered the cylinder. When O_2 sensors are used on NG-fueled engines (Cummins, Mack, Caterpillar), it functions as a close loop driver.

   Construction A typical O_2 sensor is constructed of a zirconium dioxide ceramic material with gas permeable platinum electrodes exposed on one side to the exhaust gas (measuring electrode) and on the other to ambient air (reference electrode). The zirconium dioxide begins to conduct O_2 ions at a temperature value of around 300°C (550°F). At this temperature, as the O_2 proportion in the exhaust gas and that in the ambient air differ, a small voltage is generated due to the electrolytic properties of zirconium dioxide. The greater the *difference* of the oxygen proportions, the higher the voltage produced. These voltages range between 0.1 V and 1.0 V after the trigger temperature value is attained: In a majority of systems, a signal voltage of around 0.45 V-DC indicates a stoichiometric AFR. Lambda sensors are used in SI engines required to be fueled at stoichiometric ratios.

4. NO_x sensors were introduced on 2007 products (notably some Mack vocational engines) and will likely become more common in 2010 diesel engines. They function by sampling the exhaust gas and calculating total NO_x based on the sample. A NO_x sensor consists of two chambers, which can be described as:
   - Oxidation chamber: This is the first stage. Exhaust gas enters the oxidation cavity and a pair of platinum electrodes oxidize (HCs), (CO), and H_2. What remains then passes to the sampling chamber.
   - Sampling chamber: This is the second stage. After the oxidation stage, all that remains of the exhaust gas is just a few ppm of O_2 and N_2. First the O_2 is removed by catalytic reaction with the platinum electrodes, then the remaining NO_x is electrolytically reduced to O_2 and N. In other words, NO_x molecules are broken up into elemental nitrogen and oxygen. Finally, the O_2 is removed by catalytic reaction at the measuring electrode: The amount of current required to achieve this is directly proportional to the amount of NO_x that first entered the sensor.

This type of NO_x sensor requires contact with ambient O_2 in same way an O_2 sensor does. It references NO_x with respect to the variability of atmospheric O_2 in proportion with increase in altitude. The operating temperature of these devices is about 650°C (1,200°F) so they usually incorporate a heater.

MAF Sensors

Mass airflow (MAF) sensors have been introduced recently by diesel engine manufacturers although they have been used for decades in automobile applications. Their objective is to provide the ECM with accurate measurement of the air mass delivered to the engine cylinders. MAF sensors use one of a "hot wire," critical flow venturi, or vortex flow operating principles.

Hot Wire MAF A hot wire MAF is one in which a sensing element is heated then maintained at a specified temperature. The sensing element is a film or wire positioned in the intake air stream to the engine and supplied with a constant voltage. Because of the applied voltage, the element is heated by the electrical current that the voltage produces. When intake air flows across the heated element, it has a cooling effect. The more air passing across the heated element, the greater the cooling effect. The heated wire or film element has a PTC electrical characteristic so its resistance decreases in proportion to any drop in temperature. As resistance decreases in the element, more current must flow through it in order to maintain the specified programmed temperature. The current flow at any given moment of operation is measured and transduced as an input signal to the ECM to be computed as airflow. A hot wire MAF works in conjunction with an inlet temperature sensor to ensure maximum accuracy.

Critical Flow Venturi A **critical flow venturi (CFV)** MAF sensor is sometimes known as pressure differential flow sensor. Its operating principle is based on the relationship between inlet pressure and flow rate through a venturi. By definition a venturi is a gradually diminishing sectional area within a pipe designed to accelerate gas flow and decrease pressure: It is the flow principle that was used in carburetors. When the pressure is known at the entry point to the venturi and at its exit point, the differential pressures can be used by the ECM to calculate the mass of the airflow. Some OEMs refer to these as delta pressure sensors. The venture and pressure sensing points on a Detroit Diesel Series 60 are shown in Figure 30–11.

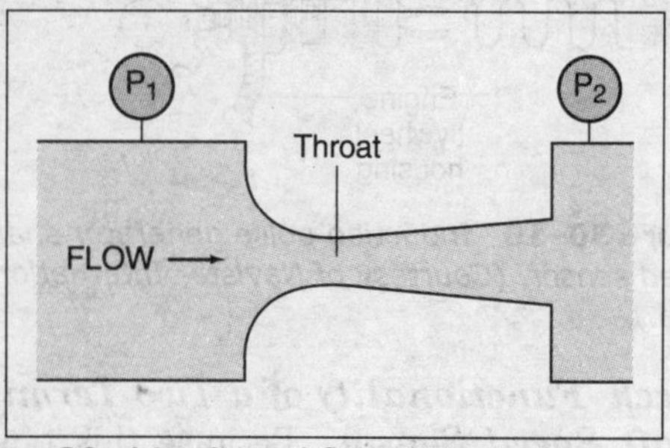

Figure 30–11 *Critical flow venturi, pressure differential MAF sensor.*

CFV Operation When air flows through the venturi, it reaches maximum velocity wherever the flow area is minimized. If you take a look at Figure 30–11 you will observe that the minimum flow area occurs at the throat of the venturi. When air (or a mixture of air and recirculated exhaust gas) is flowed through the venturi a pressure differential is established between the P1 and P2. The differential readings will change as inlet pressure increases or decreases—or if the exit pressure increases or decreases. Whatever flow occurs is proportional to the differential pressure across it while the entrance pressure (P1) remains constant. For instance, if the exit pressure (P2) is reduced the following occurs:

- Differential pressure increases.
- Flow increases.
- Air velocity at the throat increases.

These factors allow the ECM to compute the MAF with some degree of accuracy.

Vortex Airflow Sensor These are sometimes known as Karman vortex flow sensors. The upper inlet of the sensor has an obstruction in the airflow. As air passes around the obstruction, it generates a stream of vortices (like mini-tornadoes) that increase as the airflow velocity increases. By locating an ultrasonic speaker and pickup (microphone) across the stream of vortices, which spin off in opposite directions, a frequency modulated shift can be signaled to the ECM. Frequency increases in proportion to air velocity, enabling the ECM to compute mass.

Summary of MAF Operation MAF sensors are designed to work in conjunction with other temperature and pressure sensors in the intake circuit. This permits the ECM to factor variables such as ambient air temperature, humidity, barometric pressure, and manifold boost. For instance, cool air is denser than warm air, so the signal from an NTC-type ambient temperature sensor is used by the ECM to accurately compute actual airflow data. Humidity always affects air density because humid air is denser than dry air. However, because humid air has increased cooling effect on the sensing element no other compensation is required.

Water-in-Fuel Sensors

A **water-in-fuel (WIF) sensor** detects water contamination of fuel. A WIF sensor uses a couple of probes and a 12-volt supply. Because water has different electrical resistivity (known as dielectric properties) than fuel, a return signal is output from the sensor when the electrical path across the probes acts through water rather than fuel. At this point the WIF broadcasts a service alert. Note that a WIF can sometimes produce a service alert immediately after draining the water sump: The reason is that water-resident bacteria can coat the probes after draining and trigger a false signal.

Thermocouple Pyrometers

In an earlier generation of truck diesel engines, a **pyrometer** was often located at a specified distance downstream from the turbocharger and used to provide the driver with engine loading information useful for shifting, especially downshifting during prolonged loading. In post-2007 engines, one or more thermocouple pyrometers are used to accurately monitor the high temperatures produced in **diesel particulate filters (DPFs)**. DPFs will be studied in detail in Chapter 41 of this textbook, but we take a look at the electrical operating principles of thermocouple pyrometers here.

Thermocouple Principle A thermocouple pyrometer consists of two dissimilar insulated wires (often pure iron and constantin [55% copper, 45% nickel]) connected at each end to form a continuous circuit. The two junctions are known as hot end, which is located where temperature sensing is required, and reference end, which is connected to a millivolt meter. Whenever the junctions are at different temperatures, there is current flow, the voltage increasing with temperature rise. Thermocouple pyrometers and their wiring circuit should be regarded as a complete unit: Never attempt to repair the wiring and always replace as an integral unit. The practice of placing a thermocouple pyrometer in the exhaust tract of each engine cylinder is used on large diesel engines to monitor cylinder balance: It is not used as original equipment on any current truck diesel engines.

Switches

Switches complement the sensor circuit and can usually be classified as command inputs. Switches may be electromechanical or "smart," that is, they use signals rather than analog voltage values to signal a change in status. Switches can be subdivided into three groups:

1. Switches grounding a reference signal (V-Ref). A good example would be that of a coolant level sensor. This receives a reference signal from the ECM that grounds through the coolant in the upper radiator tank. Should the coolant level drop below the sensor level, the reference signal loses its ground, which after a preprogrammed time period is, for example, 8 seconds. The time lag is required to prevent a temporary loss of ground caused by braking to open the circuit and trigger an ECM response. The ECM responds with whatever action it is programmed with—electronic malfunction alert or engine shutdown. Figure 30–12 and Figure 30–13 show examples of switches used in a typical electronically managed engine circuit. The switch shown in Figure 30–12 is a driveline disengagement switch (DDS), which is located on the clutch pedal: This switch has to be closed to enable engine brake operation.
2. Manual switches control electrical circuit activity. Many versions of this type of switch are used by the operator to control vehicle functions. Some examples would be the ignition key, engine retarder mode switches, and the cruise control switches. Switches that are controlled by the driver are sometimes called command switches.

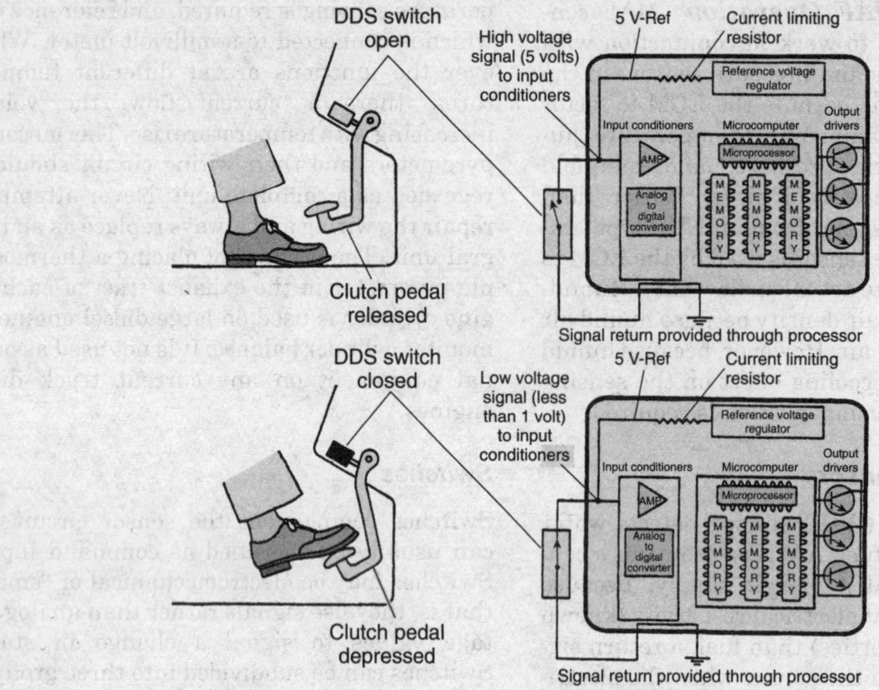

Figure 30–12 *Switch-type input sensors. (Courtesy of Navistar International Corp.)*

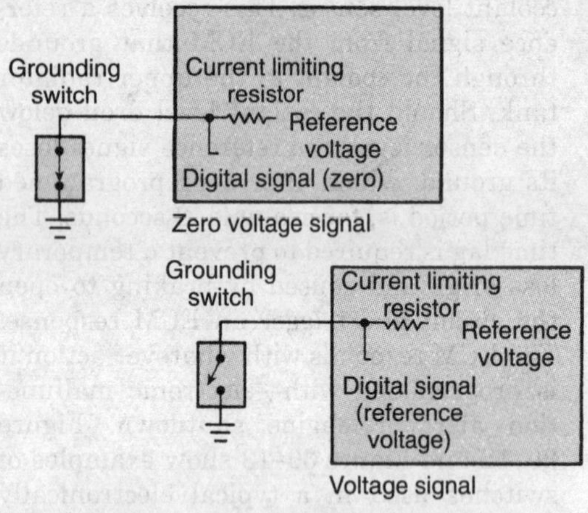

Figure 30–13 *Grounding-type switch operation. (Courtesy of Navistar International Corp.)*

3. Smart switches. Smart switches use digital signals to indicate a change in status. The signal produced by a smart switch may be automatically generated by a change in status condition or be generated by a mechanical action such as an operator toggling a switch. Smart switches are studied some detail in Chapter 33.

ECMs AND THE PROCESSING CYCLE

Figure 30–14 shows the location of an ECM and some sensors on a Navistar DT466E diesel engine. As mentioned a little earlier in this chapter a diesel engine management ECM has four basic functions:

1. Regulating reference voltage (V-Ref)
2. Input conditioning, amplification, and ADC
3. Processing
4. Managing output drivers

ECM is the generic term for the unit housing the engine management electronics. It usually incorporates the complete computer assembly and some, if not all, of the output drivers or switching apparatus. ECM is also the SAE-recommended term. However, some engine OEMs choose to use their own terms. The engine ECM receives command and monitoring data from its input circuit. Inputs are responded in the processing cycle and outputs are mapped. Among its outputs, the ECM manages V-Ref at a specific value. V-Ref is used both to power-up and benchmark input signals.

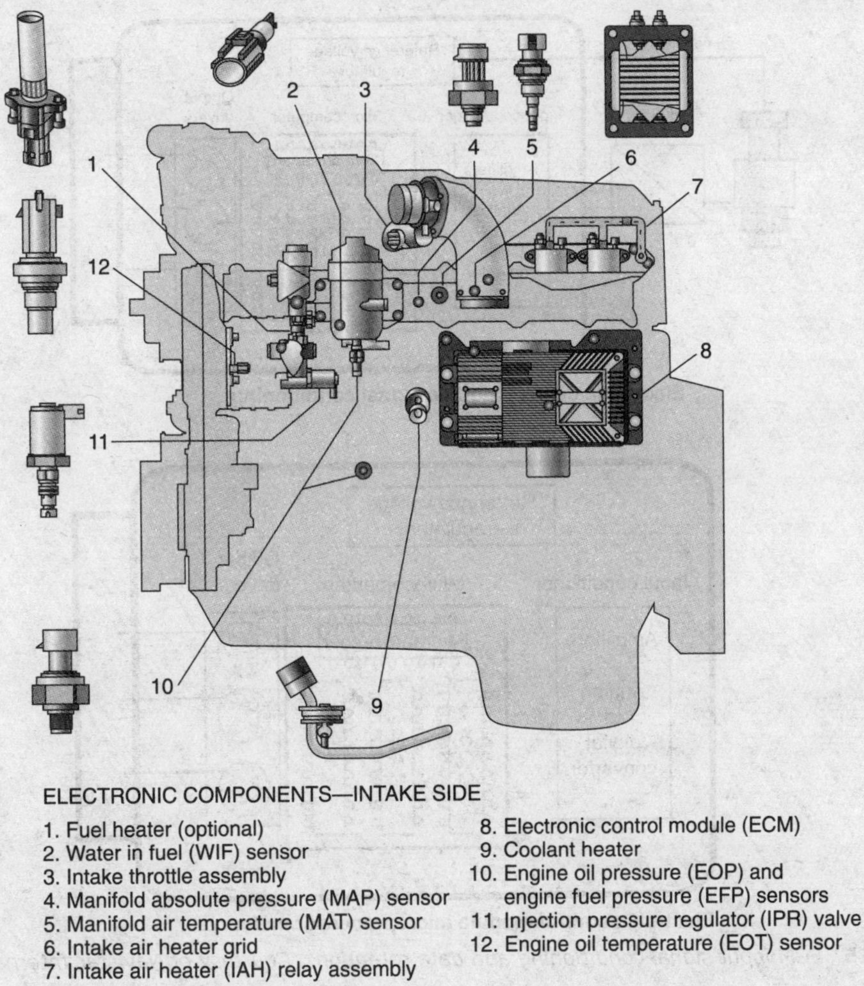

Figure 30–14 *Navistar DT466E seen from the left side and showing the location of the ECM and some sensors. (Courtesy of Navistar International Corp.)*

The electronic subcircuit consists primarily of sensors located variously on the engine and chassis. Because the data they input is mostly in the form of analog voltage values, these must be converted to digital codes for processing by an ADC (analog to digital converter) integral with the ECM. Weak signals may be conditioned or strengthened before processing. Preparing input signals for processing are ECM functions categorized as input conditioning. Figure 30–15 outlines the role of V-Ref and signal conditioning in an ECM.

Processing Cycle

Processing in the ECM involves scanning the programmed fuel map data (in ROM, complemented or finally defined in PROM and EEPROM), engine and chassis monitoring sensors (such as engine coolant temperature [ECT], VSS, and so on) and command inputs (TPS, cruise control, engine brake, and so on), and subsequently plotting an actual fuel quantity to be delivered.

Processing Speeds

ECM processing may occur at different frequencies classified as **foreground** and **background computations**. Foreground operations would include response to a critical command input such as the TPS whose signal must be processed immediately to generate the appropriate outcome. The monitoring of engine oil temperature

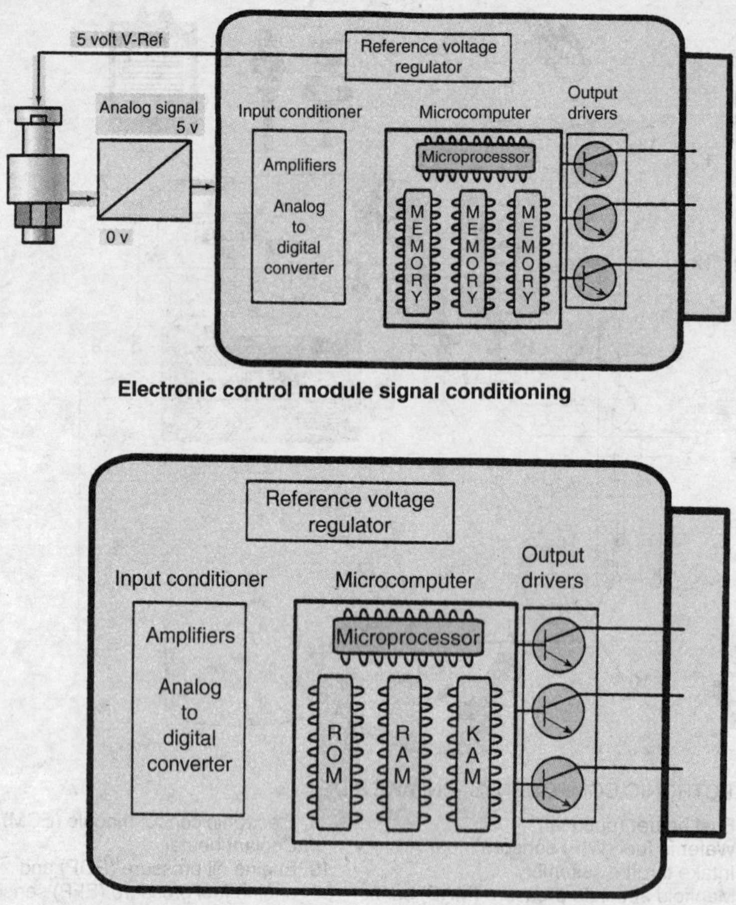

Figure 30–15 *ECM input signal conditioning and data retention. (Courtesy of Navistar International Corp.)*

is often classified as a background operation. Although obviously valid, this signal does not require as immediate an adjustment to operating strategy as that from the accelerator.

Generating Outputs

ECM outputs are switching functions. Switching functions are usually driven at system chassis voltage or, in some instances, they are spiked to higher values. In this way the ECM manages the cycles of fuel injection apparatus, exhaust gas recirculation (EGR), and diesel particulate filters (DPFs). In doing this it also communicates with other chassis electronics by sharing data. Figure 30–16 is a simplified example of the processing cycle in a three-module Navistar ECM assembly consisting of an ECM, personality module (PROM and EEPROM memory), and a separate injector driver module (IDM).

ECM Location

ECMs can be located anywhere on the vehicle chassis. Engine management ECMs are often mounted on the engine close to the devices they have to monitor and switch. While engine-mounted ECMs require better shock and vibration insulation, they have certain advantages in being close to both the sensors and actuators and, where heat is a problem, can be mounted on a heat exchanger with diesel fuel acting as the cooling medium. Another option is to locate the ECM under the dash inside the vehicle cab, reducing the requirement for shock and vibration insulation. The following are examples of some OEM acronyms used to refer to engine and fuel system controllers:

EEC—Electronic engine control
ECI—Electronically controlled injection

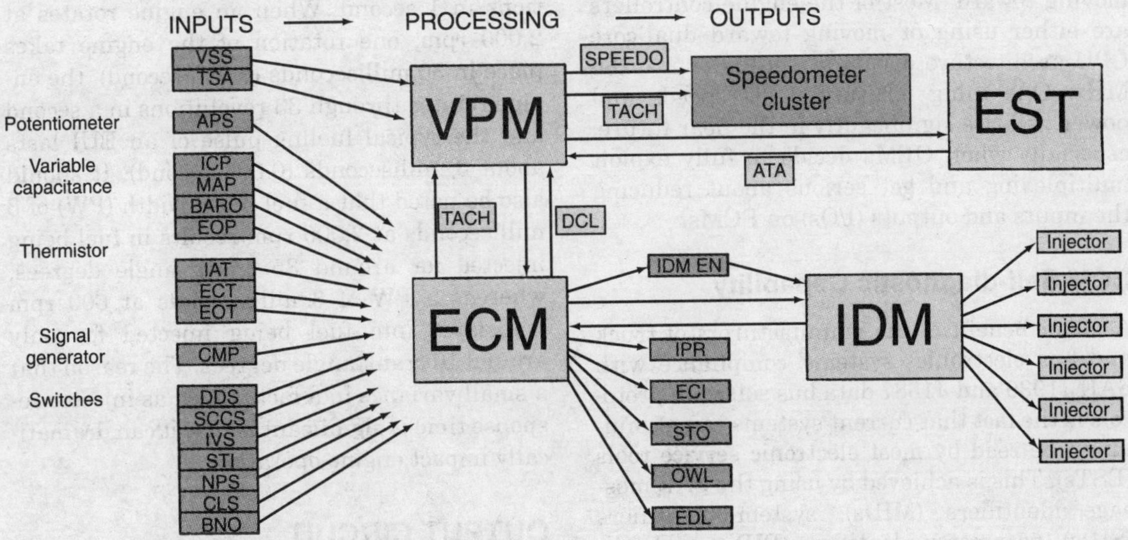

Figure 30–16 *The processing cycle of an International Trucks HEUI, three-module (3-box) controller system. (Courtesy of Navistar International Corp.)*

ECU—Electronic control unit
FIC—Fuel injection control module
MCM—Motor control module
PCM—Powertrain control module

Processing Algorithms

The term **algorithm** is used to describe the sequence of processing events in an ECM, from the arrival of a signal to the switching of an outcome. The algorithm required to produce any computer outcome is the mapping of how the processing cycle uses inputs, program instructions, and data bus (other chassis computers) data. Manufacturers use this term to describe all of the factors and preconditions required in processing to produce a specific outcome. For instance, if we say that 1 + 2 + 3 + 4 = 10, we can say that we have used the rules of addition, applied the values, and come up with an outcome, 10, the sum of the values. If any one of the values used in that example changes, the end result or outcome will change. So if the number 1 is replaced by the number 6, the outcome or sum of the values has to change: When we apply the rules of addition, the outcome changes to 15.

The control algorithm used by an engine management system uses the same means to produce outcomes, except that there are hundreds of rules and thousands of values. The values are received into the processing cycle by the input circuit, from memory, and off the chassis data bus. The rules are logged into the ECM memory components known as ROM, PROM, and EEPROM. The computer then manipulates this data in main memory or RAM and computes an outcome.

Processing Power

Controllers networked to the J1939 bus run from 8-bit CPUs that perform straightforward tasks such as managing air bags or climate control to 32-bit diesel engine management CPUs. The industry is moving toward 64-bit controllers largely because of the need to more precisely manage cylinder combustion and the complex emission control hardware. There is a disadvantage in using high-power 32- and 64-bit controllers (it is not cost, they are inexpensive today) in chassis that do not make significant use of multiplexing (many do not, at the time of writing) due to the vast number of wires required for the input and output devices. This results in a wiring harness that looks like an anaconda emerging from the ECM.

If you compare the wiring congestion on a generously multiplexed chassis such as recent versions of Navistar's Diamond Plus system with those that are not, it is easy to tell why multiplexing is the direction the industry is moving toward. Most of the engine controllers are either using or moving toward dual-core CPU architecture driven at speeds of 2 × 280 MHz. One thing for sure, CPU speeds and power will rise significantly in the near future, especially when OEMs decide to fully exploit multiplexing and get serious about reducing the inputs and outputs (I/Os) on ECMs.

ECM Self-diagnostic Capability

A major benefit of the manufacturers of truck and bus electronics systems' compliance with SAE J1939 and J1587 data bus software protocols is the fact that current systems can at minimum be read by most electronic service tools (ESTs). This is achieved by using the SAE message identifiers (MIDs), system identifiers (SIDs), parameter identifiers (PIDs), and failure mode indicators (FMIs).

Self-diagnostic capability in diesel engine electronics systems can be grouped as onboard in the form of flash codes or digital data displays, and EST-read such as reader/programmers and PC-based systems. The use of these instruments is covered in Chapter 32 and in chapters dealing with each specific system. The self-diagnostic capability of most current engines goes beyond just identifying electronic faults. Logic maps are used to identify many common hydromechanical faults.

Time and Electronically Managed Engines

Until the widespread use of electronics, engine technicians generally referred to timing and events within the engine cycle in terms of crank angle degrees. This tendency has shifted as the data readouts produced by the onboard electronics in many systems tend to use time as well as crank angle to report engine events such as EUI response time and duty cycle. It may help the technician to have a basic understanding of some of the time dimensions within which engine events take place especially when it comes to making sense of some the data displayed by electronic service tools.

Time and rpm

If an engine rotates at a typical low idle speed of 600 rpm, it turns through 10 complete rotations in 1 second. When an engine rotates at 2,000 rpm, one rotation of the engine takes place in 30 milliseconds (0.030 second), the engine rotates through 33 revolutions in a second and the typical fueling pulse of an EUI lasts about 3 milliseconds (0.003 second). It should also be noted that a fuel pulse width (PW) of 3 milliseconds at 2,000 rpm results in fuel being injected for around 35 crank angle degrees, whereas a PW of 3 milliseconds at 600 rpm translates into fuel being injected for only around 10 crank angle degrees. The reason that a small variance in a factor such as injector response time is significant is that it can dramatically impact engine operation.

OUTPUT CIRCUIT

The switching apparatus used with each electronic management system is what truly differentiates each system from any other. Although the switching and pumping apparatus in full authority engine management systems have certain similarities regardless of OEM system, it is more appropriate to address this technology by manufacturer rather than generally in this section. At the moment of writing, full authority engine management systems use EUIs, hydraulically actuated electronic unit injectors (HEUIs), EUPs, and common rail (CR) actuators that characterize each OEM system. There are indications that this is changing, and fast. Medium- and heavy-duty highway diesel engines certified for 2007 and beyond are using just four basic fuel systems:

- EUI with electrohydraulic nozzles
- CR with electrohydraulic injectors
- HPI-TP
- EUP systems

Two of the above systems will be obsolete by 2010 so this could mean that diesel technicians of the future will have to learn just two fuel systems instead of the dozen or so around at present. The output circuit always begins in the ECM or in a module bussed to it. Figure 30–17 shows the relationship of the ECM switching unit or output drivers within the processing cycle.

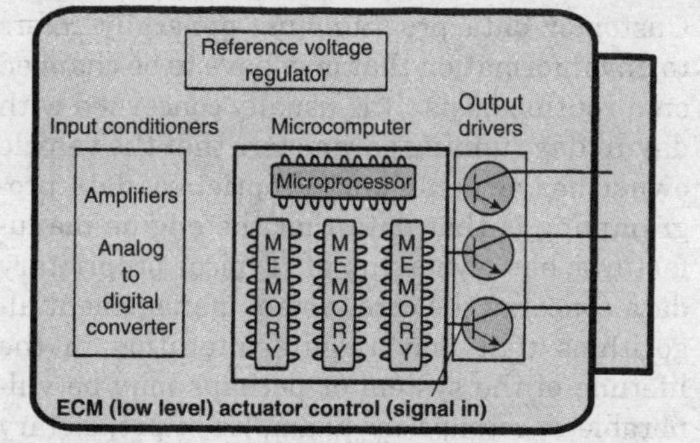

Figure 30–17 *ECM actuator control. (Courtesy of Navistar International Corp.)*

Actuators

A wide range of actuators are used on diesel engine control systems. Most are fairly simple in terms of their operating principle. We take a brief look at some of them here.

Solenoids

A simple solenoid consists of a coil and an armature. The coil is usually stationary and the armature moves within it. The armature can be integral with devices such as hydraulic poppet valves, spool valves, and levers that effect mechanical movement. A simple solenoid has two status conditions: *off* or *on*. In most cases, the armature is spring-loaded to default to the mechanical *off* position when no current is flowed through coil. When an ECM driver energizes the coil, the magnetic field that builds mechanically moves the armature. This is the principle that is used in the control cartridge of a solenoid-actuated injector. Because solenoids use electromagnetism, they respond more slowly due to the time required to build and collapse electromagnetic fields: This limits their use in applications that require fast response and frequent cycling.

Proportioning Solenoids

These are coil and armature devices that function similarly to a solenoid except that they are capable of precise linear or rotary positioning. To ensure the accuracy of the position, most proportioning solenoids are equipped with a means of signaling *actual* position while the ECM driver attempts to control current flow to the proportioning solenoid to maintain desired position. An example of a rotary proportioning solenoid would be the brushless torque motor (BTM) used in some early electronically managed Caterpillar engines. Examples of linear proportioning solenoids would be the injection actuation pressure (IAP) oil control spool valve used in HEUI-fueled engines, or the rail pressure control valve used in the high-pressure pumps on CR fuel systems.

Stepper Motors

A stepper motor is a brushless electric motor capable of precision positioning of a shaft (and whatever is connected to it). A normal direct current (DC) electric motor spins when current is flowed through its electromagnetic fields. Stepper motors use multiple toothed electromagnets arranged around a rotor gear integral with the motor shaft. All but one of the toothed electromagnets are slightly offset from the gear teeth on the rotor: This means that when the rotor gear teeth are perfectly aligned with one of the toothed electromagnets, they are slightly offset from the teeth on the others. When the next electromagnet in the rotation is energized, the rotor gear moves slightly to realign the magnetic field and completes a "step." Typically, four electromagnets are used spaced 90 degrees apart in the rotation. In this way the rotor can be precisely aligned to any position in rotation. The precision increases with the number of teeth in the rotor and, correspondingly, on the electromagnets. Figure 30–18 demonstrates the principle of a stepper motor.

ECM-controlled stepper motors are versatile positioning devices and have application in door/gate controls such as those used in EGR mixers. Controls are simple and when driven by an 8-bit controller, outputs can be divided into up to 256 micro steps, providing some precision of placement. Because a constant DC current flow creates heat (and therefore resistance variables), they are regulated by pulse width modulation (PWM) to reduce resistance-generated heat in the controller.

Piezoelectric Actuators

These have become much used in recent years in diesel fuel injection systems. They are based on the reversibility of piezoelectric effect (piezo-

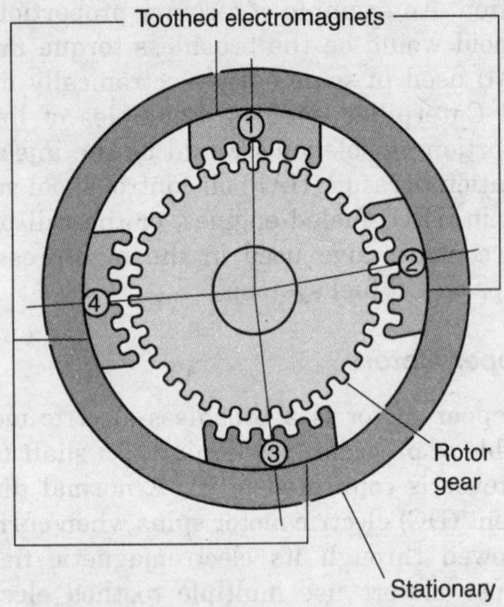

Figure 30–18 *Principle of a stepper motor.*

electricity is covered in Chapter 28). When opposing faces of a piezoelectric crystal are subjected to voltage, the crystalline structure changes almost instantly, resulting in either expansion or contraction. This change of shape produces a powerful mechanical force that can be used for actuating a valve in much the same way as a solenoid but at much faster speeds.

Piezoelectric actuators can be used to replace any solenoid but at this moment in time have the following characteristics:

- Faster response times when a control voltage is applied.
- Low electrical power consumption.
- Currently, they are bulkier than an equivalent solenoid but this will probably change.

A piezoelectric actuator is constructed of several hundred piezo crystals in stacked wafers. These piezo crystal wafer stacks have the ability not only to operate nozzle needles against the extremely high diesel fuel injection pressures but to do it quickly and repetitively. For a full explanation of piezoelectric actuators and their application, reference Chapters 27 and 37.

ECM PROGRAMMING

Vehicle ECM programming can be divided into two general categories: customer data programming and proprietary data programming. Customer data programming generally refers to any information that may have to be changed on a routine basis. It is usually concerned with day-to-day running parameters that the vehicle owner has ownership of. Proprietary data programming is that data that the engine manufacturer has ownership of. Typical proprietary data concerns the core system management algorithms that may require alteration in the lifetime of the system or perhaps may be vulnerable to corruption. Examples of proprietary data are the engine fuel map and maximum BHP.

Customer Data Programming

Most OEMs have designed their systems so that a wide range of running parameters can be altered to accommodate a change in chassis application and driver or owner preference. Usually all that is required to alter "customer data" is the ECM customer password. Fleets may sometimes not provide their drivers with the vehicle ECM password. Additionally, some fleets use an alpha password to either prevent or make difficult customer data programming changes with a generic reader programmer (ProLink 9000). Many customer data programming options are electronically toggled. This means you have no choice but to choose one option or the other. An example is the governor-type option. In this programming field you can either choose variable speed (VS) or limiting speed (LS) governing. We define this as toggle option.

Typical Customer Programming Options

Other customer programming options would define limits, such as road speed limit (RSL) or idle shutdown time limit. Some examples of typical customer data programming options offered by OEMs are:

- Governor type. Any current truck engine management system can be programmed to governor type according to the preferences of the operator. The common options, LS and VS, are normally toggled options to ensure that one or the other is selected. See Chapter 25 for an explanation of governor options.

- Power takeoff (PTO) governing/limits. Precise isochronous (zero droop) governing was, before the electronic age, an expensive governing option required in any engine driving a generator, in which application a change in engine load required an immediate governor response in fueling to maintain a constant speed to avoid variations in the electrical frequency. Isochronous governing in electronically managed engines is incorporated in the programmed software, and the term is used by some OEMs to describe the operation of the engine when driving PTO equipment such as pumps and compressors.
- Cruise control parameters. The upper and lower limits are programmable. Actual values are programmed by the driver operating the vehicle using dash switches. Smart cruise (flexible parameter) programming available in many current systems makes cruise control capable of producing superior fuel-to-mileage figures than hard (rigid parameter) cruise control. Smart cruise programming can also "reward" a driver who improves on projected vehicle fuel economy by increasing governed road speed.
- Road speed limit. Any observer of trucks driven on the highways will note that this is a popular option with large fleets that capitalize on the fuel-to-mileage and safety benefits of having a vehicle run at a lower maximum road speed. It is a programming option that most owner-operators choose not to take advantage of.
- Critical shutdown sensors. A coolant level sensor can be programmed to a number of failure strategies if its ground (through the liquid coolant) is lost for a predetermined period. The failure strategies could vary from a simple warning alert to the driver, to ramp down to a default rpm/maximum load, or to shut down the engine. The ECM could be programmed not to shut the engine down in a vehicle routed through mountainous terrain.
- Tire size. Tire size must be programmed using a tire specification manual that defines a tire's rolling radius. This is usually programmed to the ECM in revolutions per mile or revolutions per kilometer. It is important to note that two nominal 22.5" tires of different tread code ratings (even if they are manufactured by the same manufacturer) may have quite different dynamic rolling radii. Incorrectly programmed tire data will result in false road speed data signaled to the ECM and, coincidentally, inaccurate speedometer and mileage information.

Additional examples of typical customer data programming options offered by OEMs include:

- Transmission ratio
- Carrier ratio
- Progressive shifting limits
- Idle speed
- Torque rise profile
- Peak brake power
- Idle shutdown duration

Customer data programming must fall within parameters defined in the ECM's data retention media. For instance, if the maximum high-idle speed in a given engine is defined by its management ECM as 2,000 rpm, this value cannot be reprogrammed to a higher value using the preceding methods. You can reference the reprogramming procedure that appears in Chapter 31 if you want to take a closer look at reprogramming right now.

Proprietary Data Programming

The technician enables proprietary data programming but does not actually perform the reprogramming. Proprietary data includes any ECM-programmed data that the OEM does not want the customer to alter. A good example would be the fuel map. There are circumstances in which this might have to be rewritten or altered, but it would have to be accomplished by the OEM. In the event that critical ECM files become damaged or corrupted, this could require proprietary data reprogramming. Proprietary data programming normally takes place in three distinct stages: downloading, programming the ECM, and uploading verification.

Downloading

Downloading requires accessing the OEM's centrally located mainframe via a PC, the Internet, and OEM software. The vehicle is identified electronically, usually by its vehicle identification number (VIN), and the OEM software routes itself to the correct mainframe location of the system files: In some cases, the technician may be required to input passwords to enable this step. The mainframe computer can be viewed as a sort of filing cabinet. The function of the OEM software on your computer is to locate system files that pertain to the specific vehicle identified by its VIN. Having located these files, they are next downloaded to the PC's electronic memory from which they can be copied to disks or transferred directly to the vehicle ECM. This step in the process does not require that the vehicle be electronically connected to whatever computer is being used. In fact, it could be out on the highway at the time of the transaction.

Programming the ECM

The system files that were copied to computer memory can now be transferred to the vehicle ECM. A portable PC (a laptop computer would be the instrument of choice) can be connected to the vehicle ECM by means of:

- A communications adapter (CA) and the data connector, a six- or nine-pin Deutsch connector
- A wireless handshake (if both the laptop and vehicle are wireless enabled)

Now, by adhering to some menu-driven sequencing, the downloaded files can be transferred to the ECM usually by scan and overwrite so the procedure is fast. On completion, a verification file is created in computer (laptop) memory.

Before reprogramming, certain requirements usually must be met. For instance, the vehicle ignition circuit must usually be closed (switched *on*) with the engine stationary. All fault codes usually have to be cleared before reprogramming. Additionally, it would make sense to obtain a printout of all customer programmable options *before* reprogramming because these may be erased by the process depending on the system. Customer data sometimes has to be reprogrammed after a proprietary data reprogramming procedure. Some OEMs use the term *flashing* or *reflashing* ECM memory to describe programming.

Uploading

The term **upload** usually describes the transfer of data from one computer system to another, and in this sequence of events it is the final step in ECM reprogramming. The verification file created in computer memory by the vehicle ECM following reprogramming must, within a certain period of time, be uploaded to the OEM's data hub. This confirms that the reprogramming has been successful in OEM mainframe computer's files on the vehicle. Once again the Internet is used and so the verification file can be uploaded. Uploading verification files is a recordkeeping procedure.

Proprietary data reprogramming is a simple set of procedures. These require some familiarity with PCs. However, most of the sequencing is menu driven and OEMs have designed the procedures to be as user friendly as possible. As each new software version is released, reprogramming becomes progressively more simple.

ENGINE MANAGEMENT SYSTEMS

Management systems for current truck and bus fuel systems tend to be classified by the degree of control they have over the fueling pulse. It should be noted that much of the language we use to describe truck computer systems was introduced in the early days of engine computer controls. Some of this language is out of date. However, the speed of the computer revolution in vehicles has been such that in a repair shop today you will still find entirely hydromechanically managed trucks alongside those that represent the very latest in electronic management.

Full Authority Systems

For purposes of study, we will identify full authority engine management systems as those that were specifically designed for control by computer. This includes:

- EUIs
- EUPs
- HEUIs
- Time-pressure (TP) fuel systems
- CR fuel systems

The term *full authority* indicates that the system ECM has full control of the fueling pulse, but in the case of EUI, EUP, and TP systems, this necessarily occurs within a timing window established by the physical dimensions of the actuating cam profile. So in reality, it can legitimately be argued that among the preceding systems some are "more full authority" than others.

Partial Authority Systems

Partial authority systems are ECM-managed systems in which hydromechanical fueling apparatus has been adapted for computerized control. These are rapidly becoming obsolete and are not used on any current systems. The extent to which the ECM can control fueling varies considerably, ranging from a limited function as a vehicle road speed governor to complex management of vehicle functions and engine fueling. Some partial authority management systems offer comprehensive onboard diagnostics, extensive customer data programming options, and proprietary data reprogramming capability. Examples of partial authority management systems are:

- Cat PEEC (3406) based on a port-helix injection pump
- Mack V-MAC I and V-MAC II based on a port-helix injection pump
- Cummins PT PACE/PACER on N14 based on PT pump fueling
- E-control of rotary distributor pumps: used exclusively on off-highway

Some partial authority systems appeared on off-highway diesel engines up until 2004 but none appear to be used on equipment certified for 2008 and beyond.

REVIEW QUESTIONS

1. The acronym that describes the internal computer component that executes program instructions is the:
   a. CRT
   b. RAM
   c. CPU
   d. ROM
2. Which of the following data retention media is electronically retained?
   a. RAM
   b. ROM
   c. PROM
   d. EEPROM
3. To which of the following memory categories would the master program for engine management be written in a typical truck engine ECM?
   a. RAM
   b. ROM
   c. PROM
   d. EEPROM
4. To which of the following memory categories would customer data programming be written from an electronic service tool?
   a. RAM
   b. ROM
   c. PROM
   d. EEPROM
5. Which of the following components conditions V-Ref?
   a. ECM
   b. Voltage regulator
   c. EUI
   d. Personality module
6. A thermistor is responsible for signaling data concerning:
   a. Pressure
   b. Temperature
   c. Rotational speed
   d. Rotational position
7. A Hall-effect sensor is usually responsible for signaling data concerning:
   a. Pressure
   b. Temperature
   c. Fluid flow
   d. Rotational or linear position
8. Induction pulse generators are used to input data concerning:
   a. Pressure
   b. Temperature
   c. Rotational speed
   d. Altitude
9. Which of the following components is uses a voltage divider principle and can be used to signal accelerator pedal travel?
   a. Pulse generator
   b. Potentiometer
   c. Thermistor
   d. Galvanic sensor

10. Which of the following input components would more likely use a variable capacitance-type sensor?
    a. CPS
    b. TRS
    c. VSS
    d. MAP
11. In a full authority diesel engine management system, EUI switching is performed by:
    a. A personality module
    b. Injector drivers
    c. V-Ref
    d. FIC module
12. *Technician A* states that a disadvantage of a piezo actuator is that it requires a much higher current draw than comparable solenoid actuators. *Technician B* states that an advantage of piezo actuators is that they are much less bulky than comparable solenoids. Who is correct?
    a. Technician A only
    b. Technician B only
    c. Both A and B
    d. Neither A nor B
13. *Technician A* states that Hall-effect TPSs use a noncontact principle that promises to make them more reliable than potentiometer types. *Technician B* states that a Hall-effect TPS does not require an input voltage to function. Who is correct?
    a. Technician A only
    b. Technician B only
    c. Both A and B
    d. Neither A nor B
14. *Technician A* states that V-Ref and the injector driver voltage are usually managed by the ECM to the same voltage values. *Technician B* states that V-Ref is controlled by the vehicle battery and is not conditioned by the ECM. Who is correct?
    a. Technician A only
    b. Technician B only
    c. Both A and B
    d. Neither A nor B
15. *Technician A* states that in some of the earliest engine management ECMs, the only method of changing customer options, such as idle shutdown duration and transmission ratios, was to replace the ECM PROM chip. *Technician B* states that in most current vehicle management systems, customer programmable options are written to ROM. Who is correct?
    a. Technician A only
    b. Technician B only
    c. Both A and B
    d. Neither A nor B
16. *Technician A* states that in most diesel engine ECMs, RAM data is lost whenever the computer is shut down. *Technician B* states that some systems write data to a special type of RAM that is nonvolatile because it connects directly to the battery. Who is correct?
    a. Technician A only
    b. Technician B only
    c. Both A and B
    d. Neither A nor B
17. *Technician A* states that engine overspeed rpm is usually programmed to the ECM as a hard value. *Technician B* states that the engine oil/coolant overheat temperature value was usually programmed to the ECM as a soft value because it would depend on the ambient temperatures. Who is correct?
    a. Technician A only
    b. Technician B only
    c. Both A and B
    d. Neither A nor B
18. *Technician A* states that it is not unusual for the vehicle maximum speed to be programmed at a lower speed than maximum cruise speed. *Technician B* states that because fuel economy is so important in trucking operations, soft cruise is a popular option. Who is correct?
    a. Technician A only
    b. Technician B only
    c. Both A and B
    d. Neither A nor B
19. Which of the following is the correct address for an engine ECM on a chassis data bus?
    a. MID 128
    b. MID 130
    c. MID 164
    d. MID 178
20. Which of the following customer data programming fields could be described as a toggle option?
    a. Road speed limit
    b. Transmission ratios
    c. Tire rolling radii
    d. Governor type

31

ESTs and SISs

INTRODUCTION

The acronym **EST (electronic service tool)** is generally used in the trucking industry to cover a range of electronic service instruments ranging from onboard diagnostic/malfunction lights to sophisticated computer-based communications equipment. The use of generic ESTs and procedures is reviewed in this chapter; proprietary ESTs are designed to work with an original equipment manufacturer's (OEM's) specific electronics and are not introduced in this text. The proprietary EST is rapidly becoming a thing of the past, with OEMs generally opting for low-cost, powerful personal computers (PCs) whose hardware and software can be cheaply upgraded.

Data Connector Standards

ESTs capable of reading engine control module (ECM) data are connected to the onboard electronics by means of Society of Automotive Engineers (SAE) J-standard data connectors:

- J1587/J1708 six-pin Deutsch
- J1939 keyed nine-pin Deutsch

Standardized connectors and the adherence by the engine electronics OEMs to SAE data bus software protocols enable proprietary software of one manufacturer to at least read the parameters and conditions of its competitors. This means that if a Cummins-powered truck has an electronic failure in a location where the only service dealer is Detroit Diesel Corporation (DDC), some first-level problem diagnosis can be undertaken using the DDC electronic diagnostic software. Standard data bus codes can also be read by a ProLink equipped with a general heavy-duty reader card. Data connector terminal assignments are standard and are indicated in Figures 31–1 and 31–2.

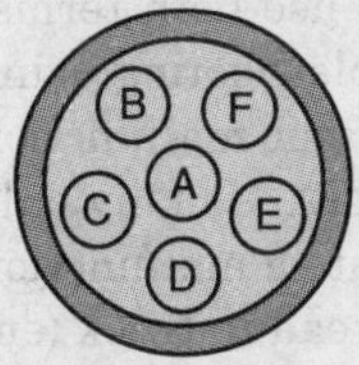

Figure 31–1 *J1587/1708 connector cavity pin assignments.*

MID Selection

Because there are usually a number of electronic systems networked to a truck data backbone, after making the EST connection to the data bus, the specific system a technician wants to work with has to be selected. Each major electronic system on the data bus is assigned a message identifier, or MID, and this must be selected before information in that system can be accessed. It actually sounds more complicated than it is. After connecting an EST to the data bus, all of the MIDs on that chassis will be displayed. If you want to work on the engine MID, use the EST to select this from the menu. The procedure for connecting to the data bus and the importance of communications adapters are covered later in this chapter.

For purposes of study, we divide ESTs into eight general categories first then study some of them in greater detail later in the chapter.

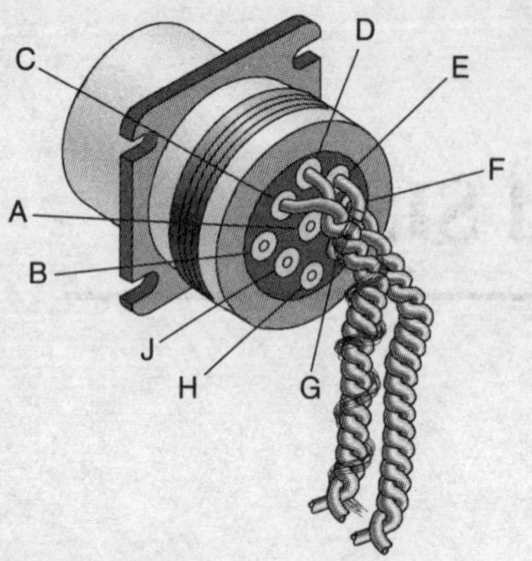

Cavity	Label
A	Battery ground
B	+12V DC
C	J1939 data link (+)
D	J1939 data link (–)
E	J1939 shield
F	J1587 data link (+)
G	J1587 data link (–)
H	Plug
J	Plug

Figure 31–2 *J1939 connector cavity pin assignments.*

Onboard Diagnostic Lights

Blink or **flash codes** (OEMs use both terms) are an onboard means of troubleshooting using a dash or ECM-mounted electronic malfunction light or check engine light (CEL). Usually only **active codes** (ones that indicate a malfunction at the time of reading) can be read in truck (engine) ECMs but some also read historic (inactive) codes.

Digital Multimeters

The **digital multimeter (DMM)** is an essential diagnostic instrument that should be part of every truck technician's tool chest. It is used to troubleshoot electronic circuits and test components. We take a close look at how to use a DMM in this chapter, including some information on what to look for when purchasing one.

Scanners

Scanners were originally read-only tools capable of reading active and **historic** or **inactive** (logged but not currently indicating a malfunction) **codes** and sometimes system parameters but little else. They are obsolete as a truck/bus diesel engine diagnostic tool, but the term is sometimes used to describe a reader-programmer. The term has lingered on in automotive technology and is often used to describe what we define as a *reader-programmer* in this book.

Graphing Meters

Graphing meters are designed to display waveforms produced by a variety of electrical signals. They are seldom used on the shop floor today because OEM diagnostic software will do the same job much more rapidly when it is important to interpret a waveform anomaly. However, they are useful instruments in a learning environment because they can help students create a visual image of an electrical waveform and background versus real time.

Generic Reader-Programmers

Generic **reader-programmers** are microcomputer-based ESTs designed to read and reprogram all proprietary systems in conjunction with the appropriate software cartridge. These are usually tough and portable. **ProLink 9000** became the industry standard and although its use is currently in decline the instrument may enjoy a resurgence when engine management diagnostics (EMD)/onboard diagnostics II (OBD II) come into effect for on-highway diesel engines.

Proprietary Reader-Programmers

These are usually PC-based test instruments packaged by the OEM for use exclusively on its system. They have the advantage of offering an optimum degree of user friendliness and the disadvantage of being system specific and high in cost. Proprietary reader-programmers still exist in truck technology such as in the example of Caterpillar's MultiTool but these tend to be limited in scope. Automotive OEMs continue to rely on this category of EST, partly as a means of limiting access to the chassis data bus and confining service diagnostics within an OEM dealer network.

Personal Digital Assistants

A handheld *personal digital assistant (PDA)* is one way of achieving potent computing capability along with good portability. Its use as an EST has increased recently and while functionality may be limited, it has proven itself as a good way of reading the chassis data bus during road testing and field service diagnosis.

Personal Computers (PCs)

All the truck engine OEMs have made the generic PC and its operating systems (Windows platform) their diagnostic and programming tool of choice. PCs are inexpensive, easily upgradable, and have vast computing power when comparedwithproprietary reader-programmers. OEMs can upgrade/update software more easily either by CD/DVD or online with access to their data hubs. PCs are connected to a vehicle ECM through the chassis data connector; a communications adapter or serial link is also required, depending on the system and the production year. Truck engine OEMs use MS Windows-driven graphical user interfaces (GUIs) to run their software on. All the OEMs offer comprehensive courses on their own management systems: Although navigating OEM software has become progressively more user friendly, technicians are advised to take OEM courses so they can get the best out of the system. Most of these courses can be taken online with OEM data hub access.

DIGITAL MULTIMETERS

A digital multimeter (DMM) is simply a tool for making electrical measurements. DMMs may have any number of special features but, essentially, they measure electrical pressure or volts, electrical current quantitative flow or amps, and electrical resistance or ohms. A good quality DMM with minimal features may be purchased for as little as $100; as the features, resolution, and display quality increase in sophistication, the price increases proportionally. A DMM is one of the most important tools in the technician's toolbox. Surprisingly, too many technicians do not understand how powerful a diagnostic instrument a good quality DMM can be. This section takes a look at DMMs with the objective of understanding how they function and what they can do for the technician.

Most electronic circuit testing *requires* the use of a DMM. Avoid using analog multimeters when attempting to diagnose circuit malfunctions: You may find analog multimeters in some older shop volts-amps testers (VATs). Reliability, accuracy, and ease of use are all factors that should be considered when selecting a DMM for purchase. Some options that technicians may wish to consider are a protective rubber holster (this will greatly extend the life of the instrument!), analog bar graphs, and enhanced resolution. This section deals with the practice of using DMMs, and a knowledge of basic electricity is assumed.

DMM Subcomponents

Figure 31–3 shows the location of the various features and subcomponents on a typical DMM. When learning how to get the best use of a DMM, you should first understand the function of each feature.

Input Terminals

The input terminals are located at the base of most DMMs as shown in Figure 31–4. The DMM shown has four input terminals into which lead jacks are inserted. The black lead is used for *all* DMM measurements and is inserted into the common jack. To make voltage

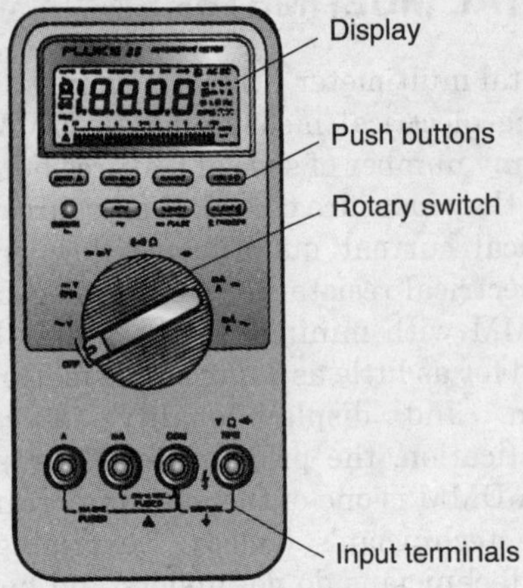

Figure 31–3 *Fluke 88 DMM. (Courtesy of International Navistar Corp.)*

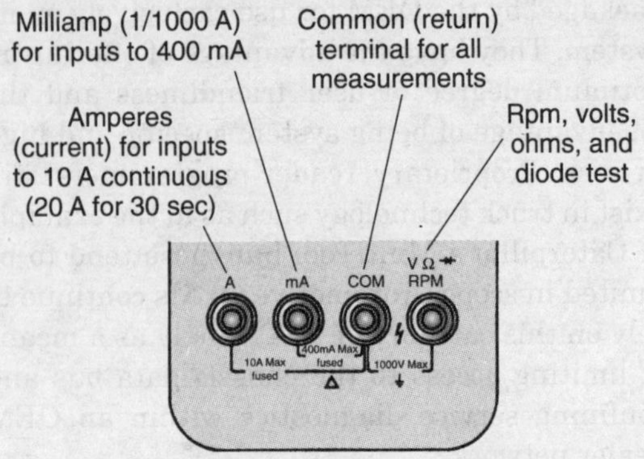

Figure 31–4 *DMM input terminals. (Courtesy of International Navistar Corp.)*

and resistance measurements, select the common and volts/rpm/ohms/diode jacks and insert the red lead into the volts/rpm/ohms/diode jack. The two jacks to the left of these are used only for series current measurements and these are self-explanatory: one for milliamps, the other for amps. Remember that mixing leads up can blow the internal DMM fuses. These fuses are expensive so blowing a couple after purchasing a DMM can be regarded as a learning experience, and after that as stupidity.

Rotary Switch

The DMM rotary switch, shown in Figure 31–5, is used to set the DMM function. The DMM can make alternating current (AC) and direct current (DC) voltage measurements measure circuit resistance and test diodes and make series amperage measurements. Most current DMMs will autorange, that is, determine what the measurement value is and select the appropriate display scale. Technicians working daily with DMMs should only consider units with autoranging capability.

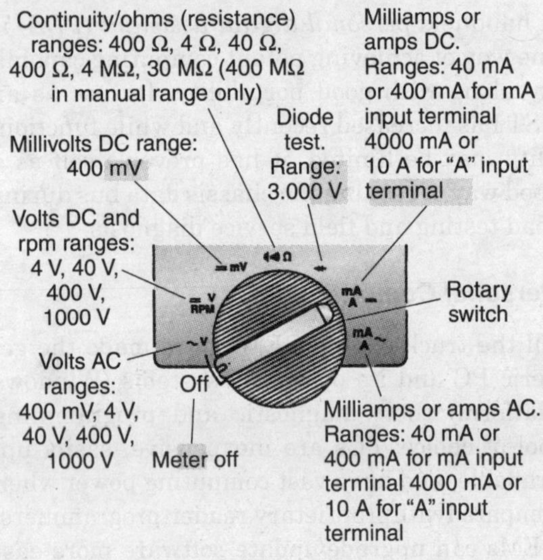

Figure 31–5 *DMM rotary switch. (Courtesy of International Navistar Corp.)*

Push Buttons

The push buttons on a DMM are used to select or fine-tune meter operations. On the DMM shown, when a push button is depressed a symbol is displayed on the data display, and in some models a beeper will sound. Figure 31–6 shows the push buttons used on a high-end Fluke DMM. In most cases, when the rotary switch setting is changed, the push-button settings will revert to default status.

Data Display

The data display is the means used to read the measurements made by the DMM. Figure 31–7 shows the data display used on a high-end DMM. Many current DMMs have the ability to display both digital and analog readings. The analog display is simply an analog representation of the digital reading made by the meter, but it is often easier to interpret. The digital display should be used for stable input values, while the analog display should be used

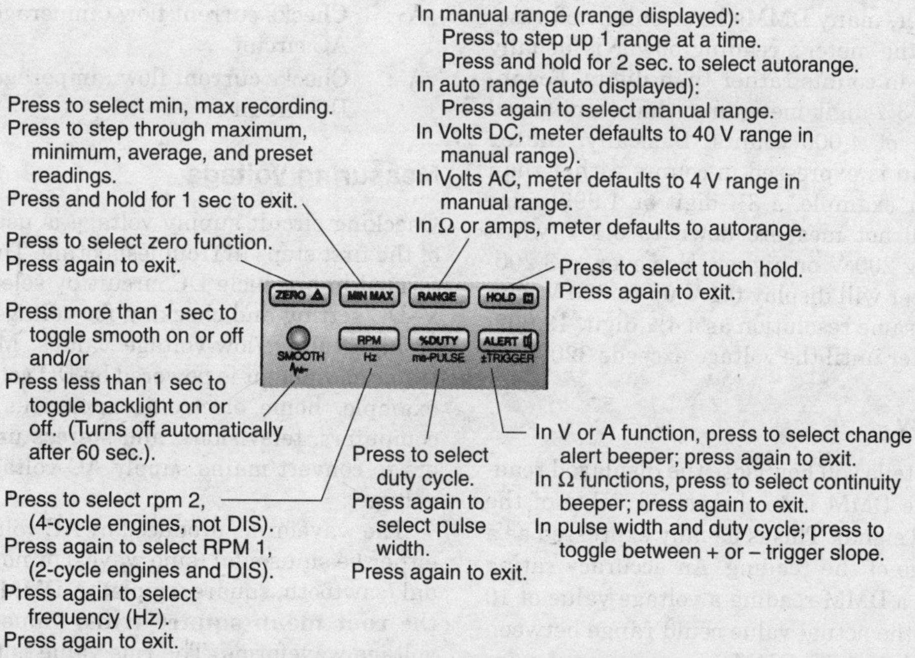

Figure 31–6 *DMM push buttons. (Courtesy of International Navistar Corp.)*

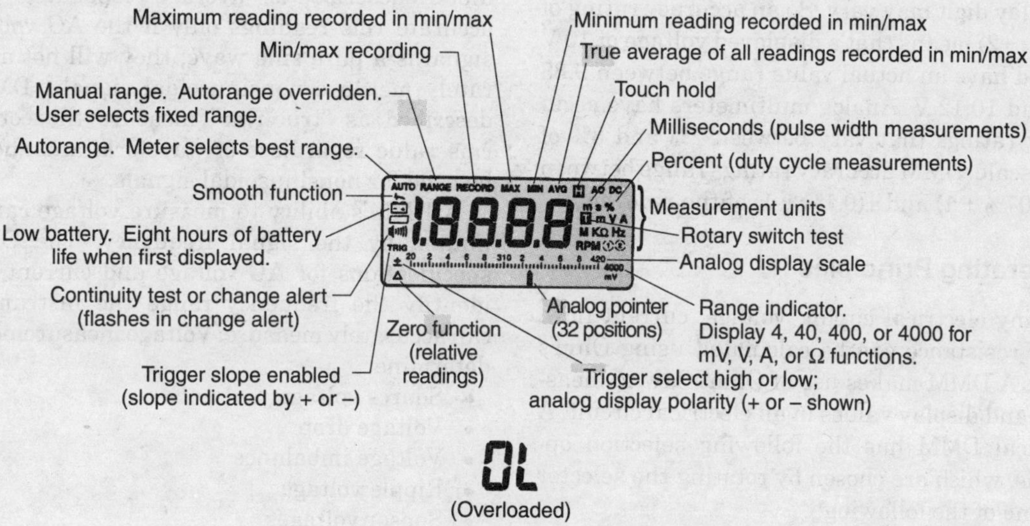

Figure 31–7 *DMM data display. (Courtesy of International Navistar Corp.)*

for frequently changing inputs. The acronym OL used in the display means overloaded or out of range.

Resolution

Resolution specifications are often misunderstood. They tell you how fine a measurement can be made with the instrument. Digits and counts are used to describe the resolution capability of a DMM. A 3½ digit meter can display three full digits ranging from 0 to 9 and one-half digit that displays either 1 or is left blank. A 3½ digit meter therefore displays 1,999 counts of resolution. A 4½ digit DMM can display up to 19,999 counts of resolution.

However, many DMMs have enhanced resolution so the meter's reading power is usually expressed in counts rather than digits. For instance, a 3½-digit meter may have enhanced resolution of 4,000 counts. Basically, **meter resolution** is expressed in counts rather than digits. For example, a 3½-digit or 1,999-count meter will not measure down to 0.1 V when measuring 200 V or higher. However, a 3,200-count meter will display 0.1 V up to 320 V, giving it the same resolution as a 4½-digit, 19,999-count meter until the voltage exceeds 320 V.

Accuracy

Accuracy tells you how close the displayed reading on the DMM is to the actual value of the measured signal. This is usually expressed as a percentage of the reading. An accuracy rating of ±1% in a DMM reading a voltage value of 10 V means the actual value could range between 9.9 V and 10.1 V. DMM accuracy can be extended by indicating how many counts the right display digit may vary. So an accuracy rating of ±(1% +2) means that a displayed voltage of 10 V could have an actual value range between 9.88 V and 10.12 V. Analog multimeters have accuracy ratings that vary between 2% and 3% of full scale. DMM accuracy ratings range between ±(0.07% + 1) and ±(0.1% + 1) of the *reading*.

Operating Principles

In any electrical circuit, voltage, current flow, and resistance can be calculated using Ohm's law. A DMM makes use of Ohm's law to measure and display values in an electrical circuit. A typical DMM has the following selection options, which are chosen by rotating the selector to one of the following:

off: Shuts down the DMM

V~: Enables AC voltage readings

V–: Enables DC voltage readings

mV-: Enables low-pressure DC voltage readings

Ω: Enables component or circuit resistance readings

α: Enables **continuity** (a circuit capable of being closed) testing. Identifies an open/closed circuit

A~: Checks current flow (amperage) in an AC circuit

A–: Checks current flow (amperage) in a DC circuit

Measuring Voltage

Checking circuit supply voltage is usually one of the first steps in troubleshooting. This is performed in a vehicle DC circuit by selecting the V–DC setting and checking for voltage present or high-voltage/low-voltage values. Most electronic equipment is powered by DC voltage. For example, home electronic apparatus such as computers, televisions, and stereos use rectifiers to convert mains supply AC voltage to DC voltage.

The waveforms produced by AC voltages can either be sinusoidal (sine waves) or nonsinusoidal (sawtooth, square, ripple). A DMM displays the **root mean square (rms)** value of these voltage waveforms. The rms value is the effective or equivalent DC value of the AC voltage. Meters described as "average responding" give accurate rms readings only if the AC voltage signal is a pure sine wave; they will not accurately measure nonsinusoidal signals. DMMs described as "true-rms" measure the correct rms value regardless of waveform and should be used for nonsinusoidal signals.

A DMM's ability to measure voltage can bc limited by the signal frequency. The DMM specifications for AC voltage and current will identify the frequency range the instrument can accurately measure. Voltage measurements determine:

- Source voltage
- Voltage drop
- Voltage imbalance
- Ripple voltage
- Sensor voltages

Figure 31–8 shows a DMM set up to read DC voltage.

Measuring Resistance

Most DMMs measure resistance values as low as 0.1 Ω and some measure high-resistance values up to 300 MΩ (megaohms). Infinite resistance or resistance greater than the instrument can measure is indicated as "OL" or flashing digits on the display. Resistance measurements

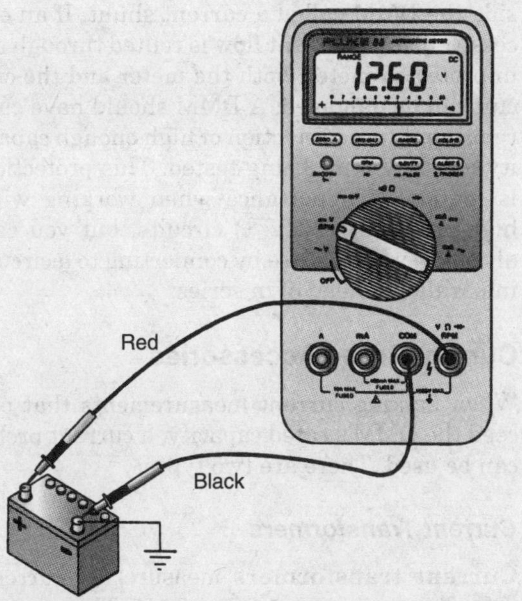

Figure 31–8 *DMM Set up for making DC voltage measurements. (Courtesy of International Navistar Corp.)*

test the continuity of a circuit when it is not energized. Resistance tests will identify an **open circuit** (one in which there is no path for current flow and would read "OL" on the display) and high circuit resistance.

Resistance and continuity measurements should only be made on unenergized circuits (Figure 31–9). Using the resistance or continuity settings to check a circuit or component that is energized will result in damage to the test instrument. Some DMMs are protected against such "accidental" abuse and the extent of damage will depend on the model. For accurate low-resistance measurement, test lead resistance, typically between 0.2 Ω and 0.5 Ω depending on quality and length, must be subtracted from the display reading. Test lead resistance should never exceed 1 Ω. Some DMMs have <zero> function that factors test lead resistance to produce a more accurate reading.

If a DMM supplies less than 0.3 V-DC test voltage for measuring resistance, it is capable of testing resistors isolated in a circuit by diodes or semiconductor junctions, meaning that they do not have to be removed from the circuit board. Resistance measurements determine:

- Resistance of a load
- Resistance of conductors

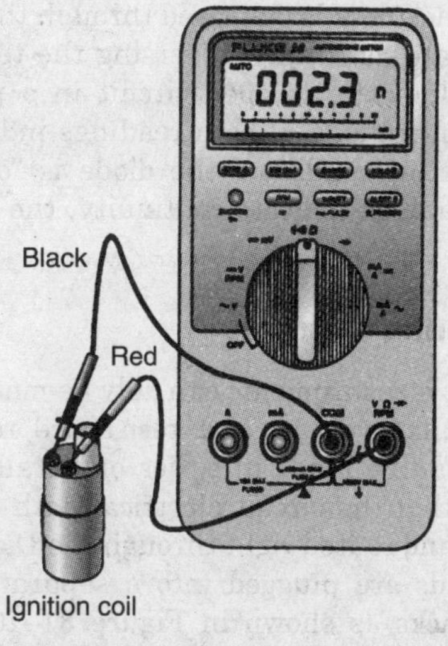

Figure 31–9 *DMM set up for measuring resistance. (Courtesy of International Navistar Corp.)*

- Value of a resistor
- Operation of variable resistors

Continuity

Continuity is a quick resistance check that distinguishes between an open and a closed circuit. Most DMMs have audible continuity beepers that beep when they detect a closed circuit, permitting the test to be performed without looking at the meter display. The actual level of resistance required to trigger the beeper varies from model to model. Continuity tests determine:

- Fuse integrity
- Open conductors
- Shorted conductors
- Switch operation
- Circuit paths

Diode Testing

A diode is an electronic switch that can conduct electricity in one direction while blocking current flow in the opposite direction. It functions like a one-way electrical check valve. Diodes are commonly enclosed in glass cylinders; a dark band identifies the cathode or blocking terminal. Current flows when the anode is more positive than the cathode. Additionally, a diode will not conduct until the forward voltage pressure reaches a certain value, which is 0.3 V in a silicon diode. Some meters have a diode test mode. When testing a diode with the DMM in this mode, 0.6 V is delivered through the device to indicate continuity; reversing the test leads should indicate an open circuit in a properly functioning diode. If both readings indicate an open circuit condition, the diode is "open." If both readings indicate continuity, the diode is shorted.

Measuring Current

Current measurements can only be made in series, unlike voltage and resistance readings, which can be made in series or parallel. This means that the circuit electrical path is interrupted and routed right through the DMM. The test leads are plugged into a separate set of input jacks as shown in Figure 31–10. Make sure you have an approximate idea of the current you are testing. Get it wrong, and you will be replacing fuses in your DMM. Current measurements determine:

- Circuit overloads
- Circuit operating current
- Current in different branches of a circuit

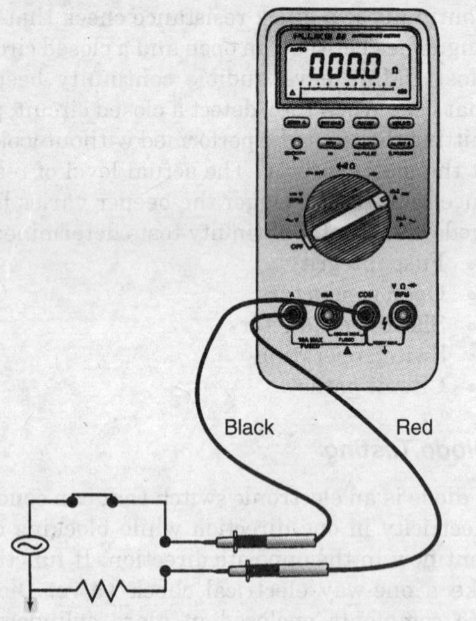

Figure 31–10 *DMM set up for measuring current flow. (Courtesy of International Navistar Corp.)*

When the test leads are plugged into the current input jacks and they are used to measure voltage, this causes a direct short across the source voltage through a low value resistor inside the DMM called a current shunt. If an excessively high current flow is routed through an unprotected meter, both the meter and the circuit can be damaged. A DMM should have current input fuse protection of high enough capacity for the circuit being tested. This protection is mainly of importance when working with high-pressure (220 V +) circuits, but you can also blow a DMM fuse by connecting to a circuit in parallel instead of in series.

Current Probe Accessories

When making current measurements that exceed the DMM's rated capacity, a current probe can be used. There are two types:

Current Transformers

Current transformers measure AC current only. Their output is 1 mA per 1 A. Therefore, a current flow of 100 A is reduced to 100 mA, which can be handled by most DMMs. The test leads would be connected to the mA and common input jacks and the meter function switch set to "mA AC." This is not a very accurate circuit test and is used for ballpark reckoning only. There are no applications for this tool in truck electrical systems.

Hall-Effect Probe

The output of a **Hall-effect probe** is 1 mV per ampere. It measures AC or DC. The test leads are connected to the V and common input jacks. The DMM function switch should be set to the V or mV scale, selecting V-AC for AC current and V-DC for DC current measurements; this is not especially accurate. The Hall-effect probe is commonly used to measure cranking motor current draw.

DMM Features

When considering a DMM for purchase, the following features should be considered:

- Fused current inputs
- Use of high potential fuses (600 V +)
- High-voltage protection in resistance mode
- Protection against high-voltage transients
- High input impedance (greater than 10 M)
- Insulated test lead handles
- Ammeter required

- Autorange capability
- Cost of replacement fuses

Some Typical DMM Tests

Always perform tests in accordance with truck and bus OEM specifications; never jump sequence or skip steps in sequential troubleshooting charts. Most DMM tests on truck and bus electronic systems will be used in conjunction with a generic reader-programmer or PC. The following tests assume the use of a Fluke 88 DMM.

Engine Position (Fuel Injection Pump Camshaft), Cam, and Crank Position Sensors

This procedure tests Hall-effect sensors so make sure to correctly identify the sensor (it could be an inductive pulse generator):

a. Cycle the ignition key, then off.
b. Switch the meter to measure V-DC/rpm.
c. Identify the ground and signal terminals at the Hall sensor. Connect the positive (+) test lead to the signal terminal and the negative (–) test lead to the ground terminal. Crank the engine. At cranking speeds, the analog bar graph should pulse; at idle speeds or above, the pulses are too fast for bar graph readout.
d. Press the duty cycle button once. Duty cycle can indicate square wave quality with poor quality signals having a low duty cycle. Functioning Hall sensors should have a duty cycle of around 50% depending on the sensor. Check to specifications.

Potentiometer-Type TPS

Resistance test:

a. Key off.
b. Disconnect the throttle position sensor (TPS).
c. Select Ω on the DMM. Connect the test probes to the signal and ground terminals. Next, move the accelerator through its stroke while observing the DMM display.
d. The analog bar should move smoothly without jumps or steps. If it steps, there may be a bad spot in the sensor.

Voltage test:

a. Key on, engine off.
b. Set the meter to read V-DC. Connect the negative lead to ground.
c. With the positive lead, check the reference voltage value and compare to specifications. Next, check the signal voltage (to the ECM) value through the accelerator pedal stroke. Check values to specification. Also observe the analog pointer; as with the resistance test, this should move smoothly through the accelerator stroke.

Inductive Pulse Generator Sensors

Inductive pulse generator-type sensors are often used as shaft speed and position sensors. They may also be known as magnetic or variable reluctance sensors function. They function similarly to a magneto. A toothed wheel cuts through a magnetic field, which induces an AC signal voltage in the signal (output) terminal: The ECM reads the frequency of the AC signal to determine a shaft speed. The AC voltage rises proportionally with rotational speed increase and range from 0.1 V up to 5.0 V-AC. Vehicle speed sensors (VSS), engine speed sensors (ESS), and antilock brake system (ABS) wheel speed sensors all use this method of determining rotational speed. Test using V-AC switch setting and locating test leads across the appropriate terminals: Then spin up the device.

Min/Max Average Test for Lambda (λ) (O_2) Sensors

a. Key on, engine running, DMM set at V-DC. Select the correct voltage range.
b. Connect the negative test lead to a chassis ground and the positive test lead to the signal wire from the lambda sensor. Press the DMM min/max button.
c. Ensure that the engine is warm enough to be in closed loop mode (100 mV–900 mV oxygen [O_2] sensor output). Run for several minutes to give the meter time to sample a scatter of readings.
d. Press the min/max button slowly three times while watching the DMM display. A maximum of 800 mV and a minimum of less than 200 mV should be observed. The average should be around 450 mV.

e. Next, disconnect a large vacuum hose to simulate a lean burn condition. Repeat steps c and d to read the average voltage. Average voltage should be lower, indicating a lean condition.
f. The same test can be performed using propane enrichment to produce a rich AFR condition and therefore higher voltage values.
g. Lambda sensor tests can be performed while road testing the vehicle; 450 mV normally indicates stoichiometric fueling ($\lambda = 1$) but check to specifications.

Thermistors

Most **thermistors** used in computerized engine systems are supplied with V-Ref (5 V) and have a negative temperature coefficient (NTC), meaning that as the sensor temperature increases, its resistance decreases. They should be checked to specifications using the DMM ohmmeter function and an accurate temperature measurement instrument.

Applying DMM Tests

OEMs seldom suggest random testing of suspect components. The preceding tests are typical procedures. Circuit testing in today's computerized engine management systems is highly structured and part of sequential troubleshooting procedure. It is important to perform each step in the sequence precisely; skipping a step can invalidate every step that follows.

Breakout Boxes and Breakout Ts

The DMM is often used in conjunction with a **breakout box** or **breakout T**. Breakout devices are designed to be teed into an electrical circuit to enable circuit measurements to be made on both closed (active) and de-energized circuits. The objective is to access a circuit with a test instrument without interrupting the circuit. A breakout T is a simpler diagnostic device that is inserted into a simple two- or three-wire circuit such as that used to connect an individual sensor, while a breakout box accesses multiple wire circuits for diagnostic analyses of circuit conditions.

Most of the electronic engine management system OEMs use a breakout box that is often inserted into the interface connection between the engine electronics and chassis electronics harnesses. The face of the breakout box displays a number of coded sockets into which the probes of a DMM can be safely inserted to read circuit conditions. Electronic troubleshooting sequencing is often structured based on the data read by a DMM accessing a circuit. A primary advantage of breakout diagnostic devices is the fact that they permit the reading of an active electronic circuit, for instance, while an engine is running.

Connector Dummies and Spoons

Diagnostic connector dummies are used to read a set of circuit conditions in a circuit that has been opened by separating a pair of connectors. The dummies are manufactured by the electrical/electronic connector manufacturer as a means of accessing the circuitry with a DMM without damaging the connector sockets and pins. Diagnostic spoons are curved and designed to be inserted behind a connector without damaging the insulation boot. When the correct dummies and spoons are used, energized circuits can be tested without inflicting damage providing some care is exercised.

DMM Circuit Testing

Many truck technicians learn to use DMMs on-the-fly, that is, they acquire their understanding on an as-needed basis, with the consequence that most do not properly understand the potential of a high-end DMM. It does not make sense to spend $500 on a complex DMM that will never be used to perform tasks that a DMM one-fifth of the cost will handle. This next section introduces some simple tasks that can be performed by low-end DMMs using components readily available in introductory electricity/electronics labs.

Testing Resistors

Most carbon resistors used in electronic circuits are color coded. Although individual resistors would seldom be tested in the real world of the truck shop, testing them in a teaching/learning environment can be a great way of getting to know the ohmmeter function of your DMM. You can grab a pile of resistors and check the color-coded resistance value against the actual measurement. First you have to understand the color codes:

Color	*Value*
Black	0
Brown	1
Red	2
Orange	3
Yellow	4
Green	5
Blue	6
Violet	7
Gray	8
White	9

In the above list, note that the darkest colors like black and brown are used for the lowest numeric values (0 and 9), moving up to white, which is given the highest numeric value. These codes are standardized by the Electronics Industries Association (EIA) and are also used for capacitors. In addition, the colors gold and silver are used to rate tolerance or the percentage amount that similarly rated resistors can differ from each other and still function within specification. Gold indicates 5% and silver 10% tolerance. Figure 31–11 shows the significance of each band used on a resistor.

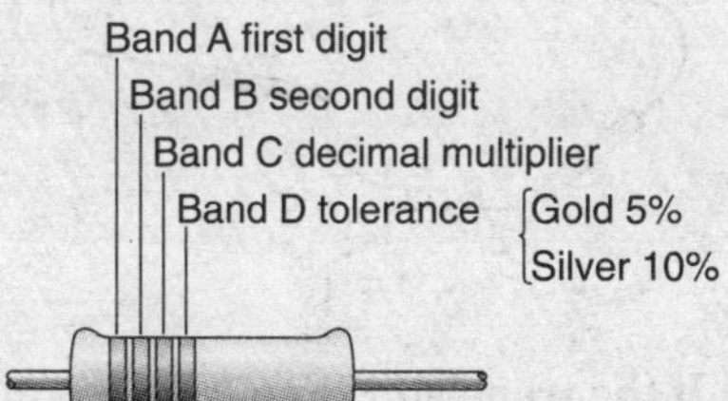

Figure 33–11 *Interpreting bands on a carbon resistor.*

Next, we can look at two actual examples shown in Figure 31–12. In Figure 31–12A the arrangement is:

1st stripe = red = 2
2nd stripe = green = 5
3rd stripe = red = multiplier to power of 2 or 10^2, so you *add* 2 zeros = 2,500 Ω

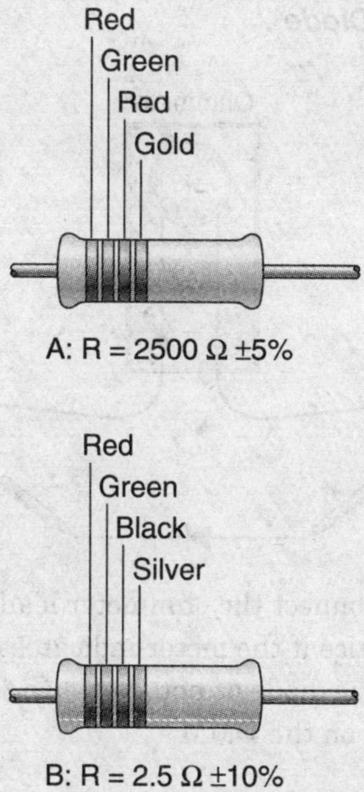

Figure 31–12 *Interpreting band resistance codes.*

4th stripe = gold = tolerance rated at within 5%

At first glance, Figure 31–12B does not appear to be unlike the previous resistor:

1st stripe = red = 2
2nd stripe = green = 5
3rd stripe = black = multiplier is zero so the value is 25 Ω
4th stripe = silver = tolerance rated at within 10%

Testing Semiconductors

Figure 31–13 and Figure 31–14 outline some methods of testing semiconductors that can be helpful in learning how to use a DMM to test circuit components and to understand how diodes and transistors operate.

GENERIC READER-PROGRAMMER

The ProLink 9000 is an example of a generic reader-programmer. This handheld tool was, for a number of years, the industry standard, electronic service tool capable of reading most current systems with the correct software car-

Testing a Diode

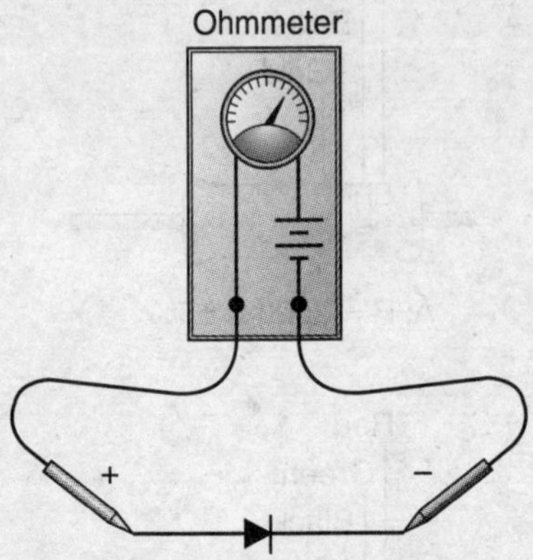

Step 1. Connect the ohmmeter leads to the diode. Notice if the meter indicates continuity through the diode or not. Now use the diode test mode on the DMM.

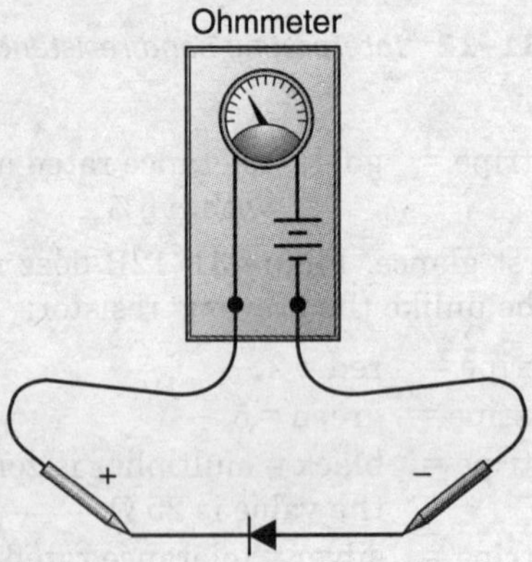

Step 2. Reverse the diode connection to the ohmmeter. Notice if the meter indicates continuity through the diode; the ohmmeter should indicate continuity through the diode in only one direction. (Note: If continuity is not indicated in either direction, the diode is open. If continuity is indicated in both directions, the diode is shorted.)

Figure 31–13 *Testing diodes. (Courtesy of Navistar International Corp.)*

Testing a Transistor

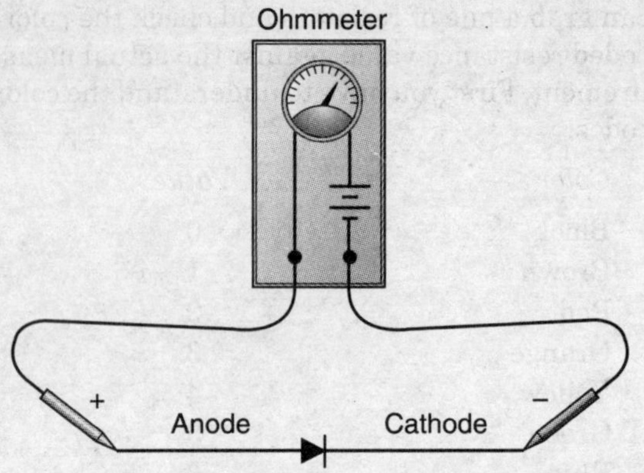

Step 1. Using a diode, determine which ohmmeter lead is positive and which is negative. The ohmmeter will indicate continuity through the diode only when the positive lead is connected to the anode of the diode and the negative lead is connected to the cathode.

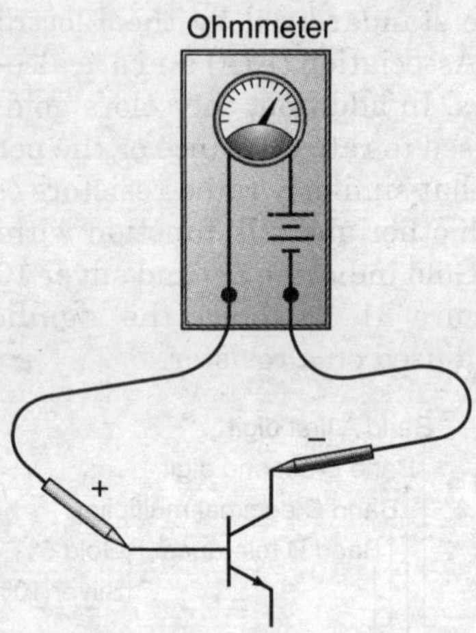

Step 2. If the transistor is an NPN, connect the positive ohmmeter lead to the base and the negative lead to the collector. The ohmmeter should indicate continuity. The reading should be about the same as the reading obtained when the diode was tested.

Figure 31–14 *Testing transistors. (Courtesy of Navistar International Corp.)*

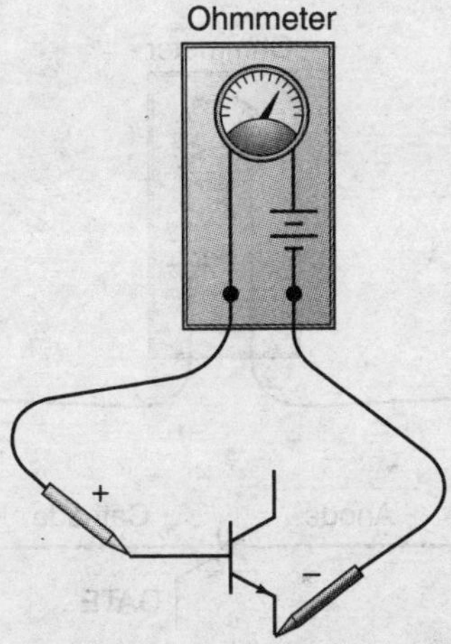

Step 3. With the positive ohmmeter lead still connected to the base of the transistor, connect the negative lead to the emitter. The ohmmeter should again indicate a forward diode junction. (Note: If the ohmmeter does not indicate continuity between the base-collector or the base-emitter, the transistor is open.)

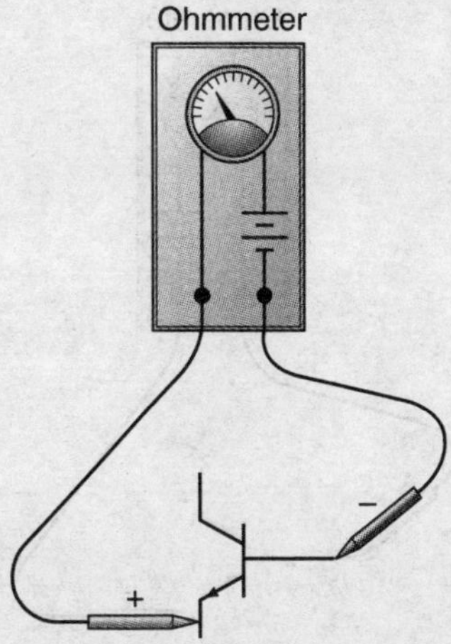

Step 5. With the negative ohmmeter lead connected to the base, reconnect the positive lead to the emitter. There should again be no indication of continuity. (Note: If a very high resistance is indicated by the ohmmeter, the transistor is "leaky" but may still operate in the circuit. If a very low resistance is seen, the transistor is shorted.)

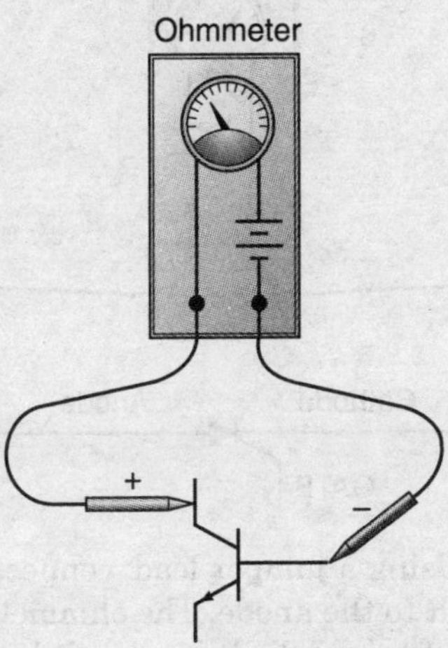

Step 4. Connect the negative ohmmeter lead to the base and the positive lead to the collector. The ohmmeter should indicate infinity or no continuity.

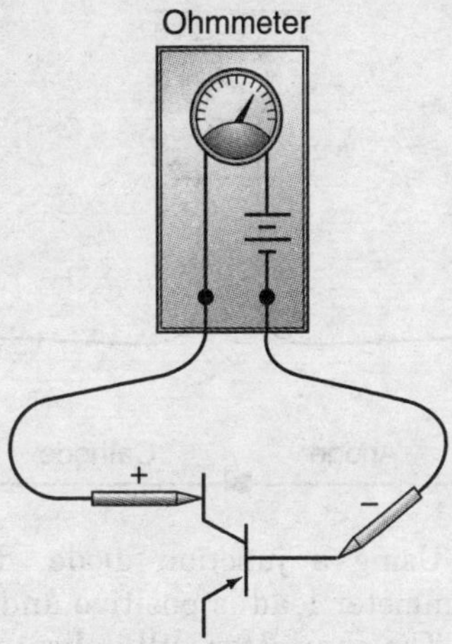

Step 6. To test the PNP transistor, reverse the polarity of the ohmmeter leads and repeat the test. When the negative ohmmeter lead is connected to the base, a forward diode junction should be indicated when the positive lead is connected to the collector or emitter.

Figure 31–14 *(Continued)*

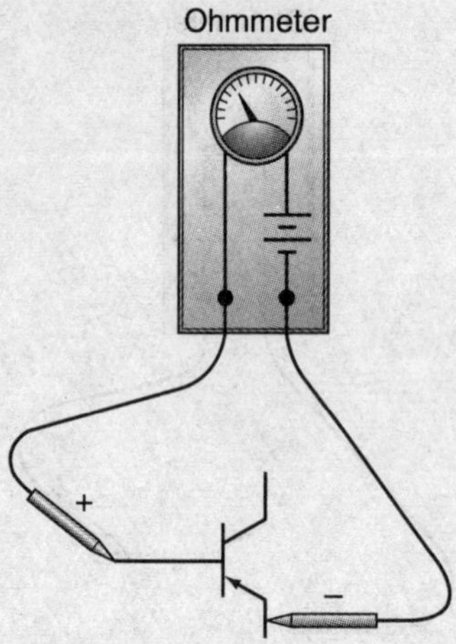

Step 7. If the positive ohmmeter lead is connected to the base of a PNP transistor, no continuity should be indicated when the negative lead is connected to the collector or the emitter.

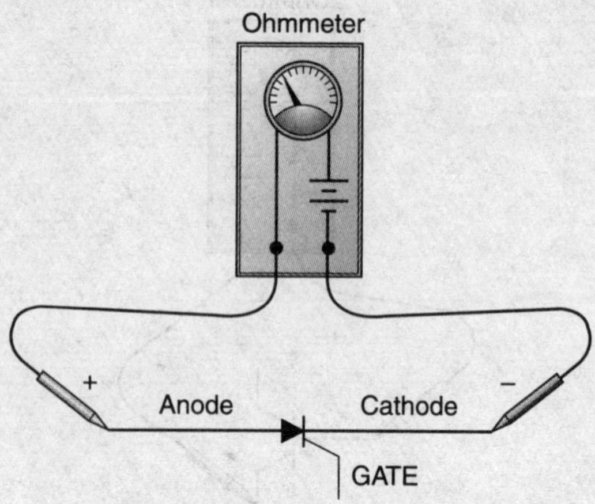

Step 2. Connect the positive ohmmeter lead to the anode of the SCR and the negative lead to the cathode. The ohmmeter should indicate no continuity.

Testing an SCR

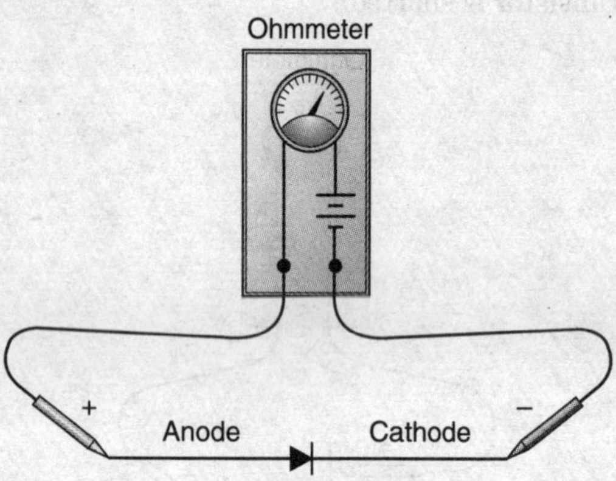

Step 1. Using a junction diode, determine which ohmmeter lead is positive and which is negative. The ohmmeter will indicate continuity only when the positive lead is connected to the anode of the diode and the negative lead is connected to the cathode.

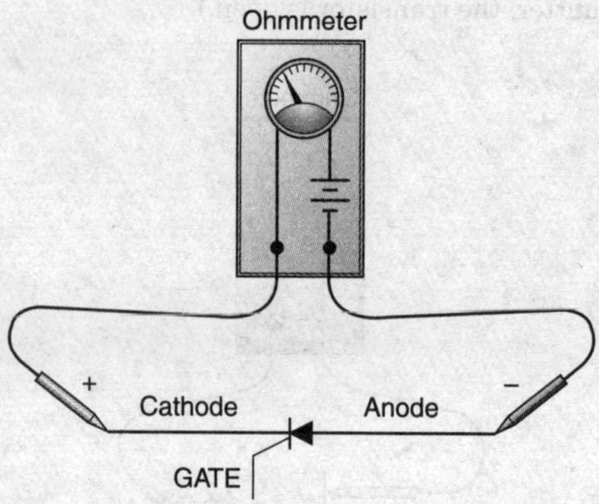

Step 3. Using a jumper lead, connect the gate of the SCR to the anode. The ohmmeter should indicate a forward diode junction when the connection is made. (Note: If the jumper is removed, the SCR may continue to connect, or it may turn off. This will be determined by whether the ohmmeter can supply enough current to keep the SCR above its holding current.)

Figure 31–14 *(Continued)*

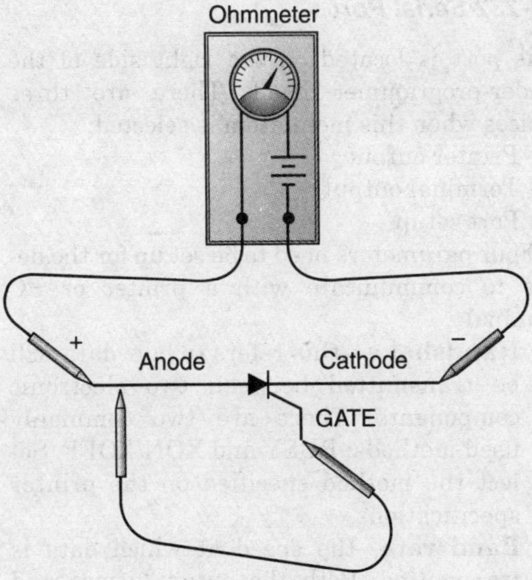

Step 4. Reconnect the SCR so that the cathode is connected to the positive ohmmeter lead and the anode is connected to the negative lead. The ohmmeter should indicate no continuity.

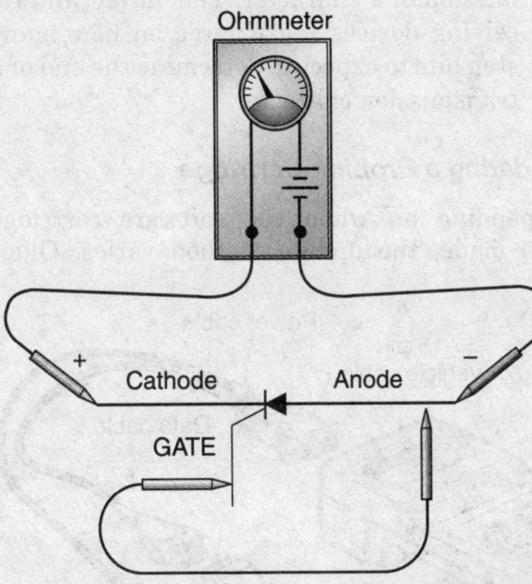

Step 5. If a jumper is used to connect the gate to the anode, the ohmmeter should indicate no continuity. (Note: SCRs designed to switch large currents [50 amperes or more] may indicate some leakage current with this test. This is normal for some devices.)

Figure 31–14 *(Continued)*

tridge or data card. ProLink 9000 reader-programmers survive today, though they are somewhat limited when it comes to working with the complex management systems we use today. These microprocessor-based test instruments will do the following (depending on the system):

- Access active and inactive/historic codes; erase **inactive historic** codes
- View all system identification data
- View data on engine operation with the engine running
- Perform diagnostic tests on system subcomponents such as EUI testing
- Reprogram customer data parameters on engine and chassis systems
- Act as a serial link to connect the vehicle ECM via modem to a centrally located mainframe for proprietary data programming (some systems only)
- Snapshot system data parameters to assist intermittent fault finding solutions

Advantages and Disadvantages

ProLink 9000 has some notable advantages:

- User friendly. Technicians require very little time to develop a working knowledge of these tools. The best method of becoming familiar with them is simply to use them. Menus can be selected, data can be scrolled through. Stuck? Simply unplug and reconnect. This can be done at any time without sustaining physical or electronic damage.
- Universality. These tools were for a while the industry standard and can be used extensively even on systems that require the use of PCs for more detailed diagnostic and programming functions. Used with a heavy-duty reader (cartridge or card) they will read most of diesel engine and other onboard systems along with automobile systems.
- Easily updated. Cartridges are updated simply by opening them up, removing and discarding out-of-date programmable read-only memory (PROM) chips or cards, and inserting new one(s). At PROM chip replacement, the random access memory (RAM) chip may have to be momentarily disconnected from the motherboard to

dump its nonvolatile content. The newer multiprotocol cartridges (MPC) are updated by data card.

- Tough. The tough plastic housing really does withstand the rough treatment some of these instruments are likely to receive in the typical service facility.

The disadvantages of generic reader-programmers are:

- Display window is limited to four lines.
- Impossible to input alpha passwords (some systems) or long-winded procedure of inputting an alpha password on the numeric keypad.
- As system management systems become more complex, the limitations of ProLink 9000 become more evident. PC-based systems are more often required to properly troubleshoot electronic problems.

Generic reader-programmers are connected to the chassis data bus by means of the data connector. Both six- and nine-pine Deutsch connectors may be used as shown in Figure 31–15.

Multiprotocol Cartridge

More recent versions of ProLink use an MPC. The MPC is designed to accept a credit card size data card with the requisite OEM or Heavy Duty Standard system software. This makes the MPC-enabled ProLink easier to use and easier to update (replace the data card) though the cost has remained unchanged despite the increased simplicity.

RS-232 Serial Port

This port is located on the right side of the reader-programmer head. There are three choices when this menu item is selected:

1. Printer output
2. Terminal output
3. Port setup

Four parameters need to be set up for the device to communicate with a printer or PC terminal:

1. **Handshake**—this refers to how data will be transmitted between two electronic components. There are two commonly used methods: BUSY and XON/XOFF. Select the method specified on the printer specifications.
2. **Baud rate**—the speed at which data is transmitted. Both the output device and the receiving device must agree on transmission speed.
3. **Parity**—this is the even or odd quality of the number of 0s and 1s. The menu options are none, odd, or even.
4. **Stop bits**—the last element in the transmission of a character. The output and receiving devices must agree on how many stop bits to expect to determine the end of a transmission character.

Updating a ProLink Cartridge

Depending on when the software cartridge was made, the update method varies. Older

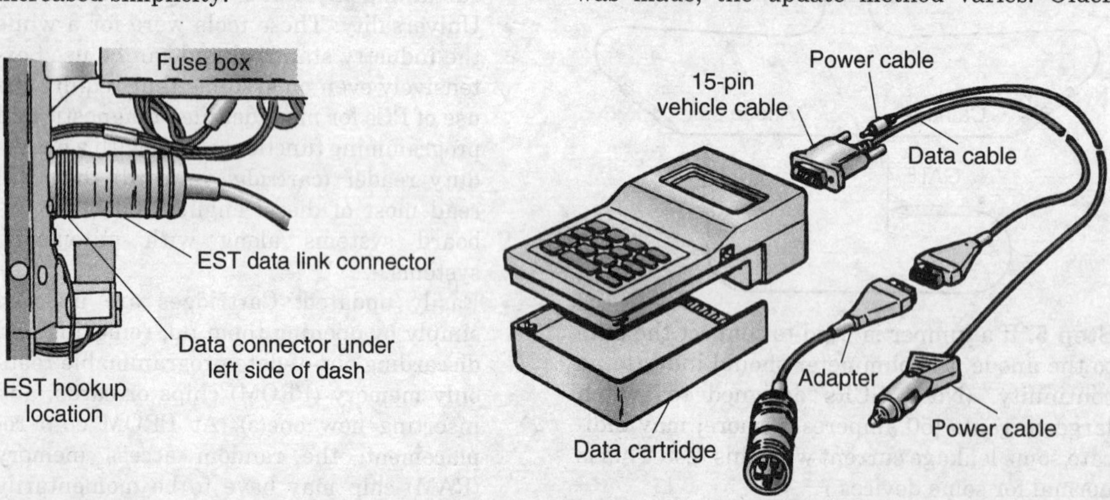

Figure 31–15 *ProLink data connection hardware. (Courtesy of International Navistar Corp.)*

cartridges are updated by replacing the PROM chip or chips using the following procedure:

1. Ensure that during this procedure, there is no danger of a static charge being unloaded into the cartridge; wear a **ground strap**.
2. Remove the cartridge from the ProLink 9000 reader-programmer. Remove the four screws fastening the cartridge cover plate, then separate the motherboard from the cartridge casing. Place on a clean wooden bench.
3. Using a chip lever, carefully remove the PROM chip(s) (one or two depending on the system) from the motherboard. Using the schematic in the update chips packaging, install the updated chips. Special care is required to avoid bending/damaging the connecting prongs of the new chips when installing them. Next, it may be required to temporarily remove the RAM chip in the same manner as the PROM chips. This should be momentarily separated from the motherboard (there is a nonvolatile memory segment in some systems) and then carefully reinstalled. On other systems, removing the battery temporarily from the motherboard will accomplish the same objective. Ensure that all the chips are fully installed in their sockets.
4. Reassemble the cartridge. Note the date of the PROM transfer and the software version somewhere on the cartridge exterior. Then install in the ProLink 9000 tool.
5. Power up the ProLink by connecting to an appropriate vehicle ECM. Variously scroll through the menus to ensure everything is working.

More recent ProLink cartridges are updated by replacing the data card, a credit card-sized data retention card that is inserted into a slot in the MCP cartridge.

Communications Adapters

Most ESTs designed to access data on a truck chassis data bus require the use of a communications adapter (CA), which is also known as a serial data link transducer. Although all CAs meet the same TMC RP 1210A communications protocols, most are electronically embedded to only function with one specific system. This means that a general service garage may be required to have a dozen or more of these nearly identical CAs.

Establishing the Connection

For purposes of illustrating this simple procedure, we are going to use the example of connecting to a chassis equipped with a Caterpillar electronic engine. We will also assume that the data bus connector is J1939 and there are multiple ECMs networked to the chassis data backbone, and your objective is to communicate with the engine ECM. The hardware required for this operation consists of:

- A PC loaded with Cat Electronic Technician (ET) software
- A Cat communications adapter (CA) meeting 171- 4401 or 275-5121 Cat and RP 1210A compatibilities
- Appropriate connection cables and a J1939 data connector

The connections should be made as indicated in Figure 31–16.

Use the foregoing sequential procedure to make and verify the CA connection.

1. Turn the dash keyswitch *off*. Boot the PC and launch ET.
2. Connect the CA (4) to a communication port on the PC by using one of the following methods:

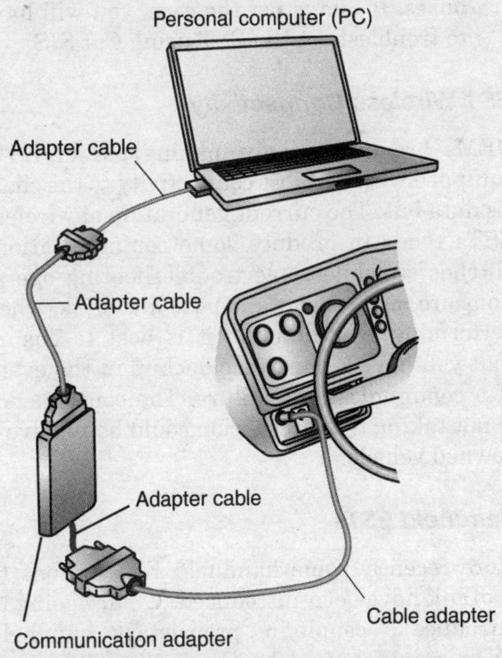

Figure 31–16 *How a communications adapter (CA) is connected as a serial link between a notebook PC and vehicle data bus. (Reprinted Courtesy of Caterpillar Inc.)*

   a. Connect cables (3) and (2) between the *computer* end of the CA (4) and a universal serial bus (USB) port on the PC (1).
   b. Connect cable (3) between the *computer* end of the CA (4) and the RS232 serial port of the PC (1).
3. Connect cables (5) and (6) between the *data link* end of the CA (4) and the service tool connector.
4. Turn the dash keyswitch *on*. First Cat ET will attempt to communicate with the chassis data bus. If this is successful, any accessible modules on the data bus network will be displayed. Because you want to communicate with the engine electronics, select the engine ECM.
5. Observe the CA and Cat ET GUI. If both the "J1939/DeviceNet" and the "**J1708**" indicator light-emitting diodes (LEDs) are flashing and Cat ET does not display a message that refers to a communication problem, Cat ET is communicating with the engine ECM on both data links. This is important: Cat electronics broadcasts to and receives messages from both data buses. If this is not the case, you will have to troubleshoot the EST using Cat SIS.

EST Wireless Connectivity

OEMs are increasingly making use of ESTs equipped with wireless connectivity to the chassis data bus. The current generation of wireless ESTs tends to produce some communications glitches so while some troubleshooting operations are made easier and pose little risk, when performing programming it is best to ensure that you have a hard-wire backup in the event of a communications failure. The consequence of not taking this precaution could be to have a downed vehicle.

Handheld ESTs

More recently some handheld EST options to ProLink have been introduced. We are going to introduce a couple of proprietary handheld ESTs manufactured by Caterpillar. When an OEM produces its own handheld EST, it does not have to hand over software and access profiles to a third party such as was the case with ProLink. Today's handheld ESTs have greater computing power driven by Palm OS, but their functionality is limited when compared to PC-based software. Figure 31–17 shows a Caterpillar's Pocket Tec PDA. There is both a driver and technician version: The technician version is can be used as a road test EST. Caterpillar has recently released an improved handheld EST known as MultiTool. This has a more shop-friendly design than Pocket Tec. An advantage of this later generation of handheld ESTs is that they can be updated online but they are limited when it comes to working with post-2007 systems. Pocket Tec and MultiTool are shown in Figure 31–17 and Figure 31–18.

Figure 31–17 *Caterpillar Pocket Tec. (Reprinted Courtesy of Caterpillar Inc.)*

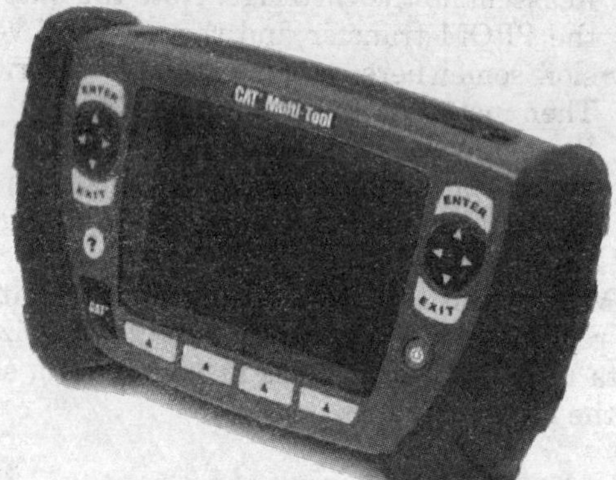

Figure 31–18 *Caterpillar MultiTool. (Reprinted Courtesy of Caterpillar Inc.)*

Reprogram Sequence

We will now switch systems and use Detroit Diesel electronic controls (DDECs) as a model to perform a reprogramming sequence with the objective of a calibration update: The procedure is similar with other OEMs.

1. Connect the DDEC communications adapter (CA) and verify J1939 handshake.
2. Select the engine MID 128.
3. From menu, select <file, print inputs, and outputs>.
4. Select <calibration>, then <ECM accumulators file print>.
5. Select <injector calibration> and record the numbers.
6. Next select <program update customer calibration at DDC>. Cycle the ignition key to log this into the laptop memory.
7. Close the window for <DDEC reprogramming>.
8. Select <DDC mainframe> to connect to the DDC data hub.
9. Select <upload history> and key in the password. This transfers ECM history to mainframe files.
10. Select <download engine serial calibration>.
11. Clear the units then key in the unit serial number, then hit <download>.
12. Close the window for the DDC mainframe.
13. Select <DDEC reprogramming>.
14. Now choose the <program ECM with mainframe data> option.
15. The display for <overwriting data> now appears: Select OK.
16. Identify the serial number, highlight, and click OK.
17. Select <no charge> or repair order # and <enter> unit is programming.
18. When complete, the display shows the serial number and I/Os. Close the window.
19. Select <update customer calibration>. Check all parameters and change if required.
20. Check the ECM accumulators and set to previous values. Check the injector settings. Close.

ELECTRONIC TROUBLESHOOTING

As we have progressed through this textbook, we have taken a look at some of the characteristics of mechanical failures in diesel engines, and technicians who have grown up in the computer age should remind themselves from time to time that diesel engines sometimes malfunction hydromechanically. Technicians must not assume that every problem is electronically based. Having said this, modern electronic troubleshooting programs are usually interactive and are capable of identifying many problems that have nothing to do with the vehicle electronics.

When undertaking troubleshooting, the approach should be methodical. We know this as sequential troubleshooting. Sequential troubleshooting takes place in stages. It is critical that the instructions in each stage are precisely undertaken before proceeding to the next; subsequent stages will generally be rendered meaningless if a stage is skipped. Most OEMs also prefer that you use PC- or online-based sequential troubleshooting programs for their systems. These simplify the sequencing of the troubleshooting stages and save a lot of page turning.

Online Service Information Systems

As mentioned a number of times earlier in this book, most OEMs prefer that the first step in any troubleshooting procedure is to log on to their online service information system whether the technician is working on an electronic engine or one of an earlier vintage. Some examples of engine OEM SIS are:

- Caterpillar Service Information System or **SIS**
- Cummins QuickServe Online or **QSOL** (or just QS)
- Detroit Diesel/Mercedes Benz **DDDL** (Detroit Diesel Diagnostic Link).
- Mack Trucks Electrical Information System (**EIS**)
- Navistar International Service Information System or **ISIS**
- Paccar **ServiceNet**
- Volvo **IMPACT**

Most OEM service information systems are updated daily. They tend to be easy to navigate. Technicians should learn how to use the Website search engines. In most cases, information searches are by engine serial number and may include searches of the parts data. It is important to note that most online SIS supports engines manufactured before the Website service existed so try to make a habit of using it.

Troubleshooting Tools

In order to effectively use electronic troubleshooting methods, the technician should be capable of using:

- The OEM online service information system (SIS)
- The OEM diagnostic software
- The appropriate electronic service tool (EST) and communication adapter (CA)
- A DMM (and have a basic understanding of electrical and electronic circuitry)
- Electrical schematics
- The correct tools, including terminal spoons, breakout Ts, and breakout boxes

Using the correct tools is often overlooked. When the use of a breakout box is mandated, make sure that it is used. Also, it should be remembered that electrostatic discharge can damage solid state components, so it is good practice to wear a ground strap when opening up any housing containing a microprocessor. When testing separated Deutsch and Weatherpac-type connectors with a DMM, ensure that the sockets have been identified before inserting test leads, terminal spoons, or breakout Ts. Use socket adapters/spoons where necessary to avoid damaging terminal cavities. Never spike wires when testing circuits.

Diagnostic Software

Each engine OEM produces its own diagnostic software, and if the technician has to do anything more than just scan a data bus, this software, along with the appropriate EST, must be used. Some examples of engine OEM diagnostic software are listed ahead; the terms normally used on the shop floor are in bold:

- Caterpillar **Electronic Technician** (ET)
- Cummins **InSite**
- Detroit Diesel /Mercedes Benz Detroit Diesel Diagnostic Link **DDDL** (Version 5.0 or higher required for both DDC and MB applications)
- Mack Trucks **Premium Tech Tool** (PTT)
- Navistar **Master Diagnostics (MD)**
- Volvo **V-Cads Pro** (Volvo computer-aided diagnostic system and programming)

It should be noted that some versions of the above software may have limited functions: Software for dealerships, body builders, and fleet applications may all be different depending on the level of programming required. For instance, a Volvo dealership uses V-cads Pro (full function) while fleets would be restricted to V-cads Elite, which cannot enable proprietary (Volvo) data programming.

Master Diagnostics

Master Diagnostics (MD) is Navistar International's diagnostic software. We use this software to show the functions and capabilities of one OEM's software. Most OEM diagnostic software share a lot in common, but the objective of using MD is to give some shape to this section. MD functions best if it is integrated into the computer software architecture and the computer is dedicated to use on Navistar International trucks and the Navistar data hub. This minimizes the chances of producing time-consuming software file clashes. If you purchase an EZ-Tech laptop computer from International Trucks, all the Navistar software, including MD, is loaded onto a Panasonic Toughbook, which is a notebook computer that adapts well to the truck shop floor. In common with all the OEMs, the functionality of the computer is designed to be as user friendly as is possible. Avoid loading competitor software onto an EZ-Tech laptop because doing so may cause software file clashes that can result in communication problems. The EZ-Tech GUI is shown in Figure 31–19.

Figure 31–19 *Navistar EZ-Tech GUI (Courtesy of Navistar International Corp.)*

MD Functions

MD software is an umbrella program that houses all the software required to troubleshoot Navistar International Trucks and its partner's subsystem electronics by directing the user into their online SIS. MD is designed to handshake directly to the vehicle data bus and access Navistar's Diamond Logic Controller (chassis control module) on the truck or bus. The MD software is designed to perform the following:

1. Display engine parameters in text format.
2. Display engine parameters in graphical format.
3. Snapshot test engine parameters.
4. Perform off-chassis diagnostics.
5. View and edit programmable parameters.
6. Display and clear **diagnostic trouble codes (DTCs)**.
7. Display engine and chassis information.

Launching MD

The procedure for launching MD is similar to other OEM diagnostic software packages. Because the software is required to crank, start, and run the engine during the tests, perform the following before beginning:

1. Block the wheels on two axles.
2. Ensure that the parking brake is engaged.
3. Place the transmission in neutral.

Then follow this sequence to connect to the chassis data bus:

1. Boot the EZ-Tech computer.
2. Connect the EZ-Tech CA to a serial or USB port on the EZ-Tech computer.
3. Connect the J1939 connector to the vehicle J1939 data connector.
4. Turn the ignition key on (do not start engine).
5. Open the COM port. When the J1939 connection is established, all of the MIDs on the chassis data bus will be identified. Select the one you want to work with. At any time the <Sniffer> menu bar will identify other MIDs networked to the data bus. We will assume it is the engine electronics and that the engine is manufactured by Navistar.
6. Three MD icon options will appear when the engine selection is made: Each addresses a different family or generation of engines. This will change with time and as new engine families are introduced.

Opening a Session

Click on the <Session> text bar to begin a session. This will immediately:

- Open a pair of parameter graphical views (blank) on the left side of the display
- Open a text display of engine parameters on the right side of the display
- Display all DTCs on the lower part of the display window

You can select parameters you want monitored during the diagnostic session by double-clicking. This can view selected parameters in graphical format and text/numeric formats. Note some of the following features:

- Double-clicking a DTC will drop down a <Help> dialog box with navigating suggestions: Close a text box by clicking on X.
- Placing the display cursor over a parameter and right-clicking will drop down advanced features: These vary with the parameter selected.
- Clicking on the <Diagnostics> pull-down menu will enable you to run diagnostic tests: Just click on the drop down radio bar to run a specific test.
- Tests are divided into <Key-Off> and <Key-On> tests.
- Left-clicking over a value will change it from English to metric and vice versa.
- You can open a graphical display at any time by clicking on the <Graphics> icon.

Graphical Display

When you select a graphical display, a black screen will be displayed until you select parameter data to be displayed. Selected display parameters will be identified at the top of the graph each assigned with a different color: This is so you can identify parameter data as it is displayed on the graph in actual specification plus real time. On the graph:

- Vertical axis: Min and Max values are shown with maximum at top.
- Horizontal axis: real time is displayed with each gradient representing one second.
- The color used to display the acronym of a selected parameter at the top of the graph is that used to plot the parameter values on the graph itself.

Using OEM Diagnostic Software The guidelines for using Navistar International's MD software are close to what you might expect to see on their competitor's diagnostic software.

Today's versions of diagnostic software are user friendly and providing you have a little knowledge of navigating OEM software, you should be able learn the ins and outs by simply experimenting. It is difficult to create problems if you use a little common sense, but be careful when entering programming fields: When altering customer data programming, it is preferable that you have received the OEM specific training.

Snapshot Tests

Most systems will accommodate a **snapshot test** readout from the ECM to facilitate troubleshooting intermittent problems that either do not generate codes or log a code with no clear reason. Snapshot mode troubleshooting can be triggered by a variety of methods (codes, conditions, manually) and record data frames before and after the trigger; these can be recorded to a reader-programmer instrument (ProLink or MultiTool) or a PC (laptop) while the vehicle is running. Because frames can be snapshot both before and after the trigger, it is possible to analyze a wide spectrum of data that may have contributed to a problem; each data frame can be examined individually after the event. The portability of handheld ESTs is ideal for this test.

DATA BUS CODES AND PROTOCOLS

The following is a partial listing of SAE codes that have been adopted by all the North American truck engine/electronics OEMs using a data bus network (explained in Chapter 33). The first-generation serial bus used in trucks adhered to SAE J1587. This slower serial bus was divided as follows:

- **SAE J1587** covered common software protocols used on the bus.
- **SAE J1708** covered common hardware protocols.

The more recent **SAE J1939** covers both hardware and software protocols. It is much faster and is built on CAN 2.0 architecture (see Chapter 33). The acceptance and widespread usage of these protocols enables the interfacing of electronic systems manufactured by different OEMs on truck and bus chassis, and additionally provides generic software with some ability to at least read competitor electronic systems. The truck technician who works with multiple OEM systems may find it easier to work using these codes rather than using the proprietary codes.

The acronym **MID (message identifier)** is used to describe a major vehicle electronic system with processing capability and a network address on the chassis data bus. Here are some of the common MIDs used on a current truck or bus.

Message Identifier (MID)	*Description*
128	Engine Controller
130	Transmission
136	Brakes—Antilock/Traction Control
137–139	Brakes—Antilock, Trailer #1, #2, #3
140	Instrument Cluster
141	Trip Recorder
142	Vehicle Management System
143	Fuel System
146	HVAC Control System
162	Vehicle Navigation
163	Vehicle Security
165	Communication Unit—Ground
171	Driver Information System
178	Vehicle Sensors to Data Converter
181	Communication Unit—Satellite
186	Actively Pressurized Coolant System
216	Lighting Control Module
232	Supplemental Restraint System
249	Bodybuilder Module
250	Steering Wheel Module

The acronym **PID (parameter identifier)** is used to code components within an electronic subsystem. The following are some PIDs used at the 128 MID address:

Parameter Identifier (Pid)	*Description*
26	Fan Speed %
27	EGR Position Sensor 1
45	Preheat Status
65	Service Brake Switch
68	Torque Limiting Factor
70	Parking Brake Switch
71	Idle Shutdown Timer Status
74	Maximum Road Speed Limit
81	Particulate Trap Inlet Pressure
83	Road Speed Limit Status
84	Road Speed
85	Cruise Control Status
89	PTO Status
91	Percent Accelerator Pedal Position
92	Percent Engine Load
93	Output Torque
94	Fuel Delivery Pressure
98	Engine Oil Level
100	Engine Oil Pressure
101	Crankcase Pressure
102	Manifold Boost Pressure
105	Intake Manifold Temperature
108	Atmospheric Pressure
110	Engine Coolant Temperature
111	Coolant Level
113	Governor Droop
121	Engine Retarder Status
153	Engine crankcase pressure
156	Injector Timing Rail Pressure
157	Injector Metering Rail Pressure
164	Injection Control Pressure
167	Charging Voltage
168	V-Bat
171	Ambient Temperature
173	Exhaust Gas Temperature
174	Fuel Temperature
175	Engine Oil Temperature
182	Trip Fuel
183	Fuel Rate
184	Instantaneous MPG
185	Average MPG
190	Engine Speed
362	EGR Position Sensor 2
412	EGR Temperature

The acronym **SID (subsystem identifier)** is used to identify the major subsystems of an electronic circuit. The SIDs listed here relate to cruise control.

Subsystem Identifiers (Sid) Common to All Mids	*Description*
242	Cruise Control Resume Switch
243	Cruise Control Set Switch
244	Cruise Control Enable Switch
245	Clutch Pedal Switch
248	Proprietary Data Link
250	SAE J1708 (J1587)
1939	Data Link

The following SIDs are used by engine and fuel system controllers:

Subsystem Identifiers (Sid) for Mid 128, 143	*Description*
01–16	Injector Cylinder #1 through #16
17	Fuel Shutoff Valve
18	Fuel Control Valve
19	Throttle Bypass Valve
20	Timing Actuator
21	Engine Position Sensor
22	Timing Sensor
23	Rack Actuator
24	Rack Position Sensor
29	External Fuel Command Input

Fault Mode Indicators

Because the diagnostic software used on trucks today is designed to be user friendly, you certainly do not have to remember specific MIDs, PIDs, SIDs, and FMIs **(fault mode indicators)** because in most cases the software is designed to route the user through the troubleshooting path without having to memorize a sequence of meaningless numeric values. Having said that, if you work on engines often enough, you will soon come to know that a 128 MID identifies the engine electronics address on the data bus, but it is more important that you are aware that MIDs identify separate and integral, onboard electronic systems networked to the data bus. Every MID has some processing capability. Every MID is therefore a computer.

When you want to communicate with any MID onboard the truck, first you have to be routed to the MID controller ECM. At that point, the subcircuit electronic troubleshooting kicks in and you are led through the PIDs and SIDs until you can identify a specific malfunctioning component.

If every engine OEM were allowed to come up with coding ways in which one of its components could fail, there would be a scary number of options, most overlapping or slightly differentiating similar failure modes. FMIs are SAE defined and common to all OEMs. That means that when the diagnostic path has been completed, the failed component must be assigned to one FMI. A FMI is a specific failure mode. Beginning in 2007, the number of FMIs has been extended so that categories of imminent failure could be identified. The following is a *current* listing of FMIs. FMIs 15 through 31 are those that were added in 2007.

Fault Mode Indicators (FMI)	*Description*
00	Data valid but above normal operating range (most severe)
01	Data valid but below normal operating range
02	Data erratic, intermittent, or incorrect
03	Voltage above normal or shorted high
04	Voltage below normal or shorted low
05	Current below normal or open circuit
06	Current above normal or grounded circuit
07	Mechanical system not responding properly
08	Abnormal frequency, pulse width, or period
09	Abnormal update rate
10	Abnormal rate of change
11	Failure mode not identifiable
12	Bad intelligent device or component
13	Out of calibration
14	Special instructions
15	Data valid but above normal (least severe)
16	Data valid but above normal (moderate severity)
17	Data valid but below normal (least severe)
18	Data valid but below normal (moderate severity)
19	Received network data in error
20	Data drifted high
31	Not available

The graph in Figure 31–20 shows how the new FMIs configure with the definition of "operating range" using the example of a temperature-sensing thermistor. Note that by adding the new FMIs some distinction can be made in the severity of a failure.

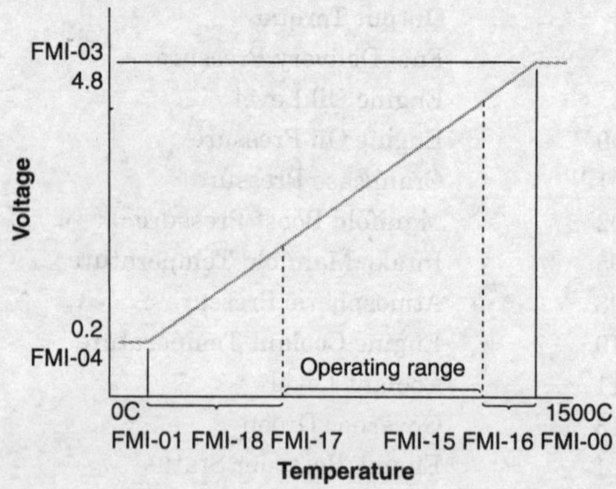

Figure 31–20 *New FMIs: distinguishing the severity of a failure in a thermistor.*

Diagnostic Display Reporting

As mentioned the way a problem is displayed by ECM diagnostics is common to all OEMs. So once a MID has been accessed, the numeric structure of the diagnostic path appears as five-digit display on the J1587 bus and as eight-digit display on the J1939 bus. Table 31–1 shows how the numeric values of SIDs, MIDs, and FMIs are sequenced in the display: five-digit on J1587, eight-digit on J1939.

EMD AND OBD II

For a decade now, automobile OEMs have been compelled to make any diagnostic codes that re-

Table 31–1 *PIDs, SIDs, and FMI Display*

Chassis Data Bus	*SIDs and PIDs*	*FMI*
J1587/1708	XXX	XX
J1939	XXXXXX	XX

late to emissions accessible for display with nonproprietary ESTs. This was covered by **on-board diagnostics** (generation) **II**, known as **OBD II**. A similar system has been proposed for all diesel-powered truck systems for 2010. This may be known as OBD II or as **engine management diagnostics (EMD)**. Whatever it ends up being called, as with the current automotive the OBD II requirement, EMD will impact only on failure conditions either directly or indirectly relating to emissions controls. After it is introduced, emissions-related failures will be required to be displayed using protocols open to generic ESTs.

Monitored Fields

Any system that either directly or indirectly impacts on engine emissions will require EMD/OBD II monitoring and reporting. The list will vary according to type of fuel system and the specific exhaust gas aftertreatment devices the vehicle is equipped with but includes:

- Misfire detection
- Fuel system efficiency
- Exhaust gas recirculation (EGR) monitor
- Catalytic efficiency
- Manifold boost pressure
- Oxides of nitrogen (NO_x) adsorber monitor
- Selective catalytic reduction (SCR) efficiency
- Urea (for SCR) level
- Glow plug monitor
- Diesel particulate filter monitor
- O_2 and NO_x reports

REVIEW QUESTIONS

1. Which EST is used to read blink/flash codes?
   a. Dash diagnostic lights
   b. Generic reader-programmer
   c. Diagnostic fork
   d. PC
2. The appropriate EST for performing a hardwire, resistance test on a potentiometer-type TPS isolated from its circuit is a:
   a. DMM
   b. Scanner
   c. Generic reader-programmer
   d. PC
3. Which of the following is usually required to hardwire couple a diagnostic PC with a truck data bus?
   a. Modem
   b. Jumper wires
   c. Data connector and CA
   d. Parallel link connector
4. When resistance and continuity tests are made on electronic circuit components, the circuit should be:
   a. Energized
   b. Not energized
5. DMM test lead resistance should never exceed:
   a. 0.2 ohm
   b. 1.0 ohm
   c. 10 ohms per inch
   d. 100 ohms per inch
6. The output (signal) of an inductive pulse generator, shaft speed sensor is measured in:
   a. V-DC
   b. Ohms
   c. V-AC
   d. Amperes
7. Which of the following cannot be performed by a ProLink 9000 reader-programmer?
   a. Erase historic fault codes
   b. Customer data programming
   c. Snapshot testing
   d. Proprietary data programming
8. Which of the following procedures cannot be performed using a PC and the OEM software?
   a. Erasing historic fault codes
   b. Customer data programming
   c. Erasing active fault codes
   d. EUI cutout tests
9. *Technician A* states that when a laptop is hardwire connected to a chassis data bus a CA is usually required. *Technician B* states that a PC with appropriate OEM software and wireless connectivity can communicate with a truck data bus. Who is correct?

a. Technician A only
b. Technician B only
c. Both A and B
d. Neither A nor B

10. When reading an open TPS circuit, which FMI would be broadcast to the data bus?
a. 0
b. 3
c. 5
d. 9

11. How many counts of resolution can be displayed by a 4½ digit DMM?
a. 1,999
b. 19,999
c. 9,995
d. 99,995

12. The specification that indicates how fine a measurement can be made by a DMM is known as:
a. rms (root mean square)
b. Percentage deviation
c. Resolution
d. Hysteresis

13. When a DMM is set to read AC, which is the term used to describe the averaging of potential difference to produce a reading?
a. rms (root mean square)
b. Percentage averaging
c. Resolution
d. Saw toothing

14. Which port on a handheld reader-programmer can be used to hardwire connect a printer?
a. ATA data connector
b. SAE J1939 connector
c. Parallel output port
d. RS–232 port

15. Which of the following components is used to test energized circuit conditions without damaging the wiring?
a. Cycling breakers
b. Alligator clips and jumper wire
c. Breakout boxes
d. Testlight with a sharp probe

16. Which electrical pressure value does a DMM output when in diode test mode?
a. 0.3 V
b. 0.6 V
c. 0.9 V
d. 1.1 V

17. *Technician A* states that the accuracy of an analog meter is usually no better than 3%. *Technician B* states that most DMMs have an accuracy factor that is within 0.1% of the reading. Who is correct?
a. Technician A only
b. Technician B only
c. Both A and B
d. Neither A nor B

18. Which of the following is the preferred source for service literature covering a repair procedure?
a. Hardcopy service manual
b. Technical service bulletin
c. Online SIS
d. DVD base service literature

19. *Technician A* states that using the snapshot test mode on an EST to troubleshoot an intermittently occurring fault code could help identify the conditions that produced the code. *Technician B* states that the snapshot test mode can only analyze historic codes on an audit trail. Who is correct?
a. Technician A only
b. Technician B only
c. Both A and B
d. Neither A nor B

20. When confirming the signal produced by a potentiometer-type TPS, the DMM test mode selected should be:
a. V-AC
b. V-DC
c. Diode test
d. Resistance

32

Electrical Wiring, Connector, and Terminal Repair

INTRODUCTION

In the past couple of years there has been a move toward standardizing the wiring used by the different truck original equipment manufacturers (OEMs). Beginning in 2007 trucks, manufacturers are bringing wiring color and codes into line with **SAE J2191** and TMC **RP146** standards. Though this has some potential to initially create grief for truck technicians who have to learn yet another system, in the long run this will be beneficial, especially for those who have to work on multiple OEM products. We begin this chapter by outlining the basics of this new wiring system for heavy-duty trucks.

One of the reasons that today's truck electronically managed engines present relatively few wiring- and terminal-related circuit problems is the overall quality of the weather-sealed components used. In a drive-by-wire chassis, it is crucial that wiring and connector repairs, when required, duplicate the original standards. Terminals may be platinum or gold coated to minimize resistance, so it is always important to observe OEM guidelines when undertaking any circuit repairs. The following instructions demonstrate general correct practices.

SAE WIRING STANDARDS

Any truck technician working in a service facility in which multiple OEM trucks are repaired will likely identify tracing electrical and electronic problems as a major challenge. Part of this challenge is that until recently, OEMs have each had their own way of coding and coloring the wiring in their electrical systems. Beginning in 2007, all truck OEMs began a process of adhering to what will probably become known as SAE wiring codes. SAE wiring codes are based on an SAE J-standard known as J2191 and a later TMC-recommended practice known as RP146. We offer a primer in SAE wiring codes in this chapter, but note that most of the OEMs provide a more detailed introduction in their SIS systems.

Color Codes

Gone are the 100 different colors packed into a wiring harness: colors that faded with age and time into a fraction of that number. SAE wiring codes use just three colors as follows:

- *Red:* Used for any unprotected (by fuse or breaker) wire. Specific wires are identified by *black* codes spaced every 10 cm (4 in.) apart.
- *White* with *orange* tracer: Used in any protected circuit. Most of the wiring on the chassis will fall into this category. Specific wires are identified by *black* codes spaced every 10 cm (4 in.) apart.
- *Grey:* Used for ground circuits. Specific wires are identified by *black* codes spaced every 10 cm (4 in.) apart.

Wire Identification

Although this may seem complicated at first, the codes used to identify each wire in a circuit make navigating wiring schematics easier because each wire is identified by an alphanumeric code in black print spaced every 10 cm (4 in.) apart. Figure 32–1 shows a typical SAE alphanumeric code, which we will decode it in this section then explain some of the terms used.

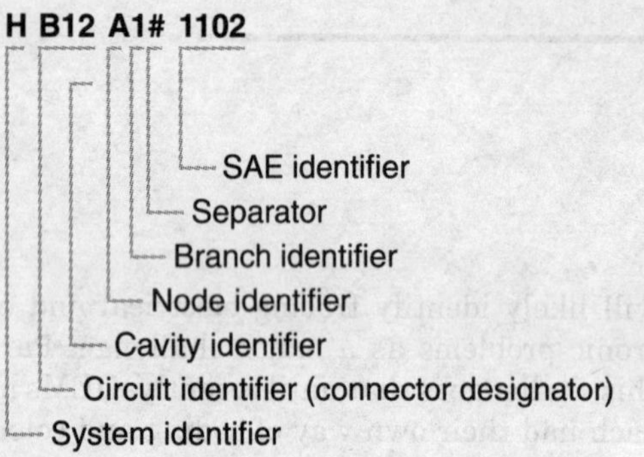

Figure 32–1 *SAE wiring code interpretation.*

System Identifier

The system identifier is the first alpha character in the code. At the time of writing 16 system identifiers are used as shown in Table 32–1.

In Figure 32–1, you will see that the system identifier alpha code is H, which tells us that the system we are working with is the instrumentation system.

Circuit Identifiers

The circuit identifier indicates what section of the system you are working with: This is often a connector. In cases where the circuit originates at a connector, the cavity number follows. Alternatively, where a circuit originates at a protection device such as a breaker or fuse, the number(s) that follow indicate the protection device number. In Figure 32–1, the circuit identifier is B12 so the B connector is at cavity number 12.

Node and Branch Identifiers

The node identifier is the first alpha character following the circuit identifier: It indicates which power node of the circuit the wire(s) form.

Table 32–1 *SAE Wiring System Identifiers*

Character Designation	*System Identifier*
A	Braking/traction control/ centralized tire inflation
B	Electrical charging circuit
C	Control and management circuits: fuel, engine, security
D	Emergency/vision
E	Engine accessories
F	Protected circuits
G	HVAC
H	Instrumentation/monitoring/ warning systems
I	Lighting systems
J	Serial data bus communications (J1939)
K	Protected power
L	Operator conveniences (GPS, WiFi, Satellite com, entertainment)
M	Trailer systems
N	Transmission, drivetrain, and drive axle carriers
U	Unprotected power
X	Ground

By definition, a node is any unswitched portion of a circuit transmitting electrical current from a driving device to a load. This means that the node identifier must change every time it passes through a switching device such as a relay, switch, or field effect transistor (FET). Each time the node id changes the next available alpha is used. An example of how this occurs is shown in Figure 32–2, which shows a continuation of the circuit shown in Figure 32–1.

The branch identifier is the number located immediately after the node id. A branch describes a wire that electrically connects to a node and permits uninterrupted current flow. The specific branch id number is the next sequential number on the node as you can see by referencing Figure 32–2. A splice or double terminal in the circuit is treated as a branch.

Separators and SAE Identifiers

The separator is simply the # symbol and it is used to keep the circuit identifier separate from

Figure 32–2 *Node id changes.*

the SAE supplementary identifier. The SAE supplementary ID is also known as the suffix. Table 32–2, Table 32–3, Table 32–4, and Table 32–5 list a sampling of some of the SAE IDs that you would be more likely to encounter while working with the engine electronics: A complete list is provided in the Student Guide that accompanies this textbook.

Table 32–2 *Power Circuit SAE Supplementary IDs*

SAE Suffix	*Subsystem Description*	*Circuit Function*
1101	Power	Battery bus feed
1102	Power	Ignition bus feed
1103	Power	Battery disconnect
1104	Power	Key switch feed
1105	Power	Cab feed
1106	Power	Accessory bus feed
1107	Power	Battery cable
1108	Power	Ignition spare
1109	Power	Battery spare
1110	Power	Accessory spare
1111	Power	Ignition bus control: chassis engine control module (ECM)

Table 32–3 *Ground Circuit SAE Supplementary IDs*

SAE Suffix	*Subsystem Description*	*Circuit Function*
1201	Ground	Instrument ground
1202	Ground	Starter ground
1203	Ground	Lighting ground
1204	Ground	Cab ground
1205	Ground	Engine ground
1206	Ground	Electronic circuit ground
1207	Ground	Alternator ground
1208	Ground	Chassis ground
1209	Ground	Battery ground

Data Bus Identification

The operating principles of chassis data buses are discussed in some detail in the next chapter so our objective here is simply to identify how the circuit appears in SAE code format. Figure 32–3 shows how this will appear on a schematic while Table 32–5 shows the SAE supplementary ids.

This should give you a general idea of how to navigate SAE wiring schematics. It is worth spending some time getting familiar with them because the objective is for all OEMs to adopt them in the near future. Figure 32–4 shows you how an SAE code appears on an actual section of wire, and Figure 32–5 shows the message identifiers (MIDs) connected to a Mack Trucks data bus using the new SAE wiring schematic architecture.

CONNECTOR ASSEMBLY AND REPAIR

Most of the wiring and connectors used on electronically managed diesel engines in North America are manufactured by a couple of manufacturers, so although the procedures outlined in this section are specific to one OEM, they are representative of those required for all the major OEMs.

Weatherproof Connectors

The following sequences demonstrate the disassembly, repair, and reassembly procedure required for the Weather-Pack and **Metri-Pack connectors** used in truck electronic circuits. This procedure is similar to that required for the assembly and repair of most electronic wiring and connectors.

Installation of Weatherproof Square Connectors

Weather-Pack and Metri-Pack connectors use a pull-to-seat design. The cable is pushed through

Table 32–4 *Control Circuit SAE Supplementary IDs*

SAE Suffix	*Subsystem Description*	*Circuit Function*
1501	Control: fuel and engine system	Engine shutdown
1502	Control: fuel and engine system	Engine retarder/brake
1503	Control: fuel and engine system	Idle shutdown timer
1504	Control: fuel and engine system	Cruise control on/off
1505	Control: fuel and engine system	Idle verification
1506	Control: fuel and engine system	rpm throttle set
1507	Control: fuel and engine system	Clutch/driveline disengage
1508	Control: fuel and engine system	Cruise set resume
1509	Control: fuel and engine system	Engine overspeed protection
1510	Control: fuel and engine system	(PTO) controls
1511	Control: fuel and engine system	Stop switch signals
1512	Control: fuel and engine system	Electronic/engine control
1513	Control: fuel and engine system	Spare relay: controlled by chassis ECU
1514	Control: fuel and engine system	Customer option: controlled by ECU
1515	Control: fuel and engine system	Air management control

Table 32–5 *Data Bus SAE Supplementary IDs*

SAE Suffix	*Subsystem Description*	*Circuit Function*
1601	Serial data communications	J1587/J1708
1602	Serial data communications	J922
1603	Serial data communications	J1939

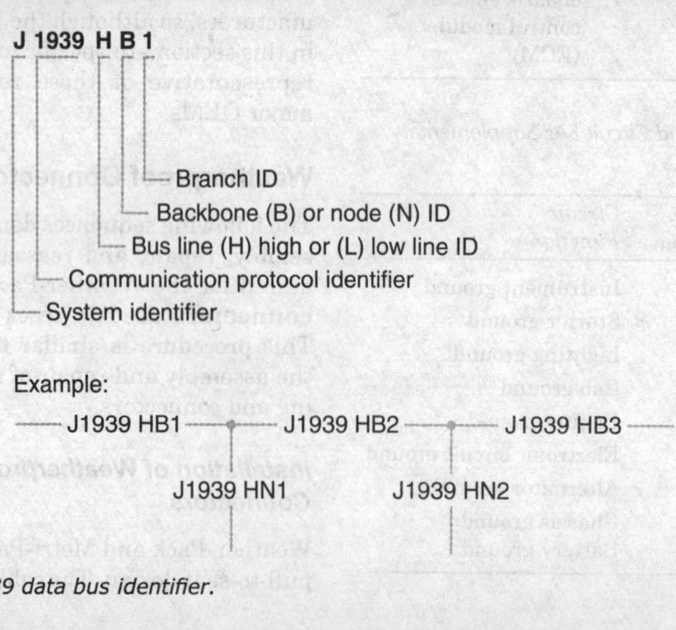

Figure 32–3 *J1939 data bus identifier.*

the seal and correct cavity of the connector before crimping the terminal to the cable. It should be stripped of insulation *after* it is placed through the seal and connector body. Use the following instructions for terminal installation:

1. Position the cable through the seal and correct the cavity of the connector (Figure 32–6).
2. Using wire strippers, strip the end of the cable to leave 5.0 ± 0.5 mm (0.2 ± 0.02 in.) of bare conductor.

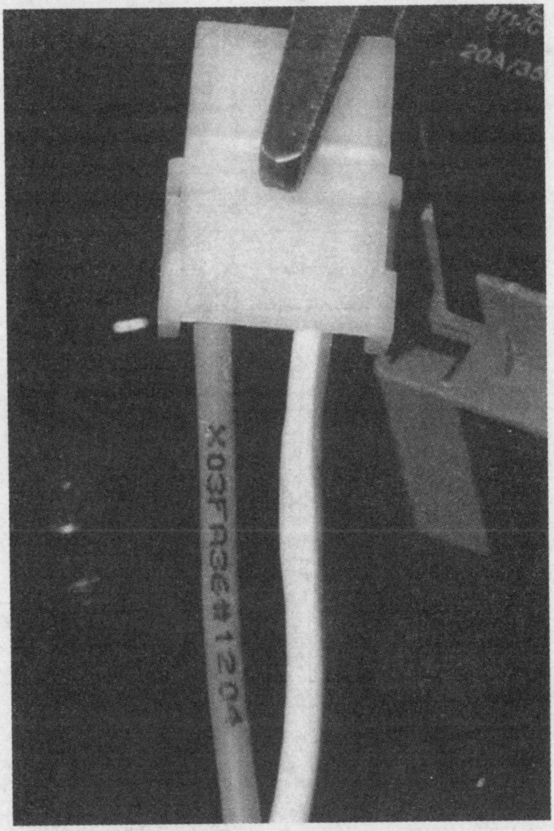

Figure 32–4 *SAE code printed on a gray protected wire. (Courtesy of Mack Inc.)*

3. Squeeze the handles of the crimping tool together firmly to cause the jaws to automatically open.
4. Hold the "wire side" facing you.
5. Push the terminal holder to the open position and insert the terminal until the wire attaching portion of the terminal rests on the 20-22 anvil of the tool. Be sure that the wire core wings and the insulation wings of the terminal are pointing toward the upper jaw of the crimping tool (Figure 32–7).
6. Insert the cable into the terminal until the stripped portion is positioned in the wire core wings and the insulation portion ends just forward of the insulation wings (Figure 32–8).
7. Compress the handles of the crimping tool until the ratchet automatically releases and the crimp is complete.
8. Release the crimping tool with the lock lever located between the handles in case of jamming.
9. Align the **locking tang** of the terminal with the lettered side of the connector.
10. Pull the cable back through the connector until a click is heard (Figure 32–9). Position the seal into the connector.

Removal and Repair

A tang on the terminal locks into a tab molded into the plastic connector to retain the cable assembly. Remove Metri-Pack 150 terminals, using the following instructions.

1. Insert the **tang release tool** into the cavity of the connector, placing the tip of the tool between the locking tang of the terminal and the wall of the cavity (Figure 32–10).
2. Depress the tang of the terminal to release it from the connector.
3. Push the cable forward through the terminal until the complete crimp is exposed.
4. Cut the cable immediately behind the damaged terminal to repair it.
5. Follow the installation instructions for crimping the terminal and inserting it into the connector.

Assembly of Metri-Pack Round Connectors

Use the following instructions for terminal installation:

1. Insert the terminal into the locating hole of the crimping tool, using the proper hole according to the gauge of the cable to be used (Figure 32–11).
2. Insert the cable into the terminal until the stripped portion is positioned in the cable core wings, and the seal and insulated portion of the cable are in the insulation wings (Figure 32–12).
3. Compress the handles of the crimping tool until the ratchet automatically releases and the crimp is complete. A properly crimped terminal is shown in Figure 32–12.
4. Release the crimping tool with the lock lever located between the handles in case of jamming.
5. Push the crimped terminal into the connector until it clicks into place. Gently tug on the cable to make sure it is secure (Figure 32–13).

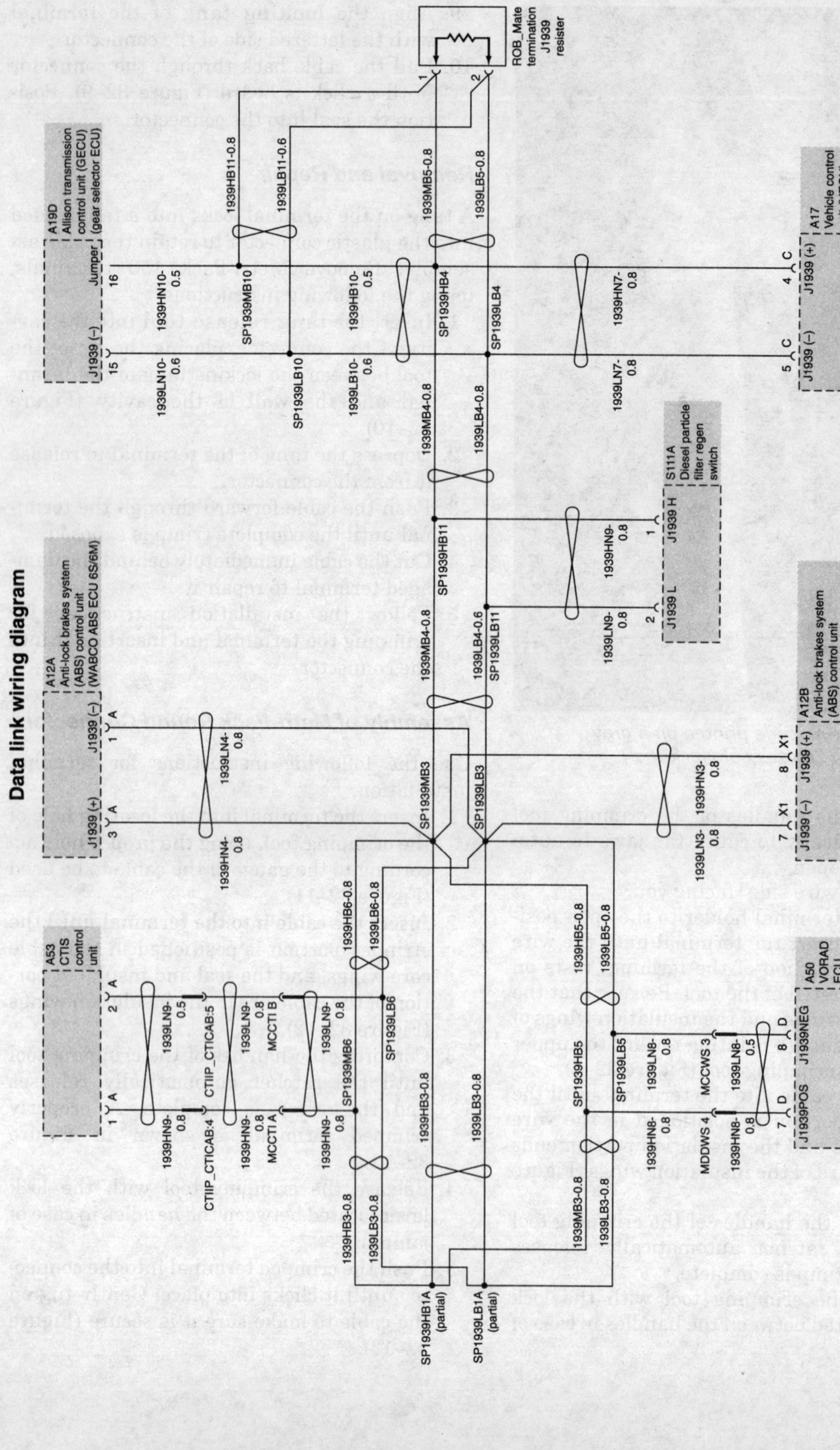

Figure 32–5 *SAE wiring schematic showing the MIDs connected to a Mack Trucks data bus. (Courtesy of Mack Inc.)*

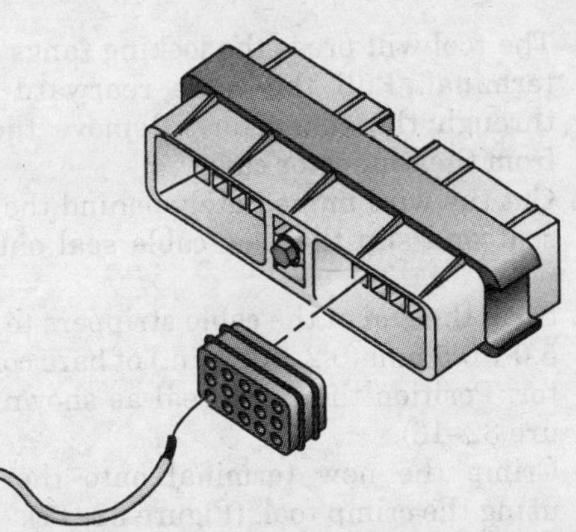

Figure 32–6 *Inserting wire in the connector.*

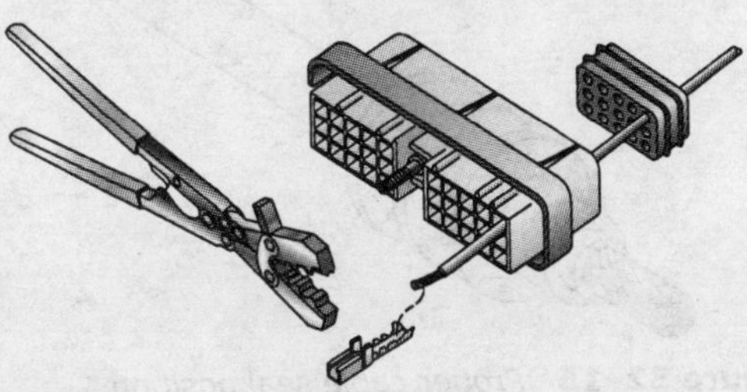

Figure 32–7 *Terminal and crimping tool position.*

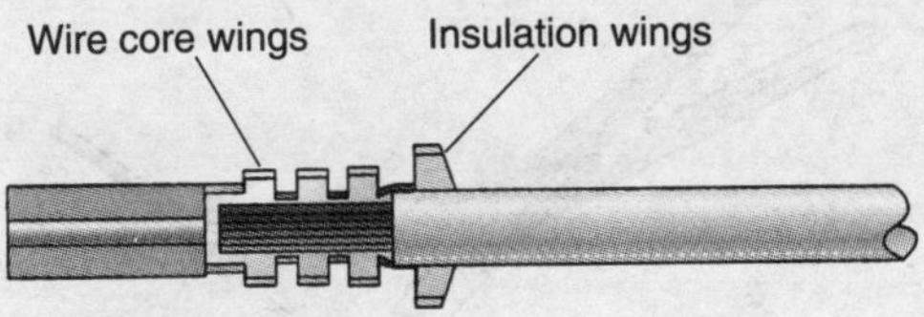

Figure 32–8 *Cable-to-terminal alignment.*

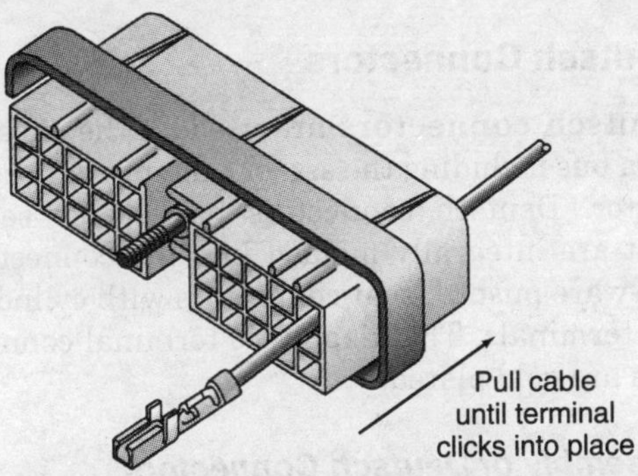

Figure 32–9 *Pulling the terminal to seat.*

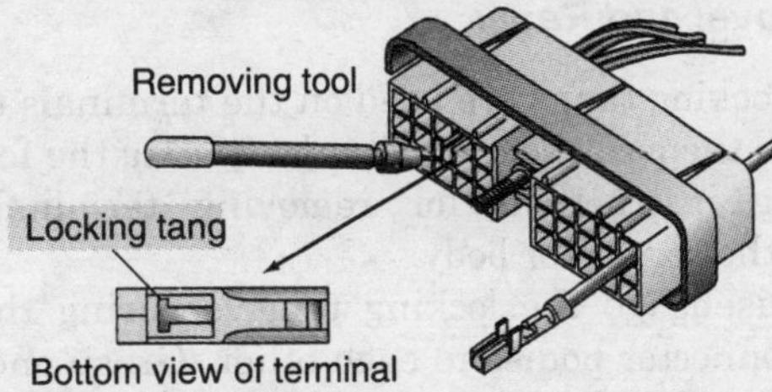

Figure 32–10 *Terminal removal.*

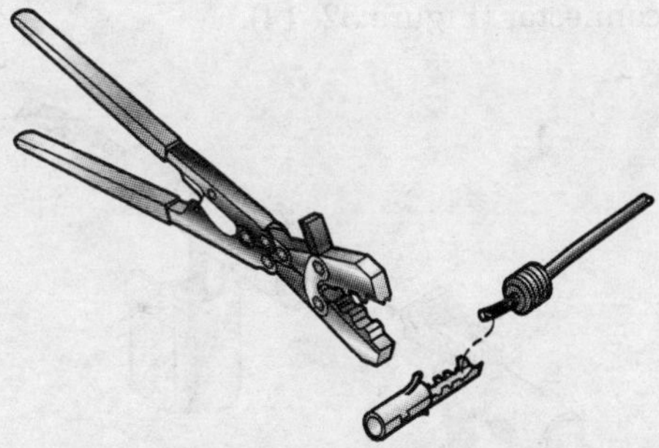

Figure 32–11 *Terminal position.*

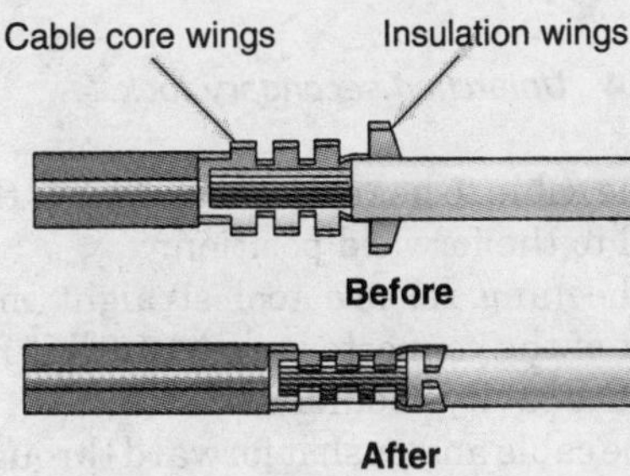

Figure 32–12 *Cable and terminal position before and after crimping.*

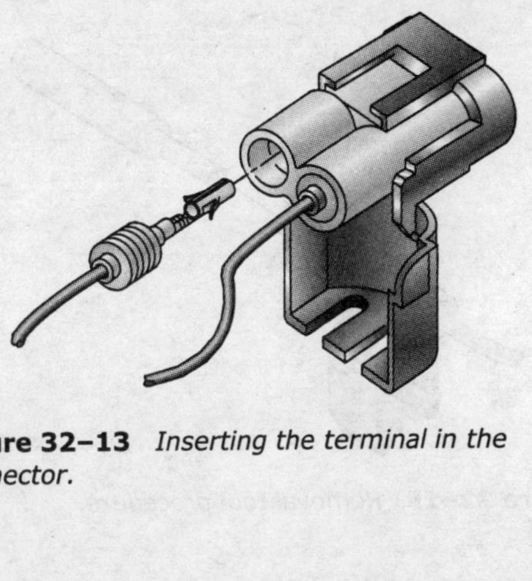

Figure 32–13 *Inserting the terminal in the connector.*

Removal and Repair

Two locking tangs are used on the terminals to secure them to the connector body. Use the following instructions for removing terminals from the connector body:

1. Disengage the locking tang, securing the connector bodies to each other. Grasp one-half of the connector in each hand and gently pull apart.
2. Unlatch and open the secondary lock on the connector (Figure 32–14).

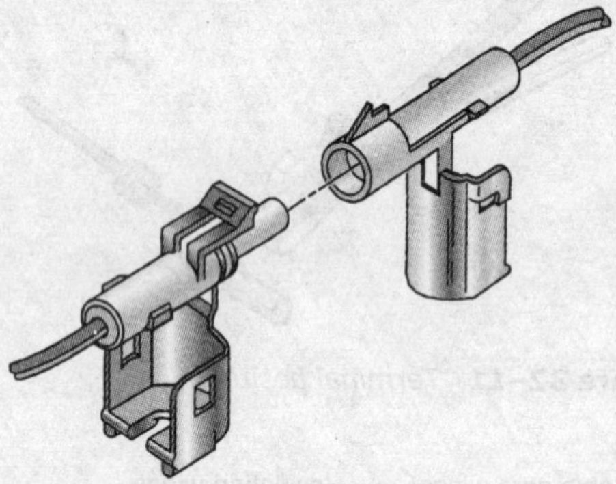

Figure 32–14 *Unlatched secondary lock.*

3. Grasp the cable to be removed and push the terminal to the forward position.
4. Insert the tang release tool straight into the front of the connector cavity until it resists on the cavity shoulder.
5. Grasp the cable and push it forward through the connector cavity into the tool while holding the tool securely in place (Figure 32–15).

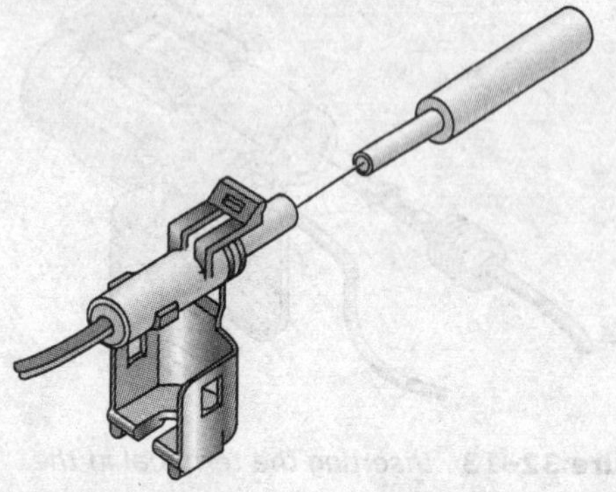

Figure 32–15 *Removal tool procedure.*

6. The tool will press the locking tangs of the terminal. Pull the cable rearward (back through the connector). Remove the tool from the connector cavity.
7. Cut the wire immediately behind the cable seat and slip the new cable seal onto the wire.
8. Strip the end of the cable strippers to leave 5.0 ± 0.5 mm (0.2 ± 0.02 in.) of bare conductor. Position the cable seal as shown (Figure 32–16).
9. Crimp the new terminal onto the wire, using the crimp tool. (Figure 32–17).

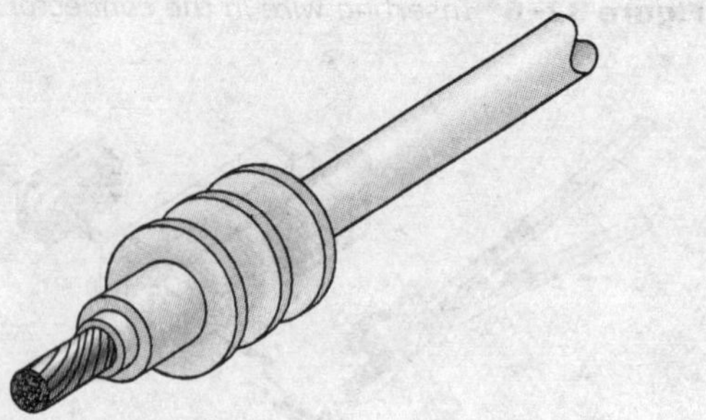

Figure 32–16 *Proper cable seal position.*

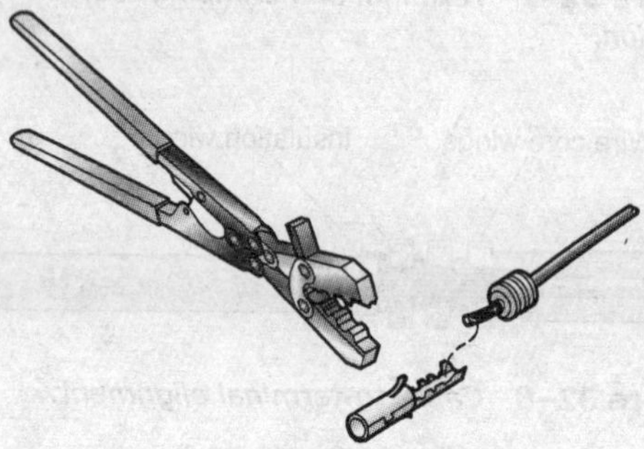

Figure 32–17 *Crimping procedure.*

Deutsch Connectors

Deutsch connectors are used on the chassis data bus including the six- or nine-pin data connector. Deutsch connectors have cable seals that are integrally molded into the connector. They are push-to-seat connectors with cylindrical terminals. The diagnostic terminal connectors are gold plated.

Assembly of Deutsch Connectors

Use the following instructions for installation:

1. Strip approximately 6 mm (¼ inch) of insulation from the cable.

2. Remove the lock clip, raise the wire gauge selector, and rotate the knob to the number matching the gauge wire that is being used.
3. Lower the selector and insert the lock clip.
4. Position the contact so that the crimp barrel is 0.75 mm (1/32) of an inch above the four indenters (Figure 32–18). Crimp the cable.
5. Grasp the contact approximately 25 mm (1 in.) behind the contact crimp barrel.
6. Hold the connector with the rear grommet facing you (Figure 32–19).
7. Push the contact into the grommet until a positive stop is felt (Figure 32–20). A slight tug will confirm that it is properly locked into place.

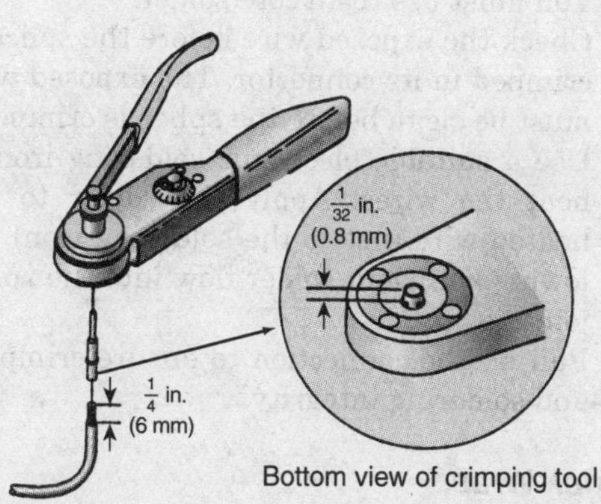

Figure 32–18 *Setting wire gauge selector and positioning the contact.*

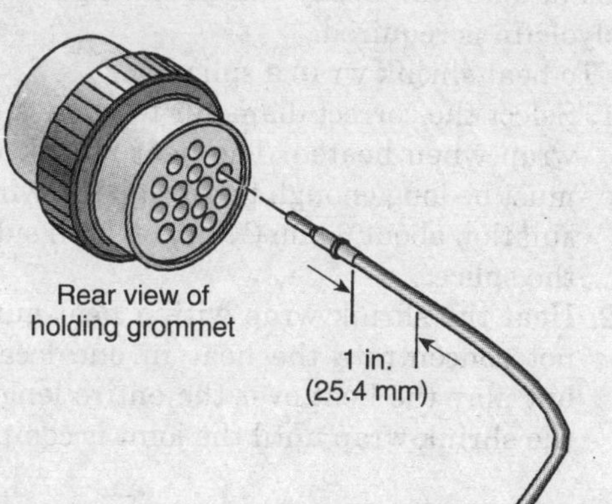

Figure 32–19 *Pushing the contact into the grommet.*

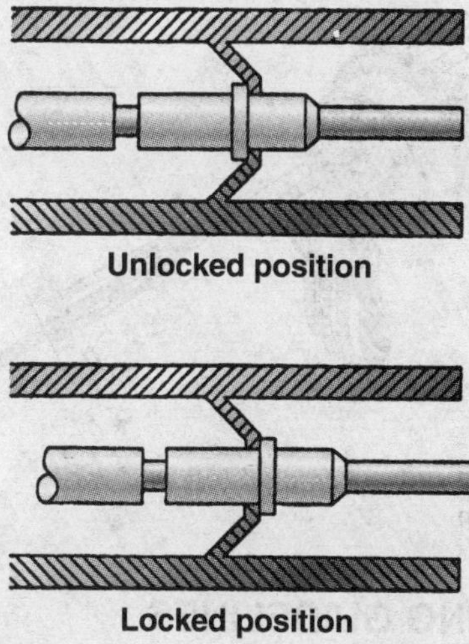

Figure 32–20 *Locking the terminal into the connector.*

Removal

The appropriate size removal tool should be used when removing cables from connectors.

1. With the rear insert toward you, snap the appropriate size remover tool over the cable of contact to be removed (Figure 32–21).
2. Slide the tool along the cable into the insert cavity until it engages and resistance is felt. Do not twist or insert tool at an angle.
3. Pull the contact cable assembly out of the connector. Keep reverse tension on the cable and forward tension on the tool (Figure 32–22).

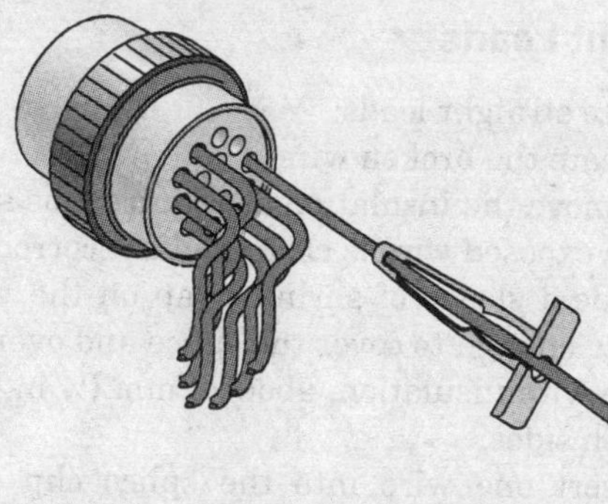

Figure 32–21 *Removal tool positioned for removal.*

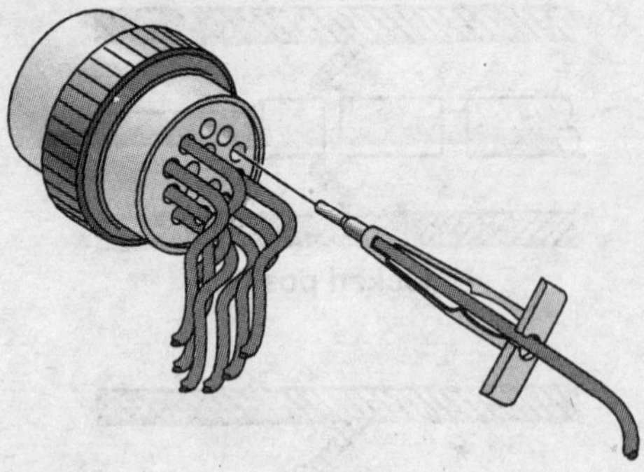

Figure 32–22 *Removal tool and wire withdrawn.*

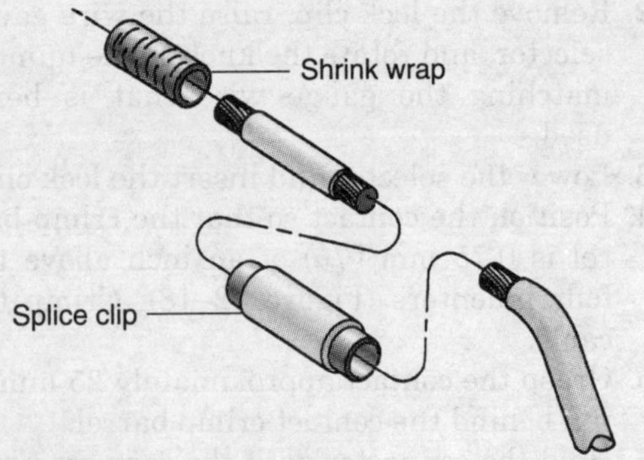

Figure 32–23 *Spliced wire.*

SPLICING GUIDELINES

Not all wiring in electronic circuits can be spliced, so always refer to the OEM service literature before attempting such a procedure. The following may be used as a general guideline. The objective is to produce a high-quality, tight splice with durable insulation that should outlast the life of the vehicle. The selection of crimping tools and splice connectors will depend on the exact repair being performed.

Tools Required:

- Soldering iron
- Rosin core solder
- Wire strippers
- Heat shrink tubing
- Splice clips
- **Crimping pliers**

Straight Leads

To splice straight leads:

1. Locate the broken wire.
2. Remove the insulation as required; be sure the exposed wire is clean and not corroded.
3. Slide a sleeve of shrink wrap on the wire long enough to cover the splice and overlap the wire insulation, about 6 mm (¼ in.) on both sides.
4. Insert one wire into the splice clip and crimp.
5. Insert the other wire into the splice and crimp (Figure 32–23).

Solder

Soldering splice connectors is optional. To solder splice connectors:

1. You must use rosin core solder.
2. Check the exposed wire before the splice is crimped in its connector. The exposed wire must be clean before the splice is crimped.
3. Use a suitable electronic soldering iron to heat the wires. Apply the solder to the heated wire (not to the soldering iron), allowing sufficient solder flow into the splice joint.
4. Pull on the connection to ensure crimping and soldering integrity.

Shrink Wrap

Shrink wrap is required. Alpha FIT-300, Raychem TAT- 125, or any equivalent heat shrink dual wall epoxy encapsulating adhesive polyolefin is required.

To heat shrink wrap a splice:

1. Select the correct diameter to allow a tight wrap when heated. The heat shrink wrap must be long enough to overlap the wire insulation about 6 mm (¼ in.) on both sides of the splice.
2. Heat the shrink wrap with a heat gun; do not concentrate the heat in one location, but play the heat over the entire length of the shrink wrap until the joint is complete.

Multiple Broken Wires

To **splice multiple** broken wires:

1. Stagger the position of each splice as illustrated in Figure 32–24.

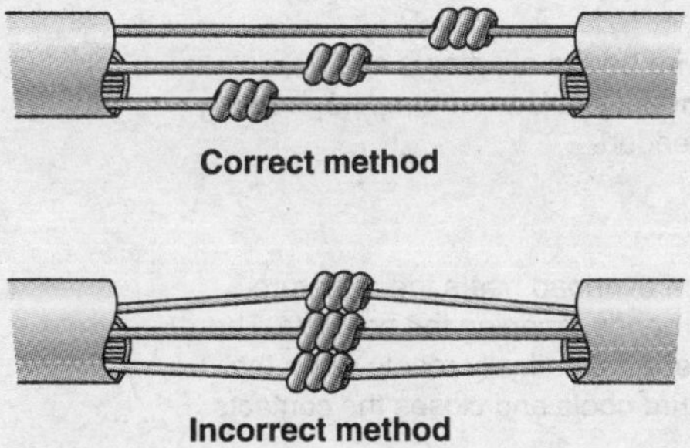

Figure 32–24 *Multiple splices.*

2. You *must* stagger positions to prevent a large bulge in the harness and to prevent the wires from chafing against each other.

Three-Wire Splice

Three-way splice connectors are commercially available to accommodate three-wire splices. The technique is the same as a single **butt splice** connector (Figure 32–25).

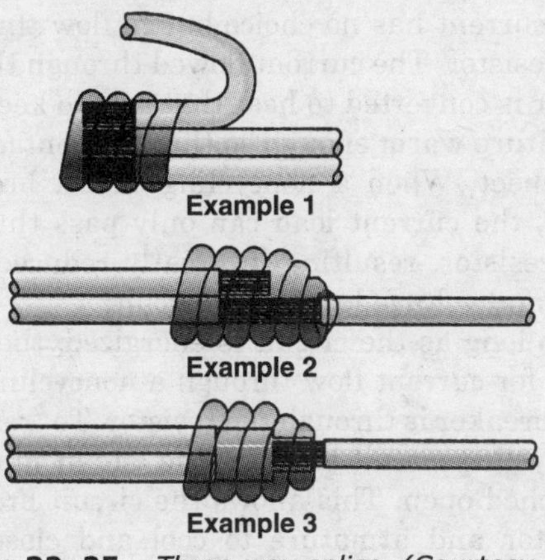

Figure 32–25 *Three-way splice. (Courtesy of Detroit Diesel Corporation)*

CIRCUIT PROTECTION DEVICES

Circuit protection components are designed to limit current flow through a circuit. Almost all truck electrical circuits are protected by either a fuse or circuit breaker. When troubleshooting electrical circuits, one of the first steps is to identify the location of the circuit protection device on any suspect wiring path. If the chassis uses SAE standard wiring codes, the color white with an orange tracer is used for any protected circuit.

Fuses

Fuses are designed to fail once, after which they have to be replaced. A typical fuse consists of a wire bridge through which circuit current is routed. The bridge is rated for specific maximum amperage. If the specified amperage is exceeded, the bridge overheats and melts. Fuses have made something of a comeback during recent years. This is due to the sensitivity of low potential electronic circuits to current overload discharge. A number of different types of fuses are used in current truck electrical circuits. Figure 32–26 shows a 400 amp mega fuse used on the ground circuit of a 2007 Mack chassis. The symbol for a fuse is shown in Figure 32–27.

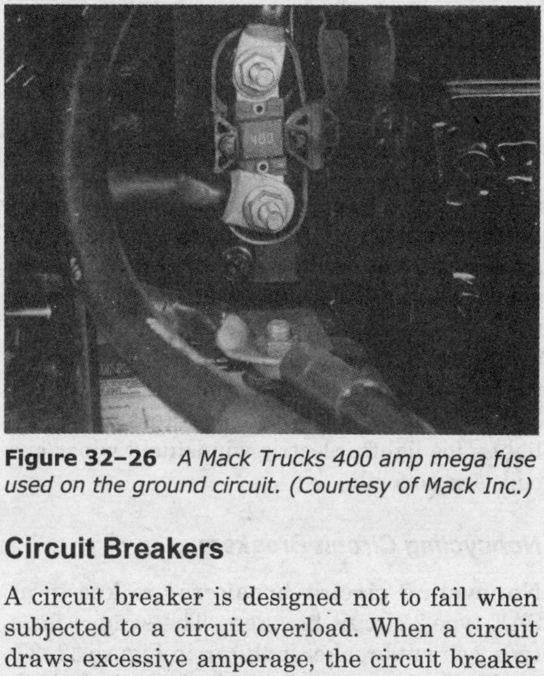

Figure 32–26 *A Mack Trucks 400 amp mega fuse used on the ground circuit. (Courtesy of Mack Inc.)*

Circuit Breakers

A circuit breaker is designed not to fail when subjected to a circuit overload. When a circuit draws excessive amperage, the circuit breaker is designed to open the circuit, after which it will reset. A typical circuit breaker consists of an armature and contacts: The contacts "break" when the circuit is overloaded. There are two types used on trucks: Reference Figure 32–28 to see how each is represented in circuit schematic.

Cycling Circuit Breakers

Cycling circuit breakers are known as SAE type 1 circuit breakers. The symbol for a type 1

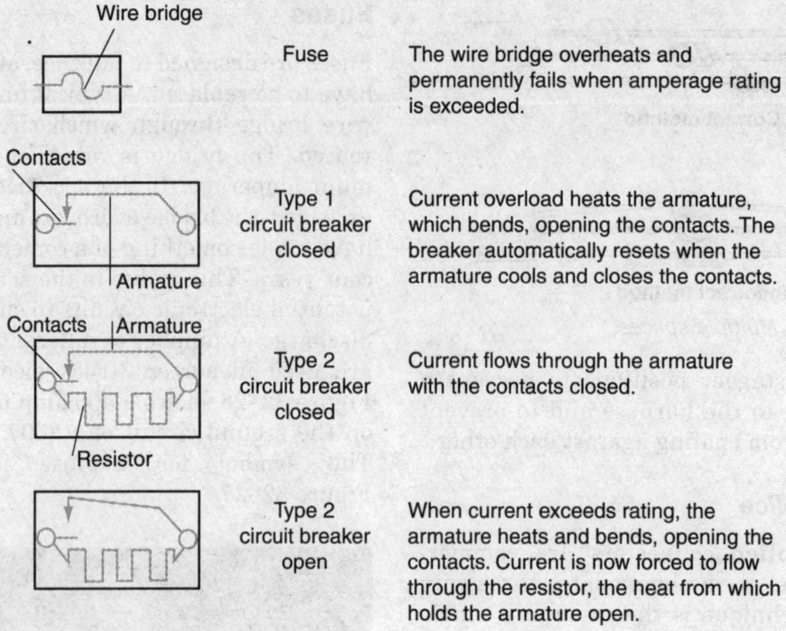

Figure 32–27 *Circuit protection devices.*

circuit is shown in Figure 32–27. When current flowed through the armature exceeds the specified rating, the armature heats up and bends away from the contact point shown by the arrow: This immediately opens the circuit. With the circuit open no current flows through the armature so it cools and as it straightens closes the circuit as the contacts connect. Typical circuits that use SAE type 1 breakers are headlights, fog lights, electric windshield wiper circuit, and so forth.

Noncycling Circuit Breakers

Noncycling circuit breakers are known as SAE type 2 circuit breakers. The symbol for a type 2 circuit breaker is shown in Figure 32–27: In this figure, you can see the breaker in both closed and tripped conditions. A type 2 circuit breaker functions similarly to the type 1 except that it has a resistor below the armature. When the circuit it is protecting is closed (energized) there are two possible current paths through the breaker, but because electricity will use the path of least resistance, most of the current will flow through the armature. If the specified amperage of the circuit breaker is exceeded, the armature heats up and bends away from the contacts, opening the circuit: At this point, circuit current has no choice but to flow through the resistor. The current flowed through the resistor is converted to heat that acts to keep the armature warm enough so that the contacts do not meet. When a noncycling circuit breaker trips, the current load can only pass through the resistor, resulting in greatly reduced current and voltage loss to the circuit.

So long as the circuit is energized, the only path for current flow through a noncycling circuit breaker is through the resistor. To "reset" a noncycling circuit breaker, the circuit must be switched open. This allows the circuit breaker resistor and armature to cool and close the breaker contacts. Noncycling circuit breakers are used extensively in truck electrical circuits.

RELAYS

A **relay** is an electromechanical switch. They are used extensively on truck electrical circuits and every technician should know exactly how one functions. A relay consists of two electrically isolated circuits. One circuit (coil) is used to electromagnetically change the switch status of the power circuit, either opening or closing it. Relays enable a low-current circuit (coil) to control high-current ("power") circuits. Take a look at Figure 32–28 and Figure 32–29 showing SAE standard relay terminal assignments when following the explanation in the text.

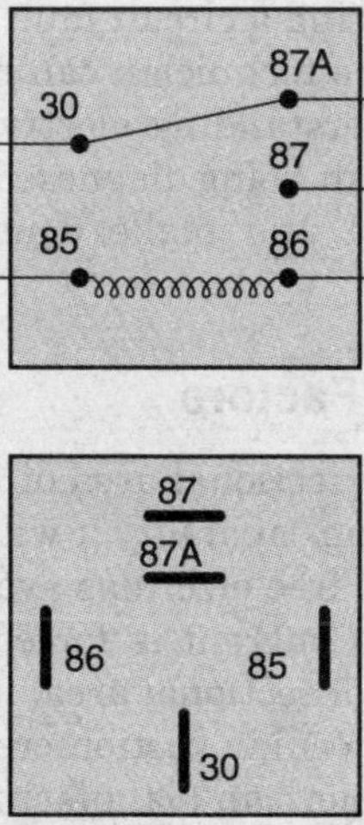

Figure 32–28 *Older style standard SAE relay terminal assignments.*

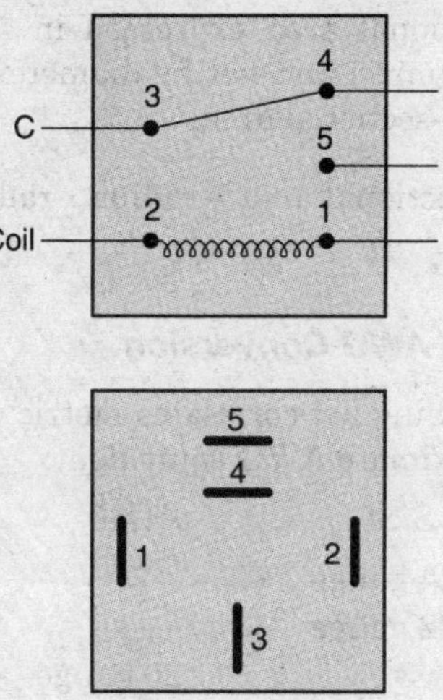

Figure 32–29 *Newer style standard SAE relay terminal assignments.*

Coil Circuit

The relays shown in Figure 32–28 and Figure 32–29 are used in all OEM electrical systems. Two sets of terminal assignment numeric codes are used. The control or coil circuit is indicated by terminals 85 (2) and 86 (1). The polarity usually does not matter but it might. For instance, Volvo Trucks uses a diode on some of its control circuits, so on these terminal 85 (2) must always be chassis ground. When no current flows through the coil circuit, the status of switching circuit is in NC, or normally closed. In most cases (depending on how the relay is being used), this will mean that the output is open. When current flows through the control coil, an electromagnetic field is created and the movable switch is pulled toward the coil, opening the 30-87a (3-4) circuit and closing the 30-87 (3-5) circuit.

Switched Circuit

Referencing Figure 32–29 and Figure 32–30 once again, you will see that the switched circuit of the relay is represented by terminals 30 (3), 87 (5), and 87a (4). Terminal 30 (3) is common. This means that a voltage reading could normally be measured at terminal 30 (3) regardless of the status of the switch. When the coil circuit of the switch is not energized, 30 (3) would close on the NC terminal 87a (4). In most cases, this would mean that the switching status of the relay is open so no current flows through the switch circuit. When the control coil is energized, the switch is moved from the 87a (4) normally closed pole, to the 87 (5) normally open (NO) pole: This action permits current to flow from common terminal 30 (3) to the NO terminal 87 (5).

Terminal Assignments

Standard SAE designations were recently revised by the SAE with the result that two numbering systems are in use. For instance, some OEMs continue to use the older numbering system even on their current product. However, we can say that one of the following methods of numbering relays are standard through the industry so it will pay you to commit the terminal designations to memory (Table 32–6).

Check these terminal assignments by referencing Figure 32–28 and Figure 32–29 again. Relays are simple switches that only cause confusion if you do not know how one operates. Take a little time and memorize the terminal assignments and it will pay off in troubleshooting time saved many times over.

Table 32–6 *SAE Relay Terminal Assignments*

Terminal Designation	*Former Number*	*Revised Number*
Coil (control current positive) often stamped +	86	1
Coil (control current ground: direct or switched)	85	2
Common (supply current in)	30	3
Normally closed (NC)	87a	4
Normally open (NO)	87	5

Relay Functionality

A good method of remembering the function of each relay terminal is shown in Table 32–7

Table 32–7 *Relay Terminal Functions*

New SAE #	*Old SAE #*	*Function*
1	86	Control current (usually positive, especially when a diode is used)
2	85	Control current out (usually to ground but could be switched)
3	30	Supply current in
4	87a	High current out—normally closed
5	87	High current out—normally open

Microrelays

Microrelays have become more commonly used in recent years. They use the same numbering codes as standard relays: Both the old and new numbering codes are used. Microrelays are compact and accordingly are specified to lower maximum current loads.

Four Terminal Relays

Four-terminal relays are now widely used in truck electrical circuits because they are more compact. A four-terminal relay retains the coil circuit (terminals 1 and 2) and eliminates the NC option on the power circuit. Operation is simple. The common terminal (3) is the power in. When the coil circuit (terminals 1 and 2) is energized, the power circuit closes so that current can flow from terminal 3 to terminal 5.

TROUBLESHOOTING TECHNIQUES

In an earlier chapter, we said that there were only three types of electrical circuit fault, namely, opens, shorts, and high resistance. This said, locating a circuit fault can often be a tough task. The technician can make this task easier by understanding electrical basics and the actual circuit being diagnosed. Learn to rely on your DMM and make double-checking a habit.

Wire Gauge Factors

The larger the sectional area of wire, the more current-carrying capability it will have because there are more free electrons available to carry it. If too much current is forced down a wire with too small a sectional area, it will overheat due to the molecular friction created. Because some North American manufacturers are using metric wire gauge specifications, you should be able to convert them to American wire gauge (AWG) standards. Metric wire is identified by cross-sectional area expressed in square millimeters (mm^2) and not by diameter. To calculate cross-sectional area:

$$\text{wire sectional area} = \text{radius} \times \text{radius} \times \pi\ (3.1416)^*$$

Metric to AWG Conversion

The following list correlates metric wire size to its approximate AWG equivalent:

Metric Size	*AWG*
00,22	24 gauge
00,35	22 gauge
00,5	20 gauge
00,8	18 gauge
01,0	16 gauge
02,0	14 gauge
03,0	12 gauge
05,0	10 gauge
13,0	6 gauge
32,0	2 gauge
62,0	0 gauge

*π is the Greek letter pi that expresses the ratio of the diameter of a circle to its circumference.

When selecting wire gauge, the sectional area is obviously a factor but so is the length of the wire. As wire length increases, so will voltage losses when current is pumped through the wire. This means that the wire gauge size must be increased if the current destination is far away from the source. Figure 32–30 shows how wire length and amperage affect voltage drop.

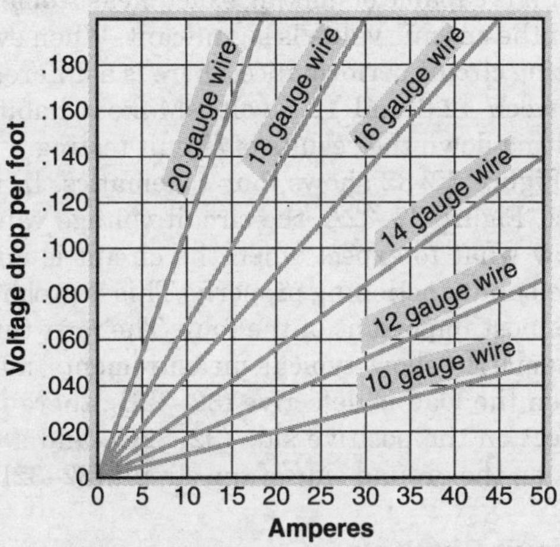

Figure 32–30 *Voltage drop and wire length chart.*

Circuit Voltage Behavior

When a discharged battery tested at 8 volts is connected to a battery charger with voltage measured at 16 volts, initially there will be high amperage flow from the charger to the battery. As the electrical pressure in the battery builds during charging, the rate of current transfer will slow due to the drop in charge differential, which is caused by battery voltage rising. The result is that as battery voltage rises to its fully charged level, the rate of amperage flow from the charger to the battery drops.

Basic circuit theory (see Chapter 28) tells us that the sum of voltage drops must always equal source voltage. In a perfect electrical circuit, voltage should be dropped only across the intended loads in that circuit. Should voltage be dropped elsewhere in the circuit due to high resistances, there will be insufficient voltage to allow the intended load in a circuit to operate properly.

Voltage Drop versus Resistance

Every technician should understand the importance of voltage drop testing and that means understanding the limitations of the ohmmeter. If a copper starter motor cable has deteriorated to the extent that only 10% of its wire strands remain intact, an ohmmeter measurement will indicate that it checks out with a resistance reading similar to that of a cable in perfect condition. This is because the ohmmeter is forcing a very small current through the test circuit. However, if a voltage drop test were to be performed on the cable while cranking the engine, the voltage drop would immediately indicate the problem.

TESlite® Circuit Testing

TESlite voltmeter leads are a set of digital multimeter (DMM) leads that have circuitry added that helps locate corrosion in some electrical circuits using digital voltmeter mode. The tool functions best when used on standard 12 V-DC circuits. The hardware consists of a set of normal DMM test leads connected through the TESlite module, which is a small plastic case with a trigger button shown in Figure 32–31. Until the button is depressed, the TESlite leads function exactly like regular DMM test leads. When the TESlite trigger button is depressed, a resistance is placed in parallel across the test leads. This means that a component can be removed from its circuit, which can then be subjected to a dynamic load test. Because of the nearly universal adoption of sealed weatherproof connectors in truck electrical circuits making access to the wires difficult without damaging them, this permits voltage drop testing for corrosion in the wiring itself.

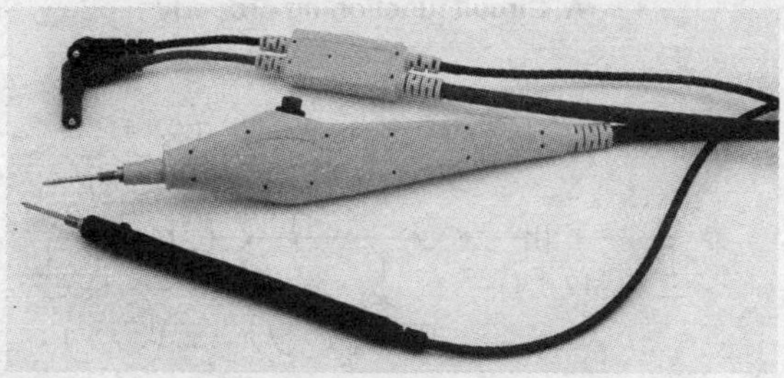

Figure 32–31 *TESlite® enhanced voltmeter leads. (Courtesy of Sullivan Systems)*

TESlite® Test Procedure

The TESlite testing sequence, which follows, is simple. The leads function as normal DMM test leads until the trigger button is depressed, so they are designed to replace the original set of leads.

1. Open the circuit at the component, for instance, a relay.
2. Connect the voltmeter in the exact place where the load component was: The negative (black) lead should be placed on the ground terminal, not to chassis ground.
3. Take a standard voltage reading with the circuit energized.
4. If nominal (system) voltage exists, depress the TESlite trigger button.
5. If voltage remains nominal (does not drop more than 0.5 V) the circuit can be regarded as clean and the test can be concluded (Figure 32–32A).
6. If the intended load in the circuit produces no voltage drop (Figure 32–32B) the load can be assumed to be defective.
7. If voltage drop exceeds 0.5 V, there is corrosion present. Proceed to the next steps in the test procedure.
8. Move the negative lead to a good chassis ground and depress the TESlite trigger.
9. If voltage remains low, the corrosion is in the positive wire (Figure 32–32C).
10. If voltage rises to nominal, the corrosion is in the negative wire (Figure 32–32D).

Circuit Analysis

Getting used to using your DMM on real world circuits should mean testing properly functioning circuits. Too often, a technician tests a circuit for a defect without knowing how that circuit would test if it were fully functional. Try to make a practice of testing good circuits as well as bad. It is important to remember that your DMM is capable of making exact measurements and the specific value is significant. When evaluating circuit performance, there is a difference between 12.6 and 12.5 volts: Make a habit of writing down the results of circuit testing.

Figure 32–32 shows four schematics. In the first, Figure 32–32A, the circuit voltage values show what to expect when the circuit is energized and functioning properly. This is probably the most important of the four. The next three schematics show typical measurements made when the load is defective (32–32B), there is a defect on the positive side (32–32C), and a defect on the ground side of the circuit (32–32D).

Circuit Symbols

While there may be similarities, most engine manufacturers like to use their own wiring schematic symbols. This may change if OEMs decide to adopt SAE standard symbols. Despite some commonality, there are enough differences to make these symbols confusing if you are working with a variety of different manu-

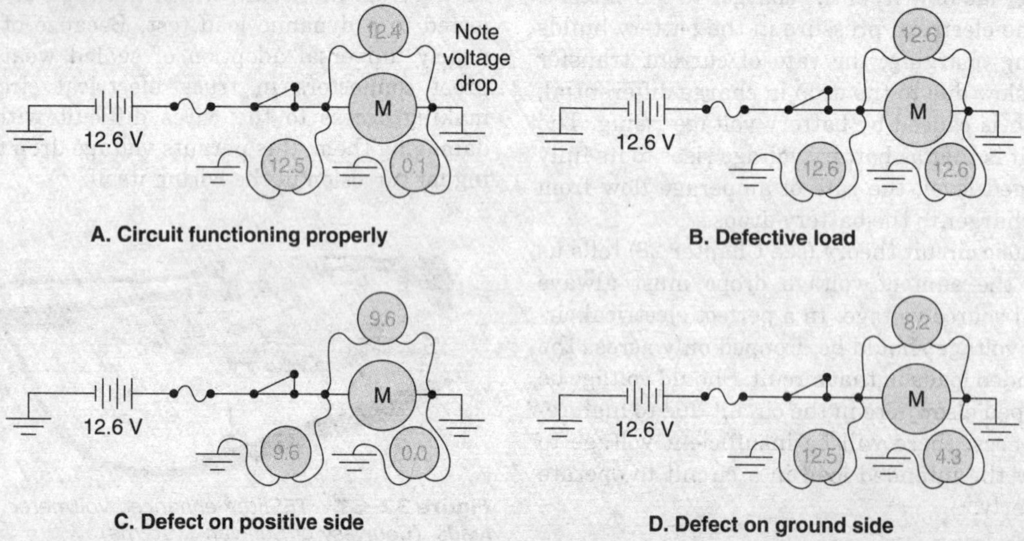

Figure 32–32 *Circuit analysis.*

facturers' product. However, you will almost always find these symbols are deciphered somewhere in the service literature or online service link. Figure 32–33 shows some examples of typical circuit symbols in current use.

Interpreting Circuit Schematics

Until the recent move to establishing an SAE standard electrical schematics (introduced at the beginning of this chapter), there was no universal set of standards for wiring schematics. This can make the interpretation of schematics challenging. Even if the new SAE standards become embraced by all OEMs, it will take a generation to obsolete the many different types of schematic used up until 2007. Figure 32–34 shows a Navistar International wiring schematic: See if you can navigate it to determine how the injector driver module (IDM) and ECM are powered up.

Virtual Schematics

Hard paper wiring schematics are not usually produced anymore by OEMs. The reason is that electrical and electronic systems have simply become too complex to represent on a single sheet of paper however large it opens up to. Most technicians today use schematics that have been downloaded off an SIS. Navigating schematics is required to identify the specific section of the electrical or electronic circuit you wish to work with. You can then display this on a laptop computer. Many OEMs today offer a variety of navigational options, including colorization and animation. This means you should get used to using them on the computer display and not waste time and paper by printing them out. Figure 32–35 shows a subsection of a schematic exiting a Mack chassis power distribution module downloaded from EIS.

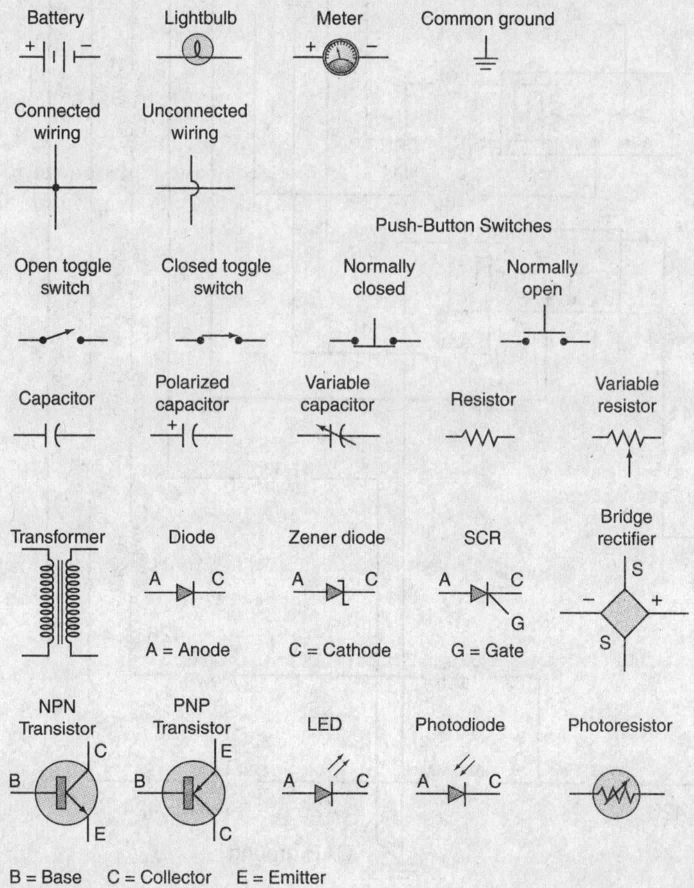

Figure 32–33 *Circuit symbols.*

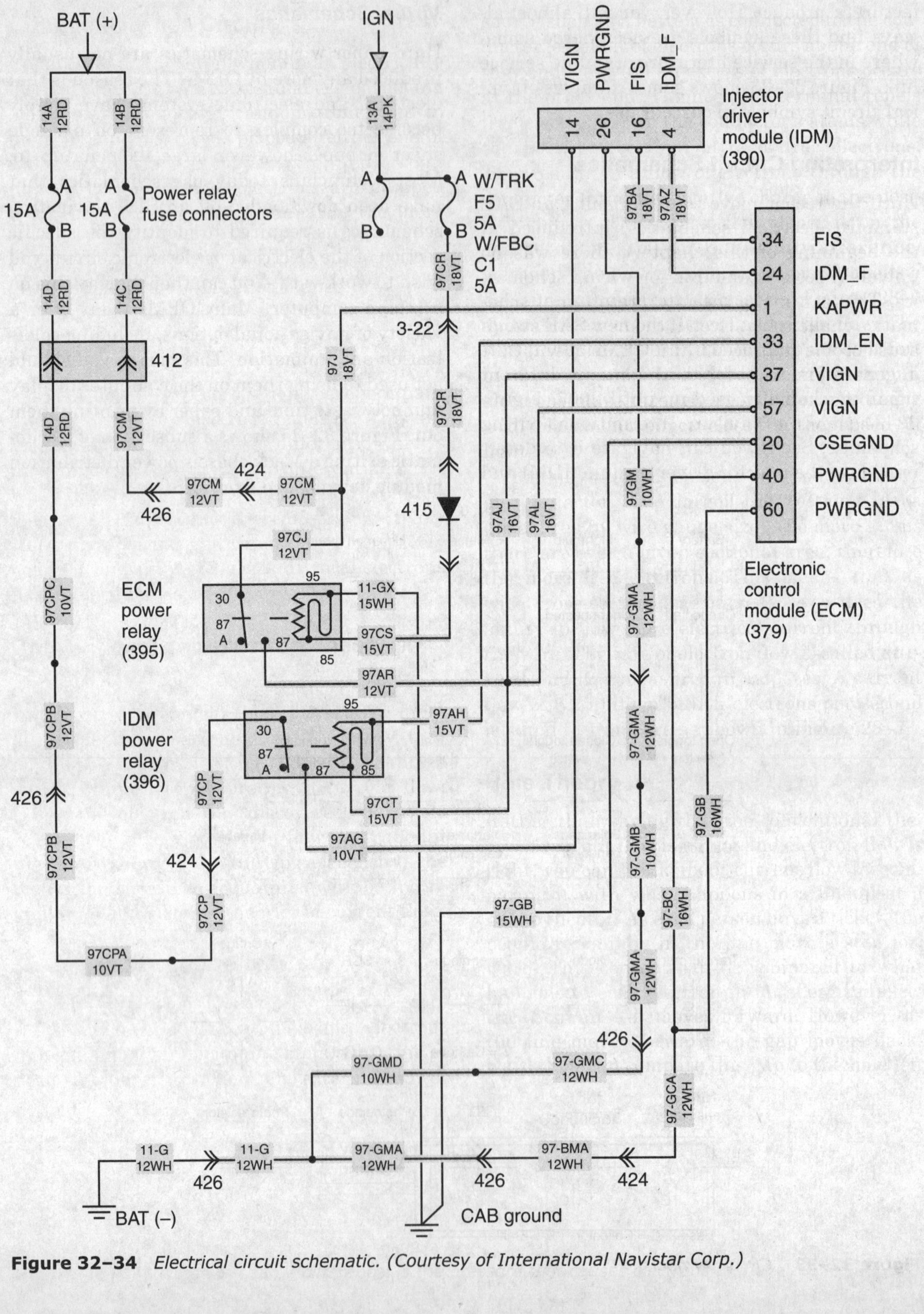

Figure 32–34 *Electrical circuit schematic. (Courtesy of International Navistar Corp.)*

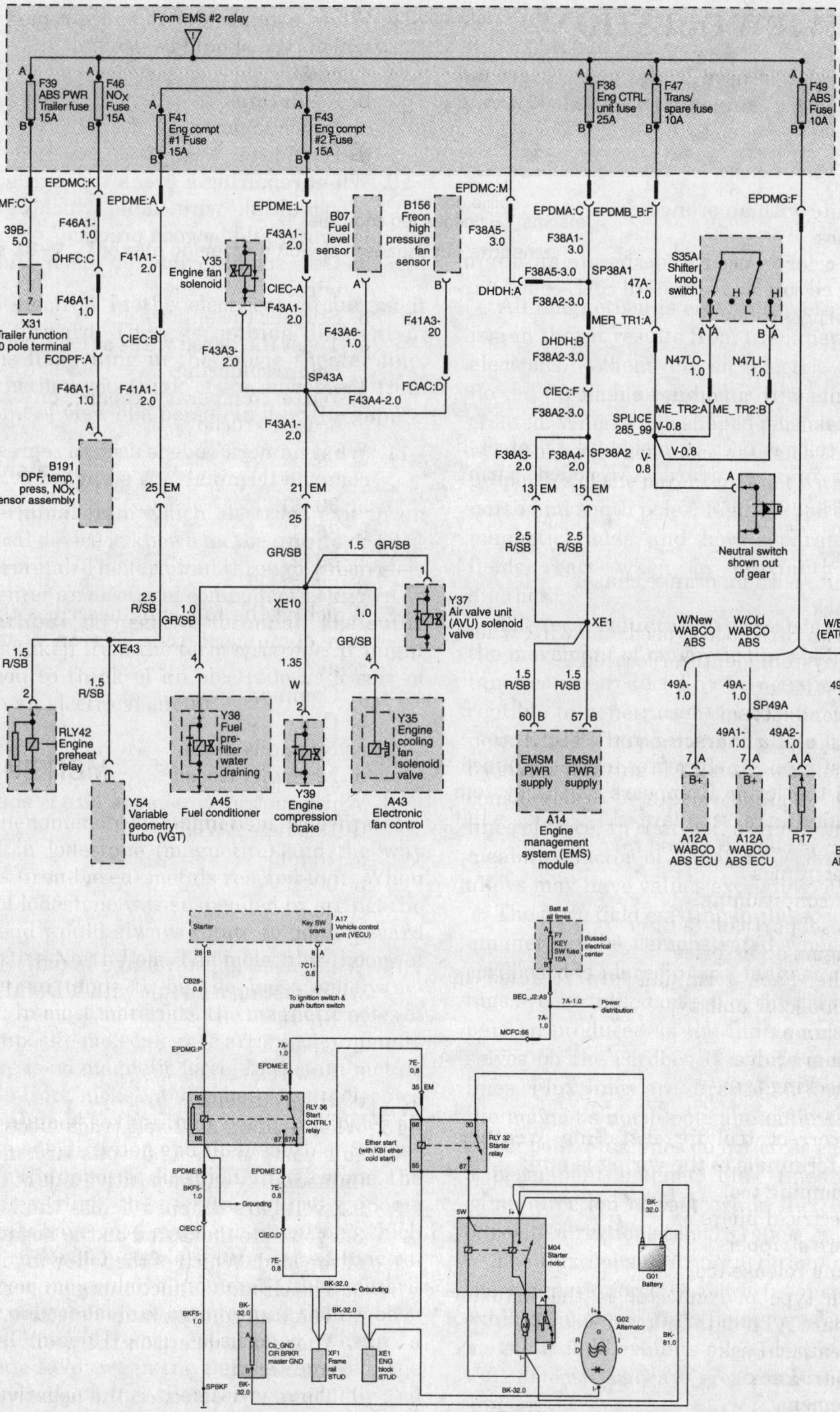

Figure 32–35 *Subsection of a schematic exiting a Mack chassis power distribution module. (Courtesy of Mack Inc.)*

REVIEW QUESTIONS

1. What color is used to designate a ground circuit in a standard SAE-coded wiring harness?
   a. Red
   b. Gray
   c. White with an orange chaser
   d. Black
2. What color is used to designate an unprotected circuit in a standard SAE-coded wiring harness?
   a. Red
   b. Gray
   c. White with an orange chaser
   d. Black
3. What color is used to designate a protected circuit in a standard SAE-coded wiring harness?
   a. Red
   b. Gray
   c. White with an orange chaser
   d. Black
4. When an SAE code is used on a wire, where is the system identifier located?
   a. First letter
   b. Second letter
   c. Last alpha character in the sequence
   d. Last numeric character in the sequence
5. When the letter E appears as the system identifier in a standard SAE code, what system is being referred to?
   a. Electronics
   b. Air conditioning
   c. Diesel particulate filter
   d. Engine accessories
6. Weather-Pack terminals are retained in the connector unit by:
   a. A clamp
   b. A jam bridge
   c. A locking tang
   d. Silicone
7. The correct tool for installing Weather-Pack terminals to the wiring is a(n):
   a. Crimping tool
   b. Electrical pliers
   c. Wire strippers
   d. Tang release tool
8. Which type of connector is the industry standard ATA data link connector?
   a. Weather-Pack
   b. Metri-Pack
   c. Siemens
   d. Deutsch
9. When using solder to splice repair wiring, which type should be used?
   a. 50/50
   b. Rosin core
   c. Silver solder
   d. Liquid set
10. When repairing a break through a section of a multiple wire cable, which of the following would be good practice?
    a. Use shrink cable on each individual wire.
    b. Stagger the splices.
    c. Twist the bared wires at each connection before soldering.
    d. Write "temporary repair" on the work order hardcopy.
11. What numeric code is used to represent the common terminal on a standard relay?
    a. 30 (3)
    b. 85 (1)
    c. 86 (2)
    d. 87 (5)
12. Which of the following describes the function of numeric code 87a on a standard relay?
    a. Common
    b. Coil
    c. Normally open
    d. Normally closed
13. Which metric wire gauge size is equivalent to 18 gauge AWG?
    a. 18 mm
    b. 00,22
    c. 00,8
    d. 05,0
14. What does ghost voltage usually indicate when testing a circuit with a DMM?
    a. Short
    b. Open
    c. Low resistance
    d. High current
15. When testing a 12-volt truck electrical circuit component, 12.5 volts are measured immediately on the positive side of the load, 9.2 volts are dropped across the load, and 3.3 volts are measured on the negative side of the load. Which of the following is *true*?
    a. The circuit is functioning properly.
    b. The load component is defective.
    c. There is a defect on the positive side of the circuit.
    d. There is a defect on the negative side of the circuit.

33

Multiplexing

INTRODUCTION

In the days of sailing ships there existed a chain of command that extended from a captain through a first mate, lieutenants, and ending up with sailors at the bottom of the chain, who would actually put into effect the instructions that began with a captain's whim. Communication was verbal. The captain delegated to subordinates. At the end of the chain of command were the sailors. They were entrusted with specific tasks such as raising a sail, turning the tiller, or dropping an anchor. Each task would probably be specialized to some extent and require some training. In this way, the critical functions of the ship were relayed and undertaken by word of mouth communications.

If we go back a couple of decades and take a look at a highway truck of that era, most of the critical functions were managed directly by the driver using mechanical, pneumatic, hydraulic, and electrical signals. An accelerator pedal required a mechanical linkage to connect the driver's boot with the engine fuel control mechanism. Likewise, every electrical function on the chassis had to be hardwired from dash switch through relay to whatever type of actuator was being controlled. Over the years, as chassis systems became more complex, more and more miles of wire were required to enable the electrical and mechanical outcomes required to manage the vehicle. That is, until **data bus communications** were introduced.

Data bus communications make managing a modern truck chassis similar to the way a large sailboat was managed a couple of hundred years ago. Instead of having a ship's captain bellow out some orders to be passed down a chain of command, a module broadcasts digital signals to a data backbone. These signals can either be read and responded to by other modules connected to the data backbone or ignored if the message does not fall within that module's realm of responsibility.

Data bus communications make use of a technology called **multiplexing** that has been used in truck computer control systems since the first electronic control modules (ECMs) were introduced on trucks in the late 1980s. In those early days, multiplexing was usually module to module and truck technicians did not have to have much of an understanding of what it was all about. For instance, one engine manufacturer used this type of communication to allow a chassis management module to "talk" to the fuel injection control module: This was a dedicated bus connection that meant the technician could not troubleshoot bus malfunctions. Today technicians regularly access the truck data bus to troubleshoot, reprogram, and read the systems within it.

Neural Network

Simply put, multiplexing means data sharing between multiple system control modules. Our definition of a control module is one that has both processing and outcome-switching capability. If you set up a chassis computer system so that all of the control modules "speak" the same language and provide a common, shared, communication path between them, then you have a multiplexed system. Today most trucks use multiplexing systems that communicate using just a couple of "language" systems. Some manufacturers use the term **neural network** to describe a comprehensively multiplexed truck chassis: This analogy makes the comparison of a multiplexed chassis with a human body's central nervous system. This is not really a fair comparison because a multiplexed truck chassis is much less complex and more logically organized than a human body.

Although the term sounds complex, multiplexing actually simplifies truck electronics. It does this by giving electronic subsystems a common communication language, and, by using a **data bus** or information highway, allows data signals to take the place of hard wire in the electronic input and output circuits. More recently, truck technicians have necessarily become more aware of the term *multiplexing* because some truck manufacturers are making increasing use of the technology in truck chassis.

Second-Generation Multiplexing

All trucks today use multiplexing. Some trucks use it much more than others. On every current truck, the chassis data bus is used to network all of the major system controllers (computers) used on the chassis. However, multiplexing is being increasingly used by a couple of original equipment manufacturers (OEMs) to do more than just get system controllers interchanging information. This new generation of trucks networks both the electronic controllers and many of the sensors and status switches used in the circuit. For instance, even the simplest electromechanical toggle switch requires two wires to signal on or off status: And those wires must run from switch to relay and from relay to control device. Multiplexed smart switches function differently. They require just one wire to signal up to five status conditions. A smart switch functions not by turning current flow on or off but instead by broadcasting a message to the data bus. Once broadcast this message can be sent through the data bus neural network and picked up by any controller that wants to use it.

Communications by digital signals rather than switched electrical circuits eliminates miles of hard wiring, reduces the number of I/O (in-out) pins on modules, and optimizes vehicle operation by giving electronic subsystems an "accountability" that extends beyond the hardware they control. Many highway/city bus technicians might already have a more advanced understanding of multiplexing because they have progressed to a higher level of sophistication in bus chassis, at least at this moment in time. An example of how the volume of hard wire can be reduced by multiplexing is demonstrated by the International Trucks Diamond Logic system that has decreased the number of wires to the instrument gauge cluster from 67 to just 7.

So what do you really have to know about multiplexing today? If you have ever used a six-pin or nine-pin **data connector** to connect a diagnostic reader or shop computer to read the electronic systems on a truck, you already have some experience of accessing a truck data bus. As every year passes, the extent to which a truck technician must be familiar with working with truck data bus systems increases. Within this chapter we reference the Freightliner M2/Cascadia and International Trucks Diamond Logic systems for examples on how multiplexing functions. Up to now, the fully multiplexed chassis has not been embraced by all the chassis OEMs. But both Freightliner and Navistar International have used it on medium-duty trucks since 2003 and have more recently released fully multiplexed Class 8 trucks.

MULTIPLEXING, CLIENTS, AND SERVERS

We have already said that multiplexing refers to transactions between a series of networked computers. In the professional world if you consult an accountant or lawyer, you become their customer and are known as their **client**. In the computer world anyone or anything that wants something is referred to as a client. For a transaction to take place there must be someone or something that fulfills that need. The fulfilling of a client need is provided by a **server**. So the data backbone serves as a path for connecting clients with servers. The messages used to make these connections are known as **packets**. There are rules about packets and how they are constructed—think of the difficulty and expense of sending an odd-shaped package by courier.

The roads used by telecommunications systems can be wires, fiber-optic cables, light beams, or radio waves. The speed limit on the highway is rated in miles per hour. The speed limit on a telecommunications channel is measured in baud and K-baud. More K-baud is better and faster. An interstate with more lanes is generally considered to be better and the number of lanes on a road can be likened to the telecommunications term **bandwidth**. They both represent volumetric capacity of the road or transmission medium. One refers to the number of cars and trucks that can travel simultaneously, the other, the number of packets or data volume that can be pumped down a channel. On the highway, because there are other cars on the road each filled with other people with their own objectives, you can closely compare the dynamics of road travel with data multiplexing—multiple packets traveling a highway heading from one place to another, using common transmission mediums, moving at pretty much the same rate.

Multiplexing modernizes electronic communications by making electronic circuits a lot more like our highway system. In doing so, miles of unnecessary hard wire and replicated components can be eliminated. Multiplexing uses electrons modulated by some strict **protocols** (rules and regulations) to simplify subsystem-to-subsystem electronic transactions. Analog inputs are converted to digital signals by the primary multiplexing processor: These signals can then be relayed digitally to slave processors and then changed back to analog format to effect actions.

MULTIPLEXING BASICS

As mentioned before, if you have worked on trucks manufactured in the 1990s, you already have some experience with multiplexing, even though it may have been unknowingly. Beginning in 2002, a new generation of truck chassis was introduced, beginning a second generation of truck multiplexing. All of the major OEMs have already introduced or plan to introduce what can be described as a fully multiplexed chassis. This replicates technology that has been used in city and highway buses since the late 1990s.

There is a good reason that buses moved ahead of trucks in their use of multiplexing. In a typical bus, the driver sits in the front and the engine, transmission, and most of the chassis electronics are located in the back. To hardwire connect the driver with these complex electronic systems located in the back of the bus required miles of wire that had to travel the length of the chassis. A data bus consisting of a twisted wire pair provides a simple solution. The driver generates commands from his position at the front of the bus, and these are sent as messages via an ECM down the communication backbone. These messages can then be read by any modules connected (networked) to the data bus. The modules process the messages and generate whatever outcomes are desired. By using switches that generate communication signals to the data bus rather than changes in electrical status, many miles of hard wire can be eliminated.

Our initial definition of multiplexing was to describe it as chassis information sharing between the different onboard computer control systems to optimize vehicle performance. For instance, some of the key input sensor signals are required by more than one of the chassis control modules. The throttle position sensor (TPS) is a good example of such a signal. The TPS signal is a required input for the engine management, transmission management, dash display module, and antilock brake system (ABS)/automatic traction control (ATC) and collision warning systems. The TPS signal is hardwire delivered to one module (usually the engine ECM) then broadcast over the data bus so that any other networked modules that require this signal can pick it up. Some electronic systems that are commonly multiplexed include the following:

- Engine control module
- Fuel injection control module
- Transmission control module
- Dash display module
- Chassis control module
- ABS
- ATC system

- Electronic immobilizer
- Collision warning system

Exchange of information between computer-controlled systems reduces the total number of sensors required and optimizes vehicle performance by making each system perform according to the requirements of the vehicle system rather than each of its subsystems.

Conventional data transmission in a vehicle required that every input and output signal had to be allocated an individual conductor (dedicated wire and terminals) because binary signals can only be transmitted using the states "1" or "0" (binary code). On/off ratios can be used to transmit continually changing parameters such as the status of an accelerator pedal-travel sensor. However, the increase in data exchange volume between vehicle electronic systems has reached such dimensions that it does not make sense to attempt to handle it by older hardwire and plug-in connectors, such as that shown in Figure 33–1. Note that each module shown requires a direct connection to any other module it needs to communicate with. Today the objective is to keep wiring-harness complexity down to a controllable level to reduce costs and downtime when there is a hard wire malfunction.

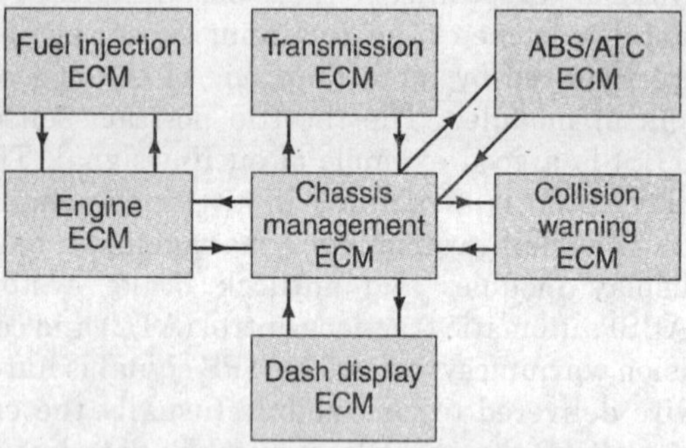

Figure 33–1 *Older style data transmission—no common multiplex bus.*

Power Line Carrier

A more primitive method of multiplexing known as **power line carrier** has been in use in trucks since 1985. Power line carrier enables communication transactions to take place on a nondedicated communication wire. Power line carrier communications were introduced when ABS became an option on highway trailers, but its use became mandatory in 2001, 3 years after the introduction of mandatory ABS. Trailer ABS requires that a warning light be illuminated in the dash of the truck in the event of a trailer ABS malfunction. Because all the wires on a standard SAE J560, seven-pin connector between truck and trailer were already dedicated, power line carrier technology was used to convert a communication signal to a radio frequency signal and then superimpose it over the 12-volt auxiliary power wire. A transducer converts the radio frequency signal back to a signal that the receiver ECMs can use. These signals use the original **J1587/J1708** data bus hardware and protocols and were diagnosed accordingly.

Types of Logic

Most current engine management systems use largely open loop (fuzzy) processing as opposed to closed loop logic so inputs from any onboard electronic circuit can influence how the engine is being managed at a given moment. This adds to volume of exchange data between the various control modules managing a chassis. To illustrate the difference between close loop and fuzzy logic, we use an example you should be familiar with if you have worked on or driven a truck using cruise control.

The older hard cruise, commonly used in most autos, required that the driver set a road speed of, say, 60 mph. After the road speed has been selected, the road speed sensor drives the management of the powertrain to maintain the desired road speed regardless of conditions. This type of close looped operation puts the road speed sensor in command of the powertrain and that can be a disadvantage. For instance, if the vehicle were traveling through mountainous terrain, attempting to maintain the input road speed at exactly 60 mph would waste fuel.

Most trucks today use "soft" or "smart" cruise. When the driver inputs a road speed cruise control command into a soft cruise system, the circuit "thinks" for itself. It uses multiple inputs and programmed instructions to manage fueling. For instance, if the soft road speed was set at 60 mph and the vehicle was traveling through undulating hills, the processing cycle uses an algorithm to "learn" from the terrain and in doing so, uses a variety of inputs to plot an actual road speed. From a fuel economy point of view, it makes sense to allow the truck to slow slightly through each uphill climb and accelerate on each downgrade to prepare for the next hill. Hard cruise is an example of closed-loop operation, while soft or smart cruise is an example of fuzzy logic.

Serial Data Transmission

So you can see now that many of the problems of data transfer volume using conventional interfaces (hard wire) can be simplified by using **bus systems** or data highways. The basis of serial data transmission is to use a single wire that delivers "instructions" rather than electrical signals to the controller modules. One example is **controller area network**, or **CAN**, a data bus system developed by Robert Bosch and Intel for vehicle applications. CAN is a serial data transmission network used for the following applications in a vehicle:

- ECM networking
- Comfort and convenience electronics
- Mobile onboard and external communications

CAN 2.0 is the basis for SAE J1939, the high-speed network in standard use on trucks and buses in North America. J1939 functions from 125 kilobits per second (kb/s) up to 1M b/s (250 kb/s is typical), making it a Class C bus: Class A and B buses use much slower speeds and no longer have applicability in trucks. In comparative terms, J1939 is about 25 times faster that the J1587/1708 (the original) data backbone. Transfer rates must be high to ensure the required real time responses are met so you can bet that transaction speeds will continue to increase in the coming years.

Multiplexing Clock Speeds

Microprocessor clock speeds of at least 16 MHz are required for J1939 transactions. A clock speed of 16 MHz translates to an ability to make up to 16 million binary "decisions" per second: Millions of these binary decisions are required to process a simple output command. Most current truck engine and transmission management processors have clock speeds of at least 16 MHz and are designed to broadcast on both the J1587/1708 and J1939 buses.

ECM Networking

CAN uses serial data transmission architecture. *Serial* means single track as opposed to multiple track, which is known as *parallel* transmission. Several system management modules such as the engine, fuel system, ABS/ATC, electronic transmission, dash display, and collision warning modules are networked on the serial data bus. Each ECM is assigned equal priority and connected using a linear bus structure as shown in Figure 33–2. One advantage of this structure is the fact that should one of the stations (subscribers) fail, the remaining stations continue to have full access to the network. The probability of total failure is therefore much lower than with other logical arrangements such as loop or star (hub) structures like that shown in Figure 33–1. With loop or star structures, failure of a single stations or the central or command ECM necessarily results in total system failure.

ISO 9141

Control module interconnectivity and data protocols fall under the International Standards Organization's **ISO 9141**. Because American-engineered and -built equipment is marketed throughout the world it makes sense that both the hardware and communication protocols are internationally common. Most OEMs manufacturing equipment on the J1939 multiplexing platform ensure that it is ISO 9141 compliant.

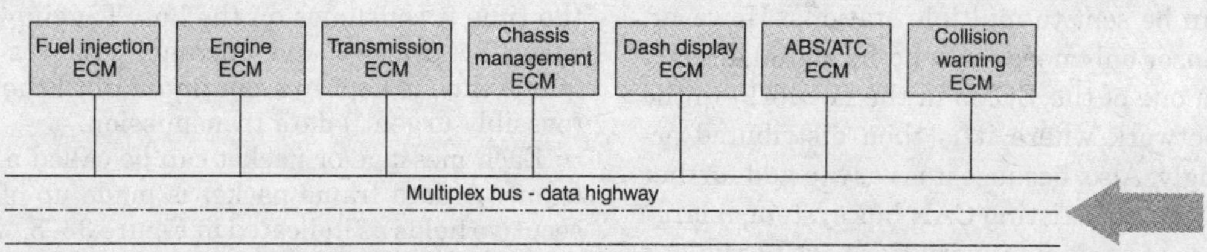

Figure 33–2 *Bus topology of J1939: linear bus structure.*

ISO 9141 specifies the requirements for the interchange of information between the various electronic/engine control units (ECUs) on vehicles to facilitate readout, diagnosis, adjustments, and data exchange. It does not apply to system-specific equipment or dedicated CAN connections.

Bus Topology

The word *topology* means an organized structural system that is unaffected by being subject to a sequence of events. The term **bus topology** is used in multiplexing to indicate that no single controller networked to the bus has more priority or status than another: No ECM/ECU is "in-charge," in other words. For instance, if a data bus was to be cut in half, theoretically all of the controllers on each side of the break should be able to communicate with each other though not with the controllers on the other side of the break. Figure 33–2 shows the bus topology used in a truck data backbone, and as you can see it is linear.

Content-Based Addressing

CAN transactions are simplified. Instead of addressing individual stations in the network, an identifier label is assigned to every message pumped out onto the data highway. Each message is coded with a unique 11- or 29-bit identifier that identifies the message contents. For instance, engine speed data is of significance to the engine, transmission and, collision warning systems but probably has no importance to the climate control module. Each station is designed to process only those messages whose identifiers are stored in its acceptance list. This means that all others messages are simply ignored.

Content-based addressing means that a signal can be sent to multiple stations. However, the sensor only needs to send its signal directly (or via one of the ECMs in the network) to the bus network where it is then distributed accordingly. Also, because it is easy to add further stations to an existing CAN bus system, a large number of equipment variations are possible.

Assigning Priorities

In a multiplex system, message identifiers can be labeled for data content and/or the priority of any message sent. In a vehicle management network, some signals are necessarily prioritized over others. An example of a high-priority signal would be the accelerator pedal angle or TPS signal. Any change in the TPS signal value must be responded to at high speed and is therefore allocated a higher priority than a signal that changes relatively slowly, such as that from the coolant-temperature sensor. A brake request signal would have the highest priority status, and a critical signal such as this can effectively shut down the data bus for a nanosecond (one-billionth of a second) or so.

Bus Arbitration

Handling traffic on the data highway requires rules and regulations. Just as every driver on an interstate is taught to cede to emergency vehicles, so must there be rules to handle data traffic. When the bus is free any station can transmit a message. However, if several stations start to transmit simultaneously, bus arbitration awards first access to the message with the highest priority with no loss of either time or data bits. Lower-priority messages are shuffled by pecking order to automatically switch to receive and repeat their transmission attempt the moment the bus is freed up.

Message Tagging

In a busy airport with insufficient runways like Chicago's O'Hare, there is an orderly line-up at the beginning of each runway for takeoff. The protocol (rule) on the airport runway is usually to sequence takeoffs in line-up order. The protocol in the truck data bus is necessarily different. Each message is assigned a tag, a **data frame** of around 100–150 bits for transmission to the bus. The tag codes a message for sequencing transmission to the data bus and also limits the time it consumes on the bus. Tagging messages by priority and currency ensures that queue time is kept to a minimum until the next (possibly urgent!) data transmission.

Each message or packet can be called a data frame. A data frame packet is made up of consecutive fields as indicated in Figure 33–3, which shows how a typical packet is constructed.

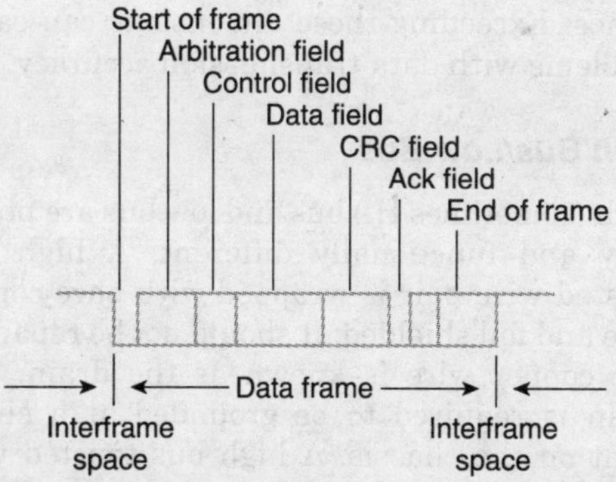

Figure 33–3 *CAN message format.*

Start of Frame Start of frame announces the start of a message and synchronizes all stations on the data bus.

Arbitration Field This contains the message identifier and an additional control bit. When this field is broadcast, the transmitter tags the transmission of each bit with a check to ensure that no higher-priority station is also transmitting. The control bit functions to classify the message as a data frame or remote frame.

Control Field The control field simply indicates the number of data bytes in the data field.

Data Field The data field is sized between 0 and 8 bytes. A message with data length 0 can be used for synchronizing distributed processes.

Cyclic Redundancy Check (CRC) Field The CRC proofs a message to identify possible transmission interference.

Ack Field The Ack or acknowledgment field contains acknowledgment signals by which the receiver stations on the bus indicate they have received a noncorrupted message.

End of Frame Field The end of frame (EOF) indicates the end of a message.

Message Bit Encoding

Figure 33–4 shows the number of bits dedicated to each segment of a data packet. Remember that there are 8 bits to a byte.

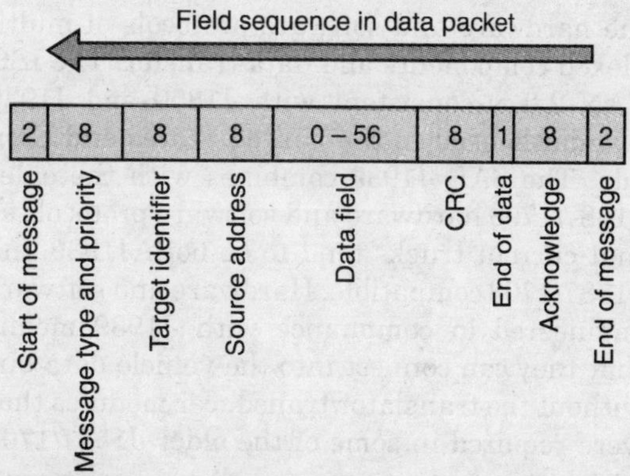

Figure 33–4 *Message bit encoding of a data packet.*

Integrated Diagnostics

A vehicle multiplexing bus system must be provided with monitoring capability so that transmission, transaction, and reception errors can be detected. This would include the check signal in the data frame and monitoring components in a transmission. It is a feature of a multiplex bus that each transmitter receives its own transmitted message again and by file comparison can detect any deviation or corruption. If a station identifies an error, it sends out an error flag that halts the current transmission. This prevents other addresses networked to the bus from receiving a faulty message.

In the event of a major station defect, it is possible that all messages, both good and bad, would be terminated with an error flag. To protect against this, a multiplex bus system has a built-in function that enables it to distinguish between intermittent and permanent errors. Furthermore, this feature can sometimes localize station failures and route the data network to an appropriate failure strategy. The process of localizing a station-type failure is based on statistical evaluation of error factors so it is not entirely foolproof.

Multiplexing Standardization

Multiplexing standardization for vehicles has been orchestrated by the ISO internationally and by the Society of Automotive Engineers (SAE) in North America. In North America SAE J standards, **J1850** (light-duty vehicles) and **J1939** (heavy-duty highway vehicles), define the hardware and software protocols of multiplexed components and data transfer. The ISO CAN 2.0 is consistent with J1850 and J1939 standards used in the United States and Canada. The SAE J1939 combines with the older J1587/1708 hardware and software protocols so that current trucks tend to be both J1939 and J1587/1708 compatible. Hardware and software engineered in compliance with J1939 means that they can connect into the vehicle data bus without the translator/transducer modules that were required in some of the older J1587/1708 systems.

The J1939 Backbone

A multiplexing data bus must be stable and be protected against unwanted radiation interference that could corrupt the data signals. The data bus or backbone used in a J1939 multiplexed chassis is a pair of twisted wires, color coded yellow and green. The reason for using a **twisted wire pair** is to minimize the surface area on which outside low-level radiation or **electromagnetic interference (EMI)** can act on the bus. The idea is to prevent the data backbone from acting like a radio antenna. Reducing EMI (electromagnetism, radar, microwave) susceptibility is critical to protecting data from transmission corruption.

The SAE standard for the J1939 data bus requires that the data bus wires twist through a full cycle once per centimeter—that is, 2½ times per inch. At either end of the data bus, a **terminating resistor** is used. The function of the terminating resistors is to prevent the twisted pair from attracting signal interference by acting like a giant antenna. Terminating resistors test at 120 ohms: Their function is explained in a little more detail later in this section. Also there cannot be any open T-connector cavities. At points where the backbone has to be spliced, special Deutsch T connectors with gold-plated contacts are used. Gold plating minimizes resistance because the nominal voltage on the data bus is low, which is usually around 2.5 volts or lower.

The J1939 data backbone uses stubs to make the physical connection between the backbone and each controller networked to it. The maximum length of a J1939 data backbone is 130 feet and each stub should be no longer than 40 inches. Exceeding these dimensions can cause problems with data transmission accuracy.

High Bus/Low Bus

High (sometimes Hi) bus and low bus are physically and functionally different. A high bus twisted wire pair is wrapped with bare copper wire and foil shielded: It should not be repaired. The copper wire is known as the drain: The drain is required to be grounded at a single point on the chassis. A high bus twisted wire pair transacts at the highest speeds (up to 800 kb/s to 1 Mb/s) and is used for real time data signaling such as would be required by accelerator pedal and electronically controlled transmission system initiated input/output transactions. A low bus twisted wire pair has no foil shielding. Low bus signaling in older systems included such things as audio system transactions, for instance, an increase in audio volume as road/engine speed increases. However, the use of low bus systems is increasing on late-model vehicles.

The high speed and volume of data pumped through a truck chassis data bus requires wiring that is a little more complex. There are two reasons for twisting the two data wires used in the bus:

- To provide immunity to magnetic fields: The voltage induced into any conductor by a magnetic field depends on distance so by twisting the data wires, induced voltage effect will be about equal on both wires. In digital signaling using binary pulses, signal transceivers (any modules networked to the bus) are designed to look at the difference between CAN H (or CAN–) and CAN L (or CAN+), not voltage levels referenced to ground.
- To provide consistent capacitance values: The distance between the two conductors on the bus alters the capacitance between them, with capacitance decreasing as the wires become further apart. The insulation material the conductor wires are wrapped in also impacts on capacitance because it acts like the dielectric discussed in Chapter 28. On the data bus, the dielectric properties of insulation should be constant: Substituting ordinary wire for J1939 cable can corrupt communications on the bus by altering the impedance.

Impedance

If, like many persons growing up today, your first experience of electronic systems was setting up audio and visual equipment, you will know that high-frequency cables are specified in ohms. For instance, coaxial TV cable has a center conductor wrapped in Teflon with the other conductor braided around it. This type of cable has what is known as a *characteristic impedance* of around 75 ohms. You cannot test characteristic impedance with an ohmmeter; it is just a measure of how the conductor appears (regardless of length) to a high-frequency signal.

A truck J1939 data bus has a characteristic impedance of 120 ohms. This is achieved by maintaining the dielectric constant (wire insulation) and keeping the physical spacing between the two conductor wires constant by having a consistent twist pattern. Once again, because this is characteristic impedance, you cannot measure it with an ohmmeter. Do not confuse the 120-ohm characteristic impedance with the 120-ohm terminating resistors used at the end of data backbone: This can be measured with an ohmmeter.

Terminating Resistors

Terminating resistors are used to absorb signal energy, leaving no energy for reflections or echoes that result in electronic noise. Electronic noise causes scrambling of signals: It is a little like listening to an outdoor public address system where audio input is being received from several different speakers different distances away.

Besides minimizing electronic noise, terminating resistors also provide a low-resistance path for current to flow between CAN H and CAN L. This permits capacitance to discharge and cancel rapidly. The length of time for a capacitor to discharge is proportional to the resistance of the conductor it is discharging through. If system capacitance cannot rapidly discharge when a device is trying to transmit at a low level, then the voltage differential between CAN H and CAN L remains high, which can shut down the data bus.

Missing Terminating Resistors

If only one terminating resistor is missing from a J1939 data backbone, it will likely function problem-free, though if you scoped the waveform, it could be noticed due to longer capacitive discharge times and increased signal reflections. However, if both terminating resistors are missing, no communication can take place. When this happens, the length of time required for each signal pulse pumped down the bus to neutralize increases and the resulting signal reflections scrambles the data bus so no transactions are possible. Figure 33–5 compares a pulsed scope signal on a J1939 data bus equipped with both terminating resistors with one that has had both terminating resistors removed.

Loss of J1939

In most trucks, if the J1939 connection is lost, the driver will be alerted by dash display and often a chassis signal such as flashing the hazard lights. Low bus wires are unshielded and may be repaired under some circumstances. Strictly observe the OEM instructions when repairing twisted wire pairs: This will include maintaining the existing-twisting cycles, gauge size, and using the correct solder.

As truck chassis management electronics become increasingly dependent on the data bus, the consequences resulting from a data bus malfunction increase in severity. Physical damage to any data bus described as high bus (shielded) warrants replacement, not repair. High bus lines are shielded. A J1939 high bus consists of the twisted wire pair (communication wires) already discussed, a shield wire, and often a couple of filler wires that function to keep the communication wires separate. The harness is usually wrapped with tin foil-like shielding.

Summary of Multiplexing Basics

Although the definition of multiplexing is fairly specific when applied to vehicle electronics, if we broaden the definition somewhat, it is easy to see how it can be effective. In organizational theory, a company president tends not to do very much of the actual work himself. He dele-

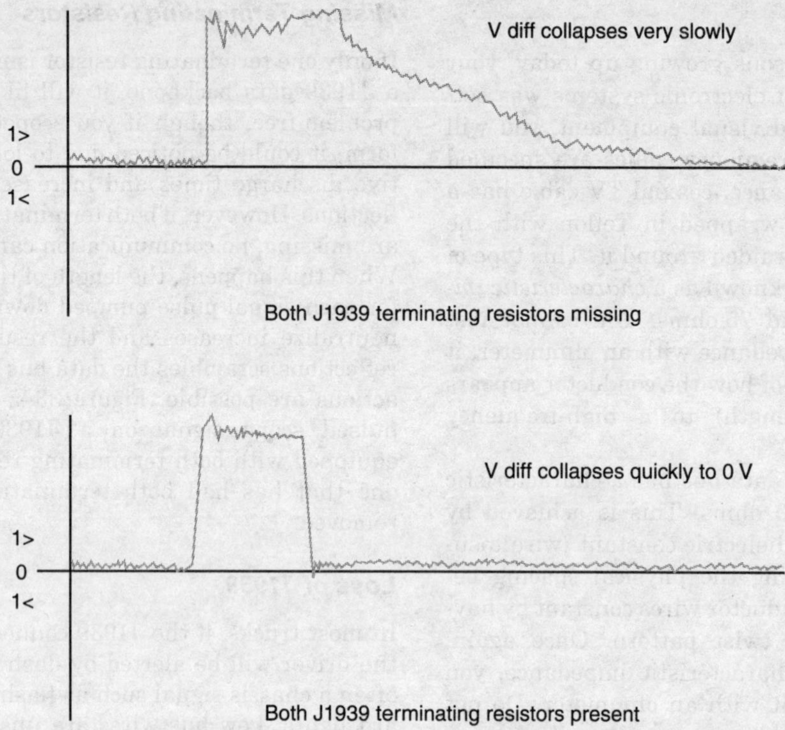

Figure 33–5 *Comparison of J1939 scope patterns with and without 120-ohm terminating resistors.*

gates those responsibilities to those who report to him. Communications within the company flow through a chain of command. In an effective company, reports also flow back up the chain of command. Sure, there are going to be occasional problems in the two-way flow of information but in a good company there are checks and balances built into communication systems that quickly identify problems and rectify them. Using the example of a truck electrical system, rather than control a 20-amp blower motor directly through a switch that could overheat, we use a low potential current through the control switch, which acts to control a relay that carries the high current load.

In computer technology for many years electronics have been used to simulate relays in devices known as **programmable logic controllers (PLCs)**. These small computers can simulate thousands of relays in a fraction of the space using elaborate stacking of transistors and **silicon controlled rectifiers (SCRs)**. Now some of the principles used in PLCs are moving beyond the ECM housing. A key factor that has moved us into a second generation of multiplexing is this "smart" switching technology that we will take a closer look at in the next section of this chapter.

Standard Switches

Every truck on the road today is equipped with many different switches, and multiplexed trucks will not reduce this to any great extent. Some examples of typical standard switches are cruise control on/off, cruise control set/resume, hazard warning switch, and ignition switch. While a switch may be hardwired, making it standard by classification, its status can also be broadcast on the data bus.

Smart Switches

Smart switches are used to describe two distinct types of switch used in multiplex circuits. A smart switch may have some processing capability and use that to broadcast switch status onto the data bus. Alternatively, the term *smart switch* can be used to describe a **ladder switch**, named because it contains a ladder of resistors, usually five per switch, known as a ladder bridge. The processor that receives data from the ladder switches has a library of resistor values that enables it to identify switch status and its commands. The Freightliner M2 truck uses a ladder bridge-type, smart switch in the dash whenever possible, and also has capability of monitoring the switch status that can be broadcast on the data bus. Each smart switch has a light emitting diode (LED) designed to indicate (to the driver) that a switch request has actually been effected. Sometimes the LED will flash while a particular action is in the process of being effected and stay on when completed. Smart switches can be toggle (two position on/off), multiposition, or spring-loaded, momentary operation.

The module responsible for monitoring data bus negotiations is designed to self-check smart switch operation and signal a fault if one is detected. It does this by analysis of the ladder bridge resistances. If a wire loosens or a terminal corrodes (high resistance), the system will not only know it, it can probably locate it. Ladder switches are simple in construction with little to go wrong and eliminate the processing capability found in other types of OEM multiplex circuit switches.

Field Effect Transistors

Okay, now it is time to take a look at what is happening at the other end of the processing circuit. First, we should examine the role of field effect transistors (FETs). These are becoming commonplace in truck chassis electronics because they are inexpensive to manufacture and function reliably. FETs are electronic relay switches. In their simplest formats, there are two types of FETs: N channel and P channel. The channel behaves as a resistor that can conduct current from the source side to the drain side. The gate controls the resistance and therefore the operation of the semiconductor device by saying whether or how much current flows through the device. So, depending on the gate voltage, the FET can be used either as a straightforward relay switch (most current truck applications) or as an amplifier. As a "relay," the FET has some great advantages in that gate-channel resistance is so high that, firstly, there is almost no current flow in the gate-current circuit, and, secondly, there is all but no effect on external components connected to the device. Figure 33–6 shows a schematic of an FET.

Wake-up

In the same way that the electronics in your automobile "wake up" (usually on door open/dome light), the multiplexing electronics in a truck are designed to reactivate following a period of inactivity. Wake-up signals are sourced from the ignition switch, dome light, brake lights, and headlights.

Simple Multiplexing Transaction

To show how a simple multiplexing transaction takes place, we take a look at a simple switch operation from switch to signal to outcome. The suspension dump, smart switch is used to illustrate the execution of a data bus enabled command. The mechanical objective (the desired

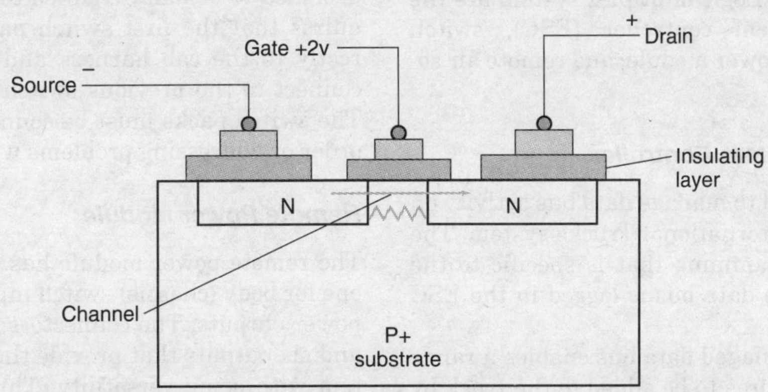

Figure 33–6 *FET operation: An FET uses a positive charge to the gate, which creates a capacitive field to permit electron flow. It acts like a relay with no moving parts.*

"outcome") of this operation is to exhaust air from the chassis air suspension. The operator actuates the suspension dump, ladder switch. This causes the dash suspension dump switch to start flashing and alters the resistance status on the ladder bridge so a packet is broadcast to the data bus. The data on this packet means nothing to any of the modules on the data bus until it is delivered to the module with the algorithm designed to process this input.

Using the example of a Freightliner multiplexed chassis, this would be the bulkhead module (BHM). The BHM then broadcasts a command signal over the J1939 data bus to another module with some switching capability: the chassis control module (CHM). The CHM in turn generates an output signal to the FET that acts as a relay for the suspension dump solenoid. The signal puts the FET in forward bias, completing the solenoid ground circuit and energizing the coil. Source power for the solenoid is the power distribution module. At the point the action is completed, that is, the suspension dump has been effected, and air drains from the suspension air bags, the dash LED on the suspension dump switch ceases to flash and remains on. This is a simple data bus transaction that will be used increasingly in the next few years. Expect to see a lot of FETs in truck electrical circuits in the near future.

Diamond Logic

The International Trucks Diamond Logic multiplexing system functions a little differently from the Freightliner system. The components of the Diamond Logic multiplex system are the electronic system controller (ESC), switch packs, remote power module, and remote air solenoid module.

Electronic System Controller

The ESC is used to manage data bus activity on the Navistar International Trucks system. The logic and programming that is specific to the operation of the data bus is logged in the ESC module.

The ESC-managed data bus enables a range of optional features to be added to the truck by checking boxes on a computer screen or by dragging and dropping graphics on a virtual electronic wiring diagram displayed on the Navistar EST (known as EZ-Tech) screen. After adding options to the chassis electronics, International requires that users access its data hub and upload the modified data files. As with any other type of modification to the vehicle files, the purpose is to make any new or modified files available to other service facilities and technicians.

International Trucks, in common with other OEMs, is currently working on enabling satellite communications to network its data hub with the chassis data bus electronics. This could eliminate the hard wire requirement for upload and download transactions with the International data hub: The system would function similarly to the fleet satellite communications in current use.

Switch Packs

The switch packs contain long-life microswitches located behind what looks like a typical truck dash rocker switch. The ESC communicates the switch packs monitoring the switch status. Switch status is therefore monitored by the ESC and then broadcast over the data bus to be picked up by other modules networked to the data backbone such as engine, transmissions, brakes, and the remote air solenoid modules.

Each truck may have up to two 6-switch packs and one 12-switch pack. Some low-optioned, late-model trucks may also use a three-switch pack that has the same physical dimensions as a six-switch pack. The switch packs are designed to be daisy-chained together: This requires that the first switch pack connects directly to the cab harness and the remainder connect to the previous switch pack in series. The switch packs must be connected in correct order or addressing problems will result.

Remote Power Module

The remote power module has two connectors: one for body (chassis) switch inputs and one for power outputs. The connectors have six inputs and six outputs that provide the multiplex system with some versatility. Three power modules are available, which can provide up to 18 20-ampere outputs.

A real advantage of the International Trucks system is that, using its Diamond Logic software and an EST, electronic gauges and rocker switches can be relocated by dragging and dropping images on the EST computer screen. This means that the physical locations of switches is completely soft and can be arranged according to user preference. Aftermarket equipment can be installed and powered up by either a rocker switch in the cab or by running a single wire into a remote switch input. Connecting the selected remote switch input to battery voltage will instruct the module to enable 12-volt power to the selected power module output. Biasing the switch to ground opens the electrical supply to the added circuit.

Remote Air Solenoid Module

In much the same way as the Freightliner system, the air modules are a means of controlling air-operated accessories such as differential locks, suspensions, or transaxles off the data bus. A system air supply is routed to one input on the air module. Modular air solenoids are supplied by this single air source. The on/off status of the solenoids is controlled by the ESC using the data bus.

Virtual Fusing

Navistar FETs are designed to feed back the current values they conduct to the controller that actuates them. This allows a microprocessor to shut off an FET if it is conducting more current for which it is specified. For instance, if too many lights were installed on a circuit, causing an excess current draw, the controller could shut down that circuit. Virtual fusing is designed to have identical performance characteristics to circuit breakers. The difference is that processor logic is used to determine the manner in which virtual circuit breakers cycle so we classify them into type 1 (cycling) and type 2 (noncycling).

Type 1 Virtual Fusing

This is used on headlamp and wiper circuits. In a current overload condition on the circuit, the ESC first turns the circuit off, then after ½ second it retries the circuit. If the current overload no longer exists, the circuit functions normally. However, if the overload condition continues, the ESC doubles the length of time between each retry, 1 second, 2 seconds, 4 seconds, up to 512 seconds. If the 512 recycle is achieved, the ESC shuts down the circuit until the ignition key is cycled.

Type 2 Virtual Fusing

Type 2 virtual circuit breaking logic is used on all the remaining FETs. Should a circuit be subject to a current overload, the ESC shuts off the circuit, then retries after ½ second. However, using type 2 virtual breaker logic, if the short or overload still exists after the initial retry, the ESC shuts the FET off and does not initiate a retry until the ignition key is cycled. If three successive key cycles take place with an overloaded circuit in memory, the ESC will only briefly (100 ms) permit the FET to be switched on after each key cycle just to test the overload condition.

Benefits of Networking

There are four primary benefits of networking a truck electronic system:

1. Greatly decreased hard wire requirement, reducing the size of the wiring harnesses. This impacts cost, weight, reliability, and serviceability.
2. Sensor data such as vehicle speed, engine temperature, and throttle position is shared, eliminating the need for redundant sensors.
3. Networking allows greater vehicle content flexibility because functions can be added by making simple software changes. Existing systems would require additional modules and additional I/O pins for each function added.
4. Many additional features can be added at little or no additional cost. For instance, driver preference data once installed in memory can be routed on the data bus to multiple processors to share such diverse information as seat preference, mirror positions, radio station presets, and engine governor type (LS or VS).

ACCESSING THE DATA BUS

Truck technicians are required to access the chassis data bus to read, troubleshoot, and reprogram the systems networked to it. The photo sequence in this section illustrates through the procedure required. A number of different generic and proprietary tools may be used to make the connection, but commonly today a handheld, laptop, or portable computer system is required. Whatever kind of computer is used, it must be loaded with either generic reader or OEM read/diagnose/reprogram software. The connection is made using one of two types of ATA connector.

J1587/1708 Connector

A J1587 data bus was the original truck data backbone: It is accessed using a six-pin **Deutsch connector**. J1587 is an SAE standard that governs the communications protocols used for data transactions. J1708 governs all the hardware standards used on a J1587 data bus. Figure 33–1 shows the cavity pin assignments used on a J1708 ATA connector.

J1939 Connector

The J1939 data bus is more recent and it covers both hardware and software standards. Because a current truck chassis equipped with a J1939 data bus is also equipped with the older J1587/1708 data bus, the data connector used has some extra pins. A J1939 connector is a nine-pin connector with the cavity pin assignments shown in Figure 33–7. High bus connections within the data bus backbone are made on a three-pin connector (also known as *stubs*) with cavity pin assignments as shown in Figure 33–7, while low bus connections use a two-pin connector.

Making the Connection

You will need an EST loaded with the OEM software plus the appropriate connectors and cabling. Most transactions can be undertaken without network access. Examples of ESTs are:

- ProLink equipped with Heavy-Duty Reader cartridge
- ProLink equipped with OEM specific cartridge
- ProLink with MPC (multiprotocol) cartridge and OEM specific data card
- Handheld (Palm OS) computer and OEM software
- Laptop/notebook personal computer (PC) and OEM software
- PC station and OEM software

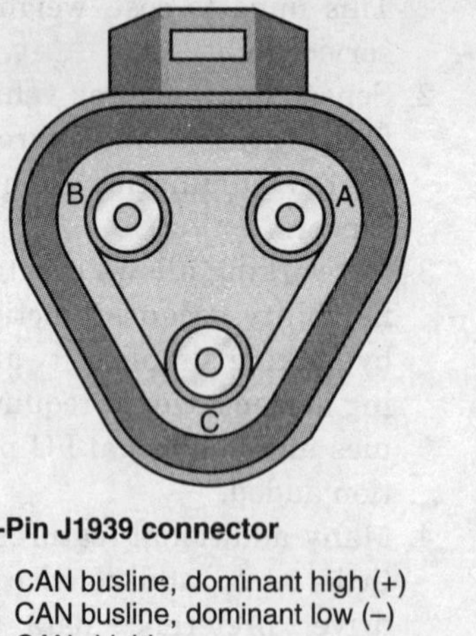

3-Pin J1939 connector

A CAN busline, dominant high (+)
B CAN busline, dominant low (–)
C CAN shield

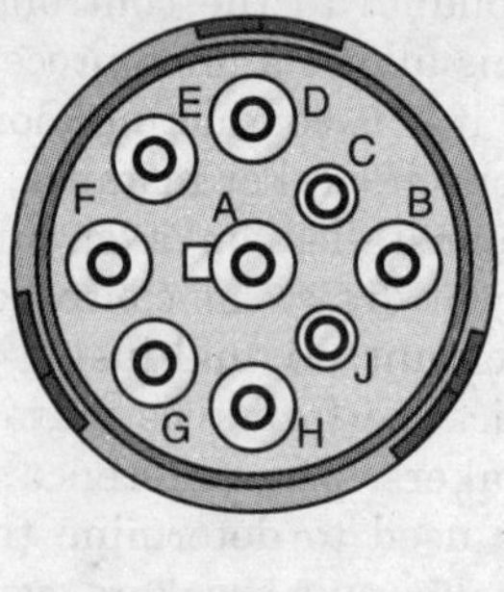

9-Pin J1939 connector

A Battery negative
B Battery positive
C J1939 CAN busline, dominant high (+)
D J1939 CAN busline, dominant low (–)
E CAN shield
F J1587 busline, dominant high (+)
G J1587 busline, dominant low (–)

Figure 33–7 *J1939 connector cavity pin assignments: three-pin Deutsch for high bus connections within the bus backbone, nine-pin Deutsch to connect an EST to the data bus.*

Identify MID

After connecting the data connector, you will be required to identify the message identifier (MID) you want to communicate with. All this means is the actual electronic system on the data bus you want to talk to. If you reference Figure 34–5, this shows a data link wiring diagram of a post-2007 Mack data bus equipped with multiple MIDs: See if you can identify the MIDs, and the symbols for twisted wire pairs, and terminating resistor. The following are some examples of common MIDs used on a truck data bus:

- Engine controller MID 128
- Transmission controller MID 130
- Tractor ABS and traction control MID 136
- Trailer #1 ABS controller MID 137
- Trailer #2 ABS controller MID 138
- Trailer #3 ABS controller MID 139
- Instrument cluster controller MID 140
- Trip recorder controller MID 141
- Vehicle (chassis) management system MID 142
- Fuel system controller (when separate from engine ECM) MID 143
- Vehicle navigation controller MID 162
- Vehicle security controller MID 163
- Ground communications controller MID 165
- Driver information systems controller MID 171
- Vehicle sensors to data converter MID 178
- Satellite communications controller MID 181

All the MIDs connected to the chassis data will be displayed as menu choices on the EST you are using to read the system. If you wish to access the transmission electronics, you will scroll to the 130 MID and select it. Once you have entered the transmission electronics, everything displayed on the EST display screen will only relate to the transmission. Menu choices will route you through the subsystems (SIDs) and parameters (PIDs) specific to the transmission. Figure 33–8 shows the location of the MIDs (modules) on the Freightliner M2 chassis.

Identify FMIs

If you were troubleshooting a fault in the transmission, you would be routed through the appropriate PIDs and SIDs to locate the FMI that had triggered the fault code. There are 4 limited number of malfunction categories that are assigned as FMIs and these are listed in Chapter 31 of this book.

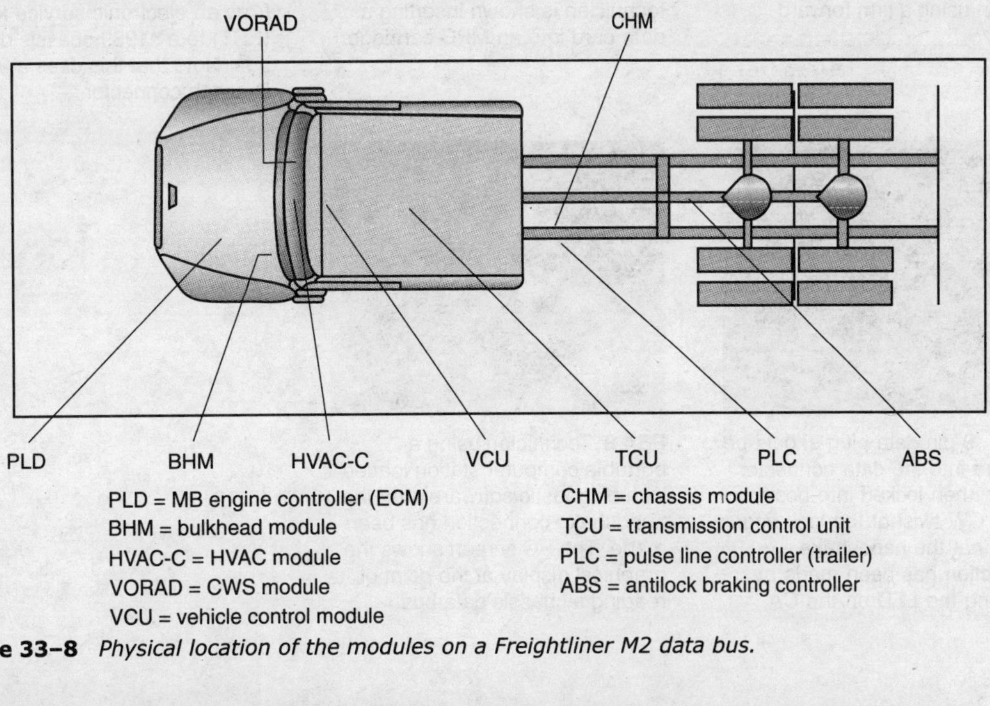

Figure 33–8 *Physical location of the modules on a Freightliner M2 data bus.*

Photo Sequence 4

Chapter 33 Accessing a Truck Data Bus Using an ATA Connector

PS4-1 Location of a 6-pin data connector (J1587/J1708) on a Mack Trucks CH chassis: lower left side of dash.

PS4-2 Location of a 9-pin data connector on a Volvo Class 8 chassis: underside of the dash, left of the steering column.

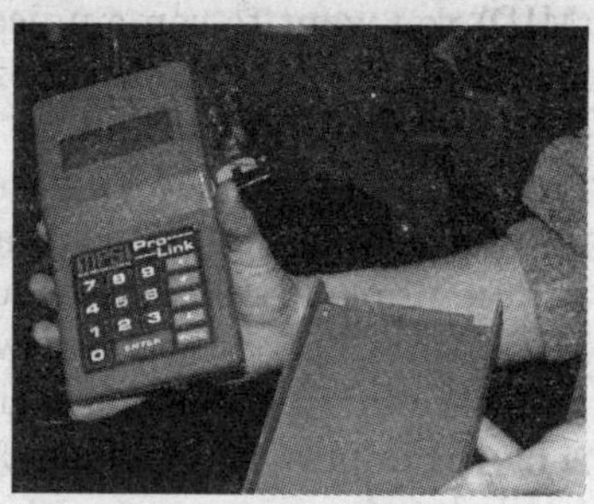

PS4-3 Data bus can be accessed using a ProLink 9000 and the appropriate cartridge.

PS4-4 To install the cartridge into the ProLink head, snap it into position using a firm forward thrust.

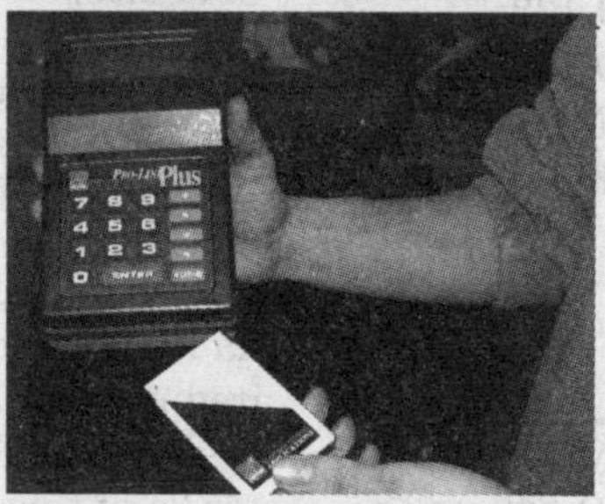

PS4-5 Alternatively, a ProLink MPC may be used: The technician is shown inserting a data card into an MPC cartridge.

PS4-6 Technician preparing to make a handshake connection from an electronic service tool (EST) to a J1939 chassis data bus: Note that this uses a 9-pin Deutsch connector.

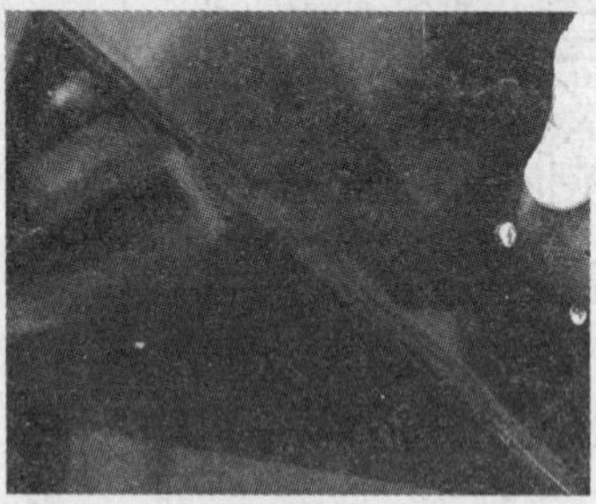

PS4-7 9-pin data plug should be inserted into the data connector socket, then locked into position with a CW twist of the lock ring. Varify that the handshake connection has been made by checking the LED on the CA.

PS4-8 Technician using a portable computer station loaded with diagnostic software after the handshake connection has been made. The PC screen shows the graphical display at the point of reading a chassis data bus.

Connect to Chassis Data Bus

Connecting to the chassis data bus is easy, and to read every MID networked to the bus you need only have generic heavy-duty reader software. Photo Sequence 4 shows where to look for the connector and how to make the connection to a typical truck chassis data bus.

REVIEW QUESTIONS

1. What is used to rate the speed of transmission on a data bus?
   a. Bandwidth
   b. K-baud
   c. Protocol
   d. CAN
2. Which of the following terms is used to rate the data volume that can be transmitted on a data bus?
   a. Bandwidth
   b. K-baud
   c. Frequency
   d. Bus arbitration
3. How many fields are used in a data packet in a J1939 data transaction?
   a. One
   b. One or two
   c. Five
   d. Seven
4. Which of the fields in a data packet contains the message identifier?
   a. Start of frame
   b. Arbitration
   c. Control
   d. Cyclic redundancy
5. Which field in a data packet is used by receiver stations to indicate that they have received an uncorrupted message?
   a. Arbitration
   b. Control
   c. Cyclic redundancy
   d. Ack
6. What is used to indicate the end of a data packet message?
   a. CRC field
   b. Ack field
   c. EOI
   d. EOF field
7. What current SAE J standard is used as the backbone of a multiplex data bus in current medium- and heavy-duty highway trucks?
   a. J1850
   b. J1939
   c. J1930
   d. J1667
8. Which of the following describes the role of an FET in a smart switch transaction?
   a. Transformer
   b. Electronic noise suppression
   c. Relay or amplifier
   d. Signal filter
9. What will result if a smart switch is disconnected at the connector?
   a. The vehicle will stop.
   b. The data bus will shut down.
   c. Operation is in default mode.
   d. A fault code will be logged.
10. How many pins are used in the J1939 Deutsch connector used to connect an electronic service tool (EST) to a truck data bus?
    a. One
    b. Four
    c. Six
    d. Nine
11. *Technician A* states that after connecting a diagnostic reader to the chassis data connector, you must first identify the MID you want to work with. *Technician B* states that FMIs are used to categorize the different types of component failure. Who is correct?
    a. Technician A only
    b. Technician B only
    c. Both A and B
    d. Neither A nor B
12. *Technician A* states that a J1708 data connector uses nine-pin assignments. *Technician B* states that a current J1939 data connector also has a J1587/1708 data backbone. Who is correct?
    a. Technician A only
    b. Technician B only
    c. Both A and B
    d. Neither A nor B
13. How many resistors are contained in a typical ladder switch?
    a. One
    b. Two
    c. Four
    d. Five

14. Which of the following would best describe the means used to identify a twisted wire pair on an electrical schematic?
    a. Circle
    b. Triangle
    c. Figure eight
    d. Trapezoid
15. *Technician A* states that voltage values on a data bus are usually 2.5 volts or lower. *Technician B* states that the twist frequency on a J1939 data bus is one complete cycle per 1 centimeter of length. Who is correct?
    a. Technician A only
    b. Technician B only
    c. Both A and B
    d. Neither A nor B

34

Bosch Electronic Distributor Systems

INTRODUCTION

The subject matter covered by this chapter applies mainly to small-bore diesel engines. The truck technician who specializes in lighter duty highway engines will be familiar with Bosch **Electronic Diesel Controls (EDCs)** because they have appeared on a number of popular fuel systems in recent years, sometimes as a standalone system, other times integrated by multiplexing into an existing engine management system such as with older versions of Cummins popular 5.9-liter, ISB series engines. Because the hydromechanical version of these injection pumps was studied in Chapter 24, it will be assumed that the operation of a Bosch VE pump is fully understood. Most of the operating principles of the hydromechanical and electronically controlled versions of these pumps are nearly identical. EDC management of rotary distributor pumps is a **partial authority** diesel engine management system. It should be noted that the EDC acronym is also used to describe other Bosch engine management systems, some of which are full authority.

ELECTRONIC DIESEL CONTROL

When used to manage a rotary distributor pump system, Bosch EDC can be defined as a partial authority, engine management system. Typically it includes drive-by-wire features but is otherwise fairly primitive in the way it adapts the popular Bosch rotary distributor injection pumps to computer control. EDC-managed rotary distributor pumps are rapidly becoming obsolete due to stringent emissions controls and the superior ability of common rail (CR) systems to meet them.

In an EDC-controlled rotary distributor pump fueled engine, the driver has no direct influence through the accelerator pedal on injected fuel quantity because it is drive by wire. Injected fuel quantity is managed by the **electronic/engine control unit (ECU)** based on a range of input variables (sensor status) and output requirements. Sensor status inputs include accelerator pedal angle, operating temperatures, nozzle **needle-motion sensor**, and ambient/engine fluid pressures. EDC monitors ambient, chassis, and engine conditions and is capable of detecting faults and, depending on their severity, can strategize countermeasures such as limitation of torque and limp-home operating modes in addition to exchanging data with other chassis electronic systems in the vehicle.

Bosch describes the electronic controls that manage a VE pump as system blocks, simply input, processing, and output. Bosch usually refers to its ECM as an ECU, so we use that acronym in this chapter. The specific components that appear in each system block are identified in Figure 34–1.

Input Circuit

The sensor circuit inputs data to the EDC microprocessor by means of protective circuitry and, where necessary, via signal transducers

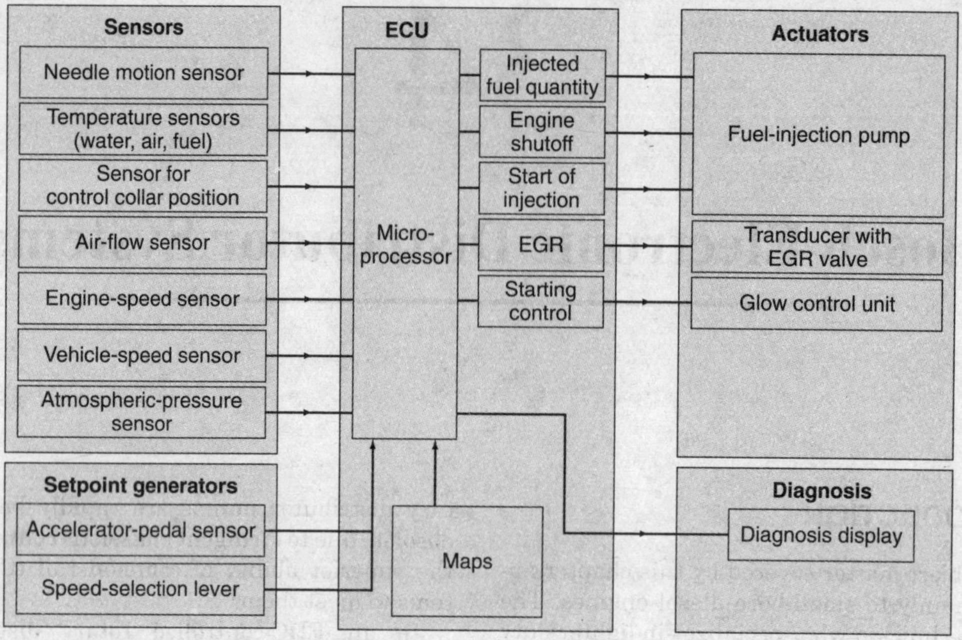

Figure 34–1 *EDC blocks. (Courtesy of Robert Bosch LLC)*

and amplifiers. The specific components used to input data to the ECM are no different from other engine management systems and are covered in Chapter 30. Both analog and digital signals are input to the ECU processing cycle.

Manifold air quantity, engine fluid and intake-air temperatures, engine pressure values, and battery voltage are examples of analog signals that must be converted to digital values by an analog to digital (A/D) converter in the ECM microprocessor. Digital input signals include on/off switching signals and digital sensor signals such as rotational speed/position pulses from Hall-effect sensors and these bypass the A/D unit to be processed directly by the microprocessor.

Nozzle Valve Motion Sensor

EDC-managed VE pumps use pintle-type nozzles with a nozzle valve motion sensor (NVMS) shown in Figure 34–2. It consists of a coil and pressure pin. Pressure pin movement through the coil winding changes the coil reluctance, inducing a signal voltage proportional to the valve's speed of movement. The NVMS enables the ECU to determine the precise moment of start of injection, which helps build the algo-

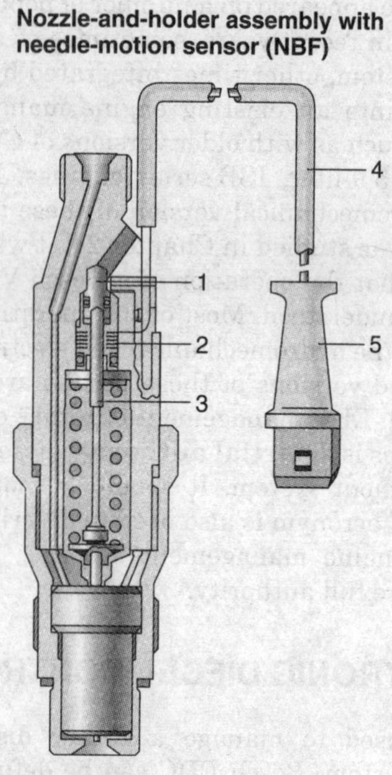

FIGURE 34–2 *PINTLE-TYPE NOZZLE EQUIPPED WITH AN NVMS. (COURTESY OF ROBERT BOSCH LLC)*

rithms required for injection timing and exhaust gas recirculation (EGR) management. It is designed to detect and signal nozzle valve movement (at nozzle opening) because the pressure pin moves with the nozzle valve. In Figure 34–3, tracking waves #1 and #2 show a graphic representation of the NVMS waveforms, unconditioned (#1) and conditioned (#2).

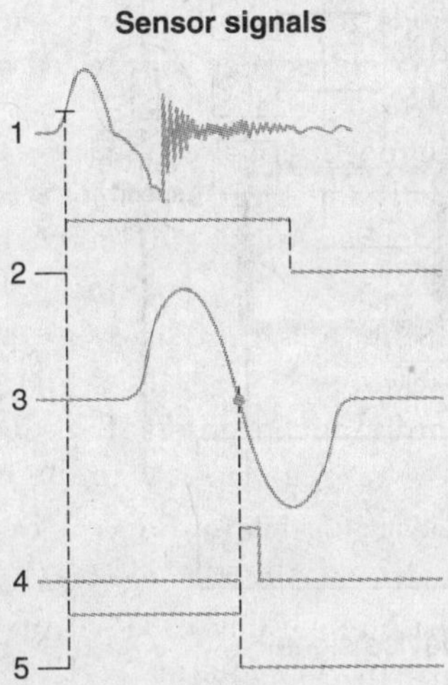

1. Untreated signal from the needle-motion sensor (NBF)
2. Signal derived from the NBF signal
3. Untreated signal from the engine-speed signal
4. Signal derived from untreated engine-speed signal
5. Evaluated start-of-injection signal

Figure 34–3 *NVMS tracking waves #1 and #2 show a graphic representation of the NVMS waveforms, unconditioned (#1) and conditioned (#2). (Courtesy of Robert Bosch LLC)*

Signal Conditioning

In order to suppress interference pulses and other electronic noise, pulse-shaped input signals from inductive sensors, which carry information on shaft speed and position, are conditioned by a special circuit in the ECM and converted to square-wave form. Signal conditioning essential squares up electronic noise as you can see by the representation of the tracking waves in Figure 34–3.

Depending on the level of integration, some signal conditioning can take place either completely or partially in the sensor. Protective circuitry is used to limit the incoming signals to a maximum voltage level. Inputted signals are freed of superimposed interference signals by filtering and then amplified to match it to the ECM working voltage.

Processing Cycle

Because of the range of fuel system types and engine applications of EDC, the programming and switching apparatus within the ECM varies considerably: Each specific ECU is provided with a variant code. Using this code, a selection of operational maps are stored in ECU memory (programmable by the manufacturer or in the workshop), with the objective of providing the specific functions required for the vehicle variant in question. Engine-specific curves and engine management maps are also written to ECU memory and would include such data as immobilizer strategies, calibration, and manufacturing data. Current ECUs also have a write-to-self capability that will log fault codes and audit trails to electronically erasable, programmable read-only memory (EEPROM). The ECU is designed to interface with other chassis electronic systems. For a more detailed look at vehicle engine management computers and how a logic processing cycle functions, refer to Chapter 30.

Output Circuit

With their output signals, Bosch microprocessors effect the results of the processing cycle into actions. When output commands are triggered within the ECM, the signals produced are usually powerful enough for direct connection to the actuators. While there is much in common in the input and processing hardware in both an electronically managed, rotary distributor pump and a CR fuel system, the output hardware is necessarily very different; therefore, the triggering of individual actuators is dealt with later in the chapter. The output circuit is proofed for common electrical failures such as short-circuit to ground, short to battery voltage, and electrical overload. Such faults are identified and reported to the microprocessor. Figure 34–4 shows a schematic of how output algorithms are mapped in a typical Bosch EDC circuit.

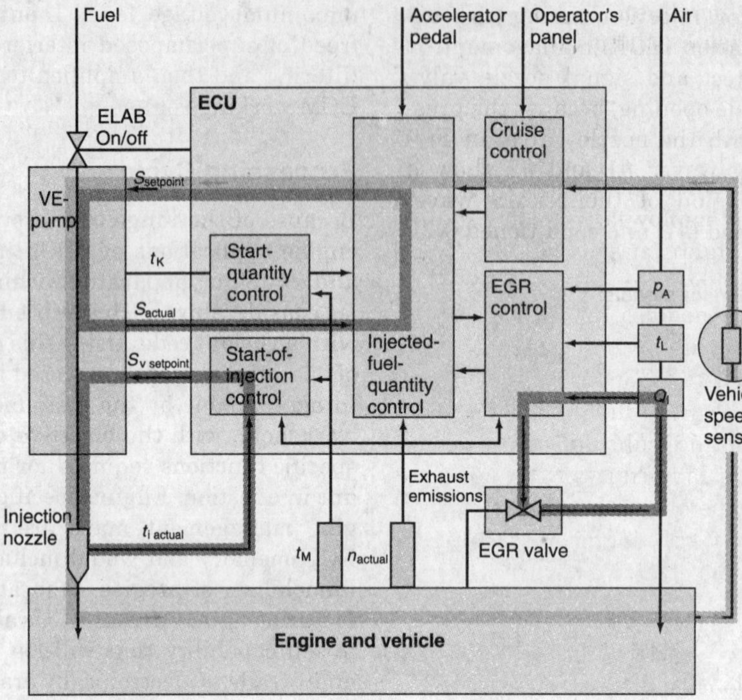

Q	Air-flow quantity	$s_{V\ set}$	Timing-device signal (setpoint)
n_{act}	Engine speed (actual)	t_K	Fuel temperature
p_A	Atmospheric pressure	t_L	Intake-air temperature
s_{set}	Control-collar signal (setpoint)	t_M	Engine temperature
s_{act}	Control-collar position (actual)	$t_{i\ act}$	Start of injection (actual)

Figure 34–4 *Schematic of how output algorithms are mapped in a typical Bosch EDC-managed VE pump. (Courtesy of Robert Bosch LLC)*

Multiplexing

Bosch EDC systems are capable of CAN 2.0 multiplexing, usually using the J1850 (see Chapter 33) data backbone. Most current engine management systems use largely open loop (fuzzy) processing as opposed to closed loop logic so inputs from any onboard electronic circuit can influence how the engine is being managed. Some systems that can be multiplexed connected to Bosch EDC include the following:

- Transmission control module
- Antilock braking system (ABS)
- Traction control system (TCS)
- Electronic immobilizer (EWS)
- Chassis control module
- Dash control module

Exchange of information between computer-controlled systems reduces the total number of sensors required and optimizes vehicle performance by making each system perform according to the requirements of the vehicle system rather than each of its subsystems.

Conventional data transmission in a vehicle required that every signal had to be allocated an individual conductor (dedicated wire and terminals) because binary signals can only be transmitted using the states "1" or "0" (binary code). While on/off ratios can be used to transmit continually changing parameters such as the status of an accelerator pedal-sensor, miles of unnecessary wiring and component replication was required in a typical truck chassis.

Perhaps most important to the technician troubleshooting vehicle electronic systems, multiplexing permits integrated diagnostics that monitor hardware functions and the data stream on the data bus, because each transmitter receives its own transmitted message back again and can therefore detect any deviations. So when a subsystem detects an error, it transmits "error flag," which stops the current transmission and prevents other subsystems from receiving a possibly faulty message. If a whole subsystem was defective, it could happen that all messages, good and bad, would be terminated with an error flag. To prevent this from happening, the multiplexing system incorporates a function that can distinguish between intermittent and permanent errors and often localize station failures.

ELECTRONIC CONTROL OF SLEEVE METERING, SINGLE PLUNGER DISTRIBUTOR PUMPS

In order to understand this section, you must have a working knowledge of the Bosch sleeve metering, single plunger distributor pumps introduced in the second part of Chapter 24. Because EDC adapts this hydromechanical injection pump for management by computer, it can be classified as a partial authority engine management system. As such, most of the hydromechanical components are retained in the electronic version and are simply adapted for computer control. The operating principles of the hydromechanical version of the pump are not repeated in this section. Figure 34–5 shows a sectional view of an electronically controlled VE pump highlighting those components that differ from its hydromechanical predecessor.

Difference between Electronic and Hydromechanical Versions

Most EDC capabilities described in the first section of this chapter are integrated in the ECM used to control the electronic version of a sleeve metering, single plunger distributor pump. By applying an electronic input circuit, highly flexible electronic data processing, replacing mechanical governing with electronic, and incorporating output switching circuits

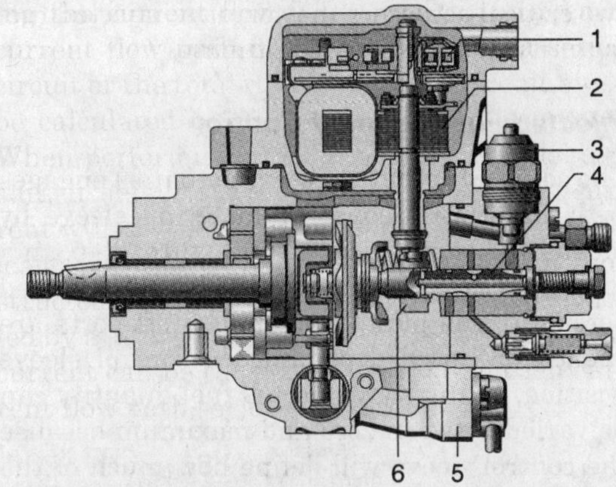

1. Control-collar position sensor
2. Solenoid actuator for the injected fuel quantity
3. Electromagnetic shutoff valve
4. Delivery plunger
5. Solenoid valve for start-of-injection timing
6. Control collar

Figure 34–5 *Partial authority VE distributor injection pump adapted for electronic control. (Courtesy of Robert Bosch LLC)*

that replace some of the mechanical controls, the versatility and performance of this injection pump is greatly enhanced. Most important, electronic controls enabled this diesel fuel injection pump to meet statutory emissions requirements that extended its life in the on-highway marketplace. The EDC also permits data exchange with other chassis electronic systems such as electronic transmission-shift control, allowing it to be integrated to the overall vehicle management system.

Because the input circuit and processing features are essentially similar to other partial authority systems, these are not examined in this chapter. The ECU output or actuator circuit is what makes the system distinct so we will take a closer look at this. The ECUs used with sleeve metering, single plunger injection pumps are usually located within the passenger compartment rather than under the hood.

Output Circuit Components

The ECM processes input circuit data and plots output commands that are sent to the output circuit. The hardware that converts the results of ECU processing into mechanical action are the actuator components. In a Bosch sleeve metering, single plunger injection pump there are two critical actuators that manage injected fuel quantity and fuel injection timing.

Injected Fuel Quantity Control

The solenoid actuator (rotary actuator) engages with the control collar or metering sleeve by means of a shaft as shown in Figure 34–5. In a manner similar to the mechanically governed version of the pump, the spill/cutoff ports are opened or closed depending on control sleeve position. In this way injected fuel quantity can be varied between zero and maximum because the control sleeve will define how much of the plunger stroke will be effective. By using a potentiometer-type sensor, the position of the control sleeve or collar is continually reported back to the ECM. In this way the ECM has complete control over plunger effective stroke and, therefore, engine output. When no voltage is applied to the sleeve actuator, return springs position it to reduce injected fuel quantity to zero. In other words, in the event of a system electrical malfunction or solenoid failure, the fuel management system is designed to default to no-fuel and shut the engine down.

Start-of-Injection Control

The pump is charged by a positive displacement-type transfer pump unloaded into a defined flow area, so pump internal pressure is dependent on engine speed. Similar to the timing device used on the mechanical version of the pump, this pressure is applied to the timing-device piston as shown in Figure 34–5. In the EDC-controlled version, fuel pressure acting on the timing device, pressure side, is modulated by a linear proportioning solenoid valve.

When the solenoid valve is electrically open (pressure reduction phase), the start of injection is at its most retarded location. When fully energized (pressure increase) the timing solenoid will locate start of injection to the most advanced position. Because solenoid position can be proportioned by ECU, any timing location between the two extreme values can be achieved by the EDC. If the solenoid was to fail, the system is designed to default to the most retarded timing value.

EDC Performance Outcomes

Improved control over injected fuel quantity influences vehicle starting, idling, power output, and driveability characteristics, but perhaps most significantly, improves engine particulate emissions. Driver input is delivered to the ECM by means of the accelerator sensor and the ECM will plot outcome responses in terms of speed, power, and torque based on pedal angle. Factoring programmed fuel map data, and the actual input values from the sensors, a setpoint is calculated for locating the control sleeve actuator in the pump. Check-back signaling ensures that the control sleeve/collar is correctly positioned in any given moment of operation.

Injection timing (start of injection) has a decisive influence on startup, engine noise, fuel consumption, and exhaust emissions. Start-of-injection maps programmed to the ECM take these interdependencies into account to optimize engine operation. The NVMS in the nozzle assembly signals the actual start of injection to the ECM and compares this with the desired or programmed start of injection. Any deviation in actual and desired timing values will result in a change to the on/off ratio of the timing solenoid valve.

ECU Failure Strategies

Consistent with early generation vehicle computer-controlled systems, EDC can be programmed with a limited range of failure compensation strategies. Some of these are shown in Table 34–1.

Table 34–1 *EDC VE Failure Responses*

Failure	*Monitoring*	*Output Response*	*Warning Lamp*	*Fault Code*
Correction sensors	Signal range	Engine derate	No	Yes
System sensors	Signal range	Limp home mode	Yes	Yes
ECU	Self-test	Limp home mode	Yes	Yes
Fuel quantity actuator	Permanent deviation	Engine shutdown	Yes	Yes

REVIEW QUESTIONS

1. *Technician A* states that Bosch Electronic Diesel Control (EDC) is used on later versions of the VE sleeve metering, rotary distributor pump. *Technician B* states that when a VE pump is managed electronically, the system is classified as a partial authority management system. Who is correct?
   a. Technician A only
   b. Technician B only
   c. Both A and B
   d. Neither A nor B
2. When a VE injection pump is managed by an ECU, which of the following is no longer required?
   a. Control sleeve/collar
   b. High-pressure pipes
   c. Timing advance
   d. Governor flyweights
3. On a Bosch electronically controlled, sleeve metering, rotary distributor injection pump (VE), what would most likely happen if the electrical signal from the atmospheric pressure sensor failed?
   a. Default to hydromechanical operation
   b. Default to limp-home mode
   c. Engine derate
   d. Engine shutdown
4. How many pump plungers are there on a VE pump used to fuel a six-cylinder engine?
   a. One
   b. Three
   c. Six
   d. Eight
5. Which of the following terms best describes the means by which signal voltage is generated by an NVMS?
   a. V-Ref
   b. Induction
   c. Hall effect
   d. Temperature differential
6. Which type of nozzle is commonly used on EDC-managed VE injection pumps?
   a. Poppet
   b. Pintle
   c. Multi-orifii
   d. Electrohydraulic injector (EHI)
7. Which of the following would be more likely to occur if the EDC-controlled VE pump, fuel quantity actuator failed?
   a. Fault code only
   b. Engine derate
   c. Limp-home mode
   d. Engine shutdown
8. *Technician A* states that an NVMS induces a signal proportional to its speed of movement. *Technician B* states that an NVMS induces a signal proportional to its distance traveled. Who is correct?
   a. Technician A only
   b. Technician B only
   c. Both A and B
   d. Neither A nor B
9. What moves the pressure pin in a pintle injector equipped with NVMS?
   a. Nozzle valve
   b. Fuel pressure
   c. Solenoid
   d. Pulse width modulation (PWM) signal
10. *Technician A* states that the control sleeve solenoid on an EDC-VE injection pump controls the fuel pump effective stroke. *Technician B* states that the start-of-injection solenoid valve on an EDC-VE injection pump controls injection timing. Who is correct?
    a. Technician A only
    b. Technician B only
    c. Both A and B
    d. Neither A nor B

35

Mack V-MAC I and II

INTRODUCTION

Mack Trucks **vehicle management and control (V-MAC)** versions I and II are both classified as partial authority, electronic engine (chassis) management systems because the system is built around a hydromechanical, port-helix metering injection pump, adapted for computer control. V-MAC was introduced in the late 1980s, evolved into V-MAC II, and the system was used until 1998 when it was replaced by V-MAC III. The fuel management system used in V-MAC I and II is distinct from V-MAC III, and the current V-MAC IV (see Chapter 37) systems. However, the electronically controlled fuel pumps described in this chapter were used by off-highway diesel engines up until around 2003.

The actual metering and pumping apparatus used with these electronically controlled pumps differs little from that on the hydromechanical fuel pumps they replaced. The computer controls focus primarily on the means used to control injected fuel quantity and beginning of delivery timing. In addition, V-MAC electronics comprehensively manages vehicle and chassis systems, plots the fueling logic to meet a variety of chassis and running conditions, and permits a wide range of customer and proprietary (Mack) programming options. Mack Trucks was also the first of the truck original equipment manufacturers (OEMs) to adopt the personal computer (PC) as its primary diagnostic, system reader, and reprogramming instrument. It is assumed that the reader has a sound knowledge of inline, port-helix metering injection pumps before reading this chapter.

SYSTEM OVERVIEW

The V-MAC I system layout incorporated two modules bussed (electronically connected to each other) to manage engine fueling, diagnostics, and chassis functions (Figure 35–1). The **fuel injection control (FIC) module** was essentially the Bosch Electronic Diesel Control (EDC) module used on other electronic applications of the PE 7000/8000 series of port-helix metering, injection pumps (used previously by John Deere/Volvo and other OEMs). The FIC module sensed and switched most of those functions directly associated with engine fueling. The FIC module was bussed to the **V-MAC module** by a dedicated CAN, which "managed" the FIC module from the broader perspective of the specific chassis application of the engine, performance, noxious emissions management, and command inputs such as power takeoff (PTO) and cruise controls. V-MAC II eliminated the separate FIC module, incorporating its functions in a single V-MAC II module, which was manufactured by Motorola until the system became obsolete in 1998. Both versions of V-MAC had the module(s) located in the cab dash compartment on the passenger side of the vehicle. In the case of V-MAC I, the FIC module was piggyback located on top of the V-MAC module.

V-MAC modules can be accessed for fault code display, reading system parameters, and

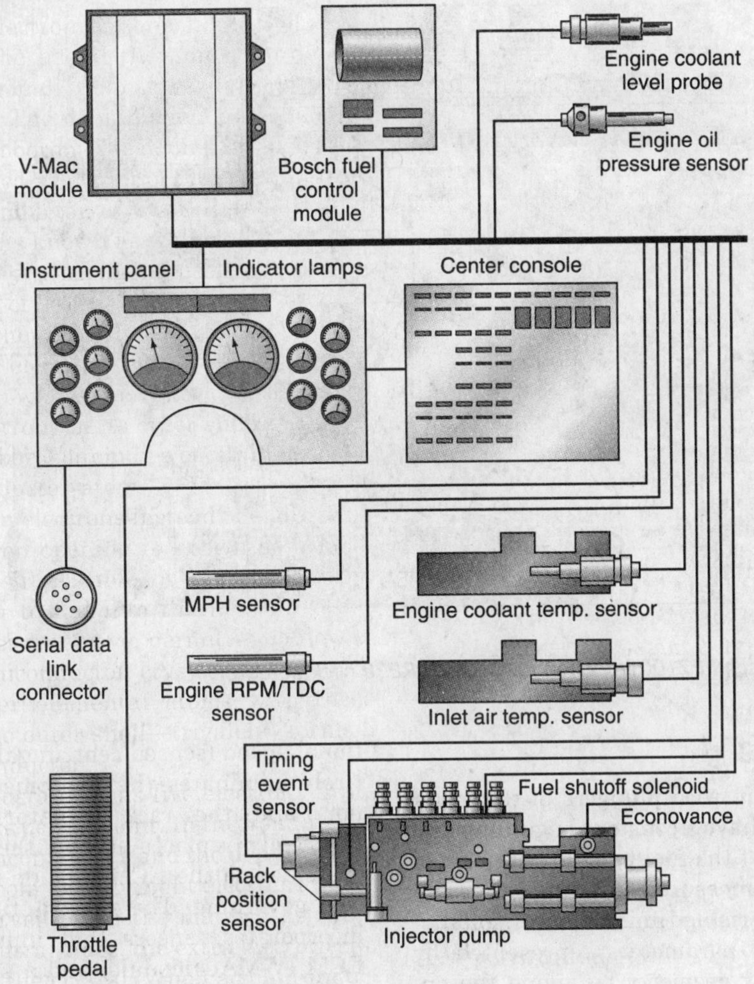

Figure 35–1 *V-MAC I system overview. (Courtesy of Mack Inc.)*

customer and proprietary data programming via the standard six-pin data bus connector (SAE J1708) located under the dash on the driver's side of the vehicle. No trucks with V-MAC I or II use a nine-pin J1939 data connector. Active system faults can also be read by blinking them through the dash "electronic malfunction" light.

BOSCH PE7100/PE8500 INJECTION PUMPS

These are essentially "beefed-up" versions of the Bosch PE3000 series injection pump. Both can be managed with hydromechanical governors. The PE7100 was used with the first V-MAC I system (Figure 35–2). The injection pump itself was designed to produce peak injection pressures of 1,050 bar (106.4 MPa or 15,500 psi) using 10 mm plungers with lower helix geometry and a cam lift of 12 mm. The pumping elements were fitted with funnel spill deflectors and a reinforced housing. The more recent version, PE8500, is closely related to the PE7100 in most of its critical specifications but is capable of peak injection pressures of 1,150 bar (116.5 MPa or 16,900 psi). Both pumps are driven off the engine timing geartrain at camshaft speed and use an electronically controlled, electric-over-hydraulic variable timing device, **Econovance**, which is located between the pump and the engine accessory drive gear.

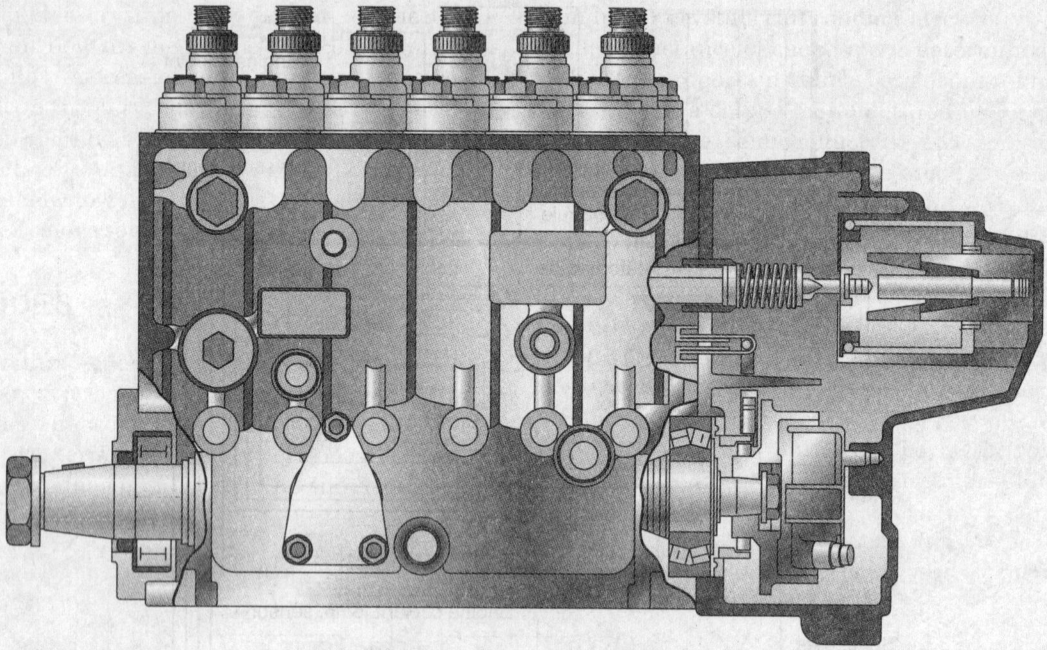

Figure 35–2 *Bosch PE7100 injection pump with RE30 rack actuator.*

Variable Timing

An injection pump with plungers of the lower helix design will have a constant beginning and variable ending of the fuel pulse, the length of which would be measured in crank angle degrees. Using a variable timing device "softens" the hard timing parameters represented by plunger and helix geometry by giving the engine control module (ECM) a window within which it can select effective stroke. The constant beginning, variable ending timing characteristic of the typical lower helix geometry is adapted into a variable beginning, variable ending of the delivery pulse, which is entirely managed by ECM processing logic. All V-MAC system injection pumps use Econovance, and the extent to which the port closure value can be moderated depends on the application; this is described in detail later in this section.

Electronic Governing

The function of the hydromechanical governor was to manage fueling on the basis of inputted fuel demand (accelerator/speed control lever), manifold boost (puff limiter), and engine rotational speed (sensed centrifugally). V-MAC entirely eliminates the hydromechanical governor, and the rack actuator housing that physically replaces it and the software programmed to the ECM provide fully computerized governing. The **rack actuator housing** incorporates sensors that input data to the ECM (V-MAC module) and a switched output from the ECM capable of precisely controlling rack position and, thus, injected fuel quantity.

RE24/30 Rack Actuator Housings

It was stated previously that although its location positions it at the rear of the fuel injection pump, the rack actuator housing should not be confused with a governor. The governor is a self-contained unit that, after calibration by the technician, sets hard parameters (rigid values) for fueling the engine through its performance range. The V-MAC I and II rack actuator housing is a unit that contains mechanisms both for reporting data to the system module(s) (that is, sensors) and for effecting the fueling commands that are the result of ECM processing logic. The fueling parameters can be described as "soft" or nonrigid.

Truly, the programmed **ECU** (Mack Trucks uses ECU or **electronic control unit**, rather than the acronym ECM) software is the "governor" in a V-MAC management system. The rack actuator housing is not as a rule field repaired but its operation must be understood by the technician seeking to effectively troubleshoot V-MAC systems. Subcomponent testing is outlined in Mack technical support literature and detailed in Robert Bosch diagnostic literature.

The rack actuator housing consists of a **rack actuator** mechanism controlled electrically by a V-MAC II module or by an FIC module in V-MAC I. Within the housing there are also several critical sensors that input engine data to the ECU(s). An RE30 rack actuator housing is shown in Figure 35–3.

Rack Actuator

The function of the "rack" in an inline, port-helix metering injection pump is to rotate the plungers in unison within their barrels when it is moved linearly. When plungers with lower helices are used, port closure ("static" timing) determines the beginning of the fueling pulse at a constant timing location of the engine, while port opening sets the end of the pulse, which will vary according to the amount of fuel to be pumped.

The fueling window is therefore defined precisely by where the plunger helices register with the barrel spill ports. Plunger rotation is effected by linearly moving the rack, which is tooth meshed to plunger control sleeves. In Bosch rack actuator housings, the control rack extends into the rack actuator housing and is attached to the rack actuator mechanism (Figure 35–4); the rack actuator can be electrically defined as a **linear magnet** or **proportional solenoid**. The rack actuator is also known as the **fuel control actuator**. The rack assembly is spring loaded to the rack "no-fuel" position. This no-fuel rack position rotationally positions all the pump plungers so their vertical slots register with the barrel spill ports throughout their cam-actuated stroke. This plunger position permits them to do no more than displace fuel as the plungers are actuated.

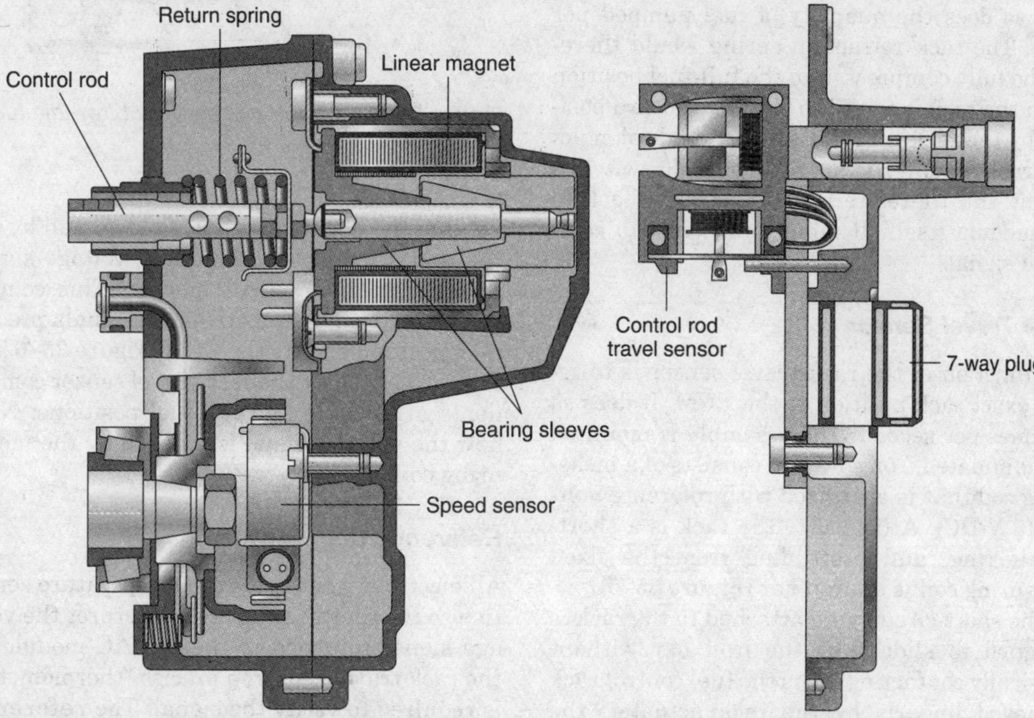

Figure 35–3 *Sectional view of Bosch RE30 rack actuator housing.*

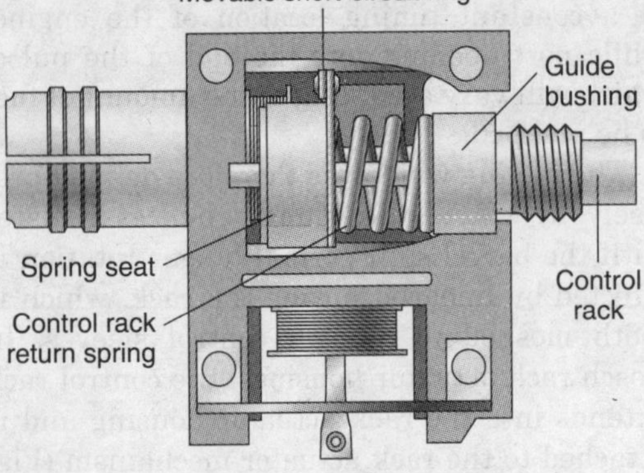

Figure 35–4 *Rack actuator assembly components.*

When the rack actuator is energized by V-MAC module switching, the rack moves linearly in opposition to this spring pressure, thereby altering the rotational position of the plungers to increase plunger effective stroke. As the amount of current flowed through the rack actuator coil is increased, the further inboard the rack is forced. As the rack is forced inboard against the spring pressure, the length of the plunger effective stroke increases and, therefore, so does the quantity of fuel pumped per cycle. The rack retraction spring would therefore be fully compressed in the full-fuel position when the rack is forced to its fully inboard position by the rack actuator proportional solenoid. The rack actuator is an ECU module switched output and therefore returns no data back to the module itself. It simply responds to command signals.

Rack Travel Sensor

The function of the rack travel sensor is to report exact rack position to the ECM. It does so 60 times per second. The assembly is mounted in a laminated iron core and consists of a measuring coil that is energized with reference voltage (5 V-DC). Attached to the rack is a short-circuit ring, and protruding from the fixed measuring coil is an iron bar (Figure 35–5).

The short-circuit ring attached to the rack is designed to slide over the iron bar without physically contacting it. As the fuel control rack is moved linearly by the rack actuator, the short-circuit ring will either move closer to or further away from the measuring coil, thereby

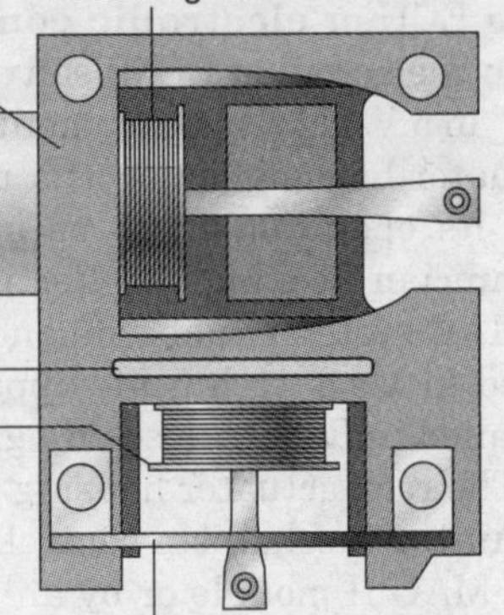

Figure 35–5 *Rack travel sensor components.*

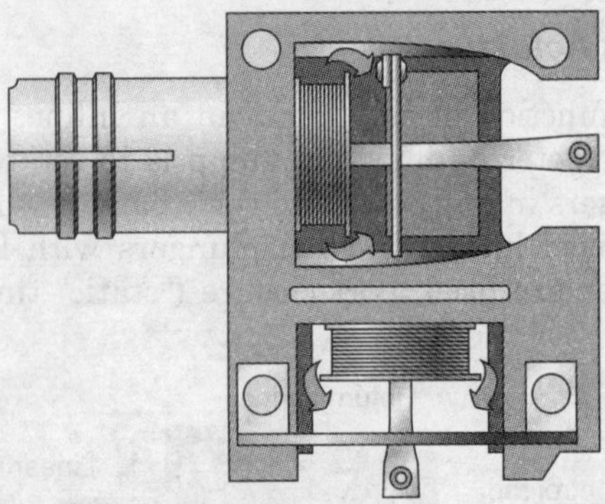

Figure 35–6 *Rack travel sensor at low-idle fuel rack position.*

varying the electromagnetic field spread by the coil and, therefore, the induced voltage signal returned to the V-MAC module. This component is extremely sensitive and signals precise rack position data to the ECM. Figure 35–6 and Figure 35–7 show the rack travel sensor components at low-idle and full-fuel positions: Note how the voltage signal is induced by the measuring coil.

Reference Coil

All electromagnet fields are temperature sensitive so in order to accurately interpret the voltage signal returned to the V-MAC module by the rack travel sensor, a precise "thermometer" is required to verify the signal. The **reference coil** is located close to the rack travel sensor coil; its location is shown in Figure 35–5. It is

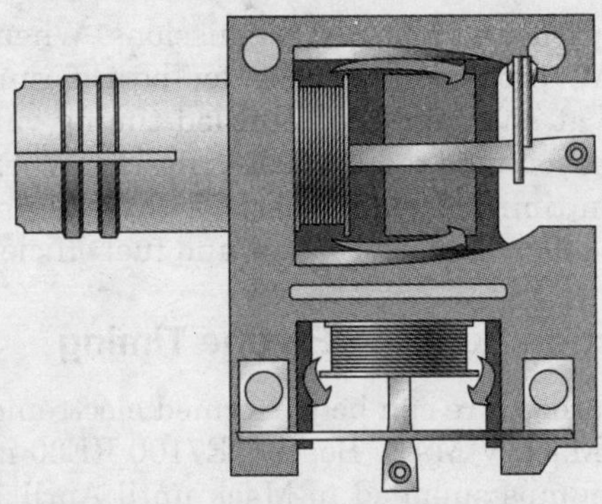

Figure 35–7 *Rack travel sensor at maximum rack (peak fuel per cycle) position.*

wound identically to the rack travel sensor coil. When it is energized, this coil provides a fixed magnetic field. The signal sent to the ECM by the **rack position sensor** will be modulated by the actual temperature conditions in the rack actuator housing so that when the rack is in a certain linear position, the signal outputted by the rack position sensor will be different when the engine is cold compared to that when the engine is at running temperatures.

To enable the ECM to accurately interpret the signal from the rack actuator, the reference coil is used. For instance, as temperature in the rack actuator housing rises, so does the temperature in both electromagnets used in the rack position sensor and the reference coil. The result of the temperature rise is that resistance increases in each coil winding and the signal returned to the ECM varies. The opposite occurs when the temperature decreases. The signal returned by the reference coil is varied only by temperature change. By comparing the signals from both electromagnets, the ECM is able to accurately evaluate the voltage signal returned from the measuring coil on the rack position sensor and thereby determine exact rack position.

Pulse Wheel

The pulse wheel is the only rotating component in the rack actuator housing. The pulse wheel is a toothed impeller located at the rear of the injection pump camshaft and driven within the rack actuator housing. It is a sintered steel component whose total indicated runout (TIR) deviation must not exceed 0.03 mm (0.001"). Straightening of impeller teeth should never be attempted, and the impeller should be replaced if measured (or observed) to be defective. The impeller or pulse wheel (tone wheel) is responsible for two inputs to the V-MAC module.

The first is engine speed (it is the primary reference for engine rpm) and the next is engine location reported by the **timing event marker (TEM)**. There are 16 protruding teeth on the pulse wheel impeller that cut through the rpm sensor's magnetic field to produce an alternating current (AC) voltage signal to the ECM, whose frequency increases proportionally with an increase in rotational speed. The TEM sensor consists of a single notch that cuts the magnetic field of the TEM sensor, producing a signal that works on the same principle as the rpm sensor, but whose input is used by the ECM as an engine position reference. The TEM notch is also used for the electronic static timing of the pump to the engine (this procedure is covered later in this chapter). Both these sensors can be electrically classified as induction pulse generators that feed a small AC analog signal to the ECM.

Econovance

The PE7100 or PE8500 injection pump is driven off the engine crankshaft gearing at camshaft speed on Mack engines. The Econovance unit serves as an intermediary between the engine-driven pump drive gear and the injection pump drive. Econovance provides an advance on the static timing, port closure location of 8-degree, 10-degree, or 20-degree crank angle depending on the engine model. The last generation versions of V-MAC tended to use a 20-degree Econovance, which is perhaps better described as a variable timing. Econovance establishes a port closure timing window within which the ECM can select port closure. For example, if static timing was set at 7 degrees before top dead center (BTDC) on a V-MAC engine with 20-degree Econovance, V-MAC could select port closure occurrence at any value between 7 degrees BTDC and 27 degrees BTDC.

Econovance alters the position of the injection pump camshaft relative to that of the engine. The device is controlled electronically by the V-MAC module. The engine's injection pump drive gear rotates a hub within which is a sliding sleeve machined with a helical spline; this is spring loaded to a nonadvanced position. To advance timing, engine oil used as the hydraulic medium acts to move the sliding sleeve along the axis of rotation of the helical splines. The following components are critical in actuating Econovance:

- Proportional solenoid. The proportional solenoid used is known as the **timing actuator**. It is a linear magnet that can be smoothly moved to any position within its range of travel by varying current flow through the coil. The proportional solenoid is switched by the V-MAC module, and its function is to control the Econovance hydraulic spool valve.
- Hydraulic spool valve. This controls oil flow (engine lube) to the Econovance sliding sleeve. The hydraulic spool valve is directly controlled by the proportional solenoid, and it establishes the extent of advance by managing oil flow. Actual engine oil pressure values will not affect its operation (so long as the oil pressure is above the V-MAC "failure" value) because the operation of Econovance is flow dependent. The ECM monitors the advance location of the injection pump by reading the TEM signal.

Econovance Operation

Static timing values as retarded as 4 degrees BTDC seem to make little sense mechanically because the firing pulse and cylinder pressure rise would appear to be completely out of synchronization with the crank throw vector angles and act to "inefficiently" load the engine powertrain. However, it should be remembered that static timing the engine at 4 degrees BTDC with a 20-degree Econovance simply provides the ECU with the option of selecting a port closure value anywhere from 4 to 24 degrees BTDC, and an actual 4-degree BTDC port closure setting would seldom be used.

Emissions control in truck diesel engines is often a case of maintaining a tricky narrow cylinder temperature window of values in which the consequences of being slightly under optimum value would be excess hydrocarbon (HC) emissions and the consequences of being slightly over produce higher NO_x emissions. When the V-MAC programming opts for port closure to occur at substantially retarded timing values, the reason is to reduce NO_x emission. When injection timing is this retarded, there is an adverse effect on performance and fuel efficiency.

Injection Pump-to-Engine Timing

This procedure can be performed electronically with Mack V-MAC. Bosch PE7100 RE30 injection pumps supplied to Mack until April 1992 were fitted with **ELAB**, a German acronym for *fuel shutoff solenoid.* This apparatus was located near the front of the pump, requiring that it be timed on the #6 cylinder. ELAB actually provided a secondary no-fuel method of shutting the engine down by gating off the supply fuel when the ignition was turned off. On later PE7100 and the more recent PE8500, fuel shutdown is achieved by simply moving the rack to the no-fuel position. However, all V-MAC PE7100 injection pumps are timed to the engine referencing #6 cylinder. All V-MAC PE 8500 pumps are timed on the #1 cylinder.

Electronic timing is performed by electromechanically positioning the protruding TEM notch using a Mack timing device. Essentially, this consists of positioning the pump so that a pair of lights on an electronic service tool (EST) illuminates simultaneously, ensuring a timing accuracy within ½ degree crank angle if the gear installation procedure has been performed accurately.

Static Timing Setting on an E-7 Engine with V-MAC Electronics

The following procedure is outlined according to Mack Trucks recommended procedure; the assumption here is made that the engine is timed to the #1 engine cylinder. The static timing procedure requires the use of Mack Trucks fixed timing tool; this consists of a display box with a pair of lights marked A and B, a ground wire and clamp, and a sensor. Attempting to static time a V-MAC engine without this tool will result in out-of-specification timing.

1. Using an engine barring device, preferably the toothed Mack barring ratchet adapter, turn the engine counterclockwise (CCW) as

viewed from the front to a location around 45 degrees BTDC of the #1 engine cylinder.

2. Next, rotate the engine clockwise (CW) until the timing pointer aligns with the calibration scale on the flywheel or vibration damper.
3. Remove the TEM sensor from the fixed timing port located on the side of the rack actuator housing, and install the fixed timing sensor tool. Take care to ensure that the sensor (fixed timing tool) is properly aligned with the locating groove in the TEM aperture.
4. Attach the ground clamp from the fixed timing tool sensor to the engine; ensure that there is continuity between the two points. A poor ground will render the timing tool inoperative.
5. Using the hub rotation tool, rotate the engine timing gear hub CW until both lights on the fixed timing tool illuminate. It is important that the inner shaft nut not be used to rotate the hub or incorrect static timing will result.
6. Install the injection pump-driven gear on the Econovance outer shaft hub so that the screw holes in the hub are properly centered in the gear slots.
7. Install the timing gear hub screws and install finger tight. This should permit the hub gear to be held in place while allowing relative movement between the Econovance hub and the pump-driven gear.
8. Using the hub rotation tool, rotate the timing gear hub CCW until the screws contact the stops at the end of the slots in the injection pump drive gear. At this point, both lights on the static timing tool should not be lit.
9. Rotate the timing gear hub CW until both lights on the fixed timing tool illuminate. If the hub is rotated too far CW so that the A light extinguishes, back up and repeat step 8.
10. Torque the fasteners that clamp the injection pump-driven gear to the Econovance hub to the specified value. During the torquing procedure, it is normal for one of the two lights to extinguish, and this can be expected.
11. Next bar the engine CCW to a minimum of 45 degrees BTDC; this will cause both the A and B lights to extinguish.
12. Validate the timing by rotating the engine CW until both the A and B lights illuminate. Ensure that the engine is only rotated CW during this procedure; once again, if the engine is barred past the point where both lights are lit, back up and repeat from step 11. When both lights are illuminated, check the engine calibration scale and pointer. The specification should be within ½ degree of the specified static timing (port closure) value; if out of specification the timing procedure will have to be repeated.

V-MAC ELECTRONICS

While somewhat primitive by today's standards, V-MAC I and II incorporated comprehensive monitoring, engine management, customer and proprietary data programming, and fault/data retention consistent with a full option, partial authority management system. The sensor circuit was as comprehensive as most of the full authority systems manufactured in the late 1990s. The primary sensors on which these early generation V-MAC systems relied on to plot fuel ratio and timing logic are the two temperature thermistors, engine coolant, and ambient air. Figure 35–8 shows the V-MAC II ECU housing; Figure 35–9 is a schematic of inputs and outputs to the module.

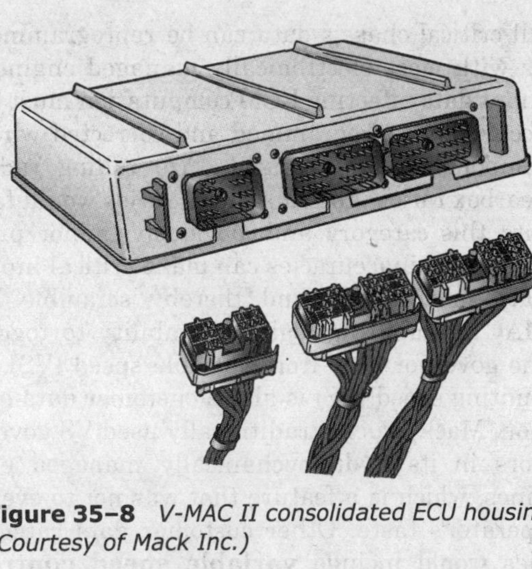

Figure 35–8 *V-MAC II consolidated ECU housing. (Courtesy of Mack Inc.)*

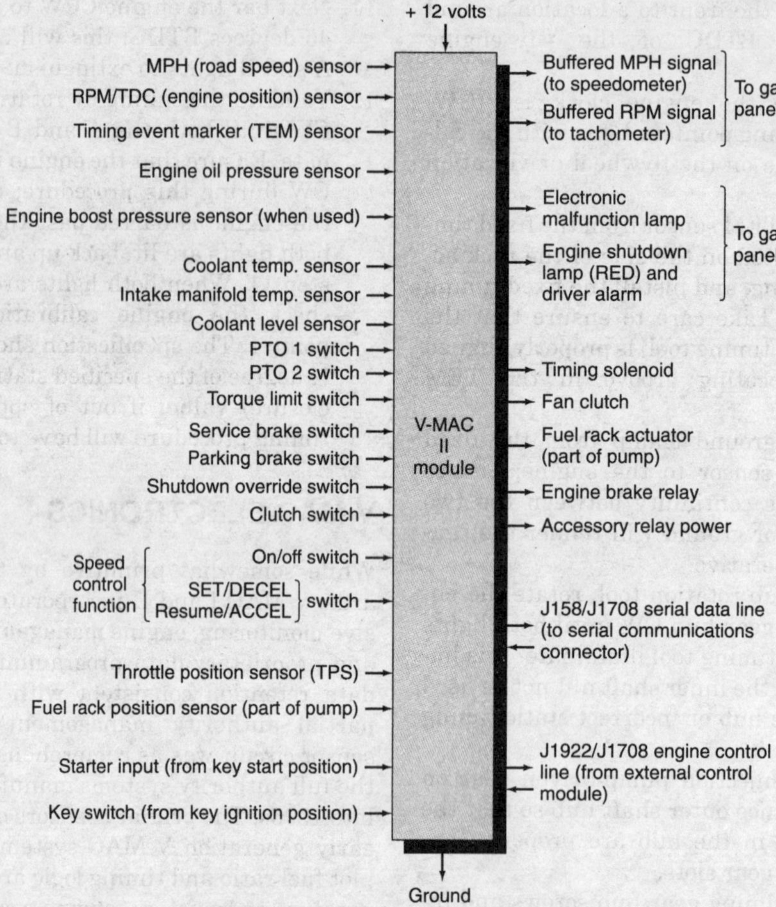

Figure 35–9 *V-MAC II ECU inputs and outputs. (Courtesy of Mack Inc.)*

Customer Data Programming

All critical chassis data can be reprogrammed as with most electronically managed engines. Input data affecting ECM computation must be accurately reprogrammed and corrected when components are changed. Tire rolling radii, gearbox ratios, and final drive ratios would fall into this category and, obviously, minor programming inaccuracies can make critical input data meaningless and thereby scramble V-MAC's computing logic. The ability to toggle the governor type from variable speed (VS) to limiting speed (LS) is also a customer data option. Mack Trucks traditionally used VS governors in its hydromechanically managed engines, which is a feature that was not to every operator's taste. Other customer data categories would include **variable speed control (VSC)**, the setting of rpm using the dash cruise control switches, and **variable speed limit (VSL)**, which limits rpm while moving a vehicle in PTO mode.

In early V-MAC software versions, fuel trip data could be logged into ECU files for subsequent readout; more recent versions of V-MAC offer CoPilot, a comprehensive driver data display tool described in detail later in this section. V-MAC customer data programming would also include a number of options designed to enhance the operating efficiency and response strategy to a critical monitored failure condition.

VSC parameters including the **road speed limit (RSL)** cruise window and PTO management would fall into the first category. The RSL is a hard parameter that may be programmed to the ECM and require a password to alter. In common with most electronically governed engines, V-MAC in PTO mode provides **single speed control (SSC)** or isochronous governing; that is, as engine load changes, the engine rpm will be maintained with zero droop during the load changes. V-MAC may also be programmed with either alert/shutdown or alert-only failure strategy on the following sensor inputs:

CTS - coolant temperature sensor
OPS - oil pressure sensor
CLS - coolant level sensor

It is strongly recommended by Mack Trucks that the shutdown strategy (rather than alert-only) be programmed, especially in fleet applications. A V-MAC-managed engine shutdown can be overridden by an operator in 30-second increments any number of times, but it should be noted that override incidents will be logged to the ECM.

Resetting Low-Idle Speed

This procedure is outlined as being an example of setting a typical customer data programming option. The V-MAC system permits the setting of the engine low-idle speed between a range of 500 and 750 rpm. This procedure may be undertaken using the dash speed control switches using the following method:

1. V-MAC must have the Low-Idle Adjust option enabled in customer data options; if not, a ProLink or PC (Figure 35–10) with the appropriate Mack Trucks software must be used. This function can, in other words, be disabled by the customer. Additionally the PTO minimum engine speed must be programmed to 475 rpm, or once again a ProLink or PC will be required to perform the procedure.
2. Ensure that no active fault codes are logged into V-MAC, that the vehicle is stationary, that the parking brake is applied, and that the throttle pedal is in the low-idle position.
3. The Speed Control *on/off* switch must be turned on/off three times within a 2-second time period. This will cause the idle speed to automatically drop to 500 rpm. V-MAC is now ready to accept a new speed. Ensure that the switch is left in the *on* position.
4. Depress and hold the accelerator pedal until the desired idle speed is produced. Then press and release the Set switch.

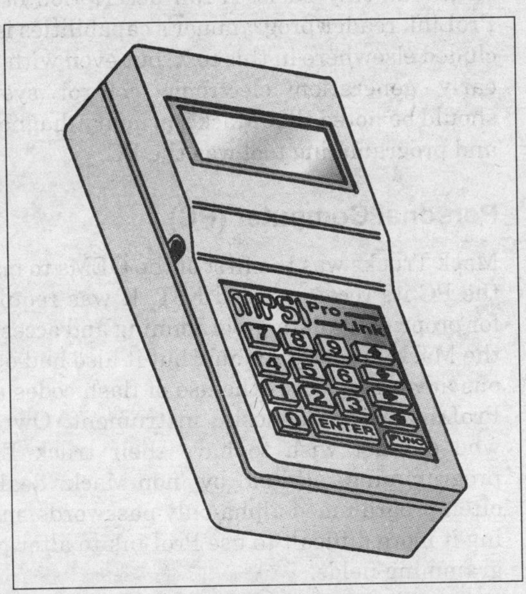

ProLink 9000

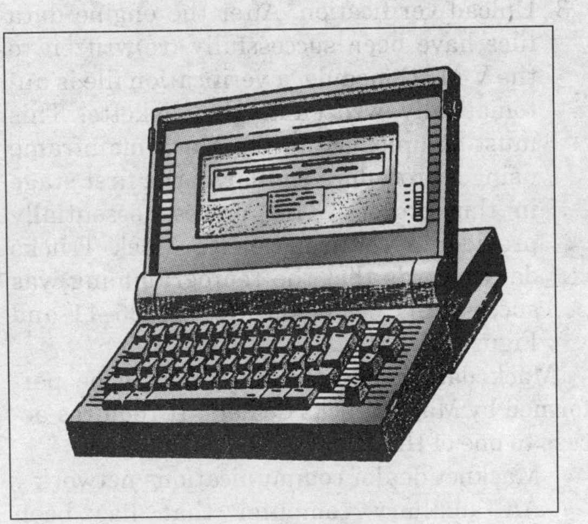

Diagnostic computer
(any IBM PC-compatible computer)

Figure 35–10 *V-MAC electronic service tools. (Courtesy of Mack Inc.)*

5. At the desired engine rpm, depress the clutch pedal and V-MAC will respond by locking the new speed as the idle speed.

ProLink or a PC may be used for customer data programming. However, only a PC may be used for proprietary (Mack Trucks) data programming.

Mack Trucks Data Programming

Proprietary data reprogramming is seldom required. However, when an ECU module is replaced for whatever reason, V-MAC module files are suspected of being corrupted or, exceptionally, the fuel map requires modification, the V-MAC module must be reprogrammed. The process required to perform this is undertaken in three distinct stages:

1. Download. This process electronically locates the appropriate file in the Mack Trucks Allentown mainframe via PC/modem using the vehicle identification number (VIN) and downloading it to a diskette.
2. Reprogram the engine data file. The engine data files on the diskette are next transferred to the V-MAC module using a PC, serial link, and a standard SAE J1708/J1939 connector.
3. Upload verification. After the engine data files have been successfully (re)written to the V-MAC module, a verification file is automatically written to the diskette. This must be uploaded to the Mack mainframe using a procedure similar to the first stage in this process. This process essentially provides a confirmation to Mack Trucks data records that the reprogramming was successfully completed (Figure 35–11 and Figure 35–12).

Mack data programming can only be performed by Mack Trucks dealers. It requires access to one of the following:

- Macknet dealer communications network
- An auxiliary computer that has been equipped with MACKCOMM communications software

Reading V-MAC ECUs

V-MAC I and II ECUs may be read in three different ways, the most simple of which is using the onboard diagnostic capability of the system. The onboard diagnostics are necessary to read a condition when more appropriate tooling is not available, such as during an on-highway breakdown. ProLink and PCs loaded with the appropriated Mack Trucks software should be used when troubleshooting the electronics in the service facility.

Flash Codes

These will blink out active numeric codes using the "electronic malfunction" dash light. Multiple active codes may be blinked but only one is signaled per demand.

ProLink

ProLink can be connected to the truck chassis and the V-MAC module(s) by means of the standard six-pin J1708 data link. A dedicated V-MAC cartridge is required to perform customer data programming or to use the unit as PC serial link. This was upgradable through each software versions progressed by replacing the two programmable read-only memory (PROM) chips in the cartridge circuit board. If a Heavy Duty "general" ProLink cartridge is used, a portion of the system can be scrutinized on a read-only basis. A full description of the ProLink reader/programmer's capabilities is included elsewhere in this text, but even with this early generation electronic control system should be noted that Mack's primary diagnostic and programming tool was the PC.

Personal Computer (PC)

Mack Trucks was the first of the OEMs to make the PC its recommended EST. It was required for proprietary data programming and access to the Mack Trucks data hub, but it also had obvious advantages over the use of flash codes and ProLink as a diagnostic instrument. Owners who did not wish to have their truck ECU programming altered by non-Mack dealers often programmed alpha-only passwords, making it more difficult to use ProLink to alter programming fields.

Troubleshooting and customer data programming are simplified using a PC and the appropriate linkage hardware and software.

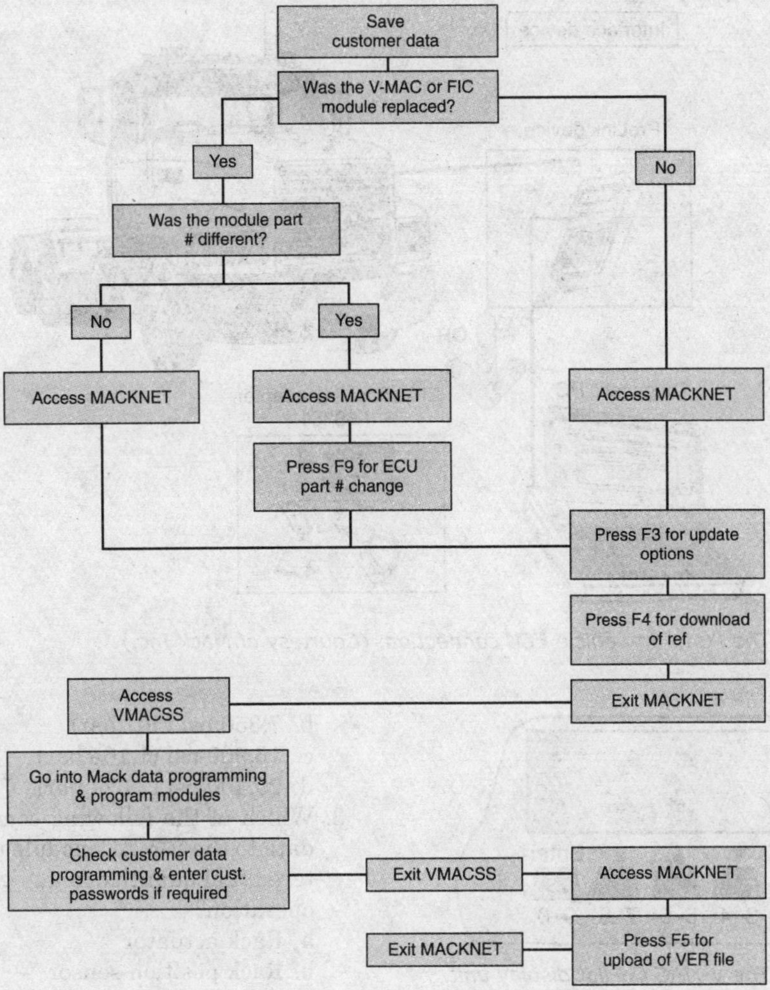

Figure 35–11 *Mack data programming flowchart. (Courtesy of Mack Inc.)*

Proprietary data programming must be downloaded from an Allentown, Pennsylvania-based mainframe accessed by modem to PC memory using the three-stage process described previously. Mack programming manuals tend to be user friendly and the reprogramming procedure is simplified by clearly denoted input fields and menu screens.

CoPilot Display

Consistent with other engine OEMs of its era, Mack Trucks's dash-located, digital displays known as **CoPilot** enable the driver to monitor specific ECU data relating to engine performance, fuel economy, trip information, time, and active and logged fault codes while the vehicle is in operation (Figure 35–13). It consists of a vacuum fluorescent screen located in the upper center of the dash display. The screen displays two rows of alphanumeric characters, and five buttons permit the operator to input display requests.

One of the real benefits of driver data displays (especially when used by a fleet) is the fuel economy readout, which has proved itself to be a cheap and highly effective driver training tool. CoPilot simply enables driver access to a portion of V-MAC running data that can enhance the driver's ability to efficiently operate the vehicle (Figure 35–14). The system is user friendly to the extent that it can be mastered

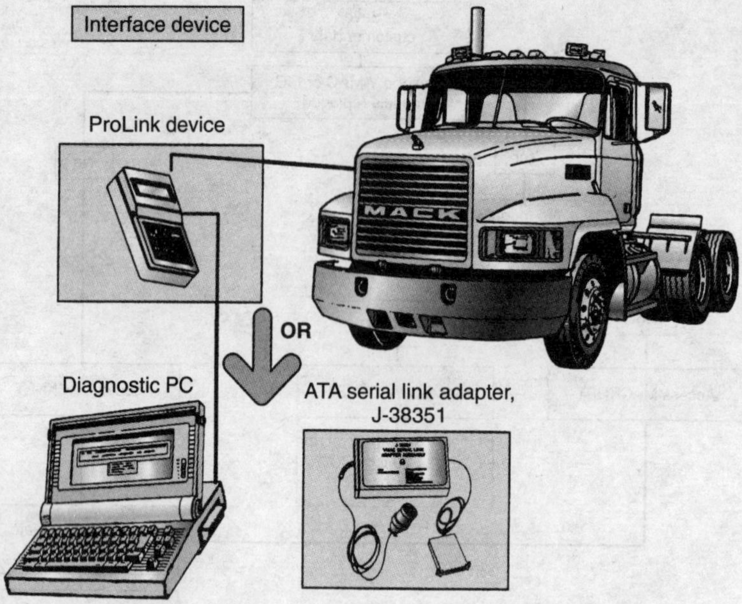

Figure 35–12 *The PC to on-vehicle ECU connection. (Courtesy of Mack Inc.)*

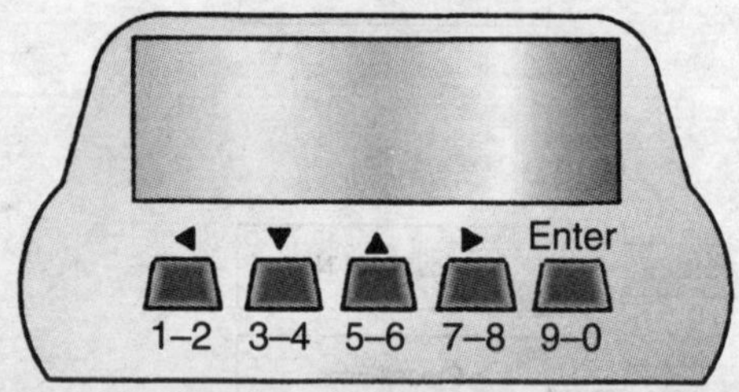

Figure 35–13 *The V-MAC CoPilot display unit. (Courtesy of Mack Inc.)*

with a few minutes of instruction. CoPilot also provides coded access to the ignition circuit, which is a theft protection feature.

REVIEW QUESTIONS

1. Which version of V-MAC used an ECU and a separate FIC module to manage an inline injection pump?
   a. V-MAC I
   b. V-MAC II
   c. V-MAC III
   d. V-MAC IV
2. Peak injection pressures in V-MAC Bosch PE 8500 injection pumps can reach:
   a. 2,200 psi (152 bar)
   b. 7,350 psi (507 bar)
   c. 16,900 psi (1,165 bar)
   d. 29,400 psi (2,027 bar)
3. Which of the following components inputs data to the V-MAC module that defines metered fuel quantity in any given moment of operation?
   a. Rack actuator
   b. Rack position sensor
   c. Reference coil
   d. Timing event marker
4. Which of the following components could be described as a linear magnet?
   a. Rack actuator
   b. Rack position sensor
   c. Reference coil
   d. Pulse wheel
5. Which is the moving component on the Bosch rack position sensor assembly?
   a. Reference coil
   b. Short-circuit ring
   c. Proportional solenoid
   d. Pulse wheel
6. Which of the following sensors is responsible for signaling engine position data to the V-MAC module?
   a. RPM/TDC sensor
   b. TEM sensor

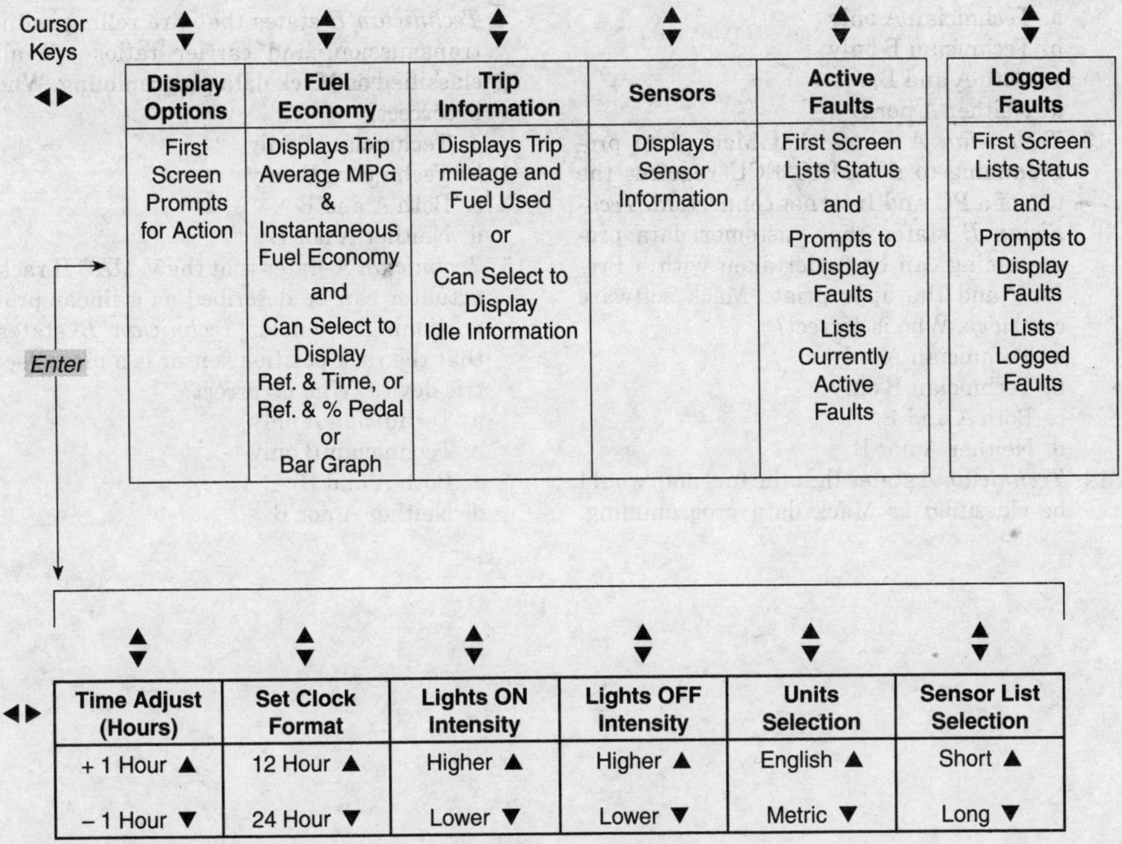

Figure 35–14 *V-MAC CoPilot quick reference. (Courtesy of Mack Inc.)*

c. CTS
d. OPS

7. How many fuel pump elements are required on a Bosch 8500 injection pump fueling an inline six-cylinder engine?
   a. One
   b. One per cylinder head
   c. Three
   d. Six
8. When performing the procedure required to reprogram a V-MAC module with Mack data, which of the following is the correct sequence?
   a. Upload, reprogram, download
   b. Download, reprogram, upload
   c. Reprogram, upload, download
9. Which of the following correctly describes CoPilot?
   a. A reader/programmer instrument
   b. A dash-located data display
   c. A customer data, programming instrument
10. Which of the following programmed data types would be classified as Mack data?
    a. Tire rolling radius
    b. Transmission ratio
    c. Road speed limit
    d. Fuel map
11. *Technician A* states that the component that signals rack position data to the V-MAC ECU is called a rack actuator. *Technician B* states that rack position data is signaled to the ECU in a digital format. Who is correct?
    a. Technician A only
    b. Technician B only
    c. Both A and B
    d. Neither A nor B
12. *Technician A* states that the V-MAC II module is mounted on the right side of the E-7 engine. *Technician B* states that early versions of V-MAC used a two-module electronic management system. Who is correct?

a. Technician A only
b. Technician B only
c. Both A and B
d. Neither A nor B

13. *Technician A* states that Mack data programming to a V-MAC ECU requires the use of a PC and Internet connection. *Technician B* states that customer data programming can be undertaken with a Pro-Link and the appropriate Mack software cartridge. Who is correct?
a. Technician A only
b. Technician B only
c. Both A and B
d. Neither A nor B

14. *Technician A* states that the fuel map would be classified as Mack data programming. *Technician B* states that tire rolling radii, transmission, and carrier ratios are all classified as Mack data programming. Who is correct?
a. Technician A only
b. Technician B only
c. Both A and B
d. Neither A nor B

15. *Technician A* states that the V-MAC II rack actuator can be described as a linear proportioning solenoid. *Technician B* states that the rack position sensor is a piezoelectric device. Who is correct?
a. Technician A only
b. Technician B only
c. Both A and B
d. Neither A nor B

36

Detroit Diesel Electronic Controls (DDEC) for EUI Systems

INTRODUCTION

Detroit Diesel Corporation (DDC) was the first diesel engine manufacturer to offer an electronically managed diesel engine in the North American truck and bus markets. DDC electronics are described by the acronym **DDEC**, which stands for **Detroit Diesel Electronic Controls**. DDEC I was introduced in 1985 and made generally available in 1987 on two-stroke cycle, **Series 92** engines (displacement of 92 cubic inches per cylinder). DDEC I was a full authority management system with mechanically actuated (camshaft) **electronic unit injectors (EUIs)** that were computer controlled by a dash-located engine control module (ECM) and an engine-mounted, **injector driver** unit called an **electronic distributor unit (EDU)**. The injector drivers are the ECM actuators responsible for effecting ECM-plotted fueling commands by switching the EUI duty cycle.

DDC Series 60

Shortly after the introduction of the full authority EUI system on the Series 92 engine, Detroit Diesel launched what was to become the most successful diesel engine of the 1990s: the Series 60. The **Series 60** engine is an inline, six-cylinder, four-stroke cycle engine fueled by EUIs and managed by full authority electronics. The engine was the first truck diesel engine to be sold in North America that was designed specifically for computer controls. Electronic management was by DDEC II electronics. DDEC II combined the EDU and ECM assemblies into a single, engine-mounted module. Initially, the Series 60 engine was available in 11.1- and 12.7-liter displacements. DDC added a 14-liter displacement version in 2002.

The 14-liter displacement Series 60 engine was not available from 2004 through 2007, but it is the only Series 60 option for 2007. The **Series 50** engine is a four-cylinder version of Series 60, which shares a high degree of component commonality with the six-cylinder version and most of its operating principles. It is basically a Series 60 with two cylinders chopped off and has become a popular transit bus engine. Figure 36–1 shows a 2007 Series 60 engine.

Corporate Changes

DDC was purchased by the DaimlerChrysler Corporation in 2001 and became allied with Freightliner trucks. This move meant that Detroit Diesel lost its "independent" identity, which resulted in Freightliner competitors to no longer consider spec'ing it in their trucks. The merging of DDC and Mercedes-Benz (MB) diesel engines in Detroit resulted in a common platform engine family scheduled for phased introduction from 2008 to 2010. Changes to the Series 60 for the 2007 model year include moving to a single 14-liter option and adopting a

Figure 36–1 *DDEC VI managed, 2007 Series 60 14-liter engine. (Courtesy of Detroit Diesel Corporation)*

single-piece, drop forged steel trunk piston. The current Series 60 piston is a Mahle Monotherm piston similar to that adopted by most DDC competitors. The merging of DDC and MB engine development and production led to the Series 60 engine being designated as linehaul and the MB 4000 as a vocational engine. The DDC family of EUI-fueled engines past and present includes:

DDC Series 60	11.1 liters	Transit bus and fleet applications
DDC Series 60	12.8 liters	Fleet and linehaul applications
DDC Series 60	14.0 liters	Linehaul applications
DDC Series 50	8.5 liters	Transit bus and shorthaul applications

DDEC Electronics

DDEC electronics have been upgraded through the years, and the current version, DDEC VI, was introduced for the 2007 model year. DDEC electronics and the fuel system hardware used on DDC engines have remained relatively unchanged over the years until 2007, so this chapter primarily references the earlier versions of DDEC and occasionally references current versions. The management system used on current DDC EUI-fueled engines (and MB EUP-fueled engines) is DDEC VI. The DDEC VI management system has also been adapted to manage the DD series of common rail (CR) fuel systems described in Chapter 40. Note that DDEC VI is a two-module system.

DDEC VI Twin Controllers

Because the sixth generation of DDEC was tailored to run DD EUI, Mercedes Benz EUP, and the DD family of CR-fueled engines, a two-module system was required. The common module with a 128 MID address on the data bus is known as the **common powertrain controller (CPC)**. The CPC uses a proprietary bus connection to the second controller known as the **motor control module (MCM)**. The MCM computes and drives those engine functions that are specific to the Series 60 engine such as the EUI injectors. Only the CPC is networked directly to the data bus. This means that data bus access to the MCM is routed through the CPC: These two modules are connected by a proprietary data bus. Figure 36–2 shows how DDC represents the CPC electronics. The overall functions of the two-module system are not dissimilar to the single module system so in most cases we do not distinguish between the two.

DD EUI-fueled diesel engines have used C-EGR since the 2004 model year and for 2007 adopted a multiple stage exhaust gas aftertreatment system with a diesel particulate filter (DPF). In common with most other original equipment manufacturers (OEMs), DDEC-managed engines using ULS fuel and multistage exhaust gas aftertreatment have not compromised fuel economy to any notable extent. The Series 60 established itself a reliable fleet engine during the 1990s and continues to rely on the linehaul market today. In this chapter, we primarily reference the Series 60 engine and attempt not to distinguish between different versions of DDEC except where necessary. Figure 36–3 shows a pre-2007 schematic representation of a typical DDEC system, which is the layout reference in this chapter.

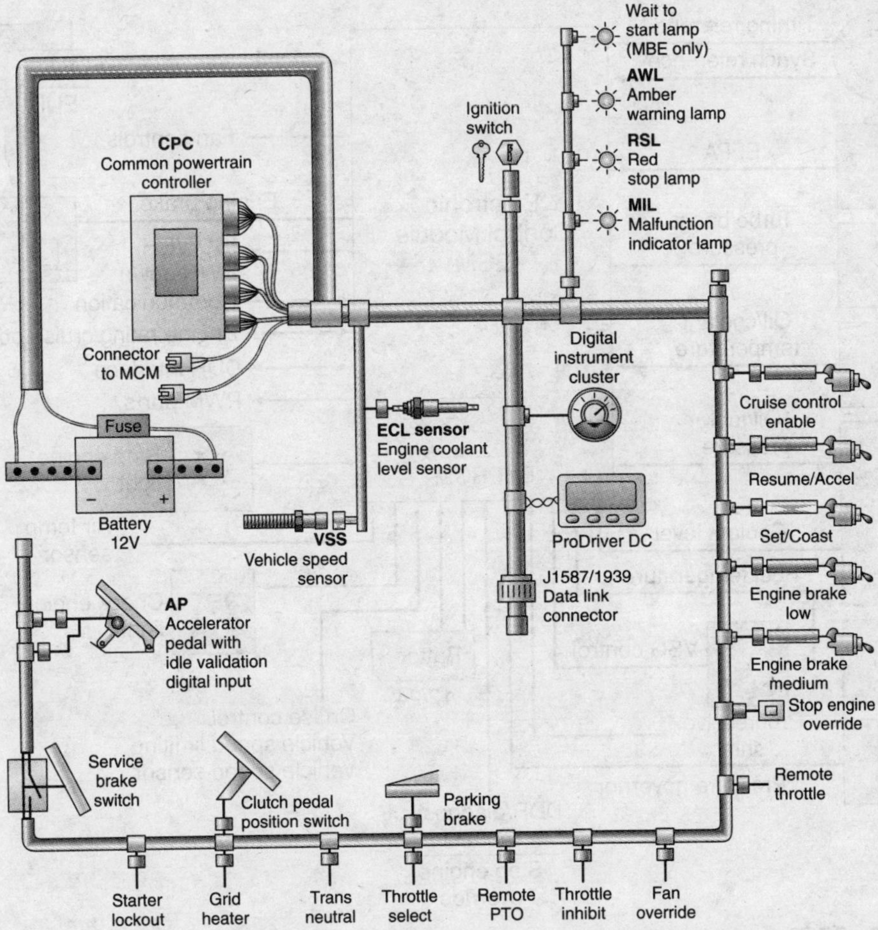

Figure 36–2 *Schematic showing DDEC VI CPC electronics. (Courtesy of Detroit Diesel Corporation)*

Heavy-Duty Engine Platform

DDC and Mercedes Benz introduced their Heavy-Duty Engine Platform (HDEP) family beginning in 2008. These engines will be badged as DDC for the NAFTA geographic area and as Mercedes Benz elsewhere in the world. The engine platform will include displacements of 15.6, 14.8, 12.8, and 10.6 liters. Engine management and fueling will use a second generation of CR diesel fuel injection in which rail pressures are amplified within an electrohydraulic injector (EHI) immediately prior to injection. Fuel-amplified (CR) fuel injection is covered at the end of Chapter 40 in this textbook. DDC will use very similar electronics to those on current DDEC VI engines to manage this new family of engines, so technicians familiar with the current system should experience few problems adapting to the new system. It is important that the EUI system discussed in this chapter is not confused with the DDC CR system.

SYSTEM OVERVIEW

The Series 60 and 50 DDEC engine management system uses cam-actuated, electronically controlled EUIs. DDEC electronics have offered progressively more comprehensive engine monitoring, audit trails, self-diagnostics, and customer and proprietary (DDC) data programming with each generation. For purposes of study, the system can be divided as follows:

1. Fuel subsystem
2. ECM
3. Input circuit—sensors
4. Output circuit—EUIs and others

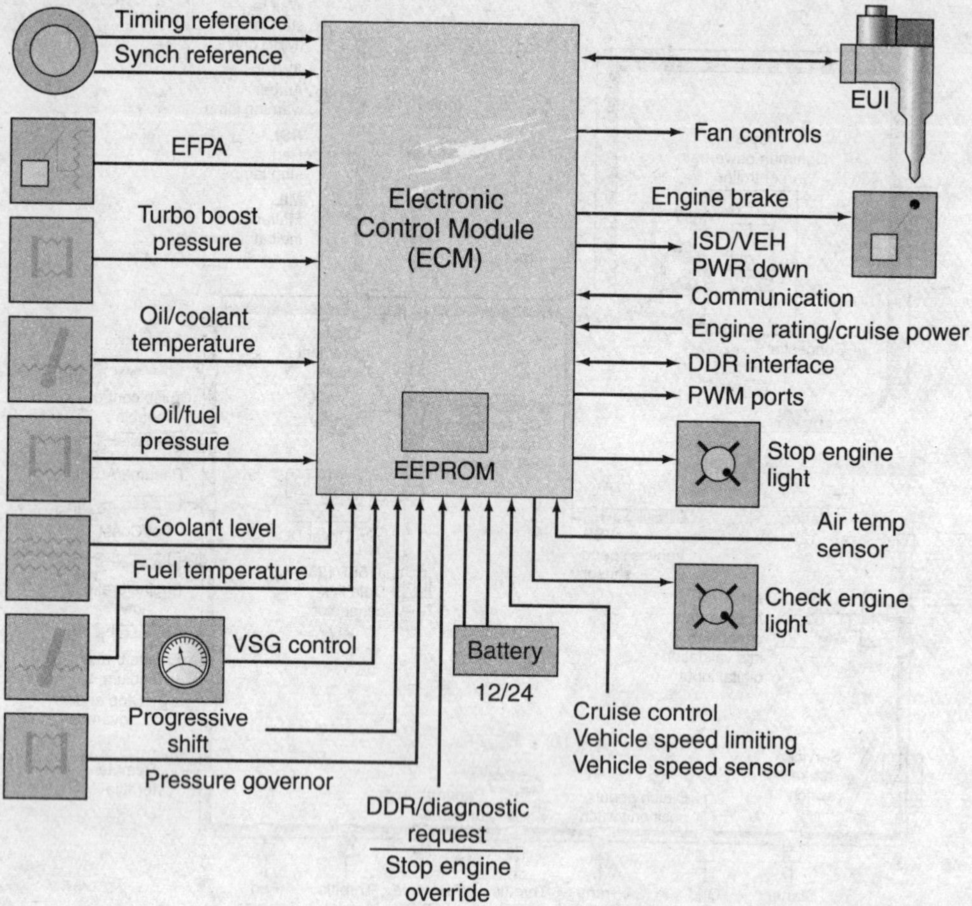

Figure 36–3 *DDEC management system layout showing input and output circuits. (Courtesy of Detroit Diesel Corporation)*

FUEL SUBSYSTEM

Fuel is drawn from the fuel tank by a gear pump, which is responsible for all movement of fuel through the fuel subsystem. A primary filter is usually but not always plumbed in series between the fuel tank and the transfer pump. Fuel discharged under pressure by the transfer pump is routed through a secondary filter, in some cases an ECM heat exchanger, and then to the fuel supply and return galleries located in the engine cylinder head. The DDC-owned Davco Fuel Pro system is recommended by DDC as a single filter system on older engines, or as a primary filter assembly to be used in conjunction with a DDC secondary filter.

Fuel Routing in the Cylinder Head

The cylinder head fuel galleries charge the cylindrically shaped EUIs. Series 60 EUIs are manufactured with two external annular recesses in the injector body, the lower of which is exposed to the charging gallery in the cylinder head. The upper annular recess in the EUI discharges fuel to the return gallery from which it is sent back to the fuel tank. In this way, fuel is continually flowed through the EUIs. In the older Series 92 engines (many of which are still around), fuel is transferred to and from the EUIs by means of jumper pipes in much the same manner as in mechanical unit injectors. The fuel constantly circulated through the EUIs while the engine is running is at a charging pressure typically around 50 psi (334 kPa).

Fuel Filters

The latest version of DDEC EUIs, the N3 and E3 units, are susceptible to fuel contaminants and problems associated with prolonged service intervals. For this reason, DDC will void warranty for injectors if the specified DDC filters are not used. In all cases, a DDC secondary spin-on filter must be used and this filter should be serviced within 15,000 (24,000 km) linehaul miles. The secondary filter can be used in conjunction with a Davco Fuel Pro 382 primary filter or aftermarket primary filter. DDC warns that the consequence of not using the specified secondary filters is plugged filters and damage to N3 and E3 injectors.

ELECTRONIC CONTROL MODULE (ECM)

The first-generation DDEC I computer was located within the vehicle cab, with the injector drivers (switching apparatus) or EDU mounted on the engine close to the EUIs. DDEC I rapidly evolved into DDEC II, so there are few examples of the first-generation electronics still in service. Versions of DDEC II through V were single module systems that housed the computer and injector drivers in a single, engine-mounted unit. The physical size of DDEC ECM housings has decreased with each progressive generation while computing capability and speed has increased. With the introduction of DDEC VI, DD reverted to a two-module system. The CPC is engineered to run the different MCMs required for the Series 60, the MB electronics for its EUP-fueled 4000 and 900 series engines, and the recently introduced DD family of engines. Figure 36–4 shows a DDEC VI CPC module.

Functions

DDEC ECMs are microprocessor-driven management and switching modules. Whether single or dual module, the controllers house the microprocessor and output actuators. DDEC electronics are responsible for engine governing, fuel timing logic, self-diagnostics, comprehensive system monitoring, **audit trails** (data tracking), **tattletales** (covert writing of electronically monitored events or conditions to EEPROM), and customer and DDC data programming. Earlier generations of DDEC ECMs could manage up to eight EUIs so in off-highway engines with more than eight cylinders, multiple controllers were required. In multi-ECM-managed engines (off-highway applications only), one of the ECMs masteres engine outcomes, receiving all inputs and processing all data, while the other (or others) act as receivers, being used primarily as injector drivers. However, in the event of master ECM communication failure, receiver ECMs can temporarily enable operation of each portion of the engine.

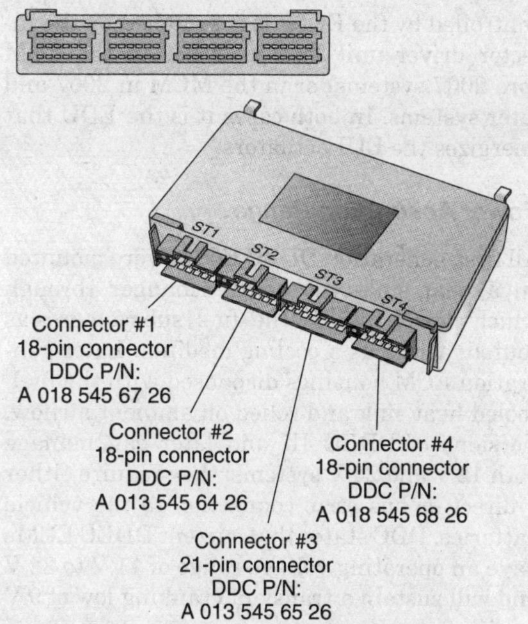

Figure 36–4 *DDEC VI common powertrain control (CPC) module. (Courtesy of Detroit Diesel Corporation)*

Injector Drivers and Duty Cycle

DDEC electronics are programmed with application specific detailing such as engine governing, torque profile shaping, cold-start logic, and engine protection strategy. The results of ECM arithmetic, logic, and fetch-and-carry computations in the processing cycle are converted to actuator or control signals that switch the EUIs. Command pulses can be referred to as **duty cycle** or **pulse width (PW)** and are measured in milliseconds: DDEC electronics can optionally display duty cycle in crank angle degrees. Duty cycle or PW is the time period within the EUI cam-actuated downstroke in which fuel is actually being injected. PW is calculated and controlled by the ECM. It is switched by the injector driver unit located either in the ECM (pre-2007 systems) or in the MCM in 2007 and later systems. In both cases it is the EDU that energizes the EUI actuators.

Power Absorption Range

All first-generation DDEC ECMs were mounted on a heat sink-type heat exchanger through which diesel fuel in the fuel subsystem was routed, acting as a cooling medium. Later-generation ECM housings dispensed with the fuel-cooled heat sink and relied on ambient airflow. Versions of DDEC III and later can manage both 12 V and 24 V systems; they require either a direct or bus strip connection to the vehicle batteries. DDC states that current DDEC ECMs have an operating voltage range of 11 V to 32 V and will sustain a transient cranking low of 9 V minimum.

Data Bus Address

All generations of DDEC II through version V used a single module ECM networked to the chassis data bus using the MID 128 address. The current DDEC VI system uses an ECM, which DDC choose to call a CPC, resident at the MID 128 address on the data bus. The CPC communicates by means of a proprietary bus to the MCM. The MCM has some processing capability but it primarily houses the drivers required to switch the EUI actuators. Technician data communication with the MCM is only through MID 128, that is, the CPC module. Figure 36–5 shows the networking relationship between the MCM, CPC, and the two chassis data buses. Bus networking is primarily through the faster J1939 with the J1587/1708 providing redundancy.

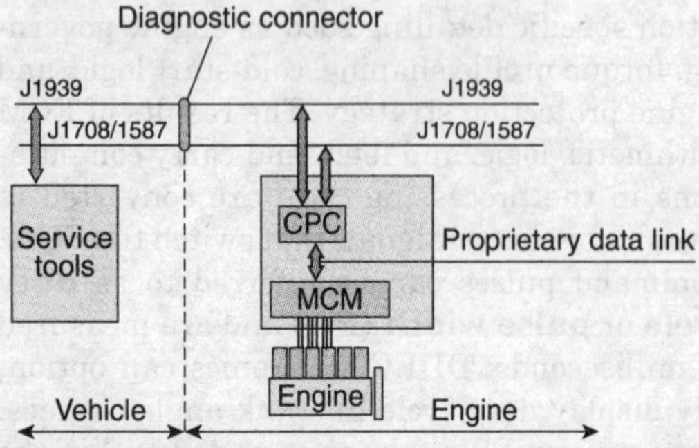

Figure 36–5 *Networking relationship between the DDEC VI MCM, CPC, and the two chassis data buses. (Courtesy of Detroit Diesel Corporation)*

INPUT CIRCUIT

As DDEC has evolved, the sensor circuit has become more comprehensive. The following is a description of some current DDEC sensors without detailing the electrical operating principles, which are covered in Chapter 30.

Shaft Speed Sensors

DDEC shaft speed sensors use an alternating current (AC) pulse generator principle to input a frequency signal to the ECM. The signal frequency increases proportionally with shaft speed. The two primary engine shaft speed sensors are described next but speed sensing, whether turbo speed or transmission tailshaft speed, is critical for factoring engine outcomes.

Timing Reference Sensor

The **timing reference sensor**, or **TRS**, signals crank position to the ECM. An AC pulse generator is used. A 6-tooth pulse wheel (reluctor) is used on DDEC II, and this was upgraded to a 36-tooth pulse wheel on more recent versions. In DDEC versions II through VI, the pulse wheel is fitted to the front of the crankshaft (Figure 36–6).

Synchronous Reference Sensor

The **synchronous reference sensor**, or **SRS**, signals the ECM to indicate a specific engine position and acts to confirm the TRS signal. An AC pulse generator is used. Both TRS and SRS must work for the engine to run; if either is not functional in DDEC II, a code is logged after 10 seconds cranking; in later versions of DDEC, this code is produced immediately. The SRS consists of a single dowel that acts as a reluctor tooth located on the cam gear and a stationary magnetic pickup in Series 60/50.

Electronic Foot Pedal Assembly

In common with most truck/bus electronically managed engines, DDEC is "drive by wire."

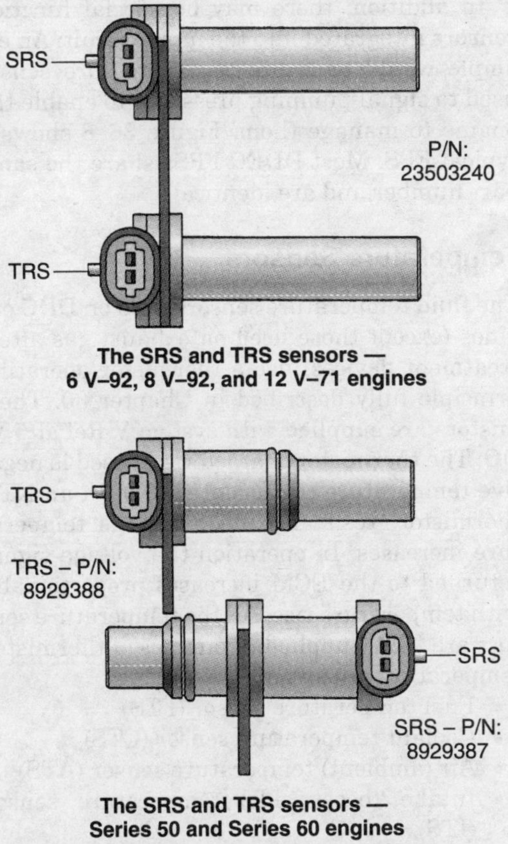

Figure 36–6 *DDEC SRS and TRS sensors on two-stroke cycle and four-stroke cycle engines. (Courtesy of Detroit Diesel Corporation)*

DDEC VI can be enabled for either a potentiometer or Hall-effect-type accelerator pedal. The critical sensor in the potentiometer type **electronic foot pedal assembly (EFPA)** is the **throttle position sensor (TPS)**. A potentiometer-type TPS receives a 5 V reference voltage (V-Ref) and returns a portion of it proportional to pedal mechanical travel. Alternatively, a noncontact Hall-effect TPS can be used to signal accelerator pedal angle. An EFPA also contains an idle validation sensor (IVS). The electrical principles of both potentiometer and Hall-effect TPS are explained in Chapter 30. In the DDEC system, the actual voltage signal returned to the ECM is converted to counts, which can be read with an electronic service tool (EST). TPS counts can be interpreted as follows:

Zero travel, idle: 100–130 TPS counts

Full travel: 920–950 TPS counts

48 counts or less: low-voltage code generated

968 counts or higher: high-voltage code generated

The EFPA is supplied by the chassis manufacturer but must be consistent with DDC specifications (Figure 36–7).

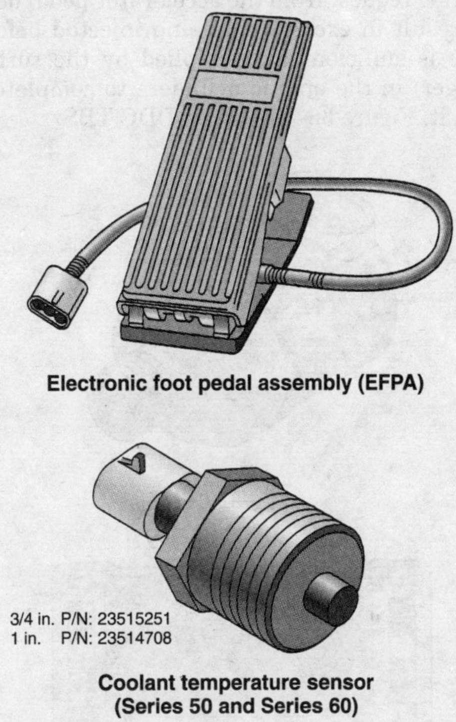

Figure 36–7 *The EFPA and CTS. (Courtesy of Detroit Diesel Corporation)*

Turboboost Sensor

The DDEC **turboboost sensor (TBS)** is responsible for signaling the ECM with both manifold boost and barometric pressure data. Early versions of DDEC TBS used a variable capacitance principle. Pressures from 0 to 45 psi (3 bar/304 kPa) can be measured. Recent versions of DDEC use a piezoresistive (Wheatstone bridge) sensor to signal manifold boost to the ECM. This sensor is the same as that used by General Motors (GM) on some of its Gen III and IV managed engines.

These sensors are more accurate and less susceptible to tampering. Whichever type is used, the TBS is a DDC-provided sensor. The electronic principles of piezoresistive pressure sensors are covered in Chapter 30.

TBS data is required by the ECM to plot air-fuel ratio (AFR). By monitoring atmospheric pressure, oxygen density is known, so ECM fuel quantity programming is automatically derated at altitude. Turboboost pressure monitoring indicates actual engine load and prevents transient overfueling. This means that a sudden full-fuel request from the accelerator pedal does not result in excess fuel being injected before there is sufficient air (supplied by the turbocharger) in the engine cylinders to completely burn it. Figure 36–8 shows a DDC TBS.

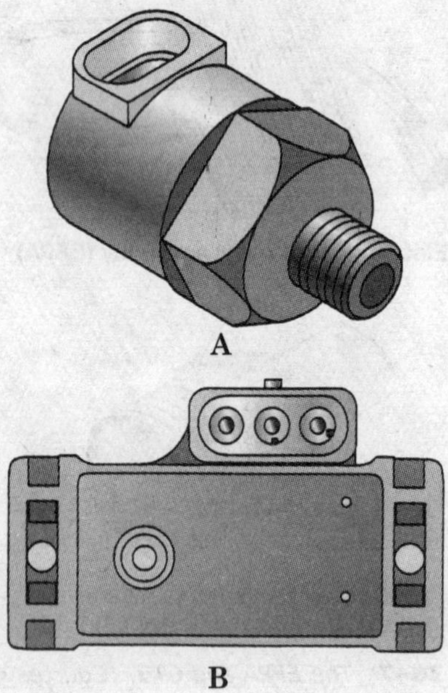

Figure 36–8 *A. Oil, fuel, and coolant pressure sensors; B. Turboboost sensor. (Courtesy of Detroit Diesel Corporation)*

Fluid Pressure Sensors

Most of the fluid pressure sensors used on DDEC-managed engines use a variable capacitance principle. They are DDC supplied. The electronic principles of this category of sensors are discussed in Chapter 30. Fluid pressure sensors include:

- **Fuel pressure sensor (FPS)**. Location: charge side of the transfer pump
- Oil pressure sensor (OPS). Location: lube gallery in the cylinder block
- Coolant pressure sensor (CPS). Location: water manifold

In addition, there may be special function sensors integrated into the input circuit: An example would be a fire truck pressure sensor used to signal pumping pressures to enable the engine to manage them. Figure 36–8 shows a typical FPS: Most DDEC FPSs share the same part number and are identical.

Temperature Sensors

The fluid temperature sensors used on DDC engines (except those used on exhaust gas aftertreatment devices) use a thermistor operating principle fully described in Chapter 30. Thermistors are supplied with system V-Ref at 5 V-DC. The thermistor characteristic used is negative temperature coefficient (NTC). In an NTC thermistor, resistance decreases as temperature increases. In operation the voltage signal returned to the ECM increases proportionally with temperature rise. All the temperature sensors are DDC supplied. Examples of thermistor temperature sensors are:

- Fuel temperature sensor (FTS)
- Coolant temperature sensor (CTS)
- Air (ambient) temperature sensor (ATS)
- Intake (boost side) temperature sensor (ITS)
- Oil temperature sensor (OTS)

Mass Airflow Sensors

DDEC uses a pressure differential-type mass airflow (MAF) sensor in its post-2004 EGR engines. The DDEC MAF sensor uses a critical flow venturi (CFV) with variable capacitance sensors positioned up- and downstream from the venturi. The operating principle is based on the relationship between inlet pressure and flow rate through a venturi. When the pressure is known at the entry point to the venturi and at its exit point, the differential pressures can be used by the ECM to calculate the mass of the airflow. Differential airflow-type MAF sensors are described in more detail in Chapter 30.

Coolant Level Sensors (CLS)

A coolant level sensor is a switch-type sensor consisting of an integrated resistor that grounds through the engine coolant and is located in the top radiator tank. If the ground is interrupted, engine protection strategy is initiated after a prescribed time period if programmed. The CLS is OEM supplied.

Multiplexing

Where DDEC electronics are required to interface with other chassis system ECMs, both the J1587 and J1939 data communication buses can be used but the faster J1939 is used for critical chassis communications. Multiplexing is used by DDEC to eliminate the need for duplicating simple chassis hardware such as TPS and **vehicle speed sensor (VSS)** to more complex, multi-MID communication transaction required for anti-yaw-rollover electronic interventions.

OUTPUT CIRCUIT

Output circuit devices are components controlled by the ECM; they are the set of components that effect the results of the programmed software and the processing computations. Any component that is switched or can be controlled by the ECM can be called an output.

Electronic Unit Injector (EUI)

DDEC EUIs are integral pumping, metering, and atomizing devices. In common with the mechanical unit injectors (MUIs) used by DDC in the past, they are cam actuated but with that the similarity ceases. The EUI plunger has no cross and center drillings or helical recesses but is a simple cylindrical plunger that reciprocates within the pump stationary element or bushing (see Figure 36–9). Actual plunger travel does not vary because this is dictated by cam profile. The EUI tappet spring holds the plunger in a raised position for most of the cycle until the actuating cam on the engine camshaft ramps toward the cam nose and the EUI plunger is driven downward into the pump chamber. DDEC EUIs can be identified by the number of actuators:

- Single actuator: a solenoid control cartridge used on N3, N2, and earlier EUIs
- Dual actuator: use a control cartridge and integrated hydraulic nozzle assembly; known as E3 injectors introduced in 2007

We will focus on single-actuator EUIs first then study dual-actuator E3 units.

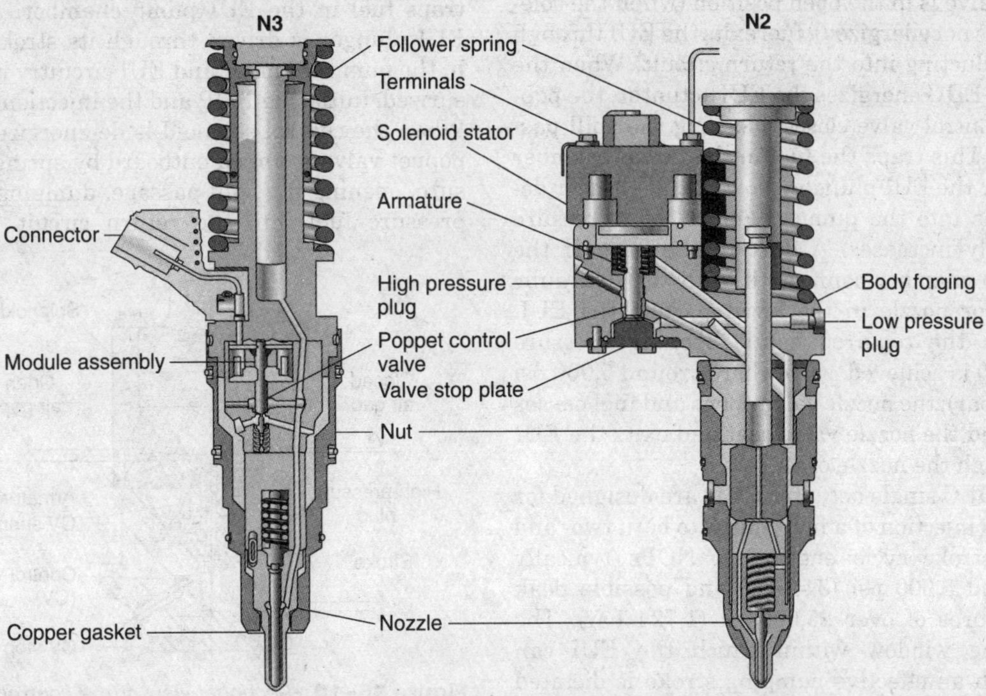

Figure 36–9 *Component identification on single-actuator DDEC N2 and N3 EUIs. (Courtesy of Detroit Diesel Corporation)*

Single Actuator EUIs

There are two types of single actuator EUIs:

- N2, using an external control cartridge
- N3, using an internal control cartridge

The DDEC N3 is lighter in weight than the N2 EUI it replaced, and its control cartridge is integrated into the EUI body so it is less bulky. However, the operating principles of both N2 and N3 injectors are nearly identical so we will study them under a single heading: However, it should be noted that they are timed using different procedures. Figure 36–9 is a sectional view of both N2 and N3 EUIs indicating the main components.

N2 and N3 Operation

Fuel from the fuel subsystem is circulated through EUI internal ducting whenever the engine is running. This charging pressure fuel is 50 psi (3.3 bar). Fuel entering the EUI internal circuit is routed through the EUI pump chamber, nozzle assembly, and EUI spill circuit. The spill circuit is opened or closed by a control solenoid. The control solenoid armature is integral with a poppet control valve. Whenever the poppet valve is in the open position (when the solenoid is not energized), fuel exits the EUI through spill ducting into the return circuit. When the ECM EDU energizes the EUI actuator, the poppet control valve closes, blocking the spill passage. This traps the fuel in the pump chamber under the EUI plunger. Now as the plunger descends into the pump chamber, fuel pressure rapidly increases. A duct in the base of the pump chamber connects it with the hydraulic injector nozzle in the lower part of the EUI. When the required nozzle opening pressure (NOP) is achieved, somewhere around 5,000 psi (340 bar), the nozzle valve opens and fuel passes around the nozzle valve seat and exits the EUI through the nozzle orifii.

DDEC single actuator EUIs are designed for direct injection of a fuel charge to both two- and four-stroke cycle engines at NOPs typically around 5,000 psi (344 bar) and possible peak pressures of over 25,000 psi (1,724 bar). The fueling window within which the EUI can switch an effective pumping stroke is dictated by cam profile because the EUI plunger must be in the act of descending for this to occur. Plunger mechanical stroke is constant. Plunger *effective stroke* must occur within the mechanical stroke hard window, beginning when the EUI solenoid is energized and ending when it is de-energized. Unlike many of its competitors that spike the actuation voltage at the beginning of injection, DDEC EUIs are switched at an electrical pressure of 12 V: Because of the lower electrical pressure, DDEC electronics are required to monitor injector response time. The actual time the EUI control cartridge remains energized is displayed as pulse width (PW) in milliseconds.

EUI Actuator

The DDEC EUI actuator used in N2 and N3 EUIs is a solenoid that opens (de-energized) and closes (energized) the EUI spill passage. In this way, it controls the duration of the effective stroke of the EUI pumping plunger. Figure 36–10 shows a sectional schematic of the EUI control cartridge, and the poppet control valve is labeled control valve (CV) in this diagram. The DDEC ECM (specifically the EDU) energizes the solenoid using a pulse width modulation (PWM)-type signal, closing the CV. This action traps fuel in the EUI pump chamber. As the EUI plunger is driven through its stroke, fuel in the pump chamber and EUI circuitry is pressurized, initiating NOP and the injection pulse. When the control solenoid is de-energized, the poppet valve is forced outboard by spring pressure, opening the spill passage, dumping high-pressure fuel into the return circuit. When

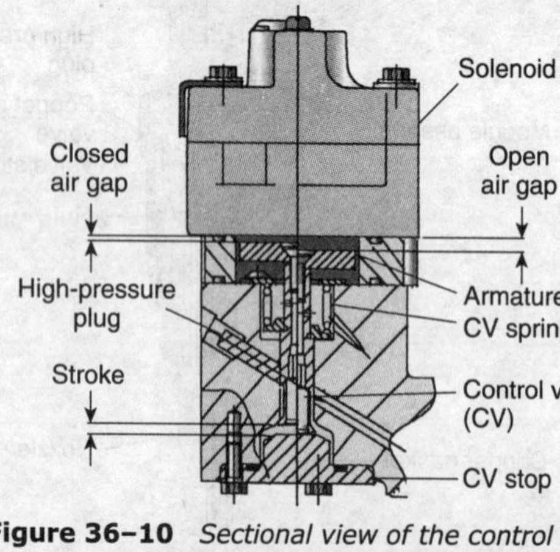

Figure 36–10 *Sectional view of the control solenoid and poppet valve from a DDEC N2 EUI. (Courtesy of Detroit Diesel Corporation)*

pressure drop is sufficient to allow the nozzle valve to close, the nozzle spring seats the nozzle valve, ending the injection pulse.

Calibration Codes

DDEC single actuator EUIs are graded with a calibration code that ranges from 00 to 99, enabling the ECM to be programmed with data concerning how a specific EUI flows fuel. The **digital diagnostic reader (DDR)** is used to program this data to the ECM and enables highly accurate and balanced fueling of the engine. The programming of calibration codes is achieved in the customer data programming field. The injector cylinder number has to be selected first, after which digital calibration code is input. Earlier DDEC II calibration codes used alpha fuel flow codes, A, B, or C, with A denoting the least fuel flow and C the most fuel flow. This critical procedure helps the ECM balance cylinder fueling.

Injector Response Time

DDEC EUIs are switched at 12 V, unlike most other EUI systems that spike the application voltage to values around 100 V. This means that the time lag between **beginning of energizing (BOE)** or the instant that the EDU initiates current flow to the EUI solenoid and **beginning of injection (BOI)** is to some extent variable. DDEC electronics are programmed to measure the response time between BOE and BOI by studying the actuation voltage wave of the previous two actuations of each EUI on a continual basis. **Injector response time (IRT)** is essentially the time lag between the output of the actuating signal at the EDU and the moment the EUI poppet control valve actually closes. It is measured in milliseconds and displayed on a DDR. Because IRT is monitored, DDEC is capable of adjusting to minor electrical circuit resistance variations within a certain window to maintain balanced fueling and timing.

It should be noted also that there is a fractional lag between EUI control valve closure (BOI) and the actual opening of the EUI hydraulic nozzle valve (NOP), which truly begins injection. Similarly, at the completion of the switched duty cycle or PW known as **ending of energizing (EOE)**, there is not an immediate cessation of injected fuel because some fraction of time is required to drop the pressure in the EUI pump chamber to the **valve closing pressure (VCP)** value at which point the injection pulse truly ceases—**ending of injection (EOI)**. DDC claims that this collapse phase is of shorter duration in N3 injectors than in N2s.

Graphing BOI, IRT, and PW

If you are not intimidated by graphs, maybe the best way of understanding IRT is to look at what happens using a graph. Figures 36–11 and 36–12 show the time and electrical waveforms that define IRT and PW, while 36–13 shows real time representations of IRT.

Split-Shot Injection

DDEC electronics, in conjunction with N2 and N3 single actuator EUIs, could manage **pilot injection**, which DDC commonly referred to as **split-shot injection**. This technology was a forerunner of the multipulse capability we accept in current EUIs. Initially, it was an ECM cold-start strategy designed to eliminate diesel knock and minimize startup smoke emission. Pilot injection essentially breaks up the fueling pulse by switching the EUI at high speeds. Using this technology under cold-start conditions would mean delivering a short pulse of fuel to the engine cylinder, calculating the moment of ignition, and at that point resuming the EUI fueling pulse. Diesel knock is a cold-start detonation condition caused by delayed ignition and excess fuel in the cylinder at the point of ignition. When pilot injection is controlled by the ECM, it can be used to help manage the combustion process: Besides cold-start fueling, it is also used at low-speed, high-load lug. In addition to reducing engine wear, pilot injection also greatly lowers cold-start emissions.

E3 Injectors

The DDEC VI system on post-2007 Series 60 engines uses a double actuator EUI manufactured by Delphi. This four-terminal EUI is used by some DDC competitors, and it boasts two significant advantages over single actuator EUIs and any current CR injection systems:

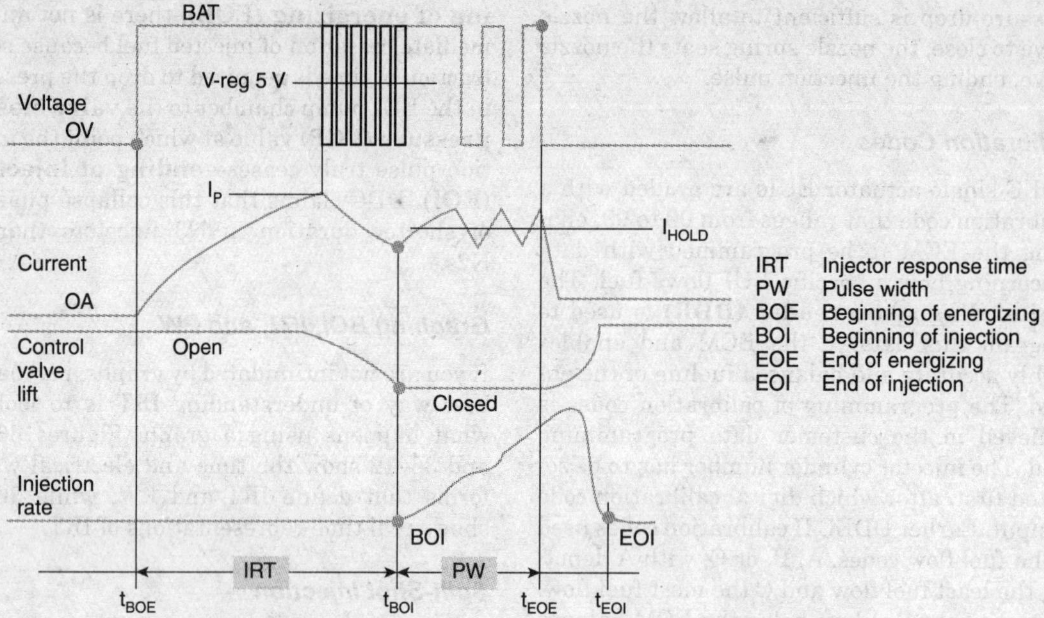

Figure 36–11 *DDEC injection cycle time graph. (Courtesy of Detroit Diesel Corporation)*

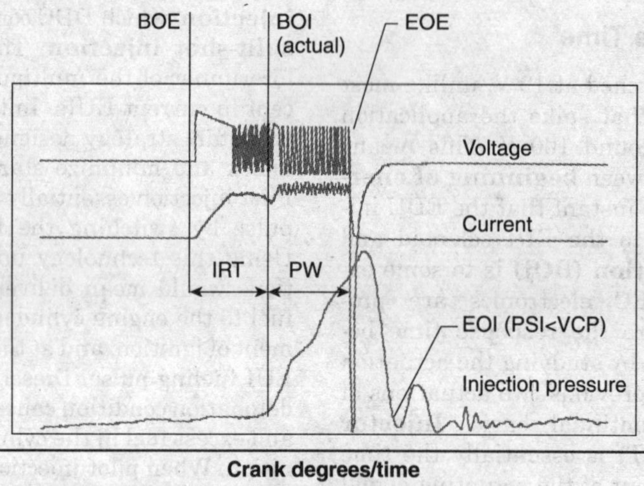

Figure 36–12 *DDEC injection cycle electrical waveform to time graph. (Courtesy of Detroit Diesel Corporation)*

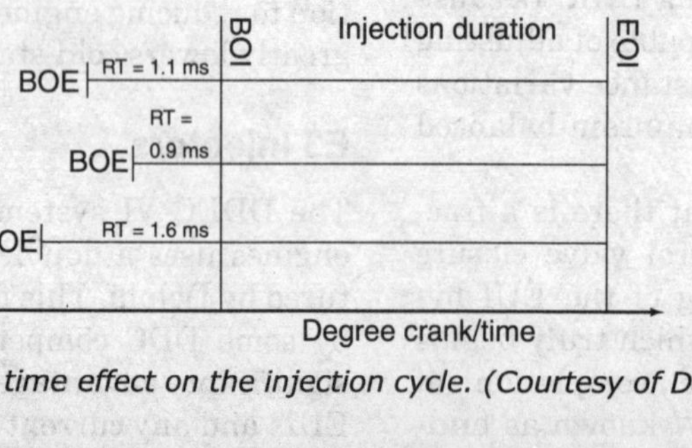

Figure 36–13 *Response time effect on the injection cycle. (Courtesy of Detroit Diesel Corporation)*

- Higher peak pressures than any current CR diesel fuel injection system: Maximum pressures exceed 32,000 psi (2,250 bar).
- Soft (ECM) control over NOP and rapid response to multipulse injection switching.

The amperage required to actuate the EUI control valves has become reduced with each DDC generation. The current required to actuate an E3 is 9 amps peak, with a 3 amp hold specification. Figure 36–14 shows an external view of a DDC Delphi E3 injector.

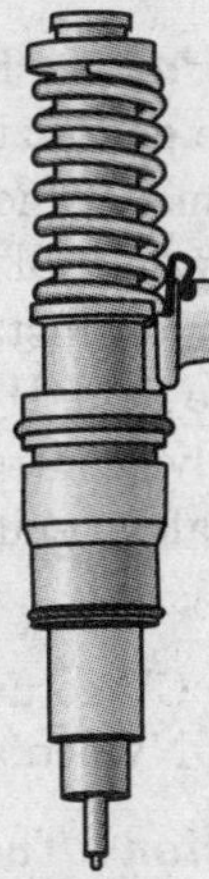

Figure 36–14 *DDC Series 60 Delphi E3 injector.*

DDEC E3 injectors are lightweight and have a sleek design similar to the N3 injectors they replace. The functions of the two actuators used are:

- **Control valve (CV) solenoid:** The CV is an MCM-switched actuator that does exactly what the control cartridge solenoid did in earlier N2 and N3 injectors. The CV opens and closes the EUI pump chamber spill circuit. Energizing and closing the CV solenoid traps fuel in the EUI pump chamber under the plunger, creating pressure rise.
- **Nozzle valve (NV) actuator:** The Delphi E3 uses an electrohydraulic nozzle valve actuator. The NV actuator is integrated into the EUI body just above the nozzle assembly as shown in Figure 36–15. The NV actuator is an electrohydraulic nozzle that enables a soft NOP value; in other words, NOP is controlled by the ECM, or more specifically, by the MCM, which contains the DDEC VI injector drivers. Review electrohydraulic nozzles in Chapter 19 if you cannot remember how these function.

Delphi E3 Construction

Dual actuator DDEC EUIs eliminate three of the major disadvantages of first-generation EUIs:

- Bulky assembly with protruding control cartridge
- Hard value NOP due to the hydraulic injector nozzle spring
- Hard value nozzle closing pressure with a prolonged collapse phase at the end of injection

Despite some increased complexity and higher peak pressures, Delphi E3 EUIs weigh about one-half as much as the older N2 EUIs. In addition, the EUI is slim and the obstruction to the injector and valvetrains created by the protruding control solenoid cartridge on the older N2 EUIs has been significantly reduced.

Delphi E3 Operating Principles

In terms of general operating principles, the four-terminal Delphi E3 EUI is essentially a "two-terminal EUI" fitted with an MCM-controlled electrohydraulic injector nozzle. We reference Figure 36–15 with the assumption that you remember how electrohydraulic injectors operate. The following description divides injector operation into six phases:

At Rest Fuel at the charge pressure flows into the EUI and circulates through the pump chamber and nozzle assembly, exiting through the spill circuit. Both actuators are de-energized and the EUI pump plunger is in its raised position because the injector cam is on its inner base circle (IBC). In this phase, the nozzle valve is held seated by spring force.

Actuator status: CV actuator: de-energized.
NV actuator: de-energized.

Pressurizing Pressurizing begins when the injector cam profile begins to ramp off its IBC: This actuates the EUI plunger, driving it downward in the bushing bore. When the MCM energizes the CV, the spill circuit closes, trapping fuel in the pump chamber under the plunger. As pressure rise is created by the descending

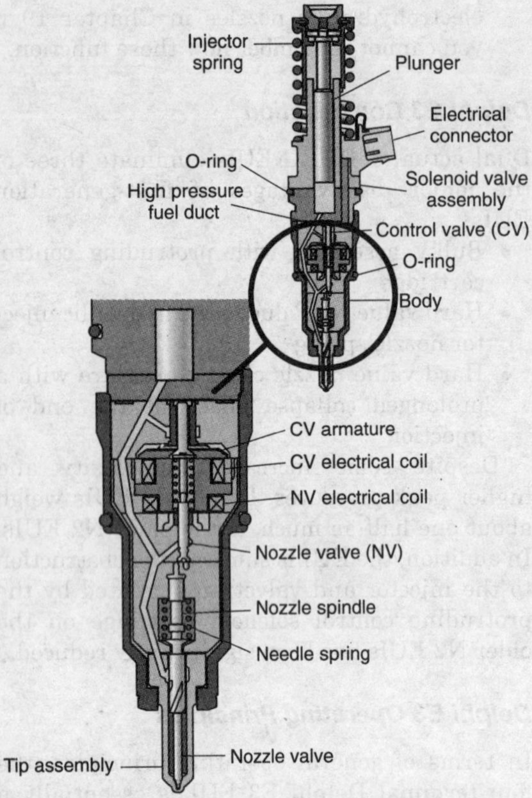

Figure 36–15 *Sectional view of a four-terminal (two-actuator) EUI featuring an electrohydraulic injector nozzle.*

plunger, identical pressure acts on both the upper and lower pressure fields: However, because the sectional area of the upper pressure field (acting on the nozzle valve piston) exceeds that of the lower pressure field (acting on the nozzle valve pressure chamber sectional area), the nozzle valve remains firmly seated. In this phase the nozzle valve is held seated by a combination of spring force and hydraulic force acting on the nozzle valve control piston.

Actuator status: CV actuator: energized.
NV actuator: de-energized.

Injection When desired NOP is achieved within the EUI circuit, the MCM energizes the NV actuator. This pulls the nozzle valve into the nozzle actuator, collapsing the upper pressure chamber field. Because the upper pressure chamber field collapses and the lower pressure chamber field remains unchanged, hydraulic force lifts the nozzle valve, beginning injection. While injection continues, the nozzle valve remains lifted, which means no pressure can build in the upper pressure chamber: The exit flow area spilling fuel from the upper nozzle pressure chamber is very small so spilling fuel from the upper pressure field has little effect on the pressure in the remainder of the EUI circuit.

Actuator status: CV actuator: energized.
NV actuator: energized.

End of Injection Pulse The MCM ends an injection pulse by de-energizing the NV actuator. Spring force slams the nozzle piston downward, closing the nozzle spill passage, and consequently permits almost instant reestablishing of the upper pressure field. Hydraulic force acting on the nozzle valve control piston abruptly closes the nozzle valve, ending the injection pulse.

Actuator status: CV actuator: energized.
NV actuator: de-energized.

Multipulse Injection The Delphi E3 EUI lends itself to multipulse injection events. For multipulse injection to take place, the NV actuator is energized more than once during a single mechanical downstroke of the EUI plunger. The advantages of multipulse injection relate to better combustion efficiency and lower emissions. EUIs equipped with electrohydraulic nozzles better suit multiple pulse injection events because the NOP and injection pressure can be made to rise incrementally with each successive pulse, producing the smallest droplets later on in the combustion phase.

During multipulse injection the NV actuator is rapidly cycled. This means that the nozzle valve has to open and close while injecting fuel to the engine cylinder at much higher opening pressures than were possible with older hydraulically actuated nozzles. In fact, as indicated earlier opening and injection pressures can be incrementally increased as each injection event occurs progressively later in the injection cycle, producing smaller atomized droplets, ensuring complete combustion.

Actuator status: CV actuator: energized.
NV actuator: cycling on and off.

End of Injection Phase The injection phase ends when both the CV and NV actuators are de-energized by the MCM drivers. This necessarily must occur before peak lift is attained by the injector actuation cam. Because Delphi E3 EUIs are mechanically actuated by cam profile, the MCM can only switch effective strokes while the EUI plunger is descending. So in cases where the CPC algorithm requires multiple injection pulses during one cycle (that is, one injection phase), these must occur within a hard window defined by cam profile. So at the completion of the injection phase the actuating cam profile ramps down to IBC, and with both actuators de-energized, the "at rest" status resumes.

Actuator status: CV actuator: de-energized.
NV actuator: de-energized.

GOVERNING

All versions of DDEC engine management offer both limiting speed governing (LSG) and variable speed governing (VSG). Although managed electronically, both LSG and VSG are designed to function similarly to mechanical LSG/VSG used in older DDC two-stroke cycle engines. The LSG option fuels the engine based on percent throttle (pedal position) selected by the operator. The VSG option will hold engine speed at a specific requested value and attempt to maintain it as the load on the engine fluctuates; it is used for PTO operation and electronic cruise control.

The ECM (DDEC versions before 2007) or CPC (DDEC VI 2007 and later) are programmed with a detailed fuel map that manages engine fueling through both normal and transient abnormal conditions. Smoke control and cold-start fueling are managed by selecting optimum injection timing and injected fuel quantity based on such factors as ambient temperature, barometric pressure, manifold boost, and throttle demand. Cold-start strategy on DDEC-managed engines may produce an automatic increase in engine idle speed and advanced injection timing to minimize cold operation white smoke emission; this would return to normal as the engine approaches operating temperatures.

Cruise Control

Cruise control can be operated in any gear when engine speed is above 1,100 rpm and providing the road speed is above 20 mph (32 km/h). Cruise control can be programmed to master engine brake operation so that if the vehicle accelerates above set cruise speed on a downgrade, the engine brake will automatically actuate.

Progressive Shifting

Progressive shifting is a governing feature designed to achieve better fuel economy by encouraging the driver to upshift before the engine reaches governed speed. When the speed of the engine is limited, the driver remains in the higher torque range of torque rise (that is, the lower rpm range) longer where the engine operates at better fuel efficiency.

Engine Protection

DDEC electronics monitors a range of vehicle and chassis functions and is programmed with engine protection strategy that may derate or shut down the engine when a running condition could result in catastrophic failure. The operator is alerted to system problems by illuminating dash lights, the **check engine light (CEL)**, and the **stop engine light (SEL)**. The CEL is illuminated when a system fault is logged to alert the operator that DDEC fault codes have been generated. The SEL is illuminated when the ECM detects a problem that could result in a more serious failure. DDEC may also be programmed to shut down or ramp down to idle speed after such a problem has been logged. A shutdown override switch known as an **SEO**, or **stop engine override**, switch will permit a temporary override of the shutdown command.

ENGINE DIAGNOSTICS

Faults that occur are stored in ECM memory and may be recalled by any of the following means:

Blink Codes

The CEL and SEL will flash codes when the diagnostic request switch is toggled. The CEL flashes inactive codes, and the SEL flashes active codes.

Detroit Diesel Diagnostic Link

Detroit Diesel Diagnostic Link (DDDL) is the online service information system and troubleshooting software program required to communicate with DDEC. All troubleshooting and even routine servicing should be routed through DDDL. It functions similarly to other OEM interactive diagnostic and analytical software with full-function online service information capability.

Chassis OEMs can program engines on their assembly lines using a DDC software package called *Vehicle Engine Programming Station (VEPS)* and a PC with a Windows environment. A PC interface (serial link adapter) is required to translate the datalink signal of the ECM to the VEPS software. VEPS PC software can also be used to rewrite some of the parameters programmed to the DDEC III ECM.

Diagnostic Data Reader (DDR)

Although DDC recommends that a PC with DDDL software and online access be used to diagnose, read, and reprogram DDEC-managed engines, DDEC can be accessed using a diagnostic digital reader (DDR) (J38500-E). This is a ProLink 9000 with the appropriate level multiprotocol card (MPC) card or dedicated cartridge. The DDR connects to the chassis data bus by means of an SAE/ATA J1708 six-pin/J1939 nine-pin Deutsch connector in the same way that a serial link adapter for a PC connects. This provides two-way communications for reading system parameters, calibration data, diagnostics, data programming, recalling faults, and audit trails. Chapter 31 provides detailed information on using ESTs and service information systems (SIS).

Fault Codes

DDEC logs two types of codes: active and inactive.

1. Active Codes—An active code is a fault that is present at the moment codes are checked. Active codes are flashed to the SEL when the diagnostic request switch is toggled.
2. Inactive Codes—Inactive codes are codes that have occurred sometime previously but are not active at the moment of request. Every code has an audit trail (tattletale) in DDEC and is tagged with the following data:
   a. First occurrence of each diagnostic code logged in engine hours
   b. Last occurrence of each diagnostic code logged in engine hours
   c. Total time in seconds that the engine code was active

Injector Cutout Test

There is no appropriate way to mechanically short out a DDEC EUI, so when a cylinder misfire has to be diagnosed, the ECM/PCM electronics perform this task by electronically cutting out the EUIs in sequence and analyzing the performance effect. When an ECM commands an engine to run at a specific rpm and an injector cutout test is performed, first the average duty cycle of the EUI in milliseconds is displayed. If the engine rpm is to be maintained during the test (this test is normally run at idle or 1,000 rpm depending on software generation), as each EUI is cut out electronically by the ECM, the PW of the remainder of the injectors will have to be lengthened if engine rpm is held at the test value. This would be true until a defective EUI was cut out; in this case, there would be no increase in the average PW of the operating EUIs, because average PW was initially established with that one dead cylinder. If the example of a Series 60 engine is used with a dead EUI, the following would occur if an injector cutout test sequence were set with the engine running at 1,000 rpm. To run the engine at 1,000 rpm, the five functioning EUIs would produce fueling values in PW milliseconds, which would be averaged on the display.

When one of the five functioning EUIs is electronically cut out by the ECM, only four EUIs would be available to run the engine at the test speed of 1,000 rpm, causing the average PW to increase. As each functioning EUI is cut out in sequence, the average PW would have to increase to maintain the test rpm. However, when the defective EUI was cut out, there would be no change in the average PW because it is based on the engine running on five cylinders. Figure 36–16 shows the DDDL screen display following an injector cutout test that has identified a defective #2 EUI.

Engine Protection Strategy

Engine protection strategy can be programmed to three levels of protection in most versions of DDEC. The levels begin at a low level of protection, essentially a driver alert, and progress to a complete shutdown of the engine. The level of protection is selected by the vehicle owner and programmed to DDEC as a customer option. Because it is a customer option, the failure strategy can be changed at any time by a technician equipped with the access password and an EST with the DDEC software. The parameters monitored are:

Low coolant level
Low coolant pressure
High coolant temperature
Low oil pressure
High oil temperature
Auxiliary digital input

When DDEC detects a fault that may result in catastrophic engine damage, it can be programmed to react to the problem in three ways:

1. *Warning only.* The CEL and SEL will illuminate but it is left up to the operator to take action to avoid potential engine damage. No power or speed reduction will occur. Engine protection is therefore left entirely to the discretion of the operator.
2. *Ramp down.* The CEL and SEL will illuminate when a fault is detected and the ECM will reduce power (fuel quantity injected) over a 30-second period after the SEL illuminates. The maximum power value (fuel quantity) for this initial reduction is set by the fuel rate just before the logging of the fault condition. A diagnostic request/stop engine override switch is required when this engine protection option is selected.
3. *Shutdown.* Both the CEL and SEL will illuminate and the engine will shut down 30 seconds afterward. Once again, a diagnostic request/stop engine override switch is required if this option is selected.

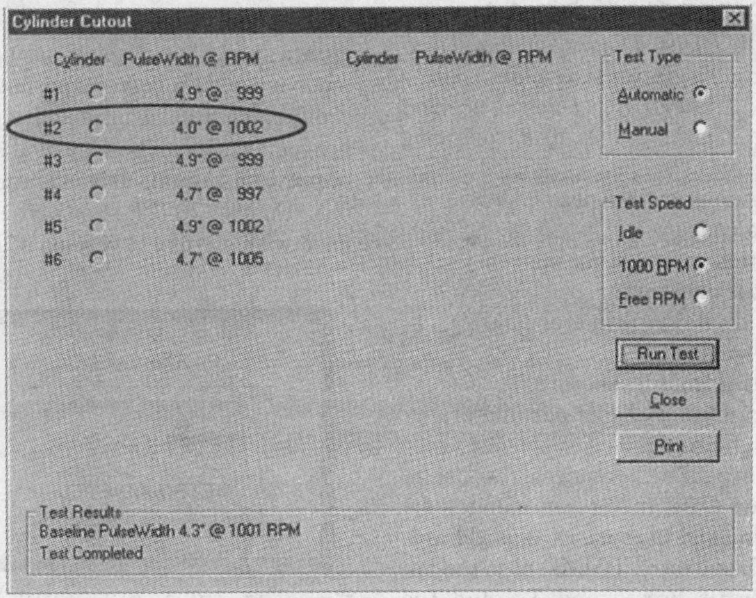

Figure 36–16 *DDDL screen display identifying the results of an injector cutout test. (Courtesy of Detroit Diesel Corporation)*

Stop Engine Override Options

Three types of stop engine overrides are available:

1. Momentary override—This resets the 30-second shutdown timer, restoring engine power to the value at the moment the SEL was illuminated. To obtain subsequent overrides the switch must be recycled after 5 seconds. DDEC will record the number of overrides actuated after the initial fault occurrence.
2. Continuous override, option 1—This stop engine override option is used when a vehicle needs full power during a shutdown sequence. Full power capability is maintained for as long as the override switch is pressed. It is intended for coach applications only.
3. Continuous override, option 2—This stop engine override option is used primarily for industrial and construction applications. It essentially disables the engine protection system until the ignition key is cycled.

Troubleshooting DDEC Electronics

The correct approach to diagnosing DDEC problems is to use DDDL. DDEC troubleshooting requires a basic knowledge of electrical circuits and schematics, DDEC operating fundamentals, and the DDC *DDEC Diagnostic and Troubleshooting Guide*. The latter is available online through DDDL. The *DDEC Diagnostic and Troubleshooting Guide* covers the following areas:

- Reading diagnostic codes
- Clearing codes
- Calibration reprogramming
- Connector checkout
- Using a digital multimeter (DMM)
- System testing
- Sequential troubleshooting
- Reprogramming engine parameters
- DDEC schematics

Troubleshooting DDEC requires the use of ESTs such as the DDR (not recommended for DDEC VI versions and higher), PC-based hardware with DDDL software, DMMs, plus the appropriate **breakout boxes** and Ts, and the DDEC service literature.

Calibration Reprogramming

Depending on the software version a DDR or PC can be used to make the following calibration changes:

- Rewrite DDR calibration configuration password.
- Add/delete 5-minute idle shutdown.
- Change droop (governor speed deviation from set speed).
- Set PTO droop (isochronous if required).
- Enable/disable cruise control.
- Add/delete a VSS.
- Change vehicle speed limit.
- Change cruise control speed limit.
- Add/delete engine shutdown.
- Set PTO maximum speed.
- Switch between available engine ratings.
- Enter injector calibration values.
- Set progressive shift configuration.

Prodriver

DDEC III versions and later feature as an option ProDriver, a dash-mounted data display that can provide the operator with immediate feedback on idle time, fuel economy, and other performance factors. Current versions of ProDriver have a graphical user interface and user-friendly, five-key controls and will write to a data card. The display provides instant data feedback to the driver and has been proved to produce an almost immediate performance improvement. It can be regarded as low-cost (compared to a human driver trainer) driver training. Figure 36–17 shows the display screen used with a current version of ProDriver.

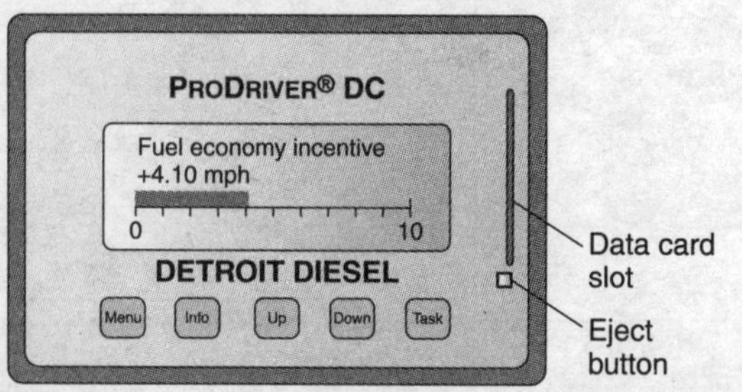

Figure 36–17 *ProDriver display screen. (Courtesy of Detroit Diesel Corporation)*

Promanager

ProManager is a PC software program that permits data to be downloaded from the DDEC ECM for analysis. Fleets interested in improving vehicle and operator performance have found ProManager a useful tool. The latest DDEC versions are driven from MS Windows or Vista software, and remote data interface (RDI) technology (see next section) is used.

Remote Data Interface

The **remote data interface**, or **RDI**, is a communications device that enables data exchange between the vehicle-based electronics, data hub products, and a fleet's computer network. System operation is automatic. The driver parks the vehicle adjacent to the RDI and connects the extraction cable to the ATA connector on the vehicle. RDI includes status lights to update the driver on the progress of the communications. The latest versions of RDI transfer are wireless.

SERIES 60 AND 50 TUNE UP

Consistent with any other section of this textbook in which a hands-on procedure is outlined, the technician is reminded that the information presented in this text is no more than an outline and that DDDL service literature be consulted before performing a tune up. Only two basic settings are checked or adjusted. The basic settings are the valve lash dimension and the EUI follower height dimension. The procedure for setting DDC valves is consistent with that used to set valves on most other engines, and that used to adjust the EUI height dimension correlates directly with the injector height dimension setting of older DDC-MUI engines.

Both Series 60 and 50 engines use a four-valve, parallel port valve configuration. DDC has eliminated the need for yokes/bridges to actuate the pairs of valves by using a distinct rocker arrangement in which each of the four valves is actuated by an individual rocker arm. The exhaust valves are set on one linear plane and the intake on another. This arrangement permits equal and uninterfered cylinder gas flow through each valve. A fifth rocker located at the center of the cylinder in between the intake and exhaust valve planes is responsible for actuating the EUI. Each rocker is adjusted for lash/load by means of an Allen head screw and locknut. Table 36–1 may be used as a guide for correlating engine position with setting location on a six-cylinder Series 60 engine.

Table 36–1 *Series 60 with N2 EUIs Tune-Up Chart*

Valve Overlap	*Set Valves on Cylinder Number*	*Set EUI Height Dimension on Cylinder Number*
#6	1	5
#2	5	3
#4	3	6
#1	6	2
#5	2	4
#3	4	1

Valve Set and Injector Timing

A detailed account on adjusting valves is given elsewhere in this textbook. As with its MUI-fueled engines, DDC prefers the use of go-no-go gauges when adjusting valve lash. The procedure used to set EUI follower height is a variation on that used in DDC MUI engines. However, it is important to identify the injector generation because the methods vary with each:

- N2 and earlier: Injector height set with follower up.
- N3 (2004) and E3 (2007): Injector set with follower down.

We now describe the procedures required for adjusting and servicing the N2 and earlier EUIs.

N2 and Earlier EUI Families

This procedure essentially synchronizes the actuating cam geometry with the EUI plunger. The correct height adjustment gauge must be selected by consulting the DDC specifications. Timing dimension tools of 78.2 mm, 78.8 mm, 80.3 mm, and 81.0 mm are commonly used with engines using N2 injectors so ensure that you select the specified tool: The engine must next be located in the correct position as shown in Table 36–1.

1. Insert the dowel of the EUI height adjustment tool into the gauge hole in the EUI. Place a drop of the same engine oil used in the engine on top of the EUI follower; the lower ledge of the T on the height adjustment tool should just wipe the oil as it is rotated.
2. If the follower flange-to-EUI body height is incorrect, loosen the locknut on the injector rocker and, using a 3/16 Allen socket, adjust the rocker train to obtain the exact specified height dimension. Torque the locknut to specification and recheck the EUI body-to-follower height. Because this procedure sets a dimension, in an emergency, that is, when the correct EUI height dimension tool is not available, an accurate electronic Vernier caliper can be used to check or adjust this setting.

EUI Removal

When the troubleshooting sequencing indicates that an EUI must be removed, it is essential that fuel be drained from the cylinder head fuel supply gallery before attempting the removal using the following procedure:

1. First, clean off the exterior of the valve housing cover and remove it.
2. Drain the cylinder head fuel gallery by removing the supply hose from the inlet fitting at the rear of the cylinder head. Collect the fuel that drains from the cylinder head and then remove the outlet hose from the fitting at the cylinder head.
3. Use regulated shop air and an air nozzle to blow out any remaining fuel from the fuel supply gallery.
4. Remove the two rocker shaft through bolts and remove the rocker assembly from the engine.
5. Loosen the EUI terminal screws. The EUI connectors have keyhole slots that permit them to fit over the terminal screw caps. Avoid attempting to remove the terminal screws because this will damage the solenoid housing threads.
6. Use an injector prybar to lift the EUI from its seated position.
7. Cap the EUI bore to prevent dirt from entering the engine. Handle the EUI with care.

EUI Installation

1. Lightly lubricate the O-rings with engine oil and carefully install them on the EUI body.
2. Insert the EUI into the injector tube. Visually align the EUI body between the valve springs. The EUI is seated by pressing it firmly into position using the heel of one hand.
3. The EUI is retained by a hold-down crab; position the hold-down crab and install the hold-down fastener. Ensure that the hold-down crab does not interfere with either the EUI follower spring or the valve springs. Torque the hold-down fastener to the required specification.
4. Install the EUI wires by positioning the keyhole terminal over the terminal screw; next, position the terminal so that the narrow slot is immediately below the terminal screw and torque to the DDC specification.
5. Install the rocker shaft assemblies.
6. Set the valve lash and EUI height dimension to specification.
7. Reinstall the fuel inlet and outlet lines to the cylinder head fittings and replace the valve housing cover.

N2 EUI Solenoid Replacement

DDEC N2 injector solenoids are serviceable without removing the EUI assembly from the cylinder head.

1. Loosen the EUI terminal screws. The EUI connectors have keyhole slots that permit them to fit over the terminal screw caps. Avoid attempting to remove the terminal screws because this will damage the solenoid housing threads.
2. Remove the four solenoid hex capscrews to remove the solenoid. The solenoid, load plate, follower retainer, and capscrews should be discarded. DDC stresses that the capscrews should always be replaced when undertaking this procedure.
3. Remove the spacer and seals from the EUI body. The seals may be discarded but the spacer *must* be reused. The spacer is matched to the EUI armature and cannot be interchanged.

4. Install a new seal in the spacer groove and position the spacer on the body with the seal facing downward. Lightly coating the seal with grease will help retain it during assembly.
5. Install a new seal in the solenoid groove and assemble the solenoid on the spacer.
6. Install the new solenoid capscrews through the load plate, follower retainer, solenoid, and spacer, threading the screws into the EUI body.
7. Torque the capscrews in the correct sequence to the DDC specification.
8. Using an electric etching pencil, write the final four digits of the injector part number on the load plate.
9. Pressurize the fuel system and visually check for leaks.

Timing N3 and E3 Injectors

The N3 EUIs were introduced for the Tier 2 generation (EPA model year 2004) of DDC Series 60 engines. The E3 EUIs described earlier in this chapter replaced the N3 EUIs for Tier 3 (EPA model year 2007). The general procedure for timing both N3 and E3 EUIs is similar and is outlined here, but remember to consult the DDC service procedure before performing this task.

1. Disable engine startup.
2. Remove the rocker housing covers.
3. Bar the engine in its correct direction of rotation. Identify a cylinder in which an injector actuating rocker is just beginning to depress the EUI follower: At this point and on this cylinder, both intake and exhaust valves should be closed, which is indicated by valve lash.
4. Fit a magnetic base dial indicator so that it can monitor the upward lift of the injector cam profile on that cylinder. You should set the pedestal of the dial indicator on top of the injector cam roller and adjust so that it is capable of measuring the full cam lift dimension.
5. Now continue to bar the engine in its correct direction of rotation until the indicator stops registering any further lift. At this point, the injector can be set.
6. Now beginning at the cylinder you have been working on, use Table 36–2 to adjust both the valves and injectors.
7. Set the injector. Loosen the jam nut on the adjusting screw by backing off two turns.
8. Tighten the adjusting screw until the injector plunger bottoms out: Torque to 40 lb-in. (4.50 N·m).
9. Next, back off the adjusting screw ¾ turn. Now torque the jam nut to 30–35 lb-ft. (41–47 N·m).
10. The EUI is now adjusted. Next set the valves on the corresponding cylinders identified in Table 36–2.

Table 36–2 *Setting Valves and Injectors on N3 and E3 Series 60 Engines*

Maximum EUI Cam Lobe Travel on Cylinder #	*Adjust EUI on Cylinder #*	*Adjust Valves on Cylinder #*
6	6	2
2	2	4
4	4	1
1	1	5
5	5	3
3	3	6

Injector Cups

Injector cups provide the means for seating and cooling DDEC EUIs. The injector cup has direct contact with the water jacket in the cylinder head. As a consequence, this can be a source of coolant leakage to the combustion chamber. Injector cup sealing integrity should be hydrostatically checked (see Chapter 8) when cylinder heads are serviced. Figure 36–18 shows a sectional view of a Series 60 injector cup.

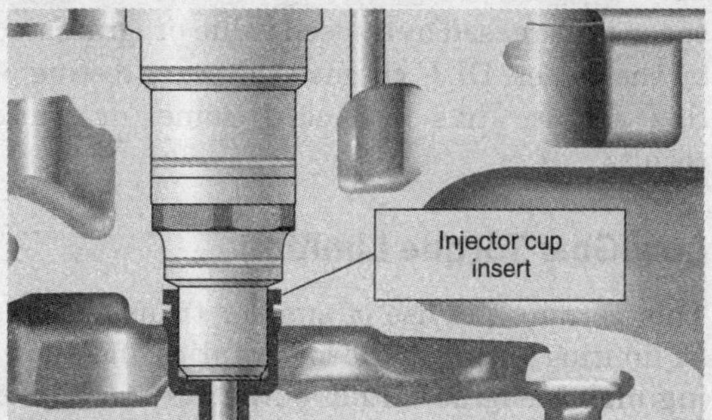

Figure 36–18 *DDC Series 60 injector cup. (Courtesy of Detroit Diesel Corporation)*

E3 E-Trim Cal Codes

E-trim programming is required to phase injection timing and balance fuel delivery for a set of injectors in an engine. E-trim essentially allows the DDEC fueling algorithm to adjust for minor fueling differences between injectors. It trims the injector driver sourced waveform to each EUI control cartridge that begins and ends each effective stroke to compensate for variations between injectors. Each E-trim cal code is determined on the OEM calibration test bench and then inscribed on each EUI in dot-matrix and alphanumeric form. Cal codes are programmed to the CPC via the MCM. The MCM is programmed with logic that can identify incorrectly input codes and prevent them from being logged. The cal codes are inscribed above the EUI terminal socket and should be programmed to the customer data programming field.

DDEC FEATURES

Because vehicle management electronics are upgraded on a continual basis, it is difficult to identify the significant changes that mark a numeric graduation from one generation of electronics to the next. This section outlines some of the features that DDEC provides extending through the current DDEC VI.

Fuel Economy Driver Incentive

The vehicle owner can set a fuel economy goal and program a maximum vehicle speed. If the driver fails to meet the goal, nothing changes. However, for each 0.10 mpg increase over the fleet economy target, the maximum vehicle speed is increased by a speed value programmed by the owner. DDC claims that early testing of this feature has produced some dramatic results.

Low Gear Torque Limiting

This permits the use of smaller transmissions by limiting torque in the lower ratios and allowing higher engine torque output in the higher gears.

Internal Compression Brake

DDEC engine internal engine compression brakes usually permits three levels of braking. Engine braking is controlled electronically and may be optioned into cruise control strategy using the DDEC platform or with collision warning systems (CWS) and vehicle stability control using the chassis data bus.

Optimized Idle

Versions DDEC IV and later can be programmed to an Optimized Idle mode that operates like a home heating thermostat when the vehicle is parked in jurisdictions where this is legal. The objective is to automatically stop and start the engine to:

- Maintain a comfortable cab/bunk temperature.
- Keep the vehicle batteries charged.
- Keep the engine warm.

Ether Start System

This feature is available in versions of DDEC III and later. Ether Start provides automatic ether injection during startup. The ECM is programmed to use the Ether Start option only when the temperature conditions require its use. By eliminating driver-controlled ether injection, the chances of abusing an engine with excess ether are reduced. A low ether supply indicator is included with the package.

Data Analysis

As the computing power of vehicle ECMs increases, more parameters are monitored and the data recorded can be used for progressive analysis of operating conditions as, for instance, in the driver fuel economy incentive. The extent to which the data can be analyzed can be greatly increased by download to a PC-based system programmed with software designed to study every performance detail of the vehicle. Data analysis is useful to the service technician. First-generation electronic systems identified what was malfunctioning in a system but offered few clues as to why the problem was occurring. Current troubleshooting packages can produce precise performance profiles that tell the diagnostic technician exactly how the vehicle has been driven and handled and the conditions at the exact moment a fault code was logged.

ProDriver (described earlier in this section) data can be downloaded by a Fleet Manager for analysis and using DDC software produce a profile of exactly how a driver operates a piece of equipment. ProDriver data may be accessed with ProDriver Reports software (requires MS Windows software) or the RDI system. RDI is weatherproofed so it can be mounted in a convenient location such as a fuel island for hardwire downloads. ProDriver can now also be programmed for wireless extraction of data from the vehicle electronic system.

REVIEW QUESTIONS

1. What force creates injection pressure values in all DDEC EUIs?
   a. Electrical
   b. Hydraulic
   c. Spring
   d. Mechanical
2. What defines the hard limit window available to the DDEC ECM when it computes EUI effective stroke?
   a. Engine rpm
   b. Accelerator position
   c. Road speed
   d. Cam profile
3. Which of the following is true regarding a DDEC N3 injector?
   a. One actuator
   b. Two terminals
   c. Multipulse capable
   d. All of the above
4. What type of transfer pump is used to move fuel through the fuel subsystem to charge the DDEC EUIs?
   a. Diaphragm
   b. Plunger
   c. Swash plate
   d. Gear
5. What is the DDC EUI system charging pressure on most generations of DDEC?
   a. 15 psi (1 bar)
   b. 50 psi (3.3 bar)
   c. 735 psi (50 bar)
   d. 5,880 psi (400 bar)
6. Which of the following versions of DDEC uses a two-module system?
   a. DDEC VI
   b. DDEC V
   c. DDEC IV
   d. DDEC III
7. What is the primary function of the SRS?
   a. Signal engine position data to the ECM
   b. Signal engine rpm data to the ECM
   c. Signal vehicle road speed data to the ECM
   d. Signal ambient temperature data to the ECM
8. How many teeth does the TRS pulse wheel have on a current DDEC system?
   a. 6
   b. 12
   c. 24
   d. 36
9. Which of the following features makes a DDEC E3 injector distinct from all previous DDEC EUIs?
   a. It is mechanically actuated.
   b. It uses an electrohydraulic nozzle.
   c. It uses a single actuator.
   d. CV actuator controls the spill circuit.
10. *Technician A* states that DDEC N3 EUIs feature an external control solenoid. *Technician B* states that DDEC N2 injectors are larger and more bulky than the later N3 versions. Who is correct?
    a. Technician A only
    b. Technician B only
    c. Both A and B
    d. Neither A nor B
11. How is DDEC effective stroke on a single-actuator EUI controlled?
    a. Varying plunger mechanical stroke
    b. Switching the EUI solenoid
    c. Altering the charging rail values
    d. Modulating the actuation voltage
12. The hydraulic injector nozzles in single-actuator DDEC EUIs usually have NOPs around:
    a. 2,200 psi (152 bar)
    b. 3,300 psi (228 bar)
    c. 5,000 psi (345 bar)
    d. 25,000 psi (1,724 bar)

13. Peak injection pressures in DDEC EUIs may exceed:
    a. 3,300 psi (228 bar)
    b. 5,000 psi (345 bar)
    c. 25,000 psi (1,724 bar)
    d. 50,000 psi (3,448 bar)
14. Which dash diagnostic light is used to flash active DDEC fault codes in systems without ProDriver?
    a. Oil pressure warning
    b. SEL
    c. Electronic malfunction
    d. CEL
15. What is the primary objective when the progressive shifting option is selected?
    a. To extend the service life of the transmission
    b. To improve fuel economy
    c. To frustrate the driver
    d. To ensure that engine oil temperature remains stable
16. *Technician A* states that DDEC E3 injectors allow the engine electronics some control over the NOP. *Technician B* states that the E3 injector is not suited to multipulse injection events. Who is correct?
    a. Technician A only
    b. Technician B only
    c. Both A and B
    d. Neither A nor B
17. When a DDEC engine programmed for *ramp down* as a failure strategy detects a low oil pressure condition, which dash light(s) will illuminate?
    a. CEL
    b. SEL
    c. CEL and SEL
    d. Ignition
18. The reason for programming DDEC ECMs with calibration codes is to:
    a. Enable the ECM to balance cylinder fueling
    b. Maintain charging pressure values
    c. Synchronize EUI solenoid valve closure
    d. Maintain the same IRT for each injector solenoid
19. *Technician A* states that one of the reasons DDEC VI electronics uses a two-module system is that the electronic management system is also used on the 2007 MB 4000. *Technician B* states that in DDEC VI the injector drivers are located in the MCM. Who is correct?
    a. Technician A only
    b. Technician B only
    c. Both A and B
    d. Neither A nor B
20. In the DDEC VI management system, which module uses the MID 128 address on the chassis data bus?
    a. ECM
    b. CPC
    c. MCM
    d. EDU

37

Volvo and Mack EUI Systems

INTRODUCTION

Volvo launched its first electronically controlled, mechanically actuated, unit-injected engine with the introduction of the first generation of VECTRO-controlled VDE-12 E in the mid-1990s. This family of engines was fueled by **electronic unit injectors (EUIs)**. Following the merger of Volvo and Mack Trucks, a new generation of **common platform** engines was introduced beginning in the 2007 model year. Volvo has used Delphi electronic unit injectors (EUIs) (Delphi-Lucas/CAV: referred to in this text as Delphi) to fuel its VDE-12 engines since their introduction to North America. The 2007 and Volvo and Mack engine families have few real differences. The primary differences are those of paint color and name. In this chapter we reference the earlier generations of Volvo EUI-fueled engines managed by VECTRO I and II, plus the current, and nearly identical, Volvo VECTRO III and Mack V-MAC IV common platform engine families.

Pre-2007 engines:

Volvo VE D12	1995	12-liter displacement
Volvo V-Pulse D-12	2002	12-liter displacement (EGR)

2007 and later common platform engines:

Volvo VE D11 (11-liter)	Mack Trucks MP7 (700 cu. in.)
Volvo VE D13 (13-liter)	Mack Trucks MP8 (800 cu. in.)
Volvo VE D16 (16.1-liter)	Mack Trucks MP10 (1000 cu. in.)

Volvo entered the North American truck market with its purchase of White Trucks and the GM heavy trucks division a generation ago. The Volvo VE D12 engine evolved from a similar 12-liter platform that used hydromechanical PLN fueling. This first generation of Delphi EUI-fueled Volvo engines was introduced in the 1990s and underwent few changes until the October 2002-introduced, V-pulse version. This exhaust gas recirculation (EGR) generation of Volvo engines was minimally reengineered to meet Environmental Protection Agency (EPA) 2004 emissions requirements, which were achieved with a constant geometry turbocharger and addition of EGR.

Current versions (post-2007) of both Volvo and Mack Trucks engines have moved away from two-piece articulating pistons, adopting steel trunk (Monotherm) pistons. Internal and external (exhaust) compression brakes are standard. At the top end of the power range of the common platform engines, the 16-liter engine can produce 625 BHP (465 kW) and up to 2,250 lb-ft. (3,350 N·m) of torque. The electronics used to manage the EUI fueling on the Volvo and Mack Trucks common platform engines are nearly identical but, probably for purposes of marketing, are badged as different systems. The Volvo engine and chassis management system is known as **VECTRO**, a loose acronym for *Volvo electronic controls*: The post-2007 version is known as **VECTRO III** or **Electronic Management System (EMS)** to describe its chassis electronics. The Mack Trucks engine and chassis management system is known as **V-MAC**, an acronym for **Vehicle management and control**: The post-2007 version is known as **V-MAC IV**. While VECTRO III shares a high degree of commonality with its previous versions, it is important to remember that V-MAC IV has almost nothing in common with previous versions of V-MAC. Figure 37–1 shows an exploded view of a Volvo VE D12 engine.

For whatever reasons, Volvo and Mack Trucks have attempted to maintain some degree of corporate identity when describing each original equipment manufacturer (OEM) badged set of engines. This may be effective from a marketing viewpoint but it makes it a little confusing for technicians learning the technology. It means that separate terminology and acronyms are required to describe exactly the same components depending on whether the engine is called a Volvo or a Mack. I guess we can begin by stating that Volvo-badged engines have their usual forest green paint job. On the other hand, if the engine is Mack Trucks badged, the engine gets a brand new scarlet paint job. Engine color is the easy part. In dealing with this subject matter in this book, there is little option but to use both the Mack and Volvo terms and hope that it does not create confusion.

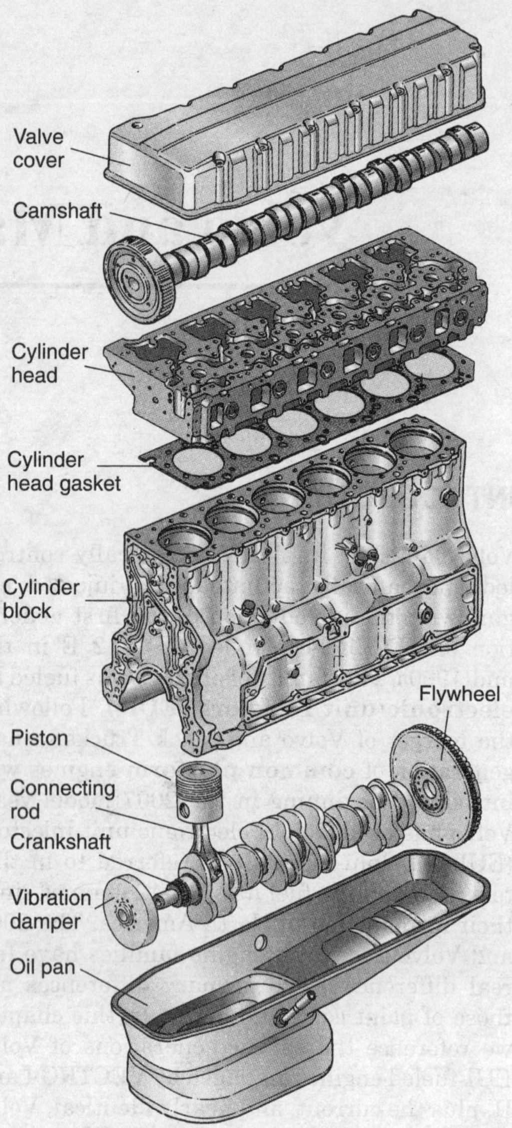

Figure 37–1 *Exploded view of a VE-D12 engine. (Courtesy of Volvo Trucks NA)*

SYSTEM OVERVIEW

Consistent with other EUI-fueled, full authority electronic engines, the mechanically actuated EUIs have an effective pumping stroke managed and switched by the electronic control module (ECM). Also in common with other EUI-fueled engines, the actual EUI plunger stroke is cam actuated. This means that the ECM is confined to a hard limit window (represented by cam profile) within which it can select an effective stroke. In this chapter, we focus on the features that make two- and four-terminal Delphi EUI-fueled engines distinct from other EUI-fueled engines. We also focus on the electronics used to manage these families of engines.

Both Volvo and Mack Trucks built post-2007 engines can be divided into the following subsystems for purposes of studying fueling control:

- Fuel subsystem
- Electronic input circuit
- Management electronics (electronic engine control unit [EECU] and vehicle control unit [VCU])
- Output circuit

FUEL SUBSYSTEM

The fuel subsystem incorporates those components that enable the transfer of fuel from the fuel tank to the EUIs. However, because the fuel subsystem incorporates a return circuit, it is best introduced by looking at the schematic shown in Figure 37–2. If you follow the directional flow arrows in this figure it will make it easier to understand the description in the text.

Callouts for Figure 37–2:

1. Supply pump (external gear type)
2. Fuel tank
3. Electronic engine control unit (MID 128) mounted on fuel-cooled heat sink
4. Fuel filter housing pad
5. Pre-filter
6. Primary fuel filter and water separator (under suction)
7. Secondary filter (under charge pressure)
8. Fuel gallery (under charge pressure: supplies and returns fuel from EUIs)
9. Electronic unit injectors (EUIs)
10. Charging pressure control valve: defines charging pressures at ± 60 psi
11. Safety check valve: if tripped, routes fuel back to tank
12. Non-return valve: a siphon break to check fuel drain back during hand priming
13. Hand primer pump
14. WIF sensor
15. Dash-mounted water drain switch
16. Water separator drain valve
17. Bleeder valve (used in event of loss of prime)
18. Charging pressure test measuring point
19. Fuel pressure sensor (signals *actual* charge pressure to EECU)
20. Optional fuel heater

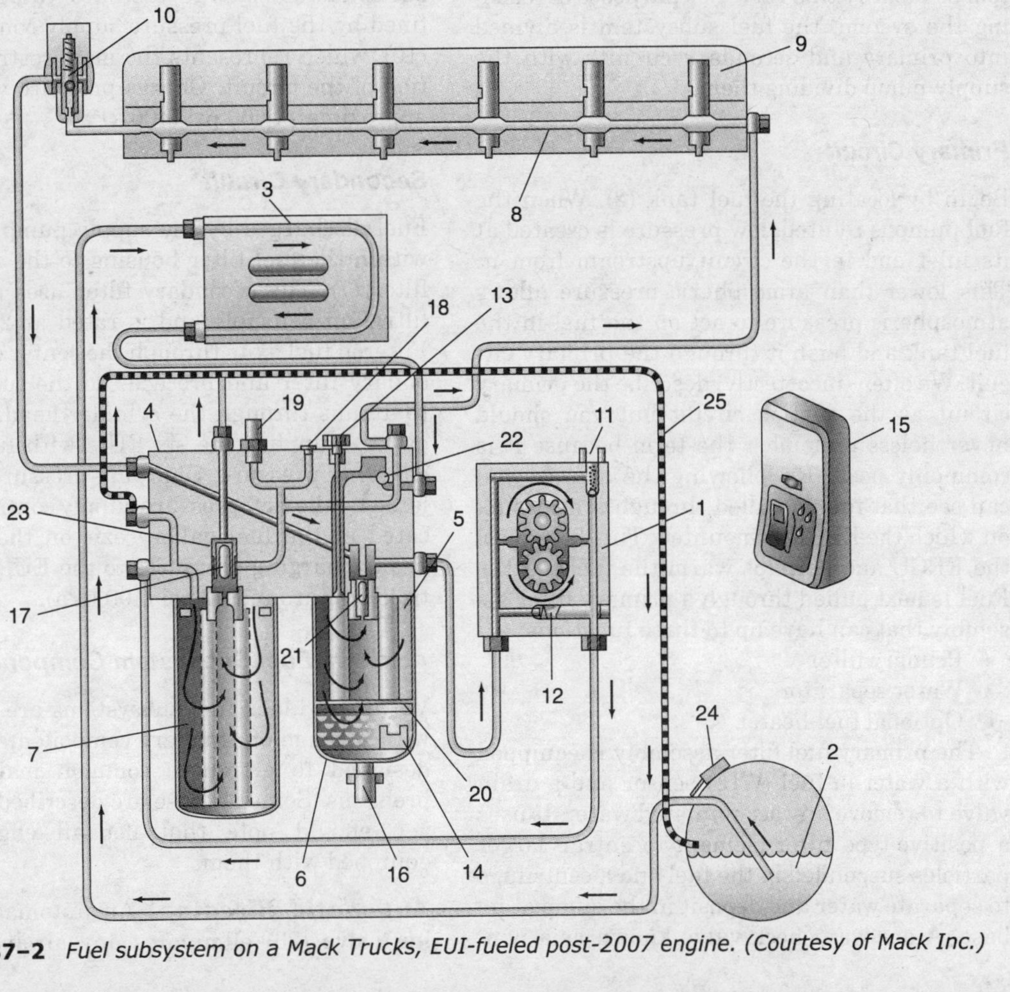

Figure 37–2 *Fuel subsystem on a Mack Trucks, EUI-fueled post-2007 engine. (Courtesy of Mack Inc.)*

21. Valve peg (lock-off for servicing)
22. Non-return valve: a siphon break that traps fuel in the primary circuit on shutdown
23. Bleeding valve (used in event of loss of prime)
24. Breather vent (aspirates fuel tank as fuel is pulled out)
25. Fuel return pipe

Fuel Subsystem Routing

For purposes of this description, we are going to use a post-2007 engine: Earlier versions of Volvo engines have minor differences. We will also assume that you are familiar with Chapter 17 in this textbook. Use the callouts in Figure 37–2 to follow the fuel from the tank, through the fuel circuit, and back to the tank again. A fixed clearance gear pump (1) is responsible for all movement of fuel through the fuel subsystem so identify this first. For purposes of teaching the system, the fuel subsystem is divided into primary and secondary circuits with the supply pump dividing them.

Primary Circuit

Begin by locating the fuel tank (2). When the fuel pump is rotated, low pressure is created at its inlet and in the circuit upstream from it. This lower than atmospheric pressure allows atmospheric pressure to act on the fuel in the fuel tank and push it through the primary circuit. We often (incorrectly) describe the primary circuit as the suction circuit but you should nevertheless remember the term because it is commonly used. By following the arrows, you can see that fuel is pulled through a heat sink on which the EECU is mounted: This helps cool the EECU and perhaps warm the fuel a little. Fuel is next pulled through a primary filter assembly that can have up to three functions:

- Primary filter
- Water separator
- Optional fuel heater

The primary fuel filter assembly is equipped with a water-in-fuel (WIF) sensor and a drain valve to remove any accumulated water. It uses a positive-type filter element to entrap larger particles suspended in the fuel and a centrifuge to separate water and deposit in the sump at its base. A one-way check valve known as a non-return valve (22) prevents fuel siphon from the primary circuit when the engine is shut down. The downstream circuit from the primary filter connects to the supply pump (1).

Supply Pump

The supply pump (1) is an external gear pump. The supply pump is usually flange mounted to the power steering hydraulic pump, which is driven by an auxiliary drive gear in the engine timing geartrain. Providing the supply pump is primed, when rotated it pushes fuel to its exit outlet and in doing so creates lower than atmospheric pressure in the circuit upstream from its inlet. Atmospheric pressure acting on the fuel stored in the fuel tank pushes fuel toward the inlet of the supply pump. Because fuel from the supply pump is discharged to the secondary circuit, you can say that it produces charging pressures. Actual charge pressure values are defined by the fuel pressure supply control valve (10), which represents the most restricted portion of the circuit. Charge pressure values are approximately 60 psi (400 kPa).

Secondary Circuit

Fuel discharged by the supply pump is routed within the fuel filter housing to the secondary filter (7). The secondary filter uses a positive filtration principle and is rated at 2 microns. Filtered fuel exits through the center of the secondary filter and proceeds to the fuel gallery that runs through the cylinder head. The fuel gallery supplies the six EUIs with fuel at the charging pressure. Charging pressure is regulated by the fuel pressure supply control (10) located at the fuel gallery exit on the cylinder head. Charging pressures to the EUIs are controlled at around 60 psi (400 kPa).

Auxiliary Fuel Subsystem Components

Volvo and Mack fuel subsystems are equipped with a range of auxiliary components that are designed to eliminate common maintenance problems. Some of these are described here but you should note that not all engines are equipped with them.

Automatic Bleeding An automatic bleeding valve (23) eliminates any air in the fuel charge circuit on startup. During bleeding, any air in the charge circuit is routed to the return line (25).

Pressure Safety Valve A pressure safety valve (11) is a relief valve located in the supply pump. It is designed to trip in the event of a blockage or partial restriction in the secondary circuit. When the pressure safety valve trips, it diverts the fuel back to the pump inlet.

Fuel Pump Non-Return Valve The fuel pump non-return valve (12) is designed to act as a one-way check valve any time the hand primer pump is used. If you take a look at the circuitry surrounding the transfer pump you will be able so see why this is required: When the hand primer pump is actuated, it is creating low pressure in a portion of the circuit that is normally under charge.

Water Drainage Valve The water drainage valve (16) is located in the water separator (6). When water is detected by the WIF sensor (WIF sensor operation is explained in Chapter 17) the dash water drainage switch (15) illuminates and an alert is posted on the driver information display (DID). When this occurs, the driver may activate the water drainage switch in the dash providing the following conditions are met:

- The ignition key is *on.*
- The engine is not running.
- The parking brake is applied.

When the dash switch is depressed, the drain valve (16) opens for about 15 seconds to drain any water in the sump of the water separator. In cases where additional drainage is required, it is necessary to wait for a minimum of 6 minutes before attempting to repeat the drain cycle.

Manual Hand Primer A manually actuated, hand primer pump (13) is located on top of the combination filter housing assembly. This should only be used when the engine is not running and is intended for use following repair work in which the fuel system has been drained or when the system has run dry. When priming the circuit using the hand primer, the inlet side of the charge fuel line to the cylinder head should be cracked loose.

Filter Pad Valve Pegs Filter pad valve pegs are used on both the primary (21) and secondary (17) filter mounts. The valve pegs are spring-loaded lockoffs that prevent fuel siphoning when the filters are removed for servicing.

INPUT CIRCUIT

Command and monitoring sensors and switches in the Volvo and Mack Trucks applications are consistent with other comparable full authority management systems. Current engine platforms require a standard potentiometer-type throttle position sensor (TPS) supplied with V-Ref. This type of TPS returns an analog signal representing a portion of the V-Ref that can be correlated with accelerator pedal travel. However, the systems are designed to accommodate the new style Hall-effect TPS (explained in Chapter 30), which have no wiper contacts and output a digital signal. Most input circuit components used by Volvo and Mack Trucks electronics are similar and mostly generic. In common with most diesel engine OEMs the number of input sensors has grown exponentially as each level of emissions legislation has come into effect. The operating principles of input circuit components are described in detail in Chapter 30 and you may want to reference this. Figure 37–3 shows the location of some of the input circuit sensors on a Mack Trucks MP-7 engine and Figure 37–4 shows the engine harness from a Volvo VE-D16 showing the location of engine sensors and actuators.

MANAGEMENT ELECTRONICS

With the objective of minimizing confusion, we examine post-2007 systems only in this chapter. Pre-2007 versions of the system described here and used on Volvo VD12 engines can be regarded as its predecessors: These earlier systems tended have more simple electronics. It is important that you understand the contents of Chapters 30 and 33 before beginning this section.

The architecture and organization of the chassis data bus and functions of the controllers networked to it in both Volvo and Mack Trucks chassis are nearly identical. Both Mack and Volvo use a **vehicle control unit (VCU)**

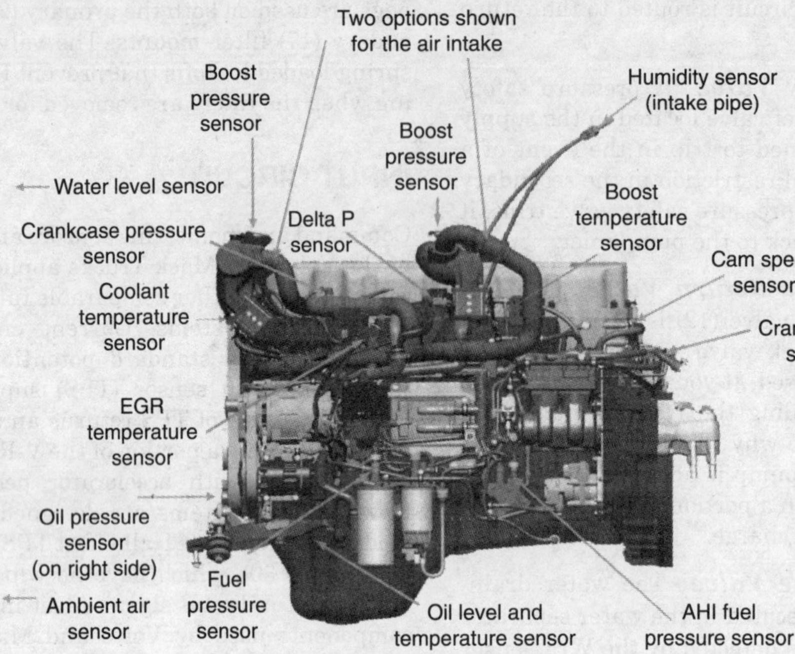

Figure 37–3 *Location of input sensors on a Mack Trucks MP-7 engine. (Courtesy of Mack Inc.)*

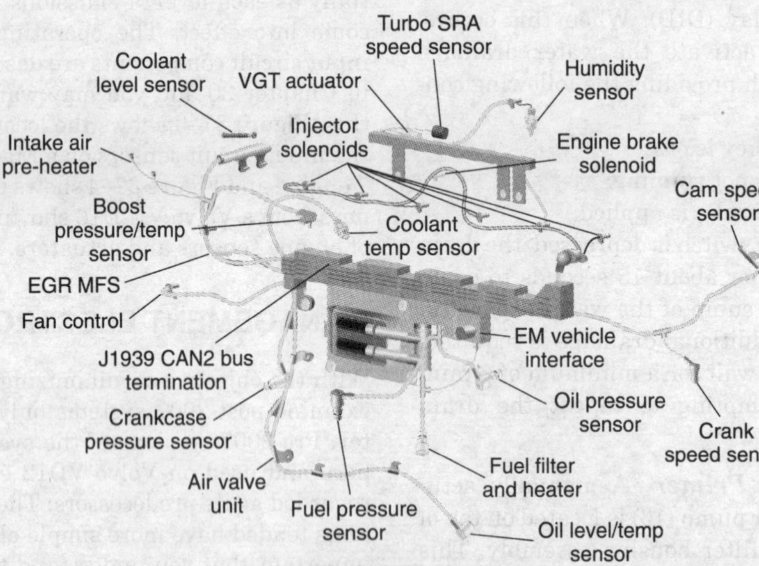

Figure 37–4 *Engine wiring harness from a Volvo VE-D16 engine. (Courtesy of Volvo Trucks NA)*

(MID 144) to "master" powertrain operations on the data bus and a separate **electronic engine control unit (EECU)** (MID 128). Sharing of input data takes place over the chassis data bus. Some Mack chassis equipped with noncatalyzed diesel particulate filters (DPFs) use a separate processor mounted on the exhaust stanchion: This processes and manages DPF active cycles and is explained in Chapter 41.

How MIDs 144 and 128 Share Input Data

The operating principles of chassis computers are explained in Chapter 30 and there is nothing especially different in the Volvo or Mack Trucks units. Here we take a brief look at how the input circuit routes signals to the EECU and the VCU, and identify the PID or SID (see Chapter 31 for the definitions of PID and SID). The lists shown in Table 37–1 and Table 37–2 are not comprehensive; they are just a sample to give you an idea how the "responsibilities" of each ECU are shared.

Table 37–1 *Inputs to EECU (MID 128)*

PID 26	Fan speed %
PID 27	EGR position 1
PID 84	Vehicle speed
PID 91	Accelerator position
PID 94	Fuel supply pressure
PID 97	WIF sensor
PID 100	Engine oil pressure
PID 105	Boost pressure
PID 158	Battery voltage
PID 171	Ambient temperature
SIDs 1–6	SIDs 1–6, EUIs 1–6
SID 22	Engine rpm, flywheel sensor
SID 211	V-Ref
SID 230	Idle validation switch
SID 231	J1939 control link

Table 37–2 *Inputs to VCU (MID 144)*

PID 84	Vehicle speed
PID 91	Accelerator position, percentage position
PID 150	PTO output
PID 191	Output shaft speed (tailshaft of transmission)
PPID 61	Retarder switches Volvo engines only
PPID 69	Idle validation switch, Volvo engines only
PPID 71	Cruise control and engine brake supply voltage
PPID 74	VECU power supply
SID 230	Idle validation switch circuit
SID 231	SAE J1939 control data link
SID 240	Program memory
SID 243	Cruise control set switch
SID 250	Data exchange link to J1587/1708
PSID 4	Engine brake stalk for six-stage Cummins ISX
PSID 200	Engine timeout

EECU and VCU Programming

Both the engine and vehicle management ECUs are programmable with customer and proprietary (Volvo or Mack Trucks) data. The same processors are used regardless of whether badged as Volvo or Mack Trucks. Current ECUs are resistant to radio frequency (RF), electromagnetic interference, and other low-level radiation experienced during normal driving. The EECU is additionally responsible for outputting system reference voltages and receiving inputs both directly from engine monitoring sensors and those broadcast on the J1587 and J1939 data backbones. Their output capability includes the ability to broadcast on the data buses and control engine and chassis actuators. Communication between the two ECUs is on J1939 with the slower J1587 providing system redundancy. The EECU also incorporates the injector driver units within the housing.

The computing power and memory capability of the ECUs has greatly increased with each successive generation while their physical size has generally decreased. Figure 37–5 shows the location and a sectioned view of the EECU used by Volvo on its VE-D12 engine.

Processing Transactions

The control **algorithm** contains a comprehensive fuel map. The EECU analyzes data input from the command and monitoring sensors, and referencing the instructions programmed to its data retention media, plots a fueling profile. Startup fueling strategy, failure modes, and all the system default values are programmed into the ECU software. The EECU is also programmed to run self-diagnostic tests on input and output circuits and identify specific component and circuit faults. Mack Trucks and Volvo troubleshooting shooting software is known as Premium Tech Tool (PTT), which runs on a MS-Windows environment and can perform interactive and guided diagnostics. Some PTT functions require the use of specialty breakout boxes.

EECU Injector Drivers

Volvo and Mack Trucks state that the trigger pulses delivered to the EUI control cartridges

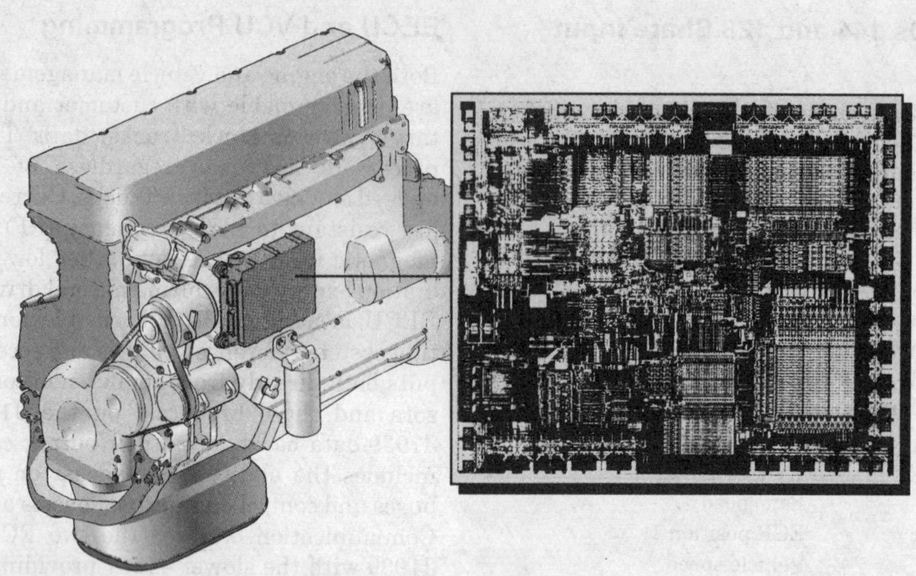

Figure 37–5 *Volvo VE-D12 EECU. (Courtesy of Volvo Trucks NA)*

are driven at 100 V. The 100 V is induced by coils and the EUIs draw about 10 amps current during the initiation spike. Once the control solenoid has closed to put the EUI into effective delivery stroke, current draw by the cartridge drops to around 4 amps. The inductive kick produced on collapse is suppressed by a capacitor. Though not required for any kind of troubleshooting, you can use training or test engines to observe the pulse width modulator (PWM) voltage profile by connecting a lab scope to the actuator terminals, then run the engine. If you plan to do this, actuation frequency should be set for a maximum of 5 ms and voltage peaks set for a maximum of 130 V-DC.

OUTPUT CIRCUIT

The Volvo and Mack Trucks output circuit devices effect the results of the computer processing cycle into action. In an EUI fuel system, the primary engine outputs are the EUIs and exhaust aftertreatment devices. Other outputs are V-Ref and any information broadcast to the data backbone such as instrument cluster displays and powertrain management.

Electronic Unit Injectors

The Delphi EUIs used by Volvo have recently undergone some changes since their introduction in the early 1990s. We divide our description of EUIs into:

- Two-terminal: single actuator units used by Volvo VECTRO I and II
- Four-terminal: dual actuator units used in current Volvo and Mack Trucks EUIs

Both **two-terminal EUI** and **four-terminal EUI** are mechanically actuated by an injector train in the same way cylinder valves are actuated. In an engine with an overhead camshaft, a rocker is used to actuate the EUI as shown in the Volvo VE-D12 example in Figure 37–6. Effective stroke of the EUI is electronically controlled by the EECU, specifically by the **injector drivers** discussed earlier in this chapter under EECU components.

Two-Terminal EUI Operation

The EUI is fitted to a cylindrical bore in the engine cylinder head. In terms of its functions, the EUI can be divided into three subsystems:

- Pumping circuit: mechanically actuated by cam profile
- Control circuit: EECU switched to manage effective stroke
- Injector circuit: hydraulic multi-orifice nozzle

Charging pressure fuel from the fuel supply manifold (drilled passages in the cylinder head) enters the EUI through the fill port and circu-

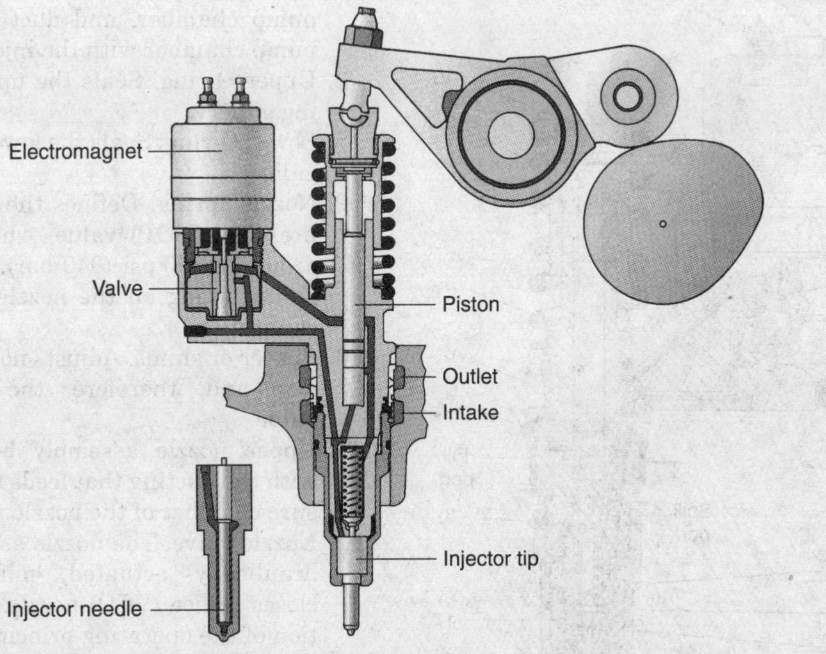

Figure 37–6 *Volvo VE-D12 EUI actuated by an overhead cam. (Courtesy of Volvo Trucks NA)*

lates through the EUI. This circuit is responsible for routing charge pressure fuel from the fuel subsystem through the EUI to feed the pumping chamber in the EUI barrel pump element. The EUI pumping chamber is formed by the cam-actuated plunger and the stationary barrel. When the injector train actuating cam is ramped off inner base circle (IBC) toward the cam nose, the EUI plunger is driven downward and acts on the fuel in the pump chamber. As the plunger descends, its leading edge first closes off the fill passage and next displaces fuel in the EUI pump chamber through the internal passages leading to the spill circuit. This displacement of fuel continues until the EUI control cartridge is energized by the injector driver circuit in the EECU.

An effective stroke can only occur when the EUI solenoid is energized. Energizing the EUI control cartridge moves the EUI control valve to close off the spill circuit. Closing the spill circuit traps fuel in the EUI pumping chamber. Now as the EUI plunger continues to descend, pressure rise takes place. Located below the EUI pump chamber and connected by a passage is a hydraulic, multi-orifice injector nozzle. Pressure rise caused by trapping fuel in the EUI pump chamber acts on the nozzle valve pressure chamber and when the required NOP is achieved, it unseats, beginning the injection pulse. Pressure continues to rise after NOP, and the peak injection pressure achieved depends on the length of the duty cycle. Effective stroke ends when the EECU switches the EUI solenoid open, allowing fuel to spill from the EUI circuit and initiating pressure collapse.

Two-Terminal EUI Subcomponents

Figure 37–7 shows a sectional view of a two-terminal Delphi EUI with the subcomponents labeled. Reference the callout codes in the description that follows.

1. Terminals. Connect to the injector drivers in the ECM.
2. Control cartridge. Solenoid consisting of a coil and armature with an integral poppet control valve (4). A spring loads the armature open. Energizing the solenoid closes the armature/poppet control valve.
3. EUI tappet spring. Loads the EUI tappet upward. This enables the tappet/ plunger actuation train to ride the cam profile and return the tappet after a mechanical stroke.

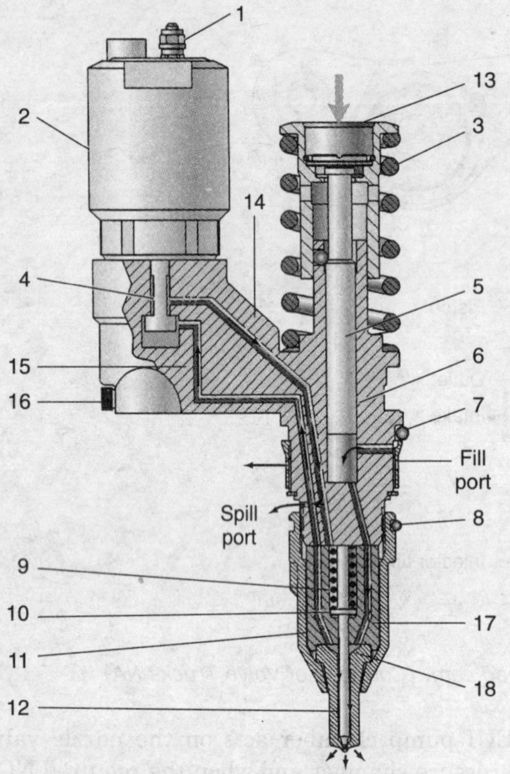

1. Solenoid connection (to the multiplex enable circuit)
2. Solenoid valve assembly
3. Spring
4. Valve (shown in the closed position)
5. Plunger
6. Barrel
7. O-ring
8. O-ring
9. Spring
10. Spacer
11. Body
12. Nozzle valve
13. Tappet
14. Spill duct
15. Spill control circuit
16. Calibration port
17. Fuel duct
18. Pressure chamber

Figure 37–7 *Sectional view of a two-terminal Delphi EUI identifying the subcomponents. (Courtesy of Delphi)*

4. Poppet control valve. A valve integral with the solenoid armature. In Figure 37–7, the control valve is shown closed (solenoid energized): This prevents fuel from exiting through the spill port, trapping it in the EUI circuitry, enabling effective stroke.
5. Plunger. The reciprocating member of the pump element, the plunger, is lugged to the tappet, so it reciprocates with it. In Figure 37–7, the plunger is shown in its upward position.
6. Barrel. The stationary member of the EUI pumping element containing the fill port, pump chamber, and duct connecting the pump chamber with the injector nozzle.
7. Upper O-ring. Seals the upper fuel charging gallery.
8. Lower O-ring. Seals the lower fuel charging gallery.
9. Nozzle spring. Defines the nozzle opening pressure (NOP) value, which is typically around 5,000 psi (340 bar). NOP is set by shims acting on the nozzle spring during calibration.
10. Spacer or shims. Adjusts nozzle spring tension and, therefore, the specific NOP value.
11. Upper nozzle assembly body. Machined with the ducting that feeds fuel to the pressure chamber of the nozzle valve.
12. Nozzle valve. The nozzle assembly is a hydraulically actuated, multi-orifii valve closes orifice (VCO) nozzle. A full description of the operating principles of this type of nozzle is provided in Chapter 19.
13. Tappet. Cam-actuated by the injector train.
14. Spill duct. The exit path of fuel from the control circuit.
15. Spill control circuit. This is open when the control solenoid is de-energized, allowing fuel to spill from the EUI. When the EECU energizes the control solenoid, the spill circuit is closed off at the control plunger trapping fuel in the EUI: This is how plunger effective stroke and injection timing are effected.
16. Calibration port. Used only for bench calibration to determine the fuel flow values.
17. Fuel duct. Connects the EUI pump chamber with the injector circuit.
18. Pressure chamber. The sectional area of the nozzle valve over which pressurized fuel acts to open the nozzle valve and begin injection.

Two-terminal EUI NOP values are typically around 5,000 psi (345 bar) and by factoring nozzle valve pressure differential ratio, closing pressure values are typically 3,800 psi (262 bar). When the injector actuating cam has attained peak lift, the injector train is unloaded, as the tappet ramps down the cam flank toward cam base circle. Simultaneously, the EUI spring lifts the plunger within the barrel, returning it to its raised position and thereby exposing the fill port. Charging pressure fuel is then permitted to circulate throughout the EUI passages for purposes of cooling the assembly, exiting the spill port until the effective cycle is repeated.

Four-Terminal EUIs

Current Volvo and all Mack Truck EUIs use four-terminal EUIs such as the Delphi E3. Volvo engines using four-terminal EUIs first appeared following the 2004 model year but are universal in all post-2007 engines. Although it uses a sleeker and lighter design, a four-terminal EUI has a pair of actuators with quite different functions:

- Spill valve (SV) actuator: EECU switched control cartridge that fulfills the same role as the control solenoid in the two-terminal EUI: In other words, it opens and closes the spill circuit to manage EUI effective stroke.
- Nozzle control valve (NCV) actuator: This is integrated into the EUI body as shown in Figure 37–8. The NCV actuator is an electrohydraulic nozzle that allows a soft NOP value.

Four-Terminal EUI Construction

Four-terminal EUIs have eliminated three of the major disadvantages of first generation EUIs:

- Bulky assembly with protruding control cartridge
- Hard value NOP due to the hydraulic injector nozzle

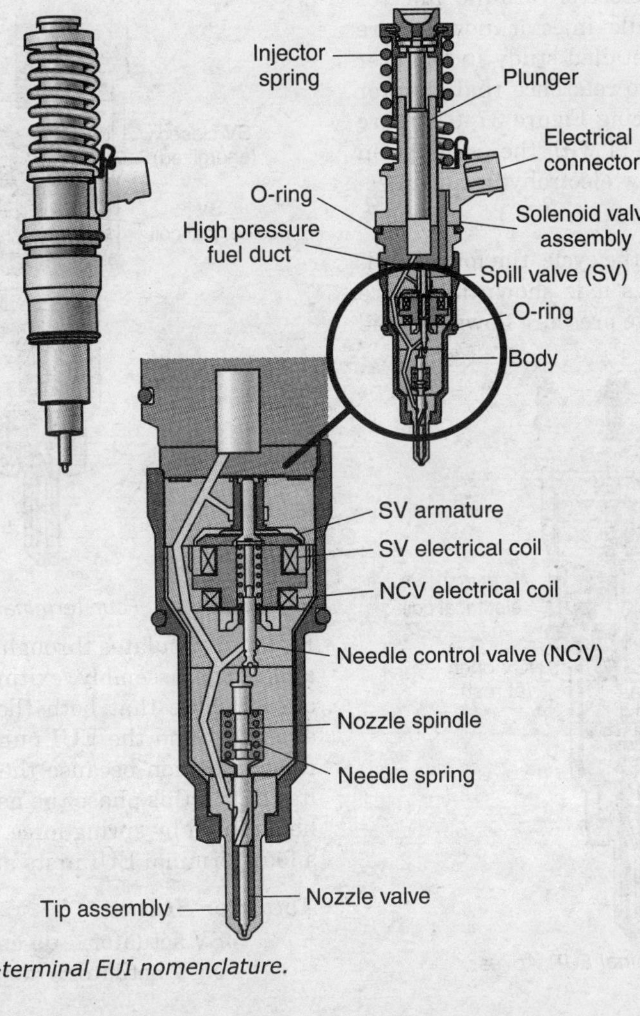

Figure 37–8 *Four-terminal EUI nomenclature.*

- Hard value nozzle closure at the end of injection, resulting in prolonged pressure collapse phase at the end of injection

Despite their increased complexity, four-terminal EUIs weigh about one-half of an equivalent two-terminal EUI. In addition, the obstruction to the injector and valve trains created by the protruding control solenoid cartridge has been significantly reduced. Figure 37–8 shows the nomenclature Volvo uses to describe its four-terminal EUI: Use this to guide you through the following description of its operating principles.

Four-Terminal EUI Operating Principles

In terms of general operating principles, four-terminal EUIs such as the Delphi E3 injectors are essentially a "two-terminal EUI" fitted with an EECU-controlled electrohydraulic injector nozzle. Electrohydraulic injector nozzles are the subject of some detailed study in Chapter 19 so you may want to reference that chapter again. We are referencing Figure 37–9, Figure 37–10, and Figure 37–11 with the assumption that you remember how electrohydraulic injectors operate.

At Rest For most of the cycle, the four-terminal EUI will appear as it is shown in Figure 37–9. Fuel at the charge pressure flows into the EUI and circulates through the pump chamber and nozzle assembly, exiting through the spill circuit. Note that both the actuators are de-energized and the EUI pump plunger is in its raised position because the injector cam is on its IBC. In this phase the nozzle needle valve is held seated by spring force. Figure 37–9 shows a four-terminal EUI in its at rest phase.

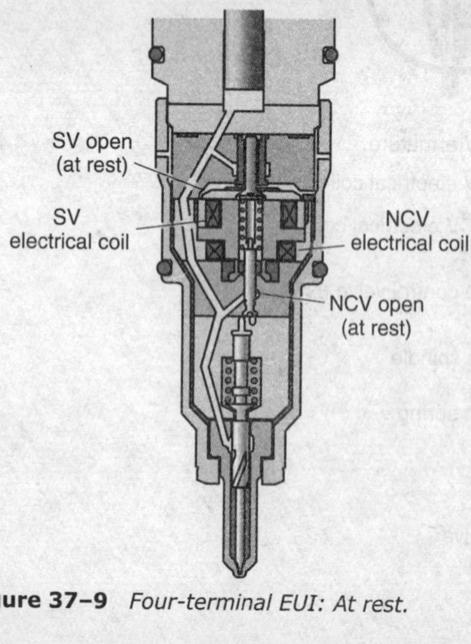

Figure 37–9 *Four-terminal EUI: At rest.*

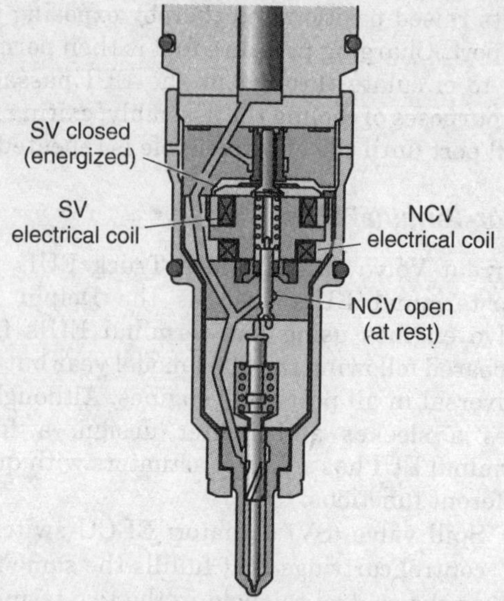

Figure 37–10 *Four-terminal EUI: Pressurizing.*

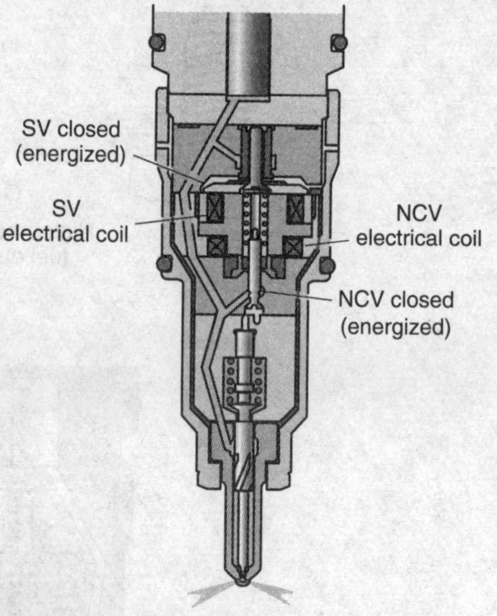

Figure 37–11 *Four-terminal EUI: Injection.*

Actuator Status:

SV actuator: de-energized.
NCV actuator: de-energized.

Pressurizing Pressurizing begins when the injector cam profile begins to ramp off its IBC: This actuates the EUI plunger, driving it downward in its bore. When the EECU energizes the SV actuator, the spill circuit closes, trapping fuel in the pump chamber underneath the plunger. As pressure rise is created by the descending plunger, identical pressure acts on both the upper and lower pressure fields: However, because the sectional area of the upper pressure field (acting on the nozzle control piston) exceeds that of the lower pressure field (acting on the needle pressure chamber sectional area), the nozzle needle valve remains firmly seated. In this phase the nozzle needle valve is held seated by a combination of spring force and hydraulic force acting on the needle control piston. Figure 37–10 shows a four-terminal EUI during the pressurizing phase.

Actuator Status:

SV actuator: energized.
NCV actuator: de-energized.

Injection When the desired NOP is achieved within the EUI circuit, the EECU energizes the NCV actuator. This pulls the NCV valve into the NCV actuator, collapsing the upper pressure chamber field. Because the upper pressure chamber field collapses and the lower pressure chamber field remains the same, hydraulic force lifts the nozzle needle valve, beginning injection. While injection continues, the NCV valve remains lifted, which means no pressure can build in the upper pressure chamber: The exit flow area spilling fuel from the upper chamber is very small so spilling fuel from the upper pressure field has little effect on the pressure in the remainder of the EUI circuit. Figure 37–11 shows a four-terminal EUI during injection: Note that both actuators are energized.

Actuator Status:

SV actuator: energized.
NCV actuator: energized.

End of Injection Pulse The EECU ends an injection pulse by de-energizing the NCV. Spring force slams the NCV downward, closing the NCV spill passage, and consequently permits almost instant reestablishing of the upper pressure field. Hydraulic force acting on the nozzle needle control piston abruptly closes the nozzle needle valve, ending the injection pulse.

Actuator Status:

SV actuator: energized.
NCV actuator: de-energized.

Multipulse Injection The four-terminal EUI lends itself to multipulse injection, especially when fast response, piezo-type NCV actuators are used. For multipulse injection to take place, the NCV actuator is energized more than once during a single mechanical downstroke of the EUI plunger. The advantages of multipulse injection relate to better combustion efficiency and lower emissions. Four-terminal EUIs better suit multiple pulse injection events because the NOP can be made to rise incrementally with each successive pulse, producing the smallest droplets late on in the combustion phase.

Actuator Status:

SV actuator: energized.
NCV actuator: cycling ON and OFF.

End of Injection Phase The injection phase ends when both the SV and NCV actuators are de-energized by the EECU drivers. This necessarily must occur before peak lift is attained by the injector actuation cam. Because EUIs are mechanically actuated by cam profile, the EECU can only switch effective strokes while the EUI plunger is descending. So in cases where the EECU algorithm requires multiple injection pulses during one cycle (that is, one injection phase), these must occur within a hard window defined by cam profile. So at the completion of the injection phase, the actuating cam profile ramps down to IBC and both actuators are de-energized: The "at-rest" status resumes.

Actuator Status:

SV actuator: de-energized.
NCV actuator: de-energized.

Advantages of the Four-Terminal EUI

The key advantage of the four-terminal EUI over the older two-terminal version is that it incorporates an electrohydraulic nozzle. This means that the EECU has control over the NOP. Remember that pressure at the nozzle's orifii determines the size of the fuel droplets that exit the nozzle: Because the EECU can control the pressure, it removes the disadvantages associated with the hard opening and closing values in the two-terminal version. Additionally, the four-terminal version uses more advanced manufacturing technology to make it lighter and much less bulky.

Fuel Calibration Codes In common with other EUIs, Volvo and Mack Trucks four-terminal injectors have a fuel flow code printed on side of the unit. The code numeric value is established when an EUI is factory bench-tested and a precise determination is made of how the injector flows fuel. This data is required by the EECU to ensure that the fueling to each engine cylinder is balanced by compensating for minor hydraulic differences between injectors. A new fuel calibration code for an EUI must be programmed to the EECU when it is changed out.

DRIVER INFORMATION DISPLAY

The digital information displayed by the **driver information display (DID)** is for the benefit of both the driver and the technician. The display data is read off the chassis data bus and can be used to read all the message identifiers (MIDs) networked to it. This means that fault routing through PIDs and SIDs through to the identification of the specific fault mode indicator (FMI), including the frequency of occurrence, can be achieved using the dash display electronics. In common with most manufacturers, the first indications of a failure are the warning lamps, buzzer, and driver digital information display if equipped.

Warning Lamps and Buzzer Alerts

In the Volvo and Mack Trucks chassis electronics systems, driver alerts can be classified as follows:

- Yellow lamp. Illuminating the yellow lamp indicates that an engine fault of a nature not likely to result in engine damage has been detected. The cause should be diagnosed as soon as possible because it could limit certain engine functions and affect vehicle driveability. It is equivalent to the *Check Engine* lamp used by other OEMs.
- Red lamp. Illuminating the red lamp indicates that an engine fault of a serious nature has been detected. Depending on the specific fault detected engine power output may be derated and may result in engine shutdown. It is equivalent to the *Stop Engine* lamp used by other OEMs.
- Buzzer. This sounds when the red dash lamp is illuminated.

DID Navigation

The DID can be used as a first-level diagnostic tool but it also provides the operator with comprehensive chassis management analytical tools. Some of its features include:

- Fuel economy analytical data
- Trip information
- Message in/out functions using wireless/satellite to base communications
- Driver id and theft deterrence
- Display of DPF cycles and status
- Diagnostics of any MIDs on the data bus

Figure 37–12 shows a routing map of the menu options used on a 2007 Mack Trucks DID. It should be noted that some of the features such as those appearing under Road Connect require enabling by subscription to satellite or WiFi providers.

Using DID to Identify an Active Fault

If you take a look at Figure 37–12, you should be able to identify the path to display a fault code. From the *Main Menu* select the *Diagnostics* option, then scroll through *Electronic Faults* to select the *Active* option. This will display any active faults and the MID within which it occurred. Figure 37–13 shows how a DID displays an active code. Use the information on MIDs, PIDs, SIDs, and FMIs in Chapter 31 to identify the problem.

TECH TOOL

Tech Tool™ (TT) replaces the pre-2007 VCADS software for Volvo and Mack Trucks. In a Mack Trucks application, this software is known as **Premium Tech Tool (PTT)**. Loaded onto a Mack or Volvo laptop it is designed for wireless interface with both the truck database and the Volvo or Mack data hub via the Inter-

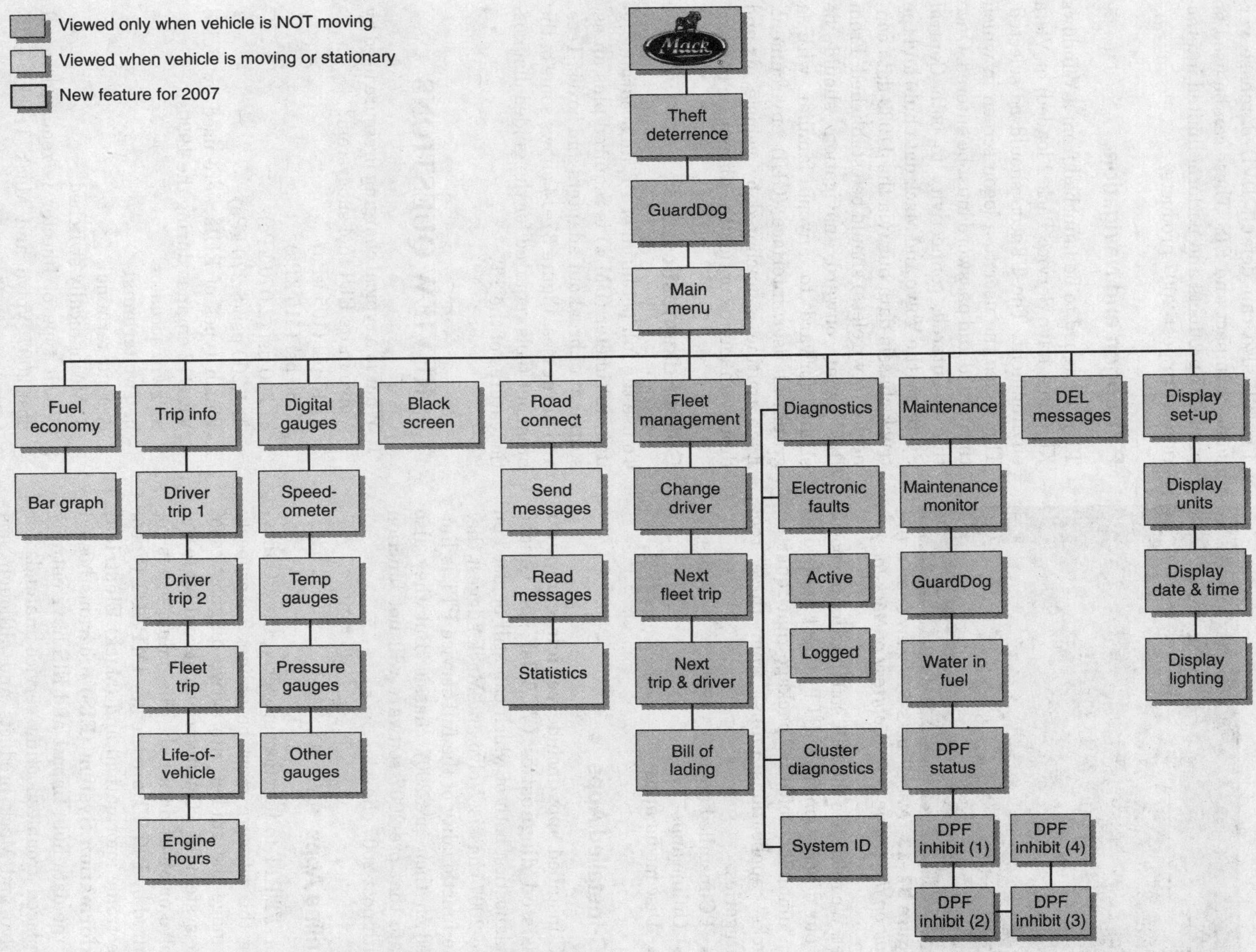

Figure 37–12 *Mack Trucks DID menu map. (Courtesy of Mack Inc.)*

MACK Fault Codes Display
(Active Faults)

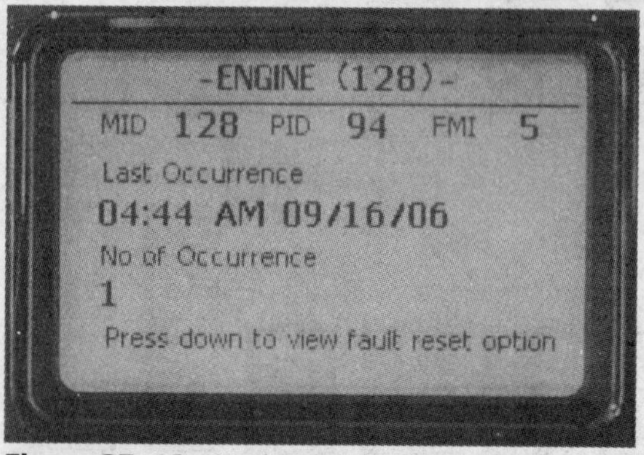

Figure 37–13 *Mack Trucks DID view of an active fault on the engine MID. (Courtesy of Mack Inc.)*

net. We reference the PTT software for purposes of our description here. PTT is best defined as an "umbrella" software package that "quarterbacks" applications (apps) from three root categories:

- PC-installed apps
- Online apps
- Disc media apps

PC-Installed Apps

PC-installed apps include **vehicle computer assisted diagnostics (VCADs)** and Guided Diagnostics, both of which can still be used on pre-2007 chassis. Volvo and Mack suggest that the functionality of the full service PTT is limited in their pre-2007 chassis. But they also state that pre-2007 software will not function with post-2007 chassis.

Online Apps

For TT/PTT to function properly, the PC driving the software must be online. To properly manage the guided interactive diagnostics of the system the PC software must connect to the Volvo or Mack data hub. The online facets required by PTT/TT are the key **IMPACT** (Volvo electronic information) and (Mack) **Electrical Information System (EIS)** programs hosted at the data hub. Impact and EIS have entirely replaced troubleshooting service manuals for Volvo and Mack products and although versions of both are available on DVD, service updates are embedded into the online version.

Disc Media Apps

TT/PTT also can default DVD database versions of Impact and EIS. These can be used on pre-2007 products and perform limited interactions with post-2007 products.

PTT Logon and Navigation

TT is designed to be launched from a Windows XP (must have Service Pack 2 loaded) or Vista environment. The TT icon should be selected. This takes the user to a logon screen in which the user id and password must be entered. Once this is entered, connectivity to both Central Systems (the Volvo and Mack data hub) and the truck chassis data bus (via the J1939 data connector or wireless) should be established. From this point onward, navigation should be straightforward to anyone familiar with a graphical user interface (GUI) environment. Each routine is anchored by vehicle id, and dropdown menus option the diagnostic path.

Guided diagnostics

To use the full potential of TT, an accurate digital multimeter (DMM) is required plus an assortment of breakout adapters and boxes, Tees, and harnesses. Figure 37–14 shows some of the breakout tools required for the guided diagnostic features of the TT.

REVIEW QUESTIONS

1. What is a typical charging pressure used by Volvo in its EUI fuel subsystem?
   a. 15 psi (105 kPa)
   b. 60 psi (415 kPa)
   c. 300 psi (2,070 kPa)
   d. 1,200 psi (8,270 kPa)
2. Which types of EUIs were used by Volvo VE-D12 engines during the 1990s?
   a. Twin actuator
   b. Two terminal
   c. Four terminal
   d. Mechanically switched
3. Which type of injector is required in all post-2007 Volvo and Mack Trucks EUI-fueled engines?
   a. Two-terminal EUI

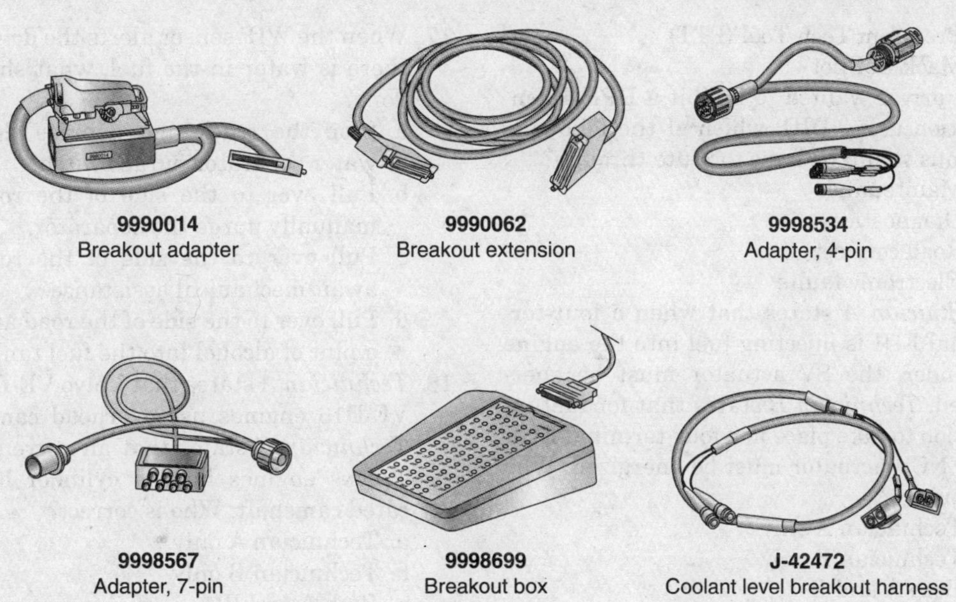

Figure 37–14 *Tech Tool™ breakout tools.*

b. Two-terminal EUP
c. Four-terminal EUI
d. Four-terminal EUP

4. How many actuators are used in a four-terminal EUI?
   a. None
   b. One
   c. Two
   d. Four
5. Which of the following is the closest common platform relative of the Volvo VE-D16 engine?
   a. Volvo VE-D11
   b. Volvo VE-D13
   c. Mack Trucks MP-8
   d. Mack Trucks MP-10
6. Which of the following metric values would be closest to the engine displacement of a Mack MP-7 engine?
   a. 7 liters
   b. 10 liters
   c. 11 liters
   d. 13 liters
7. Which of the following values would represent a typical NOP for a two-terminal EUI equipped with a hydraulic nozzle?
   a. 60 psi (415 kPa)
   b. 120 psi (830 kPa)
   c. 3,700 psi (255 bar)
   d. 5,000 psi (345 bar)
8. Where is the fuel pressure sensor located in a Mack Trucks MP-8 engine?
   a. In the fuel tank
   b. Upstream from the primary filter
   c. Upstream from the secondary filter
   d. Downstream from the secondary filter
9. What does the fuel pressure sensor on a MP-8 measure?
   a. Fuel inlet restriction
   b. Charging pressure to the EUIs
   c. Injection pressure
   d. EUI pump chamber pressure.
10. Post-2007, dual actuator EUIs are switched by the injector drivers at:
    a. 5 V
    b. 12 V
    c. 24 V
    d. ± 100 V
11. Which of the following values would represent a typical NOP for a four-terminal EUI equipped with an electrohydraulic nozzle?
    a. This generation of EUIs uses soft value NOPs.
    b. 3,800 psi
    c. 5,000 psi
    d. 22,000 psi
12. What is required to diagnose system problems on any post-2007 Mack Trucks engine?
    a. Mack service manual
    b. VCADS

c. Premium Tech Tool (PTT)
d. Mack CoPilot

13. If a driver wanted to inhibit a DPF regeneration using DID, which of the following menus would he have to route through?
a. Maintenance
b. Diagnostics
c. Road connect
d. Electronic faults

14. *Technician A* states that when a four-terminal EUI is injecting fuel into the engine cylinder, the SV actuator must be energized. *Technician B* states that for fuel injection to take place in a four-terminal EUI, the NCV actuator must be energized. Who is correct?
a. Technician A only
b. Technician B only
c. Both A and B
d. Neither A nor B

15. *Technician A* states that four-terminal EUI calibration codes must be reprogrammed to the EECU whenever one is changed out. *Technician B* states that the four-terminal EUI NCV control cartridge is located outside the cylindrical body of the injector. Who is correct?
a. Technician A only
b. Technician B only
c. Both A and B
d. Neither A nor B

16. *Technician A* states that when the dash warning yellow light illuminates on a 2007 Mack Trucks chassis, the driver should pull over to the side of the road. *Technician B* says that when the dash warning red light illuminates, a buzzer is usually activated. Who is correct?
a. Technician A only
b. Technician B only
c. Both A and B
d. Neither A nor B

17. When the WIF sensor alerts the driver that there is water in the fuel, what should be done?
a. Stop the truck and depress the dash water separator purge switch.
b. Pull over to the side of the road and manually purge the separator.
c. Pull over to the side of the road and await mechanical assistance.
d. Pull over to the side of the road and pour a pint of alcohol into the fuel tanks.

18. *Technician A* states that Volvo VE-D12 and VE-D16 engines use overhead camshafts. *Technician B* states that all current Mack Trucks engines have a cylinder block-located camshaft. Who is correct?
a. Technician A only
b. Technician B only
c. Both A and B
d. Neither A nor B

19. *Technician A* states that 2007 Mack Trucks EECUs are mounted on the right side of the engine. *Technician B* states that 2007 Mack Trucks EECUs are mounted on a heat sink through which fuel is routed. Who is correct?
a. Technician A only
b. Technician B only
c. Both A and B
d. Neither A nor B

20. *Technician A* states that if the NCV actuator fails on a four-terminal EUI, fuel injection can still take place because the SV actuator provides redundancy. *Technician B* states that the EUI pump chamber cannot produce injection pressures until the SV actuator is energized. Who is correct?
a. Technician A only
b. Technician B only
c. Both A and B
d. Neither A nor B

38

Cummins CELECT Plus

INTRODUCTION

Cummins introduced **Cummins electronic engine control (CELECT)**, its full authority electronic management system, in 1991 on L10 and N14 engines, and later on its M11 engine family when it was introduced in 1994. CELECT Plus electronics were released into the marketplace in 1996 and are used to manage current M11 Plus and N14 Plus engines (Figure 38–1). The M11 engine still survives today but with a different management system: The engine series is now known as ISM and actually displaces 10.8 liters. Cummins claims that the ISM engine produces better fuel economy than any competitor's 11-liter engine.

Figure 38–1 *CELECT Plus managed engine. (Courtesy of Cummins, Inc.)*

It should be noted that L10 (10-liter), M11 (11-liter), and N14 (14-liter) engines could have PT, CELECT, or CELECT Plus fuel management with each engine incorporating different hardware. For instance, the M11 engine with PT fueling has been available until recently as an off-highway engine. CELECT Plus represents an enhancement of the CELECT engine management system, providing more comprehensive engine monitoring, faster processing speed, and greater memory retention and programmability. The CELECT Plus system is generally used as the basis for this description of Cummins electronic management systems.

SYSTEM OVERVIEW

CELECT Plus is a full authority, computerized engine management system that uses cam-actuated, electronically controlled **CELECT injectors** that can be compared to the EUIs on other full authority systems. As indicated earlier, this fuel system survives on the ISM engine. IS is an acronym for Interact System, the management system used on the entire range of Cummins highway diesel engines. In addressing this Cummins fuel system we are going to call the system CELECT Plus whether we are referencing the current ISM or earlier versions of the system. The location of the key components used in a CELECT-managed engine is shown in Figure 38–2.

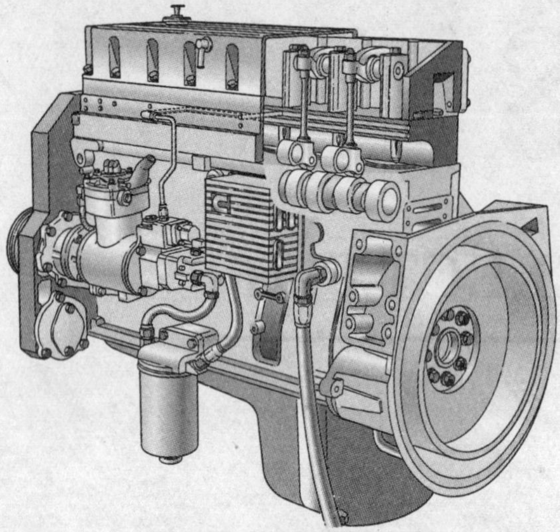

Figure 38–2 *Location of key components on a CELECT-managed engine. (Courtesy of Cummins, Inc.)*

The CELECT Plus system can be divided as follows:

1. Fuel subsystem
2. Input circuit—sensors
3. ECM
4. Output circuit—CELECT injectors, C-brake, and so forth

FUEL SUBSYSTEM

Fuel movement in the fuel subsystem is provided by a gear pump driven by an accessory drive that rotates both the air compressor and gear pump at engine speed. The gear pump is flange mounted behind the compressor. It is fitted with a pressure regulator responsible for maintaining a 150 psi (11 bar) charge pressure to the CELECT injectors. An electrical shutdown valve is positioned on top of the fuel pump at the fuel outlet. It is energized to run. The fuel pump is also fitted with a small filter consisting of a mesh and magnet to prevent wear metals from the gear pump from entering the charge pipe.

Fuel is drawn from the fuel tank by the gear pump. It is pulled through the cooling plate (heat sink) on which the CELECT **electronic control module (ECM)** is mounted; it circulates through this heat exchanger, helping to dissipate heat from the ECM. From the ECM cooling plate, fuel passes through the fuel filter, which has a high-capacity 10-micron filtering medium. Both the ECM cooling plate and the filter are therefore under suction. From the filter head, the fuel is piped to the gear pump, which pressurizes then discharges fuel at a constant charge pressure of 150 psi (11 bar) to tubing that delivers fuel to the cylinder head galleries and makes it available to the CELECT injectors. After circulating through the CELECT injectors, fuel exits by means of a fuel drain and enters the cylinder head return gallery from which it is piped back to the fuel tank (Figure 38–3).

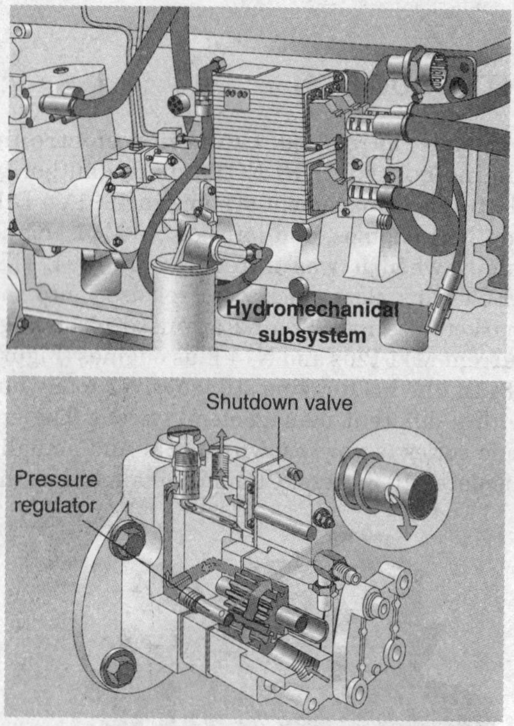

Figure 38–3 *The hydromechanical fuel subsystem on a CELECT engine. (Courtesy of Cummins Inc.)*

INPUT CIRCUIT

In the CELECT Plus electrical subsystem, components are connected to the ECM by a single wiring harness, integrating the separate sensor and actuator harnesses of the CELECT system. CELECT Plus sensors are defined by type; the electrical principles on which they are based are explained in Chapter 30.

Engine Monitoring Sensors

Reference Figure 38–4 to locate the sensors on the harness:

1. Ambient pressure sensor—piezoelectric type—for altitude fuel deration.
   M11—located in front of the ECM
   N14—located behind the ECM
2. Oil pressure sensor—piezoelectric type—located in the main oil rifle in the block.
3. Engine position sensor—a magnetic pulse generator-type sensor. The cam gear located pulse wheel has 24 evenly spaced recesses to feed back rotational speed data and a small dowel to locate top dead center (TDC) #1 cylinder for engine position data.
4. Oil temperature sensor—thermistor—located in the main oil rifle in the block.
5. Intake manifold temperature sensor—thermistor— located centrally on the intake manifold.
6. Boost pressure sensor—piezoresistive type—located in the intake manifold.
7. Coolant temperature sensor—thermistor—located in the thermostat housing.
8. Coolant level sensor—a switch-type sensor that grounds through the engine coolant with dual probes to indicate normal or low coolant levels. This sensor is original equipment manufacturer (OEM) supplied.

Command Sensors and Switches

1. Throttle pedal assembly—OEM supplied, accelerator pedal that incorporates a 5 V throttle position sensor (potentiometer) and an idle validation switch (open–close circuit). CELECT Plus is a drive-by- wire system with no mechanical backup. An accelerator interlock is optional.
2. **Jacobs C-brake**—A Cummins designed, Jacobs manufactured, engine compression brake managed by the CELECT ECM. The current version permits engine braking down to 900 rpm (previously 1,500 rpm) and can be automatically actuated by the ECM in cruise control mode.
3. Cruise control—Managed by the CELECT Plus ECM with some advanced features: (a) Governor tailoring (see section on governor tailoring); (b) Cruise auto resume after shift; (c) OEM cruise switch configuring; (d) Automatic engine braking in cruise mode. The minimum cruise control speed limits the top vehicle speed while in cruise control mode, and this cannot exceed the maximum vehicle speed setting.
4. Power takeoff (PTO) control—This feature controls engine rpm at a constant speed selected by the operator. Either a cab or a remote switch can be used, and PTO mode can be selected up to 6 mph (10 km/h) vehicle speed.

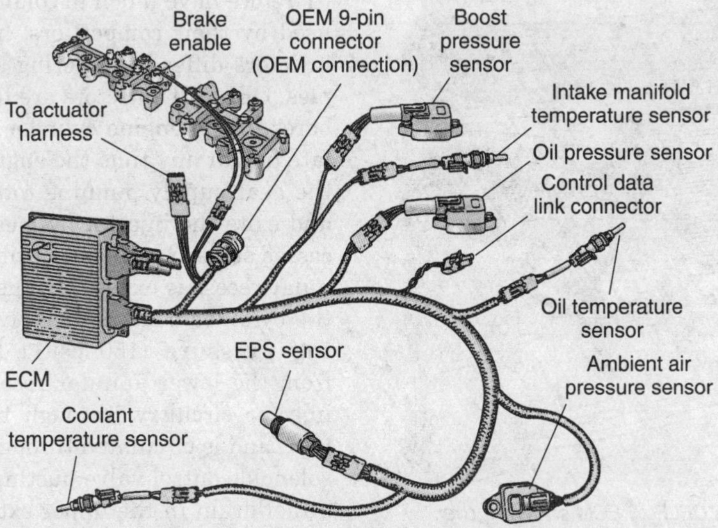

Figure 38–4 *The CELECT sensor harness. (Courtesy of Cummins Inc.)*

ELECTRONIC CONTROL MODULE (ECM)

The ECM receives and processes information from the various engine and chassis sensors and switches, plots fueling and timing commands, and outputs signals to the injector driver circuitry (switching apparatus). A more detailed accounting of ECM functions is covered elsewhere in this text. The CELECT Plus injector driver unit uses coils to spike the EUI solenoid voltage to 78 V and is the primary output of the ECM.

Consistent with other manufacturers, the CELECT Plus ECM performs comprehensive self-diagnostics, which can be read by flash codes or electronic service tools (ESTs) and can be programmed to adopt a variety of failure mode strategies (Figure 38–5 and Figure 38–6). Cummins claims that CELECT Plus offers significant advantages over CELECT, including:

- Faster processing speed
- Altitude fuel deration
- Cylinder pressure monitoring
- Faster, more accurate throttle response
- Clutch engagement validation
- Multilevel programming security
- Engine warm-up protection
- Accelerator interlock (i.e., bus door, vehicle brake, or PTO switched)

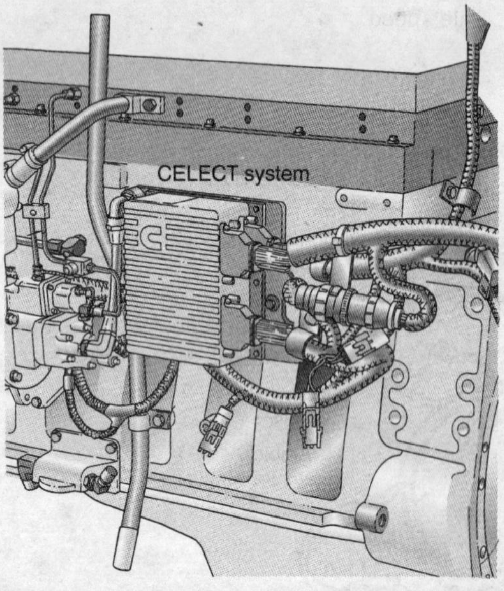

Figure 38–5 *The CELECT ECM showing the sensor and actuator harnesses. (Courtesy of Cummins Inc.)*

OUTPUT CIRCUIT

The primary ECM output function is control of the CELECT injectors whose operation is described in detail. However, the Jacobs C-brake, PTO apparatus, and all ECM readouts can also be described as being part of the output circuit. The CELECT ECM drives output circuit functions through the actuator harness shown in Figure 38–7. A brief description of C-brake and PTO operation is given in the input circuit description.

The primary diagnostic and programming tool for use with CELECT Plus systems is the personal computer (PC) with **Windows (Microsoft)**-driven, **INSITE** software, and the appropriate serial links. **Echeck** and **Compulink** (proprietary ESTs) software packages will also be made available.

The Celect Plus Injector

The CELECT injector is substantially different from the Cummins PT injector, and probably the only real resemblance is that they both are cam actuated (Figure 38–8). CELECT and CELECT Plus injectors use similar operating principles but the CELECT Plus injector is referenced in the following description.

CELECT injectors are cam-actuated, electronically controlled injector assemblies that use multi-orifii, hydraulic injector nozzles and therefore have much in common with the EUIs used by their competitors; however, CELECT injectors differ in metering and timing principles. CELECT injectors are fitted to cylindrical bores in the engine cylinder head. They circulate fuel at any time the engine is running and the gear supply pump is rotating. Fuel enters and exits the injector by means of annular recesses sealed by rubber O-rings. The lower annular recess is exposed to the supply gallery in the cylinder head and receives fuel at **charging pressure** (150 psi/11 bar). Fuel passes from the lower annular recess to the internal injector circuitry, through the metering spill port, and is circulated through the injector and solenoid control valve ducting. It exits through a fuel drain in the upper external annulus. In order to properly understand the operation of

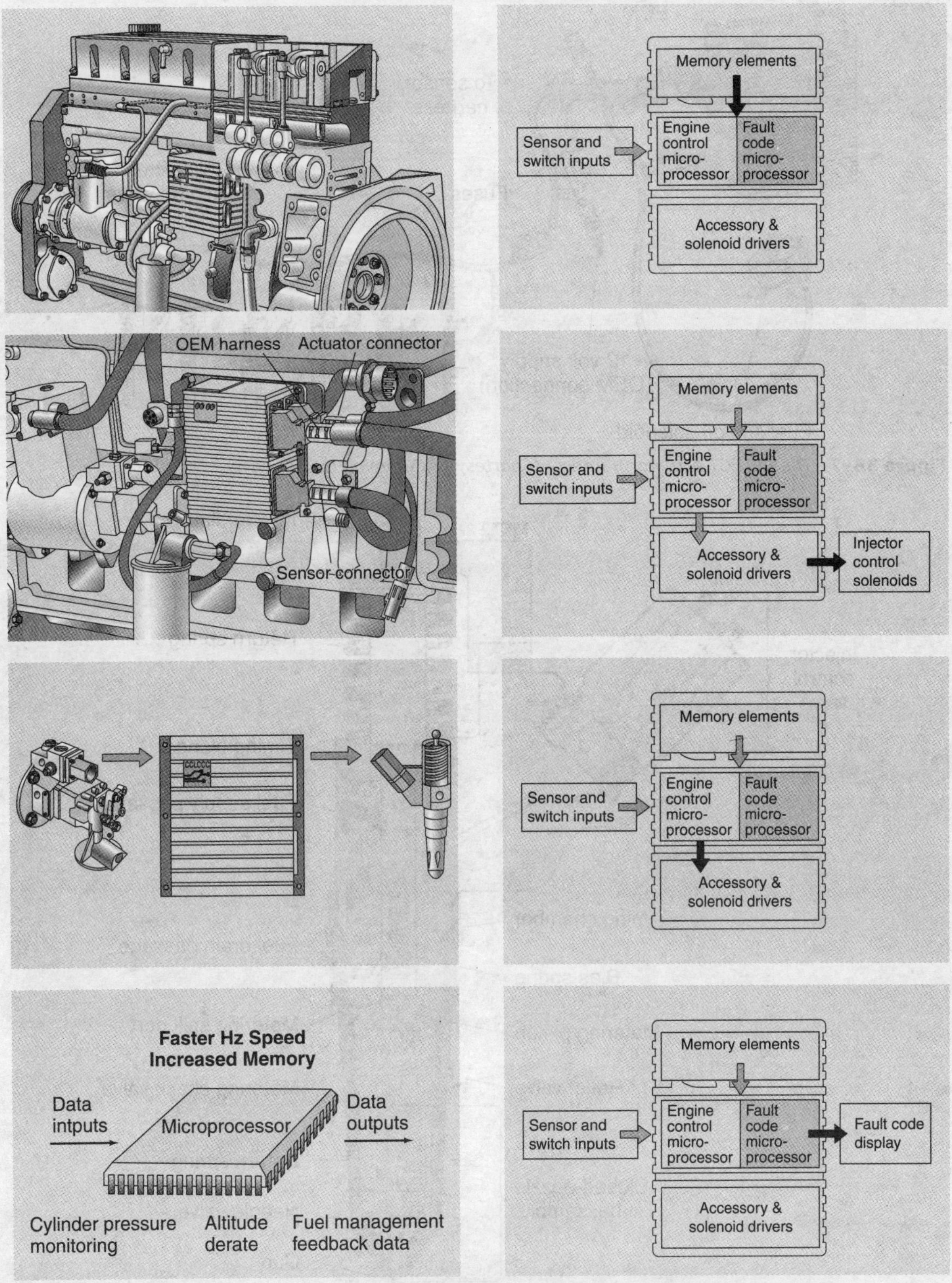

Figure 38–6 *CELECT Plus processing cycle graphics. (Courtesy of Cummins Inc.)*

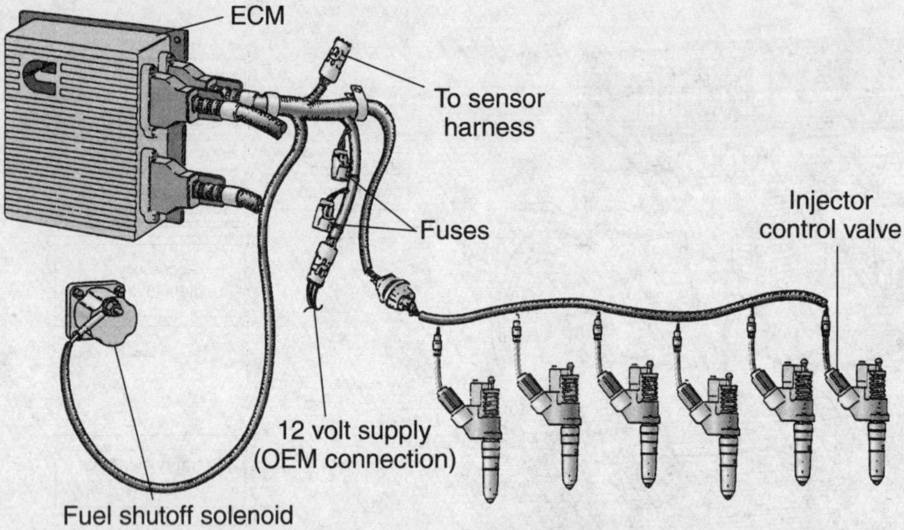

Figure 38–7 *The CELECT actuator harness. (Courtesy of Cummins Inc.)*

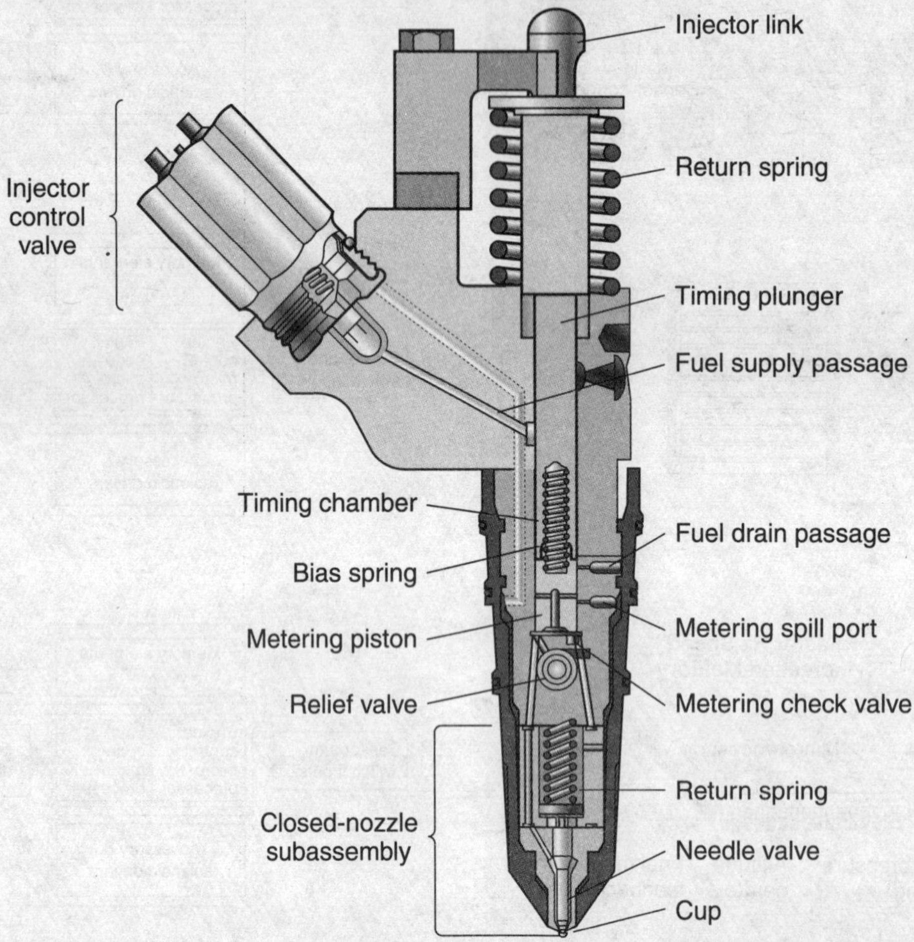

Figure 38–8 *Sectional view of the key CELECT injector components. (Courtesy of Cummins Inc.)*

this injector assembly it is necessary to refer to the following sequence of diagrams as shown in Figure 38–9. Figure 38–9 references the procedure on an M11 in which the camshaft rotates clockwise (CW) orientated from the front the engine. The procedure for an N14 is similar but you should note that the N-series camshaft rotates counterclockwise (CCW).

The ECM controls the CELECT injector by varying the time the injector control valve remains open and closed. CELECT and CELECT Plus injectors, while sharing operating principles, are *not* interchangeable, the latter incorporating increased orifii sizing, modified spray angles, and other performance enhancements.

Celect Plus Engine Protection System

The CELECT Plus ECM monitors critical engine sensors and will log fault codes when an out-of-normal condition is determined. If an out-of-range condition is detected in a critical sensor, an engine derate action may be initiated. The operator will be alerted by flashing of the appropriate dash light. The engine protection system monitors:

- Coolant temperature
- Oil temperature
- Intake manifold temperature
- Oil pressure
- Coolant level
- Engine overspeed

At engine derate, the operator is alerted by the flashing light, and engine power and speed deration will occur incrementally depending on the severity of the condition. The engine will not actually shut down unless the ECM is programmed with the engine shutdown option. The engine protection shutdown option automatically shuts down the engine when a fluid pressure or temperature out-of-normal range problem occurs that could lead to catastrophic damage. Should this occur, the dash engine protection lamp will flash for 30 seconds before the shutdown. Key off is required to restart the engine but if the shutdown trigger persists, the engine will be shut down again after 30 seconds.

Celect Plus Programmable Features

CELECT Plus offers a full range of programmable features consistent with other full authority management systems, including:

Cruise control
Automatic engine brake in cruise control
PTO governing
Gear down protection
Progressive shifting
Limiting speed/variable speed governing
Disable air conditioner
Low idle adjustment
Idle shutdown option
Engine protection shutdown
Vehicle speed monitoring
Disable engine fan
Automatic transmission bussing
Maintenance monitor options

These features and their options are programmable as customer data using Compulink, Echeck, or INSITE ESTs.

Road Speed Governing Tailoring

Road speed governing or **RSG** adjusts or moderates the upper and lower droop values. Droop is governed speed deviation from a given set speed. Moderating droop can tailor the torque curve (by managing injected fuel quantity) to optimize engine fuel economy as road and speed conditions vary. Upper droop will tailor the torque curve upward (by providing extra fuel) before maximum vehicle speed is reached under conditions of high engine load.

Lower droop tailors the torque curve upward in a downhill or light load condition, permitting increased speed before cutting off fueling. This allows a vehicle to accelerate above the governed road speed, permitting an increase in momentum before going up the next hill. RSG upper and lower droop can be set between 0 and 3 mph (0 and 5 km/h). For instance, if road speed droop were set at zero, this would represent a hard value. However, if it were programmed at 5 km/h, this would be a *soft* value because it would permit 5 km/h of latitude

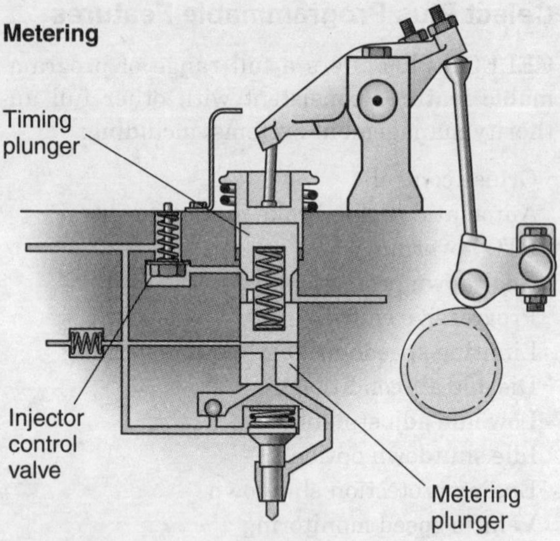

1. At the start of metering, both the metering plunger and the timing plunger are at their lower travel limit as their actuating cam is at peak lift. The injector control valve closes, actuated by a 78-V induction coil derived spike delivered from the CELECT injector driver unit.

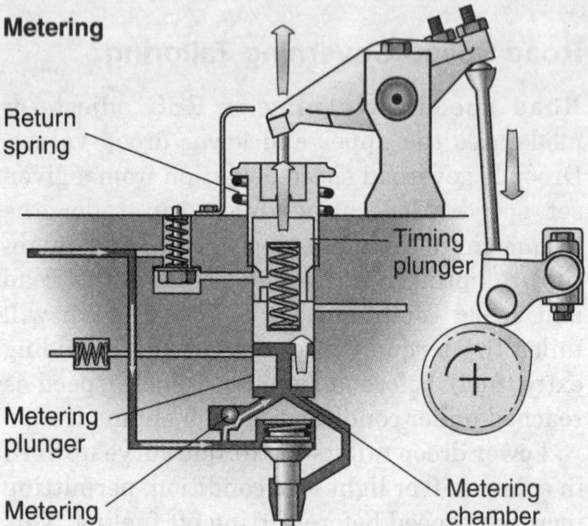

2. As the cycle continues, the cam ramps off the nose toward base circle, unloading the injector train and permitting the timing plunger return spring to lift the timing plunger. This enables fuel to flow past the metering checkball into the metering chamber. This flow continues as long as the timing plunger is moving upward and the injector control valve is closed. Supply pressure acting on the bottom of the metering piston forces it to maintain contact with the timing plunger.

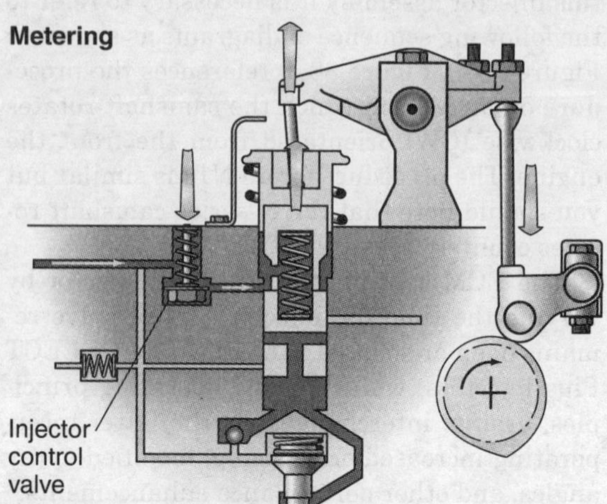

3. The ECM will determine the end of metering by switching the injector control valve to its open position. This action causes the metering checkball to seat and permits fuel to pass around the injector control valve.

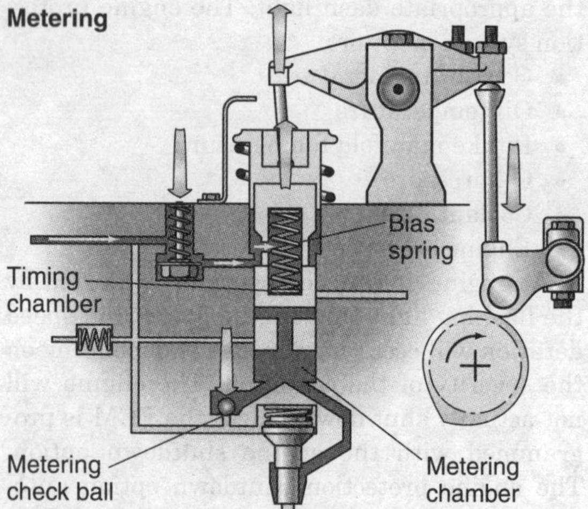

4. Fuel at supply pressure then flows into the timing chamber stopping metering piston travel. The bias spring ensures that the metering plunger remains stationary, preventing it from drifting upward as the timing plunger moves upward. This same force against the metering plunger results in sufficient fuel pressure below the metering piston to keep the metering check-ball seated. The result is a precisely metered quantity of fuel in the metering chamber.

Figure 38–9 *The CELECT Plus injector operating cycle on an M11 engine.*

Timing

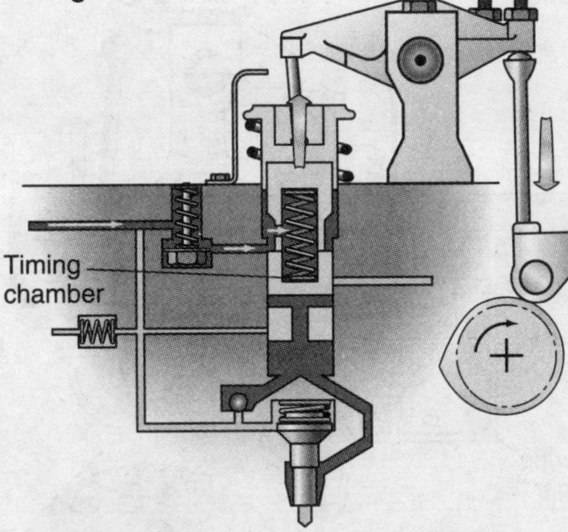

5. As the cycle continues, the injector train continues to ride toward cam inner base circle permitting the timing chamber to fill with fuel.

Timing

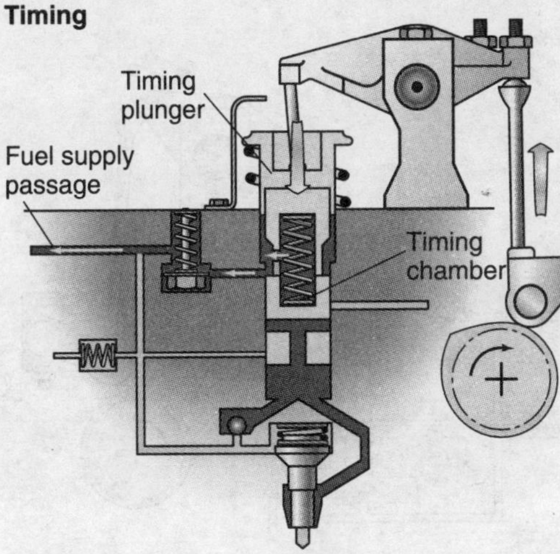

6. Next, the injector cam passes over the inner base circle location and begins ramping toward outer base circle. This action loads the injector train and consequently the timing plunger begins its downstroke. Initially the injector control valve remains open, allowing fuel to spill from the timing chamber and reverse flowing it through the fuel supply passage.

Timing

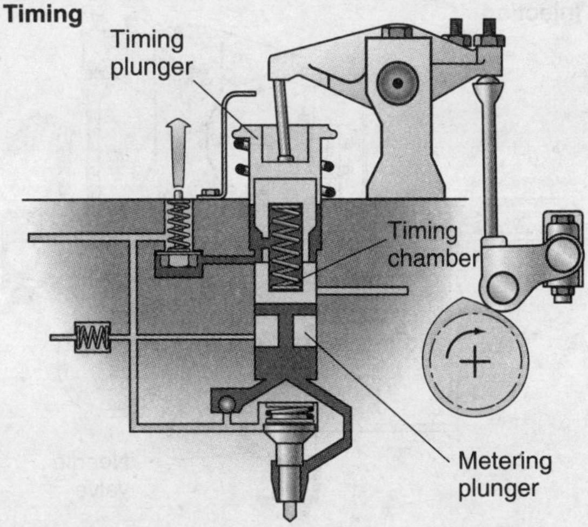

7. The delivery sequence begins when the ECM switches the injector control valve to its closed position, trapping fuel in the timing chamber. This trapped fuel acts as a hydraulic link between the timing plunger and the metering plunger; this forces the metering plunger downward with the timing plunger.

Injection

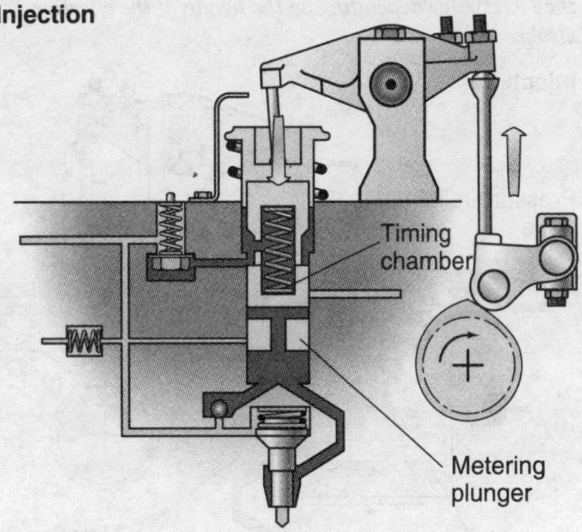

8. Because the metering plunger is being driven downward (hydraulically), rapid pressure rise begins in the metering chamber.

Figure 38–9 *(Continued)*

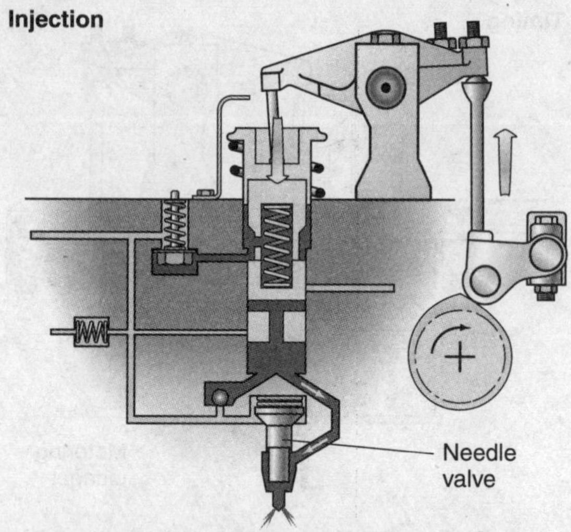

9. Ducting connects the metering chamber with the pressure chamber of the hydraulic, multi-orifii injector nozzle located at the base of the CELECT injector. When the pressure in the metering chamber (and therefore in the nozzle pressure chamber) reaches the NOP value, approximately 5,000 psi (340 atms/34.442 MPa), the nozzle valve opens and injection begins. Fuel is forced through the nozzle orifii and is atomized directly into the engine cylinder. The minute size of the nozzle orifii means that they are unable to relieve the pressure as fast as it is created, and peak pressure is capable of rising well above the NOP value depending on the length of the effective stroke.

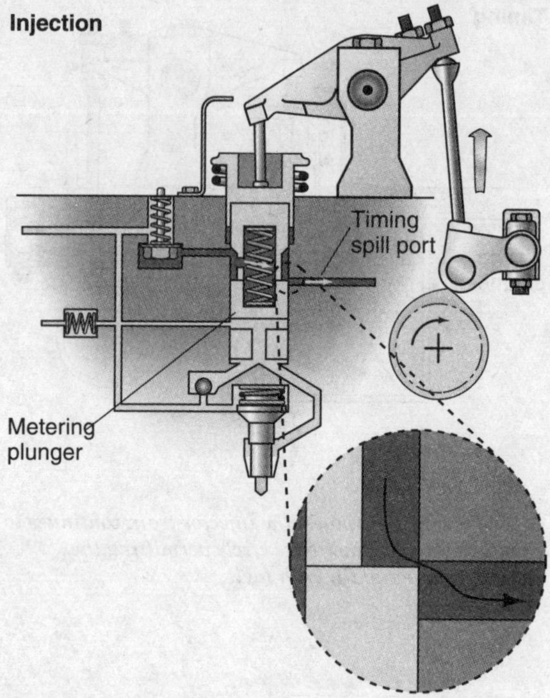

11. Immediately after the metering spill port is exposed by the downward travel of the metering plunger, its upper edge exposes the timing spill port.

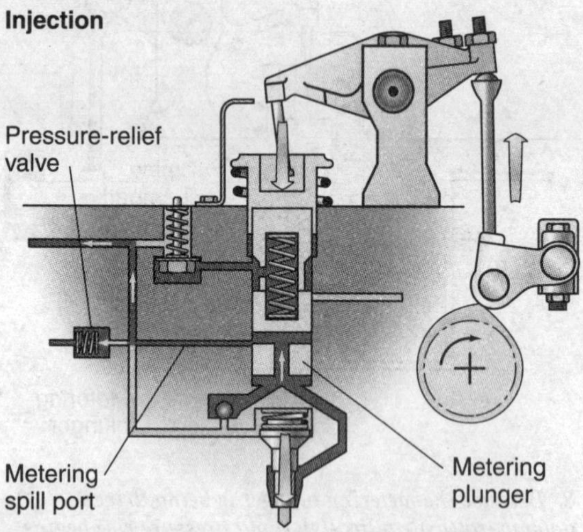

10. Injection continues until the metering plunger passes the spill passage. This action causes a collapse of metering chamber pressure, permitting abrupt nozzle valve closure. At this moment, the pressure relief valve will relieve, minimizing the effect of the high pressure spike that occurs at metering spill: The relief valve passage connects to the fuel drain line.

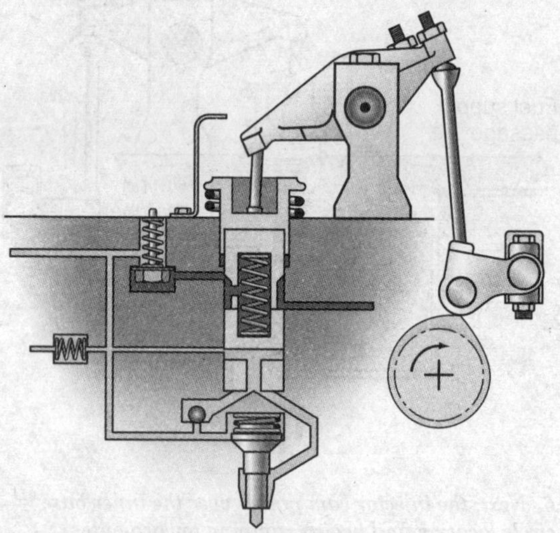

12. This action permits the fuel in the timing chamber that was used as hydraulic medium to spill to the fuel drain. This completes the cycle.

Figure 38–9 *(Continued)*

around the set cruise speed. Using increased droop settings enables the ECM to strategize fueling to road conditions and engine load to optimize fuel economy.

Electronic Smart Power

Electronic smart power (ESP) engine ratings improve driveability and fuel economy in hilly terrain. In these conditions ESP results in more consistent road speed, faster trip times, and less shifting. It is essentially a cruise option that "learns" terrain and manages vehicle speed using soft limits. On level or near level terrain, ESP engines operate at a base power rating determined by the ESP calibration programmed into the ECM. The ECM continually monitors the average speed the operator is attempting to maintain and logs this as "learned speed." The engine then automatically switches to the high torque ESP mode if all ESP operating conditions have been met.

ESP Operating Conditions

1. Engine must be operating in cruise control or between 90% and 100% throttle (a momentary throttle position change such as required by a gear shift sequence will not disable ESP).
2. The transmission must be in a drive ratio of 2.62:1 or higher.
3. Vehicle speed must not exceed the ECM programmed maximum cruise control speed.
4. Using the foot brake to reduce vehicle speed will not activate ESP mode, but momentary use will not disable it.
5. Any active fault codes that relate to the **throttle position sensor (TPS)**, **vehicle speed sensor (VSS)**, or **engine position sensor (EPS)** will prevent ESP operation.

The foregoing conditions met, the ECM will switch to the ESP high-torque mode any time the vehicle speed drops more than a calculated value (3 mph or 5 km/h) below "learned speed." Once "learned speed" has been attained and the load reduced, the engine rating will return to the base power mode.

RoadRelay

RoadRelay is a dash-mounted driver data display that keeps the operator informed of conditions on board. All RoadRelay data are recorded and may be downloaded to a PC database for analysis. The system will track the following data:

Trip length and average mpg
Trip running time and fuel consumed
Idle fuel consumed
Cruise control time
Top road speed

SYSTEM TROUBLESHOOTING AND ENGINE TESTING

Troubleshooting the CELECT Plus system requires the use of *Cummins Troubleshooting and Repair Manual 3666130–00 (9/95)* volumes 1 and 2, a digital multimeter (DMM), and the appropriate EST—preferably a PC with INSITE software. Troubleshooting is sequential and the stepped procedures must be undertaken precisely; skipping a step usually invalidates subsequent steps. Customer data programming is fully covered in this manual. In common with other full authority, electronic management systems, CELECT Plus will perform self-diagnostic procedures. Some of these are briefly outlined in the remainder of this section with an emphasis on tests and procedures that have least in common with other OEM tests.

Setup for Dynamometer Test

This test mode enables some advance diagnostic tests to run, which are structured by the CELECT ECM. The maximum engine speed without VSS, maximum vehicle speed in top gear ratio, and the maximum vehicle speed in lower gear are all set to their maximum values. At the same time, progressive shift programming and idle shutdown are disabled for the duration of the dynamometer test run. To properly enable the "setup for dynamometer test," some chassis electronic subsystems must be disabled for the duration of the test. Dynamometer testing is covered in detail in Chapter 14.

Automated Cylinder Performance Test

The objective of the cylinder performance test is a thorough test of cylinder performance that is displayed by reporting cylinder load contribution in percentage terms. The driving software for the test is contained in the EST, which must be connected.

The following requirements must be met to run this test:

1. Ensure that the area around the engine and fan are clear and the exhaust properly vented.
2. Ensure that the engine oil temperature is at a minimum of 170°F (77°C).
3. Shut down the engine.
4. Lock the fan clutch to the engaged position, disable the air-conditioning (A/C) system and any other auxiliary devices that could represent a parasitic load on the engine.
5. With the vehicle stationary, start the engine and allow it to idle.
6. Engage the PTO (the test cannot be performed unless the PTO feature is programmed to the ECM).

The test profile instructions should be run according to Cummins CELECT service literature. The test produces a *pass/fail* message for each engine cylinder.

In the event of cylinder performance outside the specification window, the percent contribution value will be displayed with the *fail* message. When the test is complete, the following sequence should be run:

1. Shut down the engine.
2. Return the fan clutch, A/C system, and all other disabled devices to normal operation.

Cylinder Cutout Test

The cylinder cutout test removes individual injectors from the cylinder firing cycle to monitor engine parameters while a selected cylinder is electronically disabled. The electronic service tool displays the percent load and engine rpm value while the cylinder cutout test is being run. Again, the test is performed while the vehicle is stationary and the EST is connected to the ECM. During the cylinder cutout test, only one cylinder can be electronically shorted out at a time in current versions of CELECT management.

REVIEW QUESTIONS

1. Which of the following CELECT Plus input sensors could be categorized as a "command" input?
   a. Ambient pressure
   b. Oil pressure
   c. Coolant level
   d. Throttle position
2. Cummins CELECT injectors are charged at a pressure of:
   a. 15 psi (103 kPa)
   b. 50 psi (345 kPa)
   c. 70 psi (482 kPa)
   d. 150 psi (1,034 kPa)
3. What type of fuel transfer pump is used in the Cummins CELECT system?
   a. Vane
   b. Diaphragm
   c. Gear
   d. Plunger
4. What operating principle does a Cummins CELECT intake manifold temperature sensor use?
   a. Piezoresistive
   b. Potentiometer
   c. Variable capacitance
   d. Thermistor
5. The CELECT boost pressure sensor would be described electrically as:
   a. Piezoresistive
   b. Potentiometer
   c. Variable capacitance
   d. Thermistor
6. At what electrical pressure are the CELECT injectors switched?
   a. 5 V
   b. 12 V
   c. 78 V
   d. 100 V
7. The Cummins software for accessing its data hub, programming CELECT ECMs, and performing system diagnosis is known as:
   a. Compulink
   b. Windows XP
   c. INSITE
   d. DOS

8. The hydraulic injector nozzle used in the CELECT injector is set to have a nozzle opening pressure (NOP) of:
   a. 150 psi (11 bar)
   b. 5,000 psi (345 bar)
   c. 25,500 psi (1,759 bar)
   d. 30,000 psi (2,069 bar)
9. The "smart" road speed governing option on CELECT Plus is called:
   a. Echeck
   b. ESP
   c. INSITE
   d. RSG tailoring
10. Which of the following will prevent ESP operation?
    a. Any logged fault code
    b. Any active fault code
    c. TPS, VSS, or EPS active code
    d. 1:1 transmission drive ratio
11. *Technician A* states that a CELECT Plus injector is energized with a 78 V spike. *Technician B* states that the CELECT Plus injector voltage increases proportionally with the duty cycle. Who is correct?
    a. Technician A only
    b. Technician B only
    c. Both A and B
    d. Neither A nor B
12. *Technician A* states that CELECT engine position sensors use a magnetic pulse generator principle. *Technician B* states that the engine position sensor is located on the engine crankshaft. Who is correct?
    a. Technician A only
    b. Technician B only
    c. Both A and B
    d. Neither A nor B
13. *Technician A* states that the CELECT engine position sensor signal is generated by the engine camshaft. *Technician B* states that the CELECT engine position sensor is fitted with 24 evenly spaced recesses with a dowel indicating the #1 piston, TDC position. Who is correct?
    a. Technician A only
    b. Technician B only
    c. Both A and B
    d. Neither A nor B
14. *Technician A* states that the ambient pressure sensor in a CELECT M-11 is located in front of the ECM housing. *Technician B* states that the ambient temperature sensor in a CELECT N-14 is also located in front of the ECM housing. Who is correct?
    a. Technician A only
    b. Technician B only
    c. Both A and B
    d. Neither A nor B
15. *Technician A* states that the coolant and oil temperature sensors are both of the thermistor type. *Technician B* states that both sensors must be supplied with a reference voltage value. Who is correct?
    a. Technician A only
    b. Technician B only
    c. Both A and B
    d. Neither A nor B
16. *Technician A* states that a CELECT injector has a single internal plunger that defines timing and metering. *Technician B* states that injection pressures in the CELECT injector are created by varying the supply pressure. Who is correct?
    a. Technician A only
    b. Technician B only
    c. Both A and B
    d. Neither A nor B
17. *Technician A* states that the CELECT and CELECT Plus injectors use shared electronic operating principles. *Technician B* states that CELECT and CELECT Plus injectors are interchangeable so long as the new fuel code is correctly programmed to the ECM. Who is correct?
    a. Technician A only
    b. Technician B only
    c. Both A and B
    d. Neither A nor B
18. *Technician A* states that the CELECT ECM is mounted on a heat sink and cooled by fuel. *Technician B* states that both the ECM and the filter are charged by the gear pump at a pressure of 150 psi (11 bar). Who is correct?
    a. Technician A only
    b. Technician B only
    c. Both A and B
    d. Neither A nor B

19. *Technician A* states that the function of the CELECT coolant level sensor is to shut the engine down should the sensor fail to ground through the coolant. *Technician B* states that the CELECT coolant level sensor has a pair of probes that can read high and low coolant levels in the radiator. Who is correct?
   a. Technician A only
   b. Technician B only
   c. Both A and B
   d. Neither A nor B

20. *Technician A* states that the actuating force of a CELECT injector is mechanical and determined by cam geometry and engine rpm. *Technician B* states that the CELECT injector metering plunger is actuated by a hydraulic column of fuel. Who is correct?
   a. Technician A only
   b. Technician B only
   c. Both A and B
   d. Neither A nor B

39

Cummins CAPS

INTRODUCTION

Introduced in 1998, the **Cummins Accumulator Pump System**, or **CAPS**, was initially used on the ISC engine, then on the ISL engine. Cummins abandoned using the CAPS system on both the ISC and ISL engines for the 2004 model year, replacing it with a Bosch common rail fuel system. CAPS is best described as a medium-pressure, **accumulator** system suitable for use on direct-injected diesel engines classified as small- to medium-bore in truck and bus applications. In common with other recently developed diesel engine fuel management systems, CAPS fuel injection pressures can be controlled entirely independently from engine speed. This permits much higher fuel injection pressures at low engine rpm. When fuel injection pressure can be accurately controlled through a wide range unrestricted by engine speed, much faster responses to sudden changes in accelerator input requests are possible.

CAPS-fueled engines share many of the engine control module (ECM) software and hardware advantages of Cummins' larger engines because they are **Interact System (IS)** managed. IS uses common software protocols and hardware architecture for the full family of Cummins engines and, although the output circuit for a CAPS-fueled engine is necessarily distinct, it has many electronic features previously associated with large-bore truck diesel in a small-bore package.

The CAPS pump uses a modular design and therefore is easy to field repair. The Cummins procedure used to repair CAPS pumps is referenced in this text, but technicians should understand that Cummins training is required prior to undertaking repairs. Figure 39–1 shows an external view of a CAPS fuel pump.

Figure 39–1 *CAPS pump (Courtesy of Cummins Inc.)*

FUEL SYSTEM COMPONENTS

We begin with an overview of the CAPS fuel system components. The components are simply reintroduced in this chapter and because most of them will be familiar to readers, they are not re-examined in great detail. The distinct component in this system is the CAPS accumulator injection pump, and its operation is presented in more detail later in the chapter. Fuel flow in the ISC is nearly double that on the earlier C-series engine (600 lb/h versus 375 lb/h), so much larger supply and drain plumbing is required. You can follow the fuel routing using Figure 39–2.

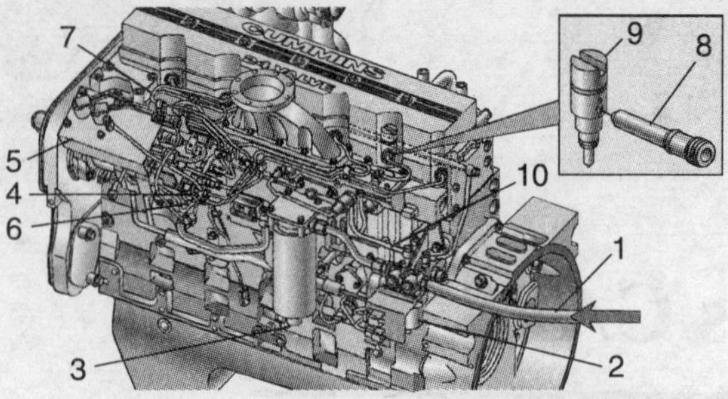

1. Fuel suction line
2. Electric lift pump
3. Fuel filter WIF sensor
4. Regulator valve
5. Accumulator module
6. Distributor snubber valve
7. High-pressure pipe
8. Fuel injector connector and edge filter
9. Hydraulic injector
10. Return circuit

Figure 39–2 *CAPS fuel flow routing. (Courtesy of Cummins Inc.)*

Electric Lift Pump

A 12 V-DC electric lift pump is used to initiate the movement of fuel through the fuel subsystem. The pump operates for 30 seconds when the key switch is first closed, whether the engine is running or not. The lift pump produces a minimum pressure of 10 psi (689 kPa). If additional priming is required, cycling the key will trigger another 30-second priming sequence.

The lift pump is used only for starting the engine. After startup, the lift pump is shut off and bypassed, leaving the gear pump to supply fuel from the fuel tank.

Gear Pump

After the CAPS fuel subsystem has been primed by the lift pump, a gear-type transfer pump is used to move fuel through the system. The priming pump uses an external gear pumping principle.

Combination Filter/Water Separator

A combination filter/water separator is used, located upstream from the gear pump. Cummins claims that the filter functions at 97% efficiency using Stratopore™ filtering media effective to 10 μ particulate. The assembly is fitted with a water-in-fuel (WIF) sensor that alerts the operator when it is immersed in water.

CAPS Pump

The heart of the CAPS fuel system is the accumulator injection pump, which is driven by a camshaft drive gear to which it is timed. This pump is presented in some detail later in the chapter. Its function can be briefly described by saying that it uses a pair of camshaft-actuated plungers to charge fuel to an accumulator: The pressure in the accumulator is precisely managed by the ECM, which can therefore control the exact pressure delivered to the injectors during each cycle. Injection pressures are consistent with accumulator pressures. The accumulator delivers fuel to the distributor head, which responds to ECM output switching and routes fuel to the injectors in engine firing sequence.

High-Pressure Circuit

Steel high-pressure lines connect the CAPS distributor head with fuel connectors located in the cylinder head. The internal dimensions and elasticity of the high-pressure pipes are critical requirements for maintaining engine fueling balance. The fuel connectors are transversely mounted in the cylinder head, and a flared nipple seats each one to a recessed angled seat in the injector. The fuel connectors are fitted with an integral edge filter that screens anything that might have somehow escaped the extensive upstream filtering processes.

Hydraulic Orifice-type Injectors

Bosch cylindrical style, hydraulic injectors are used in the CAPS fuel system. They use a multi-orifii nozzle, and shims are used to define nozzle opening pressure (NOP) at 4,500 psi (310 bar), resulting in a nozzle closing pressure of around 2,700 psi (186 bar). The injector nozzle is a seven-orifice, concentric geometry type. A leak-off circuit routes nozzle valve leakage back to the fuel tank. NOP is bench set in the usual way with shims used to trim the exact value. The injectors are field serviceable, but replacement of nozzle assemblies is recommended over reconditioning. Hydraulic injector servicing is covered in Chapter 19.

It is important to remove the transverse fuel connector in the cylinder head before attempting to remove an injector. A 1.5 mm copper alloy washer seals the nozzle from the engine cylinder and precisely positions nozzle protrusion into the cylinder. The center location of the injector and design of the cylinder head permit removal and reinstallation of the injectors without disturbing the rockers.

CAPS PUMP

The CAPS accumulator pump uses a modular design that facilitates field diagnosis and repairs. It is the heart of this fuel system and technicians should have some understanding of how it operates so that they can perform accurate first-level diagnoses in conjunction with Cummins INSITE software. Field repairs are not difficult but should only be attempted by Cummins trained technicians.

The five modules that comprise a CAPS unit are:

1. Gear pump module
2. Camshaft and housing module
3. Accumulator housing module
4. Injection control valve module
5. Distributor module

The operation of each of these modules will now be studied in detail. An exploded view of the modules is shown in Figure 39–3.

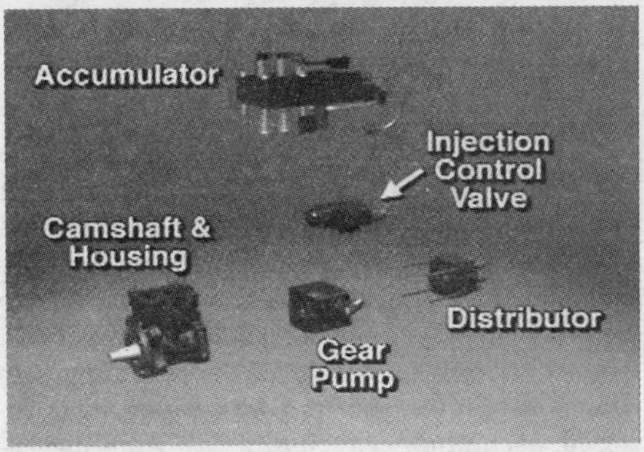

Figure 39–3 *CAPS modules. (Courtesy of Cummins Inc.)*

Gear Pump Module

The subcomponents of the CAPS gear pump module are a drive shaft, a pair of external pumping gears, front and rear couplings, and a piston-type pressure regulator. The gear pump module is located behind the CAPS camshaft. The gears are driven by the CAPS camshaft. Fuel flow through the gear pump module begins when fuel enters the inlet: As the gears rotate, fuel passes between the teeth and liner to the outlet. Once the system is initially primed, the gear pump is responsible for movement of fuel through the subsystem. Pump pressures maximize at 1,100 rpm engine speed and stay consistent as engine rpm rises above this value. A piston-type regulator manages fuel pressure at a maximum of 165 psi (11.4 bar) with excess fuel spill rerouted back to the inlet side of the gear pump. Fuel is discharged from the gear pump to the **pressure control valves (PCVs)** located in the accumulator module. Figure 39–4 shows the CAPS gear pump module separated from the main pump housing.

Figure 39–4 *CAPS gear pump assembly.*

Camshaft and Housing Module

The subcomponents of the CAPS camshaft (Figure 39–5) and housing module (Figure 39–6) are the main housing, a camshaft, camshaft main bearings, and two tappet assemblies.

The camshaft is supported by taper roller bearings at both ends and is driven by the main CAPS pump drive at half engine speed. The camshaft has a pair of three lobe cam profiles. Each cam profile is identically dimensioned. A pair of roller tappets is located above each cam profile in bores so that they ride the cam profiles. As the cam turns through one revolution, each tappet is linearly actuated three times. The rear of the camshaft drives the gear pump

Figure 39–5 *CAPS Camshaft: Each cam profile actuates one of the two pumping elements.*

Figure 39–7 *CAPS: main housing, camshaft, and piston.*

Figure 39–6 *CAPS main housing module.*

Figure 39–8 *Accumulator housing module.*

module. By actuating the pair of tappets, the camshaft provides the reciprocating force required by the accumulator module to develop injection pressures. Figure 39–7 shows the disassembled camshaft and housing module components.

Accumulator Housing Module

The accumulator housing module contains the high-pressure pumping elements required to develop injection pressures. The subcomponents include a pair of plungers and barrels, barrel retainers, retraction springs, a pressure control valve (PCV), rate shape tube and snubber, pressure and temperature sensors, and an air bleed valve. The accumulator housing module is shown in Figure 39–8.

High-Pressure Pumping

Internal drillings in the CAPS housing direct fuel from the gear pump to the PCVs located above each of the two plunger and barrel pumping elements. The plungers are ceramic with no recess geometry: Each is lugged to a tappet and the tappet retraction springs ensure they ride the cam profiles. As the CAPS camshaft is rotated, each plunger reciprocates within its stationary barrel.

When the PCVs are de-energized, fuel is allowed to flow into the pump chambers formed by the plunger and barrels at the gear pump pressure of 165 psi (11.4 bar). When a PCV is energized, it closes, trapping fuel in the pump chamber and subjecting it to pressure rise as the plunger rises through its stroke. As the pump plunger is driven into the slug of fuel trapped in the pump chamber, when the pressure exceeds that in the accumulator, a check valve opens, forcing pressurized fuel into the accumulator. The effective pump stroke continues until the plunger reaches the top of its travel when the accumulator check valve closes.

Managing Accumulator Pressure

The timing of PCV valve opening and closing is managed by the ECM that senses accumulator pressure. Accumulator pressure is therefore precisely controlled by the ECM prior to each injection event. A bleed valve located at the top of the accumulator permits any air in the system to be removed. When fuel enters the accumulator, it is stored in six cylindrical accumulator passages awaiting injection: Accumulator volume is designed to store a sufficient volume of fuel so that a maximum pressure drop of 10% will follow each injection event.

Fuel leaves the accumulator by means of the rate shape snubber valve. The rate shape snubber limits fuel injected in the initial portion of the fuel injection pulse, resulting in a pressure dip as fuel is unloaded. In some ways the CAPs fuel system can be compared to first-generation diesel common rail (CR) systems because the accumulator feeds the injectors in the high-pressure circuit in much the same way as the rail does in a CR system.

Injection Control Valve Module

The **injection control valve (ICV)** module is an ECM-controlled electrohydraulic solenoid. The subcomponents are a housing and pressure regulator assembly made up of a stator, stator housing, and outer (armature) and inner pins. When the ICV stator is de-energized, spring pressure forces the outer pin onto its seat, creating a flow path between the distributor port and drain port. Energizing the ICV causes the outer tube to be pulled toward the stator, shutting off the drain port and permitting pressure to build on the distributor port. Fuel now flows through one of the six drillings to the injection snubber valves in the distributor module. De-energizing the stator allows spring pressure to force the outer pin off its seat with the drain port, creating a flow path between the distributor port and drain, collapsing the injection pressure, and permitting the injector nozzle to close rapidly. The regulator allows fuel to spill to the drain until the pressure drops to 330 psi (22.4 bar), then closes, retaining a residual line pressure of about that value until the next injection cycle. At the completion of each injection pulse, the outer pin seats, closing off source fuel.

By controlling the energizing and de-energizing of the stator, the ECM manages both injected fuel quantity and timing. Both mechanical and electronic variables and lags are built into the ECM software algorithms (processing rules/logic). A **transorb diode** in the harness protects the ECM from the inductive kick created by the ICV when its magnetic field collapses. If the transorb diode fails either by shorting or opening, an ECM fault code will log.

Distributor Module

The distributor module functions a little like the distributor in your car except that its medium is high-pressure fuel. Its components are the distributor housing, rate shape inlet, rotor coupling, rotor, and six (one for each engine cylinder) distributor snubber valves. The rotor is rotated at camshaft speed, that is, through one complete rotation per full effective cycle of the engine (two crankshaft rotations). The distributor module is shown in Figure 39–9.

Figure 39–9 *CAPS distributor: Note the location of the distributor rotor.*

There are six injection ports in the distributor housing, each connected to an injection line. The port in the rotor (Figure 39–10) is of sufficient size so that it does not play any role in controlling either the beginning or ending of injection. It merely identifies the window within which the ECM can select an injection pulse for any engine cylinder. The beginning, duration, and ending of injection is managed by the ECM by energizing and de-energizing the ICV. The six injection high-pressure pipes are connected to the CAPS distributor module to provide the 1-5-3-6-2-4 firing sequence of the engine cylinders. The injection pumping cycle described in this section is repeated for each cylinder.

Figure 39–10 *CAPS distributor rotor.*

CAPS TROUBLESHOOTING

Troubleshooting problems in all electronically managed engines should always be performed by technicians with OEM training using guided software and service literature, specifically INSITE and QSOL when working on pre-2004 ISC and ISL engines. The following data is intended to give technicians some guidelines for diagnosing performance complaints.

QuickServe Online

Cummins **QuickServe Online** (**QSOL** or just QS) is required to work on all Cummins electronically controlled engines. QSOL is updated daily. Learn how to use the Website search engine. Information searches on QSOL are by engine serial number and include searches of the parts data if you use the full text search feature (in the upper right-hand corner of the screen). If you add double quotes on either side of a search term such as "piston ring," QS will locate all documents containing both words. QSOL also supports engines manufactured before the Website existed so make a habit of using it.

Cummins CAPS Specs on ISC and ISL

The following specifications apply to Cummins ISC and ISL engines manufactured before model year 2004. Model year 2004 was introduced in October 2002. Be careful when working with ISC and ISL engines because these have undergone significant generational changes in 2004 and 2007 model years.

Minimum lift pump pressure:
10 psi (69 kPa) at 25 secs/key-on

Fuel filter restriction: max. 6 in. (152 mm) Hg new filter
max. 10 in. (254 mm) Hg used filter

Gear pump pressure: Idle: 20 psi (138 kPa) min.
High idle: 120 psi (138 kPa) min.

Max. fuel drain restriction:
10 in. (254 mm) Hg

Max. fuel inlet temperature: 160°F (71°C)

NOP: 4,500 psi (310 bar)

CAPS accumulator pressure: 4,800 psi (331 bar) min at idle
15,000 psi (1,034 bar) min at peak torque

Maximum air intake restriction:
25 in. (635 mm) H_2O at rated power

Maximum exhaust restriction:
3 in. (76 mm) Hg at rated power

REVIEW QUESTIONS

1. For how long does the CAPS electric lift pump run following startup?
   a. During cranking only
   b. Until the engine starts
   c. 30 seconds or until the engine starts
   d. 30 seconds whether the engine is running or not
2. What types of injectors are used in a CAPS fuel system?
   a. Hydraulic, multi-orifii
   b. Electronic
   c. Mechanically actuated
   d. Pintle
3. How many lobes are machined onto each cam profile in a CAPS fuel pump camshaft?
   a. One
   b. Two
   c. Three
   d. Four
4. Which of the following would best describe the high-pressure pumping elements used in a CAPS fuel pump?
   a. Port helix
   b. Plunger and barrel
   c. External gear
   d. Vane type
5. What results when a CAPS water-in-fuel (WIF) sensor is immersed in water?
   a. Alarm buzzer sounds
   b. Engine shutdown
   c. Driver alert
   d. Power ramp-down
6. *Technician A* states that a WIF sensor is used to alert the driver that the fuel filter needs to be replaced. *Technician B* states that the WIF sensor triggers an alert when it detects water in the combination filter/separator assembly. Who is correct?
   a. Technician A only
   b. Technician B only
   c. Both A and B
   d. Neither A nor B
7. *Technician A* states that a CAPS pump is divided into five modules, any one of which can be replaced if diagnosed defective. *Technician B* states that CAPS pumps are simple enough that they can be field repaired without any Cummins technical training? Who is correct?
   a. Technician A only
   b. Technician B only
   c. Both A and B
   d. Neither A nor B
8. *Technician A* states that to pull a CAPS injector, the rocker assembly must be backed off. *Technician B* states that transverse fuel connector nut torque is responsible for defining the sealing force of the connector nipple to injector seat. Who is correct?
   a. Technician A only
   b. Technician B only
   c. Both A and B
   d. Neither A nor B
9. After a CAPS engine has been run for some time, the engine is shut down and the transorb diode in the harness is too hot to touch. *Technician A* states that the diode is shorted and should be replaced. *Technician B* states that a hot transorb diode indicates that the ICV is about to fail. Who is correct?
   a. Technician A only
   b. Technician B only
   c. Both A and B
   d. Neither A nor B
10. What is the function of a transorb diode in the harness of a CAPS fuel system?
    a. It protects the ECM from inductive voltage spikes caused by the ICV.
    b. It protects the ICV from inductive spikes caused by ECM output drivers.
    c. It isolates the CAPS electronics from the chassis electrical system.
    d. It isolates CAPS ECM from the CAPS sensor and actuator circuit.
11. What is the maximum inlet restriction permitted measure downstream from the fuel filter and upstream from the CAPS gear pump?
    a. 25 in. (635 mm) H_2O
    b. 10 in. (254 mm) H_2O
    c. 10 in. (254 mm) Hg
    d. 6 in. (152 mm) Hg
12. What is the minimum permissible fuel pressure measured at a CAPS accumulator when the engine is run at idle speed?
    a. 4,500 psi (310 bar)
    b. 4,800 psi (331 bar)
    c. 15,000 psi (1,034 bar)
    d. 24,000 psi (1,655 bar)

40

Common Rail Systems

INTRODUCTION

Electronically controlled, **common rail (CR)** diesel fuel injection systems were introduced on small-bore and automobile diesel engines in the late 1990s. These systems were manufactured by Robert Bosch and Delphi Lucas, and both had similar operating principles. The first-generation CR systems typically had lower peak pressures than those we associate with medium- and large-bore diesels. Maximum pressures were typically just over 20,000 psi (1,379 bar).

However, since the earliest generation of CR diesel fuel systems, peak system pressure potential has crept upward and today can exceed 30,000 psi (2,068 bar) in some systems. By 2010, almost every major diesel original equipment manufacturer (OEM) will be using a CR system on at least one of their engine families. There are industry experts who predict that by 2014, there will be just one diesel fuel system and that it will be some form of CR injection. This will be good news to students of diesel technology who for years have had to study nearly a dozen different fuel systems while their automotive counterparts contended with just two.

What is CR?

For purposes of defining CR diesel fuel injection, the term *CR* is used to refer to only those systems in which fuel injection pressure values are held in the **rail** that directly feeds the injectors. For instance, some texts refer to the hydraulically actuated electronic unit injector (HEUI) system as a CR system but because injection pressures are not developed in the manifold supplying the HEUIs (this is held relatively constant at a low charge pressure), this cannot be considered a CR system at least in the way we use the term. Beginning in 2008, a second generation of CR fueling in which rail pressures are amplified at the injector prior to injection was introduced: This system is described at the end of this chapter. Figure 40–1 shows a current Bosch CR system, which you should reference during this introductory explanation.

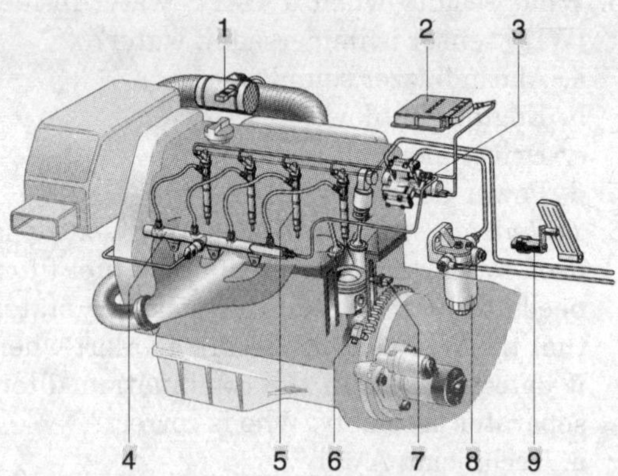

1. Air-mass meter
2. ECU
3. High-pressure pump
4. High-pressure accumulator (rail)
5. Injectors
6. Crankshaft-speed sensor
7. Coolant-temperature sensor
8. Fuel filter
9. Accelerator-pedal sensor

Figure 40–1 *Common rail injection system on a four-cylinder diesel engine. (Courtesy of Robert Bosch LLC)*

Function of the Rail

In this text, the term *rail* is used to describe the supply manifold or gallery that directly feeds all of the diesel fuel injectors. The basic principle of diesel CR can be compared to the way in which automotive gasoline fuel injection (GFI) systems operate. The major differences between GFI and diesel CR is that the diesel systems:

- Operate at much higher pressures, some exceeding 30,000 psi (2,068 bar)
- Precisely manage rail pressure within a wide range of values

In terms of managing and switching the injectors, there are many commonalities between diesel and gasoline CR injection. The similarities increase when the comparison is made between diesel CR and current direct injection (DI) gasoline fuel injection systems.

Engines Using CR Diesel Systems

As indicated earlier, CR diesel fuel systems have been used since the end of the 1990s. This means that diesel engine manufacturers have already progressed through a first generation of the technology. Some current applications of diesel CR fuel systems are shown in Table 40–1.

Table 40–1 does not include some well-established pre-2007 engines such as the 5.9-liter Cummins ISB and the Mack Trucks E-3 engine: These engines are no longer in production. You might have noticed in Table 40–1 that most of the engines listed are classified as medium to small bore. However, the MaxxForce 11 and 13 engines (produced jointly by Navistar International and MAN of Germany), which were introduced in 2007, break this trend. MaxxForce 11 and 13 engines are using the Bosch CR system. Cummins has already announced that its 2010 version of the ISX Series (15 liter) will use a CR fuel system, and the Detroit Diesel/Mercedes-Benz common platform engine family introduced in 2008 (DD15) through to 2010 (11-, 13-, and 16-liter versions) uses the amplified CR fueling described at the end of this chapter. Figure 40–2 shows a schematic layout of a Bosch CR fuel system.

Table 40–1 *Examples of Post-2007 Engines with CR Fuel Systems*

OEM	*Series*	*Displacement*	*Peak Power BHP (kW)*	*Maximum Torque lb-ft. (N·m)*
Audi	Q7-V12	6.0 liters	490 BHP (365 kW)	730 lb-ft. (990 N·m)
Caterpillar	C7 ACERT	7.2 liters	330 BHP (246 kW)	860 lb-ft. (1,166 N·m)
Caterpillar	C9 ACERT	9.3 liters	350 BHP (260 kW)	1,250 lb-ft. (1,695 N·m)
Cummins	ISB	6.7 liters	325 BHP (242 kW)	750 lb-ft. (1,017 N·m)
Cummins	ISC	8.3 liters	325 BHP (242 kW)	1,000 lb-ft. (1,356 N·m)
Cummins	ISL	8.9 liters	325 BHP (242 kW)	750 lb-ft. (1,017 N·m)
Deere	Power Tech 6.8	6.8 liters	200 BHP (150 kW)	500 lb-ft. (678 N·m)
GM Isuzu	6600 V-8	6.6 liters	360 BHP (268 kW)	650 lb-ft. (880 N·m)
GM Isuzu	7800	7.8 liters	300 BHP (224 kW)	860 lb-ft. (1,156 N·m)
Hino	J08E TB	7.7 liters	260 BHP (193 kW)	585 lb-ft. (794 N·m)
Navistar	MaxxForce 7 (V-8)	6.4 liters	350 BHP (260 kW)	650 lb-ft. (880 N·m)
Navistar	MaxxForce 9	8.6 liters	330 BHP (246 kW)	950 lb-ft. (1,288 N·m)
Navistar	MaxxForce 10	9.3 liters	350 BHP (260 kW)	1,150 lb-ft. (1,560 N·m)
Navistar	MaxxForce 11	10.5 liters	390 BHP (290 kW)	1,450 lb-ft. (1,966 N·m)
Navistar	MaxxForce 13	12.4 liters	475 BHP (354 kW)	1,700 lb-ft. (2,305 N·m)
Paccar	PX-6	6.7 liters	325 BHP (242 kW)	750 lb-ft. (1,017 N·m)
Paccar	PX-8	8.3 liters	325 BHP (242 kW)	1,000 lb-ft. (1,356 N·m)

NOTE: All the engines are I-6 configured unless otherwise indicated.

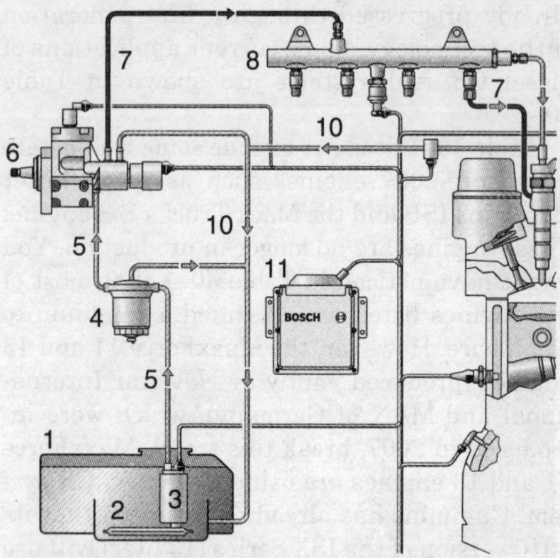

1. Fuel tank
2. Prefilter
3. Presupply pump
4. Fuel filter
5. Low-pressure fuel lines
6. High-pressure pump
7. High-pressure fuel lines
8. Rail
9. Injector
10. Fuel-return line
11. ECU

Figure 40–2 *Schematic layout of a Bosch common rail diesel fuel system. (Courtesy of Robert Bosch LLC)*

CR System Manufacturers

CR fuel system technology was launched by Bosch and Delphi Lucas but these companies have been joined by others. In most cases, diesel engine OEMs have opted to use one of the CR systems manufactured by a specialty diesel fuel system manufacturer rather than develop their own. Current systems are manufactured by:

- Bosch
- Caterpillar
- Delphi Lucas
- Denso
- Siemens

Fuel-Amplified CR

A second generation of CR diesel fuel injection systems was introduced in 2008 in preparation for the 2010 Environmental Protection Agency (EPA) model year, initially in Detroit Diesel Corporation (DDC) branded engines. This family of engines is the result of the Mercedes-Benz (MB) and DDC heavy-duty engine platform (HDEP). This engine family will be branded as DDC in North American Free Trade Agreement (NAFTA) countries and as MB elsewhere. The engine platform will include displacements of 15.6, 14.8, 12.8, and 10.6 liters. Fuel-amplified CR fuel injection is covered at the end of this chapter.

Advantages of CR Diesel Fuel Systems

If you have studied the many different diesel fuel management systems that appear in this textbook up to this chapter, you might conclude that many of them were pretty complicated. One of the primary advantages of CR diesel fuel systems is their simplicity. A technically less complex fuel system results in less potential for failures as the system ages. Not only is CR diesel fueling less complex, it also allows the engine control electronics much more precise management of cylinder combustion. This precise management of combustion results in:

- Lower emissions
- Improved fuel economy
- Lower engine noise levels
- Optimization of cylinder pressures

In fact, the CR system is in many ways so much superior to its predecessors you might well ask why it has taken so long for diesel engine OEMs to adopt it. Two key innovations have made the current generation of CR diesel fuel systems possible:

- Metals technology allowing containment of large volumes of ultra-high injection pressure fuel in the rail and rail-to-injector lines
- Electrohydraulic injectors (EHIs) permitting super-fast open and close events at the injector nozzle

CR SUBSYSTEMS AND COMPONENTS

A typical diesel CR system consists of the following key components:

- Fuel subsystem: Stores and supplies fuel to the high-pressure rail pump.
- High-pressure pump: An engine-driven pump capable of producing flow sufficient to produce injection pressures up to, and exceeding, 30,000 psi (2,000 bar). Either a radial piston or multicylinder inline piston pump is used.

- ECM-controlled **rail pressure control valve**: A linear proportioning solenoid with an integral spool valve is typically used. The linear proportioning solenoid is an ECM output. Its function is detailed a little later in this chapter.
- A V-Ref supplied **rail pressure sensor**: A variable capacitance electronic device that signals "actual" rail pressure to the ECM at any given moment of operation.
- CR: Stores fuel at injection pressures. Distribution lines or pipes connect the rail with EHIs. This can also be known as the high-pressure **accumulator** (Bosch).
- EHIs: These function as ECM-switched hydraulic valves that inject fuel directly into the engine cylinders at pressure values close to the rail pressure value.

CR Features

The reasons for the rapid and nearly universal acceptance of CR diesel fuel systems by engine OEMs are:

- Their ability to achieve high injection pressures entirely independent of engine speed.
- Their ability to control the emitted droplet size to optimize engine performance under almost any temperature, load, or speed condition.
- Their ability to switch EHIs at high speed. In current systems, there can be up to seven injection "events" during a single power stroke.

These features translate into much better management of fuel, providing improved fuel economy, reduced exhaust emissions, and improved response to a request for a change in engine output. CR diesel fuel systems also provide superior cold start performance and dramatically reduced, cold start emissions because of their capability of producing finely atomized fuel during cranking and cold idle conditions.

Engine Controller Acronym

Because diesel CR systems are used by the smallest to the largest diesel engines on our highways, some of the automobile diesel OEMs use acronyms such as ECU (electronic control unit) or PCM (powertrain control module) in preference to **ECM (engine control module)**. In addition, some of these fuel systems are manufactured offshore and, for whatever reasons, continue to use a foreign acronym on their product marketed here. For the sake of consistency when dealing with this system, we use the acronym ECM when referring to engine system controllers except when making reference to a specific OEM that prefers one of the other terms.

CR MANAGEMENT ELECTRONICS

CR fuel systems feature full authority electronic controls that really do not differ too much regardless of OEM. While attempting to be somewhat generic in our approach to diesel CR systems, we nevertheless reference the Bosch CR system used by Cummins, Paccar, and others, as well as making some references to the other OEM systems.

In all CR fuel systems, injected fuel quantity is defined by the engine management computer based on a range of input variables and output requirements that include accelerator pedal angle, operating temperatures, and emissions requirements. As in all full authority engine management systems, the ECM monitors ambient, chassis, and engine conditions and is capable of detecting both electronic and hydromechanical faults. Depending upon the severity of any faults identified, the ECM can strategize countermeasures such as torque limitation, power-derate, and limp-home operating modes. All the CR systems identified in this chapter are networked to the J1939 data bus and can therefore interface with other powertrain and chassis components. Some also interface with J1850 light duty data bus.

Input Circuit

The sensor circuit inputs data to the ECM by means of dedicated circuitry and most current systems are equipped with electronic signal transducers, electronic signal amplifiers, and fluid pressure transducers. The specific components used to input data to the ECM are no different from other engine management systems and are covered in Chapter 30. Both analog and digital signals are inputted to the processing cycle.

Analog Inputs

Mass airflow, engine fluid and intake-air temperatures, engine pressure values, diesel particulate filter (DPF) pyrometers, and battery voltage are examples of analog signals that have to be converted to digital values by an A/D converter in the ECM. When fluid pressure values are transduced to electrical values, these are signaled in an analog format but some systems transduce directly to digital.

Digital Inputs

Digital input signals include *on/off* switching signals and digital sensor signals such as rotational speed/position pulses from Hall-effect sensors and these bypass the A/D unit to be processed directly by the microprocessor. Depending on the manufacturer, the throttle position sensor (TPS) signal may use either an older style analog signature or produce a digital signal using one of two types of devices:

- Hall-effect (noncontact) TPS broadcasting pulse width modulated (PWM) signal
- Potentiometer with built-in integrated circuit (I/C) broadcasting PWM signal

Signal Conditioning

In many current systems where alternating current (AC) inductive pulse generators are used to measure shaft speeds (such as crank and camshaft speed/position sensors) the signal is transduced to a digital format to suppress interference pulses and other electronic noise. In engines designed to run at higher speeds (especially automotive diesels) the likelihood that Hall-effect shaft speed/position sensors are used increases: These are entirely digital devices. In most cases, shaft speed sensor inputs are delivered to the ECM in a square-wave form in most current systems.

This means that depending upon the level of integration, some signal conditioning can take place either completely or partially in the sensor. Protective circuitry is used to limit the incoming signals to a maximum voltage level. Inputted signals are freed of superimposed interference signals by filtering, and then amplified to match it to the ECM working voltage.

Processing Cycle

The ECMs used to manage CR fuel systems function using a selection of operational maps (algorithms) stored in memory that can be modified within certain limits by the OEM or by the technician. Engine-specific curves and engine-management maps are also written to ECM memory and include data such as failure strategies, calibration data, and power trim/derate programming. All OEM ECMs used on post-2007 products have a write-to-self capability that will log fault codes and audit trails to electronically erasable, programmable read-only memory (EEPROM). In addition, the ECM is networked to the J1939 data bus, which permits it to interface with other chassis electronic systems.

Fueling Algorithm

The key processing transaction that concerns engine fueling is the "closed" loop cycle between *desired* rail pressure and *actual* rail pressure. This can be summarized as follows:

- Desired rail pressure: Computed by the ECM based on sensor inputs, emissions monitoring, and stored instructions in memory. The ECM then attempts to maintain desired rail pressure by controlling the rail pressure control valve.
- Actual rail pressure: Measured by pressure sensor (a piezoresistive or variable capacitance type) and signaled to the ECM.

A major function of the ECM is to monitor and manage engine emissions so every fueling algorithm is mapped around the requirement to minimize emissions. The operation of emission controls is studied in Chapter 41: The CR systems used in post-2007 products use EGR, catalytic converters, and DPFs. For a more detailed look at vehicle engine management computers and how a logic processing cycle functions, refer to Chapter 30.

Output Circuit

The output driver system insofar as the fuel system management is concerned is not that complex. However, in keeping with most post-2007 engines, the driver apparatus for EGR and DPF systems tends to be a little more complicated: These are addressed in Chapter 41. The two ECM driver outputs that manage fueling are the rail management pressure control valve(s) and the injector drivers.

Rail Pressure Management Control (RPMC)

The rail pressure management control (RPMC) valve, as its name suggests, manages rail pressure. Though each OEM has its own terms to describe this valve and some use two of them (Caterpillar), they all perform the same function. It is necessarily located downstream from the high-pressure pump and upstream from the rail. RPMC valves are PWM to control its linear position: The valve is spring loaded to default to no-fuel status. Depending on the current flowed through its coil, an RPMC spool options fuel from the high-pressure pump to either the rail or the fuel return circuit. In order to precisely maintain desired rail pressure, it is typically capable of linear correction at high frequency (60 kHz).

Injector Drivers

ECM-located injector drivers are used to switch the electrohydraulic injectors. They function similarly to drivers used to switch other comparable EUI, electronic unit pump (EUP), or HEUI injectors. The ECM outputs a PWM control pulse spiked to around 100 V-DC. Current draw during actuation depends on whether solenoid or piezoelectric actuators are used, with the latter requiring less current draw.

Multiplexing

Communication between the CR ECM and the vehicle data bus uses serial CAN 2.0 protocols. However, these are not necessarily J1939 and depend on the weight classification of the vehicle. The CAN 2.0 version used on light-duty vehicles is SAE J1850 (light duty). We have said little about the J1850 data bus in this textbook because it relates exclusively to light-duty automotive applications, but we can best summarize its operation by stating that it is nearly identical to J1939. However, unlike J1939 it tends to be less accessible. This is because automotive OEMs do their utmost to keep anyone without their proprietary software off their data bus: In general, the only J1850 data that can be read is that falling under onboard diagnostics II (OBD II). OBD II relates specifically to emissions and must be "read-only" accessible using generic ESTs. For more detailed information on multiplexing, see Chapter 33.

CR FUEL ROUTING CIRCUIT

The CR fuel routing system is modular so for purposes of study, we can divide it up as follows:

- Fuel subsystem
- High-pressure pump
- Pressure accumulator or rail
- High-pressure distribution system
- Electrohydraulic injectors

Figure 40–3 shows the layout of the Caterpillar CR system used on its post-2007 C7 and C8 engines.

Fuel Subsystem

The fuel subsystem varies by OEM but the Bosch example shown in Figure 40–2 is typical. All CR systems use a transfer pump, which Bosch calls a presupply pump and may be located in the fuel tank. Caterpillar's transfer pump is a gear-type pump flange mounted to the high-pressure pump.

Fuel Tank

The fuel tank is an OEM responsibility but Bosch recommends that it be manufactured of noncorroding materials and designed to be free from leaks at double the normal operating pressure. Normal operating pressure according to Bosch is 2.5 psi (0.15 bar/17 kPa) so the required tank test pressures are 5 psi (0.3 bar/34 kPa). The tank should be properly vented (at the moment of writing, venting to atmosphere is permitted) and equipped with appropriate fittings for cycling fuel through the system.

Transfer Pump

The transfer or presupply pump may be located within the fuel tank or inline. At present, there are two possible versions:

- Electric roller-cell fuel pump located in the fuel tank. It functions similarly to the fuel supply pump in a gasoline-fueled automobile.

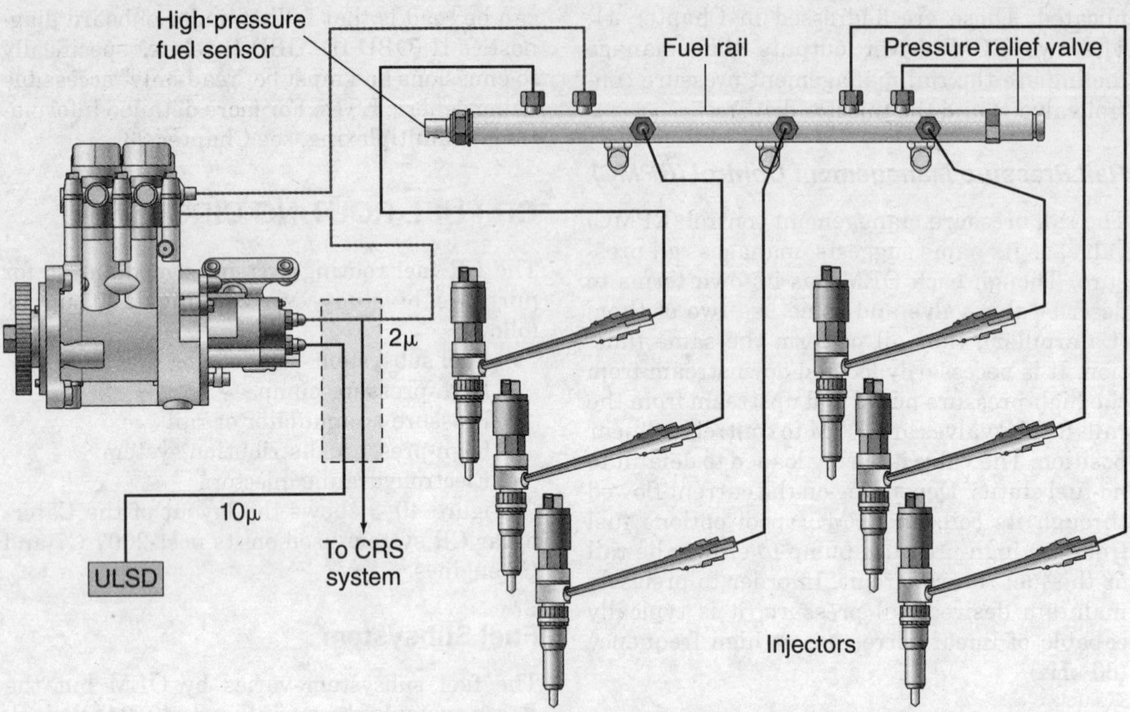

Figure 40–3 *Key components of the Caterpillar CR system used on C7 and C9 engines: The micron ratings shown are the fuel filtration requirements of the system. (Reprinted Courtesy of Caterpillar Inc.)*

- External gear pump mechanically driven by the engine. The gear pump is often located behind, and driven by, the CR high-pressure pump.

In both cases, the fuel transfer pump is positive displacement and functions to push fuel through the fuel subsystem to deliver it to the high-pressure stage.

Fuel Filter(s)

The fuel filter is an OEM responsibility but it should meet the specifications of the CR fuel system manufacturer. In a Bosch CR application, the fuel filter should have a nominal entrapment capability of 8 microns. Caterpillar requires the use of its high-efficiency secondary filter with a nominal entrapment capability of 2 microns. All fuel filtration must occur upstream from the high-pressure pump. Similar to other injection systems, CR fuel systems should use either a fuel filter with a water separator or integrate a separate water separator into the fuel subsystem. Collected water should be drained at regular intervals in systems equipped only with manual drains. Some water separators are equipped with a water-in-fuel (WIF) sensor that triggers a warning lamp, indicating that water should be drained.

High-Pressure Pumps

There are some differences between the type of high-pressure pump used by CR manufacturers so we first describe the radial piston pumps used by Bosch and Denso, and follow with a description of the inline piston pumps used in Caterpillar systems. The high-pressure pumps used in a CR diesel fuel system are simple in terms of the role they play. All they are required to do is create sufficiently high flow volume so that when subject to flow restriction, they achieve the specified window of rail pressures. Pump output is unloaded to the rail. The high-pressure pump plays no role in metering and timing of the fuel delivery. Figure 40–4 shows all the high-pressure stage components in a typical CR diesel fuel injection system.

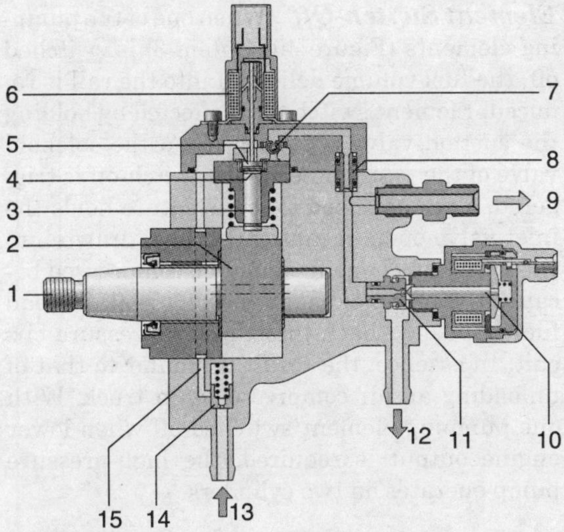

1. Driveshaft
2. Eccentric cam
3. Pumping element with pump plunger
4. Pumping element chamber
5. Suction valve
6. Element shutoff valve
7. Outlet valve
8. Seal
9. High-pressure connection to the rail
10. Pressure control valve
11. Ball valve
12. Fuel return
13. Fuel inlet from the presupply pump
14. Safety valve with throttle bore
15. Low-pressure passage to the pumping element

Figure 40–4 *Radial piston, high-pressure pump (schematic, longitudinal section). (Courtesy of Robert Bosch LLC)*

Radial Piston Pumps

Bosch and Denso use a three-cylinder, **radial piston pump** to produce the injection pressures required for system operation. They are responsible for developing rail pressures. Pressurized fuel from the high-pressure pump is unloaded by means of a high-pressure line into a tubular, high-pressure fuel accumulator that we call the rail. The high-pressure pump continually attempts to maintain the ECM-desired pressure to the rail, with the result that in contrast to conventional systems, fuel does not have to be specially pressurized for each individual injection pulse. The advantage of using a radial piston pump is its compact design.

The high-pressure pump is installed on a diesel engine so that it can be driven by the engine timing geartrain. It may be driven at either camshaft or crankshaft speed, depending on the system, but camshaft speed is more common. The high-pressure pump is usually drive coupled to the timing geartrain using a dedicated gear, but both chain and toothed belt drives may be used in light-duty applications. The pump is internally lubricated by the diesel fuel it pumps. Figure 40–5 shows a schematic cross-section of a radial piston, high-pressure pump: Note how the pumping elements are actuated by the central camshaft.

Radial Piston Pump Components

Pressure rise is created in the high-pressure pump by three radially arranged piston-type, pump elements that are evenly offset at an angle of 120 degrees around the pump drive shaft. This means that each pump element is actuated once during a single pump rotation, meaning that the stress on the pump drive remains uniform. Each pump piston is actuated by eccentric cams machined to the drive shaft (see Figure 40–5) and discharged through an outlet valve.

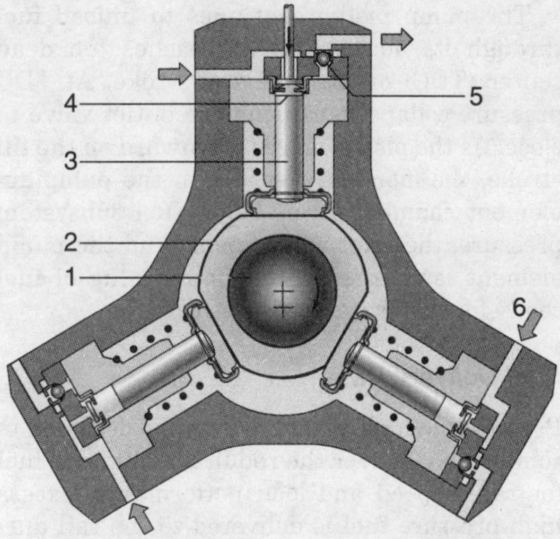

1. Driveshaft
2. Eccentric cam
3. Pumping element with pump piston
4. Inlet valve
5. Outlet valve
6. Inlet

Figure 40–5 *Radial piston, high-pressure pump (schematic, cross-section). (Courtesy of Robert Bosch LLC)*

Operating Principle

Fuel forced through the fuel subsystem is delivered to each pump element through the fuel inlet (see Figure 40–4, item 13) and the safety valve. Fuel passes through the safety valve throttle bore (14) and enters the high-pressure pump lubrication and cooling circuit, charging the pump elements. As the drive shaft (1) machined with its eccentric cams (2) rotates, it sequentially actuates the three pump plungers (3). These, therefore, reciprocate in their bores because they ride the cam profiles.

Any time delivery pressure (created in the fuel subsystem) exceeds the safety valve opening pressure, usually from 5 to 25 psi (0.5 to 1.5 bar), the pumping-element chamber (4) is charged with fuel as the pump piston moves downward on its suction stroke. The pump element inlet valve closes when the piston passes through bottom dead center (BDC), and at this moment fuel is trapped in the pump chamber. As the piston is driven into the pump chamber pressure rise is created and when pump chamber pressure exceeds the rail pressure, the outlet valve (7) opens and compressed fuel is charged to the high-pressure circuit.

The pump piston continues to unload fuel through its outlet until it reaches top dead center (TDC) on the delivery stroke. At TDC pressure collapses, causing the outlet valve to close. As the piston moves downward on the fill stroke, the moment pressure in the pumping-element chamber drops below fuel subsystem pressure, the inlet valve opens, and the pump element is charged with another slug of fuel ready for the next effective stroke.

Fuel Delivery Rate

Because the high-pressure pump is designed to comfortably deliver the required volume of fuel for rated speed and load performance, excess high-pressure fuel is delivered to the rail during idle and low-load operation. This excess fuel is returned to the tank by means of the rail pressure control valve, which routes the fuel back to the tank. This results in wasted energy because of the mechanical effort required to actuate the pumping elements. Some of this parasitic loss can be recovered by switching off one of the pumping elements.

Element Switch-Off When one of the pumping elements (Figure 40–4, item 3) is switched off, the fuel volume delivered into the rail is reduced. Element switch-off is effected by holding the suction valve (5) open. When the solenoid valve of the pumping-element switch-off is triggered, a pin attached to its armature holds the inlet valve open, meaning that fuel drawn into the pumping element cannot be compressed because the pump element does not seal. Instead fuel is pushed back into the low-pressure circuit: In essence, the result is similar to that of unloading an air compressor on a truck. With one pumping element switched off when lower engine output is required, the high-pressure pump operates on two cylinders.

Inline Piston Pumps

Caterpillar uses a two-cylinder inline piston pump to produce rail pressures in its post-2007 C7 and C9 engines. The pump is flange mounted and direct driven by an accessory drive gear. The pump is lubricated with engine oil. Fuel from the secondary filter is routed to an inlet on the upper outboard face of the pump. This fuel is then routed to the two high-pressure fuel pump elements. The transfer pump that moves fuel through the fuel subsystem is a gear-type pump coupled directly to the high-pressure pump. This low-pressure pump also supplies the fuel for dosing the Cat Regeneration System (CRS). Figure 40–6 shows and external view of the Caterpillar CR system high-pressure pump: Note how the transfer pump is coupled to, and driven by, the high-pressure pump.

Rail Pressure Control Valves

The rail pressure control valve (RPCV) is PWM actuated by the ECM. The ECM uses it to define the "desired" pressure in the rail at any given moment of operation. As discussed earlier, it is a linear proportioning solenoid and spool valve that can option fuel unloaded by the high-pressure pump either to the rail or into the return circuit. Its operation can be summarized as follows:

- If *actual* rail pressure (signaled to the ECM by the rail pressure sensor) is higher than *desired* rail pressure, the RPCV opens to

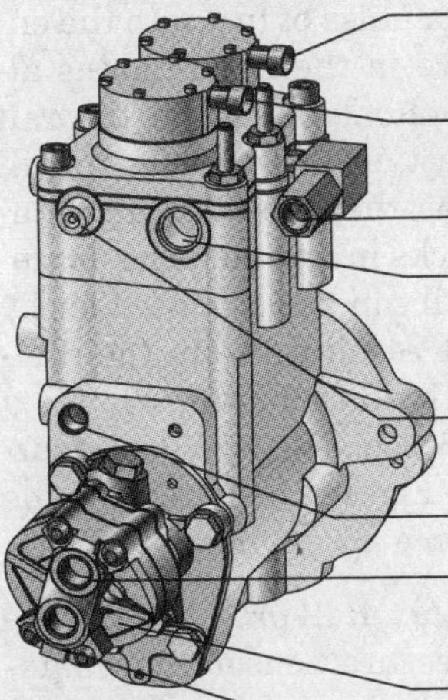

Figure 40–6 *Caterpillar CR system high-pressure pump. (Reprinted Courtesy of Caterpillar Inc.)*

divert fuel from the rail, optioning it to the return circuit.

- If *actual* rail pressure is lower than *desired* rail pressure, the pressure control valve closes, sealing the rail and permitting pressure rise.

A sectional view of the ECM-controlled rail pressure control valve is shown in Figure 40–7.

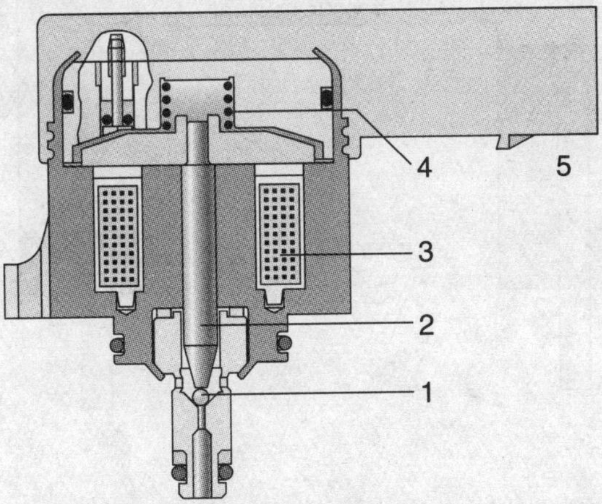

1. Valve ball
2. Armature
3. Electromagnet
4. Spring
5. Electrical connection

Figure 40–7 *Bosch type rail pressure control valve. (Courtesy of Robert Bosch LLC)*

The Bosch rail pressure control valve is located on the high-pressure pump (see Figure 40–5, item 10) and is flange mounted to it. Some systems use a pressure control valve located at the rail inlet. To seal the high-pressure rail from the return circuit, the control valve armature forces a ball into a seat to create a seal. Two forces act on the armature. Mechanical force is provided by a spring, and opposing this, electromagnetic force is created by the solenoid coil when energized. Two control loops are used to manage pressure control valve operation:

- A slow-response electrical control loop for setting (variable) mean pressure in the rail
- A fast-response mechanical control loop to compensate for the high-frequency pressure fluctuations

Pressure Control Valve Nonenergized When desired rail pressure is higher than actual rail pressure, the rail pressure control valve must drop rail pressure. High pressure at the rail or at the high-pressure pump outlet acts on the pressure control valve via the high-pressure input. Because the nonenergized electromagnet in the control valve exerts no force, high-pressure fuel exceeds the spring force, opening the control valve and spilling rail fuel to the return circuit.

Pressure Control Valve Energized When desired rail pressure is lower than actual rail pressure, the pressure control valve must allow rail pressure to rise. If the pressure in the high-pressure circuit is to be increased, the force of the electromagnet must combine with the mechanical force of the spring. The ECM energizes the pressure control valve, causing it to close and remain closed until equilibrium is established between desired and actual rail pressures. This means that a balance is reached between the high-pressure fuel forces on the one side and the combined forces of the spring and the electromagnet on the other. The valve then remains open and maintains rail pressure constant. Any change in the pump delivery quantity/engine load is compensated for by the valve assuming a different setting.

Caterpillar Rail Pressure Controls

Rail pressure control in the Caterpillar high-pressure pump is achieved by a pair of solenoids located above the pump elements. The operating principle is simple. Depending on the solenoid status, fuel is optioned either to charge the rail or routed to the low-pressure fuel return circuit. Figure 40–8 shows how the pump elements work in conjunction with the Caterpillar high-pressure pump.

Common Rail

The CR or accumulator receives fuel from the high-pressure pump, and by means of dedicated lines makes it available to the ECM-controlled, electrohydraulic injectors. The accumulator feature of the rail means that even after an injector has discharged a pulse of fuel into the engine, the fuel pressure in the rail remains almost constant. The volume of fuel in the rail has a dampening effect on the changes in rail pressure that occur as the injectors are actuated and the RPCV kicks in and out. The dampening ability of the rail can be accounted for by its accumulator effect resulting from the compressibility factor of the fuel. Figure 40–9 is a line drawing of the Bosch CR used on a four-cylinder engine, and Figure 40–10 shows the Caterpillar CR used on a C7 engine.

Rail Pressure Limits Rail pressure is monitored by the rail pressure sensor and main-

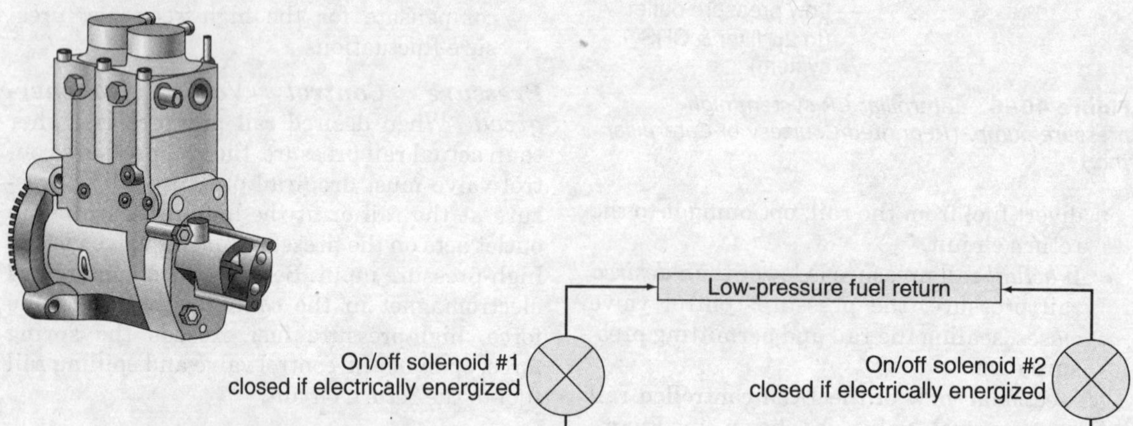

Figure 40–8 *Sectioned view of a Caterpillar CR high-pressure pump and the solenoids that act as rail pressure control valves. (Reprinted Courtesy of Caterpillar Inc.)*

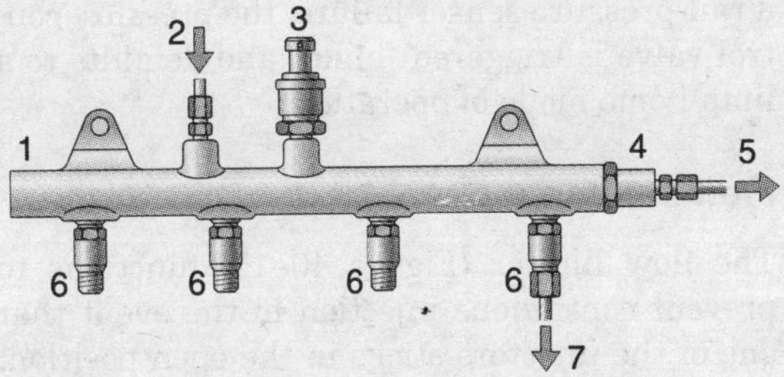

1. Rail
2. Inlet from the high-pressure pump
3. Rail pressure sensor
4. Pressure limiter valve
5. Return from the rail to the fuel tank
6. Flow limiter
7. Line to the injector

Figure 40–9 *Bosch common rail. (Courtesy of Robert Bosch LLC)*

tained at the desired value by the ECM pressure control valve. A pressure limiter valve acts to limit the maximum fuel pressure in the rail to a specification that usually exceeds the intended peak rail pressure by a small margin. For instance, in the current Caterpillar CR system the highest specified rail pressure is 27,550 psi (1,900 bar) so the pressure limiter valve is designed to trip at 33,300 psi (2,300 bar). Fuel in the rail is made available to the injectors by means of flow limiters, which prevent excess fuel from being injected. The rail and its critical components on a Bosch CR system are shown in Figure 40–9.

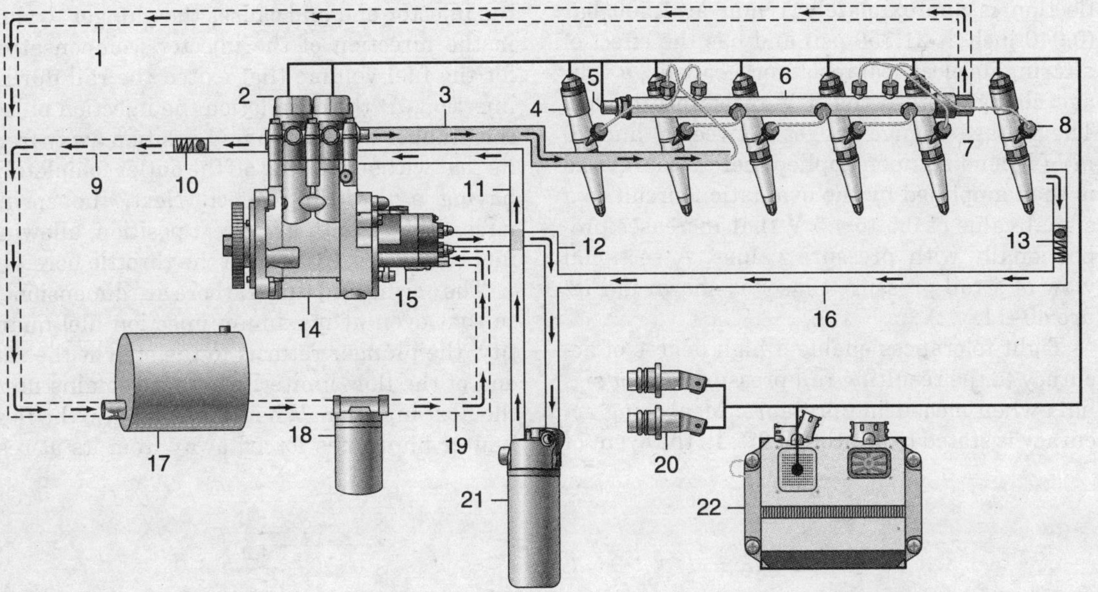

FUEL SYSTEM DIAGRAM

1. Return line from pressure relief valve to tank
2. Solenoid for the fuel pump
3. High pressure supply line
4. Electrohydraulic injector (EHI)
5. Fuel pressure sensor
6. Fuel rail
7. Pressure relief valve for the fuel rail
8. Quill tube
9. Return line from fuel pump to tank
10. Pressure relief valve for the fuel pump
11. Fuel line from secondary filter to fuel pump
12. Fuel line from transfer pump to fuel filter
13. Pressure regulator for the drain line
14. Fuel pump
15. Transfer pump
16. Fuel line for the drain back to filter
17. Fuel tank
18. Primary fuel filter
19. Fuel line from the primary filter to the transfer pump
20. Speed/timing sensor
21. Secondary fuel filter
22. Engine control module (ECM)

Figure 40–10 *Common rail system layout on a Caterpillar C7 engine. (Reprinted Courtesy of Caterpillar Inc.)*

Rail Pressure Sensor

The rail pressure sensor consists of a sensor housing, integral printed circuit, and sensor element. A rail pressure sensor can use either a piezoresistive, Wheatstone bridge operating principle (most common) or a variable capacitance-type sensor: Both these types of sensors are usually supplied with a 5-volt V-Ref and are fully explained in Chapter 30. Here we review the more commonly used piezoresistive sensor. Pressurized fuel acts on the sensor diaphragm through a blind hole and the sensor element (semiconductor device) is mounted on this diaphragm. When the diaphragm shape changes as a result of pressure acting on it, the electrical resistance of the layers attached to the diaphragm also changes. Sensing diaphragm deflection is approximately 1 mm at 1,500 bar (0.040 inch at 21,750 psi) and has the effect of altering its electrical resistance, causing a voltage change across the 5 V resistance bridge. The voltage change ranges between 0 and 70 mV (depending upon applied fuel pressure) and is then amplified by the evaluation circuit to a signal value of 0.5 to 4.5 V that increases proportionally with pressure values. A sectional view of a rail pressure sensor is shown in Figure 40–11.

Tight tolerances enable a high degree of accuracy to the resulting rail pressure sensor signals when measuring pressure. Measuring accuracy is stated to be around 2%. In the event of a rail-pressure sensor failure, the pressure control valve is triggered "blind" and defaults to a limp-home mode of operation.

Flow Limiter

The flow limiter (Figure 40–12) functions to prevent continuous injection in the event that one of the injectors sticks in the open position. Once the fuel quantity leaving the rail exceeds a predetermined volume, the flow limiter shuts off the line to the problem injector. The flow limiter consists of a metal housing with external threads for installation into the rail at one end and external threads at the other for connecting the injector lines. A plunger inside the flow limiter is forced toward the rail end by a spring. When fuel is injected, pressure drops at the injector end and causes the plunger to shift in the direction of the injector, compensating for the fuel volume that exited the rail during injection. At the completion the injection pulse, the plunger moves to a mid-position away from its seat without closing off the outlet completely, having a throttling effect. Next, the spring forces it back onto its at-rest position, allowing fuel to flow freely through the throttle bore.

The spring and throttle bore are dimensioned so that even at maximum injection fuel quantity, the plunger returns to the stop at the rail end of the flow limiter, where it remains until the next injection. When a leak occurs, the flow limiter plunger is forced away from its at-rest

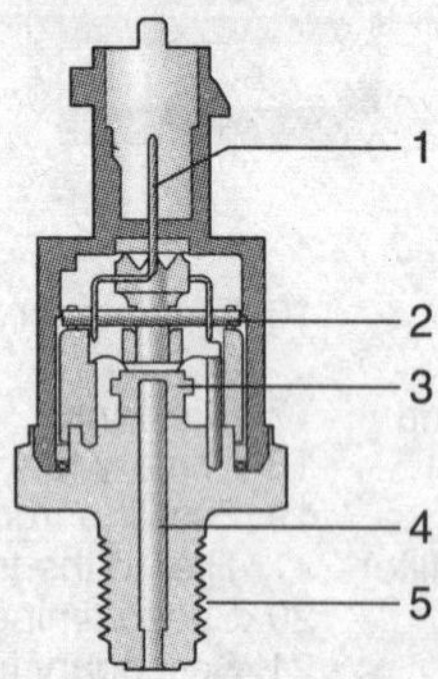

1. Electric connections
2. Evaluation circuit
3. Metal diaphragm with sensor element
4. High-pressure connection
5. Mounting thread

Figure 40–11 *Rail pressure sensor. (Courtesy of Robert Bosch LLC)*

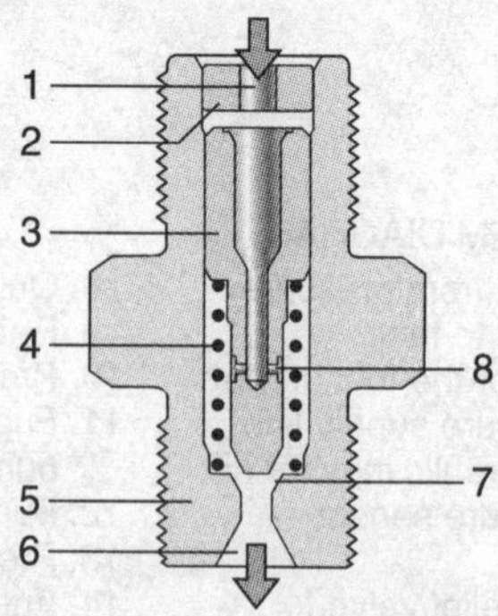

1. Connection to the rail
2. Sealing washer
3. Plunger
4. Spring
5. Housing
6. Connection to the injector
7. Seat
8. Throttle

Figure 40–12 *Flow limiter. (Courtesy of Robert Bosch LLC)* position up against the seal seat at the outlet. It remains in this position up against the stop at the injector end of the flow limiter and prevents fuel delivery to the injector.

High-Pressure Fuel Lines

The fuel lines or pipes used in CR systems have requirements similar to the high-pressure pipes used in other high-pressure diesel fuel injection systems. They must be capable of sustaining the maximum system pressures with a wide margin of safety and the pressure wave reflections that result from high-speed hydraulic switching. In common with the high-pressure pipes used with older pump-line-nozzle (PLN) systems, the injection lines tend to be of identical length and internal diameter. Should this not be the case, however, there would be no consequences that could influence injection timing. The lines used with Bosch CR systems have an outside diameter of 6 mm and an internal diameter of 2.4 mm.

Caterpillar High-Pressure Lines

The high-pressure lines used in CR diesel fuel systems may directly connect the rail with the injector or, in the case of Caterpillar applications, use a high-pressure pipe that connects with a quill tube. Caterpillar EHIs are cylindrical in shape and vertically positioned in the cylinder head. Rail fuel is routed into the EHI by machining a recess into the side of the injector: A nipple on the quill tube fits into and seals the inlet recess. You can see how this works by studying Figure 40–13 and Figure 40–14. Torque values at the quill tube nut are critical and must be observed. Figure 40–15 shows the rail, the high-pressure pipe, and the quill tube both disassembled and assembled.

Electrohydraulic Injectors

Electrohydraulic injectors (EHIs) were first introduced in Chapter 19. Two general types of EHIs are used, differentiated by the type of actuator used. First-generation EHIs used a solenoid control valve, while more recent types use piezoelectric actuators. However, the general principles used by each type remain identical; only the actuator mechanism differentiates

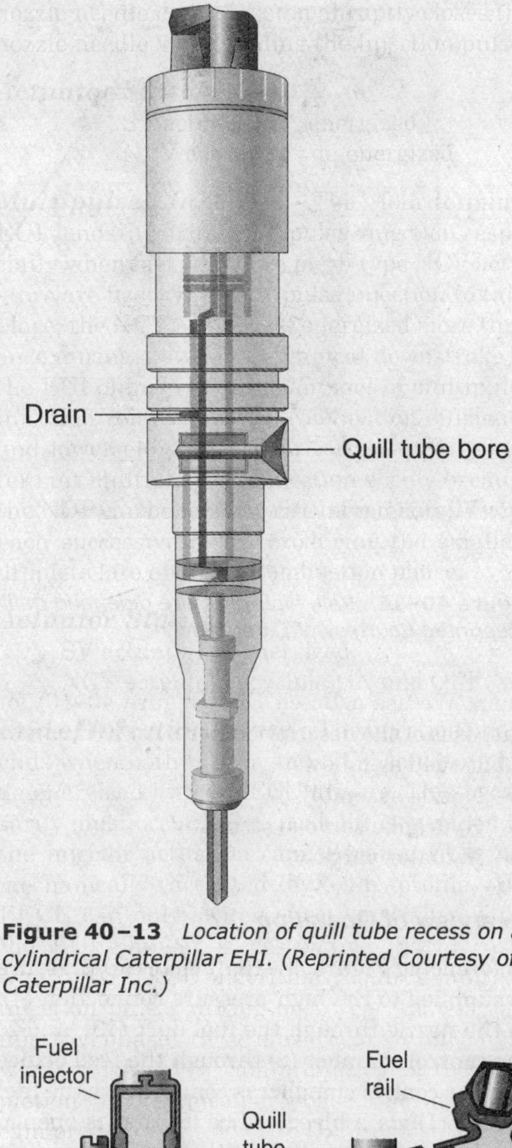

Figure 40–13 *Location of quill tube recess on a cylindrical Caterpillar EHI. (Reprinted Courtesy of Caterpillar Inc.)*

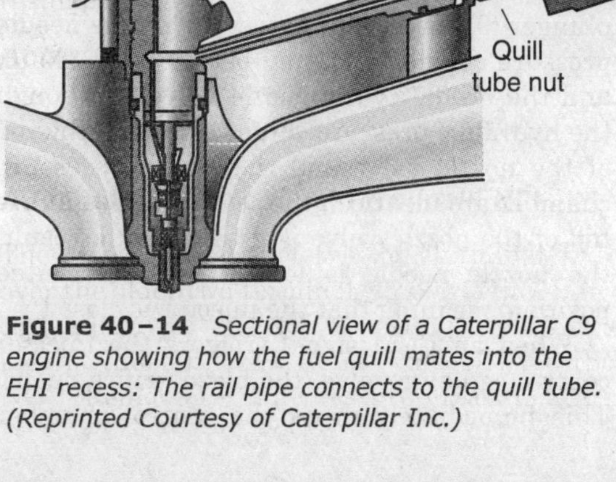

Figure 40–14 *Sectional view of a Caterpillar C9 engine showing how the fuel quill mates into the EHI recess: The rail pipe connects to the quill tube. (Reprinted Courtesy of Caterpillar Inc.)*

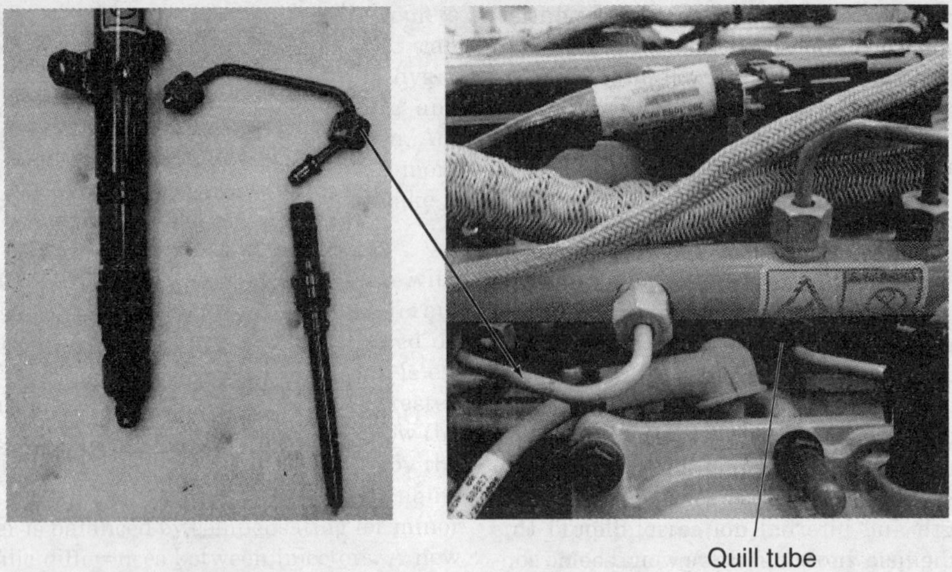

Figure 40–15 *Rail, high-pressure pipe, and quill tube disassembled and assembled on the engine. (Reprinted Courtesy of Caterpillar Inc.)*

them. We use a Bosch EHI (Figure 40–16) for purposes of the description here. An EHI can be subdivided as follows:

- Nozzle assembly
- Hydraulic servo-system
- Actuator valve

Overview of Operation

Referencing Figure 40–16, fuel at rail pressure is supplied to the high-pressure connection (4), to the nozzle through the fuel duct (10), and to the control chamber (8) through the feed orifice (7). The control chamber is connected to the fuel return (1) via a bleed orifice (6) that is opened by the actuator valve. With the bleed orifice closed, hydraulic force acts on the valve control plunger (9) exceeding that at the nozzle needle pressure chamber (located between the shank and the needle of the nozzle valve). Although the hydraulic pressure values acting on the top of the nozzle valve and that in the pressure chamber are identical, the sectional area at the top of the nozzle valve is greater. As a result, the nozzle needle is loaded into its seated position, meaning that the injector is closed.

When an ECM signal triggers the injector control actuator valve, the bleed orifice opens. This immediately drops the control chamber pressure and, as a result, the hydraulic pressure acting on the top of the nozzle valve (11) also drops. When hydraulic force acting on top of the nozzle valve drops below the force on the nozzle needle pressure shoulder, the nozzle valve retracts and allows fuel to pass around the seat to be injected through orifii into the engine cylinder. The hydraulic assist and amplification factor are required in this system because the forces necessary for rapid nozzle valve opening cannot be directly generated by an actuator valve alone. Fuel used as hydraulic media to open the nozzle valve is in addition to the injected fuel quantity, so this excess fuel is routed back to the tank. In addition to this fuel, some leak-by fuel losses occur at the nozzle valve-to-body clearance and the valve-plunger guides clearance. This leak-off fuel volume is also returned to the fuel tank via the fuel return circuit.

Injector Operating Phases

When the engine is stopped, all injector nozzles are closed, meaning that their nozzle valves are loaded onto their seats by spring pressure. In a running engine, injector operation takes place in three phases.

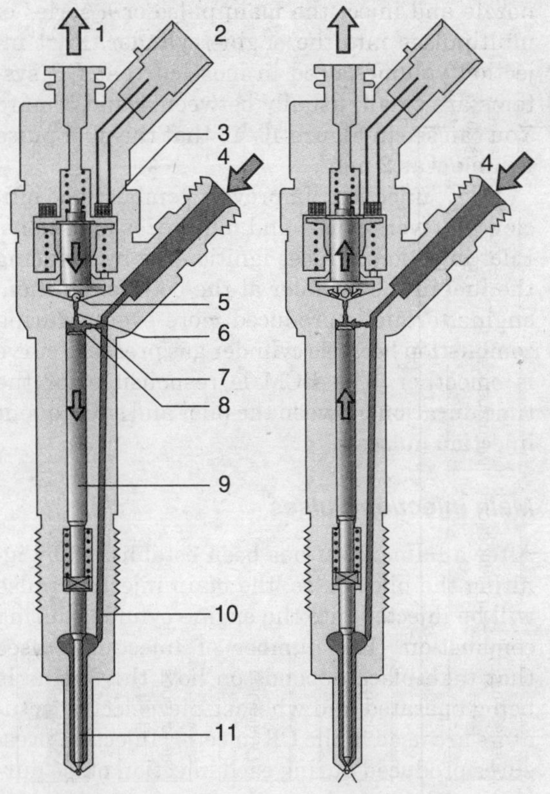

a. Injector closed (at-rest status)
b. Injector opened (injection)
1. Fuel return
2. Electrical connection
3. Triggering element (solenoid valve)
4. Fuel inlet (high pressure) from the rail
5. Valve ball
6. Bleed orifice
7. Feed orifice
8. Valve control chamber
9. Valve control plunger
10. Feed passage to the nozzle
11. Nozzle needle

Figure 40–16 *EHI. (Courtesy of Robert Bosch LLC)*

Injector Closed In the at-rest state, the actuator valve is not energized and, therefore, the nozzle valve is loaded onto its seat by the injector spring combined with hydraulic pressure (from the rail) acting on the sectional area of the valve control plunger. With the bleed orifice closed, the actuator valve spring forces the armature ball check onto the bleed orifice seat. Rail pressure builds in the injector control chamber, but identical pressure will be present in the nozzle pressure chamber. Given equal pressure acting on the larger sectional area of the nozzle control plunger (which is mechanically connected to the nozzle valve) and in the nozzle pressure chamber, this and the force of the nozzle spring combine to load the nozzle valve on its seat, holding the injector closed.

Nozzle Opening The injector actuator valve is PWM energized by the ECM injector driver. The actuation voltage varies by OEM but typically is spiked at around 100 V-DC. The current draw during injector energization depends on whether a solenoid or piezoelectric actuator is used: Current draw is lower when piezoelectric actuators are used. Force exerted by the triggered actuator now exceeds that of the valve spring and the control valve opens the bleed orifice. Almost instantly, the high-level pick-up current to the actuator drops off to a lower hold-in current flow. As the bleed orifice opens, fuel flows from the valve control chamber into the cavity above it and out to the return circuit. This collapses the hydraulic pressure acting on the valve control plunger that was helping to hold the nozzle valve closed. Now, pressure in the valve control chamber is much lower than that in the nozzle pressure chamber that is maintained at the rail pressure. The result is a collapsing of the force that was holding the nozzle valve closed, causing the nozzle valve to open and beginning the injection pulse.

The nozzle needle opening velocity is determined by the difference in the flow rate through the bleed and feed orifices. When the control plunger reaches its upper stop it is cushioned by fuel generated by flow between the bleed and feed orifices. When the injector nozzle valve has fully opened, fuel is injected into the combustion chamber at a pressure very close to that in the fuel rail.

Nozzle Closing When the actuator valve is de-energized by the ECM, its spring forces the control valve downward and the check ball closes the bleed orifice. The closing of the bleed orifice creates pressure buildup in the control chamber via the input from the feed orifice. This pressure should be the same as that in the rail, and now it exerts an increased force on the nozzle valve control plunger through its end face. This force, combined with that of the nozzle spring, exceeds the hydraulic force acting on the nozzle valve sectional area and the nozzle valve closes, ending injection. Nozzle valve closing velocity is determined by the flow through the feed orifice. The injection pulse ceases the instant the nozzle valve seats.

Nozzle Hole Geometry

CR EHIs use both sac-chamber nozzles and what are generally referred to as valve-closes-orifice (VCO) nozzles. The latter tend to be required in engines required to meet North American emissions standards. Most CR injectors use a 4 mm nozzle valve diameter. Both sac-hole and seat-hole nozzles have the edges of each orifice rounded by hydroerosive (HE) machining. HE machining helps prevent edge wear caused by the abrasive particles in the fuel and carbon coking that can reduce flow and distort the spray geometry.

Phases of Injection

Solenoid-actuated CR fuel injection systems are capable of breaking up an injection pulse in two or three shots, while piezoelectric-actuated systems can manage up to seven shots per cycle. Multipulse injections require high-speed actuator switching, and this is better achieved by the more recently introduced piezoelectric EHIs. Start of injection (base timing) is plotted by the ECM using the running conditions of the chassis and engine factored with programmed fuel maps. This will result in the pilot pulse. The pilot pulse will then be followed by main pulses: You can see what this looks like in Figure 40–17, which graphically represents the events of a five-phase multipulse injection cycle.

Pilot Injection

Pilot injection has been used by diesel engine OEMs for a generation in other fuel systems, and CR technology has adopted it because it is ideally suited to multiple pulse injection. Pilot injection requires a small shot of fuel to be injected into the engine cylinder to initiate combustion. The injector closes after injecting this initial pulse of fuel. The ECM then calculates the point at which the pilot shot of fuel ignites using sensor data such as ambient temperature, boost pressure, and engine temperature. At the moment the pilot charge ignites (combustion begins), the ECM can then reopen the nozzle and inject the main pulse or a series of multipulses into the engine cylinder. Pilot injection volumes used in most current CR systems are small, usually between 1 and 5 mm^3: You can see in Figure 40–17 that the pilot pulse volume was 2 mm^3.

Pilot injection improves combustion efficiency, lowers noise, and minimizes emissions. Pilot injection reduces ignition lag by reducing the fuel in the cylinder at the point of ignition; engine torque is produced more evenly during combustion because cylinder gas pressure curve is smoother. The ECM is responsible for the time duration between the pilot and subsequent injection pulses.

Main Injection Pulses

After a flame front has been established by igniting the pilot pulse, the main injection pulse will be injected into the engine cylinder during combustion. The number of injection pulses that take place depends on how the engine is being operated and whether piezoelectric actuators are used in the CR injector. Injection pressures produced during each injection pulse during multipulse injection events remains pretty stable during a single cycle regardless of duration or volume. These pressures directly depend on the rail pressure that is being managed by the ECM at any given moment of operation. The action of opening and closing injectors during multipulse events causes little deviation in actual rail pressures. This noticeably contrasts with comparable systems that use older hydraulic injector nozzles that open and close at fixed hard values much lower than typical CR rail pressures. Main injection volumes used in most current CR systems tend to maximize at around 8 mm^3 but it tends to be less in newer, multipulse systems.

Dosing or Afterburn Injection

Some OEMs use either the term *afterburn injection* or *dosing injection* to describe the injection of fuel into the engine cylinder that is not intended to be combusted in the cylinder. Afterburn injection refers to injection of fuel into the engine cylinder after the main pulses, so certain types of reduction-type catalytic converters (NO_x reduction) can be temporarily provided

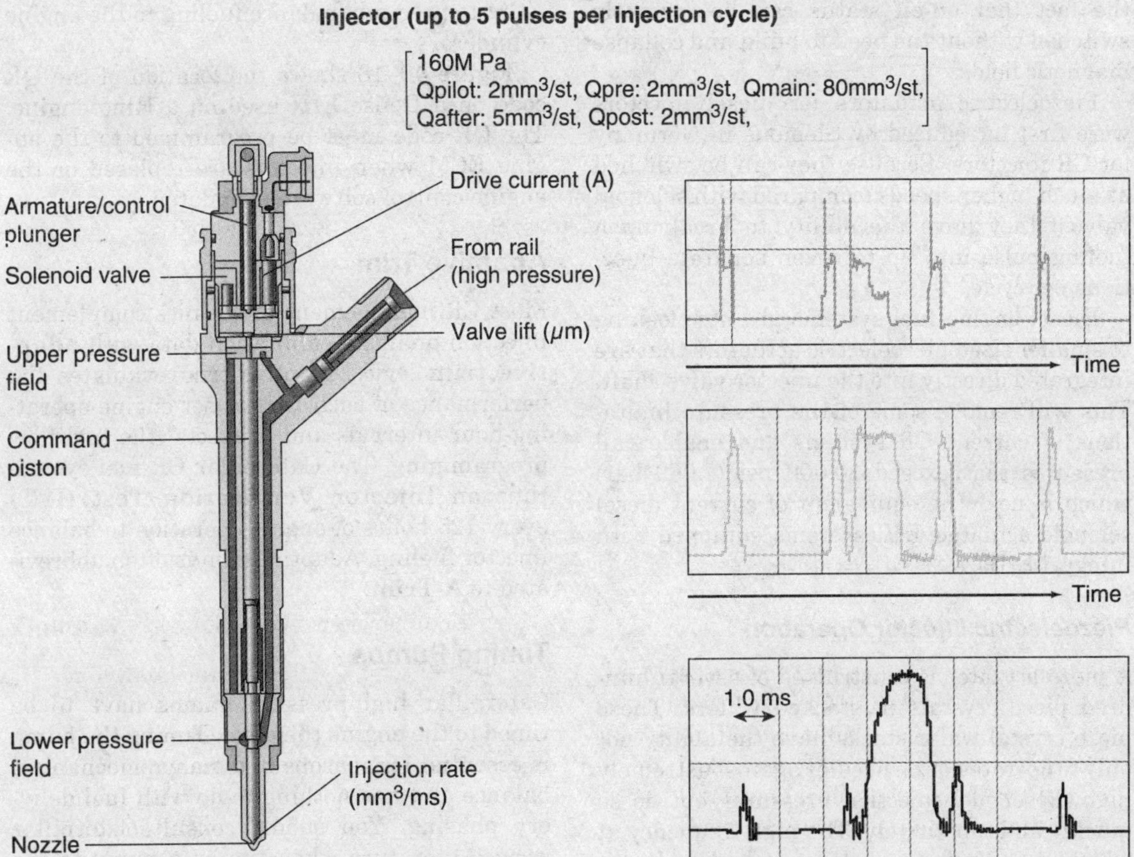

Figure 40–17 *A Denso EHI five-phase multipulse injection cycle represented graphically: Note that the rail pressure is 1,600 bar (23,225 psi). (Courtesy of DENSO Corporation)*

with a richer air-fuel ratio. In contrast to the pilot and main injection processes, afterburn phase injected fuel is not intended to be combusted but instead vaporizes due to the residual heat in the exhaust gas. When end gas is expelled from the cylinder, a mixture of exhaust gas and raw fuel exits through the exhaust valves and into the exhaust gas system. Some of this mixture may be routed back into the engine cylinder via the exhaust gas recirculation (EGR) system: This in itself can have the same effect as a very advanced pilot injection pulse. When rhodium base NO_x catalytic converters are fitted, afterburn phase fuel uses the unburned fuel in the exhaust gas as a reduction agent to lower the NO_x content in the exhaust gas.

Piezoelectric EHIs

Piezoelectric-actuated CR injectors are becoming common as the need for rapid response, multipulse injection systems increases. EHI nozzles were studied in Chapter 19 and the subject of piezoelectricity was examined in Chapter 28 so just a brief summary explanation is provided here. Piezoelectric actuators are based on reversibility of piezoelectric effect. When piezo crystals are subjected to voltage, their crystal bonds change almost instantly, resulting in either expansion or contraction. This change of shape produces a powerful force that makes it possible for them to be used as actuators. As such they function much like very fast response solenoids. The fast response is due to the fact that on-off status can be instantly switched without the need to build and collapse magnetic fields.

Piezoelectric actuators for diesel injectors were first introduced by Siemens in Germany for CR injectors. Because they can be switched at much higher speeds (compared with solenoid valves) they have the ability to break up a fueling pulse into up to seven separate injections per cycle.

Diesel engine fuel systems are now looking to smaller sized piezoelectric actuators that are integrated directly into the injector valve shaft. This will remove some of the pressure limitations of current CR systems and enable rail pressures that exceed 30,000 psi (2,069 bar) which is near the equivalent of current diesel solenoid-actuated EUI systems equipped with integral EHIs.

Piezoelectric Injector Operation

A piezo actuator is constructed of several hundred piezo crystals in stacked wafers. These piezo crystal wafer stacks have the ability not only to operate nozzle needles against extremely high diesel fuel injection pressures but do so quickly and repetitively. The high frequency at which the injectors can be switched makes it possible to reduce the intervals between each injection pulse and divide the fuel injected in one cycle into up to seven separate injections. Piezoelectric actuator technology has been a key to improving fuel economy, emissions, and noise levels in the latest generation of diesel engines.

CR Injector Calibration Programming

Most CR fuel systems require their ECMs to be programmed with a fuel flow code specific to each EHI. The fuel flow code is the **injection quantity calibration data**. The injection quantity calibration data is a specification determined when the EHI is flow tested on a calibration bench: This specification is then printed on the injector for programming. Caterpillar uses the term **E-trim**, and Denso the term **quick response (QR) code** to describe injector flow rates. Injection quantity calibration data or E-trim tells the ECM exactly how a specific injector flows fuel, and that allows the ECM to precisely balance fueling to the engine cylinders.

Figure 40–18 shows the location of the QR code on a Denso EHI used on a Hino engine. The QR code must be programmed to the engine ECM when injectors are replaced or the engine control software is updated.

Adaptive Trim

Most CR management electronics complement injection quantity calibration data with **adaptive trim** logic. Adaptive trim evaluates the performance of each EHI at set engine-operating-hour intervals and "corrects" the fuel flow programming. The Caterpillar CR fuel system runs an **Injector Verification Test (IVT)** every 125 hours of engine operation to balance injector fueling. Adaptive trim is often abbreviated to **A-Trim**.

Timing Pumps

Caterpillar high-pressure pumps have to be timed to the engine they fuel. Timing the pump is essential for reasons of dynamic mechanical balance and has nothing to do with fuel delivery phasing. You should consult Caterpillar service literature when timing a pump to the engine. Caterpillar requires technicians to use a variation of the timing pin method similar to that used on its new scroll port-helix metering pumps: The timing pin locks the pump into position so it can be mated up in complementary balance with the engine. Figure 40–19 shows a timing pin locked into position in the pump.

EHI Admission Valves

A typical EHI has the disadvantage of having the rail pressure fuel present at the nozzle valve seat constantly. In the event of a completely stuck open nozzle valve, a flow limiter would shut off the fuel supply to the EHI. However, because rail pressures approach 30,000 psi (2,000 bar) on more recent systems, if the nozzle valve was to remain in a partially open position insufficient to engage the flow limiter, the result would be fuel dribble to the engine cylinder. This type leakage at the injector seat usually results in costly engine damage. One way around this problem is to incorporate an admis-

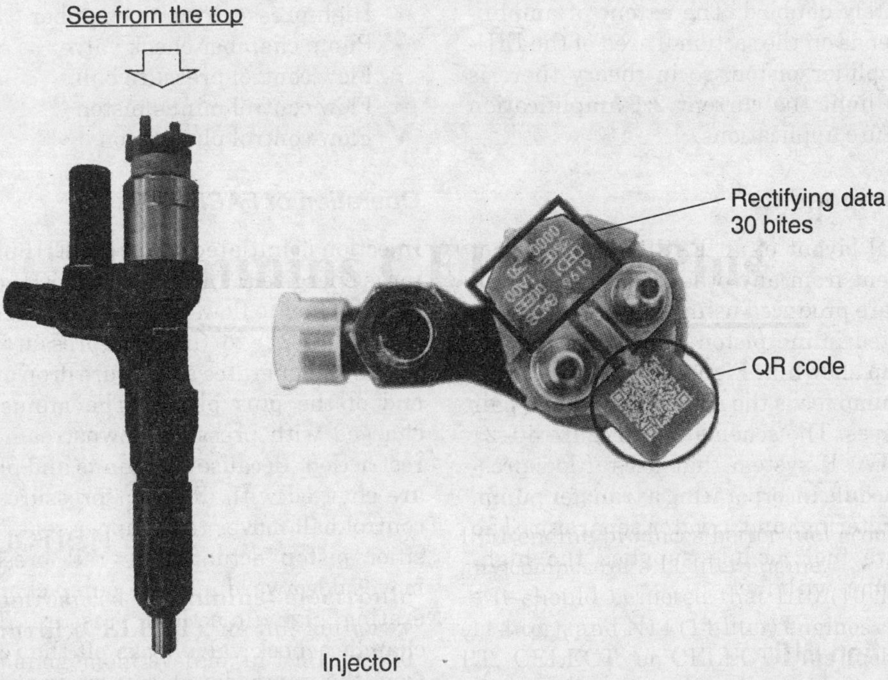

Figure 40–18 *Location of the QR code on a Denso EHI used on a Hino engine. (Courtesy of Hino Motors, Limited)*

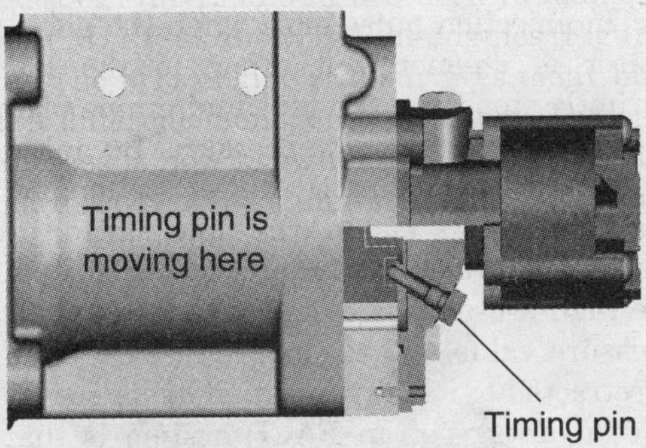

Figure 40–19 *Caterpillar high-pressure pump timing pin locked into position. (Reprinted Courtesy of Caterpillar Inc.)*

sion valve, typically a poppet valve. This valve is designed to hold rail supply fuel off the nozzle seat until energized. When integrated into EHI it reduces both internal and external leakage frequencies.

FUEL-AMPLIFIED CR SYSTEM

Fuel-amplified common rail (FACR) systems in their current format are manufactured by Bosch and Siemens. The FACR system is used on the DDC DD family of engines with engine displacements beginning at 10.6 liters and extending up to 15.6 liters. This family of engines is known in-house as the heavy-duty engine platform (HDEP) in Mercedes-Benz/Detroit Diesel Corporation. They are branded as DDC engines in the NAFTA market zone and as Mercedes-Benz (MB) engines elsewhere. In the systems currently in production, the high-pressure pump produces approximately 50% of the desired injection pressure at any given moment of operation. In fact, the FACR EHI functions somewhat similarly to an HEUI, with the actuation medium being rail fuel pressure rather than stepped oil pressure.

Amplification Principle

In the FACR EHI, a fuel pressure amplifier or intensifier piston within the injector body is actuated by CR-sourced fuel (pressure managed by ECM), and acts on fuel contained in a separate pump chamber (supplied by CR) to amplify the pressure. In current systems, the amplification ratio is 2:1, enabling the rail pressure to be approximately doubled. The extent of amplification depends on the sectional area of the EHI-located amplifier piston, so in theory there is nothing to limit the current 2:1 amplification ratio in future applications.

FACR Components

The general layout of an FACR system is not that different from any other CR system. Rail pressures are produced using an engine-driven, cam-actuated inline piston pump such as the Bosch pump shown in Figure 40–20. The high-pressure pump feeds the rail by means of a pair of outlet pipes. The schematic in Figure 40–21 shows an FACR system that uses a low-pressure fuel module incorporating a transfer pump, two-stage filtering, and a water separator. The low-pressure fuel module supplies the high-pressure pump with fuel.

Amplification EHI

The key difference in an FACR system and a standard diesel CR system is in the EHI. The FACR EHI uses a piezo actuator to control the pressure amplifier. Figure 40–22 should be referenced when following the description here, which is limited to the EHI intensifier circuit. The components of the intensifier circuit are:

- High-pressure plunger
- Plunger retraction spring
- Intensifier piston
- High-pressure pump chamber
- Pump chamber check valve
- Flow control pressure ball
- Flow control minus piston
- Flow control plus piston

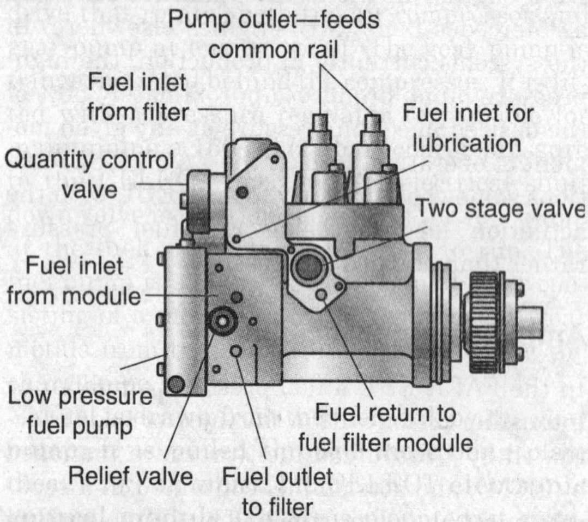

Figure 40–20 *Bosch high-pressure pump used in an FACR system.*

Operation of FACR EHI

Injection is initiated when the EHI piezo actuator is energized. This opens the EHI circuit and permits fuel to flow into the amplifier inlet (see Figure 40–22) to the high-pressure chamber. This flow generates a pressure drop at the lower end of the plus piston: The minus piston is charged with pressure downstream from this restriction. Because the minus and plus pistons are charged with different pressures, the flow control ball moves to its upper seat. The intensifier piston actuated by rail pressure then moves downward into the pump chamber, generating high (injection) pressure: The pump chamber check valve locks off the return path from the pump chamber to prevent back pressurization. High-pressure fuel is then charged to the fuel duct feeding the injector exit orifii from which it is injected to the engine cylinder.

The injection pulse ends when the piezo actuator is de-energized. Flow to the nozzle abruptly stops and the plus and minus pistons are exposed to the same pressure. Because the plus piston has a slightly larger sectional area than the minus piston, the flow control pressure ball moves to its lower seat. At this point, the high-pressure chamber returns to the rail pressure value and the high-pressure plunger is retracted to its start position by its spring. A key advantage of an FACR system is that it permits lower rail pressures while generating high injection pressures: This reduces the potential for both internal and external leakage.

REVIEW QUESTIONS

1. What type of fuel transfer pump is used on Caterpillar CR fuel systems that manage C7 and C9 engines?
   a. Reciprocating plunger
   b. Inline piston
   c. Vane
   d. External gear

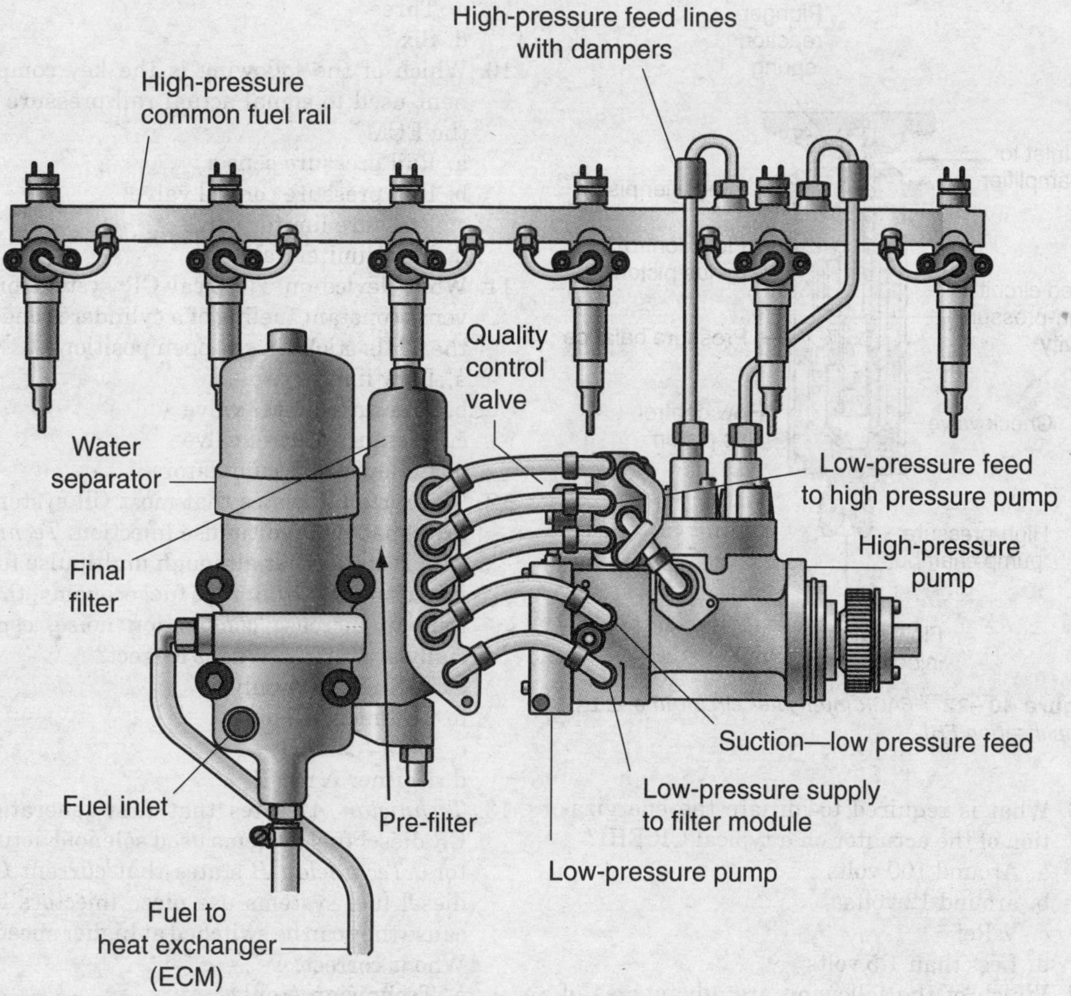

Figure 40–21 *Components of a typical FACR fuel system.*

2. Which of the following best describes the type of injectors used with CR fuel systems?
   a. Pintle injectors
   b. Electrohydraulic injectors (EHIs)
   c. Poppet injectors
   d. Electronic unit injectors (EUIs)
3. How many high-pressure pump elements are used on a Bosch radial piston pump fueling a six-cylinder engine with a CR (common rail) fuel system?
   a. One
   b. Two
   c. Three
   d. Six
4. In a Bosch CR fuel system, the rail could also be known as a(n):
   a. Charging gallery
   b. Accumulator
   c. Jumper pipe
   d. Injector tube
5. What is the normal peak rail pressure used by Caterpillar on its C7 and C9 CR-fueled engines when the system is functioning properly?
   a. 5,000 psi (345 bar)
   b. 12,000 psi (827 bar)
   c. 27,550 psi (1,900 bar)
   d. 33,300 psi (2,296 bar)

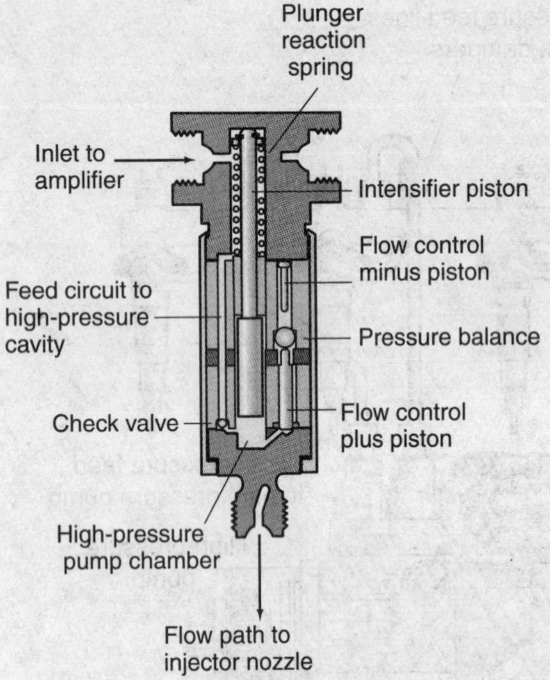

Figure 40–22 *FACR intensifier circuit in a 2:1 amplification EHI.*

6. What is required to initiate the energization of the actuator on a typical CR EHI?
   a. Around 100 volts
   b. around 12 volts
   c. V-Ref
   d. Less than 1.5 volts
7. Which of the following are advantages of multipulse injection cycles?
   a. Lower cold start emissions
   b. Lower noise levels
   c. Better fuel economy
   d. All of the above
8. Which of the following best describes the means by which the ECM drivers control the actuation of CR EHIs?
   a. V-Ref signal
   b. V-Bat signal
   c. Distributor spike
   d. Pulse width modulation (PWM)
9. How many pumping elements are used on a Caterpillar CR high-pressure pump fueling a C7 engine?
   a. One
   b. Two
   c. Three
   d. Six
10. Which of the following is the key component used to signal actual rail pressure to the ECM?
    a. Rail pressure sensor
    b. Rail pressure control valve
    c. Pressure limiter valve
    d. Flow limiter valve
11. What device on a typical CR system prevents constant fueling of a cylinder if one of the EHIs sticks in the open position?
    a. Flow limiter
    b. Pressure limiter valve
    c. Pressure control valve
    d. Collapse of accumulator
12. *Technician A* states that most CR systems are capable of multipulse injection. *Technician B* states that although multipulse fuel injection events improve fuel economy, they tend to increase combustion noise, especially at startup. Who is correct?
    a. Technician A only
    b. Technician B only
    c. Both A and B
    d. Neither A nor B
13. *Technician A* states that first-generation CR diesel fuel systems used solenoid actuators. *Technician B* states that current CR diesel fuel systems use piezo injectors because they can be switched at higher speeds. Who is correct?
    a. Technician A only
    b. Technician B only
    c. Both A and B
    d. Neither A nor B
14. What force is used to hold a CR EHI nozzle valve in its closed and seated position?
    a. Spring force only
    b. Electrical force only
    c. Combined hydraulic and electrical forces
    d. Combined hydraulic and spring forces
15. When the control plunger in an EHI is energized, what should happen to the rail pressure charged to the injector?
    a. Significant decrease
    b. Significant increase
    c. Remains almost unchanged
    d. First increases then drops off

16. What electronic principle is used in the rail pressure sensor fitted to most CR diesel fuel systems?
    a. Piezo resistive
    b. Thermister
    c. Potentiometer
    d. Differential pressure
17. *Technician A* states that to purge a typical CR system, the line nuts on the pipes connecting the rail to the EHIs should be cracked and retorqued after priming. *Technician B* states that some OEMs suggest that high-pressure pipes on CR systems be used once only because the seats deform to seal when torqued. Who is correct?
    a. Technician A only
    b. Technician B only
    c. Both A and B
    d. Neither A nor B
18. Which of the following best describes the electrical principle that makes a piezo-type actuator on an EHI operate?
    a. Electromagnetism
    b. Piezoelectric effect
    c. Reversibility of piezoelectric effect
    d. Inductive capacitance
19. Which of the following is another way of saying *injection quantity calibration data*?
    a. Fuel flow codes
    b. E-trim codes
    c. Quick-response (QR) codes
    d. All of the above
20. *Technician A* states that CR systems that use adaptive trim do not require reprogramming of injection quantity calibration data when EHIs are replaced. *Technician B* states that CR systems with adaptive trim routinely evaluate EHIs performance and can make minor adjustments to the fuel trim. Who is correct?
    a. Technician A only
    b. Technician B only
    c. Both A and B
    d. Neither A nor B

41

Emissions

INTRODUCTION

Until the late 1990s, the **Environmental Protection Agency (EPA)** focus on the **noxious emissions** from diesel engines was primarily on the engine manufacturer and testing of new engines prior to certification. The state of California has traditionally led the way for the rest of the jurisdictions in North America when it comes to defining vehicle emissions standards and enforcing them. The California agency responsible for emissions legislation and enforcement is the **California Air Resources Board**, or **CARB**. Any new engines introduced to the marketplace require testing to EPA and CARB standards prior to certification. However, until recently once an engine series had received certification, it was seldom tested in service.

Part of the emissions equation is fuel economy itself. The ATA has been proactive in lobbying for better fuel efficiency: This lobbying has included a move to increase vehicle payload ratings and trailer length. A decade ago if an organization lobbied in favor of increasing truck over-axle weights and total rig length, their argument tended to be dismissed by public sentiment that could attribute it to motives of operator profitability; that is, greed. Not so today. When every load carried on our roads is correlated to the amount of fuel required to haul it, this presents an argument for both the environment and for less dependence on imported petroleum. The EPA and United States government are currently considering mandating fuel efficiency standards for trucks, perhaps beginning as soon as 2014. Nobody can predict what will happen during the next few years but one thing you can count on is that plenty is going to change. If you are working as a truck technician today, this chapter is one of the most important in this textbook.

Importance of CARB

For many years CARB (http://www.arb.ca.gov/) was important to engine original equipment manufacturers (OEMs) because to market engines in the most populous state with the most powerful economy, CARB certification was required. Recently, the influence of CARB has increased with other states adopting CARB standards. This list of "green" states includes Maine, Vermont, Rhode Island, Massachusetts, Connecticut, Pennsylvania, Washington, Oregon, and New York—with Maryland and New Jersey planning to adopt CARB certification standards in the near future. Because this list now represents more than half of the United States economy, it is becoming unlikely that OEMs can afford to manufacture a product that is not CARB compliant.

Field Testing of Diesel Emissions

Although precertification exhaust emissions standards on diesel engines are increasingly demanding, the standards that field-tested diesel engines currently are required to meet are much less stringent than those required of gasoline-fueled, spark-ignited engines. This is because most field testing currently used in North America is based on an outdated SAE standard known as **J1667**. Published in 1991, SAE J1667 was based on a trucking world that used mostly hydromechanical diesel engines. Very few of these engines are around today.

J1667 test requirements only address visible smoke emissions, and the test procedures are entirely based on highway diesel engines commonly used in the 1980s. You might ask why this is so. While the EPA plays hardball with the diesel engine manufacturers prior to certification, field enforcement of highway diesel engine emissions is left up to state and municipal jurisdictions. Their enforcement agencies are caught in a squeeze between the voting public who are requesting tougher standards and the hard fact that aggressive enforcement and tough penalties can hurt them economically.

Economics of Emissions Compliance

By enforcing the J1667 standard, enforcement agencies can respond to environmentalists by saying that they are adhering to an accredited national standard. The truth is that the test does not begin to examine the type of emissions system shortcomings of modern, computer-controlled truck engines. However, this situation is likely to change within the next few years. As the voting public becomes more aware of the economic consequences of smog pollution and global warming, the pressure on jurisdictions to develop better ways of enforcing diesel emissions compliance that mirror EPA and CARB certification standards will increase.

Visible Smoke Emission

Because J1667 tests visible smoke emissions, the primary system being tested was that used to sense manifold boost and limit maximum fueling until turboboost pressure reached a specified pressure value. Turboboost devices are known as **aneroids**. An aneroid is a low-pressure measuring device. The aneroid devices used in hydromechanical truck engines were known by names such as turboboost sensor, manifold boost sensor, puff limiter, or air-fuel ratio control, and most of them could be disabled by a person with minimal mechanical knowledge. They were disabled because drivers believed that they were being cheated of low end power by aneroid devices. A disabled aneroid produced a telltale puff of smoke from the exhaust stack at every upshift, and it was often difficult to explain to a driver that if fuel exited the exhaust pipe in the form of smoke, then it was probably being wasted. The good news is that a J1667 opacity will identify any of these engines of a generation ago that have been tampered with.

EPA Certification

Until 2004, truck highway diesel engines relied primarily on engine management computers to meet the stringent emissions standards required for EPA certification. To meet 2004 and 2007 EPA emissions standards, simply managing combustion in diesel engine cylinders was no longer sufficient. A range of external emission control devices was added in the effort to produce much cleaner burning diesel engines. Today, truck technicians have to be aware of the emission control devices on the vehicle, including the type of fuel and engine lube used. Tampering with any emission control device is illegal. Sooner or later, the widely used SAE J1667 testing will be replaced by a testing that more accurately assesses the emission control equipment on current vehicles and the original EPA and CARB certification standards the engine had to meet. As every year passes, truck technicians will have to know more about the nature of exhaust gas emissions, how to test vehicles for compliance, and how to repair engines that fail to meet whatever standards are in place.

Scope of this Chapter

This chapter begins by providing the technician with a basic understanding of the causes of smog, how diesel engine exhaust gases contribute to fouling the environment, the components used to help reduce noxious emissions, and the methods used to test vehicle exhaust. We then briefly look at some methods that the EPA uses for testing engines for certification and, finally, a detailed look at how to perform opacity testing.

SMOG

The word **smog** comes from the words "fog" and "smoke." Throughout the world there are two main types of smog. The first type is **sulfur dioxide (SO_2)** smog produced by the burning of sulfurous fuels such as coal and heavy oils. Sulfurous smog conditions are aggravated by humidity and, therefore, fog. It is produced mostly by industry, especially in those countries that have lax standards for combusting industrial coals and heavy oils. The extent to which the burning of modern day vehicle fuels contributes to this problem is minimal due to the adoption of ultra-low sulfur fuels.

Photochemical Reaction

A **photochemical reaction** is any type of chemical process initiated by exposure to visible **ultraviolet (UV) radiation** or **infrared** radiation. For example, photosynthesis, the process by which green plants transform light energy into chemical energy and gaseous oxygen is a photochemical reaction. To say that a photochemical reaction has occurred, light energy has to be absorbed by a substance. Another natural example of a photochemical process is the production of **ozone** in the earth's upper atmosphere. Ozone (O_3) is responsible for filtering out most of the sun's harmful UV radiation. It is produced when sunlight breaks the bonds of some oxygen molecules (O_2) to form atomic oxygen (O), which then combines with O_2 to form O_3.

Photochemical Smog

Photochemical smog (also known as photosynthetic and photoelectric smog) requires neither smoke nor fog in its formative stage. It is produced mainly by the gases emitted from vehicle exhaust systems. Photochemical smog is characterized by a yellowish/light brownish haze. The effects of photochemical smog are as serious as those produced by sulfurous smog: Plant damage, eye irritation, and respiratory failure in animals (including humans) are characteristic. Photochemical smog is produced in two stages. First, **hydrocarbons (HC)** and nitrogen oxides (emitted mostly by vehicles) react with exposure to sunlight to produce O_3. O_3 then reacts with gaseous HCs to produce smog. O_3 by itself is highly toxic in low concentrations, and the smog that results from exposure to light is a major problem in America today.

Geographical Factors

Most large cities have experienced problems with photochemical smog, especially during the summer months. The location most noted for the problem is Los Angeles, California. Conditions in Los Angeles are ideal for producing photochemical smog. It is the second largest population center in the United States, per capita automobile ownership is the highest in the world, the area experiences a high ratio of sunny days, and the San Gabriel mountains act as a wall that helps entrap air masses that tend to move from the west to the east. Many years of exercising the toughest emissions standards in the world have contained, but by no means eliminated, the photochemical smog problem in Southern California. In fact, throughout North America, photochemical smog problems have worsened slightly since the first emissions control legislation several decades ago due to increased per capita ownership of vehicles and greatly increased yearly distances traveled.

Economic Factors

Doing something about this problem is not simple. Money is the problem. There is a cost attached to developing low emissions engines, and fueling them with friendlier fuels can be more costly and less fuel efficient. Environmentalists counter this argument by saying that the "hidden" costs of polluting the environment are borne by health care and the generations that follow ours.

Global Warming

The emerging issue and the one associated with global warming is **carbon dioxide** emission. Carbon dioxide **(CO_2)** is classified as a greenhouse gas and *not* as a noxious emission. Reducing CO_2 emissions will mean taking a serious look at the chemical composition of the fuels we burn, specifically reducing or eliminating the carbon content. As both gasoline and diesel fuel are composed of approximately 85% carbon, some major changes in fuel formulations are probable in the coming years.

Ingredients of Smog

Knowing a little basic chemistry helps in understanding the nature of smog. Combustion is a chemical reaction. Under optimum conditions, when an HC fuel is combusted using the oxygen available in ground-level air as the reactant, the HC fuel is oxidized to form H_2O (water) and CO_2. Neither H_2O nor CO_2 are classified as noxious products of the combustion process, so if these two gases were the only products of the reaction, the term *perfect* combustion could be applied to what has happened. The problems arise when the combustion of the HC is not "perfect" or the nitrogen that makes up 80% of air at ground level becomes involved in the combustion process.

When a hydrocarbon fuel is not completely oxidized, gaseous HC, particulate HC, and **carbon monoxide (CO)** result. When emitted from the tailpipe of a vehicle these chemicals are all classified as noxious emissions. Nitrogen can become involved in the combustion process under certain conditions. Within the combustion temperature window of a typical internal combustion engine, this tends to be when temperatures are higher. When nitrogen is oxidized, it forms a number of different compounds known collectively as **oxides of nitrogen** or **NO_x**, but of these, **nitrogen dioxide (NO_2)** presents the most problems. NO_2 is a key to the formation of ozone and acid rain. All NO_x emissions are classified as noxious emissions by the EPA.

Sulfur

In recent years, the maximum sulfur content of diesel fuels has been dramatically reduced. The current maximum permitted sulfur content in an ASTM 1D or 2D fuel is 0.0015%, (15 ppm). This standard was introduced in October 2006 and applies to 80% of the total fuel dispensed at the pumps in North America. This standard has come close to eliminating the tendency of diesel fuel to contribute to the sulfurous combustion emissions.

When sulfur is oxidized in the combustion process, it forms SO_2, an ingredient of sulfurous but not photochemical smog. Sulfur is acidic, toxic, and an ingredient of acid rain. While sulfur compound emissions from the combustion process are serious, highway trucks are no longer a major contributor. Sulfur emissions produce an unpleasant, acidic odor. Because of the acidity, high sulfur emissions from a diesel engine will usually produce tattletales such as rust-out of exhaust piping.

Hydrocarbons

The HC emitted in the combustion process of an internal combustion engine burning liquid or gaseous petroleum product may be gaseous, liquid, or solid (particulate) state. They are only observable in liquid or solid states when they may be seen as white smoke (liquid emission) or black smoke (particulate emission). A more detailed analysis of diesel engine smoke emission appears in chapter 48. In the days before rigid emissions standards, a truck diesel engine commonly emitted HC in liquid state (that is, condensing in the exhaust gas) at cold startup and in particulate state under conditions of overfueling, acceleration, or air starvation.

Particulate Matter

Anything in the exhaust gas that is in the solid state can be classified as particulate matter (PM). Carbon soot and dust form PM. Incomplete combustion is the main culprit in the formation of PM. These minute particulate compounds can cause respiratory problems in human and animal life. The EPA classifies particulates into two general categories:

1. Inhalable particulates that range between 2.5 and 10 microns in diameter
2. Respirable particulates that tend to be 2.5 microns or less in diameter

Respirable particulates are not entrapped by nasal filters and are small enough to get deep into the lungs, causing wheezing, coughing, and shortness of breath. PM emission is always visible from a diesel engine and is easily detected with smoke density-measuring equipment such as opacity meters. The objective of diesel particulate filters (DPFs), studied later in this chapter, is to minimize PM.

Volatile Organic Compounds

Volatile organic compounds or **VOCs** are HCs in the gaseous state that boil off fuels during production, distribution, and pumping. Diesel-fueled vehicles contribute VOCs in a small way: Fuel tanks exposed to the high heat of a summer's day vaporize the more volatile fuel fractions to the extent that the cetane number (CN) rating can degrade. The transportation sector accounts for approximately 30% of the VOCs found around population centers. VOCs are more likely to be an atmospheric problem in hot weather conditions when the most volatile fractions of fuels more readily boil off. VOCs react with sunlight to produce ground level O_3.

Carbon Monoxide

CO is a colorless, odorless, tasteless, and highly poisonous gas. It is a result of the incomplete combustion of an HC fuel. CO can be combusted to form harmless CO_2. Approximately two-thirds of the atmospheric levels of CO in North America can be attributed to vehicle emissions. Exposure to CO can impair brain function, cause fatigue, and may be fatal in high concentrations.

Ozone

Atmospheric oxygen is diatomic, that is, combines to form molecules consisting of two oxygen atoms (O_2). O_3 is triatomic, that is, three oxygen atoms covalently bond to form a molecule of O_3. It occurs naturally in small quantities in the earth's stratosphere where it absorbs solar UV radiation. O_3 can be produced by passing a high-voltage electrical discharge (arc) through air: You can sometimes smell it during a thunderstorm. It is manufactured commercially using an electric arc process. As a pollutant, O_3 results from the photochemical reaction between NO_x and various HCs (especially those categorized as VOCs). It is known to irritate the eyes and mucous membranes. Exposure to O_3 at low-level concentrations can be a health threat.

Oxides of Nitrogen

Despite the fact that nitrogen under usual atmospheric conditions and temperatures is **inert** (unlikely to participate in chemical reactions), when subjected to certain conditions in an engine cylinder, it reacts. That is, it can be oxidized. It forms nitrous oxide (N_2O), nitric oxide (NO), and NO_2 when oxidized. All NO_x produced in the combustion process of internal combustion engines are known collectively simply as NO_x. NO_2 is the most reactive of the NO_x compounds emitted from vehicle tailpipes. In North America on- and off-highway vehicle engines are the source of about 60% of the NO_x in our atmosphere. NO_x directly affects persons with respiratory problems, and medical opinion states that children's lungs are susceptible to concentrations of 0.1 ppm (parts per million). This accounts for the reason that the EPA has so aggressively attacked NO_x emissions from diesel engines in its certification requirements.

Air Quality Index

Most jurisdictions today monitor air quality and use the monitoring results to alert the public on those days that it rises above acceptable standards. If you suffer from asthma or other respiratory problems, you probably notice that each year that passes usually has more smog alert days when you are advised to avoid the outdoors. Table 41–1 is an air quality index (AQI) used by one jurisdiction.

Smog Summary

We can summarize the consequences of vehicle emissions as follows:

NO_x + VOCs + Sunlight = Ozone

Ozone + Particulate matter = Smog

Because the costs of doing something about vehicle emissions are upfront and levied on big business, they tend to be resisted by powerful political lobbying. However, the cost effects of pollution are delayed and impact on the health of individuals who then have to resort to the big business of our health system. Whether you are an environmentalist or antienvironmentalist probably has much to do with where you live. Residents of geographical areas that are susceptible to smog tend to advocate stronger emissions legislation, whereas those in relatively unaffected areas dismiss such controls as

Table 41–1 *Air Quality Index and Environmental Effects*

Air Quality Index and Environmental Effects	*Category*	*Carbon Monoxide (CO)*	*Nitrogen Dioxide (NO_2)*	*Ozone (O_3)*	*Sulfur Dioxide (SO_2)*	*Suspended Particles (SP)*	*SO_2 + SP*	*Total Reduced Sulfur (TRS)*
100 and over	Very poor	Increasing cardiovascular symptoms in nonsmokers with heart. disease Some visual impairment	Increasing sensitivity of patients with asthma and bronchitis.	Light exercise produces respiratory effects in patients with chronic pulmonary disease.	Increasing sensitivity in patients with asthma and bronchitis.	Increasing sensitivity in patients with asthma and bronchitis.	Significant respiratory effects in patients with asthma and bronchitis.	Sensitive individuals may suffer nausea and headache due to severe odor.
50–99	Poor	Increased cardiovascular symptoms in smokers with heart disease.	Odor and discoloration. Some increase in bronchial reactivity and asthma attacks.	Decreasing performance by athletes exercising heavily.	Odorous. Increasing vegetation damage.	Visibility decreased. Soiling evident.	Increased symptoms in patients with chronic respiratory disease.	Extremely odorous.
32–49	Moderate	Blood chemistry changes but no detectable impairment.	Odorous.	Injurious to many vegetation species (e.g., white beans, tomatoes)	Injurious to some species of vegetation.	Some decrease in visibility.	Injurious to vegetation due to sulfur dioxide.	Odorous.
16–31	Good	No effects.	Slight odor.	Injurious to some vegetation species in combination with SO_2.	Injurious to some vegetation species in combination with ozone.	No effects.	No effects.	Slight odors.
0–15	Very good	No effects	No effects	No effects	No effects	No effects	No effects	No effects

unwanted government interference. While the implementation of tougher emissions controls and their enforcement is going to be politicized and controversial, expect them nevertheless to become more aggressive by the year, simply because there are more voters in big cities than rural areas who are demanding such controls.

EPA CERTIFICATION TESTING

It is not really necessary for a truck technician to know much about the precertification testing used by the EPA, although you should recognize that it is highly exacting and goes well beyond the basic opacity or smoke density field testing used by most jurisdictions. Gaseous emission of toxic combustion by-products cannot be measured using an **opacity meter** because gaseous emissions do not show as smoke when discharged from the vehicle tailpipe. Before EPA certification, diesel engine OEMs must conduct gaseous exhaust analysis and what is known as **constant volume sampling (CVS)** test standards. CVS is a process that involves diluting exhaust gas with purified ambient air and subsequently routing the air through three tests:

1. Filtration. The gas is forced through filters that are then weighed. The increase in mass indicates particulate emission.
2. Flame ionization detection (FID). This test indicates HC concentration.
3. Exhaust sample bags. These measure CO, CO_2, and NO_x.

Noxious gas emissions can also be tested by sampling and analyzing raw exhaust gas using **gas analyzers** not unlike those used for spark-ignited (SI) engine testing and described later in this chapter.

Actual EPA Emissions Requirements by Year

Given the fact that emissions from highway truck diesel engines were unregulated until 1970, what has been accomplished since is nothing short of astounding. In 1970, a typical diesel engine emitted 16 grams per HP/h of NO_x and 10 grams of PM per HP/h. The NO_x emission standard required in 2007 is 0.2 gram per HP/h or an 8,000% reduction. Similarly, the PM standard required for 2007 has been reduced to 0.01 gram per HP/h, a 10,000% reduction. Table 41–2 outlines the progress of heavy-duty highway truck EPA emissions standards from 1990 to 2010. Figure 41–1 shows the EPA diesel emissions story over the 40-year period extending from 1970 to 2010.

Table 41–2 *EPA Emissions Requirements by Year*

Year	*HC (Grams BPH/h)**	*PM (Grams BPH/h)*	*NO_x (Grams BPH/h)*
1990	1.3	0.60	6.0
1991	1.3	0.25	5.0
1994	1.3	0.10	5.0
1996	1.3	0.10	5.0
1998	1.3	0.10	4.0
2004	0.5	0.05	0.5
2007	0.28	0.01	0.2
2010	0.14	0.01	0.2

*BPH/h = brake horsepower per hour

DIESEL ENGINE EMISSIONS CONTROLS

In the early years of legislated emissions controls designed for diesel engines, the primary emission control device on the engine was the computer that managed it. This meant that from the late 1980s, full-authority computer-controlled systems were introduced in truck engines well ahead of their automobile counterparts, which tended to rely on external controls to a larger extent. The external devices used on automobiles beginning in the 1970s included catalytic converters, air pump units, exhaust gas recirculation (EGR) systems, and positive crankcase ventilation (PCV) systems. This has changed. Now the diesel engine relies on a full range of external emission control devices that include oxidation catalysts, reduction catalysts, particulate filters, and urea injection systems. Most of the emission control additions to diesel engines are required to operate with the 2007 ultra-low sulfur (ULS) fuel standard (see Chapter 16). In this next section, we take a look at some of the devices used on today's diesel engines to limit noxious emissions.

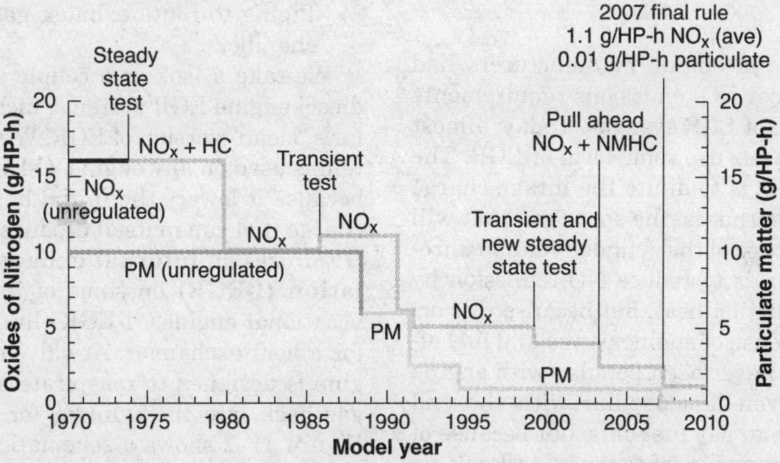

Figure 41–1 *EPA heavy-duty engine emissions standards.*

Engine Control Module

Those of you who have studied automobile technology will be familiar with the term *closed-loop fueling.* Closed-loop fueling is the term used to describe the running an SI, gasoline-fueled engine using the exhaust gas sensor signal to manage a stoichiometric air-fuel mixture. In the lean burn technology of the highway diesel engine, things are a little more complex in that the input of no single sensor is ever used to drive the **fueling algorithm** written to the electronic/engine control module (ECM) software. The fueling algorithm is the set of ECM programmed rules and procedures designed to produce the desired performance and emissions from the engine at any given moment of operation. It is based on monitored sensor data, operator commands, and programmed instruction in memory. It is also known as mapping. And in today's engines, the fueling algorithm is required to manage the engine to meet emissions standards in all operating phases.

Temperature Factors

Managing fuel injection quantity, timing, and features such as multipulse injection in diesel engines while meeting emissions standards requires high-speed processing capability. In most cases, combustion temperatures have to be managed within a narrow temperature window. HC emissions tend to increase when combustion temperatures are in the lower range and the air-fuel ratio (AFR) is richer. On the other hand NO_x emissions tend to increase when AFR leans out and the combustion temperatures are in the higher range.

Timing Factors

It is also important to note that injection timing or, more specifically, ignition timing, affects noxious emission characteristics. Retarded timing tends to reduce NO_x emission but over-retarded timing can greatly increase HC emission. Timing deviations of as little as 1 degree crank angle can increase NO_x or HC content in the **end gas** by as much as 10%. This timing sensitivity has made computer management of CI engines essential in attempting to meet emissions requirements. In general terms, advancing the engine timing will increase engine temperatures and therefore tend to increase NO_x emission, while retarding timing can lead to incomplete combustion of the fuel and therefore an increase in HC emission.

External Emission Controls

If you take a look at any truck diesel engine series manufactured in the year 2001 and compare it with its series equivalent today, you would have difficulty in identifying them as being of the same series of engine. Today's diesel engines use an extensive array of external emission control hardware. In this section we take a close look at some of these devices.

Cooled-EGR

Until 2004, diesel engine manufacturers had been able to meet NO_x emissions requirements without the use of EGR systems. Today, almost all highway diesels use some form of EGR. The objective of EGR is to dilute the intake charge with "dead" gas, that is, the spent gas that will make a percentage of the cylinder volume unreactive. The idea is to reduce NO_x emission by lowering combustion heat, but because this occurs at the expense of engine power and fuel efficiency, it has never been popular with engine OEMs. It is even less popular with the end users who have to pay fuel bills, but because of advances in engine performance technology, fuel efficiency has tended to remain fairly static over the past ten years. When rerouting engine exhaust gas back into the engine cylinders, it makes sense to cool it as much as possible. For this reason, most engine OEMs use **cooled exhaust gas recirculation**, usually known by its acronym of **C-EGR**.

C-EGR Components

A typical C-EGR system on a diesel engine has to be ECM controlled because the mixture percentage (commonly known as cut-percentage) varies according to how the engine is being operated. This ranges between 0% and 50% depending on the manufacturer. Typically a C-EGR system consists of:

- A heat exchanger (engine coolant is used)
- ECM-controlled mixing chamber (mixes boosted air with exhaust gas)
- Mass airflow sensor
- Plumbing to route engine coolant through the heat exchanger
- Piping to route exhaust gas to the mixing chamber

We take a look at a couple of variations on diesel engine EGR systems, including Caterpillar's "clean" version of EGR. When an EGR system is used on any engine, C-EGR makes sense because it lowers the initial heat of the intake charge that can in itself produce more NO_x. One OEM uses an **internal exhaust gas recirculation (I-EGR)** on some of its medium-bore, vocational engines. I-EGR eliminates the need for a heat exchanger. At valve overlap, the engine is designed to reaspirate some of the end gas back into the cylinder for the next cycle. Figure 41–2 shows a schematic of a typical C-EGR system.

Clean Gas Induction (CGI)

Caterpillar was able to meet the EPA 2004 highway diesel engine emissions standards without resorting to EGR. However, Caterpillar opted to use a modification of EGR to meet the 2007 EPA standards and beyond. Earlier, we defined the objective of an EGR system as diluting the intake charge with "dead" gas. Dead gas is unreactive gas: It enters the engine cylinder and does nothing but take up volumetric space. It is sort of like reducing the engine displacement during certain modes of operation. The objective is to reduce engine temperatures and therefore NO_x output. It has the effect of making a large displacement engine behave like a smaller displacement engine. To make the EGR principle function properly you need a source of dead gas. The fact of the matter is that there is no better source of dead gas than the engine exhaust system because it is produced any time the engine is running.

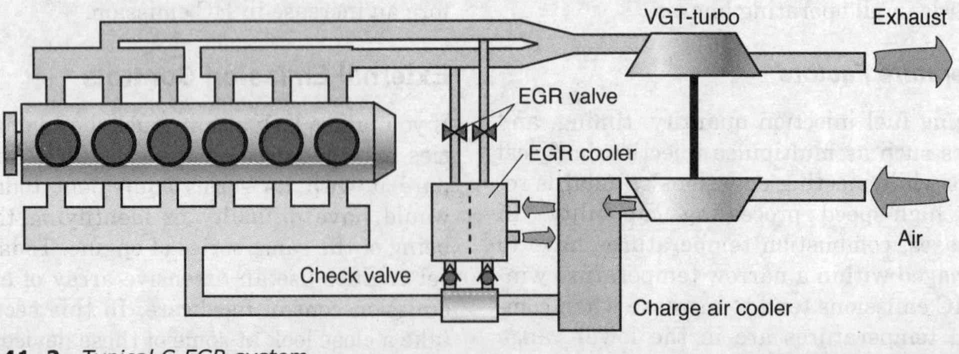

Figure 41–2 *Typical C-EGR system.*

Clean gas induction (CGI) is a variation on EGR used exclusively on Caterpillar's post-2007 ACERT family of engines. It differs from EGR in that it sources its "dead" exhaust gas downstream from the DPF. Cat likes to call its system "clean EGR." CGI is monitored by mass airflow sensor (MAF) and just like its competitors' EGR, the system objective is to keep combustion temperatures lower to reduce under certain running conditions. It is easy to identify a Cat CGI engine because the CGI pipe can be seen to exit the muffler/converter/ DPF assembly close to its exit. This pipe routes the theoretically clean dead gas back to the intake. Figure 41–3 is a Caterpillar schematic of its CGI system.

In terms of managing CGI cycles, the CGI gate is spring loaded to default to no-recirculation. Gas from the clean side of the DPF is admitted up to a 15% mixture concentricity (with EGR systems this can often be over 30%). CGI is ECM managed so it does not occur during DPF regeneration cycles.

CGI Sensors A CGI system uses three sensors. Two of these are pressure sensors, an absolute pressure sensor and a pressure differential measurement sensor (these are explained in Chapter 30), while the third is a temperature sensor. Gas exiting the CGI venturi passes into a bundle-type, coolant-cooled heat exchanger. In common with EGR engines, there is an advantage to cooling recycled exhaust gas. And Cat's heavy-duty family of engines will also use

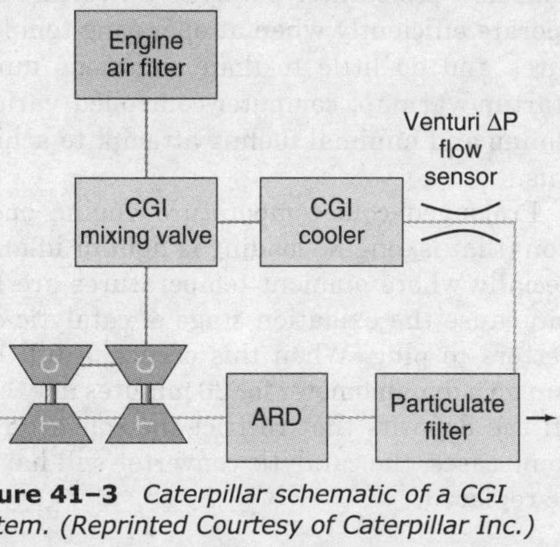

Figure 41–3 *Caterpillar schematic of a CGI system. (Reprinted Courtesy of Caterpillar Inc.)*

a precooler. The CGI pressure sensor is a delta (three-terminal), variable capacitance-type sensor. The CGI gate is controlled by a pulse width modulated (PWM) actuator, in other words, a linear proportioning solenoid, energized by V-Bat voltages.

Catalytic Converters

Some truck and bus diesel engines used **catalytic converters** before 2002, but almost all current engines use oxidation converters; some use both oxidation and reduction converters. By definition, a **catalyst** is a substance that enables a chemical reaction without itself undergoing any change. For many years, automobile catalytic converters have been two-stage, three-way converters and it is only since 2007 that truck diesels have incorporated two-stage devices. The two stages are as follows:

- Oxidation stage: attempts to oxidize HC and CO not oxidized in the engine cylinder
- Reduction stage: attempts to reduce NO_x back to elemental nitrogen and oxygen when NO_x compounds have been formed during combustion

Oxidation Stage

When cylinder end gas contains HC and CO, the oxidizing stage of the catalytic converter attempts to oxidize (combust/burn) these to harmless H_2O and CO_2. **Platinum** and **palladium** are both oxidation catalysts. They are expensive. Oxidizing catalysts enable what can be termed *catalytic afterburning,* sometimes assisted by the introduction of fresh air into the reaction; peak temperatures run in the range of 800°C to 1,000°C (1,472°F to 1,832°F). Oxidation catalytic converters first began to appear on highway diesels in the late 1990s, but today almost all diesel engines are required to use them. Because of the high operating temperatures, it is not unusual for an oxidation converter to glow red during nighttime operation, especially when engines are under load.

Reduction Stage

Where oxidized nitrogen compounds (known collectively as NO_x) are present in the exhaust gas, a reduction stage attempts to reduce these to elemental nitrogen and oxygen. **Rhodium** is the usual reduction catalyst and has been used by automobile manufacturers since the 1970s. It attempts to reduce NO_x compounds back to elemental oxygen and nitrogen. However, rhodium can only function as a reduction catalyst when the cylinder burn is managed close to stoichiometric ratios. This means that to use a rhodium catalytic converter in a lean burn diesel engine, the AFR mixture concentricity in the converter has to be close to stoichiometric. OEMs that use rhodium-based catalytic converters achieve this by **dosing**, that is, injecting excess fuel into the exhaust gas upstream from the converter.

NO_x Adsorber Catalysts

As mentioned earlier, rhodium-based reduction catalysts have been used on gasoline-fueled automobile engines for many years. These used a rhodium catalyst to "reduce" NO_x compounds back to elemental nitrogen and oxygen. Because rhodium only functions as a reduction catalyst when the air-fuel ratio is at stoichiometric or richer it could not be used on lean burn diesels that continuously unload excess (unreacted) oxygen into the exhaust. With the introduction of the **NO_x adsorber catalyst (NAC)**, diesel engine designers found a way around this problem.

Adsorption is a chemical term that is best defined as *adhesion*: When a substance is adsorbed it does not undergo any significant chemical change so do not get this confused with the term *absorb*. NO_x adsorber catalysts (NACs) use base metal oxides to initially "store" NO_x compounds then use a rhodium reduction catalyst to reduce them back to N_2 and O_2 during an ECM-managed reduction phase. In the storage phase, NO_x is adsorbed onto a metal (barium) oxide substrate: When the available NO_x storage sites within the device are occupied, the engine ECM temporarily simulates a "rich" (reduced oxygen) running condition or injects fuel directly into the exhaust system. Operation in this simulated "rich" mode releases the NO_x from its base metal storage locations and allows a rhodium catalyst to convert it to nitrogen gas (N_2), oxygen (O_2), and water vapor (H_2O). In other words an NAC functions in two stages:

- NO_x storage: NO_x is adsorbed to the base metal oxide substrate during the normal lean burn operation of the diesel engine.
- NO_x reduction: The engine ECM temporarily operates the engine operation in a rich AFR mode or injects fuel into the exhaust system to simulate a rich AFR, allowing a rhodium catalyst to reduce the NO_x to elemental nitrogen (N_2), elemental oxygen (O_2), and water vapor (H_2O).

NAC Regeneration Cycle

The key to enabling a NAC to function in its regeneration cycle is to temporarily eliminate excess oxygen from the exhaust. NACs require the use of the new ULS fuel. Sulfur at any levels can be damaging to NACs and some systems will require occasional de-sulfation regenerative modes even when only the appropriate ULS fuel has been used. A de-sulfation regenerative cycle requires temporarily raising the exhaust temperatures using fuel injected directly into the exhaust system.

Cat Converter Construction

The noble metal catalysts are thinly coated on aluminum oxide or granulate monolith **substrate** (catalytically inert material onto which active catalysts are coated), providing optimum use of catalyst surface area combined with minimal flow resistance. Catalytic converters only operate efficiently when at operating temperatures and do little to limit emissions during startup/warmup; computer-controlled variable timing and minimal fueling attempt to achieve this.

Prolonged cold temperature engine operation (that is, engine loading is light or idle), especially where ambient temperatures are low, can cause the oxidation stage of catalytic converters to plug. When this occurs, a full load run on a dynamometer for 30 minutes may burn off the deposits that restrict the converter. In some cases, the catalytic converter will have to be replaced.

Diesel Particulate Filters (DPFs)

Diesel particulate filters (DPFs) have been around in the industry for a number of years usually in special applications such as a garbage packer required to operate inside for long periods of time. On this type of application they used to be known as *particulate traps*. Things have changed. A DPF is now a requirement on every on-highway diesel engine meeting 2007 emissions and beyond.

By definition, a DPF is an exhaust aftertreatment device designed to eliminate diesel particulate matter (PM), or soot, produced by the engine cylinder combustion process. The DPF substrate is an extruded material, typically cordierite or silicon carbide, that is housed in a muffler body. DPF substrates typically capture over 95% of the diesel PM, but they will plug up unless they are cleaned or regenerated.

Regeneration Methods

There are two main regeneration approaches—passive and active. Passive DPFs are coated with precious metal catalysts, such as platinum and palladium. These enable a flameless burning or oxidizing of the diesel PM. The function of the catalysts is to reduce the oxidation temperature of diesel PM to 300°C (compared to 650°C in air). These temperatures fall within the normal operating exhaust temperature range of typical diesel engines. However, to ensure effective filter regeneration, the exhaust temperature must be monitored to make sure the low threshold temperature for passive regeneration is met.

Active regeneration events do not rely on catalysts, but instead use alternately developed heat to regenerate the DPF substrate. Active system approaches employed by engine OEMs in 2007 include fuel injection and fuel burners. Electric systems are also used in retrofit or aftermarket applications. Fuel injection systems inject diesel fuel across a diesel oxidation catalyst, which ignites releasing heat. Controlled fuel injection is designed to raise the temperatures to 650°C. At this temperature, the DPF will burn off collected soot.

Regen Chemistry

HC matter such as diesel PM converts to CO_2 and H_2O during the oxidation process. Inorganic materials, like lube oil additives, may produce CO_2 and H_2O, but they also create a residue or ash. This ash builds up in the filter over time, gradually creating a restriction and increasing exhaust backpressure. This eventually degrades engine performance and this fact means that DPF regeneration is required at intervals.

Most engine manufacturers attempted to minimize any cost increase to their 2007 engine families but this excluded DPFs, which have become universal on post-2007 products. Typically, a DPF represents something around a \$4,000 to \$12,000 additional cost. Though usually supplied by the engine manufacturer, DPFs are considered an OEM chassis additional cost. This fact makes diesel engine OEMs happy because emissions legislation forced significant increases in engine pricing in 2001 and 2004.

Types of DPF

In this section our objective is to focus primarily on the mandatory post-2007 DPFs. Determining what type of DPF is used on a particular diesel engine depends largely on average exhaust temperatures. There are three general types of DPF in use:

- Catalyzed. This more common type of DPF is integrated into a canister that also contains multistage catalytic converters and what used to be known as the muffler. It is probably best described as a comprehensive aftertreatment device. Catalyzed DPFs will be found on most highway trucks and are capable of both passive and active regeneration cycles. They are suited for applications in which *average* exhaust temperatures exceed 225°C (437°F).
- Noncatalyzed. This type of DPF is required in diesel-powered, vocational vehicles that spend prolonged periods in low engine load operation. Typical applications are cement trucks, fire trucks, and garbage packers. Noncatalyzed DPFs are only capable of active regeneration cycles and manage them with a control module that is independent of the engine ECM. The noncatalyzed DPF

control module is assigned an address on the data backbone and communicates with engine ECM via the data bus.

- **Low-temperature filter (LTF)**. The LTF is designed to remove entrapped diesel particulate emissions from applications that run at low loads. Collected ash has to be periodically removed. LTFs are designed for applications in which *average* exhaust temperatures are less than 225°C (437°F) but generally exceed 200°C (392°F).

DPF Operating Principles

The idea behind a DPF is that engine-emitted soot first collects on the walls of the device. Engine manufacturers design DPFs to function primarily in self-regeneration mode. This means that when soot collection reaches a threshold level, it is burned off in what is known as a **regeneration cycle**. In most cases, the regeneration cycle will occur when exhaust temperatures are sufficiently high during normal operation. When this regeneration takes place unassisted by additional fuel or air injection to the exhaust gas it is known as **self-regeneration**. Self-regeneration can also be known as **passive regeneration**: The terminology varies by engine OEM.

Operating temperatures of the combined oxidation converter and DPF can exceed 600°C (1,112°F) so most of these devices incorporate heat shielding. Depending on the engine and its power rating, both single and dual canister versions are used. Typically, single and dual canisters are specified by horsepower rating as follows:

- Single canister: up to 600 BHP (447 kW)
- Dual canister: 600 BHP and higher (447 kW)

When the operating environment is not conducive to a self-regeneration cycle, regeneration can also occur assisted by injection of some fuel (diesel) sometimes assisted by spark ignition. This mode of regeneration is known as **active regeneration**. Fuel for an active regeneration cycle is usually sourced from the fuel subsystem and delivered at the specified charging pressure. For instance, on a Caterpillar ACERT C15, this would be at 250 psi (17 bar) delivered by an electromechanical injector. Most diesel engine OEMs are using similar DPF devices

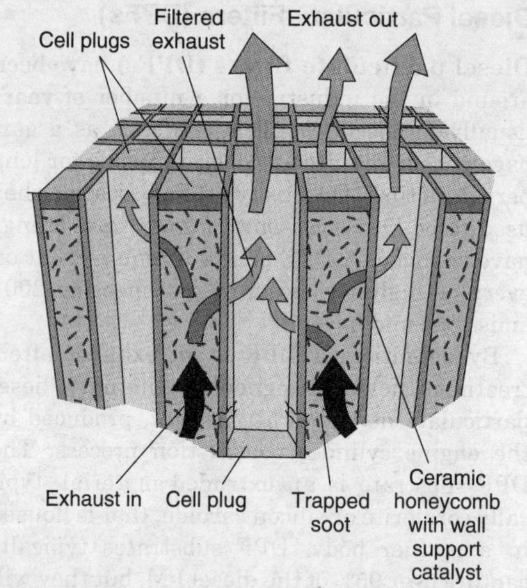

Figure 41–4 *Cutaway view of a Detroit Diesel DPF core showing gas flow. (Courtesy of Detroit Diesel Corporation)*

but they are managed in different ways. The key difference among OEMs using catalyzed DPFs is whether an active regeneration cycle requires the assistance of a spark plug. Figure 41–4 shows a cutaway core view of a Detroit Diesel DPF—a unit that regenerates without the assistance of spark ignition.

Regen Frequency

Regeneration cycles are designed to occur at set intervals. These set intervals may be as often as once every hour or as infrequent as once every 8 hours of operation depending on the engine, its power ratings, how the engine is being operated, and the type of DPF. Truck drivers should be informed about the expected intervals of the regeneration cycles of the equipment they operate. During both passive and active regeneration, there will be an increase in exhaust gas temperatures, so in all cases the driver can inhibit the regen.

Regeneration Cycles

We are going to use as reference one engine manufacturer's DPF to outline regeneration cycles: This OEM uses spark assist for active regens but is otherwise generally consistent with what is used industry-wide:

1. Passive: This occurs when exhaust temperatures are high enough for regeneration to occur without the assist of a flame burn-off cycle. This is intended to be the primary means of DPF regeneration that should usually occur when the vehicle is running down the highway with the engine under load. Typically, the normal passive cycle of regeneration is designed to occur every 8 hours or so. During this normal passive cycle, DPF temperatures can exceed 600°C (1,112°F).
2. Passive plus: This is designed to occur after a prolonged period of idling as soot buildup reaches 125% and the DPF is becoming restricted. This mode of passive regeneration occurs at a much lower temperature than the normal passive regeneration cycle. A small quantity of fuel (much less than during active transient mode) and spark ignition are used. The objective is to burn off a little soot every 2 hours or so to maintain the status quo. During passive plus regeneration, DPF outlet temperatures do not exceed 400°C (752°F).
3. Active transient: This is set to occur when soot levels are at 100% and the DPF is becoming restricted. The vehicle must be traveling at a speed exceeding 20 mph (32 km/h) for an active transient burn cycle to take place.
4. Active steady state: This is commanded by Electronic Technician (ET) using a dash button. The vehicle must be at 0 mph (km/h) and in park. Fuel is injected and a spark ignites the fuel.

Transient regeneration events will *not* occur when:

1. Ambient temperatures are less than 20°C (68°F)
2. Turbo-out temperature exceeds 550°C (1,022°F) (because passive regeneration is possible)
3. Coolant temperature is (below) 65°C (150°F)

Regeneration system (RS) alerts (broadcast on data bus by engine ECM):

When soot level exceeds 100% .... RS light illuminates

When soot level exceeds 125% .... RS light blinks

When soot level exceeds 150% .... engine derates by 25%

When soot level exceeds 175% .... engine derates by 75%

When soot level exceeds 200% .... engine shuts down

Regeneration System Hardware

The RS hardware consists of an RS head incorporating:

- A fuel injector
- A swirler plate
- A spark plug
- Cooling system ports

The fuel injector nozzle can be compared to the low-pressure injectors used in gasoline fuel injection: The fuel injector is supplied by the diesel fuel subsystem at its specified charging pressure and this is the reason some OEMs have significantly raised fuel subsystem charging pressures in their post-2007 products. A defective fuel injector is usually not serviceable and often the whole RS head must be replaced. Figure 41–5 is a schematic showing the layout of the Caterpillar regeneration system (CRS) and Figure 41–6 shows the detail of a CRS head.

The CRS spark plug consists of an electrode only (does not self-ground). It is designed to ground to a swirler plate located directly below it. Spark plug cycles are driven by the ECM.

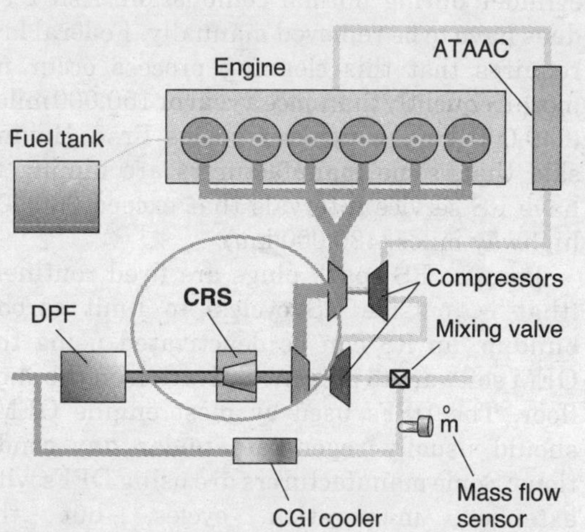

Figure 41–5 *Schematic layout of the Caterpillar CRS. (Reprinted Courtesy of Caterpillar Inc.)*

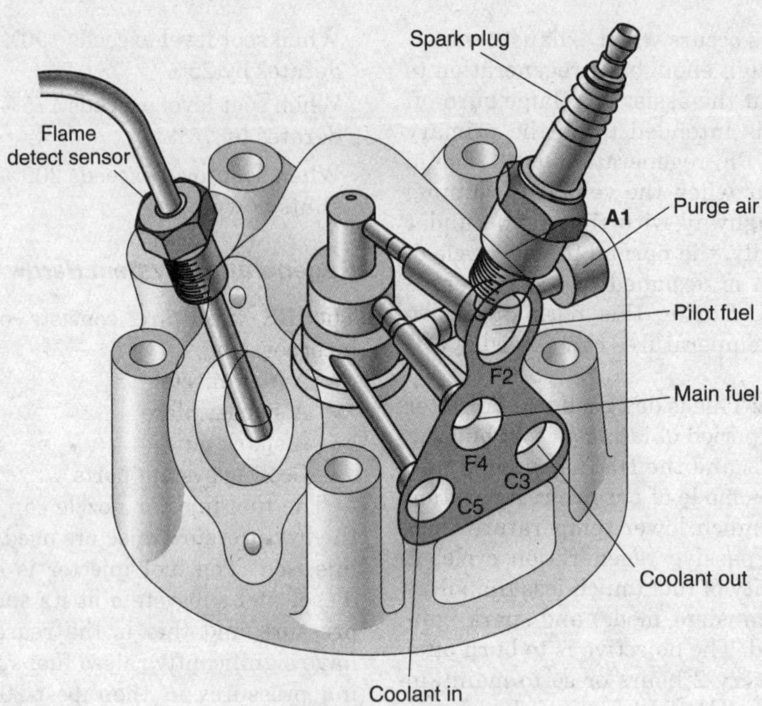

Figure 41–6 *Detail of the CRS head.*

The gasket should be replaced any time the head is removed because it functions as an insulator.

DPF Cleaning Frequency

Regeneration leaves some ash residues. These ashes primarily originate from the additive package in the engine lube that is burned in the cylinder during normal combustion. Ash residues have to be removed manually. Federal law requires that this cleaning process occur no more frequently than once a year or 150,000 miles (240,000 km), whichever comes first. Having said that, some manufacturers are aiming to have RS service intervals that exceed 300,000 highway miles (480,000 km).

Because RS spark plugs are fired routinely (that is, outside RS cycles) to limit carbon buildup, an RS can be deactivated using the OEM software to prevent activation on the shop floor. The DPFs used by most engine OEMs should usually regenerate under *any* conditions: Some manufacturers are using DPFs with extended in-operation cycles, but the disadvantage is that they must pull off the road to regenerate.

Temperature monitoring is by dual and single thermocouples (pyrometers). The operating principles of pyrometers are discussed in Chapter 32, and it is important to note that these thermocouple devices must be replaced as a complete unit when diagnosed as failed. The DPF is a complex assembly within which the regeneration cycles and temperatures have to be precisely managed. Figure 41–7 is a comprehensive schematic of the Caterpillar CRS showing the location of the key components and sensors.

Troubleshooting and Servicing DPFs

DPFs can be evaluated by measuring total exhaust system backpressure. Each engine OEM produces its own backpressure specifications but there is not too much variance between them. During a full load dyno test, typical specified backpressures (measured at the DPF outlet) are in the region of 15 to 20 in. (380 to 508 mm) H_2O. Most engine OEMs will provide a maximum backpressure specification, and if this is exceeded the DPF requires shop cleaning. DPF shop cleaning requires the use of special apparatus provided by the OEM. Attempt-

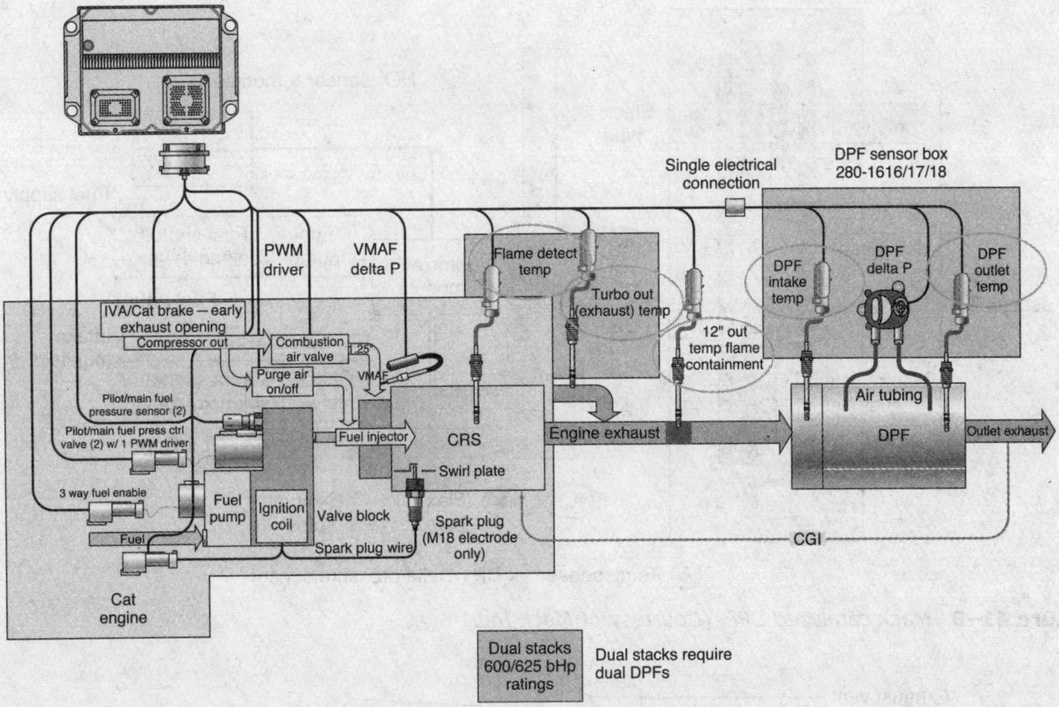

Figure 41–7 *Comprehensive schematic of the Caterpillar CRS showing the location of the key components and sensors. (Reprinted Courtesy of Caterpillar Inc.)*

ing to clean DPFs without using the specified tooling and method can result in damage. Figure 41–8 shows the Mack catalyzed DPF.

DPF-Related Codes DPF performance is monitored by the engine ECM, and typically there are three types of failure that can trigger codes:

- High restriction
- Failure to ignite
- Failure to combust

DPF Cleaning Station

Most engine OEMs state that their DPFs require routine in-shop cleaning using special equipment in the form of a DPF cleaning station. DPF cleaning is designed to remove the inorganic residues of regeneration cycles that collect in the DPF: This inorganic ash is sourced from the engine lube oil and because it does not burn, it accumulates in the filter. DPF shop cleaning stations are either OEM or aftermarket supplied. Although Caterpillar initially supplied a portable DPF cleaning cart that was designed to clean a DPF on chassis, most DPF cleaning requires that the aftertreatment canister be removed from the chassis for the cleaning procedure.

Cleaning Procedure A typical DPF cleaning station is a two-stage device capable of:

- Backpressure pulsing the DPF filter
- Thermally regenerating the filter

A typical DPF cleaning procedure would only require the pulse cleaning stage of the process. After the first stage is complete, the filter is tested for restriction. If the restriction measured falls below specification, the filter is thermally regenerated (baked) followed by a second round of pulse cleaning. Figure 41–9 shows the principles behind a Donaldson generic DPF cleaning station. It should be noted that any DPF cleaning station will only function on certain designated DPFs and that OEM service literature must be reference before connecting a cleaning station to a DPF.

Noncatalyzed DPFs

Noncatalyzed diesel particulate filters (NC-DPFs) are required to be used in highway-certi-

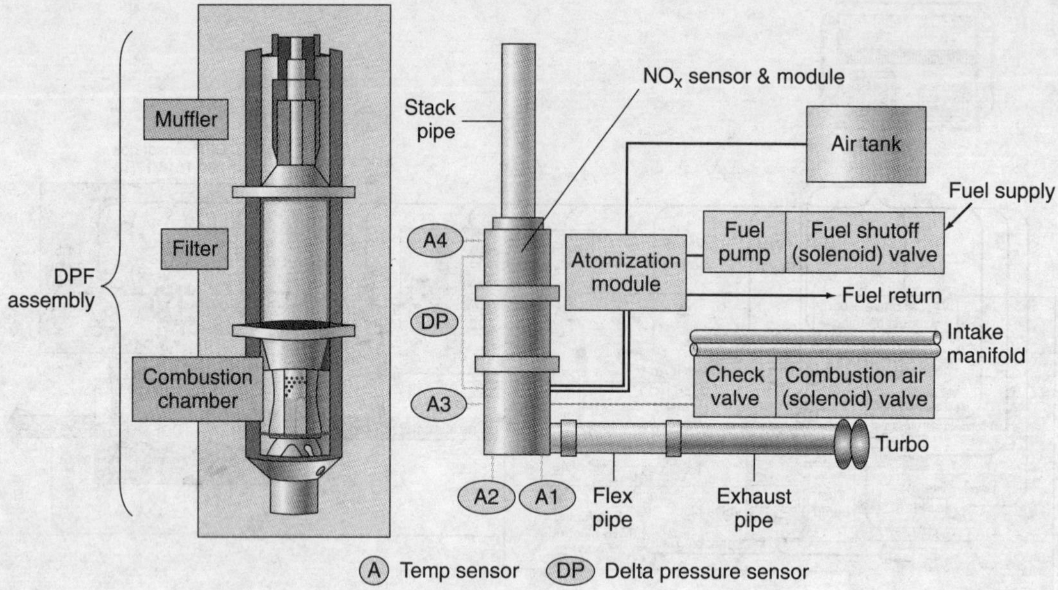

Figure 41–8 *Mack catalyzed DPF. (Courtesy of Mack Inc.)*

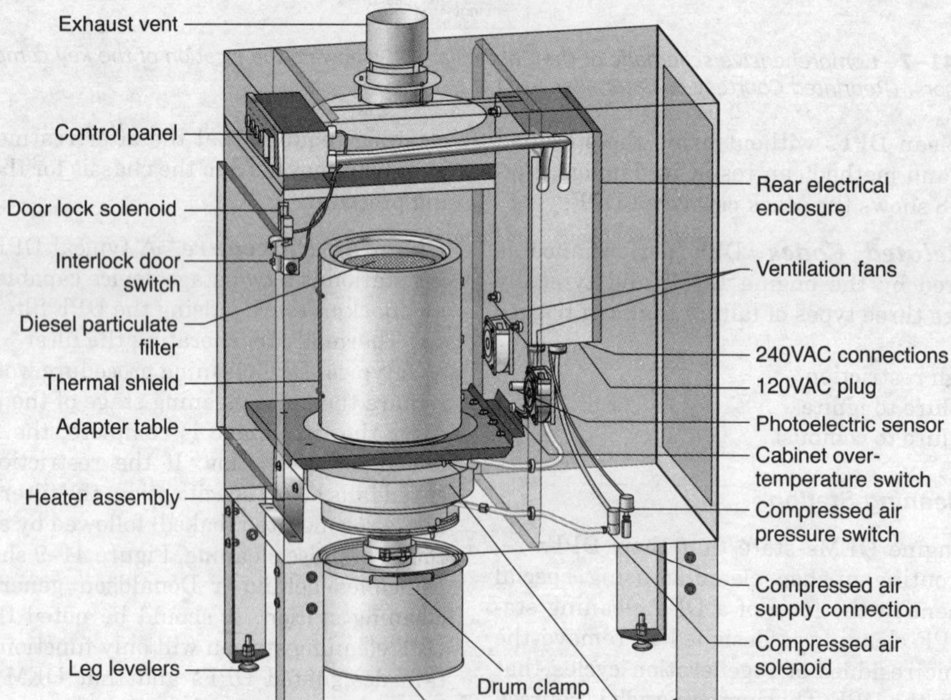

Figure 41–9 *Donaldson generic DPF cleaning cart. (Courtesy of Donaldson Company, Inc.).*

fied, vocational diesel engines that are run for prolonged periods at low engine load. This type of application is distinct in that exhaust temperatures seldom rise high enough for oxidation catalysts to function properly or for DPF passive regeneration cycles. The unit described in this section would be typical of a Mack Trucks unit used in a garbage packer. Because of the lower exhaust temperatures characteristic in this type of vocational vehicle all regeneration cycles are active and may be initiated in the following ways:

- System initiated or automatic. Requires that the vehicle be moving faster than 10 mph (16 km/h)
- Driver or EST software initiated

NC-DPF Controls

NC-DPFs are managed by a dedicated electronic control unit (ECU) known as a thermal control unit (TCU). Although there is provision for a dedicated message identifier (MID) address on the J1939 data bus the Mack Trucks NC-DPF is managed by the engine ECU using a proprietary bus connection. The input and output circuits of a typical NC-DPF are shown in Figure 41–10.

NC-DPF Components

An NC-DPF aftertreatment canister is usually mounted vertically to a stanchion behind the vehicle cab and can be subdivided into three sections as follows:

1. **Combustion chamber section.** Engine exhaust downstream from the turbocharger is piped into the base of the combustion chamber. The key components in or mounted close to the combustion chamber section of the NC-DPF are:
   - Combustion air valve (ported from intake manifold) and check valve
   - Supplementary air valve (ported from the pneumatic system: supplies atomization module)
   - Fuel pump, fuel shutoff valve, and atomization module
   - Ignition module and coil
   - Ignitors (pair of spark plugs)

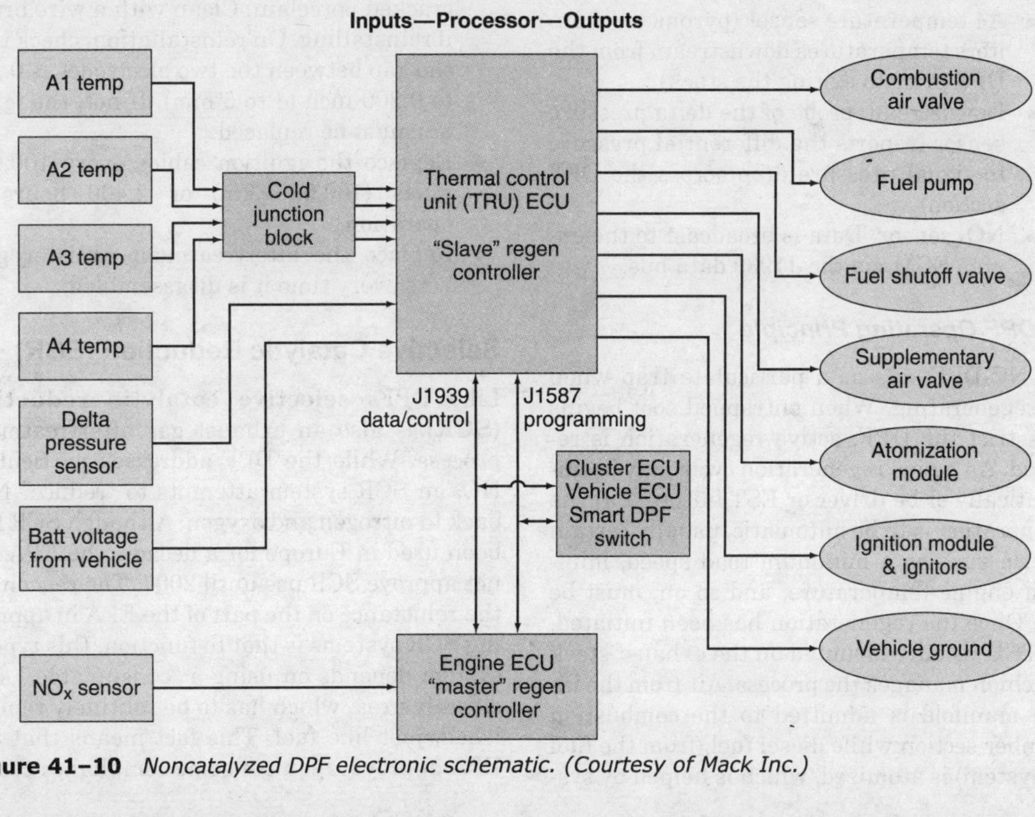

Figure 41–10 *Noncatalyzed DPF electronic schematic. (Courtesy of Mack Inc.)*

- Fuel injection nozzle
- Inlet air temperature probe (A1: exhaust temperature pyrometer)
- Flame temperature probe (A2: flame temperature pyrometer)
- Delta pressure sensor (upstream probe)

2. **DPF section.** This is the center section of the aftertreatment canister and contains the particulate filter. Exhaust gas from the combustion chamber is routed through it to be discharged to the muffler section of the canister. The key components in or mounted close to the DPF section are:
   - DPF filter core
   - Supplementary air valve (ported from the pneumatic system: supplies atomization module)
   - DPF
   - A3 temperature sensor (DPF upstream pyrometer)
3. **Muffler section.** Exhaust gas exiting the DPF section of the aftertreatment canister is routed through the muffler section. The key components in or mounted close to the muffler section of the NC-DPF are:
   - Resonator-type muffler
   - A4 temperature sensor (pyrometer monitors temperatures downstream from the DPF prior to exiting the stack)
   - Downstream probe of the delta pressure sensor (reports the differential pressure to signal pressure drop across the DPF section)
   - NO_x sensor. Data is broadcast to the engine ECM via the J1939 data bus.

NC-DPF Operating Principle

The NC-DPF acts as a particulate trap when not regenerating. When entrapped soot begins to restrict the DPF, active regeneration is required. An active regeneration cycle may be automatically or be driver or EST initiated. If the regeneration is to be automatic, usually certain criteria such as a minimum road speed, minimum engine temperature, and so on, must be met. Once the regeneration has been initiated, the TCU usually mounted on the exhaust stack stanchion manages the process. Air from the intake manifold is admitted to the combustion chamber section while diesel fuel (from the fuel subsystem) is atomized, which is helped by system air pressure and injected across a pair of ignitors (spark plugs). This produces a flame of sufficient intensity to oxidize soot particles entrapped in the DPF.

NC-DPF Maintenance

An NC-DPF must meet the EPA minimum maintenance intervals and will usually exceed them by a margin. Some of the key steps required to maintain a Mack NC-DPF are highlighted here:

- Remove the particulate filter every 250,000 miles (400,000 km) or 4,500 hours to remove accumulated ash.
- Visually inspect the particulate filter. Replace if there is evidence of soot at the outlet or if there are cracks.
- Remove and inspect the fuel injector nozzle every 100,000 miles (160,000 km). Clean with mineral solvent and use compressed air to blow out orifice holes.
- Remove and inspect the ignitors (spark plugs) every 100,000 miles (160,000 km) or 2,400 hours of operation, checking for eroded electrodes, carbon tracking, and cracked porcelain. Clean with a wire brush if reinstalling. On reinstallation, check that the gap between the two electrodes is 0.160 to 0.200 inch (4 to 5 mm). If not, the ignitors must be replaced.
- Replace the ignition cables every 100,000 miles (160,000 km) or 2,400 hours of operation.
- Replace the aftertreatment canister gaskets every time it is disassembled.

Selective Catalytic Reduction (SCR)

Like DPFs **selective catalytic reduction (SCR)** is also an exhaust gas aftertreatment process. While the DPF addresses particulate HC, an SCR system attempts to "reduce" NO_x back to nitrogen and oxygen. Although SCR has been used in Europe for a decade, the EPA did not approve SCR use until 2007. The reason for the reluctance on the part of the EPA in approving SCR systems is that to function, this type of system depends on using a "consumable," specifically urea, which has to be routinely replenished, just like fuel. This fact means that the EPA approval requires some compliance conditions. Very briefly, we will take a look at the simple chemistry behind SCRs here.

SCR Chemistry

NO_x are formed when conditions in the combustion chamber have caused the normally inert nitrogen (N_2) component of air to oxidize to compounds of oxygen and nitrogen. SCRs attempt to "reduce" these compounds back to harmless elemental oxygen (O_2) and nitrogen (N_2) that are the constituents of the air we breathe. Gasoline-fueled auto engines have used NO_x reduction catalytic converters (rhodium is the catalyst) for over 30 years, but these have the disadvantage of only being able to function when the AFR is at stoichiometric or richer. They do not function as stand-alone devices in lean burn diesel engines and, when used, are incorporated into a more complex device known as a NO_x adsorber catalyst, which is described earlier in this chapter.

Aqueous Urea SCRs use consumable **urea** to achieve what the rhodium catalyst does in the gasoline-fueled engine. Urea is composed of crystallized nitrogen compounds sourced from natural gas. The urea is in a solution with water, known as **aqueous urea**. The aqueous urea is injected into the exhaust gas stream by a computer-controlled injection system. After injection, the urea reduces to ammonia, which itself reacts with NO_x compounds, reducing them back to oxygen and nitrogen.

SCR Management

SCR is managed by the engine ECM. The urea is contained in aqueous/water form (32%) in a replenishable vessel. One manufacturer name for a 32% cut of aqueous urea is **AdBlue**. It is injected upstream from the converter/DPF/muffler assembly sometimes assisted by on-chassis compressed air. The urea injection has to be precisely metered by the ECM. Too much urea can result in ammonia discharge through the exhaust system. Too little results in NO_x emission. Figure 41–11 shows a schematic of SCR components and operation.

Urea is carried onboard the truck in tanks with capacities of 20 to 50 gallons (75 to 190 liters). An aqueous urea solution freezes at 12°F (–8°C) so it must be freeze protected in winter operation. The urea is consumed at a rate that varies between approximately 1.5% and 10% of the fuel used, the variability depending on how the engine is being operated, that is, how much NO_x has to be reduced.

EPA Compliance Guidelines

The EPA granted approval for OEMs to use SCR systems to meet NO_x reduction requirements in 2007, but stresses that OEMs must obtain certification approval for each specific system. According to the EPA, an SCR management strategy must incorporate:

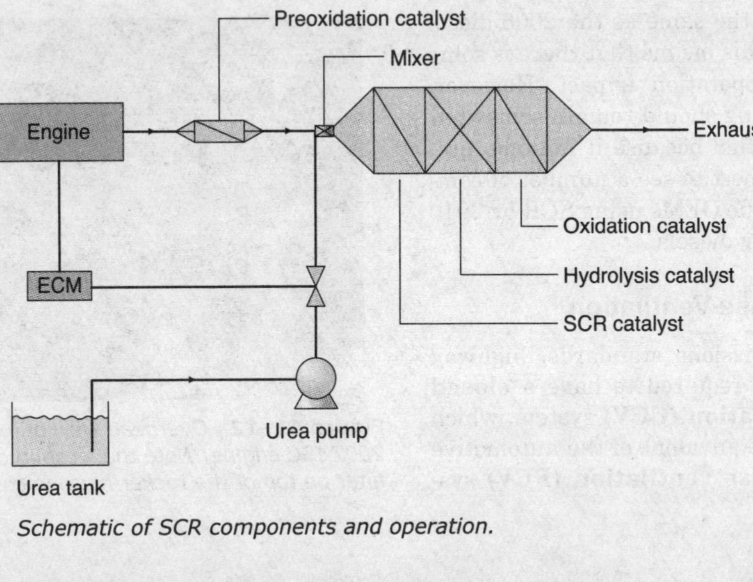

Figure 41–11 *Schematic of SCR components and operation.*

- Driver warning system: alerts when the urea reservoir is low
- Driver inducement: limited number of restarts, fuel lock-out system, or engine derate, until the reducing agent is replenished
- Monitoring system for aqueous urea quality to prevent dilution or substitutes
- Supply infrastructure: assurance to aqueous urea availability
- System reliability and durability: a warranty precondition of 435,000 miles (700,000 km) and minimum service interval of at least 150,000 miles (240,000 km) for heavy-duty highway diesels
- Tamper-resistant protection and deactivation protection

By the time these systems become commonplace in 2010, the EPA expects to receive assurances from diesel engine OEMs that its compliance guidelines can be met. Included in this will be the development of a supply infrastructure that should make aqueous urea available at truck stops, OEM dealerships, fueling stations, and repair/service operations. The EPA suggests that some last-resort persuasive alerts to the operator should be incorporated in an SCR system to minimize the chances of the system being operated without the urea reducing agent. Compliance will also require the system to be able to detect either incorrect or excessively diluted reducing agent.

In the United States a 32% aqueous urea solution costs about the same as the 2006 diesel fuel pump price. This means that there is some negative cost of operation impact. However, aqueous urea pricing should remain somewhat more stable than fuel because it is home produced. You can expect to see a number of commercial diesel engine OEMs using SCR by 2010 and most light-duty diesels.

Closed Crankcase Ventilation

To meet 2007 emissions standards, highway diesel engines are required to have a **closed crankcase ventilation (CCV)** system, which is a diesel engine equivalent of the automotive **positive crankcase ventilation (PCV)** system. The objective of a CCV system is to prevent venting to atmosphere of crankcase gases. Crankcase gases consist of:

- Blow-by gas from engine cylinders
- Boil-off gases from lubricant

Crankcase Gas Composition

The actual composition of gases in a diesel engine crankcase is a complex mix of raw fuel, partially combusted fuel, combustion by-products, and the boil-off residues of engine lube. There is agreement that it is toxic, but it is not produced in large quantities. The chemical action in the crankcase is made even more complex because of the presence of moisture-laden air and heat combined with plenty of sloshing and churning activity. However, as technicians we classify crankcase gas simply as HC emission.

CCV Operation

Diesel CCV systems can be directly compared with PCV circuits used on automobiles. The system is plumbed upstream from the turbocharger impeller housing so that some pull is exerted on the crankcase. The CCV piping is routed through a filter assembly, which in the case of the 2007 Cummins ISC engine shown in Figure 41–12, is located on top of the rocker housing cover. OEM recommendations for CCV service intervals should be observed.

Figure 41–12 *Overhead view of a Cummins 2007 ISC engine: Note the location of the CCV filter on top of the rocker housing cover.*

Auxiliary Devices

In an effort to meet ever more stringent emissions standards, the diesel engine has changed over the years. Here are some of the changes that have taken place over the years and how they have helped reduced emissions. Most of them are still used or have an equivalent in the most recent versions of diesel engines.

VCO Injectors

Valve closes orifice (VCO) injectors were introduced in Chapter 20. Elimination or reduction of nozzle sac volume in hydraulic injection injectors reduces the cylinder boil/dribble of sac fuel at the completion of injection. This nozzle design principle is used in all injectors today.

Charge Air Cooling

Effective cooling of intake air lowers combustion temperatures, making it less likely that the nitrogen in the air mixture is oxidized to form NO_x. Air-to-air charge air coolers cool air more effectively than those that rely on engine coolant. Note that anything that compromises the charge air cooler's ability to cool will result in higher NO_x emissions. This is why codes are logged when charge air heat exchangers become plugged.

Variable Geometry Turbochargers

Variable geometry turbochargers (VGTs) that perform effectively over a much wider load and rpm range can make a significant difference to both HC and NO_x emissions, especially when ECM controlled, by providing the ability to manage boost on the basis of the fueling and emissions algorithms. VGTs are used rather than constant geometry turbos in almost every U.S. 2007 compliant highway diesel engine. When a constant geometry turbocharger is used today, it is usually as one of the pair used in series turbocharged engines.

Low Headland Pistons

Low headland volume pistons raise the upper compression ring close to the leading edge of the piston crown. This keeps the headland gas volume close to minimum. Headland gas volume tends to be unreactive and can increase HC emissions. The requirement for low headland volume pistons by engine designers has resulted in some radical design changes in diesel engine pistons within a short period of time. Most diesel engine manufacturers favor trunk-type steel pistons today such as the Mahle Monotherm™, which feature low headland volumes.

Aneroids

Although most types of manifold boost measuring devices have been covered elsewhere in different chapters of this book, some of the more common methods of smoke control used on older hydromechanical engines are briefly examined here for ease of reference. The way in which each operates has dictated how the actual test compliance procedure (refer to J1667 testing later in this chapter) has been determined.

- Caterpillar AFRC (air-fuel ratio control). An aneroid device that is not effective until after the first boost application. This feature permits the engine to be full fueled for cold engine startup but not after the AFRC has been activated.
- Cummins AFC (air-fuel control). An aneroid built into the PTG pump. Manifold boost is ported directly to the AFC diaphragm, which can then modulate fueling based on actual boost values.
- Mack Trucks puff limiter. Used on many Mack engines of the 1980s, the puff limiter used an aneroid device located on the intake manifold to sense manifold boost and system air pressure to signal a plunger-type bimba valve located on the fuel injection pump and designed to block rack travel to full fuel. It functioned by cutting system air pressure to the bimba valve when boost pressure was sufficient: This allowed devices such as air brake status and transmission gear ratio to be used to limit unnecessary full fueling.
- Robert Bosch LDA. LDA is a German acronym meaning "aneroid": It became known by many Mack Truck technicians as a "lazy dog adjuster" because of its connection with lack of power complaints. The device is a

direct-acting aneroid that requires that manifold boost be ported to it. Like the puff limiter, this type of aneroid is on/off and rated with a trigger value.

Throttle Delay Cylinder

Perhaps the most primitive of smoke control devices, the Detroit Diesel throttle delay cylinder was used on its two-stroke cycle, mechanical unit injector (MUI) engines. Its function was to create a hydraulic lag between a full fuel request (from the accelerator pedal) and the act of moving the rack control tube to full fuel. It consists of a throttle delay cylinder and piston that reciprocates within it: The piston is attached by means of a lever actuated by the control rack. Engine oil is drizzled into the throttle delay cylinder so that when the control tube attempts to move from a low or idle fuel position, oil has to be displaced from the throttle delay cylinder first by means of a restriction port. You should know something about how throttle delay cylinders function if you are performing opacity compliance testing. These devices were defeated by drilling to oversize the cylinder spill port or by loosening the rack yoke.

FIELD TESTING OF SMOKE DENSITY

Field testing for emissions refers to any testing performed after an engine series has been EPA certified. In the past couple of years, most U.S. states and Canadian provinces have launched highway diesel testing programs that conform to the SAE J1667 standard. The implementation of J1667 test programs can be regarded as a first step in the development of a comprehensive field testing program that examines real world emissions of computer-controlled diesel engines. At present few jurisdictions require exhaust gas analysis.

Opacity Meters

Opacity meters are simple to use and usually require that the unit's sensor head probe be fitted to the outlet of the exhaust pipe. Two different types are used: partial flow and full flow. In the partial flow type, during the test procedure, a portion of the exhaust gas flow is diverted to a sensing chamber in which the opacity of the smoke can be read by the light sensor in the head assembly. A partial flow opacity meter is shown in Figure 41–13. The full flow type is more commonly used and makes the opacity measurement directly at the stack outlet. Figure 41–14 shows the operating principle of a full flow opacity test meter.

Light Extinction

Partial and full flow opacity meters are classified as light extinction test instruments: A beam of light from a light-emitting diode (LED)

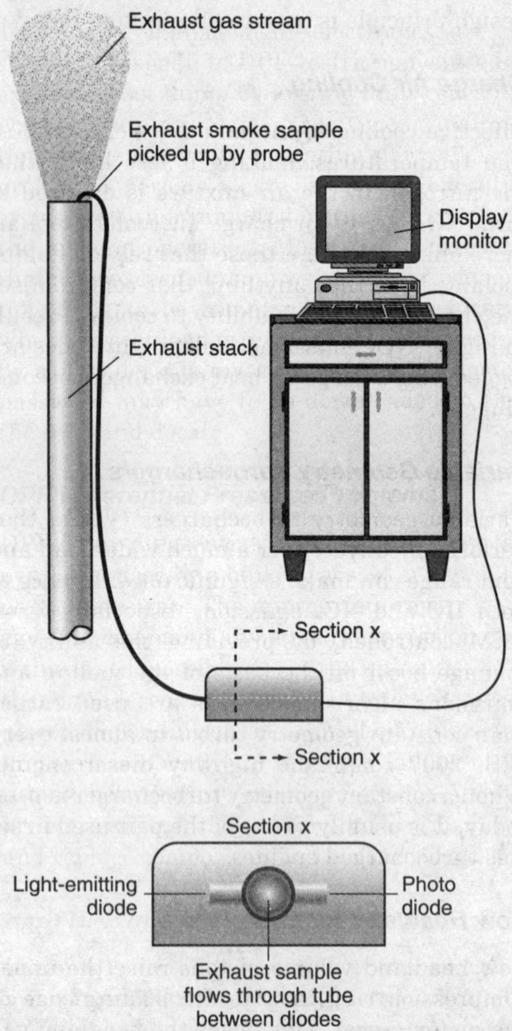

Figure 41–13 *Partial flow opacity meter.*

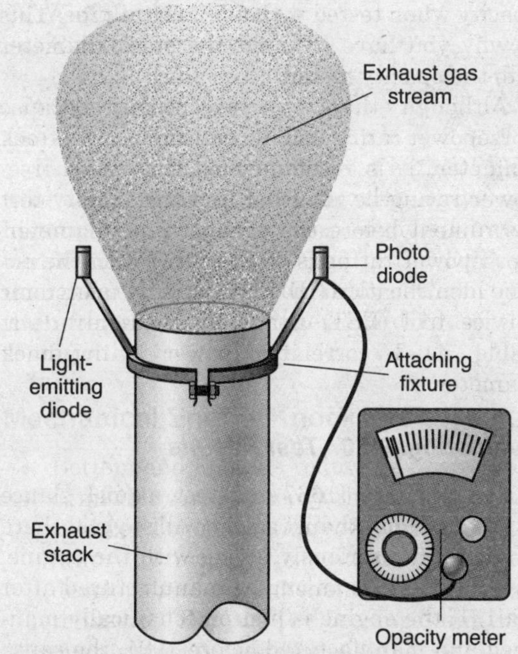

Figure 41–14 *Full flow opacity meter.*

is directed at a photo diode sensor through the exhaust gas stream. The amount of light blocked from the photo diode sensor is determined by the particulate density in the engine exhaust. The higher the particulate density, the less light is capable of penetrating it to be read by the sensor. Smoke density is expressed as a percentage reading by the opacity meter. In most cases, the opacity meter is equipped with an extension handle for the sensor head, so the device can be placed at the exhaust stack outlet by the technician working at ground level.

Opacity meter readings are displayed in percentage readings and the actual readings must always be factored with engine power output and the exhaust system outlet pipe diameter. Most should produce accurate results regardless of weather conditions and many will log the data electronically. The data logged into the opacity meter can be hardcopy printed or transferred to personal computer (PC) or Web-based systems for analysis. Manufacturers of opacity meters claim accuracy factors within 2% and programmed response times within 1 second. Any opacity meter can be used to evaluate smoke density in diesel smoke. However, those used for official testing and to enforce compliance must be approved and have PC-managed and logged test sequences.

Performing J1667 Testing

In 1997, the EPA recommended that all jurisdictions adopt the SAE J1667 procedure for the testing of heavy-duty diesel vehicles perhaps because it would promote some consistency in test standards between different jurisdictions. The test routine outlined here adheres closely to that performed by compliance enforcement agencies. While the emissions standards do vary by jurisdiction and year of manufacture of the truck, there is some consistency nationwide, and recently the maximum opacity standards have been reduced in many jurisdictions.

Ambient environmental conditions can have some impact on smoke density, and changes in air density influence the operation of any internal combustion engine. Air density is influenced by ambient temperature, and, generally, high air density will result in higher air-fuel ratios, which will tend to produce lower smoke opacity readings. For this reason, SAE J1667 imposes restrictions on the environmental conditions during the administration of an "official" snap acceleration test.

Environmental Conditions

The following environmental conditions are required when performing a J1667 test:

- Temperature in the test area must be between 35°F (2°C) and 86°F (30°C).
- Testing must not take place outdoors when there is visible precipitation such as rain, snow, or fog.
- Testing must not take place when the temperature is at or below the dew point, such as during fog or high humidity conditions.
- Vehicles using downward-directed exhaust systems should be tested over ground surface conditions that are not dusty so that dirt particulates are not combined with the exhaust gas being measured.

Snap Test

Each J1667 throttle snap consists of a three-phase cycle that must be precisely executed by the test administrator:

1. Accelerator is snapped to high idle and held for 1 to 4 seconds or until prompted by the opacity meter to release.

2. Engine rpm must be allowed to drop to the specified low idle speed.
3. Engine must run at the specified low idle for a minimum of 5 seconds and a maximum of 45 seconds before initiating the next snap as prompted by the opacity meter.

SAE J1667 Test Cycle

The SAE J1667 test cycle consists of four phases and, again, must be precisely executed by the test administrator:

1. Preliminary snaps. Three full snaps are required to clear the exhaust system of loose particles and precondition the vehicle.
2. Official snaps. Three official snaps should be undertaken as prompted by test instrument software.
3. Validation. The difference between the highest and the lowest maximum opacity readings of the three official test snaps should be within five opacity percentage points. If the difference is less than 5%, the meter software will compute the average. If the difference is greater than 5%, additional official snaps must be undertaken up to a maximum of nine.
4. Drift factor. Validation is required that the drift between three official test cycles does not exceed 2% opacity. A variation greater than ± 2% will result in an invalid test.

Stack Dimensions

Exhaust pipe diameter has an effect on the opacity reading of exhaust gas. Before undertaking a snap test, the diameter of the exhaust stack has to be known. For instance, an engine that tests at 30% opacity when tested with a 3-inch exhaust stack would only measure 20% opacity when tested with a 5-inch tailpipe. This is why you have to input the stack diameter into the opacity meter.

Although there are exceptions, because horsepower rating usually determines the stack diameter, it is recommended that the horsepower rating be inputted into the opacity test instrument before the exhaust pipe diameter. Horsepower rating is often specified on the engine identification (ID) plate or in the electronic service tool (EST) accessible read-only data. Table 41–3 correlates power with stack diameter.

Evaluating J1667 Test Results

If a truck fails a J1667 snap test, a quick glance at the emitted exhaust smoke will suggest there is something seriously wrong with the engine, especially if it is an engine manufactured after 1991. If the engine is hydromechanically managed and manufactured before 1991, the cause is frequently in the boost management system, which is a circuit that is often managed by an aneroid device.

Opacity Readings

Because J1667 is used for compliance testing diesel engine emissions, the compliance test profiles have to be simple and not vulnerable in court if offenders decide to fight fines. That is why the snap test is used. However, if you have an opacity meter, you can also look at some of the other criteria that were components of J1667. Smoke opacity readings of 5% or less are not easily observed. When using an opacity meter, it is important to meticulously observe the test procedures if any credence is to be given to the results. Remember that all of the data in Table 41–4 relates to engines manufactured before 1991.

Table 41–3 *Horsepower and Recommended Exhaust Pipe Diameter*

Rated Engine Power (BHP)	*Rated Engine Power (kW)*	*Exhaust Pipe Diameter (Inches)*	*Exhaust Pipe Diameter (Millimeters)*
Less than 101	Less than 75	2	51
101 to 200	75 to 149	3	76
201 to 300	150 to 224	4	102
301 or more	225 or more	5	127

Table 41–4 *Original (1991) J1667 Specifications*

Engine Operating Mode	*Maximum Opacity (Smoke Density)*
Idle speed	10%
High-idle speed	15% (transient response to throttle snap)
Rated speed and low (peak power)	20% (requires chassis dynamometer)
Engine lug	25% (requires chassis dynamometer)
Maximum puff on throttle snap	50%
Full acceleration	30% (requires chassis dynamometer)
Cold start at 0°F (–17°C)	Specific to OEM in original J1667

(NOTE: Current compliance specs vary by jurisdiction.)

SMOKE ANALYSIS

Any truck diesel engine that fails an opacity snap test has a *severe* smoking problem, one that could not pass unnoticed under any circumstances. Many sick engines emitting large amounts of visible smoke can still pass the opacity standards defined in J1667 testing. Chapter 48 on troubleshooting explains in some detail how to evaluate smoke emission from diesel engines, but in this section we quickly outline some examples. The appearance of exhaust smoke to the observer's eye depends on the *state* of the emission. Physical matter has three states: vapor or gaseous, liquid, and solid.

- Gaseous emission. To the observer's eye, this is "clean" exhaust smoke. And it may be clean—or contaminated with noxious pollutant in a vapor state. Because light is not affected by passing through a gas, we see nothing when we look at exhaust in a 100% gas state exiting the stack(s).
- Liquid emission. To the observer's eye, liquid condensing in the exhaust gas stream appears white. When light attempts to pass through condensing liquid droplets in the exhaust it either *reflects* or *refracts,* making it appear white to the observer.
- Solid emission. To the observer's eye, solids (we call these *particulates*) contained in the exhaust gas stream appear black. When light attempts to pass through particulate suspended in the exhaust gas it is blocked, making it appear black to the observer.

Check out the section titled "Smoke Analysis" in Chapter 48 for a more detailed look at the causes of diesel engine smoking.

REVIEW QUESTIONS

1. Which of the following compounds is not classified by the EPA as a noxious emission?
   a. Oxides of nitrogen
   b. Hydrocarbons
   c. Sulfur dioxide
   d. Carbon dioxide
2. By mass, which of the following is present in the largest quantity in the diesel engine cylinder during combustion?
   a. Oxygen
   b. Nitrogen
   c. Fuel
   d. Carbon monoxide
3. Which exhaust gas emission does a NO_x adsorber catalyst attempt to reduce?
   a. Particulate matter (PM)
   b. Oxides of nitrogen
   c. Hydrocarbons
   d. Ozone
4. Which of the following exhaust gas test instruments uses a light extinction principle to measure smoke density?
   a. CVA sampler
   b. Seven-gas analyzer
   c. Two-gas analyzer
   d. Opacity meter
5. Which of the following would properly describe atmospheric oxygen at ground level?
   a. O
   b. O_2
   c. O_3
   d. O_4
6. Which two compounds would result from theoretically "perfect" combustion of an HC fuel reacted with oxygen?
   a. Carbon monoxide and carbon dioxide
   b. Carbon dioxide and nitrogen dioxide
   c. Nitrogen dioxide and water
   d. Carbon dioxide and water

7. When a diesel engine is operated at lower-than-normal combustion temperatures, which of the following categories of noxious exhaust emissions is likely to increase?
   a. Oxides of nitrogen
   b. Carbon dioxide
   c. Ozone
   d. Hydrocarbons
8. When a diesel engine is operated at higher-than-normal combustion temperatures, which of the following categories of noxious exhaust emissions is likely to increase?
   a. Oxides of nitrogen
   b. Carbon dioxide
   c. Ozone
   d. Hydrocarbons
9. Which SAE standard defines the guidelines for testing diesel engine exhaust opacity emissions?
   a. J1587
   b. J1667
   c. J1708
   d. J1939
10. Which of the following compounds is classified as a greenhouse gas responsible for contributing to global warming?
    a. Oxides of nitrogen
    b. Hydrocarbons
    c. Sulfur dioxide
    d. Carbon dioxide
11. *Technician A* states that advancing the static timing of a fuel injection pump by one crank degree can increase NO_x emission by up to 10%. *Technician B* states that retarding static timing could produce higher HC emissions. Who is correct?
    a. Technician A only
    b. Technician B only
    c. Both A and B
    d. Neither A nor B
12. *Technician A* states that the maximum puff test when using an opacity meter can be performed with the vehicle stationary and zero load on the engine. *Technician B* states that the full acceleration test when using an opacity meter is a throttle snap test that can be performed with the vehicle stationary. Who is correct?
    a. Technician A only
    b. Technician B only
    c. Both A and B
    d. Neither A nor B
13. *Technician A* states that the combustion of diesel fuel is the major cause of ambient sulfur dioxide in North America. *Technician B* states that sulfur dioxide is the main cause of sulfurous smog. Who is correct?
    a. Technician A only
    b. Technician B only
    c. Both A and B
    d. Neither A nor B
14. Which of the following best describes what happens in an oxidation-type catalytic converter?
    a. Nitrogen in NO_x is combusted.
    b. CO and HCs are combusted.
    c. HCs are entrapped on substrate.
    d. Particulate matter is filtered out.
15. *Technician A* states that DPFs can run up to 8 hours in a linehaul diesel engine before regeneration. *Technician B* states that some DPFs must be removed from the vehicle for cleaning on a weekly basis. Who is correct?
    a. Technician A only
    b. Technician B only
    c. Both A and B
    d. Neither A nor B
16. *Technician A* states that photochemical smog requires both HC and NO_x plus a period of exposure to sunlight. *Technician B* states that the presence of ozone is a requirement for creating photochemical smog. Who is correct?
    a. Technician A only
    b. Technician B only
    c. Both A and B
    d. Neither A nor B
17. *Technician A* states that most DPF regeneration in highway diesel engines should occur as passive regeneration. *Technician B* states that DPF active regeneration cycles usually require the injection of fuel upstream from the DPF. Who is correct?
    a. Technician A only
    b. Technician B only
    c. Both A and B
    d. Neither A nor B
18. How many official throttle snaps are required when performing a J1667 opacity test?
    a. One

b. Two
c. Three
d. Four

19. A hydromechanically managed, diesel-powered truck is road tested and produces a puff of smoke at each upshift. Which of the following devices should be checked out first?
a. Shift tower air signal
b. Aneroid
c. Turbocharger wastegate
d. Governor torque trim spring

20. Which of the following is *not* an aneroid device?
a. DDC throttle delay cylinder
b. Caterpillar AFRC
c. Mack Trucks puff limiter
d. Cummins AFC

42

Troubleshooting and Diagnosis

INTRODUCTION

The term **troubleshooting** is generally used to describe noninvasive (meaning that as little of the engine is disassembled as possible) methods of determining the cause of an engine problem. These methods can vary from educated guesswork to highly structured procedures, such as are required when diagnosing electronic engine management systems. Manufacturers make troubleshooting guides available for purposes of diagnosing common engine complaints. This chapter should be used in conjunction with Chapter 33, which outlines how to use proprietary **service information systems (SIS)** and **electronic service tools (ESTs)**. We use the acronym SIS generically to describe the various proprietary original equipment manufacturer (OEM) online diagnostic software programs.

TROUBLESHOOTING TOOLS

The following tools are required to effectively troubleshoot diesel engines today:

- EST loaded with the OEM diagnostic software and SIS
- Serial link interface or **communications adapter (CA)**
- Wireless or wired Internet connection
- Breakout boxes, Tees, terminal spoons, and a digital multimeter (DMM)

Thinking Inside the Box

Troubleshooting today's diesel engines could not be more different than troubleshooting your Dad's diesel engine. Troubleshooting is driven by diagnostic software. In most cases, you are required to connect to the chassis data bus and access the engine message identifier (MID). The EST you are required to use is loaded with the OEM diagnostic and has an active Internet connection. After connecting with the engine MID, you are required to follow a structured troubleshooting path, which is often interactive. In other words, the technician could be instructed to make a voltage measurement using a DMM, then input the value measured into the diagnostic software. The next step in the diagnostic sequence depends on the value input. Troubleshooting electronically managed engines uses this type of structured procedure based on software-guided **sequential troubleshooting charts** because it reduces the time required to identify the source of a problem. Failure to observe OEM diagnostic routines can result in:

- Denial of warranty
- Excessive amounts of wasted troubleshooting time
- Trial and error diagnostics: This practice makes the whole industry look bad
- Engine damage incurred by the troubleshooting itself

You might argue that thinking inside the box does not require you to think. That is correct in a way; while you navigate diagnostic software you are really just obeying a sequence of instructions. The bottom line is that a high percentage of engine malfunctions can be quickly and accurately identified as to root cause using computer diagnostics. Another major advantage of online SIS is that a repair procedure can be tagged with all the relevant **technical service bulletins (TSBs)**. Technicians today have to learn to accept that their technical ability should only be brought into play when diagnosing a problem exceeds the capability of the software diagnostic maps.

Thinking Outside the Box

Because today's diesel engines are capable of some pretty exacting self-diagnosis, the most challenging troubleshooting problems tend to be those that fall just outside the scope of the troubleshooting paths embedded in the OEM software. This requires a troubleshooting technician who can think outside of the box, when the limitations of the diagnostic software are exceeded. The trick is to methodically work with the software-guided diagnostic routines as required by the OEM—and only when this fails to produce a result should the technician resort to that higher level of analytical skills that make the human brain a superior instrument to a computer. Thinking outside the box requires a toolbox of technical skills, some of which take some experience to develop:

- Thorough understanding of a diesel engine's mechanical systems
- Thorough understanding of the engine's computer control systems
- OEM-specific training on the engine the technician is working on
- Ability to navigate the online SIS using an EST and OEM software
- Ability to navigate OEM wiring schematics and virtual schematics

The general objective of this chapter is to help technicians to work methodically and not overlook an obvious possible cause of a problem. Although we cannot sufficiently stress the importance of online SIS, sometimes technicians forget that an engine is a fundamentally mechanical device and that some problems not only require knowledge of how those mechanical systems fail, but may not be identified by diagnostic software.

Online Service Information Systems

All truck diesel engine OEMs require that technicians use online SIS to quarterback troubleshooting routines. These are covered fully in Chapter 33 but must be also referenced here. As indicated earlier, the consequence of not using the OEM SIS can result in excessive time in identifying the cause of a problem and possible denial of warranty. The advantage of using SIS is that technology changes so rapidly that hard paper approaches to repair methods become dated as soon as they are published. Most OEMs update their online information daily.

SIS is usually oriented by engine serial number and provide the technician with user-friendly Website search engines that can even include parts data. Most truck chassis OEM SIS use an umbrella approach that permits the technician direct access to the SIS Web locations of the manufacturer of the chassis subsystems. For instance, if you log onto Navistar's ISIS, among the options you can select will be a route into Cummins QSOL because Cummins supplies engines as an option to Navistar chassis. There is a double message when it comes to using SIS. Yes, using online information systems is crucial but an expert technician also must have a good understanding of engine mechanics.

SMOKE ANALYSIS

An engine can emit from its exhaust noxious gases, water, and carbon dioxide. Providing these emissions remain in the gaseous state, they will not be observed exiting the exhaust piping. For exhaust gas to be identified as smoke, the state of the emission must be in a liquid or a solid state. When smoke appears to be black, the state of the emission is solid; specifically, particulate solids through which light will not pass at all, making it appear black to the observer's eye. The root causes of black smoke are associated with incompletely combusted fuel or engine oil.

When exhaust smoke is white in color, the emission is in the form of condensing liquid droplets from which light reflects or refracts, making it appear white to the observer's eye. The condensing liquid could be engine coolant, fuel, or when cold enough, the water that is a normal product of combustion. When smoke is blue, it is normally associated with engine oil emission. This is usually classified as a condensing liquid emission from which light is mainly reflected rather than refracted, due to the usually dark color of engine oil.

As outlined in a previous chapter, there are three states of matter—solid, liquid, and vapor. The technician analyzing the smoke emitted from a diesel engine is making an observation on the state of the emission. It is important to note that the absence of observable smoke emission does not mean that the engine is not producing noxious emissions. It means only that engine emissions are in a gaseous state. Gaseous emissions can only be analyzed using costly exhaust gas analysis equipment. Nevertheless, observable smoke emitted from an engine tells a story and the technician should have some ability to interpret the causes of a smoking engine.

In summary, exhaust emission from the stack(s) of a diesel engine can appear in one of the following four ways:

1. Clear. The exhaust gas stream is in a gaseous state.
2. White. Some liquid is condensing in the exhaust gas stream.
3. Black. Some solid matter (particulate) is contained in the exhaust gas.
4. Blue. This usually indicates presence of engine oil in exhaust gas.

Black Smoke

The term **black smoke** is used to describe anything from a greyish haze to heavily sooted exhaust gas emission. It is the result of the incomplete combustion of fuel and therefore has many causes, the most common of which are:

- Insufficient combustion air
- Restricted exhaust
- Exhaust gas recirculation (EGR) failure
- Diesel particulate filter (DPF) failure
- Excess or irregular fuel distribution
- Improper grade of fuel

Insufficient Combustion Air

The causes for insufficient combustion air are air starvation caused by any performance defect in the air intake system components from the air filter, turbocharger, boost air/heat exchanger, plugged EGR or CGI heat exchanger, EGR or CGI control valve failure, through to intake valve problems; clogged cylinder sleeve ports (two-stroke cycle engines); restricting emergency stop gate, and so on. Check for a problem that can be read electronically first in electronically managed engines. Test for intake air inlet restriction using a negative pressure gauge or manometer that reads in inches of H_2O (mm of H_2O). The specifications should always be checked to the OEM values, but typical maximum values will be close to:

15″ H_2O (380 mm H_2O) vacuum—naturally aspirated engines

25″ H_2O (635 mm H_2O) vacuum—boosted engines

Exhaust System Restriction

Exhaust system restriction problems can usually be related to turbocharger failure, collapsed exhaust system piping, internal failure of engine silencer or catalytic converter, a plugged-up DPF, or exhaust brake malfunction. Because of emissions systems electronic monitoring found in most post-2004 and all post-2007 highway diesel engines, connect to the chassis data bus to troubleshoot even if a fault code has not been logged.

Excess Fuel/Irregular Fuel Distribution

Excess fuel (overfueling) or irregular fuel distribution can be caused by out-of-spec injection timing (intentional or accidental), injector nozzle failure, variable timing control device failure, unbalanced fuel rack settings, lugging engine (operation at high loads at speeds below peak torque rpm), incorrect governor settings, inoperative or tampered with manifold boost management system, or defective barometric capsule (altitude compensator). When troubleshooting electronically managed engines, make sure you connect to the chassis data bus, select the engine MID (see Chapter 33), and perform a cylinder cutout test: This can pinpoint unbalanced cylinder fueling and usually identify the cause.

Improper Fuel Grade

When fuel is stored for prolonged periods, the more volatile fractions evaporate, altering the fuel's chemical characteristics. Additionally, fuel suppliers seasonally adjust fuel to accommodate temperature extremes. The use of used engine lube/fuel mixers often produces smoking. Heavier residual oils may not vaporize, and if they do, there is insufficient time to properly combust them. This practice may cause high acidity in the exhaust gas (high sulfur), which may rust out exhaust systems unusually quickly. Ensure that the engine lube/fuel mixer permits the engine to meet the emissions standards for the year the engine was manufactured. Check the American Petroleum Institute (API) gravity using a hydrometer as outlined in Chapter 17.

White Smoke

White smoke is caused by condensing liquid in the exhaust gas stream. Temperature usually plays a role when white smoke is observed, both ambient temperature and the engine-operating temperature. Remember that water is a natural product of the combustion of any hydrocarbon fuel and in midwinter conditions it is normal for some of this to condense in the exhaust gas.

However, when white smoke is a problem, the following are some of the possible causes.

Cylinder Misfire

Check for a problem that can be read electronically first on electronically managed engines; perform an electronic cylinder cutout test using a personal computer (PC) or other EST if that option is in the software diagnostics. With nonelectronic engines, disable each injector in sequence using a rocker claw/lever for mechanical unit injector (MUI) and pressure-time (PT) injectors, and by cracking (loosening the line nut) high-pressure pipes in hydromechanical pump-line-nozzle (PLN) systems.

Low Cylinder Compression Pressure

There are a number of possible causes of low cylinder compression. These are usually mechanical problems that prevent the cylinder from sealing properly. The procedure for checking cylinder compression pressures is outlined later in this chapter but this type of test is seldom performed today due to the speed and accuracy of electronic cylinder balance tests.

Low CN Fuel

This is more likely to be a problem experienced by operators that bulk purchase fuel in large volumes and then store it for prolonged periods. Diesel fuel cetane number (CN) is defined by its most volatile fractions and it is these that are most likely to boil off when exposed to high temperatures. Low CN fuel may produce white smoke emission: This appears as a haze in the exhaust gas stream when fueling a warm engine. Test the fuel specific gravity using a fuel hydrometer and compare to the fuel supplier's specification. Consult the fuel supplier as to the correct additive and proportions to correct a CN deficiency in a storage tank. This is seldom a problem in North America but when refueling in more remote parts of Mexico, ensure that the fuel is purchased from reputable suppliers.

Air Pumped through the High-Pressure Injection Pump Circuit

Check the fuel subsystem because this is usually the source of the air, depending on the type of system. This condition is normally accompanied by rough engine operation. Using a diagnostic sight glass can help diagnose the problem and locate it. A diagnostic sight glass is usually a section of optically clear Perspex tube coupled into dash-6, dash-8, or dash-10 hydraulic hose: The sight glass is coupled in series into the circuit to be tested.

Coolant Leakage to Cylinders

Confirm that the emission truly is coolant. Engine coolant has an acrid, bittersweet odor that is very noticeable. Locate the source. Some possibilities are injector cup failure, head gasket (fire ring) failure, cracked cylinder head, cracked wet liner flange, or **cavitation** perforation of wet liner. Coolant leakage may be difficult to locate; if the failure is not immediately evident, drop the oil pan, pressurize the cooling system, and observe. If necessary, place clean cardboard under the engine with the cooling system pressurized and leave for a while to attempt to identify the source; this may save an unnecessary engine disassembly. It is often easier to identify an internal coolant leak with the engine intact rather than disassembled, so explore all the options with the engine assembled first.

Low Combustion Temperatures

Low combustion temperatures may be the result of extreme low temperature conditions or a fault in the cooling system management system, such as a defective thermostat, fan drive mechanism, or shutters.

Blue Smoke

Blue smoke is usually caused by lube oil getting involved in the combustion process. Some possible causes are turbocharger seal failure, Roots blower seal failure, pullover of lube oil from oil bath air cleaner sump, worn valve guides, ring failure, glazed cylinder liners, high oil sump level, excessive big end bearing oil throwoff, low-grade fuel, fuel contaminated with automatic transmission fluid (ATF), or engine lube placed in fuel tanks as an additive.

Smoke Emission Measurement

For a number of years, a number of state and provincial jurisdictions have set exhaust stack maximum smoke density standards. Most of these standards are based on SAE J1667 opacity tests. An opacity test is performed using a light extinction test instrument called an opacitymeter or opacimeter. Smoke opacity is defined as that percentage of light transmitted from a source that is prevented from reaching a sensor. Opacity is expressed as a percentage value. This type of emission testing applies to the aftermarket only, that is, field inspections performed by enforcement agencies. Some jurisdictions in the United States and Canada choose to enforce these standards with heavy fines. Smoke emission testing is covered in Chapter 41.

TROUBLESHOOTING GUIDELINES

As a general rule when troubleshooting a system complaint, investigate the possible causes that can be eliminated easily (and inexpensively) before proceeding to those that require more time. Component disassembly should generally be avoided until all other investigative options have been pursued. This practice will help avoid presenting a customer with an inflated bill for diagnosing a simple complaint. It makes sense to develop some guidelines for troubleshooting engine problems in a service facility. A set of guidelines will help the technician to strategize the troubleshooting procedure. Engine OEMs produce some excellent troubleshooting charts for their own products; Figure 42–1 reproduces a Navistar chart that relates to engines fueled by its hydraulically actuated electronic unit injector (HEUI) system.

The troubleshooting strategy used by the technician will depend on whether the engine is hydromechanically or electronically managed and on the engine OEM. As we said before, the troubleshooting procedure required of most electronically managed engines is structured and sequential; it must be adhered to. Warranty may be denied by an engine OEM if its diagnostic procedures are not followed and often it requires the submitting of online forms. The subject of electronic troubleshooting is dealt with in detail in the third section of this textbook. What we are attempting to do here is outline some typical engine problems that are mechanical in origin along with possible causes and solutions.

Low Oil Pressure

Verify the problem:

1. Check the oil sump level.
2. Install a **master gauge** (an accurate, fluid-filled gauge).
3. Warm engine to operating temperature and check oil psi.
4. Investigate oil consumption history.
5. Determine the cause.

Some possible causes and suggested solutions:

- Restricted oil filter or oil cooler bundle: Change the oil and filter(s). If the problem persists, clean or replace the oil cooler bundle (core), and check or clean the filter and oil cooler bypass valves.
- Contaminated lube (fuel): Detect by oil analysis or by odor. Engine lube contaminated with fuel can have a darker appearance and feel thin to the touch. If the cause is fuel, locate the source. This may be difficult and the procedure varies with the type of fuel system and the routing of the fuel to

INTERNATIONAL	DT–466E AND THE INTERNATIONAL 530E HARD START / NO START & PERFORMANCE ENGINE DIAGNOSTICS	Date	Miles	Hours	Technician
		Eng. S/N	VIN		Unit #
		Eng. HP	Ambient Temp.	Coolant Temp.	

HARD START / NO START DIAGNOSTICS

1. SUFFICIENT CLEAN FUEL

- Free of Water–Icing and clouding
- Correct grade of fuel

Method	Check
Visual	

2. VISUAL INSPECTION

- Inspect for leaks
- Inspect for loose connections, etc.

Fuel Oil Coolant Electrical Air	
Method	Check
Visual	

3. CHECK ENGINE OIL LEVEL

- Check engine crankcase oil level
- Check for contaminants (fuel, coolant)
- Correct Grade/Viscosity
- Miles/Hours on oil, correct level
- Check oil pressure on dash gauge

Method	Check
Visual	

4. INTAKE/EXHAUST RESTRICTION

REFER TO FIGURE A ON REVERSE SIDE

- Inspect hoses and piping
- Check filter minder
- Inspect exhaust system

Method	Check
Visual	

Perform Test 7 if EST is not available or inoperative

5. EST TOOL – FAULT CODES

REFER TO FIGURE B ON REVERSE SIDE

- Install Electronic Service Tool

Active	
Inactive	

☐ See Electronic Diagnostic Form for codes

6a. EST – ENGINE OFF TESTS

- Select "Engine Off" test from diagnostic test menu

Faults Found	

☐ Repair fault codes before continuing

6b. EST–INJECTOR "BUZZ TEST"

NOTE: "Engine Off Test" must be performed first, in order to gain access to the Injector "BUZZ TEST"

- Select "Injector Test" from "The Engine Off Tests" menu

Faults Found	

☐ See Electronic Diagnostic Form for codes

Perform Test 7 if EST is not available or inoperative

7. STI BUTTON – FLASH CODES

REFER TO FIGURE F ON REVERSE SIDE

- Depress and hold "Engine Diagnostics" switch, then turn the ignition switch to the "ON" position.

Faults Found	

☐ Refer to Electronic Diagnostic form, if fault code(s) set

8. EST TOOL – DATA LIST

- Select and enter the following data as the first 3 lines in a custom data list
- Monitor the data while cranking the engine for 20 seconds minimum

Data	Spec	Actual
Bat. Voltage	7 Volts min.	
Eng. RPM	150 RPM min.	
ICP Pressure	800 PSI min.	

☐ If voltage is low, refer to ECM diagnostics
☐ If no RPM is noted, recheck fault codes
☐ If ICP pressure is low, refer to Test 10

Perform Test 9 if EST is not available or inoperative

9a. ECM VOLTAGE

REFER TO FIGURE C ON REVERSE SIDE

- Check while cranking the engine
- Measure with DVOM
- Breakout box pins 57+ & 40–

Instrument	Spec	Actual
DVOM 57+ & 40–	7 Volts minimum	

☐ If voltage is low, refer to ECM diagnostics

9b. ENGINE CRANKING RPM

REFER TO FIGURE C ON REVERSE SIDE

- Minimum 150 RPM engine cranking speed for 20 seconds
- Breakout box pins 34+ & 46– with Fluke 88

Instrument	Spec	Actual
Fluke 88 34+ & 46–	150 RPM minimum	

☐ If no RPM is noted, recheck fault codes

9c. INJECTION CONTROL PRESSURE

REFER TO FIGURE D OR E ON REVERSE SIDE

- Minimum 150 RPM engine cranking speed for 20 seconds
- Measure with breakout box: pins 27+ & 46– or breakout "Tee" signal (green) & ground (black)

Instrument	Spec	Actual
DVOM 27+ & 46–	1 Volt Minimum	

☐ If ICP pressure is low, refer to Test 10

10. LOW ICP PRESSURE TEST

NOTE: Perform this test if ICP Pressure was low in Test 8 or 9C.

REFER TO FIGURE G ON REVERSE SIDE

- Remove EOT sensor and check for oil in reservoir and reinstall EOT
- Remove high pressure hose from oil manifold
- Attach adapter and ICP sensor to hose
- Monitor pressure while cranking the engine

Instrument	Spec	Actual
EST	800 PSI min.	
DVOM	1 Volt min.	

☐ If pressure is within specifications, check for high pressure oil leakage. Refer to EGES–145 Sec. 2.2
☐ If pressure is still low, verify that pump is rotating
☐ If pressure is still low, replace IPR and retest

11. FUEL PUMP PRESSURE

REFER TO FIGURE K ON REVERSE SIDE

- Measure at bleeder valve on filter header
- Minimum 150 RPM cranking speed for 30 seconds

Instrument	Spec	Actual
0–160 PSI Gauge	20 PSI minimum	

☐ If pressure is low, replace fuel filter, clean fuel strainer and retest.

☐ If pressure is still low, perform Transfer Pump Restriction Test 2B (of Performance Diagnostics)

PERFORM TESTS IN SHADED AREAS IF EST TOOL IS NOT AVAILABLE OR ATA CODES ARE NOT TRANSMITTED

EGED–150 MARCH 1995 © NAVISTAR INTERNATIONAL TRANSPORTATION CORPORATION

Figure 42–1 *International Truck troubleshooting strategy chart. (Courtesy of Navistar International Corp.)*

Injector P/N	Turbocharger P/N
Engine Family Rating Code	
Complaint	

PERFORMANCE DIAGNOSTICS

ALL TESTS SHOULD BE PERFORMED WITH ENGINE AT OPERATING TEMPERATURE

1. CHECK ENGINE OIL LEVEL

- Check engine crankcase oil level
- Check for contaminants (fuel, coolant)
- Correct Grade/Viscosity

Method	Check
Visual	

2. SUFFICIENT FUEL/PRESSURE

REFER TO FIGURE K ON REVERSE SIDE

- Drain sample from tank(s)
- Inspect fuel for contamination
- Measure fuel pressure at fuel filter bleeder
- Measure pressure at high idle

Instrument	Spec	Actual
0–160 PSI Gauge	20 PSI minimum @ High idle	

☐ If pressure is low, replace fuel filter, clean fuel strainer and retest.

☐ If pressure still low, proceed with step 2B.

2b. TRANSFER PUMP RESTRICTION

REFER TO FIGURE L ON REVERSE SIDE

NOTE: Perform this test only if fuel pressure is low.

- Measure at fuel filter inlet @ High idle.

Instrument	Spec	Actual
0–30" Vacuum Gauge	Less than 8 " Hg.	

☐ If restriction is high, check for blockage between pump and fuel tank

☐ If restriction < 8 " Hg., refer to EGES–145 Sec. 2.3 for additional diagnostics.

3. EST TOOL – FAULT CODES

REFER TO FIGURE B ON REVERSE SIDE

- Install Electronic Service Tool

Active	
Inactive	

☐ See Electronic Diagnostic Form for codes

4a. EST – ENGINE OFF TESTS

- Select "Engine Off Test" from diagnostic test menu

Faults Found	

☐ Repair fault codes, before continuing

4b. EST–INJECTOR "BUZZ TEST"

NOTE: "Engine Off Test" must be performed first, in order to gain access to the Injector "BUZZ TEST"

- Select "Injector Test" from "The Engine Off Tests" menu

Faults Found	

☐ See Electronic Diagnostic Form for codes

Perform Test 5 if EST is not available or inoperative

5. STI BUTTON – FLASH CODES

REFER TO FIGURE F ON REVERSE SIDE

- Depress and hold "Engine Diagnostics" switch, then turn the ignition switch to the "ON" position.

Faults Found	

☐ Refer to Electronic Diagnostic form if fault code(s) set

6. INTAKE RESTRICTION

REFER TO FIGURE H ON REVERSE SIDE

- Measure at high idle and no load
- Use manometer or magnehelic gauge

Instrument	Spec	Actual
Manometer or Magnehelic Gauge	12.5" H_2O	

7a. EST–ENGINE RUNNING TEST

- Select "Engine Running" test from the diagnostic test menu

Faults Found	

☐ Refer to Electronic Diagnostic form if fault code(s) set

7b. EST TOOL–INJECTOR TEST

(CYLINDER CONTRIBUTION)

NOTE: "Engine RUNNING Test" must be performed first, in order to gain access to the "INJECTOR TEST"

- Select "Injector Test" from "Engine Running" test menu

Faults Found	

☐ Refer to Electronic Diagnostic form if fault code(s) set

Tests 8, 9 & 10 to be performed at Full load

8. FUEL PRESSURE (FULL LOAD)

REFER TO FIGURE K ON REVERSE SIDE

- Measure fuel pressure at fuel filter bleeder.
- Measure pressure at full load rated speed.

Instrument	Spec	Actual
0–160 PSI Gauge	20 PSI minimum	

☐ If pressure is low, replace fuel filter, clean fuel strainer & retest

☐ If pressure is still low, perform Test 2B.

9. ICP PRESSURE

REFER TO FIGURE D OR E ON REVERSE SIDE

- Monitor ICP pressure and engine RPM with the EST tool in data list mode

Or use breakout "TEE" and DVOM

- Refer to EGES–145 for specifications

Data	Spec	Actual
Low Idle	PSI/Volts	
High Idle	PSI/Volts	
Full Load	PSI/Volts	

☐ If pressure is low or unstable, disconnect ICP sensor and retest

☐ If problem is resolved, refer to ICP diagnostics

☐ If pressure is still low or unstable, replace IPR and retest

10. BOOST PRESSURE

REFER TO FIGURE I ON REVERSE SIDE

- Monitor boost pressure and engine RPM with the EST tool in data list mode

Or use dash tach and 0–30 PSI gauge and "T", if EST tool is not available

- Measure pressure at full load rated speed
- Refer to EGES–145 for specifications

Spec	Actual
PSI @ RPM	

11. CRANKCASE PRESSURE

REFER TO FIGURE J ON REVERSE SIDE

- Measure at road draft tube with orifice tool (ZTSE–4039)
- Measure at High Idle no load RPM

Instrument	Spec	Actual
0 to 60" H20 Magnehelic Gauge	< 6" H_2O	

STOP

IF GUIDELINE DATA WAS OBTAINED DURING THE FIRST 11 TESTS, ENGINE OPERATION IS SATISFACTORY. NO FURTHER TESTING IS REQUIRED

12. WASTEGATE ACTUATOR TEST

- Apply regulated air to actuator
- Inspect for leakage
- Inspect actuator for movement

Instrument	Spec	Actual
0 to 60 PSI Gauge	28–32 PSI	

13. EXHAUST RESTRICTION

- Visually inspect exhaust system for damage
- Measure at a point 3 to 6 inches after turbo outlet
- Measure at full load and rated speed

Instrument	Spec	Actual
Manometer or Magnehelic Gauge	0–35" H_2O	

14. VALVE CLEARANCE

- Engine off: Hot or Cold

Instrument	Spec	Actual
Feeler Gauge		

Figure 42–1 *(Continued)*

the injector. Pressure testing the fuel delivery components may be required. Porosity in the cylinder head casting, failed injector O-ring seals, leaking fuel jumper pipes, and cracked cylinder head galleries are some possible causes. Perform repairs as required, then service the oil and filters.

- Excessive crankshaft bearing clearance: Inspect the bearings to determine the cause. Visually check the crank journals to determine whether removal is required. Replace the bearings, ensuring that the clearance of the new bearings is checked.
- Excessive camshaft or rocker shaft bearing clearance: Replace the bearings.
- Pump relief valve spring stuck open or fatigued: Clean the valve and housing, replacing parts as necessary. Check the bypass/diverter valves in the oil cooler and filter mounting pad.
- Oil pump defect: Recondition or replace the oil pump.
- Oil suction pipe defect: Replace the oil suction pipe.
- Defective oil pressure gauge or sending unit: Replace the oil pressure gauge or sending unit.
- Broken-down (chemically degraded) lube oil: Change the oil and filters.

High Oil Consumption

Verify the condition by monitoring oil consumption and analyzing exhaust smoke. Some possible causes and suggested solutions:

- Excess cylinder wall lubrication: High oil sump level, excessive connecting rod and big end bearing oil throwoff, plugged oil control/wiper rings, oil pressure too high, or oil diluted with fuel. This type of problem usually requires engine disassembly to diagnose and repair.
- External oil leaks: Steam or pressure wash the engine, then load on a chassis dynamometer or road test to determine the source.
- High oil temperatures: Malfunctioning boost air heat exchanger, lug down engine loading, overfueling, incorrect fuel injection timing, or problem with the oil cooler or engine cooling system.
- Worn piston rings fit abnormality: Replace the rings.
- Piston ring failure: Determine the cause. Check for other damage, then replace the rings.
- Turbocharger seal failure: Recondition the turbocharger (recore or replace).
- High oil sump level: This causes aeration. Determine the cause of high oil level. Check for operator or service error first then for the presence of fuel and engine coolant in oil.
- Glazed cylinder liners or sleeves: Caused by improper break-in procedure or prolonged engine idling. Replace cylinder liners/sleeves or use a glaze-buster to re-establish crosshatch.
- Worn cylinder head valve guides or seals: Measure to specification and recondition the cylinder head if required.
- Improper dipstick marking: Drain the oil, fill the sump to OEM specification, then check/alter dipstick markings.

High Oil Temperature

Some possible causes and suggested solutions:

- Insufficient oil in circulation: Check the sump level, pump pressure, and lubrication circuit restrictions. Repair as required.
- High water jacket temperatures: Test cooling system performance.
- Plugged/failed oil cooler: Disassemble and inspect galleries, bundle, and bypass valve for restrictions and scaling. Service the cooling system afterward to prevent a recurrence.
- Oil badly contaminated: Submit a sample for analysis and repair cause.
- Engine lugdown: Provide some driver training and identify the transmission shift points. Drivers familiar with older engines may not have been adequately retrained to operate today's fuel efficient engines, which may be programmed with short torque rise windows.

Cooling System Problems

Cooling system problems can be grouped into the following categories:

- Overheating

- Overcooling
- Loss of coolant
- Defective radiator cap
- Defective thermostat

From the troubleshooting perspective, we can study these problems by looking at the performance problems they produce:

- High coolant temperatures
- Low coolant temperatures

High Coolant Temperature

Accurately verify the condition:

1. Check coolant level in the radiator.
2. Install a master gauge to verify the problem.
3. Observe exhaust pyrometer readings and compare to specification.

Some possible causes and solutions:

- Incorrect mixture: The correct mixture for ethylene glycol (EG) is usually an equal proportion of water and antifreeze with somewhere between 3% and 6% coolant conditioner. Increasing the antifreeze proportion with both EG and propylene glycol (PG) reduces cooling efficiency, but when high concentrations of PG are used for arctic operation, you may have no choice.
- Aerated coolant: Usually caused by combustion gases entering the cooling circuit through a defective cylinder head or failed fire ring/cylinder head gasket. Heat is not transferred efficiently through a gas. Verify the condition: Check for bubbles at the water manifold.
- Fan clutch: Test fanstat/thermal fan operation using a master gauge and loading the engine until it reaches the temperature required to cycle the fan.
- Radiator: Check for internal and external flow restrictions. Clean externally. Internal restrictions usually require the removal of the radiator and either recoring or the use of specialized reconditioning equipment.
- System does not seal: This permits the coolant to boil at a lower temperature and cause boilover. Test with a cooling system pressure tester. Check and repair/replace the defective radiator cap and/or pressure relief valves.
- Improper airflow: Incorrectly sized fan or missing/ damaged radiator shroud may significantly reduce flow through the radiator and engine compartment.
- Loose pump and fan drive belts: This will cause a reduction in coolant and/or airflow. Check visually and with a belt tension gauge.
- Coolant hoses: Coolant hoses should be changed every couple of years. They often fail internally while appearing sound externally. A collapsed hose may cause a significant flow restriction.
- Restricted air intake: May cause high engine temperatures. Test inlet restriction with a water manometer. Typical maximum specifications are 15″ (380 mm) H_2O for naturally aspirated engines and 25″ (635 mm) H_2O for boosted engines.
- Exhaust restriction: Causes high engine temperatures. Uncouple the exhaust piping from turbo or remove the exhaust manifold to see if the condition is corrected. Check the exhaust silencer and, if equipped, the catalytic converter for internal collapse.
- Shunt line failure: A restriction in the shunt line from the radiator top tank to the water pump inlet may cause a boil condition at the inlet, reducing coolant flow and causing overheating.
- Thermostat: Test out of engine for opening value and full open position using a boiler and a thermometer.
- Water pump: Remove and check for a loose or damaged impeller and check the impeller to housing clearance to specification. Rebuild or replace as required.
- Boost air heat exchanger: An airflow or internal restriction will cause a rise in engine temperatures. Test airflow through the engine compartment. Winter grille covers impede airflow through the heat exchangers and may unevenly load the fan.
- High-altitude operation: Cooling system efficiency diminishes as altitude increases. Therefore, larger cooling system capacity is required for high-altitude operation.
- Lugging: Operating an engine at high loads and lower speeds means high temperatures and reduced coolant flow. Driver training is required.

- Overfueling: This raises the amount of rejected heat and may exceed the cooling system's ability to handle it. Recalibrate engine fueling to specification.
- Fuel injection timing: Depending on the type of fuel system, both retarding and advancing fuel injection timing may result in engine overheating. Fuel injection timing may be tampered with in older engines; Set timing to OEM specification. Greatly diminished engine life is the consequence of minor injection timing adjustments, whereas major component failure is the consequence of more aggressive adjustments.

Low Coolant Temperature

Accurately verify the condition:

1. Check cooling system temperature management components such as shutter and fan operation.
2. Install a master gauge to verify the problem.
3. Observe exhaust pyrometer readings and compare to specification.

Some possible causes and solutions:

- Thermostat: A thermostat that is stuck in the open position will cause the engine to run cool. Remove the thermostat and check its start to open and fully open temperature specifications.
- Air vent valve: If stuck in the open position, this may cause low coolant temperatures when the engine is under light loads.
- Prolonged idle or light load operation: When little fuel is being used by the engine, there is less rejected heat for the cooling system to handle, and operating temperatures may be lower.
- Fan clutch: Test fanstat/thermal fan operation using a master gauge and loading the engine until it reaches the temperature required to cycle the fan: Ensure that this occurs at the specified temperature and not a lower value.
- Improper airflow: Incorrectly sized fan or missing/damaged radiator shroud may result in overcooling by creating irregular flow through the engine compartment.

Cylinder Compression Problems

Overall low or unbalanced cylinder compression can create problems of excessive blowby, low power, and vibration. Cylinder pressure may be checked using a compression tester. In diesel engines, the compression test should be performed when the engine is close to operating temperature by first removing all of the injectors, fitting the compression test gauge to the injector bore in one of the cylinders, and then cranking through five rotations. The batteries should be fully charged during this test. Test each cylinder sequentially and match the test results to specifications. The results should be within the engine OEM cylinder compression parameters and normally within 10% of each other. Compression testing will locate which cylinder(s) is/are at fault, but not the cause of the problem.

A cylinder leakage or air test kit sometimes stands a better chance of identifying the cause of a problem. The cylinder leak test consists of a pressure regulator and couplings that fit to the injector bores. Once again the engine should be close to its operating temperature at the beginning of the test. The test is performed with the piston in the cylinder to be tested at TDC and regulated air pressure is delivered to the cylinder. The test may indicate valve seal leaks, head gasket leaks, leaks to the water jacket, leaking piston rings, and defective pistons. The **cylinder leakage test** regulator delivers air at a controlled volume and pressure then measures the percentage of leakage.

High Exhaust Temperature Readings

When diagnosing high exhaust temperatures on highway diesel engines equipped with catalytic converters and DPFs, always use the engine OEMs recommended procedure. Until the universal adoption of DPFs in 2007, a pyrometer (thermocouple) was primarily a driver assist. When pyrometer readings were dash displayed in pre-2007 engines, its purpose was to alert the driver to downshift: A pyrometer was often spec'd into trucks hauling heavy loads through mountainous terrain. In post-2007 diesel engines, thermocouple-type pyrometers are used in some DPFs to help manage regeneration cycles. In pre-2007 trucks with high exhaust temperatures, first verify the condition:

1. Install a **master pyrometer** (do not rely solely on the vehicle pyrometer).
2. Chassis **dynamometer** test using a full instrumentation readout. See Chapter 15 on engine testing.

Some possible causes and solutions to high exhaust temperatures are:

- Air inlet restriction: Test inlet restriction value to specification using a manometer.
- Flow restricted, boost air heat exchanger: Check airflow through the hood intake grille. A winter grille cover can limit airflow through the heat exchangers located at the front of the chassis. Check the operation of a *tip turbine* if equipped.
- High engine load, low airflow operation: Engines using air-to-air boost air cooling *require* airflow through the heat exchanger. A dump truck with air-to-air boost cooling operating in a pit under extreme temperatures may produce high pyrometer readings even when operated skillfully.
- Fuel injection timing: Check to specification.
- Overfueling: Look for other indicators of an overfueling condition and check fuel settings to specification.

Sudden Engine Shutdown

This can result from a number of conditions. First, try to determine whether the check engine lamp (CEL) or stop engine lamp (SEL) illuminated. If this was the case, connect to the chassis data bus, and access the engine MID to route a path to the fault mode indicator (FMI). Some possible causes and solutions:

- Electrical failure: Use the appropriate electrical circuit troubleshooting to locate the problem. Check the most obvious causes first.
- No fuel: Refuel tanks and use an appropriate priming method to get the engine running again.
- Air in fuel system: Air may be pulled into the fuel subsystem from any of a number of locations; common causes are the fuel filter sealing gaskets, double gaskets (caused by failure to remove a used gasket during servicing), and a failed filter mounting pad assembly.
- Water in fuel system: Most current secondary filters will plug on small quantities of water. Locate the source of the water. Replace and prime filters.
- Plugged fuel line: Sequentially work through the fuel subsystem circuit to locate the failed component. An internally failed fuel hose can be difficult to locate because an internal flap may require moving fuel through the circuit to block flow. Similarly, something simple such as a small decal floating in the fuel tank can plug a fuel pick-up line and shut down the engine, after which it floats away.
- Lubrication failure: Caused by a failed lubrication circuit component. A lubrication circuit failure will rapidly develop into a seized engine. Attempt to manually bar the engine over. A lubrication failure often results in a complete engine disassembly.

Engine Runs Rough

This describes a condition in which the engine produces an inconsistent rpm roll (nonrhythmic), hunting (rhythmic), cylinder misfire, or surging at a specific rpm or throughout the engine-operating range.

Some possible causes and solutions:

- Air in fuel system: Often accompanied by white smoke emission. Locate the source of the air, checking the suction side of the fuel subsystem first.
- Any problems in the engine breathing circuit including heat exchangers, EGR and CGI coolers, venturis, and control valves, VGT turbos, and DPFs.
- Leak or restriction on the charge side of the fuel subsystem: Repair a leak if evident. Use a pressure gauge to check charging pressure to specification.
- Injection nozzle, PT, or MUI failure(s): This results in a cylinder misfire condition. Locate the affected cylinder by disabling each injector sequentially. Crack the high-pressure pipe nuts at the pump on hydraulic injector systems and load the injector rocker in its downstroke position on PT and MUI systems. When checking electronically managed engines, use the electronically driven diagnostics and never mechanically short out an injector.

- CN value of fuel is too low: This occurs when fuel is stored for prolonged periods either in the vehicle or base storage tanks and the fuel deteriorates chemically or biochemically.
- Fuel cloud point is high: When fuel begins to wax, it can restrict filters and pumping apparatus. Analyze the fuel and add an approved cloud point depressant as required.
- Advanced injection timing: A common tampering problem in older hydromechanical engines. It may be a problem in some EUI-fueled engines that have been intentionally advance timed by stepping a timing gauge. Check injection timing and reset to specification if required.
- Unbalanced MUI fueling: Produces a hunting condition. Balance the MUI racks according to the OEM procedure. If the condition persists after the tune up, check that the MUI fuel codes are matched.
- Bent or broken MUI pushrod: Usually a cylinder misfire accompanied by mechanical clatter. Replace defective injector train components.
- Maladjusted valves: Perform an overhead adjustment.
- Cylinder leakage: Caused by a failed fire ring(s), a crack in the cylinder head/block, or valve leakage. Perform a cylinder leakage test appropriate for the engine.

Engine Blowby

Because rings create the seal of a piston in a cylinder bore, some blowby is inevitable due to end-gap requirement. Normally, blowby is a specification determined by the OEM. Remember, piston rings seal most efficiently when cylinder pressures are highest, and engine rpm is high when there is less time for gas blowby.

Some causes and solutions of excessive engine blowby are:

- Cracked head or piston: Diagnose the cause and recondition the engine.
- Worn, stuck, or broken piston rings: Diagnose the cause and recondition the engine.
- Glazed liner/sleeve inside wall (due to improper break-in practice or prolonged idling): Diagnose the cause and recondition the engine.
- Poor quality or degraded engine lube: Change the oil and filters.

High crankcase pressures can also be caused by:

- Air compressor (plugged discharge line, worn rings, air pumped through oil return line): Test air compressor operation.
- Turbocharger (seal failure allows the impeller housing to leak pressurized air through to the turbine shaft and back through the oil return piping): Test turbocharger operation and recondition or replace as required.

Engine Will Not Crank

Determine whether the problem is related to the cranking circuit (electrical or pneumatic problem) or a mechanical engine condition. Focus on the cranking circuit first and eliminate any obvious causes; in most cases, the cause of a failure to crank condition will be found here. Attempt to bar the engine over by hand. If it cannot be turned over, check the engine externally and both the transmission and power takeoff (PTO). If engine seizure is suspected, remove all the injectors and again attempt to bar the engine over by hand before beginning to disassemble it. If disassembly is required, carefully remove and label components so that the cause of the failure can be determined.

Lack of Power

"Lack of power" probably appears more frequently on shop work orders than any others. In many cases, the lack of power is more closely associated with driver expectations than a genuine engine problem. A chassis dynamometer with a data printout is usually required to convince a skeptical truck driver that his equipment is functioning properly. Although the technician can eliminate many causes of low power, realistically, when performing this work for a customer, it is essential to test that the engine performs to specification on a chassis dynamometer when the work is completed.

It makes a lot of economic sense for shops working frequently on low-power complaints to prepare a strategy check sheet that addresses the specific equipment worked on; this ensures that the work is performed sequentially (eliminating obvious/quick to perform tasks first) and enables each technician to pursue the steps in more or less the same manner. A successful method of developing a "lack of power checklist" is to devote a technicians' shop meeting to brainstorming troubleshooting strategies and sequences for specific engine/fuel systems. Apart from the fact that the resulting checklist will incorporate the input of the entire workshop team, this exercise will have the added benefit of "educating" the less-experienced technicians. In short, the time devoted to the meeting will be paid for many times over by the time saved avoiding inconsistent and inaccurate troubleshooting practices.

There are too many variables based on engine manufacturer and fuel system type to provide definitive lack of power troubleshooting strategies. The following is a general list of some causes of low power and their solutions:

- Restricted fuel filters: Replace the filters. Check the vehicle fuel tanks for contamination.
- Restricted air inlet system: Test the inlet restriction value using a manometer. Remember, air filter elements are best tested without removal from the canister—do *not* replace unless they fail an inlet restriction test or have exceeded the in-service time limit.
- Leaks in the boost air circuit: Test for leakage using an OEM-approved method such as direct application of a soap solution with the engine under load, ether spray at idle, or others. Small leaks in charge air cooler cores can be difficult to locate and may require removal and pressure testing. When dynamometer testing, always fit instrumentation to read manifold boost.
- Low manifold boost—turbocharger problem: Visually inspect the turbocharger, checking radial play and endplay—check for rotation drag caused by carbon deposits (coking).
- Restricted exhaust system: This condition is usually accompanied by poor engine response. Check piping, engine muffler(s), DPFs, and catalytic converters when equipped. Mufflers can fail internally when baffles and resonator walls collapse; test backpressure value to specification.
- Low fuel subsystem charging pressure: Fit a pressure gauge and test fuel pressure at idle and high-idle speeds, referencing the specifications.
- Valve lash maladjustment: Often accompanied by smoking and top end/valve clatter. Adjust valves to specification.
- Defective boost pressure sensing and fuel control devices: All current engines use either a variable capacitance or piezoresistive turboboost sensors that seldom malfunction. Older engines used different types of aneroid devices, which were also known as puff limiters, LDAs (Bosch), AFC (air-fuel control) valves, air ratio control (ARC), and FARC (fuel-air ratio control) valves. The devices used on hydromechanical engines were not only more likely to malfunction, they could be easily tampered with. Most aneroids had a set trigger specification (predetermined manifold boost value), and their function was to limit fueling until there was enough air in the cylinders to properly combust it. Some aneroids permitted a graduated increase of fueling, proportional to increase in manifold boost, whereas others simply switch at a predetermined boost value, limiting fueling until it is achieved. Replacing a properly functioning boost fueling control device with one of lower trigger value will not cure a low-power complaint or increase engine power, although it will cause puff smoking at shifting, waste fuel, and contaminate the atmosphere. Use the OEM method of testing the device, and ensure that the correct one is fitted. Some use an intake manifold sensor and system air pressure to actuate a governor-located pilot plunger—these must be precisely set to specification.
- Governor maladjusted: Usually a result of repeated tampering with components that should only be adjusted on a pump comparator bench. Performance test the engine on a dynamometer and reset the governor, removing the fuel injection pump assembly, if required, if the governor is integral.
- Contaminated fuel: Fuel is abused unknowingly by both drivers and technicians whenever they add any substance to vehicle fuel tanks other than diesel fuel. Some of these

additives may create conditions that result in lack of power complaints. There are certainly seasonal conditions that mandate the addition of alcohols to fuel tanks (crossover pipe freeze-up), but drivers and technicians should be educated to understand that dumping additives into fuel tanks, especially aftermarket additives of dubious chemistry, should be avoided. Purchasing #1D fuel is much cheaper than purchasing inexpensive, summer grade #2D fuel and paying for the problems it creates when operating equipment in extreme winter conditions.

- Defective fuel: Not a common problem in North America and difficult to diagnose without the use of specialized equipment. Check with the fuel supplier. Testing specific gravity may identify a fuel in which the lighter fractions have boiled off, lowering the CN, but the exact original values must be known. Defective fuel problems are usually the consequence of storing a fuel for prolonged periods or use of fuels outside the season in which they were purchased.

Engine Vibration

While driveline vibrations are not uncommon, true engine vibrations are not often a problem. When investigating a vibration complaint, the technician should eliminate all possible causes in the driveline behind the engine first. In doing this, it should be remembered that the clutch may disengage the driveline from the transmission and beyond from the engine, but that most of the mass of the clutch is rotated with the engine engaged and disengaged.

- Cylinder misfire: See section "Engine Runs Rough."
- Loose vibration damper: Check bolts for shear damage and damper fastener holes for elongation. Visually inspect damper for other damage. Retorque and test operation.
- Defective vibration damper: Viscous-type dampers should be replaced at each engine overhaul but seldom are—visually inspect. The slightest external defect is reason to replace the unit. To dynamically test a vibration damper, a lathe and dynamic balance apparatus is required—a procedure beyond the scope of most shops. Visually inspect rubber drive ring dampers.
- Defective external driven component: Every driven engine accessory is capable of unbalancing the engine. Most air compressors marketed today are balanced units and do not require timing; but some single cylinder compressors must be timed to the engine.

Some common accessories that may cause an unbalanced engine condition when they or their bearings fail are the fan assembly (broken/damaged blades, failed bearings), water pump, accessory drive bearings, idler pulley bearings, alternator, and others.

Soot in Inlet Manifold

Some soot in the intake manifold is normal in most diesel engines; it usually indicates an engine operated at low loads and speeds for prolonged periods in engines manufactured prior to 2004. In such operating conditions, the valve overlap duration is at a maximum in real time values and manifold boost is minimal or nonexistent. Excessive soot in the intake manifold may be an indication of imminent turbocharger failure, an injection timing problem, or a problem related to the EGR circuit in post-2004 diesels. We take a brief look at EGR circuit troubleshooting in the next section.

EGR Circuit Malfunctions

When the EGR circuit fails, in most cases at least one fault code is logged because the circuit is so thoroughly monitored. When a fault code is logged, make sure you adhere to the OEM software-driven diagnostic routine. Typical diagnostic fault codes that relate to the EGR circuit are:

- Short to ground
- Dead short
- EGR valve position circuit fault

A typical OEM diagnostic routine to verify EGR PWM-actuated valve function would appear as follows:

- Check for multiple codes: When other engine codes are logged, these should be solved first.

- Disconnect the EGR valve electrical connector: Check for pin damage and corrosion and repair as necessary.
- Reconnect the EGR valve electrical connector. Key *on,* engine *off* (KOEO): Use the OEM software to monitor EGR valve actual position.
- If it reads less than the specified KOEO position (usually expressed as %): Command EGR pulse width (PW) to step to 50%. If it fails to move to within 5% of the input 50%, the EGR valve is defective and should be replaced. If it moves within 5% of the requested 50%, next command it to move to 100% PW. In this position, actual valve position should be at 95% or higher. If not, replace the valve. If it is:
  - Remove the EGR valve and check for physical damage and carbon/sludge buildup. Either condition requires the valve to be replaced, but in the case of carbon/sludge buildup, check the EGR cooler for internal failure.

Manifold Boost Problems

When manifold boost is either too high or too low, engine performance complaints result. Common causes of lower than specified manifold boost are:

- Air system restriction upstream or downstream from turbocharger impeller
- Air leakage downstream from turbocharger impeller
- Low fuel delivery
- Mismatched turbocharger

Common causes of higher than specified manifold boost are as follows:

- Mismatched turbocharger
- Sticking vane ring on VGT turbos
- Excessively high inlet air temperatures
- Deposits on turbine volute or nozzle
- Overfueling caused by governor or by calibration programming
- Advanced fuel injection timing

DPF Malfunctions

DPF malfunctions are characterized by one or more of the following:

- Black smoke emission.
- DPF collects excessive soot.
- DPF is restricted.
- DPF temperatures are high.

Begin troubleshooting a DPF malfunction by inspecting it externally for indications of dents or cracks. Remove the DPF and check for indications of fuel or engine oil. If fuel is present, perform a cylinder balance test to identify a defective injector. If oil is present, check the turbocharger(s) for oil leakage. Note that any problem upstream in the engine breathing circuit can produce excessive soot loads and black smoke: Pay special attention to the venturi for the EGR or CGI systems, EGR mixing valves, and the aftercooler.

DPF Restriction

The engine management electronics are designed to respond to DPF restriction caused by excessive soot dump by logging fault codes and power derate strategies. This type of malfunction can be caused by operating the vehicle for prolonged periods when passive regeneration cycles are either inhibited by the operator or by the mode of operation. Most have a threshold that once exceeded requires an active regeneration cycle to be commanded. In some cases, when the soot load threshold has been exceeded by a considerable margin (say 200%) a factory password may have to be downloaded to clear the code and derate.

DPF Differential Pressure Problems

DPF high differential pressure problems are usually caused by a buildup of soot or ash in the particulate trap. The appropriate repair strategy is to command an active regeneration cycle if this is still possible: If not the unit must be either laundered on-chassis or removed and replaced, depending on the OEM. Low differential pressure is usually caused by a failure of the tubing between the particulate trap and the pressure differential sensor. Check this circuit and replace components as necessary.

High DPF Intake Temperatures

High DPF intake temperatures can be caused by a range of engine circuits upstream from the DPF, including:

- Restricted boost air heat exchanger

- Air inlet restriction
- Exhaust restriction
- High altitude operation
- Overfueling
- Low coolant level

In every case, to troubleshoot high DPF intake temperatures, check for any active and historical codes: In most cases, high DPF intake temperatures are not sourced in the DPF, although an exhaust restriction downstream from the aftertreatment canister could cause this condition.

Mechanical Engine Knock

- Bottom end knock: Produces an easily recognizable low-frequency thump. The cause is big end or connecting rod journal bearing failure, which is a condition that rapidly develops into a crankshaft failure if not attended to. Replace the bearings and thoroughly inspect and measure the throw journals.
- Failed crankshaft: Diagnose the problem in-chassis by visual inspection of journals and their bearings and the critical crankshaft stress points.
- Damaged gears: Often identified by a high-frequency whine. Remove the timing gear cover and replace failed components, remembering that debris from the failure may have been pumped through the entire lubrication circuit.
- Failure of feedback circuit component: Check rocker trains and camshaft. Repair components as required, once again remembering that debris from the failure may have been pumped through the entire lubrication circuit.

Combustion Knock

Combustion knock is sometimes known as *diesel knock.* If the noise is rhythmic, it may be that it is being produced from one engine cylinder, which may be verified by comparing cylinder temperatures with an infrared thermometer. Some causes and solutions are:

- Fuel injection timing: Usually produces erratic knock, amplified at higher loads and rpm. Check fuel injection timing to specification.
- Air in fuel: Usually accompanied by white smoke emission. Check for air admission on the suction side of the fuel subsystem.
- Low grade fuel: Fuel with a low CN (caused by additive contamination, prolonged storage, nonhighway ASTM grade, etc.) will extend the ignition lag phase. This results in excess fuel in the cylinder at the time of ignition, rapid pressure rise, and detonation. Analyze the fuel quality using the fuel supplier/refiner's recommended procedure.
- MUIs out of balance: A single MUI delivering excess fuel due to a maladjusted rack can produce combustion knock. Check overhead adjustment. Check individual MUI racks for binding.

Fuel Pump/Injector Scuffing

Fuel injector pump plunger scuffing has become a problem in recent years in part due to the lower lubricity of LS and the more recent ULS diesel fuels and excessive addition of alcohol into fuel. Surface scuffing of the pump plungers can be difficult to diagnose, especially when the surface scuffing is minor. This type of failure is not common but when it appears it is usually found on MUI, PLN, EUI, HEUI, and EUP high-pressure pump plungers. Its causes are water in the fuel (usually in solution with methyl hydrate alcohols allowing it to pass through separators) or low lubricity fuel. To overcome the problem, some OEMs coat both the plunger and barrel assembly of high-pressure fuel injectors with tungsten carbide. The symptoms of fuel injector scuffing are:

- Low power: results from severe scuffing
- Slight engine miss: most notable at low load operation
- Misfire: again, most notable under low load operation
- Smoking: some high load hazing evident when multiple injectors are affected

An electronic cylinder balance (aka cylinder contribution test [CCT]) can sometimes identify an injector scuffing condition before it sets fault codes, especially when using diagnostic software capable of graphing displays. Severe cases are more likely in older electronic and hydromechanical diesel engines that lack the diagnostic software capable of identifying the condition.

QUICK REFERENCE DIAGNOSTIC CHARTS

The following sets of diagnostic charts (Tables 42-1 through 42-16) repeat some of the information already outlined and are intended as a guideline to diagnose some typical problems. As much as possible the approach is generic but where it is not, the specific engine or fuel system is referenced.

Table 42–1 *Symptom: Engine Will Not Crank*

Possible Cause	*Action*
Discharged or dead batteries	Test the batteries. Charge or replace as required.
Loose or corroded terminals	Identify the location using voltage drop (VD) testing. Clean, tighten, or replace terminals.
High resistance in cranking circuit cables	VD test cables: (1) Switch to starter. (2) Battery to starter motor. Replace as required.
Defective starter motor or starter solenoid	Check operation using a DMM. Repair as required.
Defective key switch	Test and repair as indicated by electronic troubleshooting. Replacement switch may require reprogramming.
Seized engine	Attempt to bar engine over through two revolutions. Diagnose cause of seizure.

Table 42–2 *Symptom: Engine Cranks, Will Not Start*

Possible Cause	*Action*
Slow crank	Reference Table 42–1: Engine Will Not Crank.
Check fault codes	Codes logged: Use software diagnostics. No codes logged: Continue using this chart.
No fuel to engine	(1) Check fuel tank level. (2) Check for restriction in the fuel subsystem lines and filters. Replace as required.
Defective fuel transfer pump	(1) Check fuel charging pressure to spec. (2) Check for air in fuel. Repair or replace.
Degraded fuel, water in fuel	Drain fuel from the tank. Replace fuel and fuel filters. Flush new fuel through the system before cranking.
Incorrect oil viscosity (cold weather condition)	Replace the engine oil and filters when possible: 15W-40 engine lube gels in cold weather and cannot be drained.
Low compression	Perform a cylinder compression test: Use Table 42–14 to troubleshoot.
Defective charge pressure regulator	Remove the charging pressure regulator and inspect: Replace if required.
Defective electrical connections	Perform wiggle wire tests at the ECM and critical engine sensors.

Table 42–3 *Symptom: Engine Misfire*

Possible Cause	*Action*
Codes logged	Run diagnostic routines. Check the ECM audit trails.
Degraded or contaminated fuel	Drain the fuel tanks and replace filters.
Low charging pressure	Check fuel tank level. Sequentially check the fuel subsystem for restrictions or air ingestion. Use an Hg manometer in the primary circuit, a pressure gauge in the secondary circuit. Check the charging pressure regulator valve or restrictor fitting.
Valve lash adjustment	Perform top end adjustment.
Failed camshaft	Adjust valve lash. Run the engine hard, preferably on a dyno for 30 minutes. Recheck valve lash. Check cam lift to specification. Replace the camshaft if required.
Valves not seating	Remove the cylinder head(s). Recondition valves and seats.
Defective fuel injector or pump	Condition should be identified by codes in all current engines: If the problem is intermittent, wiggle wires, examine audit trails, and perform an electronic cylinder balance test.
Air in fuel	Locate the source of air ingestion. Check the filter pads on the primary circuit first. Use an inline diagnostic sight glass to locate the source.
Blown cylinder head gasket	Locate the source of the leak by external visual inspection. Compression test each cylinder to check for blown fire rings. Check the coolant and oil for contamination.
Leakage at high-pressure pipes	Check for trace leakage. Check the pipe nut torque on high-pressure pipes. Identify the source of the leak and check all other high-pressure pipes on PLN, EUP, and CR fuel systems.

Table 42–4 *Symptom: Engine Stalls*

Possible Cause	*Action*
Codes logged	Run diagnostic routines. Check audit trails. If no active codes, road or dyno test using EST snapshot mode: Set the trigger to collect data beginning at the code log event.
Idle speed set low	Reset or reprogram idle speed.
High parasitic load cut-in events	Check the air compressor and alternator, especially on city and highway buses.
Fuel tank vent circuit plugged or restricted	Check the vent or filter. Tank vent filters can rapidly plug in construction and agricultural operations. Clean/ replace the vent/filters.
Fuel starvation	Check the fuel tank pickup tube(s). Check the fuel tank for foreign matter: A decal or piece of paper over the pickup tube inlet can shut the system down. Check the fuel lines for internal restriction such as a rubber flap. Replace the lines as required.
Low fuel supply operating on rough terrain	Check the fuel tank level and replenish if necessary.
Defective injector nozzle(s)	Identify the affected injector(s) and replace. Identify the cause of nozzle failure and remedy: If caused by water, identify the source.
Defective injection pump(s)	This condition should produce codes in current engines. In older engines, use the OEM service literature.

Table 42–5 *Symptom: Irregular Engine rpm*

Possible Cause	*Action*
Air ingested into the primary side of the fuel subsystem: results in erratic engine rpm roll	Locate the air source and repair: Check the primary fuel filter pad and fuel tank pickup tube first.
Fractured fuel transfer pump, pulsation damper: results in erratic engine rpm roll	Test the charging pressure in the secondary fuel circuit. Replace the pulsation dampening disc.
Unbalanced fueling: results in rhythmic engine rpm fluctuation known as hunting	Perform a cylinder balance test. In older MUI and PT engines, check the injection pump actuation train.
Cylinder leakage	Perform a cylinder balance test on electronically managed engines. Cylinder leakage can be caused by blown fire ring (head gasket), cylinder valves, or piston rings.

Table 42–6 *Symptom: Lack of Power (LOP)*

Possible Cause	*Action*
Codes logged	Run diagnostic routines and check the audit trails before doing anything else.
EGR mixing chamber problem	Inspect the mixing gate door and leakage. Repair as needed.
Boost side leak	Check for leaks downstream from the turbocharger impeller housing. Test the charge air cooler for leaks.
Plugged or restricted fuel tank vent(s) or filters	Clean the vents and replace the filters if necessary.
Plugged air filter	Check filter restriction using a water-filled manometer. Replace if outside of specification.
Degraded fuel	Determine the cause of fuel degradation, especially when biodiesel is used. Drain the fuel and replace the fuel filters. Refuel with fresh fuel.
Low fuel charging pressure	(1) Check the fuel tank level. (2) Check fuel filter restriction. (3) Check for air ingestion. (4) Check the charging pressure fuel pressure regulator. Fill the fuel tank and replace defective components.
Defective injector nozzle(s)	Identify the affected injector(s) and replace. Identify the cause of nozzle failure and remedy: If caused by water, identify the source.
Valve lash adjustment	Perform a valve adjustment.
Turbocharger malfunction	Remove flow piping: Inspect radial play at the turbine and impeller. Check for contact or lube leakage on both sides. Replace or recore the turbo.
Exhaust restriction	Check for restriction in the exhaust piping and DPF canister. In vehicles with low horizontal exhausts, check the exit piping from the DPF/muffler canister.
Externally or internally plugged charge air cooler (CAC)	Visually check for external plugging of exchanger fins (results in insufficient cooling). Remove the CAC and perform flow and leakage tests.
Restricted boost side ducts and pipes	Remove and visually inspect: Replace if defective.

NOTE: Caused by many factors. If possible, verify the lack of power complaint by chassis dyno testing. Never act on a driver's word alone: Over 50% of LOP complaints prove to be baseless.

Table 42–7 *Symptom: Poor Fuel Mileage*

Possible Cause	*Action*
Codes logged	Run diagnostic routines and check the audit trails before doing anything else.
Restricted air intake circuit	Use an H_2O manometer (upstream from the impeller housing) and pressure gauges (downstream from the impeller housing) to identify the location. Repair/replace as required.
Fuel leakage	Check for evidence of external fuel leakage and repair.
Defective fuel injectors	Dribbling nozzles may not trip a fault code in some systems. Usually accompanied by more frequent DPF regen cycles. Test older hydromechanical nozzles and repair/replace.
Internal engine wear	Identify the root cause. You may find that an electronic cylinder balance test helps. Worn rings may also produce other symptoms such as plugging of pressure control valve (PCV) filters and more frequent DPF regen cycles. Repair the engine as required.

Table 42–8 *Symptom: Excessive Oil Consumption*

Possible Cause	*Action*
External oil leaks	Check the engine for obvious external leakage. Check all seals and gaskets. Add ultraviolet (UV) dye to the oil then pressure wash if the source of the leakage is difficult to locate.
Turbocharger oil leakage to manifold	Check for oil in the turbocharger inlet and soot in the exhaust outlet. Check for high inlet restriction (plugged air filter). Perform repairs as necessary.
Leakage at EGR valve piston shaft	Inspect the EGR circuit for oil loss. Replace the EGR valve if necessary.
Front and rear main seal leakage	Check for oil tracking at both seals. Check the seal races on the crankshaft. Replace and realign the main seals and install wear sleeves if necessary.
Air compressor passing oil	Check the air supply circuit downstream from the air compressor. Remove the compressor and check the drain-back ports for plugging. Replace the compressor flange gasket and/or compressor.
Plugged crankcase breather or PCV circuit	Identify the cause of clogging and clean or replace components.
High exhaust backpressure	Check the exhaust backpressure. Check the exhaust circuit for internal restrictions such as collapsed resonator baffles.
Worn valve guides and seals	Check the valve guide bore size to specification and seal integrity. Repair/replace components as necessary.
Worn piston rings	You may identify this condition by running a cylinder balance test if you are familiar with the duty cycle results produced by a healthy engine. Perform a cylinder leakage test in nonelectronic engines. Rebuild the engine.
Leakage from HEUI oil actuation circuit to the fuel supply gallery	Condition is usually caused by rolled or failed center O-rings in the HEUI injector assembly.

Table 42–9 *Symptom: Engine Overheating*

Possible Cause	Action
Low coolant level	Locate the cause. Visually inspect for external leaks and pressure test the circuit. Repair and replenish the coolant.
Loose or worn fan/water pump drive belts	Adjust the belt tension or replace the drive belts.
EGR heat exchanger leak	Replace the EGR cooler: Do not attempt to repair it.
Restricted airflow through the radiator	Carefully clean bugs and road dirt from the rad cooling fins using a low-pressure hose. Test after cleaning.
Defective radiator cap	Dynamic pressure test the rad cap to ensure it seals to specified pressure. Replace if necessary.
Defective thermostat(s)	Test the thermostat opening temperature: Also check the temperature sensor accuracy. Replace components as necessary.
Defective fan cycle controls	Check the viscous fan operation by running the engine to the specified fan temperature. With on/off fan controls, check the means used to disengage/engage the fan, electric and pneumatic.
Combustion gas in coolant	Check for evidence of cylinder head gasket and injector sleeve failure. Repair as necessary.
Defective water pump	Check the water pump drive. Remove the water pump and check the impeller to shaft integrity. Repair or replace the water pump and reinstall.
Plugged oil cooler	Usually produces higher than specified oil temperatures along with normal range coolant temperatures. Remove the oil cooler and disassemble to clean/remove the restriction.
Winterfront not opened, or insufficiently open	Check the winterfront and advise the owner that winterfronts are not recommended by OEMs.
Defective shutterstat	Remove the shutterstat and test pilot temperature and on/ off cycles: Shutters should default to the open position.
Improper driver shifting practices	Check the engine audit trail if available. Refer operator to driver trainer.

Table 42–10 *Symptom: High Exhaust Temperature*

Possible Cause	Action
Codes logged	Run diagnostic routines to identify the root cause. Where separate DPF module (noncatalyzed type) is used, check for DPF codes.
Improper driver shifting practices (high load and steep terrain operation)	Check the engine audit trail if possible. Refer operator to driver trainer.
High EGR temperatures	EGR flow rate is too high or ineffective EGR heat exchanger. Check the EGR mixing gate operation and EGR heat exchanger for clogging.
Restricted air intake	Use an H_2O manometer to check for inlet restriction and repair the cause.
Leak in air intake circuit	Check hose clamps, pipes, and ducts. Repair as required.
Exhaust leak upstream from turbine housing	Check for external leaks under load. Replace gaskets and manifold coupling seals as necessary.
Exhaust restrictions	Check for exhaust circuit restrictions, including the cylinder head exhaust tract insulation, loose baffles, plugged catalytic converters, and DPFs. Replace or repair as necessary.
Overfueling or advanced timing	More common in hydromechanical engines: Check the audit trail for fuel injection system related events; run a cylinder balance test. Perform a tune up on nonelectronic engines, verifying base injection timing.
High load operation with insufficient ram air	High load operating a ram air-cooled CAC when the vehicle is moving at low velocity can produce overheat. Review the type of operation the engine has been spec'd for with the owner.

Table 42–11 *Symptom: Low Oil Pressure*

Possible Cause	*Action*
Low oil level	Add oil and check for oil leaks or an oil burning condition.
Incorrect oil viscosity	Drain the oil and replace the filters. Replenish with oil meeting the OEM-specified viscosity for the temperature conditions.
Defective oil pressure gauge or sensor	Check the operation of the oil pressure sensor first then the pressure gauge. Replace components as necessary.
Restricted full flow oil filter	Plugged full flow filters trip the bypass valve on the oil filter flange and may, or may not, cause low oil pressure. Replace the oil and filters.
Engine oil diluted with fuel	Check for fuel system leaks: This varies according to the type of fuel system used. Check the O-rings on the fuel injectors and EUPs. Repair and replace as necessary.
Oil pump failure	Check the mounting arrangement. When the oil pump is remote mounted, carefully check the pickup plumbing for air ingestion. Replace or repair components as necessary.
Oil filter and filter mounting pad problems	Remove the filter(s) and visually inspect the gaskets. Check for possible restrictions to the in/out ports of the oil filter. Repair, clean, or replace components as necessary.
Excessive clearance between crankshaft journals and bearings	Perform a bearing roll-in. Note that this is not a common problem today as friction bearing technology has advanced significantly.
EGR valve leaking excessively	Check EGR valve operation and repair as necessary.
Degraded engine oil	Low-quality engine oils, especially those with high detergent loads, can break down and thin, resulting in low oil pressure. Replace the engine oil and filters with those specified by the OEM.

Table 42–12 *Symptom: Oil in Coolant*

Possible Cause	*Action*
Defective oil cooler core or bundle O-rings	Remove and disassemble the oil cooler. Test the bundle or plate stack. Replace if defective. Rolled or failed bundle O-rings can also cause this condition. Replace as necessary. Clean out the entire cooling system using detergent.
Failed head gasket	Replace the cylinder head gasket. In cases where nonintegral grommets are used, take special care in placing these.
Cylinder head or cylinder block porosity	Identify the location of the failure and replace as necessary.

Table 42–13 *Symptom: Coolant in Engine Oil*

Possible Cause	*Action*
Cracked cylinder head	Disassemble and repair the oil cooler: This is a less common outcome of an oil cooler failure than oil in coolant.
Head gasket failure	Pressure test the cooling system and replace the cylinder head gasket.
Injector cup seal failure	Hydrostatically test the cylinder head and replace the injector cups as required.
Cylinder wet liner failure	Pressure test the system and replace the defective components as required.
Defective oil cooler core	Hydrostatically test the cylinder head to locate the failure and replace the cylinder head.

Table 42–14 *Symptom: Low Compression*

Possible Cause	*Action*
Incorrect valve lash adjustment	Perform a complete overhead tune up, making sure the valves are set to spec.
Blown fire ring(s) in the head gasket	Replace the head gasket.
Broken or fatigued valve springs	Check and replace as required: In the case of fatigued valve springs, it makes sense to replace them all.
Valves not seating	Remove the cylinder head and check the valve seat faces. Replace the valves and valve seats.
Piston ring failure	Rings may be stuck, worn, broken, or glazed. Determine the root cause of failure and replace the ring sets.
Failed camshaft and/or lifters	Determine the root cause and replace failed components, being sure to check every component in the valve train.

Table 42–15 *Symptom: Fuel in Lube*

Possible Cause	*Action*
Excessive idling in cold weather (cylinder washdown)	Program optimized idle, install auxiliary cold weather assists, and instruct the driver how to handle the truck in cold weather.
Injector nozzle malfunction	Mostly a condition on older hydromechanical engines. Remove and test the injectors. Replace the nozzle valves and reassemble the injectors.
Injector O-ring failure	Pull the injector assembly and inspect for worn, torn, or rolled O-rings. Replace all O-rings whether observed to be defective or not.
HEUI IAP circuit failures	Check the HEUIs, jumper pipes if used, and the cylinder oil and fuel manifolds. Replace components as required.

Table 42–16 *Symptom: EGR System Malfunctions*

Possible Cause	*Action*
Excessive idling—especially in cold weather, it causes carbon coking in the EGR heat exchanger that chokes down on flow	Program optimized idle and instruct the driver how to handle the vehicle in cold weather. Equip the vehicle with auxiliary power for cold weather operation.
High EGR temperature	Indicates a failed EGR heat exchanger. Replace the heat exchanger or heat exchanger core.
EGR valve housing cracked	Inspect the EGR housing and replace if necessary.
EGR gate valve sticking	Check oil supply to the valve. Check for oil coking or corrosion around the actuator bushing. Also check electrical connections to the actuator. Clean, repair, and replace components as required.
EGR valve not functioning properly due to low oil pressure	Troubleshoot the low oil pressure condition using Table 42–11.
Turbo VGT actuator does not respond	Determine whether the problem is mechanical, pneumatic, or electrical (depending on the OEM). Check for codes. If none, focus on a mechanical cause at the turbo VGT actuator or within the turbocharger.

TECHNICAL SUPPORT

It is in the business interest of every engine OEM to ensure that breakdowns occur as little as possible. It is also in their interests that when breakdowns do occur, they are repaired in a timely manner and steps are taken to avoid a recurrence. For this reason, all the OEMs provide technical support for their products. It is usually called something like Service Operations Department. In most cases, OEM service operations divide up territory geographically and provide a district service manager (DSM) for each area. While dealership managers will often contact DSMs for warranty approval, this is only one responsibility of a DSM. A more important DSM role is that of providing product technical support.

DSM Role

If you are routinely working on engine diagnostic routines it is inevitable that you will be working "outside of the box" from time to time. For this reason, you should have the contact information of the DSM pasted into your shop email address book and the phone number on your toolbox. The DSM is one of the first to take training on a new OEM engine series. Because it then becomes a DSM responsibility to provide product support most field problems are routed through him or her. A good DSM collates questions and feedback and can usually provide answers to questions before they appear in online TSBs and SIS amendments.

REVIEW QUESTIONS

1. Which of the following instruments would check dry air filter restriction with the highest accuracy?
   a. Pyrometer
   b. Trouble light
   c. H_2O manometer
   d. Dash restriction gauge
2. A typical maximum inlet restriction specification on a turboboosted diesel engine would be:
   a. 25″ (635 mm) H_2O
   b. 25″ (635 mm) Hg
   c. 25 psi (172 kPa)
   d. 250 kPa (36 psi)
3. Which of the following conditions would *least* likely cause black smoke emission?
   a. Restricted air filter
   b. Retarded injection timing
   c. Restricted diesel particulate filter
   d. Excessive use of cetane improver
4. *Technician A* states that excessive idling can result in plugging of the EGR heat exchanger. *Technician B* states that excessive idling periods can produce abnormally high intake manifold temperatures later when the engine is operated normally. Who is correct?
   a. Technician A only
   b. Technician B only
   c. Both A and B
   d. Neither A nor B
5. Removing an engine coolant thermostat would most likely result in an increase of which noxious emission?
   a. HC
   b. NO_x
   c. SO_2
   d. CO_2
6. Which of the following operating modes would be more likely to produce valve float?
   a. Overfueling
   b. Lugdown
   c. Ether abuse
   d. Overspeeding engine
7. *Technician A* states that an excessive amount of oil in the crankcase can produce a low oil pressure problem in a diesel engine. *Technician B* states that the first thing to check when oil pressures fluctuate is the crankcase oil level. Who is correct?
   a. Technician A only
   b. Technician B only
   c. Both A and B
   d. Neither A nor B
8. *Technician A* states that a small dent in a viscous-type vibration damper is not a concern unless there is visible leakage. *Technician B* states that some compressors have to be timed to the engine or a vibration will result. Who is correct?
   a. Technician A only
   b. Technician B only
   c. Both A and B
   d. Neither A nor B

9. There is evidence of soot in the intake manifold of a boosted truck diesel engine. Which of the following conditions would be most likely to contribute to this condition?
   a. Engine lugdown
   b. High-idle operation
   c. Rated speed operation
   d. Running at low speed with low loads

10. *Technician A* states that white smoke emission can be caused by air starvation to the engine. *Technician B* states that blue smoke is associated with an engine that is burning oil in its cylinders. Who is correct?
    a. Technician A only
    b. Technician B only
    c. Both A and B
    d. Neither A nor B

Index